Lecture Notes in Computer Science 5342

Commenced Publication in 1973
Founding and Former Series Editors:
Gerhard Goos, Juris Hartmanis, and Jan van Leeuwen

Niels da Vitora Lobo Takis Kasparis
Michael Georgiopoulos Fabio Roli James T. Kwok
Georgios C. Anagnostopoulos Marco Loog (Eds.)

Structural, Syntactic, and Statistical Pattern Recognition

Joint IAPR International Workshop, SSPR & SPR 2008
Orlando, USA, December 4-6, 2008
Proceedings

Springer

Volume Editors

Niels da Vitora Lobo
Takis Kasparis
Michael Georgiopoulos
University of Central Florida Orlando,Florida, 32816, USA
E-mail: {niels, kasparis, michelg } @cs.ucf.edu

Fabio Roli
University of Cagliari 09123, Cagliari, Italy
E-mail: roli@diee.unica.it

James T. Kwok
Hong Kong University of Science and Technology
Kowloon, Hong Kong, China
E-mail: jamesk@cs.ust.hk

Georgios C. Anagnostopoulos
Florida Institute of Technology Melbourne, Florida, 32901, USA
E-mail: georgio@fit.edu

Marco Loog
Delft University of Technology, Delft, The Netherlands
and University of Copenhagen, Copenhagen, Denmark
E-mail: marco.loog@gmail.com

Library of Congress Control Number: 2008939876

CR Subject Classification (1998): I.5, I.4, I.2.10, I.3.5, G.2-3

LNCS Sublibrary: SL 6 – Image Processing, Computer Vision,
Pattern Recognition, and Graphics

ISSN 0302-9743
ISBN 10 3-540-89688-0 Springer Berlin Heidelberg New York
ISBN 13 978-3-540-89688-3 Springer Berlin Heidelberg New York

Springer is a part of Springer Science+Business Media

springer.com

Typesetting: Camera-ready by author, data conversion by Scientific Publishing Services, Chennai, India
Printed on acid-free paper SPIN: 12570430 06/3180 5 4 3 2 1 0

Preface

This volume in the Springer *Lecture Notes in Computer Science* (LNCS) series contains 98 papers presented at the S+SSPR 2008 workshops. S+SSPR 2008 was the sixth time that the SPR and SSPR workshops organized by Technical Committees, TC1 and TC2, of the International Association for Pattern Recognition (IAPR) were held as joint workshops. S+SSPR 2008 was held in Orlando, Florida, the family entertainment capital of the world, on the beautiful campus of the University of Central Florida, one of the up and coming metropolitan universities in the USA. S+SSPR 2008 was held during December 4–6, 2008 only a few days before the 19th International Conference on Pattern Recognition (ICPR 2008), which was held in Tampa, only two hours away from Orlando, thus giving the opportunity of both conferences to attendees to enjoy the many attractions offered by two neighboring cities in the state of Florida.

SPR 2008 and SSPR 2008 received a total of 175 paper submissions from many different countries around the world, thus giving the workshop an international clout, as was the case for past workshops. This volume contains 98 accepted papers: 56 for oral presentations and 42 for poster presentations. In addition to parallel oral sessions for SPR and SSPR, there was also one joint oral session with papers of interest to both the SPR and SSPR communities. A recent trend that has emerged in the pattern recognition and machine learning research communities is the study of graph-based methods that integrate statistical and structural approaches. For this reason, a special session on graph-based methods was co-organized by Technical Committee TC15 to explore new research issues in this topic. Moreover, invited talks were presented by four prominent speakers: Horst Bunke, Professor of the Department of Computer Science, at the University of Bern, Switzerland; Pedro Domingos, Professor of the Computer Science Department at the Washington State University, USA; Tin-Kam Ho, leader of the Computing Research Group of Bell Labs, Alcatel-Lucent, at Murray Hill, USA, and winner of the 2008 Pierre Devijver Award; Ludmila Kuncheva, Reader of the School of Computer Science, Bangor University, UK.

We would like to take this opportunity to thank all members of the SPR and SSPR Program Committees and the additional reviewers for their professional support in reviewing the submitted papers. We thank all the Advisory Committee members, Shun-ichi Amari, Horst Bunke, Robert P.W. Duin, Anil K. Jain, Josef Kittler, Erkki Oja, Ching Y. Suen, and Tie-Niu Tan, for their invaluable advice on organizing this event. We also thank all the authors, the invited speakers, the Organizing Committee members, the sponsors, and the editorial staff of

Springer for helping us make this event a success and helping us produce this high-quality publication in the LNCS series.

December 2008

Niels da Vitoria Lobo
Takis Kasparis
James T. Kwok
Fabio Roli
Michael Georgiopoulos
Georgios C. Anagnostopoulos
Marco Loog

S+SSPR 2008

General Chair

Michael Georgiopoulos — University of Central Florida, USA

Local Chair

Georgios C. Anagnostopoulos — Florida Institute of Technology, USA

Webmaster

Marco Loog — Delft University of Technology, The Netherlands and University of Copenhagen, Denmark

Supported by

International Association for Pattern Recognition (IAPR)
The ML^2 (Machine Learning Lab) at UCF, School of EECS at UCF, and the University of Central Florida (UCF)

SSPR Committee

Co-chairmen

SPR Committee

Co-chairmen

Niels da Vitoria Lobo — University of Central Florida, USA
Fabio Roli — University of Cagliari, Italy

Program Committee

Mayer Aladjem (Israel)
George Bebis (USA)
Nitesh Chawla (USA)
Dimitrios Charalampidis (USA)
Christos Christodoulou (USA)
Zoran Duric (USA)
Francesco J. Ferri (Spain)
Venu Govindaraju (USA)
Gregory Heileman (USA)
Tin Kam Ho (USA)
Vassilis Kaburlasos (Greece)
Taskin Koçak (UK)
Konstantinos Koutroumbas (Greece)
Adam Krzyzak (Canada)
Mineichi Kudo (Japan)
Ludmila Kuncheva (UK)
Chih-Jen Lin (Taiwan)
Jana Novovicova (Czech Republic)
Longin J. Latecki (USA)
Stelios Papadakis (Greece)
Robi Polikar (USA)
Sarunas Raudys (Lithuania)
Punam Kumar Saha (USA)
Carlo Sansone (Italy)
Mubarak Shah (USA)
Paul Smith (USA)
David M.J. Tax (The Netherlands)
Sergios Theodoridis (Greece)
Francesco Tortorella (Italy)
Olcay Kursun (Turkey)
David Zhang (Hong Kong)
Zhi-Hua Zhou (China)

Reviewers

The Program Committees for both SPR and SSPR were kindly assisted by

Georgios C. Anagnostopoulos (USA)
Davide Ariu (Italy)
Nicole Maria Artner (Austria)
Arslan Basharat (USA)
Anurag Bharadwaj (USA)
Florent Brunet (Germany)
Aurelio Campilho (Italy)
Sylvie Chambon (France)
Thierry Chateau (France)
Dimitrios Charalampidis (USA)
David Cieslak (USA)
Pietro Coli (Italy)
Darcy Davis (USA)
Luca Didaci (Italy)
Martin Drauschke (Germany)
Pasquale Foggia (Italy)
Biagio Freni (Italy)
Giorgio Fumera (Italy)
Vincent Gay-Bellile (France)
Michael Georgiopoulos (USA)
Giorgio Giacinto (Italy)
Pierre Gurdjos (France)
James Haralambides (USA)
Adrian Ion (Austria)
François Jacquenet (France)
Assem Kaylani (Jordan)
Anna Koufakou (USA)
Mikhail Lange (Romania)
Wan-Jui Lee (The Netherlands)
Janaka Liyanage (USA)
Marco Loog (The Netherlands)
Zhihu Lu (China)
Mahsa Maghami (USA)
Angello Marcelli (Italy)
Gian Luca Marcialis (Italy)
Claudio Marrocco (Italy)
Weike Pan (Hong Kong)
Julien Peyras (France)
Ignazio Pillai (Italy)
Mehran Ramin (USA)
Ruben Ramirez Padron (USA)
Ajita Rattani (Italy)
Troy Raeder (USA)
John Reeder (USA)
Imran Saleemi (USA)
Carlo Sansone (Italy)
Christopher Sentelle (USA)
Paulo Simeone (Italy)
Nicholas Shorter (USA)
Karsten Steinhaeuser (USA)
Mikhail Traskin (USA)
Roberto Tronci (Italy)
Grigorios Tsoumakas (Greece)
Sergey Tulyakov (USA)

Table of Contents

Invited Talks (Abstracts)

Data Complexity Analysis: Linkage between Context and Solution in Classification 1
Tin Kam Ho

Graph Classification on Dissimilarity Space Embedding 2
Horst Bunke and Kaspar Riesen

Markov Logic: A Unifying Language for Structural and Statistical Pattern Recognition 3
Pedro Domingos, Stanley Kok, Daniel Lowd, Hoifung Poon, Matt Richardson, Parag Singla, Marc Sumner, and Jue Wang

Linear Discriminant Classifier (LDC) for Streaming Data with Concept Drift 4
Ludmila I. Kuncheva and Indre Zliobaite

SSPR

Graph-Based Methods (Special Session)

Graph Edit Distance without Correspondence from Continuous-Time Quantum Walks 5
David Emms, Richard C. Wilson, and Edwin R. Hancock

Exact Median Graph Computation Via Graph Embedding 15
Miquel Ferrer, Ernest Valveny, Francesc Serratosa, and Horst Bunke

Similarity Invariant Delaunay Graph Matching 25
Dongjoe Shin and Tardi Tjahjadi

An Inexact Graph Comparison Approach in Joint Eigenspace 35
Wan-Jui Lee and Robert P.W. Duin

A Vectorial Representation for the Indexation of Structural Informations 45
Nicolas Sidere, Pierre Heroux, and Jean-Yves Ramel

Probabilistic and Stochastic Structural Models for PR

Stochastic Text Models for Music Categorization 55
Carlos Pérez-Sancho, David Rizo, and José M. Iñesta

Efficient Pruning of Probabilistic Automata 65
Franck Thollard and Baptiste Jeudy

Position Models and Language Modeling 76
Arnaud Zdziobeck and Franck Thollard

Melody Recognition with Learned Edit Distances 86
Amaury Habrard, José Manuel Iñesta, David Rizo, and Marc Sebban

Context First 97
Daniel Heesch, Robby Tan, and Maria Petrou

Image and Video Analysis

Activity Representation in Crowd 107
Yunqian Ma and Petr Cisar

3D Object Recognition Using Hyper-Graphs and Ranked Local Invariant Features 117
Shengping Xia and Edwin R. Hancock

Image Matching with Spatially Variant Contrast and Offset: A Quadratic Programming Approach 127
Alexander Shorin, Georgy Gimel'farb, Patrice Delmas, and Jonh Morris

Natural Versus Artificial Scene Classification by Ordering Discrete Fourier Power Spectra 137
Giovanni Maria Farinella, Sebastiano Battiato, Giovanni Gallo, and Roberto Cipolla

Fuzzy ART for Relatively Fast Unsupervised Image Color Quantization 147
Nicholas Shorter and Takis Kasparis

Shape Analysis

Recognising Facial Expressions Using Spherical Harmonics 157
James Sharpe and Edwin R. Hancock

Complex Fiedler Vectors for Shape Retrieval 167
Reinier H. van Leuken, Olga Symonova, Remco C. Veltkamp, and Raffaele de Amicis

Combining Shape Priors and MRF-Segmentation 177
Boris Flach and Dmitrij Schlesinger

Measuring the Similarity of Vector Fields Using Global Distributions ... 187
H. Quynh Dinh and Liefei Xu

Consensus Graphs for Symmetry Plane Estimation 197
Andrea Albarelli, Marcello Pelillo, and Sebastiano Viviani

Kernel Methods

Graph Characteristic from the Gauss-Bonnet Theorem 207
Hewayda ElGhawalby and Edwin R. Hancock

Quantitative Evaluation on Heat Kernel Permutation Invariants 217
Bai Xiao, Richard C. Wilson, and Edwin R. Hancock

Hierarchical Bag of Paths for Kernel Based Shape Classification 227
François-Xavier Dupé and Luc Brun

Polytopal Graph Complexity, Matrix Permanents, and Embedding 237
Francisco Escolano, Edwin R. Hancock, and Miguel A. Lozano

Gesture Recognition Based on Manifold Learning 247
Heeyoul Choi, Brandon Paulson, and Tracy Hammond

Pattern Recognition and Applications (Jointly with SPR)

Graph Characteristics from the Ihara Zeta Function 257
Peng Ren, Richard C. Wilson, and Edwin R. Hancock

A Learning Approach to 3D Object Representation for Classification ... 267
Indriyati Atmosukarto and Linda G. Shapiro

Region Based Visual Object Categorization Using Segment Features and Polynomial Modeling 277
Huanzhang Fu, Alain Pujol, Emmanuel Dellandréa, and Liming Chen

Poster Papers

Graph-Based Methods

IAM Graph Database Repository for Graph Based Pattern Recognition and Machine Learning 287
Kaspar Riesen and Horst Bunke

Hybrid Genetic Algorithm and Procrustes Analysis for Enhancing the Matching of Graphs Generated from Shapes 298
Gerard Sanromà, Francesc Serratosa, and René Alquézar

Spectral Embedding of Feature Hypergraphs 308
Peng Ren, Richard C. Wilson, and Edwin R. Hancock

Clustering Using Class Specific Hyper Graphs 318
Shengping Xia and Edwin R. Hancock

Structure Is a Visual Class Invariant 329
Bai Xiao, Yi-Zhe Song, Anupriya Balika, and Peter M. Hall

Recognition and Classification

Local Fisher Discriminant Component Hashing for Fast Nearest Neighbor Classification 339
Tomoyuki Shibata and Osamu Yamaguchi

Significance Tests and Statistical Inequalities for Region Matching 350
Guillaume Née, Stéphanie Jehan-Besson, Luc Brun, and Marinette Revenu

Advanced Homology Computation of Digital Volumes Via Cell Complexes 361
Helena Molina-Abril and Pedro Real

Modeling the Model Athlete: Automatic Coaching of Rowing Technique 372
Simon Fothergill, Robert Harle, and Sean Holden

Image and Video Analysis

Optimal Solution of the Dichromatic Model for Multispectral Photometric Invariance 382
Cong Phuoc Huynh and Antonio Robles-Kelly

Ensemble Combination for Solving the Parameter Selection Problem in Image Segmentation 392
Pakaket Wattuya and Xiaoyi Jiang

Semantic Scene Classification for Image Annotation and Retrieval 402
Özge Çavuş and Selim Aksoy

Multi-angle View, Illumination and Cosmetic Facial Image Database (MaVIC) and Quantitative Analysis of Facial Appearance 411
Yen-Wei Chen, Takayuki Fukui, Xu Qiao, Takanori Igarashi, Keisuke Nakao, and Akio Kashimoto

Applications

Shape Representation Robust to the Sketching Order Using Distance Map and Direction Histogram 421
Kiwon Yun, Junyeong Yang, and Hyeran Byun

A Stochastic Approach to Median String Computation 431
Cristian Olivares-Rodríguez and Jose Oncina

Tracking by Hierarchical Representation of Target Structure 441
Nicole M. Artner, Salvador B. López Mármol, Csaba Beleznai, and Walter G. Kropatsch

Using the Shape Characteristics of Rain to Identify and Remove Rain from Video 451
Nathan Brewer and Nianjun Liu

SPR

Ensemble Methods

A Fast Approach to Improve Classification Performance of ECOC Classification Systems 459
Paolo Simeone, David M.J. Tax, Robert P.W. Duin, and Francesco Tortorella

Partially Corrective AdaBoost 469
Yan Jiang and Xiaoqing Ding

Selection and Fusion of Similarity Measure Based Classifiers Using Support Vector Machines 479
Mohammad T. Sadeghi, Masoumeh Samiei, and Josef Kittler

On Mixtures of Linear SVMs for Nonlinear Classification 489
Zhouyu Fu and Antonio Robles-Kelly

Adversarial Pattern Classification Using Multiple Classifiers and Randomisation 500
Battista Biggio, Giorgio Fumera, and Fabio Roli

Pattern Recognition

Adaptive Learning Rate for Online Linear Discriminant Classifiers 510
Ludmila I. Kuncheva and Catrin O. Plumpton

Combining Online Classification Approaches for Changing Environments 520
Juan J. Rodríguez and Ludmila I. Kuncheva

Optimal Kernel in a Class of Kernels with an Invariant Metric 530
Akira Tanaka, Hideyuki Imai, Mineichi Kudo, and Masaaki Miyakoshi

Chaotic Pattern Recognition: The Spectrum of Properties of the Adachi Neural Network ... 540
Ke Qin and B. John Oommen

On Euclidean Corrections for Non-Euclidean Dissimilarities ... 551
Robert P.W. Duin, Elżbieta Pękalska, Artsiom Harol, Wan-Jui Lee, and Horst Bunke

Feature Selection

Feature and Classifier Selection in Class Decision Trees ... 562
Kazuaki Aoki and Mineichi Kudo

Soft Feature Selection by Using a Histogram-Based Classifier ... 572
Hiroshi Tenmoto and Mineichi Kudo

Embedded Map Projection for Dimensionality Reduction-Based Similarity Search ... 582
Simone Marinai, Emanuele Marino, and Giovanni Soda

Novel Incremental Principal Component Analysis with Improved Performance ... 592
Myoung Soo Park and Jin Young Choi

Density Estimation and Clustering

An Efficient Algorithm for Optimal Multilevel Thresholding of Irregularly Sampled Histograms ... 602
Luis Rueda

A Family of Cluster Validity Indexes Based on a l-Order Fuzzy OR Operator ... 612
Hoel Le Capitaine and Carl Frélicot

Fast Multivariate Ordinal Type Histogram Matching ... 622
Sung-Hyuk Cha

Distance between Histograms with Shuffled Cost Matrix ... 633
Sung-Hyuk Cha

Using Non Local Features for 3D Shape Grouping ... 644
Antonio Adán, Miguel Adán, Santiago Salamanca, and Pilar Merchán

Computer Vision and Biometrics

A Dynamic Programming Technique for Optimizing Dissimilarity-Based Classifiers ... 654
Sang-Woon Kim and Jian Gao

Supervised Principal Geodesic Analysis on Facial Surface Normals for Gender Classification 664
Jing Wu, William A.P. Smith, and Edwin R. Hancock

Exploring Margin Maximization for Biometric Score Fusion 674
Claudio Marrocco, Maria Teresa Ricamato, and Francesco Tortorella

Biometric Template Update: An Experimental Investigation on the Relationship between Update Errors and Performance Degradation in Face Verification 684
Gian Luca Marcialis, Ajita Rattani, and Fabio Roli

Automatic Mutual Nonrigid Registration of Dense Surfaces by Graphical Model Based Inference 694
Xiao Dong and Guoyan Zheng

Pattern Recognition and Applications (Jointly with SSPR)

A Novel Video Classification Method Based on Hybrid Generative/Discriminative Models 705
Zhi Zeng, Wei Liang, Heping Li, and Shuwu Zhang

Scale-Space Kernels for Additive Modeling 714
Chandan K. Reddy and Jin-Hyeong Park

Learning Curves for the Analysis of Multiple Instance Classifiers 724
David M.J. Tax and Robert P.W. Duin

Poster Papers

Computer Vision and Biometrics

Categorizing Perceptions of Indoor Rooms Using 3D Features 734
Agnes Swadzba and Sven Wachsmuth

Template Selection by Editing Algorithms: A Case Study in Face Recognition 745
Biagio Freni, Gian Luca Marcialis, and Fabio Roli

A Novel Coding Scheme for Indexing Fingerprint Patterns 755
Aglika Gyaourova and Arun Ross

A Theoretical and Experimental Analysis of Template Co-update in Biometric Verification Systems 765
Luca Didaci, Gian Luca Marcialis, and Fabio Roli

A New Solution Scheme of Unsupervised Locality Preserving Projection Method for the SSS Problem ... 775
Yong Xu and David Zhang

High Occupancy Vehicle Detection ... 782
Alberto J. Pérez-Jiménez, Jose Luis Guardiola, and Juan Carlos Pérez-Cortés

Ensemble Methods

Automatic Classification of NMR Spectra by Ensembles of Local Experts ... 790
Kai Lienemann, Thomas Plötz, and Gernot A. Fink

Bagging, Random Subspace Method and Biding ... 801
Satoshi Shirai, Mineichi Kudo, and Atsuyoshi Nakamura

Use of Structured Pattern Representations for Combining Classifiers ... 811
Raisa Socorro and Luisa Micó

Combination of Experts by Classifiers in Similarity Score Spaces ... 821
Roberto Tronci, Giorgio Giacinto, and Fabio Roli

Improving the Performance of a NER System by Post-processing and Voting ... 831
Asif Ekbal and Sivaji Bandyopadhyay

A Low Cost Incremental Biometric Fusion Strategy for a Handheld Device ... 842
Lorene Allano, Sonia Garcia-Salicetti, and Bernadette Dorizzi

Applications

A Hidden Markov Model Approach to Classify and Predict the Sign of Financial Local Trends ... 852
Manuele Bicego, Enrico Grosso, and Edoardo Otranto

Local Metric Learning on Manifolds with Application to Query–Based Operations ... 862
Karim Abou-Moustafa and Frank Ferrie

Behavior Analysis of Volume Prototypes in High Dimensionality ... 874
Maiko Sato, Mineichi Kudo, and Jun Toyama

Pattern Recognition Approaches for Classifying IP Flows ... 885
Alice Este, Francesco Gargiulo, Francesco Gringoli, Luca Salgarelli, and Carlo Sansone

Pattern Recognition

Outlier Robust Gaussian Process Classification ... 896
Hyun-Chul Kim and Zoubin Ghahramani

2D Shape Classification Using Multifractional Brownian Motion 906
Manuele Bicego and Alessandro Trudda

A New Performance Evaluation Method for Two-Class Imbalanced Problems 917
Vicente García, Ramón A. Mollineda, and J. Salvador Sánchez

A Pruning Rule Based on a Distance Sparse Table for Hierarchical Similarity Search Algorithms 926
Eva Gomez-Ballester, Luisa Micó, and Jose Oncina

Classification and Automatic Annotation Extension of Images Using Bayesian Network 937
Sabine Barrat and Salvatore Tabbone

Selection of Suitable Set of Decision Rules Using Choquet Integral 947
Laurent Wendling, Jan Rendek, and Pascal Matsakis

Feature Selection and Clustering

Evaluating the Stability of Feature Selectors That Optimize Feature Subset Cardinality 956
Petr Somol and Jana Novovičová

A Learning Scheme for Recognizing Sub-classes from Model Trained on Aggregate Classes 967
Ranga Raju Vatsavai, Shashi Shekhar, and Budhendra Bhaduri

An Empirical Evaluation of Common Vector Based Classification Methods and Some Extensions 977
Katerine Díaz-Chito, Francesc J. Ferri, and Wladimiro Díaz-Villanueva

Invited Talks (Full Papers)

Data Complexity Analysis: Linkage between Context and Solution in Classification 986
Tin Kam Ho

Graph Classification Based on Dissimilarity Space Embedding 996
Horst Bunke and Kaspar Riesen

Author Index 1009

Data Complexity Analysis: Linkage between Context and Solution in Classification

Tin Kam Ho

Abstract. For a classification problem that is implicitly represented by a training data set, analysis of data complexity provides a linkage between context and solution. Instead of directly optimizing classification accuracy by tuning the learning algorithms, one may seek changes in the data sources and feature transformations to simplify the class geometry. Simplified class geometry benefits learning in a way common to many methods. We review some early results in data complexity analysis, compare these to recent advances in manifold learning, and suggest directions for further research.

N. da Vitora Lobo et al. (Eds.): SSPR&SPR 2008, LNCS 5342, p. 1, 2008.

Graph Classification on Dissimilarity Space Embedding

Horst Bunke and Kaspar Riesen

Abstract. Recently, an emerging trend of representing objects by graphs can be observed. In fact, graphs offer a powerful alternative to feature vectors in pattern recognition, machine learning, and related fields. However, the domain of graphs contains very little mathematical structure, and consequently, there is only a limited amount of classification algorithms available. In this paper we survey recent work on graph embedding using dissimilarity representations. Once a population of graphs has been mapped to a vector space by means of this embedding procedure, all classification methods developed in statistical pattern recognition become directly available. In an experimental evaluation we show that the proposed methodology of first embedding graphs in vector spaces and then applying a statistical classifier has significant potential to outperform classifiers that directly operate in the graph domain. Additionally, the proposed framework can be considered a contribution towards unifying the domains of structural and statistical pattern recognition.

N. da Vitora Lobo et al. (Eds.): SSPR&SPR 2008, LNCS 5342, p. 2, 2008.

Markov Logic: A Unifying Language for Structural and Statistical Pattern Recognition

Pedro Domingos, Stanley Kok, Daniel Lowd, Hoifung Poon, Matt Richardson, Parag Singla, Marc Sumner, and Jue Wang

Abstract. Effective pattern recognition requires understanding both statistical and structural aspects of the input, but in the past these have mostly been handled separately. Markov logic is a powerful new language that seamlessy combines the two. Models in Markov logic are sets of weighted formulas in first-order logic, interpreted as templates for features of Markov random fields. Most statistical and structural models in wide use are simple special cases of Markov logic. Learning algorithms for Markov logic make use of conditional likelihood, convex optimization, and inductive logic programming. Inference algorithms combine ideas from Markov chain Monte Carlo and satisfiability testing. Markov logic has been successfully applied to problems in information extraction, robot mapping, social network modeling, and others, and is the basis of the open-source Alchemy system.

N. da Vitora Lobo et al. (Eds.): SSPR&SPR 2008, LNCS 5342, p. 3, 2008.

Linear Discriminant Classifier (LDC) for Streaming Data with Concept Drift

Ludmila I. Kuncheva and Indre Zliobaite

Abstract. Simple classifiers, including LDC, have often been praised for their robustness and accuracy. Here we consider an online version of LDC applied to streaming data with concept drift. The classifier is trained on a moving window containing the latest N observations. Current approaches to determining the window size are mostly heuristical. The talk presents a framework within which theoretical relationship can be sought between the window size and the classification error. The framework is based upon two ideas. First, past literature offers formal relationships between the size of the training set and the asymptotic accuracy for several classifier models. Such a relationship can be used as a guide to determining the moving window size. Second, after a sudden change, the "old" classifier may be better than an undertrained "new" classifier that uses only the data coming after the change. We state the optimal window size for the case of LDC applied to two Gaussian classes and a sudden change in the form of rotation and translation of the feature space. Simulation results are included in order to investigate the sensitivity of the theoretical results to violation of the underlying assumptions.

N. da Vitora Lobo et al. (Eds.): SSPR&SPR 2008, LNCS 5342, p. 4, 2008.

Graph Edit Distance without Correspondence from Continuous-Time Quantum Walks

David Emms, Richard C. Wilson, and Edwin R. Hancock

Department of Computer Science, University of York, York, YO10 5DD, UK

Abstract. We consider how continuous-time quantum walks can be used for graph matching. We focus in detail on both exact and inexact graph matching, and consider in depth the problem of measuring graph similarity. We commence by constructing an auxiliary graph, in which the two graph to be matched are co-joined by a layer of indicator nodes (one for each potential correspondence between a pair of nodes). We simulate a continuous time quantum walk in parallel on the two graphs. The layer of connecting indicator nodes in the auxiliary graph allow quantum interference to take place between the two walks. The interference amplitudes on the indicator nodes are determined by differences in the two walks. We show how these interference amplitudes can be used to compute graph edit distances without explicitly determining node correspondences.

1 Introduction

The problem of determining whether a pair of graphs are isomorphic has been extensively studied in the literature. Polynomial time algorithms have been given for graphs with a number of special properties [14,1,11]. Currently the best algorithm for isomorphism testing of general graphs is 'Nauty' developed by McKay [5]. Nauty constructs a canonical labelling of the vertices in a graph and finds generators of the graph's automorphism group. For most graphs it is extremely fast but its complexity is exponential for some classes of graphs [15]. Another important algorithm that carries out subgraph isomorphism and clique detection, as well as isomorphism testing, was developed by Ullmann [17]. The algorithm uses back-tracking to successively eliminate nodes from the search space. When we wish to consider practical problems, however, we typically need to be able to deal with structural errors. For this reason, inexact graph matching must be carried out.

Quantum computing has recently attracted attention because of the potential for considerable speed-ups over classical algorithms. For instance, Grover's search algorithm [10] is quadratically faster and Shor's factorisation algorithm [16] is exponentially faster than known classical algorithms. However, quantum algorithms also have a richer structure than their classical counterparts since they use qubits rather than bits as the basic representational unit. This structure is exploited in Shor's algorithm where the Fourier transform is used to locate factors of a composite number. However, although quantum computers potentially offer computational powerful tools for solving problems of a combinatorial nature, developing algorithms that take advantage of this power has proved elusive. This is particularly the case for problems involving graphs, and in particular the subgraph isomorphism problem.

N. da Vitora Lobo et al. (Eds.): SSPR&SPR 2008, LNCS 5342, pp. 5–14, 2008.

One route is potentially furnished by Jozsa's algorithm which performs period finding and can be generalised to solving the hidden subgroup problem [12]. The hidden subgroup problem is of interest since there is a famous reduction of the graph isomorphism problem to it, see for example [12]. Beals was able to show how to efficiently carry out the quantum Fourier transform on the symmetric group [3] but, despite much attention, a quantum algorithm that uses this to efficiently solve the graph isomorphism problem has not been found. In fact, it seems unlikely that the graph isomorphism problem will succumb to a direct attack using the quantum Fourier transform on the symmetric group. In a recent paper, Kempe and Shalev [9] examine which subgroups of the symmetric group can be distinguished from the identity subgroup using the methods of quantum Fourier sampling. Ettinger and Høyer [8] propose a quantum observable for the graph isomorphism problem. However, the computational overheads are prohibitive.

In this paper we take a different approach, based on quantum walks. These have been introduced as quantum counterparts of random walks and a good summary is given by Kempe [13]. The behaviour of quantum walks is governed by unitary rather than stochastic matrices. Quantum walks provide an approach to designing quantum algorithms that lends itself more to physical intuition. By making use of the exponentially faster hitting times that are observed for continuous-time quantum walks on graphs [13], Childs et al showed that it is possible to create an oracle based algorithm using a continuous-time quantum walk that is provably exponentially faster than any possible classical algorithm [4].

Our contribution in this paper is to develop a means of computing graph edit distance based on the continuous-time quantum walk. As we will show, it is also possible to construct the continuous-time quantum walk so that the destructive interference takes place continuously rather then requiring a final interference step. As the interference takes place continuously, the interference amplitudes that we observe are reliant more directly on the differences between the two walks at all times during the walk's evolution. Finally, the quantum amplitudes are complex-valued rather then real-valued thus giving a two dimensional distribution of interference amplitudes, and therefore more information with which to calculate the assignment probabilities. We show how to compute a measure of graph similarity using the interference amplitudes and without the need to determine explicit node correspondences. This represents a development of our recent work where we have a) used continuous time quantum walks to develop non-classical extension of the commute time [7] and b) used quantum interference to compute explicit correspondences [6].

2 The Continuous-Time Quantum Walk

The continuous-time quantum walk is a natural quantum analogue of the classical random walk. Like the classical random walk on a graph, its state space is the set of vertices of the graph. However, as with the discrete-time quantum walk, the probability of being at a certain state is given by the square of the amplitude of that state, rather then just the amplitude of the state (as is the case classically). As with the discrete-time quantum walk, this allows destructive as well as constructive interference to take place.

The state space for the continuous-time quantum walk on a graph, $G = (\mathcal{V}, \mathcal{E})$, is the set of vertices, $\mathcal{V}$, as is the case for the classical random walk. In addition, transitions only occur between adjacent vertices. If the walk is at a vertex u, it moves to adjacent vertices at a rate proportional to $\frac{1}{d(u)}$. The basis states for the continuous-time quantum walk are vectors corresponding to particular vertices, as is the case for the classical random walk, and unlike the discrete-time quantum walk where basis states correspond to arcs. The basis state corresponding to the walk being at $u \in \mathcal{V}$ is written, in Dirac notation, as $|u\rangle$. A general state of the walk is a complex-linear combination of these basis states and so the state of the walk at time t is given by a vector, $|\psi_t\rangle \in \mathbb{C}^{|\mathcal{V}|}$, which we write componentwise as $|\psi_t\rangle = \sum_{u \in \mathcal{V}} \alpha_u(t)|u\rangle$.. Thus, the amplitudes $\alpha_u(t) \in \mathbb{C}$.

The probability of the walk being in a particular state is given by the square of the amplitude of that state. Let X_t be the random variable giving the location of the walk at time t. The probability of being at $u \in V$ at time t is given by $\Pr(X^t = u) = \alpha_u \alpha_u^*$, where α_u^* is the complex conjugate of α_u. As the total probability must sum to unity, we have that $|\alpha_u(t)| \in [0, 1]$ for all $u \in V$, $t \in \mathbb{R}^+$, and $\sum_{u \in V} \alpha_u(t)\alpha_u^*(t) = 1$ for all $t \in \mathbb{R}^+$. The evolution of the state vector is given by $\frac{d}{dt}|\psi_t\rangle = -iL|\psi_t\rangle$.. Since the evolution of the probability vector of the walk at time t depends on the state vector of the walk (not merely the probability vector), we note that unlike the classical walk, the quantum walk is not a Markov chain. Given an initial state for the walk, $|\psi_0\rangle$, Equation 2 can be solved to give $|\psi_t\rangle = e^{-iLt}|\psi_0\rangle$. Thus, given the initial state, we can calculate the state of the walk at an time, t.

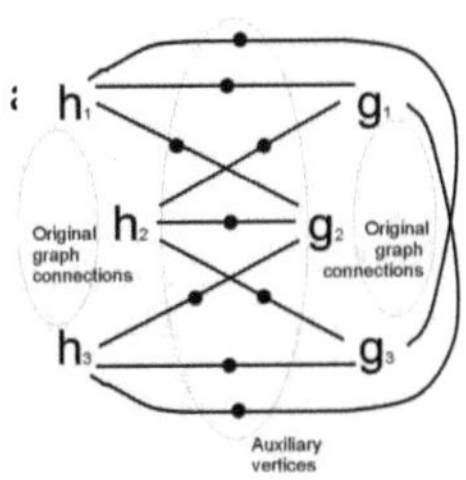

Fig. 1. The auxiliary graph, $\Gamma(G, H)$, showing the vertices $g_1, g_2, g_3 \in V_G$ and $h_1, h_2, h_3 \in V_H$ connected by way of auxiliary vertices

3 The Auxiliary Graph

Given two graphs, G and H, an auxiliary graph, $\Gamma(G, H)$, is constructed on which the walks can interfere. The auxiliary graph consists of the original two graphs, augemented by a set of auxiliary vertices that are used as correspondence indicators. Each auxiliary vertex is connected by an edge to one node in each of the original two graphs. That is, each pair of vertices, $g \in \mathcal{V}_G$ and $h \in \mathcal{V}_H$, one from each graph, are both adjacent to an auxiliary vertex labelled $v_{\{g,h\}}$. The auxiliary graph is symmetric with respect to interchanging its two arguments and with respect to permutations of the vertices in either graph. A schematic representation of its structure is shown in Figure 1.

Let $G = (\mathcal{V}_G, \mathcal{E}_G)$ and $H = (\mathcal{V}_H, \mathcal{E}_H)$, then $\Gamma(G, H) = (\mathcal{V}_\Gamma, \mathcal{E}_\Gamma)$ and the vertex and edge sets can be decomposed such that $\mathcal{V}_\Gamma = \mathcal{V}_G \cup \mathcal{V}_H \cup \mathcal{V}_A$ and $\mathcal{E}_\Gamma = \mathcal{E}_G \cup \mathcal{E}_H \cup \mathcal{E}_A$ where

$$\mathcal{V}_A = \{v_{\{g_i,h_j\}} | g_i \in \mathcal{V}_G,\ h_j \in \mathcal{V}_H\} \text{ and,}$$
$$\mathcal{E}_A = \{\{g_i, v_{\{g_i,h_j\}}\}, \{h_j, v_{\{g_i,h_j\}}\} | g_i \in \mathcal{V}_G,\ h_j \in \mathcal{V}_H\}.$$

The auxiliary vertices, $\mathcal{V}_A$, serve to connect the two graphs and act as sites on which the interference takes place. The auxiliary graph is similar to the association graph [2].

However, information about the structure of the two graphs comes from incorporating the original graphs themselves rather then through the connections between the auxiliary vertices, as is the case for the association graph. We delay our explanation of how the auxiliary graph allows graph matching to be carried out until after we have provided the mathematical details of the discrete-time quantum walk on the auxiliary graph, which we do in the next section.

4 The Continuous-Time Quantum Walk on the Auxiliary Graph

The auxiliary graph, $\Gamma(G,H)$, based on the two graphs $G = (\mathcal{V}_G, \mathcal{E}_G)$ and $H = (\mathcal{V}_H, \mathcal{E}_H)$ was described in the previous section. In this section we describe the simulation of the continuous-time quantum walk on the auxiliary graph. The state of the quantum walk on Γ at time t is given by $|\psi_t\rangle = \sum_{u \in \mathcal{V}_\Gamma} \alpha_u(t)|u\rangle$.. The starting state has amplitudes

$$\alpha_u(0) = \begin{cases} \frac{d(u)}{C} & \text{if } u \in \mathcal{V}_G; \\ -\frac{d(u)}{C} & \text{if } u \in \mathcal{V}_H; \\ 0 & \text{otherwise;} \end{cases}$$

where $C = \sum_{u \in \mathcal{V}_H \cup \mathcal{V}_H} d(u)^2$ is the normalisation constant such that the probabilities sum to unity. This starting state is used rather than the more natural starting state

$$\tilde{\alpha}_u(0) = \begin{cases} \frac{1}{\tilde{C}} & \text{if } u \in \mathcal{V}_G; \\ -\frac{1}{\tilde{C}} & \text{if } u \in \mathcal{V}_H; \\ 0 & \text{otherwise;} \end{cases}$$

where $\tilde{C} = \sqrt{|\mathcal{V}_G||\mathcal{V}_H|}$, since $|\tilde{\psi}_0\rangle = \sum_{u \in V_\Gamma} \tilde{\alpha}_u(0)|u\rangle$ is an eigenvector of the Laplacian for Γ and thus a stationary state of the walk. If we were to use this starting state then all the interference amplitudes would remain zero at all times and no information concerning the two walks would be obtained.

4.1 Behaviour of the Walk

We how consider the behaviour of the walk on the auxiliary graph, taking separately the cases when the graphs being compared are isomorphic and non-isomorphic.

Isomorphic Graphs. Let G and H be two graphs such that $\zeta : \mathcal{V}_g \rightarrow \mathcal{V}_H$ is an isomorphism between them.

Theorem 1. *If* $\zeta : \mathcal{V}_g \rightarrow \mathcal{V}_H$ *is a graph isomorphism such that* $h = \zeta(g)$ *then* $\alpha_g(t) = -\alpha_h(t)$.

Proof. Let L be the Laplacian matrix for $\Gamma = \Gamma(G,H)$ and $\tilde{L}$ be the Laplacian for $\tilde{\Gamma} = \Gamma(H,G)$. The basis states for the walk on Γ are $|u\rangle$ such that $u \in \mathcal{V}_\Gamma$ and those on $\tilde{\Gamma}$ are $|\tilde{u}\rangle$ such that $u \in \mathcal{V}_{\tilde{\Gamma}}$.

Let $\mathcal{P}$ be the permutation

$$\mathcal{P}|u\rangle = \begin{cases} |\tilde{\zeta}(u)\rangle & \text{if } u \in \mathcal{V}_G; \\ |\tilde{\zeta}^{-1}(u)\rangle & \text{if } u \in \mathcal{V}_H; \\ |\tilde{v}_{\{\zeta^{-1}(h),\zeta(g)\}}\rangle & \text{if } u \in \mathcal{V}_\Gamma \\ & \text{and } u \sim g \in \mathcal{V}_G,\ u \sim h \in \mathcal{V}_H. \end{cases}$$

The permutation $\mathcal{P}$ is such that $\tilde{L} = \mathcal{P}L\mathcal{P}^T$. The starting state for the walk on Γ is $|\psi_0\rangle = \frac{1}{C}\left(\sum_{u\in\mathcal{V}_G}|u\rangle - \sum_{v\in\mathcal{V}_H}|v\rangle\right)$ and the starting state for the walk on $\tilde{\Gamma}$ is $|\tilde{\psi_0}\rangle = \frac{1}{C}\left(\sum_{u\in\mathcal{V}_H}|u\rangle - \sum_{v\in\mathcal{V}_G}|v\rangle\right)$. We note that

$$\mathcal{P}^{-1}|\tilde{\psi_0}\rangle = \frac{1}{C}\mathcal{P}^{-1}\Big(\sum_{u\in\mathcal{V}_H}|u\rangle - \sum_{v\in\mathcal{V}_G}|v\rangle\Big) = \frac{1}{C}\Big(\sum_{u\in\mathcal{V}_G}|u\rangle - \sum_{v\in\mathcal{V}_H}|v\rangle\Big) = -|\psi_0\rangle \quad (1)$$

Consider the amplitude $\alpha_g(t)$:

$$\alpha_g(t) = \langle\tilde{g}|e^{-i\tilde{L}t}|\tilde{\psi_0}\rangle = \langle\tilde{g}|e^{-i\mathcal{P}L\mathcal{P}^{-1}t}|\tilde{\psi_0}\rangle = \langle\tilde{g}|\mathcal{P}e^{-iLt}\mathcal{P}^{-1}|\tilde{\psi_0}\rangle = -\langle h|e^{-iLt}|\psi_0\rangle = -\alpha_h(t) \quad (2)$$

□

Theorem 2. *If $\zeta : \mathcal{V}_g \rightarrow \mathcal{V}_H$ is a graph isomorphism such that $h = \zeta(g)$ then $\alpha_{v_{\{g,h\}}}(t) = 0$ for all $t \in \mathbb{R}^+$.*

Proof. Let $x(t) = \alpha_{v_{\{g,h\}}}(t)$. Equation 2 gives $d/dt|\psi_t\rangle = -i\mu L|\psi_t\rangle$. Thus, $d/dt x(t) = -i(\alpha_h(t) + \alpha_g(t) - 2x(t)) = 2i\mu x(t)$, by Theorem 1. The general solution is $x(t) = A_1 e^{2i\mu t}$. Since $x(0) = 0$ we have that $A_1 = 0$, and thus $\alpha_{v_{\{g,h\}}}(t) = x(t) = 0$ for all $t \in \mathbb{R}^+$.

□

Non-isomorphic Graphs. We now consider the case when G and H are non-isomorphic, but closely related graphs. In this case, in general, $\alpha_{v_{\{g,h\}}}(t) \neq 0$ even if $< g, h >$ is a true assignment. However, the algorithm relies upon $\alpha_{v_{\{g,h\}}}(t)$ being smaller in magnitude for true assignments than it is for false assignments. We investigate whether this assumption allows successful matching to be carried out in Section 6. In the next section we consider at what point in time the interference amplitudes for the walk should be used so that these amplitudes are most easily distinguished.

4.2 Measuring the Interference Amplitudes

The algorithm that we propose uses a set of interference amplitudes, corresponding to the set of possible vertex-vertex assignments, in order to carry out graph matching. For exact graph matching the point during the evolution of the walk at which we use the interference amplitudes, $\alpha_{v_{\{g,h\}}}(\tau)$, $g \in \mathcal{V}_G$, $h \in \mathcal{V}_H$, is not important. We know that $\alpha_{v_{\{g,h\}}}(t) = 0$ for all $t \in \mathbb{R}^+$ if $< g, h >$ is a true assignment, and it is highly improbable that if $< g, h >$ is a false assignment then $\alpha_{v_{\{g,h\}}}(\tau)$ will be exactly at some particular τ but not at other times. Thus, we base our choice of τ on how well it will allow inexact graph matching to be carried out.

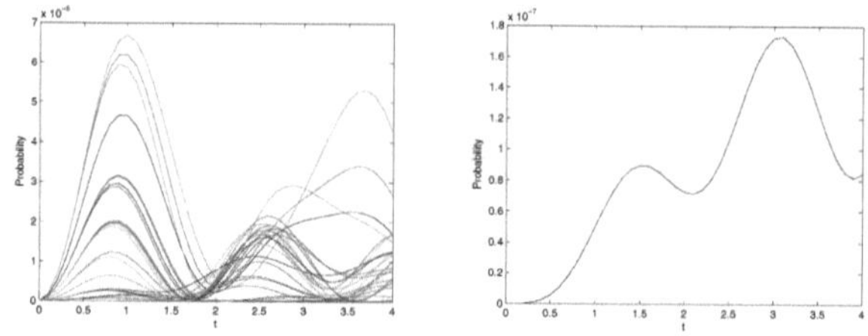

Fig. 2. The probabilities associated with the interference states for a particular vertex (left) and for just the interference state for the true assignments (right). Note the difference in scale on the y-axis.

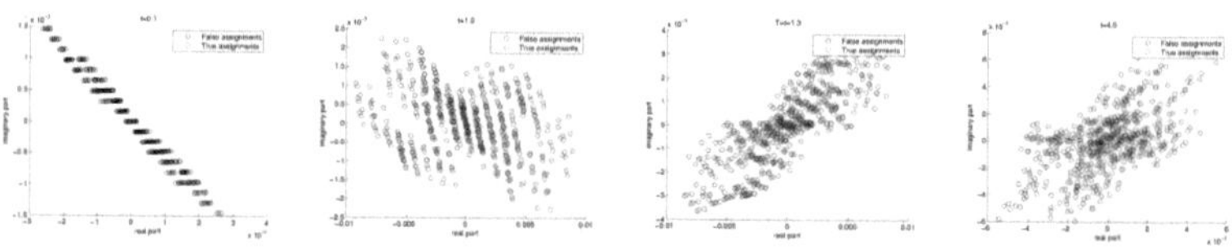

Fig. 3. The distributions of the interference amplitudes for a pair of non-isomorphic graphs, G and H, at $t = 0.1$, $t = 1.0$, $\tau = t = 1.3$ and $t = 4.0$. The probabilities for the interference states associated with a particular vertex $g \in \mathcal{V}$ are shown in Figure 2.

Let G and H be a pair of non-isomorphic, related graphs. The probabilities for the set of $|\mathcal{V}_H|$ interference states associated with a particular vertex $g \in \mathcal{V}_G$ are shown in Figure 2. The distribution of the complete set of interference amplitudes at various times during the evolution of the walk are shown in Figure 3. We note that the probabilities for the false assignments have similar periods. In comparison, the probability for the true assignment continues to increase with time as the differences in the states of the two walks increase. Thus we select τ such that the variance of the real part of the distribution of all interference amplitudes is at its first maximum. This corresponds to $t = 1.3$ in Figures 2 and 3.

We chose this value of τ since at this time the interference amplitudes for the false matches are all close to their local maxima. Thus the separation between the true and false interference amplitudes should be at its greatest. Since the periods for the probabilities for the false assignments are not exactly the same, the subsequent maxima are not as close in terms of the times at which they are obtained. Additionally, as t increases the probabilities for the true assignments continue to increase, thus they are most easily distinguished at the first maximum. We will henceforth refer to $\alpha_{v_{\{g,h\}}}(T)$ as *the* interference amplitude for $g \in \mathcal{V}_G$ and $h \in \mathcal{V}_h$ and will denote it by $\alpha^-(g,h)$, or simply α if it is unimportant to which pair of vertices it refers.

5 A Distance Measure from the Interference Amplitudes

Let $\Omega = \{|\alpha^-(g_i,h_j)|^2 : g_i \in \mathcal{V}_G, h_j \in \mathcal{V}_H\}$ be the squared magnitudes of the set of interference amplitudes for the two graphs. Further, let $\Omega_{|\mathcal{V}_G|}$ (where G is the graph with the fewest amplitudes if $|\mathcal{V}_G| \neq |\mathcal{V}_H|$) be the smallest $|\mathcal{V}_G|$ elements of Ω. We define $d(G,H)$ to be the variance of $\Omega_{|\mathcal{V}_G|}$ divided by the variance of Ω:

$$d(G,H) = \frac{\text{var}(\Omega_{|\mathcal{V}_G|})}{\text{var}(\Omega)}.$$

Theorem 3. *If $G \cong H$ then $d(G,H) = 0$*

Proof. If $G \cong H$ then there exists an isomorphism, $\zeta : \mathcal{V}_G \to \mathcal{V}_H$. Thus, for each $g \in \mathcal{V}_G$ we have, by Theorem 2, $\alpha^-(g,\zeta(g)) = 0$, for all $t \in \mathbb{R}^+$. Therefore there are at least $|\mathcal{V}_G|$ assignment amplitudes equal to zero, hence $\text{var}(\Omega_{|\mathcal{V}_G|}) = 0$ and so $d(G,H) = 0$. □

Note that we only need calculate the set of interference amplitudes in order to calculate $d(G,H)$, we need not compute the correspondence match between the graphs. As we will show in the experimental section, the variance of the smallest interference amplitudes are closely related to the edit distance between graphs.

6 Experiments

We now consider the proposed graph distance measure, firstly on synthetic graphs and secondly on graphs obtained from real-world data. Figure 4 shows plots of the distance measure as functions of the amount of noise. The left-hand panel in Figure 4 shows the distance measure averaged over the graphs with different numbers of vertices. It can be seen that the relationship between the distance measure and the amount of noise is close to linear. The right-hand panel in Figure 4 shows the distance measure as a function of noise for graphs with different numbers of vertices. The distance d increases steadily as the level of noise is increased for all the graphs.

We now explore the use of the distance measure for graphs derived from real world image data. For this we take images from the COIL database and construct Delaunay triangulations. We consider 12 different objects from the database with view angles between 0° and 45°. Firstly, we consider the distance measure's ability to form trajectories of the graphs, ordering them according to the view angle. Secondly, we analyse the effectiveness of the distance measure for graph clustering. Examples of views of each of the images used together with their Delaunay triangulations are shown in Figure

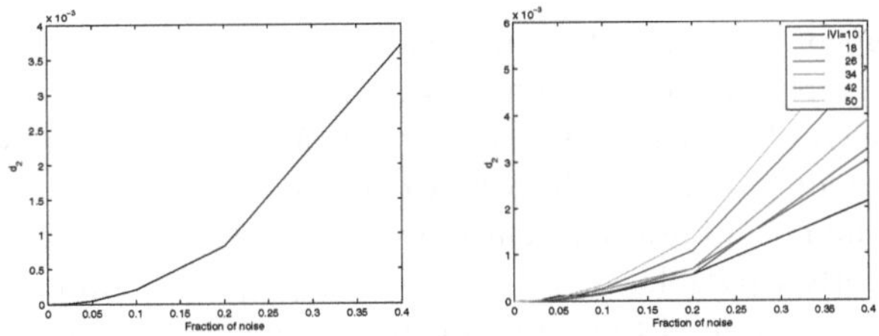

Fig. 4. [Left: The distance measure as a function of the amount of noise averaged over graphs with between 10 and 50 vertices. Right: As a function of the amount of noise for graphs with between 10 and 50 vertices.

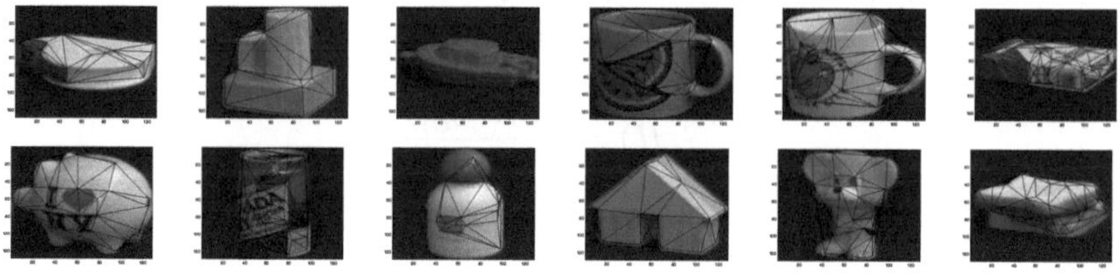

Fig. 5. Example COIL Images

5. Note that the images are represented by un-attributed, un-weighted graphs—all the information is represented in the connectivity structure of the Delaunay triangulations.

Figure 6 shows the average of our graph distance as a function of the difference in angle between the views from which the graphs were obtained. The distance between a graph and itself (that is, when the angular distance is zero) is guaranteed to be zero by Theorem 3. The difference in the distance measure between a pair of graphs with zero angular distance (between the images) and 5° is significantly larger than any other change in distance for a 5° change in angular distance. This shows that the distance measure distinguishes most strongly between pairs of isomorphic graphs and pairs of graphs that are related but non-isomorphic than it does between closely related and distantly related non-isomorphic graphs. Such behaviour reflects the understanding that there is a clear distinction between isomorphic graphs and non-isomorphic graphs but a less clear distinction (and hence a smaller change in graph distance) between 'closely' and 'less closely' related graphs.

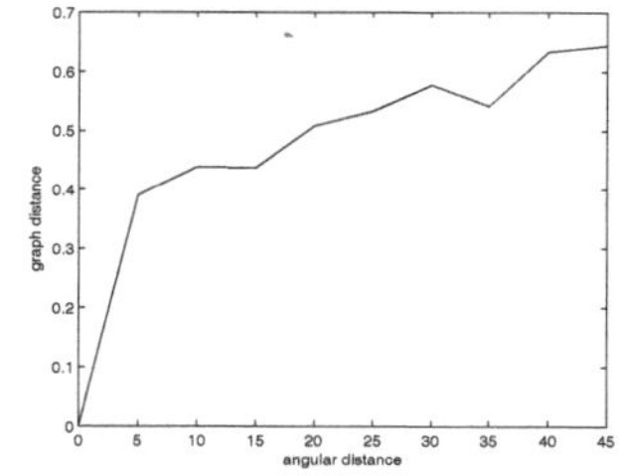

Fig. 6. Distance Measure versus Angular Distance for COIL Graphs

We now turn our attention to the ability of the distance measure to distinguish between different sets of graphs and so allow graph clustering to be carried out. We look at the ability of the distance measure to cluster two distinct sets of graphs. That is, we take two sets of graphs at a time, carry out dimensionality reduction on the matrix of distances between them and embed them in 2D space. We then look at whether the two clusters are linearly separable, and what proportion of the graphs in each set have another graph from the same set as their nearest neighbour.

We take two sets of graphs, each corresponding to views of a different object, each set containing 10 graphs. We form the matrix of distances between all these graphs to give a 20 by 20 matrix of distances. We then perform MDS on this matrix of distances to produce a 2D embedding of the graphs, where one data point corresponds to one graph. Table 1 shows the number of times the clusters of graphs for each of the 132 pairings of the objects are linearly separable when the two sets of graphs are embedded using MDS. It also shows the proportion of times that the nearest neighbour for a particular graph in the cluster is a graph derived from the same image as it was. Looking at the second column, we see that in over half of the cases the two clusters are linearly separable. The third column shows the total number of graphs in each set for which its nearest

Table 1. Clustering results of graphs derived from images from COIL. The number of clusters that are linearly separable and the mean number of graphs whose nearest neighbour is from the same set, averaged over all of the clusters.

Object	Number of clusters linearly separable (out of 11)	Nearest neighbour from same set (out of 10)
1	6	9.1
2	7	8.5
3	4	8.4
4	4	7.9
5	7	9.2
6	8	9.5
7	7	8.9
8	10	9.7
9	3	8.2
10	6	8.7
11	6	8.8
12	6	8.7
Mean	6.2	8.8

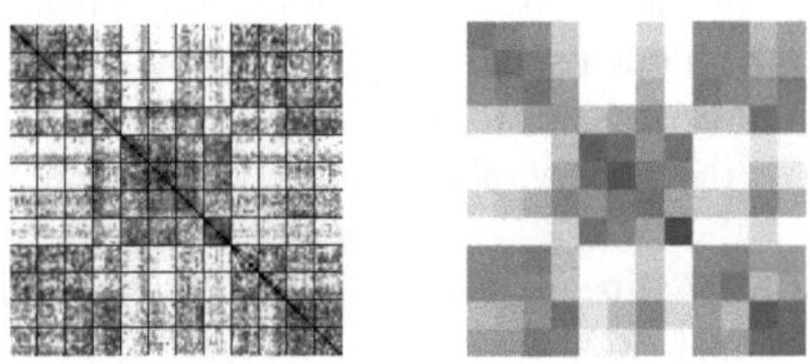

Fig. 7. The matrix of distances between individual graphs (left) and the average distances between sets of graphs (right). Completely black indicates a distance of 0, the lighter the grey the greater the distance.

neighbour is from the same set, averaged over the 11 clusterings carried out for each object. The mean for all the different sets is 8.8 out of a possible 10 indicating that the distance measure distinguishes successfully between graphs from different sets.

Figure 7 shows the matrix of distances between pairs of graphs and the mean distances between the sets of graphs. The smallest distances between graphs (left-hand plot, those closest to black in the figure) lie closest to the diagonal, these are the pairs of graphs representing images of the same objects and with a small angular distance between them.

7 Conclusions

We have suggested a distance measure based on the interference of the continuous-time quantum walk . The measure is computed by taking the variance of the smallest n interference amplitudes (where n is the number of vertices) and normalising by dividing by the variance of the whole distribution. Our experiments showed that this measure performed well.

References

1. Babai, L., Grigoryev, D.Y., Mount, D.M.: Isomorphism of graphs with bounded eigenvalue multiplicity. In: STOC 1982: Proceedings of the fourteenth annual ACM symposium on Theory of computing, pp. 310–324. ACM Press, New York (1982)
2. Barrow, H.G., Burstall, R.M.: Subgraph isomorphism, matching relational structures and maximal cliques. Information Processing Letters 4(4), 83–84 (1976)
3. Beals, R.: Quantum computation of Fourier transforms over symmetric groups. In: Proceedings of the 29th Annual ACM Symposium on the Theory of Computation (STOC), Texas, pp. 48–53. ACM Press, New York (1997)
4. Childs, A.M., Cleve, R., Deotto, E., Farhi, E., Gutmann, S., Spielman, D.A.: Exponential algorithmic speedup by a quantum walk. In: STOC 2003: Proc. 35th ACM symposium on TOC, pp. 59–68. ACM Press, New York (2003)
5. McKay, B.D.: Practical graph isomorphism. Congressus Numerantium 30, 45–87 (1981)
6. Emms, D., Hancock, E.R., Wilson, R.C.: Graph similarity using interfering quantum walks. In: Kropatsch, W.G., Kampel, M., Hanbury, A. (eds.) CAIP 2007. LNCS, vol. 4673, pp. 823–831. Springer, Heidelberg (2007)
7. Emms, D., Wilson, R.C., Hancock, E.R.: Graph Embedding Using Quantum Commute Times. In: Escolano, F., Vento, M. (eds.) GbRPR. LNCS, vol. 4538, pp. 371–382. Springer, Heidelberg (2007)
8. Ettinger, M., Høyer, P.: A quantum observable for the graph isomorphism problem (1999)
9. Ettinger, M., Høyer, P.: The hidden subgroup problem and permutation group theory (2004)
10. Grover, L.: A fast quantum mechanical algorithm for database search. In: Proc. 28th Annual ACM Symposium on the Theory of Computation, pp. 212–219. ACM Press, New York (1996)
11. Hopcroft, J.E., Wong, J.K.: A linear time algorithm for isomorphism of planar graphs. In: Proceedings of the 6th Annual ACM Symposium on the Theory of Computing (STOC 1974), pp. 172–184 (1974)
12. Jozsa, R.: Quantum factoring, discrete logarithms, and the hidden subgroup problem. Computing in Science and Engineering 03(2), 34–43 (2001)
13. Kempe, J.: Quantum random walks—an introductory overview. Contemporary Physics 44(4), 307–327 (2003)
14. Köbler, J.: On graph isomorphism for restricted graph classes. In: Beckmann, A., Berger, U., Löwe, B., Tucker, J.V. (eds.) CiE 2006. LNCS, vol. 3988, pp. 241–256. Springer, Heidelberg (2006)
15. Miyazaki, T.: The complexity of Mckay's canonical labeling algorithm (1996)
16. Shor, P.W.: Polynomial-time algorithms for prime factorization and discrete logarithms on a quantum computer. SIAM J. Comput. 26, 1484–1509 (1997)
17. Ullmann, J.R.: An algorithm for subgraph isomorphism. J. ACM 23(1), 31–42 (1976)

Exact Median Graph Computation Via Graph Embedding

Miquel Ferrer[1], Ernest Valveny[1], Francesc Serratosa[2], and Horst Bunke[3]

[1] Centre de Visió per Computador, Dep. Ciències de la Computació, Universitat Autònoma de Barcelona, Edifici O, Campus UAB, 08193 Bellaterra, Spain
{mferrer,ernest}@cvc.uab.cat

[2] Departament d'Enginyeria Informàtica i Matemàtiques, Universitat Rovira i Virgili, Av. Països Catalans 26, 43007 Tarragona, Spain
francesc.serratosa@urv.cat

[3] Department of Computer Science and Applied Mathematics, University of Bern, Neubrückstrasse, 14 CH3012 Bern, Switzerland
bunke@iam.unibe.ch

Abstract. Given a set of graphs, the median graph is defined as the graph which has the smallest sum of distances (SOD) to all the graphs in the set. It has been proposed as a tool to obtain the representative of such a set. In spite of its potential applications, the existing algorithms are computationally complex and have a very limited applicability. In this paper we propose a new approach for the exact computation of the median graph based on graph embedding in vector spaces. Graphs are embedded into a vector space and the median is computed in the vector domain. After that, the median graph is recovered from this median vector. Experiments on a synthetic database show that our approach outperforms the previous existing exact algorithms both on computation time and number of SOD computations.

1 Introduction

Given a set of graphs, the median graph [1] is defined as a graph that has the minimum sum of distances (SOD) to all graphs in the set. It can be seen as the representative of the set and, therefore, it has a large number of potential applications including many classical algorithms for learning, clustering and classification. Furthermore, it can be potentially applied to any graph-based algorithm where a representative of a set of graphs is needed. However, the computation of the median graph is exponential both in the number of input graphs and their size [2]. A number of algorithms have been reported in the past to compute the median graph. Two exact approaches have been proposed in [2] and [3]. As the computational cost of these algorithms is very high, a set of approximate algorithms have also been presented in the past based on different approaches such as genetic search [1,2], greedy algorithms [4] and spectral graph theory [5].

In this paper we propose a novel technique for the median graph computation based on graph embedding into vector spaces. It is composed of three main steps:

N. da Vitora Lobo et al. (Eds.): SSPR&SPR 2008, LNCS 5342, pp. 15–24, 2008.

the embedding of graphs in a vector space; the median vector computation and the recovering of the median graph from this median vector. In all these steps we use the graph edit distance under the assumption it is computed using a particular cost function. The use of this cost function is motivated by the fact that it permits to relate the distance between two graphs with their *mcs* [6] and to dramatically reduce the search space of the median graph as shown in [3]. Differently of the approach presented in [3], where a suboptimal method for median computation in the vector space is applied and the general edit distance is used to obtain the back-transformation, our method yields to obtain exact solutions for the median graph.

We performed a set of experiments using synthetic data and we have compared our approach with all the existing methods for the exact median computation. Results show that our method outperforms the existing methods both in computation time and the number of SOD computations, while obtaining the same or equivalent medians in terms of their SOD. With these results at hand, the use of the median graph in real (although limited) applications becomes a realistic choice.

The rest of this paper is organized as follows. In the next section we define the basic concepts and we introduce the notation we will use later in the paper. Then, in Section 3 we introduce in detail the concept of the median graph. In Section 4 the proposed method for the median computation is described. Section 5 reports a number of experiments and present results achieved with our method. Finally, in Section 6 we draw some conclusions and we point out to possible future work.

2 Basic Definitions

2.1 Graph and Subgraph

Definition 1 (Graph). *Given L, a finite alphabet of labels for nodes and edges, a graph g is defined by the four-tuple $g = (V, E, \mu, \nu)$ where V is a finite set of nodes, $E \subseteq V \times V$ is the set of edges, μ is the node labeling function ($\mu : V \longrightarrow L$) and ν is the edge labeling function ($\nu : V \times V \longrightarrow L$). The number of nodes of a graph g is denoted by $|g|$.*

Notice that L is not constrained in any way. It can be defined as a vector space (i.e. $L = \mathbb{R}^n$) or simply as a set of discrete labels (i.e. $L = \{\Delta, \Sigma, \Psi, \cdots\}$).

Definition 2 (Subgraph). *Let $g_1 = (V_1, E_1, \mu_1, \nu_1)$, and $g_2 = (V_2, E_2, \mu_2, \nu_2)$ be two graphs. The graph g_1 is a subgraph of g_2, denoted by $g_1 \subseteq g_2$ if: $V_1 \subseteq V_2$, $E_1 = E_2 \cap (V_1 \times V_1)$, $\mu_1(u) = \mu_2(u)$ for all $u \in V_1$ and $\nu_1(e) = \nu_2(e)$ for all $e \in E_1$.*

From Definition 2 it follows that, given a graph $g = (V, E, \mu, \nu)$, a subset $V' \subseteq V$ of its vertices uniquely defines a subgraph, called the subgraph *induced* by V'. That is, an induced subgraph of g can be obtained by removing some of its nodes $(V - V')$ and all their adjacent edges.

2.2 Maximum Common Subgraph and Minimum Common Supergraph

Definition 3 (Maximum Common Subgraph (mcs)). *Let* $g_1 = (V_1, E_1, \mu_1, \nu_1)$ *and* $g_2 = (V_2, E_2, \mu_2, \nu_2)$ *be two graphs. A graph* g *is called a common subgraph (cs) of* g_1 *and* g_2 *if there exists a subgraph isomorphism from* g *to* g_1 *and from* g *to* g_2*. A common subgraph of* g_1 *and* g_2 *is called maximum common subgraph (mcs) if there exists no other common subgraph of* g_1 *and* g_2 *with more nodes than* g*.*

Definition 4 (Minimum Common Supergraph (MCS)). *Let* $g_1 = (V_1, E_1, \mu_1, \nu_1)$ *and* $g_2 = (V_2, E_2, \mu_2, \nu_2)$ *be two graphs. A graph* g *is called a common supergraph* (CS) *of* g_1 *and* g_2 *if there exists a subgraph isomorphism from* g_1 *to* g *and from* g_2 *to* g*. A common supergraph of* g_1 *and* g_2 *is called minimum common supergraph (MCS) if there exists no other common supergraph of* g_1 *and* g_2 *with less nodes than* g*.*

Definition 5 (Minimum Common Supergraph of a Set of Graphs). *Let* $S = \{g_1, g_2, ..., g_n\}$ *be a set of graphs. A graph* $g_M(S)$ *(also denoted by* $MCS(S)$*) is called a minimum common supergraph of* S *if* $\{g_1, g_2, \cdots, g_n\}$ *are subgraphs of* $g_M(S)$ *and there is no other common supergraph of* $\{g_1, g_2, \cdots, g_n\}$ *with less nodes than* $g_M(S)$*.*

2.3 Graph Edit Distance

The graph edit distance [7,8], has been shown as one of the most widely used methods to compute the dissimilarity between two graphs.

The basic idea behind the graph edit distance is to define the dissimilarity of two graphs as the minimum amount of distortion required to transform one graph into the other. To this end, a number of distortion or edit operations e, consisting of the insertion, deletion and substitution of both nodes and edges are defined. Given these edit operations, for every pair of graphs, g_1 and g_2, there exists a sequence of edit operations, or edit path $p(g_1, g_2) = (e_1, \ldots, e_k)$ (where each e_i denotes an edit operation) that transforms g_1 into g_2. In general, several edit paths exist between two given graphs. This set of edit paths is denoted by $\wp(g_1, g_2)$. To quantitatively evaluate which edit path is the best, edit costs are introduced through a cost function. The basic idea is to assign a penalty cost c to each edit operation according to the amount of distortion it introduces in the transformation. The edit distance d between two graphs g_1 and g_2, denoted by $d(g_1, g_2)$, is the minimum cost edit path that transforms one graph into the other.

Graph Edit Distance and the Maximum Common Subgraph. In this work, we will use the graph edit distance under a particular cost function [6], where deletions and insertions of nodes have always a cost of 1, deletions and insertions of edges have always a cost of 0, and node and edge substitutions take the values of 0 or ∞ depending on whether the substitution is identical or not, respectively. Under this cost function, the graph edit distance between two graphs $d(g_1, g_2)$ is related to their mcs as follows:

$$d(g_1, g_2) = |g_1| + |g_2| - 2\,|mcs(g_1, g_2)| \tag{1}$$

Although the definition of the cost function might seem quite simple, this result demonstrates the intuitive idea that the more two graphs have in common, the lower is their distance. Several other distances related to the mcs have been proposed, and summarized in [9]. In the rest of the paper we will assume, that the distance between two graphs is computed according to the Eq. (1).

3 Generalized Median Graph

Definition 6 (Generalized Median Graph). *Let U be the set of graphs that can be constructed using labels from L. Given $S = \{g_1, g_2, ..., g_n\} \subseteq U$, the* **generalized median graph** *$\bar{g}$ of S is defined as:*

$$\bar{g} = arg \min_{g \in U} \sum_{g_i \in S} d(g, g_i) \tag{2}$$

That is, $\bar{g}$ is a graph $g \in U$ that minimizes the sum of distances (SOD) to all the graphs in S. Notice that $\bar{g}$ is usually not a member of S, and in general more than one generalized median graph may exist for a given set S.

Two important aspects in the median graph computation are the distance computation and the search space. In the general case, the search space becomes exponential with respect to the sum of nodes of the graphs in S. Nevertheless, in [3], it is shown that under the particular cost function introduced in Section 2.3,

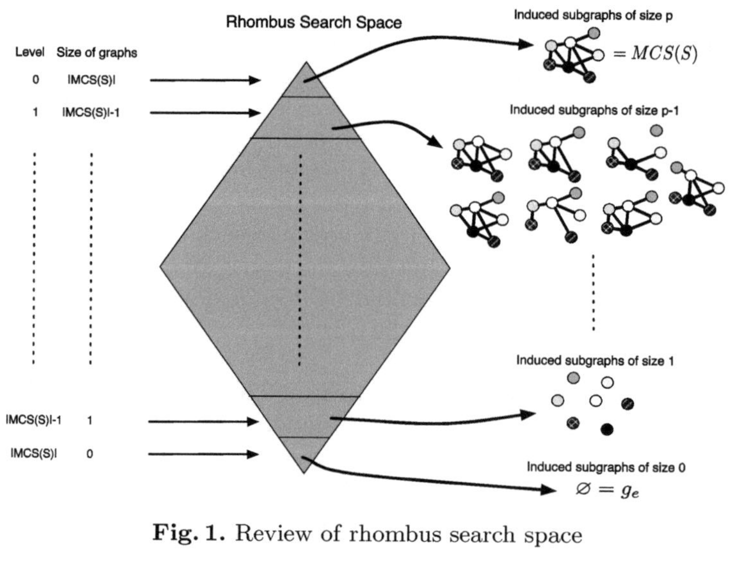

Fig. 1. Review of rhombus search space

the search space of the median graph can be drastically reduced and is composed only of the induced subgraphs of $g_M(S)$. Figure 1 shows this search space (also called Rhombus). At the top of the Rhombus there is the minimum common supergraph of S, $MCS(S)$. At the bottom, there is the empty graph $g_e = \phi$. Inside the rhombus there are all the possible induced subgraphs of $MCS(S)$. Each graph in S is located in this search space. In the following we will assume that we are working in this search space and therefore the term $MCS(S)$ have to be computed.

4 Exact Median Graph Via Graph Embedding

In this section we present a new embedding technique for the median graph computation (we will assume that a set of n graphs $S = \{g_1, g_2, \ldots, g_n\}$ for the median graph computation is given), that is composed of three main steps:

- **1. Graph Embedding in a Vector Space:** Each graph in S becomes a point in an n-dimensional space. The graph edit distance is used for this embedding.
- **2. Median Vector Computation:** The median vector is computed using the points of the first step.
- **3. Going back to the Graph Domain:** The median vector is converted back to a graph, which is taken as the median graph of S.

4.1 Graph Embedding in Vector Spaces

In order to perform the embedding step, we will use a novel procedure proposed in [10], briefly described in the following.

Assume we have a set of training graphs $T = \{g_1, g_2, \ldots, g_n\}$ and a graph similarity measure $d(g_i, g_j)$. Then, a set $P = \{p_1, \ldots, p_m\} \subseteq T$ of m prototypes is selected from T (with $m \leq n$). After that, the similarity between a given graph of $g \in T$ and every prototype $p \in P$ is computed. This leads to m dissimilarity values, $d_1, \ldots, d_m$ where $d_k = d(g, p_k)$. These dissimilarities can be arranged in a vector $(d_1, \ldots, d_m)$. In this way, we can transform any graph of the training set T into an m-dimensional vector using the prototype set P.

We perform the graph embedding step according to this definition, but we let the training set T and the prototype set P be the same, i.e, the set S for which the median graph is to be computed. So, we compute the distance between every pair of graphs in the set S. These distances are arranged in a distance matrix. Each row/column of the matrix can be seen as an n-dimensional vector representing one of the graphs in S. Figure 2 illustrates this procedure. In this example, it is assumed that the first row in the matrix corresponds to the distances of the black graph in the set to all other graphs in S.

At the end, each graph has a corresponding point (n-dimensional vector) in the vector space. What is important to remark here is the meaning of each

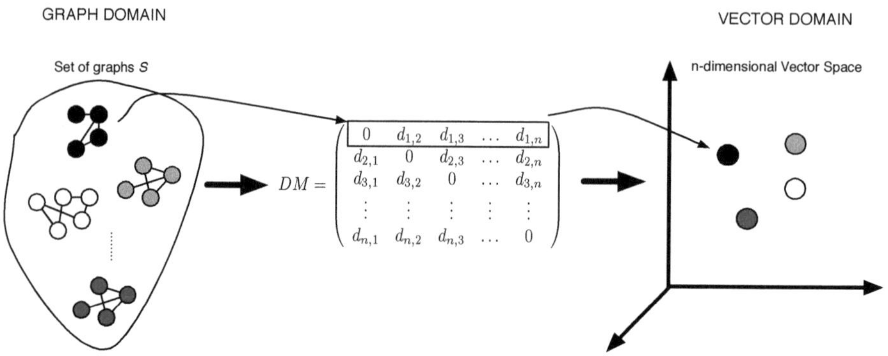

Fig. 2. Illustration of the first step (graph embedding)

position in this vector. If a vector $\boldsymbol{v}_i$ corresponds to the graph $g_i \in S$, then the coordinate j (with $j = 1 \ldots n$) of this vector is the distance from the graph g_i to the graph g_j, that is $d(g_i, g_j)$.

4.2 Median Vector Computation

The median vector is computed using the points obtained in the first step. Although we are in a Euclidean space, the nature of the cost function, in which only nodes insertions and deletions, each with cost 1 will occur [6], allows us to assume a L_1 metric can be used to compute the median vector. Using this metric, the median vector can be computed by means of the *Manhattan median*:

Definition 7 (Manhattan Median). *Given a set* $X = \{\boldsymbol{x}_1, \boldsymbol{x}_2, \ldots, \boldsymbol{x}_m\}$ *of* m *points with* $\boldsymbol{x}_i \in \mathbb{R}^n$ *for* $i = 1 \ldots m$, *the Manhattan median is defined as*

$$Manhattan\ median = arg\ \min_{y \in \mathbb{R}^n} \sum_{i=1}^{m} |\boldsymbol{x}_i - \boldsymbol{y}| \tag{3}$$

where $|\boldsymbol{x}_i - \boldsymbol{y}|$ *denotes the Manhattan distance between the points* $\boldsymbol{x}_i, \boldsymbol{y} \in \mathbb{R}^n$.

The *Manhattan median* [11] is simply the median of each coordinate of the vector, which makes the *Manhattan median* computation really simple.

4.3 Back to the Graph Domain

The median vector is used to go back to the graph domain and obtain the median graph. Every component of the median vector represents the distance between the median graph and one of the graphs in the set S (see Figure 3). In this example, the median graph would have a distance of 2 to the graph $g_1 \in S$, a distance of 1 to the graph $g_2 \in S$, and so on.

Knowing the distance from the median to one graph in S, we can draw an interval in the Rhombus search space around this graph according to the distance.

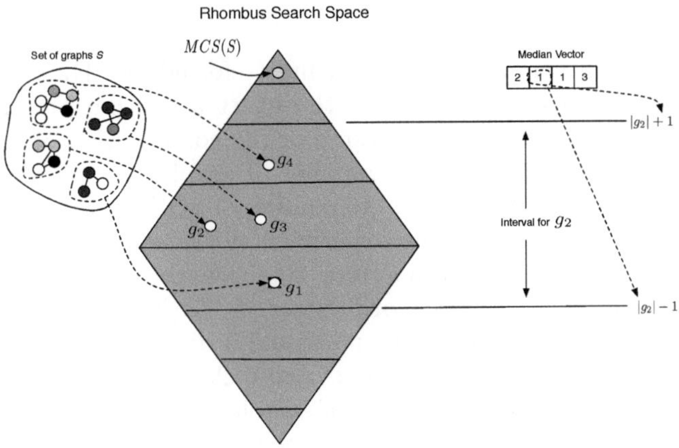

Fig. 3. Interpretation of the median vector

Since in the distance we are using only insertions and deletions of nodes, and these operations have a cost of 1, a distance k from the median to the graph g_i will imply that we would have to take into account graphs around g_i with size $|g_i| \pm k$, since a distance k will add (or remove) at most k nodes to g_i. Figure 3 shows a simple situation where the set S is composed of 4 graphs. For clarity, in this example only the interval for g_2 is shown.

Once the intervals are set, we choose only the smallest interval and its corresponding graph (this corresponds to the graphs 2 or 3 in Figure 3). The median is chosen as the graph with minimum SOD only in this interval. With this approach the search space is reduced to only this interval (grey part in Fig. 4).

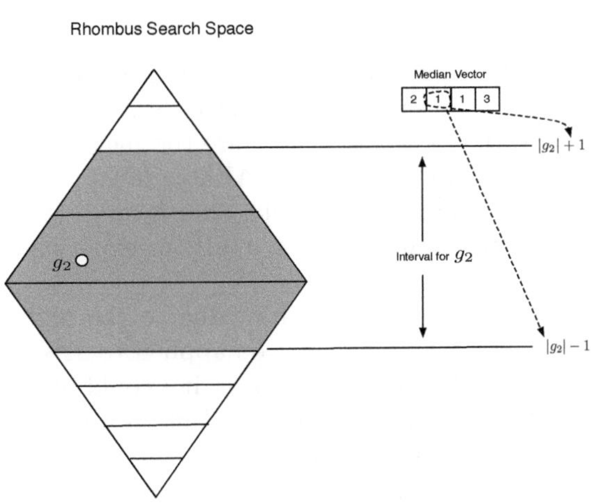

Fig. 4. Explored part of the search space (grey part)

5 Experimental Setup

In this section, we will compare our algorithm (referred as EE) with the multimatch algorithm (referred as MM) and the exact algorithm presented in [3] (referred as ER). The experiments consisted of the computation of several generalized median graphs using different number of graphs in each computation. For each median, the elapsed time and the number of SOD computations needed to compute it were recorded. In these experiments, the graph dataset was composed of graphs that represent capital letters [10]. From the original set composed by 15 letters, we only used a subset of 6 letters L, V, N, T, K and M. Then, from each original model, we manually generated 4 distorted instances. Therefore, our dataset is composed of 30 elements and 6 classes. Each class represents a letter and contains 5 different instances of each letter. Table 5 shows the original letters in the first row and the four distorted letters in the rest of the rows. Notice that, in the distorted letters, some lines have been erased or moved their position, but the number of terminal point has been kept unchanged. The letters are represented by graphs as follows. The straight lines are represented by edges and the terminal points of the lines by the nodes. Nodes are labelled by a two-dimensional attribute that represents the position (x,y) of the terminal point. Edges have no attributes.

Table 1. Original models (first row), some distorted models (other rows)

The results for both the computation time and the SOD computations required as a function of the total number of nodes in S are shown in Figures 5(a) and 5(b), respectively. Notice that the results are the mean values obtained for all sets S having the same number of nodes. Although the median vector computation yields the optimal median, the back-transformation is not proven to really give us a true median graph. However, in all our experiments we obtained optimal medians.

First of all it is important to remark that, due to the high computational requirements, the MM approach could only be applied to sets of graphs whose sum of nodes is 12, while the ER and EE approaches could be applied to all the sets. In addition, both figures show a substantial improvement of the EE and ER algorithms with respect to the MM approach. It is important to remark that the EE approach clearly outperforms the ER algorithm. The results show that while for the ER algorithm, the computation time for sets of graphs with 25 nodes is around 18 seconds, the EE algorithm only needs less than 3 seconds. Similar results can be observed for the number of SOD computations. For the

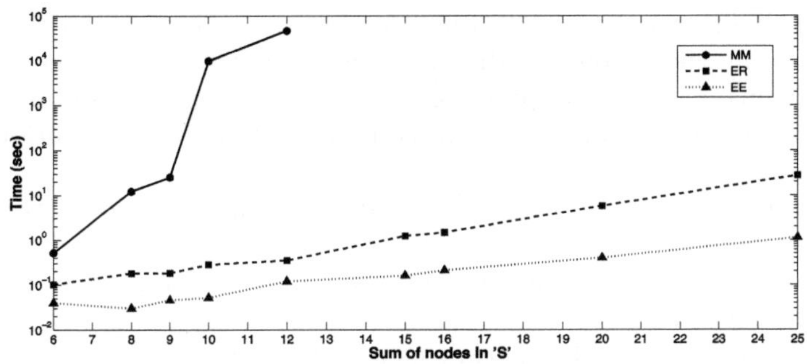

(a)

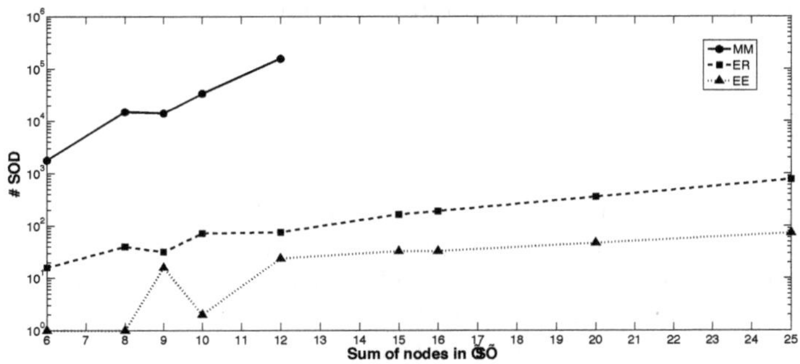

(b)

Fig. 5. Computation time (a) and number of SOD computations (b) as a function of the total number of nodes of the graphs in S

ER algorithm this number grows up to 500 SOD computations (for sets of graphs with 25 nodes), while for the EE algorithm it is only 75 in the same case.

As a final conclusion, we can say that with the embedding approach for the median graph computation we are able to obtain exact solutions for the median graph with relatively large sets of graphs with a low computation time. This suggests this method may extend the exact median graph computation to real (although limited) applications.

6 Conclusions

The median graph has been shown as a good choice to obtain a representative of a set of graphs, which has many potential applications in pattern recognition and related areas. Nevertheless, it suffers from a large complexity.

In this paper we have presented a new technique for the exact median graph computation based on graph embedding into vector spaces. This procedure is composed of three steps: the embedding of graphs into a real vector space, where each graph becomes a point in this space; the median vector computation using these points; the recovering of the median graph from the median vector. Although this technique can be used in many ways, we have particularized it to the exact median graph computation assuming the distance between graphs is carried out using the graph edit distance and a particular cost function. Under these conditions, the search space for the median graph computation can be drastically reduced and this technique is used to obtain exact solutions for the median graph.

Experiments made on a synthetic database show that our method clearly outperforms the previous existing methods for the exact median graph computation both in the computation time and the number of graph distances needed to obtain the median.

The proposed technique has permit us to extend the median graph computation to more realistic (although limited) sets of graphs. This situation opens the door to the application of the median graph to more complex problems, such as classification or clustering tasks.

References

1. Jiang, X., Münger, A., Bunke, H.: On median graphs: Properties, algorithms, and applications. IEEE Trans. Pattern Anal. Mach. Intell. 23(10), 1144–1151 (2001)
2. Münger, A.: Synthesis of prototype graphs from sample graphs. Diploma Thesis, University of Bern (in German) (1998)
3. Ferrer, M.: Theory and algorithms on the median graph. application to graph-based classification and clustering. PhD Thesis, Universitat Autònoma de Barcelona (2007)
4. Hlaoui, A., Wang, S.: Median graph computation for graph clustering. Soft Comput. 10(1), 47–53 (2006)
5. White, D., Wilson, R.C.: Mixing spectral representations of graphs. In: 18th International Conference on Pattern Recognition (ICPR 2006), Hong Kong, China, August 20-24, pp. 140–144. IEEE Computer Society, Los Alamitos (2006)
6. Bunke, H.: On a relation between graph edit distance and maximum common subgraph. Pattern Recognition Letters 18(8), 689–694 (1997)
7. Sanfeliu, A., Fu, K.: A distance measure between attributed relational graphs for pattern recognition. IEEE Transactions on Systems, Man and Cybernetics 13(3), 353–362 (1983)
8. Bunke, H., Allerman, G.: Inexact graph matching for structural pattern recognition. Pattern Recognition Letters 1(4), 245–253 (1983)
9. Schenker, A., Bunke, H., Last, M., Kandel, A.: Graph-Theoretic Techniques for Web Content Mining (Machine Perception and Artificial Intelligence) (Series in Machine Perception and Artificial Intelligence). World Scientific Publishing Co., River Edge (2005)
10. Riesen, K., Neuhaus, M., Bunke, H.: Graph embedding in vector spaces by means of prototype selection. In: Escolano, F., Vento, M. (eds.) GbRPR 2007. LNCS, vol. 4538, pp. 383–393. Springer, Heidelberg (2007)
11. Krause, E.F.: Taxicab Geometry: An Adventure in Non-Euclidean Geometry. Dover, New York (1986)

Similarity Invariant Delaunay Graph Matching

Dongjoe Shin and Tardi Tjahjadi

School of Engineering, University of Warwick, Coventry, UK, CV4 7AL
{Dongjoe.Shin,T.Tjahjadi}@warwick.ac.uk

Abstract. Delaunay tessellation describes a set of arbitrarily distributed points as unique triangular graphs which preserves most local point configuration called a clique regardless of noise addition and partial occlusion. In this paper, this structure is utilised in a matching method and proposed a clique-based Hausdorff Distance (HD) to address point pattern matching problems. Since the proposed distance exploits similarity invariant features extracted from a clique, it is invariant to rotation, translation and scaling. Furthermore, it inherits noise robustness from HD and has partial matching ability because matching performs on local entities. Experimental results show that the proposed method performs better than the existing variants of the general HD.

Keywords: Point pattern matching, Delaunay tessellation, Hausdorff distance, Similarity invariant distance.

1 Introduction

Point Pattern Matching (PPM) is a problem that searches the best point correspondences by investigating underlying point pattern and it includes many vision applications, e.g., motion estimation, image registration and object recognition [1, 2, 3, 4]. Although much research has been intensely done to address the problem, the PPM result is not reliable particularly when a point set includes noise or outliers, when whole or part of the data has been similarity transformed (e.g., rotated, translated and scaled) and when the number of matching points is different, i.e., some points should have either multiple matching opponents or none.

One classical approach to PPM is a spectral method that compares correlation of eigen vectors of a distance matrix referred to as a proximity matrix. The intra-distances between all possible pairs of points in an image are measured by a Gaussian-weighted distance metric and stored as a matrix in which eigen vectors are extracted as new features for matching [2]. Although the earlier methods show a weakness in dealing with partial matching, noise and significant image transformation, more recent methods address partial matching using a sub-matrix matching algorithm [5] and achieve robustness against noise by combining the spectral analysis with the Expectation Maximization (EM) framework [6]. Caelli et al. improve the accuracy of matching by re-normalising the eigen vectors and values used in comparison. The similarity of graph or tree of vertices is then measured by the distance of clusters in a tree or graph in the re-normalised subspace instead of the general point distance [7].

N. da Vitora Lobo et al. (Eds.): SSPR&SPR 2008, LNCS 5342, pp. 25–34, 2008.

Another approach is the Hausdorff Distance (HD) matching. The key advantage of using HD is that exact matching is not required when computing distance. Without a knowledge of exact point correspondences, HD measures distance between two sets of points. Moreover, it allows small perturbation and partial point pattern matching [8]. HD is a non-directional and nonlinear operation, and by making some changes in the distance function and combining two directional HDs, useful variants of HD can be generated, e.g., ranked HD and modified HD [9]. These methods have been extended to line and curve features in accurate recognition systems [10,11,13]. A recent algorithm which associates an affine invariant coordinate called a homogeneous barycentric coordinate with HD, has been introduced to address the affine transformed point pattern matching [14], but the barycentric coordinate relies on the convex shape of points and its mean position. Therefore, if the convex shape is deformed by an outlier, result is not reliable.

Since Euclidean distance of point features is not always sufficient for matching, some methods utilise contextual conditions derived from the point pattern.These methods structure points as a tree or a graph, and use the structure to solve the general PPM problems. In particular, weighted connection of nodes in a graph indicates the strength of connection and this concept is utilised by stochastic approaches, e.g., probabilitic relaxation methods [15]. Li et al. introduced a tree structure which is obtained by applying k-D tree algorithm to partition points, and the similarity of trees is measured for sparsely distributed point patterns [16]. Strickland et al. suggested a method that uses an iterative relaxation algorithm based on the probability defined by a separation of length and angle between connected branches in a non-rigid shape [1]. Zheng et al. considered point matching as an optimisation problem in order to preserve local neighbourhood structure, and the optimal solution is searched using relaxation labelling [3]. Andrew et al. focussed on the Delaunay graph of a set of data points and suggested a relaxation labelling solution for Delaunay graph matching which simplifies the joint probabilities defined by neighbouring nodes and reduces the computation time of the iteration [17]. The concept is extended to a transform estimation in an EM framework in their later work [18]. Despite these extensive researches, PPM still remains a challenging task: spectral methods are not robust to corruption in structure (e.g., due to noise) and the performance of relaxation methods degrade when there are significant increases in the sizes of the point sets since defining every joint probability is complicated [4].

In this paper a Delaunay graph is investigated as a method for uniquely constructing a local configuration called a clique, and propose a clique-based HD which exploits a graphical shape difference between cliques. In the proposed method, an idea of robust matching is motivated by HD and the affine invariant features from the cross ratio, and it effectively addresses noise, outliers, occlusion and inexact graph matching. The proposed distance measures inter-distance, i.e., it measures a clique distance directly from a model to a test data unlike a spectral method. Thus, when the sizes of the two graphs being compared are different, a rectangular matrix stores all the possible distances between cliques and strong correspondences are identified to define initial transform between two sets. Once the local transform is estimated initially, the guided matching increases point correspondences by collecting point pairs (called supporting pairs) that reside within an error bound of the estimated transform and repeating the estimation.

The paper is organised as follows. A Delaunay tessellation and its characteristics are explained in Sect. 2, which also introduces the theoretical representation of data points used in the proposed method. Section 3 explains how the distance between two cliques is measured and how it is related to HD. Finally, experimental results and conclusions are presented in Sect. 4 and 5, respectively.

2 Point Set Representation

A Delaunay graph is a result of Delaunay tessellation. As a dual of a Voronoi diagram which divides distinct n points according to the nearest neighbour rule, the Delaunay graph can be constructed from connecting points of Voronoi polygons that are adjacent to one another [see Fig.1(a) and (b)] and result thus forms a pattern of packed convex polygons [19, 20]. One characteristic of a Delaunay graph is that no point is allowed within a circumcircle of a triangle. In other words, noise only affects local graph where circumcircles are contaminated [see Fig.1(c)]. Furthermore, even when points are patly occluded, resulting graph is similar to that prat of the original graph (see Fig.1(d)).

Mathematically, the representation of a Delaunay graph is similar to that of an ordinary graph. Suppose that there are point sets of a model $\mathcal{V}^{\mathrm{m}} = \{\boldsymbol{v}_1, \cdots, \boldsymbol{v}_{|\mathcal{V}^{\mathrm{m}}|}\}$, where $|\cdot|$ is the cardinality of a point set and $\mathcal{V}^{\mathrm{t}}$ is a test

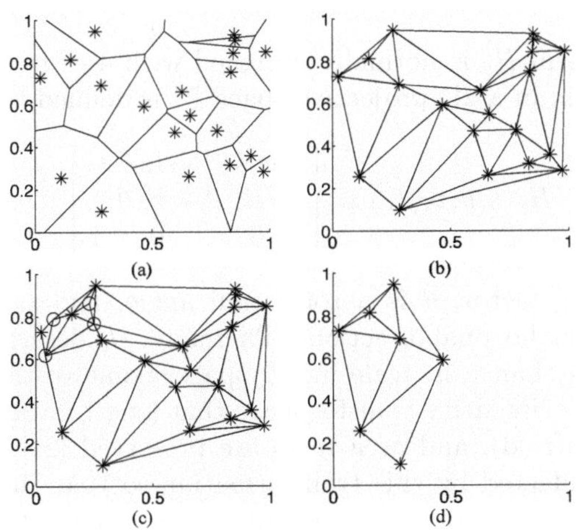

Fig. 1. (a) A Voronoi diagram of 20 feature points (denoted by '*') that are randomly generated and ranged [0 1]. (b) A Delaunay graph, the dual of the Voronoi diagram in (a). (c) The general shape of Delaunay graph in (b) is not changed by the addition of a noise point 'o'. (d) A randomly selected small portion of features does not change the local graph significantly.

set. A Delaunay graph is then defined by sets of points, edges and faces, i.e., a graph $\mathcal{G}^{\mathrm{m}} = (\mathcal{V}^{\mathrm{m}}, \mathcal{E}^{\mathrm{m}}, \mathcal{F}^{\mathrm{m}})$, where $\mathcal{E}^{\mathrm{m}}$ is the list of node connections (i.e., $\mathcal{E}^{\mathrm{m}} = \{(i,j) \mid \forall \boldsymbol{v}_i,\ \boldsymbol{v}_j \in \mathcal{V}^{\mathrm{m}}\}$), and $\mathcal{F}^{\mathrm{m}}$ is a set of triplets of indices, in which each triplet represents a triangle such as $\mathcal{F}^{\mathrm{m}} = \{(i,j,k) \mid \forall (i,j), (j,k) \text{ and } (j,k) \in \mathcal{E}^{\mathrm{m}}\}$. Local entity of a graph called a clique is then defined as a cluster of points connected by $\mathcal{E}^{\mathrm{m}}$. By using clique notation in [17], a model clique centred at a point $\boldsymbol{v}_i$ is given by

$$\mathcal{C}_i^{\mathrm{m}} = \{i\} \cup \{j | \forall (i,j) \in \mathcal{E}^{\mathrm{m}}\} \ . \tag{1}$$

A centre point $\boldsymbol{v}_i$ of a clique $\mathcal{C}_i^{\mathrm{m}}$ is especially referred to as the seed of a clique, and the indices of the other points are called neighbours, which are ordered in the clockwise direction.

One benefit of using this structured points is that it creates unique local configuration of points and the connections of nodes are not as complicated as a generic non-directional graphical model, e.g., Markov random fields. This is because the Delauany graph restricts the number of point connections as a triangle and the connections are only made within the nearest neighbours. However, a problem occurs when the sizes of the two cliques being compared are different, i.e., $|\mathcal{C}_i^{\mathrm{m}}| \neq |\mathcal{C}_j^{\mathrm{t}}|$. Thus, if a general point-pair based distance is used for matching, it is necessary to establish inexact matching of neighbour points in advance, e.g., some points should have multiple opponents, or a null point concept is required [3, 18]. However, the proposed matching framework enables distance of two data sets with different sizes to be measured.

3 Clique Distance

A similarity transform H_s is normally designed with 4 degrees of freedom, i.e., the transform matrix in a 2D projective space $\mathbb{P}^2$ is defined by

$$H_s(s, \theta, t_{\mathrm{x}}, t_{\mathrm{y}}) = \begin{bmatrix} s\cos\theta & s\sin\theta & t_{\mathrm{x}} \\ -s\sin\theta & s\cos\theta & t_{\mathrm{y}} \\ 0 & 0 & 1 \end{bmatrix} , \tag{2}$$

where s is a scaling factor, θ is a rotational angle, and t_{x} and t_{y} represent translations in two orthogonal directions. Therefore, H_s describes a rigid motion of an object with a change in scale in 2D space. Some of the useful invariant properties under the similarity transform are that length ratio, the combination of vectors (e.g., centroid), and area ratio are preserved [21]. In particular the angle ratio is not affected by this transformation so that the Delaunay graph of a transformed point pattern also remains unchanged. These properties are exploited in a new distance between cliques, which makes the new measure more reliable for a transformed point pattern.

The shape difference of a clique is measured by a graphical distance. Instead of direct use of edge length and angles, the proposed graphical distance compares a set of cross ratios of a clique. The cross ratio is the length ratio defined by four distinct points on a collinear line, e.g., the cross ratio of four points, $\boldsymbol{v}_1$, $\boldsymbol{v}_2$, $\boldsymbol{v}_3$, and $\boldsymbol{v}_4$ is

$$f_{\mathrm{cr}}(\boldsymbol{v}_1, \boldsymbol{v}_2, \boldsymbol{v}_3, \boldsymbol{v}_4) = \frac{||\boldsymbol{v}_1 - \boldsymbol{v}_3|| \cdot ||\boldsymbol{v}_2 - \boldsymbol{v}_4||}{||\boldsymbol{v}_1 - \boldsymbol{v}_4|| \cdot ||\boldsymbol{v}_2 - \boldsymbol{v}_3||} \ . \tag{3}$$

Because the cross ratio is invariant up to a projective transform, it can correctly measure the shape difference of two similarity transformed graphs. To retain four points for the cross ratio estimation, the proposed method exploits the midpoints of two sides of a triangle belonging to a clique, and the two ends of the boundary edge. Since these four points are not collinear, the two midpoints are projected onto the boundary line. Suppose a neighbour index of $\mathcal{C}_i^{\mathrm{m}}$ is denoted as $n_{i1}, \cdots, n_{i(|\mathcal{C}_i^{\mathrm{m}}|-1)}$. The boundary of a clique $\mathcal{C}_i^{\mathrm{m}}$ is then defined as

$$\mathcal{B}_i^{\mathrm{m}} = \{(n_{ij}, n_{i(j+1)}) | j = 1, \cdots, |\mathcal{C}_i^{\mathrm{m}}| - 2\} \ . \tag{4}$$

If the seed of a clique is enclosed by its neighbours, the boundary set has another element $\{(n_{i(|\mathcal{C}_i^{\mathrm{m}}|-2)}, n_{i1})\}$. This is because the face of an internal clique is normally defined by a circular permutation of two adjacent neighbours but a clique whose boundary is identical to the boundary of a graph is not, i.e., the maximum size of a boundary set is limited by $|\mathcal{B}_i^{\mathrm{m}}| \leq |\mathcal{C}_i^{\mathrm{m}}| - 1$.

All the midpoints from two side edges of a triangle in $\mathcal{C}_i^{\mathrm{m}}$ are stored in

$$M_i^{\mathrm{m}} = [[\boldsymbol{\mu}_{i1}^1 \boldsymbol{\mu}_{i1}^2][\boldsymbol{\mu}_{i2}^1 \boldsymbol{\mu}_{i2}^2] \cdots [\boldsymbol{\mu}_{ij}^1 \boldsymbol{\mu}_{ij}^2]] \ , \tag{5}$$

where $\boldsymbol{\mu}_{ij}^1 = 0.5(\boldsymbol{v}_i + \boldsymbol{v}_{n_{ij}})$ and $\boldsymbol{\mu}_{ij}^2 = 0.5(\boldsymbol{v}_i + \boldsymbol{v}_{n_{i(j+1)}})$. Consequently, each triangle in a clique is represented by the cross ratio. For example, a triangle of $(i, n_{ij}, n_{i(j+1)})$ is defined by

$$r_{ij} = f_{\mathrm{cr}}\left(f_{\mathrm{o}}(\boldsymbol{v}_{n_{ij}}, \boldsymbol{\mu}_{ij}^{1'}, \boldsymbol{\mu}_{ij}^{2'}, \boldsymbol{v}_{n_{i(j+1)}})\right) \ , \tag{6}$$

where $\boldsymbol{\mu}_{ij}^{k'}$ is a projection point of $\boldsymbol{\mu}_{ij}^k$ onto the boundary vector $\boldsymbol{v}_{n_{i(j+1)}} - \boldsymbol{v}_{n_{ij}}$, and $f_{\mathrm{o}}(\cdot)$ represents an ordering function which orders four points in the direction of the boundary vector. Figure 2 illustrates some examples of graphical distances of a Delaunay graph from 20 random points. When a triangular face has an obtuse angle, the projections of midpoints lie outside of the triangle [see Fig.2(b)]. In this case, the order of four points is different from the order of the acute triangle shown in Fig. 2(a). To make the order of points consistent, an ordering function $f_{\mathrm{o}}(\cdot)$ rearranges four points in the direction of the boundary vector and any direction of the boundary gives the same cross ratio, i.e., $f_{\mathrm{cr}}(\boldsymbol{v}_1, \boldsymbol{v}_2, \boldsymbol{v}_3, \boldsymbol{v}_4)$ is equal to $f_{\mathrm{cr}}(\boldsymbol{v}_4, \boldsymbol{v}_3, \boldsymbol{v}_2, \boldsymbol{v}_1)$. Thus, the face $\mathcal{F}^{\mathrm{m}}$ can also be represented in terms of the cross ratios of $\mathcal{C}_i^{\mathrm{m}}$, i.e., $\mathcal{F}^{\mathrm{m}} = \{R_1^m, \cdots, R_{|V^m|}^m\}$, where $R_i^m = [r_{i1} \ \cdots \ r_{i(|\mathcal{B}_i^{\mathrm{m}}|)}]$.

The graphical distance between two cliques is incorporated in the modified HD, where the size of the triangle is used for the weight, i.e.,

$$h_{\mathrm{g}}(R_k^{\mathrm{m}}, R_l^{\mathrm{t}}) = \frac{1}{|R_k^{\mathrm{m}}|} \sum_{\alpha \in R_k^{\mathrm{m}}} \min_{\beta \in R_l^{\mathrm{t}}} \{g(r_{k\alpha}, r_{l\beta})\} \ , \tag{7}$$

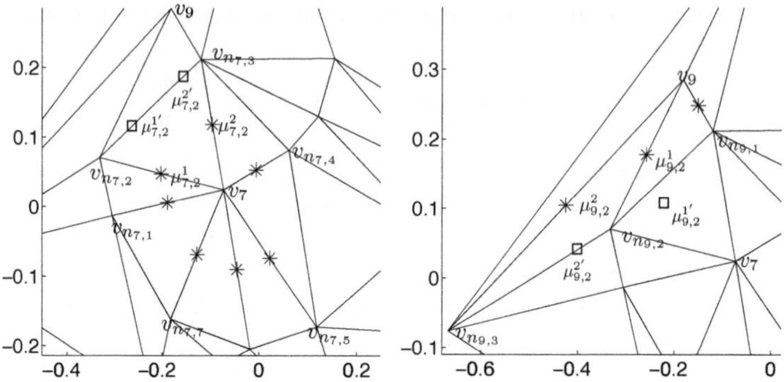

(a) (b)

Fig. 2. Illustration of a graphical distance: (a) A triangle defined by $(\boldsymbol{v}_7, \boldsymbol{v}_{n_{7,2}}, \boldsymbol{v}_{n_{7,3}})$ is a face of $\mathcal{C}_7$ and its cross ratio is $f_{\mathrm{cr}}(\boldsymbol{v}_{n_{7,2}}, \boldsymbol{\mu}^{1'}_{7,2}, \boldsymbol{\mu}^{2'}_{7,2}, \boldsymbol{v}_{n_{7,3}})$; (c) For a face $(\boldsymbol{v}_9, \boldsymbol{v}_{n_{9,2}}, \boldsymbol{v}_{n_{9,3}})$ with an obtuse angle, the projection of a midpoint is out side of a boundary

where $g(\cdot)$ is

$$g(r_{k\alpha}, r_{l\beta}) = (1 + ||\frac{a_\alpha}{a_{\mathcal{C}^{\mathrm{m}}_k}} - \frac{a_\beta}{a_{\mathcal{C}^{\mathrm{t}}_l}}||)||r_{k\alpha} - r_{l\beta}|| \ , \tag{8}$$

and $a_{\mathcal{C}^{\mathrm{m}}_k}$ is the total area of a clique $\mathcal{C}^{\mathrm{m}}_k$, and a_α represents the area of a triangle indexed by α. Thus, the graphical distance accounts for the difference of area ratio in a clique. The non-directional version of the graphical distance $h_2(R^{\mathrm{m}}_k, R^{\mathrm{t}}_l)$ is obtained by choosing the maximum of two directional distances. Finally, a graphical proximity matrix is defined as

$$\Delta_{\mathrm{g}} = \begin{bmatrix} h_2(R^{\mathrm{m}}_1, R^{\mathrm{t}}_1) & h_2(R^{\mathrm{m}}_1, R^{\mathrm{t}}_2) & \cdots & h_2(R^{\mathrm{m}}_1, R^{\mathrm{t}}_{|\mathcal{V}^{\mathrm{t}}|}) \\ \vdots & \vdots & \cdots & \vdots \\ h_2(R^{\mathrm{m}}_{|\mathcal{V}^{\mathrm{m}}|}, C^t_1) & h_2(R^{\mathrm{m}}_{|\mathcal{V}^{\mathrm{m}}|}, R^{\mathrm{t}}_2) & \cdots & h_2(R^{\mathrm{m}}_{|\mathcal{V}^{\mathrm{m}}|}, R^{\mathrm{t}}_{|\mathcal{V}^{\mathrm{t}}|}) \end{bmatrix} \ , \tag{9}$$

and the matrix is normalised by the maximum value of its elements.

4 Experimental Results

To evaluate the performance of the proposed method in partial matching, the Harris detector was used to extract 251 model points from a 348×360 grey level image of a cup. Figure 3(a) shows the Delaunay graph of the model points overlaid on the image. To create a portion of the model data, a reference line and a sweeping line are used in the experiment. The reference line is parallel to the horizontal axis and has an intersection with the mean point of the model data. The sweeping line sweeps the model points in a clockwise direction at a sweeping angle θ_s measured from the reference line. The sweeping angle determines the size of the selected portion. Figure 3(b) shows a selected test data with $\theta_s = 60°$, and its Delaunay graph. Since the general HD is able to cope with partial matching, the performance of HD and its variants in partial matching are compared with that of the clique HD from $\theta_s = 20°$ to $\theta_s = 340°$. The results are shown in Fig.3(c), where matching distances are nomalised for comparison. Despite the small portion of model data selected with $\theta_s = 60°$, the clique HD matching distance of 0.589 is the best, whilst HD, MHD, 30% Ranked HD (RHD), 50% RHD and 70% RHD scores are respectively 0.953, 0.786, 0.388, 0.782 and 0.783. This is because the local graphical information is not changed. RHDs score 0 as the size of the portion is close to the model data because they only use the best ranked score. However, this characteristic is not desirable for identification.

Images of three small objects as shown in Fig.4 are used for evaluating the identification capability of the proposed method. Two of the objects have a similar shape except for details of their decoration [see Fig.4(a) and (b)] and the last test image is generated from Fig.4(a) using a similarity transform. The

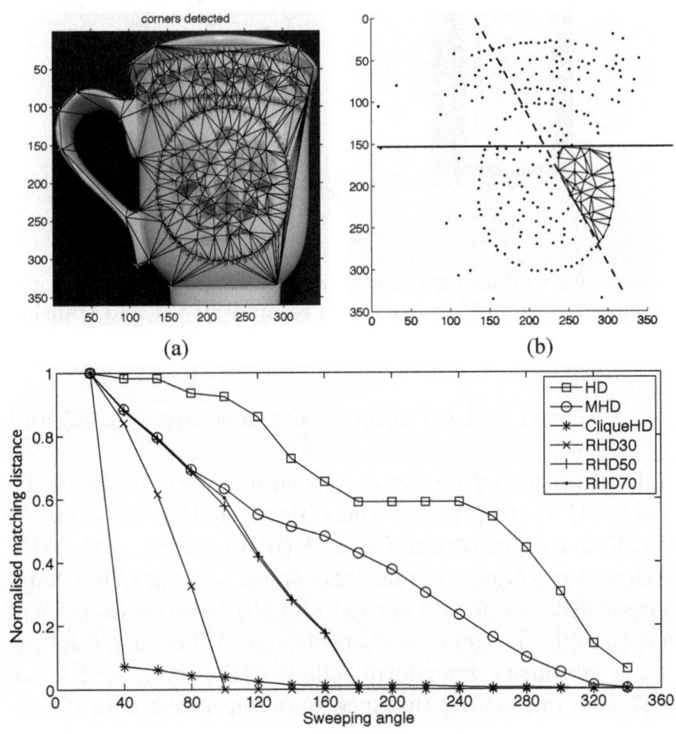

Fig. 3. (a) The 251 model points are extracted from the 348×360 image for the partial matching experiment. They are denoted by *, with their Delaunay graph overlaid. (b) The selected data is bounded by the solid reference line and the dashed sweeping line with $\theta_s = 60°$. (c) Normalised matching distances of HD, MHD, 30% RHD, 50% RHD, 70% RHD and clique HD, are respectively denoted by □, ∘, x, +, • and *.

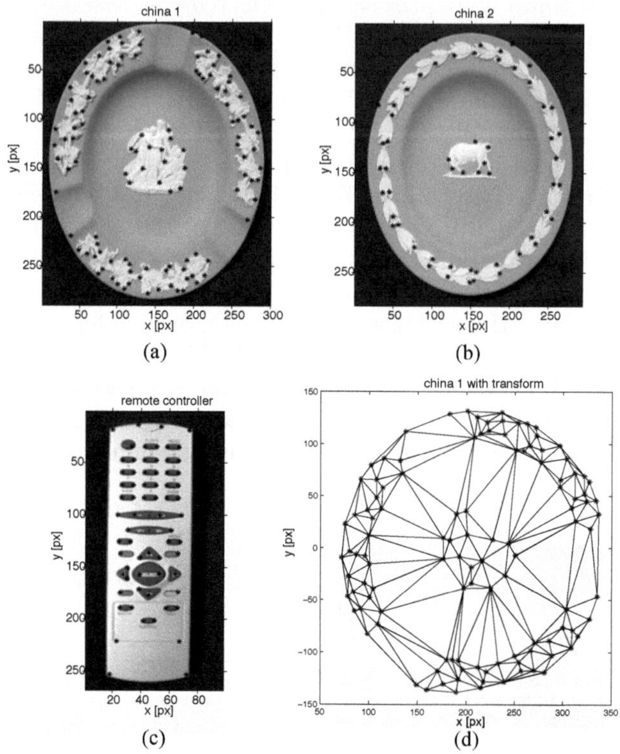

Fig. 4. Test images for evaluating the identification capability: (a) china 1; (b) china 2; (c) remote controller; and (d) image in (a) rotated by 45° and translated.

images in Fig. 4(a), (b) and (c) respectively generate 113, 57 and 68 feature points.

The normalised matching distances are summarised in Table 1 where a, b, c and d respectively correspond to the objects in Fig. 4(a), (b), (c) and (d). Since the overall shapes of object (a) and (b) are similar, the HD score (i.e., distance) for these two objects is relatively small. On the other hand, the clique HD gives a larger distance for object (a) and (b) because their local details are different, even though the general shape is not. When a model and test data are related by a similarity transform, clique HD scores 9.83 whereas general HD scores 60.7 (see mataching distance between a and d in Table 1), i.e, the HD matching score between (a) and (d) is significantly increased but clique HD gives a small distance under the transform. These results show that the clique HD achieves a much better performance in identifying an object, e.g., for model based matching. However, when general shape of two objects are considerably different, e.g., rectangular remote and circular china, traditional HD performs better and the average distances of clique HD and HD in this case are 59.73 and 85.21, respectably (see c column in the table).

Table 1. Identification test results

	clique HD				HD			
	a	b	c	d	a	b	c	d
a	0	49.41	38.61	9.83	0	11.92	81.88	60.07
b	49.41	0	100	44.12	11.92	0	73.76	59.89
c	38.61	100	0	40.58	81.88	73.76	0	100
d	9.83	44.12	40.58	0	60.07	59.89	100	0

5 Conclusions

HD-based point matching has advantageous characteristics, e.g., capable of matching without exact point pairs and partial matching, and robustness against noise and outliers. Therefore, many variants of HD have been introduced. This paper proposes a clique HD that incorporates a graphical distance in the traditional HD matching framework. The proposed method performs matching at local point sets called cliques, which are uniquely formed by the Delaunay tessellation.

To achieve similarity invariance, the proposed method uses a set of cross ratios defined by four collinear points on every boundary edge in a clique. The normalised area of a face in a clique is used for weight when matching two triangles. Experiments show that the proposed matching method is robust to noise and outliers, which normally would deteriorate the performance of graph-based matching methods. However, the method cannot account for apparent shape matching (e.g., matching by the boundary or silhouette of an object) because it is solely based on features from local entities. As future works, the proposed method can be modified to address an affine transformed data, and recent estimating techniques that are robust to noise, e.g., LMeds or RANSAC algorithm will be explored to achieve a more accurate estimation of a local transform.

References

1. Strickland, R.N., Mao, Z.: Computing correspondences in a sequence of non-rigid shapes. Pattern Recognit. 25(9), 901–912 (1992)
2. Shapiro, L.S., Brady, J.M.: Feature-based correspondence: An eigenvector approach. Image Vision Compt. 10(5), 283–288 (1992)
3. Zheng, Y., Doermann, D.: Robust point matching for nonrigid shapes by preserving local neighborhood structures. IEEE Trans. Pattern Anal. Machine Intell. 28(4), 643–649 (2006)
4. Caetano, T.S., Caelli, T., Schuurmans, D., Barone, D.A.C.: Graphical models and point pattern matching. IEEE Trans. Pattern Anal. Machine Intell. 28(10), 1646–1663 (2006)
5. Saber, E., Xu, Y., Tekalp, A.M.: Partial shape recognition by sub-matrix matching for partial matching guided image labeling. Pattern Recognit. 38(10), 1560–1573 (2005)

6. Carcassoni, M., Hancock, E.R.: Spectral correspondence for point pattern matching. Pattern Recognit. 36(1), 192–204 (2003)
7. Caelli, T., Kosinov, S.: An eigenspace projection clustering method for inexact graph matching. IEEE Trans. Pattern Anal. Machine Intell. 26(4), 515–519 (2004)
8. Huttenlocher, D.P., Klanderman, G.A., Rucklidge, W.J.: Comparing images using the hausdorff distance. IEEE Trans. Pattern Anal. Machine Intell. 15(9), 850–863 (1993)
9. Jolly, M.P.D., Jain, A.K.: A modified hausdorff distance for object matching. In: Proc. Int. Conf. Pattern Recognition, vol. A, pp. 566–568 (1994)
10. Yi, X., Camps, O.I.: Line-based recognition using a multidimensional Hausdorff distance. IEEE Trans. Pattern Anal. Machine Intell. 21(9), 901–(1999)
11. Gao, Y., Leung, M.K.H.: Line segment Hausdorff distance on face matching. Pattern Recognit. 35(2), 361–371 (2002)
12. Gao, Y., Leung, M.K.H.: Face recognition using line edge map. IEEE Trans. Pattern Anal. Machine Intell. 24(6), 764–779 (2002)
13. Yu, X., Leung, M.K.H.: Shape recognition using curve segment Hausdorff distance. In: Proc. Int. Conf. Pattern Recognition, vol. 3, pp. 441–444 (2006)
14. Gope, C., Kehtarnavaz, N.: Affine invariant comparison of point-sets using convex hulls and Hausdorff distances. Pattern Recognit. 40(1), 309–320 (2006)
15. Rosenfeld, A., Hummel, R.A., Zucker, S.W.: Scene labeling by relaxation operations. IEEE Trans. Syst., Man, Cybern. 6(6), 420–433 (1976)
16. Baihua Li, Q.m., Holstein, H.: Similarity K-d tree method for sparse point pattern matching with underlying non-rigidity. Pattern Recognit. 38(12), 2391–2399 (2005)
17. Finch, A.M., Wilson, R.C., Hancock, E.R.: Matching Delaunay graphs. Pattern Recognit. 30(1), 123–140 (1997)
18. Cross, A.D.J., Hancock, E.R.: Graph matching with a dual-step EM algorithm. IEEE Trans. Pattern Anal. Machine Intell. 20(11), 1236–1253 (1998)
19. Bowyer, A.: Computing Dirichlet tessellations. Comput. J. 24(2), 162–166 (1981)
20. Aurenhammer, F.: Voronoi diagrams: A survey of a fundamental geometric data structure. ACM Comput. Surv. 23(3), 345–405 (1991)
21. Hartley, R., Zisserman, A.: Multiple view geometry, 1st edn. Cambridge University Press, Cambridge (2000)

An Inexact Graph Comparison Approach in Joint Eigenspace

Wan-Jui Lee and Robert P.W. Duin

Faculty of Electrical Engineering, Mathematics and Computer Sciences,
Delft University of Technology, The Netherlands
W.J.Lee@tudelft.nl, r.duin@ieee.org

Abstract. In graph comparison, the use of (dis)similarity measurements between graphs is an important topic. In this work, we propose an eigendecomposition based approach for measuring dissimilarities between graphs in the *joint* eigenspace (JoEig). We will compare our JoEig approach with two other eigendecomposition based methods that compare graphs in *different* eigenspaces. To calculate the dissimilarity between graphs of different sizes and perform inexact graph comparison, we further develop three different ways to resize the eigenspectra and study their performance in different situations.

1 Introduction

Graphs are a general and powerful data structure for object representation in structural pattern recognition[11,2]. The nodes in a graph can represent different parts of a object and the relations between the parts are represented by edges. Also, labels and attributes for the nodes and edges can further be used to incorporate more information in a graph representation. But for simplicity, we only consider nodes and edges without labels and attributes in this work.

In pattern recognition and other related areas, (dis)similarity among objects is an important issue. If the graphs are used for object representation, the problem turns into determining (dis)similarity between objects, which is usually referred to as graph comparison. In the study of graph comparison, there are two main directions. One is by comparing their spatial structures and the other is by comparing their spectral structures. Early approaches to graph comparison were mainly based on the first. The best known example is maximum common subgraph [2] where the similarity of two graphs is decided by how much they spatially share in common. An alternative is the graph edit distance [8]. In graph edit distance computation, one introduces a set of graph edit operations, such as deletion, insertion or substitution, to apply on nodes as well as edges. The edit distance of two graphs is defined as the shortest sequence of edit operations that transforms one graph into the other. These methods for finding spatial (dis)similarity between graphs are guaranteed to find the optimal solutions but require exponential time and space due to the NP-completeness of the problem.

N. da Vitora Lobo et al. (Eds.): SSPR&SPR 2008, LNCS 5342, pp. 35–44, 2008.

In 1988, Umeyama [6] introduced the spectra of graphs for the exact graph comparison problem. This approach is based on spectral graph theory that is concerned with characterising the structural properties of graphs using the eigenvectors of the adjacency matrix or the closely related Laplacian matrix (the degree matrix minus the adjacency matrix). To compare two different graphs A and B, the idea is to transform graph A into the space of graph B. So there exists an approximated graph $A^{'}$ of graph A in the space of graph B. The difference between A and B is actually calculated by the difference between $A^{'}$ and B. There is another approach for inexact graph comparison based on graph spectra proposed by Caelli [1]. Instead of transforming graph A into the space of graph B, it projects graph A into its own eigenspace and the same is done for graph B. Because graph A and B could have different numbers of nodes, a renormalization step is introduced for eigenspace projections. Therefore, both graphs are projected into a unit hypersphere defining the eigenspaces. In this renormalized space, two graphs are close if their rescaled eigenvectors have similar angles and their eigenvalues are similar. The authors declare that the graph comparison was done in a common eigenspace of these two graphs, but actually it's done in their own eigenspaces but with the common unit hypersphere.

In Umeyama's method, the graphs are compared in the original space of one of the graphs, but this is not unique since there are two graphs. As a matter of fact, we further discover that the projection of graph A in the space of graph B is actually determined by how the eigenvalues of graph A live in the eigenspace of graph B. So if we want to have a symmetric dissimilarity measurement using Umeyama's method, we have to take an average or choose one of the results from two different eigenspaces, i.e. the eigenspace of A or B. In Caelli's approach, the dissimilarities of the graphs are compared by their projections in their own eigenspaces. For a fair comparison of two graphs, we will bring them into the same eigenspace and then compare them in this unique eigenspace. Inspired by Umeyama's approach, we actually can find an unique joint eigenspace (JoEig) for both graphs, which is a space with the eigenvectors of both graphs. However, the sizes of these two sets of eigenvectors could be different and therefore it is essential to adjust these two sets of eigenvectors for sharing the same number of dimensions in order to construct the joint eigenspace. We study three possibilities for setting the number of eigenvectors in this work. The first two choices are setting the number of the eigenvectors according to the size of the larger or the smaller graph, respectively. We study as well for a number smaller than both of the graphs to emphasize the importance of larger eigenvalues. We will compare these three settings and discuss how the dimensionality of eigenvectors effects the performance of classifiers in the experiments.

The rest of the paper is organized as follows. In Section 2, we introduce Umeyama's and Caelli's approaches for comparing graphs, respectively. Our graph dissimilarity measurement approach, JoEig, is described in Section 3. Simulation results are presented in Section 4. Finally, a conclusion is given in Section 5.

2 Previous Eigendecomposition Approaches for Graph Comparison

Before we introduce Umeyama's and Caelli's approaches for graph comparison, some definitions and introduction on graphs are given as in the following.

A graph is a set of nodes connected by edges in its most general form. Consider the undirected graph $G = (V, E, W)$ with the node set $V = \{v_1, v_2, \ldots, v_n\}$, the edge set $E = \{e_1, e_2, \ldots, e_m\} \subset V \times V$, and the weight function $W : E \to (0, 1]$. If the graph edges are weighted, the adjacency matrix A for the graph G is the $n \times n$ matrix with elements

$$A_{ij} = \begin{cases} W(v_i, v_j), & \text{if } (v_i, v_j) \in E; \\ 0, & \text{otherwise.} \end{cases} \tag{1}$$

Clearly since the graph is undirected, the matrix A is symmetric. The Laplacian [10] of the graph is defined by $L = D - A$, where D is the diagonal node degree matrix whose elements $D_{ii} = \sum_{k=1}^{n} A_{ik}$. The Laplacian matrix of G is positive semidefinite and singular, and it is more often adopted for spectral analysis than the adjacency matrix because of its properties.

2.1 An Eigendecomposition Approach for the Weighted Graph Matching Problem

In 1988, Umeyama [6] developed an eigendecomposition approach for the weighted graph matching problem. Let $G = (V_1, E_1, W_1)$, $H = (V_2, E_2, W_2)$ be weighted graphs with n nodes (these two graphs are with the same size). The weighted graph matching problem is the problem of finding a one-to-one correspondence Φ between $V_1 = \{v_1, v_2, \ldots, v_n\}$ and $V_2 = \{v_1^{'}, v_2^{'}, \ldots, v_n^{'}\}$ which minimizes the difference between G and H. The following criterion is used as a measure of difference:

$$J(P) = \|P L_G P^T - L_H\|^2 \tag{2}$$

where the permutation matrix P presents the node correspondence Φ and $\|\cdot\|$ is the Euclidean norm ($\|L\| = (\sum_{i=1}^{n} \sum_{j=1}^{n} l_{ij}^2)^{\frac{1}{2}}$). Thus the weighted graph matching problem is reduced to the problem of finding the permutation matrix P which minimizes $J(P)$. In general, it is not easy to find the exact solution of the weighted graph matching problem since it is purely combinatorial. Thus, an efficient method which gives a "nearly" optimum solution, i.e., a permutation matrix $P^{'}$ whose criterion value $J(P^{'})$ is very close to the optimum value. Since the optimum criterion value cannot be known in advance, the aim is to determine a permutation matrix of a small criterion value.

Let G and H be weighted undirected graphs and L_G and L_H be their Laplacian matrices, respectively. The eigendecomposition of L_G and L_H are performed as

$$L_G = V_G D_G V_G^T, \; L_H = V_H D_H V_H^T, \tag{3}$$

where V_G and V_H are orthonormal matrices and D_G and D_H are diagonal matrices of the eigenvalues (in ascending order) of G and H, respectively. Now if G and H are isomorphic,

$$PV_G D_G V_G^T P^T = V_H D_H V_H^T. \tag{4}$$

It means there exists a permutation matrix P that has the one-to-one correspondence of nodes between G and H. Also, the eigenvalues of these two graphs should be the same. That is, $D_G = D_H$, and therefore $PV_G = V_H$ which leads to $P = V_H V_G^T$. As a result, the difference between two graphs can be computed as

$$\|V_H V_G^T L_G V_G V_H^T - L_H\|^2, \text{ or } \|V_G V_H^T L_H V_H V_G^T - L_G\|^2. \tag{5}$$

Even though Umeyama's approach is only designed for graphs with the same size, we extend it in this paper to compare it with the other approaches. First of all, the Lapacian matrix is used instead of adjacency matrix. Secondly, the eigenvalue and eigenvector matrices are resized because the graphs might be with different numbers of nodes. We study three different ways for computing the resized eigenvalue diagonal matrix and the eigenvector matrix, and they will be discussed further in the next section. Finally, the difference between two graphs is the difference between two approximations instead of the one between the original graph and its approximation. This is to prevent asymmetricity which can easily be seen in Eq.(5). The difference between two graphs can be different if the original graph is different. Therefore we use a new criterion for Umeyama's approach to make the results symmetric, and that is

$$\|V_H V_G^T L_G V_G V_H^T - V_G V_H^T L_H V_H V_G^T\|^2 = \|V_H D_G V_H^T - V_G D_H V_G^T\|^2. \tag{6}$$

By observing Eq.(6), we can see that the approximation of one graph in Umeyama's approach is by relocating the other graph's eigenvalues in its eigenspace. This suggests that the difference between these two approximations is the difference derived from two different eigenspace.

2.2 An Eigenspace Projection Clustering Method for Inexact Graph Matching

Subspace projection methods are conventionally used to reduce the dimensionality of data by minimizing the number of dimensions and the amount of information loss. To solve the inexact graph matching problem, Caelli [1] proposed an approach to project graphs into their subspaces. Furthermore, to compare graphs in different subspaces, these subspaces should have the same dimensionality. This was done by setting a number k to fix the dimensionalities of these subspaces. In Caelli's approach, the graph Laplacian matrix, L, is first decomposed into the familiar eigenvalue and eigenvector matrix product $L = VDV^T$ as in Eq.(3). The original data is then projected onto a smaller number of k most important (i.e., the k largest eigenvalues) principal components.

To compare the similarities of two vertices from two graphs of different sizes, an additional renormalization step for such projections was introduced. Both eigenvector (V) and eigenvalue (D) matrices are truncated according to the chosen number of projection dimensions (k)-largest eigenvalues and their associated eigenvectors. The renormalized eigenvectors and eigenvalues become: $V_k^{'} = \frac{V_k}{\|V_k\|}$, $D_k^{'} = \frac{D_k}{\|D_k\|}$. As a result, the renormalized subspace projection of each vertex (column of L) was obtained as $L_k^{'} = D_k^{'}(V_k^{'})$. In this renormalized subspace, vertices from different graphs will be close together if their rescaled eigenvectors have similar angles and their rescaled eigenvalues are similar. Such scaling can map initially quite different dimensional vectors into similar positions in the subspace as long as their eigenspectrum components in this lower dimensional subspace share similar amounts of the matrix variance and the rescaled unit eigenvectors have similar orientations. It is easy to see that the similarities are dependent on similar truncated and rescaled eigenspectra as well as similar unit rescaled eigenvectors but they are independent of the relative sizes of the original graphs.

To be more precise, in this work, the difference between two graphs using this method was computed as

$$\|D_G^{'}(V_G^{'})^T - D_H^{'}(V_H^{'})\|^2, \tag{7}$$

where $D^{'}$ is the resized eigenvalue diagonal matrix while $V^{'}$ is the resized eigenvector matrix. The number of the k most important components are set to the size of the smaller graph.

3 JoEig: Graph Comparison in Joint Eigenspace

Unlike the previous methods, which project graphs into different eigenspaces and compare them, we project each pair of two graphs into a joint eigenspace. This joint eigenspace is expanded by both set of eigenvectors. Given Eq.(3), we can further rewrite the equations to

$$D_H = V_H^T L_H V_H,\ D_G = V_G^T L_G V_G. \tag{8}$$

If graph G and H are isomorphic, we will have $D_G = D_H$ and therefore

$$V_G^T L_G V_G = V_H^T L_H V_H. \tag{9}$$

By multiplying V_G on the left side and V_H^T on the right side of Eq.(9), we can further derive

$$L_G V_G V_H^T = V_G V_H^T L_H. \tag{10}$$

Therefore, $V_G V_H^T$ is the joint projection vector for both graphs G and H. By introducing Eq.(3) into Eq.(10),we will have

$$V_G D_G V_G^T V_G V_H^T = V_G V_H^T V_H D_H V_H^T. \tag{11}$$

After removing $V_G^T V_G$ and $V_H^T V_H$ from Eq.(11),

$$V_G D_G V_H^T = V_G D_H V_H^T \tag{12}$$

will be obtained. Therefore, the difference between two graphs using our method can be defined as

$$\|V_G D_G V_H^T - V_G D_H V_H^T\|^2. \tag{13}$$

Unlike Umeyama's method in which the approximation of one graph is defined by relocating the other graph's eigenvalues in its own eigenspace, our approach approximates the graph by relocating its eigenvalues in the joint eigenspace constructed by the eigenvectors of both graphs as shown in Eq.(13).

However, the sizes of graphs might be different, and therefore it might not be possible to make a matrix product between V_G and D_H or V_H and D_G as in Eq.(13). A feasible solution is to fix the number of eigenvectors for both graphs. In this work, we study three possibilities for setting the number of eigenvectors. First, we make full use of the eigenvectors from the larger graph (by larger graph, we mean graphs with more nodes). In this case, we have to expand the eigenvectors of the smaller graph with zero vectors and also assign zero eigenvalues to them. But this will lower the influence of the smaller graph because the newly added zero eigenvalue and eigenvectors have limit impact on the results. On the other hand, we also try to make full use of the eigenvectors from the smaller graph and keep the same number of eigenvectors and eigenvalues in the larger graph as in the smaller graph by removing less important eigenvalues and eigenvectors from the larger graph. Less important eigenvectors are those with smaller eigenvalues. This, on the other hand, is in favour of the smaller graph. Moreover, we ignore the size of these two graphs and just pick a reasonable fix small number of eigenvectors and eigenvalues for both graphs. In this way, the strength of the relationship inside a graph is more important than the size of the graph, but it is not easy to choose an appropriate number of eigenvalues and eigenvectors to use.

4 Experiments

In this section, we compare the JoEig approach with other two methods, i.e. Umeyama's and Caelli's methods, described in Section 2 in the dissimilarity space [9]. Three classifiers, i.e., linear discriminant (ldc), quadratic discriminant (qdc) and nearest mean classifier (nmc), are adopted to have a more general understanding over the performance in the dissimilarity space. Three real-world datasets, i.e., Mutag [3], House [13] and Coil-20 [7], are used in the experiments. We use 60 representative objects to construct the dissimilarity space for the House and Coil-20 datasets and 30 representative objects for the Mutag dataset. Also, the eigenvalue diagonal and eigenvector matrices are resized in three different ways. The first is to enlarge the matrices of the smaller graph to the size of the larger graph, and it is marked as *max* in the figures. The second is by

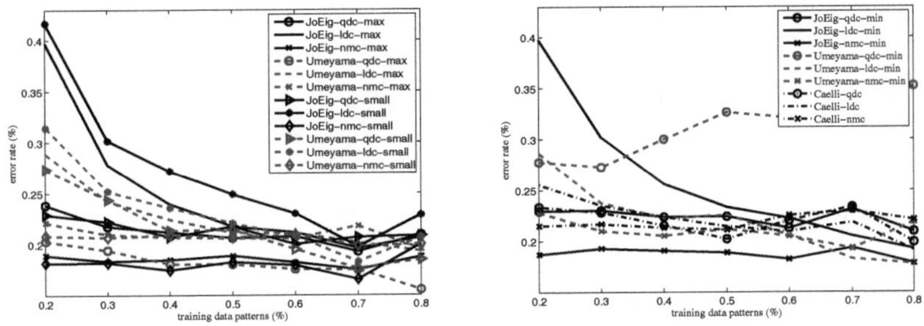

(a) (b)

Fig. 1. Mugtag Dataset: Learning curves of (a) JoEig and Umeyama's method with qdc, ldc, and nmc using max and small strategies and (b) all methods with qdc, ldc, and nmc using min resizing strategies

shrinking the matrices of the larger graph to the size of the smaller graph, and it is called *min*. The third one is called *small*, and it is done by reducing the matrices of both graphs to the size of the smaller graph minus 5. Moreover, all the results in the following are the average over 50 repetitions of experiments resulting in a very small standard deviation.

4.1 Experiment 1: Mutag Dataset

The Mutag dataset consists originally of 230 chemical compounds assayed for mutagenicity in Salmonella typhimurium. Among the 230 compounds, however, only 188 (125 positive ones, 63 negative ones) are considered to be learnable [3] and thus only 188 patterns are used in the simulations.

From Figure 1, it can be observed that under the cases of using a smaller number of eigenvalues and eigenvectors, nmc performs much better than qdc and ldc using JoEig and Umeyama's method, especially with small sample sizes. However, by using *max* to enlarge the smaller graph, Umeyama's method prefers qdc than nmc while nmc still performs better than qdc with JoEig. Which indicates that Umeyama's method using the *max* strategy brings the data points into a dissimilarity space where the global distance is not good enough for separating classes and therefore requires a more complex classifier to fulfill the task.

4.2 Experiment 2: House Dataset

The house dataset contains 101 , 111, and 182 image sequences from three different houses, respectively. We extract the feature points using the scale-invariant feature transform (SIFT) method[5] and then compute the Voronoi tessellations of the feature points to construct the region adjacency graph, i.e., the Delaunay triangulation, of the Voronoi regions.

As shown in Figure 2, all methods prefer ldc and qdc over nmc when the *max* strategy is used, but nmc is more preferable when the *min* strategy is adopted.

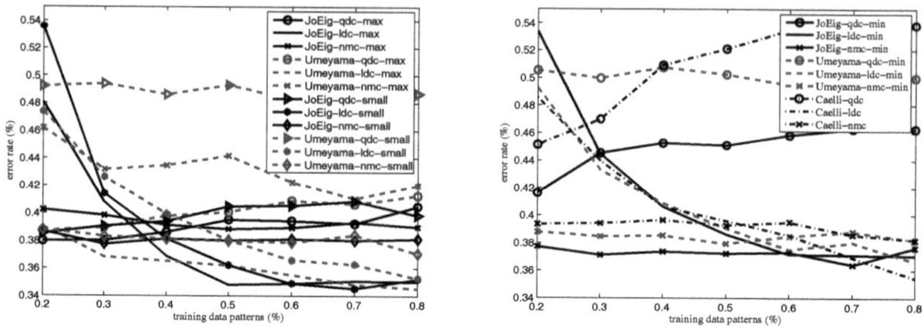

(a) (b)

Fig. 2. House Dataset: Learning curves of (a) JoEig and Umeyama's method with qdc, ldc, and nmc using max and small strategies and (b) all methods with qdc, ldc, and nmc using min resizing strategies

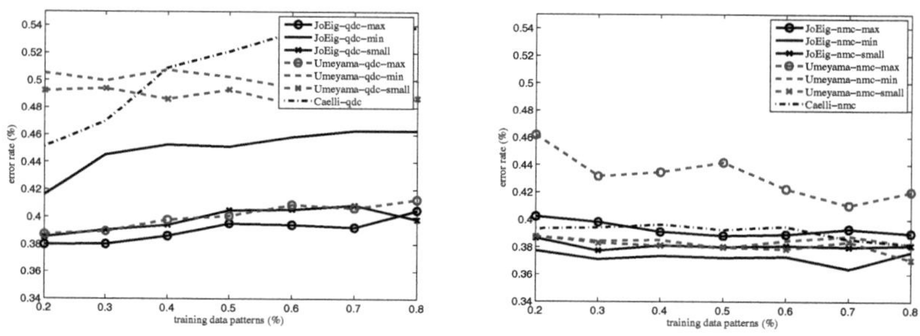

(a) (b)

Fig. 3. House Dataset: Learning curve of (a) quadratic discriminant and (b) nearest mean classifier using JoEig, Umeyama's, and Caelli's methods for graph comparison

Compared to the other two datasets, the graphs in this dataset are relatively more similar and therefore their relative distances are also closer. Which makes the global distance of not so great help in separating these three classes. However, selecting a small number of eigenvalues and eigenvectors to reconstruct the graph distances in the dissimilarity space seems to enhance the distance between graphs and makes it an easier task for nmc to handle. From Figure 3, we can also observe that nmc performs better with *small* and *min* strategies and qdc is better with the *max* strategy.

4.3 Experiment 3: Coil-20 Dataset

The Coil-20 contains multiple views of the same object in different poses with respect to the camera. There are originally 20 objects (classes) in the data, but we only use 5 objects and in total 358 images to form the dataset. Graphs are derived with the same method described in Section 4.2.

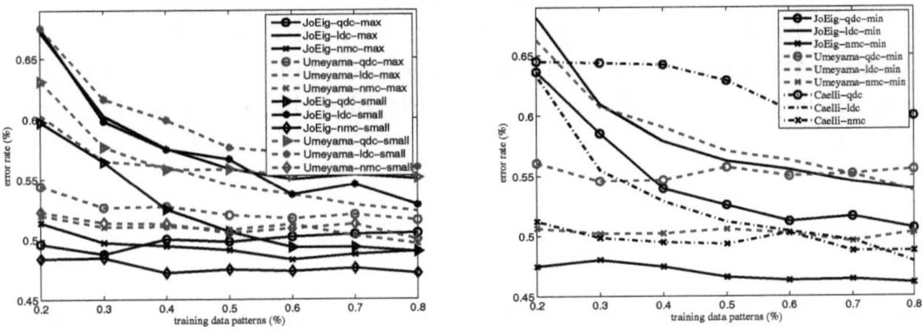

(a) (b)

Fig. 4. Coil-20 Dataset: Learning curves of (a) JoEig and Umeyama's method with qdc, ldc, and nmc using max and small strategies and (b) all methods with qdc, ldc, and nmc using min resizing strategies

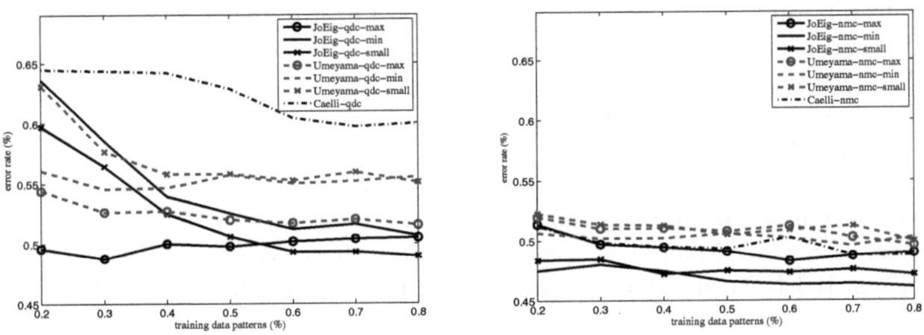

(a) (b)

Fig. 5. Coil-20 Dataset: Learning curve of (a) quadratic discriminant and (b) nearest mean classifier using JoEig, Umeyama's, and Caelli's methods for graph comparison

As given in Figure 4, again we discover that nmc performs better with JoEig and Umeyama's method using different resizing strategies despite the fact that COIL-20 dataset is with a much more complex graph structure than the Mutag dataset. From Figure 5, we derive the same conclusion as before that nmc performs better with *small* and *min* strategies while qdc is better with the *max* strategy.

5 Conclusions

We propose a graph comparison approach named JoEig in the joint eigenspace based on eigendecomposition. The main contribution of this work is introducing the joint eigenspace and comparing graphs in it. We also study three different ways for resizing the eigenvalue diagonal and eigenvector matrices to solve the inexact graph matching problem. From a series of examples presented in the experiments, we conclude that our proposed JoEig approach is in general better than Umeyama's and Caelli's methods. Furthermore, nmc performs better with *small* and *min* strategies while qdc is better with the *max* strategy. All in all, resizing graphs with *small* or *min* strategies and then using the nmc classifier in JoEig space is probably a better solution despite the original structure of the graph dataset.

References

1. Caelli, T., Kosinov, S.: An Eigenspace Projection Clustering Method for Inexact Graph Matching. IEEE Trans. Pattern Analysis and Machine Intelligence 26(4), 515–519 (2004)
2. Conte, D., Foggia, P., Sansone, C., Vento, M.: Thirty Years of Graph Matching in Pattern Recognition. International Journal of Pattern Recognition and Artificial Intelligence 18(3), 265–298 (2004)
3. Debnath, A.K., Lopez de Compadre, R.L., Debnath, G., Shusterman, A.J., Hansch, C.: Structure-Activity Relationshop of Mutagenic Aromatic and Heteroaromatic Nitro Compounds. Correlation with Molecular Orbital Energies and Hydrophobicity, Journal of Medicinal Chemistry 34, 786–797 (1991)
4. He, P.R., Zhang, W.J., Li, Q., Wu, F.X.: A New Method for Detection of Graph Isomorphism Based on the Quadratic Form. Journal of Mechanical Design 125, 640–642 (2003)
5. Lowe, D.G.: Distinctive Image Features from Scale-Invariant Keypoints. International Journal of Computer Vision 60(2), 91–110 (2004)
6. Umeyama, S.: An Eigendecomposition Approach to Weighted Graph Matching Problems. IEEE Trans. Pattern Analysis and Machine Intelligence 10(5), 695–703 (1988)
7. Nene, S.A., Nayar, S.K., Murase, H.: Columbia University Image Library (COIL-20), `http://www1.cs.columbia.edu/CAVE/software/softlib/coil-20.php`
8. Neuhaus, M., Bunke, H.: Edit Distance-Based Kernel Functions for Structural Pattern Classification. Pattern Recognition 39, 1852–1863 (2006)
9. Pekalska, E., Duin, R.P.W.: Dissimilarity Representations Allow for Building Good Classifiers. Pattern Recognition Letters 23(8), 943–956 (2002)
10. Qiu, H.J., Hancock, E.R.: Spectral Simplication of Graphs. In: Proceedings of the 8th European conference on computer vision, Czech Republic, pp. 114–126 (2004)
11. Wilson, R.C., Hancock, E.R., Luo, B.: Pattern Vectors from Algebraic Graph Theory. IEEE Trans. Pattern Analysis and Machine Intelligence 27(7), 1–13 (2005)
12. Zhao, G.X., Luo, B., Ting, J., Ma, J.X.: Using Eigen-Decomposition Method for Weighted Graph Matching. In: Huang, D.-S., Heutte, L., Loog, M. (eds.) ICIC 2007. LNCS, vol. 4681, pp. 1283–1294. Springer, Heidelberg (2007)
13. CMU Vision and Autonomous System Center's Image Database, `http://www.ius.cs.cmu.edu/idb/`

A Vectorial Representation for the Indexation of Structural Informations

Nicolas Sidere[1,2], Pierre Heroux[1], and Jean-Yves Ramel[2]

[1] Université de Rouen
LITIS EA 4108
BP 12 - 76801 Saint-Etienne du Rouvray, France
[2] Université François Rabelais Tours
Laboratoire Informatique de Tours
64 Avenue Jean Portalis
37200 Tours, France
{nicolas.sidere,jean-yves.ramel}@univ-tours.fr,
pierre.heroux@univ-rouen.fr

Abstract. This article presents a vectorial representation of structured data to reduce the complexity of dissimilarity computations in an information retrieval context. This representation enables, via a computation of an adapted measure, to approximate the distance between structural representations in both context of distance between graphs and searching occurrences of subgraphs. Preliminary results show that the proposed representation offers comparable performance with those of the literature.

Keywords: graph signature, structured information indexation and retrieval.

1 Introduction

The evolution of digitization techniques, the easiness of broadcasting by the way of the Internet and the wish to keep and access digitized collections of documents put the themes of indexation and retrieval in the heart of many research axes. With the diversity of thematics (preserving old books, archiving administrative data, automatic reading, ...) the amount of data generated grows up and up. This contributes to the multiplication of works in this domain. If some solutions come out, these are often restricted to a specific application domain or corpus. Many uses a keyword-based indexing extracted by optical recognition characters systems which are inefficient on particular documents (old books archives, graphical documents, ...) or with a manual annotation limited by the size of the corpus or by the subjectivity of the user. So, one can see a certain interest in a new querying modalities. Consequently, the nature of works are more and more aimed to a new characterization of documents with indices such as structure. This is the point we intend to study in this article. According to the situation, the structure analyses can be used in many ways:

N. da Vitora Lobo et al. (Eds.): SSPR&SPR 2008, LNCS 5342, pp. 45–54, 2008.

- the layout description. For example, the layout of a phone book, which is quite significant.
- the logical organization (title, section, paragraph, ...) can be used to differenciate some documents, a newspaper from a novel for example.
- the frequency of the same element, in a technical drawing with several occurences of a symbol.

The document retrieval consists in measuring how a structural description is relevant with respect to the user's need, which can be expressed with a structural representation of the request too. The purpose is to return k documents ranked according to their relevance. In fact, as the notion of structure of a document can be expressed by different ways according to the user, it is quite important that the user has the final choice.

Most of the times, this information is represented by graphs. Many methods can be found to label a graph to obtain this representation. Nevertheless, the computation of a graph-to-graph distance reveals to be a NP-Complex problem. This complexity grows in an exponential way with the number of nodes.

Our work aims at reducing this complexity. An interesting approach consist in extracting a numerical characteristic vector which embeds a part of the graph topological information. The comparison of two graphs is reduced to a more simple computation of a distance between two vectors in an euclidean space. More, an information retrieval context where the documents are described by structural informations, indexing can be done offline. Some works have already been done:

- A first method is presented in [1]. This representation is based on vertex degrees. The simplicity of the description combined with a comparison of two graphs reduced to a linear time allow to find topologically similar graphs in the most of cases.
- In the second approach presented in [3] where the representation of a document is focused on occurent elements (subgraphs). The document is described with a bag of symbol representation. A rich knowledge is necessary to enable the final identification.

The characterization of a graph by a vector gives, in the case presented above, advantages, but also drawbacks. For example, the description proposed by Lopresti and Wilfong ([1]) is not bijective. Thus, two non-isomorphic graphs can have the same vectorial description for which the distance is null. This ambiguity is due to the vector construction method. It is only based on the vertex degree. Several configurations from the same set of nodes with different graphs but the same signature may occur. We think that this description does integrate the informations on the graph topology in a superficial way.

Barbu's approach need a rich knowledge of the domain, that does not allow to treat a document set with a huge heterogenieity. So, we can notice that the relevance of the results are dependant of the graph vector characteritics.

Our approach relies on these works. The idea is to build a lexicon with model graphs differenciated by their topology and their size. The vector will be built according to the occurences of these figures present in the target graph. All the problematic is on the choice of the lexicon, which must answer to the genericity problem but also needs to be performant and cute in results

In the next section we present the lexicon that we chose and how it can be built. The section 3 shows the construction process of a vectorial representation of a graph. Section 4 presents two distances operator using this vectorial description, both distances correspond to two different cases of use. Section 5 presents the first experiments showing that the representation offers performance equivalent to Lopresti and Wilfong in a context of classification. Finally, Section 6 evaluates and sets out a number of prospects for the continuation of this work.

2 The Lexicon Construction

As said before, the lexicon is the basement of the construction of our graph vectorial signature. So, the lexicon content is quite determining in the relevance of the vectorial representation. We have seen ([3]) that it is possible to build it from the n most frequent subgraphs of the database, for example. Nevetheless, this is only possible when the base presents a high homogeneity. The frequency is revealing a certain semantic. Thus, in the work of Barbu, subgraphs are frequently associated with graphic symbols, entities which carry meaning in a technical documentation.

Unfortunately, there are many cases where these assumptions are not verified. For example, old books are different depending on the author, publisher, date ... Therefore, to keep a generic nature, it is preferred to use a lexicon totally independent from the database content. However, this lexicon must be sufficiently comprehensive to ensure that these terms can afford to discriminate a graph from another.

We have therefore decided to take as a baseline the non-isomorphic graphs network presented in [2]. The network presents all graphs composed of n edges up to N (N is the maximum number of edges). This network is built iteratively from a graph made up of a single vertex. At each iteration, it is possible to construct a graph of rank n adding an edge to a graph of rank $n-1$ with the ability to add a vertex if needed. All solutions are being considered which makes the network complete. A graph with rank n built from a graph with rank $n-1$ is called successor. Conversely, the graph of $n-1$ is called predecessor. A graph of this network may have several successors. Similarly, several graphs with rank $n-1$ can rise to a single successor. Ways of construction of this non-isomorphic graph network can be stored to build all predecessors and successors of a graph.

Thereafter, the term *pattern* will refer to a subgraph element of the non-isomorphic graph network. So, the lexicon is composed of all patterns until the defined rank.

For example, the figure 1 shows the non-isomorphic graph network until the fourth rank giving a lexicon of 11 patterns. The dotted arrows indicate the path of construction of the network, the arrows are directed from the predecessors towards the successors.

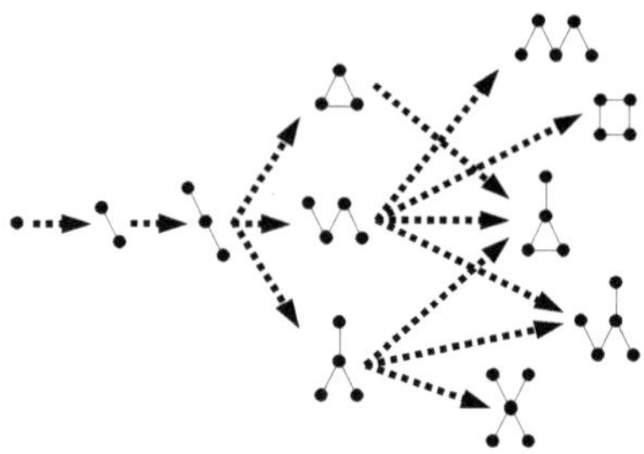

Fig. 1. The non-isomorphic graph network

Table 1 gives the number of elements in the lexicon depending on the maximum rank of the non-isomorphic graph network.

Table 1. Size of the lexicon depending on the rank of the non-isomorphic graph network

Rank	Size
0	1
1	2
2	3
3	6
4	11
5	23
6	51
7	117
8	276

We can notice that the number of patterns increases exponentially with the rank. The size of the lexicon is a parameter to determine according to several criteria. Indeed, the complexity of the transformation to a vectorial representation is directly dependent of the number of patterns. However, the more the size of the lexicon increases, the bigger the patterns it integrates are. The vectorial representation then integrates more information on topology. Therefore, it is necessary to find a trade-off between expressiveness and complexity.

3 The Construction of the Vectorial Representation

The construction of the vector consists on determining the frequency for each patterns of the lexicon in the graph to describe. The dimension of the vector is the size of the lexicon. Its construction can become very costly in time. However, the indexing phase, which is to build the vectorial representation of all graphs of the base, can be done offline. The complexity of construction of the vector is only critical when processing the query graph. Indeed, this graph can be obtained from a sample document presented by a user. The extraction method of representation will be the same that is used for the base and may be costly.

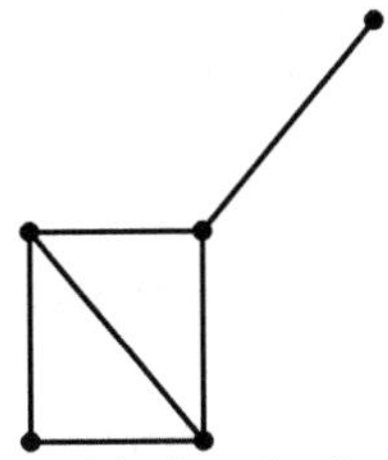

(a) Graph G

Motif						
Frq.	5	6	10	2	10	3

(b) Number of occurences of each pattern in G

Motif						
Frq.	5	6	3	1	1	1

(c) Number of edge-disjoint occurences of each pattern in G

Motif						
Frq.	5	2	1	1	1	1

(d) Number of vertex-disjoint occurences of each pattern in G

Fig. 2. A simple graph and its vectorial descriptions

The lexicon is sorted in the order of the subgraphs network, the first value of the vector describing a graph is then the number of vertices, the second the number of edges, the third the number of subgraphs with two edges, ...

The search for patterns may be different depending on constraints. The recording (or not) of patterns that share one or several components (vertices or edges) directly influences the vectorial representation of a same graph. The use of this type of constraints can be justified by the need to find a bijection between the graph and its vectorial representation. Indeed, for the vectorial description to be closest to the graph, a pattern discovered in a graph must be removed from it. Elements (edges or vertices) can not belong to another occurrence of the same pattern. However, the complexity of the extraction of the vectorial representation with these constraints increases. Following the case, it is not necessary to apply such constraints. We give details on this point in the next section.

As a didactic example, the figure 2 represents the vectorial description associated with a undirected and unlabelled simple graph (Fig. 2(a)) for a lexicon of size 6 (order maximum of patterns : 3) without constraints (Fig. 2(b)) with edge-disjoint constraint (Fig. 2(c)) and vertex-disjoint constraint (Fig. 2(d)).

In the next section, we present two ways to use the vectorial description. We will show the advantage of being able to decline the representation according to the selected use case.

4 Examples of Measure Depending on the Case of Use

A computation of distances in a grah space can quantify the difference between two graphs. However, the complexity related to the computation of the distance between graphs prohibits its use in a case of seeking information where the purpose is to order all documents of the basis (or the k nearest) depending on their proximity to a query. The vectorial representation that we propose will enable to approximate this distance by a dissimilarity measure between structural representations of documents on the one hand, and a request expressed by the same structural mean, on the other. Even if the extraction of this vectorial representation requires a important cost, this task of indexing can be done offline and can be tolerated.

The various dimensions of the vectorial representation that we offer presents a certain redundancy. Indeed, if a pattern of order n is counted, all its predecessors are also included in the statement. This redundancy is the integration of a degree of robustness we wished to provide in our presentation. It seemed appropriate to take into consideration the disturbances that could infer on structural representations often extracted automatically : appearance or disappearance of vertices or edges. Thus, if two identical graphs have strictly identical vectorial descriptions, two graphs which one is a noisy version of the other have at least some patterns below in common.

Finally, we justify our proposal of vectorial description by the fact that it can be used in two cases of use. In the first practice, it is to find in the indexed graph database the nearest to the query graph. A second application is to find graphs of the database containing the largest number of occurrences of the query graph. Both following subsections introduced measures on our vectorial description and applying to these two cases of use.

4.1 Measuring Dissimilarity Graph to Graph

In this case of use, the purpose is to find, among a graph database, those closest to a query graph. The graphs in the database and the query graph are represented by a vectorial features. The distances between graphs of the database on the one hand and the query graph on the other hand is approximated by a dissimilarity measure. This measure of dissimilarity corresponds to the distance between the vectorial representations of graphs.

This problem has already been raised in the literature. We have chosen to use a Euclidean distance. In a n dimensions space, the distance is expressed in this form:

$$D(G_1, G_2) = \sum_{i=1}^{N} \sqrt{(k_{1i} - k_{2i})^2}$$

with G_1 and G_2 two graphs to compare, k_{i1} and k_{i2} the occurrences of the pattern i in G_1 and G_2.

Depending on the application framework, this measure can evolve. In technical document retrieval, for example, experts can provide *a priori* knowledge on semantics of symbols present in the document. This can "punish"or "prime" the presence or absence of a symbol. Of course, our approach can be expanded to all types of documents where sufficient knowledge allows to consider the relevance of patterns. That is rendered by weighting α_i each pattern i.

$$D(G_1, G_2) = \sum_{i=1}^{N} \sqrt{\alpha_i (k_{1i} - k_{2i})^2}$$

The weights α_i can then be determined by optimization algorithms or artificial learning depending on case of use. Similarly, if that vectorial representation is used on a uniforme database, it is possible to apply feature selection methods aiming at the reduction of dimensionality. The purpose of this reduction of the vector is to improve performance by removing aberrant or unnecessary patterns. It is also possible to combine patterns. There are several methods in literature, such as principal components analysis or feature selection.

In this case, the use of representations without constraints give a redundancy of the information in order to increase the precision of the vectorial representation.

4.2 Finding Occurences of a Query Graph

Other applications do not need to measure a distance between two vectors to quantify the similarities between them, but investigate the presence and number of occurrences of the query graph in a graph, even in a pre-indexed database.

For example, finding a specific electrical component in a plan is a possible application. Here, the aim is to find the number u of occurrences of the query graph S in the graph G.

V_G is the vectorial representation of G and V_S for S. There is v_{Gi} (respectively v_{Si}) the occurences number of the pattern i in G (respectively in S).

$$V_G = \begin{bmatrix} v_{G1} \\ \vdots \\ v_{GN} \end{bmatrix}$$

$$V_S = \begin{bmatrix} v_{S1} \\ \vdots \\ v_{SN} \end{bmatrix}$$

So,

$$\forall i \in \mathbb{N}, 1 \leq i \leq N, v_{Gi} \geq u.v_{Si}$$

In other words, if a graph G includes u occurrences of a query graph S, then each subgraph S is present in G u times at least.

This scheme may be applied to all subgraphs of S and particularly on the patterns of the lexicon. Thus, for all i dimensions of the vectorial representation.

$$\forall i \in \mathbb{N}, 1 \leq i \leq N, u \leq u_i = \frac{v_{Gi}}{v_{Si}} \text{ avec } v_{Si} \neq 0$$

Consequently,

$$u \leq \min_{i/v_{Si} \neq 0} \left(\frac{v_{Gi}}{v_{Si}} \right)$$

Then, the quantity $\min_{i/v_{Si} \neq 0} \left(\frac{v_{Gi}}{v_{Si}} \right)$ can be used to approximate u, the occurences number of S in G.

It is noteworthy that in this case of use, a vectorial representation extracted with tight constraints, as we have seen in the third paragraph, must be used. Indeed, research occurrences of a subgraph often needs that all occurrences must be fully present in the graph. They can not share edges or node. The choice of constraint depends on the context and then requires an *a priori* knowledge of an expert.

5 Experiments

We describe in this section the first experiments on the vectorial description of graphs we conducted. During these tests, our vectorial description of graph is compared to the vectorial description proposed by Wilfong and Lopresti.

For both descriptions, we have limited the size of the vector to dimension 6. Thus, the vectorial description Lopresti and Wilfong graphs are in the vertices of degree 1 to 6 and our description are the patterns with 3 edges or less.

Both descriptions are compared on a classification task by nearest neighbours using the technique known as "leave one out", ie that each element is ranked considering all others belong to the learning database.

The first comparison was made on synthetic graphs. In this database, each class is associated to a random generating model with two parameters that are n the number of vertices of the graph and d the average degree of vertices. 20 classes have been generated for n taking values among 5, 10, 20, 50 and 100 and d taking values 0.2, 0.5, 1 and 2. Thus, 20 graphs were generated for each class, making a database of 400 graphs.

The rate of correct classification using the description of Lopresti and Wilfong reached 99.45 %. The few confusions are between classes of graphs C (n = 5, d = 0.2) and C (n = 5, d = 0.5). The classification using the proposed description given strictly identical results, the same confusion occuring on the same elements.

In a second experiment, graphs correspond to structural representations of symbols from GREC competition (recognition of symbols). We used 10 classes of symbols which have been alterated by rotation and vectorial bending. The structural representations are adjacency regions graphs corresponding to the

Fig. 3. Examples of symbols from the GREC database

black and white connected components. The classes contain between 5 and 19 elements per class, the total number of elements of the database rise to 88. Figure 3 shows examples of symbols from the database.

The rate of correct classification obtained through the representation of Lopresti and Wilfong amounts to 51.14 % while the results obtained through our representation reached 53.41 %.

Good recognition rate achieved in the first experimentation should be put into perspective considering that generative models of random graphs as they are defined clearly distinguish the classes. Indeed, classes could be classified without any confusion by considering only the number of vertices and the number of edges.

The smaller rate classification observed on the application of recognition of symbols must be relativised because the structural representations which are used are not labelled and they only reflect the topology of the adjacency regions graphs without qualifying the nature of the region.

Table 2. Summary of the first results

	Lopresti & Wilfong method	Our method
Synthetic graphs	99.45%	99.45%
GREC	51.14%	53.41%

In one case as in others, it is interesting to note that the rate classification obtained with the two vectorial descriptions are comparable. This indicates that the description we propose can be used in a search for nearest neighbour.

6 Conclusions

In this article, we have presented a new vectorial description of graphs in order to reduce the complexity of computing distances needed for information retrieval applications with structured data. This representation is proposed to be applied to two cases of different use:

1. Finding similar graphs;
2. Finding graphs containing multiple occurrences of a query graph.

The results of preliminary tests highlight that our vectorial representation is equivalent in terms of performance to other approaches proposed in the literature to approximate distance between graphs in a task of classification.

However, we have seen throughout this article that some points remain outstanding. It now seems important to measure the influence of the size of the lexicon on the construction of the description and its complexity. Indeed, table 1 shows the need to limit the description to low rank to reduce the complexity of its construction. However, initial tests have proved that the relevance of the description was better with a important number of patterns. future work will naturally focus on the compromise between expressiveness and dimensionality of our representation.

To evaluate the relevance of our vectorial representation in the context of information retrieval to order graphs based on the number of occurrences of a query graph we have used structural representations of old document images formerly extracted from the platform AGORA (cf. [4]). We intend to build a set of queries with relevant documents to assess details and reminders for the k first documents according to the value of k.

The first experiments have shown that in the case of structural representations with many edges, the extraction of representation required a consequent computation time. The algorithmic optimizations would be realised.

Acknowledgement

The works described in this article were done with the support of the ANR within the project NAVIDOMASS ANR-06-MDCA-12.

References

1. Lopresti, D., Wilfong, G.: A fast technique for comparing graph representations with applications to performance evaluation. Int. J. Doc. Anal. Recognit. 6(4), 219–229 (2003)
2. Jaromczyk, J., Toussaint, G.: Relative neighborhood graphs and their relatives. Proceedings of the IEEE (1992)
3. Barbu, E., Heroux, P., Adam, S., Trupin, E.: Clustering document images using a bag of symbols representation, ICDAR, pp. 1216–1220 (2005)
4. Ramel, J.Y., Busson, S., Demonet, M.L.: AGORA:the Interactive Document Image Analysis Tool of the BVH Project, DIAL, pp. 145–155 (2006)

Stochastic Text Models for Music Categorization

Carlos Pérez-Sancho, David Rizo, and José M. Iñesta

Department of Software and Computing Systems
University of Alicante
P.O. box 99, E-03080 Alicante, Spain
{cperez,drizo,inesta}@dlsi.ua.es
http://grfia.dlsi.ua.es

Abstract. Music genre meta-data is of paramount importance for the organization of music repositories. People use genre in a natural way when entering a music store or looking into music collections. Automatic genre classification has become a popular topic in music information retrieval research. This work brings to symbolic music recognition some technologies, like the stochastic language models, already successfully applied to text categorization. In this work we model chord progressions and melodies as n-grams and strings and then apply perplexity and naïve Bayes classifiers, respectively, in order to assess how often those structures are found in the target genres. Also a combination of the different techniques as an ensemble of classifiers is proposed. Some genres and sub-genres among popular, jazz, and academic music have been considered. The results show that the ensemble is a good trade-off approach able to perform well without the risk of choosing the wrong classifier.

1 Introduction

Organization of large music repositories is a tedious and time-intensive task for which music genre is an important meta-data. Automatic genre and style classification have become popular topics in Music Information Retrieval (MIR) research because musical genres are categorical labels created by humans to characterize pieces of music and this nature provides genre with a high semantic and cultural information to the items in the collection.

Traditionally, the research on music genre classification has been divided into the audio [1] and symbolic [2] music analysis and retrieval domains. This work focuses on the symbolic approach, where music is found as *digital scores*. Language modeling is a common practice in natural language tasks, such as speech recognition [3], and also in text categorization [4]. Our goal is to bring stochastic language models, already applied successfully in text categorization, to music genre classification. Some of those technologies, at least those working at a lexical level, can be transferred to music with some adaptation work.

In a former work [5], we studied how melodies could be coded and classified. Here, we are going to test whether chords can be represented by a model that learns probabilities from the progressions used in each genre.

N. da Vitora Lobo et al. (Eds.): SSPR&SPR 2008, LNCS 5342, pp. 55–64, 2008.

Little attention has been paid in the genre recognition literature to how harmony can help in this task. For example, in [6] the authors use pitch histograms as feature vectors (either computed from audio signals or directly derived from MIDI data). In contrast to that low-level harmony features, the representation chosen in this paper is to model higher level harmonic structures, like chord progressions, as n-grams and strings in order to compute how often those structures are found in the target genres. There are a small number of papers in the literature that poses this problem.

To test these models, we have built a classification framework with different genres in order to see whether such a model built from a genre is able to correctly identify new chord sequences belonging to it.

2 Experimental Data

Music works from different genres covering a wide range of music areas have been utilized: popular, jazz, and academic music. Popular music data have been separated into three sub-genres: *pop*, *blues*, and *celtic* (mainly Irish jigs and reels). For jazz, three styles have been established: a *pre-bop* class grouping swing, early, and Broadway tunes, *bop* standards, and *bossanovas* as a representative of latin jazz. Finally, academic music has been categorized according to historic periods: *baroque*, *classicism*, and *romanticism*. All these categories have been defined with the help and advice of music experts that have also collaborated in the task of assigning meta-data tags to the files and rejecting outliers in order to have a reliable ground-truth for the experiments.

The number of files eventually used for each genre is displayed in Table 1. The total amount of pieces was 856, providing around 60 hours of music data.

Table 1. Number of files per genre and subgenre in the different training sets

Academic	235	**Jazz**	338	**Popular**	283
Baroque	56	Pre-bop	178	Blues	84
Classical	50	Bop	94	Pop	100
Romanticism	129	Bossanova	66	Celtic	99

The Academic and Jazz corpora have been obtained from sample files for the software 'Band in a Box'[1]. The Popular music files were downloaded from the Internet[2]. These data can be obtained upon request to the authors. In order to focus on the chord progressions instead of the chords themselves, all the songs have been transposed to C Major / A minor and the chords are encoded as their degrees for the tonality the score is in. For example, the sequence Cm – B♭ – E♭ in Fig. 1-left is coded as VIm – V – I, as it corresponds for the E♭ major key.

[1] `http://www.pgmusic.com` (PG Music)
[2] `http://www.alisdair.com/gearsoftware/biablinks.html`

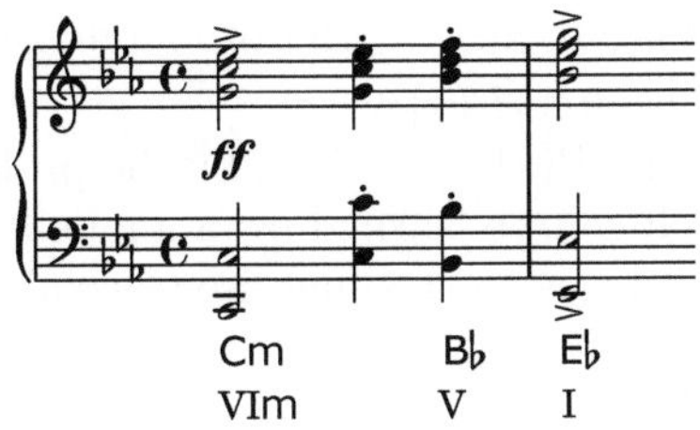

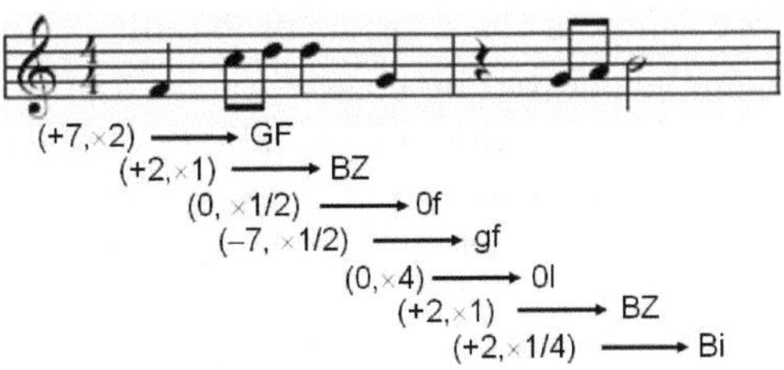

Fig. 1. Example of chord encoding (left) and melodic words (right)

The other source of information is the melody, which is represented as a sequence of intervals and duration ratios between consecutive notes (silences are not considered, being like an extension of the former note), that are mapped into sequences of alphanumeric characters (named 2-words, since they are extracted from pairs of notes). See Fig. 1-right for an example.

3 Classification Methods

For the experiments in section 4 two statistical classification methods have been used: a naïve Bayes classifier and n-grams. Both of them have been widely used in text classification tasks and, more recently, in music genre classification by melody [5,7].

Both techniques allow the construction of a statistical model for each genre from a set of documents. While n-grams make use of context information to compute the probabilities for each word, naïve Bayes computes the probability of each word on its own[3].

3.1 Naïve Bayes Classifier

The naïve Bayes classifier, as described in [8], has been used. It assumes that all words in a document are independent of each other, and also independent of the order in which they are generated. This assumption is false in our problem and also in the case of text classification, but naïve Bayes can obtain near optimal classification errors in spite of that [9].

We have a set of classes $\mathcal{C} = \{c_1, c_2, \ldots, c_{|\mathcal{C}|}\}$. A music piece is represented as a vector $\mathbf{x} \in \{0,1\}^{|\mathcal{V}|}$, where each component x_t represents whether the word w_t appears in the document or not, and $|\mathcal{V}|$ is the size of the vocabulary. A document $\mathbf{x}$ is assigned to the class $c_j \in \mathcal{C}$ with maximum a posteriori probability, in order to minimize the probability of error:

$$P(c_j|\mathbf{x}) = \frac{P(c_j)P(\mathbf{x}|c_j)}{P(\mathbf{x})} \,. \tag{1}$$

[3] We are using "document" in a metaphoric way for "music piece", while a "word" can be either a chord or a short melody fragment.

where $P(c_j)$ is the a priori probability of class c_j, $P(\mathbf{x}|c_j)$ is the probability of $\mathbf{x}$ being generated by class c_j, and $P(\mathbf{x}) = \sum_{j=1}^{|\mathcal{C}|} P(c_j)P(\mathbf{x}|c_j)$.

The class-conditional probability of a document $P(\mathbf{x}|c_j)$ is given by the probability distribution of words w_t in class c_j, which can be learned from a labelled training set of melodies, $\mathcal{X} = \{\mathbf{x}_1, \mathbf{x}_2, \ldots, \mathbf{x}_{|\mathcal{X}|}\}$, using a supervised learning method.

Each class follows a multivariate Bernoulli distribution:

$$P(\mathbf{x}|c_j) = \prod_{t=1}^{|\mathcal{V}|} x_t P(w_t|c_j) + (1 - x_t)(1 - P(w_t|c_j)) \tag{2}$$

where $P(w_t|c_j)$ are the class-conditional probabilities of each word in the vocabulary, and these are the parameters to be learned from the training set.

Bayes-optimal estimates for probabilities $P(w_t|c_j)$ can be easily calculated by counting the number of occurrences of each word in the corresponding class:

$$P(w_t|c_j) = \frac{1 + M_{tj}}{2 + M_j} \tag{3}$$

where M_{tj} is the number of documents in class c_j containing word w_t, and M_j is the total number of documents in class c_j. Also, a Laplacean prior has been introduced in the equation above to avoid zero probabilities. Prior probabilities for classes $P(c_j)$ can be estimated from the training sample using a maximum likelihood estimate:

$$P(c_j) = \frac{M_j}{|\mathcal{X}|} \tag{4}$$

Classification of new documents is performed then using Equation 1, which is expanded using Equations 2 and 4.

Feature selection. A common practice in text classification is to reduce the dimensionality of those vectors, through a feature selection process. This is done by selecting the words which contribute most to discriminate the class of a document, using a ranked list of words extracted from the training set. A widely used measure to rank these words is the *average mutual information* (AMI) [10], which gives a measure of how much information about a class is provided by each single word. Informally speaking, we can consider that a word is informative when it is very frequent in one class and less in the others.

AMI is calculated between 1: the class of a document and 2: the absence or presence of a word in the document. We define C as a random variable over all classes, and F_t as a random variable over the absence or presence of word w_t in a document, F_t taking on values in $f_t \in \{0, 1\}$, where $f_t = 0$ indicates the absence of word w_t and $f_t = 1$ indicates its presence. The AMI is calculated for each w_t as[4]:

$$I(C; F_t) = \sum_{j=1}^{|\mathcal{C}|} \sum_{f_t \in \{0,1\}} P(c_j, f_t) \log \frac{P(c_j, f_t)}{P(c_j)P(f_t)} \tag{5}$$

[4] The convention $0 \log 0 = 0$ was used, since $x \log x \to 0$ as $x \to 0$.

where $P(c_j)$ is the number of documents for class c_j divided by the total number of documents; $P(f_t)$ is the number of documents containing the word w_t divided by the total number of documents; and $P(c_j, f_t)$ is the number of documents in class c_j having a value f_t for word w_t divided by the total number of documents.

3.2 *n*-Grams

In this case, our language model is a probability distribution that assigns a probability to a progression of words $P(w_1, \ldots, w_k)$, so that the probability of each word in the sequence is dependent on its *context* $P(w_i|w_1, \ldots, w_{i-1})$.

Estimating the probabilities of such a model can be an arduous task, and maybe computationally unaffordable, when dealing with long sequences. This is why language models are often approximated using n-gram models. An n-gram is a sequence of n words in which the first $n-1$ words are considered as the context. Thus, the estimated probability of a word w_i given a context is computed as $P(w_i|w_{i-n+1}, \ldots, w_{i-1})$. Note that now we are modeling the probability of finding a given word after a context of $n-1$ words, while formerly we were coding words as symbols in a vocabulary, and probabilities were then estimated for those symbols.

In order to perform genre classification with the n-grams, a different language model must be constructed for each genre in the data set. Each sequence (song) in the data set is decomposed in n-grams of a fixed length n (see Fig. 2). Then, the probability of each different n-gram is computed as the probability of the last word given its context. This probability can be easily calculated by dividing the number of occurrences of the n-gram by the number of occurrences of its context in the given data set:

$$P(w_i|w_{i-n+1}, \ldots, w_{i-1}) = \frac{\mathcal{N}(w_{i-n+1}, \ldots, w_i)}{\mathcal{N}(w_{i-n+1}, \ldots, w_{i-1})} \quad . \tag{6}$$

Once a language model is constructed for each genre, the probability that a new progression $w = w_1, \ldots, w_k$ has been generated by model c is:

$$P_c(w) = \prod_{i=1}^{k} P_c(w_i|w_{i-n+1}, \ldots, w_{i-1}) \quad . \tag{7}$$

Thus, a test sample can be classified by following the risk minimization criterion, i.e. given the set of classes $\mathcal{C} = \{c_1, \ldots, c_{|\mathcal{C}|}\}$, each test sample is assigned to the class $\hat{c}$ of the model which holds $\hat{c} = \arg\max_c P_c(w)$.

Progression: I – V – VI – V – I

Bigrams:	I – V	V – VI	VI – V	V – I
Trigrams:	I – V – VI	V – VI – V	VI – V – I	

Fig. 2. Decomposition of a chord progression in bigrams and trigrams

Parameter smoothing. Even when the training set is big enough to build a good language model, there can be situations where we can find words in a test sample that have not been seen previously. When such situation occurs, the probability of the n-grams containing those words is zero, thus causing the probability of the whole sequence being zero by the application of Eq. (7).

To avoid this problem, it is common to use a procedure known as *smoothing*, in which a small probability is substracted from the set of known words, and then shared out among all unseen words. There are several techniques to calculate the optimal amount of probability that must be taken off, and what percentage of it must receive every unseen word.

A similar problem happens when a new sequence of words is found. This can be solved using a process known as backing-off in which the probability of a previously unseen sequence of words can be estimated using a lower order model built using $(n-1)$-grams. In this work the *Witten-Bell* discounting method [11] combined with backing-off has been used.

3.3 Classifier Ensembles

The results for different similarity models may differ substantially. This feature can be exploited by combining the outcomes of individual comparisons. Two weighted voting methods have been utilized. These approaches are described below.

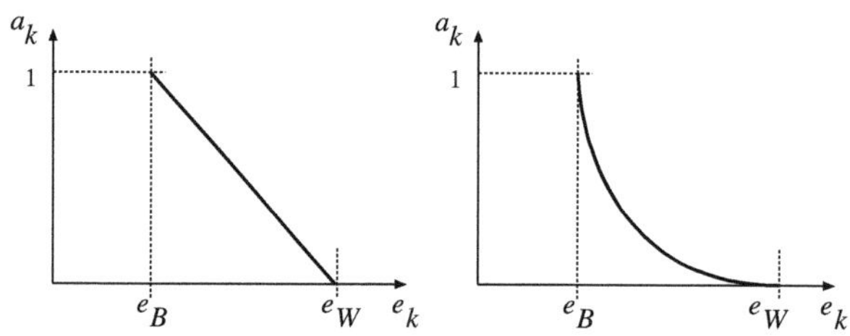

Fig. 3. Different models for giving the authority (a_k) to each classifier in the ensemble as a function of the number of errors (e_k) made on the training set

Best-worst weighted vote (BWWV). In this ensemble, the best and the worst classifiers C_k in the ensemble are identified using their estimated accuracy. A maximum authority, $a_k = 1$, is assigned to the former and a null one, $a_k = 0$, to the latter, being equivalent to removing this classifier from the ensemble. The rest of classifiers are rated linearly between these extremes (see figure 3-left). The values for a_k are calculated using the number of errors e_k as follows:

$$a_k = 1 - \frac{e_k - e_B}{e_W - e_B} ,$$

where

$$e_B = \min_k\{e_k\} \quad \text{and} \quad e_W = \max_k\{e_k\}$$

Quadratic best-worst weighted vote (QBWWV). In order to give more authority to the opinions given by the most accurate classifiers, the values obtained by the former approach are squared (see figure 3-right). This way,

$$a_k = (\frac{e_W - e_k}{e_W - e_B})^2 \quad .$$

Classification. For these voting methods, once the weights for each classifier decision have been computed, the class receiving the highest score in the voting is the final class prediction. If $\hat{c}_k(\mathbf{x}_i)$ is the prediction of C_k for the sample $\mathbf{x}_i$, then the prediction of the ensemble can be computed as

$$\hat{c}(\mathbf{x}) = \arg\max_j \sum_k w_k \delta(\hat{c}_k(\mathbf{x}_i), c_j) \quad , \tag{8}$$

being $\delta(a, b) = 1$ if $a = b$ and 0 otherwise.

4 Experiments

Experiments were performed at two levels of the genre hierarchy. The first level corresponds to the three music domains considered: academic, jazz, and popular; and the second level to the 9 sub-genres shown in Table 1. In order to build ensembles of classifiers, we need to evaluate first the performance of each single classifier so that they can be properly weighted. For this reason, the data set was randomly split in two: 80% (684 songs) were used in section 4.1 to evaluate the single classifiers and compute their weights, and the remaining 20% (172 songs) were used in section 4.2 to evaluate the performance of the ensembles.

4.1 Evaluation of the Single Classifiers

In this preliminary stage we wanted to evaluate the performance of the classification methods on two different dimensions of the data set: 1) using the two representations (harmony vs. melody), and 2) looking at the two levels of the genre hierarchy. Besides the 3- and 9-class problems, we also considered the intradomain problems, i.e. classifying among the 3 sub-genres within each domain. For this, an experiment was run for each data set, classifier, and encoding. All these experiments were validated using a 10-fold cross-validation scheme. Values of $n = \{2, 3, 4\}$ were used for the n-gram models.

Table 2 shows the results obtained in these experiments. Note that the best results were obtained when classifying among the three broad domains. In this case the use of harmony is significantly better than using the melodies. When descending to the second level in the genre hierarchy (9-class problem) there were significant differences between the use of harmony or melody: using melody helps to make better decisions. Although the performance for the 9-class problem is poorer, a 62% in the best case, note that the baseline of the recognition rate is now around 21%. Note also that the deviations for the three sub-problems in the second level of the general hierarchy are much higher than those for the 3-class problems. This shows less consistency in the performance against the different

Table 2. Average classification percentages and deviations from the 10-fold sub-experiments obtained with n-grams and naïve Bayes

	Chords (harmony)				Melodic words			
	2-grams	3-grams	4-grams	MB	2-grams	3-grams	4-grams	MB
3 classes	86 ± 3	85 ± 3	85 ± 3	86 ± 4	76 ± 5	80 ± 6	79 ± 5	76 ± 6
9 classes	36 ± 10	38 ± 12	38 ± 15	62 ± 6	55 ± 8	50 ± 8	50 ± 9	55 ± 12
academic	40 ± 20	40 ± 20	40 ± 20	54 ± 14	60 ± 20	60 ± 20	60 ± 20	62 ± 10
jazz	62 ± 7	63 ± 8	63 ± 6	72 ± 7	70 ± 10	54 ± 11	54 ± 11	73 ± 10
popular	56 ± 12	60 ± 9	60 ± 9	85 ± 6	81 ± 6	80 ± 5	81 ± 5	79 ± 6

partitions that suggests the presence of outliers. The validity of the ground-truth should be addressed in further works.

When these results are studied in more detail (see Fig. 4) one sees that the errors mainly occur within the broad domains. Misclassifications among different domains are much less frequent. We can profit from the good results achieved at the 3-classes level in order to improve the performance, by pruning the errors made between different domains using a hierarchical classifier.

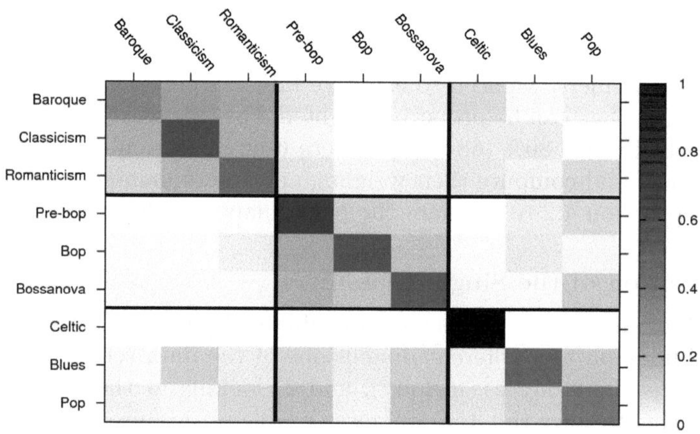

Fig. 4. Confusion matrix for the 9-class problem using classification by Bernoulli with naïve Bayes. Grey levels represent the classification percentages. Rows are the ground truth and columns the actual system output. The parts of the matrix corresponding to the different music domains have been highlighted.

4.2 Hierarchical Classification

From the analysis of the results obtained in the previous section, we propose a two-level classification scheme. At the first level of the hierarchy, decisions will be made based on harmonic information to distinguish between the academic, jazz, and popular domains. At the second level, three different classifiers will be used, each one trained with the sub-genres belonging to their respective domains. At this level we decided to use melodic information because it consistently performed better than using harmony.

At both levels, decisions were made by an ensemble of classifiers, built as explained in section 3.3. This is better than choosing just the classifier which performed the best in the previous step because there is a risk that it will not perform so well with a new data set, given the standard deviations observed in the previous experiment, specially in the intradomain problems.

The weights for each classifier in the ensembles were computed using the results shown in Table 2 and the two proposed weighting methods. As a result, we have two hierarchical classifiers, one made up of ensembles using the BWWV voting method, and another using the QBWWV method. Both classifiers were trained using the whole training set, and then evaluated using the remaining 20% of previously unseen files. We also compared this approach with the one using just the classifier which performed the best at each level. Finally, the performance of the individual classifiers was assessed with this validation set, in order to check whether there is a real improvement when using the hierarchical approach. Results of these experiments are shown in Table 3. As it can be seen, all hierarchical classifiers performed slightly better that the best single classifier, with the best results obtained when using ensembles with the QBWWV voting method.

Table 3. Classification percentages at the two levels of the hierarchy using single and hierarchical classifiers

	Single classifiers				Hierarchical		
	2-grams	3-grams	4-grams	MB	Best	BWWV	QBWWV
3 classes	87.8	88.4	88.4	87.8	88.4	90.1	90.1
9 classes	54.7	48.8	47.7	58.7	60.5	62.8	63.4

5 Conclusions and Future Work

In this paper, the feasibility of classifying music in genres using harmonic and melodic information has been tested. For this, we applied n-gram models and a naïve Bayes classifier on a corpus of music files from three broad genres (academic, jazz, and popular music), and nine sub-genres. Full chord names encoded as degrees with respect to the tonality have been used for encoding chord progressions, whereas pairs of intervals and duration ratios between consecutive notes have been used for melodies.

When classifying the three broad genres we obtained an 86% recognition rate using harmonic information, and comparable results were observed when using both the naïve Bayes classifier and n-gram models. For the nine-subgenre experiment the performance was poorer, but the system was still able to correctly classify close to a 60% of the target samples. Classification errors were mainly made between close sub-genres, inside each of the three domains.

When comparing the performance of the system using harmonic and melodic information better results were obtained by using harmony for distinguishing between the three broad domains, while the use of melodies was of better use for the intradomain problems. Thus, we have proposed a hierarchical (two-level) classification scheme in which decisions are made using harmony and melody in the first and second levels, respectively. At each level, an ensemble of classifiers is used in order to increase system stability. The results obtained with the hierarchical ensembles equalled or outperformed those obtained with the best single classifier in the nine-subgenre problem.

Acknowledgments

The authors want to thank Pedro J. Ponce de León and José L. Heredia for their help and advice. This work is supported by the Spanish PROSEMUS project (TIN2006-14932-C02) and the research programme Consolider Ingenio 2010 (MIPRCV, CSD2007-00018).

References

1. Tzanetakis, G., Cook, P.: Musical genre classification of audio signals. IEEE Transactions on Speech and Audio Processing 10, 205–210 (2002)
2. McKay, C., Fujinaga, I.: Automatic genre classification using large high-level musical feature sets. In: Proceedings of the ISMIR, pp. 525–530 (2004)
3. Jelinek, F.: Statistical Methods for Speech Recognition. MIT Press, Cambridge (1998)
4. Cavnar, W.B., Trenkle, J.M.: N-gram-based text categorization. In: Proceedings of SDAIR 1994, 3rd Annual Symposium on Document Analysis and Information Retrieval, Las Vegas, US, pp. 161–175 (1994)
5. Pérez-Sancho, C., Iñesta, J.M., Calera-Rubio, J.: Style recognition through statistical event models. Journal of New Music Research 34, 331–340 (2005)
6. Tzanetakis, G., Ermolinskyi, A., Cook, P.: Pitch histograms in audio and symbolic music information retrieval. In: Proceedings of the ISMIR, Paris, France (2002)
7. Cruz-Alcázar, P.P., Vidal, E., Pérez-Cortes, J.C.: Musical style identification using grammatical inference: The encoding problem. In: Sanfeliu, A., Ruiz-Shulcloper, J. (eds.) CIARP 2003. LNCS, vol. 2905, pp. 375–382. Springer, Heidelberg (2003)
8. Mccallum, A., Nigam, K.: A comparison of event models for naive bayes text classification (1998)
9. Domingos, P., Pazzani, M.: Beyond independence: conditions for the optimality of simple bayesian classifier. Machine Learning 29, 103–130 (1997)
10. Cover, T.M., Thomas, J.A.: Elements of Information Theory. John Wiley, Chichester (1991)
11. Witten, I., Bell, T.: The zero-frequency problem: Estimating the probabilities of novel events in adaptive text compression. IEEE Transactions on Information Theory 37, 1085–1094 (1991)

Efficient Pruning of Probabilistic Automata*

Franck Thollard and Baptiste Jeudy

Université de Lyon,
Université Jean-Monnet,
Laboratoire Hubert Curien UMR CNR 5516
{franck.thollard,baptiste.jeudy}@univ-st-etienne.fr

Abstract. Applications of probabilistic grammatical inference are limited due to time and space consuming constraints. In statistical language modeling, for example, large corpora are now available and lead to managing automata with millions of states. We propose in this article a method for pruning automata (when restricted to tree based structures) which is not only efficient (sub-quadratic) but that allows to dramatically reduce the size of the automaton with a small impact on the underlying distribution. Results are evaluated on a language modeling task.

1 Introduction

Probabilistic automata have proved to be very useful in many fields. Among these, we can note natural language processing, e.g. machine translation [1], character recognition [2], ... Unfortunately, some particular tasks – such as language modeling for speech recognition – cannot be achieved because of the time complexity of the algorithms. For example, alergia [3], acyclic-infer [2], MDI [4], DDSM [5], multinomial-infer [6] have a worst case quadratic complexity.

In these grammatical inference algorithms, an automaton representing the data is built as a first step. It is then generalized using state merging operations. In this second step, pairs of states are considered and merged if they are sufficiently close to be considered equivalent. The algorithms differ in the strategy they use to choose the pair of states to consider and by the equivalence criteria.

We propose in this article to prune the probabilistic automaton obtained in the first step by deleting some – carefully selected – states. Our goal is to get a much smaller automaton on which the inference will be efficient.

Of course, the pruned automaton is only an approximation of the data. The deleted states are therefore chosen to minimize the distance between the distributions before and after the pruning.

The next section provides the notations. Section 3 describes in more details the problem we address in Sect. 4. A new algorithm is presented in Sect. 5 which is evaluated in Sect. 6. We then conclude.

* This work was supported by the BINGO2 project (ANR-07-MDCO 014-02).

N. da Vitora Lobo et al. (Eds.): SSPR&SPR 2008, LNCS 5342, pp. 65–75, 2008.

2 Definitions and Notations

Let Σ be a *finite alphabet* and $\Sigma^\star$ (resp. Σ^+) be the set of all strings that can be built from Σ, including (resp. not including) the empty string denoted by λ. We denote $\Sigma' = \Sigma \cup \{\lambda\}$.

A *language* is a subset of $\Sigma^\star$. By convention, symbols in Σ will be denoted by letters from the beginning of the alphabet $(a, b, c, ...)$ and strings in $\Sigma^\star$ will be denoted by end of the alphabet letters $(..., x, y, z)$.

A *stochastic language* $\mathcal{D}$ is a probability distribution over $\Sigma^\star$. We denote by $P_{\mathcal{D}}(x)$ the probability of a string $x \in \Sigma^\star$ under the distribution $\mathcal{D}$ or $P_{\mathcal{D}_{\mathcal{A}}}(x)$ if the distribution is modeled by a syntactic machine $\mathcal{A}$. The distribution must verify $\sum_{x\in\Sigma^\star} P_{\mathcal{D}}(x) = 1$.

A *sample* S is a multi-set of strings: as samples are usually built through sampling, one string may appear more than once. The number of times a string x appears in S is its *multiplicity* and is denoted $|S|_x$. The *cardinality* of sample S is the total number of strings in the sample (each counted with its multiplicity): $|S| = \sum_{x\in S} |S|_x$. The empirical finite-support distribution associated with S will be denoted as $\mathcal{D}_S$ and is defined by: $P_{\mathcal{D}_S}(x) = |S|_x/|S|$.

2.1 Probabilistic Automata

Definition 1. *A Deterministic Probabilistic Finite state Automaton* DPFA *is a tuple $\mathcal{A}=\langle Q_{\mathcal{A}}, \Sigma, \delta_{\mathcal{A}}, q_0, p_{\mathcal{A}}\rangle$, where $Q_{\mathcal{A}}$ is a finite set of* states; *$q_0 \in Q_{\mathcal{A}}$ is the* initial state; *Σ is* the alphabet; *$\delta_{\mathcal{A}} : Q_{\mathcal{A}} \times \Sigma \to Q_{\mathcal{A}}$ is a* transition function; *$p_{\mathcal{A}} : Q_{\mathcal{A}} \times \Sigma' \to \mathbb{R}^+$ are* transition probabilities *such that*

$$\forall q \in Q_{\mathcal{A}}, \qquad \sum_{a\in\Sigma'} p_{\mathcal{A}}(q, a) \;=\; 1. \tag{1}$$

For each state q, the probability $p_{\mathcal{A}}(q, \lambda)$ is not associated with a transition. It represents the probability that the string ends at q. Functions $\delta_{\mathcal{A}}$ and $p_{\mathcal{A}}$ are extended recursively from Σ to $\Sigma^\star$.

The probability of a string s according to a DPFA $\mathcal{A}$ is then defined as: $P_{\mathcal{A}}(s) = p_{\mathcal{A}}(q_0, s) \times p_{\mathcal{A}}(\delta_{\mathcal{A}}(q_0, s), \lambda)$ if $s \neq \lambda$ and $P_{\mathcal{A}}(\lambda) = p_{\mathcal{A}}(q_0, \lambda)$ else. If (1) holds, these probabilities define a probability distribution $\mathcal{D}_{\mathcal{A}}$ over $\Sigma^\star$.

For a state q, $\Gamma^+(q)$ is the set $\{a \in \Sigma : p(q, a) \neq 0\}$ of the letters labeling the outgoing transitions. We also define the set of the descendants of q as $\mathrm{desc}(q) = \{p : \text{there is a path from } q \text{ to } p \text{ in } \mathcal{A}\} \bigcup \{q\}$ (we consider that $q \in \mathrm{desc}(q)$). This definition can be extended to a set of states D: $\mathrm{desc}(D) = \bigcup_{q\in D} \mathrm{desc}(q)$. Finaly, the size of a DPFA $\mathcal{A}$, noted $|\mathcal{A}|$, is defined as its number of states.

2.2 Probabilistic Prefix Trees

Definition 2. *A* probabilistic prefix tree automaton PPTA *is a particular case of* DPFA *where the underlying graph (defined by the states and the non zero probability transitions) is a tree rooted at the initial state q_0.*

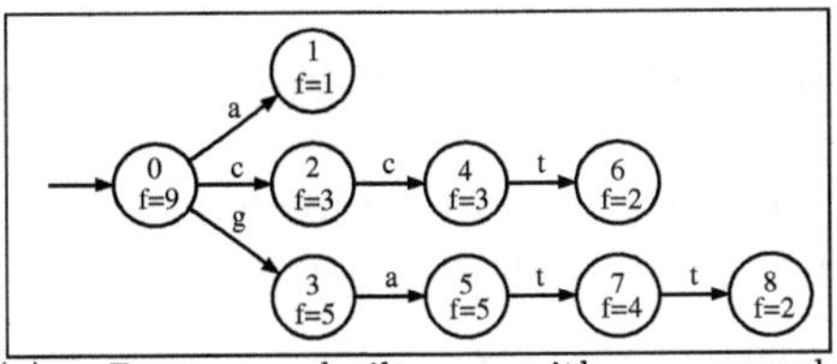

(a) PPTA built with sample $S = \{a, cct, cct, cc, gatt, gat, gat, ga, gatt\}$

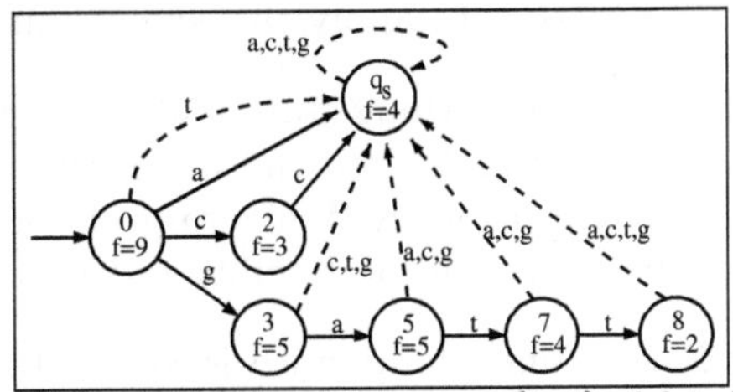

(b) Same PPTA with states $\{1, 4\}$ deleted and a smoothing state q_s added.

Fig. 1. A PPTA and a pruned smoothed PPTA. States are labeled by their numbers and their frequencies freq_S. Dashed arrows represent smoothed transitions.

A PPTA built from a sample S (denoted PPTA_S) is the prefix tree built on S. Transitions which are not in the prefix tree are added as loops with zero probabilities. Other probabilities are estimated using state frequencies: the state frequency of $q \in Q$ with respect to a sample S is defined as $\text{freq}_S(q) = \sum_{x \in \{uv \in S : \delta(q_0, u) = q\}} |S|_x$. The transition probability $p(q, a)$ is then estimated by $\text{freq}_S(\delta(q, a))/\text{freq}_S(q)$ and $p(q, \lambda)$ is computed in order to satisfy (1). Given this construction, the distribution induced by PPTA_S and $\mathcal{D}_S$ are identical. More details on probabilistic automata can be found in [7].

Figure 1(a) presents a PPTA built from the sample $S = \{a, cct, cct, cc, gatt, gat, gat, ga, gatt\}$.

2.3 Smoothing Probabilistic Automata

Smoothing discrete distributions is a widely studied field (see [8] for a survey). It has been addressed mainly in real world applications (speech recognition [8], information retrieval [9], ...) as it has shown to dramatically improve the estimate of discrete distributions. Moreover, smoothing must be taken into account from the theoretical point of view (see [2,10] for some links between smoothing and learnability of probabilistic automata). We will therefore consider here smoothed automata.

When the main model cannot parse the input string, a back-off transition is dynamically built that goes to a back-off model (an unigram model) in which the end of the parsing is done. The probability $P_{\mathcal{U}}(a), a \in \Sigma \cup \{\lambda\}$ is estimated by $|a|/||S||$ where $|a|$ is the frequency of the letter a in S and $||S||$ the total number of symbols in S. This smoothing technique uses a parameter $0 < \varepsilon < 1$. The smoothed transition probability p_{smooth} is estimated by:

$$p_{\text{smooth}}(q, a) = \begin{cases} \frac{\text{freq}(\delta(q,a)) - \varepsilon}{\text{freq}(q)} & \text{if } a \in \Gamma^+(q) \\ K.P_{\mathcal{U}}(a) & \text{else} \end{cases} \tag{2}$$

$$p_{\text{smooth}}(q, \lambda) = \begin{cases} \frac{\left(\text{freq}(q) - \sum_{a \in \Gamma^+(q)} \text{freq}(\delta(q,a))\right) - \varepsilon}{\text{freq}(q)} & \text{if } p(q, \lambda) \neq 0 \\ K.P_{\mathcal{U}}(\lambda) & \text{if } p(q, \lambda) = 0 \end{cases} \tag{3}$$

where value K is chosen to ensure that (1) holds.

In practice, to apply this smoothing, the PPTAis modified in three steps:

1. a new smoothing state (called q_s) is added. For all a in Σ, $\delta(q_s, a) = q_s$ and for all a in Σ', $p(q_s, a) = P_{\mathcal{U}}(a)$;
2. all transitions with a zero probability in the PPTA are redirected to q_s;
3. the transition probability p is replaced by the smoothed one p_{smooth}.

If this smoothing state is added to a PPTA, the resulting automaton is no longer strictly speaking a PPTA. We call such an automaton a *smoothed* PPTA. In the following, all PPTAs are assumed to be smoothed.

Fig. 1(b) presents the smoothed automaton obtained after pruning states 1 and 4.

2.4 Deleting States

We will prune a given automaton $\mathcal{A} = \langle Q, \Sigma, \delta, q_0, p\rangle$ by removing a state q_d ($q_d \neq q_s$). In order to maintain a complete automaton, the incoming transitions of q_d are redirected to q_s. After the deletion of q_d, several states may become unreachable from the initial state and are also deleted. This set of unreachable states is denoted $U_{\mathcal{A}}(q_d)$. The formal definition of the automaton $\mathcal{A}_{\overline{q_d}} = \langle Q', \Sigma, \delta', q_0', p'\rangle$ obtained from $\mathcal{A}$ by deleting q_d is then: $Q' = Q \setminus U_{\mathcal{A}}(q_d)$; $\delta'(q, a) = q_s$ if $\delta(q, a) = q_d$, else $\delta'(q, a) = \delta(q, a)$; $q_0' = q_0$; p' is the restriction of p to Q'. Note that the obtained automaton still respects equation (1) and thus defines a stochatic language.

The deletion operation is commutative (*i.e.*, removing state q and then state q' is equivalent to removing state q' and then state q) and we can therefore define $\mathcal{A}_{\overline{D}}$, the automaton obtained from $\mathcal{A}$ by deleting a set of states D.

Given an automaton $\mathcal{A}$ and two sets of states D and D', if they have the same set of descendants (*i.e.*, $\text{desc}(D) = \text{desc}(D')$), then $\mathcal{A}_{\overline{D}} = \mathcal{A}_{\overline{D'}}$ (for instance, the PPTA of figure 1(b) can be obtained by deleting $D = \{1, 4, 6\}$ or $D = \{1, 4\}$). In a PPTA, there is a minimal set D among those having the same descendants ($\{1, 4\}$ in our example). We call this set a *cut set of state* and is formally defined as a set D such that for every q, q' in D, $q \notin \text{desc}(q')$.

Next section presents the dissimilarity measure used to quantify the modification of the distribution induced by the pruning method.

2.5 Kullback-Leibler Divergence

The dissimilarity between two distributions will be evaluated through the Kullback-Leibler divergence [11]:

$$KL(\mathcal{D}, \mathcal{D}') = -\sum_{x \in \Sigma^\star} \mathcal{D}(x) \log \frac{\mathcal{D}(x)}{\mathcal{D}'(x)} \tag{4}$$

Although this divergence is not a metric, some nice properties hold, *e.g.* it is positive and it bounds the L_1 distance from above [12]. The divergence is not symmetric. One of the distribution thus takes the role of a reference one. In term of information theory, $KL(\mathcal{D},\mathcal{D}')$ represents the number of bits one must pay by coding messages drawn according to $\mathcal{D}$ using an optimal code derived from $\mathcal{D}'$. In our case, the distribution represented by the PPTA will serve as the reference.

3 Problems Setting

Depending on the application, the PPTA can be very large (*e.g.*, several million states). When processing these models (*e.g.*, applying a grammatical inference algorithm, or using such model *as is*), the size of the model can be problematic (most of inference algorithms have a quadratic complexity). A reduction of the size of the model is thus needed.

We follow an approach consisting in doing lossy compression: we want to find a trade-off between the size of the automaton and divergence. We define the loss function between two automata $\mathcal{A}$ and $\mathcal{A}'$ by:

$$\mathcal{L}(\mathcal{A},\mathcal{A}') = \frac{KL(\mathcal{A},\mathcal{A}')}{|\,|\mathcal{A}| - |\mathcal{A}'|\,|}. \tag{5}$$

In our case, the automaton $\mathcal{A}'$ is obtained from $\mathcal{A}$ by deleting some states. We will focus on two optimization problems:

1. Find the cut set of states D such that $\mathcal{L}(\mathcal{A},\mathcal{A}_{\overline{D}})$ is minimal.
2. In the previous problem, there is no control on the number of deleted states. In practice, we want to delete a significant number of states. The second problem is thus to find the optimal cut set of states D with the constraint that the total number of deleted states $|\text{desc}(D)|$ is above a given threshold.

4 Problems Solutions

In this section, the goal is to find efficiently the optimal set of states D. Given a sample S, the algorithm will proceed in two steps: first build the PPTA_S and then find the set D. In order to be efficient even with large automata, the numerator of (5) must be computed efficiently.

4.1 Computing the Divergence between Two Automata

Carrasco [13] proposed an efficient way to compute the divergence between two distributions when they are represented by two DPFA $\mathcal{A} = \langle Q,\Sigma,\delta,q_0,p\rangle$ and $\mathcal{A}' = \langle Q',\Sigma,\delta',q_0',p'\rangle$. The computation of $KL(\mathcal{A},\mathcal{A}')$ is made by summing state divergences:

$$KL(\mathcal{A},\mathcal{A}') = \sum_{q\in Q}\sum_{q'\in Q'} kl(q,q') \quad \text{with} \quad kl(q,q') = c_{q,q'} \sum_{a\in\Sigma'} p(q,a)\log\frac{p(q,a)}{p'(q',a)} \tag{6}$$

and

$$c_{q,q'} = \sum_{x \in L_{q,q'}} p(q_0, x) \text{ where } L_{q,q'} = \{x \in \Sigma^\star : \delta(q_0, x) = q \wedge \delta'(q'_0, x) = q'\} \quad (7)$$

If $\mathcal{A}$ and $\mathcal{A}_{\overline{\{q_1,q_4\}}}$ are the automata of the running example, we have $L_{7,7} = \{gat\}$, $L_{6,q_s} = \{cct\}$, $L_{5,2} = \emptyset$, $c_{7,7} = p_{\mathcal{A}}(q_0, gat) = \frac{5-\epsilon}{9} \times \frac{5-\epsilon}{5} \times \frac{4-\epsilon}{5}$ (where ϵ is the smoothing parameter, see (2), Sect. 2.3).

Equation (6) is very general and applicable to any automata. Its computation complexity is $O(|\mathcal{A}|.|\mathcal{A}'|.|\Sigma|)$ times the complexity of the computation of the coefficients $c_{q,q'}$. In his paper, Carrasco gives an iterative method that converges to the values of these coefficients.

4.2 Pruning Automata

In this section, (6) is simplified in the case where $\mathcal{A}' = \mathcal{A}_{\overline{D}}$ for some set of states D. The automaton $\mathcal{A} = \langle Q, \Sigma, \delta, q_0, p\rangle$ is a complete DPFA with a smoothing (or unigram) state denoted by q_s. We consider the deletion of a set of states D of $\mathcal{A}$ ($q_s \notin D$). The resulting DPFA is $A_{\overline{D}} = \langle Q', \Sigma, \delta', q'_0, p'\rangle$.

Lemma 1

$$KL(\mathcal{A}, \mathcal{A}_{\overline{D}}) = \sum_{q \in desc(D)} kl(q, q_s) \quad (8)$$

The (omitted) proof of this lemma is based on the fact that most of the $kl(q, q')$ are zero because either $c(q, q') = 0$ or $p(q, a) = p'(q', a)$. The complexity of computing $KL(\mathcal{A}, \mathcal{A}_{\overline{D}})$ is therefore less than $O(|\mathcal{A}|.|\Sigma|)$ times the complexity of the computation of the coefficients c_{q,q_s}. However, we still need to find the optimal set of states D, and it is of course not possible to test all subsets of Q.

Another important point is that, in a PPTA, c_{q,q_s} does not depend on the set of deleted states D and therefore neither does $kl(q, q_s)$. It means that it is possible to compute $kl(q, q_s)$ for all q and then consider several possibilities for the cut set of states D without recomputing the kl values.

4.3 Solutions of the Optimization Problems

Given a cut set of states D and following (8), the loss function is now

$$\mathcal{L}(\mathcal{A}, \mathcal{A}_{\overline{D}}) = \frac{KL(\mathcal{A}, \mathcal{A}_{\overline{D}})}{|\mathcal{A}| - |\mathcal{A}_{\overline{D}}|} = \frac{\sum_{q \in \text{desc}(D)} kl(q, q_s)}{|\text{desc}(D)|} \quad (9)$$

which is the average value of $kl(q, q_s)$ on all deleted states $q \in \text{desc}(D)$.One can easily shows that the optimal cut set of states D (for our first problem) consists of only one state: the state q_{opt} that minimizes the average kl of its descendants. The whole set of deleted states is therefore limited to the branch $\text{desc}(q_{\text{opt}})$.

Preliminary experiments showed that this branch is often reduced to only one state in practice. Since our goal is to delete a significant number of states, we will focus on the second problem in which we want to delete at least nb states. It can be solved with a recursive algorithm. For each subtree of the PPTA (rooted in q) and for all $0 \leq i \leq nb$, the algorithm computes the optimal cut set of states $D(q,i) \subseteq \mathrm{desc}(q)$ to delete with the constraints that $|\mathrm{desc}(D(q,i))| > i$ and $\mathcal{L}(\mathcal{A}, \mathcal{A}_{\overline{D(q,i)}})$ is minimal. However, this algorithm has a complexity $O(nb.|\mathcal{A}|.|\Sigma|)$ which is too high for large PPTA and large values of nb (we want to be able to have nb of the order of $|\mathcal{A}|$). We present a heuristic in the next section.

5 Algorithm and Complexity

We focus here on our second problem, i.e., to find a cut set of states $D = \{q_1, .., q_k\}$ such that:

- $|\mathrm{desc}(D)| > nb$;
- D minimizes $\mathcal{L}(\mathcal{A}, \mathcal{A}_{\overline{D}}) = \frac{\sum_{1 \leq i \leq k} |\mathrm{desc}(q_i)|.v(q_i)}{|\mathrm{desc}(D)|}$ with $v(q_i) = \frac{\sum_{q \in \mathrm{desc}(q_i)} kl(q, q_s)}{|\mathrm{desc}(q_i)|}$ which is the weighted average of the $v(q_i)$ with weights $|\mathrm{desc}(q_i)|$.

If the q_i could be chosen without constraints in Q, this problem could be solved by:

1. For each state $q \in Q$, compute $v(q)$ and $|\mathrm{desc}(q)|$ (these values can be computed in time $O(|\mathcal{A}|)$ using a recursive traversal of the PPTA);
2. Construct a list $L =< q_1, ... >$ by sorting the states q in increasing order of $v(q)$ in time $O(|\mathcal{A}|.\log|\mathcal{A}|)$ (notice that the first element of this list is q_{opt}, the solution of the first problem);
3. Take the shortest prefix $< q_1, ..., q_k >$ of L such that the sum of the weights of q_1,...,q_k is greater than nb.
4. Take $\{q_1, ..., q_k\}$ as D.

This algorithm is optimal and has complexity $O(|\mathcal{A}|.\log|\mathcal{A}|)$. However, in our problem, the set D must be a cut set of states, *i.e.*, there cannot exist two states p and q in D such that $p \in \mathrm{desc}(q)$. Our heuristic is to replace the third and fourth steps by:

3'. Take the shortest prefix $< q_1, ..., q_k >$ of L such that, when considering in this prefix only the states that are not descendant of one another, the sum of their weights is greater than nb. This can be done in time $O(|\mathcal{A}|)$ by marking each descendant of each state q_i and summing only the weights of the non marked states.
4'. Take the non marked states of this prefix as D.

The complexity of this algorithm is therefore $O(|\mathcal{A}|.\log|\mathcal{A}|)$.

6 Experimentations

We will evaluate the pruning first by analyzing the increment of the KL w.r.t. the number of states pruned. Then we measure its impact on grammatical inference by comparing the result of the inference when started from a pruned PPTA and from a non-pruned PPTA.

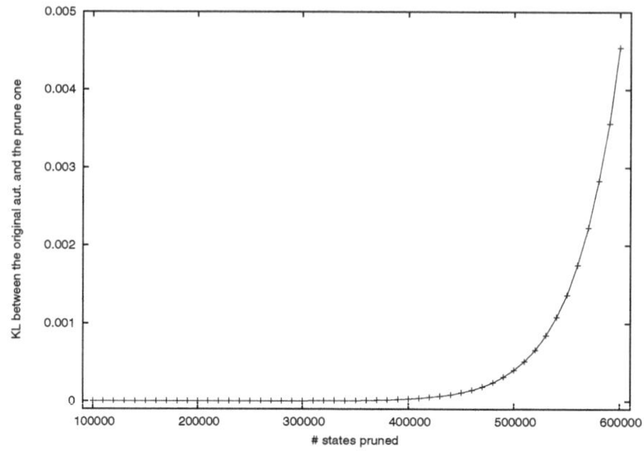

Fig. 2. KL w.r.t. the number of states pruned

6.1 KL Behavior with Respect to the Number of States Pruned

In order to evaluate the quality of the pruning procedure, the second algorithm is used to compute a cut set of state D for different pruning thresholds nb (minimal number of states to delete).

Figure 2 presents the divergence $KL(\mathcal{A}, \mathcal{A}_{\overline{D}})$ with respect to nb. The input PPTA $\mathcal{A}$ has 868,851 states (built on the wall street journal task, see section 6.3). We expect the KL to increase monotonically. It is however noticeable that it is almost flat up to 400,000 states pruned (that is 46% of the automaton's states). The pruning is in practice efficient as it takes less than 20 seconds on a recent PC machine[1] (including Input/Output time). This value is negligible with respect to the 91 hours needed for the inference on the whole learning set (see table Sect. 6.3).

This means that the distribution $\mathcal{D}_{\mathcal{A}_{\overline{D}}}$ represented by the pruned automaton is close to the distribution represented by $\mathcal{A}$ even when a large number of the states are deleted. We thus expect to be able to use the pruned model $\mathcal{A}_{\overline{D}}$ as a replacement for the PPTA $\mathcal{A}$ built with the full training sample.

6.2 Evaluation Measure

We will evaluate the pruning method in the context of language modeling. In many probabilist grammatical inference algorithms [3,4,5,6], a PPTA is built from a sample as a first step of the inference. It is then generalized by merging equivalent states. The algorithms differ in the way equivalence of states is considered.

These algorithms are evaluated through the ability of the resulting automata to parse new strings, that is by the *per symbol log-likelihood* of strings x belonging to a test sample S according to its underlying distribution:

[1] The machine used is a linux box with a 3 Ghz processor and 1 GB of memory.

$$LL(S) = \left(-\frac{1}{\|S\|} \sum_{x \in S} |S|_x \sum_{i=1}^{|x|} \log p_{\mathcal{A}}(\delta_{\mathcal{A}}(q_0, x_1 x_2 ... x_{i-1}, x_i) \right)$$

where x_i is the i-th symbol of x and $|S|_x$ its multiplicity in S. This average log-likelihood is related to the KL divergence between an unknown target distribution and the hypothesis by considering the test sample as the empirical estimates of the unknown distribution [11].

The *test sample perplexity* $PP(S)$ is most commonly used for evaluating language models. It is given by $PP(S) = 2^{LL(S)}$. The minimal perplexity $PP = 1$ is reached when the next symbol x_i is always predicted with probability 1 from the current state q_i (*i.e.* $p(\delta(q_0, x_1...x_{i-1}), x_i) = 1$) while $PP = |\Sigma|$ corresponds to random guessing from an alphabet of size $|\Sigma|$.

6.3 The Wall Street Journal Task

The data used are drawn from the Wall Street Journal database, a large syntactically annotated corpus subdivided in 25 sections. We followed the preprocessing used by [14]: Sections 0 to 20 were used as the training set (962,612 words), sections 21 and 22 were used as a development set (48,024 words) and sections 23 and 24 serve as the test (101,189 words). Digit numbers were replaced by a unique character. Following [14], the 10,000 most frequent words were kept and the remainder were transformed in a unique symbol `unknown`. In order to make the data more realistic for a speech to text task, the punctuation was removed and the words transformed in their lower case form. The average length of the sentences is 22 words and the size of Σ is around 10000.

We applied the pruning algorithm on the PPTA built on the whole training set (868,851 states) with different levels of pruning (number of deleted states ranging from 100K to 600K). The MDI [4] algorithm is then applied on each of these pruned PPTAs. The MDI algorithm depends on a parameter α that controls the generalization rate. To choose the value of this parameter, several inferences were performed on each PPTA with α ranging from 0.00008 to 0.01. The value that gives the best perplexity on the development set is chosen. The following table gives the best value of α and the total time spent to compute all automata for all values of α.

Then, the best automaton (corresponding to the best value of α) is evaluated by its perplexity computed on the test set (see table). The size of this best automaton is also given (# states post inference).

Test set results of the inference using pruned PPTA

% states pruned	0	11%	23%	35%	46 %	57%	69%
# states pruned (×1000)	0	100	200	300	400	500	600
α ($\times 10^{-4}$)	2	2	2	2	2	1	1
Total time (h)	91	72	61	47	60	20	8
$PP_{MDI}(test)$	549	550	564	547	564	567	568
# states post inference	451	463	458	422	434	1989	1863

7 Discussion

As can be seen in the above table, the test set perplexity of the inferred model does not change significantly up to a pruning level of 35% of the states. This is consistent with Fig. 2. We can also notice that the pruning has little influence over the size of the inferred automaton (at least for pruning levels below 46%).

It is worth stressing that the optimal value of the parameter α at different pruning thresholds does not change a lot. The pruning can thus be used to estimate the optimal parameter α in a less demanding situation. Only one inference will then be done on the full PPTA using the optimal setting of the pruned one.

Related work : language modeling is classically done using n-grams. As the size of the corpora increases, increasing the size of n improves the prediction. Unfortunately, the size of the model dramatically increases in n. In [15] Stolcke proposed to prune the n-gram model using an entropy based criterion similar to ours. As it is applied on n-gram, such a pruning will remove the estimation of part of sentences and not end of sentences as we do. Another main difference is the length of the removed chunk. In our case, the size of the chunk we can remove is not bounded as with n-gram. If a very long sentence appears once, our pruning method will remove the whole branch, which will not be the case with n-gram.

Regarding the prediction performances, we need to mention that the MDI algorithm is outperformed by the unpruned trigram model that obtains a test set perplexity of 165[2]. Many improvements can nevertheless be achieved at the grammatical inference level: use another algorithm [5,6,16], or use preprocessing techniques (*e.g.* word clustering [17], bagging [18]) or post-processing techniques (*e.g.* combining
automata [5]).

8 Conclusion and Further Work

We presented in this article an efficient way to prune a smoothed PPTA. We showed that we can dramatically reduce the size of the automaton keeping a great similarity between the distributions represented by the original automaton and the pruned one. We then showed, on a reasonable language modelling task, that the pruning can be seen as a preprocessing technique before applying grammatical inference algorithm. The performance of the learning algorithm is not changed up to a pruning level of 35% of the states of the initial automaton.

From a theoretical view point, we intend to study pruning of general automata, which leads to both computational and theoretical problems, since the efficient computation of $\mathcal{A}_{\overline{q}}$ is more complicated to achieve in the general case. From the practical point of view, it is of great interest since the pruning could be applied as a post-processing step after the inference, leading to smaller models that perform similarly.

[2] The trigram considered here used a Kneyser-Ney smoothing [8].

References

1. Casacuberta, F., Vidal, E.: Machine translation with inferred stochastic finite-state transducers. Computational Linguistics 30(2), 205–225 (2004)
2. Ron, D., Singer, Y., Tishby, N.: On the learnability and usage of acyclic probabilistic finite automata. In: Proceedings of COLT 1995, pp. 31–40 (1995)
3. Carrasco, R.C., Oncina, J.: Learning stochastic regular grammars by means of a state merging method. In: Carrasco, R.C., Oncina, J. (eds.) ICGI 1994. LNCS, vol. 862, pp. 139–152. Springer, Heidelberg (1994)
4. Thollard, F., Dupont, P., de la Higuera, C.: Probabilistic DFA inference using Kullback-Leibler divergence and minimality. In: Langley, P. (ed.) ICML. Morgan Kaufmann, San Francisco (2000)
5. Thollard, F.: Improving probabilistic grammatical inference core algorithms with post-processing techniques. In: ICML 2001, pp. 561–568. Morgan Kaufmann, San Francisco (2001)
6. Kermorvant, C., Dupont, P.: Stochastic grammatical inference with multinomial tests. In: Adriaans, P.W., Fernau, H., van Zaanen, M. (eds.) ICGI 2002. LNCS, vol. 2484, pp. 149–160. Springer, Heidelberg (2002)
7. Vidal, E., Thollard, F., de la Higuera, C., Casacuberta, F., Carrasco, R.C.: Probabilistic finite-state machines – Part I and II. IEEE trans. on PAMI 27(7) (2005)
8. Goodman, J.: A bit of progress in language modeling. Technical report, Microsoft Research (2001)
9. Zhai, C., Lafferty, J.: A study of smoothing methods for language models applied to information retrieval. ACM Trans. on Information Systems 22(2), 179–214 (2004)
10. Clark, A., Thollard, F.: Pac-learnability of probabilistic deterministic finite state automata. JMLR 5, 473–497 (2004)
11. Cover, T.M., Thomas, J.A.: Elements of Information Theory. Wiley Series in Telecommunications. John Wiley & Sons, Chichester (1991)
12. Abe, N., Warmuth, M.: On the computational complexity of approximating distributions by probabilistic automata. Machine Learning 9, 205–260 (1992)
13. Carrasco, R.C.: Accurate computation of the relative entropy between stochastic regular grammars. RAIRO TIA 31(5), 437–444 (1997)
14. Charniak, E.: Immediate-head parsing for language models. In: 10th Conf. of the Association for Computational linguistic, ACL 2001 (2001)
15. Stolcke, A.: Entropy-based pruning of backoff language models. In: DARPA Broadcast News Transcription and Understanding Workshop, pp. 270–274 (1998)
16. Callut, J., Dupont, P.: Learning partially observable markov models from first passage times. In: Kok, J.N., Koronacki, J., Lopez de Mantaras, R., Matwin, S., Mladenič, D., Skowron, A. (eds.) ECML 2007. LNCS, vol. 4701, pp. 91–103. Springer, Heidelberg (2007)
17. Dupont, P., Chase, L.: Using symbol clustering to improve probabilistic automaton inference. In: Honavar, V.G., Slutzki, G. (eds.) ICGI 1998. LNCS (LNAI), vol. 1433, pp. 232–243. Springer, Heidelberg (1998)
18. Thollard, F., Clark, A.: Shallow parsing using probabilistic grammatical inference. In: Adriaans, P.W., Fernau, H., van Zaanen, M. (eds.) ICGI 2002. LNCS, vol. 2484, pp. 269–282. Springer, Heidelberg (2002)

Position Models and Language Modeling*

Arnaud Zdziobeck and Franck Thollard

Université de Lyon
Université Jean Monnet, Saint-Étienne
Laboratoire Hubert Curien, UMR CNRS 5516
arnaud.zdziobeck@bvra.univ-st-etienne.fr,
thollard@univ-st-etienne.fr

Abstract. In statistical language modelling the classic model used is n-gram. This model is not able however to capture long term dependencies, *i.e.* dependencies larger than n. An alternative to this model is the probabilistic automaton. Unfortunately, it appears that preliminary experiments on the use of this model in language modelling is not yet competitive, partly because it tries to model too long term dependencies. We propose here to improve the use of this model by restricting the dependency to a more reasonable value. Experiments shows an improvement of 45% reduction in the perplexity obtained on the Wall Street Journal language modeling task.

1 Introduction

In statistical language modelling, the n-gram model is broadly used. This model allows to model dependencies of size n. By using non shadowing smoothing, the n-gram model will also capture dependencies smaller than its order n.

With the increase of the size of the corpora, researchers attempt to increase the size of the history in n-gram models. Unfortunately, the data sparseness problem is getting more an more important as the size of the order of the model (*i.e.* the history) augments. Moreover the size of the model dramatically increases, even if some pruning techniques can be applied with a small impact on the prediction power of the model [15,16].

As the improvements with the n-gram model tend to be harder and harder to obtain, another strategy could be to consider other models. An alternative is to use probabilistic automata, which, in general, have a greater expressive power. On the contrary to the non probabilistic case, non-deterministic automata have a greater power of expression than the deterministic one. See [20] for a presentation of probabilistic finite state machines and their associated algorithms. As the use of non-deterministic machines is much more complex (*e.g.* parsing for example is much efficient in the non deterministic case), most of the work has been concentrated on deterministic ones.

Even if many algorithms exist for inferring deterministic probabilistic finite state automata (DPFA for short), their successful application mainly appears in

* This work was supported by the BINGO2 project (ANR-07-MDCO 014-02).

N. da Vitora Lobo et al. (Eds.): SSPR&SPR 2008, LNCS 5342, pp. 76–85, 2008.

contexts with small vocabularies and short sentences (*e.g* character recognition [14], shallow parsing [18]). Unfortunately, work is still needed in the domain of statistical language modelling as size of the vocabulary is generally large, and more importantly, sentences length are, on average, quite long.

We think that DPFA can be accurate up to a moderate length, that is on part of sentences. We follow here a windowing approach in which the sentences to model are split into windows of a given size, and a model is learned on each window. The parsing is then made using all the automata.

Next section presents the notations used. We then present some preliminary experiments that demonstrate the relevance of modeling strings by chunks. Our experimental set up will follow together with the results. We then conclude.

2 Definitions and Notations

Let Σ be a *finite alphabet* and $\Sigma^\star$ (resp. Σ^+) be the set of all strings that can be built from Σ, including (resp. not including) the empty string denoted by λ. We denote $\Sigma' = \Sigma \cup \{\lambda\}$. A *language* is a subset of $\Sigma^\star$.

A *stochastic language* $\mathcal{D}$ is a probability distribution over $\Sigma^\star$. We denote by $P_\mathcal{D}(x)$ the probability [1] of a string $x \in \Sigma^\star$ under the distribution $\mathcal{D}$. The distribution must verify $\sum_{x\in\Sigma^\star} P_\mathcal{D}(x) = 1$. If the distribution is modeled by some syntactic machine $\mathcal{A}$, the probability of x according to the probability distribution defined by $\mathcal{A}$ is denoted $P_\mathcal{A}(x)$.

A *sample* S is a multi-set of strings: as samples are usually built through sampling, one string may appear more than once. The number of times a string x (resp a symbol x_i) appears in S is denoted $|S|_x$ (resp. $|S|_{x_i}$). The number of symbols $x_i \in \Sigma$ in S is denoted by $||S||$.

The goal of a statistical language model consists in providing a probability to strings of $\Sigma^\star$. If the length of x is know, this can be estimated using the chain rule:

$$P(x = x_1x_2\dots x_m) = P(x_1) \times \prod_{l=2}^{m} P(x_l \mid x_1 \dots x_{l-1}) \quad (1)$$

This decomposition requires the estimate of a long dependency term $P(x_l \mid x_1 \dots x_{l-1})$.

In the following, we recall the n-gram model in which a bound on the history is assumed, and the probabilistic automata which can model unbound dependencies.

2.1 The n-Gram Model

The n-gram assumption considers that

$$P(x_l \mid x_1 \dots x_{l-1}) \sim P(x_l \mid x_{l-(n-1)} \dots x_{l-1})$$

[1] As usual, we denote $P(x)$ as $P(X = x)$ for any discrete random variable X.

that is that the probability of a symbol depends only on the $n-1$ preceding ones.

Each probability is estimated using maximum likelihood:

$$P(x_l \mid x_{l-(n-1)} \dots x_{l-1}) = \frac{|S|_{x_{l-n+1}x_{l-n+2}\dots x_l}}{|S|_{x_{l-n+1}x_{l-n+2}\dots x_{l-1}}}$$

A special case of the n-gram model is the unigram model ($n = 1$), noted $\mathcal{U}$: for each symbol $x_i \in \Sigma$, $P_{\mathcal{U}}(x_i) = \frac{|S|_{x_i}}{||S||}$. This model can be represented as a one state automaton, with a looping transition for each symbol of the alphabet.

When dealing with large vocabularies, the denominator of the previous equation can have a null estimate. Hence a smoothing strategy needs to be set in order to provide a non null probability on each n-gram.

2.2 Probabilistic Automata

Definition 1. *A Deterministic Probabilistic Finite state Automaton* DPFA *is a tuple* $\mathcal{A}=\langle Q_{\mathcal{A}}, \Sigma, \delta_{\mathcal{A}}, q_0, p_{\mathcal{A}}\rangle$, *where:*

- $Q_{\mathcal{A}}$ *is a finite set of* states*;*
- $q_0 \in Q_{\mathcal{A}}$ *is the* initial state*;*
- Σ *is* the alphabet*;*
- $\delta_{\mathcal{A}} : Q_{\mathcal{A}} \times \Sigma \rightarrow Q_{\mathcal{A}}$ *is a* transition function*;*
- $p_{\mathcal{A}} : Q_{\mathcal{A}} \times \Sigma' \rightarrow \mathbb{R}^+$ *are* transition probabilities *such that*

$$\forall q \in Q_{\mathcal{A}}, \qquad \sum_{a \in \Sigma'} p_{\mathcal{A}}(q, a) \;=\; 1. \qquad (2)$$

For each state q, the probability $p_{\mathcal{A}}(q, \lambda)$ is not associated with a transition. It represents the probability that the string ends at q.

Functions $\delta_{\mathcal{A}}$ and $p_{\mathcal{A}}$ can be extended recursively from Σ to $\Sigma^\star$.

The probability of a string s according to a DPFA $\mathcal{A}$ is then defined as: $P_{\mathcal{A}}(s) = p_{\mathcal{A}}(q_0, s) \times p_{\mathcal{A}}(\delta_{\mathcal{A}}(q_0, s), \lambda)$ if $s \neq \lambda$ and $P_{\mathcal{A}}(\lambda) = p_{\mathcal{A}}(q_0, \lambda)$ else. If (2) holds, these probabilities define a probability distribution $\mathcal{D}_{\mathcal{A}}$ over $\Sigma^\star$ [20].

This definition of DPFA considers only *complete* automata, *i.e.* functions p and δ are defined for all pairs of state and symbol. If an automaton is not complete, then transitions with zero probabilities can be added to get a complete automaton without changing its distribution $\mathcal{D}_{\mathcal{A}}$.

The automaton considered being deterministic and the size of the vocabulary being constant through the paper, the size of a DPFA $\mathcal{A}$ is defined as its number of states and will be noted $|\mathcal{A}|$.

Figure 1 presents a smoothed DPFA, smoothing being drawn using dash lines. The probability of string cb according to this automaton is $\frac{8}{32} \times \frac{16}{32} \times \frac{16}{16}$.

3 Building Language Models

As seen in section 2.1, the n-gram model is estimated by counting term frequencies in a training corpus. We address in this section the building of probabilistic

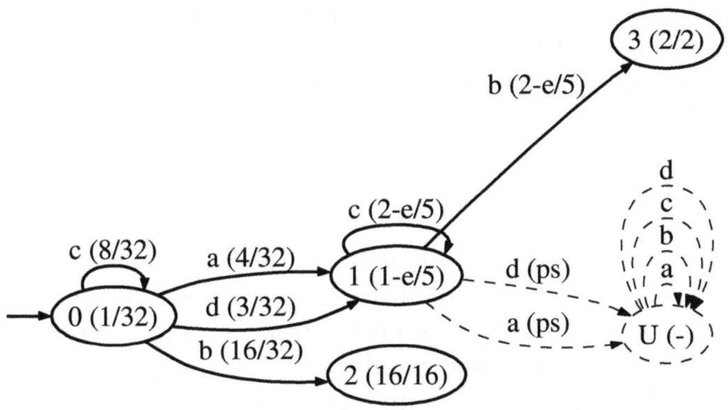

Fig. 1. Probabilistic automaton with a smoothed state

automata. In most of the probabilistic grammatical inference algorithm a first tree shape model is built that represents the maximum likelihood estimate of the training data. A second step is then used to generalize this model. This is done by merging states that are considered close according to some distance and some tolerance parameter α [2,10,19]. The algorithms differ in the distance used. We will consider here the MDI algorithm [19] as i) it uses a global criterion for comparing state and ii) the merging criterion favors small models. The automata produced are thus quite compact as compare to the ones output by the other algorithms. For our purpose, the MDI algorithm can be seen as a black box that takes as input a training set and a tuning parameter α and provides, as the output, a probabilistic automaton.

4 Smoothing Language Models

As mentioned above, smoothing language models is needed to handle the sparse data problem. This is mandatory either from a practical point of view [4,7] or from the theoretical one [13,14] . The main strategy consists in removing some probability mass from the seen events and redistribute them on the unseen ones. A back-off model, usually less accurate but more general, is used to provide a redistribution of the probability mass more realistic than, say, a uniform one. The classic smoothing scheme for n-gram is the Kneyser-Ney smoothing [11] while smoothing probabilistic automata is still an widely open problem (see [6,17] for some attempts).

In [17] smoothing a probabilistic automaton is done at a state level. A value ϵ is removed from each transition. This leaves a probability mass that can be redistributed, according to the unigram model, to estimate the probability of the symbol for which no transition exists. In order to keep the model compact, this smoothing scheme is done dynamically at parsing time. As an illustration, state 1 of figure 1 is explicitly smoothed: two smoothed transitions are added and goes to the unigram state $\mathcal{U}$. The ps values are estimated with the unigram adequately renormalized by $\varepsilon/5 + \varepsilon/5 + \varepsilon/5$, as this represents the probability mass saved respectively from the stopping probability, and the outgoing original transitions "c" and "b".

5 Motivations

We made some preliminary experiments on the "*Air Travel Information System*" (ATIS) corpus[8]. Even though it is now considered a small corpus it is large enough to make relevant preliminary work. In Fig. 2 we compute the average probability estimated by the unigram model, a probabilistic automaton inferred by the MDI algorithm [19] and a trigram model smoothed by Kneyser-Ney smoothing [11]. As can be seen, apart from the very beginning and the very end of the sentences[2], it appears that, the unigram and the trigram performances are not much influenced by the position in the sentence. This is not the case however for the probabilistic automaton as its prediction tends to the unigram one as the position increases. What can be noticed however is the fact that both the trigram and the automaton have a peak of performance at the same point (namely at position 3). The curve labelled "*raw aut*" is obtained by averaging the non null probabilities provided by a non smoothed automaton, that is the probabilities obtained when no smoothing is required. It is interesting to see that this predictions are, on average, rather good.

This means that, when it can parse, the automaton is rather good. An interpretation of the difference between this curve and the one of the smoothed automaton, is that the automaton suffers from under-generalization, and the parsing at that point is usually done by the unigram model. Unfortunately, tuning the grammatical inference parameter in order to build a more general automaton will decrease the prediction performances at the beginning of the sentence. In order to circumvent this problem, we propose here to build a model for which the generalization will depend on the position. As the models inferred by MDI are quite compact (few hundred of states), we can combine several DPFA without fear of facing any storage problem.

6 Experiments

The analysis above were conduct on the small ATIS task. In order to evaluate the position model on a more up to date database, we now consider the Wall Street Journal database.

6.1 The Wall Street Journal Task

The data used are drawn from the Wall Street Journal database [12], a large syntactically annotated corpus subdivided in 25 sections. We followed the

[2] Before position 3, the trigram does not have information to model properly conditional probabilities. Above position 23, very few strings exists (namely 34) that makes any interpretation suspicious.

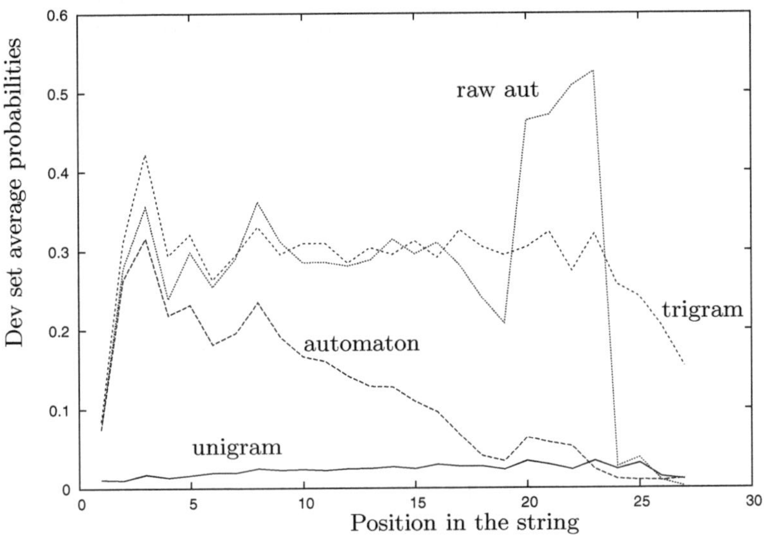

Fig. 2. Prediction power w.r.t the position

preprocessing used by [3]: Sections 0 to 20 were used as the training set (962,612 words), sections 21 and 22 were used as a development set (48,024 words) and sections 23 and 24 serve as the test (101,189 words). Digit numbers were replaced by a unique character. As in [3], the 10,000 most frequent words were kept and the remainder were transformed in a unique symbol unknown. In order to make the data more realistic for a speech to text task, the punctuation was removed and the words transformed in their lower case form. The average length of the sentences is 22 %(± 10) words and the size of Σ is around 10000.

6.2 Experimental Protocol

We present here the experimental protocol we followed. According to the Fig. 2 we consider that probabilistic automata are not able to handle cleverly histories of size greater than 10. Each string is thus split in different chunk (position 1 to 10, 8 to 15, ...). A model is inferred on each chunk and the final parsing will change the model as the position changes to another chunk. As can be seen on fig 2, the parsing at the very beginning of the sentence do not have enough evidence to provide good estimate. Moreover, as the position increases, smoothing is more likely to appear. For theses reasons, we propose to make an overlap between the chunks.

Defining the size of the overlap is not easy, as there is no reason that a good overlap size at the beginning of the string will be relevant at the end of the string. We therefore propose to have a lot of concurrent overlaps and to learn which one is relevant and which one is not.

The strings are thus split using different overlapping chunks (*e.g* 1-5, 1-6, 1-7, ..., 1-15, 3-11, 3-12, ...). For each chunk we inferred a probabilistic automaton[3]. A given position (say 8) will be covered by different chunks (*e.g.* 6-10 and 7-12). We make chunks of different size (ranging from 5 to 15 words) at different position (ranging from 0 to 58). Given the small amount of data at far position, a unique automaton is inferred from position 61. We will note C^i (resp. A^i) the set of chunks that covers position i (resp. the set of automata inferred on chunk C^i). Obviously, all the automata in A^i do not have the same accuracy in their probability estimate: as seen on Fig. 2 depending on the position in the automaton during the parsing, the probability estimated by the model do not have the same accuracy. The probability estimated at position i by our position model will be an interpolation of the probabilities provided by the automata in A^i. We note λ^i_j the weight of an automaton $A^i_j \in A^i$

More formally, the probability of a string will be computed as :

$$P(x = x_1..x_n) = \prod_{pos=i}^{n} \sum_{A^i_j \in A^i} \lambda^i_j \times P_{A^i_j}(x_i | x_1...x_{i-1})$$

The values of the λ^i_j coefficients are evaluated using an EM procedure.

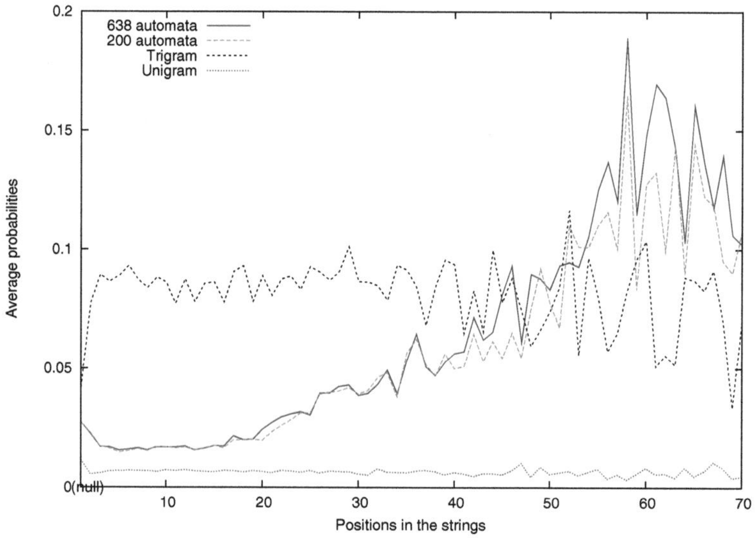

Fig. 3. Prediction power w.r.t the position

Fig. 3 shows the average probability of what we call the position model (curve named "638 auts") at different positions. As a point of comparison, the average probability is also provided for the trigram model and the automaton inferred

[3] Here again, a set of inferences is done by varying the MDI parameter α from 0.0008 to 0.1. The best automaton according to its dev set performance is selected.

on the full strings. It is interesting to notice that the position model starts to be relevant from position 20 and outperforms the trigram from position 40. Unfortunately, as the number of strings of length greater than 40 is quite small, this performance does not compensate the bad results obtained at the beginning: the position model is outperformed on average by the trigram. Another point should nevertheless be considered: despite the fact we used many automata, the overall size of our model is still quite reasonable as it is far smaller than the size of the trigram. Moreover, it appears that some of the λ_j^i are quite small as compare to other ones. We thus propose to reduce the size of the overall position model by iteratively removing the automaton that has the smallest weight.

Fig. 4 shows the behavior of the position model when we remove some automata. Note that on the contrary to the over figures, the quality criterion used is the perplexity: the smaller the perplexity on a given data set, the better the model predicts the strings of this set. This measure is linked, on the one hand, with the Kullback-Leibler divergence between the model and the development set, and on the other hand, on the inverse of the (geometric) average of the probabilities assigned by the model. See *e.g.* [20] for more detail on this measure and its links to the average on the test set string probabilities.

On Fig. 4 we can see that the our pruning strategy can save up to 50% of the size of the position model with a small impact on the development set perplexity (curve "200 auts" on fig 3).

For sake of comparison, the table 1 presents the *test set* perplexity of the position model as compare with the Trigram model (with a Kneyser-Ney smoothing).

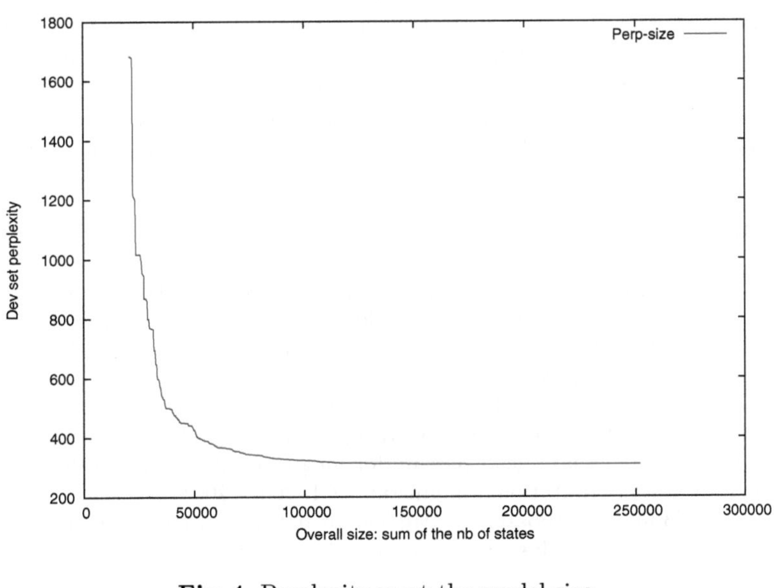

Fig. 4. Perplexity w.r.t the model size

We include the size of the model [4] and see first the that more compact model is the automaton inferred on the whole data set (name "full aut"). This model however performs rather badly as its perplexity is around 550 (as compare to the 150 of the trigram). The "All aut" model is the one in which all the 638 automata were kept. Using this model we obtain 45% of perplexity reduction, the size of the model being reasonable. When concern with the size of the model, we can see that the 200 auts model is a good tradeoff between size and prediction as the gain in size (w.r.t the All aut model) is interesting as regard to the prediction performances.

Table 1. Test set perplexity on the WSJ task

Model	#states	Perp (test)
3-grams	1,589,844	150
Full Aut	451	556
100 auts	32,609	399
200 auts	54,083	381
All aut	252,026	364

7 Conclusion and Further Work

We presented in this article a position model for statistical language modeling using DPFA. This position model is evaluated on the Wall Street Journal task. It achieve a test set perplexity reduction of 45% at the expense of having a bigger model. This model is still smaller than the trigram one.

In the near future, we want to consider the tradeoff between the size of the model and the prediction power. To this end, some trigram compression method (*e.g.* [9]) or trigram pruning methods (such as [16,15]) will be compare to the automaton's size, adequately compacted by compressing method [5]).

Another improvement can come from the use of other more recent probabilistic grammatical inference algorithms such as [1] and other smoothing schemes [6].

References

1. Callut, J., Dupont, P.: Learning partially observable markov models from first passage times. In: Kok, J.N., Koronacki, J., Lopez de Mantaras, R., Matwin, S., Mladenič, D., Skowron, A. (eds.) ECML 2007. LNCS, vol. 4701, pp. 91–103. Springer, Heidelberg (2007)
2. Carrasco, R.C., Oncina, J.: Learning stochastic regular grammars by means of a state merging method. In: Second ICGI, pp. 139–152 (1994)
3. Charniak, E.: Immediate-head parsing for language models. In: 10th Conf. of the Association for Computational linguistic, ACL 2001 (2001)
4. Chen, S.F.: Building Probabilistic Models for natural Language. PhD thesis, Harvard University Cambridge Massachusetts (May 1996)

[4] The size of the trigram model is computed as the sum of the number of entries of the different hash tables.

5. Daciuk, J., van Noord, G.: Finite automata for compact representation of language models in NLP. In: Watson, B.W., Wood, D. (eds.) CIAA 2001. LNCS, vol. 2494, pp. 65–73. Springer, Heidelberg (2003)
6. Dupont, P., Amengual, J.C.: Smoothing probabilistic automata: an error-correcting approach. In: Oliveira, A.L. (ed.) ICGI 2000. LNCS, vol. 1891, pp. 51–64. Springer, Heidelberg (2000)
7. Goodman, J.: A bit of progress in language modeling. Technical report, Microsoft Research (2001)
8. Hirschman, L.: Multi-site data collection for a spoken language corpus. In: DARPA Speech and Natural Language Workshop, pp. 7–14 (1992)
9. Kenneth, C., Ted, H., Jianfeng, G.: Compressing trigram language models with Golomb coding. In: Joint EMNLP-CoNLL, pp. 199–207 (2007)
10. Kermorvant, C., Dupont, P.: Stochastic grammatical inference with multinomial tests. In: Adriaans, P.W., Fernau, H., van Zaanen, M. (eds.) ICGI 2002. LNCS, vol. 2484, pp. 149–160. Springer, Heidelberg (2002)
11. Kneser, R., Ney, H.: Improved backing-off for m-gram language modeling. In: Intl. Conf. on Acoustic, Speech and Signal Processing, pp. 181–184 (1995)
12. Marcus, M., Santorini, S., Marcinkiewicz, M.: Building a large annotated corpus of English: the Penn treebank. Computational Linguistics 19(2), 313–330 (1993)
13. McAllester, D., Shapire, R.: On the convergence rate of the good-turing estimators. In: Conf. on Computational Learning Theory, pp. 1–66 (2000)
14. Ron, D., Singer, Y., Tishby, N.: On the learnability and usage of acyclic probabilistic finite automata. In: Proceedings of COLT 1995, pp. 31–40 (1995)
15. Siivola, V., Hirsimäki, T., Virpioja, S.: On growing and pruning kneser-ney smoothed n-gram models. IEEE Transactions on Audio, Speech and Language Processing 15(5), 1617–1624 (2007)
16. Stolcke, A.: Entropy-based pruning of backoff language models. In: DARPA Broadcast News Transcription and Understanding Workshop, pp. 270–274 (1998)
17. Thollard, F.: Improving probabilistic grammatical inference core algorithms with post-processing techniques. In: ICML 2001, pp. 561–568. Morgan Kaufmann, San Francisco (2001)
18. Thollard, F., Clark, A.: Shallow parsing using probabilistic grammatical inference. In: Adriaans, P.W., Fernau, H., van Zaanen, M. (eds.) ICGI 2002. LNCS, vol. 2484, pp. 269–282. Springer, Heidelberg (2002)
19. Thollard, F., Dupont, P., de la Higuera, C.: Probabilistic DFA inference using Kullback-Leibler divergence and minimality. In: ICML (2000)
20. Vidal, E., Thollard, F., de la Higuera, C., Casacuberta, F., Carrasco, R.C.: Probabilistic finite-state machines – Part I and II. PAMI 27(7) (2005)

Melody Recognition with Learned Edit Distances*

Amaury Habrard[1], José Manuel Iñesta[2], David Rizo[2], and Marc Sebban[3]

[1] Laboratoire d'Informatique Fondamentale, Université de Provence, 13453 Marseille cedex 13, France
[2] Dept. Lenguajes y Sistemas Informáticos, Universidad de Alicante, E-03080 Alicante, Spain
[3] Laboratoire Hubert Curien, Université de Saint-Etienne, 42000 Saint-Etienne, France

Abstract. In a music recognition task, the classification of a new melody is often achieved by looking for the closest piece in a set of already known prototypes. The definition of a relevant similarity measure becomes then a crucial point. So far, the edit distance approach with a-priori fixed operation costs has been one of the most used to accomplish the task. In this paper, the application of a probabilistic learning model to both string and tree edit distances is proposed and is compared to a genetic algorithm cost fitting approach. The results show that both learning models outperform fixed-costs systems, and that the probabilistic approach is able to describe consistently the underlying melodic similarity model.

Keywords: Edit distance learning, music similarity, genetic algorithms, probabilistic models.

1 Introduction

In a music recognition task, the classification of a new melody is often achieved by looking for the closest piece in a set of already known prototypes. The definition of a relevant similarity measure becomes then a crucial point. Several approaches for the musical similarity computation can be found in the literature, such as geometric approaches [9], language models [6] or edit distance based-methods [2,7]. This last category has encountered a great success during the past few years. Actually, musical pieces can be naturally represented by symbolic representations such as strings or trees containing the sequence of notes in the melody. In this context, the edit distance is a performing similarity measure because it is able to directly work on the structure of the data and thus the melody structure itself.

Let us recall that the basic edit distance between two strings corresponds to the minimum number of primitive edit operations allowing us to transform one

* This work was funded by the French *ANR Marmota project*[1,3], the Spanish *PROSEMUS project* (TIN2006-14932-C02)[2], the research programme *Consolider Ingenio 2010* (MIPRCV, CSD2007-00018)[2], and the *Pascal Network of Excellence.*[1,2,3]

N. da Vitora Lobo et al. (Eds.): SSPR&SPR 2008, LNCS 5342, pp. 86–96, 2008.

input string into an output one. Three edit operations are usually allowed: the deletion of a symbol in the input string, the insertion of a symbol in the output one, or the substitution of a symbol in the input string by a letter in the output string. To take into account the relative importance of each edit operation, one usually assigns a cost to each of them. Therefore, the computation of the edit distance boils down to looking for the less costly transformation. String edit distance in music recognition has already been applied in former works (e.g. [3]) and it has been extended to the context of tree-structured data [7] where a tree representation of the melodies is utilized to compare the instances with tree edit distances, using fixed edit operation costs. In fact, whatever the edit distance we use, the problem of assigning relevant costs to achieve a performing music recognition task remains the crucial point. To tackle this problem, two solutions can be considered. First, one can use background knowledge and human expertise to fix them. For instance, Mongeau and Sankoff [2] compared melodies by using an extended string edit distance (with consolidation and fragmentation edit operations), where the edit costs are tuned according to musicological knowledge. The second strategy consists to automatically tune or learn the edit costs. One of the main motivations for learning the edit costs from a training set is to define models able to better estimate similarities between structured objects resulting in a better accuracy while performing pattern recognition tasks (as also done by approaches which learn Mahalanobis distances in the context of numerical vectors). In other words, such learned edit costs capture background knowledge that enable an improvement of recognition performances.

For example, in the music recognition field, Gómez et al. [3] posed the problem of edit cost tuning in string edit distances given a training set through an exhaustive space searching. Another approach was used in [4] by running a genetic algorithm to fit the edit costs. However, as it will be confirmed in our experiments, this method does not seem to provide an interpretable model from a music recognition standpoint, performing as a blackbox. To overcome this drawback, we suggest in this paper the use of recent probabilistic learning algorithms of edit distance that have not been exploited in music recognition so far, but that have already proved their efficiency in other pattern recognition tasks [14,16,17]. This type of models has another interesting advantage. It also allows the representation of understandable knowledge. For example, in music analysis, one can wonder, in variations of some musical tunes, what are the most common changes that makes the tune still recognizable. Our hypothesis is that those changes will have small edit costs, in opposition to high valued edit operations.

One solution for both learning efficiently edit costs and keeping the understandability of learned weights is to use a probabilistic framework. In this context, rather than fixing edit costs, one aims to learn a matrix of edit probabilities which provides higher meaning than absolute weights. These edit probabilities can be estimated by learning the parameters of *memoryless* joint transducers as initially introduced by Ristad and Yianilos [17]. However, this joint model has the known disadvantages of generative approaches [1]. To overcome this drawback, Oncina and Sebban [16] have proposed to learn a discriminative *memoryless* transducer where conditional probabilities are directly estimated. This approach has shown its efficiency in handwritten digit recognition [16] and has been recently extended to tree-structured data [10,14]. Note that non memoryless models have been proposed during the past few years [11,13,15]. While these approaches allow us to model more complex phenomena, their understandability is more complicated because the knowledge is captured in several matrices of finite state machines. This explains why we will only deal in our following experiments with memoryless models.

In this paper, our objective is to apply the very efficient learning algorithms presented in [10,16], respectively on string and tree-based representations of musical pieces [7]. The quality of the system is assessed in terms of the two aforementioned qualities: classification accuracy and understandability of the induced model. The performance improvement is measured using the classical edit distances with a-priori fixed costs as baseline. The ability to extract significant information in the form of edit cost matrices is compared to a version of the classical edit distance algorithm where the costs have been fitted using a genetic algorithm.

The rest of this paper is organized as follows: The framework of edit distance learning is introduced in Section 2. Section 3 is devoted to the presentation of the musical corpus, the experimental setup and the analysis of the results. Finally, we conclude in Section 4.

2 Edit Distance and Learning

2.1 Notations and Classical Edit Distance

For sake of simplicity, we mainly focus on the string case but the notions presented here can be extended to trees. Strings are defined over a finite alphabet Σ and the empty symbol is denoted by $\lambda \notin \Sigma$. In this paper, we use letters in lowercase for symbols of $\Sigma \cup \{\lambda\}$ while letters in uppercase correspond to strings built from Σ.

The classical edit distance between two strings, also called the Levenshtein distance, is the minimal cost to transform by edit operations one input string into an output one. Three kinds of basic edit operations can be used in the transformation process: The insertion of a letter, the deletion of a letter and the substitution of a letter by another one. To each operation a so-called *edit cost* is assigned. Suppose, for example, that strings are built from an alphabet of two letters a and b. The costs for computing an edit distance can be represented by a 3×3 matrix c where the rows describe the input alphabet and the columns the output one. Therefore, $c(a, b)$ denotes the cost of applying the edit operation (a, b) where a is an input letter while b belongs to the output alphabet; If $a = \lambda$, the operation denotes an insertion; If $b = \lambda$ the operation is a deletion. Note that the operation (λ, λ) is not allowed.

Definition 1. *An edit script* $e = e_1 \cdots e_n$ *is a sequence of edit operations* $e_i = (a_i, b_i) \in (\Sigma \cup \{\lambda\}) \times (\Sigma \cup \{\lambda\})$ *allowing the transformation of a string* X *into* *a string Y. The cost of an edit script $\pi(e)$ is the sum of the costs of the edit operations involved in the script: $\pi(e) = \sum_{i=1}^{n} c(e_i)$.*

Let $S(X,Y)$ be the set of all the scripts that enable the emission of Y given X, the edit distance between X and Y is defined by: $d(X,Y) = min_{e \in S(X,Y)} \pi(e)$.

The computation of such an edit distance can be computed in quadratic time using dynamic programming techniques.

Note that many extensions have been presented in the context of tree-structured data [18,19]. We will consider in this paper the algorithm of Selkow [18] which takes into account three basic edit operations: The substitution of a node by another one, the deletion of an entire subtree and the insertion of an entire subtree. The interested reader will find more details in [18].

2.2 Learning in a Stochastic Framework

Learning the parameters of an edit distance requires the use of an inductive principle. A possible approach is to use the maximum likelihood paradigm as used in [10,16,17]. In this context, the algorithms do not learn a true metric between strings but rather an edit similarity estimated from the probability of a transformation between two strings. To overcome the bias of generative models such as [17], a relevant strategy consists to directly learn a conditional model [16]. Therefore, the similarity between two strings X and Y is assessed from the estimation of $p(Y|X)$ corresponding to the probability of generating by edit operations an output Y given an input X.

Let $\delta(b|a)$ be the conditional probability to change the symbol a into an output letter b (note that $\delta(b|a)$ is in some way equivalent to the cost $c(a,b)$ in the stochastic framework). To learn a conditional distribution over the edit operations, the δ function must fulfill the following condition of a statistical distribution:

$$\forall a \in \Sigma, \sum_{b \in \Sigma \cup \{\lambda\}} (\delta(b|a) + \delta(b|\lambda)) = 1.$$

Let us now extend the notion of edit script in the probabilistic framework.

Definition 2. *A stochastic edit script $e = e_1 \cdots e_n$ is a sequence of edit operations $e_i = (b_i|a_i)$ allowing the transformation of a string X into a string Y. The probability $\pi_s(e)$ of such a script corresponds to the product of the probabilities of the edit operations involved in the script: $\pi_s(e) = \prod_{i=1}^{n} \delta(e_i)$.*

Definition 3. *The probability $p(Y|X)$ of generating a string Y given an input string X is the sum of the probabilities of all the edit scripts transforming X into Y: $p(Y|X) = \sum_{e \in S(Y|X)} \pi_s(e)$.*

One can remark that in this stochastic framework, we consider all the edit scripts to estimate $p(Y|X)$. Given $p(Y|X)$, one can compute an edit similarity between X and Y.

Definition 4. *The edit similarity between two strings X and Y, conditionally to X, is defined as: $sim(X,Y|X) = -\log p(Y|X)$.*

(a) Original melody. The figures below the score indicate the Σ_p (interval from tonic).

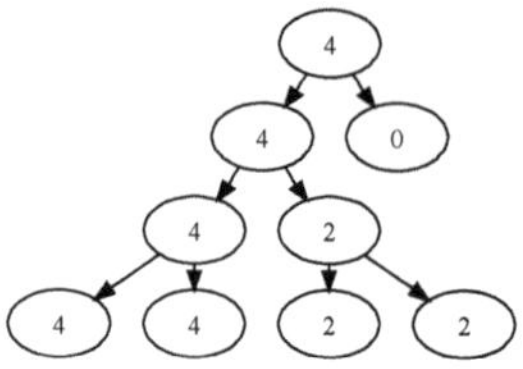

(b) Corresponding tree.

Fig. 1. Original tune

The learning of the parameters (*i.e.* the δ matrix) is done by an adaptation of the well known *expectation-maximization* algorithm [12] that aims to learn the hidden parameters of a probabilistic model by maximization the likelihood of a learning sample composed of pairs of $(input, output)$ strings.

3 Experiments and Analysis in Melody Recognition

3.1 Representation of Music and Database

In our series of experiments, we use both string and tree representations of music pieces [7]. For representing a monodic melody s as a string, symbols σ_i from a pitch description alphabet Σ_p are used: $s \in \Sigma_p^*, s = \sigma_1\sigma_2...\sigma_{|s|}$. In this paper, the interval from the tonic of the song is utilized as pitch descriptor (Fig. 1a): $\Sigma_p = \{p \mid 0 \leq p \leq 11\}$. In the tree representation (Fig. 1b), the node labels are symbols from the same alphabet Σ_p. Initially, only leaf nodes are labelled. Then, a melodic analysis [8] is used to decide which notes are more important in order to implement a bottom-up propagation process until all the inner nodes are labelled.

In our experiments, we used a corpus consisting of a set of 420 monophonic 8-12 bar incipits of 20 worldwide well known tunes of different musical genres [1]. For each song, a canonic version was created by writing the score in a musical notation application and exported to MIDI and MP3 format. The MP3 files were given to three amateur and two professional musicians who listened to the songs (mainly to identify the part of the tune to be played) and played with MIDI controllers the same tune several times with different embellishments and capturing performance errors. This way, for each of the 20 original scores, 20 different variations have been built.

3.2 Fitting the Edit Costs by Genetic Algorithms

One of the objectives of the paper is to compare the probabilistic edit distance learning methods to other existing cost fitting approaches that have already been

[1] The MIDI data is available upon request to the authors.

used in music recognition. In this context, a conventional genetic algorithm, as introduced in [4], has been used. The objective of this algorithm is to find the best insertion, deletion and substitution costs to be used in the classical edit distance. In our series of experiments, these costs can be represented by a matrix of size $13 \times 13 - 1 = 168$ (the operation (λ, λ) is not allowed) that represents the cost of the substitution of each of the 12 pitches for any other pitch, plus the insertion and the deletion of each pitch. Individuals containing the cost matrix in the form of vector are represented as chromosomes with 168 genes, one for each cost to be optimized, where the genes are real-valued in the range [0,2].

The fitness of each chromosome is evaluated experimentally by training and testing a classifier using the edit costs contained in the chromosome, where better classifier performance yields higher fitness. The fitness function has been computed as the averaged *precision-at-n* for all the melodies in the training set. The *precision-at-n* corresponds to the number of relevant hits among the n closest melodies, n being the number of correct prototypes in the learning set. The system has been implemented with an open-source robust GA software named JGAP [5]. We used the JGAP default settings that include mutation rate and selection method. Note that the algorithm does not stop until all the specified evolutions are finished. In the experiments, the number of evolutions has been empirically set to 100 from the observation of the fitness function evolution curve.

3.3 Melody Classification Accuracy

The main goal of our first series of experiments is to identify a melody from the set of all the different variations played by the musicians. For a melody, the edit distances to all the other prototypes are computed and the 1-NN rule was applied. A 3-fold cross-validation scheme has been followed to perform the experiments. This evaluation process has been achieved from both string and tree representations of the data.

The average error rates are shown in Fig.2. These results show that both the stochastic and the genetic methods outperform the results of the classical edit distance with insertion and deletion costs fixed to 1, and substitution fixed to 2, in the case of strings. The improvement is even much more significant for trees, while comparing with a standard algorithm with all the costs set to 1. We can note that the genetic algorithm seems to be a little bit more efficient than the stochastic method from the only accuracy point of view.

However, to improve the music recognition systems, we have to develop methods that are able not only to improve the classification accuracy of standard edit distance algorithms, but also to provide understandable knowledge explaining the distortions of the melodies. This is the aim of our second series of experiments.

3.4 Analysis of the Learned Edit Matrices

To analyze the understandability of the inferred models, we compared the learned cost matrices obtained with both genetic and probabilistic approaches. In order

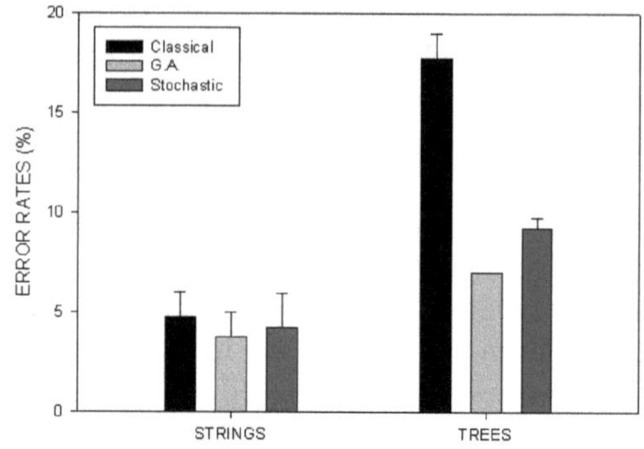

Fig. 2. Error rates from both string and tree representations

to represent graphically these matrices, they have been averaged and normalized over the three folds. To allow the comparison, the conditional probabilities of the stochastic model have been translated into costs using negative logarithms. Our aim here, is to see if the learned edit costs are able to model the changes and mistakes that can be found in the corpus. By analyzing it, the mistakes correspond in general to note elisions, grace notes insertions, the substitution of the intended note by another at a distance of a semitone or a tone, and note duration and onset changes.

The matrices learned with the genetic approach are shown in Fig. 3. From the lighter gray diagonal in the plot in Fig. 3a it can be drawn that the smallest costs are found in the substitution of a pitch for itself as it could be expected. No other pattern can be found that explains the most common differences between the original songs with its variations. In the case of the learned costs for trees (Fig. 3b), nor even the diagonal can be identified. From these results, it can be stated that the genetic approach to fit costs is useful to improve the success rates but not to learn a readable model of the melodic similarity phenomena.

The learned edit probabilities in the form of edit costs have been plotted in Fig. 4. In the case of strings (Fig. 4a), both the diagonal with the identity substitution cost and the insertion row (top) and deletion column (left-most) are

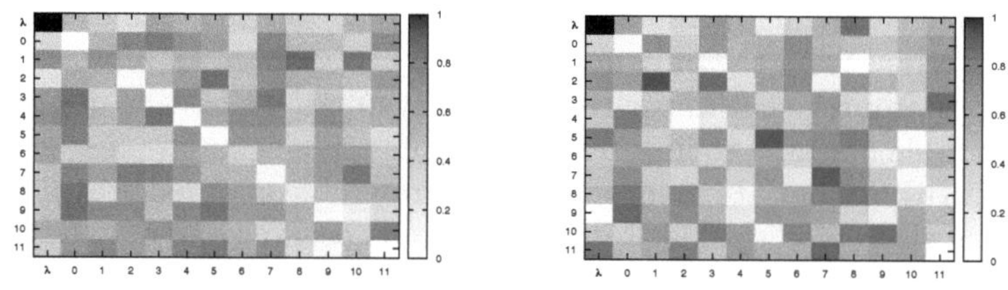

(a) String edit costs. (b) Tree edit costs.

Fig. 3. Learned edit costs by a genetic algorithm

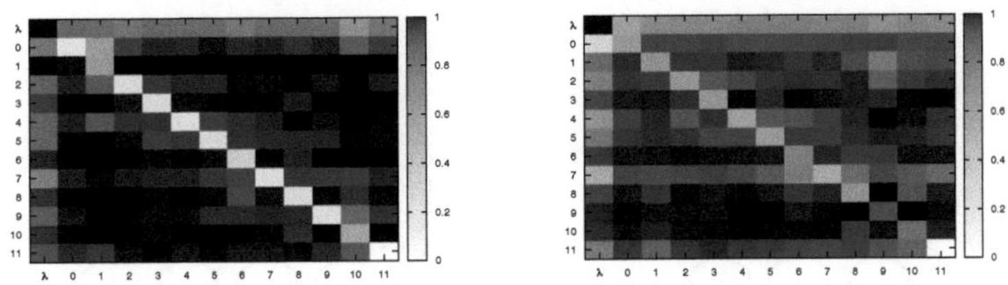

(a) String edit costs. (b) Tree edit costs.

Fig. 4. Learned edit costs by a stochastic algorithm

evident. These insertions and deletion reflect changes in note onset and durations. Moreover, the values close to the diagonal are also given higher probabilities, showing that the stochastic system has the ability of capturing changes of semitones made in the playing of the songs as pointed above in Sec. 3.1. The tree learned costs in Fig. 4b contain more or less the same scheme even if the diagonal is less dark. This behavior can be easily explained by the restriction imposed by Selkow's learning algorithm [10] used in the tree case: It allows only the insertions and deletions of whole subtrees which implies that deletion and insertion operations cover a larger part of the probability density of edit operations.

Based on the previous observations, it is much more difficult to understand the mapping from the original version to the variation using the matrices generated by the genetic algorithm. The reason why the genetic algorithm is unable to produce an understandable set of costs in opposite to the probabilistic approach seems to be due to the fact that in the classical edit distance computations, only the best path in the dynamic programming matrix is retained, and the genetic algorithm just tries to minimize the cost of one path for each prototype in the training set. This is only a local minimum in the parameter space. On the other hand, being the stochastic distance a sum of probabilities of all possible paths, all those probabilities must remain high (*i.e.* with low costs), being more likely to assess the importance of each operation. Thus, this method is more capable to capture relevant knowledge even with little corpora.

(a) Variation of tune in Fig.1a.

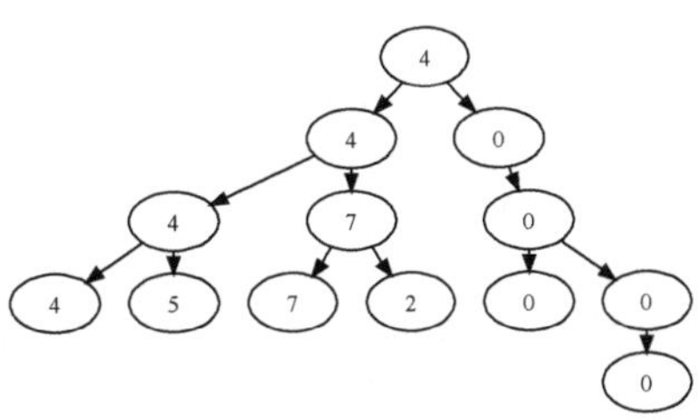

(b) Corresponding tree.

Fig. 5. Variation of tune in Fig.1

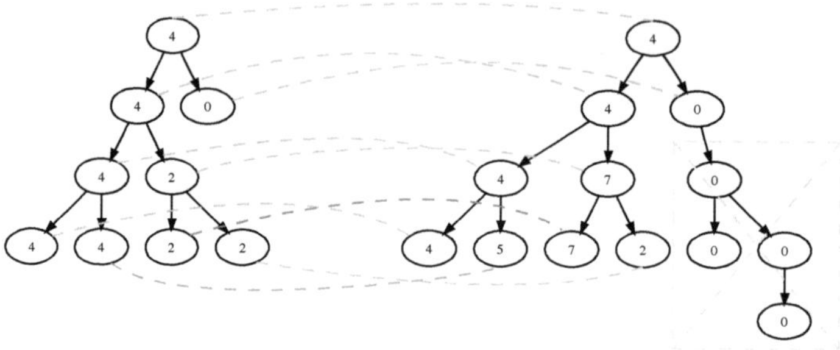

Fig. 6. Edit operations between the original tune and one of its variations. The deleted nodes are inside a rectangle. Gray levels of edit operations follow values in the costs matrix in Fig. 3b.

To illustrate the adequacy of the proposed approach to the melody similarity problem, two different variations of the same bar of the same song are shown in Figs. 1 and 5. The most probable edit script is indicated in Fig. 6 with dashed lines. The gray level of these dashed lines is set according to the gray levels in the matrix of Fig. 3b. Note that the structure of the tree is preserved, the substitution costs and the deletion of the nodes with a label "0" (tonic note) are the most probable option given the cost matrix. The string edit script, that can be deduced easily from the strings indicated below the scores in Figs. 1a and 5a, follow the same substitution and deletion pattern. Although not all the examples in the dataset are so easy to describe, in general the same process as the one we have just exposed can be followed.

4 Conclusions and Future Work

The ability for machine learning approaches to provide interpretable models is a desirable quality not frequently found. In this paper a recent algorithm of edit distance learning has been applied to the melodic similarity problem. It has been compared to both edit distance systems with a-priori fixed costs and to a genetic algorithm that fits the edit weights through the exploration of the cost space. It has been shown that both proposed methods to learn the edit costs, genetic and probabilistic schemes, are better than using initially fixed costs.

Moreover, the probabilistic approach has been shown to be more adequate for explaining the variation or noise introduction process being able to produce a readable cost matrix coherent with the training corpus. With a bigger corpus maybe the genetic approach will generate more understandable cost matrices, but also the parameters used in stochastic distance learning approaches will be more precisely assessed.

In the future, we plan to consider more pitch representations such as contour, high definition contour, pitch classes, and intervals to evaluate their adequacy to the melody matching problem. Another perspective is to work on the more frequent polyphonic music. Thus, one of the most interesting lines of work to be followed is the extension of the presented method to the polyphonic similarity computation. In this problem, the tree representation is the most suited to accomplish the task which brings us to focus on tree edit distance learning.

References

1. Bouchard, G., Triggs, B.: The trade-off between generative and discriminative classifiers. In: COMPSTAT 2004. Springer, Heidelberg (2004)
2. Mongeau, M., Sankoff, D.: Comparison of musical sequences. Computers and the Humanities 24, 161–175 (1990)
3. Gómez, C., Abad-Mota, S., Ruckhaus, E.: An Analysis of the Mongeau-Sankoff Algorithm for Music Information Retrieval. In: Proc. of the 8th International Conference on Music Information Retrieval (ISMIR), Vienna, Austria, pp. 109–110 (2007)
4. Grachten, M., Arcos, J., López de Mántaras, R.: Melody Retrieval using the Implication/Realization Model. In: Proceedings of the 6th International Conference on Music Information Retrieval, ISMIR 2005 (2005)
5. Rotstan, N., Meffert, K.: JGAP: Java Genetic Algorithms Package (accessed June 7, 2008), `http://www.jgap.sf.net`
6. Doraisamy, S., Rüger, M.: Robust Polyphonic Music Retrieval with N-grams. J. Intell. Inf. Syst. 21, 53–70 (2003)
7. Rizo, D., Moreno-Seco, F., Iñesta, J.M.: Tree-structured representation of musical information. In: Perales, F.J., Campilho, A.C., Pérez, N., Sanfeliu, A. (eds.) IbPRIA 2003. Lecture Notes in Computer Science (LNAI), vol. 2652, pp. 838–846. Springer, Heidelberg (2003)
8. Illescas, P.R., Rizo, D., Iñesta, J.M.: Harmonic, melodic, and functional automatic analysis. In: Proc. International Computer Music Conference, pp. 165–168 (2007)
9. Aloupis, G., Fevens, T., Langerman, S., Matsui, T., Mesa, A., Nuñez, Y., Rappaport, D., Toussaint, G.: Algorithms for Computing Geometric Measures of Melodic Similarity. Computer Music Journal, Fall 2006 30(3), 67–76 (2006)
10. Bernard, M., Habrard, A., Sebban, M.: Learning stochastic tree edit distance. In: Fürnkranz, J., Scheffer, T., Spiliopoulou, M. (eds.) ECML 2006. LNCS, vol. 4212, pp. 42–53. Springer, Heidelberg (2006)
11. Bernard, M., Janodet, J.-C., Sebban, M.: A discriminative model of stochastic edit distance in the form of a conditional transducer. In: Sakakibara, Y., Kobayashi, S., Sato, K., Nishino, T., Tomita, E. (eds.) ICGI 2006. LNCS, vol. 4201, pp. 240–252. Springer, Heidelberg (2006)
12. Dempster, A., Laird, M., Rubin, D.: Maximum likelihood from incomplete data via the EM algorithm. J. R. Stat. Soc. B(39), 1–38 (1977)
13. Durbin, R., Eddy, S., Krogh, A., Mitchison, G.: Biological sequence analysis. Cambridge University Press, Cambridge (1998)
14. Boyer, L., Habrard, A., Sebban, M.: Learning metrics between tree structured data: Application to image recognition. In: Kok, J.N., Koronacki, J., Lopez de Mantaras, R., Matwin, S., Mladenič, D., Skowron, A. (eds.) ECML 2007. LNCS, vol. 4701, pp. 54–66. Springer, Heidelberg (2007)

15. McCallum, A., Bellare, K., Pereira, P.: A conditional random field for disciminatively-trained finite-state string edit distance. In: Proc. of UAI 2005, pp. 388–400 (2005)
16. Oncina, J., Sebban, M.: Learning stochastic edit distance: application in handwritten character recognition. J. of Pattern Recognition 39(9), 1575–1587 (2006)
17. Ristad, S.V., Yianilos, P.N.: Learning string-edit distance. IEEE Transactions on Pattern Analysis and Machine Intelligence 20(5), 522–532 (1998)
18. Selkow, S.M.: The tree-to-tree editing problem. Information Processing Letters 6(6), 184–186 (1977)
19. Zhang, K., Shasha, D.: Simple fast algorithms for the editing distance between trees and related problems. SIAM Journal of Computing, 1245–1262 (1989)

Context First

Daniel Heesch, Robby Tan, and Maria Petrou

Imperial College London, SW7 2AZ, UK

Abstract. We propose a probabilistic model that captures contextual information in the form of typical spatial relationships between regions of an image. We represent a region's local context as a combination of the identity of neighbouring regions as well as the geometry of the neighbourhood. We subsequently cluster all the neighbourhood configurations with the same label at the focal region to obtain, for each label, a set of configuration *prototypes*. We propose an iterative procedure based on belief propagation to infer the labels of regions of a new image given only the observed spatial relationships between the regions and the hitherto learnt prototypes. We validate our approach on a dataset of hand segmented and labelled images of buildings. Performance compares favourably with that of a boosted, non-contextual classifier.

1 Introduction

Object recognition in general scenes remains a formidable task for artificial systems. Part of the difficulty stems from the way humans categorise things: it is first and foremost shared functional or causal characteristics that define most classes of interest, not similarity in appearance [19]. The problem is compounded by the observation that even very similar objects may look very different under different viewing angles and partial occlusion. Amongst the most successful approaches is that of modelling objects in terms of large sets of discriminant keypoints in conjunction with representations that are invariant with respect to several types of transformation (e.g. [14], [15], [16], [22], [8], [25] and [20]). These appearance-based models have in common that the number of classes to be distinguished is very small and that images contain objects only of one class. It is not clear how to scale to the several thousands of categories humans discriminate without effort.

Promising alternatives are hierarchical models in which features are allowed to be shared between different classes thus reducing the computational burden over flat models, e.g. [9]. Another route is to employ information about other objects of the same scene or information about the type of scene. Such contextual information may reduce the space of plausible object hypotheses and suggest a smaller set of dedicated non-contextual classifiers. It is known that the gist of the scene or the relationships between objects can be captured by the low-frequency content of an image [18]. It has also been shown in neuro-physiological studies that low-frequency information is processed relatively early during visual recognition [2]. Combining these two observations suggests that context may play a pivotal role as an early facilitator during visual recognition.

N. da Vitora Lobo et al. (Eds.): SSPR&SPR 2008, LNCS 5342, pp. 97–106, 2008.

While other authors have explored the use of scene information for object recognition [23] or region information for scene classification [4], [5], [10], our work investigates the question how and to what extent local geometric and topological relationships between objects can be exploited for object classification. Our approach is motivated by the discovery in [1] of cortical context networks that have been implicated in the storage of typical configurations of objects. Spatial relationships can arguably be extracted more easily than specific details of individual regions. Their extraction is also virtually insensitive to photometric variation. It may therefore not be surprising that these spatial networks also exhibit early activation during visual recognition tasks [1].

In contrast with other contextual models, we do not believe that contextual information should only be used to resolve tension between conflicting non-contextual evidence, e.g. [12] and more recently [21] in the context of probabilistic relaxation. Rather, we believe that context on its own can get us a long way and indeed may be the crucial ingredient to make object recognition scalable.

This paper makes three contributions: (i) we propose a fuzzy representation of the local neighbourhood of a region and a method to obtain *typical* neighbourhood configurations or *prototypes*; (ii) we propose a way to use these prototypes in the formulation of a random field over image regions; (iii) we provide an optimisation technique based on belief propagation to relax the random field.

The paper is structured as follows. Section 2 describes related work. In Section 3 we formulate the graphical model, formalise the set of spatial relationships used and describe how representative configurations, or prototypes, are obtained from a training set of annotated images. Section 4 explains how inference is performed. Section 5 describes our experiments. Section 6 ends the paper with a discussion.

2 Related Work

Several contextual models have been formulated that are concerned with dependencies between objects (as opposed to hierarchical dependencies). Amongst the probabilistic models, Markov random fields are the most popular, e.g. [17], [6], [13], [11], [20]. The authors in [11] and [20] define a conditional random field over individual pixels. In [20], contextual information is incorporated by using the joint boosting algorithm [24] for learning potential functions. Neither work explicitly considers spatial relationships, although [11] includes the absolute position of a site in the potential function.

In [6], each image is assumed to be associated with a bag of words and the precise term-region associations have to be learnt from training data. The Markov random field is specified through single and pair-wise clique potential functions which are learnt on the assumption that they are symmetric. The model therefore does not capture asymmetries in the dependency relationship. The model also does not take into account spatial relationships and thus is indifferent to whether, for example, a blue patch is above (sky) or below (sea) another.

In [17], a graphical model is defined over image regions by specifying the clique functions for all types of single and pair-wise cliques. The potential functions are weighted sums of basis functions with the parameters being set manually. Our work has the same objectives as those of [6] and [17]. What sets it apart is that we allow neighbouring regions to influence each other differently depending on their relative spatial positions and topological relationships. This added complexity is best handled by specifying the field in terms of local conditional probability distributions which are obtained empirically from a training set.

3 Spatial Context Model

Let $S = \{1, \ldots, N\}$ index a set of regions in an image. We assume that each region is associated with a random variable x_i which takes its value from a discrete set of class labels. The neighbourhood configuration of the ith region, $\mathcal{N}_i$, comprises the labels and spatial relationships of regions that are within some radius r of the focal region. We define the probability with which label l is assigned to region i as

$$P(x_i = l|\mathcal{N}_i) \equiv \frac{1}{Z}\exp(-\psi(\mathcal{N}_i, R_l)), \tag{1}$$

where Z is a normalising constant, and $\psi(\mathcal{N}_i, R_l)$ is a function that measures the distance between neighbourhood configuration $\mathcal{N}_i$ and the set of prototypes with label l at their focal region. The next section will describe the set of relations we use. The subsequent section explains how ψ is defined and how prototypes R_l are obtained.

3.1 Pairwise Relations

Spatial relationships are often modelled such that only one relationship holds between any two objects, e.g. [21] and [5] in the context of scene understanding. The representational convenience of crisp relations comes at the cost of increased sensitivity to errors with respect to the spatial localisation and geometry of the input data. We believe that much can be gained by modelling relationships as fuzzy concepts [3]. A fuzzy relation holds to a variable *degree* determined by a membership function associated with that relation. We here consider five relations between region pairs. These are their relative vertical orientation, their relative horizontal orientation, their containment relation, and the ratio of their widths and heights, respectively. These are defined as follows.

Vertical and Horizontal Relationships. Let p_{cnt_i} and p_{n_i} be points from each of the two regions. We measure the angle ϕ_i between vector $(p_{n_i} - p_{cnt_i})$ and the unit vector $(-1, 1)^T$. The degree of aboveness (or belowness) of p_n with respect to p_{cnt} is then computed as

$$f_{v_i}(p_{n_i}, p_{cnt_i}) = \sin\phi_i \tag{2}$$

where f_{v_i} represents the vertical relationship of a point pair. Similarly, we represent the horizontal relationship between point pairs as

$$f_{h_i}(p_{n_i}, p_{cnt_i}) = \cos\phi_i. \tag{3}$$

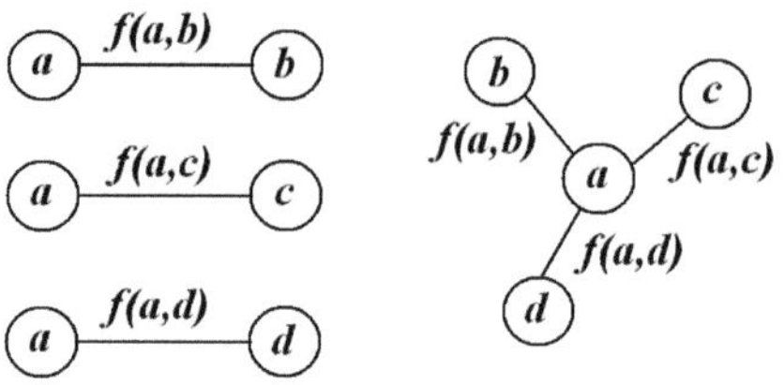

Fig. 1. The pictorial description of pairwise connections and a configuration, where a, b, c, d may correspond to wall, sky, roof and door, respectively. $f(a, b)$ in the diagram is a vector that consists of all components of the relationships between regions a and b.

To represent the vertical and horizontal relationship between two regions, we compute the average over point-wise membership values: $f_v = \frac{1}{N}\sum_i^N f_{v_i}$ and $f_h = \frac{1}{N}\sum_i^N f_{h_i}$. To be computationally efficient, we generate (p_{n_i}, p_{cnt_i}) randomly within the respective regions.

Containment Relationships. To measure whether region r_n includes region r_{cnt}, we are guided by the following decision rule:

$$f_{ct}(r_n, r_c) = \begin{cases} -1 & \text{if } (r_n) \text{ contains } (r_{cnt}) \\ +1 & \text{if } (r_n) \text{ is contained in } (r_{cnt}) \\ 0 & \text{otherwise} \end{cases} \tag{4}$$

Width and Height Relationships. We define these as the ratios between the widths and heights, respectively, of region r_n and those of region r_{cnt}. In our particular application domain, these relationships are useful to distinguish between, for example, roofs and chimneys, which are indistinguishable under all other relations. Formally, the width ratio is

$$f_{wr}(r_n, r_c) = \begin{cases} 1 - w_{cnt}/w_n & \text{if } w_n/w_{cnt} \geq 1 \\ w_n/w_{cnt} - 1 & \text{otherwise} \end{cases} \tag{5}$$

where w represents the width of a region with respect to its principal axis. The height ratio is defined analogously.

The spatial relationship between two regions can then be modeled as a vector with values in $[-1, 1]$ each component of which is the membership value for the corresponding relation.

3.2 Configurations and Prototypes

A neighbourhood configuration consists of the labels of the neighbours and their spatial relationships with respect to the focal region. Formally, it is an ordered set of relationship vectors with each vector being associated with a particular label. See Figure 1 for an illustration.

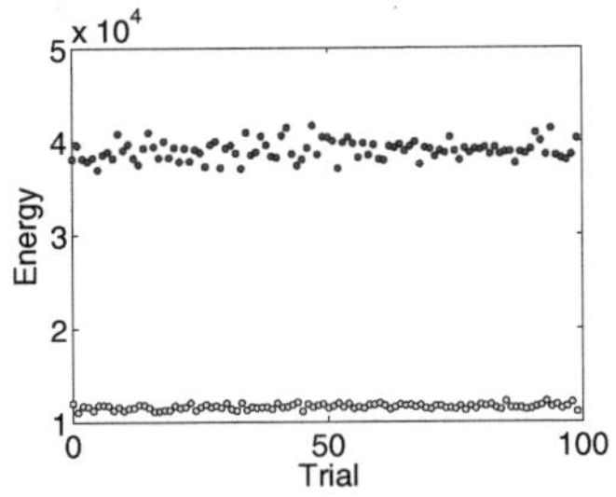

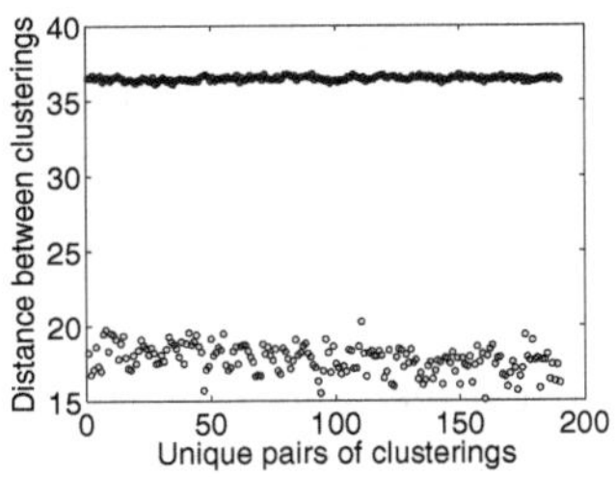

Fig. 2. Left: Value of the energy function at initialisation (top) and after convergence (bottom). Right: distances between clusterings for random clusters (top) and those obtained from the real data (bottom).

Prototypes. The purpose of the next step is to identify for each region label a small set of typical neighbourhood configurations, or prototypes. This is accomplished by clustering all those configurations that have the same label at the focal region. Clustering is based on the pair-wise distances between the configurations' respective matrix representations as described below.

Let P and Q denote the relation matrices of two configurations A and B of size M and N, respectively. Let the labels of the regions be represented by vectors p and q, respectively. For each region of configuration A, we determine its distance from all those regions of configuration B that bear the same label. This distance is computed by applying the l_1 metric to the respective row vectors of P and Q. We consider the closest region as the best match to the region of configuration A, add the distance to our overall cost and exclude the matching region from all subsequent comparisons. If configuration B does not have any region of that label, a fixed cost is applied to penalise label discrepancies. This is repeated for all other configurations of region A. The overall cost reflects both differences in the labels as well as differences in the geometry and topology of regions carrying the same label.

We employ the k-medoid algorithm to cluster configurations. Like k-means, the algorithm is guaranteed to converge because the sum of the distances between all points and their respective cluster centroid cannot increase and is also bounded from below. However, like any gradient descent algorithm, the final solution depends on the initialisation and is thus not guaranteed to be the global optimum. To assess the stability of the solution, we run the algorithm several times and compare the energy before and after convergence. As Figure 2 indicates, the final energy remains within narrow bounds and suggests that the final solutions come close to the global optimum.

A similar energy upon convergence does not imply, however, that the clusterings are the same or similar. Let $\mathcal{A}=\{A_1, A_2, \ldots, A_m\}$ and $\mathcal{B}=\{B_1, B_2, \ldots, B_n\}$ denote two clusterings. The members of each clustering are themselves sets of indices denoting a particular configuration. To assess the quality of the result of the k-medoid algorithm, we run it several times on the real data and compute a distance measure for all pairs of clusterings thus obtained. We then generate random clusterings consisting of the same number of clusters and the same cluster size distribution as the real clustering, and measure their pair-wise distances. We compute the distance between two clusterings $\mathcal{A}$ and $\mathcal{B}$ as

$$\sum_{i=1}^{m} d(A_i, \mathcal{B}) = \sum_{i=1}^{m} \left(|A_i| - \max_j |A_i \cap B_j| \right). \tag{6}$$

The distances between the random clusterings and the true clusterings are plotted in Figure 2 on the right. The plot demonstrates that the set of clusterings obtained by running k-medoid repeatedly on the same distance matrix are more similar to each other than a set of random clusterings. We take this as circumstantial evidence that k-medoid does capture intrinsic structure in the space of configurations.

Cluster centroids are those configurations for which the sum of the distances to all other members of the respective cluster is minimal. Prototypes correspond to the cluster centroids and thus are themselves configurations.

4 Inference

Given a set of regions in an image, we intend to label them using the prototypes we have generated. To arrive at correct labellings, we define a cost function that is based on the distance between the observed configurations and their closest prototype. Formally, we define the potential function as

$$\psi(\mathcal{N}_i, R_l) = \min_{R \in R_l} d(\mathcal{N}_i, R) \tag{7}$$

where $d(N_i, R)$ is the distance between a configuration and a prototype R as defined in Section 3.2. We intend to obtain the closest distance of all configurations from the corresponding prototypes, that is we want to minimise

$$E(\mathbf{x}) \equiv E(x_1 = l_1, ..., x_N = l_N) = \sum_{i \in S} \psi(\mathcal{N}_i, R_l) \tag{8}$$

In order to apply the technique of belief propagation, we change the undirected configurations into directed ones and generate a factor graph. Figure 3 shows an example of the transformation. We may subsequently use the following equations to optimise the cost function.

$$b(x_i) = \prod_{c \in N_i} m_{c \to i}(x_i) \tag{9}$$

$$m_{c \to i}(x_i) = \sum_{\{x_c\} - x_i} \psi \prod_{i' \in N_{c-i}} n_{i' \to c}(x_{i'}) \tag{10}$$

$$n_{i \to c}(x_i) = \prod_{c' \in N_{i-c}} m_{c' \to i}(x_i) \tag{11}$$

where $b(x_i)$ is the belief that x_i has a particular label. $m_{c \to i}(x_i)$ is the message from the neighbouring nodes of i with respect to the label of x_i. In Figure 3,

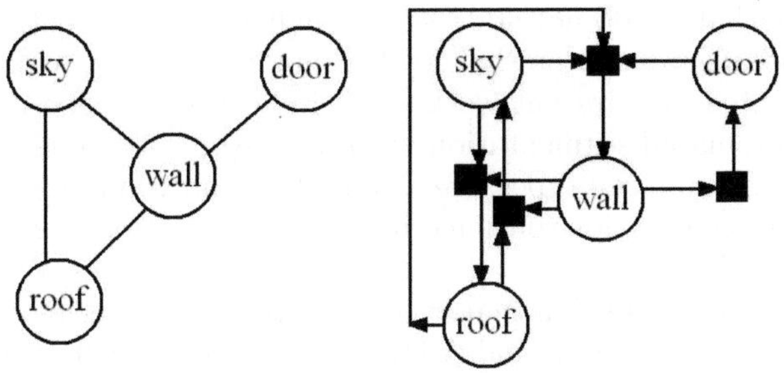

Fig. 3. The pictorial description of the transformation from an undirected graph into a directed graph

Fig. 4. Examples of manually segmented images

$m_{c \to i}(x_i)$ is represented by the black boxes that connect the regions or nodes. Since the configurations in the factor graph are directed configurations, we may simplify the equations thus

$$b(x_i) = m_{c \to i}(x_i) \tag{12}$$

$$m_{c \to i}(x_i) = \sum_{\{x_c\} - x_i} \psi \prod_{i' \in N_{c-i}} n_{i' \to c}(x_{i'}) \tag{13}$$

$$n_{i \to c}(x_i) = m_{c' \to i}(x_i) \tag{14}$$

5 Evaluation

The image collection used for training and testing consists of photographs depicting buildings from different cities of several countries, mostly taken from street-level. Each image was manually segmented and labeled with one of the following nine classes: 'window', 'chimney', 'roof', 'door', 'wall', 'stairs', 'pipe', 'sky', and 'vegetation'. Figure 4 shows two examples. The training set contains 197 images with a total of 3,675 regions. The test set comprises 80 images with a total of 1,372 regions. Images were randomly assigned to one of the two sets.

We shall note that we do not address the challenge of segmenting regions automatically. Instead, we assume that for learning and testing regions have been manually segmented. The recent work of [5] indicates that one may profitably start with an automated segmentation when the regions to be separated are visually distinct (e.g. material types like water, sky, sand). In our case, however, a general-purpose segmentation routine is unlikely to achieve sufficient accuracy for our contextual model.

Table 1 shows the confusion matrix for the proposed method. We compare performance with a non-contextual AdaBoost classifier which is trained to find the optimal combination of unary attributes pertaining to the shape of regions including the ratio of principal axes, compactness, variances, and elliptical variances. The confusion matrix for the labelling based on the adaboost algorithm is shown in Table 2. The total error is 576 (42%) compared to 473 (34%) for the contextual labelling.

Table 1. Confusion matrix for contextual classification (rows: true labels; columns: hypothesised labels)

	Win	Chi	Roo	Doo	Wal	Dor	Sta	Pip	Sky	Veg
Window	487	11	9	77	2	11	1	1	23	17
Chimney	10	70	1	0	0	24	0	0	2	0
Roof	18	0	31	0	0	2	1	0	31	8
Door	27	3	0	84	0	9	1	0	1	10
Wall	30	8	2	5	45	3	1	5	10	17
Dormer	7	1	0	1	0	2	0	0	3	1
Stairs	0	0	0	0	0	0	19	0	0	1
Pipes	12	16	0	2	3	0	0	48	1	1
Sky	11	2	1	0	0	0	0	1	79	0
Vegetation	7	2	2	3	1	1	6	5	1	35

Table 2. Confusion matrix for boosting on shape descriptors (rows: true labels; columns: hypothesised labels)

	Win	Chi	Roo	Doo	Wal	Dor	Sta	Pip	Sky	Veg
Window	435	56	9	49	4	78	4	0	1	4
Chimney	71	21	2	7	0	5	0	0	1	0
Roof	5	1	45	0	0	2	0	1	37	0
Door	71	3	0	53	3	5	0	0	0	0
Wall	2	9	3	1	83	10	1	2	14	1
Dormer	7	3	3	0	0	2	0	0	0	0
Stairs	15	1	2	0	0	2	0	0	0	0
Pipes	0	0	0	0	0	4	0	79	0	0
Sky	7	1	6	0	5	1	0	0	72	2
Vegetation	10	0	10	3	19	4	0	3	6	8

6 Discussion

In most computer vision applications that take into account contextual information, regions are initially assigned labels on the basis of their unitary attributes, and the label assignment is subsequently refined through the use of contextual information (e.g. [7]). Inspired by neuro-physiological findings about human visual processing, we here advocate the view that the order in which information is processed ought to be reversed. When humans view a scene, they first view it as a whole before focussing on particular details that merit further interpretation. Anecdotal evidence supporting this idea is the preference people show for seats in trains that look forward over seats that look backward in relation to the travelling direction. When looking backwards, one sees first the detail and then the context of the object; when looking forward, the global picture is captured first.

We proposed a probabilistic model in which a region's local neighbourhood is represented in the form of fuzzy relationship matrices. Because of the inability to define cliques in directed graphical models, the joint label probability over all regions cannot be expressed as a Gibbs distribution. Instead, we defined local conditional probabilities of a region's label that depend on the neighbourhood through a potential function that takes into consideration differences in the geometry of and the labels present in the neighbourhood. Central to our approach is the idea of typical neighbourhood configurations or prototypes which are obtained from a training set through clustering. Every label is associated with a small number of prototypes and inference aims to find a labelling of all regions such that the observed configurations are close to the labels' closest prototypes. Comparison with a non-contextual Adaboost classifier trained on a variety of shape features support the view that contextual information can provide powerful information for labelling structured scenes.

Acknowledgements. This work was supported by the EU project eTRIMS.

References

1. Bar, M., Aminoff, E.: Cortical analysis of visual context. Neuron 38, 347–358 (2003)
2. Bar, M., Kassam, K., Ghuman, A., Boshyan, J., Schmidt, A., Dale, A., Hämäläinen, M., Marinkovic, K., Schacter, D., Rosen, B., Halgren, E.: Top-down facilitation of visual recognition. Proc. National Academy of Sciences 103(2), 449–454 (2006)
3. Bloch, I.: Fuzzy Representations of Spatial Relations for Spatial Reasoning. John Wiley & Sons, Chichester (2007)
4. Boutell, M., Brown, C., Luo, J.: Learning spatial configuration models using modified Dirichlet priors. In: Workshop on Statistical Relational Learning (2004)
5. Boutell, M., Luo, J., Brown, C.: Factor graphs for region-based whole-scene classification. In: Proc. IEEE Conf. Computer Vision and Pattern Recognition, Semantic Learning Workshop (2006)
6. Carbonetto, P., de Freitas, N., Barnard, K.: A statistical model for general contextual object recognition. In: Proc. European Conf. Computer Vision, pp. 350–362 (2004)

7. Christmas, W.J., Kittler, J., Petrou, M.: Structural matching in computer vision using probabilistic relaxation. IEEE Transactions on Pattern Analysis and Machine Intelligence 17(8), 749–764 (1995)
8. Csurka, G., Bray, C., Dance, C., Fan, L.: Visual categorization with bags of keypoints. In: Proc. European Conf. Computer Vision (2004)
9. Fidler, S., Boben, M., Leonardis, A.: Similarity-based cross-layered hierarchical representation for object categorization. In: Proc. Int'l. Conf. Computer Vision and Pattern Recognition (2008)
10. Gökalp, D., Aksoy, S.: Scene classification using bag-of-regions representations. In: Proc. IEEE Conf. Computer Vision and Pattern Recognition, Beyond Patches Workshop (2007)
11. He, X., Zemel, R., Ray, D.: Learning and incorporating top-down cues in image segmentation. In: Proc. European Conf. Computer Vision (2006)
12. Kittler, J., Hancock, S.: Combining evidence in probabilistic relaxation. Journal of Pattern Recognition and Artificial Intelligence 3, 29–51 (1989)
13. Kumar, S., Hebert, H.: Discriminative random fields: a discriminative framework for contextual interaction in classification. In: Proc. Int'l. Conf. Computer Vision (2003)
14. Lowe, D.: Object recognition from local scale-invariant features. In: Proc. Int'l. Conf. Computer Vision, pp. 1150–1157 (1999)
15. Lowe, D.: Distinctive image features from scale-invariant keypoints. Int'l. Journal of Computer Vision 60, 91–110 (2004)
16. Matas, J., Chum, O., Urban, M., Pajdla, T.: Robust wide baseline stereo from maximally stable extremal regions. In: British Machine Vision Conference, pp. 384–393 (2002)
17. Modestino, J., Zhang, J.: A Markov random field model-based approach to image interpretation. IEEE Trans. Pattern Analysis and Machine Intelligence 14(6), 606–615 (1992)
18. Oliva, A., Torralba, A.: Modelling the shape of the scene: a holistic representation of the spatial envelope. Int'l. Journal Computer Vision 42(3), 145–175 (2001)
19. Petrou, M.: Learning in Computer Vision: some thoughts. In: Rueda, L., Mery, D., Kittler, J. (eds.) CIARP 2007. LNCS, vol. 4756, pp. 1–12. Springer, Heidelberg (2007)
20. Shotton, J., Winn, J., Rother, C., Criminisi, A.: Textonboost: Joint appearance, shape and context modeling for multi-class object recognition and segmentation. In: Proc. European Conf. Computer Vision (2006)
21. Singhal, A., Luo, J., Zhu, W.: Probabilistic spatial context models for scene content understanding. In: Proc. IEEE Conf. on Computer Vision and Pattern Recognition (2003)
22. Sivic, J., Zisserman, A.: Video google: a text retrieval approach to object matching in videos. In: Proc. Int'l. Conf. Computer Vision, pp. 1–8 (2003)
23. Torralba, A.: Contextual priming for object detection. Int'l. Journal of Computer Vision 52(2), 169–191 (2003)
24. Torralba, A., Murphy, K., Freeman, W.: Sharing fatures: efficient boosting procedures for multiclass object detection. In: Proc. IEEE Conf. Computer Vision and Pattern Recognition, pp. 762–769 (2004)
25. Winn, J., Criminisi, A., Minka, T.: Object categorization by learned universal visual dictionary. In: Proc. Int'l. Conf. Computer Vision, pp. 1800–1807 (2005)

Activity Representation in Crowd

Yunqian Ma[1] and Petr Cisar[2]

[1] Honeywell Labs, 1985 Douglas Drive, Golden Valley, MN 55422, USA
yunqian.ma@honeywell.com
[2] Honeywell Prague Laboratory, V Parku 2326/18, 14800 Prague, Czech Republic
petr.cisar@honeywell.com

Abstract. Video surveillance of large facilities, such as airports, rail stations and casinos, is developing rapidly. Cameras installed at such locations often overlook large crowds, which makes problems such as activity and scene understanding very challenging. Traditional activity recognition methods which rely on input from lower level processing units dealing with background subtraction and tracking, are unable to cope with frequent occlusions in such scenes. In this paper, we propose a novel activity representation and recognition method that bypasses these commonly used low level modules. We model each local spatio-temporal patch as a dynamic texture. Using a Martin distance metric to compare two patches based on their estimated dynamic texture parameters, we present a method to temporally stitch together local regions to form activity streamlines and represent each streamline by its constituent dynamic textures. This allows us to seamlessly perform activity recognition without explicitly detecting individuals in the scene. We demonstrate our method on multiple real data sets and show promising results.

1 Introduction

Activity representation and understanding in video sequences has been extensively studied over the last few years. For simple activity recognition, people use shape features to perform human movement analysis [1]. Our previous activity recognition works can identify people walk, run, skip, etc. [2][3]. For complexity activity recognition, people use hidden markov model, dynamic Bayesian networks method to inference [4][5]. However, most activity recognition methods show poor results when a large number of people are seen occluding each other in the scene. Surveillance videos in public areas such as airports and railway stations often contain crowded environments causing problems for most computer vision applications. Many of these problems stem from the inability of low-level vision algorithms (such as background subtraction, tracking, feature extraction etc.) to deal with crowded video data set. Thus commonly seen activities such as people moving in different directions, crowd formation and dispersal, etc. become very hard to detect and represent reliably.

In this paper, we present a motion texture activity representation to capture dynamic content that is not well represented by individual objects, or small numbers of interacting objects. The types of activities can include people moving

N. da Vitora Lobo et al. (Eds.): SSPR&SPR 2008, LNCS 5342, pp. 107–116, 2008.

in different direction, crowd formation and dispersal, etc. To represent it, we use dynamic textures for characterizing the underlying activities without requiring the inference of objects trajectories. Dynamic texture can be viewed as sequences of images of moving scenes that exhibit certain stationary properties in space and time [6]. [6] showed that linear dynamic textures can be characterized as the output of an auto-regressive moving average model. They obtain a closed form solution for learning the parameters of a linear dynamic texture. They demonstrated being able to extrapolate dynamic texture sequences as well as classify them based on the learnt model. Saisan et. al [7] further extended this work to model and recognize over 50 different dynamic textures. They evaluated three distance measures - the Geodesic distance metric, the Martin distance metric and Subspace angles.

Fujita et. al [9] extended the notion of a single dynamic texture spanning the whole image to multiple dynamic textures in different regions of an image. They partitioned the image into several regions, each region having its own dynamic model. They use a distance metrics for a non-linear space, so the features of dynamic textures are based on the impulse responses of state variables. These responses were shown to be efficient to compute and effective for matching purposes.

Chan et. al [8] proposed using mixtures of dynamic textures to model and segment video sequences. They used an expectation maximization algorithm to learn the parameters of the model and demonstrated their approach on natural scenes and highway traffic videos. Zhao et. al [10] extended the Local Binary Pattern operator commonly used in texture analysis to combine appearance as well as motion information. This operator was rotation invariant and robust to gray-scale variations making and was used to recognize dynamic textures. Our previous works in dynamic texture is for spatial segmentation [11] for video surveillance data.

In this paper, we first partition the whole image to multiple regions. Each region (patch) has its own dynamic models. Second, we link multiple spatio-temporal patches in time to form video streamlines. This consists of three components: temporal association of patches, stop criterion for the temporal association and streamline editing to filter out noisy links. We then aggregate multiple streamlines in a video sequence using the Karcher mean concept [12] to form an activity model. An extension of the Martin distance serves as a distance metric between two activities. This allows us to recognize activities of people and cars in different video scenarios. We demonstrate our approach on a wide variety of scenes such as an airport data set, a subway station data set, a retail store data set and a traffic intersection data set.

The rest of the paper is organized as follows. Section 2 gives a brief overview of dynamic texture modeling and presents the distance metrics that between two dynamic texture models. Section 3 discusses activity representation and recognition using dynamic texture. We present experimental results in Section 4. Section 5 is conclusion.

2 Dynamic Textures

In this section, we begin by providing a brief overview of the research in dynamic textures. The interested reader can refer the detail in [6]. For a sequence of images, we use $y(t) \in R^m, t = 1, ..., \tau$ to represents pixel intensity. Doretto et al. [6] view $y(t)$ as a noise observation on a linear dynamic texture. Therefore, they represent this linear dynamic texture by an auto-regressive, moving average process with unknown input distribution:

$$\begin{aligned} x(t+1) &= Ax(t) + v(t), \; v(t) \sim N(0,Q); x(0) = x_0 \\ y(t) &= Cx(t) + w(t), \qquad\quad w(t) \sim N(0,R) \end{aligned} \tag{1}$$

where $x(t) \in R^n$ is the hidden state, $A \in R^{n \times n}$ is the state transition matrix, representing the dynamics. $C \in R^{m \times n}$ is the output matrix, $v(t)$ is the driving input to the system, which is assumed to be Gaussian white noise, $w(t)$ is measurement noise.

A system identification method can be used to estimate the dynamic texture parameters. Let's form a matrix $Y_1^\tau = [y(1), ..., y(\tau)] \in R^{m \times \tau}$ for the sequence of τ frames. To make the estimation of the dynamic texture unique, [6] let $m \gg n$ and $rank(C) = n$, and use canonical model $C^T C = I_n$ (I_n is the $n \times n$ identity matrix) to give a closed form solution by singular value decomposition $Y_1^\tau = U \Sigma V^T$. The estimated dynamic texture parameters of matrix C and A are

$$\begin{aligned} \hat{A} &= \Sigma V^T \begin{bmatrix} 0 & 0 \\ I_{r-1} & 0 \end{bmatrix} V (V^T \begin{bmatrix} I_{r-1} & 0 \\ 0 & 0 \end{bmatrix} V)^{-1} \Sigma^{-1}, \\ \hat{C} &= U \end{aligned} \tag{2}$$

The discrimination between two dynamic models can be calculated through parameters A and C, since the dynamic texture is viewed as ARMA model. Several distance matrices between dynamic textures were proposed [7][9]. For example, the subspace angle between two dynamic textures parameters $[A_1, C_1]$ and $[A_2, C_2]$ can be computed. Using these subspace angles $(\theta_i, i = 1, ..., n)$, [7] used the Martin distance as the distance between the two dynamic texture parameters.

$$d = \ln \prod_{i=1}^{n} \frac{1}{\cos^2(\theta_i)} \tag{3}$$

3 Activity Representation

Our proposed activity representation framework is based on the analysis of motion flow in different parts of the video sequence. This allows us to recognize video sequence that correspond to different activities.

3.1 Motion Flow Forming

We begin by partitioning an image into overlapping rectangular regions. We group temporally neighboring regions to form a spatio temporal volume (which we call a *patch*) and consider each patch volume to be a dynamic texture. For example, we can use five consecutive frames to form the patch volume. The dynamic texture parameters of the patch volume are estimated as in Section 2 and the Martin distance metric is used as a distance measure between two dynamic textures. Patches are then connected temporally to form streamlines, as shown in Figure 1. There are three components to streamline formation: temporal association of the patches, a stop criterion for the temporal association and streamline editing to remove noise.

The temporal association of patches is carried out as follows. Each patch's dynamic texture parameters are estimated independent of other patches in the neighborhood. The estimation of these parameters is carried out using a five frame sequence in the immediate future of the current frame. Temporal association is to connect a patch at time t to a patch at time $t+1$ based on the distance between the corresponding dynamic texture parameters. If the motion of objects in the video sequence is sufficiently small, one can assume that the patch at time $t+1$ that is connected to a patch $B_i(t)$ at time t lies in its immediate spatial neighborhood. Thus we compute the Martin distance between patch $B_i(t)$ and the nine patches in the immediate spatial neighborhood at time $t+1$. A patch $B_{\hat{j}}(t+1)$ is temporally associated with $B_i(t)$ if,

$$\hat{j} = \text{argmin } d(DT(B_i(t)), DT(B_j(t+1))) \tag{4}$$

This procedure is continued until a stopping criterion is satisfied, resulting in a streamline.

Next we present two stopping criteria for the streamline formation procedure. First, if all nine neighboring patches at time $t+1$ do not exhibit any motion, one can assume that the object has either stopped moving or has left the scene. Thus the corresponding streamline is ceased. Second, if objects overlap leading to the occlusion of the object in consideration, the dynamic texture parameters of all nine neighborhood patches will differ greatly with the dynamic texture parameters of the current patch $B_i(t)$. In particular, if $d(DT(B_i(t)), DT(B_j(t+1))) > (\mu + 3\delta)$, where μ is the mean and δ is the standard deviation of the Martin distances calculated during the streamline formation, the corresponding streamline is ceased.

Finally we perform streamline editing. We filter out three types of noise. Firstly, we filter out the streamlines whose starting patch is on the boundary. Such streamlines may join neighboring patches out of the boundaries of the image in the next few frames, and are thus removed. Secondly, streamlines with no significant change of coordinates are considered noisy streamlines and are removed. These streamlines correspond to isolated movement observed in the background. Thirdly, spurious streamlines characterized by lengths shorter than a predefined threshold are filtered out. The resulting set of streamlines is a robust set of streamlines that can be used for the purpose of activity recognition.

Figure 2 presents an example of streamlines formed in an occlude situation. The upper images in Figure 2 shows three image frames at three time instants to represent a image sequence. Two persons move towards, get occluded and then separate. The lower image in Figure 2 present the streamlines corresponding to

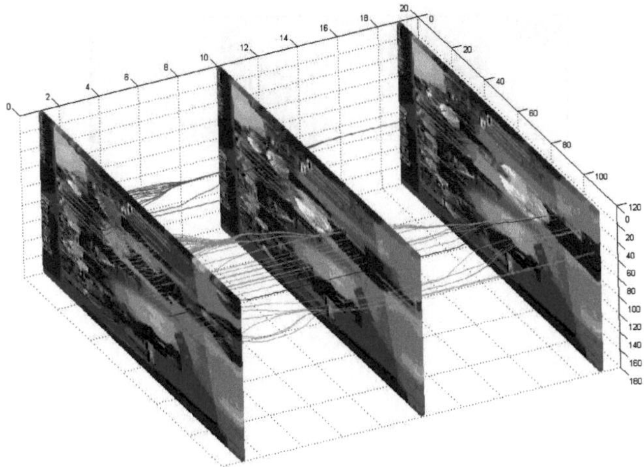

Fig. 1. Flow forming by connecting patches using the similarity between the dynamic texture parameters of the patches

the occluded person get divided at the time of the occlusion using the stopping criteria described above.

3.2 Activity Recognition

The proposed activity representation uses the motion flow formed in the above section. For example, a query sreamline S^q composes DT_t^q, where $DT, t = 1, ..., n$ means dynamic texture parameters, including A and C. That is, the estimated dynamic texture parameters of the patches forming a streamline can be used as features to describe the streamline. An extension to the Martin distance serves as a distance metric between any two streamlines. This metric is used to compare a query streamline S^q with a streamline S^p representing activities in the training data as follows,

$$d(S^q, S^p) = \frac{1}{n}\sum_{t=1}^{n} d(DT_t^q, DT_t^p) \tag{5}$$

We perform activity recognition by comparing the distance of a query activity (streamline) to all activities in the training database. Then we use a k-nearest neighbor classification method to select the best match.

There may be many streamlines labeled with the same activity category in the training data. Instead of performing an exhaustive search over every training data, which is time consuming, we present a method to aggregate multiple streamlines in each class of the training database to form one representative streamline. Thus the number of comparisons that must be made to classify a query streamline will equal the number of categories in the training dataset. Since the dynamic texture parameters are not defined in a Euclidean space, a simple mean calculation cannot be used to aggregate multiple streamlines. Instead, in

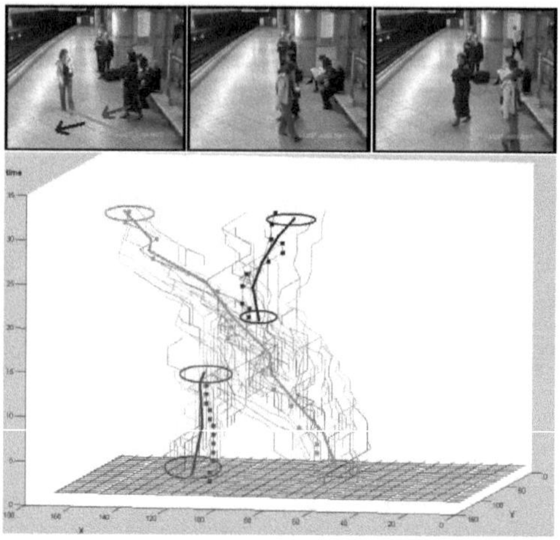

Fig. 2. upper images are an example of overlapping movement; lower image is the formed streamlines, the overlapped object is divided into two parts according to the stop criteria

this paper we use the Karcher mean concept to aggregate multiple streamlines into a representative set of dynamic texture parameters S^{rep} given by,

$$rep(t) = \underset{k}{\operatorname{argmin}} \sum_j d(DT(B_k(t)), DT(B_j(t))) \quad (6)$$

$$\text{for } j \neq k, t = 1, ..., n \quad (7)$$

where $d(\cdot)$ denotes the Martin distance given by Equation (3).

We perform activity recognition using the nearest neighbor classification by comparing the distance between the query streamline S^q with the representative streamline S^{rep} for each activity category in the training data.

4 Experimental Results

In this section, we present experimental results for the proposed activity representation and recognition method. All the experiments were conducted on a PC with an Intel Core2, 2.4 GHz processor and 2GB RAM.

Dataset: the experiments were conducted on video sequences from various real dataset. First, we have airport dataset collected at an airport. This dataset is also used in our previous paper [12]. Second, we downloaded the public dataset, i-Lid dataset[1] [13], which consists of video sequences collected from surveillance cameras overlooking a subway station. Third, we downloaded the UCF dataset from [14]. We used two sets of videos from this dataset: the retail escalator video sequences and the vehicle traffic video sequences.

[1] http://www.elec.qmul.ac.uk/staffinfo/andrea/avss2007_ss_challenge.html

For each dataset, we selected n video sequences with k different activities. Activities in every video sequence were manually labeled to form the ground truth. Half the sequences were used for training, and the other half for testing. The training video sequences were utilized to obtain representative streamlines for the different activities as given in Section 3.2. The performance of the system is given by the classification accuracy of all streamlines in the video sequences used for testing.

Table 1 gives the performance of the proposed method on the airport dataset. This consists of two major activities: people moving towards the camera (down) and away from the camera (up). The training data includes 12 *down-streamlines* and 18 *up-streamlines*. The test data includes 74 down-streamlines and 161 up-streamlines. Table 1 shows the classification accuracy for the two activities.

Table 1. Classification accuracy - Airport dataset

Activity	Down	Up
Down	97%	3%
Up	0%	100%

Table 2. Classification accuracy - Subway dataset

Activity	Down	Up	Left	Right
Down	97%	0%	3%	0%
Up	0%	76%	24%	0%
Left	15%	0%	85%	0%
Right	0%	0%	0%	100%

Fig. 3. Activity recognition shown for sample frames from the subway dataset

Fig. 4. Activity recognition shown for sample frames from the UCF retail escalator dataset. The red and green arrows denote the two major motion patterns observed in the video sequences. The corresponding color coded patches denote the labels outputted by our system.

Fig. 5. Activity recognition shown for sample frames from the UCF traffic dataset. The red and green arrows denote the two major motion patterns. The corresponding color coded patches denote the outputted activity labels. The small resolution of objects in the scene causes some patches to be erroneously classified.

There are four major activities involved in the video sequences of the subway dataset- moving left, moving right, moving up, and moving down. Figure 3 shows a few frames from the subway dataset with the recognized activities. The training data included 17 down-streamlines, 18 up-streamlines, 19 left-streamlines and 13 right-streamlines. The test data included the following streamlines: 38 down, 153 up, 100 left and 44 right. Table 2 shows the classification accuracy matrix using the subway dataset.

The UCF data set is a challenging dataset with a larger number of object in the scene. Figure 4 shows a few sample frames of the activity recognition results on the retail escalator set of videos. Figure 5 presents the activity recognition results on the vehicle traffic set of videos from the dataset.

5 Conclusion

Activity representation and recognition in video surveillance is a very important area which has been received a lot attention from researchers. The high density environment is very common in video surveillance data, where the trajectory based activity representation which needs track individual person first work poorly in this environment. We present a novel activity recognition method using dynamic textures. We first partition image sequences into patches. Then we form motion flow by temporally connecting the patch in the current frame to the patch in the next frame. Activity representation is from the dynamic texture features extracted from the streamline. We use various real data set to test the proposed method. The experimental results show good performance for activity recognition. In the future, we are going to perform more complex activity representation and recognition under this direction.

References

1. Veeraraghavan, A., RoyChowdhury, A.K., Chellappa, R.: Matching Shape Sequences in Video with Applications in Human Movement Analysis. IEEE Transactions on Pattern Analysis and Machine Intelligence 27 (December 2005)
2. Ma, Y., Damelin, S.B., Masoud, O., Papanikolopoulos, N.: Activity Recognition Via Classification Constrained Diffusion Map. In: International Symposium on Visual Computing 2006, pp. 1–8 (2006)
3. Ma, Y., Miller, B., Buddharaju, P., Bazakos, M.: Activity Awareness: From Predefined Events to New Pattern Discovery. In: IEEE International Conference on Vision Systems, New York, NY, USA, January 5-7 (2006)
4. Park, S., Aggarwal, J.K.: A Hierarchical Bayesian Network for Event Recognition of Human Actions and Interactions. Multimedia Systems; Special issue on Video Surveillance 10(2), 164–179 (2004)
5. Muncaster, J., Ma, Y.: Hierarchical Model-Based Activity Recognition With Automatic Low-Level State Discovery. Journal of Multimedia 2(5), 66–76 (2007)
6. Doretto, G., Chiuso, A., Wu, Y.N., Soatto, S.: Dynamic Textures. International Journal of Computer Vision 51(2), 91–109 (2003)
7. Saisan, P., Doretto, G., Wu, Y.N., Soatto, S.: Dynamic Texture Recognition. In: IEEE Conference on Computer Vision and Pattern Recognition, vol. 2, pp. 58–63 (2001)
8. Chan, A., Vasconcelos, N.: Modeling, Clustering and Segmentating Video with Mixtures of Dynamic Textures. IEEE Trans. on Pattern Analysis and Machine Intelligence 30(5), 909–926 (2008)
9. Fujita, K., Nayar, S.K.: Recognition of Dynamic Textures using Impulse Response of State Variables. In: 3rd International Workshop on Texture analysis and synthesis (2003)
10. Zhao, G., Pietikainen, M.: Dynamic Texture Recognition Using Volume Local Binary Patterns. In: 9th European Conference on Computer Vision, Graz, Austria, May 7 - 13 (2006)
11. Ma, Y., Cisar, P.: Motion Analysis Using Dynamic Texture in Crowd Environment. Image Analysis - From Theory to Applications, Research Publishing, pp. 49–54 (2008)

12. Ma, Y., Miller, B., Cohen, I.: Video Sequence Querying Using Clustering of Objects' Appearance Models. In: Bebis, G., et al. (eds.) International Symposium on Visual Computing 2007, pp. 328–339 (2007)
13. i-lids dataset for avss (2007)
14. Ali, S., Shah, M.: A Lagrangian Particle Dynamic Approach for Crowd Flow Segmenta-tion and Stability Analysis. In: IEEE Conference on Computer Vision and Pattern Recognition, Minneapolis, MN 2007 (2007)

3D Object Recognition Using Hyper-Graphs and Ranked Local Invariant Features

Shengping Xia[1,2,⋆] and Edwin R. Hancock[2]

[1] ATR Lab, School of Electronic Science and Engineering, National University of Defense Technology, Changsha, Hunan, P.R. China 410073
[2] Department of Computer Science, University of York, York YO1 5DD, UK

Abstract. Local invariant feature-based methods such as SIFT have been proven highly effective for object recognition. However, they have made either relatively little use or too complex use of geometric constraints and are confounded when the detected features are superabundant. Here we make two contributions aimed at overcoming these problems. First, we rank the SIFT points (R-SIFT) using visual saliency. Second, we use the reduced set of R-SIFT features to construct a class specific hyper graph (CSHG) which comprehensively utilizes local SIFT and global geometric constraints. Moreover, it efficiently captures multiple object appearance instances. We show how the CSHG can be learned from example images for objects of a particular class. Experiments reveal that the method gives excellent recognition performance, with a low false-positive rate.

1 Introduction

Recently, local invariant feature extraction (LIFE) methods[5] [8][10], such as SIFT, PCA-SIFT,GLOH and SURF, have proven successful and have been widely used for object recognition [5] [9], scene classification [1][2] [7] and video retrieval [12][13]. According to comprehensive experimental analysis on several large image databases[9], SIFT appears to be the most effective descriptor for practical uses, and has become a standard of comparison. SIFT is scale invariant, rotation invariant, and is also illumination invariant for a limited range of light source directions. However, there are two basic drawbacks to its use in object representation. Firstly, in most cases the features delivered are super-abundant and exceed the number that can be effectively used for the purposes of matching or recognition. As a result the exhaustive search of features is time consuming and too many false positive matches are often produced. One way of overcoming these problems is to use visual saliency [6] to select a subset of the features for analysis. Secondly, in the Bag-of-words[7] approach local features usually play the role of "visual words" that are predictive of a certain object class.

However, this representation overlooks both geometric and structural constraints on feature arrangement. By discarding this information recognition performance is compromised. A more versatile and expressive tool for representing structured data are attributed graphs (hereinafter simply referred to as graph). When recognition is attempted using graph-based abstractions, it is difficult to account for natural structural variations.

⋆ Corresponding author.

N. da Vitora Lobo et al. (Eds.): SSPR&SPR 2008, LNCS 5342, pp. 117–126, 2008.

One way of overcoming the problem is to construct a class prototype by merging example structures together. For instance, Torsello and Hancock [14] have constructed the class-prototype through tree-union and have performed clustering using a mixture of tree-unions controlled by a description length criterion [15]. These ideas are taken further by Escolano and Lozano [16] who extended the methodology to graphs rather than trees, and used an EM algorithm for clustering. However, although this work provides an elegant way of encapsulating within-class structural variations it is relatively impoverished in terms of the attribute information encoded. For this reason in this paper we turn to hypergraphs. Our aim is to develop a prototype structure through the merging of salient graphs. This structure encapsulates both variations in structure, and edge and node attributes.

In this paper we make two contributions to appearance based object recognition. Firstly, we rank SIFT-points using visual saliency and select only the most salient points to construct object models. Secondly, we use a class specific hypergraph (CSHG) to model objects compactly. The hypergraphs are based on multiple Delaunay graphs. Each of these is constructed from the selected feature points for a single prototype image of an object. In this way, the object models can be constructed with a minimum of object views.

2 Preliminaries

Before we detail our method, we commence by providing some preliminary definitions.

2.1 Graph and Class Specific Hyper Graph

Def. 2.1 ***Attributed Graph*** **G [4]:** An attributed graph is a 2-tuple $G = (V, E)$, where V is the (finite) set of vertices (also called nodes), $E \subseteq V \times V$ is the set of edges.

Def. 2.2 ***Spectral Eigenvalue Vector*** **(S_G):** Let A be the adjacency matrix of G and D be the diagonal degree matrix. The spectral decomposition of the normalized Laplacian matrix $\hat{L} = D^{-1/2}(D - A)D^{-1/2}$ [4] is $\hat{L} = \Phi\Lambda\Phi^T$, where $\Lambda = diag(\lambda_1, \lambda_2, ..., \lambda_{||G||})$, $\lambda_1 \leq \lambda_2 \leq ... \leq \lambda_{||G||}$ is the diagonal matrix with the ordered eigenvalues as elements and $\Phi = (\phi_1, \phi_2, ..., \phi_{||G||})$ is the matrix with the correspondingly ordered eigenvectors as columns. If $e = (1, 1, 1, .., 1)^T$ is the all-ones-vector, then the ordered eigenvalues can be represented by the spectral vector $S_G = \Lambda e$.

Def. 2.3 ***Maximum Common Subgraph*** (MCS)[4]**:** G is a maximum common subgraph of G_i and G_j, denoted by $MCS(G_i, G_j)$, if it is a common subgraph of G_i and G_j and there is no other common subgraph having more nodes than G. We denote any subgraph of $MCS(G_i, G_j)$ as $CS(G_i, G_j)$.

Def. 2.4 ***Correct Matching of Graph*****(CMG):** $CMG(G_i, G_j), G_i, G_j \in HV$, is defined as:

$$CMG(G_i, G_j) = \begin{cases} 1, & ||MCS(G_i, G_j)|| \geq n_\tau, & (1) \\ 0, & else. & (1') \end{cases}$$

where n_τ is threshold on the number of correctly matched nodes. If $||MCS(G_i, G_j)|| \geq n_\tau$, then the two graphs are a CMG pair.

Using ideas from basic graph theory [4] and hypergraph theory [3], we construct a class specific hyper graph (CSHG), which we use to capture structural variations for objects of a particular class.

Def. 2.5 ***Class Specific Hyper Graph*** **(CSHG):** A CSHG is defined as 2-tuple $CSHG = (HV, HE)$, where 1) the hyper vertex $HV = \{ G_i, i = 1, 2, ..., M \}$ is a finite set of graphs forming the nodes of the CSHG; 2) the hyper-edge HE is the edge set of the CSHG. There is an edge between two nodes G_i and G_j, denoted by $Edge(G_i,G_j)$, if and only if $CMG(G_i, G_j) = 1$.

Def. 2.6 ***Simple Class Specific Hyper Graph*****:** Given $G_i \in HV$, $\forall\, G_j\ (G_j \neq G_i) \in HV$, if $Edge(G_i, G_j) \equiv \emptyset$, G_i is isolated. If all G_i are isolated, then the graph is a simple CSHG.

Def. 2.7 ***Siblings of a Graph($\mathscr{S}$)*****:** The siblings of G_i are defined as $\mathscr{S}\{G_i\} = \{ G_j | CMG(G_i, G_j) = 1, G_j \in HV\}$. For a graph G_i, there may be more than 2 siblings, and any two siblings may not form a CMG pair.

Def. 2.8 ***A Path of CSHG*****:** A path between G_i and G_j is defined as a set of graphs

$$\mathscr{P}\{G_i, G_j\} = \{G_i, G_{j1}, G_{j2}, ...G_{jK}, G_j\}. \tag{2}$$

where $CMG\{\ G_i\ ,G_{j1}\ \}$=1, $CMG\{\ G_j\ ,G_{jK}\ \}$ =1 and for any two consecutive graphs $CMG\{\ G_{j(l)}, G_{j(l+1)}\ \}$ =1, $l = 1, 2, ..., K-1$. If $K = 0$, the path degenerates to a CMG pair.

Def. 2.9 ***Redundant Graph*****:** A graph G_t is redundant for an existing CSHG model if

$$\exists\, G_i, G_j, s.t. \mathscr{P}\{G_i, G_j\} \neq \emptyset, \mathscr{S}\{G_t\} \subseteq \bigcup \mathscr{S}\{G_l \mid G_l \in \mathscr{P}\{G_i, G_j\}\}. \tag{3}$$

The special case is that $\mathscr{S}\{G_t\} = \{G_i, G_j\} = \mathscr{P}\{ G_i, G_j \}$.

2.2 SIFT and Its Ranking for Visual Saliency

Local Invariant Feature Extraction. The SIFT algorithm is described in detail in [5]. Let $\overrightarrow{X}^t$=$(x^t_1, x^t_2)^T$ be the coordinates of the t-th SIFT point, $\overrightarrow{R}^t = (r^t, \alpha^t)^T$ be a vector with the corresponding gradient magnitudes and orientations as components, and $\overrightarrow{U}^t$=$(u^t_1, u^t_2, ..., u^t_n)^T$ be the descriptors of SIFT feature. The three vectors are concatenated to form a SIFT feature-vector $V^t = ((\overrightarrow{X}^t)^T, (\overrightarrow{R}^t)^T, (\overrightarrow{U}^t)^T)^T$ and the set of feature-vectors is $\mathcal{V} = \{V^t, t = 1, 2, ..., m\}$.

Def. 2.10 ***Positive Match between SIFT Descriptors*****:** Given two sets of SIFT points $\mathcal{V}_i = \{\ V_i^{t_i},\ \ t_i = 1, 2, ..., m_i\ \}$ and $\mathcal{V}_j = \{\ V_j^{t_j},\ \ t_j = 1, 2, ..., m_j\ \}$, for a descriptor $U_i^{t_i}$ of $\mathcal{V}_i$, $\exists\, U_j^{t_{j1}} \in \mathcal{V}_j$, subject to

$$U_j^{t_{j1}} = \arg\min\ d(U_j^{t_j},\ U_i^{t_i}),\ U_j^{t_j} \in \mathcal{V}_j\ \ and\ \ d(U_j^{t_j},\ U_i^{t_i}) \leq \varepsilon_\tau. \tag{4}$$

$$U_j^{t_{j2}} = \arg\min\ d(U_j^{t_j},\ U_i^{t_i}),\ U_j^{t_j} \in \mathcal{V}_j - V_j^{t_{j1}}\ . \tag{5}$$

$$and\ \gamma\ =\ d(U_j^{t_{j1}},\ U_i^{t_i})\ /\ d(U_j^{t_{j2}},\ U_i^{t_i})\ \leq\ \gamma_\tau. \tag{6}$$

where $d(\bullet) = \| \bullet \|_2$ is Euclidean distance between two vectors. ε_τ is a threshold of Euclidean distance and γ_τ is a threshold of the ratio of γ defined above. When the condition above is satisfied, then $U_i^{t_i}$ and $U_j^{t_{j1}}$ are said to have a positive match.

In our graph based model, the coordinates are used to construct Delaunay graph $G = (\mathcal{V}, E)$, with node-set $\mathcal{V}=\{V^t, t = 1, 2, ..., m\}$. It is important to emphasize that in our CSHG model the nodes are attributed with information such as locations and descriptors of SIFT points. We now expand our definition of $CS(G_i,G_j)$ to include the role of SIFT points.

Def. 2.11 ***Common subgraph*:** $CS(G_i,G_j)$ satisfies the conditions:

1) For each node in $CS(G_i,G_j)$, the corresponding descriptors of G_i and G_j are positive matches.

2) We obtain two weighted Delaunay graphs G_i' and G_j' using the feature coordinations of G_i and G_j. For $CS(G_i,G_j)$, the Euclidean distance of spectral vectors $S_{G_i'}$ and $S_{G_j'}$ must be less than the threshold λ_τ.

Ranking Local Invariant Features through Boosting. The method to rank SIFT features, referred to as R-SIFT, is outlined in the schematic diagram given in Figure 1. The method uses an image synthesis method to locate feature points that are robust to random image perturbations

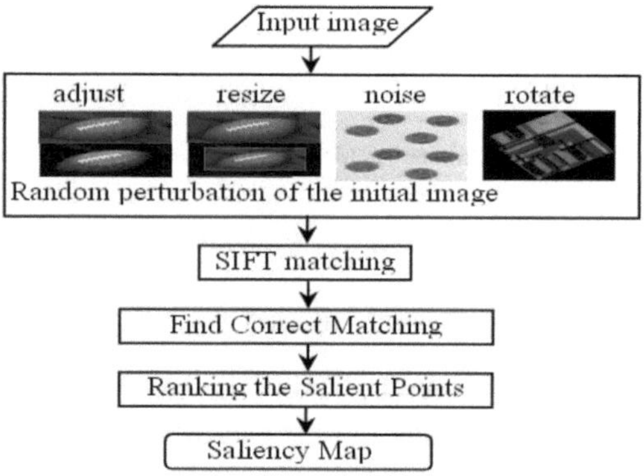

Fig. 1. Diagram of R-SIFT algorithm

We input an initial image to the image synthesis system. For the initial image the feature set is $\mathcal{V}_{ini}$ and the points belonging to the set, i.e. V_{ini}^t, are ordered according to the gradient magnitudes of $\vec{R}_{ini}^t$. We synthesized new images using a Monte Carlo method which resizes, rotates and adds noise to the original image.

These synthesized images are then processed to extract features and the features are matched with those in the initial input image. The fraction of correctly matching points (FCM) for each SIFT point in the original image is η^t. and the gradient magnitudes of $\vec{R}^t = (r^t, \alpha^t)^T$ is unified according to the maximum-minimum transform in the group of $\vec{R}_{ini}$, which is denoted as ζ^t. Our measure used for ranking the feature points is

$$\rho^t = \eta^t \times \zeta^t. \tag{7}$$

3 Constructing a CSHG from R-SIFT Features

We select the first T R-SIFT points from an image with known class identity. We denote the points by the set $\mathcal{V}=\{V^t, t = 1,2,...,T\}$ which we use to construct a Delaunay graph G. In our experiment, T is set as 40. If there are less than T points, then all the available SIFT points are selected. In this way, a set of graphs G_l, $l = 1,2,...,N$ are obtained and these are used to construct the hyper-vertex set HV of a CSHG.

According to Def. 2.9, a redundant graph means that there are sufficient similar local regions that it may be discarded without significant loss of information. In practice redundant graphs are iteratively discarded from the set G_l, $l = 1,2,...,N$, until an irreducible subset is located and used to construct a CSHG. To bootstrap the process, prior to training, a group of graphs can be randomly selected without checking for redundancy.

For convenience, we use the notation $CSHG\{\bullet\}$ for the item set in a CSHG. Hence, $CSHG\{G_l\}$ is the set of graphs and $CSHG\{\overrightarrow{U^t}\}$ is the set of R-SIFT feature descriptors. Given a new graph $G_l = (\mathcal{V}_l, E_l)$, $\mathcal{V}_l=\{V_l^t \mid V_l^t=((\overrightarrow{X_l^t})^T, (\overrightarrow{R^t}_l)^T, (\overrightarrow{U_l^t})^T)^T, t = 1,2,...,T_l\}$, and a trained CSHG model. For each descriptor $\overrightarrow{U_l^t}$, the ε nearest neighbors, denoted as $NN_\varepsilon\{U_l^t\}$, are those descriptors $\overrightarrow{U_q^j}$ in a CSHG which satisfy the following restrictions:

$$NN_\varepsilon\{U_l^t\} = \{U_q^j \mid \|U_l^t - U_q^j\|_2 \leq \varepsilon_\tau,\ U_q^j \in CSHG\{\bullet\}\}. \tag{8}$$

where ε_τ (normally less than 1) is the Euclidean distance threshold for accepting two descriptors as neighbors. If the the similarity of two the descriptors is defined as

$$\mathcal{S}(U_q^j, U_l^t) = \exp\{-\beta \cdot \|U_l^t - U_q^j\|_2\}. \tag{9}$$

where $\beta \in [0,1]$. So a modified set of nearest neighbors is

$$NN_\varepsilon\{U_l^t\} = \{U_q^j \mid \mathcal{S}(U_q^j, U_l^t) \geq \exp\{-\varepsilon_\tau\},\ U_q^j \in CSHG\{\bullet\}\}. \tag{10}$$

The ε nearest neighbor graphs of G_l in a CSHG denoted by $NNG_\varepsilon\{G_l\}$, are

$$NNG_\varepsilon\{G_l\} = \{G_q \mid \forall U_l^t \in G_l,\ U_q^j \in NN_\varepsilon\{U_l^t\},\ U_q^j \in G_q \text{ and } G_q \in CSHG\{\bullet\}\}. \tag{11}$$

The cardinality of the common subset of G_l and G_q is a measure of how well the two graphs are matched, denoted as $\|CS(G_l,G_q)\|$. In order to find the best matched graphs, combining $\mathcal{S}(U_q^j, U_l^t)$, a ranking-oriented similarity of two graphs G_t and G_q, denoted as $\mathcal{R}(G_l,G_q)$, is defined as

$$\mathcal{R}(G_l,G_q) = \|CS(G_l,G_q)\| \times \prod_{t=1,2,...,T_l} \mathcal{S}(U_q^j, U_l^t). \tag{12}$$

Each graph $G_q \in NNG_\varepsilon\{G_l\}$ can be placed in descending order according to $\mathcal{R}(G_l,G_q)$, and this order is denoted by $ONNG_\varepsilon\{G_l\} = \{G_{tq},\ tq = 1,2,...,\|NNG_\varepsilon\{G_l\}\|\}$. Hence, the first K graphs of the ordered set $ONNG_\varepsilon\{G_l\}$ are the $K-$ nearest neighbor graphs of G_l, and these are denoted as $NNG_K\{G_l\}$. If the number of ε nearest neighbor graphs

$NNG_{\varepsilon}\{G_l\}$ is large, we may select $NNG_K\{G_l\}$ instead for the purposes of recognition, clustering, retrieval or tracking.

Up to now, geometric constraints have not been utilized. As a result $ONNG_{\varepsilon}\{G_l\}$ may contain serious errors due to mismatched feature points. In fact, the false positive rate becomes rather high if we make decisions according to just $NNG_K\{G_l\}$ or $ONNG_{\varepsilon}\{G_l\}$.

To overcome this problem suppose that for a matched graph pair G_l and $G_q \in NNG_K\{G_l\}$, the coordinates of the matched features are as X_l and X_q. To match these points we use a Procrustes alignment procedure [11]. To this end we construct the orthogonal matrix

$$R = \arg\min \|X_l \cdot \Omega, X_q\|_F, \; subject\ to\ \Omega^T \cdot \Omega = I. \tag{13}$$

where $\|\bullet\|_F$ denotes the Frobenius norm. The norm is minimized by the nearest orthogonal matrix

$$R^* = \Psi \cdot \Upsilon^*, \; subject\ to\ X_l^T \cdot X_q = \Psi \cdot \Sigma \cdot \Upsilon^*. \tag{14}$$

where $\Psi \cdot \Sigma \cdot \Upsilon^*$ is the singular value decomposition of matrix $X_l^T \cdot X_q$. The goodness-of-fit criterion is the root-mean-squared error, denoted as $e(X_l, X_q)$. The best case is $e(X_l, X_q) = 0$. The error e can be used as a measure of geometric similarity between the two groups of points. Suppose there exists mismatched pairs of feature points, then $e(X_l, X_q)$ will be greater than 0. If we discard one pair of points from X_l and X_q, denoted $X_l - X_l^i$ and $X_q - X_q^i$, a set $e(X_l - X_l^i, X_q - X_q^i)$, $i = 1, 2, ..., \|CS(G_l, G_q)\|$ can be obtained. The maximum decrease of $e(X_l - X_l^i, X_q - X_q^i)$ is defined as

$$\triangle e(\|CS(G_l, G_q)\|) = e(X_l, X_q) - \min\{e(X_l - X_l^i, X_q - X_q^i)\} \tag{15}$$

if $\triangle e(\|CS(G_l, G_q)\|)/\, e(X_l, X_q) > \epsilon$, e.g. $\epsilon = 0.1$, the corresponding pair X_l^i and X_q^i is discarded as a mismatched feature pair. This leave-one-out procedure can be proceed iteratively, and is referred as the iterative Procrustes matching of G_l and G_q.

After the iterative Procrustes method has been applied, the similarity ranking of G_l and G_q can be redefined as

$$\mathscr{R}^*(G_l, G_q) = \mathscr{R}(G_l, G_q) \times (\exp(-e(X_l, X_q)))^{\kappa}. \tag{16}$$

where κ is a parameter used to amplify the influence of the geometric dissimilarity of X_l and X_q, and is adaptively determined according to the number of mismatched feature pairs discarded by iterative Procrustes matching. Similarly, according to the ranking index $\mathscr{R}^*(G_l, G_q)$, ordered graph sets $ONNG_{\varepsilon}^*\{G_l\}$ and $ONNG_K^*\{G_l\}$ can be obtained. Moreover, we define a highly ranked subset $ONNG_{K\tau}^*\{G_l\}$ of $ONNG_K^*\{G_l\}$ as follows,

$$ONNG_{K\tau}^*\{G_l\} = \{\, \mathscr{R}^*(G_l, G_q) \geq \mathscr{R}_{\tau}^*, \; G_q \in ONNG_K^*\{G_l\} \,\}. \tag{17}$$

only those $G_q \in ONNG_{K\tau}^*\{G_l\}$ will be utilized. In Section 4, we will show how $\mathscr{R}_{\tau}^*$ is determined.

To summarize the above, suppose a CSHG is constructed and we wish to locate the best matched graph for a new image so as to incrementally refine the CSHG. This involves three steps. First, construct a graph G_l based on R-SIFT features. Second, find the best matched graph set $ONNG^*_{K\tau}\{G_l\}$ from the CSHG. Third, check whether G_l is redundant or not. If not, G_l will be incremented to the CSHG.

For multiple objects, CSHG's are trained separately and can be aggregated together as a multi-object CSHG.

4 Experiments and Discussions

This section is divided into two parts. We commence in Section 4.1 by illustrating the effectiveness of the R-SIFT method. Section 4.2 reports results on using the CSHG for object recognition for single 3D object.

4.1 Experiments on R-SIFT

Figure 2 consists of a series of panels each containing three images. From left to right the leftmost is the initial image, the middle images shows the first 20 R-SIFT points (deep blue circles) and their Delaunay graph (red lines), the rightmost images shows the complete set of detected SIFT points. In the first two panels for human faces, the first 10 R-SIFT correspond to the mouth, eyes, nose, face, jowl and hair. In the remaining examples (e.g. the dog, the deer, the polar bear and the flower) the R-SIFT features are also meaningful. The result is encouraging, since most of the highly ranked points correspond to those known to be visually salient, and hence the ranking process seems to be effective.

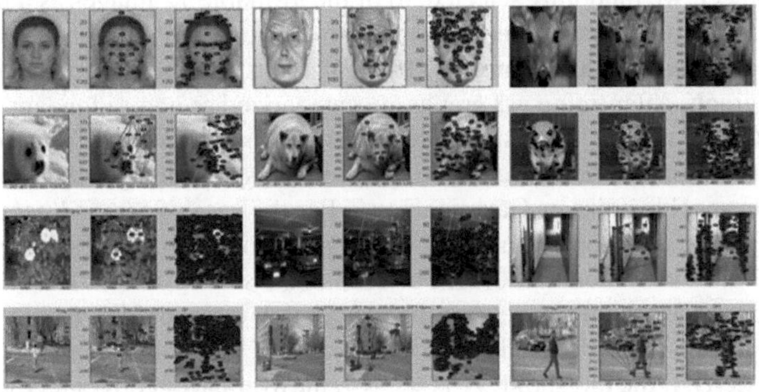

Fig. 2. Ranked salient features of various images

4.2 CSHG Modeling and Recognition for Single 3D Object

The CSHG differs from alternative recognition methods in two important respects. First, negative samples are not prerequisites in the training of the CSHG model. Second, object models can be constructed individually for different objects. In this section, we will demonstrate how the CSHG's can be constructed for different objects and then used for object recognition. We will use object 100 (the small car) from the COIL-100 data set, shown in Figure 3, to illustrate some of the steps of the process.

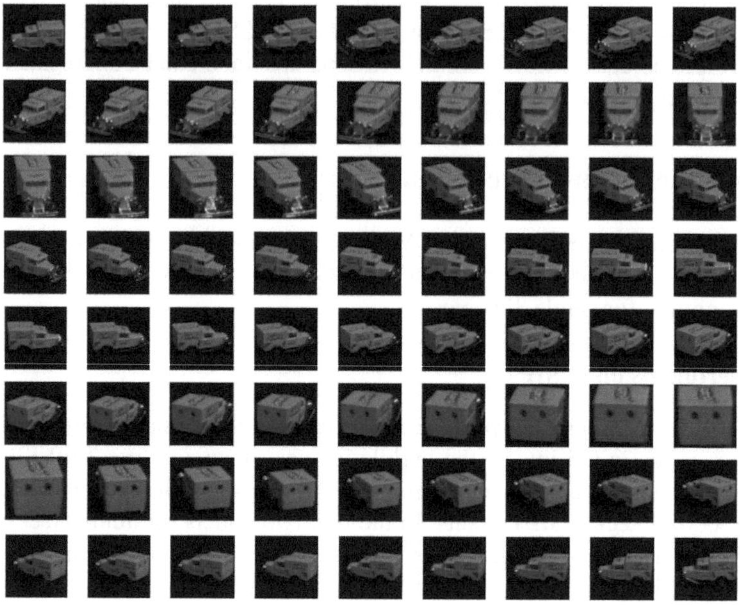

Fig. 3. Images of different views of the small car(obj100) in COIL data set

For the object, we randomly selected 56 of the available 72 images as a training set. Each of the 56 training images were processed to extract R-SIFT points. For each of the 56 training images, there are at most 40 R-SIFT points and these are used to construct a Delaunay graph. After traing, 42 graphs are categorized as non-redundant and used to construct a CSHG model. Their sibling relations are also learned and encoded.

Each of the 72 object views are used to generate noisy test images by adding salt and pepper noise, speckle noise or Gaussian noise. For the case of salt and pepper noise, a noise density d (i.e. that approximately $d*100\%$ pixels of an image affected) varies from 0.01 to 0.12 with increments of 0.01. For each noise setting 360 different images are generated. In our experiments we vary the recognition threshold $\mathscr{R}_\tau^*$ from 0 to 15. The correct recognition rates (denoted by r_g) for the case of salt and pepper noise are shown in Figure 4 (Left). If the threshold $\mathscr{R}_\tau^*$ is set as 7, the average of correct recognition rates ($\overline{r}_g$) is greater than 0.95. If, on the other hand, $\mathscr{R}_\tau^*$ is set as 6 then $\overline{r}_g$ is greater than 0.97, and if $\mathscr{R}_\tau^*$ is set as 8, then $\overline{r}_g$ decreases to less than 0.9.

The second noise process considered is pure speckle. Here multiplicative uniformly distributed random noise with mean 0 and variance v is added to the initial images. The multiplicative noise variance v is varied from 0.05 to 0.16. The recognition results obtained are shown Figure 4 (Right). The third noise process considered is an additive mixture of salt and pepper noise, speckle noise, and Gaussian noise. The variances of the all three processes are set to 0.03 , 0.04 or 0.05, and the recognition results obtained are shown in Figure 5(Left). The results are very similar to those shown in Figure 4, and hence the nature of the noise process does not appear to be a critical issue in determining performance.

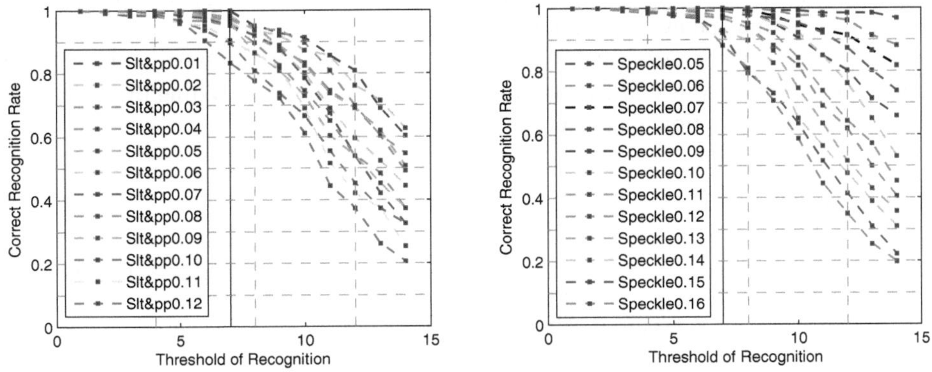

Fig. 4. Recognition results of different generated image data sets

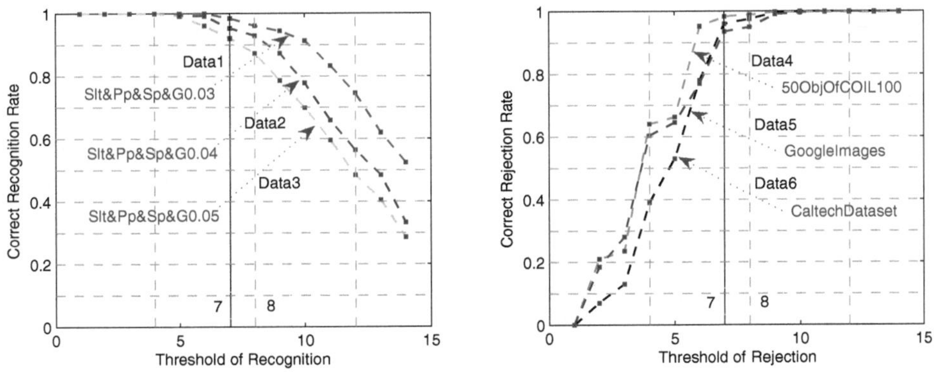

Fig. 5. Recognition results for different data sets

We have also explored how the true rejection rate (denoted by r_j) depends on the threshold. We randomly selected negative test samples from the COIL100 data set, Caltech101 data set and Google images. For COIL100, we randomly selected 36 images for 50 random objects (giving 1800 images). Similarly, we randomly selected 500 images from Caltech101 and 500 images from Google images. The rejection results are shown in Figure 5(Right). If $\mathscr{R}_\tau^*$ is set as 7, the average correct rejection rate ($\overline{r}_j$) for the three cases is greater than 0.95. On the other hand, if $\mathscr{R}_\tau^*$ is set to 8, $\overline{r}_j$ is greater than 0.98 and if $\mathscr{R}_\tau^*$ is set to 6, then $\overline{r}_j$ decreases to 0.83. Hence to compromise, we set $\mathscr{R}_\tau^*$ to 7.

5 Conclusions

We have shown how CSHG models can be obtained separately for multiple objects. Objects are efficiently represented with R-SIFT points. The CSHG model comprehensively utilizes both SIFT and geometric constraints, and hence combines both global and local information. The framework of this paper can be extended to pose estimation, action recognition and event recognition.

At the moment the construction of our class prototype is purely structural. In the future we aim to pose its construction in an information theoretic setting by casting the process into a description length setting.

References

1. Bosch, A., Zisserman, A., Muoz, X.: Scene Classification Using a Hybrid Generative/Discriminative Approach. IEEE Trans. PAMI 30(4), 1–16 (2008)
2. Bosch, A., Muoz, X., Marti, R.: Which is the best way to organize/classify images by content? Image and Vision Computing 25, 778–791 (2007)
3. Berge, C.: Hypergraphs. North-Holland, Amsterdam (1989)
4. Chung, F.: Spectral Graph Theory. American Mathematical Society (1997)
5. Lowe, D.G.: Distinctive image features from scale-invariant key points. IJCV 60(2), 91–110 (2004)
6. Elazary, L., Itti, L.: Interesting objects are visually salient. Journal of Vision 8(3), 1–15 (2008)
7. Li, F.F., Perona, P.: A Bayesian hierarchical model for learning natural scene categories. CVPR 2, 524–531 (2005)
8. Bay, H., Tuytelaars, T., Gool, L.V.: SURF: Speeded Up Robust Features. In: Leonardis, A., Bischof, H., Pinz, A. (eds.) ECCV 2006. LNCS, vol. 3951, pp. 404–417. Springer, Heidelberg (2006)
9. Zhang, J., Marszablek, M., Lazebnik, S., Schmid, C.: Local Features and Kernels for Classification of Texture and Object Categories. IJCV 73(2), 213–238 (2007)
10. Yan, K., Sukthankar, R.: PCA-SIFT: A more distinctive representation for local image descriptors. CVPR (2), 506–513 (2004)
11. Schonemann, P.: A generalized solution of the orthogonal Procrustes problem. Psychometrika 31, 1–10 (1966)
12. Sivic, J., Zisserman, A.: VideoGoogle: A text retrieval approach to object matching in videos. ICCV 2, 1470–1477 (2003)
13. Sivic, J., Russell, B.C., Efros, A.A., Zisserman, A., Freeman, W.T.: Discovering objects and their location in images. ICCV 1, 370–378 (2005)
14. Torsello, A., Hancock, E.R.: Graph Embedding using Tree Edit Union. Pattern Recognition 40, 1393–1405 (2007)
15. Torsello, A., Hancock, E.R.: Learning Shape Classes using a Mixture of Tree Union. IEEE Trans. PAMI 28, 954–967 (2006)
16. Bonev, B., Escolano, F., Lozano, M.A., Suau, P., Cazorla, M.A., Aguilar, W.: Constellations and the Unsupervised Learning of Graphs. In: Escolano, F., Vento, M. (eds.) GbRPR 2007. LNCS, vol. 4538, pp. 340–350. Springer, Heidelberg (2007)

Image Matching with Spatially Variant Contrast and Offset: A Quadratic Programming Approach

Alexander Shorin, Georgy Gimel'farb, Patrice Delmas, and Jonh Morris

Department of Computer Science, University of Auckland
Private Box 92019, Auckland 1142, New Zealand
al@cs.auckland.ac.nz, {g.gimelfarb,p.delmas,j.morris}@auckland.ac.nz

Abstract. Template matching is widely used in machine vision, digital photogrammetry, and multimedia data mining to search for a target object by similarity between its prototype image (template) and a sensed image of a natural scene containing the target. In the real-world environment, similarity scores are frequently affected by contrast / offset deviations between the template and target signals. Most of the popular least-squares scores presume only simple smooth deviations that can be approximated with a low-order polynomial. This paper proposes an alternative and more general quadratic programming based matching score that extends the conventional least-squares framework onto both smooth and non-smooth signal deviations.

1 Introduction

Template matching is a crucial part of autonomous computer vision systems that detect, classify or retrieve natural objects under varying imaging conditions and operating environments. Inherent inflexibility of many matching algorithms is often exacerbated by poor expressiveness of their underlying models of noise (here 'noise' is an umbrella term embracing all sources that cause differences between corresponding target and template signals). Common models of image noise relate pixel-wise signal differences to interdependent spatially invariant or variant contrast factors and offsets and independent additive random differences with a known and typically central-symmetric probability distribution.

Most of the state-of-the-art matching methods support only smooth (i. e. mildly non-linear) contrast and offset deviations originating from varying illumination conditions [9,2]. Although such models are more or less appropriate for matching images of planar surfaces, they are inadequate as far as most three-dimensional objects are concerned. 3D geometry of those objects can create shadows following highly non-smooth contrast functions. To appreciate the sheer complexity of noise modelling, let us consider a rather simplistic case when additive noise for each individual pixel is a linear function of signals in the two adjacent neighbours. An exact (lossless) model of this complex noise function would require approximately $3N$ parameters where N is the total number of pixels. Comparing this model to the one in [9] where only five parameters are used clearly shows that the low-order polynomial model may not be satisfactory.

N. da Vitora Lobo et al. (Eds.): SSPR&SPR 2008, LNCS 5342, pp. 127–136, 2008.

The model proposed in this paper deals with a much larger search space of noise parameters than any currently existing template matching algorithm. Specifically, it involves approximately $6N$ parameters that appear in a constrained least-squares matching problem with up to six constraints per pixel-wise signal to be solved with quadratic programming (QP). The resulting matching algorithm can successfully handle inhomogeneous contrast deviations over the target image that are introduced by a variety of noise sources including shadowing of complex non-planar objects.

The rest of the paper is structured as follows. Section 2 describes previous work done in this area and shows how our proposal fits into the existing body of knowledge. Section 3 presents the quadratic programming based matching algorithm. Experimental results and conclusions are covered in Sections 4 and 5.

2 Previous Work

In order to better understand main strengths and limitations of known contributions and novelty of the proposed approach, let us introduce a somewhat arbitrary dichotomy of contrast / offset deviations between the target and template images: *smooth* (or homogeneous) versus *non-smooth* (non-homogeneous). The former may be encountered due to inhomogeneities of an image sensor or because of illumination variations over planar objects. Three possible smooth functions are shown in Fig. 1. The existing matching methods [8,9] can handle this type of the deviations. The latter (non-homogeneous) signal deviations frequently arise when dealing with non-planar real-world objects because even a slight difference in the lighting angle can produce highly non-smooth contrast patterns like those shown in Fig. 2.

As a delineation criterion for these two broad categories of noise, a polynomial function of some low order p such that $p \ll N$ is adopted as the upper bound for the complexity which can be categorised as homogeneous. The choice is due to the fact that most of the known template matching methods have $p \leq 2$ as

(a) Pattern 1 (b) Pattern 2 (c) Pattern 3 (d) Template (e) Target

Fig. 1. Low-order polynomials as idealised contrast factors: three out of a host of possible 2D polynomial patterns of the 2nd degree (1a–1c), the template (1d), and the target image (1e) obtained by using the scaled pattern (1a) for the template (1d). The pattern (1e) was computed from (1d) by applying the transformation $f(x, y) = (0.035x - 0.00021x^2)t_{x,y} + 28$.

(a) 0° (frontal) (b) 15°R 0°U (c) Difference (d) 90°R 75°U (e) Difference

Fig. 2. Inhomogeneous contrast deviations: even a slight change in the light source orientation results in complex non-smooth contrast patterns. The contrast pattern (2c) for the target (2b) with respect to the template (2a) can hardly be approximated by a low-order polynomial. The same holds for more extreme illumination angles as shown in (2d) and (2e).

the upper bound and therefore would not be able to handle non-homogeneous contrast / offset deviations of higher order.

The interest in computationally efficient template matching has been around for quite some time. First attempts were done with various correlation-based methods [1] assuming the constant contrast and offset ($p = 0$). More recently, in response to the failure of traditional correlation techniques to handle the task [4], methods based on robust statistics and low-order polynomial models of contrast and offset such as proposed by Lai [8] made a step forward and branched out into a family of related algorithms [3,15,14]. In these algorithms, the traditional squared-error kernel is replaced by a statistical M-estimator being more robust in the presence of large signal differences and contrast and offset deviations are modelled with a low-order polynomial of pixel x- and y-coordinates.

Let $g : \mathcal{R} \rightarrow \mathcal{Q}$ denote a greyscale digital image $\mathbf{g}$ on a finite arithmetic lattice $\mathcal{R} = [(x, y) : x = 0, \ldots, m-1; y = 0, \ldots, n-1]$ where $\mathcal{Q} = \{0, 1, \ldots, Q-1\}$ is a finite set of grey levels. Let $t : \mathcal{R} \rightarrow \mathcal{Q}$ be a template $\mathbf{t}$ to be matched to $\mathbf{g}$.

Both the contrast factor $\alpha_{x,y}$ and offset $\beta_{x,y}$ for a pixel-wise signal $g_{x,y}$ in the target image are assumed to be polynomial functions of its co-ordinates (x, y) on the supporting pixel lattice so that the target signal depends on the respective signal $t_{x,y}$ in the template image as follows:

$$\begin{aligned} g_{x,y} &= \alpha_{x,y} t_{x,y} + \beta_{x,y} \\ &= (\alpha_0 + \alpha_1 x + \alpha_2 y + \ldots) t_{x,y} + (\beta_0 + \beta_1 x + \beta_2 y + \ldots) \end{aligned} \tag{1}$$

The matching algorithms use the basic Newton-Raphson descent method for iterative suboptimal minimisation of the Lorentzian or similar function chosen as the error kernel. Although the number of parameters could be expanded to a higher number (more than customary used five [8]), clearly there is a limit to what can be done under this formulation mainly due to high computational cost of its kernel and the kernel's influence function.

Other attempts to increase the number of parameters and thus improve the richness of the parameter space were done too. One approach is to avoid dealing with the entire N-dimensional space of images and resort to some selective heuristics-based sampling of the pixel space [14]. Unfortunately, this method relies on manually crafting a heuristic requiring human expert intervention. Additionally, much of pixel space is completely ignored thus resulting in a rather limited applications of the approach.

Another alternative is to pre-process input data with edge detection and contouring [15]. Again as with earlier approaches, this direction managed to account for a very special low-order polynomial model of contrast / offset deviations but it fails when non-smooth variations such as shadows are introduced. Clearly, for all methods based on robust statistics additional computational costs associated with the robust estimators (error kernels) force these methods to tackle only a tiny subset of parameters preventing them from coping with more complex functions. As the result, adequate modeling of non-homogeneous noise becomes computationally intractable.

Alternative methods which do not use robust statistics have been also proposed. Unfortunately, they suffer from similar inability to handle rich parametric spaces and thus fail to capture non-homogeneous signal deviations. Examples include template matching handled in the frequency domain [6], modifications to correlation-based statistics [13,16], conquer-and-divide approaches to pixel space by breaking down the entire pixel space into smaller subspaces and subjecting them to localised matching [17], or even methods of matching stereo images where richer problem spaces are considered [2]. More recent alternative global or subglobal methods [11,12] rely on a number of restrictive assumptions and thus cannot handle more general non-homogeneous deviations. At present, to the best of our knowledge, no matching score for loosely constrained non-homogeneous contrast and offset deviations has been proposed[1].

3 Our Approach

Our approach is to capitalise on the computational convenience of the least-squares error function and formulate the matching problem as a global quadratic programming one (QP) with a simple set of linear constraints being sufficiently descriptive to handle arbitrary non-homogeneous contrast and offset deviations of target images. The re-formulation of the QP task in terms of the Karush-Kunh-Tucker (KKT) conditions leads to an elegant and relatively fast implementation of the proposed new matching score using the Hildreth-D'Esopo algorithm [7,5].

The resulting non-parametric noise model combining spatially variant contrast and offset deviations with a central-symmetric independent noise and the derived on its basis matching score are far more general than its existing rivals and thus can successfully overcome the above limitations of the latter.

Let admissible contrast and offset deviations over the target image $\mathbf{g}$ with respect to the template $\mathbf{t}$ result in changes of the adjacent signals being constrained to a certain multiplicative range $E = [e_{\min}, e_{\max}]$ such that $0 < e_{\min} \leq e_{\max}$

[1] This paper deals only with photometric deviations, but the gradient-based search for geometric transformations of a template can be readily added much as in the other algorithms, e.g. in [8,11,12].

and with $e_{\min} < 1$ and $e_{\max} > 1$. Then contrast and offset of a transformed template can both decrease and increase. If $h : \mathcal{R} \rightarrow \mathcal{Q}$ is an image $\mathbf{h}$ obtained by the admissible changes of the template $\mathbf{t}$, then the local constraints on $\mathbf{h}$ are as follows:

$$\begin{array}{rcccl} \Delta_{\min;x,x-1;y} & \leq & h_{x;y} - h_{x-1;y} & \leq & \Delta_{\max;x,x-1;y} \\ \Delta_{\min;x;y,y-1} & \leq & h_{x;y} - h_{x;y-1} & \leq & \Delta_{\max;x;y,y-1} \\ 0 & \leq & h_{x;y} & \leq & Q-1 \end{array} \qquad (2)$$

for all the adjacent pairs $((x,y);\,(x-1,y))$ and $((x,y);\,(x,y-1))$ in $\mathcal{R}$ where

$$\begin{array}{ll} \Delta_{\min;x,x-1;y} = \min\limits_{e \in E}\{e\ (t_{x,y} - t_{x-1,y})\}; & \Delta_{\max;x,x-1;y} = \max\limits_{e \in E}\{e\ (t_{x,y} - t_{x-1,y})\} \\ \Delta_{\min;x;y,y-1} = \min\limits_{e \in E}\{e\ (t_{x,y} - t_{x,y-1})\}; & \Delta_{\max;x;y,y-1} = \max\limits_{e \in E}\{e\ (t_{x,y} - t_{x,y-1})\} \end{array}$$

Assuming the central-symmetric random noise, the least-squares matching score minimises the Cartesian distance between the target image $\mathbf{g}$ and the template $\mathbf{t}$ under the constrained signal deviations of Eq. (2):

$$d(\mathbf{g}, \mathbf{t}|E) = \min_{\mathbf{h} \in \mathcal{H}(\mathbf{t};E)} \left\{ \sum_{(x,y) \in \mathcal{R}} \left(h_{x;y} - g_{x;y}\right)^2 \right\} \qquad (3)$$

where $\mathcal{H}(\mathbf{t}; E)$ denotes a set of all images $\mathbf{h}$ with the admissible signal deviations of Eq. (2) from the template $\mathbf{t}$. To solve this QP problem, it is converted to a dual problem that is solved efficiently with the Hildreth-d'Esopo QP algorithm [7,5] that is guaranteed to converge to a solution that is arbitrary close to the desired optimal solution. Since our constraint matrix is very sparse, this iterative implementation is efficient and converges quickly in spite of the large dimensionality $6N$ of the search space. There could be different stopping criteria, e.g. processing time (the number of iterations), convergence to the goal score (a threshold for its changes at successive iterations), or a magnitude of residual violations of the constraints. Since we found that suboptimal solutions can be adequate for image matching purposes as described in the next section, the latter criterion is not generally required.

Geometrically speaking, the algorithm starts the search for the solution $\mathbf{h}$ from the target image $\mathbf{g}$, i.e. $\mathbf{h}^{[0]} = \mathbf{g}$ so that the matching score is zero at the very beginning and the initial location of $\mathbf{h}^{[0]}$ is most likely to be outside the feasible area specified by Eq. (2). Nevertheless the algorithm then quickly moves closer to the feasible area as the constraints are being negotiated. First several hundred iterations produce considerable improvement in meeting the constraints in the matched image (the score grows the fastest during the first iterations) and then the steps along the search space gradually level off.

4 Experimental Results

We have implemented the proposed QP-based matching algorithm and performed experiments with images having various non-homogeneous contrast /

offset deviations introduced by transforming templates to targets as follows (see Fig. 3 illustrating these experiments):

1. **Uniform contrast / offset:** A potentially lossy transformation for the high- and low-value signals because the transformed signals are truncated to fit the range [0..255]. This simple linear relationship corresponds to a frequently occurring situations such as variations due to software or hardware brightness and gain adjustments.
2. **Quadratic contrast:** Most likely it is even more lossy than the uniform one and may arise e.g. due to different angles of illumination of planar objects. See Fig. 1 for an example.
3. **Occlusions:** This scenario assumes that the original image signals are either partially or completely suppressed in a large area due to e.g. occlusions (a "step-function" contrast).
4. **Natural non-homogeneous deviations:** Shadowing of 3D object due to varying illumination angles is one of its examples. In practice it is one of the most relevant to computer vision types of noise because the exact image capturing conditions are almost never reproducible. See Fig. 2.
5. **Swapped uniform contrast / offset:** Finally, a transformation from less rich to more expressive images is modelled by swapping the target and template images; in our experiments, Case 1 has been used to swap the target and the template).

The main challenge to template matching of non-planar objects is posed by the natural non-homogeneity (Case 4) caused by shadows and varying illumination under different angles. It has never been addressed in the literature before. One of most suitable datasets for testing whether a matching algorithm can cope with this problem is the well-known MIT facial database [10], which we have used for the experiments below.

The dataset contains 360 images of 10 persons captured with the dominant light rotating such that its vertical orientation changes from 0° (direct) to 75° (top), and its horizontal orientation – from 0° (direct) to 90° (right). The additional ambient light source was used for making all facial features visible regardless of the dominant light position. All images are invariant to facial expression, geometry and backgrounds so that the experiments can be conducted in controlled conditions. The resulting complex shadows were the ultimate test for the proposed QP based approach (see Fig. 2c for an example of the deviations involved).

In order to benchmark the proposed algorithm, it was compared to the state-of-the-art alternative, namely, the gradient-descent based minimisation of the M-estimator of mismatches presented in [8] and capable to handle the contrast / offset deviations represented by simple polynomial models. We implemented a faster version [9] of this algorithm using the Newton-Raphson method.

Five tests were designed, corresponding to the above five scenarios of transforming templates to their target images. In our QP based matching the multiplicative range constants were set to $E = [0.25, 4]$ and the stopping rule was 7.500 iterations independently of whether other criteria were or not satisfied.

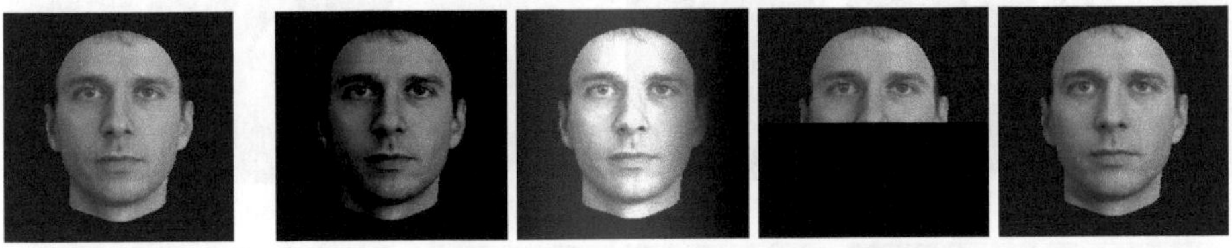

(a) Template (b) Uniform (c) Quadratic (d) Occlusion (e) Natural

Fig. 3. Test cases used in our experiments: the template (a) and the target images for the above scenarios. In Case 5 the target (a) and template images and (b) are simply swapped. In the case of natural non-homogeneous deviations (e) due to illumination, the target may visually appear the same as (a), but in fact it does have mild shadowing variations e.g. under the subject's left cheek bone.

For the benchmark counterpart, we use a 10-parameter polynomial model, i.e. 2nd-order polynomials without cross products for spatially variant contrast and offset. The stopping criterion was based on 20 iterations without improvement of the matching score.

For the uniform (or global) contrast / offset model, the target $f = \alpha t_{x,y} + \beta$ was obtained from its template with $\alpha = 1.3$, and $\beta = -40$ (see Fig. 3b), and the values $f(x,y)$ outside the range [0...255] were truncated to their respective bounds. The contrast factor in Fig. 3c is a 2nd-degree polynomial along the horizontal axis only $f(x,y) = (0.035x - 0.00021x^2)t_{x,y} + 28$ (the value truncation was done after that transformation). For the step function, the lower half of the target image was set to $f_{x,y} = 0$ as in Fig. 3d. Targets with natural non-homogeneous deviations were chosen by randomly selecting another image of the same individual but with different lighting orientation (see Fig. 3e). Finally, for Case 5 the template and target images used in Case 1 were swapped. Results of experiments in comparison to the benchmark algorithm are shown in Fig. 4 and Table 1.

The QP-based matching algorithm displays clear superiority for the uniform contrast / offset, the natural shadows, and the swapped uniform case. It also is

Table 1. Scaled matching scores (total squared differences $d(\mathbf{g}, \mathbf{t}) \times 10^{-5}$ between the targets and the transformed templates) for the five test scenarios (mean, std, min, and max denote the mean score value, the standard deviation, and the minimum and maximum values of the matching scores in our experiments)

Test scenario	Benchmark				Our algorithm			
	Mean	Std	Min	Max	Mean	Std	Min	Max
1. Uniform contrast / offset	23	3.7	15	29	**1.7**	**0.4**	**0.4**	**7.1**
2. Quadratic contrast	**6.1**	**7.6**	**0.8**	**28**	64	1.6	49	85
3. Occlusion	210	59	110	350	**87**	**4.5**	**46**	**134**
4. Natural deviation	62	39	39	150	**4.4**	**0.6**	**0.2**	**11**
5. Swapped uniform case	74	27	37	150	**6.6**	**0.8**	**1.5**	**17**

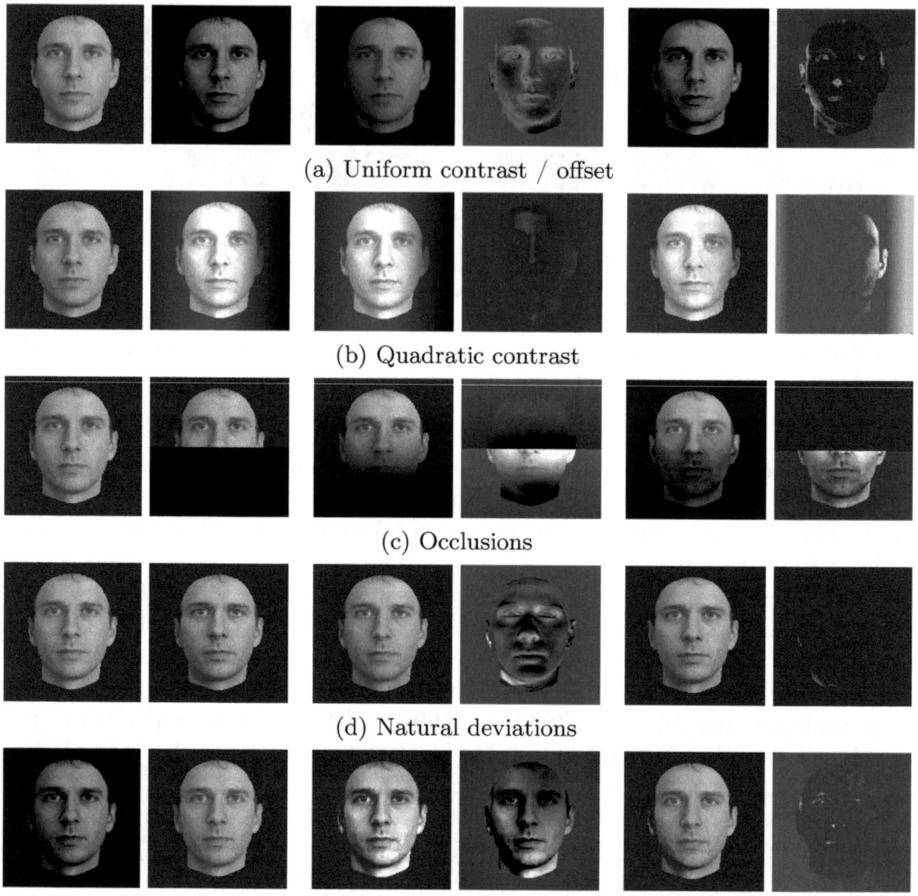

(a) Uniform contrast / offset

(b) Quadratic contrast

(c) Occlusions

(d) Natural deviations

(e) Swapped uniform contrast / offset

Fig. 4. Experimental results for the five test scenarios: the benchmark [8,9] vs. our QP-based matching: templates and targets are in columns 1 and 2, respectively; the matched image and its residual absolute difference from the target (scaled up for visualising purposes) for the benchmark is columns 3 and 4, respectively, and the like results for our matching are shown in columns 5 and 6. Note that the less the residual (i.e., the darker the residual), the closer (better) pixel-wise matching.

better in the case of occlusions, but is worse in the case of quadratic contrast function. The weaker performance of the benchmark algorithm [8,9] for the uniform contrast / offset and excellent performance for the quadratic contrast are worthy to note. The benchmark method presumes a second degree polynomial model so it is not surprising that it behaves nicely with the quadratic function. It is better than the more general QP based approach because of this perfect fit between the model and the function to be modeled. However, when handling simpler surfaces such as given in Case 1 the polynomial algorithm is just too numerically unstable due to its assumed higher (quadratic) polynomial factors.

In fact the best match in this case is achieved after just the first iteration, and then it begins to deteriorate.

The superior performance of our method has been demonstrated in all cases except the quadratic case when our approach behaves worse than the benchmark. It was quite expected because the benchmark polynomial model allows for different signs of local signal changes in adjacent pixels that correspond to one another in the transformed and original templates whereas in our model these signs have to remain the same. The other notable case relates to occlusions. Just as with the quadratic function, there were reasons to suspect that the benchmark could outperform our model since a single step function can be approximated by a 2nd-order polynomial. Nonetheless, our model gave better results in this case too. Statistical results for the 30 target–template pairs for each scenario are given in Table 1 and confirm the intuitive results visible on the output images.

5 Conclusions

This work focuses on template matching under spatially variant (inhomogeneous) contrast and offset deviations and presents the new theoretical framework and implementation of a general non-parametric QP-based algorithm. In a number of cases it shows superior results comparing to the current state-of-the-art polynomial approach. The latter benchmark algorithm had outperformed the proposed QP-matching only when the noise could be accurately approximated with the quadratic function assumed in the former method.

In the practical sense, the most important improvement of the QP approach is its ability to handle natural non-homogeneous variations resulting from changing lighting source orientation. This requirement is crucial when image recognition of non-planar objects is involved since it is almost impossible to reproduce the exact lighting conditions under which the template image was taken. Shadows are quite non-homogeneous constructs, and they can hardly have accurate low-order polynomial approximations.

The proposed noise model is more general and has greater expressive power. Its adaptability to signal deviations lies in the boundaries of the multiplicative range which were used to define the constraints. The greater the range, the more variations the model can successfully capture. This is true since the magnitude of the constraints represents the size of the admissible deviations the algorithm is aiming at. The selection of the multiplicative range should be done carefully which is a goal for future work. In this paper the authors set up a conservative range which nevertheless resulted in impressive improvement over the benchmark in four out of five test cases.

Finally, the authors are currently working on extending the current approach to include a mechanism for handling geometric transformations and masks for outliers. This will improve flexibility and power of template matching in a range of tasks where up to now matching has not been successful.

References

1. Aschwanden, P., Guggenbuhl, W.: Experimental results from a comparative study on correlation-type registration algorithms. In: Förstner, W., Ruwiedel, S. (eds.) Robust Computer Vision, pp. 268–289. Karlsruhe, Wichmann (1992)
2. Basri, R., Jacobs, D., Kemelmacher, I.: Photometric stereo with general, unknown lighting. International Journal of Computer Vision 72(3), 239–257 (2007)
3. Chen, J., Chen, C., Chen, Y.: Fast algorithm for robust template matching with m-estimators. IEEE Transactions on Signal Processing 51(1), 230–243 (2003)
4. Crowley, J., Martin, J.: Comparison of correlation techniques. In: Proc. International Conference on Intelligent Autonomous Systems (IAS-4), Karlsruhe, Germany, March 27–30, pp. 86–93. IOS Press, Amsterdam (1995)
5. D'Esopo, D.: A convex programming procedure. Naval Research Logistics Quarterly 6, 33–42 (1959)
6. Fitch, A., Kadyrov, A., Christmas, W., Kittler, J.: Fast robust correlation. IEEE Transactions on Image Processing 14(8), 1063–1073 (2005)
7. Hildreth, C.: A quadratic programming procedure. Naval Research Logistics Quarterly 4, 79–85 (1957)
8. Lai, S.: Robust image matching under partial occlusion and spatially varying illumination change. Computer Vision and Image Understanding 78(1), 84–98 (2000)
9. Lai, S., Fang, M.: Method for matching images using spatially-varying illumination change models. US patent 6, 621, 929 (2003)
10. MIT face database (accessed August 24, 2006), `http://vismod.media.mit.edu/pub/images/`
11. Pizarro, D., Peyras, J., Bartoli, A.: Light-invariant fitting of active appearance models. In: Proc. IEEE Conf. on Computer Vision and Pattern Recognition (CVIP 2008), Anchorage, Alaska, USA, pp. 1–6 (June 2008)
12. Silveira, G., Malis, E.: Real-time visual tracking under arbitrary illumination changes. In: Proc. IEEE Conf. on Computer Vision and Pattern Recognition (CVIP 2007), Minneapolis, USA, pp. 1–6 (June 2007)
13. Tombari, F., Di Stefano, L., Mattoccia, S.: A robust measure for visual correspondence. In: Proc. 14th Int. Conf. on Image Analysis and Processing (ICIAP), Modena, Italy, 2007, pp. 376–381 (September 2007)
14. Wei, S., Lai, S.: Robust and efficient image alignment based on relative gradient matching. IEEE Trans. on Image Processing 15(10), 2936–2943 (2006)
15. Yang, C., Lai, S., Chang, L.: Robust face image matching under illumination variations. EURASIP Journal on Applied Signal Processing 2004(16), 2533–2543 (2004)
16. Zhu, G., Zhang, S., Chen, X., Wang, C.: Efficient illumination insensitive object tracking by normalized gradient matching. IEEE Signal Processing Letters 14(12), 944–947 (2007)
17. Zou, J., Ji, Q., Nagy, G.: A comparative study of local matching approach for face recognition. IEEE Trans. on Image Processing 16(10), 2617–2628 (2007)

Natural Versus Artificial Scene Classification by Ordering Discrete Fourier Power Spectra

Giovanni Maria Farinella[1], Sebastiano Battiato[1],
Giovanni Gallo[1], and Roberto Cipolla[2]

[1] University of Catania, IT
{gfarinella,battiato,gallo}@dmi.unict.it
[2] University of Cambridge, UK
cipolla@eng.cam.ac.uk

Abstract. Holistic representations of natural scenes is an effective and powerful source of information for semantic classification and analysis of arbitrary images. Recently, the frequency domain has been successfully exploited to holistically encode the content of natural scenes in order to obtain a robust representation for scene classification. In this paper, we present a new approach to *naturalness* classification of scenes using frequency domain. The proposed method is based on the ordering of the Discrete Fourier Power Spectra. Features extracted from this ordering are shown sufficient to build a robust holistic representation for *Natural* vs. *Artificial* scene classification. Experiments show that the proposed frequency domain method matches the accuracy of other state-of-the-art solutions.

1 Introduction and Motivations

Humans are able to recognize complex visual scenes at a single glance, despite the number of objects with different poses, colors, shadows and textures that may be contained in the scenes. Recent studies suggest that the human visual system achieves its discrimination power using global information about the overall structure of the scene [1].

Many computer vision researchers [2,3,4,5] have proved that holistic approaches can be efficiently used to solve the problem of rapid and automatic scene classification. In particular holistic approaches are able to recognize a scene bypassing the recognition of the objects and the details inside the scene. All of the proposed holistic approaches share the same basic structure that can be schematically summarized as follows:

1. A suitable features space is built (e.g. textons vocabulary). This space must emphasize specific image cues such as, for example, corner, oriented edges, etc.
2. Each image under consideration is projected into this space. A descriptor, as a whole entity, of the image projection in the feature space is built (e.g. textons histograms).

N. da Vitora Lobo et al. (Eds.): SSPR&SPR 2008, LNCS 5342, pp. 137–146, 2008.

3. Scene classification is obtained by using a probabilistic model on the new holistic representation of the images.

A wide class of classification algorithms based on the above scheme work extracting features on perceptually uniform color spaces (e.g. CIELab). Typically, filter banks [3] or local invariant descriptors [5] are employed to capture image cues and to build the visual vocabulary to be used in a bag of visual words model [2]. An image is considered as a distribution of visual words and this holistic representation is used to perform classification. Eventually local spatial constraints are added in order to capture the spatial layout of the visual words within images [2].

Alternatively, as Torralba and Oliva have shown in several papers, the frequency domain can be a useful and effective source of information to encode holistically an image for scene understanding. The statistics of natural images on frequency domain [6] reveal that there are different spectral signatures for different image categories. In particular by considering the shape of the spectrum of an image it is possible to address scene category [4], scene depth [7], and object priming [8] such as identity, scale and location.

This paper propose a new holistic representation obtained in the discrete Fourier frequency domain. This representation is used to perform classification at the superordinate level of description of scenes [4]. Specifically we are interested in discriminating *Natural* vs. *Artificial* scenes[1].

The new and main contribution of the present work is to demonstrate that the ordering of the frequencies is a useful feature for *naturalness* classification. Ordering the frequencies indeed enables capture the overall shape of the scene in frequency domain. Similar to [4] we infer the category of an image from the shape of the scene spectrum in the frequency domain, but differently from [4], we use the relative position of pre-selected ordered frequencies as a global holistic cue.

This paper is organized as follows: Section 2 describes the model we have used and how the *discriminative frequencies* are selected. Section 3 illustrates the dataset, the setup of our experiments and the results obtained using the proposed method. Finally in Section 4 we conclude and describe future works.

2 Ordering Discrete Fourier Power Spectra

The classifier proposed in this paper has been obtained taking the following statements as starting points:

Statement-I: The energy distribution over the spectrum space, as captured by the spectra *signature* or *shape*, is very distinctive for different scene categories. This statement is strongly supported by the results in [4,6].

[1] In this work the term *Artificial* refers to images in which are depicted man-made environments (cities, buildings, streets, etc) whereas *Natural* refers to images in which natural landscapes are represented (open country, mountain, forest, coast, etc).

Statement-II: It is possible to experimentally observe [4,6] that the spectrum of *Natural* scenes is quite isotropic with no preferred direction, whereas the spectrum of *Artificial* scenes have strong "vertical" and "horizontal" axis. Thus the "diagonal[2]", "horizontal[3]" and "vertical[4]" frequencies should be particularly powerful in scene *naturalness* discrimination. Specifically, *Artificial* scenes have well marked "horizontal" and "vertical" structure in spatial domain, so "vertical" and "horizontal" frequencies are stronger than other frequencies. On the other hand in *Natural* scenes the "diagonal" frequencies are stronger than "vertical" and "horizontal" frequencies. A classic exception to the isotropy of *Natural* scene spectrum is related to the scenes in which "landscapes" are depicted. In these scenes there is a well marked "horizontal" structure in spatial domain, so "vertical" frequencies are stronger than "horizontal" frequencies. Taking into account this last specific case the "diagonal" and the "horizontal" frequencies should be particularly powerful in classifying a scene as *Natural* or *Artificial.*

Statement-III: Shape descriptors that take into account the relative positions of shape contour points have proved very powerful for shape recognition [9]. Although this property has been demonstrated in the spatial domain it is straightforward to adopt this kind of strategy to recognize *shape* or *signatures* in the frequency domain.

A key point of the proposed classification scheme indeed is to look at each specific frequency $f_{x,y}$ within its *context*, i.e. according to the relationship that $f_{x,y}$ has with the other frequencies in the Fourier space. To capture the shape signature of images in the frequency domain we use the position of the frequencies after ordering them according to their magnitudes. The proposed holistic image representation is then inspired by the following rationale.

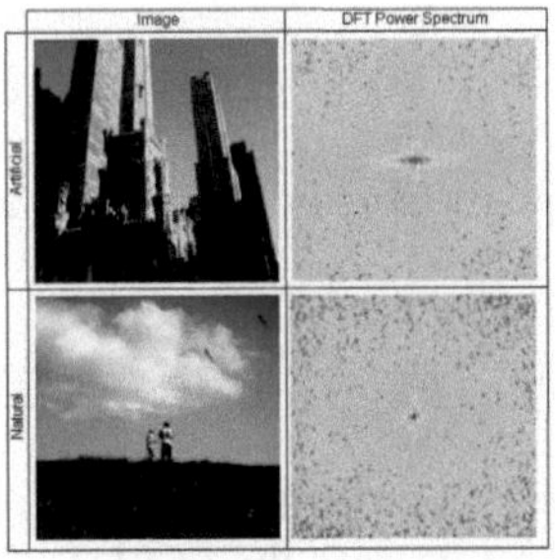

Fig. 1. Examples of two images from Natural (top-left) and Artificial (Bottom-Left) classes and their corresponding Discrete Fourier Power Spectrum (Right column)

[2] "Diagonal" frequencies correspond to any kind of structure without a preferred "horizontal" or "vertical" direction on spatial domain.

[3] "Horizontal" frequencies correspond to "vertical" structure on spatial domain.

[4] "Vertical" frequencies correspond to "horizontal" structure on spatial domain.

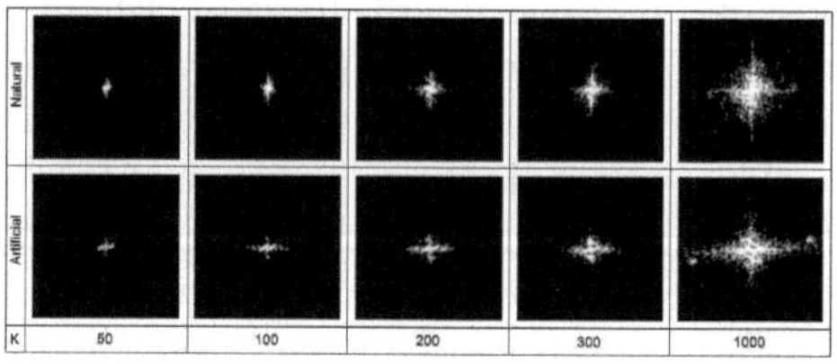

Fig. 2. The first K∈{50,100,200,300,1000} frequencies in the magnitude ordering of the power spectrum corresponding the two images reported in Figure 1

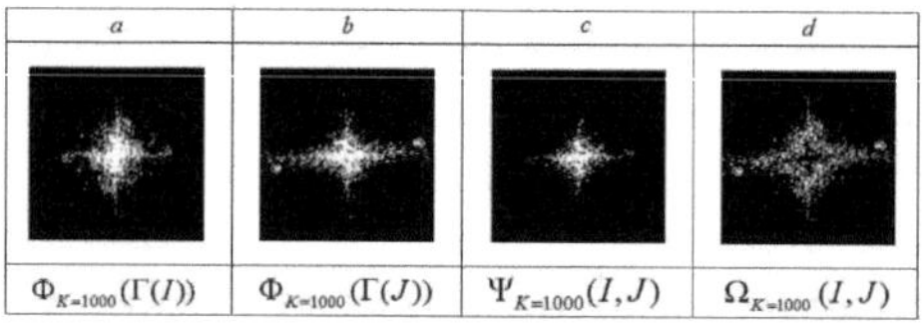

Fig. 3. The K-bounded descending ordering of Γ for the two images reported in Figure 1 allows to capture the shape of their related category ((*a*) *Natural*, (*b*) *Artificial*). The frequencies present in both ((*c*) not Discriminative) and just in one ((*d*) Discriminative) of the K-bounded ordering are shown.

Let I be an image and let $\Gamma(I)$ be the corresponding Discrete Fourier Power Spectrum (Figure 1). Ordering the frequencies in the Discrete Power Spectrum by their magnitude and selecting the first K frequencies in the descending order, it is possible to capture the specific shape signature of the image class (Figure 2).

Let $\Phi_K(\Gamma(I))$ be the set of the top K frequencies when $\Gamma(I)$ is ordered by decreasing magnitude. Figure 2 shows $\Phi_K(\Gamma(I))$ sets for different K values relative the images in Figure 1. If I is a *Natural* scene image and J is an *Artificial* scene image, the frequencies in

$$\Psi_K(I,J) = \Phi_K(\Gamma(I)) \bigcap \Phi_K(\Gamma(J)) \quad (1)$$

are likely to be not discriminative for our classification task. On the other hand the frequencies in

$$\Omega_K(I,J) = (\Phi_K(\Gamma(I)) \bigcup \Phi_K(\Gamma(J))) \setminus \Psi_K(I,J) \quad (2)$$

are likely to be useful for the recognition of the signature of each class because they belong only to one of the two sets $\Phi_K(\Gamma(I))$, $\Phi_K(\Gamma(J))$.

Figure 3 shows the $\Psi_K(I,J)$ and $\Omega_K(I,J)$ sets (K=1000) relative to the image pair in Figure 1. The value K=1000 has been choosen only as an example. Automatic selection of this parameter is thoroughly discussed in Section 2.1.

Frequencies in $\Omega_K(I,J)$ are called *disciminative frequencies*.

It is straightforward to generalize the $\Omega_K(I,J)$ to capture the shape of the whole classes of *Natural* and *Artificial* scenes by an averaging process. Let $\boldsymbol{N}$ and $\boldsymbol{A}$ be two ensembles of images respectively in the *Natural* and *Artificial*

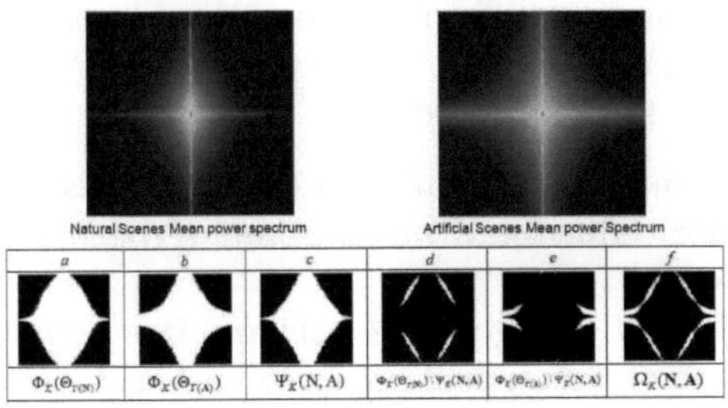

Fig. 4. Discriminative frequencies: (d) the frequencies selected from the *Natural* power spectrum span mostly "diagonally" whereas (e) the frequencies selected from the *Artificial* power spectrum span mostly "horizontally"

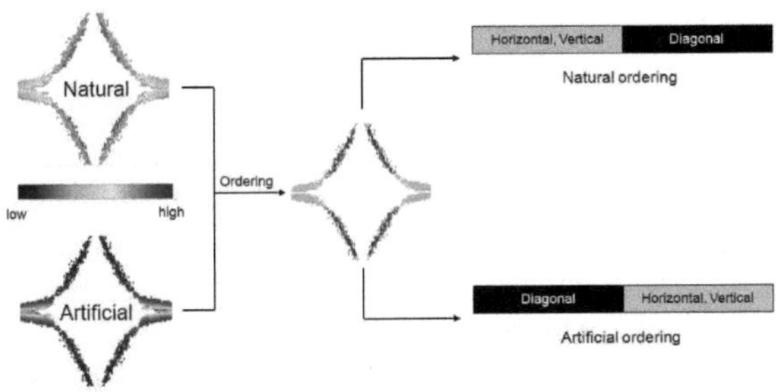

Fig. 5. Discriminative frequencies are reported on the left side. The magnitudes are color coded. In the middle, "diagonal" frequencies are represented in black while "horizontal" frequencies are represented in gray. On the right the *discriminative frequencies*, coded with the same colors than in the middle, are ordered. Separation properties characterizing the *Natural* vs. *Artificial* classes are evident in this schematic graphic diagram. The probability to find a "diagonal" frequency in the lower part of the ordering is low (high) for the *Natural* (*Artificial*) class. The probability to find a "horizontal" and "vertical" frequency in the lower part of the ordering is high (low) for *Natural* (*Artificial*) class.

classes. We indicate with $\Theta_{\Gamma(\boldsymbol{N})}$ and $\Theta_{\Gamma(\boldsymbol{A})}$ the mean power spectra obtained by respectively averaging over the power spectra of the two ensembles $\boldsymbol{N}$ and $\boldsymbol{A}$. $\Psi_K(\boldsymbol{N},\boldsymbol{A})$ and $\Omega_K(\boldsymbol{N},\boldsymbol{A})$ may hence be defined as:

$$\Psi_K(\boldsymbol{N},\boldsymbol{A}) = \Phi_K(\Theta_{\Gamma(\boldsymbol{N})}) \bigcap \Phi_K(\Theta_{\Gamma(\boldsymbol{A})}) \tag{3}$$

$$\Omega_K(\boldsymbol{N},\boldsymbol{A}) = (\Phi_K(\Theta_{\Gamma(\boldsymbol{N})}) \bigcup \Phi_K(\Theta_{\Gamma(\boldsymbol{A})})) \setminus \Psi_K(\boldsymbol{N},\boldsymbol{A}) \tag{4}$$

In Figure 4 the mean power spectra of the two classes, *Natural* and *Artificial* are reported. The mean power spectrum of the *Natural* class has been obtained averaging the spectra of images within the basic classes *Forest*, *Coast*, *Mountain* and *Open Country*. The mean power spectrum of the *Artificial* class has been obtained averaging the spectra of images within the basic classes *Building*, *Street*, *Highway* and *City*. Figure 4 confirms the *Statement-II* of this section.

The classification scheme can benefit from the corresponding relative order between the selected *discriminative frequencies* (Figure 5). In particular the following two cases can be considered:

- *The image is Natural*: "horizontal" and "vertical" *discriminative frequencies* typically rank before "diagonal" *discriminative frequencies* in the magnitude ascendent ordering;
- *The image is Artificial*: "diagonal" *discriminative frequencies* typically rank before "horizontal" and "vertical" *discriminative frequencies* in the magnitude ascendent ordering.

Our classification model should capture the fact that in the ascending ordering of the power spectrum of a *Natural* (*Artificial*) image there is high (low) probability that the ascendent ordering position of a "diagonal" frequency is higher (lower) than the ascendent ordering position of "horizontal" and "vertical" frequencies.

The above rationale suggest that the ordered *discriminative frequencies* provide an effective feature set for *Natural* vs. *Artificial* classification, indeed it is able to capture the discriminative shape signature in frequency domain. In some sense the ranking order of a *discriminative frequency* $f_{x,y}$ can be thought as the context in which $f_{x,y}$ lie with respect to the other *discriminative frequencies.* This capture the essence of the *Statement-III* of this section.

Let $\boldsymbol{r}$ be the vector of the relative position order of the *discriminative frequencies* after their selection and let $\boldsymbol{s}$ be the corresponding vector of their magnitudes. We use these two feature vectors alone or in combination to holistically represent the scene.

2.1 Discriminative Frequencies Selection

A key point in the construction of our holistic representation is the selection of the discriminative frequencies. This selection, as pointed above, is parameterized by the number K of the highest ranking frequencies in the mean power spectra. The selection process is hence reduced in choosing a suitable value for K. The idea is to select a maximal set of *discriminative frequencies.* We summarize the selection process as follows:

1. Compute the mean power spectrums $\Theta_{\Gamma(\boldsymbol{N})}$ and $\Theta_{\Gamma(\boldsymbol{A})}$ of the two classes using respectively the *Natural* and *Artificial* training images;
2. Sort $\Theta_{\Gamma(\boldsymbol{N})}$ and $\Theta_{\Gamma(\boldsymbol{A})}$ and to each frequency $f_{x,y}$ in the mean spectrums assign the ranking positions $Pos(f_{x,y},Natural)$ and $Pos(f_{x,y},Artificial)$;
3. Let $\Phi_n(\Theta_{\Gamma(\boldsymbol{N})})$ and $\Phi_n(\Theta_{\Gamma(\boldsymbol{A})})$ be the set of the first n frequencies in the descending ordering. Compute the set of the best *discriminative frequencies* $\Omega_K(\boldsymbol{N}, \boldsymbol{A})$ such that $K = \arg\max_n(\mid \Omega_n(\boldsymbol{N}, \boldsymbol{A}) \mid)$.

In Figure 6 the selection process relative to the mean power spectra of Figure 4 is illustrated. The discriminative set of frequencies, corresponding to the global maximum of the function $\mid \Omega_n(\boldsymbol{N}, \boldsymbol{A}) \mid$, is selected. Clearly, $\mid \Omega_K(\boldsymbol{N}, \boldsymbol{A}) \mid$, which is the number of *discriminative frequencies*, is smaller than K (which is the

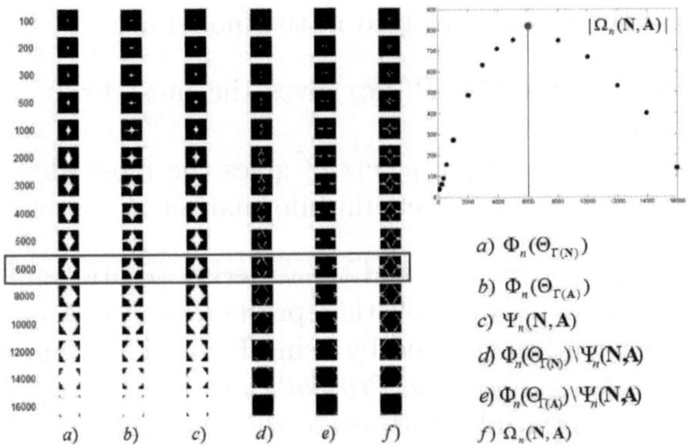

Fig. 6. Discriminative frequencies selection

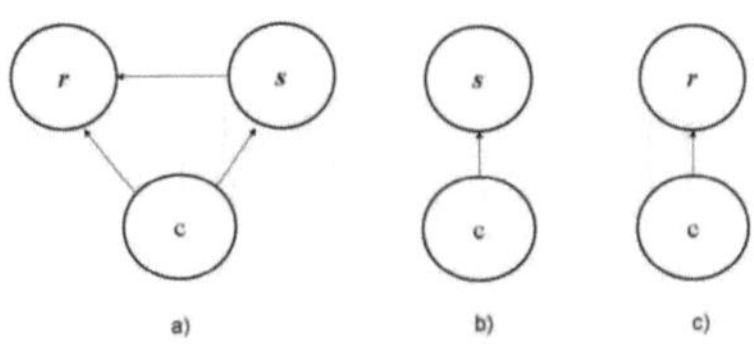

Fig. 7. The Generative Models used for the classification task. In the cnfiguration a) the power spectrum of the *discriminative frequencies* (node $\boldsymbol{s}$) and their ordering (node $\boldsymbol{r}$) are depending from the class of an image (node $\boldsymbol{c}$). Moreover, the ordering (node $\boldsymbol{r}$) depends from the magnitude of the power spectra (node $\boldsymbol{s}$). The configuration b) takes into account only the magnitude of the power spectrum, whereas the configuration c) takes into account only the ordering position.

maximum order considered). In Figure 6, we can see that $\mid \Omega_n(\boldsymbol{N}, \boldsymbol{A}) \mid$ is less than 850 for $K= 6000$.

2.2 Classification Model

Given the holistic representation of a new unclassified image I, a simple generative model together with a MAP classification rule can be used as a framework to establish which category c$\in\{$*Natural*, *Artificial*$\}$ the image I belongs to. Specifically, we use three generative models as shown in Figure 7. Let be $f_{x,y}^{(1)}...f_{x,y}^{(H)}$ the H pre-selected frequencies in $\Omega_K(\boldsymbol{N}, \boldsymbol{A})$. For each $f_{x,y}^{(i)}$ the values $\boldsymbol{s}_i$ and $\boldsymbol{r}_i$ refer respectively to the power spectrum value and its ordering position.
When the spectrum magnitudes and the ordering positions of *discriminative frequencies* are used in combination as holistic representation (Figure 7.a), the classification problem requires the evaluation of the class likelihood function P(c$|\boldsymbol{s}$,$\boldsymbol{r}$). This posterior probability can be factorized as follows:

$$P(c|\boldsymbol{s}, \boldsymbol{r}) = \frac{P(\boldsymbol{r}|\boldsymbol{s}, c)P(\boldsymbol{s}|c)}{\sum_c P(\boldsymbol{r}|\boldsymbol{s}, c)P(\boldsymbol{s}|c)} \tag{5}$$

where equiprobability is assumed for the prior P(c).

The meanings of the factors involved in the model are:

- *Power Spectrum Probability*: $P(\boldsymbol{s}|\text{c})$ gives the most likely spectrum values given a scene category.
- *Relative Position Probability* : $P(\boldsymbol{r}|\boldsymbol{s},\text{c})$ gives the most likely ordering for a scene category given power spectrum information;

When the spectrum magnitudes and the ordering positions of *discriminative frequencies* are used separately as holistic representation (Figure 7.b and 7.c) the posterior probability can be obtained by using Bayes Theorem. In such case we use respectively the *Power Spectrum Probability* or the *Relative Position Probability* when spectrum magnitudes features or the ordering positions are involved.

In the proposed classification frameworks, the *Power Spectrum Probability* and the *Relative Position Probability* are modeled as H-dimensional multivariate gaussian distribution:

$$P(\boldsymbol{s}|c) = \prod_{i=1}^{H} \frac{1}{\sqrt{2\pi\sigma_{i,c}}} \exp\left[-\frac{1}{2}\left(\frac{s_i - \mu_{i,c}}{\sigma_{i,c}}\right)^2\right] \tag{6}$$

$$P(\boldsymbol{r}|\boldsymbol{s},c) = \prod_{i=1}^{H} \frac{1}{\sqrt{2\pi H}} \exp\left[-\frac{1}{2}\left(\frac{r_i - Pos(f_{x,y}^{(i)}, c)}{H}\right)^2\right] \tag{7}$$

For the distribution of *Power Spectrum Probability* $\sigma_{i,c}$ are obtained considering the frequency $f_{x,y}^{(i)}$ of all the training images belonging to the class c. For the distribution of the *Relative Position Probability* the i-th gaussian is centered at the relative position $Pos(f_{x,y}^{(i)}$,c) that assumes the frequency $f_{x,y}^{(i)}$ in the ordering of the mean power spectrum $\Theta_{\Gamma(c)}$. In $P(\boldsymbol{r}|\boldsymbol{s},c)$ all the standard deviations are set to be equal to H, which is the maximum absolute difference that can be observed between r_i and $Pos(f_{x,y}^{(i)}$,c).

3 Dataset, Experimental Setup and Results

The experiments described in this section are aimed to demonstrate that the relative order of the discrete Fourier power spectra can be a useful feature for *Natural* vs. *Artificial* scene classification. The database used in our tests is composed of eight basic scene categories collected by Oliva and Torralba [4]: *Coast*, *Forest*, *Open Country*, *Mountain*, *Highway*, *Building*, *City*, *Street*. The first four basic classes have been considered as *Natural* scenes whereas the other classes have been considered *Artificial*. The overall database contains 2360 labeled images of 256×256 pixel in size. covering a large variety of *Natural* and *Artificial* outdoor places. As in [4], strongly ambiguous scenes in term of *naturalness* were discarded. Images have been converted in gray scale after performing histogram equalization on color domain. A logarithmic transformation of the discrete Fourier power spectrum of each image have been applied normalizing the spectrum in [0,1]. All experiments have been repeated ten times with different randomly selected training (75%) and test images (25%). The parameters involved in the classification models have been learned from the training set at each run (see Section 2) and the measures of Accuracy, Recall and Specificity were recorded at each run.

In Table 1, 2, 3 the measures of classification performances obtained at each run are reported. The results are shown with respect to the three proposed holistic representation: magnitude $\boldsymbol{s}$, relative position order $\boldsymbol{r}$, magnitude and relative position order $(\boldsymbol{r},\boldsymbol{s})$.

Table 1. Percentage of test images that were classified correctly

ACCURACY	1	2	3	4	5	6	7	8	9	10	Average
s	81.44	83.26	84.46	86.16	83.46	81.10	84.65	80.80	84.55	84.72	**83.46**
r	93.50	92.97	90.77	93.09	93.92	92.38	94.14	91.08	92.65	92.28	**92.68**
(r,s)	92.57	91.27	92.53	91.44	92.81	90.72	93.00	90.61	93.90	93.05	**92.19**

Table 2. Percentage of *Natural* test images that were classified correctly

RECALL	1	2	3	4	5	6	7	8	9	10	Average
s	97.98	96.87	98.54	98.20	98.67	98.98	97.78	97.53	97.70	98.96	**98.12**
r	93.46	92.31	88.46	92.76	93.73	92.38	93.03	88.84	90.80	90.10	**91.59**
(r,s)	96.47	94.88	95.66	93.58	98.53	96.23	94.61	95.42	98.09	96.68	**96.02**

Table 3. Percentage of *Artificial* test images that were classified correctly

SPECIFICITY	1	2	3	4	5	6	7	8	9	10	Average
s	55.01	63.02	62.54	67.55	58.25	53.88	63.21	55.57	60.53	62.47	**60.20**
r	93.57	93.95	94.38	93.60	94.19	92.38	95.97	94.46	95.41	95.68	**94.36**
(r,s)	86.34	85.90	87.65	88.13	84.42	82.33	90.36	83.35	87.66	87.38	**86.35**

First, let us examine the performance when the magnitude of the *discriminative frequencies* is considered as holistic representation. The average accuracy rate is 83.46%, which is no higher than the best results obtained by the state of the art methods working on frequency domain. As shown in Table 2 and Table 3, by using the magnitude of the *discriminative frequencies* there is an average of 40% of false positive whereas the percentage of false negative is very low. This means that the classifier is not robust in recognizing *Artificial* scenes.

Next, let us examine the performances obtained by considering the order position of the *discriminative frequencies.* As shown in Table 1, the accuracy is greater than 90% in all tests. The average accuracy is 92.68%. In Figure 8 some images correctly classified by using the order position as holistic representation are reported according their increasing posterior probability. As shown by recall and specificity measures (Table 2 and Table 3) in this case the classifier show low percentage of misclassified test images with respect to a specific class. The average classification accuracy of the proposed method (92.68%) close match the results of state of art approaches working in the frequency domain (93.5%) [4].

Fig. 8. Posterior Probability: On the left side images correctly classified as *Artificial* scenes. On the right side images correctly classified as *Natural* scenes. The images are ordered respect to the probability of belonging to the *Artificial* or *Natural* Class.

Finally, it is interesting to observe that a combination of the magnitude and the order position of the *discriminative frequencies* do not improve the classification performances. Indeed, recognition of the *Artificial* scenes decreases respect to considering the case when the order position is used alone (Table 1,2,3).

4 Conclusion

This paper has presented a new approach for *Naturalness* Classification of scenes based on ordering Discrete Fourier Power Spectra. The proposed method works by selecting a set of *discriminative frequencies*, building an holistic representation mainly based on the frequencies ordering and using a simple probabilistic model for classification. The experiments prove that the achieved performance match the state-of-art methods working on frequency domain. Future work on this technique requires the addressing of the problem related to the classification where image resolution is not fixed and other classes are taken into account.

References

1. Oliva, A., Torralba, A.: Building the gist of a scene: The role of global image features in recognition. Visual Perception, Progress in Brain Research, 251–256 (2006)
2. Lazebnik, S., Schmid, C., Ponce, J.: Beyond bags of features: Spatial pyramid matching for recognizing natural scene categories. In: IEEE Conference on Computer Vision and Pattern Recognition, vol. II, pp. 2169–2178 (2006)
3. Renninger, L.W., Malik, J.: When is scene recognition just texture recognition? Vision Research 44, 2301–2311 (2004)
4. Oliva, A., Torralba, A.: Modeling the shape of the scene: a holistic representation of the spatial envelope. International Journal of Computer Vision 42, 145–175 (2001)
5. Bosch, A., Zisserman, A., Munoz, X.: Scene classification via pLSA. In: Proceedings of the European Conference on Computer Vision (2006)
6. Torralba, A., Oliva, A.: Statistics of natural image categories. Network: Computing in Neural Systems 14, 391–412 (2003)
7. Torralba, A., Oliva, A.: Depth estimation from image structure. IEEE Trans. Pattern Anal. Mach. Intell. 24(9), 1226–1238 (2002)
8. Torralba, A., Pawan, S.: Statistical context priming for object detection. In: Internation Conference on Computer Vision (2001)
9. Belongie, S., Malik, J., Puzicha, J.: Shape context: A new descriptor for shape matching and object recognition. In: Neural Information Processing Systems (2000)

Fuzzy ART for Relatively Fast Unsupervised Image Color Quantization

Nicholas Shorter and Takis Kasparis

University of Central Florida
School of Electrical Engineering and Computer Science, Orlando, FL, USA
nshorter@mail.ucf.edu, kasparis@ucf.edu

Abstract. The use of Fuzzy Adaptive Resonance Theory (FA) is explored for the unsupervised color quantization of a color image. The red, green and blue color component values of a given color image are passed as input instances into FA which then groups similar colors into the same class. The average of all of the colors in a given class then replaces the pixel values whose original colors belonged to that class. The FA unsupervised clustering is capable of realizing color quantization with competitive accuracy and arguably low computation time.

Keywords: Image Color Quantization, Fuzzy ART, Clustering, Unsupervised.

1 Introduction

Color quantization (CQ) is the process of reducing the number of colors in a given image by replacing a group of similar colors with a single color. The replacement should be such that the color quantized image looks as similar as possible to the original image. Color quantization is often used as a preprocessor to various digital image processing applications such as segmentation. The JSEG [1] and B-JSEG [2] algorithms both rely on image CQ preprocessing. We intend to segment high resolution aerial images (4000x3000 pixels at 15cm pixel resolution) and therefore needed a computationally efficient color quantizing technique. Because of Fuzzy Adaptive Resonance Theory's (Fuzzy ART or FA) [3] ability to realize one shot fast learning and because of its competitive unsupervised learning performance; it is chosen as the clustering algorithm to group the original image's color palette into a smaller subset of colors.

2 Problem Formulation

2.1 Fuzzy ART Parameters

Carpenter et al's FA has three user defined parameters: the vigilance parameter $\rho \in [0,1]$, the learning rate $\beta \in [0,1]$, and the choice parameter $\alpha > 0$. For fast learning, the choice parameter is chosen close to 0 (actually set to $\alpha = 0.0001$)

N. da Vitora Lobo et al. (Eds.): SSPR&SPR 2008, LNCS 5342, pp. 147–156, 2008.

and the learning rate is set to 1 ($\beta = 1$). These settings therefore correspond to one shot stable learning where no weight change or search occurs after each item of an input set is presented just once.

The vigilance parameter controls how resilient FA is to accepting new input instances to existing pattern templates. Therefore, the higher (or closer to 1) the vigilance parameter is set, the more categories (or quantized colors) are created. The lower (or closer to 0) the vigilance parameter is set, the less categories (or quantized colors) are created.

2.2 Input Representation and Preprocessing

An input image is defined as matrix $\bar{\mathbf{I}}_{\mathbf{in}}$, $i \in \{1, 2, ..., R\}$, $j \in \{1, 2, ..., C\}$, with $R \times C$ elements. Each element in the matrix $\bar{\mathbf{I}}_{\mathbf{in}}$ is a 3 dimensional vector corresponding to a given pixel's color RGB components:

$$\bar{\mathbf{I}}_{\mathbf{in}}(i, j) = \{I_{in,r}(i, j), I_{in,g}(i, j) I_{in,b}(i, j)\} \tag{1}$$

Assume the input image is 24 bits per pixel. The inputs are then presented to FA as an array of colors:

$$\bar{\mathbf{I}}_{ary}(k) = \bar{\mathbf{I}}_{\mathbf{in}}(i \cdot j) \text{ where } k = \{1, 2, ..., R \times C\} \tag{2}$$

The input image is read into memory via the cimg header file [4].

In [5], Moore describes a problem in FA where a large number of inputs erode the norm of weight vectors causing FA to create more categories than necessary to encode a given pattern (known as category proliferation). Carpenter et. al in [3] note that normalizing the input by compliment encoding it can enable FA to avoid the category proliferation problem. Assuming 8 bits, each color component input is scaled to exist between 0 and 1:

$$a_r(k) = I_{ary,r}(k) / 255 \tag{3}$$

Equation (3) shows how the red color component is scaled. Similar equations can be derived for the blue and green color scalings. The complement of 'a' is calculated as follows:

$$a^c(k) = 1 - a(k) \tag{4}$$

The compliment encoding of the input then is formulated as follows:

$$\bar{\mathbf{X}}(k) = \{a_r(k), a_g(k), a_b(k), a_r^c(k), a_g^c(k), a_b^c(k)\} \tag{5}$$

2.3 Fuzzy ART Classification

The inputs (array of colors) are presented to FA which then clusters similar colors into the same class. Therefore groups of pixels are now assigned membership to a given class:

$$\overline{\mathbf{X}}^{\mathbf{p}}(k)=\left\{a_r^p(k),a_g^p(k),a_b^p(k),a_r^{p,c}(k),a_g^{p,c}(k),a_b^{p,c}(k)\right\} \quad p\in\left\{1,2,\ldots,C_l\right\} \tag{6}$$

where C_l is the total number of classes detected by FA and p are the pixel class labels.

The color components of all of the pixels belonging to a single class are averaged. Then for all of those pixels in that class, their colors are replaced by the class color component average:

$$A_r^p=\frac{1}{N_p}\cdot\sum_p I_{in,r}^p(i,j) \tag{7}$$

Equation (7) is the class average for the red color component. Similar equations can be derived for the blue and green color class averages. Note that N_p is the total number of points in a given class. The output, O(i,j), for a given pixel located at (i,j) and belonging to class p is therefore the red, blue and green class color component averages:

$$\overline{\mathbf{O}}(i,j)=\left\{A_r^p,A_g^p,A_b^p\right\} \tag{8}$$

3 Experiments

3.1 Performance Metrics

Several metrics were devised to evaluate the performance of FA's ability to color quantize a given input image. The algorithm's execution time was noted. Execution time is obviously relative to the machine which runs the algorithm. Our tests were benchmarked on an AMD 3700+ Sandiego Single Core Processor – 2.2 GHz. The computer also had 2 gigabytes of DDR400 (PC3200) ram with 2-3-2-5 timings.

In order to evaluate the quality of the color quantized image, the root mean squared error (RMSE) between the original color image and the color quantized image is used. The RMSE is also used in both [6] and [7] where they have benchmarked their algorithms on the Lenna and Mandrill images as well. The MSE for the red color component is calculated as follows:

$$MSE_r=\frac{1}{R\cdot C}\cdot\sum_{i=1}^{R}\sum_{j=1}^{C}\left(O_r(i,j)-I_{in,r}(i,j)\right)^2 \tag{9}$$

Substituting the Output term for the color class average:

$$MSE_r=\frac{1}{R\cdot C}\cdot\sum_{i=1}^{R}\sum_{j=1}^{C}\left(A_r^p-I_{in,r}(i,j)\right)^2 \tag{10}$$

Similar equations can be written for the MSE's for green and blue colors (denoted MSE_g and MSE_b respectively). The total RMSE then is the average of the three color component mean square errors:

$$RMSE = \sqrt{\frac{1}{3} \cdot \left(MSE_r + MSE_g + MSE_b \right)} \tag{11}$$

3.2 Experimental Results

Several experiments were performed to benchmark FA's clustering ability. First, three sets of 9 experiments were conducted for the entire range of the vigilance parameter. In the first set, $\beta = 1$ and $\alpha = 0.0001$ for one shot stable learning. Fuzzy ART was only allowed to execute for a single epoch (the Input was only presented to FA once). In the second set, the learning rate was empirically set to $\beta = 0.8$ and the choice parameter set to $\alpha = 0.1$. These values were found to give the best results for when FA was allowed to execute for three epochs. Finally, the learning rate was set to $\beta = 0.8$ and the choice parameter set to $\alpha = 0.01$. Now FA was allowed to run for as many epochs until it converged. Convergence for FA was defined when two conditions were met. The first condition was that no new categories were created during the current and last epoch. The second condition was that the magnitudes of the weight changes for all dimensions were less than a predefined value:

$$\forall l, \forall p \Rightarrow \left| w_{l,p}^{new} - w_{l,p}^{old} \right| \leq 0.0001 \tag{12}$$

where l is the dimension index and its range is as follows: $l \in \{1, 2, ..., 6\}$. The p is the class index and its range is as follows: $p \in \{1, 2, ..., C_l\}$. These three sets of 9 experiments were conducted on two images: the Lenna image (labeled as '-L') and the Mandrill image (labeled as '-M') with results shown for both images in [Fig. 1].

Then, six additional sets of experiments (testing 3 methods on both the Lenna and Mandrill images) were performed using the one shot stable learning parameters. The vigilance parameters were fine tuned so that the color quantized image had the following number of colors: 16, 32, 64, 128, 256 and 512.

First, as described in section 2.3, all R,G,B color components in a single class (as clustered by FA) were averaged and then the average replaced the pixels belonging to that class in the output image.

The vigilance parameters (Vig.), the RMSE, and the execution time in seconds (Sec) for the corresponding number of colors (Col) for both the Lenna image and the Mandrill image are presented in [Table 1] (shown on page 8).

Second, instead of using the average, we experimented with representing a given class in the output image with the median of the class' spectral components. We sorted

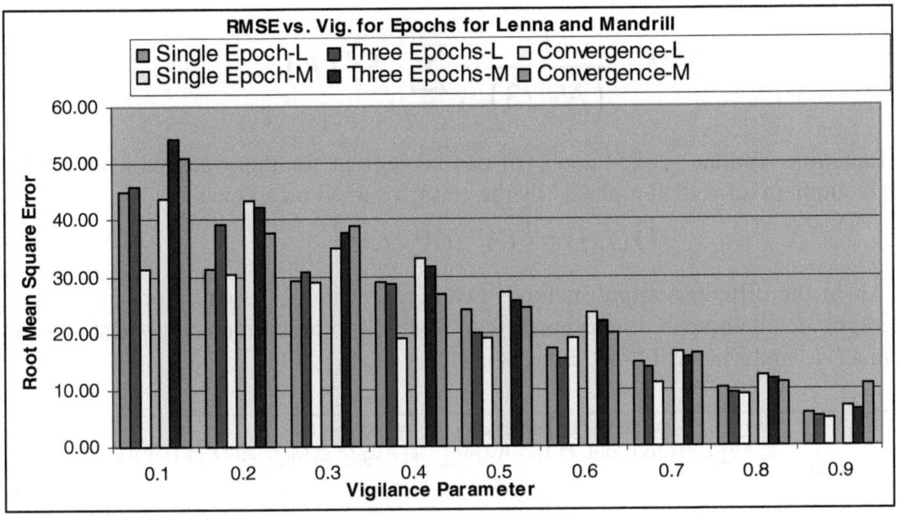

Fig. 1. RMSE for Different Epochs for Lenna

each of the R, G and B spectral components of pixels in a single class in ascending order and then found the median of each of those pixel spectral components for each class:

$$M_r^p = Median\left\{\bar{\mathbf{I}}_{in,r}^p(i,j)\right\} \tag{13}$$

where M_r^p is the median of all the red components of all pixels in class p. Similar equations can be written for the median of all blue and green pixels in class p. Then all pixels belonging to class p are replaced in the output image with the median of their class spectral components:

$$\bar{\mathbf{O}}(i,j) = \left\{M_r^p, M_g^p, M_b^p\right\} \tag{14}$$

Third, a trimmed mean median was used to represent the pixels of a given class in the output image. First, all of the spectral components (RGB colors) for all pixels in a given class were sorted:

$$\bar{\mathbf{S}}_r^p = sort\left(\bar{\mathbf{I}}_{in,r}^p(i,j)\right) \tag{15}$$

Then, the mean was calculated for only samples from $N_p/3$ to $2 \cdot N_p/3$ for each color component in a given class p:

$$TA_r^p = \frac{1}{\left(N_p/3\right)} \cdot \sum_{k=N_p/3}^{2 \cdot N_p/3} S_r^p\left(k\right) \tag{16}$$

Equations similar to (15) and (16) can be written for the green and blue colors. Each output pixel is then replaced by the class' trimmed mean median:

$$\bar{\mathbf{O}}(i, j) = \left\{TA_r^p, TA_g^p, TA_b^p\right\} \tag{17}$$

All of the different output methods (average, median of each color, and trimmed average) are all compared in terms of their RMSE for the various colors tested for the Lenna ('-L')and Mandrill ('-M') images in [Fig. 2].

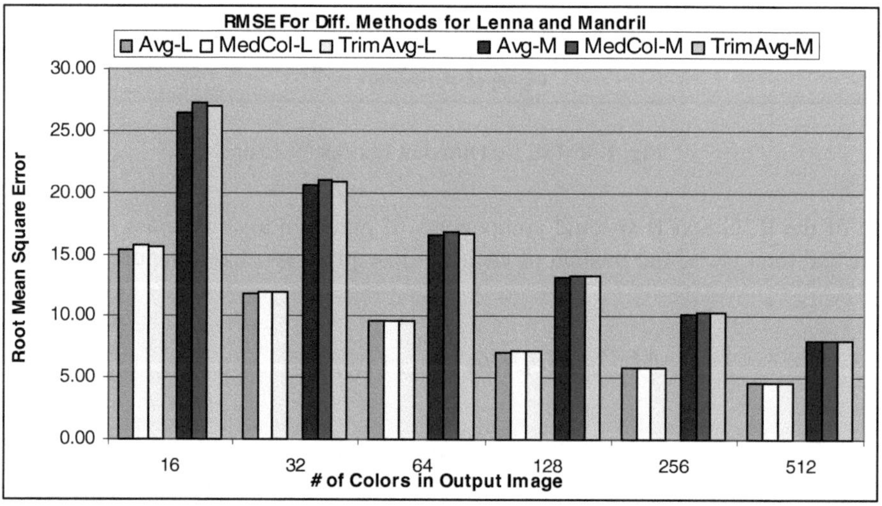

Fig. 2. RSME for Diff. Methods for Lenna

Fig. 3. Lenna as 64 Colors with FA CQ

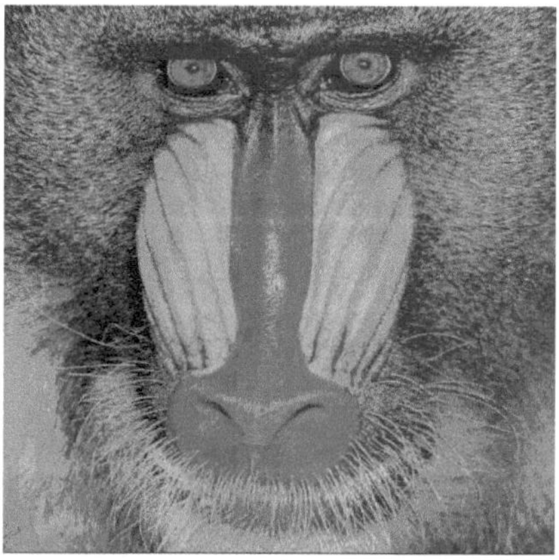

Fig. 4. Mandrill as 64 Colors with CQ

The 64 color quantized Lenna and Mandrill images (RMSE listed in [Table 1]) are presented as [Fig. 3] and [Fig. 4] respectively. Both of these figures were generated by substituting the output pixel with the class average. The original version of the standard Lenna image exists as a 512x512 pixel image with 148,279 colors while the Mandrill image is 512x512 pixels with 230,427 colors.

Finally, two additional sets of experiments were conducted where the FA color quantization algorithm was tested on two real aerial images depicting commercial and residential buildings existent in Fairfield, Australia. The aerial images have a 15 cm/pixel spacing resolution.

For these two images, only the average of each of the R,G,B spectral components of inputs belonging to the same class, as described in section 2.3, were used to represent the color clusters in the output image. The results for Scene 1 (partially depicted in [Fig. 5]) and Scene 2 (partially depicted in [Fig. 6]) are shown in [Table 2].

Table 1. Results for Average Each Color

Average of Each Color in Class						
	Lenna Image			Mandril Image		
Col	Vig	RMSE	Sec	Vig	RMSE	Sec
16	0.6257	15.36	3	0.49	26.46	3.1
32	0.72	11.83	4.2	0.6085	20.59	4.2
64	0.795	9.58	6.9	0.705	16.52	7.6
128	0.849	7.09	12	0.775	13.15	11
256	0.886	5.80	22	0.83	10.22	23
512	0.9152	4.59	40	0.8719	7.95	42

Table 2. Results for Scenes 1 and 2

Average of Each Color in Class						
	Scene 1			Scene 2		
Col	Vig	RMSE	Sec	Vig	RMSE	Sec
16	0.5	22.24	34	0.5	20.66	32
32	0.61	15.92	51	0.64	12.33	50
64	0.705	11.44	89	0.73	10.16	85
128	0.76	9.80	142	0.79	9.02	139
256	0.825	8.35	295	0.84	7.41	259
512	0.87	6.81	583	0.885	5.86	576

Fig. 5. Scene 1 Cropped

Fig. 6. Scene 2 Cropped

Fig. 7. Scene 1 Quantized to 64 Colors Cropped

Fig. 8. Scene 2 Quantized to 64 Colors Cropped

Cropped sections of the 64 color quantized Scene 1 (1510x1973 pixels with 698,843 colors) and Scene 2 (1595x1878 with 519,513 colors) are shown in [Fig. 7] and [Fig. 8] respectively. Both pictures were about 8.5 megabytes in size. The cropped sections shown were selected where there existed a rather large amount of colors so that the color quantization errors were somewhat visible.

This algorithm's executable is publicly available at [8].

4 Conclusions

Because the original Mandrill image has about 80,000 more colors than the Lenna image, it makes sense that in [Fig. 1] all of the RMSE values are higher for the Mandrill than that of the Lenna image for the listed, quantized colors.

Observe how letting FA execute for 1 to as many epochs as needed for convergence for both the Lenna and the Mandrill does not result in a significant reduction in the RMSE for vigilance values of 0.5 (which in most cases correspond 16 or more resultant quantized colors) and higher (see [Fig. 1]). Therefore, we recommend the use of one shot stable learning ($\beta = 1, \alpha = 0.0001$, and only one epoch or input list presentation) for FA input parameters as the execution time is significantly lower and the RMSE is on par with that of letting the algorithm run for additional epochs.

We have experimented with several different methods for representing a given cluster of colors existent in the input image in the output image. These methods are as follows: averaging all the colors in a given class; taking the median of the colors in a given class; and taking a trimmed mean median of all the colors in a given class. It was observed that simply averaging all the colors in a given class and replacing the pixels in that class with those color averages worked the best, yielding not only the lowest RMSE, for the tested Lenna and Mandrill images as shown in [Fig. 2] but the lowest execution time as well (as opposed to using the other two median based output representation methods).

In addition to testing the algorithm on the Lenna and Mandrill images, tests for various color quantizations were also done on real images captured over Fairfield, Australia. The proposed implementation produced acceptable color quantized versions of the images with an arguably low execution time and RMSE.

The use of FA therefore has several advantages for CQ. One of which is its low execution time (on the order of less than a minute for both the Lenna and Mandrill images when the number of colors desired is 512 or less, see [Table 1]) and on the order of minutes for the larger aerial images (see [Table 2]). The RMSE yielded between the original and color quantized image, as shown in [Fig. 2], is competitive with that of other methods in the literature [6], [7]. One unfortunate disadvantage of using FA for CQ is one cannot directly specify the number of colors that results in the color quantized image.

5 Future Work

One possible future direction in which to take this work would be to explore the use of additional features to enable the FA clustering to account for the pixel's location and context in the image in addition to the pixel's RGB value. This could possibly further improve the algorithm's performance.

Acknowledgments

This research effort has been funded by a grant from Harris Corporation. The authors would like to thank the anonymous reviewers for their critiques. The authors would also like to thank Dr. Simone Clode and Dr. Franz Rottensteiner for donating the Fairfield, Dataset which was used in this research effort.

References

1. Deng, Y., Manjunath, B.S.: Unsupervised segmentation of color-texture regions in images and video. IEEE Transactions on Pattern Analysis and Machine Intelligence 23(8), 800–810 (2001)
2. Wang, Y.-G., Yang, J., Chang, Y.-C.: Color-texture image segmentation by integrating directional operators into JSEG method. Pattern Recognition Letters 27 (2006)
3. Carpenter, G.A., Grossberg, S., Rosen, D.B.: Fuzzy ART: Fast Stable Learning and Categorization of Analog Patterns by an Adaptive Resonance System (1991)
4. CImg Header File, `http://cimg.sourceforge.net/`
5. Moore, B.: ART1 and Pattern Clustering. In: Proceedings of the 1998 Connectionist Models Summer School, pp. 174–185. Morgan Kaufmann Publishers, San Mateo (1989)

6. Ashutosh, D., Bose, N.S.C., Kandula, P., Kalra, P.K.: Modified Forward Only Counterpropogation Network (MFOCPN) for Improved Color Quantization by Entropy based Subclustering. In: International Joint Conference on Neural Networks, IJCNN 2007, pp. 1865–1870 (2007)
7. El-Mihoub, T., Nolle, L., Schaefer, G., Nakashima, T., Hopgood, A.: A Self-Adaptive Hybrid Genetic Algorithm for Color Clustering. In: IEEE International Conference on Systems, Man and Cybernetics, SMC 2006, October 8-11, vol. 4, pp. 3158–3163 (2006)
8. Nicholas Shorter's research site, `http://www.nshorter.com/`

Recognising Facial Expressions Using Spherical Harmonics

James Sharpe and Edwin R. Hancock

Department of Computer Science, University of York

Abstract. This paper explores whether facial expressions can be recognised by using the distribution of surface normal directions in the extended Gaussian image (EGI). We work with range images and extract surface normals using a mesh fitting technique. Our representation of the surface normals is based on the co-efficients of spherical harmonics extracted from the EGI. We explore whether the co-efficients can be used to construct shape-spaces that capture variations in facial expression using a number of manifold learning techniques. Based on a comparison of various alternatives, the best results are given by the diffusion map.

1 Introduction

Recently, it has been shown that statistical models based on the distribution of surface normals can offer a powerful means of representing an recognising facial shape. The reasons for this are two-fold. First, the needle map (or Gauss map) offers a representation that is rich in terms of differential geometry, and hence can potentially be used to capture subtle variations in facial shape. Secondly, surface reflectance is determined by the relative orientation of the surface and the light source. Hence, given a description of skin reflectance then a surface normal model can be fitted to image brightness data to recover 3D shape from a 2D facial image.

There are a number of approaches to capturing the statistics of surface normal direction. For instance, Smith and Hancock [1] project the surface normals into a tangent space to construct a statistical model using principal geodesic analysis. This work has recently been extended to gender recognition [2], but has proved too cumbersome for expression recognition. Bronstein, Bronstein and Kimmel [3] develop a spherical embedding, that allows faces to represented in a manner that is invariant to expression. Parameterising the distribution of surface normals, Kazhdan et al. [4] use the fact that the spherical harmonics of a given frequency form a subspace which is a rotationally invariant and which can be applied to the extended gaussian image (EGI). to create a rotationally invariant shape descriptor.

The aim in this paper is to take this work one step further and to explore whether spherical harmonics can be used for expression recognition. Specifically, we aim to explore whether the co-efficients can be used to construct shape spaces that allow different individuals and their expressions to be distinguished. To this

N. da Vitora Lobo et al. (Eds.): SSPR&SPR 2008, LNCS 5342, pp. 157–166, 2008.

end, we explore different ways of embedding the spherical harmonic co-efficients into a low dimensional pattern space. The most effective of these is the diffusion map [5], but we also provide comparison with kernel principal components analysis [6], local linear embedding [7], heat kernel embeddings and [8], commute time embeddings [9,5].

2 Representing Variations in Surface Normal Direction

In this section we describe our method. We commence from the extended Gaussian image which we parameterise using spherical harmonics. Using the co-efficients of the spherical harmonics, we construct shape-spaces using the diffusion map.

Extended Gaussian Image: The Extended Gaussian Image(EGI) [10] is a shape descriptor that represents the distribution of surface normals as data on a Gaussian Sphere. The Gaussian image is the mapping of surface normal data onto a unit sphere. The EGI is then an extension of this idea where an additional weight is assigned to each point on the sphere equal to the area of the surface that has the given normal. Computationally the EGI is formed by constructing a histogram from the surface normal data. This is achieved by representing the normal data as (θ, ϕ) pairs and then constructing a histogram.

Spherical Harmonics: Spherical Harmonics [11] arise as part of the general solution for the heat equation in spherical coordinates and are expressed in terms of Legendre polynomials. The Legendre polynomials are a set of special functions that are obtained when solving the heat equation. Specifically, they emerge as a result of solving associated Legendre equation:

$$(1-s^2)\frac{d^2\Theta}{ds^2} - 2s\frac{d\Theta}{ds} + \left(\mu - \frac{m^2}{1-s^2}\right)\Theta = 0 \tag{1}$$

Since this is a second-order differential equation, the power series is of the form $\Theta(s) = \sum_{n=0}^{\infty} a_n s^n$. where $s = cos\theta$. The solution $\Theta(s)$ is a polynomial of degree l, referred to as the Legendre Polynomial of degree k, denoted by $P_l(s)$. The Legendre Polynomials solve the Legendre equation (Eqn 1) for the case $m = 0$. To solve for $m \neq 0$ we can use the associated Legendre functions:

$$P_l^m(\cos\theta) = \sin^m \theta P_l^(\cos\theta) \qquad m = 1, 2, \ldots, l \tag{2}$$

With these ingredients the spherical harmonics are defined as

$$Y_l^m(\theta, \varphi) = \sqrt{\frac{2l+1}{4\pi}\frac{(l-m)!}{(l+m)!}} P_l^m(\cos\theta)\exp[im\varphi]$$

A Basis for the 2-Sphere: A complex valued function $f(\theta, \varphi)$ can be decomposed as a set of coefficients $a_{l.m}$ using the series expansion

$$f(\theta, \varphi) = \sum_{l=0}^{\infty} \sum_{m=-l}^{l} a_{l.m} Y_m^l(\theta, \varphi)$$

This is an infinite sum over the set of all spherical harmonics. A band limited function is a function that can be approximated by the finite set of terms

$$f(\theta,\varphi) = \sum_{l=0}^{b} \sum_{m=-l}^{l} a_{l.m} Y_m^l(\theta,\varphi)$$

Using the band limited representation we can calculate the coefficients a_{lm} by multiplying both sides by the complex conjugate $Y_m^{l*}(\theta,\varphi)$ and integrating over the solid angle Ω to give:

$$a_{lm} = \int_{\Omega} f(\theta,\varphi)(Y_m^l)^*(\theta,\varphi)d\Omega$$

Over the unit 2-sphere this becomes:

$$a_{lm} = \int_0^{\pi} d\varphi \int_0^{2\pi} f(\theta,\varphi)(Y_m^l)^*(\theta,\varphi)\sin(\theta)d\theta$$

Spherical Embedding: Recent work on expression-invariant face recognition using spherical harmonic embeddings [3] has been based on the application of the work of Kazhdan et al. [4]. Kazhdan et al. use the fact that the spherical harmonics of a given frequency l, form a subspace $V_l = \text{Span}(Y_l^{-l}, Y_l^{-l+1}, \ldots, Y_l^{l-1}, Y_l^l)$ which is a representation for the rotation group. It is this property which ensures the description is rotationally invariant. By taking each set of basis functions, summed over a given frequency, the L2-norm for the coefficients is invariant to rotation [4]. This parameterisation can be applied to the extended gaussian image (EGI) to create a rotationally invariant shape descriptor. There is however a degree of information loss which means that in general for two values with the same energy for a given frequency ($l > 2$), there are multiple representations that do not have a rotation defined between them. Kazhdan et al use this representation for expression invariant recognition, and so the information that is lost by the transform may be salient. Moreover, of the transform that is used to construct shape-spaces via manifold embedding, then the modes of shape variation may not be reliably captured.

Diffusion Map: Diffusion maps are coordinates constructed from the eigenfunctions of Markov matrices. Coifman and Lafon recognised that the majority of the existing manifold learning techniques are simply special cases of diffusion processes[5]. They provide examples of how diffusions relate to the Laplace-Beltrami operator on manifolds which Levy [12] shows as a good basis for functions of geometry and topology of objects. Coifman et al. also shows how the diffusion map is robust to noise, a useful feature for a pattern recognition algorithm, since noise is an inherent problem.

Suppose that the jth range-image can be paramterised using the vector of spherical harmonic co-efficients A_j. From the T range images available to use we compute the $T \times T$ similarity weight matrix W with elements $W(i,j) = \exp[-\sigma(A_i - A_j)^T(A_i - A_j)]$. The associated Laplacian matrix is $L = D - W$, where D is the $T \times T$ diagonal degree matrix with elements $D(i,i) = \sum_{j=1}^{T} W(i,j)$. The diffusion map commences from the random walk on a graph which has transition probability matrix $P = D^{-1}W$. Although P is not symmetric, it does have a right eigenvector matrix Φ, which satisfies the equation

$$P\Phi = \Lambda\Phi \tag{3}$$

Since $P = D^{-1}W = D^{-1}(D-L) = I - D^{-1}L$, $L\Phi = (I-\Lambda)D\Phi$. The embedding co-ordinate matrix for the diffusion map is $X = \Lambda^t \Psi^T$, where t is real. For the embedding, the diffusion distance between a pair of nodes is $d_t^2(u,v) = \sum_{i=1}^{m} (\lambda)_i^{2t} (\phi_i(u) - \phi_i(v))^2$.

The diffusion map algorithm has two input parameters; σ, which is used as the variance value when constructing the Markov chains from a Gaussian distribution and t which is the number of steps of the diffusion process. The larger the value of σ, the more weight is given to distant points. When t is set to 0 then the diffusion map is equivalent to the Graph Laplacian, when set to $\frac{1}{2}$ it is equivalent to the Fokker-Plank propagator and when set to 1 the diffusion map is equivalent to the Laplace-Beltrami operator.

3 Experiments

Range Data Collection: We have collected expression data using a Cyberware 3030 whole-head scanner. The Cyberware 3030 scanner is mounted on a PS motion platform and rotates around the subject whilst shining a low intensity laser to create the digitized points. The scan process takes about 30 seconds per scan, during which time the subject is required to remain stationary. For some expressions this is difficult to achieve and so multiple attempts were required for these scans. The scans begin at the back of the head so the effect of any slight movements of the subject during the scan are minimized as we are not interested in the area at the rear of the head where the join occurs. Movement of the subject can be seen visually in the output by stretching of features or distortion of the face. Another obstacle to overcome was the area under the chin often had holes in the scan due to the chin occluding the area from the laser during the scan. This was not a problem with all subjects but for those where this occurred the subject was asked to tilt their head back so that the laser was not obstructed by the chin.

To simplify the identification process the collected range data was edited to remove hair, ears and neck. Subjects were chosen without glasses or facial hair. Ten subjects were collected in five different facial poses: anger, neutral, smile, sad and surprise, however one subject's data was not usable due to their hair covering too much of the face and thus not producing usable meshes after the editing process. The nine subjects that were used are shown in their edited forms in figure 1. Each row in the figure corresponds to a facial expression; anger, neutral, smile sad and surprise; and each column, an individual subject.

The EGI can then be used as the input for the spherical decomposition described above. The resulting shape descriptors reconstructed from their spherical

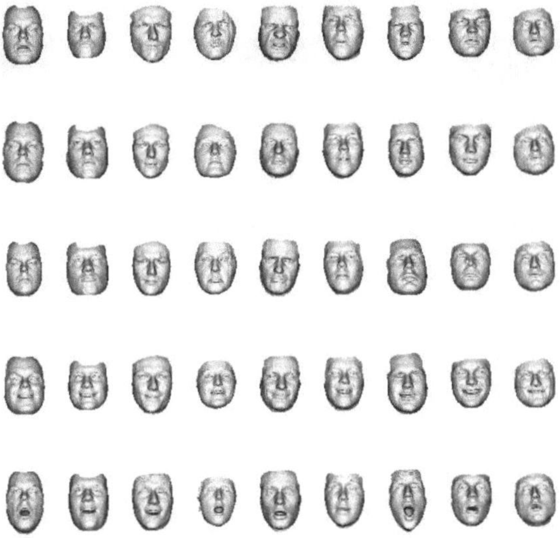

Fig. 1. Collected facial expressions

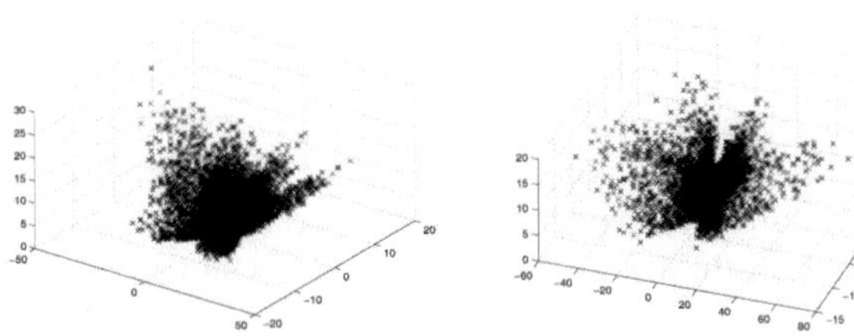

(a) (b)

Fig. 2. Shape Descriptors from histogram data

decompositions can be seen in Figure 3. As the spherical harmonic frequency is increased, higher frequency data can be represented as can be visually observed. It is important to verify that this representation, represents the original data so we will begin by looking at the accuracy of the produced shape descriptor.

Shape Descriptor Accuracy: The average error rates for the reconstruction of the shape descriptor from its spherical decomposition can be seen in Figure 4. The error value is calculated as the squared difference between the reconstructed shape descriptor and the actual shape descriptor, normalised by the number of samples. The shape descriptor was sampled with 10000 bins, with both θ and φ quantised into 100 bins each. The data was averaged over the 45 scans that were collected. The error shows an approximately logarithmic reduction in error as the frequency is increased.

Spherical Harmonic Decomposition: The surface normals for the mesh are converted from Cartesian format to spherical angles in the ranges $0 < \theta < 2\pi$ and

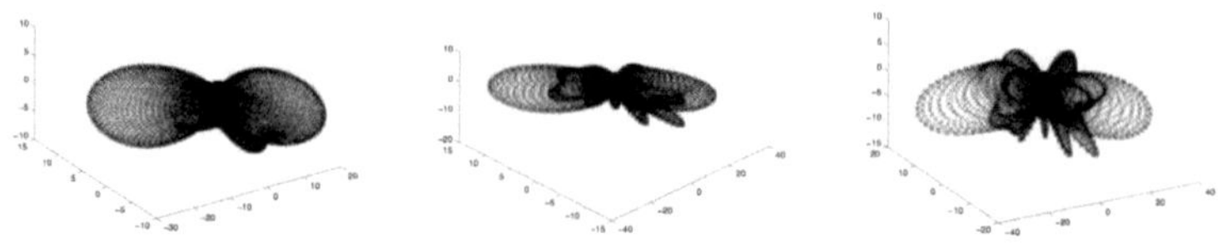

(a) 10 (b) 20 (c) 30

Fig. 3. Spherical Harmonic shape descriptors

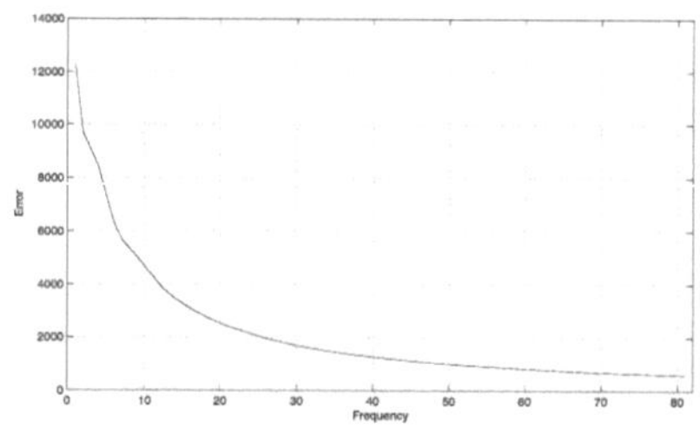

Fig. 4. Error in shape descriptor

$0 < \varphi < \pi$. An Extended Gaussian Image is calculated from these angles which is then used as the descriptor to decompose into spherical harmonics. Calculation of the Spherical Basis functions $Y_m^l : l < n, -l < m < l$ involves evaluating the function at each of the sample points (θ, φ) taken to be the centre point of each of the histogram bins calculated as part of the creation of the shape descriptor. The coefficients calculated from the decomposition are then arranged into a vector to create a vector representation of the spherical harmonic decomposition.

Embeddings: The embeddings were constructed using spherical harmonics decomposed with a bandwidth of 60. Above this the computational time becomes prohibitive within Matlab. The dimensionality of the embeddings can be determined by examining the magnitudes of the eigenvalues produced. By computing the cumulative percentage represented by the eigenvalues, a value for the dimensionality of the data can be determined. In the case of PCA (figure 5) we can see that 90% of the variation is accounted for in the first 5 eigenvalues. However, we observe a poor level of clustering for this embedding.

The degree of clustering can be visualised using the pairwise distance matrix with elements $DD(i, j) = (A_i - A_j)^T (A_i - A_j)$. In the ideal case, there should be a block structure along the diagonal when the vectors of the same class are grouped close to each other. The clustering for the spherical harmonics is shown in figure 6. The vectors are grouped by expression to determine if there is any visible block structure. There is no diagonal block structure present in this matrix which suggests a lack of clustering for the raw spherical harmonics.

We have explored the effect of varying the parameters of the diffusion map on the resulting embedded distance matrix. The value of σ controls effect of distance. Calculating the distance matrix for values of σ from 1 to 15 and varying t between 0 and 1 for each sigma value produces an array of varying distance

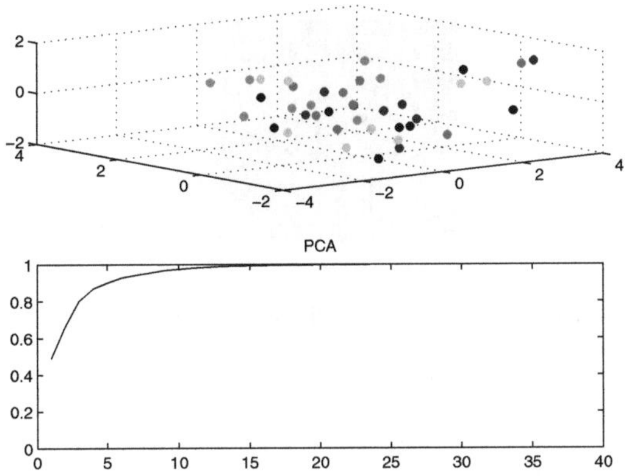

Fig. 5. Data embedded using PCA (top). Cumulative percentage of eigenvalues (bottom).

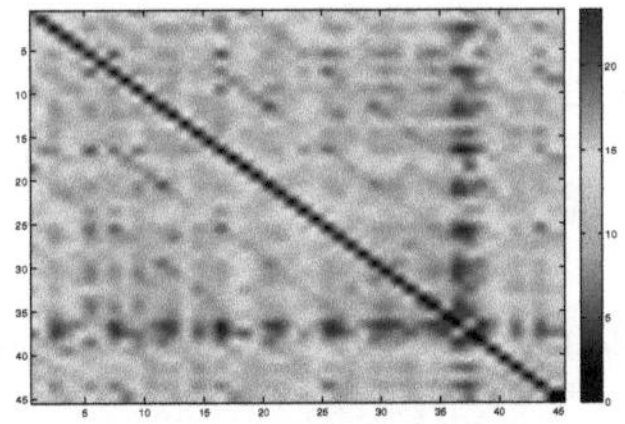

Fig. 6. Pairwise distance matrix for spherical harmonic coefficient vectors, ordered by expression

matrices of differing quality. Most are similar in structure to figure 9(e), although the result from figure 7 with t set to 0.8 and σ to 10 gives a better block diagonal structure.

Figure 8 shows the embedded points after applying the diffusion map, each colour represents a different facial expression. There is some clustering of points, as expected from the distance matrix, but there is more than one cluster for each class.

We have also applied Kernel PCA (using a polynomial kernel), heat kernel embedding, commute time embedding and LLE to the expression data. Figure 9 shows the distance matrices for the embedded points. These all give poorer results than the diffusion map. Although within the blocks the distances increase with each subject, and so would not be useful for recognition, but the diffusion map algorithm has parameters which can be adjusted to produce different results. It is interesting to note that in the KPCA distance matrix (Figure 9(a)) there is a regular pattern of banding of larger distances, these bands all belong to one subject, if the matrix was grouped by subject this would be more apparent. This suggests that this subject has some particular features that are not present in the other subjects. This additional structure may reveal itself with a larger sample size, as more subjects are likely to exhibit the characteristic,

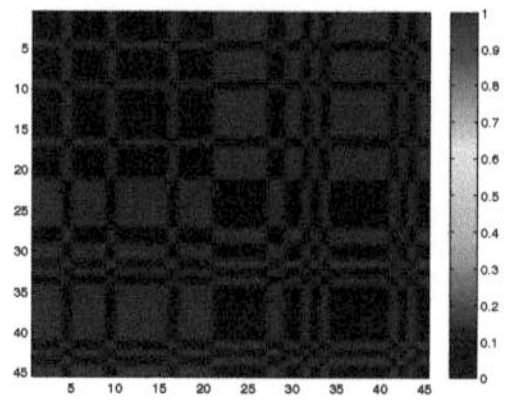

Fig. 7. Diffusion map distance matrix with alpha = 0.8 and sigma = 10

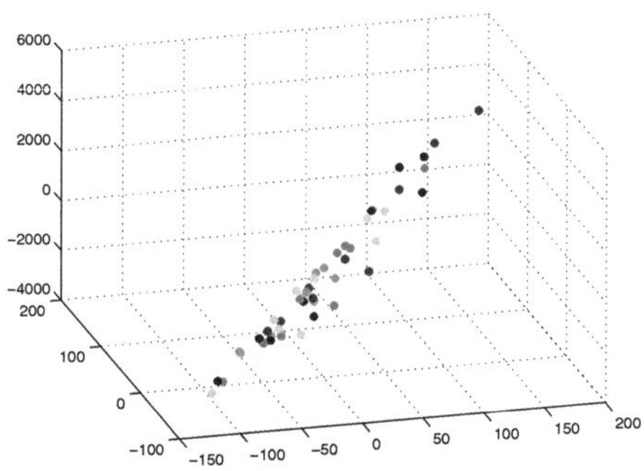

Fig. 8. Embedded data using diffusion map

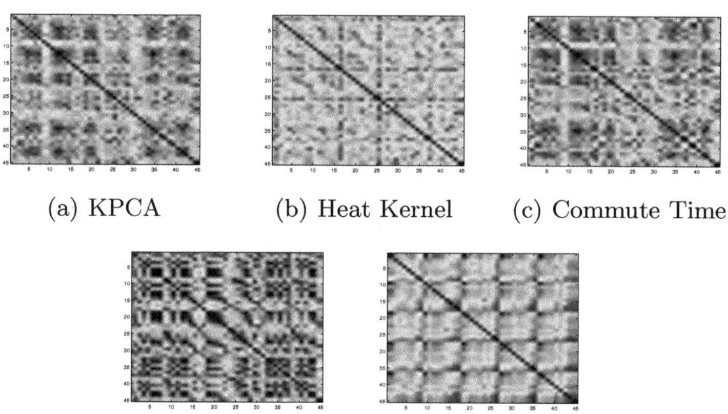

(a) KPCA (b) Heat Kernel (c) Commute Time

(d) LLE (e) Diffusion Map

Fig. 9. Pairwise distance matrices for various embedding methods

enabling the algorithm to construct a better manifold embedding due to the extra information.

Classifier: A k-nearest neighbour algorithm is used to classify a given sample point. For each point to be tested, the embedding is applied and then the nearest k neighbours are found. A k-nearest neighbour classifier is suitable for this task

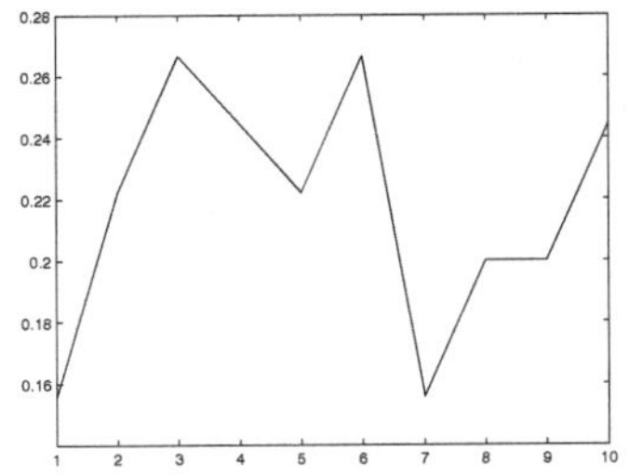

Fig. 10. Classifier accuracy as k varies

because there is a small training set and so storage for the classifier is reasonable. To test the performance of the classifier on the diffusion map embedding, each point was removed from the dataset and then the k-nearest neighbour algorithm applied. The percentage of correct classifications is then recorded. The results can be seen in figure 10 for values of k ranging from 1 to 10. The best performance is for 3 nearest neighbours which by observing the clustering in figure 8 is sensible since the clusters are generally at least 3 elements. The sample size is not large enough to do more sophisticated error analyses of the classifier.

4 Conclusions and Further Work

In this paper we have explored using the set of spherical harmonics coefficients of a model's normal map as a shape descriptor to be used in manifold learning techniques, specifically attempting to identify differences in facial expressions. We have shown that application of the diffusion map to the spherical harmonic decomposition of the surface normals of the model provides some block structure which can be utilised in classifying facial expressions. For a given expression there may be multiple clusters, although determining the differences between these clusters is an open question. A larger database for training the algorithm to create a classifier would enable a more accurate classifier. This paper only considers a static 3D representation as source data. As facial expressions are temporal in nature, more success may be achieved by looking at the time derivatives of the statistics of the normal maps.

References

1. Hancock, E.R., Smith, W.A.P.: Recovering facial shape using a statistical model of surface normal direction. IEEE Transactions on Pattern Analysis and Machine Intelligence 28(12), 1914–1930 (2006)
2. Wu, J., Smith, W.A.P., Hancock, E.R.: Weighted principal geodesic analysis for facial gender classification. Progress in Pattern Recognition, Image Analysis and Applications, 331–339 (2007)
3. Bronstein, A., Bronstein, M., Kimmel, R.: Expression invariant face recognition via spherical embedding (2005)

4. Kazhdan, M., Funkhouser, T., Rusinkiewicz, S.: Rotation invariant spherical harmonic representation of 3d shape descriptors (2003)
5. Coifman, R.R., Lafon, S.: Diffusion maps. Applied and Computational Harmonic Analysis, 5–30 (2006)
6. Kim, K.I., Jung, K., Kim, H.J.: Face recognition using kernel principal component analysis. IEEE Signal Processing Letters 9(2) (2002)
7. Rowels, S., Saul, L.K.: Nonlinear dimensionality reduction by locally linear embedding. Science 290, 2323–2326 (2000)
8. Bai, X., Wilson, R.C., Hancock, E.R.: Manifold embedding of graphs using the heat kernel. Mathematics of Surfaces, 34–49 (2005)
9. Qiu, H., Hancock, E.R.: Graph embedding using commute time. Structural, Syntatic and Statistical Pattern Recognition, 441–449 (2006)
10. Horn, B.K.P.: Extended gaussian images. Proceedings of the IEEE 72(12), 1671–1686 (1984)
11. Hobson, E.W.: The Theory of Spherical and Ellipsoidal Harmonics. Chelsea Publishing Company (1965)
12. Levy, B.: Laplace-beltrami eigenfunctions: Towards an algorithm that understands geometry (2006)

Complex Fiedler Vectors for Shape Retrieval

Reinier H. van Leuken[1], Olga Symonova[2], Remco C. Veltkamp[1], and Raffaele de Amicis[2]

[1] Universiteit Utrecht, The Netherlands
[2] Fondazione Graphitech, Trento, Italy

Abstract. Adjacency and Laplacian matrices are popular structures to use as representations of shape graphs, because their sorted sets of eigenvalues (spectra) can be used as signatures for shape retrieval. Unfortunately, the descriptiveness of these spectra is limited, and handling graphs of different size remains a challenge. In this work, we propose a new framework in which the shapes (3D models in our test corpus) are represented by multi-labeled graphs. A Hermitian matrix is associated to each graph, in which the entries are defined such that they contain all information stored in the graph edges. Additional constraints ensure that this Hermitian matrix mimics the well-studied spectral behaviour of the Laplcian matrix. We therefore use the Hermitian Fiedler vector as shape signature during retrieval. To deal with graphs of different size, we efficiently reuse the calculated Fiedler vector to decompose the graph into a limited number of non-overlapping, meaningful subgraphs. Retrieval results are based on both complete matching and subgraph matching.

1 Introduction

Although more and more 3D models populate the Internet nowadays, content based searching that is driven by shape matching remains a challenge. This research area can be divided in to areas: shape descriptors and corresponding matching algorithms at one hand, and indexing algorithms that facilitate efficient querying and database pruning on the other hand. In this paper we propose a shape descriptor that can be used as input for indexing algorithms directly. For more information about shape descriptors in general, see [1,2].

Among all shape descriptors, graph based descriptors have an advantage of providing the possibility to detect both partial and overall similarity between shapes. In particular we would like to mention Reeb graphs, which represent the topological structure of 3D model and can store local geometrical characteristics. Reeb graphs for shape description and retrieval were proposed in [3,4,5,6,7]. However, it is still a challenge to perform (sub)graph matching. Different heuristics are used to reduce the NP-hard problem of finding (sub)graph isomorphism, e.g. bipartite graph matching through minimizing the distance between node labels [6], matching based on graph hierarchy [4,7] and maximum common subgraph computation based on the node attributes [5]. An alternative approach to reduce the complexity of graph matching is to use the spectrum or eigenvectors

N. da Vitora Lobo et al. (Eds.): SSPR&SPR 2008, LNCS 5342, pp. 167–176, 2008.

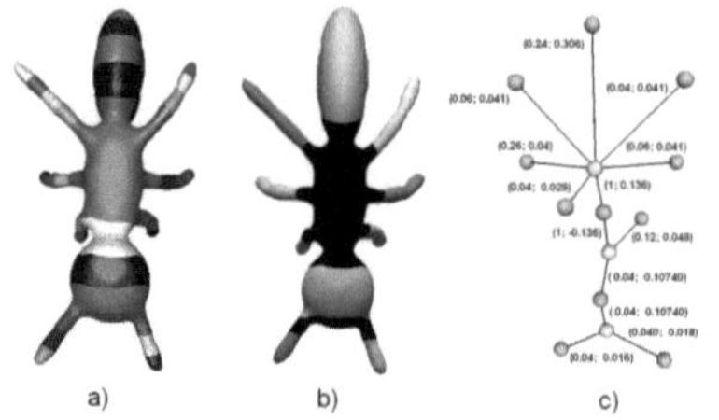

Fig. 1. ERG construction. (a) Shape segmentation. (b) Merging simple adjacent components. (c) ERG with edge attributes: shape index and segment weight.

of an associated matrix [8,9,10,11,12]. Unfortunately, the descriptiveness of the used matrices is limited, and it's often unclear how to handle graphs of different sizes.

Contributions. First, we propose a reeb-graph based shape descriptor that encodes topological and geometrical features of the model. Second, we propose to represent this descriptor by the complex-valued Fiedler vector of a Hermitian property matrix. This property matrix is constructed to contain all the features that are stored in the graph, while mimicking the well-known Laplacian matrix from a spectral point of view. Third, the calculated Fiedler vector is reused to partition the graph into non-overlapping, meaningful subgraphs that are used for partial similarity and to overcome the problem of graphs of different size. We show that our method outperforms existing spectral based indexing methods.

2 Segmentation

The Extended Reeb Graph (ERG) gives a notion about the structure of a model, and depending on the function which is used to construct the graph, it can be invariant to the position of a model in space as well as to posture variation. The ERG can be additionally enriched with geometrical shape characteristics, so that it gives a more complete shape description.

We define a mapping function f which is used to represent the manifold of a model, in our case the integral geodesic distance function [4]. The values of the mapping function are computed for each vertex of the triangular mesh.

Next, we specify N levelsets $f_i^{-1} = \{x | f(x) = f_{min} + i\frac{f_{max} - f_{min}}{N+1}\}$ for $i = 1..N$, and the intervals of the mapping function $\bigcup_{i=0}^{N} [f_i, f_{i+1}]$. These intervals define the decomposition of the model into connected components, see Figure 1a). The components can be divided into three groups, having one, two or more boundaries. In this context, a boundary is a levelset f_i^{-1} shared by two adjacent components. The components with one boundary represent minimum or maximum regions, components with two boundaries are regular regions, and components with more than two boundaries correspond to saddle regions of the mapping function [13]. We call the components with one and two boundaries *simple*. Their shape can be characterized by several geometric features [14,15]. Adjacent simple components are merged into ons segment, see Figure 1*b*.

To construct the ERG, each component obtained after the described segmentation and merging process, is represented by a graph node. There is an edge between two nodes if the corresponding components are adjacent. If two saddle components are adjacent, we insert an intermediate node between them (body part): see Figure 1*c*. Simple and intermediate nodes are connected only to saddle nodes and vice versa.

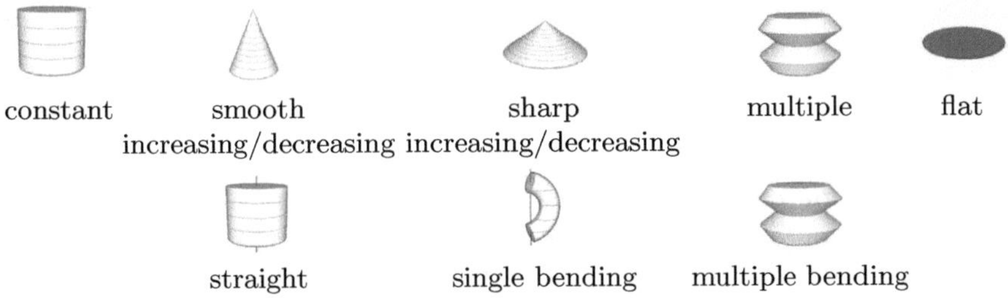

Fig. 2. Variations of the two shape criteria cross section alteration and bending

To enrich the topological structure of the ERG with geometric characteristics, we use a recently proposed shape analysis of simple components that uses two criteria [15]. The first criterion is the change in area of the cross sections. All simple components are classified in one of the following seven shape classes: constant, smooth/sharp increasing/decreasing, cross area alterations and flat components. The second shape criterion is component bending, and defines three classes: zero, single and multiple bending. See figure 2 for prototypes of shape classes; see the top row for cross area alteration and the bottom row for bending.

Combined, these two criteria define 19 different shape classes, together with the intermediate nodes there are 20 shape types. Each class is represented by a unique index. Additionally, we calculate the relative size of each segment, defined as the ratio of the surface area of the segment to the surface area of the whole model. The index of the shape classification and the segment size are stored as attributes of the edge that connects the simple node with a saddle node.

3 Graph Property Matrices

We store the graph characteristics in a Hermitian matrix because it provides space for all topological information and the edge properties. A more popular and well-studied matrix, is the Laplacian matrix. We therefore take the Laplacian matrix as our starting point, and show how this matrix can be extended to a Hermitian matrix to contain more information, without losing the validity of known theorems. The Laplacian matrix is defined as $L(G) = D(G) - A(G)$, where $D(G)$ is the diagonal matrix containing node degrees, and $A(G)$ is the adjacency matrix; entry $A_{i,j}$ is 1 if nodes i and j are connected, 0 otherwise. The spectrum of the Laplacian matrix can be used as a signature representation for the graph, and thus for efficient retrieval purposes (indexing) [8]. One of the main reasons for this is that many graph properties and invariants are reflected by the Laplacian spectrum [16]. Moreover, cospectrality for non-isomorphic graphs tends to be rare [12] and similar Laplacian matrices have similar spectra due to the interlacing theorem for two graphs where one is a slightly modified version of the other [17].

In the case of a weighted graph, $L_w(G) = D_w(G) - A_w(G)$ can be obtained, by storing edge weights in the matrices instead of the number of edges. In order to preserve the useful properties of a normal Laplacian, every edge weight $w_{a,b}$ should be positive, it may only be zero if a and b are not adjacent, and should be equal to $w_{b,a}$ (symmetry).

3.1 Hermitian Matrices

In order to encode more information about the graph in a matrix, we extend the ideas of [11] and use a Hermitian matrix to store graph characteristics. By imposing several constraints on the elements of the Hermitian matrix, we will define a complex analogue of the Laplacian matrix, such that it posesses all the properties reported in [16], [12] and [17].

A Hermitian matrix H (or self-adjoint matrix) is a square matrix with complex entries that is equal to its own conjugate transpose. Every Hermitian matrix has a real valued spectrum; the corresponding eigenvectors are complex. To impose the necessary constraints on H, it's off-diagonal elements are complex numbers written in polar form using Euler's formula, defined as $H_{a,b} = -W_{a,b}e^{iy_{a,b}}$, where each edge has the pair of properties $(W_{a,b}, y_{a,b})$. The first property, $W_{a,b}$, is used as the magnitude of the entry and should satisfy the same constraints as the edge weights of a Laplacian matrix. The second property, used as the phase of the complex matrix entry, must satisfy the following conditions:

$$y_{a,b} = -y_{b,a} \tag{1}$$

$$-\pi < y_{a,b} < \pi \tag{2}$$

The first condition (1) ensures that H is equal to its own conjugated transposed matrix. By obeying the second constraint (2), phase wrapping can be avoided. The real-valued on-diagonal entries are defined as $H_{aa} = \sum_{b \neq a} W_{a,b}$. In this way, the entries in each row of the matrix sum up to zero. We would like to stress that this is a necessary property to correctly mimic the spectral behavior of Laplacian matrices, contrary to the Hermitian matrix that is used in [11] (where additional node measurements on the diagonal are allowed). Furthermore, magnitudes should be calculated in such a way that an edge between two nodes can never be weighted 0, for it would destroy the connectivity of the graph.

As was described in section 2, an edge in the graph connects a saddle node and a simple node. As a consequence, the calculated shape index and the weight of the simple component can be stored as the attributes of the edge incident to the corresponding simple node. All 20 possible shape classifications are represented by one unique shape index. The index range is mapped to the interval $(0;1]$ to allow their use as magnitudes of complex entries of the Hermitian matrix. The positive shape index of the segment is invariant to edge direction, thus satisfying the edge weight constraints. We choose the values of shape indices based on visual shape similarity as well as on the alteration of one or both criteria used for shape analysis, assigning a greater influence to the variation of cross area than to the bending criterion. This choice can be explained by the fact that semantically equivalent subparts of articulated models frequently have different bending properties, e.g. straight and bended arms.

The relative size of the segment can be used as the phase value y_{ab} for the complex entries of the Hermitian matrix. To obey the antisymmetric condition (1) we set y_{ab} as the relative size $\text{Size}_{S_{ab}}$ of the segment S_{ab} if $\deg(a)>\deg(b)$ and as $-\text{Size}_{S_{ab}}$ if $\deg(a)<\deg(b)$. The value of $\text{Size}_{S_{ab}}$ is bounded by the interval $(0;1)$. Scaling this interval up to $(0;\pi)$ and using the above definition of y_{ab} we satisfy the constraint (2) for the phase value of the complex entries of the Hermitian matrix.

4 Retrieval through Complex Fiedler Vectors

The Hermitian matrix with attributes calculated as described above, mimics the spectral behavior of the Laplacian matrix. We propose to use the eigenvector associated to the second smallest eigenvalue (i.e. the first non-zero eigenvalue) as a signature for the graphs. These graph entities are known as Fiedler eigenvalue and Fiedler eigenvector, they encode probably the most important graph information [16].

In order to preserve permutation invariance of the representation, the entries of the Fiedler vector need to be sorted. Therefore, before evaluating the similarity between two n-dimensional Fiedler vectors, they are transformed into $2n$-dimensional vectors, by sorting them lexicographically (first based on the real part, then on the imaginary part) and interleaving the real and imaginary parts. The similarity between graphs with n nodes, can then be defined as the Euclidean distance between their $2n$-dimensional, interlaced and sorted Fiedler vectors. In practice however, models are often represented by graphs of different sizes. Furthermore, partial similarity is not taken into account by this approach. We therefore propose to base retrieval both on complete and partial graphs.

4.1 Retrieval of Complete Graphs

A common solution to the problem of comparison of differently sized graphs through their spectra or eigenvectors is padding with zeros: enlarge the smaller spectrum or eigenvector to the size of the larger spectrum by inserting zeros. This is a semantically sensible approach, since it means that the smaller graph is enlarged by inserting isolated nodes. In the context of a given database, all the graphs are brought up to the maximum size present in that database, by inserting isolated dummy nodes. When all Fiedler vectors are of the same dimension, a simple range search or nearest neighbor search among these vectors produces a ranked list of relevant database objects with respect to a query. The distance between two graphs g_1 and g_2 (based on the complete graphs) is then defined as

$$d_{\text{full}}(g_1, g_2) = \sqrt{\sum_{i=1}^{2n}(F_1(i) - F_2(i))^2} \quad (3)$$

Although this is correct semantically, it can possibly distort the retrieval results. Suppose two similar graphs are of different size as a consequence of a different level of detail that was revealed during segmentation. When dummy nodes are inserted in the coarsely segmented graph simply to enlarge the dimension of its Fiedler vector, a larger distance to the finer segmented graph is found as a consequence of matching entries related to dummy nodes with entries related to real nodes. Despite this phenomenon, the approach still produces good results for similar graphs of nearly equal size or when difference between different graphs should be magnified. We therefore propose to use it, but in combination with a method for retrieving database models based on parts of the graphs.

4.2 Subgraph Decomposition

Evaluating similarity of subgraphs is used to account for partial similarity. It also helps in solving the problem of handling graphs of different sizes, because combining similarity of parts may reduce the influence of a finer segmented region. Only subgraphs of the same size are used for similarity evaluation. The problem that remains, is to find subgraphs that are appropriate for this purpose. One proposed approach is extracting all possible subgraphs of all possible sizes, and match only subgraphs of the same size [8]. This is, however, a time consuming approach, and requires calculation of distances between many subgraphs that are not meaningful or irrelevant to the original graph. Therefore, we propose to use a graph partitioning based on [18] to find small, non-overlapping and meaningful subgraphs. The proposed partitioning approach reuses valuable information that is already calculated, i.e. the complex Fiedler vector. The approach is as follows:

1. Sort the entries of the Fiedler vectors lexicographically in decreasing order.

2. Associate a *score* with every node in the graph: $F(i) = \alpha \times \text{degree}(i) + \beta/\text{rank}(i)$, where rank$(i)$ is the position of the node in the sorted Fiedler vector, and α and β are two balancing factors to increase influence of either rank or degree. In our experiments, they are both set to 0.5.

3. Traverse through the list of the sorted Fiedler vector and select center nodes. A node is a center node if its score is higher than the scores of all its neighbors.

4. Remove the center node and its adjacent nodes from the list; together they form a subgraph. Continue processing the graph this way until all nodes are in a subgraph.

5. Decompose each subgraph s even further into every possible combination of the center node and its adjacent nodes of size 2 up to the size of s.

Steps 1-4 of the above procedure partition the graph in non-overlapping meaningful subgraphs. For an example, see Figure 3, where a segmented model of a goat is displayed, together with a view of its 3D graph and extracted meaningful subgraphs (back part, middle part and head part).

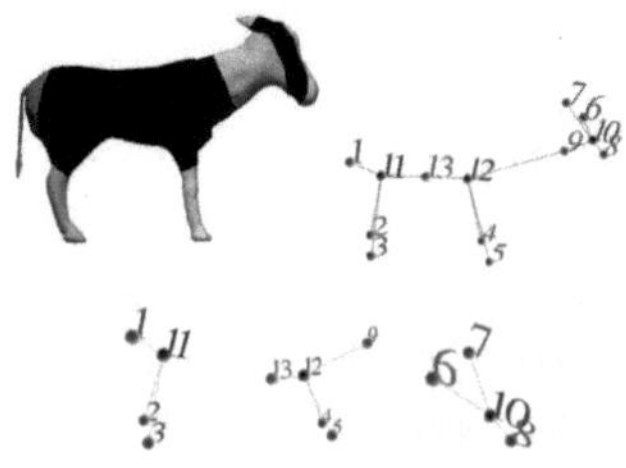

Fig. 3. Subgraph decomposition of a model of a goat

To be even more robust against small alterations in the graph structure that are caused by segmentation and/or analysis differences, step 5 of the above procedure decomposes the subgraphs even further. To calculate the subgraph-based distance between two graphs g_1 and g_2, all their subgraphs are represented by their Fiedler vectors as described in the beginning of Section 4. The distance between g_1 and g_2, based on their subgraphs, is then defined as the normalized and weighted sum of all pairwise subgraph distances, where the distance between two subgraphs is calculated only if they are of the same size. For a given subgraph size s, the average is taken of all pairwise distances of subgraphs of this size:

$$d_{\mathrm{sg\_s}} = \frac{1}{n \times m} \sum_{i=1}^{n} \sum_{j=1}^{m} d_{\mathrm{full}}(g_1^i, g_2^j) \qquad (4)$$

where g_1^i and g_2^j are the $i-th$ and $j-th$ subgraphs of size s of graphs g_1 and g_2, and where n and m are the number of subgraphs of size s that were extracted from g_1 and g_2.

A subgraph distance contributes more to the total similarity if the size of the subgraphs is larger. The total subgraph based distance between g_1 and g_2 is

$$d_{\mathrm{sg}}(g_1, g_2) = \frac{1}{\mathrm{MaxS} \times (|g_1| + |g_2|)} \sum_{d=1}^{\mathrm{MaxS}} s \times d_{\mathrm{sg\_s}} \qquad (5)$$

where $|g_1|$ and $|g_2|$ are the number of nodes in g_1 and g_2 respectively, and where MaxS is the size of the smallest maximal subgraph that could be decomposed from either g_1 or g_2.

Using this distance function, a ranked list of all the database objects can be produced for a given query. Together with the ranked list based on the comparison of complete graphs (see Section 4.1), two complementary ranked lists are produced for one query. These ranked lists need to be aggregated into one. For this purpose, each database object is assigned a score i, calculated as

$\text{score}(i) = \log(\text{r}_{\text{sg}}(i)) + \log(\text{r}_{\text{full}}(i))$, where $\text{r}_{\text{sg}}(i)$ and $\text{r}_{\text{full}}(i)$ are the ranks of object i in the ranked lists according to subgraph matching and complete matching respectively. Database objects are then re-ordered based on the scores. By taking the logarithms, a higher priority is given to the top of both ranked lists. By looking at ranks rather than distances while calculating the scores, the aggregation doesn't suffer from a difference in distance range or scale.

5 Experimental Results

We tested our framework on the dataset that was used in the Watertight 3D Models track of the SHape REtrieval Contest 2007 (SHREC '07) [19]. This dataset consists of 400 models, divided over 20 classes of 20 models each. All models were preprocessed using the described segmentation and shape analysis software. After preprocessing, the implemented framework ranks the complete database of 400 models with respect to one query in around 0.2 seconds on a 3.00 Ghz Intel Xeon processor. Each model was used as a query once; the goal is to to see the complete class of the query at the top of the ranking.

However, before running the experiments on the whole dataset, we investigated the ability of the approach to discriminate between topologically similar, but geometrically and semantically different models. To this end, we have chosen the class of human and the class of Teddy bear models. Figure 4 shows retrieval results for some sample queries when only this subset of 40 models is considered. As can be seen from these results, the Teddies can be separated from the topologically similar humans, by means of the encoded shape analysis.

To be able to compare the results to other participants of this contest, the same performance measures are used. These are precision and recall rates when the scope is set to 20, 40, 60 or 80 items; Average Dynamic Recall [1]; success rates for the first (PF) and second (PS) items; Average Ranking and Last Place Ranking [2] Unfortunately, our method does not perform as well as most of the SHREC participants. Table 1 displays the above performance values for our method (abbreviated as *CFV*, for Complex Fiedler Vectors) and the method that performed best in SHREC 2007, a method by Tung et al. [7]. At this point however we would like to emphasize that our method is an indexing method rather than a matching method; all participating methods in SHREC are matching methods encoding a large amount of shape information. This means that with our method, parts of the database can be discarded upon querying time, whereas the matching methods need an exhaustive search through the database to obtain the ranking. Our method does improve with respect to other spectral based indexing methods; we compared it to using the Fiedler vectors of adjacency and Laplacian matrices.

---

[1] $ADR = \frac{1}{20}\sum_{i=1}^{20}\frac{RI(i)}{i}$, where $RI(i)$ is the number of relevant retrieved items in the top i of the ranking. $ADR \in [0, 1]$, where 1 is the best retrieval result.

[2] $LPR = 1 - \frac{\text{Rank}-20}{380}$, where Rank is the rank of the last relevant item. $ADR \in [0, 1]$, where 1 is the ideal retrieval result.

Fig. 4. Retrieval results when only the subset of human and Teddy bear models is considered: top 20 retrieved items are displayed. Colored border means successful hit.

Table 1. Performance comparison of the proposed approach (CFV), using Laplacian matrices (Lapl), using Adjacency matrices (Adj) and the approach by Tung et al.

Measure	CFV	Lapl	Adj	Tung et al.	Measure	CFV	Lapl.	Adj.	Tung et al.
P20	0.42	0.22	0.20	0.71	R20	0.42	0.22	0.20	0.71
P40	0.29	0.16	0.16	0.41	R40	0.57	0.33	0.31	0.83
P60	0.22	0.14	0.14	0.29	R60	0.66	0.43	0.41	0.87
P80	0.18	0.12	0.12	0.23	R80	0.72	0.50	0.48	0.90
PS1	1	1	1	1	ADR	0.58	0.36	0.31	0.86
PS2	0.65	0.33	0.17	0.98	AVG	77	125	132	31
					LPR	0.48	0.25	0.17	0.74

6 Concluding Remarks

In this paper we presented a new shape retrieval method using complex Fiedler vectors of Hermitian property matrices. These matrices represent the graph based shape descriptor, in which both topology and geometry of the models is encoded, such that the matrix mimics the Laplacian matrix from a spectral point of view. By combining partial and complete retrieval, the framework shows strong performance on a benchmark of 400 watertight 3D models with respect to other spectral based retrieval techniques.

It is one of our interests in the future to test the proposed approach on other domains. After the Hermitian matrix has been constructed with features of a certain database object (regardless its modality), the framework is fixed. It is interesting to see how this approach extends to retrieval of other multimedia objects, such as music or 2D images. Furthermore, we intend to investigate the how this method can be used as an efficient prefiltering step in combination with other methods to in order to increase performance and save computing time.

Acknowledgements. This research was supported by FP7 ICT project 214993 FOCUS K3D and FP6 IST projects 511572-2 PROFI and 506766 AIM@SHAPE.

References

1. Biasotti, S., Falcidieno, B., Frosini, P., Giorgi, D., Landi, C., Marini, S., Patane, G., Spagnuolo, M.: 3D shape description and matching based on properties of real functions. In: Eurographics - Tutorials (2007)
2. Tangelder, J.W., Veltkamp, R.C.: A survey of content based 3d shape retrieval methods. In: Shape Modeling International (2004)
3. Biasotti, S., Giorgi, D., Spagnuolo, M., Falcidieno, B.: Reeb graphs for shape analysis and applications. Theoretical Computer Science (2007) (to appear)
4. Hilaga, M., Shinagawa, Y., Kohmura, T., Kunii, T.L.: Topology matching for fully automatic similarity estimation of 3d shapes. In: SIGGRAPH (2001)
5. Marini, S., Spagnuolo, M., Falcidieno, B.: From exact to approximate maximum common subgraph. In: Graph-based Representations in Pattern Recognition (2005)
6. Tierny, J., Vandeborre, J.P., Daoudi, M.: Reeb chart unfolding based 3D shape signatures. In: Eurographics (2007)
7. Tung, T., Schmitt, F.: Augmented reeb graphs for content-based retrieval of 3d mesh models. In: Shape Modeling International (2004)
8. Demirci, M.F., van Leuken, R.H., Veltkamp, R.C.: Indexing through laplacian spectra. Computer Vision and Image Understanding 110(3), 312–325 (2008)
9. Sengupta, K., Boyer, K.L.: Modelbase partitioning using property matrix spectra. Computer Vision and Image Understanding 70(2), 177–196 (1998)
10. Shokoufandeh, A., Macrini, D., Dickinson, S., Siddiqi, K., Zucker, S.: Indexing hierarchical structures using graph spectra. Pattern Analysis and Machine Intelligence 27(7) (2005)
11. Wilson, R.C., Hancock, E.R., Luo, B.: Pattern vectors from algebraic graph theory. Pattern Analysis and Machine Intelligence 27, 1112–1124 (2005)
12. Zhu, P., Wilson, R.C.: A study of graph spectra for comparing graphs. In: British Machine Vision Conference (2005)
13. Biasotti, S., Falcidieno, B., Spagnuolo, M.: Extended reeb graphs for surface understanding and description. In: Discrete Geometry for Computer Imagery (2000)
14. Mortara, M., Patane, G., Spagnuolo, M., Falcidieno, B., Rossignac, J.: Blowing bubbles for multi-scale analysis and decomposition of triangle meshes. Algorithmica 38(1), 227–248 (2004)
15. Symonova, O., De Amicis, R.: Shape analysis for augmented topological shape descriptor. In: Eurographics (2007)
16. Mohar, B.: The laplacian spectrum of graphs. Graph Theory, Combinatorics and Applications 2, 871–898 (1991)
17. Merris, R.: Laplacian matrices of graphs: a survey. Linear Algebra and its Applications 197(1), 143–176 (1994)
18. Qiu, H., Hancock, E.R.: Graph partition for matching. In: Graph-based Representations in Pattern Recognition (2003)
19. Veltkamp, R.C., ter Haar, F.B.: SHREC2007: 3D Shape Retrieval Contest. Technical Report UU-CS-2007-015, Utrecht University (2007)

Combining Shape Priors and MRF-Segmentation

Boris Flach and Dmitrij Schlesinger

Dresden University of Technology*

Abstract. We propose a combination of shape prior models with Markov Random Fields. The model allows to integrate multiple shape priors and appearance models into MRF-models for segmentation. We discuss a recognition task and introduce a general learning scheme. Both tasks are solved in the scope of the model and verified experimentally.

1 Introduction

The last decade's progress in image analysis and image interpretation is strongly influenced by a model driven attitude. In particular this applies for image segmentation, whereas the goal is to partition the image domain into meaningful areas.

In particular discrete structural models and approaches were strongly influenced by the breakthrough achieved in the area of $(\max, +)$-Labelling problems also known as Soft Constraint Satisfaction problems. Findings on polynomially solvable classes of these problems [1,2,3] and deduction of well-founded approximation algorithms for rather general classes [4,5,6] have led to considerable progress for image segmentation (and other areas of image analysis as well).

The adaption of probabilistic shape priors from level-set based models [7,8] and their further development and combination with simple segmentation models [9] and later on with Markov Random Fields and Conditional Random Fields [10,11] marks another leap in the segmentation research.

A closer examination of particular aspects like model complexity, recognition and learning reveals however, that many of the above mentioned approaches concentrate on some of these aspects and leave out others at the same time. Some of these approaches combine shape prior models with simple segmentation models – i.e. without lateral interactions – and solve the corresponding segmentation and learning tasks [9]. Other approaches combine very complex shape models and CRF-s and solve corresponding recognition tasks as well as supervised learning. Partially unsupervised learning for CRF-s (without shape priors) is considered in [12]. However, CRF-s have some principal drawbacks. They cannot be learned completely unsupervised. The same holds, if the observation is partially hidden. Moreover, it is yet unclear how to measure their generalisation capability (e.g. VC-Dimension). Finally, the majority of papers propose the MAP-criterion for segmentation, which is not natural – using the Hamming distance for loss is much more appropriate.

* Supported by Deutsche Forschungsgemeinschaft, Grant No. FL307/2-1.

N. da Vitora Lobo et al. (Eds.): SSPR&SPR 2008, LNCS 5342, pp. 177–186, 2008.

In the present paper we propose a probabilistic segmentation model based on Markov Random Fields and object/segment specific shape priors. The latter range midscale in complexity. The model combines the previously mentioned benefits and avoids unnecessary restrictions at the same time. It allows to pose *all* variants of recognition as well as *all* variants of learning in a closed fashion and to solve them approximatively.

2 The Model

For the sake of convenience we consider throughout this paper a relatively simple model for the shape priors – an expected average shape defined up to pose parameters like position and scale. Denoting the discrete image domain by $R \subset \mathbb{Z}^2$, the sample space Ω of the probabilistic model is $\mathcal{X} \times \mathcal{S}$, where $\mathcal{X}$ is the set of all images

$$x \colon R \to F, \quad F \text{ - colour space} \tag{1}$$

and $\mathcal{S}$ is the set of all segmentations into K regions

$$s \colon R \to K, \quad K \text{ - label/segment set.} \tag{2}$$

A shape prior function $\phi_k(r, \mu)$ is assigned to each object/segment, where

$$\phi_k(., \mu) \colon \mathbb{R}^2 \to \mathbb{R} \tag{3}$$

is the signed distance to the average object contour defined up to a set of pose parameters μ. The better a segmentation $s \in \mathcal{S}$ fits the prior shape, the more probable it is. The same is required with respect to "compactness". Modelling the latter by Potts-interactions, gives the following prior model for segmentations

$$P(s) = \frac{1}{Z} \exp\Big[\alpha \sum_{\{rr'\} \in E} \delta(s_r, s_{r'}) + \lambda \sum_{r \in R} \sum_{k \in K} \delta(s_r, k) \phi_k(r, \mu_k)\Big] , \tag{4}$$

where $\delta(.,.)$ denotes the Kronecker symbol and E is a neighbourhood relation on R. Z denotes the partition function and depends on α, λ, $\bar{\mu} = \{\mu_k, k \in K\}$ and the choice of $\phi_k(.,.)$.

The second part of our probabilistic model – the conditional probability to observe a certain image given a segmentation – is assumed to be pixelwise independent:

$$P(x \mid s) = \prod_{r \in R} q(x_r \mid s_r) , \tag{5}$$

where $q(f \mid k)$, $f \in F$, $k \in K$ are arbitrary conditional p.d-s describing the appearance of the object segments.

Let us remark here, that the idea to adopt signed distance functions (from level set approaches) for discrete probabilistic models is not new and was proposed e.g. in [9]. Our model is nevertheless much more complex due to the presence of Potts-interactions. On the other hand, even more complex models where proposed and used in the context of Conditional Random Fields (see e.g. [10]). Unfortunately, the good recognition results achievable by CRF-s have their price – the latter have drawbacks e.g. with respect to learning. The proposed model, ranging midscale with respect to complexity, is *generative* and allows to pose arbitrary learning and recognition tasks.

3 Recognition

In this section we study the task of pose and segmentation estimation for a given image $x \in \mathcal{X}$. We begin with pose estimation. We consider it as an unknown model parameter and estimate it by the Maximum Likelihood principle[1]

$$\bar{\mu}_* = \arg\max_{\bar{\mu}} P(x; \bar{\mu}) = \arg\max_{\bar{\mu}} \sum_{s \in \mathcal{S}} P(x, s; \bar{\mu}) \,. \tag{6}$$

At this point we would like to explain, why we advocate this approach instead of the usually used

$$\bar{\mu}_*, s_* = \arg\max_{\bar{\mu}, s} P(x, s; \bar{\mu}) \,. \tag{7}$$

We assume for simplicity, that we are interested in the pose parameters only; i.e. the segmentation does not matter. The latter is not known, since only the image x is observed. Hence, it is reasonable to sum over all segmentations, weighting them by their probabilities $P(x, s; \bar{\mu})$, in order to evaluate a certain choice of $\bar{\mu}$.

Currently there are no efficient algorithms for the calculation of the required sum over the vast number of segmentations in (6). It is reasonable to use the EM-scheme here, what gives the iteration

$$\begin{aligned} \text{E:} \quad & \beta^{(n)}(s) = P\big(s \mid x; \bar{\mu}^{(n)}\big) \\ \text{M:} \quad & \bar{\mu}^{(n+1)} = \arg\max_{\bar{\mu}} \sum_{s \in \mathcal{S}} \beta^{(n)}(s) \log P(x, s; \bar{\mu}) \,. \end{aligned} \tag{8}$$

Substitution of the used model in M-step gives

$$L = \lambda \sum_{r \in R} \sum_{k \in K} \beta^{(n)}(s_r = k) \phi_k(r, \mu_k) - \log Z(\bar{\mu}) \to \max_{\bar{\mu}} \,. \tag{9}$$

Currently it is impossible to perform this maximisation directly, because it is unknown how to calculate the partition sum. Interestingly, it is possible to derive its gradient with respect to $\bar{\mu}$, what gives:

$$\frac{\partial L}{\partial \mu_k} = \lambda \sum_{r \in R} \Big[P\big(s_r = k \mid x; \bar{\mu}^{(n)}\big) - P\big(s_r = k; \bar{\mu}\big) \Big] \frac{\partial \phi_k(r, \mu_k)}{\partial \mu_k} \,. \tag{10}$$

Thus it is necessary to calculate the difference of marginal a-posteriori and a-priori probabilities for the segment label $k \in K$ in every pixel $r \in R$ in order to perform the gradient step.

---

[1] Hereby and in what follows, we adopt the convention that parameters of a p.d. are separated from events by a semicolon.

Let us now return to the question how to estimate an optimal segmentation. We formulate this question as a task of Bayesian decision. Thereby we adopt the Hamming distance between the true and the estimated segmentation for the loss function. The resulting optimal Bayes decision strategy is

$$s_r^* = \arg\max_{k \in K} P(s_r = k \mid x; \bar{\mu}) . \tag{11}$$

Again, all we need for the estimation of the optimal segmentation are the posterior marginal probabilities for the segment labels.

It is of course clear that the attainable precision of the whole recognition scheme depends on the remaining two model parameters α and λ. Moreover, we expect a certain interplay of these parameters because the shape priors ϕ_k are assumed to be signed distance functions for the expected contours (or a sign preserving monotone function of the latter). In order to study this effect and to evaluate the learning results given in the next section, we conducted the following series of experiments. Assuming a circular shaped object

$$\phi(r) = \frac{\rho^2 - \|r - z\|^2}{2\rho} \tag{12}$$

where ρ is the radius of the circle and z is its centre position, seven images were created as follows. The contour of the circle was deformed and the resulting two-valued images were distorted by replacing the greyvalue in 80% of randomly chosen pixels by uniformly distributed noise. Figure 1 shows two of these pairs. We applied the whole recognition scheme including the estimation of the pose and segmentation to all seven images for different combinations of λ and α. The initial poses were chosen so as to have a substantial overlap of the circles – the centres were initialised randomly inside the true circle. For each α, λ combination and each image we get the estimated pose parameters as a result of the EM-algorithm. At the same time the segmentation is estimated according to (11). In addition, MAP-segmentation was determined after pose convergence for comparison. Except the case of $\lambda = 0$ the pose parameters were estimated correctly. Table 1 shows the averaged Hamming distance between the ground truth and the segmentation obtained by (11) and by MAP-decision (upper and lower number in each cell respectively). First off all, it is seen, that for $\alpha = 0$ (i.e. absence of the Potts-prior), the result is the better, the "stronger" the prior shape

Fig. 1. Two of the images pairs used for experiments. Each is shown with ground truth on the left and the distorted image on the right. (See Section 5 for the meaning of the red circle).

Table 1. Average Hamming distance between the estimated segmentations and ground truth

$\lambda\backslash\alpha$	0	0.7	1.4	1.8	2.1	2.3	3.0
0.00	1.e4 1.e4	1015 360	107 162	81 155	82 155	86 162	92 158
0.03	1405 1366	189 186	**85** 124	83 129	**82** 139	92 143	95 159
0.06	1075 1081	151 147	86 110	**80** 120	87 122	**83** 125	92 136
0.10	647 627	126 132	93 110	86 114	90 112	84 112	**89** 121
0.30	214 206	**125** 126	113 120	101 113	104 115	98 113	103 110
0.60	187 186	139 142	130 132	123 128	126 135	129 132	136 139
1.00	**183** 183	152 154	146 148	145 144	148 148	147 153	148 152

model i.e. the bigger the λ is chosen. This behaviour changes for $\alpha > 0$ – for each such α there is an optimal λ-interval inside the analysed range. The overall optimum is reached at $\alpha > 0$, what demonstrates the usefulness of the Potts-prior. And finally, the Bayes decision (11) clearly outperformes the MAP decision.

4 Learning

This section is devoted to learning. Given a learning sample, the task is to estimate the remaining model parameters α and λ and/or the appearance characteristics $\bar{q}$. Especially in the context of structural pattern recognition it is highly desirable to utilise for learning not only completely segmented examples but also partially segmented examples as well as example images without segmentation and even partially observed images with or without (partial) segmentation. Learning of model parameters from partially segmented images (without shape priors) was considered in [12] in the context of CRF-s. However, in order to recognise or learn, CRF-s require *complete* observations and at least partial segmentations for learning, whereas generative models like ours can cope with partial observations as well as with completely absent segmentations.

We propose to consider a learning sample T as a set of *events* $A_j \subset \Omega = \mathcal{X} \times \mathcal{S}$, $j = 1, 2 \ldots, \ell$. The learning task, if considered according to the Maximum Likelihood Principle, reads then

$$L = \log P(T; \alpha, \lambda, \bar{q}) = \sum_{j=1}^{\ell} \log \sum_{(x,s) \in A_j} P(x, s; \alpha, \lambda, \bar{q}) \to \max_{\alpha,\lambda,\bar{q}} . \tag{13}$$

Clearly, all previously mentioned situations are absorbed in this formulation.[2]

[2] Please note, that an event A_j might typically consist of a huge number of elementary events.

Because the log-likelihood in (13) cannot be calculated and maximised directly, we again use the EM-scheme here. We introduce non negative coefficients

$$\beta_j(x,s) \geqslant 0 \;\; \forall (x,s) \in A_j \quad \text{where} \sum_{(x,s)\in A_j} \beta_j(x,s) = 1 \tag{14}$$

for each example and write the log-likelihood L as

$$L(m) = \sum_{j=1}^{\ell} \sum_{(x,s)\in A_j} \beta_j(x,s) \log P(x,s;m) \\ - \sum_{j=1}^{\ell} \sum_{(x,s)\in A_j} \beta_j(x,s) \log P(x,s \mid A_j; m) \tag{15}$$

where m subsumes all model parameters $m = \{\alpha, \lambda, \bar{q}\}$ or a subset of them. Choosing now the coefficients β according to

$$\beta_j^{(n)}(x,s) = P(x,s \mid A_j; m^{(n)}) \tag{16}$$

in each E-step of the EM-iteration, ensures that the subtrahend in (15) has a global maximiser at $m = m^{(n)}$. Maximising (or at least increasing) the minuend, ensures an increase of the log-likelihood. The task to be solved in the M-step is thus to maximise the minuend $\sum_j L_j$ in (15) with respect to m. Substituting our model (4), (5) and denoting

$$\beta_j^{(n)}(s_r = s_{r'}) = P(s_r = s_{r'} \mid A_j; \alpha^{(n)}, \lambda^{(n)}, \bar{q}^{(n)}) \\ \beta_j^{(n)}(s_r = k) = P(s_r = k \mid A_j; \alpha^{(n)}, \lambda^{(n)}, \bar{q}^{(n)}) \, . \tag{17}$$

gives the following results. The new appearance characteristics are (up to a straightforward normalisation)

$$q^{(n+1)}(f \mid k) \sim \sum_j \sum_{r \in R_j(f)} \beta_j^{(n)}(s_r = k) \, , \tag{18}$$

where $R_j(f)$ denotes the set $\{r \in R \mid x_r^j = f\}$.

Similar as for recognition, it is impossible to maximise $\sum_j L_j$ directly, with respect to α and λ because it is unknown how to calculate the partition sum efficiently. Nevertheless, it is possible to derive its gradient, what gives

$$\frac{\partial L_j}{\partial \alpha} = \sum_{\{rr'\}\in E} \Big[\beta_j^{(n)}(s_r = s_{r'}) - P(s_r = s_{r'}; \alpha, \lambda, \bar{q}) \Big] \\ \frac{\partial L_j}{\partial \lambda} = \sum_{r\in R} \sum_{k \in K} \Big[\beta_j^{(n)}(s_r = k) - P(s_r = k; \alpha, \lambda, \bar{q}) \Big] \phi_k(r, \mu_k) \, . \tag{19}$$

In order to perform the M-step in its gradient version, we need a-priori and a-posteriori marginal probabilities conditioned by the corresponding events A_j.

In order to verify the reliability of the derived learning scheme with respect to α and λ, we accomplished a series of experiments. The true deformed circular segmentations (see Sec. 3) were used as a learning sample. This corresponds to the supervised version of learning. Because each A_j consits of only one segmentation s in this case, the posterior marginal probabilities β in (17) are known in advance. Learning of λ was performed for certain fixed α-values. The results are shown in Table 2. Comparison with the recognition results given in Sec. 3 shows an almost perfect agreement.

Table 2. Values of λ learned for different α

α	0	0.7	1.4	1.8	2.1	2.3	3.0
λ	0.91	0.32	0.07	0.05	0.048	0.046	0.091

Completely unsupervised learning of the appearance characteristics $\bar{q}$ (on a per image basis) was used and thus verified in experiments with real images (see next section).

5 Other Experiments

We have used rather good pose initialisations in all experiments described so far. In order to study the ability to cope with wrong initial pose estimates, we made the following experiment. Using one of the previously described images, the recognition scheme for pose estimation and segmentation was combined with unsupervised learning of the appearance characteristics $\bar{q}$ and the shape parameter λ. The latter was fixed to a small value at the beginning and started to learn after some delay. The initial pose was located in the upper left corner (shown as the red circle in Fig. 1), thus having no overlap with the true segment. The results of the experiment are shown in Fig. 2. At the beginning of the experiment (λ is fixed to a small value) the appearance characteristics $\bar{q}$ are learned

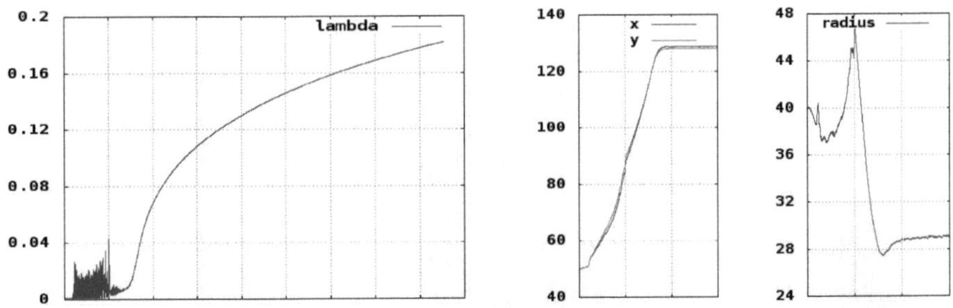

Fig. 2. Results of recognition and simultaneous learning of λ: learning curves for λ, the centre position and the radius respectively (the constant parts of the latter two were clipped)

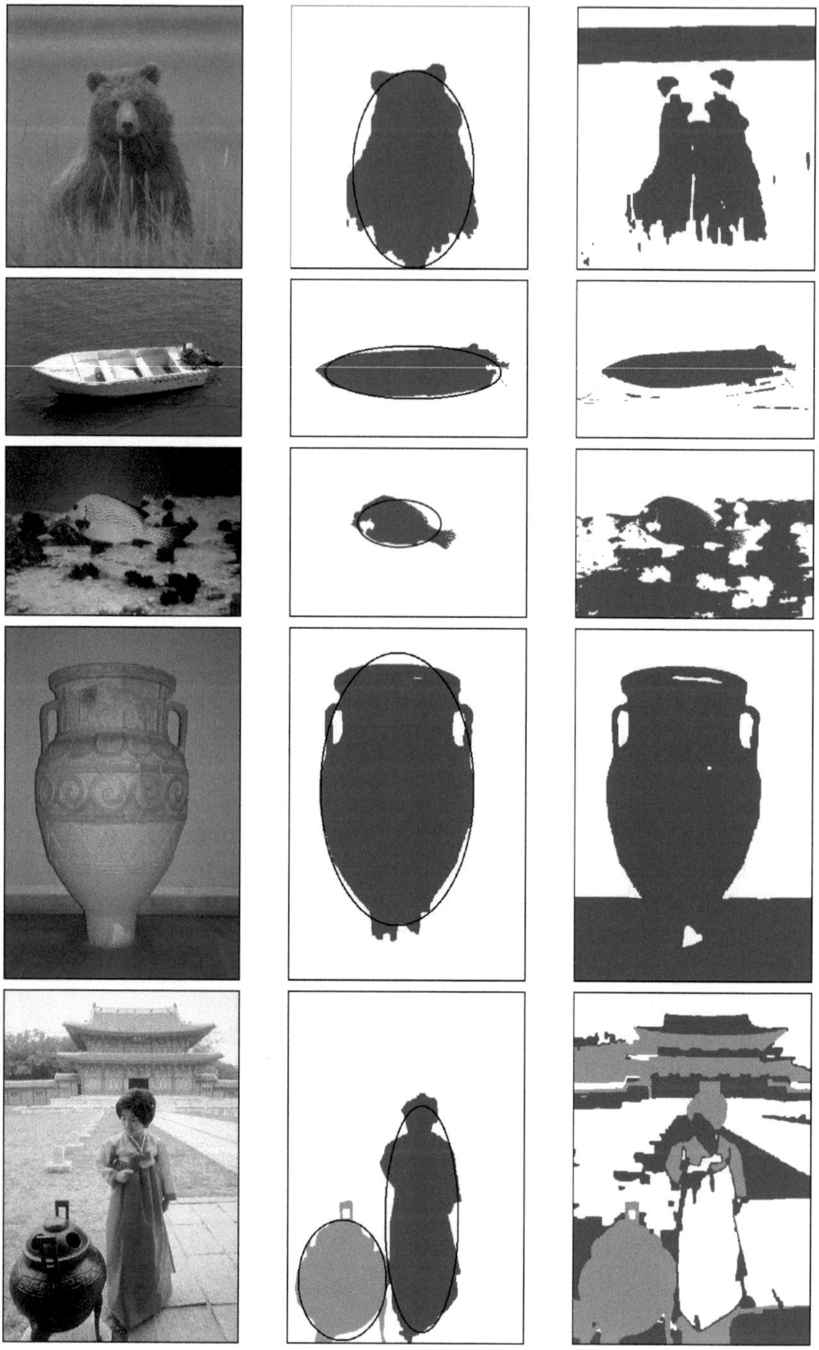

Fig. 3. Original images, segmentations with overlaid shapes, segmentations without shape prior

approximately and the segmentation shows two blobs - located at true position and at the actually estimated (wrong) pose respectively. A rather interesting phenomenon is observed when λ starts to learn: The blobs "compete" with each other and λ oscillates. At the same time the pose starts to approach the true one ($x = 128$, $y = 128$ and $\rho = 30$). Finally, the pose is correctly estimated and λ approaches a reasonable value ($\alpha = 0.7$ was used in this experiment).

The next examples (Fig. 3) show segmentation results for real images. A rather good pose initialisation was used in all experiments. The appearance characteristics $\bar{q}$ were assumed to be a mixture of multivariate Gaussians and learned completely unsupervised. The number of segments and the numbers of Gaussians for each segment (typically from 2 to 4) as well as the parameters α and λ were estimated experimentally. The appearance model includes additional a-priori slowly varying shading fields for each segment (see [13] for details). These shadings were learned simultaneously with $\bar{q}$ in a fully unsupervised manner. These experiments clearly demonstrate the usefullness of shape priors, especially for learning.

6 Conclusions and Open Questions

We introduced a combination of shape prior models with Markov Random Fields. The model allows to integrate multiple shape priors and appearance models into MRF-models for segmentation. At the same time it allows to pose various recognition tasks and a wide spectrum of learning tasks. Our main goal was to pay attention to concise and reasonable formulation of these tasks as opposed to increasing the model complexity. As demonstrated in experiments, segmentation of real images is greatly improved, even if relatively simple shape priors are used.

Yet there are several open questions which we are currently investigating or want to investigate in the future:

In the current model we considered the pose of a shape as a parameter. Moreover, only "templates" were used for the shape. It seems to be beneficial to extend these models towards statistical shape models (like e.g. in [9]). At the same time this would mean to consider other types of recognition tasks, based on Bayes decision for the pose and possibly for the shape.

It is yet unclear how to cope with wrong pose initialisations. The experiment presented in the paper is a first trial at most. In case of statistical shape models, the pose and the shape can be sampled (like e.g. in [10]), but it is still unclear, whether such a replacement of the gradient scheme will solve the problem efficiently.

References

1. Greig, D.M., Porteous, B.T., Scheult, A.H.: Exact maximum a posteriori estimation for binary images. Journal of the Royal Statistical Society B 51(2), 271–279 (1989)
2. Kolmogorov, V., Zabih, R.: What energy functions can be minimized via graph cuts? In: Heyden, A., Sparr, G., Nielsen, M., Johansen, P. (eds.) ECCV 2002. LNCS, vol. 2352, pp. 65–81. Springer, Heidelberg (2002)

3. Schlesinger, D., Flach, B.: Transforming an arbitrary minsum problem into a binary one. Technical Report TUD-FI06-01, Dresden University of Technology (April 2006)
4. Boykov, Y., Veksler, O., Zabih, R.: Fast approximate energy minimization via graph cuts. In: Proc. of the 7. Intl. Conf. on Computer Vision, vol. 1, pp. 377–384 (1999)
5. Kovtun, I.: Partial optimal labeling search for a NP-hard subclass of (max,+) problems. In: Michaelis, B., Krell, G. (eds.) DAGM 2003. LNCS, vol. 2781, pp. 402–409. Springer, Heidelberg (2003)
6. Kolmogorov, V.: Convergent tree-reweighted message passing for energy minimization. IEEE Transactions on Pattern Analysis and Machine Intelligence 28(10), 1568–1583 (2006)
7. Leventon, M.E., Grimson, W.E., Faugeras, O.: Statistical shape influence in geodesic active contours. In: Proc. IEEE Conf. Comp. Vision and Patt. Recog. (2000)
8. Cremers, D., Sochen, N., Schnörr, C.: A multiphase dynamic labeling model for variational recognition-driven image segmentation. International Journal of Computer Vision 66(1), 67–81 (2006)
9. Pohl, K.M., et al.: Shape based segmentation of anatomical structures in magnetic resonance images. In: Liu, Y., Jiang, T., Zhang, C. (eds.) CVBIA 2005. LNCS, vol. 3765, pp. 489–498. Springer, Heidelberg (2005)
10. Kumar, M.P., Torr, P.H.S., Zisserman, A.: An object category specific MRF for segmentation. In: Ponce, J., Hebert, M., Schmid, C., Zisserman, A. (eds.) Toward Category-Level Object Recognition. LNCS, vol. 4170, pp. 596–616. Springer, Heidelberg (2006)
11. Levin, A., Weiss, Y.: Learning to combine bottom-up and top-down segmentation. In: Leonardis, A., Bischof, H., Pinz, A. (eds.) ECCV 2006. LNCS, vol. 3954, pp. 581–594. Springer, Heidelberg (2006)
12. Verbeek, J., Triggs, B.: Scene segmentation with CRFs learned from partially labeled images. In: Advances in Neural Information Processing Systems, vol. 20 (2008)
13. Schlesinger, D., Flach, B.: A probabilistic segmentation scheme. In: Rigoll, G. (ed.) DAGM 2008. LNCS, vol. 5096, pp. 183–192. Springer, Heidelberg (2008)

Measuring the Similarity of Vector Fields Using Global Distributions

H. Quynh Dinh and Liefei Xu

Department of Computer Science
Stevens Institute of Technology
quynh@cs.stevens.edu, lxu1@stevens.edu

Abstract. Sensors such as video surveillance and weather monitoring systems record a significant amount of dynamic data which are represented by vector fields. We present a novel algorithm to measure the similarity of vector fields using global distributions that capture both vector field properties (e.g., vector orientation) and relational geometric information (e.g., the relative positions of two vectors in the field). We show that such global distributions are capable of distinguishing between vector fields of varying complexity and can be used to quantitatively compare similar fields.

Keywords: vector field matching, shape distribution, geometric histogram.

1 Introduction

Comparing vector fields is becoming increasingly relevant as sensor systems now record a significant amount of dynamic data. For example, vector fields may come from video surveillance data (via tracking and optical flow), from weather phenomena (wind speed and direction and wave currents), and from medical imaging (phase-contrast magnetic-resonance angiography captures 3D velocity fields of moving tissue [1]). In video analysis, comparing these vector fields may improve the recognition of activities, faces, and facial expressions [2,3]. For weather data, measuring the similarity between vector fields can help scientists understand how wind and wave currents have changed over time and to compare with simulations. In medical imaging, measuring the difference between vector fields enable doctors to analyze changes in tissue over time and study the variation of particular organs or tissue groups over large populations.

We present a novel algorithm to measure the similarity of vector fields using global distributions (one for each vector field) that can be quantitively compared by computing the difference between them. Our key contribution is the development of global distributions that capture both vector field properties such as vector orientation and relational information (*e.g.*, the relative positions of two vectors in the field). Previous work in 3D shape matching has proven that relational information is effective in shape classification, and we show that this is also true for comparing vector fields.

2 Related Work

Vector field analysis identifies and locates singularities (critical points where the magnitude of the vector vanishes), separatrices emanating from saddle points, and periodic

N. da Vitora Lobo et al. (Eds.): SSPR&SPR 2008, LNCS 5342, pp. 187–196, 2008.

orbits. Critical points are further classified into sinks, sources, saddles, focus, and center points using the eigenvalues and eigenvectors of the Jacobian matrix. There is significant research in the area of vector field analysis and singularity classification, including [4,5,6,7,8,9]. The algorithm we present does not depend on locating critical points in vector fields because many naturally occuring vector fields do not actually contain singularities and because locating such points is computationally intensive. Instead, we collect statistics on the vector field in the form of discrete distributions, or histograms, and measure the similarity between vector fields by computing the difference, or norm, between the histograms. As results show, our approach is able to differentiate between vector fields with different configurations of singularities even though we do not explicitly locate them.

Geometric distributions (also called *shape histograms* [10] or *shape distributions* [11]) provide a way of comparing shapes based on statistical properties. These approaches record the distribution of a selected feature in a histogram for efficient storage, indexing, and comparison via norms on the histogram. A single histogram may be generated for the entire field (global histograms), or one for each surface point (local histograms) to be used as a local descriptor. Global histograms store the statistics of the entire shape and are used to compare whole objects. To compare two objects, a norm is computed between the histograms. Standard methods include the Minkowski L_N norms, the χ^2 distance, the Battacharyya distance, Earth Mover's distance, and the correlation coefficient. Geometric distributions are known for their invariance to rotation and translation of the shape and robustness to noise due to their statistical nature.

Statistics on vector fields from optical flow have been explored in [12] where the goal was to obtain a statistical consensus for use as a prior in optical flow computation, not to find a disciminating distribution. We focus on global distributions that enable us to efficiently compare two vector fields by computing the difference between a single pair of histograms.

3 Global Distributions for Vector Fields

In developing a global distribution for vector fields, our goal is to find the vector field property that will result in discriminating distributions that can act as *signatures* for vector fields. We now describe different global distributions we have developed, and in Section 4, show results of measuring the similarity between test vector fields of varying complexity – from few to many singularities, complex fields that contain no singularity, and real vector fields designed for painterly rendering [9] and from flow through a diesel engine cylinder [13]. We generate a single global distribution for each vector field. The distribution is a discrete histogram in which each bin covers a range of values for a given vector field property, and each bin is a count of the number of times its range was encountered in the vector field.

3.1 A Simple One-dimensional Distribution

An intuitive statistic for 2D vector fields is to bin the vectors in the field according to orientation, resulting in a 1D distribution as shown in the 2nd column of Figure 1.

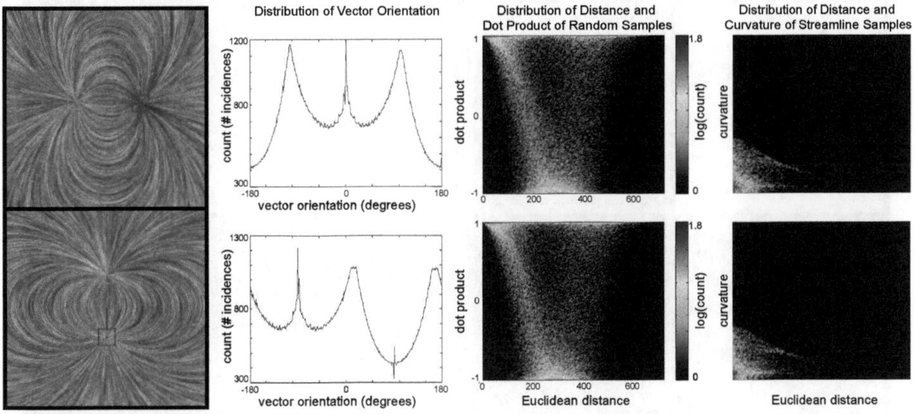

Fig. 1. Left to right: Two vector fields related by a rotation (green indicates a source, red a sink) overlayed with streamlines; histogram of vector orientation; histogram of distance and vector dot product between pairs of randomly selected points; and histograms of distance and curvature between points on streamlines

The 1D distributions are clearly equivalent up to a phase shift (due to the rotation of the singularities), and are thus sensitive to rotations in the vector field. To find the minimum difference between the histograms requires shifting one plot with respect to the other.

The 1D histogram also fails to account for geometric, or positional, relationships between vectors. In 3D shape matching, Osada *et al.* found that histograms which preserve positional relationships are more effective for shape matching [11]. They showed that a 1D distribution based on binning the Euclidean distance between pairs of randomly selected surface points was particularly successful in 3D shape classification. Ohbuchi *et al.* extend this concept to a 2D distribution by binning 3D surface points based on distance and the dot-product between surface normal vectors [14]. We now describe a geometry-preserving 2D distribution for vector fields that captures both statistics on vector properties (*e.g.*, vector dot product) and positional relationships between points in the vector field using the Euclidean distance.

3.2 Geometry-Preserving Distributions

We present a novel 2D geometry-preserving distribution for vector fields. The distributions are generated by randomly selecting pairs of points, computing the distance between them and the dot product between the vectors located at the points, and then binning the pairs in a 2D histogram based on these two computed values. The 3rd column in Figure 1 shows the resulting histograms for the vector fields on the left. A difference between the distributions can be directly computed (we use χ^2 distance) without requiring any shifts, unlike the 1D distributions based on vector orientation. Although computing the difference between 2D histograms is equivalent to shifting to find the min difference between 1D histograms, the 2D histograms are more informative and discriminating. In particular, Figure 1 shows that the 2D histograms are invariant to rotations in the vector field. The 2D geometry-preserving distributions (3rd and 4th

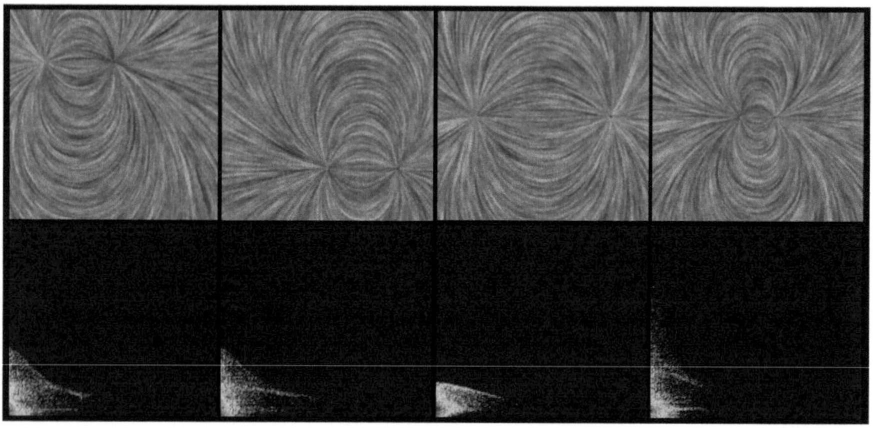

Fig. 2. Top: translation (columns 1 and 2) and scaling (columns 3 and 4) of the vector field shown in Figure 1 (top-left). Bottom: histograms of distance and curvature between points on streamlines show that the distributions are relatively invariant to translation and scaling. The χ^2 distance between the above distributions and the streamline distribution of the vector field in Figure 1 (top-left) are, from left to right: 0.077, 0.066, 0.125, 0.052 out of a max of 1.0.

columns) for the top and bottom vector fields are equivalent even though the position of the singularities are rotated by 90 degrees.

Note that we do not compute the distance and vector dot-product between every pair of points in the domain because it is computationally expensive (for 512×512 vector fields, there are 512^4 pairs). Instead, we use a large number of samples – 10,000 pairs of points are randomly selected. Figure 4 (left) shows the χ^2 distance between the vector field distributions based on the distance and dot-product between randomly selected pairs of points for 14 different data sets (10 of which are shown in Figure 3).

3.3 Distributions over Streamlines

The 2D distribution we describe above can distinguish between simple and complex vector fields (*e.g.*, those with one singularity versus many). However, it cannot distinguish between different complex fields or different simple fields. For example, the plot of Figure 4 (left) shows that simple fields 1 through 4 are equally different from each other. The same is true of complex fields 5 through 12 and 14 in the plot.

To compute a distribution that further distinguishes between vector fields, we implicitly record additional structural information by extracting streamlines from the vector field (as shown in Figure 1). Streamlines are formed by randomly seeding points in the field and allowing them to move through the field guided by the vectors. The paths taken by the points form streamlines. We uniformly distribute streamlines by preventing them from growing into regions where streamline density is high, similar to [16].

Streamlines enhance the structure present in vector fields and provide a framework for recording positional relationships between vectors. Instead of randomly selecting pairs of points from across the entire vector field as in the previous section, we randomly

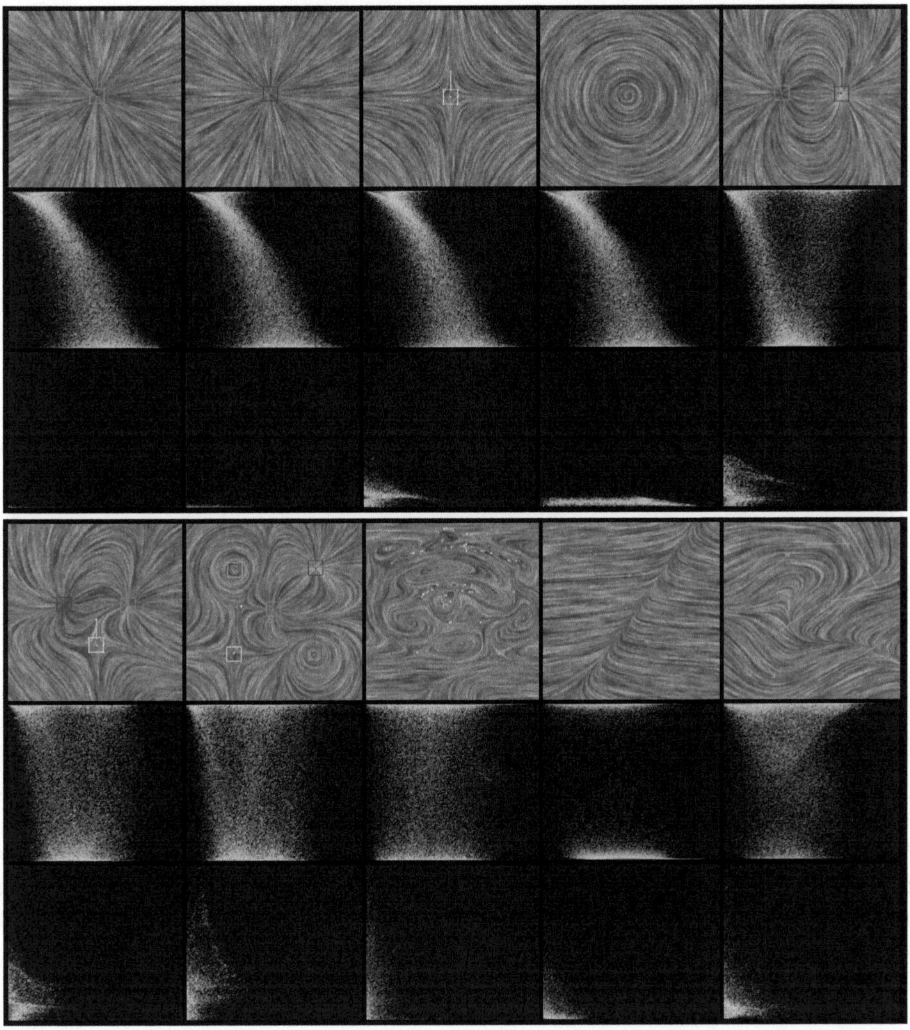

Fig. 3. Top row: vector fields corresponding to 1, 2, 3, 4, 5, 8, 9, 12, 13, 14 in plots of Figure 4. Middle row: distance and dot-product distributions from random samples over entire field. Bottom row: distance and curvature distributions from streamline samples. Due to space limitations, we have omitted four vector fields which are rotated versions of some of the above. They can be found at our website [15].

select pairs of points from the *same* streamline. The intuition is that points on the same streamline have a stronger geometric relationship than randomly selected pairs of points which may be very far apart. The resulting distribution now only collects data between strongly related points that are along the same path in the vector field.

With streamlines available, we can record curvature as the second property (rather than vector dot product) in the 2D histogram. Given two randomly selected points on

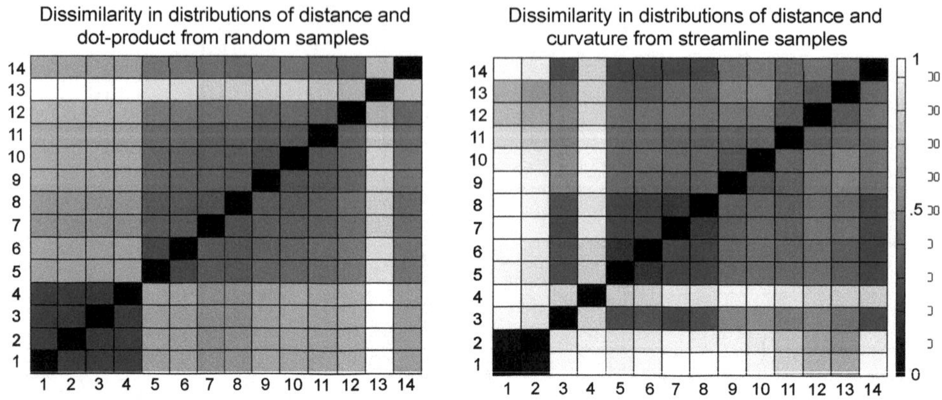

Fig. 4. χ^2 distance between 2D distributions of 14 different vector fields. Darker (closer to 0) means more similar. Vector fields are shown in Figure 3.

the same streamline, we compute the mean of the two points' discrete curvature and bin them based on distance and mean curvature. For a point (x, y) on the streamline, the discrete curvature $k(x, y)$ is:

$$k(x,y) = \frac{|x'y'' - y'x''|}{(x'^2 + y'^2)^{3/2}} \tag{1}$$

In the equation above, x' and x'' are the discrete first and second derivatives of x, respectively (and similarly for y). Each streamline is an ordered list of uniformly distributed points, making it easy to compute the discrete derivatives via finite differencing. The last column of Figure 1 are distributions of distance and curvature between points sampled on the same streamline. Figure 3 (bottom rows) are distributions of distance and curvature for the 10 vector fields shown above them. The color coding in Figure 1 is used for all distributions.

In Figure 1, we showed the rotational invariance of our geometry-preserving distributions. We further test the translational and scale invariance by moving the pair of sink and source singularities and by scaling them to be farther apart and closer together. In comparing the distributions of Figure 2 to that of Figure 1 (4th column, top), we can see that the distributions are invariant to translation (1st two columns) and partially invariant to scaling (last two columns). When the vector field is scaled-up, the singularities are farther apart and cover a larger region of the field. The original vector field is essentially cropped, and results in a distribution that, when compared to the original, measures a difference. Scale invariance does hold when the vector field is scaled down.

4 Results and Conclusions

We now present results of using our geometry-preserving distributions to measure simularity between a set of 14 test vector fields and on the flow through a diesel engine cylinder, a real-world dataset.

4.1 Measuring Similarity on Test Vector Fields

We use Zhang *et al.*'s vector field design tool [9] to generate a test dataset of 14 vector fields of varying complexity – from few to many singularities (1, 2, 3, 5, and >5), complex fields that contain no singularity, and two real vector fields designed for painterly rendering [9] (fields 11 and 12). Figure 3 shows 10 of the vector fields. Source singularities are green; sinks are red; saddles are yellow; centers are pink (attracting) and cyan (repelling); and regular vectors are cyan arrows. Squares surrounding singularities are control knobs for rotation. All vector fields are 512x512, and histograms 200x200 in size. The χ^2 distance between N-bin normalized histograms, f and g, is:

$$\chi_2 : D(f,g) = \frac{1}{2}\sum_{i=1}^{N} \frac{(f[i]-g[i])^2}{f[i]+g[i]} \tag{2}$$

Results are plotted in Figure 4 for all 14 vector fields (darker means more similar). The plots show that streamline distributions (right plot) provide a finer measure of similarity (or dissimilarity) between the vector fields than our first geometry-preserving distribution (left plot) which samples the entire field and captures broad differences between them. This also bears out in the distributions themselves (Figure 3) which reveal streamline distributions to be more distinct than those generated from randomly sampling the entire vector field. We found the computed differences between streamline distributions of distance and dot-product to be similar to results using curvature.

Note that vector fields 1 and 2 are essentially identical – 1 contains a sink in the center, whereas 2 contains a source. The only difference between the vector fields is the vector direction (inward for a sink, outward for a source). The streamline distributions show that these two fields are very similar, and the and χ^2 distance is close to 0. Intuitively, all samples on the same streamline with have a dot-product of 1.0 and a curvature near 0 for both vector fields. Randomly sampling from the entire field does not preserve this relationship, and the computed χ^2 distance is greater.

4.2 Measuring Similarity on Engine Cylinder Slices

We tested our algorithm on a real-world dataset – the simulated flow within the combustion chamber of a diesel engine. Engineers typically slice through the cylinder along its length to analyze the simulation results, resulting in the circular flow patterns seen in Figure 5. The ideal flow pattern is a spiral helix extending the length of the cylinder. This motion results in an optimal mixing of air and fuel leading to a more efficient combustion process [13]. The 2D flow pattern obtained from slicing the ideal helix is a swirl motion (Figure 5 left). By comparing the slices of the vector field along the cylinder to the ideal swirl, we can provide a quantitative analysis of how close the flow through a designed cylinder is to ideal, and order the slices based on similarity to the ideal. This ordering provides engineers with information on where to improve their design.

We computed the similarity of the ideal flow to 10 slices through the cylinder using streamline distributions and computed the χ^2 distance between them. In Figure 5,

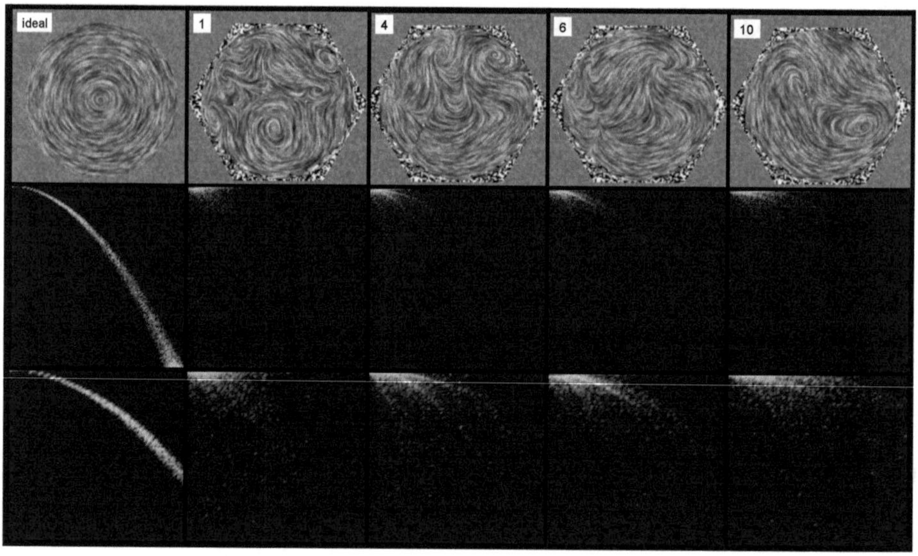

Fig. 5. Left to right: ideal flow in a slice of the engine cylinder, and slices 1, 4, 6, and 10 from the simulated flow in a designed combustion chamber. 1 and 10 are near the cylinder entrance and exit, respectively, and are where the most turbulence occurs. Slice 6 is closest to ideal. Top: planar vector fields from cylinder data. Middle: streamline distributions of distance and dot-product for each slice. Bottom: close-up of top-left corner of the histograms show how they differ. Note that the data is on a hexagonal grid, leading to the hexagonal boundaries around the circular slices.

we show 4 representative slices and their streamline distributions of distance and dot-product. The numbering of the slices corresponds to their positions along the length of the cylinder from where the fluid enters at slice 1 to where it exits at slice 10. For completeness, we show all 10 slices and distance and curvature distributions in Figure 6.

The simulation data is embedded in a hexagonal grid, and hence, are essentially aligned. We can directly compare vectors of a slice to the ideal slice in a point-wise manner at each grid point using the dot-product. To do so, we sum up 1 minus the normized vector dot-product for each grid point pair, so that similar vectors will contribute 0, while vectors pointing in opposite directions will contribute 2 towards the sum. Similar slices will have a sum closer to 0. For a consistent comparison between this direct approach using the dot-product and our streamline approach, we compute the streamline distributions of distance and dot-product. The results of both methods are similar and consistent with the engineers' expectations. Slices close to the top of the cylinder where the fluid enters and slices near the bottom where the fluid exits tend to be most turbulent and far from ideal. Slices in the center where the flow approaches a stable steady-state is closer to ideal. Using streamline distributions based on dot-product, the order of the 10 slices from most to least similar to ideal is: 6, 7, 5, 8, 9, 10, 4, 3, 2, 1. Using the sum of 1 minus normalized dot-product between corresponding pairs, the order is: 6, 7, 5, 8, 1, 4, 3, 2, 9, 10. The point-wise comparison is, of course, more efficient than computing and comparing the streamline distributions.

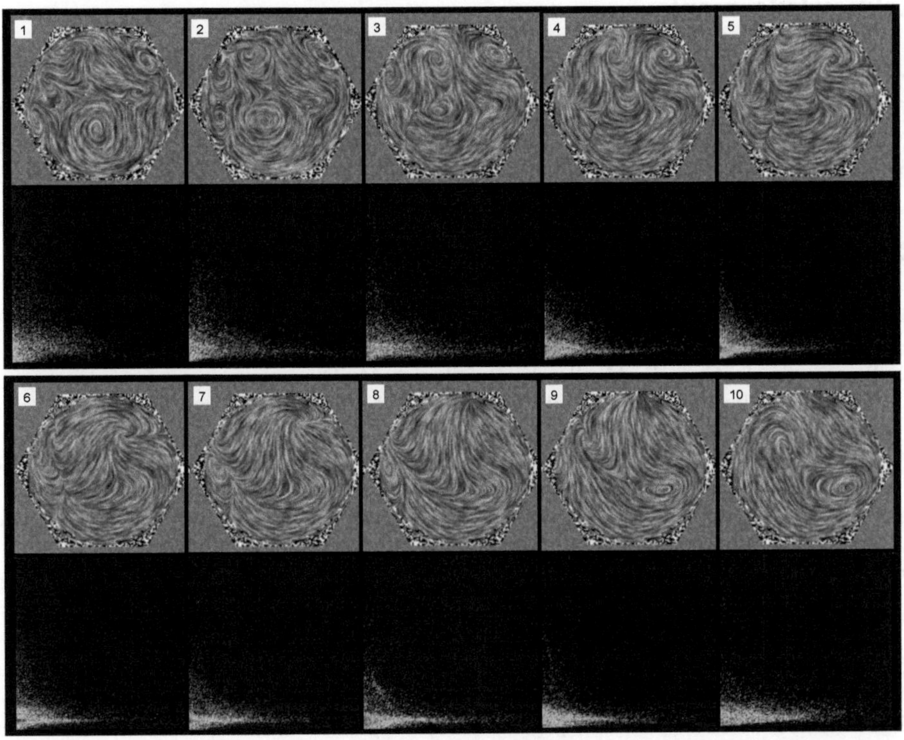

Fig. 6. 10 slices of the simulated flow within a designed combustion chamber and histograms of distance and curvature between randomly selected pairs of points on the same streamline. Note that the distribution for the ideal slice (not shown) is a horizontal line near the bottom.

Although we are not taking advantage of the rotational invariance of our distibutions (since the cylinder slices are aligned), we have found that the positional, or contextual, information implicitly stored in the distributions result in a better ordering of slice similarity than the point-wise comparison. In the point-wise comparison, slice 1 falls in the middle of the ordering based on similarity to ideal even though it is the slice with the most visual turbulence, and slices 2 and 3 are closer to ideal than slices 9 and 10 near the exit even though they appear more turbulent than 9 and 10.

5 Future Work

We have shown that geometry-preserving distributions – in particular, streamline distributions – enable us to quantitatively compare complex vector fields and distinguish between them. They are invariant to rotation and translation of the vector field, and have limited invariance to scaling. In future work, we will apply our algorithms to vector fields generated from optical flow and dense tracking of video.

Acknowledgements

We thank Robert S. Laramee of Swansea University for the original diesel engine data, and Eugene Zhang and Lin Zhongzang of Oregon State for extracting the planar slices from the data, for the vector field design tool, and for the vector fields from painterly rendering. This work was funded in part by NSF CreativeIT Award# IIS-0742440.

References

1. Tovar, M.: Vector-field classification in magnetic-resonance angiography. In: Proc. of AMIA Symposium, pp. 926–930 (1998)
2. Pamudurthy, S., Guan, E., Mueller, K., Rafailovich, M.: Dynamic approach for face recognition using digital image skin correlation. Audio- and Video-based Biometric Person Authentication (2005)
3. Lu, X., Jain, A.: Deformation modeling for robust 3d face matching. In: Proc. of CVPR (2006)
4. Helman, J., Hesselink, L.: Surface representations of two- and three- dimensional fluid flow topology. In: Proc. of IEEE Visualization, pp. 6–13 (1990)
5. Scheuermann, G., Krüger, H., Menzel, M., Rockwood, A.: Visualizing nonlinear vector field topology. IEEE Transactions on Visualization and Computer Graphics (TVCG) 4(2), 109–116 (1998)
6. Tricoche, X., Scheuermann, G., Hagen, H.: Continuous topology simplification of planar vector fields. In: Proc. of IEEE Visualization, pp. 159–166 (2001)
7. Tricoche, X., Scheuermann, G., Hagen, H.: Topology-based visualization of time-dependent 2d vector fields. In: Data Visualization, Proc. of IEEE TVCG Symposium on Visualization, pp. 117–126 (2001)
8. Ebling, J., Scheuermann, G.: Clifford convolution and pattern matching on vector fields. In: Proc. of IEEE Visualization, pp. 193–200 (2003)
9. Zhang, E., Mischaikow, K., Turk, G.: Vector field design on surfaces. ACM Transactions on Graphics (TOG) 25(4), 1294–1326 (2006)
10. Ankerst, M., Kastenmuller, G., Kriegel, H., Seidl, T.: 3d shape histograms for similarity search and classification in spatial databases. In: Proc. of 6th International Symposium on Spatial Databases (1999)
11. Osada, R., Funkhouser, T., Chazelle, B., Dobkin, D.: Shape distributions. ACM Trans. on Graphics 21(4), 807–832 (2002)
12. Roth, S., Black, M.: On the spatial statistics of optical flow. Int'l. J. of Computer Vision 74, 33–50 (2007)
13. Laramee, R., Weiskopf, D., Schneider, J., Hauser, H.: Investigating swirl and tumble flow with a comparison of visualization techniques. In: Proc. of IEEE Visualization, pp. 51–58 (2004)
14. Ohbuchi, R., Minamitani, T., Takei, T.: Shape-similariy search of 3d models by using enhanced shape functions. In: Proc. Theory and Practice of Computer Graphics, pp. 97–104 (2003)
15. `http://www.cs.stevens.edu/~quynh/vfield_global.html`
16. Verma, V., Kao, D., Pang, A.: A flow-guided streamline seeding strategy. In: Proc. of IEEE Visualization, pp. 163–170 (2000)

Consensus Graphs for Symmetry Plane Estimation

Andrea Albarelli[1], Marcello Pelillo[1], and Sebastiano Viviani[2]

[1] Dipartimento di Informatica, Universita' Ca' Foscari di Venezia
[2] Evolvenda s.r.l.

Abstract. Finding the most relevant symmetry planes for an object is a key step in many computer vision and object recognition tasks. In fact such information can be effectively used as a starting point for object segmentation, noise reduction, alignment and recognition. Some of these applications are strongly affected by the accuracy of symmetry planes estimation, thus the use of a technique that is both accurate and robust to noise is critical. In this paper we introduce a new weighted association graph which relates the main symmetry planes of 3D objects to large sets of tightly coupled vertices. This technique allows us to cast symmetry detection to a classical pairwise clustering problem, which we solve using the very effective Dominant Sets framework. The improvement of our approach over other well known techniques is shown with several tests over both synthetic data and sampled point clouds.

Keywords: symmetry, dominant sets, shape registration, clustering.

1 Introduction

Among all the features that can be extracted from a 2D image or 3D object symmetry is one of the most important. For instance symmetry axes can be used to set a canonical orientation, which in turn can be useful for matching different objects, computing partial registration of surfaces or simply to offer a comparable point of view in visual application. Other applications exploit symmetries as a descriptor for searching in large databases or as a starting point for locating other features. In addition partial local symmetries can be very effective as a cue in segmentation. Finally, due to its abundance in both natural and crafted object, symmetries play an important role in visual perception and Gestalt theory. Because of this wide range of applications, symmetry estimation has been a popular research field for long time. Most seminal papers appeared in the eighties and were focused on perfect symmetries, mainly because this kind of analysis required less computational power. Atallah[1] and Wolter[2] solved the problem very efficiently by casting it into substring matching. Some more recent approaches use orientation histograms[3], singular value decomposition[4], extended Gaussian image[5] and generalized complex moments[6]. Also principal component analysis[7] can be used with good results when dealing with strongly symmetric objects. All these approaches rely on the existence of a perfect plane

N. da Vitora Lobo et al. (Eds.): SSPR&SPR 2008, LNCS 5342, pp. 197–206, 2008.

of symmetry, thus they are very sensitive to noise and cannot be applied to objects where partial or local symmetries are present.

In order to deal with partial or imperfect symmetries some measures of the distance between a given object and the nearest perfectly symmetric shape have been proposed[8,9]. Such a measure could be used to search in the full plane space for the solution that minimizes the error, however since it can be applied to just one plane at a time it would not be an efficient solution.

A more practical approach, that has been proved effective in a wide range of contexts, is to use some kind of transform. That is a mapping from a function defined over the space of the object data to a function defined over the space of the feasible solutions, where local and global maximum are usually associated to local and global solutions of the problem itself. The most well known transform in computer vision is the Hough transform, which maps points on a plane to the space of parametrized straight lines. Over the last decade several symmetry transform have been proposed[10,11,12,13,14]. Recently the Planar-Reflective Symmetry Transform (PRST) has been presented in[15,16] and extended in[17]. This is a transform that maps the points of three dimensional objects to a real-valued function in the three dimensional space of parametric planes. As for any transform technique in practical implementations it is computed in a discrete form: that is, the function is evaluated only on a large set of randomly selected points and the results are used to vote for the best solutions in a discrete set of bin that are uniformly distributed over the space of planes. This approach leads to coarse errors, due to the discrete size of the bins. However a continuous refinement technique is proposed that greatly enhances the solution by searching for a more accurate symmetry plane inside the winning bin by using an error minimization technique inspired by the Iterative Closest Point algorithm[18].

In this paper we propose a novel approach which casts the problem to the search for pairwise clusters in a newly defined weighted association graph that we named Symmetry Consensus Graph. In such a graph each vertex is associated to a candidate symmetry plane and the similarity between vertices expresses the consensus between planes. This way each large cluster in the graph captures a set of highly similar plane candidates and leaves out all the unwanted outliers, according to the high internal homogeneity and external inhomogeneity properties of clusters. In section 2 we define the Symmetry Consensus Graph in both a binary and a weighted form. In section 3 we suggest a very effective framework in order to find large cluster in a Consensus Graph. Finally, in sections 4 and 5 we compare our approach with PCA and the very accurate PRST technique.

2 The Symmetry Consensus Graph

Given a large set of 3D points describing an object our goal is to find the largest subsets that share a common symmetry plane. The most widely used approach is to select a large number of point pairs and thus collect votes for the plane between them in an appropriate parameters space. As for any other voting technique, concrete implementations of this approach need to choose a quantization of the parameters space into discrete bins. In many scenarios the choice of the size of the bins can be very critical as a too coarse quantization can lead to a poor precision in the estimation of the symmetry, while a too fine one could suffer of vote dispersion due to the dimensionality curse. In [16] the authors suggest to adopt a quite coarse parameters space quantization ad thus apply a continuous refinement step in order to minimize the local symmetry error among points that voted for the winning bin. This step enhances the quality of the solution found, but still is quite sensible to the presence of local outliers in the rather large winning bin. In order to obtain the best results an optimal bin size should be used, that is a quantization that allows to collect just the most relevant votes, leaving out local outliers. Unfortunately it is very hard to know in advance which size to use as it depends on many factors such as noise and quality of the symmetry. By contrast, the main idea which underpin our method is to avoid the quantization by using a more global approach able to weight all the voting point pairs at once.

2.1 Binary Symmetry Consensus Graph

In this section we focus our attention on objects that, although not being symmetric as a whole, contain one or more subset of points that are perfectly specular by a plane. Obviously this scenario is not very realistic from a practical point of view, nevertheless it permits us to define an useful structure that we will expand in section 2.2 to be more suitable for real world applications:

Definition 1. *Given a set of points in space* $P = \{p_1, p_2 \dots p_n\}$*, their Binary Symmetry Consensus Graph (BSCG) is the undirected graph* $G_B = (V_B, E_B)$ *where* $V_B = \{(p_a, p_b) \in P | a < b\}$ *is the set of pairs of points in* P *and where* $((p_a, p_b), (p_c, p_d)) \in E_B$ *if and only if the plane between* p_a *and* p_b *is exactly the same plane between* p_c *and* p_d*.*

Each vertex in such a graph represents a pair of points in the original object and is attributed with the plane that separates such points in a symmetric way. The key property of this graph is that two vertices are connected if and only if they share the same plane. As this is an equivalence relation the transitive property holds and it is easy to see that each set of vertices that share the same symmetry plane with each other forms a complete subgraph (clique). By building the BSCG over a set of points we can thus cast the search for the most relevant symmetry planes into the search for the largest cliques over the Consensus Graph. Although this is known to be a NP-complete problem many powerful heuristics have been suggested for it.

In figure 1 we illustrate a simple example of a BSCG. On the left an object with two partial perfect symmetry planes is shown. On the right the BSCG is built over all the pairs of point of the object. Only two cliques are found in the BSCG, respectively between pairs ag, bf, ce, which are symmetric with respect to the vertical axis, and between pairs bc, ef, which are symmetric with respect to the horizontal axis. All other pairs do not share their symmetry plane with any other pair in the graph, thus all of them are singleton.

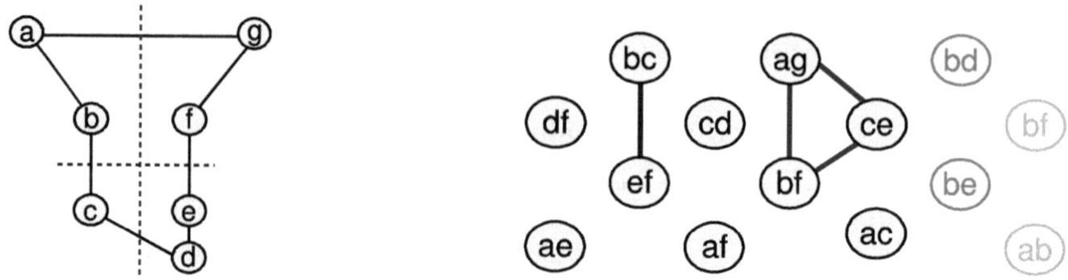

Fig. 1. A very simple object with perfect partial symmetries (left) and its Binary Symmetry Consensus Graph (right). Only pairs of vertices that share the same symmetry plane are connected by an edge. For practical reasons only a part of the graph is shown.

2.2 Weighted Symmetry Consensus Graph

The assumptions taken in section 2.1 are very strict in practical scenarios. In fact noise, numerical errors and not perfect symmetries in the object itself can all make very difficult for many pairs of points to share exactly the same symmetry plane. One possible solution to this problem could be the use of a threshold in order to state that two plane are equals. Unfortunately this approach introduces two more hurdles: the first is the correct choice for such threshold, which is not different from the choice of the correct bin size, the other is that this way the transitive property would be no more valid, thus breaking the correspondence between cliques in the Consensus Graph and symmetries.

Our proposal is to extend the BSCG presented in 2.1 to a weighted graph where each edge is weighted according to the similarity between the planes associated to the vertices connected:

Definition 2. *Given a set of points in space* $P = \{p_1, p_2 \dots p_n\}$, *their Weighted Symmetry Consensus Graph (WSCG) is the edge weighted undirected graph* $G_W = (V_W, E_W, \omega)$ *where* $V_W = \{(p_a, p_b) \in P | a < b\}$ *is the set of pairs of points in* P, $E_W = V_W \times V_W$ *is the set of edges and* $\omega : E_W \rightarrow \mathbb{R}^+$ *is a weight function that assigns to each edge a positive real value* $\omega((p_a, p_b), (p_c, p_d))$ *proportional to the similarity between the symmetry plane of* p_a, p_b *and the symmetry plane of* p_c, p_d.

From a topological point of view a WSCG is by construction a complete graph, so it would not be interesting to search for cliques in it as we made for the BSCG. Instead we are interested in finding large sets of vertices that are all associated to very similar symmetry planes. In other words, since the edges reflect the similarity between planes we are interested in finding sets of vertices that present a high pairwise similarity among them and a low pairwise similarity with respect to vertices external to the set itself. These two properties (high internal similarity and low external similarity) are commonly used to define the notion of a cluster, which indeed generalizes to the weighted graphs context the notion of clique in binary graphs.

In figure 2 we illustrate a simple example of a WSCG. The object shown is a slightly perturbed version of the one shown in figure 1. In this case there are

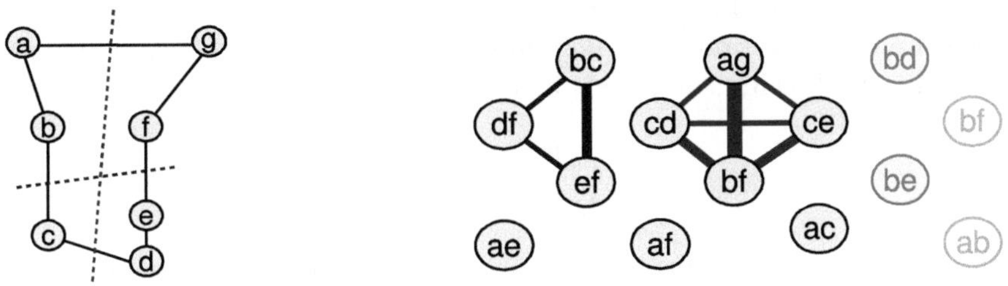

Fig. 2. A very simple object without perfect symmetries (left) and its Weighted Symmetry Consensus Graph (right). The weight of the edges is proportional to their thickness. For practical reasons negligible edges are not drawn and only a part of the graph is shown.

no subset of points that exhibit a perfect symmetry, nevertheless the separation plane between a, g is very similar to the one between b, f, which in turn is just a little less similar to the ones between c, d and c, e, and so on (note that only the edges with relevant weight are drawn). The two largest clusters correspond to the two most relevant symmetry planes in the model. It is worth noting than a cluster could assign to the same point in the model two or more different specular points, as in the case of point c which is symmetric to both e and d. This many-to-many property of the assignment is critical when dealing with data coming from sampling, where artifacts and random noise make very unlikely a direct correspondence between each point and its specular instance, even in the presence of perfect simmetry in the sampled object.

3 Dominant Sets Cluster Extraction

In section 2.2 we shown how the problem of finding optimal symmetry planes can be casted into the search for large clusters in a set of vertices over which we defined a pairwise similarity measure ω. In order to perform this search we propose to use the Dominant Set framework [19] which is a very effective pairwise clustering technique.

Given an edge weighted undirected graph $G = (V, E, \omega)$, a subset of vertices $S \subseteq V$ and two vertices $i \in S$ and $j \notin S$ the following function measures the coherence between nodes j and i, with respect to the average coherence between node i and its neighbors in S

$$\phi_S(i, j) = \omega(ij) - \frac{1}{|S|}\sum_{k \in S} \omega(ik) \qquad (1)$$

And overall weighted coherence between i and other nodes in S is defined as:

$$w_S(i) = \begin{cases} 1 & \text{if } |S| = 1 \\ \sum_{j \in S \setminus \{i\}} \phi_{S\setminus\{i\}}(i, j) w_{S\setminus\{i\}}(j) & \text{otherwise} \end{cases} \qquad (2)$$

Intuitively, $w_S(i)$ will be high if i is highly coherent with vertices in S. Given this measure $S \subseteq V$ is said to be *dominant* if $\sum_{i \in T} w_T(i) > 0$ for any nonempty $T \subseteq S$ and the following conditions hold:

$$w_S(i) > 0, \forall i \in S \tag{3}$$

$$w_{S \cup \{i\}}(i) < 0, \forall i \notin S \tag{4}$$

The conditions above correspond to the two main properties of a cluster: namely internal homogeneity and external inhomogeneity, which are exactly the properties that we expect to hold for the sets we are searching for.

Another compelling reason to use the Dominant Set framework is that, despite the complexity of their recursive definition, they are very easy to find in practice: in fact a single set (corresponding to a symmetry plane) can be found in polynomial time and a convenient technique exists in order to enumerate all the Dominant Sets in a graph.

Pavan and Pelillo [19] have shown that dominant sets correspond to local maximizer over the standard simplex (i.e. the sets of vectors with components that sum up to 1) of the quadratic function

$$f(\mathbf{x}) = \mathbf{x}^t A \mathbf{x} \tag{5}$$

where A is the weighted adjacency matrix of the graph (thus $A_{ij} = \omega(i,j)$).

These maximums can be found by exploiting the convergence properties of the payoff monotonic replicator dynamic

$$x_i(t+1) = \frac{(Ax(t))_i}{x(t)^t Ax(t)} \tag{6}$$

which is guaranteed to converge to a local maximum when the association graph is undirected and, thus, the matrix A is symmetric[20]. At convergence the value of the function f is a measure of the global coherence of the extracted set, which in turn express the strength of the symmetry found.

Since an object could express more than a single symmetry plane it would be useful to be able to enumerate at least the most relevant Dominant Sets present in the graph. In addition, although replicator dynamics are guaranteed to converge it does not need to stop on the largest Dominant Set, thus by enumerating them we could choose always the most relevant ones. To obtain such enumeration we used the technique proposed in [21].

4 Experimental Results

In order to show the effectiveness of the proposed approach we performed both qualitative tests on a database of real world objects and quantitative validation over a larger set of synthetic data generated under different conditions of noise and occlusion. In both cases our technique has been compared with the PRST[15] algorithm which we implemented with 64 bins for each dimension of the space of planes parametrization and with ISP refinement. As an optimization meant to keep the size of the Symmetry Consensus Graph reasonable we filtered the vertex set V_W by sampling a fixed number of random vertices.

The qualitative tests have been made against a database of about 40 3D scans of sunglasses models acquired as point clouds with a structured light scanner. Each model counts between 10.000 and 20.000 points which have been sampled with a metric error which is guarantee to be less than 10 micrometers. Our goal is to find for each model the main symmetry plane, that is, in this context, the plane that split each sunglasses pair in two specular halves. Despite the high quality of the sampling process it is difficult to obtain an accurate result because of slight differences in the two halves of the model, incomplete sampling and parts of the object that are not fixed and can move during the sample process (for instance nose pads). These hurdles strongly limit the use of traditional global techniques, such as PCA, and the presence of non perfect symmetry can easily lead to some slight imprecision also in more refined methods, such as PRST.

In figure 3 we show the performance of these latter methods in comparison to WSCG. The central blue plane is the symmetry plane found, the two lateral yellow planes are added reference planes that are parallel and equidistant to it. The rows in the figure show respectively the plane obtained with PCA, PRST and WSCG. The plane found by PCA is clearly skewed because of the asymmetric sampling of the object during range scanning. The result obtained by the PRST algorithm is far better than PCA, as expected, while WSCG gives another improvement over it.

Qualitative results offer a visual hint about the kind of improvement obtained with the WSCG technique, but as PRST is in itself very accurate this enhancement could appear minimal. In order to give a more quantitative insight we

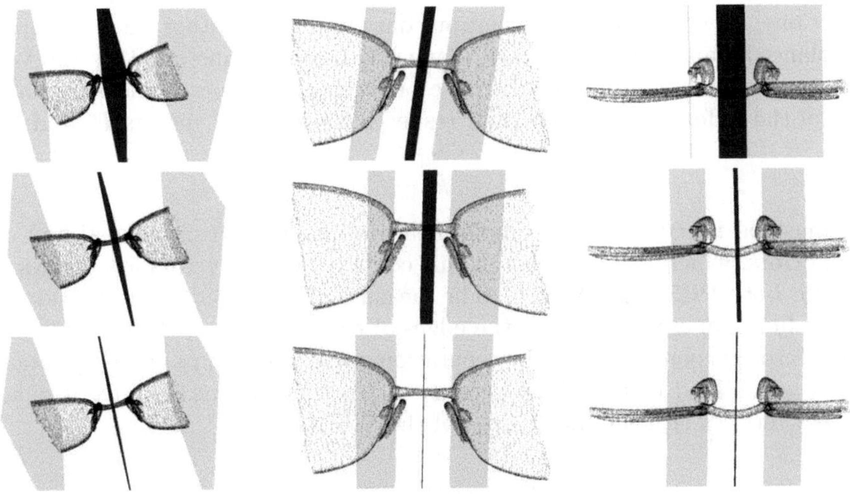

Fig. 3. Qualitative results on 3D scans. In the first row the symmetry plane obtained with PCA is shown from different angles and details. The second row contains the results of PRST and the third the results of WSCG.

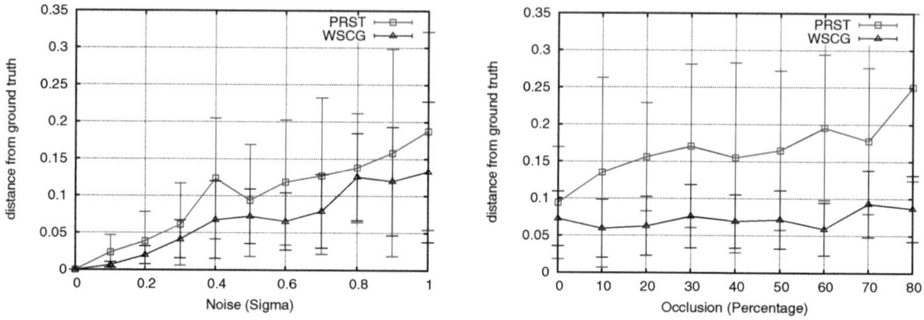

(a) (b)

Fig. 4. Accuracy of PRST and WSCG with different level of noise and occlusion

performed a set of experiments with synthetic data which allowed us to compare the symmetry planes obtained with respect to an accurate ground truth plane. We generated a random distribution of 2000 points within a range of 400 millimeters perfectly symmetric with respect to a known plane. To evaluate the performance of WSCG in different conditions we added gaussian random noise along all three axes with a value of the parameter σ ranging from 0.1 to 1 millimeters. Then we repeated the experiments by setting $\sigma = 0.5$ and simulating occlusion with the deletion of a growing percentage of random points from one side of the symmetry plane. Each experimental condition has been tested with 20 independent runs. In each run the two algorithms are applied and the distance between the obtained principal symmetry plane and the ground truth is measured. While several symmetry measures have ben proposed [8,9], we are not really interested in knowing how much the found symmetry diverge from a perfect one, as we already know that our data are noisy: rather we are interested in a plane distance measure. Each plane has been parametrized by the vector P perpendicular to it and the distance between two planes is measured as the norm of the difference between their respective associated vectors P_1 and P_2:

$$d = \| P_1 - P_2 \| \tag{7}$$

It must be noted that this simple measure mixes up translational and rotational errors, which is reasonable in this context because errors are always very small for both PRST and WSCG. The measure of equation 7 is also used to define the function ω that weights the edges of the WSCG in definition 2.2. If P_{ab} is the vector perpendicular to the separation plane between points p_a, p_b and P_{cd} the one associated to the plane between p_c, p_d then we measure the similarity between vertices (p_a, p_b) and (p_c, p_d) of the WSCG as:

$$\omega((p_a, p_b), (p_c, p_d)) = e^{-\|P_{ab} - P_{cd}\|} \tag{8}$$

In figure 4(a) the performance of WSCG is shown for several levels of gaussian noise added to the test points and in figure 4(b) the robustness with respect

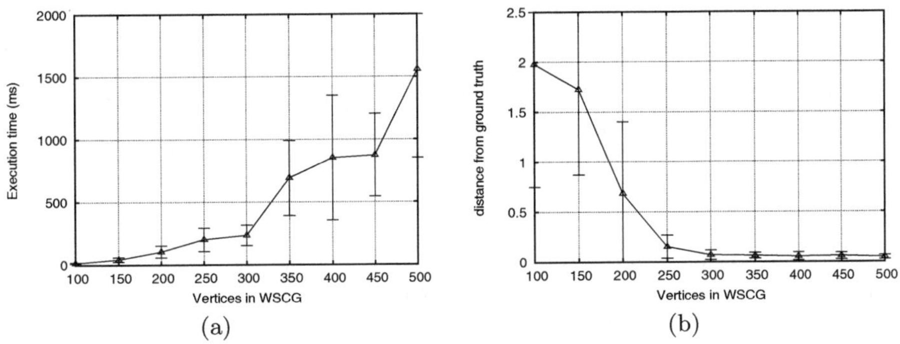

Fig. 5. Running time and accuracy of WSCG with respect to the number of points

to occlusion is shown for different percentages of sample points removal. The comparison with PRST shows that both techniques offer comparable very good performance and that WSCG is more accurate on average. In particular WSCG seems to be more tolerant to occlusion, which is expected as PRST relies on a voting algorithm. In figure 5(a) we show the convergence time of the WSCG algorithm with respect to the number of vertices that are kept in the Consensus Graph. Each test has been performed 20 times with gaussian noise of parameter $\sigma = 0.5$ and an occlusion of 50% and has been executed on a 2Ghz Athlon CPU. These result show that by keeping the number of sampled points low, WSCG can be executed in less than a second. This optimization does not imply a lower quality in the symmetry plane estimation, in fact figure 5(b) shows that the error becomes very small when keeping as few as 300 vertices.

5 Conclusions

The Weighted Symmetry Consensus Graph is a new approach to planar symmetry detection that exploits pairwise clustering to find a solution that is maximally consistent from a global perspective. The main features of WSCG are that it does not require any quantization of the solutions space and is adaptive with respect to outliers. Experimental validation shows that this approach gives better results than another state-of-the-art algorithm. Finally WSCG is quite general as the compatibility measure needs not to be related to planar symmetry. In fact future work will aim to extend the approach to radial symmetry and other features that can be estimated by exploiting a global consensus.

Acknowledgments. We thank Luxottica Group S.p.A. for funding this research and providing the experimental data. We also acknowledge the financial support of the 7th Framework Programme, under FET-Open project SIMBAD grant no. 213250.

References

1. Atallah, M.J.: On symmetry detection. IEEE Trans. Computers 34(7), 663–666 (1985)
2. Wolter, J.D., Woo, T.C., Volz, R.A.: Optimal algorithms for symmetry detection in two and three dimensions. The Visual Computer 1(1), 37–48 (1985)
3. Sun, C.: Symmetry detection using gradient information. Pattern Recogn. Lett. 16(9), 987–996 (1995)
4. Shah, M.I., Sorensen, D.C.: A symmetry preserving singular value decomposition. SIAM J. Matrix Anal. Appl. 28(3), 749–769 (2006)
5. Sun, C., Sherrah, J.: 3d symmetry detection using the extended gaussian image. IEEE Trans. Pattern Anal. Mach. Intell. 19(2), 164–168 (1997)
6. Shen, D., Ip, H.H.S., Cheung, K.K.T., Teoh, E.K.: Symmetry detection by generalized complex (GC) moments: A close-form solution. IEEE Transactions on Pattern Analysis and Machine Intelligence 21(5), 466–476 (1999)
7. Lu, C., Zhang, C., Wen, F., Yan, P.: Principle component analysis-based symmetry detection. Tien Tzu Hsueh Pao/Acta Electronica Sinica 27(5), 25–28 (1999); Cited By (since 1996): 4
8. Zabrodsky, H., Peleg, S., Avnir, D.: Symmetry as a continuous feature. IEEE Trans. on Pattern Analysis and Machine Intelligence 17(12), 1154–1166 (1995)
9. Kulkarni, P., Dutta, D.: An investigation of techniques for asymmetry rectification. Journal of Mechanical Design 117(4), 620–626 (1995)
10. Reisfeld, D., Wolfson, H., Yeshurun, Y.: Context free attentional operators: the generalized symmetry transform. Int. J. of Computer Vision, Special Issue on Qualitative Vision (1994)
11. Choi, I., Chien, S.I.: A generalized symmetry transform with selective attention capability for specific corner angles. Signal Processing Letters, IEEE 11(2), 255–257 (2004)
12. Bigün, J.: Pattern recognition in images by symmetries and coordinate transformations. Computer Vision and Image Understanding 68(3), 290–307 (1997)
13. Gesú, V.D., Valenti, C., Strinati, L.: Local operators to detect regions of interest. Pattern Recogn. Lett. 18(11-13), 1077–1081 (1997)
14. Loy, G., Zelinsky, A.: Fast radial symmetry for detecting points of interest. IEEE Trans. Pattern Anal. Mach. Intell. 25(8), 959–973 (2003)
15. Podolak, J., Shilane, P., Golovinskiy, A., Rusinkiewicz, S., Funkhouser, T.: A planar-reflective symmetry transform for 3d shapes. In: SIGGRAPH 2006: ACM SIGGRAPH 2006 Papers, pp. 549–559. ACM, New York (2006)
16. Podolak, J., Shilane, P., Golovinskiy, A., Rusinkiewicz, S., Funkhouser, T.: A planar-reflective symmetry transform for 3d shapes. ACM Trans. Graph. 25(3), 549–559 (2006)
17. Rustamov, R.M.: Augmented symmetry transforms. In: SMI 2007: Proceedings of the IEEE International Conference on Shape Modeling and Applications 2007, Washington, DC, USA, pp. 13–20. IEEE Computer Society, Los Alamitos (2007)
18. Besl, P.J., McKay, N.D.: A method for registration of 3-d shapes. IEEE Trans. Pattern Anal. Mach. Intell. 14(2), 239–256 (1992)
19. Pavan, M., Pelillo, M.: Dominant sets and pairwise clustering. IEEE Trans. Pattern Anal. Mach. Intell. 29(1), 167–172 (2007)
20. Pelillo, M.: Replicator equations, maximal cliques, and graph isomorphism. Neural Comput. 11(8), 1933–1955 (1999)
21. Rota Bulò, S., Torsello, A., Pelillo, M.: A continuous-based approach for partial clique enumeration. In: Escolano, F., Vento, M. (eds.) GbRPR 2007. LNCS, vol. 4538, pp. 61–70. Springer, Heidelberg (2007)

Graph Characteristic from the Gauss-Bonnet Theorem

Hewayda ElGhawalby[1,2] and Edwin R. Hancock[1]

[1] Department of Computer Science, University of York, YO10 5DD, UK
[2] Faculty of Engineering, Suez Canal university, Egypt
{howaida,erh}@cs.york.ac.uk

Abstract. In this paper we explore how to characterise graphs using the Gauss-Bonnet theorem. Using the Gaussian curvatures estimated from first-order cycles we compute a global estimate of the Euler index using the Gauss-Bonnet theorem. We commence by embedding the nodes of a graph in a manifold using the heat-kernel mapping. From this mapping we are able to compute the geodesic and Euclidean distance between nodes, and these can be used to estimate the sectional curvatures of edges. Assuming that edges reside on hyper-spheres, we use Gauss's theorem to estimate the Gaussian curvature from the interior angles of geodesic triangles formed by first-order cycles in the graph. From the Gaussian curvatures we make a global estimate of the Euler index of the manifold using the Gauss-Bonnet theorem. Experiments show how the Gaussian curvatures and the Euler characteristics can be used to cluster Delaunay triangulations extracted from real world images.

Keywords: Manifold embedding, Heat kernel, Hausdorff distance, Gaussian curvature, Graph matching.

1 Introduction

The task of characterising the structure of graphs for the purposes of matching, clustering or indexing is a challenging one for which no definitive solution exists. If the characterisation is attempted without recourse to edge weight or node attribute information, then the regularity of many graph structures means that local characteristics are ambiguous, and only global structural information can be used. Examples of the available global characteristics include Cheeger numbers, perimeters, diameters and edge densities. This problem can be particularly severe for k-nearest neighbor graphs, Delaunay triangulations and strongly regular graphs.

Of course, one solution to this problem is to use graph-spectral information. Although the spectrum of a graph (i.e. the set of eigenvalues of the Laplacian matrix) is a global permutation invariant characterisation, the set of eigenvectors are not permutation invariant and can only be used to characterise graphs on a node-by-node basis. Unfortunately, eigenvectors are notoriously unstable,

N. da Vitora Lobo et al. (Eds.): SSPR&SPR 2008, LNCS 5342, pp. 207–216, 2008.

and small perturbations in graph structure can lead to significant variations in eigenvector directions. Moreover, the estimation of eigenvectors is subject to directional ambiguity.

Another solution to the problem is to make use of spectral geometry [5,2,9]. This allows the nodes of graph to be embedded on a manifold, and the graph to be characterised using the local differential structure of the manifold. Hence, local graph structure can be characterised using invariant quantities such as curvature or torsion [8]. This is the approach adopted by Bai and Hancock [12] who use the heat kernel to construct, embed and characterise the edges of a graph using sectional curvature. This curvature is computed using the difference between the geodesic (edge length) and Euclidean internode distance under the embedding. Of course, the sectional curvature is one of many possible differential invariants that can be computed.

The aim in this paper is therefore to look in greater depth at the diversity of quantities that are furnished by the differential geometry of the manifold. We turn our attention to geodesic triangles. These are formed when a first-order cycle in a graph is embedded. By making a locally hyper-spherical approximation to the geometry of the manifold and the turning angles of the geodesics, we are able to estimate the Gaussian curvature of the triangle. Furthermore, the Gauss-Bonnet theorem allows us to estimate a local Euler characteristic from abutting geodesic triangles. We demonstrate how the estimated curvature information can be used for matching and clustering graphs, extracted from real world image data.

2 Manifold Embeddings

We commence by establishing a matrix representation for graph−structure. Let the graph under study be denoted by $G = (V, E)$ where V is the set of nodes and $E \subseteq V \times V$ is the set of edges. The elements of the adjacency matrix A of the graph G are:

$$A(u,v) = \begin{cases} 1 & if(u,v) \in E \\ 0 & otherwise \end{cases} \tag{1}$$

To construct the Laplacian matrix of graph G we first establish a diagonal degree matrix D with elements

$$D(u,u) = \sum_{v \in V} A(u,v) = d_u \tag{2}$$

The Laplacian matrix $L = D - A$, is the degree matrix minus the adjacency matrix. Elementwise, we have

$$L(u,v) = \begin{cases} d_u & if u = v \\ -1 & if(u,v) \in E \\ 0 & otherwise \end{cases} \tag{3}$$

The normalized Laplacian is given by $\widehat{L} = D^{-1/2}LD^{-1/2}$ and has spectral decomposition $\widehat{L} = \Phi\Lambda\Phi^T = \sum_{i=1}^{|V|} \lambda_i \phi_i \phi_i^T$ where $|V|$ is the number of nodes,

$\Lambda = diag(\lambda_1, \lambda_2, ..., \lambda_{|V|}), (0 < \lambda_1 < \lambda_2 < ... < \lambda_{|V|})$ is the diagonal matrix with the ordered eigenvalues as elements and $\Phi = (\phi_1|\phi_2|...|\phi_{|V|})$ is the matrix with the corresponding ordered eigenvectors as columns. We follow Bai and Hancock [12] and make use of the heat kernel embedding. The heat kernel plays an important role in spectral graph theory. It encapsulates the way in which information flows through the edges of graph over time under the heat equation, and is the solution of the partial differential equation

$$\frac{\partial h_t}{\partial t} = -\widehat{L}h_t \tag{4}$$

where h_t is the heat kernel and t is the time. The solution is found by exponentiating the Laplacian eigenspectrum as follows

$$h_t = \exp[-\widehat{L}t] = \Phi \exp[-t\Lambda]\Phi^T \tag{5}$$

For the heat kernel, the matrix of embedding coordinates Y (i.e. the matrix whose columns are the vectors of node coordinates) is found by performing the Young-Householder [13] decomposition $h_t = Y^T Y$ as a result the matrix of node embedding coordinates is

$$Y = (y_1|y_2|...|y_{|V|}) = \exp[-\frac{1}{2}t\Lambda]\Phi^T \tag{6}$$

where y_u is the coordinate vector for the node u. In the vector space, the Euclidean distance between the nodes u and v of the graph is

$$d_e^2(u,v) = (y_u - y_v)^T(y_u - y_v) = \sum_{i=1}^{|V|} \exp[-\lambda_i t](\phi_i(u) - \phi_i(v))^2 \tag{7}$$

When a pair of nodes are connected by an edge, then $d_G(u,v) = 1$. We assume that the geodesic between the pair of nodes can be approximated by a segment of circle whose arc-length is the geodesic distance and whose chord-length is the Euclidean distance. If the radius of the circle is $r_s(u,v)$ and the tangent vector to the circle arc undergoes a change in direction of $2\theta(u,v)$, then we have

$$d_g(u,v) = 2r_s(u,v)\theta(u,v) \tag{8}$$

and

$$d_e(u,v) = 2r_s(u,v)\sin\theta(u,v) \tag{9}$$

3 The Gauss-Bonnet Theorem

The Gauss-Bonnet Theorem links the topology and geometry of a surface in an elegant and compact manner. Spivak [10] and Stillwell [11] give accounts of the early history of its development and application.

For a smooth compact oriented Riemannian 2-manifold M, let $\triangle_g$ be a triangle on M whose sides are geodesics, i.e. paths of shortest length on the manifold.

Further, let α_1, α_2 and α_3 denote the interior angles of the triangle. According to Gauss's theorem, if the Gaussian curvature K (i.e. the product of the maximum and minimum curvatures at a point on the manifold) is integrated over $\triangle_g$, then

$$\int_{\triangle_g} KdM = \sum_{i=1}^{3} \alpha_i - \pi \tag{10}$$

where dM is the Riemannian volume element.

3.1 Geometric Preliminaries

To estimate the Gaussian curvature from the above, we must determine the interior angles α_i of the geodesic triangle. To this end we assume that T is a triangulation of a smooth manifold M, $\triangle_g$ be a geodesic triangle on M with angles $\{\alpha_i\}_{i=1}^3$ and geodesic edge lengths $\{d_{gi}\}_{i=1}^3$. Further suppose that $\triangle_e$ be the corresponding Euclidean triangle with edge lengths $\{d_{ei}\}_{i=1}^3$ and interior angles $\{\varphi_i\}_{i=1}^3$. We assume that the geodesic index i is a great arc on a sphere with radius r_i, $i = 1, 2, 3$. By averaging over the constituent geodesic edges, we treat the geodesic triangles as residing on a hyper−sphere with radius $r = \frac{1}{3}\sum_{i=1}^3 r_i$. To compute the interior angle of the geodesic triangle, we make use of two formulae. First, from the geometry of the sphere

$$\sin(\frac{\alpha}{2}) = \frac{r}{\sqrt{4r^2 - d_e^2}} \tag{11}$$

Second, when using the double integral to compute the area of a triangle on a sphere

$$\frac{\alpha}{2} = \frac{\pi r^2}{6r^2 - d_e^2} \tag{12}$$

Substituting from (11) and (12) in the Taylor expansion for $\sin(\alpha/2)$, we have

$$\sin(\alpha/2) \simeq (\alpha/2) - \frac{1}{3!}(\alpha/2)^3 \tag{13}$$

$$\frac{\sin(\alpha/2)}{(\alpha/2)} \simeq 1 - \frac{\alpha^2}{24} \tag{14}$$

So we can deduce that

$$\alpha^2 \simeq 24(1 - \frac{6r^2 - d_e^2}{\pi r\sqrt{4r^2 - d_e^2}}) \tag{15}$$

3.2 The Euler Characteristic

We can now use the Gauss Bonnet Theorem to estimate a global topological characteristic for the manifold. We commence by triangulating M so that each face is a geodesic triangle.

Summing (11) over all the triangular faces gives that the integral of K over all M is 2π times the Euler characteristic of M, i.e.

$$\int_M K dM = 2\pi\chi(M) \tag{16}$$

4 Hausdorff Distance

Once sets of Gaussian curvatures for the geodesic triangles of a graph are to hand, we require a means of comparing different graphs so that we can compute graph similarity. Here to avoid the need to compute explicit node correspondences using the Hausdorff distance between curvature sets. The Hausdorff distance has been used for a number of matching and recognition problems. It provides a means of computing the distance between sets of unordered observations when the correspondences between the individual items are unknown. In its most general setting, the Hausdorff distance is defined between compact sets in a metric space. Given two such sets, we consider for each point in one set the closest point in the second set. Hausdorff distance is the maximum over all these values. More formally, the classical Hausdorff distance (HD) [6] between two finite point sets A and B is given by

$$H(A, B) = \max(h(A, B), h(B, A)) \tag{17}$$

where the directed Hausdorff distance from A to B is defined to be

$$h(A, B) = \max_{a \in A} \min_{b \in B} \|a - b\| \tag{18}$$

and $\|.\|$ is some underlying norm on the points of A and B (e.g., the L2 or Euclidean norm). Regardless of the norm, the Hausdorff metric captures the notion of the worst match between two objects. The computed value represents the largest distance between a point in one set and a point in the other one. Several variations of the Hausdorff distance have been proposed as alternatives to the maximum of the minimum approach in the classical one. These include Hausdorff fraction, Hausdorff quantile [6] and Spatially Coherent Matching [1]. A robust modified Hausdorff distance (MHD) based on the average distance value instead of the maximum value was proposed by Dubuisson and Jain [3], in this sense they defined the directed distance of the MHD as

$$h(A, B) = \frac{1}{N_A} \sum_{a \in A} \min_{b \in B} \|a - b\| \tag{19}$$

For this paper we will consider the classical and the modified Hausdorff distances. Using these ingredients we can describe how Hausdorff distances can be extended to graph-based representations. To commence let us consider two graphs $G_1 = (V_1, E_1, T_1, k_1)$ and $G_2 = (V_2, E_2, T_1, k_2)$, where V_1,V_2 are the sets of nodes, E_1, E_2 the sets of edges, T_1,T_2 are the sets of triangles, and k_1,k_2 the sets of Gaussian curvatures for geodesic triangles defined in the previous section. We can now write the distances between two graphs as follows:

1) The classic Hausdorff distance (HD) is

$$h_{HD}(G_1, G_2) = \max_{i \in V_1} \min_{j \in V_2} \|k_2(j) - k_1(i)\| \tag{20}$$

2) The modified Hausdorff distance (MHD) is

$$h_{MHD}(G_1, G_2) = \frac{1}{|V_1|} \sum_{i \in V_1} \min_{j \in V_2} \|k_2(j) - k_1(i)\|) \tag{21}$$

5 Experiments

In our experiments we use the standard CMU, MOVI and chalet house sequences as our data set [7]. These data sets contain different views of model houses from equally spaced viewing directions. From the house images, corner features are extracted, and Delaunay graphs representing the arrangement of feature points are constructed. Our data consists of ten graphs for each of the three houses. Each node in a Delaunay graph belongs to a first order cycle, and as a result the graph is a triangulation.

To commence, we obtain the manifold embedding for each of the thirty graphs following the steps mentioned in Section 2. We compute the Euclidean distances between the nodes in each graph based on the heat kernel with the values of $t = 10.0, 1.0, 0.1$ and 0.01. Using the Delaunay graphs to triangulate the manifold, for each geodesic triangle we compute the interior angles deduced using the procedure outlined in Section 3.1. For each triangle we then compute the Gaussian curvature as described in Section 3. We use the Gaussian curvatures of the geodesic triangles as features for the purposes of gauging the similarity of graphs. Using the set of Gaussian curvatures for pairs of graphs, we compute a matrix of Hausdorff distances the thirty different graphs. Here we explore the use of both the classical Hausdorff distance and the modified Hausdorff distance. Finally, we apply multidimensional scaling (MDS) to the distance matrices to embed the individual graphs into a 2D space. Here each graph is represented by a single point. Figure 1 shows the results obtained using the classical Hausdorff distance. The subfigures are ordered from left to right, using $t = 10.0, 1.0, 0.1$ and 0.01 in the heat kernel. Figure 2 shows the corresponding results obtained when the modified Hausdorff distance is used.

Of course, there are alternatives to the heat-kernel embedding, and the procedure for computing Gaussian curvatures can be applied to these too. When the Euclidean distances are computed directly from the Laplacian embedding instead (i.e. $Y = \sqrt{\Lambda}\Phi^T$), then the results shown in Figure 3 are obtained. Here the left-hand subfigure is obtained when using the HD and the right-hand one when using MHD.

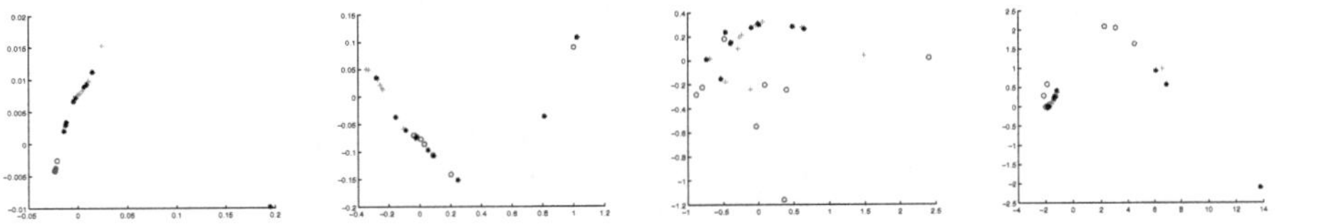

Fig. 1. MDS embedding obtained using HD for the houses data

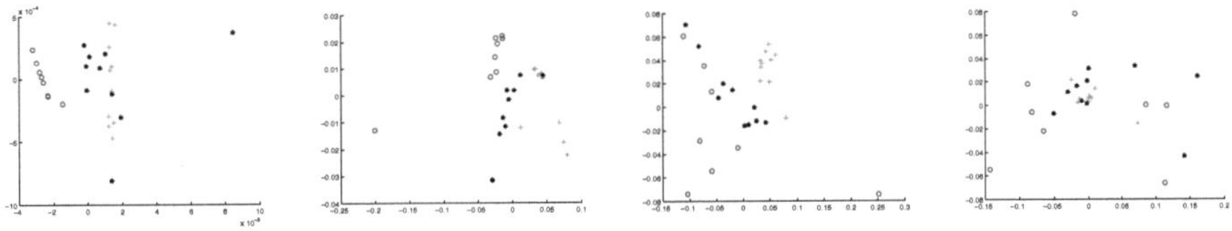

Fig. 2. MDS embedding obtained using MHD for the houses data

To investigate the data in more detail Table 1 shows the rand index for the data as a function of t. This index is computed as follows: We compute the mean for each cluster, we then compute the distance from each point to each mean. If the distance from the correct mean is smaller than those to remaining means,then the classification is correct, if not then classification is incorrect. The rand index is R= (♯ incorrect)/ (♯ incorrect + ♯ correct).

Table 1. A rand index vs. t

	lap	t=10	t=1.0	t=0.1	t=0.01
HD	0.2667	0.2333	0.3000	0.5333	0.5333
MHD	0.1333	0.1667	0.0333	0.1333	0.4000

In previous work [4], we used the sectional curvature associated with the edges of a graph under the heat kernel embedding as a feature for the purposes of gauging the similarity of graphs, and hence clustering them. Table 2 shows the rand index for the results obtained as a function of t.

A comparison shows that the curvature attributes associated with the edges give a slightly better clusters than those obtained using the attributes derived

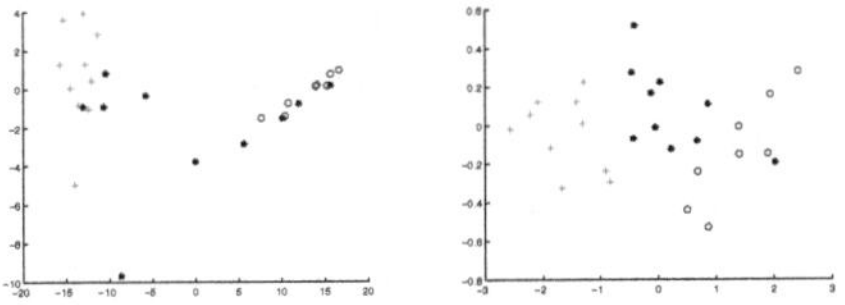

Fig. 3. MDS embedding obtained using curvatures computed from the Laplacian eigenmap and distances computed using HD (left) and MHD (right) for houses data

Table 2. A rand index vs. *t*. induced when the sectional curvature of the edges was used.

	lap	t=10	t=1.0	t=0.1	t=0.01
HD	0.0.2333	0.1000	0.1667	0.4333	0.0333
MHD	0.1000	0.1333	0.2333	0.1333	0.0333

Table 3. A rand index vs. *t*. induced when the sectional curvature of the edges was used.

lap	t=10	t=1.0	t=0.1	t=0.01
0.0.3333	0.4667	0.2333	0.3000	0.4000

from the triangles. To take this study one step further we have constructed normalised histograms of the Gaussian curvature associated with the geodesic triangles on each graph. The histograms can be treated as vectors, and can be used as input to principal components analysis. Using the leading two principal component projections we repeat the Rand index computation, The error deduced is shown in Table 3, which ranges between 0.23 and 0.47. Next, we consider the Euler characteristic whose computation was detailed in Section 3.2. In Table 4 we list the mean and variance for each group of graphs (for each house).

Table 4. The mean and variance for the Euler characteristic of the manifold embedding of each graph

		lap	t=10	t=1.0	t=0.1	t=0.01
1st	Mean	0.0217	-0.1725	2.2361	1.2835	5.1177
house	Variance	0.0099	0.0001	0.1830	4.8437	1.7395
2nd	Mean	0.0850	-0.7514	4.0221	49.1613	8.3172
house	Variance	0.1344	0.0003	0.1077	12.4022	10.2801
3rd	Mean	0.0545	-0.3970	3.5168	16.1415	6.1704
house	Variance	2.4059	0.0120	0.8116	150.9340	17.5878

Finally, we investigate how the Gaussian curvatures of the geodesic triangles are distributed over the Delaunay graph. Figures 4, 5 and 6 show distribution for sample embeddings computed from the heat kernel at (t=0.1).

From the sequence it is clear that the Gaussian curvature distribution over the different views of each house is stable, moreover it moves smoothly from positive (elliptical) to negative (hyperbolic) regions. This suggests that the arrangement of triangles and their Gaussian curvatures could be used as the basic of a matching algorithm.

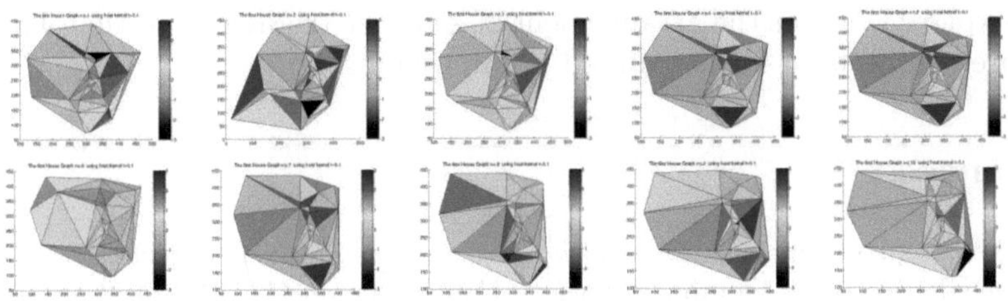

Fig. 4. The distribution of the Gaussian curvatures of the geodesic triangles for the ten graphs of the 1st house

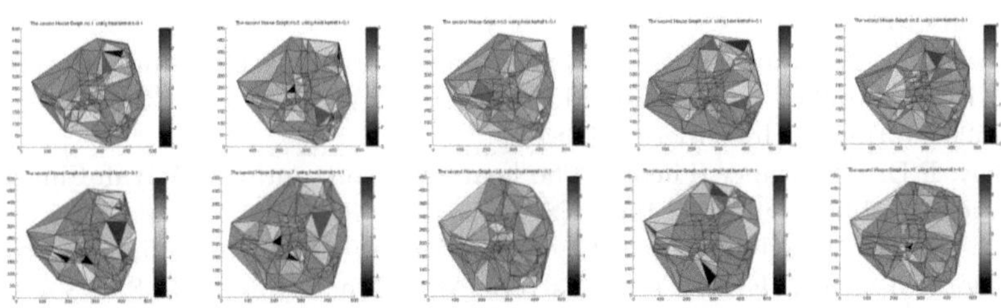

Fig. 5. The distribution of the Gaussian curvatures of the geodesic triangles for the ten graphs of the 2nd house

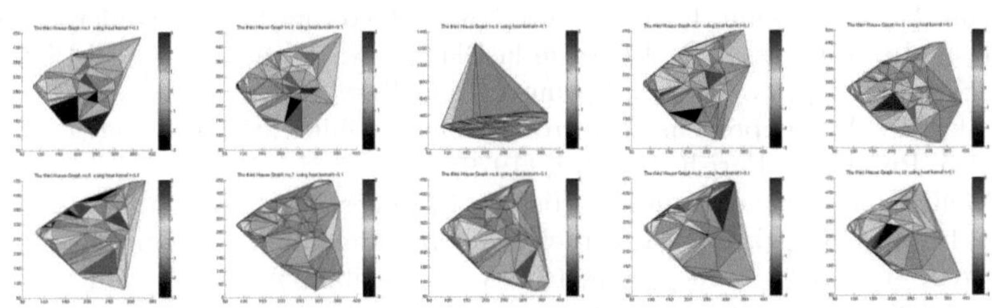

Fig. 6. The distribution of the Gaussian curvatures of the geodesic triangles for the ten graphs of the 3rd house

6 Conclusion and Future Work

We have shown how to compute the Gaussian curvatures of geodesic triangles and the Euler characteristic for graphs embedded on a manifold using the heat kernel. Experiments show that these two quantities provide effective features for graph clustering, but perform marginally worse than the sectional curvatures of the edges. However, they do lead to a deeper representation of the differential structure of embedded graphs. As future work we have a number of different lines of research in mind. First we aim to explore whether our characterisation can lead to the construction of Grassmanian manifolds for graphs. Second, we aim to explore whether the arrangement of geodesic triangles and their Gaussian curvature can be used for attributed graph matching.

References

1. Boykov, Y., Huttenlocher, D.: A new bayesian framework for object recognition. In: Proceeding of IEEE Computer Society Conference on CVPR, vol. 2, pp. 517–523 (1999)
2. Brooks, R.: Geometry, Spectral theory, Groups, and Dynamics. Amer. Mathematical Society (2005)
3. Dubuisson, M., Jain, A.: A modified hausdorff distance for object matching, pp. 566–568 (1994)
4. ElGhawalby, H., Hancock, E.R.: Measuring graph similarity using spectral geometry. In: Campilho, A., Kamel, M.S. (eds.) ICIAR 2008. LNCS, vol. 5112, pp. 517–526. Springer, Heidelberg (2008)
5. Gilikey, P.B.: Invariance Theory, The heat equation, and the Atiyah-Singer index Theorem. Publish or Parish Inc., USA (1984)
6. Huttenlocher, D., Klanderman, G., Rucklidge, W.: Comparing images using the hausdorff distance. IEEE. Trans. Pattern Anal. Mach. Intell. 15, 850–863 (1993)
7. Luo, B., Wilson, R.C., Hancock, E.R.: Spectral embedding of graphs. Pattern Recogintion 36, 2213–2230 (2003)
8. Rosenberg, S.: The Laplacian on a Riemanian manifold. Cambridge University Press, Cambridge (2002)
9. Saucan, E., Appleboim, E.: Curvature based clustering for DNA microarray data analysis. In: Marques, J.S., Pérez de la Blanca, N., Pina, P. (eds.) IbPRIA 2005. LNCS, vol. 3523, pp. 405–412. Springer, Heidelberg (2005)
10. Spivak, M.: A Comprehensive Introduction to Differential Geometry, 2nd edn., vol. 1-5. Publish or Parish, Houston (1979)
11. Stillwell, J.: Mathematics and its History. Springer, New York (1974)
12. Xiao, B., Hancock, E.R.: Heat kernel, riemannian manifolds and graph embedding. In: Fred, A., Caelli, T.M., Duin, R.P.W., Campilho, A.C., de Ridder, D. (eds.) SSPR&SPR 2004. LNCS, vol. 3138, pp. 198–206. Springer, Heidelberg (2004)
13. Young, G., Householder, A.S.: Disscussion of a set of points in terms of their mutual distances. Psychometrika 3, 19–22 (1938)

Quantitative Evaluation on Heat Kernel Permutation Invariants

Bai Xiao[1], Richard C. Wilson[2], and Edwin R. Hancock[2]

[1] Department of Computer Science, University of Bath, Bath,BA2 7AF, UK
xb202@cs.bath.ac.uk
[2] Department of Computer Science, University of York, York, YO10 5DD, UK
erh@cs.york.ac.uk

Abstract. The Laplacian spectrum has proved useful for pattern analysis tasks, and one of its important properties is its close relationship with the heat equation. In this paper, we first show how permutation invariants computed from the trace of the heat kernel can be used to characterize graphs for the purposes of measuring similarity and clustering. We explore three different approaches to characterize the heat kernel trace as a function of time. These are the heat kernel trace moments, heat content invariants and symmetric polynomials with Laplacian eigenvalues as inputs. We then use synthetic and real world datasets to give a quantitative evaluation of these feature invariants deduced from heat kernel analysis. We compare their performance with traditional spectrum invariants.

1 Introduction

The Laplacian spectrum of a graph has proved to be a convenient summary of graph structure that can be relatively easily manipulated to solve the problems such as bi-partitioning [5] and routing[1]. However, the extraction of of permutation invariants for the purpose of characterising graphs has proved to be more elusive. Although the spectrum of a graph is a permutation invariant, its eigenvectors are not. To overcome this problem Luo, Wilson and Hancock [10] have used symmetric polynomials to construct a rich family of permutation invariants from the spectral matrix (the eigenvector matrix premultiplied by the square root of the eigenvalue matrix). Moreover, in prior work [12] [11] and motivated by the work of Yau [13], we have explored related invariants that can be extracted from the heat-kernel.

The invariants described above have been described in a piecemeal way and their relationship to each other has not been fully explored. Neither has their performance on large-scale object recognition tasks been fully explored. The aims in this paper are therefore twofold. First, we aim to explain the relationships between the different Laplacian invariants in a consistent and consolidated manner. Second, we aim to explore their performance on a demanding object recognition problem furnished by a large image database.

N. da Vitora Lobo et al. (Eds.): SSPR&SPR 2008, LNCS 5342, pp. 217–226, 2008.

We commence from one of the most important permutation invariants of the heat kernel, namely the trace which is found by summing and exponentiating the Laplacian eigenvalues with time. In our previous paper[12], we showed how to characterise the shape of the heat-kernel trace using moments with respect to time. An analysis based on the Mellin transform shows that the required moment generating function is the zeta function. An alternative means of extracting invariants is to make use of heat content [6]. We show how to expand the heat content as a polynomial series in time, and demonstrate that the coefficients can be related to the eigenvalues and eigenvectors of the Laplacian. Finally, we explore the derivative of the zeta function at the origin. This is linked to the determinant of the Laplacian, and we show that it is also one of the invariants proposed by Wilson al et[10].

The outline of this paper is as follows. In Section 2 we explain the relationship between the Laplacian eigensystem and the heat kernel trace. Section 3, we develop the three sets of invariants deduced from graph heat kernel which can be used to characterize the graphs. Section 4 offers experiments, and illustrates the utility of the different permutation invariants for graph clustering. Finally, Section 5 offers some conclusions and directions for future research.

2 Heat Kernel on Graphs

To commence, suppose that the graph under study is denoted by $G = (V, E)$ where V is the set of nodes and $E \subseteq V \times V$ is the set of edges. Since we wish to adopt a graph spectral approach we introduce the adjacency matrix A for the graph, which has elements $A(u, v) = 1$ if $u, v \in E$ and 0 otherwise. We also construct the diagonal degree matrix D, whose elements are given by $D(u, u) = deg_u = \sum_{v \in V} A(u, v)$. From the degree matrix and the adjacency matrix we construct the Laplacian matrix $L = D - A$, i.e. the degree matrix minus the adjacency matrix. The normalized Laplacian is given by $\hat{L} = D^{-\frac{1}{2}} L D^{-\frac{1}{2}}$. The spectral decomposition of the normalized Laplacian matrix is $\hat{L} = \Phi \Lambda \Phi^T$, where $\Lambda = diag(\lambda_1, \lambda_2, ..., \lambda_{|V|})$ is the diagonal matrix with the ordered eigenvalues $(0 = \lambda_1 < \lambda_2 < ... < \lambda_{|V|})$ as elements and $\Phi = (\phi_1 | \phi_2 | .... | \phi_{|V|})$ is the matrix with the corresponding ordered unit-norm eigenvectors as columns.

We are interested in the heat equation associated with the Laplacian, and this is given by $\frac{\partial h_t}{\partial t} = -\hat{L} h_t$ where h_t is the heat kernel and t is time. The heat kernel is the fundamental solution of the heat equation. It can be viewed as describing the flow of information across the edges of the graph with time. The rate of flow is determined by the Laplacian of the graph. The solution to the heat equation is $h_t = e^{-t\hat{L}}$. We can compute the heat kernel on a graph by exponentiating the Laplacian eigenspectrum(see e.g. Chung[2]), i.e. $h_t = \Phi \exp[-\Lambda t] \Phi^T = \sum_{i=1}^{|V|} \exp[-\lambda_i t] \phi_i \phi_i^T$. The heat kernel is a $|V| \times |V|$ matrix, and for the nodes u and v of the graph G the resulting element is

$$h_t(u, v) = \sum_{i=1}^{|V|} \exp[-\lambda_i t] \phi_i(u) \phi_i(v) \tag{1}$$

3 Heat Kernel Invariants

In this section, we will describe a number of invariants that can be computed from the heat kernel matrix of the graph. These include the zeta function, heat content invariants, and the unary attributes from symmetric polynomials.

3.1 Zeta Function and Heat Kernel Trace Moments

The trace of the heat kernel is the sum of the diagonal elements of the heat kernel matrix of the graph.

$$Z(t) = Tr[h_t] = \sum_{i=1}^{|V|} \exp[-\lambda_i t] \tag{2}$$

The purpose in this part is to use the shape of the heat kernel trace function as a means of characterizing graph structure. Our characterization is found by taking moments of trace function over time.

To commence our development, we consider the zeta function associated with the Laplacian eigenvalues. The zeta function is defined by $\zeta(s) = \sum_{\lambda_i \neq 0} \lambda_i^{-s}$. In other words, it is the result of exponentiating and summing the reciprocal of the non-zero Laplacian eigenvalues.

To establish the link between the zeta function and the trace of the heat kernel we make use of the Mellin transform. The Mellin Transform of the function $f(t)$ is $w(s) = \int_0^\infty t^{s-1} f(t) dt$. Hence, the heat kernel trace has Mellin transform

$$\int_0^\infty Tr[h_t] t^{s-1} dt = \int_0^\infty \sum_{i=1}^{|V|} e^{-\lambda_i t} t^{s-1} dt \tag{3}$$

Suppose that the zero eigenvalue $\lambda_1 = 0$ has multiplicity C. Removing this eigenvalue from the sum under the integral, we have $\int_0^\infty Tr[h_t] t^{s-1} dt = \int_0^\infty [C + \sum_{\lambda_i \neq 0} e^{-\lambda_i t}] t^{s-1} dt$. To develop this expression for the heat kernel trace, we note that the gamma function $\Gamma(s)$ is defined to be $\Gamma(s) = \int_0^\infty t^{s-1} e^{-t} dt$. As a result $\int_0^\infty t^{s-1} e^{-\lambda_i t} dt = \frac{\Gamma(s)}{(-\lambda_i)^s}$. Hence, we can write $\int_0^\infty (Tr[h_t] - C) t^{s-1} dt = \sum_{\lambda_i \neq 0} \frac{\Gamma(s)}{(-\lambda_i)^s}$. So finally, the zeta-function is given by $\zeta(s) = \sum_{\lambda_i \neq 0} (-\lambda_i)^{-s} = \frac{1}{\Gamma(s)} \int_0^\infty (Tr[h_t] - C) t^{s-1} dt$. In other words, the zeta function is related to the Mellin moments of the heat kernel trace. It is hence a way to characterizing the shape of the heat kernel trace with time. In our experiments, we take zeta functions with different integer input to construct the feature vector.

3.2 Heat Content Invariants

We now turn our attention to heat content invariants. In a recent paper McDonald and Meyers [6] have described a set of differential invariants that can be derived from the heat content of the heat kernel. The heat content is the sum of the entries of the heat kernel over the nodes of the graph and is given by

$$Q(t) = \sum_{u \in V} \sum_{v \in V} \sum_{k=1}^{|V|} \exp[-\lambda_k t] \phi_k(u) \phi_k(v) \tag{4}$$

McDonald and Meyers [6] have shown that the heat content can be expanded as a polynomial in time, i.e. $Q(t) = \sum_{m=0}^{\infty} q_m t^m$. We expand $\exp[-\lambda_k t]$ by using MacLaurin series and substitute it with Equation 4, we have $Q(t) = \sum_{m=0}^{\infty} \sum_{u \in V} \sum_{v \in V} \sum_{k=1}^{|V|} \phi_k(u) \phi_k(v) \frac{(-\lambda_k)^m t^m}{m!}$. As a result the coefficients q_m appearing in the polynomial expansion of $Q(t)$ are given by

$$q_m = \sum_{k=1}^{|V|} \Big\{ (\sum_{u \in V} \phi_k(u))^2 \Big\} \frac{(-\lambda_k)^m}{m!} \tag{5}$$

As a result, the time-series coefficient q_m does not depend on the permutation order of the nodes of the graph.

The time-series coefficients q_m are a set of invariants which can be used for graph characterization. Since they depend on both eigenvalues and eigenvectors, they are more likely to be unique than invariants purely based in eigenvalues which are likely to be affected by problems of co-spectrality [2]. We will explore the use of the time-series polynomial coefficients for the purpose of graph clustering. To do this we construct a feature vector $\vec{B} = (q_1, q_2, ..., q_k)^T$ from the k leading coefficients of the heat content polynomial.

3.3 Unary Attributes with Symmetric Polynomials

In this part, we illustrate the relationship between the derivative of the zeta function at the origin and the symmetric polynomial invariants of Wilson, Hancock and Luo [10].

In their recent paper Wilson, Hancock and Luo [10] have reported a family of permutation invariants that can be computed by applying symmetric polynomialsS_r, to the elements from the graph spectral matrix $\sqrt{\Lambda}\Phi$.

There are of course alternative invariants that can be computed from the spectrum of the Laplacian. The most obvious of these is the sum of eigenvalues or the trace of the normalized Laplacian eigenvalue matrix of the graph. Then if we use the eigenvalues of the normalized Laplacian matrix as input for the symmetric polynomial S_r, the first symmetric polynomial is also the trace of the normalized Laplacian and the last, which is $S_1(\lambda_1, \lambda_2, ..., \lambda_n) = \prod_{i=1}^{n} \lambda_i$, is related to the derivative of the zeta function at the origin which is $\zeta'(0) = \sum_{\lambda_i \neq 0} \{-\ln \lambda_i\} = \ln \Big\{ \prod_{\lambda_i \neq 0} (\frac{1}{\lambda_i}) \Big\}$. Moreover, in the paper [10], Wilson, Hancock and Luo found it necessary to perform logarithmic squashing of the polynomials to bring them into a convenient dynamic range. This property arises naturally through the analysis of the zeta function.

In our experiments, we will use the symmetric polynomial to construct unary attributes for graph characterization. The input for the symmetric polynomial are the columns of the spectral matrix and the eigenvalues of the normalized Laplacian matrix.

4 Experiment

In this section, we will apply our methods to synthetic and real world image databases. We first use synthetic data to quantitatively investigate heat kernel invariants behaviors. This study is divided into two parts. We commence by comparing the three new methods as ways for graph characterization, and then evaluate the stability of them. In the second experiment, we will apply our methods to shape analysis. Here we use a shock tree database which composed of 150 silhouettes of 10 different types of object [9].

4.1 Graph Characterization

In the first experiment, we test our methods for graph characterization on synthetic graphs. To this end, we generate five different seed graphs with different structure, i.e. number of nodes and number of edges. We then apply random edit operations to the five seed graphs to simulate the effects of noise. The edit operations are edge deletion, edge insertion and node deletion. These operations are used to generate five structurally related graph clusters, which we would hope to separate using our methods. In Figure 1, we show the original seed graphs. For each seed graph, we randomly perform 3 to 10 edit operations to obtain 70 graph variants.

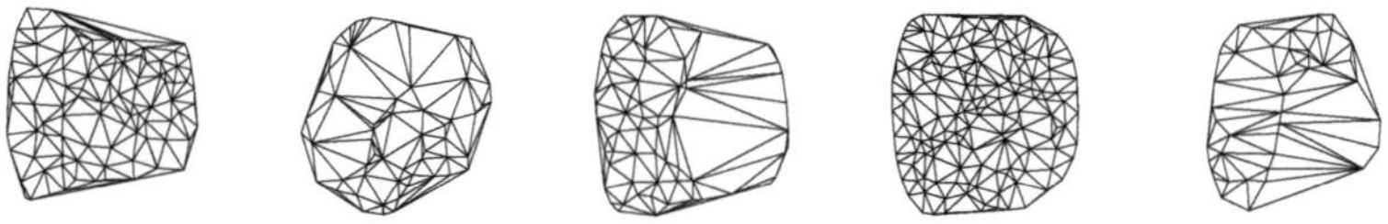

Fig. 1. Example of the seed graphs

We commence by illustrating the behavior of the zeta function. ¿From left-to-right and top-to-bottom in Figure 2 we show the values of $\zeta(1)$, $\zeta(2)$, $\zeta(3)$ and $\zeta(4)$ for the 70 randomizations of each seed graph. In each plot the different curves are for the different seed graphs, and the x-axis is the index number of the randomization. The main feature to note is that the curves for the different graph clusters are well separated, and that the individual values of the zeta function do not vary significantly with randomization index. When $s = 1$ and $s = 2$, the fluctuations of curves $\zeta(s)$ are generally small. However, the fluctuation is larger when $s = 4$. This is because when s becomes large the small eigenvalues will play a dominant role in the ζ function.

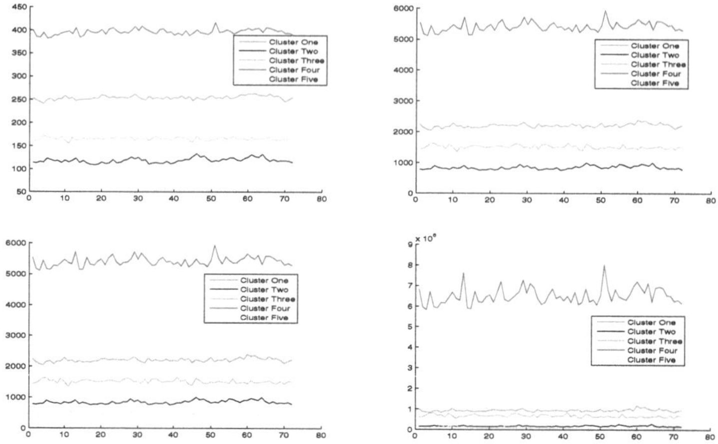

Fig. 2. Zeta function $\zeta(s)$ for randomizations of the five seed graphs. From left to right, and top to bottom, $s = 1, 2, 3$ and 4 respectively.

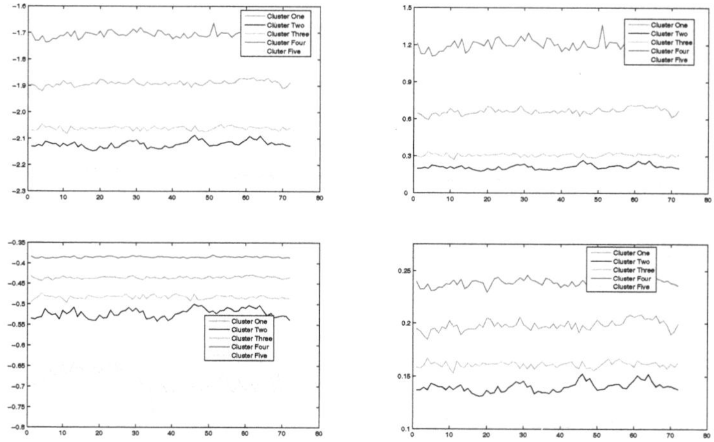

Fig. 3. Individual heat content invariants q_n for five cluster random graphs(from left to right, and top to bottom, $n = 1, 2, 3$ and 4 respectively).

In Figure 3, we plot the four leading heat content invariants coefficients q_1, q_2, q_3 and q_4 separately as a function of the randomization index for the five seed graphs. The coefficients are relatively stable under randomized edit operations. The curves are well separated for the different objects, and the fluctuations are relatively small.

We turn our attention to the unary features from symmetric polynomials. In the left of Figure 4, we plot the mean values of the symmetric polynomials for the five graph clusters, as a function of polynomial order. Here the arguments of the polynomials are the elements of the spectral matrix $\sqrt{\Lambda}\Phi$. The different curves in

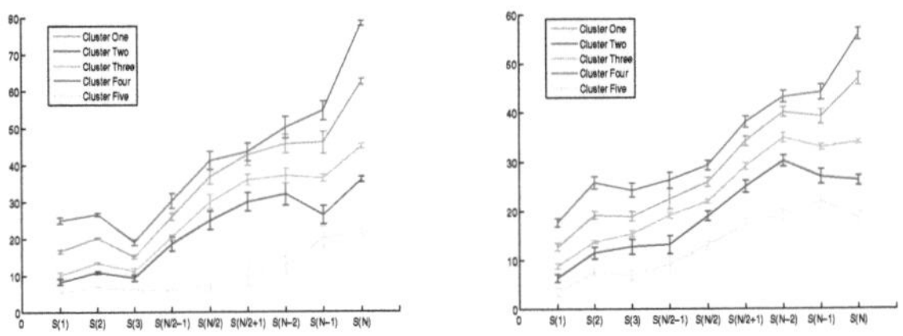

Fig. 4. Results for symmetric polynomials using spectral matrix elements for five cluster random graphs(left) and symmetric polynomials using normalized Laplacian eigenvalues(left) for five cluster random graphs

the plot are for the different seed graphs. We select 9 different symmetric polynomials for investigation. These are the first three symmetric polynomials S_1, S_2, S_3, the middle three symmetric polynomials $S_{fix(\frac{N}{2})-1}, S_{fix(\frac{N}{2})}, S_{fix(\frac{N}{2})+1}$ and last three symmetric polynomials S_{N-2}, S_{N-1}, S_N, where N is the number of node in the graph. The error bars show the standard deviation over the different sets of graphs of the same cluster. The right plot of Figure 4 repeats the analysis using the normalized Laplacian spectrum as the arguments of the polynomials. It is clear that the first three and last three symmetric polynomials computed using the normalized Laplacian eigenvalues and spectral matrix elements as inputs can be used for graph clustering. We also get a better separation of objects when using the polynomials from the normalized Laplacian spectrum rather than the elements of the spectral matrix.

4.2 Stability Evaluation

We now evaluate the stability of different graph characterization methods. We study five different methods. These are a) heat content invariants, b) the zeta function values, c) symmetric polynomials with Laplacian eigenvalues as arguments, d) symmetric polynomials with spectral matrix elements as arguments and e) the Laplacian spectrum. We generate a random graph with thirty five vertices and one hundred and twenty eight edges. We compute the Euclidean distance between the coordinate vector from the principal component analysis (PCA) projection of the graph feature vector and its modified variant obtained by deleting between 3 and 27 edges from the random graph. The edge to be deleted is chosen at random. For each deletion, we perform 30 trails in order to obtain the average distance together with its standard deviation.

In Figure 5, the x-axis in each plot shows the number of deleted edges, while the y-axis shows the mean Euclidean distance between the original random graph and the modified variant graph together with the standard deviation. These plots show that the five characterizations we proposed in this chapter have an approximately linear relationship between the Euclidean distance and the number of

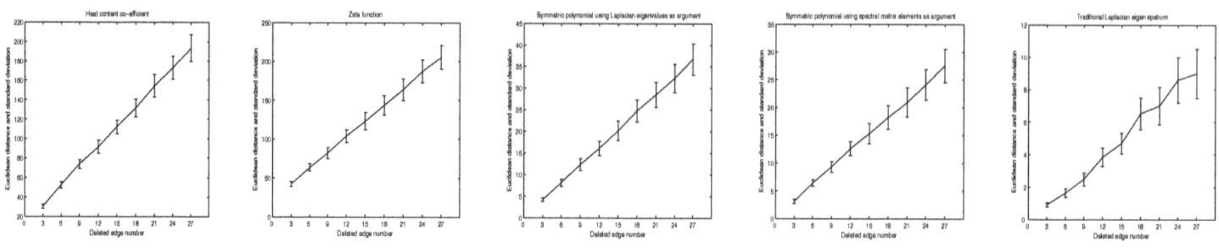

Fig. 5. Euclidean distance and standard deviation of the random graph experiments

deleted edges. The traditional characterization based on the Laplacian spectrum seems marginally less linear.

4.3 Applications on Shape Analysis

We have also explored the application of our new graph characterization methods to shape analysis. Here graph based methods have provided a powerful means of characterizing the underlying variations in shape due to deformation, articulation or natural class variability [7]. We have applied our methods to shock trees. This is a tree structure based representation of the differential structure of the boundary of a 2D shape. It is obtained by locating the shape skeleton, and examining the differential behavior of the radius of the bitangent circle from the skeleton to the object boundary, as the skeleton is traversed [8]. Recently, the shock tree extraction process has been further improved by Torsello and Hancock[9].

In this paper we use these methods in[9] to extract a tree representations from the silhouettes of 2D shape. A representative view of each object studied is shown in Figure 6.

In the database, there are over 15 shapes from 10 classes. For each shape we can construct a tree representation of the skeleton. We compute the different invariants described earlier. A GMM(Gaussian Mixture Model)[3] supervised learning process is applied to the feature vectors constructed from the invariants for the purposes of classification. In Figure 7, we plot the proportion of shapes from the database correctly classified as the number of shape classes is increased.

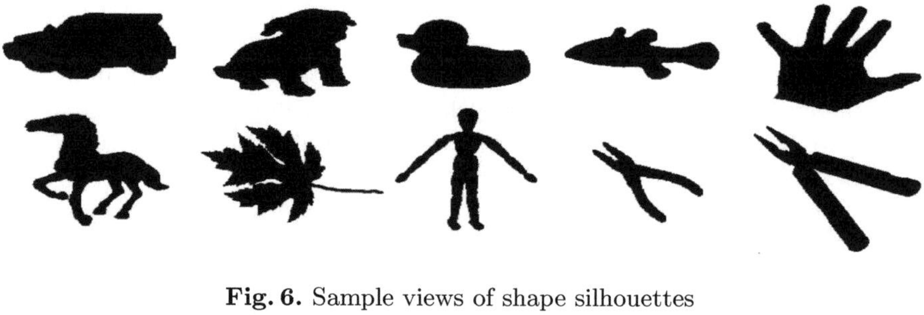

Fig. 6. Sample views of shape silhouettes

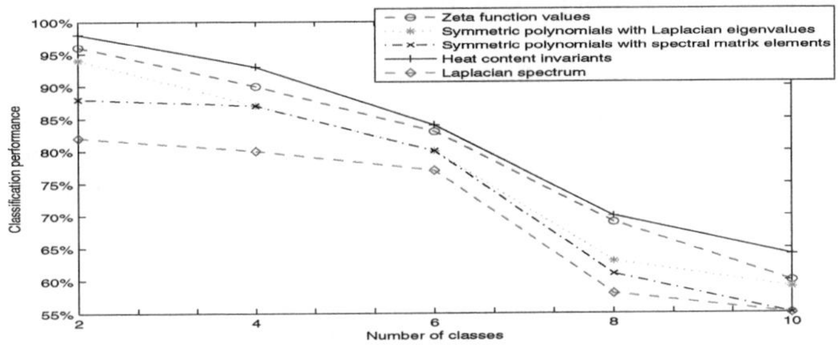

Fig. 7. Shape silhouettes classification performance as a function of silhouette numbers

In this experiment, we choose 5 shapes from each class as training data to generate the object classifier. With more classes been inserted, the classifier contains more trained Gaussian classes. All five methods explored decreases in classification accuracy as the size of the database increases. The best performance is given by the heat-content invariants, closely followed by the feature vector of zeta-function values. The poorest performance is given by the spectrum of Laplacian eigenvalues.

5 Conclusion

In this paper we have explored the use of several invariants computed from the heat kernel to characterize graphs. We have experimented with each of the newly proposed graph characterizations on synthetic and real world databases. The new methods show advantages over traditional methods in our experiments.

There are a number of ways in which the work reported in this paper can be extended. First, we aim to explore in more depth the links with spectral geometry, and the use of differential invariants computed from the spectrum of the Laplace-Beltrami operator [4]. One interesting route here may be to explore the use of the Gauss-Bonnet theorem or Atiyah-Singer index theorem to compute topological invariants. A second line of investigation will be to explore in more depth the link between the heat kernel invariants and symmetric polynomials. Finally, we intend to explore different application domains. For instance, we will explore whether the invariants described in this paper can be used to characterize molecular structures and small world networks.

References

1. Atkins, J.E., Bowman, E.G., Hendrickson, B.: A spectral algorithm for seriation and the consecutive ones problems. SIAM J. Comput. 28, 297–310 (1998)
2. Chung, F.R.K.: Spectral graph theory. American Mathematical Society (1997)
3. Figueiredo, M.A.F., Jain, A.K.: Unsupervised learning of finite mixture models. IEEE Trans. PAMI 24(3), 381–396 (2002)
4. Chaval, I., Farkas, H.M.: Extremal kahler metrics ii. Differential Geometry and Complex Analysis, 95–114 (1985)

5. Shi, J., Malik, J.: Normalized cuts and image segmentation. IEEE Trans. PAMI 22, 888–905 (2000)
6. McDonald, P., Meyers, R.: Diffusions on graphs, poisson problems and spectral geometry. Transactions on American Mathematical Society 354, 5111–5136 (2002)
7. Mori, G., Belongie, S.J., Malik, J.: Efficient shape matching using shape contexts. IEEE Trans. Pattern Anal. Mach. Intell. 27(11), 1832–1837 (2005)
8. Sebastian, T.B., Klein, P.N., Kimia, B.B.: Recognition of shapes by editing their shock graphs. IEEE Trans. Pattern Anal. Mach. Intell. 26(5), 550–571 (2004)
9. Torsello, A., Robles-Kelly, A., Hancock, E.R.: Discovering shape classes using tree edit-distance and pairwise clustering. International Journal on Computer Vision 72(3), 259–285 (2007)
10. Wilson, R.C., Luo, B., Hancock, E.R.: Pattern vectors from algebraic graph theory. IEEE Trans. PAMI 27, 2220–2237 (2003)
11. Xiao, B., Hancock, E.R.: Clustering shapes using heat content invariants. In: ICIP (1), pp. 1169–1172 (2005)
12. Xiao, B., Hancock, E.R.: Trace formula analysis of graphs. In: SSPR/SPR, pp. 306–313 (2006)
13. Yau, S.T., Schoen, R.M.: Differential geometry. Science Publication (1988)

Hierarchical Bag of Paths for Kernel Based Shape Classification

François-Xavier Dupé* and Luc Brun

GREYC UMR CNRS 6072,
ENSICAEN-Université de Caen Basse-Normandie,
14050 Caen France
{francois-xavier.dupe,luc.brun}@greyc.ensicaen.fr

Abstract. Graph kernels methods are based on an implicit embedding of graphs within a vector space of large dimension. This implicit embedding allows to apply to graphs methods which where until recently solely reserved to numerical data. Within the shape classification framework, graphs are often produced by a skeletonization step which is sensitive to noise. We propose in this paper to integrate the robustness to structural noise by using a kernel based on a bag of path where each path is associated to a hierarchy encoding successive simplifications of the path. Several experiments prove the robustness and the flexibility of our approach compared to alternative shape classification methods.

Keywords: Shape, Skeleton, Support Vector Machine, Graph Kernel.

1 Introduction

The skeleton of a $2D$ shape is defined as the location of the singularities of the signed distance function to the border of the shape. This structure has several interesting properties: it is thin, homotopic to the shape, invariant under rigid transformations of the plane and most importantly it has a natural interpretation as a graph. The representation of a shape by a skeletal (or shock) graph has become popular owing the good properties of this representation inherited from the properties of the skeleton. However, beside all this good properties, the skeletonization is not continuous and small perturbations of the boundary insert structural noise within the graph encoding the shape.

Several graph based methods have been proposed to compute a distance between shapes robust to such a structural noise. Sharvit et al. [1] propose a graph matching method based on a graduated assignment algorithm. Siddiqi [2] proposes to transform the shock graph into a tree and then applies a tree matching algorithm. Pellilo [3] uses the same tree representation but transforms the tree matching problem into a maximal clique problem within a specific association graph.

* This work is performed in close collaboration with the laboratory Cycéron and is supported by the CNRS and the région Basse-Normandie.

N. da Vitora Lobo et al. (Eds.): SSPR&SPR 2008, LNCS 5342, pp. 227–236, 2008.

All the above graph methods operate directly on the space of graphs which contains almost no mathematical structure. This lack of mathematical structure forbids the use of basic statistical tools such as the mean or the variance. Graph kernels provide an elegant solution to this problem. Using appropriate kernels, graphs can be mapped either explicitly or implicitly into a vector space whose dot product corresponds to the kernel function. All the "natural" operations on a set of graphs which were not defined in the original graph space are now possible into this transformed vector space. In particular, graph kernels may be combined with the kernelised version of robust classification algorithms such as the Support Vector Machine (SVM).

A Graph kernel used within the shape representation framework should take into account the structural noise induced by the skeletonization process. Bunke [4] proposes to combine edit distance and graph kernels by using a set of n prototype graphs $\{g_1, \ldots, g_n\}$. Given a graph edit distance $d(.,.)$, Bunke associates to each graph g the vector $\phi(g) = (d(g, g_1) \ldots, d(g, g_n))$. The kernel $k(g_1, g_2)$ between the two graphs g_1 and g_2 is then defined as the dot product $< \phi(g_1), \phi(g_2) >$.

Neuhaus [5] proposes a similar idea by defining for a prototype graph g_0, the kernel: $k_{g_0}(g, g') = \frac{1}{2}\left(d^2(g, g_0) + d^2(g_0, g') - d^2(g, g')\right)$, where $d(.,.)$ denotes the graph edit distance. Several graph prototypes may be incorporated by summing or multiplying such kernels. Using both Neuhaus [5] and Bunke [4] kernels two close graphs should have close edit distance to the different graph prototypes. The metric induced by such graph kernels is thus relative both to the weights used to define the edit distance and to the graph prototypes. This explicit use of prototype graphs may appear as artificial in some application. Moreover, the definite positive property of these kernels may not in general be guaranteed.

Suard [6] proposes to use the notion of bag of paths of finite length for shape matching. This method associates to each graph all its paths whose length is lower than a given threshold. The basic idea of this approach is that two close shapes should share a large amount of paths. A kernel between these sets should thus reflect this proximity. However, small perturbations may drastically reduce the number of common paths between two shapes (Section 3.2). Moreover, the straightforward definition of a kernel between set of paths does not lead to a definite positive kernel (Section 2.1).

This paper proposes a new definite positive kernel between set of paths which takes into account the structural noise induced by the skeletonization process. We first present in Section 2 the bag of paths approach for shape similarity. Our contributions to this field are then presented in Section 3. The effectiveness of our method is demonstrated through experiments in Section 4.

2 Kernels on Bag of Paths

Let us consider a graph $G = (V, E)$ where V denotes the set of vertices and $E \subset V \times V$ the set of edges. As mentioned in Section 1 a bag of paths P of length s associated to G contains all the paths of G of length lower than s. We denote by $|P|$ the number of paths inside P. Let us denote by K_{path} a generic kernel between paths. Given two graphs G_1 and G_2 and two paths $h_1 \in P_1$ and $h_2 \in P_2$ of respectively G_1 and G_2, $K_{path}(h_1, h_2)$ may be interpreted as a measure of similarity between h_1 and h_2 and thus as a local measure of similarity between these two graphs. The aim of a kernel between bags of paths consits to agregate all these local measures between pairs of paths into a global similarity measure between the two graphs.

2.1 The Max Kernel

This first method, proposed by Suard [6], uses the kernel K_{path} as a measure of similarity and computes for each path $h_1 \in P_1$ the similarity with its closest path in $P_2(\max_{h_j \in P_2} K_{path}(h_1, h_j))$. A first global measure of similarity between P_1 and P_2 is then defined as:

$$\hat{K}_{max}(G_1, G_2) = \hat{K}_{max}(P_1, P_2) = \frac{1}{|P_1|} \sum_{h_i \in P_1} \max_{h_j \in P_2} K_{path}(h_i, h_j). \tag{1}$$

The function $\hat{K}_{max}(G_1, G_2)$ is however not symmetric according to G_1 and G_2. Suard obtains a symmetric function interpreted as a graph kernel by taking the mean of $\hat{K}_{max}(G_1, G_2)$ and $\hat{K}_{max}(G_2, G_1)$:

$$K_{max}(G_1, G_2) = \tfrac{1}{2} \left[\hat{K}_{max}(G_1, G_2) + \hat{K}_{max}(G_2, G_1) \right]. \tag{2}$$

This kernel is not positive definite in general. However as shown by Haasdonk [7], SVM with indefinite kernels have in some cases a geometrical interpretation as the maximization of distances between convex hulls. Moreover, experiments (section 4, and [6]) show that this kernel usually leads to valuable results.

2.2 The Matching Kernel

The non definite positiveness of the kernel K_{max} is mainly due to the max operator. Suard [6] proposes to replace the kernel K_{path} by a kernel which decreases abruptly when the two paths are different. The resulting kernel is defined as:

$$K_{matching}(G_1, G_2) = K_{matching}(P_1, P_2) = \frac{1}{|P_1|} \frac{1}{|P_2|} \sum_{h_i \in P_1} \sum_{h_j \in P_2} \exp\left(\frac{-d^2_{path}(h_i, h_j)}{2\sigma^2} \right). \tag{3}$$

where d_{path} is the distance associated to the kernel K_{path} and defined by: $d^2_{path}(h_1, h_2) = K_{path}(h_1, h_1) + K_{path}(h_2, h_2) - 2K_{path}(h_1, h_2)$.

The resulting function defines a definite positive kernel. This kernel relies on the assumption that using a small value of σ, the couple of paths with the smallest distance will predominate the others in equation 3. This kernel may thus lead to erroneous results if the distance are of the same order of magnitude than σ or if several couples of paths have nearly similar distances.

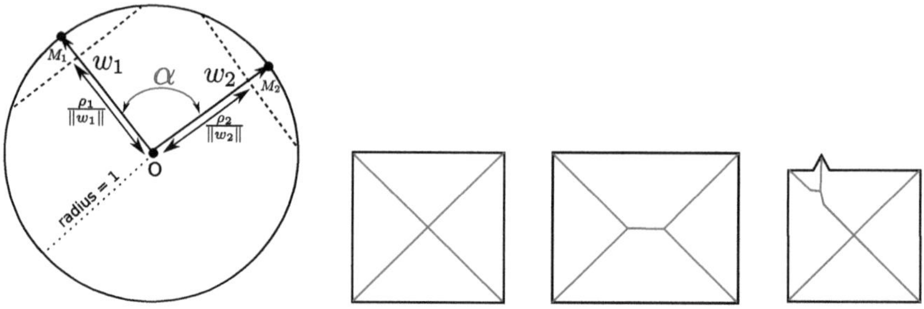

(a) Sets on the unit sphere (b) original (c) edge protrusion (d) node insertion

Fig. 1. Separating two sets using one-class SVM (a). The symbols (w_1, ρ_1) and (w_2, ρ_2) denote the parameters of the two hyperplanes which are represented by dashed lines. Influence of small perturbations on the bag of paths ((b), (c) and (d)).

2.3 The Change Detection Kernel

Desobry [8] proposed a general approach for the comparison of two sets which has straightforward applications in the design of a kernel between bags (sets) of paths. Desobry models the two sets as the observation of two sets of random variables in a feature space and proposes to estimate a distance between the two distributions without explicitly building the pdf of the two sets.

The feature space considered by Desobry is based on the normalised kernel $(K(h, h') = K_{path}(h, h')/\sqrt{(K_{path}(h, h)K_{path}(h', h'))})$. Using such a kernel we have $\|h\|_K^2 = K(h, h) = 1$ for any path. The image in the feature space of our set of paths lies thus on an hypersphere of radius 1 centered at the origin (Fig. 1). Desobry defines a region on this sphere by using a single class ν-SVM. This region corresponds to the density support estimate of the unknown pdf of the set of paths [8].

Using Desobry's method, two set of vectors are thus map onto two regions of the unit sphere and the distance between the two regions corresponds to a distance between the two sets. Several kernels based on this mapping have been proposed:

1. Desobry proposed [8] to define the distance between the two spherical arcs as a contrast measure defined by: $d^2_{Desobry}(P_1, P_2) = \frac{\arccos\left(\frac{w_1 K_{1,2} w_2}{\|w_1\|\|w_2\|}\right)}{\arccos\left(\frac{\rho_1}{\|w_1\|}\right) + \arccos\left(\frac{\rho_2}{\|w_2\|}\right)}$.
This distance is connected to the Fisher ratio (see [8, Section IV]). However, the definite positiveness of the Gaussian RBF kernel based on this distance remains to be shown.
2. Suard [6] proposed the following kernel: $K_{Suard}(G_1, G_2) = K_{Suard}(P_1, P_2) = \rho_1\rho_2 \sum_{h_i \in P_1} \sum_{h_j \in P_2} \alpha_{1,i}\ K_{path}(h_i, h_j)\alpha_{2,j}$ with $w_1 = (\alpha_{1,1}, \ldots, \alpha_{1,|P_1|})$ and $w_2 = (\alpha_{2,1}, \ldots, \alpha_{2,|P_2|})$.
This kernel is definite positive, but does not correspond to any straightforward geometric interpretation.

2.4 Path Kernel

All the kernels between bags of paths defined in Section 2 are based on a generic kernel K_{path} between paths. A kernel between two paths $h_1 = (v_1, \ldots, v_n)$ and $h' = (v'_1, \ldots, v'_p)$ is classically [9] built by considering each path as a sequence of nodes and a sequence of edges. This kernel denoted $K_{classic}$ is then defined as 0 if both paths have not the same size and as follows otherwise:

$$K_{classic}(h, h') = K_v(\varphi(v_1), \varphi(v'_1)) \prod_{i=2}^{|h|} K_e(\psi(e_{v_{i-1}v_i}), \psi(e_{v'_{i-1}v'_i})) K_v(\varphi(v_i), \varphi(v'_i)) \tag{4}$$

where $\varphi(v)$ and $\psi(e)$ denote respectively the vectors of features associated to the node v and the edge e. The terms K_v and K_e denote two kernels between respectively nodes and edge's features. For the sake of simplicity, we have used Gaussian RBF kernels between the attributes of nodes and edges (Section 4).

3 Hierarchical Kernels

Since the main focus of this paper is a new kernel method for shape classification, the construction of skeletal graphs from shapes has been adressed using classical methods. We first build a skeleton using the method proposed by Siddiqi [2]. However the graph we build from the skeleton does not correspond to the shock graph proposed by Siddiqi. Indeed, this graph provides a precise description of the shape but remains sensitive to small perturbations of the boundary. We rather use the construction scheme proposed by Suard [6] and Ruberto [10] which consists to select as node all the pixels of the skeleton which correspond to end points or junctions. These nodes are then connected by edges, each edge being associated to one branch of the skeleton. Given a skeletal graph G we valuate each of its edge by an additive weight measure and we consider the maximal spanning tree T of G. The bag of path associated to G is built on the tree T. Note that, the skeletonization being homotopic we have $G = T$ if the $2D$ shape does not contain any hole.

3.1 Bag of Path Kernel

None of the bag of path kernels proposed by Desobry or Suard (Section 2) is both definite positive and provides a clear geometrical interpretation. We thus propose a new kernel based on the following distance:

$$d^2_{change}(P_1, P_2) = \arccos\left(\frac{w_1 K_{1,2} w_2}{\|w_1\| \, \|w_2\|}\right). \tag{5}$$

This distance corresponds to the angle α between the two mean vectors w_1 and w_2 of each region (Fig. 1). Such an angle may be interpreted as the geodesic distance between two points on the sphere and has thus a clear geometrical interpretation. Based on this distance we use the Gaussian RBF kernel:

$$K_{change}(G_1, G_2) = K_{change}(P_1, P_2) = \exp\left(\frac{-d^2_{change}(P_1, P_2)}{2\sigma^2}\right). \quad (6)$$

This kernel is definite positive since the normalized scalar product is positive definite and arccos is bijective on $[0, 1]$. The Gaussian RBF kernel based on this distance is thus definite positive (see [11] for further details).

3.2 Hierarchical Kernel between Paths

A mentioned in Section 1, the use of kernels between bags of paths within the shape matching framework relies on the assumption that the graphs associated to two similar shapes share a large amount of similar paths. This assumption is partially false since a small amount of structural noise may have important consequences on the set of paths. Let us for example, consider the small deformation of the square (Fig. 1(b)) represented on Fig. 1(c). This small deformation transforms the central node in Fig. 1(b) into an edge (Fig. 1(c)). Consequently graphs associated to these two shapes only share two paths of length 2 (the ones which connect the two corners on the left and right sides). In the same way, a small perturbation of the boundary of the shape may add branches to the skeleton(Fig. 1(d)). Such additional branches i) split existing edges into two sub edges by adding a node and ii) increase the size of the bag of path either by adding new paths or by adding edges within existing paths.

The influence of small perturbations of the shape onto an existing set of paths may thus be modeled by node and edge insertions along these paths. In order to get a path kernel robust against structural noise we associate to each path a sequence of successively reduced paths, thus forming a hierarchy of paths. Our implicit assumption is that, if a path has been elongated by structural noise one of its reduced version should corresponds to the original path.

The reduction of a path is performed either by node removal or edge contraction along the path. Such a set of reduction operations is compatible with the taxinomy of topological transition of the skeleton compiled by Giblin and Kimia [12]. Note that, since all vertices have a degree lower than 2 along the path these operations are well defined. In order to select both the type of operation and the node or the edge to respectively remove or contract we have to associate a weight to each node and edge which reflects its importance according to the considered path and the whole graph.

Let us consider a skeletal graph G, its associated maximal spanning tree T and a path $h = (v_1, \ldots, v_n)$ within T. We valuate each operation on h as follows:

Node removal: Let us denote by v_i, $i \in \{2, \ldots, n-1\}$ the removed node of the path h. The node v_i has a degree greater than 2 in T by construction. Our basic idea consists to valuate the importance of v_i by the total weight of the additional branches which justify its existence within the path h. For each neighbor v of v_i not equal to v_{i-1} nor v_{i+1} we compute the weight $W(v)$

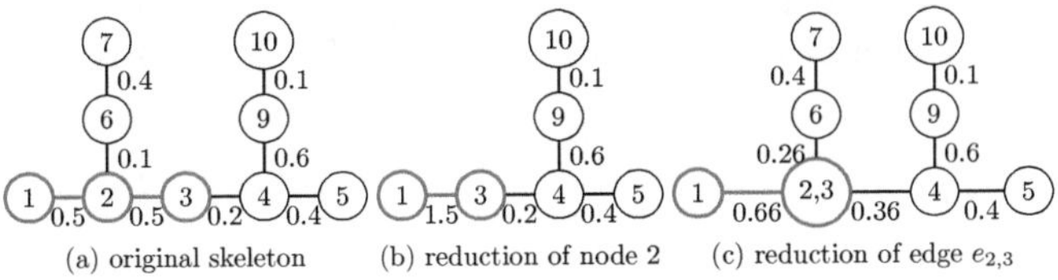

Fig. 2. Different reductions of a path (in gray) within a skeletal tree

defined as the addition of the weight of the tree rooted on v in $T - \{e_{v_i v}\}$ and the weight of $e_{v_i v}$. This tree is unique since T is a tree. The weight of the node v_i (and the cost of its removal) is then defined as the sum of weight $W(v)$ for all neighbors v of v_i (excluding v_{i-1} and v_i).

After the removal of this node the edges $e_{v_{i-1} v_i}$ and $e_{v_i v_{i+1}}$ are concatenated into a single edge in the new path h'. The weight of this new edge is defined as the sum of the weight of the edges $e_{v_{i-1} v_i}$, $e_{v_i v_{i+1}}$ and the weight of the node v_i (Fig. 2(a) and (b)).

Edge contraction: The cost of an edge contraction is measured by the relevance of the edge which is encoded by its weight. Let us denote by $e_{v_i v_{i+1}}$, $i < n$ the contracted edge of the path $h = (v_1, \ldots, v_n)$. In order to preserve the total weight of the tree after the contraction, the weight of the edge $e_{v_i v_{i+1}}$ is equally distributed among the edges of T incident to v_i and v_{i+1}:

$$\forall e \in \iota(v_i) \cup \iota(v_{i+1}) - \{e_{v_i v_{i+1}}\} \quad w'(e) = w(e) + \frac{w(e_{v_i v_{i+1}})}{d(v_i) + d(v_{i+1}) - 2}$$

where $\iota(v)$ and $d(v)$ denote respectively the set of edges incident to v and the cardinal of this set (the vertex's degree). The symbol $w(e)$ denotes the weight of the edge e.

For example, the contraction of the edge $e_{2,3}$ in Fig. 2(a) corresponds to a cost of $w(e_{2,3}) = .5$. The contraction of this edge induces the incrementation of the edge's weights $w(e_{2,1})$,$w(e_{2,6})$,$w(e_{3,4})$ by $.5/3 \approx .16$

Any additive measure encoding the relevance of a branch of the skeleton may be used as a weight. We choose to use the measure defined by Torsello [13] which associates to each branch of the skeleton (an thus to each edge) the length of the boundaries which contributed to the creation of this branch. Such a measure initially defined for each pixel of the skeleton is trivially additive.

Let us denote by κ the function which applies the cheapest operation on a path. The successive applications of the function κ associate to each path h a sequence of reduced paths $(h, \kappa(h), \ldots, \kappa^D(h))$ where D denotes the maximal number of reductions. Using $K_{classic}$ for the path comparison, we define the kernel K_{edit} as the mean value of kernels between reduced paths of equal length. Given two paths h and h', this kernel is thus equal to 0 if $||h| - |h'|| > D$. Indeed, in this case the maximal reduction of the longuest path remains longuer than the shortest one. Otherwise, $||h| - |h'|| \leq D$, and $K_{edit}(h, h')$ is defined as:

$$K_{edit}(h, h') = \frac{1}{D+1} \sum_{k=0}^{D} \sum_{l=0}^{D} K_{classic}(\kappa^k(h), \kappa^l(h')) \tag{7}$$

This kernel is proportional (by a factor $D + 1$) to a sum of R-convolution kernels [14, Lemma 1] and is thus definite positive.

Since $K_{classic}$ is equal to 0 for paths of different lengths, K_{edit} is indeed equal to a sum of kernels between reduced paths of equal length. For example, given two paths h and h' whose respective length is equal to 4 and 3 we have for $D = 2$:

$$K_{edit}(h, h') = \frac{1}{3} \left[K_{classic}(\kappa(h), h') + K_{classic}(\kappa^2(h), \kappa(h')) \right]$$

4 Experiments

We used the following features for our experiments: Each node is weighted by its distance to the gravity center of the shape and each edge is assocated to a vector of two features: The first feature corresponds to the edge's weight (section 3.2). The second feature is the angle between the straight line passing through the two nodes of the edge and the principal axis of the shape. These experiments are based on the LEMS [15] database which consists of 99 objects divided into 9 classes.

We defined three kernels for these experiments: The kernel $K_{max,classic}$ based on a conjoint use of the kernels K_{max} (equation 2) and $K_{classic}$ (equation 4) has been introduced by Suard [6]. The kernel $K_{change,classic}$ based on a conjoint use of the kernels K_{change} (equation 6) and $K_{classic}$ allows to evaluate the performances of the kernel K_{change} compared to the kernel K_{max}. Finally, the kernel K_{new} is based on a conjoint used of the two kernels K_{edit} and K_{change} proposed in this paper. The kernel $K_{classic}$ is defined by the two parameters σ_{edge} and σ_{vertex} respectively used by the Gaussian RBF kernels on edges and vertices. The kernel $K_{max,classic}$ does not require additional parameters while $K_{change,classic}$ is based on a ν-SVM and requires thus the parameter ν. It additionally requires the parameter $\sigma_{change}^{classic}$ used by the RBF kernel in equation 6. The kernel K_{edit} requires the two parameters σ_{edge} and σ_{vertex} used by $K_{classic}$ together with the maximal number of edition (D). Finally, the kernel K_{new} requires as K_{change} the two additional parameters ν and σ_{change}^{new}(equation 6). These parameters have been fixed to the following values in the experiments described below: $D = 2$, $\sigma_{edge} = \sigma_{vertex} = 0.1$, $\nu = 0.9$, $\sigma_{change}^{new} = 0.3$ and $\sigma_{change}^{classic} = 1.0$. The parameters σ_{edge} and σ_{vertex} are common to all kernels. The remaining parameters have been been set in order to maximize the performances of each kernel on the experiments below.

Our first experiment compares the distance induced by each kernel k and defined as $d^2(x, x') = k(x, x) + k(x', x') - 2k(x, x')$. The mean number of matches for each class is defined as follows: For each shape of the selected class we sort

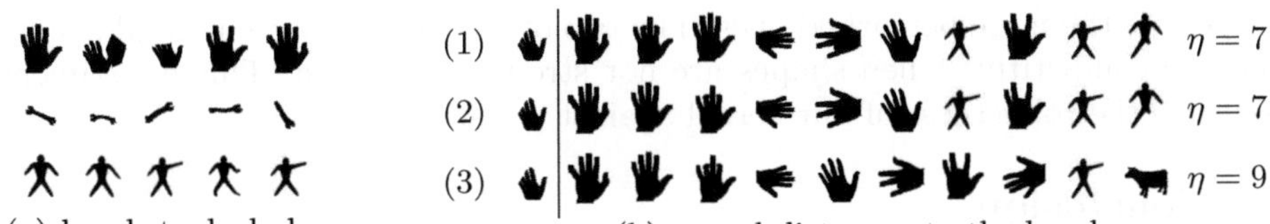

(a) hands,tools,dudes

(b) sorted distances to the hand

Fig. 3. Five representative shapes of the classes hands, tools and dudes of the LEMS database (a), and (b) the 10 closest shapes from an hand using the distances induced by the kernels $K_{max,classic}$ (1), $K_{change,classic}$ (2) and K_{new} (3)

all the shapes of the database according to their distances to the selected shape using an ascending order. The number of good matches of the input shape is then defined as the number of shapes ranked before the first shape which belongs to a different class than the selected one. For example, the 10 nearest neighbors of a hand sorted in an ascending order are represented in Fig. 3(b), the number of good matches of each shape is indicated on the right of the figure. Note that the greater number of good match being obtained for the kernel K_{new}. The mean number of good matches of a class is defined as the mean value of the number of good matches for each shape of the class. The different values represented in Tab. 1(a) represent the mean values of these number of good matches for the classes: hands, tools and dudes (Fig. 3(a)). As indicated by Tab. 1(a), the kernel $K_{max,classic}$ provides stable results but is sensitive to slight perturbations of the shapes as the ones of the class dudes and cannot handle the severe modifications of the hands. The kernel $K_{change,classic}$ leads to roughly similar results on the different classes. Though not presented here, the kernel $K_{matching,classic}$ (equation 3) gives worst results than the others kernels. This result may be explained by the drawbacks of this kernel (Section 2.2). The kernel K_{new} always provides the best results with a good robustness to perturbation on dudes and hands.

Our second experiment evaluates performances of each kernel within a classification framework. To this end, we have trained a SVM on 5 shapes of each of the three classes: dudes, hands, tools on one side and one model of each of the 6 remaining classes on the other side. The SVM margin parameter was selected in order to maximize the number of true positive while having no false positive. Tab. 1(b) shows the number of well classified shapes for each class. The kernel K_{new} gives the best performances especially for the hands where the two missing

Table 1. Kernels evaluation based on distance (a) and classification (b) criteria

	Hands	Tools	Dudes
$K_{max,classic}$	4.81	6.18	6.36
$K_{change,classic}$	5.27	5.45	6.36
K_{new}	7.09	9.82	6.36

(a) Mean number of good matches.

	Hands	Tools	Dudes
$K_{max,classic}$	7	11	10
$K_{change,classic}$	7	10	10
K_{new}	9	11	11

(b) Number of recognized shapes in one class.

shapes are the more perturbed ones. The two others kernels present good results and are competitive when shapes are not strongly deformed. This experiment confirms the robustness of our kernel against perturbed shapes.

5 Conclusion

The bag of path approach is based on a decomposition of the complex graph structure into a set of linear objects (paths). Such an approach benefits of recent advances in both string and vectors kernels. Our graph kernel based on a hierarchy of paths is more stable to small perturbations of the shapes than kernels based solely on a bag of paths. Our notion of path's hierarchy is related to the graph edit distance through the successive rewritings of a path. Our kernel is thus related to the ones introduced by Neuhaus and Bunke.

References

1. Sharvit, D., Chan, J., Tek, H., Kimia, B.: Symmetry-based indexing of image databases. Journal of Visual Communication and Image Representation 9(4), 366–380 (1998)
2. Siddiqi, K., Shokoufandeh, A., Dickinson, S.J., Zucker, S.W.: Shock graphs and shape matching. Int. J. Comput. Vision 35(1), 13–32 (1999)
3. Pelillo, M., Siddiqi, K., Zucker, S.: Matching hierarchical structures using association graphs. IEEE Trans. on PAMI 21(11), 1105–1120 (1999)
4. Bunke, H., Riesen, K.: A family of novel graph kernels for structural pattern recognition. In: Rueda, L., Mery, D., Kittler, J. (eds.) CIARP 2007. LNCS, vol. 4756, pp. 20–31. Springer, Heidelberg (2007)
5. Neuhaus, M., Bunke, H.: Edit distance based kernel functions for structural pattern classification. Pattern Recognition 39, 1852–1863 (2006)
6. Suard, F., Rakotomamonjy, A., Bensrhair, A.: Kernel on bag of paths for measuring similarity of shapes. In: European Symposium on Artificial Neural Networks, Bruges-Belgique (April 2007)
7. Haasdonk, B.: Feature space interpretation of svms with indefinite kernels. IEEE PAMI 27(4), 482–492 (2005)
8. Desobry, F., Davy, M., Doncarli, C.: An online kernel change detection algorithm. IEEE TSP 53(8), 2961–2974 (2005)
9. Kashima, H., Tsuda, K., Inokuchi, A.: Marginalized kernel between labeled graphs. In: Proc. of the Twentieth International conference on machine Learning (2003)
10. Ruberto, C.D.: Recognition of shapes by attributed skeletal graphs. Pattern Recognition 37(1), 21–31 (2004)
11. Berg, C., Christensen, J.P.R., Ressel, P.: Harmonic Analysis on Semigroups. Springer, Heidelberg (1984)
12. Giblin, P.J., Kimia, B.B.: On the local form and transitions of symmetry sets, medial axes, and shocks. In: Seventh Internat. Conf. on Computer Vision, pp. 385–391 (1999)
13. Torsello, A., Handcock, E.R.: A skeletal measure of 2d shape similarity. CVIU 95, 1–29 (2004)
14. Haussler, D.: Convolution kernels on discrete structures. Technical report, Department of Computer Science, University of California at Santa Cruz (1999)
15. LEMS: shapes databases, `http://www.lems.brown.edu/vision/software/index.html`

Polytopal Graph Complexity, Matrix Permanents, and Embedding

Francisco Escolano[1], Edwin R. Hancock[2], and Miguel A. Lozano[1]

[1] University of Alicante, Departamento de Ciencia de la Computación e Inteligencia Artificial, Spain
[2] University of York, Department of Computer Science, UK

Abstract. In this paper, we show how to quantify graph complexity in terms of the normalized entropies of convex Birkhoff combinations. We commence by demonstrating how the heat kernel of a graph can be decomposed in terms of Birkhoff polytopes. Drawing on the work of Birkhoff and von Neuman, we next show how to characterise the complexity of the heat kernel. Finally, we provide connections with the permanent of a matrix, and in particular those that are doubly stochastic. We also include graph embedding experiments based on polytopal complexity, mainly in the context of Bioinformatics (like the clustering of protein-protein interaction networks) and image-based planar graphs.

1 Introduction

The quantification of the *intrinsic complexity* of undirected graphs has attracted significant attention due to its fundamental practical importance, not only in pattern recognition but also other areas such as control theory and network analysis. Such a quantification not only allows the complexity of different graph structures to be compared, but also allows the complexity to be traded against goodness of fit to data when a structure is being learned. Previous complexity characterizations include: a) the number of spanning trees, its connections with the Laplacian spectrum, b) methods based on the path-length chromatic decomposition, c) the number of Boolean operators necessary to build the graph from generator graphs, and more recently, d) the notion of linear complexity of any of the associated adjacency matrices.

In pattern recognition and machine learning the problem of quantifying graph complexity is not only deeply related to embedding methods [1][2] for structural classification or indexing [3], but is also key to the process of constructing graph prototypes [4][5]. In this paper we explore the connection between convex polytopes (and those of the Birkhoff type in particular), heat kernels in graphs and graph structure. Our contribution is to propose and characterize a novel multi-dimensional measure of complexity. This measure is motivated by the notion of *graph entropy* introduced by Körner in [6], which provides a useful framework based on convex polytopes. Unfortunately, the graph entropy is defined for graphs with a probability distribution function over nodes. In this paper we

N. da Vitora Lobo et al. (Eds.): SSPR&SPR 2008, LNCS 5342, pp. 237–246, 2008.

focus on more general unweigthed graphs. We commence with a characterisation based on the heat kernel of a graph, and show how the heat kernel can be decomposed into Birkoff polytopes. Next, we use the work of Birkhoff and von Neuman to characterize the complexity of the decomposition, and then establish links to the permanents of a graph. We experiment with the complexity measure for the problem of graph-embedding, and provide examples on data provided by bioinformatics and computer-vision. A preliminary version of the ideas presented here appear in [7]. Here we provide a deeper theoretical and experimental analysis of the proposed method.

2 The Birkhoff Polytope

The $n-$th Birkhoff polytope $\mathcal{B}_n$ is the set of doubly stochastic matrices of dimension $n \times n$:

$$\mathcal{B}_n = \left\{ B = [b_{ij}]_{n\times n} \;:\; \begin{array}{c} \sum_i b_{ij} = 1,\ \forall j \\ \sum_j b_{ij} = 1,\ \forall i \\ b_{ij} \geq 0,\ \forall i \forall j \end{array} \right\} . \tag{1}$$

Such a polytope is convex, has dimension $(n-1)^2$ and its extreme points are the set of $n!$ permutation matrices $P = [p_{ij}]_{n\times n}$ (entries are 0 except for a single 1 in each row and column). The *Birkhoff-von Newmann (BvN) theorem* [8] states that every doubly stochastic matrix (*DSM*) can be expressed as a convex combination of permutation matrices (*PM*):

$$B = \sum_\alpha p_\alpha P_\alpha,\ \forall B \in \mathcal{B}_n \text{ and } \begin{array}{c} \sum_\alpha p_\alpha = 1 \\ p_\alpha \geq 0\ \forall \alpha \end{array} . \tag{2}$$

Thus $\mathcal{B}_n$ is the convex hull of the set of the $n \times n$ permutation matrices. However, the representation of a *DSM* in terms of many *PMs* is not unique because $\mathcal{B}_n$ is not a simplex. The barycenter of $\mathcal{B}_n$ is the van der Waerden constant matrix B_* with all entries equal to $1/n$.

3 The Link: DSMs and Heat Kernels

The *BvN* decomposition has been applied to traffic problems [9](where the P_α are crossbar switch configurations and p_α are the probabilities of choosing such configurations). Moreover, Birkhoff polytopes are special cases of *transportation* polytopes. Consider an undirected graph $G = (V, E)$ where $|V| = n$, and adjacency matrix A, as the model of a network. The so called $n \times n$ *heat/diffusion kernel* of the graph is found by the matrix exponentiation $K(G) = exp(-\beta\mathcal{L})$, (with $\beta > 0$) of the graph Laplacian $\mathcal{L} = A - D$ where $D = diag(d_i = \sum_{j\in V} A_{ij}, i \in V)$ is the degree matrix. It turns out that K is a *DSM* where K_{ij} encodes the probability of reaching vertex j from vertex i, and viceversa, through lazy random walks [10]. Therefore, the decomposition $K(G) = \sum_\alpha p_\alpha P_\alpha$ will not only map a

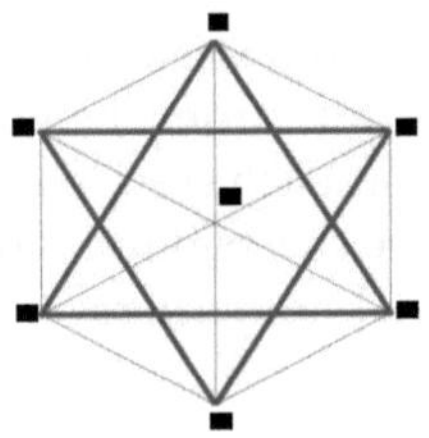

$$P_1 = \begin{bmatrix} 1&0&0\\0&1&0\\0&0&1 \end{bmatrix}\ P_2 = \begin{bmatrix} 1&0&0\\0&0&1\\0&1&0 \end{bmatrix}\ P_3 = \begin{bmatrix} 0&1&0\\1&0&0\\0&0&1 \end{bmatrix}\ P_4 = \begin{bmatrix} 0&1&0\\0&0&1\\1&0&0 \end{bmatrix}\ P_5 = \begin{bmatrix} 0&0&1\\0&1&0\\1&0&0 \end{bmatrix}\ P_6 = \begin{bmatrix} 0&0&1\\1&0&0\\0&1&0 \end{bmatrix}.$$

$$\triangle_1 = p_1P_1 + p_3P_3 + p_5P_5 = \begin{bmatrix} p_1 & p_3 & p_5\\ p_5 & p_1 & p_3\\ p_3 & p_5 & p_1 \end{bmatrix} \quad p_1 + p_3 + p_5 = 1$$

$$\triangle_2 = p_2P_2 + p_4P_4 + p_6P_6 = \begin{bmatrix} p_2 & p_4 & p_6\\ p_4 & p_6 & p_2\\ p_6 & p_2 & p_4 \end{bmatrix} \quad p_2 + p_4 + p_6 = 1$$

Fig. 1. The $\mathcal{B}_3$ polytope. It has $6 \cdot 5/2 = 15$ edges (all of them extremal, which is not the usual case). There are 6 long edges with $D^2 = Tr(A - B)(A^T - B^T) = 6$ of the triangles sides and 9 short ones with length $D^2 = 4$. All the 2-faces of $\mathcal{B}_3$ are triangles with one long and two short edges. Finally $Tr(\triangle_1 - B_*)(\triangle_2^T - B_*^T) = 0$

graph into $\mathcal{B}_n$ but also the probability distribution induced by the convex combination will *weight with each* p_α *the interactions between vertices encoded by the corresponding* P_α. An interesting example is the complete graph of n vertices C_n whose diffusion kernel is the barycenter $K(C_n) = B_*$, that is, the van der Waerden matrix. In addition, the probability distribution, with exactly n coefficients, associated to the *BvN* decomposition is the uniform one $\mathcal{U}_n$ where $p_\alpha = 1/n$, $\forall\alpha$ (see for instance $\mathcal{B}_3$ in Fig. 1).

4 Polytopal Complexity

4.1 The *BvN* Constructive Decomposition

The most intuitive way of computing the *BvN* is to exploit the constructive proof of the theorem [11] follows: (i) Initially pose $K(G)$, which is a *DSM*, as a convex combination of a permutation matrix and another *DSM* $K(G) \equiv K_1 = \lambda_1 P_1 + (1 - \lambda_1)K_2$; and (ii) Iteratively decompose $K_2 = \lambda_2 P_2 + (1 - \lambda_2)K_3$ until the final *DSM* is a permutation. Let γ be the number of permutations needed to encode K, satisfying $2 < \gamma < n^2$, and let i be the step number of the iterative process. With these ingredients, the decomposition is

$$K(G) = \lambda_1 P_1 + (1 - \lambda_1)\lambda_2 P_2 + \ldots = \sum_{\alpha=1}^{\gamma} \underbrace{\{\prod_{i=1}^{\alpha-1}(1 - \lambda_i)\}\lambda_\alpha}_{p_\alpha} P_\alpha\,, \tag{3}$$

if $\gamma = 1$ or it is even, and it is the expression above from which we remove the last λ_α otherwise. What is more important, from the point of view of graph theory, is the way permutations are selected at each step. Let K be the *DSM* which we wish to decompose. Such a matrix induces a weighted bipartite graph $H = (V', E', W)$ where $V' = \{s_1, \ldots, s_n, r_1, \ldots, r_n\}$ and there is an edge $e_{ij} = (s_i, r_j) \in E'$ when $K_{ij} \neq 0$ and has weight $W_{ij} = K_{ij}$. As H is bipartite, V' is disjointly partitioned into $S = \{s_i\}$ and $R = \{r_j\}$. Thus, for any vertex s_i its neighbors satisfy the condition $N(s_i) \subset R$, and for r_j we have $N(r_j) \subset S$. Given that K is a *DSM*, we have that $\sum_{u \in N(s_i)} W_{iu} = \sum_{v \in N(r_j)} W_{vj} = 1$. Therefore, if $A \subset R$, the sum of weights connecting vertices $x \in A$ and $y \in N(A)$ is exactly $|A| = \sum_{x \in A, y \in N(A)} W_{xy} = \sum_{x \in A} \sum_{z \in N(y)} W_{zx} = \sum_{x \in A} 1$. Similarly, the size of $N(A) \subset S$ is given by the sum of the weights connecting the vertices of $N(A)$ and the vertices of $N(N(A))$ which are a superset of the vertices of A. Therefore, $|N(A)| \geq |A|$. This allows us to apply the *Hall's Marriage Theorem* on H, and this in turn ensures the existence of a *perfect matching*. This matching contains n edges, being the largest matching possible, and which can be computed with the Kuhn-Munkres algorithm. Finally, the latter match may be encoded by the permutation P, which is the one selected for decomposing K.

Then, the corresponding $\lambda > 0$ is selected among the minimum of the components determined by P, that is $\lambda = min_{i,j=1\ldots,n}\{K_{ij} : P_{ij} \neq 0\}$. Such a selection ensures the non-negativity of the matrix $M = K - \lambda P$ whose rows and columns sum to $1 - \lambda$. Therefore, $K' = \frac{1}{1-\lambda} M$ is a *DSM*. In addition, from the n components of K_{ij} satisfying $P_{ij} = 1$, those with value λ will be zero in M and, obviously, in K'. This implies that the number of non-zero components will decrease with each step and, thus, the number of steps is bounded by n^2.

For instance, the *BvN* decomposition of B_* in $\mathcal{B}_3$ using the latter algorithm is given by:

$$K = \begin{bmatrix} \frac{1}{3} & \frac{1}{3} & \frac{1}{3} \\ \frac{1}{3} & \frac{1}{3} & \frac{1}{3} \\ \frac{1}{3} & \frac{1}{3} & \frac{1}{3} \end{bmatrix} = \frac{1}{3} \begin{bmatrix} 1&0&0 \\ 0&1&0 \\ 0&0&1 \end{bmatrix} + \frac{2}{3} \frac{1}{(1-\frac{1}{3})} \begin{bmatrix} 0 & \frac{1}{3} & \frac{1}{3} \\ \frac{1}{3} & 0 & \frac{1}{3} \\ \frac{1}{3} & \frac{1}{3} & 0 \end{bmatrix} = \frac{1}{3} \begin{bmatrix} 1&0&0 \\ 0&1&0 \\ 0&0&1 \end{bmatrix} + \frac{2}{3} \begin{bmatrix} 0 & \frac{1}{2} & \frac{1}{2} \\ \frac{1}{2} & 0 & \frac{1}{2} \\ \frac{1}{2} & \frac{1}{2} & 0 \end{bmatrix} =$$

$$= \frac{1}{3} \begin{bmatrix} 1&0&0 \\ 0&1&0 \\ 0&0&1 \end{bmatrix} + \frac{2}{3} \left\{ \frac{1}{2} \begin{bmatrix} 0&1&0 \\ 0&0&1 \\ 1&0&0 \end{bmatrix} + \frac{1}{2} \begin{bmatrix} 0&0&1 \\ 1&0&0 \\ 0&1&0 \end{bmatrix} \right\} = \frac{1}{3} P_1 + \frac{1}{3} P_4 + \frac{1}{3} P_6$$

4.2 BvN Graph Complexity

Given an undirected and unweighted graph $G = (V, E)$ with diffusion kernel $K_\beta(G)$, and *BvN* decomposition $K_\beta(G) = \sum_{\alpha=1}^{\gamma} p_\alpha P_\alpha$, we define the *$\beta$-graph complexity of G* as

$$\mathcal{C}_\beta(G) = \frac{H(\mathcal{P})}{\log_2 n} = \frac{\log_2 \gamma + D(\mathcal{P} || \mathcal{U}_\gamma)}{\log_2 n}, \tag{4}$$

where $\mathcal{P} = \{p_1, \ldots, p_\gamma\}$ is the probability density function (pdf) induced by the decomposition, $H(.)$ the entropy and $D(.)$ the Kullback-Leibler divergence

$D(\mathcal{P}||\mathcal{Q}) = \sum_{\alpha} p_{\alpha} \log \frac{p_{\alpha}}{q_{\alpha}}$. The latter definition considers: a) the size of the graph, b) the number of components of the decomposition and c) the information content of the induced pdf. Moreover, as the pdf induced by the complete graph C_n is $\mathcal{U}_n$, our definition of graph complexity is actually the entropy ratio $H(\mathcal{P})/H(\mathcal{U}_n)$. Independent of n, we have $\mathcal{C}_{\beta}(C_n) = 1$, and as a result a complete graph has unit $\beta-$graph complexity. Also, $\mathcal{C}_{\beta}(I_n) = 0$ where I_n is the graph with all its n vertices isolated (without neighbors).

As we will show in the experimental section, $\beta-$graph complexity is *quasi-invariant* to graph permutations, that is $\mathcal{C}_{\beta}(G) \approx \mathcal{C}_{\beta}(Q^T A Q)$, where A is the adjacency matrix of G and Q any permutation of order n. Quasi-invariance is fulfilled despite the fact that the *BvN* decomposition does not yield such invariance in the coefficients. As we will see, the intra-class variability is also useful for discriminating graph classes. However, such a definition is $\beta-$dependent and a more complete multi-dimensional descriptor is more convenient to describe the full complexity of the graph. Thus, we define the *BvN graph complexity* as

$$\mathcal{C}(G) = \left\{ \begin{array}{l} [\mathcal{C}_{\beta_1}(G), \ldots, \mathcal{C}_{\beta_r}(G)]^T \in R^r \\ \beta_1 \equiv \beta_{max} \geq \beta_2 \geq \ldots \geq \beta_r \equiv \beta_{min} \\ \beta_{max} = \min\{\beta \in R^+ : \mathcal{C}_{\beta}(G) = 1\} \\ \beta_{min} = \max\{\beta \in R^+ : \mathcal{C}_{\beta}(G) = 0\} \end{array} \right\} . \quad (5)$$

The latter *complexity profile*, is thus constructed by considering that when it is treated as a function we have

$$\lim_{\beta \to \infty} \mathcal{C}_{\beta}(G) = 1 \text{ and } \lim_{\beta \to 0} \mathcal{C}_{\beta}(G) = 0 \;\forall G . \quad (6)$$

Moreover, β_{max} and β_{min} are computed respectively from $\Lambda_{max} = \Phi^T K(C_n) \Phi$ and $\Lambda_{min} = \Phi^T K(I_n) \Phi$, given that for $\beta = 1$ we have that $K(G) = \exp(-\mathcal{L}) \equiv \Phi \Lambda \Phi^T$ where Φ the eigenvector matrix of $-\mathcal{L}$, $\Lambda = diag(e^{-\lambda_n}, e^{-\lambda_{n-1}}, \ldots, e^{-\lambda_1} = 1)$, and where $\lambda_1 = 0 \leq \lambda_2 \leq \ldots \leq \lambda_n$ are the eigenvalues of $\mathcal{L}$. More precisely, we have that the matrix Λ_{max} is of the form $diag(e^{-\beta\lambda_n}, e^{-\beta\lambda_{n-1}}, \ldots, e^{-\beta\lambda_1})$, and thus β_{max} may be deduced from the greatest $\frac{\beta\lambda_i}{\lambda_i} \in R^+ :\ K(C_n) \approx \exp(-\beta\mathcal{L})$. A similar rationale yields the β_{min} bound. Thus, the graph complexity trace $\mathcal{C}(G)$ is a signature of the interaction between the heat diffusion process and the structure/topology of the graph as β (and thus the range of interactions between vertices) changes. It can also be interpreted as a trajectory in $\mathcal{B}_n$ between the extreme point given by the identity permutation $P_I = \mathcal{I}_n$ and the barycenter $B_* = K(\mathcal{C}_n)$. In addition, the typical signature is heavy tailed, non-monotonically increasing from β_{max} to $\beta^* \approx \arg\max\{\mathcal{C}_{\beta}(G)\}$ and monotonically decreasing from β^* to β_{min}. When $\mathcal{C}_{\beta}(G)$ is $r-$discretized, each $k \in \{1, \ldots, r\}$ is seen as a *percentile*. The interval $[\beta_{max}, \beta^*)$ reflects intra-class variability, given that it encodes global aspects of heat diffusion, whereas the interval $[\beta^*, \beta_{min}]$ reflects inter-class one, because it encodes local (more constrained as we tend to β_{min}) diffusion. Both intervals $[\beta_{max}, \beta^*) \cup [\beta^*, \beta_{min}]$ may be discretized individually with the same rate, and then fuse both fragments of the complexity function for the purpose of pattern analysis (see Fig. 2). All profiles are aligned according to Eq. 6. In addition, the value of β^* may be easily approximated by accounting for the appearing of a increasing number of $K_{ii} \approx 1$ (the decay starts just before that point). Thus, β^* represents the most significant *topological phase transition* regarding the impact of the diffusion process in the topology of the input graph.

4.3 Connections with Matrix Permanents

The observation that the interval $[\beta^*, \beta_{min}]$ is populated by an increasing number of elements of the heat kernel for which $K_{ii} \approx 1$, or equivalently an increasing number of elements for which $K_{ij} = 0$, $i \neq j$ motivates the analysis of polytopal complexity in terms of *the rate of loss of perfect matchings over this interval.* Let A be the $n \times n$ adjacency matrix associated to a given heat kernel $K_{\beta_{ij}}$ such that $A_{ij} = 1$ if $K_{\beta_{ij}} > 0$, for a given value of β. The *permanent* of A is

$$per(A) = \sum_{\pi \in \mathcal{S}_n} \prod_{i=1}^{n} A_{i\pi(i)} \ , \tag{7}$$

being $\mathcal{S}_n$ the set of permutations of $\{1, 2, \ldots, n\}$ and $\pi(i)$ is the $i-$th element of the π permutation. As each product can be only 1 when a perfect matching exists in the bipartite graph induced by A, the overall sum counts the number of perfect matchings in the such bipartite graph. As $\lim_{\beta \to \beta_{min}} K_\beta = I_n$ the minimum number of perfect matchings in the interval $[\beta, \beta_{min}]$ is 1. However, preliminary experiments show that $per(A)$ is nearly constant over the latter interval and sometimes close to its maximum value $n!$, the one for $A = C_n$ (complete graph), before falling to 1. Consequently the $per(A)$ profile is it not qualitatively related to the one of $\mathcal{C}$. However, this is not true if we consider $per(K_\beta)$ that yields the *sum of weighted perfect matchings.*

The van der Waerden conjecture states that: $\forall B \in \mathcal{B}_n : per(B) \geq per(B_*) \equiv n!/n^n$ and $per(B) = per(B_*) \Leftrightarrow B = B_*$. The proofs solving the latter conjecture (which embodies a minimization problem) reveal that the permanent is a *mixed discriminant* of diagonal quadratic forms. Mixed discriminants are considered generalizations of the permanent. For the sake of polytopal complexity it is useful to note that $\forall B \in \mathcal{B}_n : 0 < per(B) \leq 1$ and $per(B) = 1 \Leftrightarrow B$ is a permutation matrix. Thus, an alternative definition of graph complexity relies on $\mathcal{C}^{'}_\beta = 1 - per(K_\beta)$ defining the $\mathcal{C}^{'}(G)$ profile in $[\beta_{max}, \beta_{\min}]$. However, the latter profile is typically flat and equal to 1 along $[\beta_{max}, \beta^*]$ and it is only informative at second half of $[\beta^*, \beta_{min}]$ see (Fig. 3). In addition, the computation of the permanent is a $\#-$P problem and one of the best approximate algorithms [12] takes $O(n^{23})$ on average only for generating a sample of the Markov chain in which it it relies. However, when small graphs are available or complex

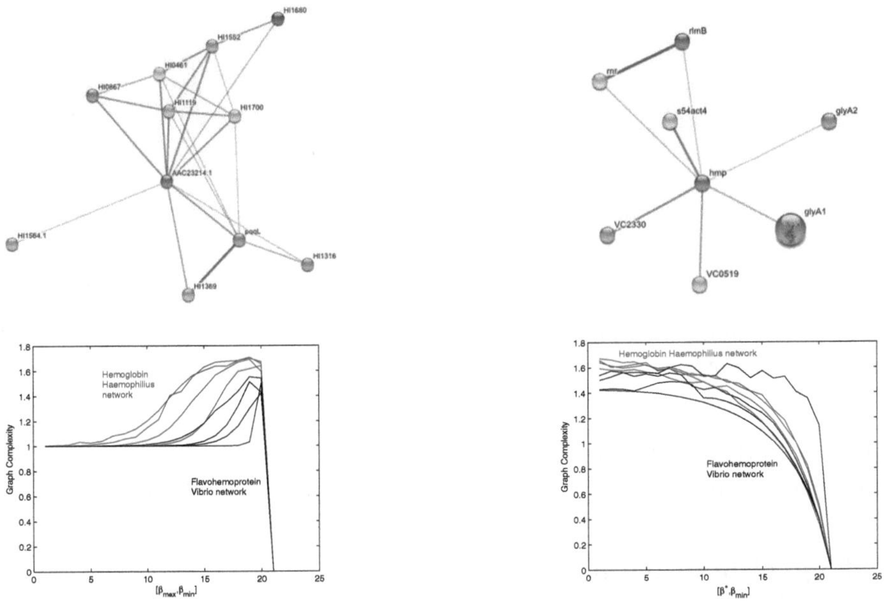

Fig. 2. Top: Examples of protein-protein interaction networks (STRING [14]) for hemoblobin in Haemophilus (left) and flavohemoprotein in Vibrio (right). Bottom: Complexity profiles for both nets in the ranges $[\beta_{max}, \beta_{min}]$ (left) and $[\beta^*, \beta_{min}]$ (rigth).

graphs are simplified (see below) we may use the Ryser's algorithm which takes $\Theta(n2^n)$ operations.

4.4 Non-metric Graph Simplification

For a graph $G = (V, E)$ with $|V| = n$ vertices, the *BvN* decomposition process takes $O(n^2)$ iterations and in each operation solves a Kuhn-Munkres (Hungarian) search for a suitable permutation, which takes $O(n^3)$. Hence, we have a global complexity of $O(n^5)$ which is not too expensive for $n = 30$ nodes, but becomes unmanageable for $n \geq 70$, taking into account that this is the complexity for each percentile. For this reason, in this paper we extend the *stratified graph simplification* method proposed in [13] to the *non-metric* case. More precisely, G is augmented with one heat-source node τ which is connected with all $v \in Periphery(G)$. The result is the so-called *affixation graph* $\mathcal{A}(G) = (V \cup \{\tau\}, E \cup \{(\tau, u), (u, \tau), \forall u \in Periphery(G)\})$. In our case, the $Periphery(G)$ is not the convex hull of points but it is defined by computing all vertices u having as excentricity $e(u) = \max_{v \in V} d(u, v)$ (where $d(u, v)$ is the minimum path length between vertices u and v) equal to the diameter of graph $Diam(G) = max_{u \in V} e(v)$. From that point we compute the *average path lengths* $\mathcal{D}(\tau, u)$ using the normalized Laplacian and perform unsupervised clustering to

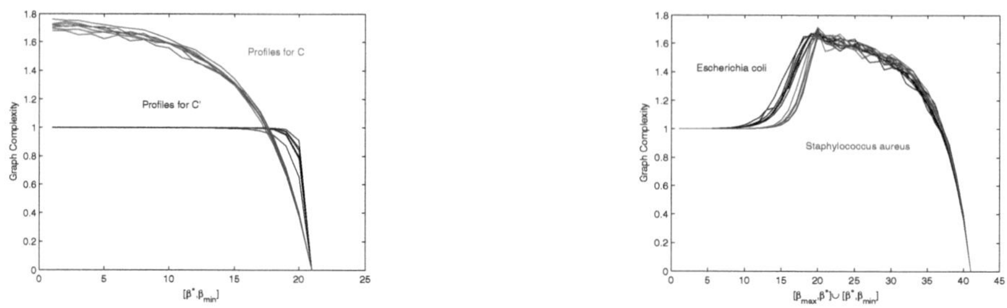

Fig. 3. Left: Polytopal vs Permanent-based complexity considering $[\beta^*, \beta_{min}]$. Rigth: Complete traces of $[\beta_{max}, \beta^*) \cup [\beta^*, \beta_{min}]$ considering same discretization.

find layers in the graph. We assign a node l to each cluster and we connect nodes with adjacency relations in $\mathcal{A}$. So, $\mathcal{S}(G)$ is the resulting *simplified graph.*

5 Experimentation

First, we report the testing of *quasi-invariance* $\mathcal{C}_\beta(G) = \mathcal{C}_\beta(Q^T AQ) \pm e(Q)$, being $e(Q)$ an error term. For the CMU, MOVI and Chalet sequences we have a *median absolute error* (for the sake of robustness) of 0.0215, 0.0167 and 0.0150 respectively, that is $O(2 \times 10^{-2})$, when considering 50 permutations for each of the 30 *simplified* graphs. In this case we set $\beta = 1$. Second, we have tested the proposed measure in the context of *graph embedding.*In Fig. 4(left) we show the results of PCA/MDS embeddings derived from the complexity and compare them with the embeddings given by the leading eigenvalues [2] showed in Fig. 4(right). In the first two rows we perform embedding over protein-protein interaction networks from STRING. More precisely, for networks emanating from the trpA (Tryptophan synthase alpha chain) in 27 sub-species corresponding to 7 species (classes). In this case, interactions account for similar genetic context, and consider two cases: networks of 10 interactors with trpA (11 nodes) and networks of 20 ones. Our complexity method (right) allows to discriminate most of the species except Bacillus in the 10 interactors case. However, they are segregated when we consider 20 interactors. In the same cases, the leading eigenvalues method confuses all classes (although it works better with 20 interactors). No simplification was applied. However, in the last row, where we show the embedding of graphs from de CMU-Chalet-MOVI sequence, graphs are simplified with the method described above. In this latter case, since simplification may yield small graphs, the number of eigenvectors used for spectral embedding, the result is a complete overlap of the three graph classes. This overlap does not appear when we consider the original graphs. So in summary, polytopal complexity seems to yield good embeddings in conditions of low structural variability where the leading eigenvectors method fails, and acceptable results after applying simplification.

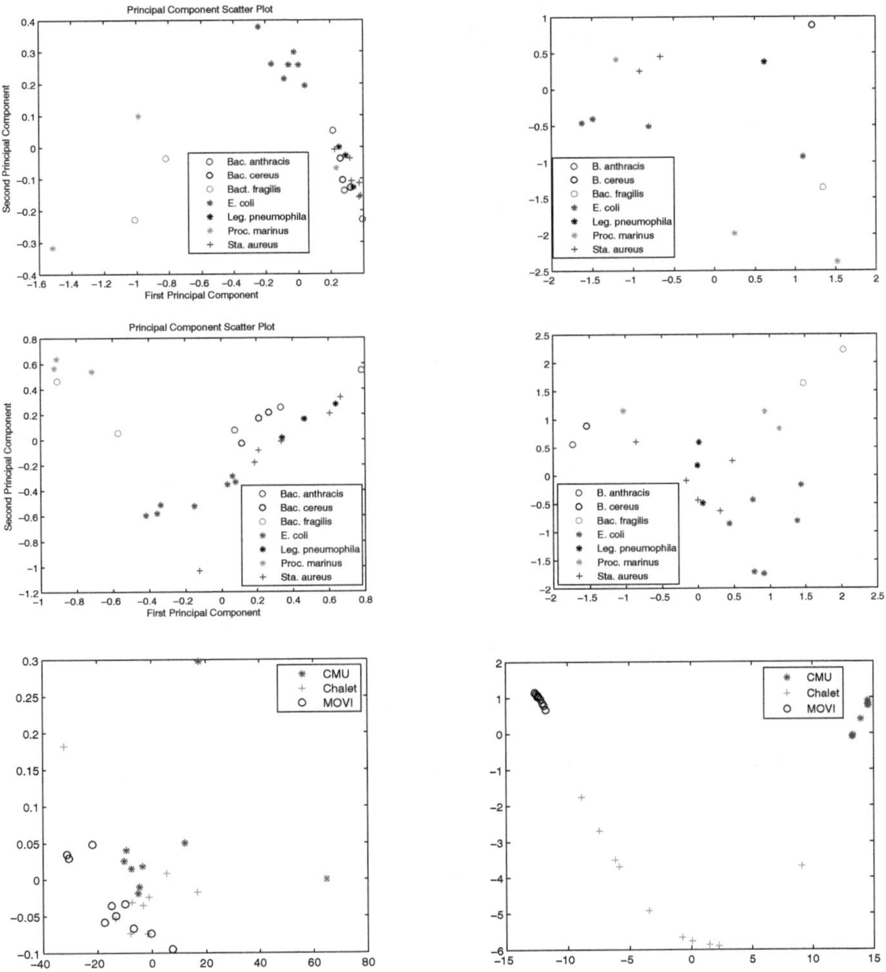

Fig. 4. Graph embedding based on $\mathcal{C}$ vs leading eigenvalues. Firs row: Protein-interaction nets with 10 interactors. Second row: with 20 interactors. Third row: for graphs of the CMU-Chalet-MOVI sequence.

6 Conclusions

In this paper, we have proposed and successfully tested a novel measure of graph complexity. We have also extended a recently reported method of graph simplification to the non-metric case. The main question for the future is the formal characterization of the main phase transition defined by the β^* value. A attempt at answering this question approach has been provided in this paper, through exploring the connections between polytopal complexity and matrix permanents. Here the progressive loss of ability to form perfect matchings seems to explain the observed behaviour. However a complete characterization of the phase transition in terms of entropy change is a topic for future research. We will also compare this method with others in the literature.

References

1. Robles-Kelly, A., Hancock, E.R.: A riemannian approach to graph embedding. Pattern Recognition (40), 1042–1056 (2007)
2. Luo, B., Wilson, R.C., Hancock, E.: Spectral embedding of graphs. Pattern Recognition (36), 2213–2223 (2003)
3. Shokoufandeh, A., Dickinson, S., Siddiqi, K., Zucker, S.: Indexing using a spectral encoding of topological structure. In: IEEE ICPR, pp. 491–497
4. Torsello, A., Hancock, E.: Learning shape-classes using a mixture of tree-unions. IEEE Trans. on PAMI 28(6), 954–967 (2006)
5. Lozano, M., Escolano, F.: Protein classification by matching and clustering surface graphs. Pattern Recognition 39(4), 539–551 (2006)
6. Körner, J.: Coding of an information source having ambiguous alphabet and the entropy of graphs. In: Trans. of the 6th Prague Conference on Information Theory, pp. 411–425 (1973)
7. Escolano, F., Hancock, E., Lozano, M.: Birkhoff polytopes, heat kernels, and graph embedding. In: ICPR (2008)
8. Birkhoff, G.D.: Tres observaciones sobre el algebra lineal. Universidad Nacional de Tucuman Revista, Serie A 5, 147–151 (1946)
9. Chang, C., Chen, W., Huang, H.: On service guarangees for input buffered crossbar switches: A capacity decomposition approach by birkhoff and von neumann. In: IEEE IWQoS, pp. 79–86 (1998)
10. Kondor, R.I., Lafferty, J.: Diffusion kernels on graphs and other discrete structures. In: Proc. ICML (2002)
11. Mirsky, L.: Proofs of two theorems on doubly stochastic matrices. Proc. Amer. Math. Soc. 9, 371–374 (1958)
12. Jerrum, M., Sinclair, A., Vigoda, E.: A polynomial-time approximation algorithm for the permanent of a matrix with nonnegative entries. Journal of the ACM 51(4), 671–697 (2004)
13. Qiu, H., Hancock, E.: Graph simplification and matching using conmute times. Pattern Recognition (40), 2874–2889 (2007)
14. von Mering, C., Huynen, M., Jaeggi, D., Schmidt, S., Bork, P., Snell, B.: String: a database of predicted functional associations. Nuc. Acid Res. 31, 258–261 (2003)

Gesture Recognition Based on Manifold Learning

Heeyoul Choi, Brandon Paulson, and Tracy Hammond

Dept. of Computer Science, Texas A&M University
3112 TAMU, College Station, TX 77843-3112, USA
{hchoi,bpaulson,hammond}@cs.tamu.edu

Abstract. Current feature-based gesture recognition systems use human-chosen features to perform recognition. Effective features for classification can also be automatically learned and chosen by the computer. In other recognition domains, such as face recognition, manifold learning methods have been found to be good nonlinear feature extractors. Few manifold learning algorithms, however, have been applied to gesture recognition. Current manifold learning techniques focus only on spatial information, making them undesirable for use in the domain of gesture recognition where stroke timing data can provide helpful insight into the recognition of hand-drawn symbols. In this paper, we develop a new algorithm for multi-stroke gesture recognition, which integrates timing data into a manifold learning algorithm based on a kernel Isomap. Experimental results show it to perform better than traditional human-chosen feature-based systems.

Keywords: Sketch Recognition, Manifold Learning, Kernel Isomap.

1 Introduction

Sketched gestures are a natural form of input for many domains, such as drawing mathematical symbols, graphs, binary trees, finite state machines, electrical circuit diagrams, military course of action diagrams and many more. To allow the computer to understand these diagrams, gesture recognition systems have been built for a large number of domain [1,2,3,4,5,6,7,8,9,10,11]. Many techniques for gesture recognition have been developed; however, current gesture recognition systems still struggle with trying to get high recognition accuracy while still providing drawing freedom. As in other recognition domains (such as image and speech), feature selection is crucial for efficient and qualified performance in gesture recognition. Previous research has provided several suggestions for good feature sets [12,13,14,11]. Rubine suggested 13 features based on stylistic drawing features and time [12]. Long modified Rubine's features by adding 11 new features and removing 2 time-based features [13]. Also, ink features were proposed in [11]. These feature sets were well designed, but they are based on manual entry of methods to extract them, which becomes tiresome.

As Mahoney says in [15], designing features manually is a tedious task and may be extended by designer's intuition, which is subjective. Moreover, feature design

N. da Vitora Lobo et al. (Eds.): SSPR&SPR 2008, LNCS 5342, pp. 247–256, 2008.

is sensitive to problems such as jitters. More interestingly, machine learning research in other domains have shown that computers are able to select their own features with similar results. As stated in [15], some researchers have tried to learn some features automatically as well as manually, which leads to a mixture of hand-chosen and machine-generated features. Human-generated features are not easily extendable, whereas it is trivial to add some dimensions in feature space for a computer. In addition, hand-chosen features are extracted based on human-observable properties, whereas machine-generated features can be optimized for clear advantages in classification.

In machine learning, the Hidden Markov Models (HMMs) [16], which use temporal and spatial structure, have been widely used for gesture recognition. Several sketch recognition systems have been built which use HMMs to help predict stroke ordering [17,18,19]. Bayesian Networks have also been applied [20] to recognize multi-stroke shapes using LADDER shape descriptions [21]. Some dissimilarity-based approaches have been proposed in [22] with image-based methods and in [23] with graph-based methods.

To our knowledge, however, manifold learning has not yet been applied to gesture recognition despite evidence that has shown manifold learning methods to be good nonlinear feature extractors. Manifold learning is an effective method for representation that works by recovering meaningful low dimensional structures hidden in high-dimensional data [24,25,26,27,28]. Previous research has attempted to use manifold learning to analyze 3D hand-gestures [29], but manifold learning has not yet been applied to gesture (or sketch) recognition.

In this work we apply the kernel Isomap manifold learning method to classify sketch data, because it has a projection property, which provides the ability to project (map) new test data into (onto) the same feature space as training data. Kernel Isomap requires a dissimilarity matrix to find a low-dimensional mapping from which it produces a feature set. We introduce a new algorithm to measure dissimilarity distance between shapes that is based on both spatial and temporal information. We also show how this algorithm can be modified to accommodate for shapes drawn with multiple strokes. For recognizing shapes, we compare our method with the widely used Rubine method [12] and the $1 recognizer [30], which is a recently proposed automatic sketch recognizer.

2 Kernel Isomap

Manifold learning involves inducing a smooth, nonlinear, low-dimensional manifold from a set of data points. Recently, various mapping methods (for example see [24,25,26,27]) have been developed in the machine learning community, and their wide applications have started to draw attention in pattern recognition and signal processing. Isomap is one of the representative isometric mapping methods, which extends metric multidimensional scaling (MDS), by using Dijkstra's geodesic distances (shortest paths) on a weighted graph, instead of Euclidean distances [25].

The geodesic distance matrix used in Isomap can be interpreted as a kernel matrix. However, the kernel matrix based on the doubly centered geodesic distance matrix is not always positive semi-definite. *Kernel Isomap* [31] exploits a constant-shifting method such that the geodesic distance-based kernel matrix is guaranteed to be positive semi-definite as a *Mercer kernel* [32] matrix. This kernel Isomap has a generalization property, enabling us to project test data points onto an associated low-dimensional manifold, using a kernel trick as in kernel PCA [33], whereas, in general, most embedding methods (including Isomap, LLE, and Laplacian eigenmap) do not have such a property. See [31] for the details.

3 Implementation

3.1 Dissimilarity from Sketch Data

To apply the kernel Isomap to sketch classification, what we need is not the data points, but a dissimilarity matrix, which is enough to find a low dimensional space. The first step in our algorithm is to scale each character to be the same width and height. Then, we need a consistent (and large) number of points in each sketch. Thus, we interpolate points between any two consecutive points that are adjacent in time, but far away in distance to ensure that each gesture has the same number of points.

Dissimilarity. To calculate the dissimilarity matrix, we calculate the sum of squared distance between two points of the same order in each sketch as in Fig. 1. This is the squared dissimilarity in our algorithm. The dissimilarity between the ith sketch data and the jth sketch data, D_{ij}, is then given by

$$D_{ij} = \sqrt{\sum_{k=1}^{S} (d_k)^2}, \tag{1}$$

where d_k is the Euclidean distance between the kth points of two sketches, and S is the number of points in one stroke.

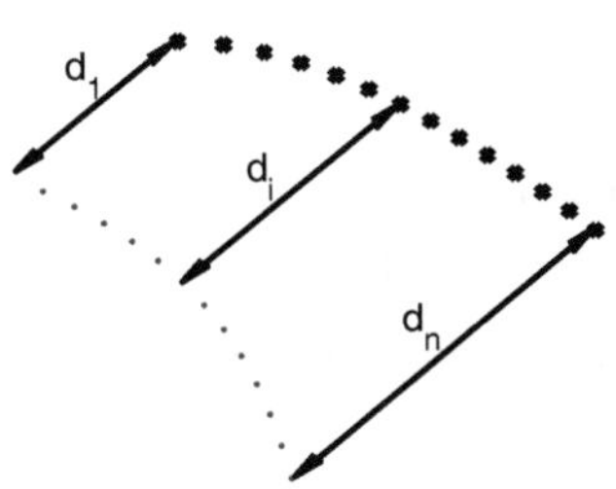

Fig. 1. Dissimilarity between two sketches

In addition to normalizing each stroke by resampling, we use different weights for different points. When we connect the two parts of '4' or '5', the connection causes confusion in calculating the distance. If we use different weights so that the fake points have less importance than real parts of the stroke, we can eliminate some of the confusion. Finally, the distance between two points is given by

$$D_{ij} = \sqrt{\sum_{k=1}^{S} (d_k)^2 w_{ik} w_{jk}}, \tag{2}$$

where w_{ik} is the normalized weight of the kth point in the ith stroke, which is given by

$$w_{ik} = \begin{cases} \frac{1}{Z_i} & \text{if the point is original in the stroke,} \\ \frac{\varepsilon}{Z_i} & \text{otherwise,} \end{cases} \tag{3}$$

where Z_i is the normalization term and ε is a constant between 0 and 1.

We call a kernel Isomap equipped with weighted distance *weighted kernel Isomap*[1]. The weighted kernel Isomap works well with multi-strokes shapes. For example, some characters such as the digit '7' might be drawn as one stroke or two strokes even by one user. In this case, weighted kernel Isomap works well since the weights make the boundary between one stroke and multi-strokes smooth so that the distance between two same characters can be closer.

There have been proposed some dissimilarities for sketch data [22]. However, those dissimilarities are based on images, while our proposed dissimilarity is based on sketch data including time information.

3.2 Classification

After getting features through kernel Isomap, we use the k nearest neighbor (kNN) method, which is a simple classifier, to check how good the features are[2] In kNN, we use Mahalanobis distance instead of Euclidean distance between points, since the variances of Rubine features are too different from each other, thus Euclidean distance does not find any meaningful structure from Rubine's features. On the other hand, in the case of kernel Isomap, the variances of the features are similar to each other, so Euclidean distance also shows good results.

4 Experiments

We carried out numerical experiments with three different data sets: (a) 26 small letters in English written by one user, where all characters are drawn by

[1] When weighted kernel Isomap and kernel Isomap work in the same way, we name (weighted) kernel Isomap

[2] Applying kNN to the projected space does not give any significant benefit in classification compared to applying it directly to the input data space. However, the results from kNN is enough to show how much our methods have the information for classification after the projection to the feature space.

one stroke and each letter has 25 data points; a third of the examples of each letter were drawn two weeks later, so there is a reasonable amount of variability in the example data, (b) 10 digits from 0 to 9, drawn by another user, where each class has 25 characters and (c) 8 different mathematical symbols ('+', '-', 'x', '/', '=', 'sin', 'cos' and 'tan'), drawn by 10 subjects, where each class of each person has 5 characters, each of which is by one stroke. In order to show the improved accuracy of our method, we compared our method with Rubine's algorithm and the $1 recognizer.

4.1 Accuracies

For more robust results, we executed 10-fold cross validation 50 times. Figs. 2 and 3 show the hit rates of classification in three data sets, respectively, by three approaches: (1) Rubine's method, (2) $1 recognizer (3) kernel Isomap and (4) weighted kernel Isomap. To make it fair, we extracted 13 features from the (weighted) kernel Isomap as Rubine's[3].

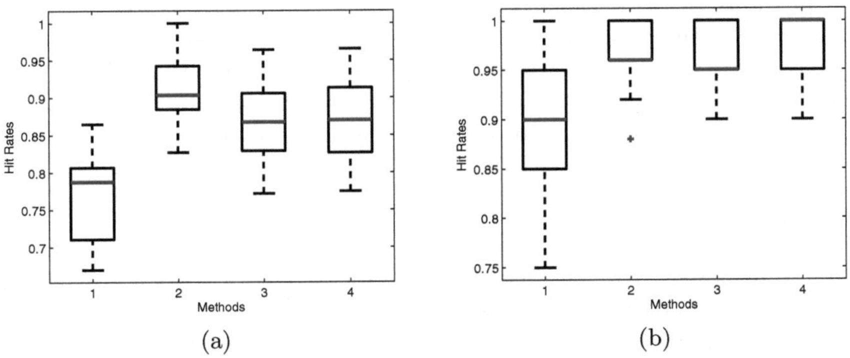

Fig. 2. For (a) alphabetical characters (b) digits, the classification accuracies of three methods: (Method 1) Rubine method (Method 2) $1 recognizer (Method 3) kernel Isomap and (Method 4) weighted kernel Isomap

Fig. 2 is the boxplots of 50 time experiments in the cases of the English letters set and digits. In these figures, weighted kernel Isomap and kernel Isomap features are better than Rubine features in the classification, but similar to each other. Moreover, we can easily extend the number of features in (weighted) kernel Isomap by taking more eigenvalues and eigenvectors of the kernel matrix, which might produce better performance, whereas the number of Rubine features is 13 and it is hard to add another feature to them. Note that in (b), some digits such as '4' and '5' are composed of 2 strokes. This verifies that the weights for points in strokes work for the classification especially when characters are composed of multiple strokes. The average hit rates for (a) are 77.31%, 90.96%, 87.19% and 87.04%, respectively, and for (b) they are 91.80%, 97.28%, 95.90% and 97.80%, respectively.

[3] $1 recognizer does not extract any features. Actually it is like using all features.

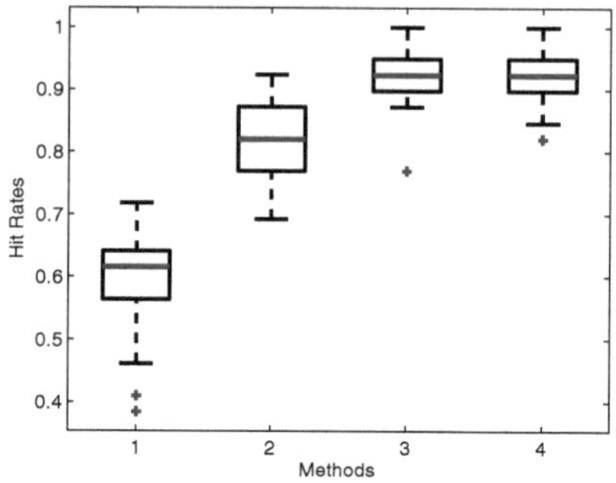

Fig. 3. For mathematical symbols ('+', '-', 'x', '/', '=', 'sin', 'cos' and 'tan'), the classification accuracies of three methods: (Method 1) Rubine method (Method 2) $1 recognizer (Method 3) kernel Isomap and (Method 4) weighted kernel Isomap. The average hit rates are 60.46%, 82.41%, 92.77% and 92.41%, respectively.

The previous two experiments used one user's data set. As for the mathematical symbols, the data set was made by 10 different subjects. However, since each character in the data set was drawn by one stroke, there is no big difference between kernel Isomap and weighted kernel Isomap as in the first experiment. In Fig. 3, the accuracies of our proposed methods are much better than Rubine's.

4.2 Feature Spaces

Figs. 4 and 5 show us how good the features are. We plotted the 4 best features from the English letters data set. In Fig. 5 which is from kernel Isomap, we can see that sketches in the same classes share similar feature values for most features, whereas in Fig. 4, which uses the Rubine feature set, just a few features are useful for classification.

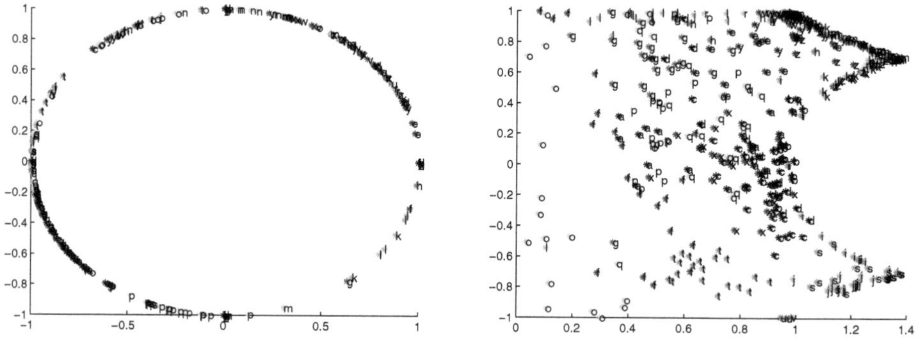

Fig. 4. Four features from Rubine's. Each axis represents one Rubine feature. Note that the features do not effectively separate the different gestures.

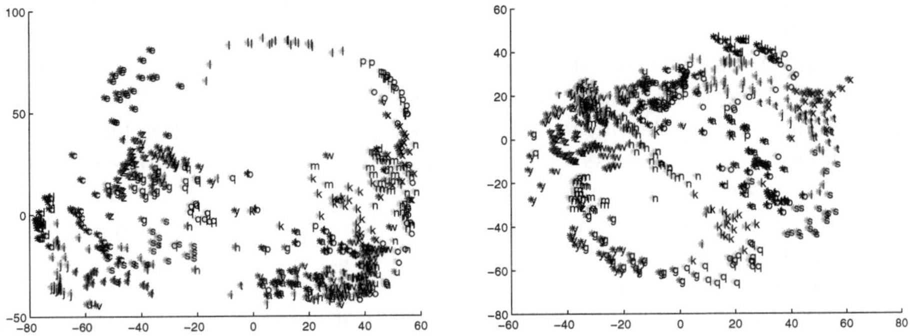

Fig. 5. Four features corresponding to the 4 largest eigenvalues of kernel matrix. Each axis represents an automatically generated feature. Note that the features effectively separate the different gestures.

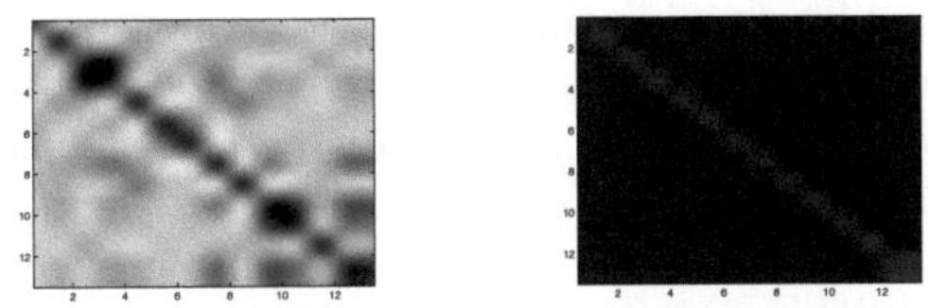

Fig. 6. Correlation matrices between 13 features (Red color represents high correlation and each axis represents 13 features): (Left) Rubine Features; (Right) (weighted) kernel Isomap Features. Note that in the figure on the right all of the automatically chosen Isomap features are highly uncorrelated, meaning that they are much more effective to include distinguishable information.

We calculated the correlation matrix of each feature set (see Fig. 6). (Weighted) kernel Isomap features are uncorrelated, which means they are more efficient than the heavily correlated and redundant Rubine features (seen particularly between features 10 and 11).

5 Discussion

As shown in the figures above, the features from (weighted) kernel Isomap are more accurate and more efficient than Rubine's for the classification. But, there are two drawbacks in the (weighted) kernel Isomap approach. To get the features, a kernel Isomap requires a lot of training data because it is assumed in manifold learning that data points should be dense. In addition, it takes some time to extract those features for training. However, after training, time does not matter when testing new data points. Also, although the Rubine's approach does not need dense data points for training, it also needs training data for the classification. More importantly, as we gather more data points, the features of a (weighted) kernel Isomap become better, whereas Rubine's features do not.

The weighted kernel Isomap works well with multi-strokes shapes. Sometimes, it is hard to say if the input character is composed of one stroke or multiple strokes. In this case, weighted kernel Isomap works well since the weights make the boundary between single stroke and multi-strokes smooth. There is some room to improve the classification hit rate by providing better weights.

The reason why (weighted) kernel Isomap is better than Rubine's for the classification is that they try to find the best features to represent the data set in an efficient way, whereas Rubine's features are already determined according to stylistic drawing properties. Therefore, the feature set of kernel Isomap reflects the domain or context dependent features.

With the letter data set which has large amount of orientation variability, the $1 recognizer is slightly better than our method because of their rotation invariant distance metric[4]. However, our method outperforms the $1 recognizer with math data set because it contained much less rotational variability. Furthermore, some shapes, such as '+' and 'x', contain rotational ambiguity which makes them hard to distinguish with $1. When a domain calls for rotation invariant features, we can improve our technique by simply using the distance measure in the $1 recognizer. Another disadvantage of the $1 recognizer is that it does not extract any features, so it can not improve anymore (it is fixed in its method of classification), whereas our method can be integrated with other types of classifiers, such as a quadratic classifier or support vector machine, which may improve overall performance. For the purposes of this work, we used a simple kNN classifier as proof that our manifold learning technique can be used to extract features from sketched data.

6 Conclusion

Gesture recognition is a natural form of human-computer interaction. As in other recognition problems, features are crucial for efficient and qualified performance in gesture recognition. In general pattern recognition problems, machines can learn effective features which may perform better than the hand-chosen ones in classification. Though, in many research areas, manifold learning methods have been shown to be good nonlinear feature extractors, only a few of them have been applied to gesture recognition. In this paper, we developed a new technique for gesture classification using a manifold learning approach that combines temporal and spatial information, while handling multiple strokes with weighted distances. Experimental results confirmed the performance to be better than the Rubine's feature set.

Acknowledgements

One of the authors was supported by StarVision Technologies's student sponsorship program and this work funded in part by NSF IIS grant 0744150: Developing Perception-based Geometric Primitive-shape and Constraint Recognizers to Empower Instructors to Build Sketch Systems in the Classroom.

[4] It is computationally so expensive to calculate rotation-invariant distances. So it takes much time even to classify a new point.

References

1. LaViola, J., Zeleznik, R.: Mathpad2: A system for the creation and exploration of mathematical sketches. ACM Transactions on Graphics (Proceedings of SIGGRAPH) 23(3) (2004)
2. Stahovich, T., Davis, R., Shrobe, H.: Qualitative rigid body mechanics. Artificial Intelligence (2000)
3. Landay, J.A., Myers, B.A.: Sketching interfaces: Toward more human interface design. IEEE Computer 34(3), 56–64 (2001)
4. Forbus, K.D., Usher, J., Chapman, V.: Sketching for military course of action diagrams. In: Proceedings of IUI 2003 (2003)
5. Do, E.Y.L.: VR sketchpad - create instant 3D worlds by sketching on a transparent window. In: de Vries, B., van Leeuwen, J.P., Achten, H.H. (eds.) CAAD Futures 2001, pp. 161–172 (July 2001)
6. Forsberg, A.S., Dieterich, M.K., Zeleznik, R.C.: The music notepad. In: Proceedings of UIST 1998, ACM SIGGRAPH (1998)
7. Igarashi, T., Matsuoka, S., Tanaka, H.: Teddy: A sketching interface for 3d freeform design. In: SIGGRAPH 1999, pp. 409–416 (August 1999)
8. Mahoney, J.V., Fromherz, M.P.J.: Interpreting sloppy stick figures by graph rectification and constraint-based matching. In: Fourth IAPR Int. Workshop on Graphics Recognition, Kingston, Ontario, Canada (2001)
9. Muzumdar, M.: ICEMENDR: Intelligent capture environment for mechanical engineering drawing. Master's thesis, Massachusetts Institute of Technology (1999)
10. Hammond, T., Davis, R.: Tahuti: A geometrical sketch recognition system for UML class diagrams. In: AAAI Spring Symposium on Sketch Understanding, March 25-27, pp. 59–68 (2002)
11. Patel, R., Plimmer, B., Grundy, J., Ihaka, R.: Ink features for diagram recognition. In: Sketch Based Interfaces and Modeling IEEE, Eurographics (2007)
12. Rubine, D.: Specifying gestures by example. Computer Graphics 25(4), 329–337 (1991)
13. Long, A.C., Landay, J.A., Rowe, L.A., Michiels, J.: Visual similarity of pen gestures. In: Human Factors in Computing Systems (2000)
14. Sezgin, T.M., Stahovich, T., Davis, R.: Sketch based interfaces: Early processing for sketch understanding. In: Proceedings of 2001 Perceptive User Interfaces Workshop (PUI 2001) (2001)
15. Mahoney, J.V., Fromherz, M.P.J.: Three main concerns in sketch recognition and an approach to addressing them. In: AAAI Spring Symposium on Sketch Understanding, Standord, CA, pp. 105–112 (March 2002)
16. Rabiner, L.R., Juang, B.H.: An introduction to hidden Markov models. IEEE Trans. Acoustics, Speech, and Signal Processing Magazine 3, 4–16 (1986)
17. Sezgin, T.M.: Sketch Interpretation Using Multiscale Stochastic Models of Temporal Patterns. PhD thesis, Massachusetts Institute of Technology (May 2006)
18. Sun, Z., Jiang, W., Sun, J.: Adaptive online multi-stroke sketch recognition based on hidden markov model. In: Yeung, D.S., Liu, Z.-Q., Wang, X.-Z., Yan, H. (eds.) ICMLC 2005. LNCS, vol. 3930, pp. 948–957. Springer, Heidelberg (2006)
19. Muller, S., Eickeler, S., Rigoll, G.: Image database retrieval of rotated objects by user sketch. In: IEEE Workshop on Content-Based Access of Image and Video Libraries, p. 40 (1998)
20. Alvarado, C., Davis, R.: Sketchread: A multi-domain sketch recognition engine. In: Proceedings of UIST 2004, pp. 23–32 (2004)

21. Hammond, T., Davis, R.: Ladder, a sketching language for user interface developers. Elsevier, Computers and Graphics 28, 518–532 (2005)
22. Kara, L.B., Stahovich, T.F.: An image-based trainable symbol recognizer for sketch-based interfaces. In: Making Pen-Based Interaction Intelligent and Natural, Menlo Park, California, October 21-24. AAAI Fall Symposium, pp. 99–105 (2004)
23. Lee, W., Kara, L.B., Stahovich, T.F.: An efficient graph-based recognizer for hand-drawn symbols. Computers & Graphics 31, 554–567 (2007)
24. Seung, H.S., Lee, D.D.: The manifold ways of perception. Science 290, 2268–2269 (2000)
25. Tenenbaum, J.B., de Silva, V., Langford, J.C.: A global geometric framework for nonlinear dimensionality reduction. Science 290, 2319–2323 (2000)
26. Saul, L., Roweis, S.T.: Think globally, fit locally: Unsupervised learning of low dimensional manifolds. Journal of Machine Learning Research 4, 119–155 (2003)
27. Belkin, M., Niyogi, P.: Laplacian eigenmaps for dimensionality reduction and data representation. Neural Computation 15, 1373–1396 (2003)
28. de Silva, V., Tenenbaum, J.B.: Global versus local methods in nonlinear dimensionality reduction. In: Advances in Neural Information Processing Systems, vol. 15, pp. 705–712. MIT Press, Cambridge (2003)
29. Jenkins, O.C., Matari, M.J.: A spatio-temporal extension to isomap nonlinear dimension reduction. In: Proc. Int'l. Conf. Machine Learning, Banff, Canada (2004)
30. Wobbrock, J., Wilson, A., Li, Y.: Gestures without libraries, toolkits, or training: A $1 recognizer for user interface prototypes. In: Proc. of the 20th Annual ACM Symposium on User Interface Software and Technology, Newport, RI, USA (2007)
31. Choi, H., Choi, S.: Robust Kernel Isomap. Pattern Recognition 40(3), 853–862 (2007)
32. Girolaini, M.: Mercer kernel-based clustering in feature space. IEEE Transactions on Neural Networks 13(3), 780–784 (2002)
33. Schölkopf, B., Smola, A.J., Müller, K.R.: Nonlinear component analysis as a kernel eigenvalue problem. Neural Computation 10(5), 1299–1319 (1998)

Graph Characteristics from the Ihara Zeta Function

Peng Ren, Richard C. Wilson, and Edwin R. Hancock

Department of Computer Science, The University of York, York, YO10 5DD, UK
{pengren,wilson,erh}@cs.york.ac.uk

Abstract. This paper shows how to extract permutation invariant graph characteristics from the Ihara zeta function. In a previous paper, we have shown that the Ihara zeta function leads to a polynomial characterization of graph structure, and we have shown empirically that the coefficients of the polynomial can be used as to cluster graphs. The aim in this paper is to take this study further by showing how to select the most significant coefficients and how these can be used to gauge graph similarity. Experiments on real-world datasets reveal that the selected coefficients give results that are significantly better than those obtained with the Laplacian spectrum.

1 Introduction

One of the bottlenecks in comparing graphs using graph edit distance[13] is that its computation requires correspondence matches. This problem can be overcome if pattern vectors composed of graph characteristics are used instead, and similarity is measured using the distance between vectors. There are a number of alternative characterizations available. For instance, one could use classical properties of the graph such as diameter, perimeter length or Cheeger number[4] or normalized quantities such as the edge or triangle densities. Several possibilities are offered by spectral graph theory, and these include the Laplacian spectrum[10] or symmetric polynomials computed from the spectral matrix[16].

Recently, however Bai and Hancock have shown that the zeta function of a graph offers an interesting means of characterizing its structure[1]. They have demonstrated that the Rosenberg zeta function[11] is related to the Mellin moments of the heat kernel trace, and have used the moments generated by sampling the Rosenberg zeta function to generate graph feature vectors. The zeta function computed by Bai and Hancock is determined by the Laplacian spectrum. However, in general the zeta function of a graph can be thought of as an analogue of the Riemann zeta function, with prime numbers replaced by prime paths[2]. In fact, the zeta function is a compact representation of information concerning the distribution of paths and path lengths on a graph.

In is this paper, we aim to explore this relationship in a greater depth. We turn to the Ihara zeta function. The Ihara zeta function was first detailed in [7] and [8]. Hashimoto subsequently deduced explicit factorizations for bi-regular bipartite graphs[6]. Bass has generalized Hashimoto's factorization to all finite

N. da Vitora Lobo et al. (Eds.): SSPR&SPR 2008, LNCS 5342, pp. 257–266, 2008.

graphs[2]. In a recent paper we have performed a preliminary study of the Ihara zeta function, and have shown how it can be sampled to give a permutation invariant means of characterizing graphs[12]. However, to be rendered tractable for real world problems the issue of how to avoid sampling the infinities at poles and how to generate stable pattern vectors from the Ihara zeta function must be addressed. We have shown how to overcome this problem by establishing pattern vectors using the polynomial coefficients of the reciprocal Ihara zeta function.

The aim in this paper is to take this work one step further. In particular we aim to study in depth the information conveyed by the individual polynomial coefficients and their relationship to the path length structure of a graph. We show how a reduced set of selected coefficients can be used to efficiently characterize graphs, and that the Euclidean distance between vectors composed of these coefficients can be used to approximate the edit distance between graphs.

2 The Ihara Zeta Function

In this section we review the theory of the Ihara zeta function used in our earlier work. The Ihara zeta function of a graph can be denoted in the form of a rational function[2]:

$$Z_G(u) = \left(1-u^2\right)^{\chi(G)} \det\left(\boldsymbol{I} - u\boldsymbol{A} + u^2\boldsymbol{Q}\right)^{-1} \tag{1}$$

Here, $\chi(G)$is the Euler Number of the graph, which is defined as the difference between the vertex number $|V(G)|$ and the edge number $|E(G)|$ of the graph, i.e. $\chi(G) = |V(G)| - |E(G)|$, and $\boldsymbol{A}$ is the adjacency matrix of the graph. The degree matrix $\boldsymbol{D}$ can be generated by placing the column sums as the diagonal elements, while setting the off-diagonal elements zeros. Finally, $\boldsymbol{Q}$ is the difference of the degree matrix $\boldsymbol{D}$ and the identity matrix $\boldsymbol{I}$, i.e. $\boldsymbol{Q} = \boldsymbol{D} - \boldsymbol{I}$.

3 Polynomial Expression

For md2 graphs, i.e. the graphs with vertex degree at least 2, it is clear-cut that (1) can be rewritten in the form of the reciprocal of a polynomial. However, it is difficult to compute the coefficients of the reciprocal of the Ihara zeta function from (1) in a uniform way, except by resorting to software for symbolic calculation. To efficiently compute these coefficients, it is more convenient to transform the form of the Ihara zeta function in (1) into a concise expression. The Ihara zeta function can also be written in the form of determinant expression[9]

$$Z_G(u) = \frac{1}{\det(\boldsymbol{I} - u\boldsymbol{T})} \tag{2}$$

where $\boldsymbol{T}$ is the Perron-Frobenius Operator[15] on the oriented line graph of the original graph, and is an $2m \times 2m$ square matrix with dimensionality m the number of the edges of the original graph. According to (2), the reciprocal of the Ihara zeta function can be rewritten as:

$$\begin{aligned} Z_G^{-1}(u) &= \det(\boldsymbol{I} - u\boldsymbol{T}) \\ &= (u)^{2m}\det\left(\frac{1}{u}\boldsymbol{I} - \boldsymbol{T}\right) \\ &= u^{2m}[c_0\left(\frac{1}{u}\right)^{2m} + c_1\left(\frac{1}{u}\right)^{2m-1} + \cdots + c_{2m-1}\left(\frac{1}{u}\right) + c_{2m}] \\ &= c_0 + c_1u + \cdots + c_{2m-1}u^{2m-1} + c_{2m}u^{2m} \end{aligned} \tag{3}$$

From (3), the coefficients of the reciprocal of the Ihara zeta function can be derived from the coefficients of the characteristic polynomial of the matrix $\boldsymbol{T}$. The calculation of the above coefficients can be converted to a summation of a series of determinants[3]:

$$c_k = \sum_{\binom{2m}{2m-k}} \begin{vmatrix} b_{1,1} & b_{1,2} & \cdots & b_{1,2m} \\ b_{2,1} & b_{2,2} & \cdots & b_{2,2m} \\ \cdots & \cdots & \cdots & \cdots \\ b_{2m,1} & b_{2m,2} & \cdots & b_{2m,2m} \end{vmatrix} \tag{4}$$

There are in total $\binom{2m}{2m-k}$ determinants in the sum. The relevant matrix in each determinant is created by replacing $(2m-k)$ of the $2m$ diagonal elements of the matrix $\boldsymbol{T}$ with -1 and the remaining elements in those corresponding rows and columns with 0.

4 Pattern Vectors Using Coefficients as Feature Components

To establish pattern vectors from the Ihara zeta function for the purposes of machine learning, one approach is to consider taking function samples as elements. However, if this strategy is adopted then there is the danger of sampling at poles, and these give rise to infinities. Hence, pattern vectors consisting of function samples are potentially unstable since the distribution of poles is unknown beforehand.

To overcome this problem, we note that the coefficients of the reciprocal of the Ihara zeta function do not give rise to infinities. These coefficients are essentially descriptors of graph structures. As long as G is a simple graph, a) the coefficients c_3, c_4, and c_5 are respectively the negative of twice the number of triangles, squares, and pentagons in G, b) c_6 is the negative of twice the number of hexagons in G plus 4 times the number of pairs of edge disjoint triangles plus twice the number of pairs of triangles with a common edge, and c) c_7 is the negative of twice the number of heptagons in G plus 4 times the number of edge disjoint pairs of one triangle and one square plus twice the number of pairs of one triangle and one square that share a common edge[14]. The highest order coefficient is associated with the number of edges incident to vertex v_i, i.e. the node degree $d(v_i)$:

$$c_{2m} = (-1)^{\chi(G)} \prod_{v_i \in V} (d(v_i) - 1) \tag{5}$$

The set of coefficients can play the role of pattern vectors for clustering, not only because of their immunity to the distribution of the poles, but also due to their ability to characterize the graph structures.

Although the full set of coefficients associated with a graph can be used to construct a pattern vector, only a subset of the coefficients contribute significantly. Some coefficients may be redundant and others may reduce the effectiveness of machine learning algorithms. We thus need to select the subset of salient coefficients, i.e. those that take on distinct values for different classes and exhibit small distinctiveness within class variation. To do this, we compute the between-class scatter $S_b = \sum_{i=1}^{M} N_i(\bar{c}_{k,i} - \bar{c}_k)^2$ and the within-class scatter $S_w = \sum_{i=1}^{M} \sum_{c_{k,i,j} \in C_i} (c_{k,i,j} - \bar{c}_{k,i})^2$ of the individual coefficients, where $\bar{c}_k$ is the mean of all the c_k samples, $\bar{c}_{k,i}$ is the mean of the c_k samples in class C_i, N_i is the number of the c_k samples in class C_i and M is the total number of classes. We then use the criterion function $J = (S_b + S_w)/S_w$ to evaluate the performance of individual coefficients. Individual coefficients which give a large contribution to the criterion function are the most significant.

5 Experiment

5.1 Synthetic Graphs

We commence by investigating the relationship between the edit distance and the Euclidean distance between vectors of Ihara coefficients. We begin with a set of randomly generated md2 graphs. The seed graph for the set has 100 vertices

(a) House Sequences

(b) COIL Datasets

Fig. 1. Datasets for Experiments

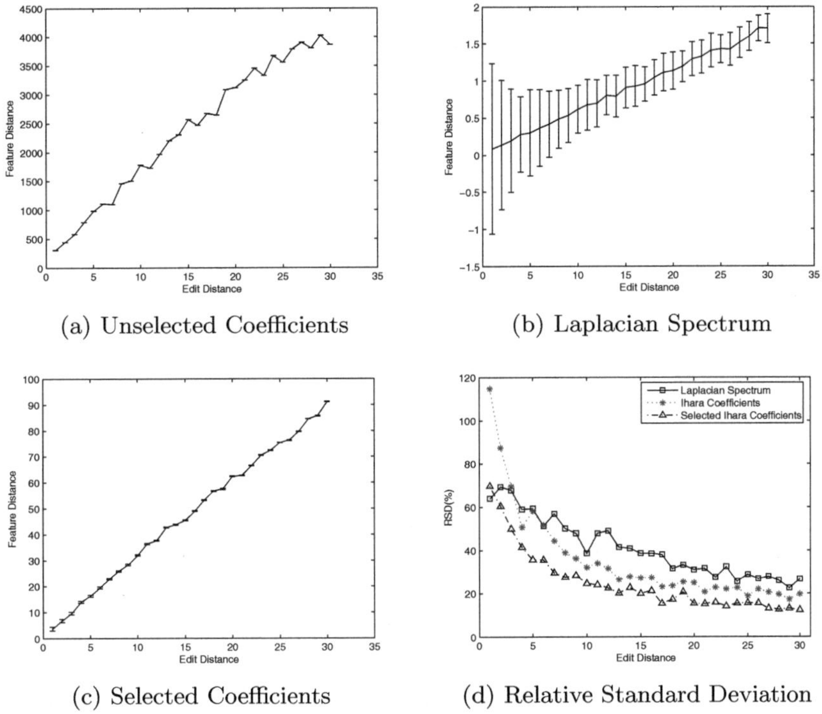

(a) Unselected Coefficients (b) Laplacian Spectrum

(c) Selected Coefficients (d) Relative Standard Deviation

Fig. 2. Feature Distance

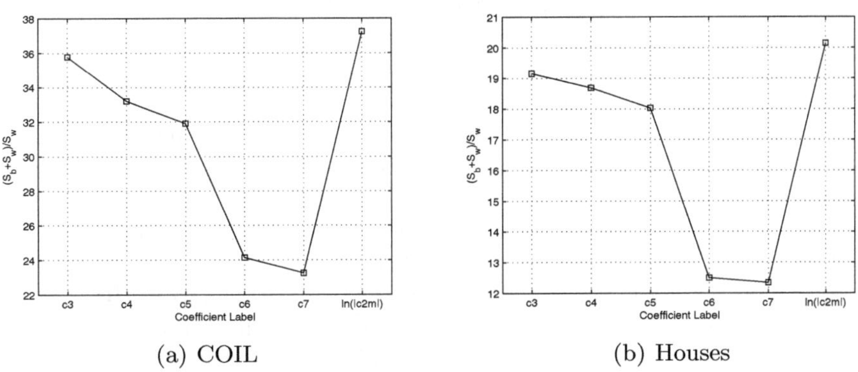

(a) COIL (b) Houses

Fig. 3. Criterion Function Value

and 300 edges. The remainder of the graphs in the set are obtained by deleting the edges of the seed graph (indexed from 1 to 30). At each level of editing, 100 trials are performed and the edges deleted are chosen randomly, subject to preserving the md2 constraint. We compare the Euclidean distance between the Ihara coefficients with the both the edit distance and the distance between Laplacian spectra. In our experiments, we compute the Ihara coefficients using (4) and

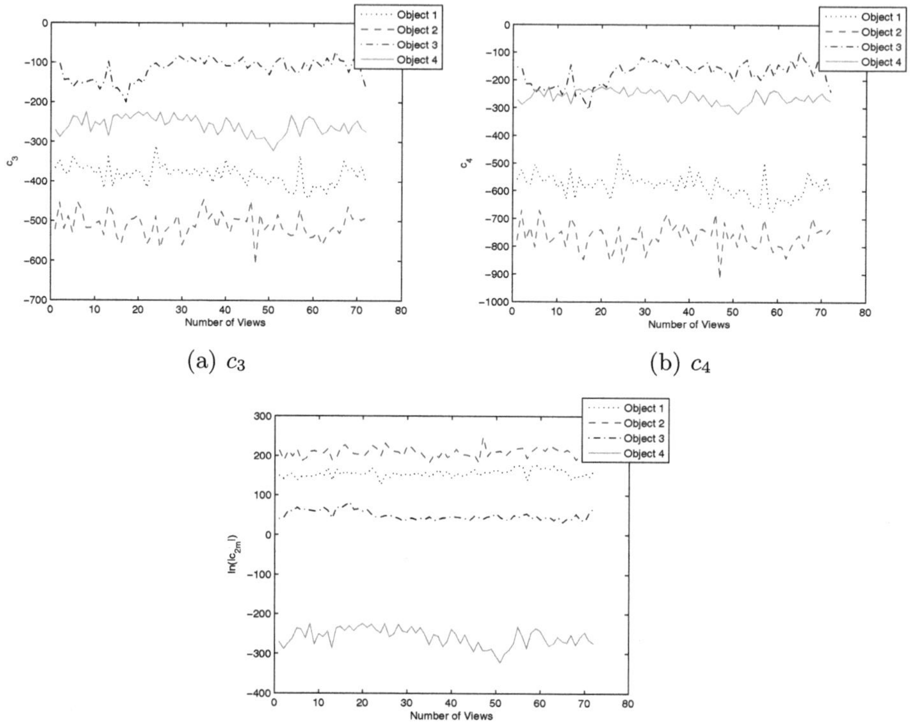

(a) c_3

(b) c_4

(c) $\ln(c_{2m})$

Fig. 4. Coefficients from COIL

(5), constructing a pattern vector in the form $v_G = [c_3\ c_4\ c_5\ c_6\ c_7\ \ln(|c_{2m}|)]^T$. The final component of the pattern vector is scaled in a logarithmic manner for the purpose of avoiding high order oscillations. The Lapalacian spectral pattern vectors are formed by taking the second smallest through to the seventh smallest eigenvalues of graph Laplacian as components. The experimental results are shown in Fig.2. Fig.2(a) shows the distance between coefficient pattern vectors from the seed graph and the edited graph as a function of the edit distance, i.e. the number of edges deleted. The relative standard deviation (RSD) is also shown as an error bar. The coefficient distance generally follows the edit distance. Fig.2(b) shows the Laplacian spectral distance as well as the corresponding RSD as a function of the edit distance (but scaled differently to Fig.2(a)). From Fig.2(a) and Fig.2(b) it is clear that the dynamic range of the coefficient feature distance is much larger than that of the spectral distance for corresponding edit distance. Thus the coefficients are more sensitive to graph edits than the Laplacian spectra. To evaluate how reliable the coefficient distance and the spectral distance predict the edit distance, we plot their RSD as a function of edit distance in Fig.2(d). The RSD of the coefficient distance is larger than that of the spectral distance. This means that the Laplacian spectra are more stable than the coefficients under graph edits.

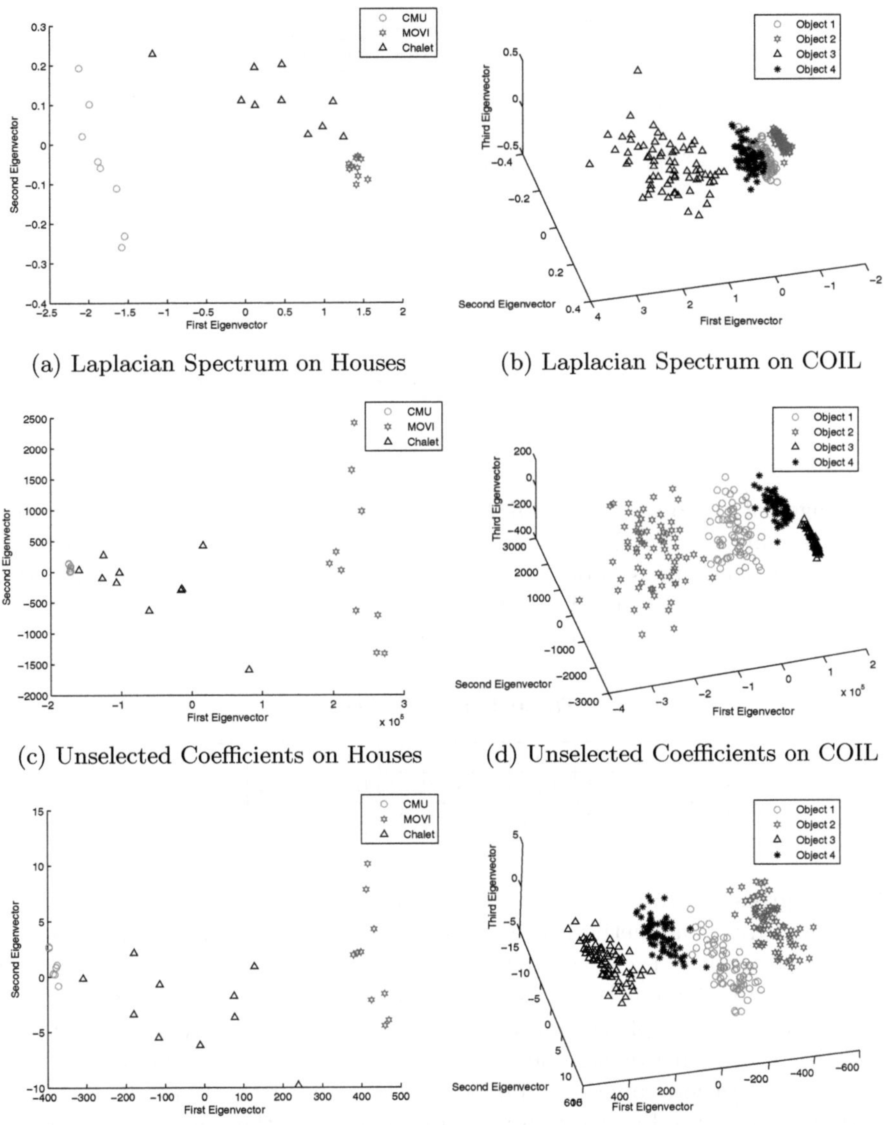

(a) Laplacian Spectrum on Houses (b) Laplacian Spectrum on COIL

(c) Unselected Coefficients on Houses (d) Unselected Coefficients on COIL

(e) Selected Coefficients on Houses (f) Selected Coefficients on COIL

Fig. 5. Clustering Performance

5.2 Visual Clustering

We apply the pattern vectors composed of Ihara coefficients to two graph datasets used previously in the work of Bai and Hancock. The first set of graphs are extracted from three sequences of images of model houses (referred to as the CMU, MOVI and chalet sequences in Fig.1(a)). The second set of graphs are extracted

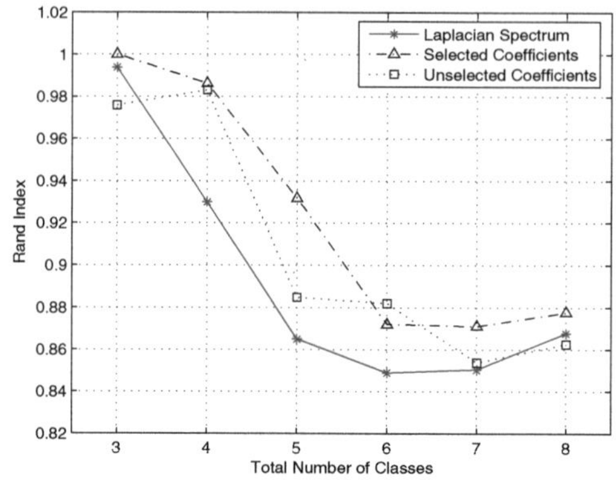

Fig. 6. Rand Index

from images of the objects in COIL database (Fig.1(b)). We first extract corner points using the Harris detector. Then we establish Delaunay graphs based on these corner points as nodes. The pattern vectors are formed as explained in Section 5.1. We perform PCA on the pattern vectors to embed them into a 3-dimensional space. We then locate the clusters using K-means method.

Table 1. Rand Indices

	(a)	(b)	(c)	(d)	(e)	(f)	(g)
House	0.82	0.71	0.83	0.73	0.73	0.83	0.40
COIL	0.99	0.84	0.98	0.79	0.79	0.80	0.66

Table 1 gives the Rand indices obtained when clustering is attempted using different features. The methods are (a) the Laplacian spectrum (second smallest eigenvalue through to the seventh samllest eignevalues), (b) the Rosenberg zeta function used by Bai and Hancock sampled at the integer values 1 to 6, (c) the Ihara zeta function sampled over the range from 0.001 to 0.006 with interval 0.001, (d) the Ihara zeta function sampled over the range from 0.01 to 0.06 with interval 0.01, (e) the Ihara zeta function sampled over the range from 0.11 to 0.16 with interval 0.01 and (f) the Ihara zeta function sampled over the range from 0.1 to 0.6 with interval 0.1. The clusters are located using the K-means method. In Table 1 we can see that the proposed method outperforms the Rosenberg zeta function, and is comparable with the Laplacian spectrum. The final three columns in Table 1 indicate the performance of pattern vectors constructed by sampling the Ihara zeta function. If the samples are appropriately chosen, then this method is comparable with the use of the coefficients. However, when function is sampled in the proximity of poles, the clustering performance deteriorates.

Finally, we explore which combination of coefficients gives the best performance. To do this, we select the coefficients according to the criterion introduced in Section 4.2. We select 3 objects for each of which 10 sample images are used as training data to compute the criterion function value. Fig.3 shows the criterion function values for the coefficients extracted from the two datasets. The first and last coefficients offer more discrimination than the middle ones. This is because the remaining coefficients provide no significantly increase in information over c_3, c_4 and c_{2m} since they are determined by the number of triangles and squares in the graph and the node degrees. This information can be fully represented by c_3, c_4 and c_{2m}. Based on this feature selection analysis we work with the pattern vector $v_{Gs} = [c_3\ c_4\ \ln(|c_{2m}|)]^T$. The three components of v_{Gs} extracted from the first 4 objects in the COIL dataset are shown in Fig.4 as a function of view numbers. Each line in a plot represents the coefficient extracted from one object. The lines in each plot are well separated thus indicating that the three coefficients are sufficient to distinguish different object classes. Fig.2(c) shows the feature distance computed using the three selected coefficients as a function of the edit distance. Compared with Fig.2(a), the selected coefficients are more linear with the edit distance than the unselected coefficients. From Fig.2(d), it is clear that although the unselected coefficients are less stable than the graph spectra, the selected coefficients offer best stability. We apply the pattern vectors of a) Laplacian spectra, b) unselected Ihara coefficients and c) selected Ihara coefficients to both the house sequences and the first four objects in COIL dataset. Fig.4 shows the clustering results obtained by performing PCA on the pattern vectors. The selected coefficients outperform both the Laplacian spectra and the unselected coefficients,and give clusters with better separation.

We then confine our attention to the COIL dataset, and evaluate the clustering performance obtained with different numbers of object classes. After performing PCA on the pattern vectors, we locate the clusters using the K-means method and calculate the Rand index. The Rand index for each pattern vector is plotted as a function of class number in Fig.4. The selected coefficients gives the best performance and the Laplacian spectra the poorest performance.

6 Conclusion

In this paper, we show how to construct pattern vectors for graph characterization using the Ihara zeta function. We compute polynomial coefficients using the reciprocal of the Ihara zeta function. We construct pattern vectors by performing feature selection on the coefficients. Unlike the samples of the Ihara zeta function, the coefficients are not prone to singularities due to poles. The method outperforms the Laplacian spectrum both in terms of stability and in clustering performance on md2 graphs.

References

1. Bai, X., Hancock, E.R.: Recent results on heat kernel embedding of graphs. In: Brun, L., Vento, M. (eds.) GbRPR 2005. LNCS, vol. 3434, pp. 373–382. Springer, Heidelberg (2005)
2. Bass, H.: The Ihara-Selberg zeta function of a tree lattice. International Journal of Mathematics 6(3), 717–797 (1992)

3. Brooks, B.P.: The coefficients of the characteristic polynomial in terms of the eigenvalues and the elements of an n×n matrix. Applied Mathematics letters 19(6), 511–515 (2006)
4. Chung, F.K.: Spectral graph theory. American Mathematical Society (1997)
5. Haemers, W.H., Spence, E.: Enumeration of cospectral graphs. European Journal of Combinatorics 25(2), 199–211 (2004)
6. Hashimoto, K.: Zeta functions of finite graphs and representations of p-adic groups. Advanced Study of Pure Mathematics 15, 211–280 (1989)
7. Ihara, Y.: Discrete subgroups of PL(2, k_{φ}). In: Proceeding Symposium of Pure Mathematics, pp. 272–278 (1965)
8. Ihara, Y.: On discrete subgroups of the two by two projective linear group over p-adic fields. Journal of Mathematics Society Japan 18, 219–235 (1996)
9. Kotani, M., Sunada, T.: Zeta functions of finite graphs. Journal of Mathematics University of Tokyo 7(1), 7–25 (2000)
10. Luo, B., Wilson, R.C., Hancock, E.R.: Spectral embedding of graphs. Pattern Recognition 36(10), 2213–2223 (2003)
11. Rosenberg, S.: The Laplacian on a Riemannian manifold. Cambridge University Press, Cambridge (2002)
12. Ren, P., Wilson, R.C., Hancock, E.R.: Pattern vectors from the Ihara zeta function. In: IEEE International Conference on Pattern Recognition (submitted, 2008)
13. Sanfeliu, A., Fu, K.S.: A distance measure between attributed relational graphs for pattern recognition. IEEE Transanctions on Systems, Man, and Cybernetics 13, 353–362 (1983)
14. Scott, G., Storm, C.K.: The coefficients of the Ihara zeta function. Involve - A Journal of Mathematics 1(2), 217–233 (2008)
15. Storm, C.K.: The zeta function of a hypergraph. Electronic Journal of Combinatorics 13(1) (2006)
16. Wilson, R.C., Luo, B., Hancock, E.R.: Pattern vectors from algebraic graph theory. IEEE Transactions on Pattern Analysis and Machine Intelligence 27(7), 1112–1124 (2005)

A Learning Approach to 3D Object Representation for Classification

Indriyati Atmosukarto and Linda G. Shapiro

University of Washington,
Department of Computer Science and Engineering,Seattle, WA, USA
{indria,shapiro}@cs.washington.edu

Abstract. In this paper we describe our 3D object signature for 3D object classification. The signature is based on a learning approach that finds salient points on a 3D object and represent these points in a 2D spatial map based on a longitude-latitude transformation. Experimental results show high classification rates on both pose-normalized and rotated objects and include a study on classification accuracy as a function of number of rotations in the training set.

Keywords: 3D Object Classification, 3D Object Signature.

1 Introduction

Advancement in technology for digital acquisition of 3D models has led to an increase in the number of 3D objects available in specific application databases and across the World Wide Web. This has motivated recent research in two major areas. The first research area is 3D object retrieval where the goal is to retrieve 3D objects from a database that are similar in shape to a given 3D object query. The second research area is 3D object classification where previously unseen objects are assigned to one of a set of previously defined object classes. Classification helps in labeling new unseen objects and can be applied to a number of fields including engineering, architecture, and medical field. The two research areas are complementary as 3D object classification helps in organizing and retrieving objects from a database. Classifying a query object before retrieving similar objects from the database helps to speed up and potentially improve the retrieval process.

Classification of 3D objects requires them to be represented in a way that captures the global and local characteristics of the object. This is achieved by creating a 3D descriptor or signature that summarizes the important shape properties of the object that can be used in classification. In this paper, we describe our method of representing a 3D object and using the representation for 3D object classification. Our 3D object signature is based on a learning approach that finds the salient points on a 3D object and represent these points in a 2D spatial map based on longitude-latitude transformation. A classifier is trained on the signatures of the objects for a number of classes and can then be used to classify new unseen objects.

N. da Vitora Lobo et al. (Eds.): SSPR&SPR 2008, LNCS 5342, pp. 267–276, 2008.

Motivated by our work on medical applications [2], we wanted to explore the performance of the signatures in classifying objects that have a similar global shape. For the experiments in this paper, we chose to classify head shapes including human, dog, cat, rabbit, horse, wild cats, and bear heads.

The rest of the paper is organized as follows. We first discuss existing classification methods and existing 3D object signatures. We then describe our data acquisition process. Next we describe our method for learning the salient points of a 3D object and then describe the longitude-latitude map signature that captures the patterns of the salient points. Finally, we explain and analyze the results results from our experiments and provide a summary and suggestions for future work.

2 Related Work

Biasotti et al. [3] presented a comparative evaluation study for 3D shape classification. In their work, they compared the performance of five similarity measures on four different shape descriptors in classifying 3D objects, and concluded that using the nearest neighbor similarity measure performed the best for all four different shape descriptors. Huber et al. [4] presented a parts-based method for 3D object classification. Their method extracted parts from an object and learned part classes and a mapping from the part classes to the object classes using probability.

Lee et al. [9] used the curvature information at every mesh vertices to find the salient points.Their method then computed the difference between mean curvature at different scales. The final saliency value was the aggregate of the saliency at all scales with a non-linear normalization. Ohbuchi et al. [11] rendered multiple views of a 3D model and extracted local features from each view image using the SIFT algorithm. The local features are then integrated into a histogram using the bag-of-features approach. Novatnack et al. [10] derived corner and edge detectors to extract salient points at each scale.

Assfalg et al. [1] captured the shape of a 3D object using the curvature map of the object's surface. The model surface is first warped into a sphere, and the curvature information is then mapped onto a 2D image using the Archimedes projection. Retrieval of 3D objects is performed by comparing the 2D map of the query object against the 2D maps of the database models using a histogram-based search as one method and a weighted walk-through of the map regions as another. Our method differs from theirs in that it does not use curvature directly, but instead uses a classifier to find salient points and labels them according to the classifier prediction score. It then uses the longitude and latitude position of the salient points on the object's surface to create a 2D map signature. A second classifier is trained on these 2D map signatures for object classification.

3 Data Acquisition

The 3D objects used in our experiments were obtained by scanning hand-made clay toys using a Roldand-LPX250 laser scanner with a maximal scanning resolution of 0.008 inches for plane scanning mode [6]. Raw data from the scanner consisted of clouds of 3D points that were further processed to obtain smooth and uniformly sampled triangular meshes of 0.9-1.0 mm resolution.

To increase the number of objects for training and testing our method, new objects were created by deforming the original scanned 3D models in a controlled fashion using 3D Studio Max [5]. Global deformations of the models were generated using morphing operators such as tapering, twisting, bending, stretching and squeezing. The parameters of each of the operators were randomly chosen from given ranges that were determined empirically. Each deformed model included at least five different deformations applied in a random sequence. Figure 1 shows an example of an original scanned human head model and two different morphed versions. The first morphed model was generated by squeezing the original model resulting in a slimmer head, while the second morphed model was generated by twisting the original model resulting in a slightly asymmetrical face.

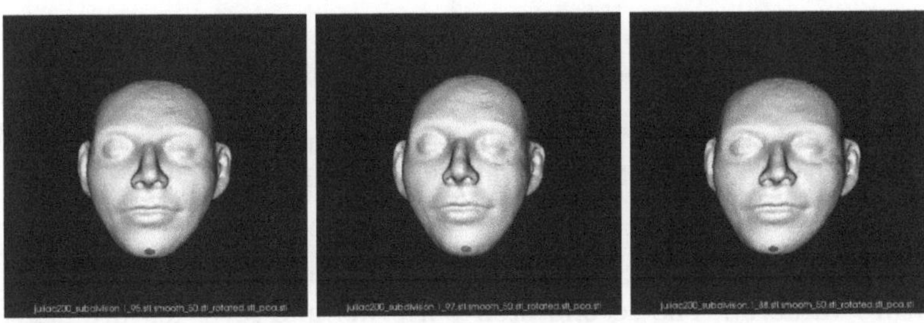

(a) (b) (c)

Fig. 1. (a) Original scanned human head model and (b,c) two different morphed versions of the model

Fifteen objects were originally scanned to create a 7-class database. In this work we focus on heads of different classes and shapes to emphasize the repeatability of the method. The seven classes are: cat head, dog head, human head, rabbit head, horse head, tiger head and bear head. Each of the fifteen original objects were randomly morphed to increase the database size. A total of 250 morphed models per original object were created. For this work, we used 75 morphed models from each of the original objects for training and testing the classifier. Points on the morphed models are in full correspondence with the original models from which they were constructed.

4 Learning Salient Points

Our methodology starts by applying a low-level operator to every point on the surface mesh. The low-level operators extract local properties of the surface points by computing a single feature value v_i for every point p_i on the mesh surface. Five different low-level operators were implemented to test the algorithm's performance. The five low-level features are Gaussian curvature, mean curvature, shape index, normal magnitude, and curvedness. In this work, we will use the absolute values of the Gaussian curvature for our experiments. The low-level feature values are convolved with a Gaussian filter to reduce the noise. Figure 2a shows an example of the absolute Gaussian curvature values of a 3D model, and Figure 2b shows the results of applying a Gaussian filter over the low-level values.

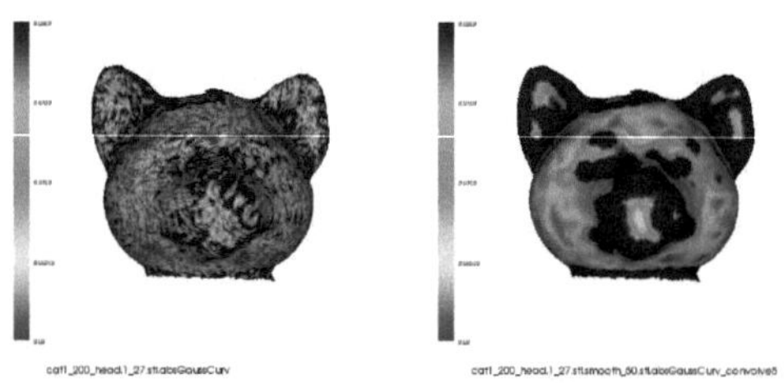

(a) (b)

Fig. 2. (a) Absolute Gaussian curvature low-level feature value and (b) Smooth low-level feature values after convolution with the Gaussian filter

After this first step, every point p_i on the surface mesh will have a low-level feature value v_i. The second step performs mid-level feature aggregation to compute an aggregate vector for a given neighborhood of every point p_i on the surface mesh. In this work, we use local histograms to aggregate the low-level feature values of each point. The histograms are computed by taking a neighborhood around each point and accumulating the low-level features in that neighborhood. The size of the neighborhood is determined by multiplying a constant c, $0 < c < 1$, with the diagonal of the object's bounding box. This ensures that the size of the neighborhood is scaled according to the object's size and that the results are comparable across objects. The aggregation results in a d-dimensional vector f_i for every point p_i on the surface mesh where d is the number of histogram bins. For our experiments, we used $d = 250$ and $c = 0.05$.

Because we were not satisfied with the salient points computed by standard interest operators (eg. Kadir's entropy operator [7] or Lowe's SIFT operator [8] applied to 3D), we chose to teach a classifier the characteristics of points that we regard as salient. Histograms of low-level features are used to train a Support Vector Machine (SVM) classifier [12] [13] to learn the salient points on the 3D surface mesh. We used the SVM implemented in WEKA for our experiments [14]. The training data for supervised learning for the classifier are obtained by manually marking salient and non-salient points on the surface of each training object. An average of 12 salient points and 12 non-salient points were marked on the training objects. Since our current database contains head shapes (human heads, wildcat heads, bear heads, etc.), salient points that were chosen included the tip of the nose, corner of the eyes, corner and midpoints of

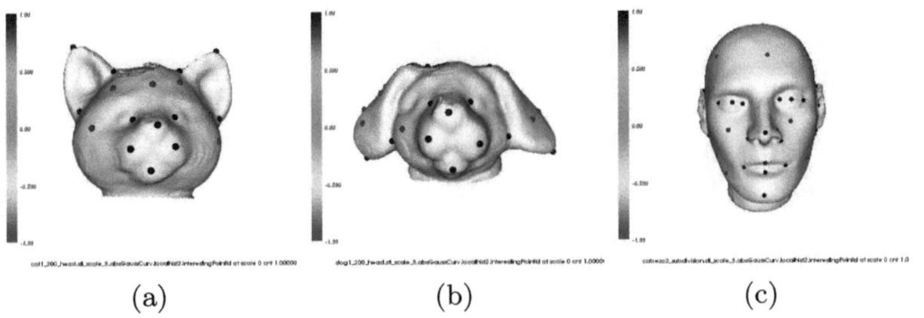

Fig. 3. Examples of manually marked salient (blue color) and non-salient (red color) points on (a) cat head model, (b) dog head model, and (c) human head model

the lips, etc. The histogram of low-level features of each of the marked points are saved and used for the training. Figure 3 shows examples of manually marked salient and non-salient points on the training data. Since the morphed models were in total correspondence with the original model, we only had to manually mark the original 15 models.

A small training set was used to train the classifier to learn the salient points. The training data for the classifier consisted of the histogram of low-level features for the manually marked salient and non-salient points of 25 instances of the cat head model, 25 instances of the dog head model and 50 instances of the human head models. The classifier was trained with 10-fold cross validation. The classifier does not learn the exact point positions, but instead learns the characteristics of the salient training points in terms of the histograms of their low-level features.

To test the classifier, the histogram of low-level features of all the points of an object are given as its input. The classifier labels each of the points of the 3D object as either salient or non-salient and provides a confidence score for its decision. A threshold T is applied to the confidence scores for the salient points.

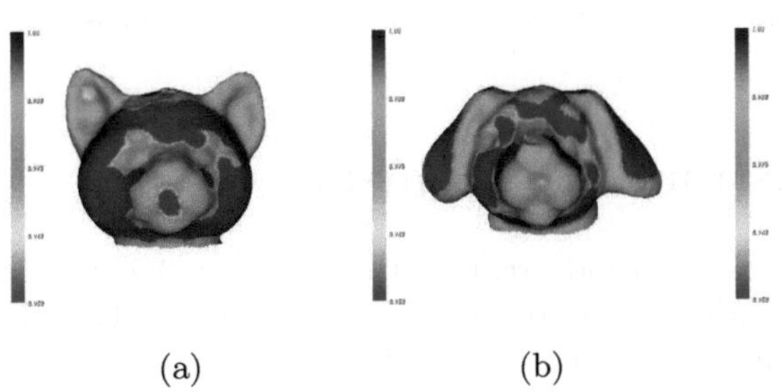

Fig. 4. Salient point prediction for (a) cat head class, (b) dog head class, and (c) human head class. The three different columns of models shows the repeatability of the salient point prediction for models of the same class. Non-salient points are colored in red, while salient points are colored in different shades ranging from green to blue, depending on the classifier confidence score assigned to the point. A threshold (T = 0.95) was applied to include only salient points with high confidence score.

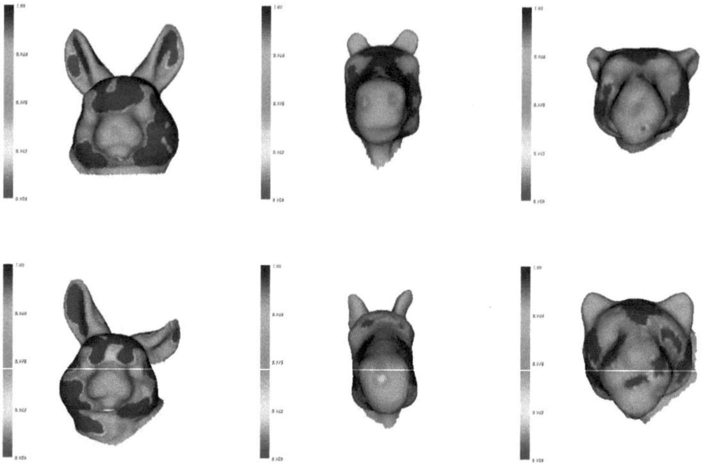

(a) (b) (c)

Fig. 5. Salient point prediction for (a) rabbit head class, (b) horse head class, and (c) leopard head class. Even though all three classes were not included in the training, the training model was able to predict repeatable salient points across the classes.

In our experiments, we used $T = 0.95$ to keep only the salient points with high confidence scores from the classifier. Figure 4 shows results of the salient points predicted on instances of the cat, dog and human head class. The salient points are colored according to the classifier confidence score assigned to the point. Non-salient points are colored in red, while salient points are colored in different shades of blue with dark blue having the highest prediction score.

The salient point prediction model was also tested using object classes that were not included in the training phase. Figure 5 shows the predicted salient points on these new object classes. The results show that the salient points predicted are repeatable across objects of the same class, and across objects of different classes.

5 2D Longitude-Latitude Map Signature

Most 3D object classification methods require the use of a 3D descriptor or signature to describe the shape and properties of the 3D objects. Our signature is based on the salient point patterns of the 3D object mapped onto a 2D plane via a longitude-latitude transformation.

Before mapping the salient point patterns onto the 2D plane, the salient points are assigned a label according to the classifier confidence score assigned to the point. The classifier confidence score range is discretized into a number of bins. For our experiments, at confidence level 0.95 and above, we chose to discretize

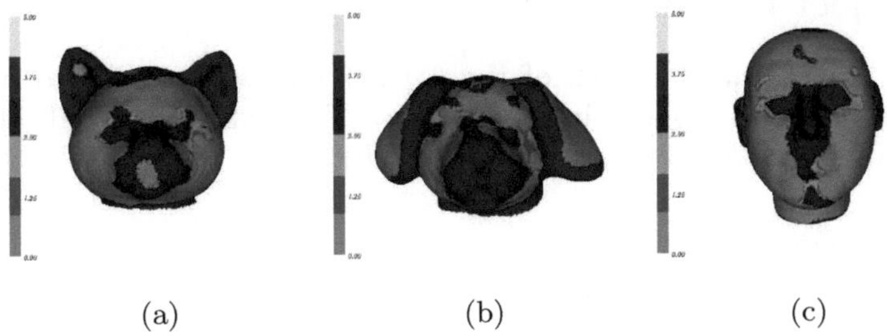

Fig. 6. Salient point patterns on 3D objects of Figure 4. Salient points with confidence 0.95 − 0.96: red, 0.96 − 0.97:green, 0.97 − 0.98: blue, 0.98 − 0.99: dark red and 0.99 − 1.0: yellow. Gray points are non-salient points.

the confidence score range into 5 bins. Each salient point on the 3D mesh is assigned a label based on the bin into which its confidence score falls. Figure 6 shows the discretized salient point patterns on the 3D objects of Figure 4.

To obtain the 2D longitude-latitude map signature for an object, we calculate the longitude and latitude positions of all the 3D points on the object's surface. Given point $p_i(p_{ix}, p_{iy}, p_{iz})$, the longitude position θ_i and latitude position ϕ_i of point p_i are calculated as follows:

$$\theta_i = \arctan(\frac{p_{iz}}{p_{ix}}) \quad \phi_i = \arctan(\frac{p_{iy}}{\sqrt{(p_{ix}^2 + p_{iz}^2)}})$$

where $\theta_i = [-\pi, \pi]$ and $\phi_i = [-\frac{\pi}{2}, \frac{\pi}{2}]$

A 2D map of the longitude and latitude positions of all the points on the object's surface is created by binning the longitude and latitude values of the points into a fixed number of bins. A bin is labeled with the salient point label of the points that fall into that bin. If more than one label is mapped to a bin, the label with the highest count is used to label the bin. Figure 7 shows the 2D map signature for the cat head, dog head, and human head model of Figure 6.

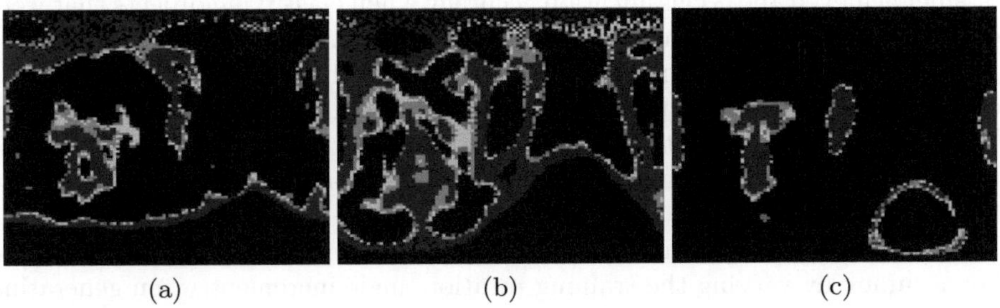

Fig. 7. 2D longitude-latitude map signatures for (a) cat head, (b) dog head, and (c) human head from Figure 6. All three objects were registered to face forward.

6 3D Object Classification

By creating a signature for each of the 3D objects, we are able to perform classification for all the objects in the database. Our experiments fall into two categories: pose normalized and rotation invariant.

6.1 Pose Normalized

For the first part of the classification experiments, all objects in the database were rotated to face forward and have the same orientation. Classification of the 3D objects in the database was performed by training a SVM classifier on the salient point patterns of each class using the 2D longitude-latitude signature of the objects in the class. We trained the classifier with the signatures of 25 objects from each class for all seven classes in the database and tested the classifier model with a new training set consisting of 50 objects per class for each of the seven classes. The classifier achieved 100% accuracy in classifying all the pose-normalized objects in the database.

6.2 Rotation Invariant

Since 3D objects may be encountered in the world in any orientation, rotation invariant classification is desirable. To achieve rotation invariance for classification, we trained the classifier with a number of rotated versions of the 2D map longitude-latitude signature for each training object. Due to WEKA space and memory constraints, we reduced the database from 7 classes to 3 classes for these tests: human head, wild cat head and bear head, and selected 5 objects per class as the base training set.

For the rotation invariance study, we ran three sets of experiments. We first tested the classification accuracy by training a classifier with rotated versions of the training data signatures at 45 degree increments for all three axes. This resulted in $8 \times 8 \times 8$ rotated signatures for each object in the database. The classifier was tested on four new objects per class. Rotated versions of the testing data signatures were generated using the same rotation degree increments. The classifier achieved 100% classification accuracy when classifying objects that were rotated in the same increments as the training data.

In the second experiment, the classification method was tested on 15 new testing instances per class that were rotated randomly. For example, a rotation of (250,4,187) was one of the random rotations that did not match any of the training rotations. The classifier was still able to achieve 100% classification accuracy.

Our third set of experiments was to explore the degradation in the classification accuracy by varying the training rotation angle increment when generating the signatures for the training data. Figure 8 shows the degradation in the classification accuracy as the training angle increment increases and the number of rotated training signature instances decreases. The graph shows that the classification accuracy steadily decreases as the number of rotated training signature

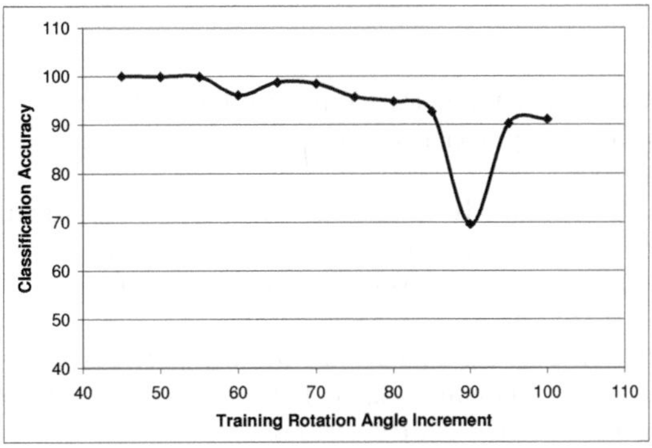

Fig. 8. Classification accuracy vs training rotation angle increment

decreases. In addition, there is a big dip in the classification performance when the training signatures are generated at 90 degree angle increments. This is because the signatures produced at 90 degree increments are not representative of angles in between the multiples of 90 degrees. Note that the classifier is still able to achieve 91% classification accuracy with training signatures generated at 100 degree increments with only $3 \times 3 \times 3 = 27$ rotated training signatures per training object, which is much better than the $8 \times 8 \times 8 = 512$ signatures that we started with.

7 Conclusions

We have described a methodology that uses a learning approach to identify salient points on 3D objects. The patterns of these salient points are then used to train a classifier for object class recognition by transforming the patterns onto a 2D map using longitude-latitude transformation. The first set of experiment results show that the classifier is able to achieve high accuracy when classifying objects that are pose normalized and are of the same orientation. We extended the approach to achieve rotation invariance by generating a number of rotated version of each of the training object signatures and trained a classifier on these rotated signatures to learn possible rotation combination and their respective maps. Experimental results show that the classifier was able to achieve 100% classification accuracy when trained on rotated signatures generated at 45 degree angle increments and can still achieve 91% classification accuracy when trained on rotated signatures generated at 100 degree angle increments.

We are investigating clustering the rotated training signatures to further reduce space usage and computation time so that we will be able to include many more models in the database. By clustering the training signatures, we will be able to reduce the training data by using only the cluster centroids. In addition, we are also working on finding a mapping function between two signatures. This will eliminate the need to generate rotated version of the signatures and will greatly improve space and computation efficiency.

References

1. Assfalg, J., Bimbo, A.D., Pala, P.: Content-based retrieval of 3d models through curvature maps: a cbr approach exploiting media conversion. Multimedia Tools Appl. 31, 29–50 (2006)
2. Atmosukarto, I., Travillian, R., Franklin, J., Shapiro, L., Brinkley, J., Suciu, D.: A unifiying framework for combining content-based image retrieval with relational database queries for biomedical applications, SIIM (2008)
3. Biasotti, S., Giorgi, D., Marini, S., Spagnuolo, M., Facidieno, B.: A Comparison Framework for 3D Object Classification Methods. MRCS, 314–321 (2006)
4. Huber, D., Kapuria, A., Donamukkala, R.R., Hebert, M.: Parts-based 3D object classification. In: Proc. IEEE CVPR (2004)
5. Autodesk 3d studio max, http://usa.autodesk.com
6. Correa, S.R., Shapiro, L., Meila, M., Berson, G., Cunnnigham, M., Sze, M.: Symbolic Signatures for Deformable Shapes. IEEE Trans. PAMI 28(1), 75–90 (2004)
7. Kadir, T., Brady, M.: Scale, Saliency and Image Description. IJCV 45(2), 83–105 (2001)
8. Lowe, D.: Distinctive image features from scale-invariant keypoints. IJCV 60(2), 91–110 (2004)
9. Lee, C.H., Varshney, A., Jacobs, D.W.: Mesh saliency. ACM Trans. Graph 24(3), 659–666 (2005)
10. Novatnack, J., Nishino, K.: Scale-dependent 3d geometric features. In: Proc. ICCV (2007)
11. Ohbuchi, R., Osada, K., Furuya, T., Banno, T.: Salient local visual features for shape-based 3d model retrieval. In: Proc. IEEE SMI (2008)
12. Scholkopf, B., Smola, A.J.: Learning with kernels. Cambridge Uni. Press, Cambridge (2002)
13. Vapnik, V.V.: Statistical Learning Theory. John Wiley and Sons, Chichester (1998)
14. Witten, I.H., Frank, E.: Data Mining: Practical machine learning tools and techniques, 2nd edn. Morgan Kaufmann, San Francisco (2005)

Region Based Visual Object Categorization Using Segment Features and Polynomial Modeling

Huanzhang Fu, Alain Pujol, Emmanuel Dellandréa, and Liming Chen

LIRIS, UMR 5205 CNRS
Ecole Centrale de Lyon - 36 Av. Guy de Collongue - 69134 Ecully Cedex - France
{huanzhang.fu,alain.pujol,
emmanuel.dellandrea,liming.chen}@ec-lyon.fr

Abstract. This paper presents a novel approach for visual object classification. Based on Gestalt theory, we propose to extract features from coarse regions carrying visually significant information such as line segments and/or color and to include neighborhood information in them. We also introduce a new classification method based on the polynomial modeling of feature distribution which avoids some drawbacks of a popular approach, namely "bag of keypoints". Moreover we show that by separating features extracted from different sources in different "channels", which are then combined using a late fusion strategy, we can limit the impact of feature dimensionality and actually improve classification accuracy. Using this classifier, experiments reveal that our features lead to better results than the popular SIFT descriptors, but also that they can be combined with SIFT features to reinforce performance, suggesting that our features managed to extract information which is complementary to the one of SIFT features.

1 Introduction

Generic visual object classification is one of the most challenging topics in computer vision. Indeed the number of real world object types as well as variations in view, imaging, lighting and occlusion pose serious problems in this domain. Furthermore we must add to this the difficulty induced by intra-class variations, typical of semantic classes of everyday objects. As such it has attracted a lot of attention in the past years [1].

1.1 Related Work

Most works in the literature make use of a "bag of features" kind of approach [2,3] which tries to adapt the "bag-of-words" representation for text categorization to "Visual Object Categorization" (VOC) problem and has shown its effectiveness, obtaining the best performance in Pascal VOC contest [1]. These methods view images as an orderless distribution of local image features, typically using the popular SIFT features [4], extracted from salient image regions,

N. da Vitora Lobo et al. (Eds.): SSPR&SPR 2008, LNCS 5342, pp. 277–286, 2008.

called interest "points" [4,5,6] or more simply from points extracted using a grid [7]. The set of these local features is then characterized by a histogram of "visual keywords" from a visual vocabulary which is learned from the training set by a hard assignment (quantization) or a soft assignment through GMMs. These distributions can thus be compared to estimate the similarities between images and categorized through a machine learning process, for instance SVM.

Although the "bag-of-local features" approach has achieved the best performance in the last Pascal VOC contests, the overall performance, with an average precision around 60% over 20 classes achieved by the best classifier, is still far from real application-oriented requirements. In particular, the size of visual vocabulary which is the basis of this approach is hard to be fixed as there are no evident similar concepts in images as compared to a textual document. The basic problem is that the "bag-of-local features" approach, while adapting the best practice from text categorization, does not necessarily correspond to a human visual perception process which is ruled by some Gestalt principles according to several studies on visual perception [8,9] and supposed to perform a holistic analysis combined with a local one through a fusion process. Moreover, the schemes so far proposed in the literature for automatic generic visual object classification also suffer from well-known machine learning problems, namely the curse of dimensionality when increasing feature vector size which leads to exponential learning complexity as well as a small and biased training dataset, in particular with an imbalanced ratio of positives versus negative samples.

1.2 Our Approach

Our basic hypothesis is that effective visual object classification should be inspired by some basic human image interpretation principles. In this paper we propose overcoming the shortfalls of the popular "bag-of-local features" approach and make use of some basic principles from the Gestalt theory, especially the well known Gestalt laws of Perceptual Organization which suggest both the grouping of pixels into homogeneous regions as well as the interaction between regions.

Desolneux et al. have given in [10] a comprehensive introduction to Gestalt theory in an image analysis perspective. Gestalt theory starts with the assumption of active grouping laws in visual perception which recursively cluster basic primitives into a new, larger visual object, a *gestalt*. These grouping laws follow criterion such as spatial proximity, color similarity. They also highlight the interaction between regions.

We feel that the popular "bag of features" like approaches, while ignoring these principles, deprive themselves of meaningful information. Thus, instead of SIFT like local features, we propose some region-based meaningful features extracted from image regions with neighborhood information. These features result from perceptually significant "Gestalts" segmented according to some basic Gestalt grouping laws. Moreover, we propose using visually meaningful features, such as color and line segment based features which we extend to provide information from neighboring regions. Our goal is to compare the use of these region based features with the popular SIFT features but also to check the efficiency of the combination of them to evaluate their complementarities.

We also propose a polynomial representation to model image feature distribution. Its interest is 3-fold. We can circumvent the difficulty of fixing the size of visual vocabulary, avoid the inaccurate assumption of Gaussian repartition of features and finally cope with a smaller number of feature vectors per image.

Finally, we also study and compare two fusion strategies, namely early fusion and late fusion. Experiments carried out show that our features can be combined with SIFT features using fusion strategy to provide better overall performance, especially in the case of using late fusion strategy.

The rest of the paper is organized as follows. In section 2 we describe our Gestalt-inspired region segmentation algorithm and the extraction of region-based features. Section 3 introduces our classification method while section 4 contains experimental results and discussion. Section 5 concludes the paper while depicting some future research directions.

2 Region Segmentation and Region-Based Features

In this section, we first introduce our Gestalt-inspired region segmentation scheme [11] and then our color and segment-based features extracted from the region map by our segmentation scheme.

2.1 Gestalt-Inspired Region Segmentation

The principle of our region segmentation algorithm is to segment an image into partial gestalts for further visual object recognition. These partial gestalts are significant regions within the image satisfying some grouping laws. The segmentation result should have some robustness against the color variation from one image to another. We thus made use of the following Gestalt basic grouping laws in our gestalt construction process: color constancy law, similarity law, vicinity

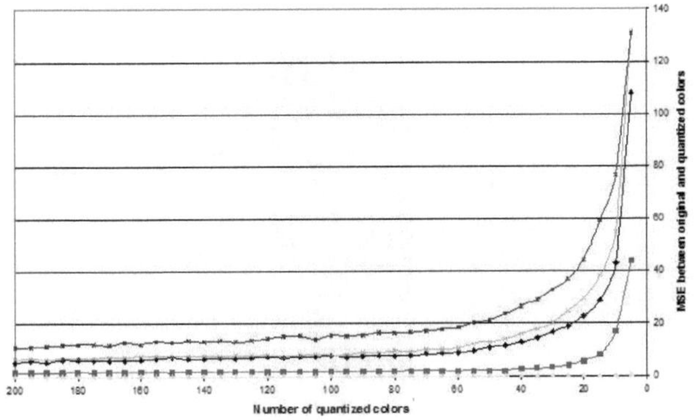

Fig. 1. Evolution of MSE between quantized colors and original colors

law and finally good continuation law. Because those laws are defined between regions and their context, at each step we assess the possibility to merge according to global information.

The segmentation is first based on color clustering making use of color constancy law but also it includes an extra post-processing step to ensure spatial consistency of the regions. In practice, our segmentation scheme is a 3-step process. First we filter the image for robustness to noise and reduce color depth by grouping perceptually similar colors. In the second step we use an iterative algorithm to determine a good color count which limits the quantization error. Indeed, quantization error can be measured by MSE between original and quantized color distribution as illustrated in Fig. 1, which clearly shows a sharp MSE rise when the number of color clusters is beyond a threshold. By performing several fast coarse clustering operations using Neural Gas algorithm [12], which is fast and less sensitive to initialization than its counterparts such as K-means, a target color cluster count is fixed at the number leading to a double MSE value of the one corresponding to 200 color clusters. We then use hierarchical ascendant clustering to achieve segmentation. The third step consists in splitting spatially unconnected regions, merging similar regions and constraining segmentation coarseness. Merging of similar regions is achieved through the use of the squared Fisher's distance as (1) (used for a similar task in [13]). Where n_i, μ_i, σ_i^2 are respectively the number of pixels, the average color and the variance of colors within region i.

$$D(R_1, R_2) = \frac{(n_1 + n_2)(\mu_1 - \mu_2)^2}{n_1\sigma_1^2 n_2\sigma_2^2} \quad . \tag{1}$$

Sample segmentation results on some Pascal challenge dataset images can be seen on Fig. 2. As we can see, we have original images at the first row having very different color distribution. The image on the left has very few colors, the middle one has very contrasted colors while the image on the right contains many colors. Our Gestalt-inspired segmentation algorithm has automatically adapted its segmentation process, producing significant partial gestalts.

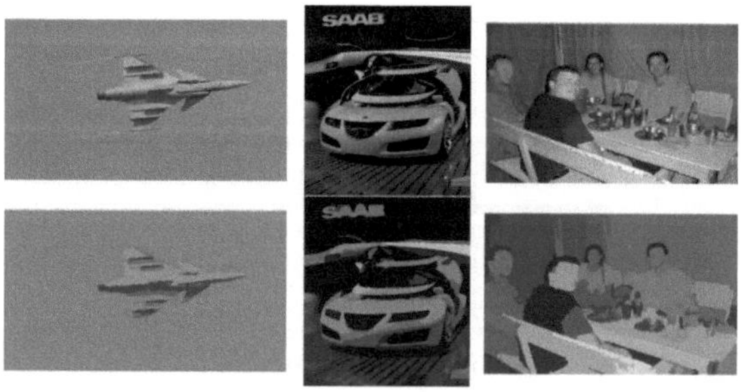

Fig. 2. Sample Segmented Images

2.2 Region-Based Features

Two kinds of features are used in this work for the purpose of image classification: color features and segment-based features. Region based color features aim at capturing a coarse perception of partial gestalts. They are defined by color moments (mean, variance and skewness) [14] for each color channel in CIELch color space. CIELch was chosen in our work as variant of Lab color space as in (2), which best fits to the human perception [15]. These features are quite compact and have proven as efficient as a high dimension histogram [16].

$$L_{Lch} = L_{Lab} \qquad c = \sqrt{a^2 + b^2} \qquad h = \arctan\frac{b}{a} \ . \tag{2}$$

The segment-based features aim on the other hand at capturing some textural and geometrical properties of partial gestalts. They rely on a fast connective Hough transform (FCHT) [17] that quickly detects segments within an image. Once all segments identified by FCHT, they are distributed to the regions. Our segment-based features, a simple histogram combining length and orientation, are then computed as follows. We first divide lengths by the longest segment's length and then compute an average orientation so that all angles can be expressed with respect to it. We therefore obtain a feature that is invariant to translation, scale and rotation. The size of the histograms was experimentally determined and set to 6 bins for orientation and 4 for length.

Finally in order to include neighborhood information, our color and segment-based features are expressed at four levels: original region, region+neighbors, region+neighbors+neighbor's neighbors, etc. Those levels are concatenated in the final feature vector. This is a basic way to integrate spatial relationship but also to include global information in each feature vector because on most images the fourth level will represent features extracted over the whole image.

3 Polynomial Image Modeling and Classification

We now turn to the problem of image modeling and classification. The basic problem is that the number of feature vectors as we propose to extract in the previous section can vary from one image to another, typically depending upon the number of segmented gestalts whereas machine-based learning schemes require a same data size for the classification. In the popular "bag-of-features" approach, one deals this problem by computing a kind of histogram for each image on the basis of a "visual vocabulary". This "visual vocabulary" is built using learning dataset either through a clustering algorithm on the set of feature vectors or the use of Gaussian Mixture Model (GMM). The drawbacks of such an approach are twofold. On the one hand, the optimal size of this visual vocabulary is hard to be fixed as there is no easy intuitive counterpart in image compared to keywords in text document. On the other hand, when a GMM is used for a soft assignment, the number of parameters along with the number of Gaussians can quickly lead to the problem of "curse of dimensionality" [18].

Instead of building a "visual vocabulary", we propose here a simple polynomial modeling to characterize the visual content represented by the set of feature vectors extracted in the previous section. Moreover, we also propose to study several fusion strategies for a better image classification accuracy.

3.1 Polynomial Modeling Based Image Representation

In order to represent the visual content of an image characterized by a set of feature vectors, we propose to consider the distribution of values in each component of these feature vectors and to model such a distribution (histogram) by a simple polynomial. The coefficients of these polynomials will then be considered as the feature vector characterizing the visual content of an image.

The polynomial model for a given feature histogram is computed as follows. Given the set D of histogram values $D = \{(x_1, y_1), ..., (x_M, y_M)\}$ (M is the number of values), a polynomial $f(x)$ of degree N, described by its set of coefficients $P = \{p_1, p_2, ...p_{n+1}\}$, is computed to interpolate the data, by fitting $f(x_i)$ to y_i in a least squares sense. Thus, vector P can be used to characterize the distribution D. An example is given in Fig. 3. Once the distribution of each component from the feature set has been modeled thanks to a polynomial, a new image feature vector Q is produced by concatenating the coefficients of all polynomials.

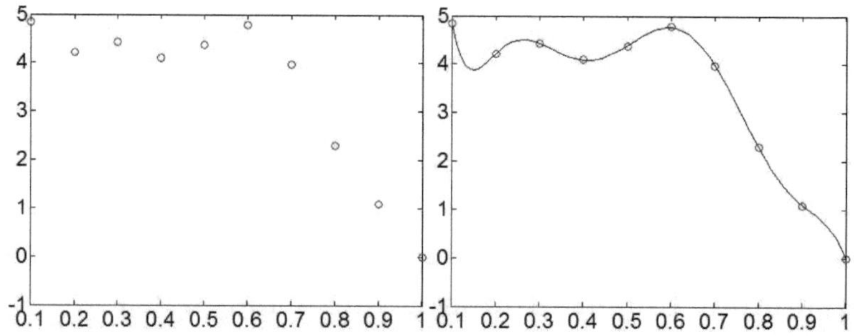

Fig. 3. (a) Histogram for one component of the image feature set. (b) A polynomial curve models the histogram in (a)

Assuming that our feature vector has L components and each component is modeled by a polynomial of degree N, then the vector Q has a dimension of $(N + 1) * L$, which generally ranges from hundreds to thousands. A vector of such high dimensionality used for classification can also lead to the "curse of dimensionality" [18]. Consequently, we further reduce the dimension using a feature selection method. While there exist several such methods in the literature [19], we have chosen the canonical discriminant analysis [20] as it is fast and enables an efficient feature dimension reduction by producing a new representation space which distinguishes the best the different classes. In most cases, we obtain

$K-1$ axes with the number of classes K. Thus with the help of this method, the overall feature vector Q becomes a much more simplified vector which is called in the subsequent *polynomial modeling based image representation.*

3.2 Classification and Fusion Strategies

Given an image to classify, we first characterize its visual content by extracting a set of feature vectors as proposed in section 2. However, other feature vectors, such as for instance SIFT, can also be used in our polynomial modeling and the subsequent classification process. Upon these feature vectors, our polynomial modeling based image representation is computed, leading to a single global feature vector for the input image. This new single global feature vector is then fed to a classifier beforehand trained or a set of such classifiers if a fusion strategy is applied, to judge whether this image contains a specified object. Any classifier, such as SVM or Neural networks, can be used for categorization of such an image representation. In our current experiments, we have chosen a simple multilayer perceptron for its ability to draw complex separating class border.

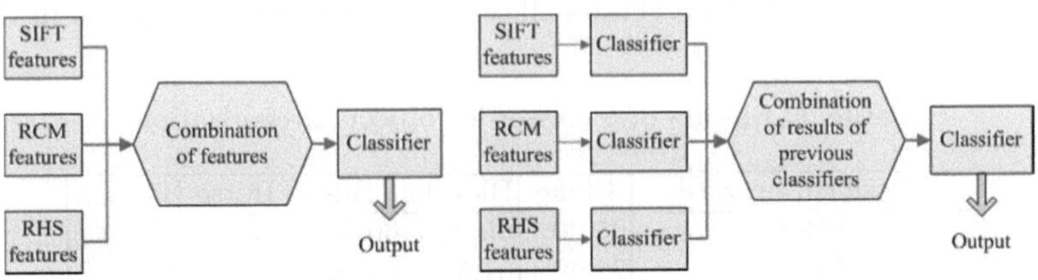

Fig. 4. (a) General scheme for early fusion (b) General scheme for late fusion

A fusion strategy can be applied when there exist several data sources in a classification process as for instance in video data analysis which can rely on both visual channel and audio one [21]. In our case, descriptors of different natures, such as SIFT, Region based Color Moments (RCM) and Region based Histogram of Segments (RHS), are extracted from an image. These descriptors can be considered as complementary modalities whose fusion can lead to a better accuracy in a classification process. Several fusion strategies can be applied: an early fusion is obtained when grouping all the features together and fed into a single classifier whereas a late fusion is to make use of "channels" with a separate classifier for each kind of features, the outputs of these classifiers being merged later [21] in a process similar to boosting [22]. Between these two strategies, there exist many intermediate strategies which consist in generating intermediate classes from different sources and to take a final decision based on these intermediate classes. In our current experiments, we have studied the two main strategies: early fusion and late fusion. Their schemes are illustrated respectively in Fig. 4(a) and 4(b).

4 Experimental Results

As a first experimental validation and evaluation of our approach, we have chosen 5 semantic representative classes namely airplane (248 images), bicycle (243 images), bus (186 images), horse (287 images) and person (2008 images) from the dataset of [1]. One versus all others multilayer perceptron was built for each class with a 4 fold-cross-validation. The structure of these perceptrons is composed of one hidden layer for all the tests, and the number of neurons in the hidden layer varies according to the number of inputs, and has 3 values: 5, 15, 2 for single channel, early fusion and late fusion respectively. Finally, the degree of polynomial for modeling the visual content of an image has been empirically set to 8.

Our region-based features, namely RCM and RHS, along with the popular SIFT features [23], were considered in our experiments. Early and late fusion strategies are also studied by interpreting these three types of features as different modalities. RCM and RHS are first merged by the strategies of Early Fusion and Late Fusion, noted as EF(RCM+RHS) and LF(RCM+RHS), and SIFT is combined afterwards, noted as EF(RCM+RHS+SIFT) and LF(RCM+RHS +SIFT). Here, we also use one versus all others multilayer perceptron in the two

Table 1. Results of 5 classes for object categorization

Classification rate	Plane	Bicycle	Bus	Horse	Person
SIFT	65,00%	55,21%	60,75%	65,49%	58,94%
RCM	72,69%	61,57%	67,90%	65,84%	62,77%
RHS	76,60%	61,98%	66,13%	62,59%	63,54%
EF(RCM+RHS)	80,34%	63,97%	70,75%	65,63%	65,17%
EF(RCM+RHS+SIFT)	81,47%	64,63%	69,30%	66,43%	65,50%
LF(RCM+RHS)	82,02%	70,95%	91,99%	79,65%	66,74%
LF(RCM+RHS+SIFT)	85,21%	72,73%	92,74%	81,54%	69,41%
Recall rate	Plane	Bicycle	Bus	Horse	Person
SIFT	68,66%	57,90%	62,58%	71,57%	60,93%
RCM	73,45%	64,10%	68,06%	66,53%	67,27%
RHS	76,55%	68,32%	71,61%	67,09%	69,11%
EF(RCM+RHS)	80,17%	65,67%	70,86%	67,09%	68,42%
EF(RCM+RHS+SIFT)	81,43%	66,67%	70,54%	70,10%	68,57%
LF(RCM+RHS)	84,20%	73,86%	89,35%	79,48%	70,01%
LF(RCM+RHS+SIFT)	85,38%	74,86%	89,78%	83,89%	72,89%
Precision rate	Plane	Bicycle	Bus	Horse	Person
SIFT	63,98%	54,90%	60,37%	63,76%	58,60%
RCM	72,35%	60,98%	67,85%	65,56%	61,72%
RHS	76,62%	60,60%	64,53%	61,49%	62,08%
EF(RCM+RHS)	80,44%	63,47%	70,71%	65,13%	64,24%
EF(RCM+RHS+SIFT)	81,50%	64,02%	68,84%	65,25%	64,61%
LF(RCM+RHS)	80,68%	69,77%	94,32%	79,71%	65,72%
LF(RCM+RHS+SIFT)	85,09%	71,77%	95,43%	80,08%	68,14%

fusion strategies. The performance of our experiments was evaluated through three classical rates, namely classification rate, recall rate and precision rate. The detailed results are presented in Table 1.

In Table 1, experimented classifiers can be categorized into 3 classes: Single Channel (SC) which means make use of only one kind of features, Early Fusion (EF) and Late Fusion(LF). As we can see, our region-based features, RCM and RHS, with an improvement of 5 points in average, perform better than SIFT features. These results tend to show the effectiveness of our RCM and RHS features using the polynomial modeling based image representation. Between RCM and RHS, we find that RHS is slightly better after comparing all 3 rates and RCM tends to favor negative side. Now focusing on EF and LF, we can note that the best classification rates are obtained when the 3 channels are merged using LF strategy which performs much better than SC and EF. The classes of bus and horse, for instance, record a classification rate increase by about 22 points and 15 points respectively compared to the second higher rate obtained in EF. This result seems to suggest that the three different channels carry complementary visual contents and their fusion helps to improve the final classification accuracy. Another reason might be that EF may have more chances to result in a conflict between different features which blur the boundary between classes. It can also explain that why EF performs only slightly better than SC and much worse than LF.

5 Conclusion and Perspectives

We have proposed in this paper a novel approach for visual object categorization, using polynomial modeling based image representation with new region based features. Two different fusion strategies, early and late, have been considered to merge information from different "channels" represented by the different types of features. Results have shown that good performance can be achieved with our approach and that our segment features carry information which is complementary to SIFT features, especially when merging feature channels according to a late fusion strategy. Encouraged by these first experimental results, we are currently experimenting our approach on the whole database in [1]. Among other perspectives under way, we are optimizing the quantification technique used for the construction of feature histograms as well as the choice of the degree of polynomial for modeling. Concerning features, we will evaluate richer representations of our region based features (e.g.: coocurrence matrix, ...) and also test more channels such as region shape and texture.

References

1. The PASCAL Visual Object Classes Challenge 2007 Results (2007), `http://www.pascal-network.org/challenges/VOC/voc2007/workshop/`
2. Dance, C., Willamowski, J., Fan, L., Bray, C., Csurka, G.: Visual categorization with bags of keypoints. In: ECCV International Workshop on Statistical Learning in Computer Vision (2004)

3. Rothganger, F., Lazebnik, S., Schmid, C., Ponce, J.: Object modeling and recognition using local affine-invariant image descriptors and multi-view spatial contraints. International Journal of Computer Vision 66(3) (2006)
4. Lowe, D.G.: Distinctive image features from scale-invariant keypoints. International Journal of Computer Vision 60(2), 91–110 (2004)
5. Lindeberg, T.: Feature detection with automatic scale selection. International Journal of Computer Vision 30(2), 79–116 (1998)
6. Mikolajczyk, K., Schmid, C.: Scale & affine invariant interest point detectors. International Journal of Computer Vision 60(1), 63–86 (2004)
7. Li, F.F., Perona, P.: A bayesian hierarchical model for learning natural scene categories. In: Proceedings of the 2005 IEEE Computer Society Conference on Computer Vision and Pattern Recognition (CVPR 2005), pp. 524–531 (2005)
8. Kaniza, G.: Grammatica del vedere. Il Mulino (1997)
9. Wertheirmer, M.: Untersuchungen zur lehre der gestalt ii. Psychologische Forschung 4, 301–350 (1923)
10. Desolneux, A., Moisan, L., Morel, J.: From Gestalt Theory to Image Analysis: A Probabilistic Approach. Springer, Heidelberg (2008)
11. Pujol, A., Chen, L.: Coarse adaptive color image segmentation for visual object classification. In: Proceedings of the 15th International Conference on Systems, Signals and Image Processing (2008)
12. Martinetz, T., Schulten, K.: A "neural-gas" network learns topologies. Artificial Neural Networks I, 397–402 (1991)
13. Zhu, S.C., Yuille, A.: Region competition: Unifying snakes, region growing, and bayes/mdl for multiband image segmentation. IEEE Trans. Pattern Anal. Mach. Intell. 18(9), 884–900 (1996)
14. Stricker, M.A., Orengo, M.: Similarity of color images. In: Storage and Retrieval for Image and Video Databases (SPIE), pp. 381–392 (1995)
15. Trémeau, A., Fernandez-Maloigne, C., Bonton, P.: Digital Color Imaging - From acquisition to Processing. Dunod (in French) (January 2004)
16. Deng, Y., Manjunath, B., Kenney, C., Moore, M., Shin, H.: An efficient color representation for image retrieval. IEEE Transactions on Image Processing 10(1), 140–147 (2001)
17. Ardabilian, M., Chen, L.: A new line extraction algorithm: Fast connective hough transform. In: Proceedings of PRIP 2001, p. 127 (2001)
18. Bellman, R.: Adaptive Control Processes: A Guided Tour. Princeton University Press, Princeton (1961)
19. Saeys, Y., Inza, I., Larranaga, P.: A review of feature selection techniques in bioinformatics. Oxford University Press, Oxford (2007)
20. Fisher, R.A.: The use of multiple measurements in taxonomic problems. Annals of Eugenics 7, 179–188 (1936)
21. Snoek, C.G.M., Worring, M., Smeulders, A.W.M.: Early versus late fusion in semantic video analysis. In: MULTIMEDIA 2005: Proceedings of the 13th annual ACM international conference on Multimedia, pp. 399–402. ACM, New York (2005)
22. Freund, Y., Schapire, R.: A short introduction to boosting. Journal of Japanese Society for Artificial Intelligence 14(5), 771–780 (1999)
23. Nowozin, S.: Libsift - scale-invariant feature transform implementation (2005), `http://user.cs.tu-berlin.de/~nowozin/libsift/`

IAM Graph Database Repository for Graph Based Pattern Recognition and Machine Learning

Kaspar Riesen and Horst Bunke

Institute of Computer Science and Applied Mathematics, University of Bern,
Neubrückstrasse 10, CH-3012 Bern, Switzerland
{riesen,bunke}@iam.unibe.ch

Abstract. In recent years the use of graph based representation has gained popularity in pattern recognition and machine learning. As a matter of fact, object representation by means of graphs has a number of advantages over feature vectors. Therefore, various algorithms for graph based machine learning have been proposed in the literature. However, in contrast with the emerging interest in graph based representation, a lack of standardized graph data sets for benchmarking can be observed. Common practice is that researchers use their own data sets, and this behavior cumbers the objective evaluation of the proposed methods. In order to make the different approaches in graph based machine learning better comparable, the present paper aims at introducing a repository of graph data sets and corresponding benchmarks, covering a wide spectrum of different applications.

1 Introduction

The first step in any system in pattern recognition, machine learning, data mining, and related fields consists in representing objects by adequate data structures. In the statistical approach the data structure is given by an n-dimensional vector $\mathbf{x} \in \mathbb{R}^n$, where each of the n dimensions represents the value of a specific feature. In recent years a huge amount of algorithms for classification, clustering, and analysis of objects given in terms of feature vectors have been developed [1].

Yet, the use of feature vectors implicates two limitations. First, as vectors represent a predefined set of features, all vectors of a set have to preserve the same length regardless of the size or complexity of the corresponding objects. Furthermore, there is no direct possibility to describe binary relationships among different parts of an object. It is well known that both constraints can be overcome by graph based representation [2]. That is, graphs allow us to adapt their size to the complexity of the underlying objects, and moreover, graphs offer a convenient possibility to describe relationships among different parts of an object.

Due to the ability of graphs to represent properties of entities and binary relations at the same time, a growing interest in graph-based object representation

N. da Vitora Lobo et al. (Eds.): SSPR&SPR 2008, LNCS 5342, pp. 287–297, 2008.

in pattern analysis can be observed [2]. That is, graphs found widespread applications in science and engineering. In the fields of bioinformatics and chemoinformatics, for instance, graph based representations have been intensively used [3, 4, 5]. Another field of research where graphs have been studied with emerging interest is that of web content mining [6]. Image classification is a further area of research where graph based representation draws the attention [7, 8, 9, 10]. Finally, we like to mention computer network analysis, where graphs have been used to detect network anomalies and predict abnormal events [11].

In statistical machine learning, the UCI Machine Learning Repository [12] is well established and widely used for benchmarking different algorithms. On the other hand a lack of standardized data sets for benchmarks in graph based machine learning can be observed. For an early discussion of benchmarking graph matching algorithms see [13]. As of today, however, there is only one standard set of graphs adequate for graph matching tasks publicly available to the knowledge of the authors, viz. the TC-15 graph database [14]. However, this data set consists of synthetically generated graphs only. Furthermore the graphs are particularly generated for exact graph matching algorithms rather than general matching tasks. In [15] benchmarks for graph problems are available. These benchmarks, however, are defined for special problems from graph theory, such as the maximum clique or vertex coloring problem, and are not related to pattern recognition and machine learning. The present paper aims at making a first step towards creating a graph repository that is suitable for a wide spectrum of tasks in pattern recognition and machine learning. These graph data sets emerged in the context of the autors' recent work on graph kernels [16] and graph embedding [17]. All graph data sets discussed in the present paper are publicly available or will be made available in the near future[1].

2 The Graph Set Repository

An attributed graph g, or graph for short, is defined as a four-tuple $g = (V, E, \mu, \nu)$, where V is the finite set of nodes, $E \subseteq V \times V$ is the set of edges, $\mu : V \to L$ is the node labeling function, and $\nu : E \to L$ is the edge labeling function. This definition allows us to handle arbitrary graphs with unconstrained labeling functions. For example, the labels can be given by the set of integers, the vector space $\mathbb{R}^n$, or a set of symbolic labels $L = \{\alpha, \beta, \gamma, \ldots\}$. Moreover, unlabeled graphs are obtained by assigning the same label l to all nodes and edges. Edges are given by pairs of nodes (u, v), where $u \in V$ denotes the source node and $v \in V$ the target node of a directed edge. Undirected graphs can be modeled by inserting a reverse edge $(v, u) \in E$ for each edge $(u, v) \in E$ with $\nu(u, v) = \nu(v, u)$.

Each of the data sets presented in the next subsections is divided into three disjoint subsets, which can be used for training, validation, and testing novel learning algorithms. Hence, the benchmarks are primarily designed for supervised learning tasks. Note, however, that the test set can be also used for benchmarking unsupervised learning algorithms. If appropriate, all three or two out of

[1] `http://www.iam.unibe.ch/fki/databases/iam-graph-database`

the three subsets can be merged. For each data set the classification result of a k-nearest neighbor classifier (k-NN) in conjunction with graph edit distance [18] on the test set is given. These results can serve as a first reference system to compare other algorithms with. Note that the meta parameters for graph edit distance (node and edge insertion/deletion cost) and the number k of graphs considered by the classifier are determined on the independent validation sets. A summary of the graph data sets together with some characteristic properties appears in Table 1. In the following subsections, each data set will be described in greater detail.

2.1 Letter Database

The first graph data set involves graphs that represent distorted letter drawings. We consider the 15 capital letters of the Roman alphabet that consist of straight lines only (*A, E, F, H, I, K, L, M, N, T, V, W, X, Y, Z*). For each class, a prototype line drawing is manually constructed. These prototype drawings are then converted into prototype graphs by representing lines by undirected edges and ending points of lines by nodes. Each node is labeled with a two-dimensional attribute giving its position relative to a reference coordinate system. Edges are unlabeled. The graph database consists of a training set, a validation set, and a test set of size 750 each. The graphs are uniformly distributed over the 15 classes. In order to test classifiers under different conditions, distortions are applied on the prototype graphs with three different levels of strength, viz. *low*, *medium* and *high*. Hence, our experimental data set comprises 6,750 graphs altogether. In Fig. 1 the prototype graph and a graph instance for each distortion level representing the letter A are illustrated. The classification rates achieved on this data set are 99.6% (*low*), 94.0% (*medium*), and 90.0% (*high*).

(a) (b) (c) (d)

Fig. 1. Instances of letter A: Original and distortion levels *low*, *medium* and *high* (from left to right)

2.2 Digit Database

The digit data set consists of graphs representing handwritten digits [19]. The original version of this database includes 10,992 handwritten digits from classes 0 to 9. For our data set a randomly selected subset of totally 3,500 digits is used. This set is split into traininig set of size 1,000, a validation set of size 500, and a test set of size 2,000. The digit graphs are uniformly distributed over the 10 classes. During the recording of the digits, the position of the pen was determined with constant time intervals. The resulting sequences of (x, y)-coordinates were converted into graphs by inserting nodes in regular intervals between the starting and ending points of a line. Successive nodes are connected by undirected edges. Each node is labeled with a two-dimensional attribute giving its position relative to a reference coordinate system. The edges are attributed with an angle denoting the orientation of the edge with respect to the horizontal direction. In Fig. 2 one particular graph instance for each digit class is illustrated. The classification rate achieved on this data set is 91.0%.

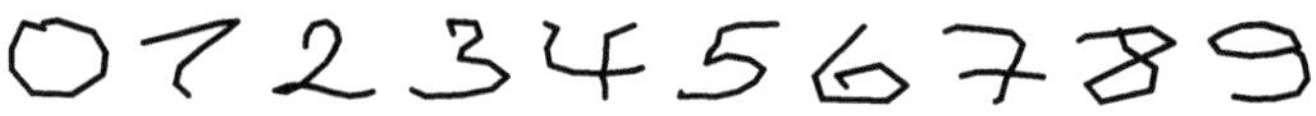

Fig. 2. A graph instance of each of the ten digit classes

2.3 GREC Database

The GREC data set consists of graphs representing symbols from architectural and electronic drawings. The images occur at five different distortion levels. In Fig. 3 for each distortion level one example of a drawing is given. Depending on the distortion level, either erosion, dilation, or other morphological operations are applied. The result is thinned to obtain lines of one pixel width. Finally, graphs are extracted from the resulting denoised images by tracing the lines from end to end and detecting intersections as well as corners. Ending points, corners, intersections and circles are represented by nodes and labeled with a two-dimensional attribute giving their position. The nodes are connected by undirected edges which are labeled as *line* or *arc*. An additional attribute specifies the angle with respect to the horizontal direction or the diameter in case of arcs. From the original GREC database [20], 22 classes are considered. For an adequately sized set, all graphs are distorted nine times to obtain a data set containing 1,100 graphs uniformely distributed over the 22 classes. The resulting set is split into a traininig and a validation set of size 286 each, and a test set of size 528. The classification rate achieved on this data set is 95.5%.

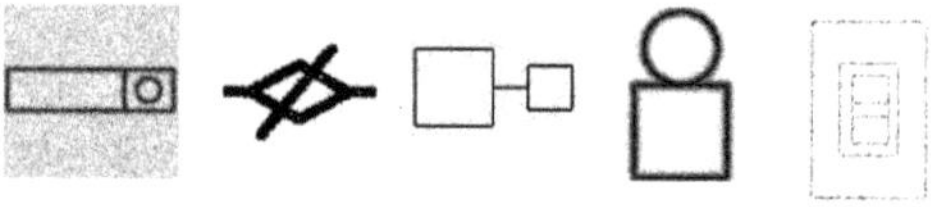

Fig. 3. A sample image of each distortion level

2.4 Fingerprint Database

Fingerprints are converted into graphs by filtering the images and extracting regions that are relevant [21]. In order to obtain graphs from fingerprint images, the relevant regions are binarized and a noise removal and thinning procedure is applied. This results in a skeletonized representation of the extracted regions. Ending points and bifurcation points of the skeletonized regions are represented by nodes. Additional nodes are inserted in regular intervals between ending points and bifurcation points. Finally, undirected edges are inserted to

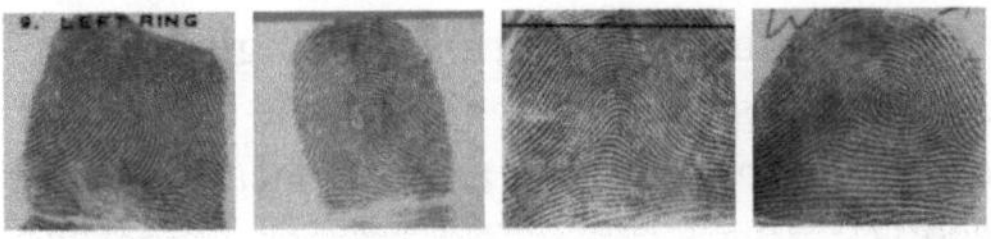

(a) *Left* (b) *Right* (c) *Arch* (d) *Whorl*

Fig. 4. Fingerprint examples from the four classes

link nodes that are directly connected through a ridge in the skeleton. Each node is labeled with a two-dimensional attribute giving its position. The edges are attributed with an angle denoting the orientation of the edge with respect to the horizontal direction.

The fingerprint database used in our experiments is based on the NIST-4 reference database of fingerprints [22]. It consists of a training set of size 500, a validation set of size 300, and a test set of size 2,000. Thus, there are 2,800 fingerprint images totally out of the four classes *arch*, *left*, *right*, and *whorl* from the Galton-Henry classification system. Note that in our benchmark test only the four-class problem of fingerprint classification is considered, i.e. the fifth class *tented arch* is merged with the class *arch*. Therefore, the first class (arch) consists of about twice as many graphs as the other three classes (left, right, whorl). For examples of these fingerprint classes, see Fig. 4. The classification rate achieved on this data set is 76.6%.

2.5 COIL-RAG Database

The COIL-100 database [23] consists of images of 100 different objects. Images of the objects are taken at pose intervals of 5 degrees. Fig. 5 shows an example image of each class. We first segment images into regions of homogeneous color using a mean shift algorithm [24]. Segmented images are transformed into region adjacency graphs by representing regions by nodes, labeled with attributes

Fig. 5. COIL images of 100 different objects

specifying the color histogram of the corresponding segment, and the adjacency of regions by edges. The edges are labeled with the length, in pixels, of the common border of two adjacent regions. For a more detailed description of the graph extraction process see [7].

The training set is composed of 12 images per object, acquired every 15 degree of rotation. From the remaining images we randomly select five images per object for the validation set, and ten images per object for the test set. This results in a training set of size 2,400, a validation set of size 500, and a test set of size 1,000. The total amount of graphs is uniformely distributed over the 100 classes. The classification rate achieved on this data set is 92.5%.

2.6 COIL-DEL Database

The same images as for the COIL-RAG database described above are used for this data set. However, a different graph extraction procedure is applied [9]. The Harris corner detection algorithm [25] is used to extract corner features from the images. Based on these corner points, a Delaunay triangulation is applied. The result of the triangulation is then converted into a graph by representing lines by undirected edges and ending points of lines by nodes. Each node is labeled with a two-dimensional attribute giving its position, while edges are unlabeled.

For this graph extraction method the training, the validation, and the test sets contain the same images as COIL-RAG. The classification rate achieved on this data set is 93.3%.

2.7 Web Database

In [6] several methods for creating graphs from web documents are introduced. For the graphs included in this data set, the following method was applied. First, all words occuring in the web document – except for stop words, which contain only little information – are converted into nodes in the resulting web graph. We attribute each node with the corresponding word and its frequency, i.e. even if a word appears more than once in the same web document we create only one unique node for it and store its total frequency as an additional node attribute. Next, different sections of the web document are investigated individually. These sections are *title*, which contains the text related to the document's title, *link*, which is text in a clickable hyperlink, and *text*, which comprises any of the readable text in the web document. If a word w_i immediately precedes word w_{i+1} in any of the sections *title*, *link*, or *text*, a directed edge from the node corresponding to word w_i to the node corresponding to the word w_{i+1} is inserted in our web graph. The resulting edge is attributed with the corresponding section label. Although word w_i might immediately precede word w_{i+1} in more than just one section, only one edge is inserted. That is, an edge is possibly labeled with more than one section label. Finally, only the most frequently used words (nodes) are kept in the graph and the terms are conflated to the most frequently occuring forms.

In our experiments we make use of a data set that consists of 2,340 documents from 20 categories (*Business, Health, Politics, Sports, Technology, Entertainment, Art, Cable, Culture, Film, Industry, Media, Multimedia, Music, Online, People, Review, Stage, Television,* and *Variety*). The last 14 catgories are sub-categories related to entertainment. The number of documents of each category varies from only 24 (Art) up to about 500 (Health). These web documents were originally hosted at Yahoo as news pages (`http://www.yahoo.com`). The database is split into a training, a validation, and a test set of equal size (780). The classification rate achieved on this data set is 80.3%.

2.8 AIDS Database

The AIDS data set consists of graphs representing molecular compounds. We construct graphs from the AIDS Antiviral Screen Database of Active Compounds [26]. This data set consists of two classes (*active, inactive*), which represent molecules with activity against HIV or not. The molecules are converted into graphs in a straightforward manner by representing atoms as nodes and the covalent bonds as edges. Nodes are labeled with the number of the corresponding chemical symbol and edges by the valence of the linkage. In Fig. 6 one molecular compound of both classes is illustrated. Note that different shades of grey represent different chemical symbols, i.e. node labels. We use a training set and a validation set of size 250 each, and a test set of size 1,500. Thus, there are 2,000 elements totally (1,600 inactive elements and 400 active elements). The classification result achieved on this data set is 97.3%.

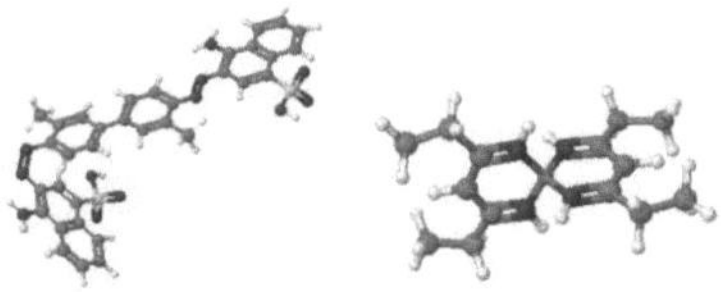

(a) Active (b) Inactive

Fig. 6. A molecular compound of both classes

2.9 Mutagenicity Database

Mutagenicity is one of the numerous adverse properties of a compound that hampers its potential to become a marketable drug [27]. In order to convert molecular compounds of the mutagenicity data set into attributed graphs the same procedure as for the AIDS data set is applied. The data set was originally prepared by the authors of [27]. The mutagenicity data set is divided into two classes *mutagen* and *nonmutagen*. We use a training set of size 1,500, a validation set of size 500, and a test set of size 2,337. Thus, there are 4,337 elements totally (2,401 mutagen elements and 1,936 nonmutagen elements). The classification rate achieved on this data set is 71.5%.

2.10 Protein Database

The protein data set consists of graphs representing proteins originally used in [3]. The graphs are constructed from the Protein Data Bank [28] and labeled with their corresponding enzyme class labels from the BRENDA enzyme database [29]. The proteins database consists of six classes (*EC 1, EC 2, EC 3, EC 4, EC 5, EC 6*), which represent proteins out of the six enzyme commission top level hierarchy (EC classes). The proteins are converted into graphs by representing the secondary structure elements of a protein with nodes and edges of an attributed graph. Nodes are labeled with their type (helix, sheet, or loop) and their amino acid sequence (e.g. *TFKEVVRLT*). Every node is connected with an edge to its three nearest neighbors in space. Edges are labeled with their type and the distance they represent in angstroms. In Fig. 7 six images of proteins of all six classes are given.

(a) *EC 1* (b) *EC 2* (c) *EC 3* (d) *EC 4* (e) *EC 5* (f) *EC 6*

Fig. 7. Protein examples of all top level classes

There are 600 proteins totally, 100 per class. We use a training, validation and test set of equal size (200). The classification task on this data set consists in predicting the enzyme class membership. We achieve a classification rate of 65.5% on the test set.

3 Conclusions

In the present paper ten graph sets with quite different characteristics are presented. They represent line drawings, gray scale and color images, HTML webpages, molecular compounds, and proteins. In Table 1 a summary of all graph datasets and their corresponding characteristics is provided. In addition to a description of the data sets, classification results achieved with a simple reference system based on a nearest-neighbor classifier are given. All data sets are publicly available or will be made available soon. We expect that the graph repository introduced in this paper provides a major contribution towards promoting the use of graph based representations and making graph based pattern recognition and machine learning algorithms better comparable against each other.

In future work we want to further expand the graph set repository. Towards this end, we highly encourage the community not only to use the available sets for developing and testing their algorithms, but also to integrate their own graph sets into our repository.

Table 1. Summary of graph data set characteristics, viz. the size of the training (*tr*), the validation (*va*) and the test set (*te*), the number of classes (#classes), the label alphabet of both nodes and edges, the average and maximum number of nodes and edges (∅/max nodes/edges), whether the graphs are uniformely distributed over the classes or not (balanced), and the recognition rate of the *k*-NN classifier (RR)

Database	size (*tr*, *va*, *te*)	#classes	node labels	edge labels	∅ nodes	∅ edges	max nodes	max edges	balanced	RR
Letter (*low*)	750, 750, 750	15	x, y coordinates	none	4.7	3.1	8	6	Y	99.6%
Letter (*medium*)	750, 750, 750	15	x, y coordinates	none	4.7	3.2	9	7	Y	94.0%
Letter (*high*)	750, 750, 750	15	x, y coordinates	none	4.7	4.5	9	9	Y	90.0%
Digit	1,000, 500, 2,000	10	x, y coordinates	Angle	11.8	13.1	32	30	Y	91.0%
GREC	286, 286, 528	22	x, y coordinates	Line type	11.5	12.2	25	30	Y	95.5%
Fingerprint	500, 300, 2,000	4	x, y coordinates	Angle	5.42	4.42	26	24	N	76.6%
COIL-RAG	2,400, 500, 1,000	100	RGB histogram	Boundary length	3.0	3.0	11	13	Y	92.5%
COIL-DEL	2,400, 500, 1,000	100	x, y coordinates	none	21.5	54.2	77	222	Y	93.3%
Web	780, 780, 780	20	Word and its frequency	Section(s) type	186.1	104.6	834	596	N	80.3%
AIDS	250, 250, 1,500	2	Chemical symbol	Valence	15.7	16.2	95	103	N	97.3%
Mutagenicity	1,500, 500, 2,337	2	Chemical symbol	Valence	30.3	30.8	417	112	N	71.5%
Protein	200, 200, 200	6	Type and aa-sequence	Type and distance	32.6	62.1	126	149	Y	65.5%

Acknowledgements

This work has been supported by the Swiss National Science Foundation (Project 200021-113198/1). For making their programs for graph extraction of the COIL images available to us, we would like to thank D. Emms (Delaunay graphs) and B. Le Saux (RAG graphs). The webgraphs and the protein graphs are due to A. Schenker and K. Borgwardt, respectively. We are very grateful to both authors.

References

1. Duda, R., Hart, P., Stork, D.: Pattern Classification, 2nd edn. Wiley Interscience, Hoboken (2000)
2. Conte, D., Foggia, P., Sansone, C., Vento, M.: Thirty years of graph matching in pattern recognition. Int. Journal of Pattern Recognition and Artificial Intelligence 18(3), 265–298 (2004)
3. Borgwardt, K., Ong, C., Schönauer, S., Vishwanathan, S., Smola, A., Kriegel, H.P.: Protein function prediction via graph kernels. Bioinformatics 21(1), 47–56 (2005)
4. Mahé, P., Ueda, N., Akutsu, T.: Graph kernels for molecular structures – activity relationship analysis with support vector machines. Journal of Chemical Information and Modeling 45(4), 939–951 (2005)
5. Ralaivola, L., Swamidass, S., Saigo, H., Baldi, P.: Graph kernels for chemical informatics. Neural Networks 18(8), 1093–1110 (2005)
6. Schenker, A., Bunke, H., Last, M., Kandel, A.: Graph-Theoretic Techniques for Web Content Mining. World Scientific, Singapore (2005)
7. Le Saux, B., Bunke, H.: Feature selection for graph-based image classifiers. In: Marques, J.S., Pérez de la Blanca, N., Pina, P. (eds.) IbPRIA 2005. LNCS, vol. 3523, pp. 147–154. Springer, Heidelberg (2005)
8. Harchaoui, Z., Bach, F.: Image classification with segmentation graph kernels. In: IEEE Conference on Computer Vision and Pattern Recognition, pp. 1–8 (2007)
9. Luo, B., Wilson, R., Hancock, E.: Spectral embedding of graphs. Pattern Recognition 36(10), 2213–2223 (2003)
10. Neuhaus, M., Bunke, H.: An error-tolerant approximate matching algorithm for attributed planar graphs and its application to fingerprint classification. In: Fred, A., Caelli, T.M., Duin, R.P.W., Campilho, A.C., de Ridder, D. (eds.) SSPR&SPR 2004. LNCS, vol. 3138, pp. 180–189. Springer, Heidelberg (2004)
11. Bunke, H., Dickinson, P., Kraetzl, M., Wallis, W.: A Graph-Theoretic Approach to Enterprise Network Dynamics. Progress in Computer Science and Applied Logic (PCS), vol. 24. Birkhäuser, Basel (2007)
12. Asuncion, A., Newman, D.: (UCI machine learning repository) University of California, Department of Information and Computer Science, `http://www.ics.uci.edu/~mlearn/MLRepository.html`
13. Bunke, H., Vento, M.: Benchmarking of graph matching algorithms. In: Proc. 2nd Int. Workshop on Graph Based Representations in Pattern Recognition, pp. 109–113 (1999)
14. Foggia, P., Sansone, C., Vento, M.: A database of graphs for isomorphism and subgraph isomorphism benchmarking. In: Proc. 3rd Int. Workshop on Graph Based Representations in Pattern Recognition, pp. 176–187 (2001)

15. Xu, K.: Bhoslib: Benchmarks with hidden optimum solutions for graph problems (maximum clique, maximum independent set, minimum vertex cover and vertex coloring),
`http://www.nlsde.buaa.edu.cn/~kexu/benchmarks/graph-benchmarks.htm`
16. Neuhaus, M., Bunke, H.: Bridging the Gap Between Graph Edit Distance and Kernel Machines. World Scientific, Singapore (2007)
17. Bunke, H., Riesen, K.: A family of novel graph kernels for structural pattern recognition. In: Rueda, L., Mery, D., Kittler, J. (eds.) CIARP 2007. LNCS, vol. 4756, pp. 20–31. Springer, Heidelberg (2007)
18. Bunke, H., Allermann, G.: Inexact graph matching for structural pattern recognition. Pattern Recognition Letters 1, 245–253 (1983)
19. Alpaydin, E., Alimoglu, F.: Pen-Based Recognition of Handwritten Digits. Dept. of Computer Engineering, Bogazici University (1998)
20. Dosch, P., Valveny, E.: Report on the second symbol recognition contest. In: Wenyin, L., Lladós, J. (eds.) GREC 2005. LNCS, vol. 3926, pp. 381–397. Springer, Heidelberg (2006)
21. Neuhaus, M., Bunke, H.: A graph matching based approach to fingerprint classification using directional variance. In: Kanade, T., Jain, A., Ratha, N.K. (eds.) AVBPA 2005. LNCS, vol. 3546, pp. 191–200. Springer, Heidelberg (2005)
22. Watson, C., Wilson, C.: NIST Special Database 4, Fingerprint Database. National Institute of Standards and Technology (1992)
23. Nene, S., Nayar, S., Murase, H.: Columbia Object Image Library: COIL-100. Technical report, Department of Computer Science, Columbia University, New York (1996)
24. Comaniciu, D., Meer, P.: Robust analysis of feature spaces: Color image segmentation. In: IEEE Conf. on Comp. Vision and Pattern Recognition, pp. 750–755 (1997)
25. Harris, C., Stephens, M.: A combined corner and edge detection. In: Proceedings of The Fourth Alvey Vision Conference, pp. 147–151 (1988)
26. DTP, AIDS antiviral screen (2004),
`http://dtp.nci.nih.gov/docs/aids/aids_data.html`
27. Kazius, J., McGuire, R., Bursi, R.: Derivation and validation of toxicophores for mutagenicity prediction. Journal of Medicinal Chemistry 48(1), 312–320 (2005)
28. Berman, H., Westbrook, J., Feng, Z., Gilliland, G., Bhat, T., Weissig, H., Shidyalov, I., Bourne, P.: The protein data bank. Nucleic Acids Research 28, 235–242 (2000)
29. Schomburg, I., Chang, A., Ebeling, C., Gremse, M., Heldt, C., Huhn, G., abd Schomburg, D.: Brenda, the enzyme database: updates and major new developments. Nucleic Acids Research 32 (2004) Database issue: D431–D433

Hybrid Genetic Algorithm and Procrustes Analysis for Enhancing the Matching of Graphs Generated from Shapes

Gerard Sanromà[1,⋆], Francesc Serratosa[1], and René Alquézar[2]

[1] Departament d'Enginyeria Informàtica i Matemàtiques, Universitat Rovira i Virgili, Spain
{gerard.sanroma,francesc.serratosa}@urv.cat
[2] Departament de Llenguatges i Sistemes Informàtics, Universitat Politècnica de Catalunya, Spain
alquezar@lsi.upc.es

Abstract. Typically, graphs generated via skeletonization of shape images are small and present low structural constraints. This fact constitutes a source of ambiguities for structural matching methods. Hybrid Genetic Algorithms have been effectively used for graph matching. This paper presents a new method which combines Hybrid Genetic Search with an enhanced model for graph matching. This enhanced model is based on the cliques model by Wilson and Hancock but introduces Procrustes Analysis over positional information in order to eliminate ambiguities. Comparative results are presented of the performance of the Hybrid Genetic algorithm with both the original cliques model and the enhanced model.

Keywords: graph matching, hybrid genetic algorithm, procrustes analysis, shape analysis, cliques model.

1 Introduction

The aims of graph matching is to associate nodes in a data graph with nodes in a model graph [1][2][3]. Nodes are associated so that a compatibility measure is maximized, usually computed using only the structural constraints (edges) provided by the graphs under match. While some authors exploit the relations of edge existence/absence between pairs of nodes [1][3], Wilson and Hancock use the overall adjacency relations of each node [2].

There is a variety of ways to represent an object by the means of a graph. For instance, given an image of a shape, we can extract the edges through a skeletonization, and place the nodes at end points, junction points and high curvature points. Graphs generated from shapes (e.g. handwritten letters) through skeletonization are not necessarily large and they also present low structural constraints (i.e. low connectivity between nodes). Note that the best connected node,

⋆ Corresponding author.

N. da Vitora Lobo et al. (Eds.): SSPR&SPR 2008, LNCS 5342, pp. 298–307, 2008.

has at most three edges, due to the nature of the skeletonization process. Structural matching methods often fail to match graphs under these circumstances. Purely structural criterions often generate ambiguous compatibility measures. Ambiguity refers to the fact of assigning the same amount of probability to a number of matching alternatives, most of them intuitively wrong.

The model by Wilson and Hancock (Section 3) provides good means for matching graphs with high or even moderate structural constraints, since it exploits the overall adjacency relations of each node [2]. It has demonstrated to be powerful when applied to large graphs of road maps and also Delaunay graphs. Delaunay triangulations, even on reduced point-sets, provide structural information enough in order to eliminate ambiguities. In the case of large graphs, local ambiguities are eliminated when evaluating the context.

Graphs generated from shapes are typically small and present low structural constraints. Delaunay triangulation is not a feasible choice for augmenting the structural constrains because the information about the shape contour provided by the edges from the skeletonization would be lost. There is a need of an extra source of information.

Landmark positions are traditionally used in Statistical Shape Analysis [4]. The main inconvenience of using positional information is that it depends on the position, orientation and scale of the point-sets. Procrustes analysis (Section 2), provide a framework to compare such point-sets in a way invariant to Euclidean similarity transformations. Our aim is to use Procrustes methods to augment the structural criterion proposed by Wilson and Hancock [2] with a measure of positional error between two point-sets and thereby, eliminate ambiguities with a more fine-grained compatibility measure.

Wilson and Hancock [2] propose an optimization procedure based on probabilistic relaxation. It consists on a gradient ascent process that updates the matching configuration through iterations. Graph-matching applications are characterized for presenting highly irregular search spaces. This fact makes gradient-ascent optimization processes, potentially able of getting trapped in local optima. Genetic Algorithms (GA) are optimization processes effectively capable of locating global optimal solutions [5]. A population of individuals is evolved in GA by the means of mutation and crossover operations in a manner similar than the natural evolution does. Convergence to the global optimum is guaranteed by selection of the most fit individuals to pass to the next generation. The fitness function, that measures the adaptation of each individual, is a key point on the genetic search.

Several works have been presented that use Hybrid GAs for optimization in graph-matching processes. Hybrid GAs arise from the combination of GAs with further optimization procedures (e.g. gradient ascent). The two ones most related with the work presented here [6][7], consist on the implementation of a hybrid GA to match graphs according with the model by Wilson and Hancock [2] (Section 3). We demonstrate that better graph-matching results are achieved applying hybrid genetic search over the augmented model presented in this paper (Section 4), than over the original one.

In Section 2 we describe Procrustes analysis, in Section 3, the cliques model by Wilson and Hancock [2] is described, in Section 4 we define our contribution to the cliques model in order to improve the matching of graphs generated from shapes, in Section 5 details on the genetic search procedure used are given, in Section 6 results are presented and finally, in Section 7 some conclusions are given.

2 Procrustes Analysis

Consider two correspondent point-sets X_1 and X_2 arranged in $k \times m$ matrices of cartesian coordinates of k points in m dimensions.

Procrustes Analysis involves the least squares superimposition of two point-sets using the similarity transformations. Estimation of the similarity parameters γ, Γ and β is carried out by minimizing the squared Euclidean distance

$$D^2_{PA}(X_1, X_2) = \inf_{\Gamma,\beta,\gamma} \| X_2 - \beta X_1 \Gamma - 1_k \gamma^T \|^2 , \tag{1}$$

where $\| X \| = \text{trace}\left(X^T X\right)^{\frac{1}{2}}$ is the Euclidean norm, $\beta \in \Re^+$ is a positive scaling factor, $\Gamma \in SO(m)$ is a $m \times m$ rotation matrix and $\gamma \in \Re^m$ is a translation m-vector.

Following Dryden and Mardia [4], to obtain the minimum of Equation (1), first we substitute X_1, X_2 by their centered versions normalized to unit size $Z_1 = \frac{CX_1}{\|CX_1\|}$ and $Z_2 = \frac{CX_2}{\|CX_2\|}$, where $C = I_k - \frac{1}{k} 1_k 1_k^T$ is the centering matrix which applies a translation that brings the centroids of X_1 and X_2 to the origin.

Afterwards, the minimizing parameters are found at

$$\hat{\gamma} = 0 , \quad \hat{\beta} = \frac{\text{trace}\left(Z_2^T Z_1\right)}{\text{trace}\left(Z_1^T Z_1\right)} , \quad \hat{\Gamma} = UV^T , \tag{2}$$

where $V \Lambda U^T = \text{svd}\left(Z_2^T Z_1\right)$ is the *singular value decomposition* of $Z_2^T Z_1$.

Alternatively, we may use the centered non-normalized versions $X_1^C = CX_1$ and $X_2^C = CX_2$ with $\hat{\beta} = 1$ if we want to preserve the original size of the point-sets when calculating the distance measure.

3 A Gradient Ascent Approach to Graph Matching

In this Section, the graph matching technique used in the hill-climbing step of the hybrid genetic algorithm is described. The aim of graph matching is to associate nodes in a data-graph $G_1 = (V_1, E_1, A_1)$, where V_1 is a set of nodes, E_1 is a set of arcs and $A_1 = \left\{ x_u^{(1)}, \forall u \in V_1 \right\}$ is a set of symbols associated with the nodes, against those in a model-graph $G_2 = (V_2, E_2, A_2)$, where $A_2 = \left\{ x_v^{(2)}, \forall v \in V_2 \right\}$. The matching is represented by a function $f : V_1 \rightarrow V_2 \cup \phi$ for the nodes in the data graph G_1 to those in the model graph G_2 augmented with a null-label.

In [2] the graphs are matched using *cliques*. For a given node indexed j from G_1, the clique $C_j = \{u_1, \ldots, u_{|C_j|}\}$ is defined as $C_j = j \cup \{i| (i, j) \in E_1\}$. The matched realization of the clique $C_j \subseteq V_1$ is denoted as

$$\Gamma_j = \{f(u_1), \ldots, f(u_{|C_j|})\} . \tag{3}$$

For each clique on the model graph S_i, a set of *structure-preserving mappings* (SPMs) are generated in order to obtain a dictionary $\Theta_j = \{S_i^k\}$ onto which we evaluate each data graph clique C_j. The generation of the SPMs is a key point consisting in two steps: (1) the model graph clique S_i is padded with dummy nodes in order to accommodate size differences with C_j due to node loss and merging and, (2) the nodes of the padded model graph relations are then permuted so as to preserve the cyclicity of the non-dummy nodes. This process effectively preserves the adjacency structure of the model graph nodes while leaving dictionary invariant to potential scene translations, scalings or rotations [2].

The final gradient-ascent rule to update the matching configuration according to Ref. [2] is

$$f(u) = \underset{v^{(2)} \in V_2 \cup \phi}{\arg\max} P\left(f(u^{(1)}) = v^{(2)} | x_u, x_v\right) P(f) , \tag{4}$$

where $P\left(f(u^{(1)}) = v^{(2)} | x_u, x_v\right)$ is the probability of match between nodes $u^{(1)} \in V_2$ and $v^{(2)} \in V_2$ given measurements only relative to the nodes under match, and $P(f)$ is the joint *prior* which gauges the overall consistency of the matching configuration. The most critical of these is a means of modeling the joint prior $P(f)$. In the following we describe the developement of this structural consistency measure.

This process commences by modelling the consistency of match of an individual clique C_j onto G_2, i.e. Γ_j. The dictionary of SPMs Θ_j represents the consistent labellings available for gauging the quality of match. As demanded by the Bayes rule:

$$P(\Gamma_j) = \sum_{S_i^k \in \Theta_j} P\left(\Gamma_j | S_i^k\right) P\left(S_i^k\right) . \tag{5}$$

Assuming independence in the matching errors between adjacent nodes of the same clique, the conditional probabilities become

$$P\left(\Gamma_j | S_i^k\right) = \prod_{r=1}^{|S_i^k|} P\left(f(u_r) | v_r\right) . \tag{6}$$

Sources of errors are assumed to occur with uniform probability P_e. Under these assumptions, the distribution rule under the product of (6) is

$$P\left(f(u_r) | v_r\right) = (1 - P_e)^{\delta_{u_r v_r}} P_e^{1 - \delta_{u_r v_r}} , \tag{7}$$

where δ_{ij} is the Kronecker delta which is a function of two variables, usually integers, which is 1 if they are equal, and 0 otherwise.

Combining this distribution with Equation (6) we obtain the following expression for the conditional probabilities:

$$P\left(\Gamma_j | S_i^k\right) = P_e^{H\left(\Gamma_j, S_i^k\right)} \left(1 - P_e\right)^{E\left(\Gamma_j, S_i^k\right)} , \tag{8}$$

where $H\left(\Gamma_j, S_i^k\right)$ is the number of conflicts between Γ_j and S_i^k (i.e. the Hamming distance), and $E\left(\Gamma_j, S_i^k\right)$ is the number of coincidences between Γ_j and S_i^k.

Assuming that each of the SPMs is equi-probable, i.e. $P\left(S_i^k\right) = \frac{1}{|\Theta_j|}$ and using the natural exponential form for expressing Equation (8), the final model for the clique matching probabilities is

$$P\left(\Gamma_j\right) = \frac{1}{|\Theta_j|} \sum_{S_i^k \in \Theta_j} \exp\left[-k_{er} H\left(\Gamma_j, S_i^k\right) - k_{eq} E\left(\Gamma_j, S_i^k\right)\right] , \tag{9}$$

where $k_{er} = \log\left[\frac{1}{P_e}\right]$ and $k_{eq} = \log\left[\frac{1}{(1-P_e)}\right]$.

Finally, the joint prior is computed by averaging the clique matching probabilities over the nodes of the data graph, i.e.

$$P(f) = \frac{1}{|V_1|} \sum_{j \in V_1} P\left(\Gamma_j\right) . \tag{10}$$

The matching configuration is updated using the MAP decision rule given in (4) and according to a Discrete Relaxation scheme [2]. The strategy is to set P_e to an initial high value to reflect a poor labelling, and reduce it through iterations.

In the next Section we describe our contributions to the model above, aimed to improve the matching of graphs generated from shapes.

4 Augmenting the Matching Criterion with Procrustes Analysis

Graphs generated from shapes are typically small and they present low structural constraints. These two facts together are a potential source of ambiguity for purely structural matching methods. Our aim is to augment the measure of structural consistency reported in [2] (Equation 10) with a measure of similarity based on the positions of the nodes. By doing so, we assume that corresponding nodes in different graphs keep similar positional relations with their neighboring nodes. Note that, although articulated shapes would break this assumption in some extent, the study of such effects is over the scope of this paper.

To start developing our model, we augment the set of unary measurements associated with the nodes with the coordinates of the feature points. Hence, $A = \{(x_i, \vec{\alpha_i}), \forall i \in V\}$, where $\vec{\alpha_i} = (a_1, \ldots, a_m)$ is a row m-vector of cartesian coordinates in m dimensions, associated with each node.

The underpinning idea is to weight the contributions of the consistently mapped nodes (i.e, those such that $f(u) = v$) with a gaussian error derived from the Euclidean distance and the expected variance-covariance, after the Procrustes superimposition. By introducing such errors, we augment the measure of structural consistence drawn up in Section 3 with a measure of shape similarity between cliques. Thus, different cliques with the same cardinality on a given graph are no longer susceptible to contribute the same amount to the energy functional.

Our expression for the marginal probabilities of (7) is

$$P\left(f(u_r), \alpha_{u_r} | v_r, \alpha_{v_r}\right) = P_{match}\left(\overrightarrow{\alpha'_{u_r}}, \overrightarrow{\alpha_{v_r}}\right)^{\delta_{u_r v_r}} \rho^{1-\delta_{u_r v_r}}\,, \tag{11}$$

where δ_{ij} is the Kronecker delta, ρ is a constant controlling the inconsistent matching process in a similar manner than P_e did in Section 3 and, P_{match} means the probability of matching nodes $u_r^{(1)}$ with $v_r^{(2)}$ according with the similarity between the shapes of the cliques they belong to. We must take care to set ρ to be always lower than the probability of matching two corresponding nodes $u_r^{(1)}$ and $v_r^{(2)}$ according to P_{match}. Notice that, while weighting the contributions of the consistently mapped nodes, we aim to preserve the idea by Wilson and Hancock [2] of penalizing structural inconsistencies with low contributions to the overall measure. Otherwise, the gradient-ascent scheme would lead to a structurally inconsistent state.

As said, P_{match} is a gaussian error function between the two points, and it is defined as

$$P_{match}\left(\overrightarrow{\alpha'_{u_r}}, \overrightarrow{\alpha_{v_r}}\right) = \exp\left[-\frac{1}{2}\left(\overrightarrow{\alpha_{v_r}} - \overrightarrow{\alpha'_{u_r}}\right) \Sigma^{-1} \left(\overrightarrow{\alpha_{v_r}} - \overrightarrow{\alpha'_{u_r}}\right)^T\right]. \tag{12}$$

In the computation of $P_{match}\left(\overrightarrow{\alpha'_{u_r}}, \overrightarrow{\alpha_{v_r}}\right)$, the transformed coordinates $\overrightarrow{\alpha'_{u_r}}$ resulting from the Procrustes Analysis are used. Therefore, a point-set X_1 is built from the cartesian coordinates of the data graph nodes that have been correctly assigned in the current mapping S_i of the matching conditional $P(\Gamma_j|S_i)$ (i.e. the data graph nodes u_r under the circumstance $f(u_r) = v_r$). Thus, $X_1 = \{\overrightarrow{\alpha_{u_r}} | f(u_r) = v_r, r = 1, \ldots, |S_i|, f(u_r) \in \Gamma_j\}$ is a $k \times m$ matrix of cartesian coordinates of k points in m dimensions. Equivalently, a point set X_2 is built from the cartesian coordinates of the model graph nodes on which the data graph nodes have been correctly mapped to. Thus, $X_2 = \{\overrightarrow{\alpha_{v_r}} | f(u_r) = v_r, r = 1, \ldots, |S_i|, v_r \in S_i\}$ is a $k \times m$ matrix as well.

Finally, the new coordinates of the data graph $\overrightarrow{\alpha'_{u_r}}$ are taken from the transformed matrix $X_1^P = \beta X_1 \Gamma - 1_k \gamma^T$, after being aligned with X_2 using Procrustes Analysis (Section 2).

Collecting terms and using the natural exponential form, our model for the clique matching probabilities becomes

$$P(\Gamma_j) = \frac{1}{|\Theta_j|} \sum_{S_i \in \Theta_j} \exp\left[-k_{er} H - \frac{1}{2}\mathrm{trace}\left(\left(X_2 - X_1^P\right) \Sigma^{-1} \left(X_2 - X_1^P\right)^T\right)\right], \tag{13}$$

where $k_{er} = \log\left[\frac{1}{\rho}\right]$. Note that the second term in Equation (8), is replaced in our model by a product of terms as the one Equation (12). This product of exponential terms can be turned into a sum of the exponents when the clique matching probabilities are expressed in the exponential form. Finally, this sum can be expressed as a squared Euclidean distance similar than the one in Equation (1) but normalized by a variance-covariance matrix (i.e. a squared Mahalanobis distance).

The joint prior is computed as in Equation (10) by averaging the clique matching probabilities over the nodes of the data graph.

Details on the implementation of the hybrid genetic search are given in the following Section.

5 Hybrid Genetic Search

Genetic search has effectively proved to be capable of finding global optimal solutions in a wide variety of applications. A population of feasible solutions is evolved by the means of mutation and crossover operations.

A match $f : V_1 \rightarrow V_2$ from a data-graph G_1 to a model graph G_2 is represented as a list of length $|V_1|$ of labels drawn from V_2.

Mutation consists on randomly reassigning the matches on individual sites with uniform probability. These reassignments are done with random labels selected from V_2.

While the works in Refs. [6] and [7] use the measure of structural consistence reported by Wilson and Hancock (Equation (10)) as fitness measure, the present work use the augmented measure of matching consistence described in Section 4.

The approaches in Refs. [6] and [7], present a genetic search augmented with the gradient ascent approach reported by Wilson and Hancock [2] (Section 3) based on discrete relaxation. Mutation and crossover operators can leave the matches into inconsistent states (e.g. two different data-graph nodes assigned to the same model-graph node). The main aims of the gradient ascent step applied after the mutation and crossover operators are, firstly to return the matches to a consistent state, and secondly to push them towards local optima via hill-climbing.

In the present work, a gradient ascent step is conducted as stated by Equation (4) using the expression developed in Section 4 as the joint *prior*.

Gradient ascent methods are capable of correcting to around 70% of the initialization errors in the type of graphs used here. The choice of population size, as stated by Myers and Hancock [7], is made so that at least a given fraction (e.g. 30%) of the nodes are correctly assigned in a random initialization. Denoted as P_s, the probability that at least some fraction s of the nodes will have at least one correct assignment can be expressed as $P_s = \sum_{s|V_1| \leq k \leq |V_2|} \binom{|V_1|}{k} P_c^k (1 - P_c)^{|V_1|-k}$, where $P_c = 1 - \left(1 - \frac{1+r}{|V_2|+1}\right)^n$ is the probability of at least one correct assignment appearing in the initial population of size n with a fraction r of nodes corrupt.

6 Experiments

We present two types of experiments, evaluating both the matching efficiency and the convergence rate. We have used 84 graphs extracted from handwritten capital letters. Samples present moderate levels of structural corruption consisting on few added extra nodes and extra arcs (notice that we are not evaluating our model under severe structural corruption). Data graphs of each class are matched against a prototypical graph of that class. Comparative results are presented between the hybrid approach with the augmented cliques model reported in this paper (HGA cliques+procrustes) and the hybrid approach with the original cliques model (HGA cliques). Population is randomly initialized in all the experiments. A known ground truth consisting on the intuitively correct assignments is available for all the matchings.

For the matching efficiency experiments we have set the mutation and crossover probabilities to 0.4 and 0.5, respectively, and the population size such that a fraction of 0.3 assignments are correct in the initial population. Correct assignments fraction is obtained by comparison with the ground truth.

Figure 1.(a) shows the correct assignments fraction at each iteration. Fraction of correct assignments is computed at each iteration by averaging this quantity among the fittest individuals. Results at the first iteration correspond to initial population with neither gradient ascent nor genetic operator steps.

The next experiment evaluates the tolerance of our method to severe noise in the coordinates of the nodes. We have applied gaussian white noise to the (x, y) coordinates of the point positions associated with the nodes. The variance of such noise ranges from zero to the total variance of the data. Therefore, in the extreme case the variance due to noise is the same quantity than due to data. Figure 1.(b) shows correct assignments fraction obtained at the end of the algorithm for a given value of noise in the positional information. The algorithm finalizes when either the correct match is found or 20 iterations are reached. Since our method is the only one sensitive to this kind of noise, we have plotted the final correct fraction of the *HGA cliques* method as a base-line.

Results show that our method evolves significantly better through iterations. A correct fraction of 85% is reached at iteration 10 of our method while, *HGA cliques* is slightly over 60%. Because of ambiguities in the purely structural criterion, a number of global optima arise when using the *HGA cliques* method. Therefore, the average correct fraction among the fittest individuals degrades the results. Although a more fine-grained measure is used in the *HGA cliques+procrustes*, it does not always find the matching corresponding to the ground-truth. Is for this reason that averaged results are below 1. It is interesting to note that the correct fraction in the randomly initialized population (around 30%) is consistent with the criterion used in the choice of the population size. Results also show that our method outperforms *HGA cliques* while positional noise is under 75% approximately. It is interesting to note that positional noise degrades the effectiveness of our method just until a given threshold (approximately 20%). After that threshold, it stabilizes.

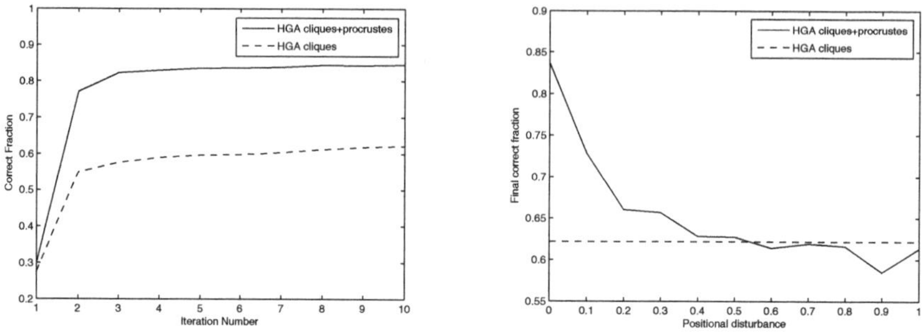

(a) Correct assignments fraction at each iteration of the algorithm. Results at the first iteration correspond to initial populaiton with no mutation, crossover or gradient ascent.

(b) Correct assignments fraction versus positional noise fraction. Since *HGA cliques* method is not sensible to this kind of noise the final correct fraction is plotted as a base-line.

Fig. 1. Matching efficiency results

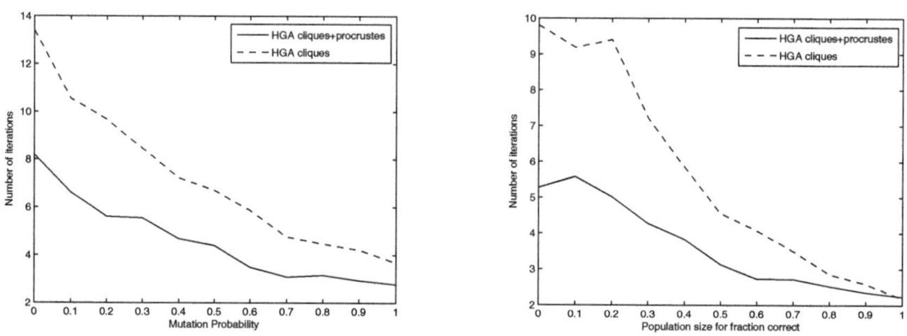

(a) Iterations until convergence versus mutation probability

(b) Iterations until convergence versus correct fraction required in the choice of population size

Fig. 2. Convergence rate results

Convergence rate experiments evaluate the number of iterations needed to find the match corresponding to the ground-truth, under mutation probability and population size variations. Otherwise, the algorithm is stopped at iteration 20. Figures 2.(a) and 2.(b) show the number of iterations needed to converge versus mutation probability and population size, respectively. The mutation probability and the correct fraction required in the choice of the population size in the cases that are fixed, are set to 0.4 and 0.3, respectively.

There are a number of matching processes that do not find the correct match, as seen in Figure 1.(a). These processes contribute to the convergence rate with a number of iterations equal to the maximum permitted (20). Despite this fact, the number of iterations needed to find the correct match tend to decrease as the mutation probability and population size increase. Anyway, our method always finds the correct match earlier.

7 Conclusions

We have presented a method for improving the matching of graphs that are small and present low structural constraints, such as those obtained from shapes (e.g. letters). We have used hybrid genetic search with an improved gradient ascent step based Wilson and Hancock's method [2] enhanced by the use of Procrustes analysis over positional information. Although we have evaluated our method with shapes it is applicable to any kind of object on which meaningful positional information can be extracted. The model presented integrates perfectly into the cliques framework. Results show a significant improvement in either matching ability and convergence rate of our method. Matching results of our method under severe positional disturbance are comparable to the ones obtained using the original hybrid approach without positional information.

Acknowledgements. This research was partially supported by Consolider Ingenio 2010, project CSD2007-00018, by the CICYT project DPI 2007-61452 and by the Universitat Rovira i Virgili (URV) through a predoctoral research grant.

References

1. Luo, B., Hancock, E.R.: Structural graph matching using the em algorithm and singular value decomposition. Pattern Analysis and Machine Intelligence 23(10) (October 2001)
2. Wilson, R.C., Hancock, E.R.: Structural matching by discrete relaxation. Pattern Analysis and Machine Intelligence 19(6) (June 1997)
3. Gold, S., Rangarajan, A.: A graduated assignment algorithm for graph matching. Pattern Analysis and Machine Intelligence 18(4) (April 1996)
4. Dryden, I.L., Mardia, K.V.: Statistical Shape Analysis. John Wiley and Sons, Chichester (1998)
5. Whitley, D., Beveridge, R., Graves, C., Mathias, K.: Test driving three 1995 genetic algorithms: New test functions and geometric matching. J. Heurist 1 (June 1995)
6. Cross, A.D.J., Wilson, R.C., Hancock, E.R.: Inexact graph matching using genetic search. Pattern Recognition 30(6) (1997)
7. Myers, R., Hancock, E.R.: Least-commitment graph matching with genetic algorithms. Pattern recognition 34 (October 1999)

Spectral Embedding of Feature Hypergraphs

Peng Ren, Richard C. Wilson, and Edwin R. Hancock

Department of Computer Science, The University of York, York, YO10 5DD, UK
{pengren,wilson,erh}@cs.york.ac.uk

Abstract. In this paper we investigate how to establish a hypergraph model for characterizing object structures and how to embed this model into a low-dimensional pattern space. Each hyperedge of the hypergraph model is derived from a seed feature point of the object and embodies those neighbouring feature points that satisfy a similarity constraint. We show how to construct the Laplacian matrix of the hypergraph. We adopt the spectral method to construct pattern vectors from the hypergraph Laplacian. We apply principal component analysis (PCA) to the pattern vectors to embed them into a low-dimensional space. Experimental results show that the proposed scheme yields good clusters of distinct objects viewed from different directions.

1 Introduction

Graph-based methods have played an important role in machine learning due to their ability to represent relational patterns. Various applications of graph-based methods can be found in data mining, natural language processing and information retrieval. In computer vision, the attractive feature of graph representations is that they concisely capture the relational arrangement of object primitives in a manner which can be invariant to changes in object viewpoint[1].

However, in many situations the graph representation for relational patterns is incomplete. This is because in real-world problems objects and their features tend to exhibit multiple relationships rather than simple pairwise ones. Since graphs are of limited use in representing complex relational patterns, hypergraphs have been adopted as a more sophisticated alternative.

Hypergraphs have certain advantages over graphs. These include their accuracy and completeness in describing feature relations and structures. As a result, hypergraphs have recently found applications in a variety of practical problems such as partitioning circuits[7] and clustering categorical data[6]. In computer vision, hypergraphs have been applied to image segmentation[3], edge detection[4] and noise removal[4]. However, most of these applications are based on low-level approximated image analysis, and thus induce a heavy computation burden.

In addition, to be easily manipulated and hence useful in machine learning, hypergraphs must be represented in an mathematically consistent way, e.g. as matrices or vectors. Since Chung's definition on the Laplacian matrix of a k-uniform hypergragh[5], there have been several investigations of the matrix representation of hypergraphs[9][11] [14]. However, none of the proposed mathematical representations has been generally accepted as a basis for the spectral manipulation of

N. da Vitora Lobo et al. (Eds.): SSPR&SPR 2008, LNCS 5342, pp. 308–317, 2008.

hypergaphs. In this paper, we develop a spectral hypergraph embedding method. The starting point is to show how to encode hyperedge structure in a improved version of the hypergraph Laplacian in [14]. The pattern vectors are then established by using the Laplacian eigenmodes, which is proved to be permutation invariant to both the vertex labels and hyperedge labels. We apply the proposed method to clustering objects based on a high-level feature-based analysis of images. The proposed method gives good object clusters.

This paper is organized as follows. In Section 2, we review the relevant background on hypergraphs. In Section 3, we develop our hypergraph model for representing the object structure. In Section 4, we introduce a Laplacian matrix representation of hypergraphs. In Section 5, we develop a spectral embedding scheme for hypergraphs by using the associated Laplacian matrix. Section 6 gives experimental results for the proposed scheme on clustering visual information. Finally, conclusions are presented in Section 7.

2 Hypergraph Fundamentals

A hypergraph is a generalization of a graph. Unlike the edge of a graph, which can connect only two vertices, the edge of a hypergraph, i.e. hyperedge, can connect any number of vertices. A hypergraph is normally defined as a pair $H(V, E)$ where V is a set of elements, called nodes or vertices, and E is a set of non-empty subsets of V called hyperedges.

An example of a hypergraph is shown in Figure 1(a), where the vertex set $V = \{v_1, v_2, v_3, v_4, v_5, v_6\}$ and the hyperedge set $E = \{e_1 = \{v_1, v_2, v_3\}, e_2 = \{v_3, v_4, v_5\}, e_3 = \{v_5, v_6\}\}$. The representation of a hypergraph in the form of sets, concretely captures the relationship between vertices and hyperedges. However, it is difficult to manipulate this form in a computationally uniform way. Thus one alternative representation of a hypergraph is in the form of a matrix. For a hypergraph with m hyperedges and n vertices, we establish a $m \times n$ matrix $\boldsymbol{H}$ with element $h_{i,j}$ as follows:

$$h_{i,j} = \begin{cases} 1 \text{ if } v_i \in e_j \\ 0 \text{ otherwise} \end{cases} \tag{1}$$

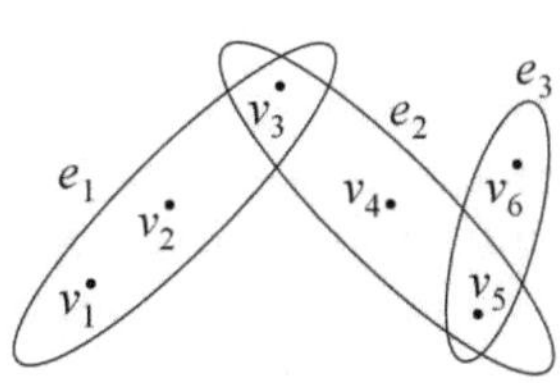

(a) Hypergraph

$$\boldsymbol{H} = \begin{array}{c} \begin{matrix} e_1 & e_2 & e_3 \end{matrix} \\ \begin{bmatrix} 1 & 0 & 0 \\ 1 & 0 & 0 \\ 1 & 1 & 0 \\ 0 & 1 & 0 \\ 0 & 1 & 1 \\ 0 & 0 & 1 \end{bmatrix} \end{array} \begin{matrix} v_1 \\ v_2 \\ v_3 \\ v_4 \\ v_5 \\ v_6 \end{matrix}$$

(b) Incidence Matrix

Fig. 1. An Example of Hypergraphs

Here, $\boldsymbol{H}$ is referred to as the incidence matrix of the hypergraph $H(V, E)$. The incidence matrix can be more easily manipulated than its equivalent set representation. The incidence matrix of the example hypergraph is shown in Figure 1(b), where each column is associated with the corresponding hyperedge, and each row with the corresponding vertex.

3 Hypergraph Representation of Objects

There has been a considerable body of research aimed at how to represent visual information using relational graphs[10][13]. However, these methods mainly focus on the representation of the relational arrangement of object primitives but neglect the intrinsic characteristics of the feature points (such as their colors and intensities). On the other hand, the hypergraph-based image representation proposed by Bretto *et al* incorporates the intensity levels of picture elements into the neighbourhood description for edge detection and denoising[4]. However, the low-level representation is pixel-based it is computationally burdensome.

To overcome these problems, in this paper, we develop a high-level hypergraph representation, i.e. feature hypergraph, based on feature points, rather than the pixel contents of an image. We exploit the intensities of feature points to characterize the similarities between different image locations. The hypergraph associated with one object is established with m vertices and m hyperedges, where m is the cardinality of the feature point set $V = \{v_1, v_2 \cdots v_m\}$ extracted from the object. Each feature point is represented by one vertex in the hypergraph and therefore the feature point set is identical to the hypergraph vertex set (denoted by V). Each hyperedge is specific to one vertex, which we define as the seed of the hyperedge. To determine the elements of each hyperedge in addition to the seed, we measure the similarities between the intensities of the seed and of the other vertices which lie in the spatial neighbourhood of the seed to confine the extent of the hyperedge. Let $\boldsymbol{c}(v_i)$ denote the spatial coordinate of the feature point v_i in an image, $I(v_i)$ denote the intensity of v_i, Th_{j1} be a fixed value which represents the distance threshold for neighbourhood, and Th_{j2} be the similarity threshold between the vertex intensities. The threshold Th_{j2} could be either a fixed value or calculated in an adaptive way[4]. The association of the vertex v_i to the hyperedge $E(v_j)$ derived from the seed v_j is determined as follows:

$$v_i \begin{cases} \in E(v_j) \text{ if } \|\boldsymbol{c}(v_i) - \boldsymbol{c}(v_j)\| \leq Th_{j1} \text{ and if } \mid I(v_i) - I(v_j) \mid \leq Th_{j2} \\ \notin E(v_j) \text{ otherwise} \end{cases} \tag{2}$$

The first condition confines one set of candidate vertices to the spatial neighbourhood of the seed. The second condition selects a set of candidate vertices satisfying the similarity constraint. The intersection of the two sets constitute one hyperedge. We hereby refer to the hypergraph established in (3) as the feature hypergraph of an object.

The incidence matrix of the feature hypergraph is hence of size $m \times m$. The ith column represents the hyperedge derived from v_i, and the jth row corresponds to vertex v_j. Thus the element of incidence matrix is denoted as:

$$h(i,j) = \begin{cases} 1 \text{ if } \|\boldsymbol{c}(v_i) - \boldsymbol{c}(v_j)\| \leq Th_{j1} \text{ and if } \mid I(v_i) - I(v_j) \mid \leq Th_{j2} \\ 0 \text{ otherwise} \end{cases} \tag{3}$$

4 Laplacian Matrix of a Hypergraph

Although the incidence matrix can fully describe the characteristics of the constructed hypergraph, it poses some drawbacks. Firstly, the elements of the incidence matrix represent a vertex-to-hyperedge relationship rather than a vertex-to-vertex relationship. Secondly, it is not possible to perform spectral analysis on the incidence matrix of a hypergraph, since the incidence matrix is normally not square (although that of our hypergraph is square). For these reasons, we establish the hypergraph adjacency matrix and Laplacian matrix.

4.1 Adjacency Matrix

For a hypergraph with a incidence matrix $\boldsymbol{H}$, the associated adjacency matrix $\boldsymbol{A}$ is:

$$\boldsymbol{A} = \boldsymbol{H}\boldsymbol{H}^T - \boldsymbol{D}_v \tag{4}$$

where $\boldsymbol{H}^T$ is the transpose of $\boldsymbol{H}$ and $\boldsymbol{D}_v$ is the diagonal vertex degree matrix whose diagonal element $d(v_i)$ is the summation of the ith row of $\boldsymbol{H}$. The (i,j)th entry of $\boldsymbol{H}\boldsymbol{H}^T$ records the number of hyperedges which contain both v_i and v_j.

An interpretation of (4) can be given in terms of the associated bipartite graph of the hypergraph[12]. To establish the associated bipartite graph, we use a dual representation in which each hyperedge is represented by a new vertex. The new vertex is incident to every original vertex in the corresponding hyperedge. The new vertex set and the original vertex set therefore constitute the associated bipartite graph with the new vertices corresponding to hyperedges on one side and the original vertices on the other side. The (i,j)th entry of the adjacency matrix $\boldsymbol{A}$ is the number of directed paths with length 2 from v_i to v_j in the bipartite graph.

The adjacency matrix of a hypergaph defined in (4) characterizes it in a way that is essentially analogous to that of a graph, since it encodes the incidence

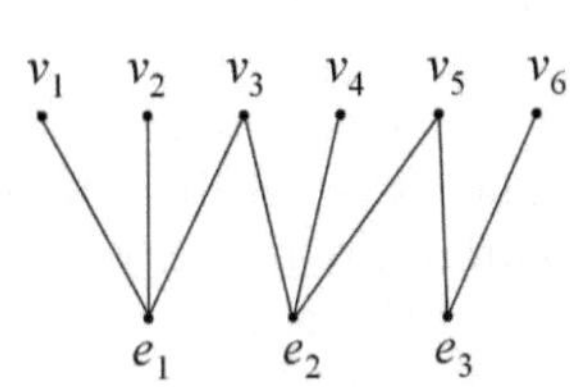

(a) Bipartite Graph

$$A = \begin{array}{c} \begin{matrix} v_1 & v_2 & v_3 & v_4 & v_5 & v_6 \end{matrix} \\ \begin{bmatrix} 0 & 1 & 1 & 0 & 0 & 0 \\ 1 & 0 & 1 & 0 & 0 & 0 \\ 1 & 1 & 0 & 1 & 1 & 0 \\ 0 & 0 & 1 & 0 & 1 & 0 \\ 0 & 0 & 1 & 1 & 0 & 1 \\ 0 & 0 & 0 & 0 & 1 & 0 \end{bmatrix} \end{array} \begin{matrix} v_1 \\ v_2 \\ v_3 \\ v_4 \\ v_5 \\ v_6 \end{matrix}$$

(b) Adjacency Matrix

Fig. 2. The Bipartite Graph and The Adjacency Matrix of Figure 1(a) relations between vertices. In the extreme situation of all hyperedges with cardinality 2, the adjacency matrix of a hypergraph reduces to the adjacency matrix of a graph.

The associated bipartite graph representation of the example hypergraph is shown in Figure 2, together with its associated adjacency matrix. The adjacency matrix can be computed either from (4) or by using the bipartite graph method. As an example of the bipartite graph method, the $(1,3)$th entry of the adjacency matrix is 1 because there is only one directed path $v_1 e_1 v_3$ with length 2 from v_1 to v_3.

4.2 Laplacian Matrix

Recently, Zhou *et al* have detailed a hypergraph Laplacian that is suitable for spectral analysis[14]. However, there are some drawbacks with their Laplacian when it is used to deal with vision problems. One of these is that the definition involves the diagonal hyperedge degree matrix to guarantee the positive semi-definiteness of the hypergraph Laplacian. This is necessary to ensure that the embedding coordinates of vertices obtained from the spectrum of the Laplacian is real. However, the diagonal hyperedge degree matrix contains redundant information. Another problem with the definition of Zhou *et al* is that it involves normalizing the hypergraph Laplacian in order to ensure that the Fiedler vector has the properties that lend it to bi-partitioning. However, in our preliminary experiments we found that the normalized Laplacian is not effective in distinguishing relational data[15]. To overcome these drawbacks, we define the hypergraph Laplacian by analogy with the graph Laplacian, as follows:

$$\boldsymbol{L} = \boldsymbol{D}_v - \boldsymbol{A} = 2\boldsymbol{D}_v - \boldsymbol{H}\boldsymbol{H}^T \tag{5}$$

This definition does not involve the diagonal hyperedge degree matrix, which as noted above contains redundant information. Furthermore, we do not normalize our Laplacian since the unnormalized Laplacian has been shown to be superior in distinguishing relational data[15]. Therefore, the hypergraph Laplacian in (5) can be regarded as an improvement over Zhou's hypergraph Laplacian, which is more concise in characterizing the structures of hypergraphs. It is also can be regarded as a star expansion representation of a hypergraph according to [2].

5 Spectral Embedding of Hypergraphs

The structure of a hypergraph is invariant under permutations of the node and hyperedge labels. However, the incidence matrix is modified by the node order and hyperedge order since the rows and columns are respectively indexed by the node order and hyperedge order. If we relabel the nodes, the incidence matrix undergoes a permutation of rows, while if we relabel the hyperedges, the incidence matrix undergoes a permutation of columns. The associated adjacency matrix and Laplacian matrix, which represent the incident relationship between nodes, might thus also undergo the permutation of rows and columns when there is a reordering of nodes or hyperedges. Hence, the problems to be solved here are a) how to establish a set of feature vectors that are permutation invariant to the change of the node and hyperedge order and b) how to embed them into a low-dimensional space.

5.1 Pattern Vectors

Suppose that $\boldsymbol{H}$ and $\boldsymbol{H_P}$ are the incidence matrices associated with two isomorphic hypergraphs $H(V,E)$ and $H_P(V,E)$ respectively. Let the matrix $\boldsymbol{P_V}$ be the node permutation matrix representing the change in node order between $H(V,E)$ and $H_P(V,E)$ and $\boldsymbol{P_E}$ the hyperedge permutation matrix representing the change in hyperedge order. Both $\boldsymbol{P_V}$ and $\boldsymbol{P_E}$ are orthogonal matrices and as a result $\boldsymbol{P_V P_V^T} = \boldsymbol{I_V}$ and $\boldsymbol{P_E P_E^T} = \boldsymbol{I_E}$, where $\boldsymbol{I_V}$ and $\boldsymbol{I_E}$ are identity matrices of the same size with $\boldsymbol{P_V}$ and $\boldsymbol{P_E}$ respectively. The permutation relationship between the two incidence matrices can be denoted as $\boldsymbol{H_P} = \boldsymbol{P_V H P_E}$. Thus, the relationship between the adjacency matrices of $H_P(V,E)$ and of $H(V,E)$ is:

$$\begin{aligned}\boldsymbol{A_P} &= \boldsymbol{H_P H_P^T} - \boldsymbol{D_{vP}} = \boldsymbol{P_V H P_E}(\boldsymbol{P_V H P_E})^T - \boldsymbol{P_V D_v P_V^T} \\ &= \boldsymbol{P_V H P_E P_E^T H^T P_V^T} - \boldsymbol{P_V D_v P_V^T} = \boldsymbol{P_V}(\boldsymbol{H H^T} - \boldsymbol{D_v})\boldsymbol{P_V^T} = \boldsymbol{P_V A P_V^T} \end{aligned} \quad (6)$$

where $\boldsymbol{D_{vP}}$ is the diagonal vertex degree matrix of $H_P(V,E)$. The relationship between the Laplacian matrix of $H_P(V,E)$ and of $H(V,E)$ is:

$$\begin{aligned}\boldsymbol{L_P} &= 2\boldsymbol{D_{vP}} - \boldsymbol{H_P H_P^T} = 2\boldsymbol{P_V D_v P_V^T} - \boldsymbol{P_V H P_E}(\boldsymbol{P_V H P_E})^T \\ &= 2\boldsymbol{P_V D_v P_V^T} - \boldsymbol{P_V H P_E P_E^T H^T P_V^T} = \boldsymbol{P_V}(2\boldsymbol{D_v} - \boldsymbol{H H^T})\boldsymbol{P_V^T} = \boldsymbol{P_V L P_V^T}\end{aligned} \quad (7)$$

From (6) and (7) it is clear that the hyperedge order has no effect on either the permutation of the adjacency matrix or the permutation of the Laplacian matrix, which are both solely controlled by the node permutation matrix.

Furthermore, let $\boldsymbol{e}$ be a normalized eigenvector of $\boldsymbol{L}$ with associated eigenvalue λ and let $\boldsymbol{e_P} = \boldsymbol{P_V e}$. With these ingredients, we have that:

$$\boldsymbol{L_P e_P} = \boldsymbol{P_V L P_V^T P_V e} = \boldsymbol{P_V L e} = \lambda \boldsymbol{P_V e} = \lambda \boldsymbol{e_P} \quad (8)$$

From (8) $\boldsymbol{e_P}$ is an eigenvector of $\boldsymbol{L_P}$ with eigenvalue λ. Hence, the isomorphic hypergraphs $H(V,E)$ and $H_P(V,E)$ with changes in hyperedge order and node order both share the common Laplacian spectrum, i.e. the same set of eigenvalues of the Laplacian. Spectra from the adjacency matrices exhibit the same property and this can be proved in a similar way as (8).

We construct our pattern vectors by using the hypergraph Laplacian spectrum. The pattern vector of a hypergraph which is permutation invariant to node labels and hyperedge labels can thus be established by sorting the eigenvalues of the hypergraph Laplacian in an ascending order and taking the second smallest up to the $(N+1)$th smallest ones as vector components. Although the hypergraphs studied might be of different sizes, thus inducing different dimensionalities of the eigenspaces, we can fix the number of eigenvalues used to restrict the dimension of pattern vectors to N. As a result, the pattern vectors of hypergraphs with different sizes can be formulated in the common form of $\boldsymbol{v} = [\lambda_2, \lambda_3 \cdots \lambda_{N+1}]^T$ without distinctions in length.

5.2 Embedding Using PCA

In Section 5.1, we construct the pattern vectors which both characterize the structures of the hypergraphs and that are invariant to label permutations. However, it is inconvenient to apply these pattern vectors to clustering objects and visualization when the vector dimension N is high. For this reason, we embed the pattern vector into a low-dimensional space by using principal component analysis (PCA). The purpose of this technique is to find a linear projection which describes the class differences rather than the overall variance of the data. Therefore the coordinates representing the original indvidual hypergraphs in the embedding space tend to be massed into clusters according to the class labels.

6 Experiments

In this section, we experiment with our new spectral representation of hypergraphs for the problem of clustering object views. All the experiments are undertaken based on the feature points extracted from objects by Harris detector[8]. We commence by comparing the proposed scheme with the traditional graph-spectral method. We embark on the experiments on a set of house images previously used in [10] and [13]. There are three classes of houses in the image set,

(a) House Sequences

(b) COIL Datasets

Fig. 3. Datasets for Experiments

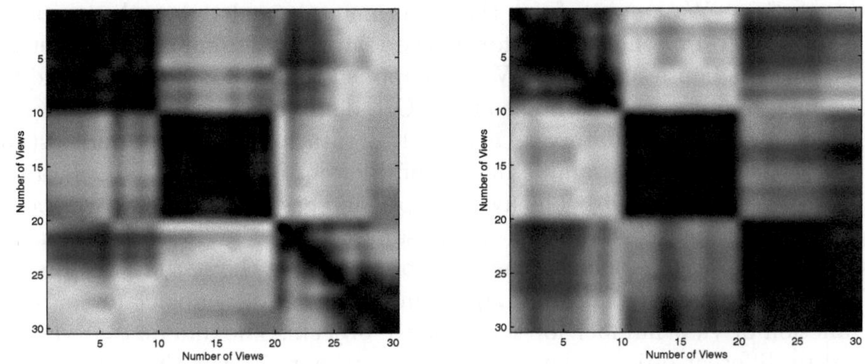

(a) Hypergraph (b) Graph

Fig. 4. Distance Map of Houses

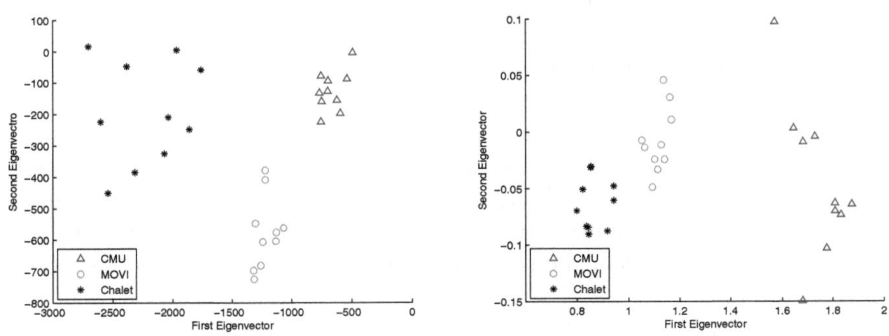

(a) Hypergraph (b) Graph

Fig. 5. Results of Clustering Houses

which are from the CMU database, the MOVI database and the chalet sequence. For each object, we have a sequence with different camera viewing directions. From the left to right, Figure 3(a) shows the sample image from the CMU database, the MOVI database and the chalet sequence. We establish the feauture hypergraphs of objects as reported in Section 3 based on the feature points as vertices. The neigbourhood threshold Th_{j1} is set to $1/4$ the size of the image. The similarity threshold Th_{j2} is computed in the adaptive way. Then we establish the pattern vector for each feature hypergraph as presented in Section 5.1. Here we set the dimensionality of the pattern space $N = 6$. Figure 4(a) is the distance map of the proposed pattern vectors, and Figure 5(a) is the clustering results of the proposed scheme in the space spanned by the first two principal linear projections in PCA. The distance map and clustering results from the graph-spectral method are illustrated in Figure 4(b) and Figure 5(b) respectively. The graphs here used are Delaunay graphs of which the feature points are the vertices. We can see from the experimental results that the proposed hyergraph-spectral method outperforms the traditional graph-spectral method in clustering the house sequences. We then applied the proposed method and the traditional graph-spectral method to the COIL database. There are 72 views for each object in the COIL database, acquired by rotating the object under study about a vertical axis. The sample images of different objects in COIL database are illustrated in Figure 3(b). The hypergraphs and graphs and their corresponding pattern vectors are established in the same way as the experiments on the house sequences. Figure 6 shows the first component of the pattern vectors, i.e. λ_2, from the proposed hypergraph Laplacian and the graph Laplacian separately as a function of view number for the first five objects. The different curves correspond to different objects. The main feature to note is that the curves in the top plot, which is the result of our hypergraph-based method, is better separated than those in the bottom plot, which is the result of the graph-based method. All the above experiments imply that the hypergraph-spectral method is generally superior to the graph-spectral method in clustering visual information. This is because the graph based method mainly focuses on the relational arrangement of objects and neglects the similarity between feature points. The proposed hypergraph method, on other hand, incorporates a similarity measure between feature points as well as considering the arrangement of feature points in terms of neighbourhood relationship. Therefore, the hypergraph model represents richer sources of scene information about objects than the graph model.

Finally, we compare the performance of our hypergraph Laplacian and of Zhou's hypergraph Laplacian[14] in clustering views of objects. We first establish the Laplacian matrices of the feature hypergraphs as reported in Section 4.2 and as reported in [14] separately. We then perform PCA on the spectral pattern vectors from the two kinds of Laplacian and embed them into a 3-dimensional space. We locate the clusters by using the K-means method and calculate the Rand index to evaluate the clustering performance. The Rand indices, about the two methods, of the clustered objects with different numbers of classes are illustrated in Figure 6. From the point of view by using Rand index for evaluation, the proposed hypergraph Laplacian is superior to Zhou's hypergraph Laplacian

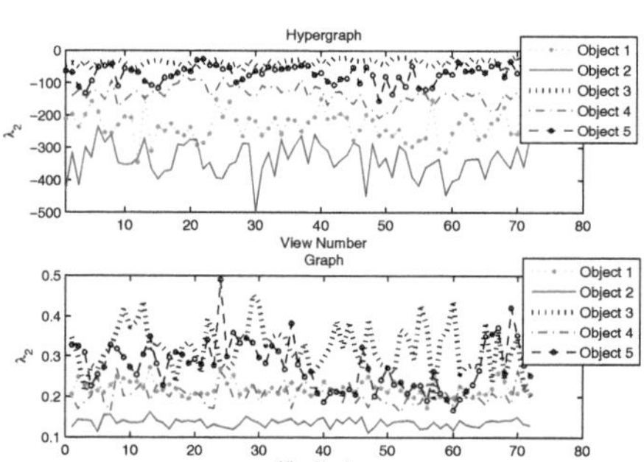

Fig. 6. Eigenvalue λ_2

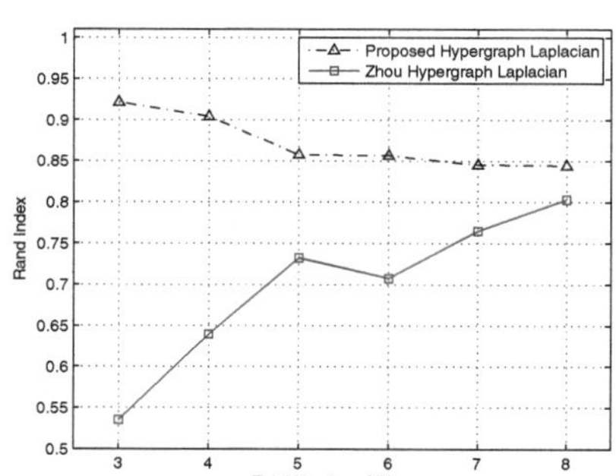

Fig. 7. Rand Index

in clustering visual information. This indicates that although Zhou's hypergraph Laplacian performs well in classifying categorical data, it is not suitable for visual clustering problems.

7 Conclusion

In this paper we have investigated how spectral pattern vectors extracted from the hypergraph Laplacian can be embedded in low-dimensional spaces for the purpose of clustering. There are two main contributions in this paper. The first is that we develop a hypergraph representation based on the locations of attributed feature points, i.e. feature hypergraph, to represent objects. The second is that we introduce an improved hypergraph Laplacian and develop one embedding method by applying PCA to the pattern vectors generated from it. The hypergraphs studied in our analysis are of different sizes and we do not need to locate correspondences. Experimental results indicate that the proposed scheme can provide useful pattern vectors for objects which perform well the in the domain of object view clustering.

References

1. IEEE TPAMI: Special Section on Graph Algorithms and Computer Vision, 23 10401151.1 (2001)
2. Agarwal, S., Branson, K., Belongie, S.: Higher-Order Learning with Graphs. In: ICML, pp. 17–23 (2006)
3. Agarwal, S., Lim, J., Zelnik-Manor, L., Perona, P., Kriegman, D., Belongie, S.: Beyond pairwise clustering. CVPR 2, 838–845 (2005)
4. Bretto, A., Cherifi, H., Aboutajdine, D.: Hypergraph imaging: an overview. Pattern Recognition 35(3), 651–658 (2002)
5. Chung, F.K.: The Laplacian of a Hypergraph. AMS DIMACS Series in Discrete Mathematics and Theoretical Computer Science 10, 21–36 (1993)
6. Gibson, D., Kleinberg, J., Raghavan, P.: Clustering categorical data: An approach based on dynamical systems. The VLDB Journal 8(3-4), 222–236 (2000)
7. Hagen, L., Kahng, A.B.: New spectral methods for ratio cut partitioning and clustering. IEEE TCAD 11(9), 1074–1085 (1992)
8. Harris, C.G., Stephens, M.J.: A combined corner and edge detector. In: Proceedings of Fourth Alvey Vision Conference, pp. 147–151 (1994)
9. Li, W., Sole, P.: Spectra of Regular Graphs and Hypergraphs and Orthogonal Polynomials. European Journal of Combinatorics 17, 461–477 (1996)
10. Luo, B., Wilson, R.C., Hancock, E.R.: Higher-Order Learning with Graphs. Pattern Recognition 36(10), 2213–2223 (2003)
11. Rodriguez, J.A.: On The Laplacian Eigenvalues and Metric Parameters of Hypergraphs. Linear and Multilinear Algebra 51, 285–297 (2003)
12. Storm, C.K.: The Zeta Function of a Hypergraph. EJC 13 (2006)
13. Wilson, R.C., Hancock, E.R., Luo, B.: Pattern Vectors from Algebraic Graph Theory. IEEE TPAMI 27(7), 1112–1124 (2005)
14. Zhou, D., Huang, J., Scholkopf, B.: Learning with Hypergraphs: Clustering, Classification, and Embedding. NIPS 19, 17–24 (2007)
15. Zhu, P., Wilson, R.C.: A Study of Graph Spectra for Comparing Graphs and Trees. Pattern Recognition (to appear, 2008)

Clustering Using Class Specific Hyper Graphs*

Shengping Xia[1,2,**] and Edwin R. Hancock[2]

[1] ATR Lab, School of Electronic Science and Engineering, National University of Defense Technology, Changsha, Hunan, P.R. China 410073
[2] Department of Computer Science, University of York, York YO1 5DD, UK

Abstract. The aim in this paper is to develop a method for clustering together image views of the same object class. Local invariant feature methods, such as SIFT, have been proven effective for image clustering. However, they have made either relatively little use or too complex use of geometric constraints and are confounded when the detected features are superabundant. Here we make two contributions aimed at overcoming these problems. First, we rank the SIFT points (R-SIFT) using visual saliency. Second, we use the reduced set of R-SIFT features to construct a specific hyper graph (CSHG) model of holistic-structure. Based on the CSHG model, a two stage clustering method is proposed. In which, images are clustered according to the pairwise similarity of the graphs, which is a combination of the traditional similarity of local invariant feature vectors and the geometric similarity between two graphs. This method comprehensively utilizes both SIFT and geometric constraints, and hence combines both global and local information. Experiments reveal that the method gives excellent clustering performance.

1 Introduction

Clustering uses features to organize data in a number of groups. One can use clustering to organize large collections of images. In this paper we aim to form clusters of images that are similar - similar in semantics as well as similar in visual appearance. Generally, there are two modalities for image clustering. The first is appearance feature based, the second is graph based.

1.1 Appearance Based Clustering

Traditionally, images are represented using feature vectors extracted from global appearance. Many kinds of central clustering methods have been proposed by using these feature vectors. These clustering methods can be categorized into partitioning methods, hierarchical methods, density-based methods, grid-based methods. The major difficulty in applying such techniques lies in obtaining feature vectors, namely building appropriate representations of images. For high dimensional vectors, the more scattered the data distribution is, the poorer the performance of clustering. A common problem is that the different views of an object present very different appearance. An example is furnished by the images of the cat from the COIL100 data set shown in Fig.1. More examples of

* Supported by ATR Key Lab Foundation project 91408001020603.
** Corresponding author.

N. da Vitora Lobo et al. (Eds.): SSPR&SPR 2008, LNCS 5342, pp. 318–328, 2008.

Fig. 1. Example images of small cat in COIL100 data set

Fig. 2. Example images of objects from COIL 100

various objects in this data set are shown in Fig.2. A global appearance based feature vector might be inappropriate in these cases. It is difficult to build a global feature for clustering these into their individual categories by using global appearance based feature vector methods.

Hence, local invariant feature based methods[3][6][8], e.g. SIFT, PCA-SIFT, GLOH and SURF, have recently been developed and have been successfully and widely used for object classification [3][7], scene classification [2][5] and video clustering or retrieval [9]. According to a comprehensive test on several large image databases, SIFT appears to be the most appealing descriptor for practical uses, and has become a standard of comparison[7][2].

Local features play the role of "visual words" predictive of certain "topic" or object class. For example, an eye is highly predictive of a face being present in the image. If our visual dictionary contains words that are sufficiently discriminative when taken individually, then it is possible to achieve a high degree of success for whole image classification, i.e. identification of the object class contained in the image without attempting to segment or localize that object, simply by looking which visual words/features are present, regardless of their spatial layout.

The success of the orderless feature model may be explained with the help of an analogy to bag-of-words models for text document classification [10]. From this point, some Bayesian text models, such as probabilistic Latent Semantic Analysis (pLSA) [5], Latent Dirichlet Analysis (LDA)[9] and transformed Dirichlet Process (TDA)[11] have been adapted and used to model scene categories. Where images are represented by the frequency of "visual words". The *Bag-of-Features*(BOF) models, which represent an image as an orderless collection of local features, have recently demonstrated impressive levels of performance[7]. In all of these methods central clustering methods can be utilized to find distinguishable visual features.

However, there are two basic drawbacks for the local invariant feature based methods. Firstly, in most cases too many orderless feature points are extracted from each image. As a result the exhaustive search over features is time consuming and too many false positives are produced. One way of overcoming these problems is to use visual saliency [4] to select a subset of the salient points for analysis. Secondly, in the BOF model, local features play the role of "visual words" in an orderless way. This representation overlooks geometric and structural constraints on feature arrangement. This is likely to compromise recognition, clustering or retrieval performance. A more versatile and expressive tool for representing structured data are attributed graphs (hereinafter simply referred to as graphs). Hence there has been a recent interest in modeling and learning structural relationships using graph based method.

1.2 Graph Based Clustering

Graph clustering is an important, yet relatively under-researched topic in computer vision [14][15]. The importance of the topic stems from the fact that it can be used as a tool for learning the class-structure of data abstracted in terms of relational graphs. Problems of this sort are posed by a multitude of unsupervised learning tasks in knowledge engineering, pattern recognition and computer vision. The process can be used to structure large data-bases of relational models or to learn equivalence classes. Broadly speaking, there are two approaches to the problem.

The first of these is to use center clustering methods. Here a class prototype is maintained, and the aim is to find clusters of graphs that are close to the prototype. This prototype may be either an explicit representation of full graph-structure or a proxy representation based on features that succinctly express the graph characteristics. The learning of explicit cluster prototypes is a difficult task due to (1) missing powerful analytical techniques for data analysis and (2) high computational complexity when measuring structural similarities between graphs.

Considering the former problem, it is not clear how to embed dynamic data structures such as graphs into a vector space without loss of structural information. Therefore it is not straightforward to provide well defined characterizations of basic concepts such as the mean or variance of a set of graphs. The latter problem of measuring the structural similarity is more generally referred to as the graph matching problem. However, taking into account the practical importance, investigations on clustering of graphs has been recently revitalized[19][20] [21][22]. However, It is well known that the graph matching problem is NP-complete[18]. These algorithms are computationally intensive and have hence only been demonstrated on rather small graphs.

The lack of analytical methods in the domain of graphs led to the formulation of clustering problem as pairwise clustering [23][17]. The process of pairwise clustering is somewhat different to the more familiar one of central clustering. Whereas central clustering aims to characterize cluster membership using the cluster mean and variance, in pairwise clustering it is the relational similarity of pairs of images which is used to establish cluster membership. Although less well studied than central clustering, there has recently been renewed interest in pairwise clustering.

Recently, Torsello and Hancock [12][17] have constructed the class-prototype through tree-union and have performed clustering using a mixture of tree-unions controlled by a description length criterion [16]. These ideas are taken further by Escolano and Lozano [13] who extended the methodology to graphs rather than trees, and used an EM algorithm for clustering. When posed in a pairwise setting, the graph-clustering problem involves measuring the pairwise distances of the graphs. Provided that the data set consists of N samples, pairwise clustering requires the solution of $O(N^2)$ NP-complete graph matching problems which might be intractable for large data sets. Thus, clustering large set of graphs is still widely unexplored and one of the most challenging problems in structural pattern recognition.

To summarize, local invariant feature based methods, which represent an image as an orderless collection of local features, have demonstrated impressive performance, though overlooking geometric and structural constraints upon feature arrangement. While the graph based method provides an elegant way of encapsulating within-class structural variations, it is relatively impoverished in terms of the attribute information encoded. For this reason, we have proposed a class specific hyper graph (CSHG) model to combine both SIFT and geometric constraints in a companion paper [1] submitted to the same workshop . The CSHG model includes two basic parts.

Firstly, we rank SIFT-points using visual saliency for each image or ROI. The highly ranked salient points are used to construct attributed Delaunay graphs. Secondly, the CSHG model is used to develop a prototype structure through the merging of salient graphs. This structure encapsulates both variations in structure, and edge and node attributes, including the information contained in the ranked local invariant feature. In the CSHG model, a new method on measuring the pairwise graph similarity is proposed.

Based on the CSHG model, we develop a new two stage clustering method in this paper. In the first stage a sample of the graphs in a trained CSHG model are clustered to form initial clusters using constraints based on the pairwise similarity. The pairwise graph matching is sketched using the SIFT matching, which is not NP-complete. Because it utilizes an RSOM [24] (a SOM-based incremental clustering tree) central clustering method in learning a CSHG model, the clustering procedure requires the solution of $O(N \lg N)$ pairwise SIFT matching problem. Hence our model can be used to deal with large datasets. The remaining graphs are either recognized as one of the existing clusters or rejected from consideration.

The structure of this paper is as follows. In Section 2 we present some preliminaries. In Section 3 we describe the framework for modeling and clustering. Section 4 provides some experimental results. This paper finishes in Section 5 with the conclusions.

2 Preliminaries

See [1] Section 2.

3 CSHG Modeling and Clustering

3.1 CSHG Modeling

See [1] Section 3.

3.2 Clustering Based on a Trained CSHG

Suppose a CSHG model has been trained from a batch of images. In the CSHG model, the relationships of the graphs $ONNG^*_{K\tau}\{G_l\}$ have been encoded and will be updated with additional graphs. The number of the items in $ONNG^*_{K\tau}\{G_l\}$ is largely influenced by the threshold $\mathscr{R}^*_\tau$. In our system, $\mathscr{R}^*_\tau$ is set as 5, and the number K of the highly ranked graphs in $ONNG^*_{K\tau}\{G_l\}$ is set as 14. We expand the definition of the siblings of a graph (Def. 2.7 in [1]) to include the role of $\mathcal{R}^*_\tau$ as follows.

Def. 3.1 ***Siblings of a Graph($\mathcal{S}$):*** Siblings of G_l is re-defined as:

$$\mathcal{S}\{G_l\} = \{G_q \mid CMG(G_l, G_q) = 1,\ G_q \in ONNG^*_{K\tau}\{G_l\} \triangleq \mathcal{S}_{\mathcal{R}^*_\tau}\{G_l\}. \tag{1}$$

For any graph $G_l \in$ CSHG, we can obtain its siblings within $ONNG^*_{K\tau}\{G_l\}$. For each graph $G_q \in \mathcal{S}\{G_l\}$, their corresponding siblings can also be obtained. In this way, we can iteratively obtain a series of graphs according to the sibling relationships. This is called a family tree of a graph G_l.

Def. 3.2 ***Family Tree of a Graph within g Generations*(FTOG):** Given a CSHG model, the family tree of a graph(FTOG) G_l within g generations, denoted as $\mathcal{F}\{G_l, g\}$, is defined as:

$$\mathcal{F}\{G_l, g\} = \mathcal{F}\{G_l, g-1\} \bigcup_{G_q \in \mathcal{F}\{G_l, g-1\}} \mathcal{S}\{G_q\}. \tag{2}$$

where G_l is non-redundant graph of the CSHG model, and g is the number of the generations of a family tree. If $g = 1$, $\mathcal{F}\{G_l, 1\} = \mathcal{F}\{G_l, 0\} \bigcup \mathcal{S}\{G_l\}$ and $\mathcal{F}\{G_l, 0\} = \{G_l\}$. If $\lim_{g\to\infty} \mathcal{F}\{G_l, g\} \triangleq \mathcal{F}\{G_l, \infty\}$ =CSHG, CSHG is called all-linked.

Def. 3.3 ***Clan of a CSHG*:** Given a CSHG model, $\exists G_{l_i}$, $i = 1, 2, ..., \mathbb{c}$, subject to $\mathcal{F}\{G_{l_i}, g\} \not\subseteq \mathcal{F}\{G_{l_j}, g\}$, $i \neq j, i, j \in \{1, 2, ..., \mathbb{c}\}$, and for each family tree, there are at least $\mathcal{N}_\tau$ graph members, e.g. $\mathcal{N}_\tau = 3$. A clan of CSHG within g generations is defined as

$$\mathcal{C}\{g\} = \{\ \mathcal{F}\{G_{l_i}, g\},\ i = 1, 2, ..., \mathbb{c}\ \}. \tag{3}$$

We can regard a family tree of a graph within g generations ($\mathcal{F}\{G_l, g\}$) as a cluster if the number of graphs is greater than or equal to $\mathcal{N}_\tau$. The clan of a CSHG is a cluster separation of the whole graph set. Corresponding to $\mathcal{R}^*_\tau$, $\mathcal{F}\{G_l, g\}$ and $\mathcal{C}\{g\}$ are separately denoted as $\mathcal{F}_{\mathcal{R}^*_\tau}\{G_l, g\}$ and $\mathcal{C}_{\mathcal{R}^*_\tau}\{g\}$.

If $\mathcal{R}^*_\tau$ =0, it is easy to obtain that $\mathcal{F}\{G_l, \infty\}$ =CSHG. If $\mathcal{R}^*_\tau$ is set as the number of the nodes in a graph, namely the selected number of R-SIFT points for an image, there might always be only one item in a family tree $\mathcal{F}\{G_l, g\}$. Especially, in the training stage, the threshold $\mathcal{R}^*_{\tau_0}$ is set to 20 to check for the redundancy. Consequently, the selection of the value of $\mathcal{R}^*_\tau$ is a key point for a CSHG model. Here we will continue to present a two stage clustering method.

Firstly, if we set a relatively large $\mathcal{R}^*_\tau$, e.g. $\mathcal{R}^*_{\tau_1} = 12$, $\forall\ G_l \in$ CSHG, $\mathcal{F}_{\mathcal{R}^*_{\tau_1}}\{G_l, \infty\}$ normally converges to a stable graph set and each $\mathcal{F}_{\mathcal{R}^*_{\tau_1}}\{G_l, \infty\}$ is regarded as the seed members of a cluster. In this way, a clan of a CSHG model can be obtained, denoted as $\mathcal{C}_{\mathcal{R}^*_{\tau_1}}\{g\}$. Suppose there are N_c family trees obtained and each of which are uniquely labeled. These are the initial seed clusters in a CSHG model, which is denoted as follows:

$$\mathcal{C}_{\mathcal{R}^*_{\tau_1}} = \{\ \mathcal{F}_{\mathcal{R}^*_{\tau_1}}\{G_l, \infty\ \} \triangleq \{\ \mathcal{F}_l \mid l \in \{L_1, L_2, ..., L_{N_C}\}\ \}. \tag{4}$$

After the first stage, generally there are still some graphs which have not been clustered. As a second stage, the un-clustered graphs will be either recognized as one of the clusters in $\mathcal{C}_{\mathcal{R}^*_{\tau_1}}$ or discarded from being clustered using the following method.

For an un-clustered graph G_l,

$$if\ \exists\ only\ one\ clustered\ graph\ G_q,\ s.t.\ \mathscr{R}^*(G_l,G_q) \geq \mathscr{R}^*_{\tau_2},\ then\ L_{G_l} = L_{G_q}. \tag{5}$$

where $G_q \in ONNG^*_{K\tau}\{G_l\}$, L_{G_l} is the cluster label of G_l, and L_{G_q} is the cluster label of G_q. Otherwise, G_l remains un-clustered.

Furthermore, for any graph G_l of the remaining un-clustered graphs, we can obtain its K-clustered nearest neighbor graphs $ONNG^*_{K\tau}\{G_l\}$ with a threshold $\mathscr{R}^*_{\tau_3}$, e.g. $\mathscr{R}^*_{\tau_3}$ =5.

$$ONNG^*_{K\tau_3}\{G_l\} = \{\ G_q \mid \mathscr{R}^*(G_l,G_q) \geq \mathscr{R}^*_{\tau_3},\ and\ \exists\ \mathscr{F}_{L_i}\ s.t.\ G_q \in \mathscr{F}_{L_i}\}. \tag{6}$$

According to the K graphs G_{q_i}, $i = 1,2,...,K$, we use a weighted voting method to determine an FTOG $\mathscr{F}_{L_{i0}}$ that the graph G_l belongs to.

$$L_{i0} = \arg\max_{\mathscr{F}_{L_i}} \{\ \overline{W}^*\{\ \mathscr{F}_{L_i}\ \} = \frac{W^*\{\ \mathscr{F}_{L_i}\ \}}{\sum_{\mathscr{F}_{L_i}} W^*\{\ \mathscr{F}_{L_i}\}} \geq 2/3\ \}. \tag{7}$$

where $W^*\{\ \mathscr{F}_{L_i}\ \} = \sum_{G_q \in \mathscr{F}_{L_i}}\{\mathscr{R}^*(G_l,G_q)\ \}$, $\mathscr{F}_{l_i} \in \mathscr{C}_{\mathscr{R}^*_{\tau_1}}$ and $\overline{W}^*\{\ \mathscr{F}_{L_i}\ \}$ satisfy the following conditions

$$E_{\overline{W}^*} = -\sum_{\mathscr{F}_{L_i}} \{\ \overline{W}^*\{\ \mathscr{F}_{L_i}\} \cdot ln\{\overline{W}^*\{\ \mathscr{F}_{L_i}\}\}\ < E_\tau. \tag{8}$$

where E_τ is set as 0.4 according to the experiments. We borrowed this equation from the information entropy theorem. If no such family tree exists, or if $\forall\ G_q \in$ CSHG, $\mathscr{R}^*(G_l,G_q) < \mathscr{R}^*_{\tau_3}$, G_l is abstained from being clustered, which is also called a rejection.

In an incremental clustering process, for a new graph G_l, if Equation (9) holds, G_l will be incremented to the CSHG model. For clustering, $\mathscr{F}_{L_{i1}}$ and $\mathscr{F}_{L_{i2}}$ are merged into the family tree $\mathscr{F}_{\mathscr{R}^*_{\tau_1}}\{G_l,\infty\ \}$of graph G_l. G_l is called a joint graph of the cluster-merging.

$$\exists G_{q1}, G_{q2} \in \mathscr{S}_{\mathscr{R}^*_{\tau_1}}\{G_l\},\ \ s.t.\ \ G_{q1} \in \mathscr{F}_{L_{i1}}, G_{q2} \in \mathscr{F}_{L_{i2}}, and\ \mathscr{F}_{L_{i1}}, \mathscr{F}_{L_{i2}} \in \mathscr{C}_{\mathscr{R}^*_{\tau_1}} \tag{9}$$

where $\mathscr{S}_{\mathscr{R}^*_{\tau_1}}\{G_l\} = \{\ G_q \mid CMG(G_l,G_q) = 1,\ G_q \in ONNG^*_{K\tau_1}\{G_l\}\ \}$ are the clustered siblings with threshold $\mathscr{R}^*_{\tau_1}$ of graph G_l, and G_q is a graph that has been clustered in the S1 of Algorithm 1.

To summarize, the principal steps for CSHG based clustering are listed in Table 1, referred as Algorithm 1.

Table 1. CSHG based matching and incremental training

Algorithm 1: CSHG Based Clustering	
Input: a CSHG model	
Output: updated CSHG model, Clan of the CSHG model	
S1	Obtaining kernel clusters $\mathscr{C}_{\mathscr{R}^*_{\tau_1}}$, namely the clan of the CSHG with threshold $\mathscr{R}^*_{\tau_1}$;
S2	Recognizing un-clustered graphs G_l according to Equation (5).
S3	Recognizing un-clustered graphs G_l according to Equation (7). <End>

4 Experiments

We use all 72 images of 50 objects from the COIL100 (The COIL set contains objects placed on a black background with no clutter, the view angle changes from 0° to 355° with a step size of 5°, and there are 72 images for each object.). For each image, 6 noised images are generated with multiple noise-parameters. There are 25200 images. We aim to cluster these images into corresponding classes.

For this clustering task, firstly, the 3600 initial noise free images are modeled with a CSHG model and then clustered. Secondly, the remaining 21600 noisy images are incrementally modeled and clustered using the CSHG model obtained from the noise free images.

Table 2. Clustering of 3600 initial images of 50 objects in COIL100 dataset

ID of Cluster	ID of Object in COIL	Num. of Ini. Images	Num. of 1st stage Clustered	Num. of Finally Abstained
31 clusters	1,6,7,8,14,15,16,19,24, 28,30,37,39,46,49,53,55, 58,60,62,66,69,72,79, 81,87,89,91,93,96	72 images per calss	totally 2232	0
32	13	72	33	3
33	13	72	28	3
34	23	72	71	1
35	26,71	144	133	6
36	27	72	70	2
37	29,32	144	144	0
38	31,54	144	144	0
39	44,59	144	144	0
40	57	72	71	1
41	61,92	144	144	0
42	67,84,98	216	216	0
43	90	72	36	2
44	90	72	29	2
45	99	72	69	1
	Total	3600	3564	21
	Percent	100%	99.42%	0.58%

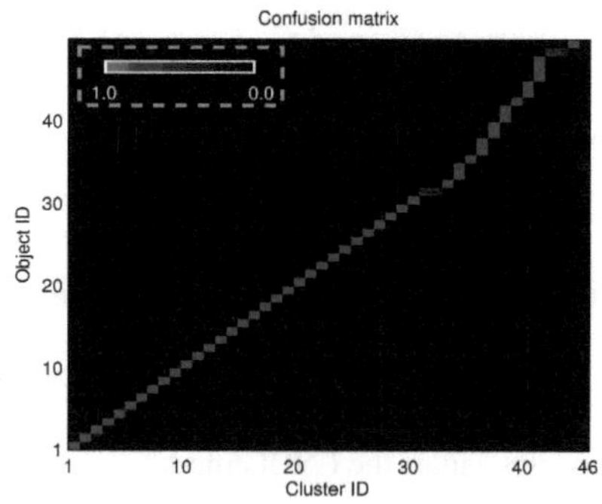

Fig. 3. Confusion matrix of clustering of 3600 initial images of 50 objects from the COIL100 dataset

For the 50 objects, the 3600 initial images are clustered into 45 clusters as shown in Table 2. The confusion matrix for the clustering result is shown in Fig.3, where the Y-axis is the order of the object ID in COIL 100. The X-axis is the cluster ID and 46 corresponds to the rejected images.

There are 35 clusters that separately correspond to 35 different objects without confusion. Example images from these 35 object-classes are shown in Fig. 2. In addition, 13 objects are assigned to 6 clusters as shown in Fig. 4. Each group of objects in Fig. 4 (A)(B)(C)(E) are actually identical in shape but color. Since it only uses gray scale information in SIFT, our method fails in this case. We hence regard these objects in the four groups as being correctly clustered according to shape.

For 2 groups of overlapping-clustered objects shown in Fig. 4(D)(F), there is only one joint graph. Finally 2 objects are assigned into 4 clusters, shown in Fig. 5. The graphs of the front view images are very different form the graphs of the back view images, while the graphs of the side view images are too simple to be joint graphs of the two clusters of the front and the back. However these clusters will be merged into one object in the supervised learning stage while assigning a same name to the front cluster and the back cluster. Finally, 0.42% of all these initial images are discarded from being clustered, namely the rejection rate is 0.42%.

Fig. 4. 6 groups of objects are overlapping-clustered into 6 clusters

Fig. 5. Identical objects are assigned into separated clusters

Table 3. Clustering of various images with multiple threshold restriction

Image source	$\mathscr{R}^* \geq \mathscr{R}^*_{\tau_0}$ ($\mathscr{R}^*_{\tau_0} = 20$)	$\mathscr{R}^* \geq \mathscr{R}^*_{\tau_1}$ ($\mathscr{R}^*_{\tau_1} = 12$)	$\mathscr{R}^* \geq \mathscr{R}^*_{\tau_2}$ ($\mathscr{R}^*_{\tau_2} = 8$)	$\mathscr{R}^* \geq \mathscr{R}^*_{\tau_3}$ ($\mathscr{R}^*_{\tau_3} = 5$)	$\mathscr{R}^* < \mathscr{R}^*_{\tau_3}$ ($\mathscr{R}^*_{\tau_3} = 5$)
Initial	40.27%	84.32%	94.00%	99.42%	0.58%
Noise0.01	31.86%	79.55%	92.61%	99.17%	0.83%
Noise0.03	22.55%	72.51%	90.53%	98.94%	1.06%
Noise0.05	15.21%	59.56%	88.58%	98.65%	1.35%
Noise0.07	9.31%	38.68%	86.83%	98.36%	1.54%
Noise0.09	6.17%	27.75%	84.61%	98.25%	1.75%
Noise0.11	3.37%	19.16%	81.32%	97.80%	2.20%

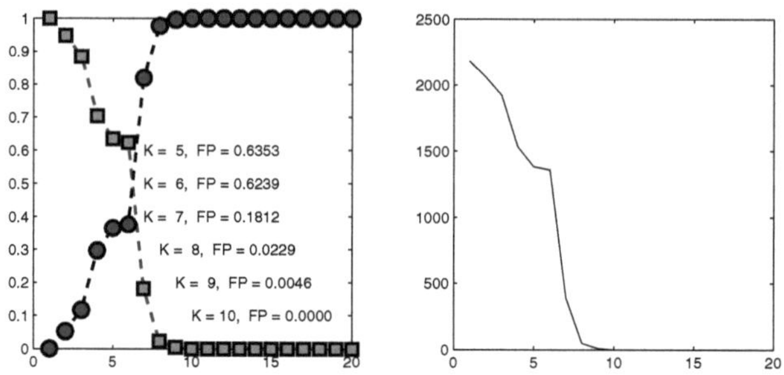

Fig. 6. Clustering test result of 2300 various negative sample images. These images are randomly selected from three data sets: Caltech-101, Caltech-256 and Google images. We use the CSGH model of 50 objects obtained above to test how these negative sample images are clustered. For the left sub-figure, Y-axis corresponds to clustering rate or rejection rate and X-axis corresponds to the threshold of $\mathscr{R}^*_\tau$. The green-face-square marker corresponds to the clustering rate, the red-face-circle marker corresponds to the rejection rate. For the right sub-figure, the plot is the number of images with different threshold of $\mathscr{R}^*_\tau$.

Our second experiment involves adding multiple kinds of noise to each image. The noise processes consists of a mixture of salt and pepper noise, speckle noise, and Gaussian noise. The fraction of added noise increases from 0.01 to 0.11 with a step 0.02. For each case, 3600 noisy images are generated. Based on the CSHG model and clustering result obtained from the initial images, the noisy images are incrementally clustered. The results are shown in Table 4. The average correct clustering rate for all these 21600 images is 98.45%, The remaining 1.55% were rejected and were not correctly clustered. This is a significant property of our method. If a graph of an image is not similar enough to the graphs of the CSHG model, it should not be clustered.

After clustering, we have constructed a compact multiple-object example CSHG model for all of these 50 objects. Hence we have also explored how negative images are recognized using this CSHG model. We randomly selected 2300 images from three data sets: Caltech-101, Caltech-256 and Google images. In Fig. 6, we investigate the false classification rate and correct rejection rate with varying threshold $\mathscr{R}^*_\tau$ for these negative images . From Fig. 6, there are 2.29% of all the 2300 images that are false clustered or recognized according to S1 of Algorithm 1. However the average false clustering rate for each class is only 0.046%. And there are 36.37% of the images that are correctly rejected according to the restriction that is $\mathscr{R}^* < \mathscr{R}^*_{\tau_3}$. The remaining 61.34% are rejected according to S3 of Algorithm 1 since Equation (7) poses a very strict condition especially on the negative samples.

5 Conclusions

In paper [1], we demonstrated how to construct a CSHG model. This model comprehensively captures both global and local information through utilizing both SIFT and geometric constraints. Based on this model, we proposed a two stage generalized pairwise clustering method, which combines the recognition process into the clustering process, to capture a more holistic and compact representation of the objects. Experiments reveal that the method gives excellent clustering performance. The model can be extended to object recognition and content based image retrieval.

References

1. Xia, S.P., Hancock, E.R.: 3D Object Recognition Using Hyper-Graphs and Ranked Local Invariant Features. In: S+SSPR 2008 (to appear, 2008)
2. Bosch, A., Zisserman, A., Muoz, X.: Scene Classification Using a Hybrid Generative/Discriminative Approach. IEEE Trans. PAMI 30(4), 1–16 (2008)
3. Lowe, D.G.: Distinctive image features from scale-invariant key points. IJCV 60(2), 91–110 (2004)
4. Elazary, L., Itti, L.: Interesting objects are visually salient. Journal of Vision 8(3), 1–15 (2008)
5. Li, F.F., Perona, P.: A Bayesian hierarchical model for learning natural scene categories. CVPR 2, 524–531 (2005)
6. Bay, H., Tuytelaars, T., Gool, L.V.: SURF: Speeded Up Robust Features. In: Leonardis, A., Bischof, H., Pinz, A. (eds.) ECCV 2006, Part I. LNCS, vol. 3951, pp. 404–417. Springer, Heidelberg (2006)
7. Zhang, J., Marszablek, M., Lazebnik, S., Schmid, C.: Local Features and Kernels for Classification of Texture and Object Categories. IJCV 73(2), 213–238 (2007)
8. Yan, K., Sukthankar, R.: PCA-SIFT: A more distinctive representation for local image descriptors. CVPR 2, 506–513 (2004)
9. Sivic, J., Zisserman, A.: VideoGoogle: A text retrieval approach to object matching in videos. ICCV 2, 1470–1477 (2003)
10. Nigam, K., Lafferty, J., Mccallum, A.: Using maximum entropy for text classification. In: IJCAI Workshop on Machine Learning for Information Filtering, pp. 61–67 (1999)
11. Sudderth, E.B., Torralba, A., Freeman, W.T., Willsky, A.S.: Describing Visual Scenes Using Transformed Objects and Parts. Int. J. Comput. Vis. 77, 291–330 (2008)
12. Torsello, A., Hancock, E.R.: Graph Embedding using Tree Edit Union. Pattern Recognition 40, 1393–1405 (2007)
13. Bonev, B., Escolano, F., Lozano, M.A., Suau, P., Cazorla, M.A., Aguilar, W.: Constellations and the Unsupervised Learning of Graphs. In: Escolano, F., Vento, M. (eds.) GbRPR 2007. LNCS, vol. 4538, pp. 340–350. Springer, Heidelberg (2007)
14. Rizzi, S.: Genetic operators for hierarchical graph clustering. Pattern Recognition Letters 19, 1293–1300 (1998)
15. Segen, J.: Learning graph models of shape. In: Laird, J. (ed.) Proceedings of the Fifth International Conference on Machine Learning, pp. 29–35 (1988)
16. Torsello, A., Hancock, E.R.: Learning shape-classes using a mixture of tree-unions. IEEE Trans. PAMI 28(6), 954–967 (2006)
17. Torsello, A., Robles-Kelly, A., Hancock, E.R.: Discovering Shape Classes using Tree Edit Distance and Pairwise Clustering. International Journal of Computer Vision 72, 259–285 (2007)
18. Garey, M., Johnson, D.: Computers and intractability: A guide to the theory of NP-completeness. W.H. Freeman and Company, New York (1979)
19. Günter, S., Bunke, H.: Self-organizing map for clustering in the graph domain. Pattern Recognition Letters 23, 401–417 (2002)
20. Lozano, M.A., Escolano, F.: Em algorithm for clustering an ensemble of graphs with comb matching. In: Rangarajan, A., Figueiredo, M.A.T., Zerubia, J. (eds.) EMMCVPR 2003. LNCS, vol. 2683, pp. 52–67. Springer, Heidelberg (2003)

21. Luo, B., Wilson, R.C., Hancock, E.R.: Spectral embedding of graphs. Pattern Recognition 36(10), 2213–2223 (2003)
22. Jain, B.J., Wysotzki, F.: Central Clustering of Attributed Graphs. Machine Learning 56, 169–207 (2004)
23. Hofmann, T., Buhmann, J.M.: Pairwise data clustering by deterministic annealing. IEEE Transactions on Pattern Analysis and Machine Intelligence 19(1), 1–14 (1997)
24. Xia, S.P., Liu, J.J., Yuan, Z.T., Yu, H., Zhang, L.F., Yu, W.X.: Theory and Algorithm of Machine Learning Based on RSOM Tree Model. ACTA Electronica sinica 33(5), 937–944 (2005)

Structure Is a Visual Class Invariant

Bai Xiao, Yi-Zhe Song, Anupriya Balika, and Peter M. Hall

Computer Science Department, University of Bath, U.K.

Abstract. The problem of learning the class identity of visual objects has received considerable attention recently. With rare exception, all of the work to date assumes low variation in appearance, which limits them to a single depictive style usually photographic. The same object depicted in other styles — as a drawing, perhaps — cannot be identified reliably. Yet humans are able to name the object no matter how it is depicted, and even recognise a real object having previously seen only a drawing.

This paper describes a classifier which is unique in being able to learn class identity no matter how the class instances are depicted. The key to this is our proposition that topological structure is a class invariant. Practically, we depend on spectral graph analysis of a hierarchical description of an image to construct a feature vector of fixed dimension. Hence structure is transformed to a feature vector, which can be classified using standard methods. We demonstrate the classifier on several diverse classes.

1 Introduction and Background

The work in this paper was motivated by the desire to classify objects no matter how they are depicted. For example, it is perfectly normal to label a picture of a bird as "bird", whether it is seen for real, is photographed, is a drawing, or is depicted in any other way. Moreover, it is common human experience to be able to recognise a real instance of some object after being shown only a drawing of it. For example, flori — catalogues of plants — allow botanists to identify plants; flori support text descriptions using drawings rather than photographs, possibly because drawings capture the class as a whole more faithfully than a single photograph of one plant. Clearly, learning to identify object categories through pictures of all kinds is an important problem. It is this problem we address, in contrast to most previous work which assumes a single depictive style, most commonly photographs.

Consider a thought experiment using four images. Three images show faces, one little more than a scribble by a young child, one the face of a clown with crosses for eyes, the other a photograph. The fourth image is a photograph of a car. Which is the odd one out? We have run this experiment in practice, and were not at all surprised to find the car was always selected as the odd one out. Yet the variance between the faces is profound. Eyes can be any shape, so can the mouth. The draw face may or may not have an outline, and if it does exist can be any shape. Children tend to draw the eyes at the top of the head when

N. da Vitora Lobo et al. (Eds.): SSPR&SPR 2008, LNCS 5342, pp. 329–338, 2008.

in fact they are in the middle. Given so much variation in shape, in position, in color, it is difficult to see any invariant except the structural arrangement of facial parts. The natural question following these observations is *what class property is invariant across depictions?* Our answer, which we believe is novel, is stated as a proposition: *the topology of an object's parts is a class property invariant to depiction.* It is this proposition that this paper sets out to test.

We learn objects classes, and so our work relates to is preceded by a considerable body of research that also learns object classes. Recent work, *circa* 2000 to date, uses localized image patches; there is too much to review here, but [2,8,14,13,19] provide typical examples. Many problems have been addressed, including recognition under occlusion [5] and from different points of view, eg [16], and learning from just a few examples [11]. Despite the many successes of this work it is all premised on the assumption that corresponding patches exhibit low variation.However, drawings and photographs would raise the variance above acceptable limits. Indeed, some deliberately filter non-photographs from their database used to build models of appearance [7]. Yet an object's class does not depend on its depiction; faces can be drawn, photographed, painted and so on.

By comparison, there is little research on classifying objects independently of depiction type. Some use curves [15], which does not necessarily assume photographic depiction, but does require low shape variance which we have observed cannot be assumed when using artwork. Schectman and Irani [18] show it is possible to match pairs of images in different depictive styles. They use the spatial relation between different patterns, which they call local self-similarities. For example, flowers might represent eyes in one picture, but photographed in another. Provided matching is consistent, Schectman and Irani [18] can match them. It might seem that matching many pairs would allow classes to be built, but as the number of representational instances grow so does the variation in depective style. This means there is no satisfactory charactisation in terms of appearance alone.

Fidler and Leonardis [6] uses a large data base of image to learn very simple primitives, which are combined into ever more specific objects until the highest layers are category-specific. The system is then able to detect categories across depictive styles. In contrast we need just a few examples of each class in order to learn. Also, Fidler and Leonardis [6] do not explicitly consider the topological relationship, as do we and Ahuja and Todorovic [1]. But they [1] cluster parts on the basis of similar geometric, photometric and topological properties. For them, topological similarity is defined via tree isomorphisms.

By using spectral graph theory [4] to encode structure as vectors we not only avoid the graph isomorphism problem and gain robustness to noise, but also can form clusters. In practice we must extract an object's structure with a method robust to wide variations in depictive style. The outline of the paper is as follows. In Section 2, we introduce a hierarchical image description by building a tree description through agglomerative merging of image primitives. Section 3 explains how a a modified form of graph energy is used to segment the tree into sub-trees, yielding a rough segmentation of objects. The objects are subject to exactly the same form of analysis, yielding their parts. The topological relation between parts is then characterized with a feature vector. These features are clustered using standard methods, in Section 4. Section 5 shows the experiments. Finally, we give the conclusion in Section 6.

2 A Hierarchical Image Description

Forming a hierarchical image description is important to us, but our claim should not depend on a particular algorithm. We can use any algorithm to produce an image description, provided that: (i) gives a covering or partition of the image, so the whole is described; and (ii) operates effectively over images of all depictive styles, so any image can be described. Unfortunately, for reasons given below, these criteria, but especially the second, rule out well known approaches. Unfortunately, we were therefore forced to develop an approach of our own. Although this exhibits some novelty, we would not wish the reader to be distracted by it. We emphasis that we make just two claims for it: (a) it operates over different dipictions; (b) it is strong enough to allow us to demonstrate our proposition that structure is class invariant.

Maximally stable extremal regions [12] and sieves [3] both naturally produce a hierarchical image description. But neither can be used because they build no tree for binary images — but line drawings are often binary images. The patch based approaches common to many visual object classifiers (see Introduction, above) depend on low variation in appearance, but using line drawings, paintings, photographs ensures wide appearance variation. Additionally patches are usually filtered to keep just the most "salient". But there is a problem with databases which include drawings and photographs, as shown in the first column of Figure 1. In the second column of Figure 1 potential patches are located via the first extremum of a difference of Gaussian (DoG) filter, which is reported to give excellent results [10]. The size of a DoG patch is set to just cover a local feature. Figure 1 filters DoG patches using a threshold on the DoG signal strength. This threshold is not set absolutely, but is set as a fraction of the strongest signal. This must be set high to include on salient feature in a photograph, but the same threshold leads to missed features in the drawing – the second column in Figure 1. Lowering the threshold includes the missing features in the drawing, but at the expense of including all patches in the photograph – the third column in Figure 1.

We initially solved this problem using watershed regions as image primitives, which we then grouped to form a hierarchical description. This paper, though, uses image primitives which are DoG patches. In this case we filter DoG patches on the basis of size, using a greedy algorithm. We keep the largest DoG patch and reject any other patch whose center falls inside it. We then keep the largest remaining patch, and simply iterate until all patches are kept or rejected. The result is a set of DoG patches, each one a circular patch. These patches can overlap, as seen in the fourth column of Figure 1. The next step is to merge neighboring image primitives to create a hierarchical tree description.

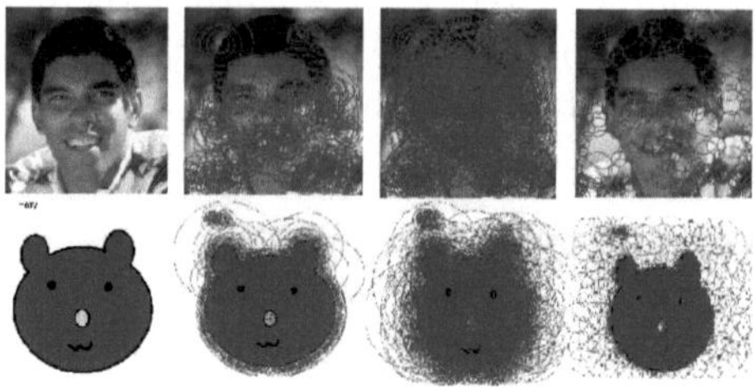

Fig. 1. Interest patches (oriented to local gradient): the first column are original images, the second column are top 50% of patches, the third column are top 90%. These patches can give poor coverage of some images, unless many are used. Notice that the eyes of the drawing are not covered until 90% of patches are used — but the photographed face is then swamped with patches. For comparison, the fourth column shows greedy filtering – the primitive we use.

At each merging iteration we merge the most similar pair of neighboring regions. The similarity measure must allow for the fact that regions can be of arbitrary shape, so we use simple statistical properties. Specifically, the distribution of color and intensity gradient of a region is modeled as a single Gaussian, written $\Omega = (N, \mu, \mathbf{C})$, where N is the number of pixel in the region, μ is the mean of the property vectors at each pixel, and $\mathbf{C}$ the covariance. When two regions, i and j, are merged it is easy to compute a new representation for them. If the regions do not overlap then the new model, $\Omega = \Omega_i + \Omega_i$, is

$$\Omega = \left(N_i + N_j, \frac{N_i \mu_i + N_j \mu_j}{N_i + N_j}, \frac{N_i \mathbf{C}_i + N_j \mathbf{C}_j}{N_i + N_j} + e \right) \tag{1}$$

The "error" term, e, compensates the eigenspace for the difference between the means of the model and is easily shown to be

$$e(i,j) = \frac{N_i N_j}{N_i + N_j} (\mu_i - \mu_j)^T (\mu_i - \mu_j) \tag{2}$$

We allow for spatially overlapping regions simply by partitioning the pixels into sets "i alone", "j alone" and "i and j", then adding the results as above. Agglomeration continues until there is just one region left. Finally note that the adjacency matrix is weighted using the error measure, so if primitives i and j are connected the weight is $w(i,j) = \exp(-e(i,j))/e_{\max}$, in which $e_{\max}$ is the maximum weight over all connected pairs. We are now ready to analyze this tree.

3 Object and Object Part Identification

Our aim now is to partition a hierarchical description of an image into objects, and then partition objects into parts. This process can not depend in any way on the visual representation, because that is assumed to be unknown. Rather, the process must depend only on the structure of tree and the graph of primitives. Figure 2 exemplifies what we mean by object structure. This figure shows structure can be successfully obtained in a simple way and one same structure is general to many depictive styles. Appearance within a class varies significantly, because of depictive style, yet images from the same object class share roughly similar structures. For example, all the flowers contain a center; faces comprise eyes and a nose; for cows and horses we can extract the legs, the body and the head. Intuitively, these structures are a property of the object; which is as it should be, for we have proposed that structure is a property of an object's class.

Fig. 2. Examples objects and their parts, found using graph energy analysis, taken from our experimental training and test sets. Facial parts are identified; four legged animals have their head and legs separated from their body; flowers have their centers extracted.

Consider first the problem of identifying a salient object in the tree. As with graph cuts [17], we want high connectivity within an object and low connectivity between objects. But we do not use graph cuts, we use graph energy instead. We make this choice because: (i) graph cuts cannot decide the number of objects there are in an image, whereas our graph energy based method can decide this quantity; and (ii) the graph energy we use shows minima for graphs of homogeneous degree. This includes circuits (polygons) and cliques (fully connected objects), both of which are visually salient. Any node in the tree can be used to identify a collection of image primitives, which in turn correspond to a subgraph the graph of neighbors. The root node corresponds to the whole image, and we compute the graph energy of this. Then we can remove the root node to reveal the lowest branches, which form a covering of the image and which partition the adjacent matrix into subgraphs. The energy of each subgraph is computed and their sum is defined as the energy of the whole. We continue in this way, moving down through the tree, computing a graph energy at each step.

We use borrow the idea of the Laplacian graph energy [9]. The Laplacian matrix is defined as $L = D - A$, in which D is a degree matrix, and A an adjacency matrix. Laplacian graph energy has the following standard definition: for a general graph $G = (V, A)$, with arc weights $w(i, j)$ the Laplacian energy is

$$\mathcal{E}(G) = \sum_{i=1}^{|V|} \left| \lambda_i - 2\frac{m}{|V|} \right| \tag{3}$$

In which: the λ_i are eigenvalues of the Laplacian matrix; m is the sum of the arc weights over the whole graph, or is half the number of edges in an unweighted graph; $|V|$ is the number of nodes in graph. Note that $2m/|V|$ is just the average (weighted) degree of a node. Now, the Laplacian energy of a graph can rise or fall; our tests show that this rise and fall is strongly correlated with the variance in the degree matrix D. This means local minima tend to occur when the graph is regular.

The standard definition of Laplacian graph energy makes no explicit account for a change in the number of connected components. We would like this to happen, because connected components are partitions of a general adjacency matrix. Now the adjacency matrix we have has just one connected component, it is constructed that way. But the tree nodes do partition our adjacency matrix into subgraphs that are connected only with arcs weaker in strength than any arc in the subgraph, so it is reasonable to consider each tree branch (object) as a connected component in an adjacency matrix, Figure 3 shows a typically energy curve, as a function of the height of the tree.

We compute graph energy $\mathcal{E}(G)$ for each distinct object. We combine these by defining the *normalized* Laplacian energy for a graph of N connected components over n nodes (image primitives/leaves)

$$\xi = \frac{n}{N} \sum_{i=1}^{N} \frac{\mathcal{E}(G_i)}{|V_i|} \tag{4}$$

The term $\mathcal{E}(G)/|V|$ is the average connection energy per node — this is rather like normalising spatial measures (such as image moments) by area. Since the partition energy, ξ, is a function of tree height, so we should write $\xi(k)$. This novel definition coincides with $\mathcal{E}(G)$ exactly if there is just one connected component, but otherwise is bounded above by $\mathcal{E}(G)$. The significant difference, though, is that ξ rises steeply when two large connected components are merged, less steeply when a large component absorbs a smaller one, and gives local minima where connected components are close to regular graphs.

As mentioned local minima of graph energy occur when (Sub)graphs exhibit homogeneous degree. Figure 3 shows that there is more than one such local minimum. We currently select the tallest tree minimum to obtain a partition

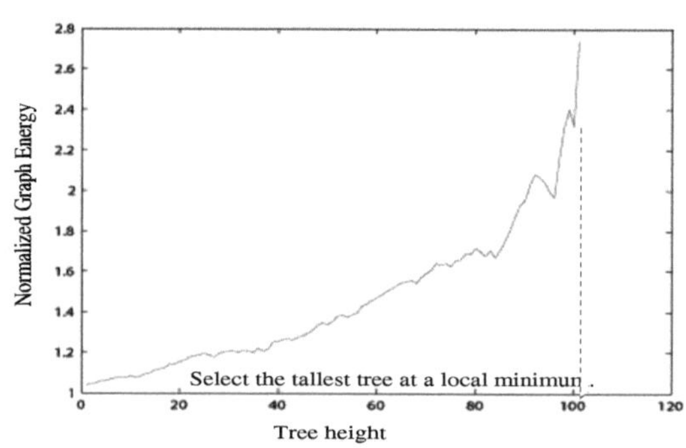

Fig. 3. An example of normalized graph energy as a function of tree height. We identify an object using the local minimum with tallest tree.

with the fewest number of large objects. We differentiate between foreground and background objects by having the user touch the foreground.

Having identified objects we turn attention to the second goal of breaking them into parts. Objects are just images, with a tree and neighbor matrix inherited from the full image. We find an object's parts by recursive application of the technique described — we just find the last local minimum in the graph energy curve for the object-image. The result is a set of objects and their parts, as seen in Figure 2. It is these structures we use to characterize the objects, as explained next.

4 Structure Vectors as Object Features, and Classification

In this section, we explain how objects are assigned feature vectors that are descriptors of their structure, and how these can be used to classify. The analysis of the tree yields an object made of parts, and the way those parts are structured. This is no more than a graph of nodes — the parts — and arcs — their connections. We follow Ping and Wilson [20], who advocate the use of eigenvalues from the signless Laplacian matrix. This is defined as $L = D + A$, where as before A is the adjacency matrix and D is the degree matrix. So, we form a feature vector, $\mathbf{f}$, as the eigenvalues of $D + A$.

We fixed $\mathbf{f}$ into a space of eight dimensions, padding with zeros if necessary. We used eight dimensions because that was the largest graph we observed, so we were sure to retain all information. The clusters visualized in Figure 5 are in two dimensions, yet remain largely separable; how best to choose the dimension of structural feature space remains an open question.

A structure vector $\mathbf{f}$ is no more than a particular kind of feature vector, which allows us to use any standard classification algorithm. After visually examining a two dimensional scatter plot of twelve different kinds of object we decided to

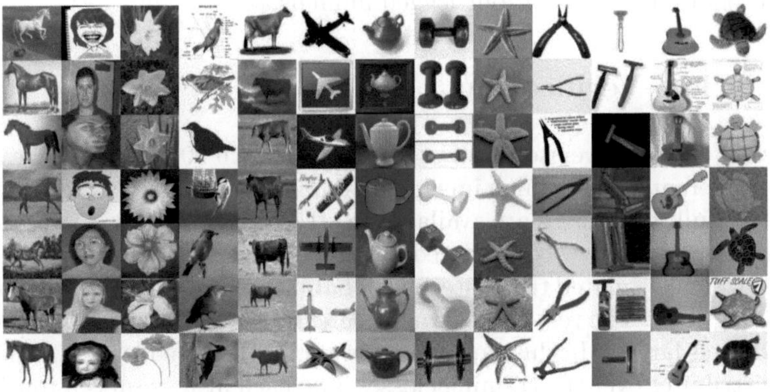

Fig. 4. The set of training images. Each column is a class (but horses and cows are both four legged animals). Many depictive styles are included in each class.

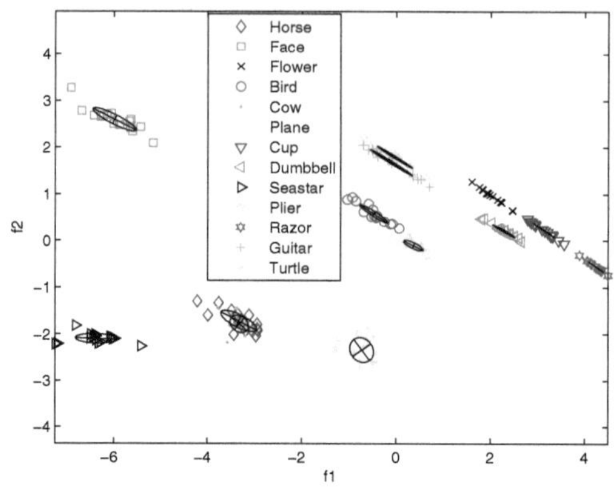

Fig. 5. Classes as components in a Gaussian Mixture Model

opt for supervised training. An eigenmodel comprising the number of point, their mean and covariance was fitted to the features of each training class, which was manually labeled hence the supervision. The individual eigenmodels were then collected into a single Gaussian Mixture Model (GMM). Figure 4 shows some of the pictures in our training set. We need only a few images per class, we trained with 10 examples per class. All sets contains drawings and paintings as well as photographs, yet form discernible clusters, as Figure 5 shows. The spread of the cluster is explained by noise that can arise from slightly different segmentations.

Notice we include both horses and cows in the training set, but these form a single cluster. This shows that the equivalence classes generated from structure are broad. We included these "four legged animals" because we predicted they would classify as one — but "four legged animals" is a perfectly legitimate class, and horses and cows can be separated other means, if the application demands. Notice that a system that always discriminates cows from horses might never be able to recognise the class of four legged animals.

5 Empirical Verification

To verify our classifier we used a set of test images, with 50 images from each class which are different from training images. Test images are highly varied in their depictive style(full test images are available at https://cs.bath.ac.uk/ xb202/data). For a given test object we constructed a feature vector, which was then input to our GMM which estimated the posterior probability of the test object belonging to each training class. The test object was assigned the class for which the posterior probability was largest. In this experiment, we constructed a histogram of how the test set distributes over the training classes, and each histogram is a row in Table 1. It is clear that most objects can be correctly classified with high probability, suggesting our GMM is a reasonable discriminative model.

Table 1. A confusion matrix; each row shows the probability a test class is classified as a given training class. The classes are: (1) Four legged animals; (2) Faces; (3) Flowers; (4) Birds; (5) Plane; (6) Cup; (7) Dumbbells; (8) Star fish; (9) Pliers; (10) Razors; (11) Guitars; (12) Turtles.

	1	2	3	4	5	6	7	8	9	10	11	12
1	0.94								0.03			0.03
2		0.94		0.06					0.03			0.03
3			0.86				0.14					
4		0.06		0.91		0.03						
5					0.86	0.14						
6			0.06			0.89					0.05	
7							1.00					
8	0.06							0.94				
9									0.97		0.03	
10						0.06				0.94		
11							0.09			0.03	0.88	
12	0.03											0.97

6 Discussions and Conclusions

This paper claims structure can be used to classify, just as other features like texture, shape are used. We have provided experimental evidence in support of this claim. Our means of forming hierarchical descriptions requires relatively simple images, but does operate over a wide range of depictive styles. Because of this we eschew any other claims regarding our method, except that is is strong enough to allow empirical support to our claim.

The characteristic that structure brings into classification is that of broad classes — sufficiently broad to cover objects depicted in a wide range of styles: photographs, drawings, and so on, yet sufficiently discriminative to be meaningful. The broadness of the classes we generate are illustrated not just by the inclusion of many depictive styles, but also by the fact cows and horses classify as one; this is not necessarily a disadvantage — an application might require "four legged animals" as a class, and subdivision on other measures is always possible.

Changing the segmentor and the classifier will not affect our conclusion, although it will make the system more robust. Similarly, there is more work to do to answer questions of scale, point of view, occlusion and so on. Again, answering these questions will not change our conclusion, but will act to make the system more robust, which provides amply future work. Also, comparative evaluation against other systems is for future work: exactly how to compare a classifier based on structure against one based on appearance is an open issue. Here we have shown that structure can be used as a feature, which was our intention and the contribution of this paper.

References

1. Ahuja, N., Todorovic, S.: Discovering hierarchical taxonomy of categories and shared subcategories in images. In: ICCV (2007)
2. Leibe, B., Schiele, B.: Interleaved object categorization and segmentation. In: BMVC
3. Andrew Bangham, J., Moravoc, K., Harvey, R., Fisher, M.: Scale-space trees and applications as filters, for stereo vision and image retrieval. In: Pridmore, T., Elliman, D. (eds.) British Machine Vision Conference (August 1999)
4. Chung, F.R.K.: Spectral graph theory. American Mathematical Society (1997)
5. Lowe, D.: Distinctive image features from scale-invariant keypoints. International Journal of Computer Vision 1, 91–110 (2004)
6. Fidler, S., Leonardis, A.: Towards scalable representations of object categories: Learning a hierarchy of parts. In: ICCV (2007)
7. Schroff, F., Criminisi, A., Zisserman, A.: Havesting image databases from the web. In: ICCV (2007)
8. Grauman, K., Darrell, T.: Unsupervised learning of catagories from sets of partially matching image features. In: CVPR (2006)
9. Gutman, I., Zhou, B.: Laplacian energy of a graph. Linear Algebra and its Applications 44, 29–37 (2006)
10. Mikolajczyk, K., Schmid, C.: A performance evaluation of local descriptors. TPAMI 2005 27(10), 1615–1630 (2005)
11. Fei-Fei, L., Fergus, R., Perona, P.: A bayesian approach to unsupervised one-shot learning. In: ICCV (2003)
12. Matas, J., Chum, O., Urban, M., Pajdla, T.: Robust wide baseline stereo from maximally stable extremal regions. In: BMVC (2002)
13. Mikolajczyk, K., Leibe, B., Schiele, B.: Multiple object class detection with a generative model. In: CVPR (2006)
14. Mutch, J., Lowe, D.: Multiclass object recognition with sparse, localized features. In: CVPR (2006)
15. Pinz, A., Opelt, A., Zisserman, A.: A boundary-fragment-model for object detection. In: ECCV (2006)
16. Yan, P., Khan, S.M., Shah, M.: 3d model based object detection in an arbitrary view. In: ICCV (2007)
17. Shi, J., Malik, J.: Normalized cuts and image segmentation. In: International Conference on Computer Vision and Pattern Recognition (1997)
18. Schechtman, S., Irani, M.: Matching local self-similarities across images and videos. In: CVPR (2007)
19. Weber, M., Welling, M., Perona, P.: Unsupervised learning of models for visual object class recognition
20. Zhu, P., Wilson, R.C.: Stability of the eigenvalues of graphs. In: Gagalowicz, A., Philips, W. (eds.) CAIP 2005. LNCS, vol. 3691, pp. 371–378. Springer, Heidelberg (2005)

Local Fisher Discriminant Component Hashing for Fast Nearest Neighbor Classification

Tomoyuki Shibata and Osamu Yamaguchi

Corporate Research and Development Center, Toshiba Corporation,
1, Komukai-Toshiba-cho, Saiwai-ku, Kawasaki, 212-8582, Japan
{tomoyuki1.shibata,osamu1.yamaguchi}@toshiba.co.jp

Abstract. This paper presents a novel approximate nearest neighbor classification scheme, *Local Fisher Discriminant Component Hashing* (LFDCH). Nearest neighbor (NN) classification is a popular technique in the field of pattern recognition but has poor classification speed particularly in high-dimensional space. To achieve fast NN classification, Principal Component Hashing (PCH) has been proposed, which searches the NN patterns in low-dimensional eigenspace using a hash algorithm. It is, however, difficult to achieve accuracy and computational efficiency simultaneously because the eigenspace is not necessarily the optimal subspace for classification. Our scheme, LFDCH, introduces Local Fisher Discriminant Analysis (LFDA) for constructing a discriminative subspace for achieving both accuracy and computational efficiency in NN classification. Through experiments, we confirmed that LFDCH achieved faster and more accurate classification than classification methods using PCH or ordinary NN.

Keywords: approximate nearest neighbor classification, high-dimensional space, hash, dimensionality reduction.

1 Introduction

Nearest neighbor (NN) classification [1] is a popular technique in the field of pattern recognition. Given a query pattern, a NN classifier classifies the pattern by simply finding the nearest pattern among training ones. However, the NN classification is not regarded as a practical, because of its inefficient memory use and poor classification speed.

To accelerate classification speed, fast NN search algorithms have been proposed. These search algorithms introduce the following two ideas:

- Reducing the number of comparisons: The number of distance calculations between a query and each training patterns is minimized by using the triangle inequality. [10][11][12][13]
- Aborting the number of distance calculation: Distance calculation are not necessary for those training patterns that can not possibly be the nearest neighbor of the query pattern. [2]

N. da Vitora Lobo et al. (Eds.): SSPR&SPR 2008, LNCS 5342, pp. 339–349, 2008.

However, the previous methods have the limitation that search cost does not reduce in high-dimensional space (more than a few tens of dimensions). In this case, NN search is no better than brute force. Thus, recently, approximate NN search methods have been gathering interest. These approximate methods do not guarantee to find the nearest pattern among training patterns.

Principal Component Hashing (PCH) [3] is one of the approximate NN search methods. PCH projects training patterns into an eigenspace obtained using principal component analysis (PCA) [4]. The eigenspace is generated by the maximizing the variance of training patterns in the lower-dimensional space. We refer to this eigenspace as PCA space. In the experiments of [3], PCH is a faster and more accurate search method than normal fast NN search methods and previous approximate methods.

However, PCH has the following barriers to achieving faster classification speed while keeping accuracy of classification.

(1) PCA space is not necessarily the optimal subspace for classification.
(2) PCH has a significant trade-off relationship between classification speed and classification accuracy.

To solve these problems, we propose a novel approximate nearest neighbor classification scheme, *Local Fisher Discriminant Component Hashing* (LFDCH). For problem (1), above, we introduce Local Fisher Discriminant Analysis (LFDA) [7], which is a supervised dimensionality reduction method, for obtaining a discriminative subspace by maximizing between-class and minimizing the within-class variance. We refer to this discriminative subspace as LFDA space. LFDCH in low dimensional LFDA space is a more accurate NN classifier than PCH in low dimensional PCA space. For problem (2), above, we recognize that the ultimate goal of NN classification is to classify the class label of the query pattern, rather than simply to find the NN pattern. Thus, LFDCH can abort the classification process when all training patterns have the same class label. LFDA space helps to increase the probability of this event compared with PCA space. As a result, classification speed on LFDCH is further accelerated, compared with the classification method based on PCH.

The remainder of this paper is organized as follows: Section 2 and Section 3 explain PCH and LFDA respectively. We propose LFDCH in Section 4 and show experimental results in Section 5. Finally Section 6 presents conclusions.

2 Principal Component Hashing

Principal Component Hashing (PCH) [3] is a search method using the hash function in one of the approximate NN search algorithms. Other approximate NN search algorithms (e.g. Locality Sensitive Hashing [5][6]) have the following drawbacks.

- Hash may fail to find the corresponding bucket when the query lies in a low-density area of training patterns. The PCA space is divided into equal-sized buckets by hashing, for each axis.

- The number of NN candidates will not decrease drastically when the query is given at a high-density area of training patterns.

Those problems are caused by the hash function which produces a disjoint decomposition of the space by equal-sized buckets.

For solving these problems, PCH searches by dividing the space into finite buckets that each contain the same expected number of training patterns. We describe below how to decompose the space into buckets in this way.

2.1 Bucket Decomposition

Previous approximate NN search methods using hash functions, such as Locality Sensitive Hashing, divided PCA space into equal intervals, with each interval referred to as a bucket, and made training patterns within buckets to hash to the same index. These methods construct the hash function without any relation to training pattern distribution, or do not allocate buckets to cover the whole pattern space. As a result, if a query falls into a high-density area of training patterns, NN search becomes less efficient. Conversely, if a query falls into an area with no training patterns, there may be no training patterns that hash to the same index as the query hash index.

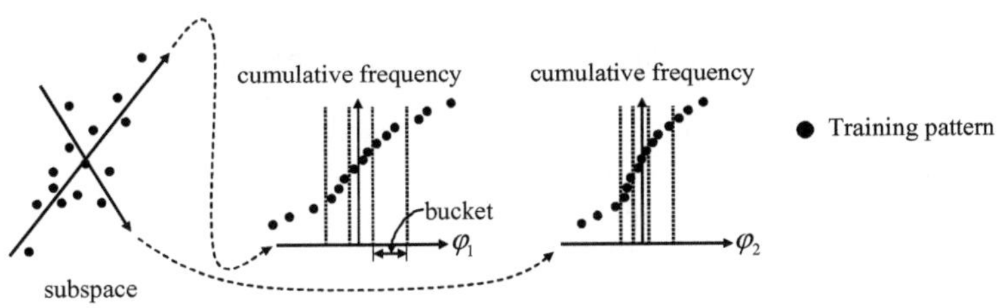

Fig. 1. Each axis is divided depending on the cumulative frequency of training patterns in PCA space. The divided interval is called a bucket, and training patterns in the same bucket hash to the same index.

The hash function constructed for PCH depends on the training pattern distribution, and each bucket contains a fixed number of training patterns as shown in Fig. 1.

Let $x_i \in \Re^d$ $(i = 1, 2, \ldots, n)$ be d-dimensional training patterns and $\varphi^T(x - \bar{x})$ be the training patterns projected into the subspace, where matrix φ is made up of the A components of the subspace, forming an orthonormal basis. Let ^{T} denotes the transpose of a matrix or vector, and $\bar{x}$ is an average vector of all training patterns. The bucket is decomposed to contain a fixed number of training patterns on φ_i. This process, which is called bucket decomposition, represents threshold processing on φ_i, and the bucket that contained the query can be determined by binary search.

2.2 Query Search

The PCH's search algorithm steps are shown in the following. **Step1.** and **Step2.** are preprocessing.

Step1. Obtain the PCA space for training patterns using PCA.
Step2. Let A be the number of dimensions of the PCA space, each bucket is decomposed, and training patterns are ascribed to the corresponding buckets.
Step3. On each φ obtained in preprocessing, the bucket that contained query q is determined.
Step4. For each training pattern, let it's *overlapped number* be the number of occasions when the patterns and the query pattern fall into the same backet. Let candidate NN patterns $C(q)$ be the training patterns that correspond to the highest $b\%$ of the calculated overlapped numbers, where $b\%$ is called *cutoff rate.*
Step5. After the approximate NN pattern of the query is determined by distance calculation from among $C(q)$, the process finishes.

Because complex processing to refer to neighboring buckets was necessary for **Step4.** in the original PCH, **Step4.** was simplified as shown above.

2.3 Refinement by Aborting the Distance Calculation

Since the approximate NN pattern of a query is determined from among $C(q)$ as shown above, the training pattern that has the most overlapped number is x_{max}, and the distance from query q to x_{max} is $D(q, x_{max})$. This distance is called provisional distance z, which means that

$$z = D(q, x_{max}), \quad where\ x_{max} \in C(q) \tag{1}$$

The distance calculation is able to be aborted using this provisional distance z. In the following discussions, this paper uses L_p distance $D_p(x_1, x_2)$. L_p distance is given by

$$D_p(x_1, x_2) = \sqrt[p]{\sum_{i=1}^{M} (x_{1i} - x_{2i})^p} \tag{2}$$

in M dimensional space.

To be based on Parseval's theorem, L_p distance can be calculated between projected vector and discretional orthonormal basis, and let L_p distance in subspace be

$$D_p(x_1, x_2) = \sqrt[p]{\sum_{i=1}^{M} \left|\varphi_i^T x_{1i} - \varphi_i^T x_{2i}\right|^p}. \tag{3}$$

Consequently, candidate x that satisfies the following conditions is not able to be NN pattern.

$$\sum_{i=1}^{m} \left|\varphi_i^T q - \varphi_i^T qx\right|^p > z^p, \quad where\ m \leq M. \tag{4}$$

Based on this property, **Step5.** (see 2.2) is modified using refinement by aborting the distance calculation by the following steps. Let A be the number of dimensions in PCA space, N be the number of NN pattern candidates, provisional NN pattern be x_t, and provisional distance be z^p.

Step5-1. Let x_{max} be the pattern that has maximum overlapped number among $C(q)$, and calculate the provisional distance z^p.
Step5-2. $n := 1$
Step5-3. Select x_n among $C(q)$ without x_t. If $n = N$, NN search is finished as NN pattern is x_t.
Step5-4. $m := 1$
Step5-5. If x_n satisfies Eqs.4, x_n is deleted, $n = n + 1$, and go to **Step5-3.**. If $m = A$, go to **Step5-7.**.
Step5-6. $m = m + 1$ and go to **Step5-5.**.
Step5-7. Calculate $D_p(q, x_n)$. If $D_p^p(q, x_n) < z^p$, $x_t = x_n$ and $z^p = D_p^p(q, x_n)$. Go to **Step5-3.**.

These steps are processed with φ vectors in decreasing order of training pattern variance. Therefore, aborting the number of distance calculation becomes efficient because more training patterns satisfy Eqn.(4). This ensured that the true NN pattern is not rejected by above steps.

3 Local Fisher Discriminant Analysis

Fisher Discriminant Analysis (FDA) [8] is a popular method for linear supervised dimensionality reduction. FDA seeks an embedding transformation such that the between-class scatter is maximized and the within-class scatter is minimized.

Let W and B be the within-class scatter matrix and the between-class scatter matrix and $\{\varphi_k\}_{k=1}^{d}$ be the generalized eigenvectors associated with the generalized eigenvalues $\lambda_1 \geq \lambda_2 \geq \cdots \geq \lambda_d$ of the following generalized eigenvalue problem:

$$B\varphi = \lambda W\varphi. \tag{5}$$

The FDA transformation matrix T_{FDA} is given by

$$T_{FDA} = (\varphi_1 \mid \varphi_2 \mid \cdots \mid \varphi_r), \tag{6}$$

where r is rank(T_{FDA}). The between-class scatter matrix B has at most rank $c-1$. This implies that the multiplicity of $lambda = 0$ is at least $d-c+1$. Therefore,

FDA can find at most $c-1$ meaningful features; the remaining features found by FDA are arbitrary. This is an essential limitation of FDA for dimensionality reduction and is very restrictive in practice. Additionally, FDA can perform poorly if patterns in a class form several separate clusters (i.e., multimodal).

To overcome these problems, Local Fisher Discriminant Analysis (LFDA) evaluates the levels of the between-class scatter and the within-class scatter in a local manner. Namely, according to the affinity matrix defined by similarity from pattern pairs, LFDA weights the values for the pattern pairs in the same class. This means that far apart pattern pairs in the same class have less influence on the within-class scatter matrix W and the between-class scatter matrix B. This allows LFDA to attain between-class separation and within-class local structure preservation simultaneously. Additionally, because the between-class scatter matrix B in LFDA is always full rank for various data sets, thanks to the affinity matrix, LFDA can find at most d meaningful features.

4 Local Fisher Discriminant Component Hashing

For faster and more accurate classification, Local Fisher Discriminant Component Hashing (LFDCH) introduces the following two ideas into PCH.

- LFDA obtains the subspace that maximizes between-class and minimizes the within-class variance.
- NN classification does not necessarily need to find the NN pattern, but only needs to classify its class label.

LFDCH projects training patterns into LFDA space and searches by hash function, which produces a disjoint decomposition of the space by finite buckets that each contain the same numbers of training patterns. LFDCH aborts the process when all training patterns have the same class label.

In contrast to PCH aimed at NN search, LFDCH is aimed at NN classification and does not need to store the vector of training patterns in the original space. LFDCH only stores the projected vectors to memory. As a result, LFDCH can use memory more efficiently than PCH can.

The LFDCH's classification algorithm steps are shown in the following. **Step1.**, **Step2.**, and **Step3.** are preprocessing.

Step1. Obtain the LFDA space for training patterns using LFDA.

Step2. Let A be the number of dimensions of the LFDA space. All training patterns are projected to the LFDA space. LFDCH only stores training patterns after dimensionality reduction.

Step3. For A components φ, each bucket is decomposed, and training patterns are ascribed to corresponding bucket as well as PCH.

Step4. On each φ obtained in preprocessing, the bucket that contained query q is determined.

Step5. Let the candidate NN patterns $C(q)$ be the patterns corresponding to the highest b% of overlapped numbers.

Step6. Refine $C(q)$ by aborting the distance calculation where possible, the same as with PCH. LFDCH can also abort the process and classify *query* when all $C(q)$ have the same class label.

5 Experimental Results

In the experiment, we compared the following methods with Correct Match Rate (CMR) and in terms of classification time:

LFDCH: Proposed algorithm.
PCH: Query is classified into the same class with NN pattern searched by PCH.
NN: Query is classified into the same class with NN pattern searched by naive NN search engine.

The classification time is the elapsed time taken to classify one query pattern. All experimental results described here are obtained on Intel Xeon PC, 3.0 GHz CPU, 16 GB memory using Microsoft Visual Studio 2005. We use the following settings as default for LFDCH and PCH:

Number of PCA and LFDA special dimensions: 20
Number of buckets per axis: 500
cutoff rate: 20%

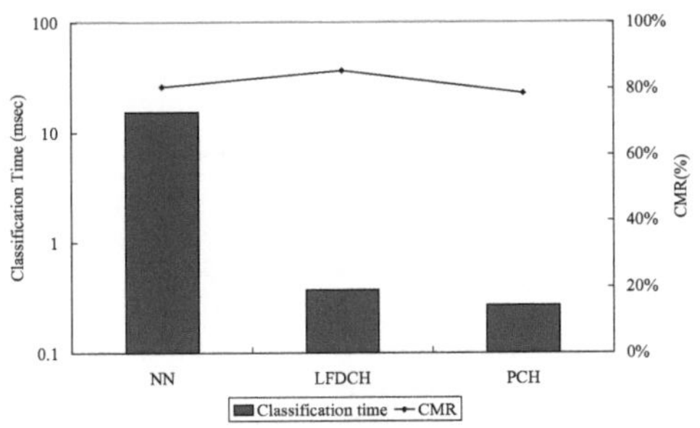

Fig. 2. This figure shows the classification time per query pattern and the CMR. The left vertical axis shows a logarithmic scale of classification time and the right vertical axis shows the CMR. The bar graph shows the classification time, and the line graph shows the CMR, for each method.

5.1 Gender Classification Using Face Image

We use multiple face image data sets and illustrate how LFDCH classifies a two-class classification problem. We classify gender using a normalizetion method that uses a generic 3D face model [9]. We describe the in-depth data as follows.

Distance: Euclidean Distance
Number of classes: 2 (male, female)

Number of training patterns: 18239 (11453 male, 6786 female)
Number of query patterns: 21573 (15991 male, 5582 female)
Dimension of patterns: 1024

Fig. 2 shows the classification time per query pattern and the corresponding CMR. LFDCH and PCH are much faster than NN. LFDCH is about 0.1 msec slower than PCH, but the CMR of LFDCH is about 7 points higher than that of PCH. Additionally, the CMR of LFDCH is about 5 points higher than that of NN. Shown on the left in fig. 3 are the classification times per query pattern and the CMR for LFDCH and PCH, changing the number of both LFDA and PCA spatial dimensions from 1 to 100. The CMR of LFDCH is always higher than that of PCH in each dimension. CMR of PCH becomes higher by adding the PCA spatial dimensions, but that of LFDCH becomes maximum in low LFDA spatial dimensions. As a result, LFDCH can achieve faster more and accurate classification than PCH. For example, LFDCH needs only 2 LFDA spatial dimensions to achieve the same CMR as NN, but PCH needs 100 PCA spatial dimensions. In this case, LFDCH is about 100 times faster than PCH, and 50 times smaller than PCH in terms of memory use. Shown on the right in fig. 3 are the classification times per query pattern and the CMR for LFDCH and PCH, changing the number of bucket decompositions from 50 to 1000. The CMR of LFDCH is more independent fo the number of bucket decompositions than PCH. Because both methods become fast by increasing the number of bucket decompositions, LFDCH speed can be further imprved while maintaing a good CMR.

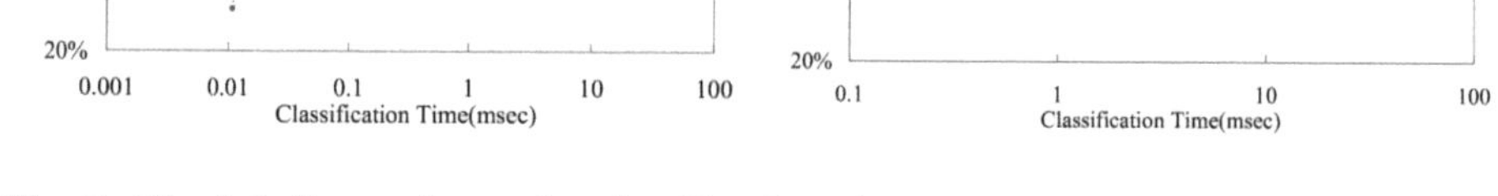

Fig. 3. The left figure shows the classification time per query pattern and the CMR on LFDCH and PCH, changing both LFDA and PCA spatial dimensions from 1 to 100. The right figure shows the classification time per query pattern and the CMR of LFDCH and PCH, changing the number of bucket decompositions from 50 to 1000. In both figures, the horizontal axis shows logarithmic scale of classification time and the vertical axis shows the CMR.

5.2 Handwriting Character Recognition

We use orientation component density of 256 dimensional features for uniquely collected handwritten characters and illustrate how LFDCH classifies the multi-class classification problem. We describe the in-depth data as follows.

Distance: Euclidean Distance
Number of classes: 62 (case-sensitive alphabet (52), digit 0 to 9 (10)
Number of training patterns: 20150 (325 patterns per character)
Number of query patterns: 20150 (325 patterns per character)

Shown on the left in fig. 5 are the classification times per query pattern and the CMR for LFDCH and PCH, changing both LFDA and PCA spatial dimensions from 1 to 100.

Fig. 4 shows the classification time per query pattern and the CMR. LFDCH and PCH are much faster than NN. LFDCH is about 0.1 msec slower than PCH, but the CMR of LFDCH is about 7 points higher than that of PCH. Additionally, the CMR of LFDCH is about 1 point higher than that of NN.

The CMR of LFDCH is always higher than that of PCH in every case. The CMR of PCH becomes higher by increasing the number of spatial dimensions,

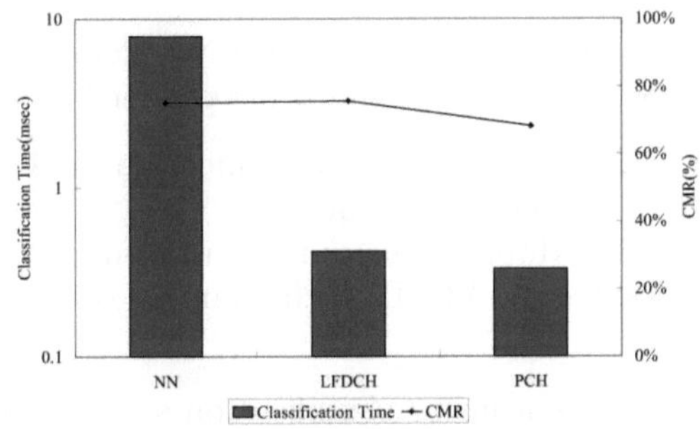

Fig. 4. This figure shows the classification time per query pattern and the CMR. The left vertical axis shows a logarithmic scale of classification time and the right vertical axis shows the CMR. The bar graph shows the classification time, and the line graph shows the CMR, for each method.

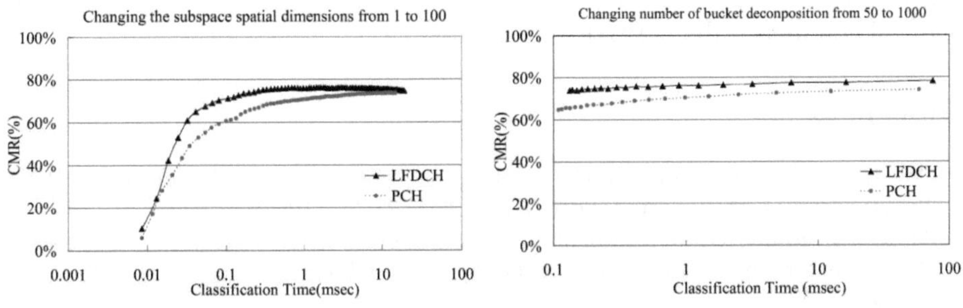

Fig. 5. The left figure shows the classification time per query pattern and the CMR on LFDCH and PCH, changing both LFDA and PCA spatial dimensions from 1 to 100. The right figure shows the classification time per query pattern and the CMR on LFDCH and PCH, changing the number of bucket decomposition from 50 to 1000. In both figures, the horizontal axis shows a logarithmic scale of classification time and the vertical axis shows the CMR.

but that of LFDCH exhibits a maximum at low LFDA spatial dimensions. As a result, LFDCH can achieve faster and more accurate classification than PCH which is the same as the situation described in Subsection 5.1. For example, LFDCH needs only 34 LFDA spatial dimensions to achieve the same CMR as NN, but PCH needs more than 100 PCA spatial dimensions. In 100 PCA spatial dimensions case, LFDCH is about 10 times faster than PCH, and 3 times smaller than PCH in terms of memory use. Shown on the right in fig. 5 are the classification times per query pattern and the CMR for LFDCH and PCH, changing the number of bucket decompositions from 50 to 1000. The CMR of LFDCH is more independent of the number of bucket decompositions than PCH. Because both methods become fast by increasing the number of bucket decompositions, LFDCH speed can be further improved while maintaining a good CMR.

6 Conclusions

In this paper, we proposed a fast and accurate NN classifier, LFDCH. The experimental results have confirmed the following properties.

- LFDCH is faster than NN search and classification methods using PCH and ordinary NN in high-dimensional space.
- The CMR of LFDCH exhibits a maximum in low-dimensional LFDA space.
- LFDCH can maintain a good CMR if the number of bucket decompositions is large.

LFDCH was able to achieve a faster classification speed than the classification method based on PCH while maintaining accuracy of classification. Future work is to automate the optimization of the number of dimensions of the LFDA space.

Acknowledgment

I am deeply grateful to Prof. Toshikazu Wada whose enormous support and insightful comments about PCH were invaluable.

References

1. Cover, B.K.B.T.M., Hart, P.E.: Nearest neighbor pattern classification. IEEE Transactions on Information Theory IT-13(1), 21–27 (1967)
2. Arya, S., Mount, D.M., Netanyahu, N.S., Silverman, R., Wu, A.Y.: An optimal algorithm for approximate nearest neighbor searching. Journal of the ACM 45, 891–923 (1998)
3. Matsushita, Y., Wada, T.: Principal Component Hashing. Meeting on Image Recognition and Understanding (MIRU), pp. 127–134 (July 2007) (in Japanese)
4. Hotelling, H.: Analysis of a complex of statistical variables into principal components. Journal of the American Statistical Association 24, 441 (1933)
5. Indyk, P., Motwani, R.: Approximate Nearest Neighbors: Towards Removing the Curse of Dimensionality. In: Proceedings of the 30th ACM Symposium on Theory of Computing (STOCf 1998), pp. 604–613 (May 1998)

6. Datar, M., Indyk, P., Immorlica, N., Mirrokni, V.: Locality-Sensitive Hashing Scheme Based on p-StableDistributions. In: Proceedings of the 20th Annual Symposium on Computational Geometry (SCG 2004) (June 2004)
7. Sugiyama, M.: Dimensionality reduction of multimodal labeled data by local fisher discriminant analysis. JMLR 8, 1027–1061 (2007)
8. Fisher, R.A.: The Use of Multiple Measurements in Taxonomic Problems. Annals of Eugenics 7, Part II, 179–188 (1936)
9. Kozakaya, T., Yamaguchi, O.: Face Recognition by Projection-based 3D Normalization and Shading Subspace Orthogonalization. FGR, pp. 163–168 (2006)
10. Vidal, R.: An algorithm for finding nearest neighbor in (approximately) constant average time. Pattern Recognition Letters 4, 145–158 (1986)
11. Mico, L., Oncina, J., Vidal, E.: A new version of the nearest-neighbor approximating and eliminating search algorithm (AESA) with linear preprocessing time and memory requirements. Pattern Recognition Letters 15, 9–17 (1994)
12. Brin, S.: Near neighbor search in large metric spaces. In: Proc. of 21st Conf. on very large database (VLDB), Zurich, Switzerland, pp. 574–584 (1995)
13. Yianilos, P.Y.: Data structures and algorithms for nearest neighbor search in general metric spaces. In: Proc. of the Fourth Annual ACM-SIAM Symp. on Discrete Algorithms, Austin, TX, pp. 311–321 (1993)

Significance Tests and Statistical Inequalities for Region Matching

Guillaume Née[1,2,⋆], Stéphanie Jehan-Besson[1], Luc Brun[1], and Marinette Revenu[1]

[1] GREYC Laboratory - Bd du Maréchal Juin. 6 - 14050 Caen - France
[2] General Electric Healthcare - Velizy - France
{gnee,jehan,brun,revenu}@greyc.ensicaen.fr

Abstract. Region matching - finding conjugate regions on a pair of images - plays a fundamental role in computer vision. Indeed, such methods have numerous applications such as indexation, motion estimation or tracking. In the vast literature on the subject, several dissimilarity measures have been proposed in order to determine the true match for each region. In this paper, under statistical hypothesis of similarity, we provide an improved decision rule for patch matching based on significance tests and the statistical inequality of McDiarmid. The proposed decision rule allows to validate or not the similarity hypothesis and so to automatically detect matching outliers. The approach is applied to motion estimation and object tracking on noisy video sequences. Note that the proposed framework is robust against noise, avoids the use of statistical tests and may be related to the *a contrario* approach.

1 Introduction

The notion of similarity between regions is a key feature for a wide range of applications in image analysis and pattern recognition such as clustering, indexation, image segmentation, shape matching or registration. This notion is usually indirectly defined through a distance or a measure of dissimilarity. In many methods proposed for matching, the most similar pattern in a set of patterns is usually chosen. In this paper we are interested in the significance of the matching. One straightforward way to decide if two connected set of pixels (regions) are significantly similar consists in comparing the dissimilarity of both regions to a fixed threshold [18]. However, such a decision rule does not usually provide a clear interpretation of the threshold which is often difficult to set and which should be adapted to each data-set.

Deciding if two regions are similar may alternatively be stated as the decision between the two opposite hypothesis: "$\mathbf{H_0}$: both regions are similar" and "$\mathbf{H_1}$: both regions are dissimilar". Given a dissimilarity measure, one must measure the probability that an observation of this dissimilarity is greater than a given threshold under both hypothesis $\mathbf{H_0}$ and $\mathbf{H_1}$. The decision may then be stated using a likelihood ratio test. The main problem of such a decision scheme is that

⋆ This work is funded by a grant co-financed by General Electric Healthcare and the Region Basse Normandie.

N. da Vitora Lobo et al. (Eds.): SSPR&SPR 2008, LNCS 5342, pp. 350–360, 2008.

both hypothesis $\mathbf{H_0}$ and $\mathbf{H_1}$ are not symmetric. Indeed the set of regions similar to a given region is usually much simpler than the set of its dissimilar regions. Consequently, the design of a general model for $\mathbf{H_0}$ is usually tractable while the design of a model for $\mathbf{H_1}$ may be quite difficult or even impossible. This last point limits the use of likelihood ratio test.

The *a contrario* approach provides a way to solve this difficulty by setting the decision only on the null hypothesis $\mathbf{H_0}$. This approach has been used in several distinct fields such as detection of gelstat [4,6,5], motion detection [16], shape matching and recognition [12,13], object matching [1], parameter estimation [7], blotch detection for digitized film restoration [15] and segmentation [14,8]. The a contrario approach comes from the perception theory and particularly the grouping law of the Wertheimer's theory. This grouping law states that "objects having a quality in common get perceptually grouped". The Helmholtz principle [4] which states that "an event is meaningful if its number of occurrences is very small in a random model" is a quantitative version of the previous law. More formally, let us consider an event E which probability under $\mathbf{H_0}$ is bounded by a low threshold δ, the *a contrario* approach leads to reject the initial hypothesis if such an event occurs. Using such a decision scheme, δ may be interpreted as an upper bound of the probability of a false alarm (rejection of $\mathbf{H_0}$ while $\mathbf{H_0}$ is actually true). The probability δ may be fixed *a priori* in which case the test $P(E|H_0) < \delta$ is called a significance test [3]. Desolneux [5] proposed to set δ according to the expected number of false alarm. Such a method provides an elegant way to fix the threshold but reduces the adaptability of the method to user requirements.

Our approach is based on a conjoint use of the *a contrario* approach and a general formulation of the McDiarmid's theorem [10]. McDiarmid's theorem allows to design statistical tests for a wide variety of problems and provides a natural robustness to noise and to outliers. Within our framework this theorem is used to bound the probability that the dissimilarity between two regions is greater than a given threshold. Such a bound is incorporated into the *a contrario* framework in order to provide an automatic threshold on the dissimilarity between two regions. This general framework is illustrated by a dissimilarity measure between patches which corresponds to the Sum of Squared Difference (SSD) involved in Block Matching algorithms. This dissimilarity measure is used in conjunction with the McDiarmid Theorem within the *a contrario* approach in order to design a robust Variable Size Block Matching (VSBM) algorithm that can be used for motion estimation and tracking in noisy video sequences.

We first present our *a contrario* approach for region matching in section 2. We then illustrate this method using a first dissimilarity measure between patches in section 3. Several experiments presented in section 4 demonstrate the effectiveness of our approach for variable size block matching (VSBM) and tracking in noisy video sequences.

2 Problem Statement and Statistical Tools

We present in this section our approach for region matching and the statistical tools which constitute the basis of our approach.

2.1 *A Contrario* Approach for Region Matching

Due to the random part in image acquisition systems, an image I is classically considered as an observation of a perfect statistical image I^*. Using such an image model, an ideal region is defined as a vector $\mathbf{X} = (X_1, \ldots, X_n)$ of n random variables (r.v.) representing the pixel intensities. A "real" region is then considered as an observation of this random vector which takes its values in $\prod_{k=1}^{n} A_k$. In natural images, the set of admissible values A_k usually corresponds to $[0; M]$ where $M = 255$. However, in medical images (e.g. : MRI, Echography), the set A_k may be larger.

Given two statistical regions $\mathbf{X}_1$ and $\mathbf{X}_2$ and a dissimilarity criterion $d(.,.)$, let us consider two observations R_1 and R_2 of respectively $\mathbf{X}_1$ and $\mathbf{X}_2$ and the event E: *the observed value* $d(R_1, R_2)$ *of the statistic* $d(\mathbf{X}_1, \mathbf{X}_2)$ *is greater than a threshold* T. As mentioned in section 1, the *a contrario* approach is based on the estimation of the probability of this event under the similarity hypothesis $\mathbf{H_0}$. Let us consider an upper bound δ of this probability:

$$\mathbf{P}\left\{d(\mathbf{X}_1, \mathbf{X}_2) \geq T | \mathbf{H_0}\right\} \leq \delta \tag{1}$$

We can remark that the probability δ and the threshold T are dependent. Indeed if the threshold T is set to a high value, this corresponds to a non probable event under $\mathbf{H_0}$ and δ should then be small. On the contrary if the threshold T is set to a small value, this corresponds to a probable event under $\mathbf{H_0}$ and so δ must be large. More generally, one may usually assume that the threshold T is a decreasing function of δ which may be denoted as $T(\delta)$.

Using the *a contrario* approach, if we take δ as a low probability value, the event E is considered as not probable under the similarity hypothesis $\mathbf{H_0}$ and this hypothesis is then rejected. Our decision rule for region matching is then defined as:

$$\text{if} \quad d(R_1, R_2) \geq T(\delta) \quad \text{then} \quad \mathbf{H_0} \text{ is rejected} \tag{2}$$

2.2 Computation of the Threshold

The main difficulty of the above approach lies in the computation of the threshold $T(\delta)$. In this work, we propose to use the extension of the McDiarmid theorem [10] which allows to bound the probability of a large class of events. Let us remind this theorem:

Theorem 1. *Let* $\mathbf{Y} = (Y_1, \ldots, Y_n)$ *be a family of random variables with* Y_k *taking values in a set* A_k*, and let* f *be a bounded real-valued function defined on* $\Omega = \prod_{k=1}^{n} A_k$*. If* μ *denotes the expectation of* $f(\mathbf{Y})$ *we have for any* $\alpha \geq \mu$*:*

$$\mathbf{P}\left\{f(\mathbf{Y}) \geq \alpha\right\} \leq \exp\left(\frac{-2(\alpha-\mu)^2}{r^2}\right) + \mathbf{P}\left\{\mathbf{Y} \in C\right\} \tag{3}$$

Where C *is a subset of* Ω *and* r^2 *is the maximal sum of squared range (section A.1) defined on* $\overline{C} = \Omega - C$*.*

The set C in the above theorem corresponds to a set of outliers for $\mathbf{Y}$. Within our framework, we define $f(Y)$ as our dissimilarity measure $d(\mathbf{X}_1, \mathbf{X}_2)$ and Y as an appropriate combination of the two vectors $\mathbf{X}_1$ and $\mathbf{X}_2$. An example will be given in the next section. Theorem 1 then provides a bound for $P(d(\mathbf{X}_1, \mathbf{X}_2) \geq \alpha)$. The hypothesis $\mathbf{H_0}$ is introduced through the parameters r and μ. Some assumptions on the noise model can be used to estimate those parameters (see section 3.1).

A simplified version of theorem 1 has already been used [14,8] within the segmentation framework. In this simplified version $\overline{C} = \Omega$ and $\mathbf{P}\{\mathbf{Y} \in C\} = 0$. To our knowledge, the general theorem 1 has never been used for region matching whereas it provides a natural tool for taking into account outliers which may occur in a statistical image. Indeed, the set $\overline{C}$ can be used to define the set of admissible values for a given type of images or regions.

Let us denote by $\Delta(\alpha)$, the probabilistic bound given by the McDiarmid's theorem:

$$\mathbf{P}\{f(\mathbf{Y}) \geq \alpha\} \leq \Delta(\alpha) \tag{4}$$

The threshold $T(\delta)$ introduced in (1) can then be computed by setting $\delta = \Delta(\alpha)$ and so $\alpha = \Delta^{-1}(\delta)$.

3 A First Dissimilarity Measure for Patch Matching

We consider two statistical regions defined in two images. In order to match heterogeneous regions we restrict our definition of region to rectangular patches. Such a restriction allows us to assume a one to one correspondence between the pixels of two matched regions.

Let us denote the considered regions by $\mathbf{X}_1 = \{X_i\}_{1 \leq i \leq n}$ and $\mathbf{X}_2 = \{X_i\}_{n+1 \leq i \leq 2n}$. The correspondence between the pixels of both regions is implicitly encoded by the index, since we suppose that a pixel X_i, $i \in \{1, \ldots, n\}$ of $\mathbf{X}_1$ corresponds to the pixel X_{i+n} of $\mathbf{X}_2$. We measure the similarity between corresponding pixels by the following dissimilarity measure which corresponds to the classical Sum of Square Differences (SSD):

$$d(\mathbf{X}_1, \mathbf{X}_2) = \frac{1}{n} \sum_{i=1}^{n} (X_i - X_{n+i})^2 \tag{5}$$

For the sake of simplicity, we introduce the random vector $\mathbf{Y} = (Y_1, \ldots, Y_n)$ where $Y_i = (X_i - X_{n+i})^2$, $i \in \{1, \ldots, n\}$. If $[0; M]$ is the interval of variation of the r.v. (X_i), then Y_i varies within $[0; M^2]$.

Our goal is to compute a decision rule that indicates if two observations R_1 and R_2 of $\mathbf{X}_1$ and $\mathbf{X}_2$ are similar or not. According to our approach (section 2), we have to upper bound the probability that the function $d(\mathbf{X}_1, \mathbf{X}_2)$ is greater than a given threshold α using the McDiarmid's theorem (theorem 1). Such an upper bound is provided by the following proposition:

Proposition 1. *Using the previously defined notations, we have for any couple $(\mathbf{X}_1, \mathbf{X}_2)$ of statistical regions and any threshold $\alpha > 0$:*

$$\mathbf{P}\{d(\mathbf{X}_1, \mathbf{X}_2) \geq \alpha\} \leq \exp\left(-\frac{2n(\alpha - \mu)^2}{N^4}\right) + K \tag{6}$$

with $K = \mathbf{P}\{\mathbf{Y} \in C\}$ *where* $C = [N^2; M^2]^n \subset \Omega$*, is defined by the parameter* N *and* $\mu = \mathbf{E}(f(\mathbf{Y}))$ *with* $f(Y) = \frac{1}{n}\sum_{i=1}^{n} Y_i = d(\mathbf{X}_1, \mathbf{X}_2)$*.*

Proof. see Appendix, section A.2.

The exact values of K and μ should be determined according to a specific model of similarity between regions (section 3.1). The parameter N determines the size of our set of outliers. Intuitively, if each pixel X_i, $i \in \{1, \ldots, n\}$ of $\mathbf{X}_1$ corresponds to the pixel X_{i+n} of $\mathbf{X}_2$, observations of the random variables Y_i should be around 0 with a large probability. We thus define *a priori* the set $[0; N^2]$ on which we expect a large number of observations and consider the set $C =]N^2; M^2]$ as our set of outliers. The parameter N should be set for each application and each dissimilarity measure. However, experiments, not reported here shown us that the optimal value of N remains stable on a wide class of images. The value of N as been fixed to 65 in all experiments reported below.

As described in section 2, given the parameter N, the probability K remains fixed and the right side of equation 6 defines a function Δ of α such that:

$$\mathbf{P}\{d(\mathbf{X}_1, \mathbf{X}_2) \geq \alpha\} \leq \Delta(\alpha).$$

The function Δ being bijective this last equation is equivalent to:

$$\mathbf{P}\left\{d(\mathbf{X}_1, \mathbf{X}_2) \geq \Delta^{-1}(\delta)\right\} \leq \delta.$$

For a threshold δ.

According to our decision rule (section 2.1), we thus reject the hypothesis $\mathbf{H_0}$ that both regions are similar if $d(\mathbf{X}_1, \mathbf{X}_2)$ is greater than:

$$T(\delta) = \Delta^{-1}(\delta) = \mu + N^2 \sqrt{\frac{1}{2n} \ln\left(\frac{1}{\delta - K}\right)} \tag{7}$$

3.1 Statistical Prior for Parameter Estimation

In order to explicitly compute the parameters μ and K given in equation 7, we need to formalize our similarity hypothesis $\mathbf{H_0}$. This task is achieved by the notion of σ-similar regions which relies on a Gaussian noise model:

Definition 1. *(σ-similarity)*
Two statistical regions $\mathbf{X}_1 = (X_i)_{i \in \{1,\ldots,n\}}$ *and* $\mathbf{X}_2 = (X_i)_{i \in \{n+1,\ldots,2n\}}$ *are called* σ*-similar if the r.v.* $(X_i - X_{i+n})$ *are independent and identically distributed (iid) and follow a normal distribution* $\mathcal{N}(0, \sigma)$ *for any* $i \in \{1, \ldots, n\}$*.*

Let us consider two σ-similar regions $\mathbf{X}_1$ and $\mathbf{X}_2$. As all r.v. $X_i - X_{i+n}$ are iid, the r.v. Y_i are iid and follow $\mathbf{P}\{Y_i = y\} = \frac{1}{\sigma\sqrt{2y\pi}} \exp\left(-\frac{y}{2\sigma^2}\right)$. We have thus:

$$\mu = \mathbf{E}(f(\mathbf{Y})) = \mathbf{E}\left(\frac{1}{n}\sum_{i=1}^{n} Y_i\right) = \frac{1}{n}\sum_{i=1}^{n} \mathbf{E}(Y_i) = \sigma^2$$

$$K = \mathbf{P}\left\{\mathbf{Y} \in C = [N^2; M^2]^n\right\} = \prod_{i=1}^{n} \mathbf{P}\left\{Y_i \in [N^2; M^2]\right\}$$

$$= \left(\frac{1}{\sigma\sqrt{2\pi}} \int_{N^2}^{M^2} \frac{1}{\sqrt{y}} \exp\left(-\frac{y}{2\sigma^2}\right) dy\right)^n = \left(\operatorname{erf}\left(\frac{M}{\sigma\sqrt{2}}\right) - \operatorname{erf}\left(\frac{N}{\sigma\sqrt{2}}\right)\right)^n$$

Note that our threshold (equation (7)) takes into account the amount of noise through the two terms $\mu = \sigma^2$ and K.

3.2 Patch Matching

In order to illustrate our approach, let us give some basic results for patch matching between two frames of a video sequence.

Given a reference patch from frame #1, we present the set of accepted/rejected patches of frame #2 in Fig. 1. The frame #2 is considered with or without the addition of a Gaussian noise $\mathcal{N}(0, 15)$. The candidate patches are presented with their associated value of dissimilarity. To compute the threshold $T(\delta)$, the ratio δ of authorised false alarms is set to 0.1 and N is set to 65 for both experiments (with or without noise on the frame #2). The value of σ is estimated by extracting homogeneous 20x20-patches from the two images. We find $\sigma = 5.91$ between the non-noisy frames #1 and #2, and $\sigma = 16.5$ between the frame #1 and the noisy frame #2. The rejected patches are the ones which present significant translations of the car hood inside the patch. We can notice that the computed threshold adapts itself automatically to the additional noise in frame #2. Indeed, using the same threshold for non-noisy sequences would lead to the rejection of all patches in noisy sequences. With this adaptive threshold, the discrimination between rejected and accepted patches is still valid.

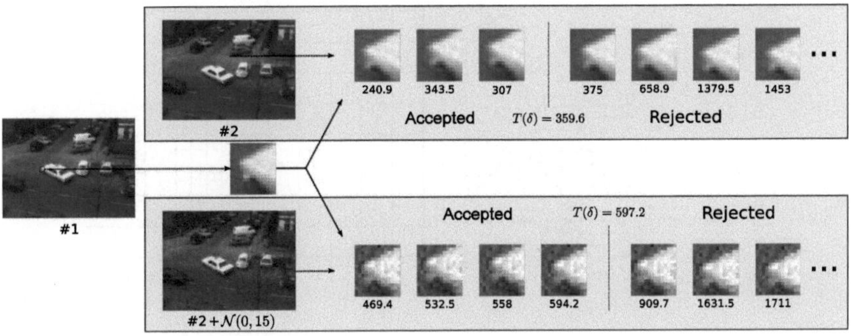

Fig. 1. Examples of acceptation/rejection of 13x15-patch matching. The candidate patches from frame #2 are selected around the position of the reference patch (from frame #1).

4 Application to Motion Estimation and Tracking

The framework developed above can be efficiently applied for motion estimation and tracking (especially in the case of noisy images). This leads us to the design of a robust variable size block matching (VSBM) method that is then used for rigid object tracking. For both methods, we present some experimental results on real video sequences and we focus our attention on two main points: robustness to noise and automatic detection of false matching.

4.1 Motion Estimation Using Variable Size Block Matching

Block matching (BM) algorithms were originally described by Jain and Jain [9] and are now widely adopted for motion estimation especially in video compression (MPEG-2,MPEG-4). Each frame is divided into a fixed number of usually square blocks. For each block in the frame, we search for the "best" matching block in the next frame. The dissimilarity measure is usually taken as the SSD measure defined in equation 5. Since a fixed size for the blocks is not adapted to the granularity of both motion and objects, VSBM algorithms have been introduced and are now widely used in new compression standards (H264, MPEG-4 AVC) [18]. If the best matching error for a block is above some threshold, the

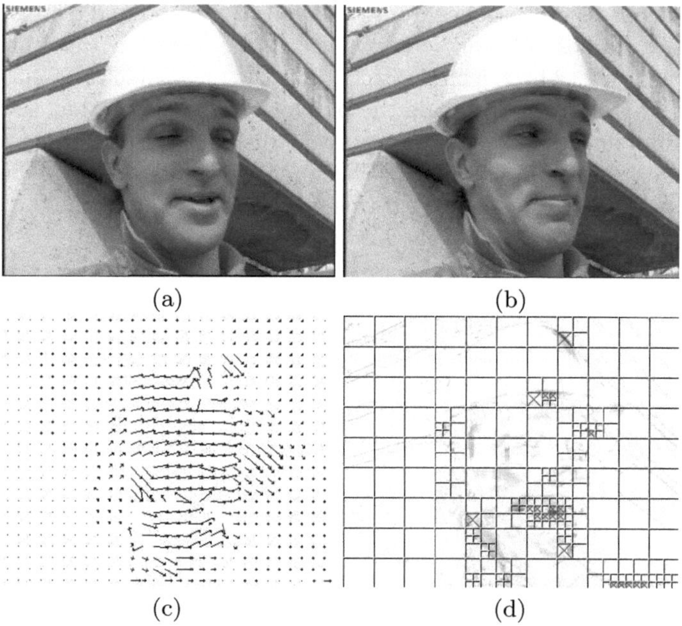

Fig. 2. "Foreman" sequence (a) and (b), the computed motion estimation (c) by our VSBM. The variable size block division, the residual between the reconstruction of (b) from (a) and rejected blocks represented by a cross are superposed on (d).

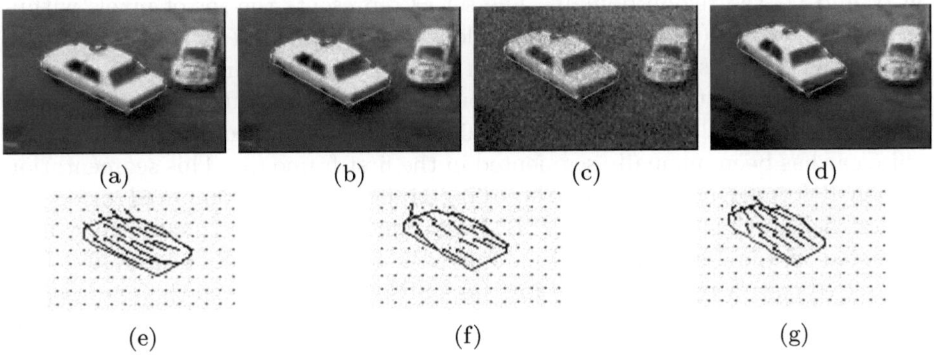

Fig. 3. First line - tracking of the white car in a sequence which presents a noisy frame (c). Second line - the motion between frames is estimated by our VSBM algorithm (section 4.1).

block is divided into four smaller blocks, until the maximum number of blocks or locally minimum errors are obtained, see [2] for example.

Using *a contrario* approaches, we have designed a statistical decision rule for the dissimilarity measure used in BM. Our decision rule allows us to detect false matching and is robust to noise. It is thus adapted to the design of VSBM algorithms. Fig. 2 presents the results of our VSBM between frames (a) and (b). Motion estimation errors are visualized through the inverse difference (d) between (b) and its reconstruction. Such a reconstruction is computed using (a) and the estimated motion field (c) obtained by our VSBM algorithm. These results have been obtained by choosing 32x32-blocks which can be recursively cut in half size blocks while the SSD is above the threshold $T(\delta)$ and until 8x8-blocks are reached. The threshold T given by (7) is computed by fixing N to 65 and δ to 0.1. We can remark that the algorithm gives smaller blocks when the motion is more complex (the mouth or the helmet). Rejected blocks (represented by a cross on (d)) are also (qualitatively) well detected. Such blocks correspond to strong and complex motions, such as the mouth which is opened on frame (a) and closed on frame (b).

4.2 Application to Tracking in Noisy Sequences

The validity of our algorithm can be further attested for rigid object tracking in noisy video sequences. In this application, we assume as in [11] that the object of interest undergoes an affine deformation from one frame to another which is estimated by minimizing the following convex criterion according to the six parameters of the affine transformation $\mathbf{A}$:

$$E(\mathbf{A}) = \sum_{\mathbf{p}\in\Omega_I/\Theta} ||\mathbf{d}(\mathbf{p}) - \mathbf{A}(\mathbf{p})||^2$$

where $\mathbf{p} = [x_1, x_2]^T$ is the pixel, $\mathbf{d}(\mathbf{p}) = [u(\mathbf{p}), v(\mathbf{p})]^T$ the motion vector estimated as explained in the previous section, $\mathbf{A}$ the affine transformation applied on $\mathbf{p}$, and Ω_I the region domain. The set Θ represents the set of pixels within the rejected blocks (section 4.1). Such blocks are represented by a cross in Fig. 2. The estimated transformation $\mathbf{A}$ is then used to deform an initial segmentation.

Fig. 3 illustrates the effectiveness of our method on the video sequence "Taxi" where an additional Gaussian noise $\mathcal{N}(0, 15)$ has been added on frame (c). The white car has been manually segmented in the first frame (a). This segmentation is then deformed according to the affine transformation computed using the motion field estimated by our VSBM (section 4.1). The estimated motion is presented on the second line of Fig. 3 and the tracking results on the first line. The set of parameters used here for the VSBM is the same as the one used in the previous experiment (section 4.1) and is kept all along the sequence. This result shows that we can track the car with the same set of parameters all along the sequence even in its noisy part.

5 Conclusion

In this paper, we propose to combine *a contrario approaches* and the recent statistical inequality of McDiarmid to design significance tests for region matching. This approach is illustrated by designing an improved decision rule for patch matching that is robust to noise. This decision rule has been applied to motion estimation with a VSBM algorithm and tracking. Our experimental results show the effectiveness of our approach on noisy video sequences. Our on going research is directed towards the design of other decision rules for region merging algorithms so as to extend the work proposed in [8]. Moreover, we plan to extend our method to other noise models such as Rayleigh or Poisson which are encountered in several application domains (e.g. medical image).

References

1. Burrus, N., Bernard, T.M., Jolion, J.M.: Bottom-up and top-down object matching using asynchronous agents and *a contrario* principles. In: Gasteratos, A., Vincze, M., Tsotsos, J.K. (eds.) ICVS 2008. LNCS, vol. 5008, pp. 343–352. Springer, Heidelberg (2008)
2. Chan, M.H., Yu, Y.B., Constantinides, A.G.: Variable size block matching motion compensation with applicationsto video coding. IEEE Communications, Speech and Vision 137(4), 205–212 (1990)
3. Coupier, D., Desolneux, A., Ycart, B.: Image denoising by statistical area thresholding. Journal of Mathematical Imaging and Vision 22(2-3), 183–197 (2005)
4. Desolneux, A., Moisan, L., Morel, J.-M.: Meaningful alignments. International Journal of Computer Vision 40(1), 7–23 (2000)
5. Desolneux, A., Moisan, L., Morel, J.-M.: Computational Gestalts and perception thresholds. Journal of Physiology 97(2-3), 311–324 (2003)
6. Desolneux, A., Moisan, L., Morel, J.-M.: Maximal meaningful events and applications to image analysis. Annals of Statistics 31(6), 1822–1851 (2003)
7. Felip, R.L., Binefa, X., Diaz Caro, J.: A new parameter estimator based on the helmholtz principle. In: International Conference on Image Processing, pp. 1306–1309 (2005)

8. El Hassani, M., Jehan-Besson, S., Brun, L., et al.: Time-consistent video segmentation algorithm designed for real-time implementation. VLSI Design (2008)
9. Jain, J.R., Jain, A.K.: Displacement measurement and its application in interframe image coding. IEEE Transactions on Communications, 1799–1808 (1981)
10. McDiarmid, C.: Concentration. In: Habib, M., McDiarmid, C., Ramirez-Alfonsin, J., Reed, B. (eds.) Probabilistic Methods for Algorithmic Discrete Mathematics. Springer, Heidelberg (1998)
11. Meyer, F., Bouthemy, P.: Region-based tracking using affine motion models in long image sequences. CVGIP: Image Understanding 60(2), 119–140 (1994)
12. Musé, P., Sur, F., Cao, F., Gousseau, Y.: Unsupervised thresholds for shape matching. In: International Conference on Image Processing, pp. 647–650 (2003)
13. Musé, P., Sur, F., Cao, F., Gousseau, Y., Morel, J.-M.: An a contrario decision method for shape element recognition. International Journal on Computer Vision 69(3), 295–315 (2006)
14. Nock, R., Nielsen, F.: Statistical region merging. IEEE Pattern Analysis and Machine Intelligence 26(11), 1452–1458 (2004)
15. Tilie, S., Laborelli, L., Bloch, I.: Blotch Detection for Digital Archives Restoration based on the Fusion of Spatial and Temporal Detectors. In: Fusion, Florence, Italy (2006)
16. Veit, T., Cao, F., Bouthemy, P.: An a contrario decision framework for region-based motion detection. International Journal on Computer Vision 68(2), 163–178 (2006)
17. Werthmer, M.: Untersuchungen zur Lehre der Gestalt, ch. 4, vol. II, pp. 301–350. Psychologische of Forschung (1923)
18. Wiegand, T., Sullivan, G.J., Bjntegaard, G., Luthra, A.: Overview of the H.264/AVC video coding standard. IEEE Transactions on Circuits and Systems for Video Technology 13(7), 560–576 (2003)

A Appendix

A.1 Statistical Notations and Definitions

Let $\mathbf{Y} = (Y_1, \ldots, Y_n)$ be a family of r.v. with Y_k taking values in a set A_k, and let f be a real-valued function defined on $\prod A_k$. Let $y_i \in A_i$ for each $i = 1, \ldots, k$ and let B_k denotes the set of events $\{Y_i = y_i\}_{i=1,\ldots k}$. For $y_k \in A_k$, let:

$$g(y_k) = \mathbf{E}\left(f(\mathbf{Y})|B_{k-1}, Y_k = y_k\right) - \mathbf{E}\left(f(\mathbf{Y})|B_{k-1}\right) \tag{8}$$

be the function that measures how much the expected value of $f(\mathbf{Y})$ changes if it is revealed that the r.v. Y_k takes the value y_k. By these notations, the range of $g(Y_k|B_{k-1})$ is defined as follows:

$$ran(y_1, \ldots, y_{k-1}) = \sup\left\{|g(y) - g(x)| : x, y \in A_k\right\} \tag{9}$$

For $\mathbf{y} \in \prod A_k$, the sum of squared range and its maximum are defined as:

$$R^2(y) = \sum_{k=1}^{n} (ran(y_1, \ldots, y_{k-1}))^2, \quad \hat{r}^2 = \sup_{y \in \prod A_k} (R^2(y)) \tag{10}$$

A.2 Proof of Proposition 1

The r.v. $\{Y_k\}_{k=1\ldots n}$ are independent and so the function g defined in equation (8) that measures how much the expected value of SSD changes can be written as:

$$g(y) = \mathbf{E}\left(f(y_1, \ldots, y_{k-1}, y_k, Y_{k+1}, \ldots, Y_n)\right)$$
$$- \mathbf{E}\left(f(y_1, \ldots, y_{k-1}, Y_k, Y_{k+1}, \ldots, Y_n)\right)$$
$$g(y) = \frac{1}{n}\left(y_k - \mathbf{E}\left(Y_k\right)\right) \tag{11}$$

Now, let us consider $C = \left\{\mathbf{Y} = \mathbf{y} : \mathbf{y} \in \left[N^2; M^2\right]^n\right\}$ the subset of all the outliers such that the vector of r.v. $\mathbf{Y} = \mathbf{y}, \mathbf{y} \in C$ is not acceptable, then the range defined by equation (9) becomes:

$$ran(y_1, \ldots, y_{k-1}) = \sup\left\{\frac{1}{n}\left|y_k - y_k' - \mathbf{E}\left(Y_k - Y_k'\right)\right| : y_k, y_k' \in \left[0; N^2\right]\right\} = \frac{N^2}{n} \tag{12}$$

where y_k' is another observation of the r.v. Y_k. Consequently:

$$\mathbf{E}\left(Y_k - Y_k'\right) = \mathbf{E}\left(Y_k\right) - \mathbf{E}\left(Y_k'\right) = 0$$

So the *maximum sum of squared ranges* defined in (10) is equal to: $\hat{r}^2 = \frac{N^4}{n}$. By applying equation (3) and from theorem 1, we obtain the proposition 1:

$$\mathbf{P}\left\{f(\mathbf{Y}) \geq \alpha\right\} \leq \exp\left(-\frac{2n(\alpha - \mu)^2}{N^4}\right) + K \tag{13}$$

where $K = \mathbf{P}\left\{\mathbf{Y} = \mathbf{y} : \mathbf{y} \in C\right\}$ and $\mu = \mathbf{E}\left(f(\mathbf{Y})\right)$.

Advanced Homology Computation of Digital Volumes Via Cell Complexes*

Helena Molina-Abril and Pedro Real

Dpto. Matematica Aplicada I, E.T.S.I. Informatica, Universidad de Sevilla, Avda. Reina Mercedes, s/n 41012 Sevilla (Spain)
{habril,real}@us.es

Abstract. Given a 3D binary voxel-based digital object V, an algorithm for computing homological information for V via a polyhedral cell complex is designed. By homological information we understand not only Betti numbers, representative cycles of homology classes and homological classification of cycles but also the computation of homology numbers related additional algebraic structures defined on homology (coproduct in homology, product in cohomology, (co)homology operations,...). The algorithm is mainly based on the following facts: a) a local 3D-polyhedrization of any $2 \times 2 \times 2$ configuration of mutually 26-adjacent black voxels providing a coherent cell complex at global level; b) a description of the homology of a digital volume as an algebraic-gradient vector field on the cell complex (see Discrete Morse Theory [5],AT-model method [7,5]) . Saving this vector field, we go further obtaining homological information at no extra time processing cost.

1 Introduction

One possible way for developing techniques for volume recognition can be through relevant structural descriptors, mainly based on topological-geometrical properties. Topology deals with connectivity and separability features and although the global nature of topological properties makes their computation difficult, topology seems to be an essential part of the vocabulary by which human visual system represent and characterize objects [2]. Homology gives us the simplest topological version of a volume in terms of information about connected components, tunnels or holes and cavities on it. Homological numbers commonly used in Pattern Recognition as robust descriptors are Betti numbers (number of connected components, holes and cavities), local topological characterization of voxels and Euler characteristic. Nowadays, the current interactions between Pattern Recognition and Computational Topology are mainly at graph-based level and it includes issues such that topological skeletonization and Reeb graphs [1]. Examples of image analysis and pattern recognition problems that benefit from the use of topological considerations include, for example, the use of homology invariants to compare

* This work has been partially supported by "Computational Topology and Applied Mathematics" PAICYT research project FQM-296, "Andalusian research project PO6-TIC-02268) and Spanish MEC project MTM2006-03722.

N. da Vitora Lobo et al. (Eds.): SSPR&SPR 2008, LNCS 5342, pp. 361–371, 2008.

objects, for image classification, or shape recognition. At present, there is an intensive research on homology computation in the context of the image ([9] for cubical homology,[15,16] for persistent homology,[3] for combinatorial maps,[12] for matrix methods,[13] for graph pyramids,[7] for algebraic-topological models).Even so, the amount of efficient numerical tools combining homological with geometrical information could be increased using and saving additional algebraic data structures at no extra computational time.

In this paper, we establish a method for computing homology numbers for a binary 26-digital volume V. 3D digital homology is developed here in a cell (more precisely, polyhedral) ambiance. Roughly speaking, a continuous analogous $K(V)$ of V is a cell complex having the same "homology" than it. In particular, the homology of V and $K(V)$ are isomorphics and all the numerical descriptors derived from homological and geometrical information in both objects are equals (Betti numbers, Euler characteristic,...) or extremely similar (size and orientation of geometrically well-adapted representative cycles of homology generators,...). Due to the fact that we encode homology in terms of an algebraic gradient vector field on the cell complex, it is possible to deduce not only Betti numbers, representative cycles of homology classes and homological classification of cycles but also homology numbers related to additional algebraic structures defined on homology (coproduct in homology, product in cohomology, (co)homology operations,...). In discrete context, this idea of encapsulating "homology" under an algebraic-topological format has already been used in theories like AT-model [6] and Discrete Morse theory [5]. For calculating the homology gradient on $K(V)$, it is possible to adequately modify classical algorithms (Smith Normal Form, incremental, etc) for computing homology. We deal with here the incremental Delfinado-Edelsbrunner technique.

The remainder of this paper is organised as follows. In section 2 we construct the cell complex $K(V)$. Definition and computation of the homology gradient vector field is given in section 3. Then we briefly comment some advantages of saving this algebraic information in section 4. We focus our interest in the canonical coproduct in homology [11].

2 3D Polyhedrization

In algebraic topology, specifically homotopy theory, shapes are often constructed out of contractile cells (having the homotopy type of a point) of different dimension which are coherently attached one to each other. Intuitively, a contractile cell is one that can be continuously shrunk to a point. In that way, cell complexes appear (see [11] for a formal definition) and constitute a class of spaces that is broader than simplicial complexes. They retain a combinatorial nature, so that computational consideration can still be applied in that context. In fact, the cell complex generalizes both the notion of graph adding cells of dimension greater than 1 as well as the notion of a simplicial complex in that triangles become polygons and tetrahedra become polyhedra. Note also that the cells need not be straight, such that a polygon may contain as few as a single edge connecting a

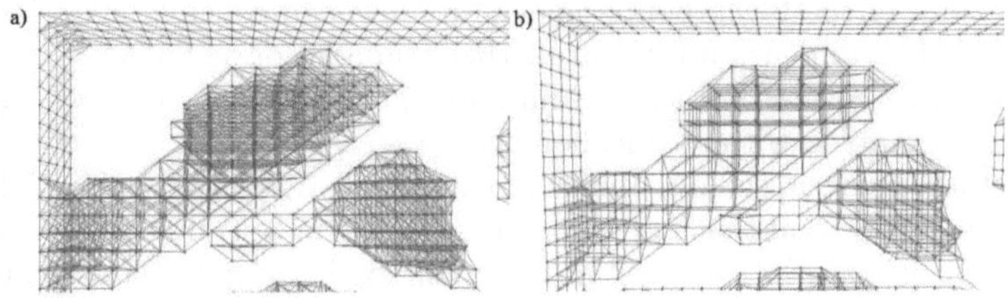

Fig. 1. A simplicial complex (a) and a cell complex (b)

single vertex to itself, and a polyhedron may contain as few as a single face, a single vertex and no edges.

A cell complex or simply complex in $\mathbf{R}^d$ is a set $K = \{K_q\}_{q \geq 0}$ of convex polyhedra (called cells). A face of a cell consist of each lower dimension cell, which make it up, in satisfying two conditions: (1) Every face of a cell is a cell (i.e. in K), and (2) If σ and σ' are cells, then their intersection is a common face of both. A simplicial complex is a cell complex whose cells are all simplices.

We associate here a three-dimensional cell complex $K(V)$ to a digital object V with 26-adjacency relationship. The complex $K(V)$ and the cubical complex canonically associated to V consisting in the set of geometric realizations of the voxels forming V are homotopically equivalent, and consequently, present the same homological information. To obtain the cell complex $K(V)$ we do as follows. Each black voxel can be seen as a point (0-cell) of our complex. The algorithm consist on dividing the volume into overlapped (its intersection being a "square" of four voxels mutually 26-adjacent) unit cubes formed by eight voxels mutually 26-adjacents, and to associate each unit cube configuration with its corresponding cell. We scan the complete volume, always taking as elementary step a unit cube.

The cell associated to a unit cube configuration is a 0-cell if there is a single point. If there are two points, the complex is a 1-cell which is the edge conecting both of them. With three or four coplanar points on the set, the 2-cell associated is a polygon. If there are four non coplanar points or more, the 3-cell is a polyhedra. In other words, the cell associated to a unit cube configuration is just

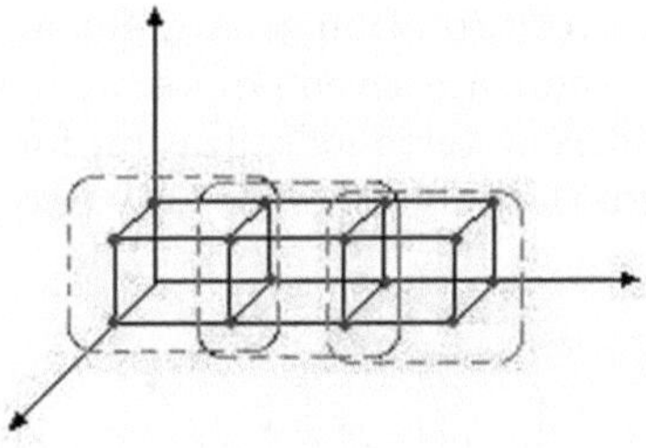

Fig. 2.

the convex hull of the black points and all its lower dimension faces. Note that for 3-cells, their 2-dimension faces are either triangles or squares.

Once we have covered all the volume and joined all the cells, we build the complete cell complex without incoherences.

Our method gives us the possibility of canonically associate a simplicial complex equivalent to V and this fact allows us to define in a straightforward manner the boundary operator of each cell in the complex. For doing that, we limit ourselves to say that we here slightly modify the idea proposed in [10] to create a new triangulation method on the surface of a digital volume. Because a $2 \times 2 \times 2$ volume includes eight points, each of which can be black or white, there are 256 possible patterns in a unit cube. If we ignore the congruent patterns differing only by the rotations of the center of the unit cube, the 256 patterns can be reduced to 23.

On a simplicial complex, a 0-simplex (v0) is a point; a 1-simplex (v0, v1) is a line connecting vertex v0 and v1; a 2-simplex (v0, v1, v2) is the triangle defined by these three vertex; and a 3-simplex (v0, v1, v2, v3) is a tetrahedra. The aim is to divide each cell into points, edges, triangles or tetraedra to build the simplicial complex.

Using 26-adjacency to divide each cell, we will find crossing simplex when there are four coplanar points into the cube. In all these cases we will choose one direction we will give preference, to avoid simplex crossing.

The problematic situations arise only when there are four or more black points into a 2x2x2 set, so we have to study each possibility separately.

As is shown in figure 3, given a cube with five black points, there are three possible positions of this points not considering cube rotations.

In figure 3.a we have to choose a direction, such that no incoherencies will appear between one cube and their neighbours. That is why all possible rotations of this point configuration must be considered separately, because depending on the cube's face we find the four coplanar points, we should choose a different tetrahedrization, according to the decision adopted for the four points. We have then six different configurations to consider for this case.

The other two situations (3.b, 3.c) are easier to carry out, because we can apply the tetrahedrization showed in the picture, to all of the possible rotations in each case without incoherencies.

With six black points, there are also three possible configurations.In two of them we have to consider every rotation separately, and there are twelve possibilities for each one. When there are seven points, we have just eight possibilities, and we should build a different tetrahedralization for each one. Finally, when all the points are black into the cube, there is only one possible tetrahedrization

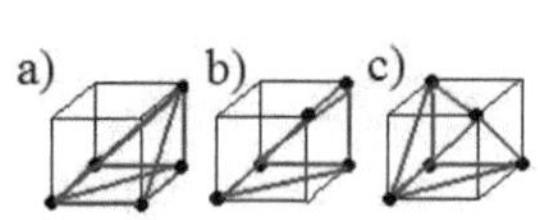

Fig. 3. Five black points

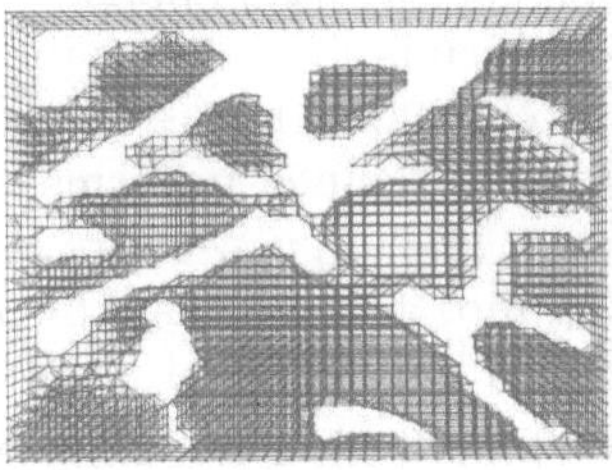

Fig. 4. Cell complex $K(V)$ of a trabecular bone volume with 21938 cells

according to the direction we chose for each cube's face. Once we have the entire cube's volume and their simplicial complex, we have to build a general simplicial complex for the total volume.

If we just join the different complexes, taking into account not to take two times the same simplex, we will not find incoherences obtaining the simplicial final complex.

3 Homology Gradient Vector Field for a 3D Digital Object

In this section we extend the Discrete Morse Theory notion of gradient vector field of a Morse function [5]. Homology is encoded here in terms of an algebraic vector field (so-called homology gradient vector field) from which it is possible to compute not only Betti numbers (number of connected components, holes and cavities), representative cycles of homology classes and homological classification of cycles but also homology numbers related to additional algebraic structures defined on homology (coproduct in homology, product in cohomology, (co)homology operations,...). In discrete setting, this idea of considering a linear map for defining homology has been extensively developed in Discrete Morse and AT-model [7] theories. In the first one, this map is a vector field and in the second one, it is a chain homotopy operator determining a strong algebraic relationship between the 3D object and its homology. Here, we suitably combine concepts and techniques of both theories in order to specify an adequate framework for efficiently computing homological information.

To explain the notion of homology gradient vector field, we first describe some elementary concepts from Homological Algebra (see [8]).

Since the objects considered in this paper are embedded in $\mathbf{R}^3$ then the homology groups vanish for dimensions greater than 3 and they are torsion–free for dimensions 0, 1 and 2. Throughout the paper, we consider that the ground ring is a finite field $\mathbf{F}$. Let K be a three-dimensional cell complex. A *q–chain* a is a formal sum of simplices of $K^{(q)}$. We denote $\sigma \in a$ if $\sigma \in K^{(q)}$ is a summand of a. The q–chains form a group with respect to the component–wise addition; this group is the *qth chain group* of K, denoted by $C_q(K)$. There is a chain group for every integer $q \geq 0$, but for a complex in $\mathbf{R}^3$, only the ones for $0 \leq q \leq 3$ may be non–trivial.

The boundary map $\partial_q : C_q(K) \to C_{q-1}(K)$ applied to a q–cell σ gives us the collection of all its $(q-1)$–faces which is a $(q-1)$–chain. By linearity, the boundary operator ∂_q can be extended to q–chains. In the concrete case of a simplicial complex, the boundary of a q-simplex defined in terms of vertices $\sigma = \langle v_0, \dots, v_q \rangle$ is defined by: $\partial_q(\sigma) = \sum \langle v_0, \dots, \hat{v}_i, \dots, v_q \rangle$, where the hat means that vertex v_i is omitted. In our case, taking into account that the 3-cells of our cell complexes can automatically be subdivided into tetrahedra, its boundary map can directly be derived from that of the component tetrahedra. It is clear that $\partial_{q-1}\partial_q = 0$. A chain $a \in C_q$ is called a *q–cycle* if $\partial\_q\ (a) = 0$. If $a = \partial\_q + 1(a')$ for some $a' \in C_{q+1}$ then a is called a *q–boundary.* Define the *qth homology group* to be the quotient group of q–cycles and q–boundaries, denoted by $H_q(\mathcal{C})$. Let $\mathcal{C} = \{C_q, \partial_q\}$ and $\mathcal{C}' = \{C'_q, \partial'_q\}$ be two chain complexes. A *chain map* $f : \mathcal{C} \to \mathcal{C}'$ is a family of homomorphisms $\{f_q : C_q \to C'_q\}_{q \geq 0}$ such that $\partial'_q f_q = f_{q-1}\partial_q$.

Now, we are ready for giving an algebraic version of the Discrete Morse Theory notion of gradient vector field, which is not necessarily associated with a Morse function. ([5]).

Definition 1. *[5] Let K be a finite cell complex. A linear map of chains $\phi : C_*(K) \to C_{*+1}(K)$ is a gradient vector field on K if the following conditions hold:*

1. *For any cell $a \in K_q$, $\phi(a)$ is a $q+1$-cell.*
2. *$\phi^2 = 0$*

The first condition means that ϕ is a combinatorial function and the second that V induces the following partition on the q-cells of K:

$$\{q - cells\ a \text{ of } K\} = \{a \text{ such that } \phi(a) \neq 0\}$$
$$\bigcup \{a \text{ such that } a \in \mathrm{Im}\phi\} \bigcup \{a \text{ such that } \phi(a) = 0 \text{ and } a \notin \mathrm{Im}\phi\}$$

Let us note that 0 is an specimen of $C_*(K)$. The cells constituting the last set are called *critical cells.*

In [14], elementary algebraic operations are used for designing an algorithm for computing homology which can be seen as a kind of algebraic-topological thinning. These elementary operations are called integral operators and they can be defined as algebraic versions of the notion gradient vector field (in the sense that ϕ applied on a cell gives us not necessarily a cell of higher dimension but possibly a sum of cells).

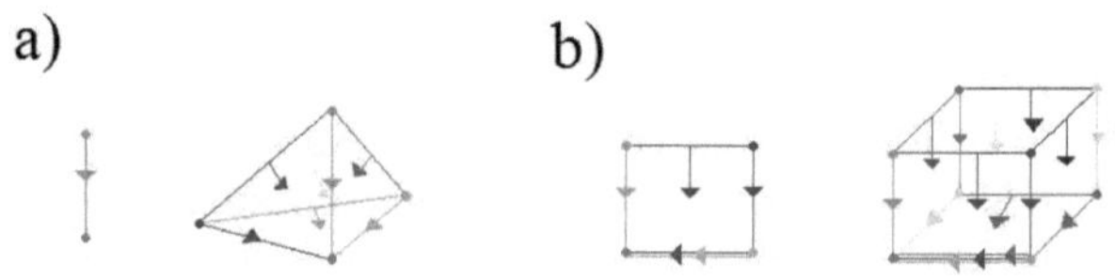

Fig. 5. Combinatorial gradients (a) and algebraic gradients (b)

The notion of homology gradient vector field of K generalizes that of gradient vector field and integral operator and it condenses all the significant homological information of the cell complex.

Definition 2. *A homology (or optimal) gradient vector field for a filtered cell complex K is a linear map of chains $\phi : C_*(K) \to C_{*+1}(K)$ such that $\phi\phi = 0$, $\phi\partial\phi = \phi$ and $\partial\phi\partial = \partial$, where ∂ is the boundary operator in $C(K)$.*

In fact, using homological algebra arguments, it is possible to deduce that ϕ determines a strong algebraic relationship connecting $C(K)$ and its homology vector space $H(K)$.

Let $\pi = id_{\mathcal{C}(\mathcal{K})} - \partial\phi - \phi\partial$. This chain map satisfies that $\pi^2 = \pi$. If $\text{Im}\pi = \{x \in C(K)$, such that $x = \phi(y)$ for some $y\}$ and $\text{Ker}\pi = \{x \in C(K)$ such that $\phi(x) = 0\}$, then $\mathcal{C}(\mathcal{K}) = \text{Im}\pi \oplus \text{Ker}\pi)$. Let $f : C(K) \to \text{Im}(\pi)$ be the corestriction of π to $\text{Im}(\pi)$ (that is, $\pi : C(K) \to \text{Im}(\pi)$) and $g : \text{Im}(\pi) \to C(K)$ be the inclusion. Let $\tilde{d}$ be the boundary operator of $\text{Im}(\pi)$. We now prove that $\tilde{d} = 0$. Taking into account that $id_{C(K)} + gf = \phi d + d\phi$, $\partial\partial = 0$ and $\partial\phi\partial = \partial$, we then obtain $\partial - \partial gf = \partial$. Therefore, $\partial gf = g\tilde{d}f = 0$. Since f is onto and g is one- to-one, we deduce that $\tilde{d} = 0$. That means that the Morse complex $M_\phi = Im\phi$ is a graded vector space with null boundary operator isomorphic to the homology $H(K)$.

We obtain a homology gradient vector field for K adequately modifying classical algorithms for computing homology (boundary matrix reduction based on Smith Normal Form [11] , incremental technique [4], ...) existing in the literature. We describe here a homology gradient version of the incremental homology method of Delfinado-Edelsbrunner [4]. For designing such an algorithm it is necessary to impose an order on the set of cells of K. We call such an ordering a *filter*. In other words, given a cell complex K, $\mathcal{K} = \{c_1, \ldots, c_m\}$ is a filter if c_i is a face of c_j for $i < j$. It is possible to "filter" K by first considering all the 0-cells in a certain order, then an order on all the 1-cells, and so on.

Algorithm 1. *Computing a homology gradient vector field for a finite filtered cell complex K (with filter) over a field.*

INPUT: The filter $\mathcal{K}_m = \{c_0, \ldots, c_m\}$ considered on the chain complex $(C(K), \partial)$, and the differential ∂_i for the basis $\mathcal{K}_i$, $i = 0 \ldots m$.

$\mathcal{C}_0 := (\{c_0\}, \partial_0)$, $\phi_0(c_0) := 0$.
For $i = 1$ to m do
  $\mathcal{C}_i := (\mathcal{C}_{i-1} \cup \{c_i\}, \partial_i)$,,
  If $(\partial_i - \partial_{i-1}\phi_{i-1}\partial_i)(c_i) = 0$, then
    $\phi_i(c_i) := 0$.
    For $j = 0$ to $i - 1$ do,
      $\phi_i(c_j) := \phi_{i-1}(c_j)$.
  If $(\partial_i - \partial_{i-1}\phi_{i-1}\partial_i)(c_i)$ is a sum of a kind $\sum_{i=1}^{r} \lambda_i u_i \neq 0$ $(u_i \in \mathcal{C}_{i-1})$, then:
    $\tilde{\phi}(u_1) := -\lambda_1^{-1} c_i$ and zero in the rest of elements of $\mathcal{C}_{i-1}$, .
    then For $j = 0$ to i do,
      $\phi_i(c_j) = (\phi_{i-1} + \tilde{\phi} - \tilde{\phi}\phi_{i-1}\partial_{i-1} - \tilde{\phi}\partial_{i-1}\phi_{i-1})(c_j)$,

OUTPUT: a homology gradient vector field $\phi_m forK$.

The algorithm runs in time at most $\mathcal{O}(m^2)$, where m is the number of cells of K.

Concerning to the homology computation of a binary digital volume V with 26-adjacency relation, the homology gradient vector field for the canonical 26-adjacency cell representation $K(V)$ (with a given filter) of V provides us the desired solution.

This method has already been implemented and has been tested on digital volumes of trabecular bones (courtesy of the Institute of Biomechanics, Zurich, Swiss). The following table shows the Betti numbers and the computation time for the homology of digital trabecular bone volumes of different sizes.

Number of cells	Betti Numbers	Time
21938	$b_0 = 3$, $b_1 = 12$, $b_2 = 0$	0,885 s.
47448	$b_0 = 2$, $b_1 = 13$, $b_2 = 0$	2,055 s.
117780	$b_0 = 2$, $b_1 = 406$, $b_2 = 13$	5,31 s.

4 Advanced Homological Information

In this section, we briefly comment some advantages of saving a homology gradient vector field ϕ for a binary digital volume V.

First, considering homology in terms of a linear map ϕ increasing the dimension by one allows to give a positive answer to problems so important in object recognition as detecting null-homological cycles (that is, cycles which become zero in homology), homologically classifying cycles at any dimension, to deform a cycle guaranteeing a homological control in the process, reversible topological skeletonization,... In fact, all these question can be handled using as main tool the triple of maps (π, inc, ϕ) derived from ϕ previously described in the last section. These maps connect in an algebraic strong way the chain complex $C(K(V))$ canonically associated to the polyhedral complex $K(V)$ to its homology $H(V)$. For instance, if we are interested in homologically classify a 1-cycle c ($d(c) = 0$), we only need to compute $\pi(c)$ for determining the homology class it belongs. If $\pi(c) = 0$, then c is a boundary and the following equality hold $c = d(\phi(c))$. If we are interested in comparing two different 1-cycles c and c' belonging to the same homology class $\pi(c) = \pi(c')$, we can obtain that $c - c' = d\phi(c - c')$. This means in particular that the 2-chain $\phi(c - c')$ is a "surface" in V which has as part of its boundary c and c'. Given a homology class $\pi(c)$, if we are interested in obtaining a specific cycle c' belonging to this homology class presenting some properties of geometrical nature (for example, to be "well-adapted" to the geometry of the object), we only need to develop the specific geometric algorithm for getting c' and finally to check that $\pi(c') = \pi(c)$. It is possible to generate a (curve, curve-surface, ...) topological skeleton $Sk(K(V))$ which is reversible in the sense that we can reconstitute the cell complex $K(V)$ and consequently, the volume V from the skeleton. In fact, it is possible to design an algorithm f or computing a chain contraction from $C(K(V))$ to $C(Sk(K(V))$ which allows us to ensure the reconstruction of a subvolume of V starting from any part of the skeleton.

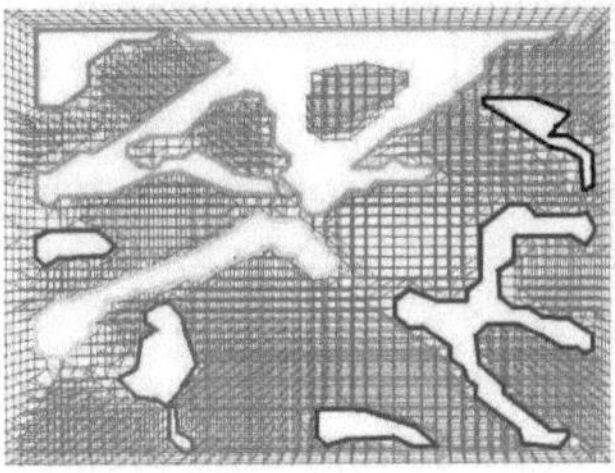

Fig. 6. Geometrically well-adapted 1-cycles

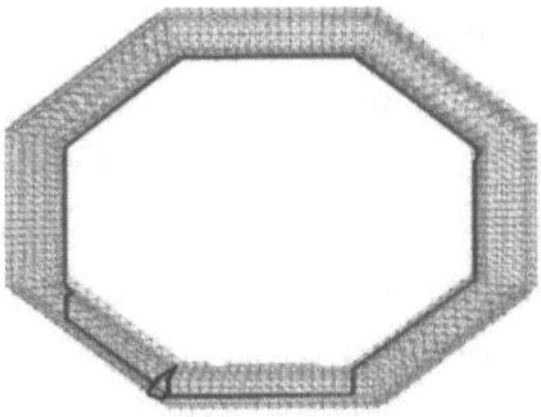

Fig. 7. The hollow torus and the representative cycles of homology classes of dimension 1 and 2

Now, we focus our interest in the computation of the coproduct in homology for simplicially subdivided cell complexes ($K(V)$ is an example of this). There is a canonical coproduct $\Delta : H(V) \rightarrow H(V) \otimes H(V)$ (also called Alexander-Whitney coproduct) in homology which is well defined in the context of Simplicial Topology. This coproduct operation at homology level can be seen as a proper decomposition of 2-cycles into 1-cycles factors. For example in the case of a hollow ring (torus, see the figure), the homology coproduct applied on the global cavity gives us the two fundamental 1-cycles. The formula for this operation at complex chain level applied to a generic triangle (v_0, v_1, v_2) is the following:

$$AW(v_0, v_1, v_2) = (v_0, v_1, v_2) \otimes (v_2) + (v_0) \otimes (v_0, v_1, v_2) + (v_0, v_1) \otimes (v_1, v_2)$$

This map can be translated to homology level using the triple (π, inc, ϕ). Let $\Delta : H(V) \rightarrow H(V) \otimes H(V)$ be $\Delta([x]) = (\pi, \pi)AW(x \times x))$ where $[x]$ means the homology class with representative 2-cycle x.

In our case, a cell 2-cycle c we must handle is a sum of cubes and polyhedra, and they can be automatically putted under the form s of sum of the component tetrahedra of them. We use this simplicial description s for the 2-cell cycle in the previous formula for the coproduct.

5 Conclusions and Future Work

This paper proposes a polyhedral-based solution to advanced homological computation, opening a door to the design of efficient algorithmic tools which are suitable for volume recognition. The fundamental tool in our approach is that of homology gradient vector field which condenses all the relevant homological information of the digital object V. This notion is interpreted in a Homological Algebra setting as a chain homotopy equivalence. This strong algebraic "connection" between V and its homology allows to go further in homological computation.

The main theorem of Discrete Morse Theory establish that a cell complex K with discrete Morse function is simple homotopy equivalent to another cell-complex with exactly one cell of dimension p for each critical cell of dimension p. Then, the fundamental idea here is to reduce the number of cells in a complex without changing the homotopy type. In this paper, we construct directly from the boundary operator the homology gradient vector field in an incremental way. If, for each digital object V we want to analyze, it would be possible to coherently install an initial (non-optimal) gradient vector field ϕ_0 in the cubic grid ($2 \times 2 \times 2$ unit cubes) "containing" all the cells of $K(V)$, then we reduce the complexity of the homology calculus, by computing the homology on the Morse complex $M_{\phi_0} = \text{Im}\phi_0$, exclusively composed of critical cells. The computational costs of the method can drastically be reduced if the number of critical cells of the corresponding Morse complex is small.

References

1. Biassoti, D., Giorgi, M., Falcidieno, B.: Reeb Graphs for Shape Analysis and Applications. Theoretical Computer Science 392(1-3), 5–22 (2008)
2. Chen, L.: Topological structure in the perception of apparent motion. Perception 14(2), 197–208 (1985)
3. Damiand, G., Peltier, S., Fuchs, L., Lienhardt, P.: Topological Map: An Efficient Tool to Compute Incrementally Topological Features on 3D Images. In: Reulke, R., Eckardt, U., Flach, B., Knauer, U., Polthier, K. (eds.) IWCIA 2006. LNCS, vol. 4040, pp. 1–15. Springer, Heidelberg (2006)
4. Delfinado, C.J.A., Edelsbrunner, H.: An Incremental Algorithm for Betti Numbers of Simplicial Complexes on the 3–Sphere. Comput. Aided Geom. Design 12, 771–784 (1995)
5. Forman, R.: A Discrete Morse Theory for Cell Complexes. In: Yau, S.T. (ed.) Geometry, Topology & Physics for Raoul Bott. International Press (1995)
6. Gonzalez-Diaz, R., Medrano, B., Real, P., Sanchez-Pelaez, J.: Algebraic Topological Analysis pf Time-sequence of Digital Images. In: Ganzha, V.G., Mayr, E.W., Vorozhtsov, E.V. (eds.) CASC 2005. LNCS, vol. 3718, pp. 208–219. Springer, Heidelberg (2005)
7. Gonzalez-Diaz, R., Real, P.: On the Cohomology of $3D$ Digital Images. Discrete Applied Math. 147, 245–263 (2005)
8. Hatcher, A.: Algebraic Topology. Cambridge University Press, Cambridge (2001)
9. Kaczynski, T., Mischaikow, K., Mrozek, M.: Computational Homology. Springer, New York (2004)
10. Kenmochi, Y., Imiya, A., Ichikawa, A.: Boundary extraction of discrete objects. Computer Vision and Image Understanding 71, 281–293 (1998)
11. Munkres, J.R.: Elements of Algebraic Topology. Addison–Wesley Co. (1984)

12. Peltier, S., Alayrangues, S., Fuchs, L., Lachaud, J.O.: Computation of homology groups and generators. Computers & Graphics 30(1), 62–69 (2006)
13. Peltier, S., Ion, A., Haxhimusa, Y., Kropatsch, W.G., Damiand, G.: Computing Homology Group Generators of Images Using Irregular Graph Pyramids. In: Escolano, F., Vento, M. (eds.) GbRPR 2007. LNCS, vol. 4538, pp. 283–294. Springer, Heidelberg (2007)
14. Real, P., González-Díaz, R., Jimenez, M.J., Medrano, B., Molina-Abril, H.: Integral Operators for Computing Homology Generators at Any Dimension. In: Ruiz-Shulcloper, J., Kropatsch, W.G. (eds.) CIARP 2008. LNCS, vol. 5197, pp. 356–363. Springer, Heidelberg (2008)
15. Zomorodian, A., Carlsson, G.: Localized Homology. Computational Geometry: Theory & Applications (to appear)
16. Zomorodian, A.: Topology for Computing. Cambridge University Press, New York (2005)

Modeling the Model Athlete: Automatic Coaching of Rowing Technique

Simon Fothergill, Robert Harle, and Sean Holden

Computer Laboratory, University of Cambridge,
Cambridgeshire, CB3 0FD, U.K.
{jsf29,rkh23,sbh11}@cam.ac.uk

Abstract. Watching athletes allows coaches to provide both vital feedback on how well they are performing and on ways to improve their technique without causing or aggravating injuries. The thoroughness and accuracy of this traditional observation method are limited by human ability and availability. Supplementing coaches with sensor systems that generate accurate feedback on any technical aspect of the performance gives athletes a fall back if they do not have enough confidence in their coach's assessment.

A system is presented to model the quality of arbitrary aspects of rowing technique found to be inconsistently well performed by a set of novice rowers when using an ergometer. Using only the motion of the handle, tracked using a high-fidelity motion capture system, a coach trains the system with their idea of the skill-level exhibited during each performance, by labeling example trajectories. Misclassification of unseen performances is encouragingly low, even for unknown performers.

Keywords: Novel Applications, Sports coaching, Quality, Rowing Technique, Body motion, Intelligent Sensing Systems, Spatiotemporal Pattern Recognition, Shape Analysis.

1 Introduction

When practising a physical skill, reassurance that our technique is good or feedback on how to correct it, are crucial for motivation and improvement. Learning a technique so we perform correctly without thinking requires lengthy (re)training of "muscle memory" as well as studying the underlying biomechanical principles, especially if the technique is complex or we need to break old habits. Real-time feedback can help athletes when learning what good and bad technique feels like. As they may not be able to feel whether the whole performance is correct and can not stop to observe themselves, feedback is required from a third party such as a coach. In this paper, the authors choose to investigate generating feedback to supplement coaches of the Olympic sport of rowing. The multi-faceted technique requires a very high level of consistency and precision at high speed between multiple rowers. On the water, a few centimeters difference in a stroke can unbalance and decelerate a boat so there is real need for continuous personal feedback. Rowing can be practiced very realistically in a laboratory environment by using a stationary ergometer.

N. da Vitora Lobo et al. (Eds.): SSPR&SPR 2008, LNCS 5342, pp. 372–381, 2008.

Good coaches will always be in demand but automated analysis of body movements may offer advantages over human coaching: world-class coaches are often busy and expensive. They often coach in squads to instill competition, giving each athlete a fraction of their time during which they might give unrepresentative and damaging feedback. Even experienced coaches can not simultaneously concentrate on all aspects of a technique, can be biased towards athletes they know, can fail to notice differences in a performer they are used to watching and can become impaired by factors such as fatigue.

An automated, surrogate system could analyze athletes continuously whenever they train, objectively focusing on any number of aspects simultaneously. As motion capture systems continue to develop, this hardware will become much more affordable than a real person. On-athlete sensors could also offer a more accurate perspective from inside the boat rather than from a distance. When athletes are interested in semantically sophisticated descriptions of a performance, informal discussions between coaches and the authors showed they may be less comfortable representing a performance as digitized trajectories of points, rather than watching their moving flesh. By automating analysis of the trajectories, athletes could instead be presented with comprehendible and familiar ontological elements from the domain of rowing, which could also be used as automatically constructed, low bandwidth indices into recorded performances.

The authors have circumstantial evidence from the sports science community that biomechanical theory can not always justify coaches' behavior and that the latter find difficultly in explaining their judgments. Supervised learning provides a framework for directly associating coach-level feedback with performances recorded using a few sensibly, but not precisely placed sensors, whereas a traditional (bio)-mechanical approach would require explicit and precise formulation of rules for every facet of the technique.

The physical sensation of rowing drives the terminology to include words such as "relaxed", "fluid", "too", "sufficient" or "that looks right!" Their amorphous and complex nature is hard to capture using explicit biomechanical formulae, especially when the authors tried to use some coaches' explanations. Approximations may have detrimental or unobvious effects on classification performance through including or missing out certain factors; when rowing even small movements are important.

Measurements of performances of a similar quality still have inherent noise from the sensors and differences in where they are worn. Characterization of what technique works best for different athletes also adds to variation that is not suited to exact mechanical formulae. Even for more tangible aspects such as "overreaching" or "leaning back too far", rules would need to be personalized and require expert biomechanical advice to be trustworthy. The authors found that a large number of biomechanical measurements are typically taken and instrumenting athletes is time consuming. Rules that rely on specific sensors become dependable on those sensors functioning. By using such rules the system is customized to the domain of rowing and the same system could not be taught different sporting techniques.

The highly complex nature of rowing where large intra-class variations are present but important differences between classes are made due to multiple, occasionally unobvious aspects changing by large or very small amounts, makes this an apt challenging problem for machine learning, especially when using a limited set of body movements. How well a learning system might recognize the quality of an individual aspect of technique when performed by different people and with different levels of skill, whilst ignoring other changing aspects of the technique, is a question that has not been addressed for the domain of rowing as far as the authors are aware.

The process of coaching athletes includes observation and analysis of their technique to determine the quality of their performance, deciding what encouragement or corrections are necessary and formulating how to communicate this so that athletes appreciate it. This involves explaining and demonstrating correct technique whilst the athletes rest so they can be prompted to recall it as necessary during a later performance. The quality of a performance can be described at various levels of precision from an overall judgment of the whole performance, to the quality of individual aspects of the technique, to a highly detailed description of each muscle. This work focuses on determining the quality of individual aspects of rowing technique using binary classification as this provides a simple yet informative representation of common judgments.

2 Previous Work

Much research exists into the capture and analysis of human body motion, including recognition of everyday activities to specific gesticulations. However, judgment of the quality of a physical performance is less common and work that exists tends to focus on a few direct biomechanical measurements that are somehow related to quality within the medical domain.

Work in [2] uses HMMs to recognize a vocabulary of gestures that could make up a surgical procedure. Using different observations of the performance, a naïve judgment of skill is based simply on the number of different gestures used in a performance, or the percentage of time spent on each gesture. No rigorous validation is completed. Authors of [5] admit that the objective measurements of surgeons' dexterity as recorded by a motion capture system within an operating theatre does not capture the complexity and differences in styles involved in assessing the skill of different surgeons. The authors of [4, 10] use HMMs to obtain values for overall quality of a complete surgical performance, rather than individual aspects. The method used in [10] splits the performances into sub-sections with semantic significance and used the distance from a template performance as a measure of quality. The authors of [4] calculated the log-likelihood of a performance by applying one model to the whole performance, disregarding any internal structure and achieved a recognition of performance that correlated highly with an expert ($r = 0.93$ $p<0.001$).

Recognition of sporting activities is the focus of work in [3], where elements of a technique such as the tennis serve are recognized in video footage. However this is done by high-level scene analysis. The work of [1] evaluates overall quality in Karate using many markers and requiring models of prototype performances that are morphed to new performances. Results are presented for one person performing 20 different Karate moves. Recognition accuracy varies over technique but is on average 85%. The author of [6] discussed the need for individual aspects of a performance to be assessed, but proposed that an automated rule-based system would struggle with the flexible or fuzzy nature of the existing rubrics.

3 Description of the System

The system provides binary classification of a population of strokes taken from a number of different performances. Strokes are separated based on the quality of an arbitrary aspect of the rowing technique used. It is able to learn the judgment rule from a set of labeled performances and provide validation of how well it generalizes over an unseen population of strokes.

3.1 Motion Capture

The VICON motion capture system [11] was used with 10 cameras and default parameter values to track the position of single points on multiple objects augmented with 3 or more retro-reflective markers (see Fig. 1). The sample rate was set to 200Hz. The objects are automatically identified using unique markers topologies.

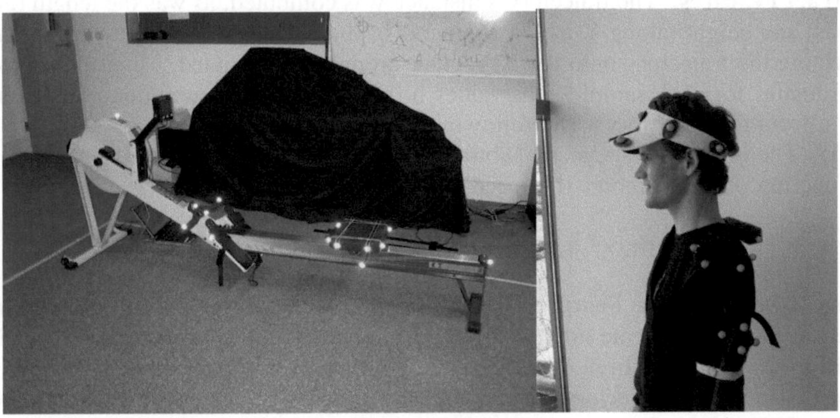

Fig. 1. Ergometer and rower augmented with VICON markers. The erg, its seat and its handle were 3 objects. Parts of the body were captured for future use.

Some markers on body parts and the seat were occluded for short periods of time no longer than half a second, which is approximately a quarter of the period of a stroke. Conventional video was also used to record the performances for review later.

3.2 Preprocessing

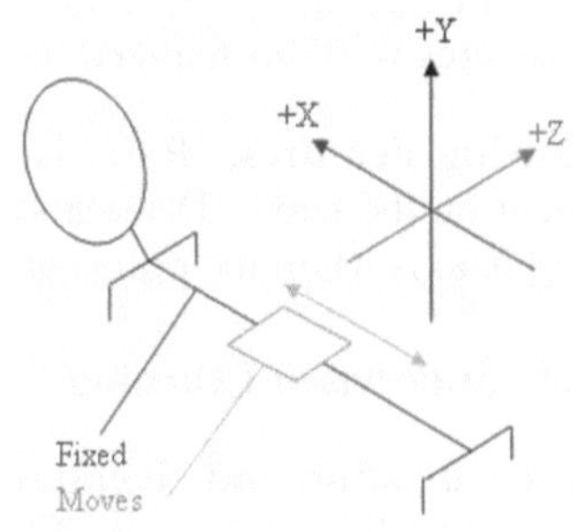

Fig. 2. Erg co-ordinate system

Occluded portions of the trajectories are recovered using linear interpolation. All trajectories are transformed from the motion capture co-ordinate system to an erg-coordinate system (see Fig. 2) to make analysis of the data more intuitive.

Markers of the seat were used to define the X-axis as the seat moves along it, and the X-Y plane from which the normal gives the Z-axis. The seat exhibited a measured deviation of less than one mm from the averaged X-axis. This work only uses markers on the handle, seat and erg-frame. This trajectory is segmented into separate strokes using an algorithm that detects the main troughs in the handle's X-coordinate. The first and last strokes were discarded in case they were unrepresentative as the athletes were warming up/down.

3.3 Feature Extraction

The tempo-spatial features listed below are used to represent the handle trajectory for a stroke. The stroke rate (number of strokes every minute) is made invariant as it is taken to be independent of the aspects of technique used in this study.

• Distance travelled	• Length	• Height	• Shape moments
• Speed moments	• Wobble	• Drive smoothness	• Recovery smoothness
• Shape smoothness	• Ratio	• Drive angle	• Recovery angle

Abstract Features. The trajectory's distance was computed, as was the length (along X-axis) and height (along Y-axis). Moments of the 2D shape formed by orthogonally projecting the trajectory onto the X-Y plane were also computed, using the projected co-ordinates for each sample s, as values $h_x(s)$ and $h_y(s)$ and summing over all samples s, from the trajectory. The following were computed for shape: λ^{11}, λ^{12}, λ^{21}, λ^{02}, λ^{20} (1). The instantaneous, mean-subtracted speed at each sample point s, was used as the function $\psi(s)$ to compute the speed moments: μ^{11}, μ^{12}, μ^{21}, μ^{02}, μ^{20} (1).

$$\lambda^{pq} = \sum (h_x(s)^p \, h_y(s)^q) \,, \qquad \mu^{pq} = \sum (h_x(s)^p \, h_y(s)^q \, \psi(s)) \,, \tag{1}$$

Physical Performance Features. The following were chosen based on preliminary experiments to investigate how well they disambiguated a small set of performances of differing overall quality. Wobble (lateral variance) was computed as the variance in the distance each trajectory sample is from the best fit line when they are projected into the X-Z plane. Smoothness of the drive and recovery in time were computed by low-pass filtering the mean-subtracted, instantaneous speed of the handle at 3Hz (as the raw motion data was too noisy), identifying large accelerations by taking the second derivative and summing the absolute values of this signal. Smoothness of the shape was computed by forming a signal from the distances of each subsequent trajectory sample point, to the centroid of the trajectory and low-pass filtering at 6Hz, followed by measuring the flatness of the absolute value of the second differential of this signal by counting the number of samples that are less than 0.4 ms^{-2}. All smoothness measures were normalized to the length of the stroke.

Rowing Features. Ratio is the ratio of the time spent on the recovery to the time spent on the slide. Drive and recovery angles are the angles between the best fit line and X axis when the points are projected into the X-Z plane.

3.4 Supervised Learning

Normalization and Negation. Each feature used in the model has parameters computed to adjust its values to be roughly within the same range as all other features in order to weight and compare them more easily. This is done for each stroke by subtracting the minimum and dividing by the difference between the maximum and minimum values of the feature, in the non-normalized training set. If the values from the training set for a single feature correlate highly but negatively with each strokes' labeled scores, the normalized feature's complement is used instead of the original feature, i.e. the original feature value is subtracted from 1. The labeled score for each stroke is normalized to 0 (bad) and 1 (good).

Classifiers. Each stroke from a population is automatically scored using a linear combination of a weighted bias and the weighted features. The weights of the model are learnt using one of two different methods: The first solves the system of simultaneous equations formed from the feature vectors of each stroke as rows of a matrix multiplied by a column vector of weights, which is made equal to a corresponding column vector of labeled scores for each stroke. It is solved using the Moore-Penrose pseudo-inverse of the feature matrix. The second method uses gradient descent to adjust the weights, which are initially set to zero after each iteration through the whole training set. A learning rate of 0.001 is used for 750 iterations to minimize the sum of the square of the differences between the labeled and machine scores of the training set. The training is repeated for a number of times using different training sets formed by leaving out a different set of strokes from the whole original population. These unseen stroke(s) are classified using the trained model and their scores recorded. Each stroke is left out exactly once. When considering multiple performers all strokes from one performer are left out at once. This "leave-one-out" validation (whether it be one stroke or one athlete) gives machine scores for each stroke in the population, ensuring it hasn't been used to train the classifier that scores it. The scores are summarized using Pearson's coefficient for the correlation between the coach and the machine scores and by using the percentage of strokes misclassified. The latter is computed by thresholding the two classes for the machine scores at a point which minimizes the misclassification. This value is calculated exhaustively.

Sensitivity Analysis. For the second training method the optimum number of iterations is chosen by repeatedly calculating the percentage of misclassified strokes, having trained using a different number of iterations, increasing by 50 each time from 50 to 750. For both methods, an optimum feature set is chosen that minimizes the percentage of misclassified strokes by repeating the sensitivity analysis for the number of iterations, for different feature sets. Each set is formed using exhaustive backwards-selection by removing the least important features until only 1 is left. The initial set is all the features and importance is measured by summing the weights over all training sets when using the optimum number of iterations. The class threshold, number of iterations and feature set that give the smallest misclassification error are assumed when presenting the results for different represented populations of strokes.

4 Empirical Validation of System

These experiments gave an indication of the system's capabilities in real-world coaching scenarios. Each stroke was scored by using the score a single amateur coach gave to the whole performance it was from and by reviewing the videos the coach considers at least 95% of the strokes of a performance to be representative of the quality they assigned to the whole performance. The coach (lead author) has been rowing for 2 years and was confident of assessing basic technique in novice rowers. Strokes by different rowers are differentiated as a person's physique and skill level effect the set of strokes recorded as does their conscious decisions for how to row and how much attention to pay the coach. Six novice, male rowers in their mid-twenties, between 60kg and 90kg were used with very little or no rowing experience. They were not initially fatigued and rowed at a comfortable rate in an uncontrived manner. The amount of input from the coach during a performance affects the consistency of a the technique during a performance and can increase how much of the stroke must be ignored or generalized over when only interested in scoring one aspect.

Each rower was given a basic explanation of how to row and gave an initial performance. A number of things were usually wrong with this performance so a coaching process was then repeated until no obvious faults existed or the rower was exhausted: the coach evaluated the last performance and taught the rower how to improve an aspect. The rower then gave another performance during which the coach helped them to maintain the improved technique for the increasing number of corrected aspects. Each performance lasted for approximately 30 strokes and the corrected aspects were maintained for at least 95% of the strokes.

The final features chosen for each experiment are recorded to investigate their usefulness. For either training method all features are used in at least 60% of the final features sets. Shape and speed moments λ^{02}, λ^{20} μ^{02} and μ^{20} were the only four used in at least 90% of the final feature sets for both algorithms.

4.1 Coaching Single Aspects for Individuals

Coaches often try to improve a single aspect of technique at a time in order to simplify the training process. This requires the coach to judge the quality of this aspect and issue prompts to the rower when necessary as they try to correct only this aspect. As only one aspect was focused on for each consecutive performance during the data collection, the system effectiveness at this task is shown by using pairs of consecutive performances from each novice as sample populations, see Table 1.

4.2 Coaching Multiple Aspects Simultaneously for Individuals

It can be more efficient to issue coaching calls on multiple aspects depending on what the rower does wrong. This requires a coach to make judgments about one aspect, no matter what the qualities of other aspects are. As data were collected on each rower whilst several aspects of the stroke were performed at different qualities, realistic combinations of different qualities for several aspects were observed as were realistic amounts of variation in strokes with the same level of quality for one aspect. Using all performances from each individual, results of the system judging the quality of each coached aspect for strokes where multiple aspects are changing are shown in Table 1.

Table 1. Percentages of misclassified strokes for all experiments when coaching single and multiple aspects of the technique of individual rowers

Rower	Coached aspect (chronological order)	Moore-Penrose training		Gradient Descent training	
		Single aspect	All aspects	Single aspect	All aspects
1	Separate arms/legs	0	0	0	0
	Overreaching	0	0	0	1
2	Separate arms/legs	0	-	3	-
3	Separate arms/legs	0	0	0	0
	Overreaching	0	0	2	1
4	Overreaching	0	0	0	0
	Shins vertical	0	0	0	0
	Early open back	0	0	2	1
5	Leaning back	0	3	0	6
	Quick hands	0	4	0	7
	Rushing slide	0	2	0	6
	Early open back	3	0	5	0
6	Overreaching	0	0	0	0
	Separate arms/legs	0	0	1	3
	Quick hands	0	0	1	1

4.3 Coaching Single Aspects across Different Individuals

In order to coach a new athlete who has not been seen before the system can only base a decision of how well they row on performances of other people. Being able to do this would provide more evidence of being able to recognize the quality of an aspect when other parts of the stroke are performed differently as in §4.2. By having enough examples, representative populations of strokes can be formed using consecutive pairs of performances form multiple people, all of whom are improving the same aspect. Table 2 shows the recognition results and the number of rowers used for each aspect.

Table 2. Percentages of misclassified strokes for aspects across a number of different athletes

Rowers	Aspect	Moore-Penrose training	Gradient Descent training
2	Quick hands	9	5
2	Early open back	33	29
3	Separate arms/legs	21	21
4	Overreaching	12	12

5 Discussion, Conclusions and Further Work

Analysis of the final feature sets shows it is wise to make use of all the features. The common choice of the four moment features suggests they may capture fundamental

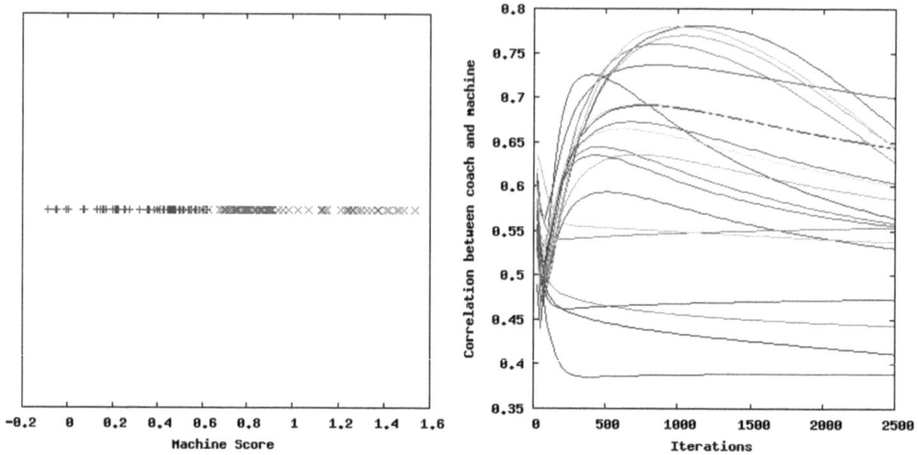

Fig. 3. Judging strokes from unseen people for the aspect "Quick Hands": machine score (left), correlation values using gradient descent up to 2500 iterations, for different feature sets (right)

and important characteristics of quality of technique and analysis of relative weights would confirm their suitability as good candidates for recognizing other sports.

Given the small amount of evidence for coaching higher numbers of aspects, when coaching up to four, for an single athletes, very low misclassification was found. The Moore-Penrose solution gave slightly better results. The minimum possible training error does minimize the generalized misclassification results and gradient descent does not find it in time at this learning rate. Recognition of strokes from unseen people gave quite low (5-21%) misclassification for three of the experiments. The fourth training set was only one person. The gradient descent method was slightly better this time: Tightening the model to fit the training data too closely by using Moore-Penrose had adverse effects on the generalization to unseen performances as shown in Fig. 3 where too many iterations caused the highest correlating feature sets to correlate less well, even though the training error was observed never to increase. Inter-variation of data from different athletes is observed to be generally greater than an athlete's intra-variation, suggesting why the results are significantly worse for unseen athletes. Further work including scoring each stroke separately, using more markers and estimating the weights differently or using other widespread classifiers. Semantically sophisticated analysis of physical performance would be possible if ontological elements from a domain of sports techniques were identifiable. Encouraging classification results are achieved for qualitative assessment of rowing technique but the limits of the recognition algorithms are not yet fully characterized.

Acknowledgments. Thanks are due to EPSRC, U.K. and the Digital Technology and Rainbow groups and Professor Andy Hopper at the Computer Laboratory, University of Cambridge U.K. for funding, support, equipment and volunteers.

References

1. Ilg, W., Mezger, J., Giese, M.A.: Estimation of Skill Levels in Sports Based on Hierarchical Spatio-Temporal Correspondences. In: Michaelis, B., Krell, G. (eds.) DAGM 2003. LNCS, vol. 2781, pp. 523–531. Springer, Heidelberg (2003)
2. Murphy, Vignes, Yuh, Okamura: Automatic Motion Recognition and Skill Evaluation for Dynamic Tasks. In: EuroHaptics 2003 (2003)
3. Zhong, Chang: Structure Analysis of Sports Video Using Domain Models. In: International Conference on Multimedia & Expo, Tokyo, Japan, pp. 920–923 (August 2001)
4. Leong, J.J.H., Nicolaou, M., Atallah, L., Mylonas, G.P., Darzi, A.W., Yang, G.-Z.: HMM Assessment of Quality of Movement Trajectory in Laparoscopic Surgery. In: Larsen, R., Nielsen, M., Sporring, J. (eds.) MICCAI 2006. LNCS, vol. 4190, pp. 752–759. Springer, Heidelberg (2006)
5. Dosis, Aggarwal, Bello, Moorthy, Munz, Gillies, Darzi: Synchronized Video and Motion Analysis for the Assessment of Procedures in the Operating Theater. Arch. Surg. 140, 293–299 (2005)
6. Gordon: Automated Video Assessment of Human Performance. In: Greer, J. (ed.) Proceedings of AI-ED 1995, pp. 541–546 (1995)
7. Australian Capital Territory Rowing Association, Rowing Technique for Coaches, `http://www.rowingact.org.au/SDO/TECHNIQUE_1.html`
8. Concept2, Rowing Technique, Faults and Corrections, `http://www.concept2.co.uk/training/faults_corrections.php`
9. Kaliouby, Robinson: Generalization of a vision-based computational model of mind-reading. In: First International Conference on Affective Computing and Intelligent Interaction (2005)
10. Rosen, Solazzo, Hannaford, Sinanan: Objective Laparoscopic Skills Assessments of Surgical Residents Using Hidden Markov Models Based on Haptic Information and Tool/Tissue Interactions. In: The Ninth Conference on Medicine Meets Virtual Reality (2001)
11. VICON, a motion capture system, `http://www.vicon.com`

Optimal Solution of the Dichromatic Model for Multispectral Photometric Invariance

Cong Phuoc Huynh[1] and Antonio Robles-Kelly[1,2]

[1] RSISE, Bldg. 115, Australian National University, Canberra ACT 0200, Australia
[2] National ICT Australia (NICTA)*, Locked Bag 8001, Canberra ACT 2601, Australia

Abstract. In this paper, we address the problem of photometric invariance in multispectral imaging making use of an optimisation approach based upon the dichromatic model. In this manner, we can cast the problem of recovering the spectra of the illuminant, the surface reflectance and the shading and specular factors in a structural optimisation setting. Making use of the additional information provided by multispectral imaging and the structure of image patches, we recover the dichromatic parameters of the scene. Thus, we formulate a target cost function combining the dichromatic error and the smoothness priors for surfaces with uniform reflectance. The dichromatic parameters are recovered through minimising this cost function in a coordinate descent manner. To this end, we make use of regular albedo patches in the scene selected automatically. The algorithm is shown to be robust, and since the cost function is convex in nature, the resulting solution is globally optimal. We illustrate the effectiveness of our method for purposes of illuminant spectrum recovery and skin recognition. We also compare our results to a number of alternatives.

1 Introduction

In computer vision, the modeling of surface reflectance is a topic of pivotal importance for surface analysis and image understanding tasks. For instance, Nayar and Bolle [1] have used photometric invariants to recognise objects with different reflectance properties. This work builds on the one reported in [2], where a background to foreground reflectance ratio is introduced. In a related development, Dror *et al.* [3] have shown how surfaces may be classified from single images through the use of reflectance properties. Moreover, although shape-from-shading usually relies on the assumption of Lambertian reflectance [4], photometric correction or specularity subtraction may be applied as a preprocessing step to improve the results obtained.

In multispectral imaging, photometric invariants pose great opportunities and challenges in the areas of material identification [5] and shape analysis. This is due to the information-rich representation of the surface radiance acquired by multispectral and hyperspectral sensing devices, which acquire wavelength-indexed data in thousands of bands across a broad spectrum. As in trichromatic imaging, the multispectral reflectance

* NICTA is funded by the Australian Government as represented by the Department of Broadband, Communications and the Digital Economy and the Australian Research Council through the ICT Centre of Excellence program.

N. da Vitora Lobo et al. (Eds.): SSPR&SPR 2008, LNCS 5342, pp. 382–391, 2008.

distribution is determined not only by the light source and viewer directions, but also by the material properties of the surface under study.

The main bulk of work concentrates in the effects encountered on shiny or rough surfaces. For shiny surfaces, there are specular spikes and lobes which must be modelled. There have been several attempts to remove specularities from images of non-Lambertian objects. For instance Brelstaff and Blake [6] use a thresholding strategy to identify specularities on moving curved objects. Ragheb and Hancock [7] have developed a probabilistic framework for specularity subtraction which uses the Torrance and Sparrow model to account for the distribution of image brightness. Despite being effective, the application of these methods to multispectral imagery is somewhat limited since they rely on the closed form of the Bidirectional Reflectance Distribution Function (BRDF) to characterise the specular spike and lobe. Moreover, other alternatives elsewhere in the literature aimed at detecting and removing specularities either make the use of additional hardware [1], impose constrains on the input images [8] or require color segmentation [9] as postprocessing steps. Hence, they are not readily applicable to multispectral images, as there can be tens or even hundreds of bands for each pixel. Thus, any pre/postprocessing and local operations must be exercised with caution and in relation to neighbouring spectral bands so as to prevent spectral signature variation.

Specific to multispectral imagery, Healey and his colleagues [10] have addressed the problem of photometric invariance for material classification and mapping in aerial imaging as related to photometric artifacts induced by atmospheric effects and changing solar illumination. In [11], a photometrically invariant approach was proposed based on the derivative analysis of the spectra. This local analysis of the spectra was shown to be intrinsic to the surface albedo. Yet, the analysis in [11] was derived from the Lambertian reflection model and, hence, its not applicable to specular reflections.

2 Recovery of the Reflection Model Parameters

Here, we present a structural approach based upon the processing of smooth surface-patches whose spectral reflectance is uniform over all the pixels they comprise. As mentioned earlier, the process of recovering the photometric parameters for the image is based on an optimisation method which aims at reducing the difference between the estimate yielded by the dichromatic model and the radiance image data. In this section, we commence by providing an overview of the dichromatic model as presented by Shafer [12]. Subsequently, we formulate a target minimisation function with respect to the model in [12] and an optimisation strategy based upon the radiance structure drawn from smooth image patches with uniform reflectance. Throughout the section, we also present our strategy of selecting patches used by the algorithm and describe in detail the coordinate descent optimisation procedure used here. This optimisation strategy is such that we fix in interleaved steps either the light spectrum or the surface properties (geometry and reflectance) so as to recover the optima of the remaining dichromatic parameters.

2.1 The Dichromatic Reflection Model

Throughout the paper, we employ the dichromatic model introduced by Shafer [12] so as to relate light spectral power, surface reflectance and surface radiance. Following this model, surface radiance is decomposed into a diffuse and a specular component. Let an object with surface radiance $I(\lambda, u)$ at location u and wavelength λ be illuminated by an unknown illuminant whose spectrum is $L(\lambda)$. The dichromatic model becomes

$$I(\lambda, u) = g(u)L(\lambda)S(\lambda, u) + k(u)L(\lambda) \tag{1}$$

In Equation 1, the shading factor $g(u)$ governs the proportion of diffuse light and depends solely on the surface geometry. We aim to recover the shading factor $g(u)$, the specular coefficient $k(u)$, the light spectrum $L(\lambda)$ and the spectral reflectance $S(\lambda, u)$ at location u and wavelength λ from the spectral radiance $I(\lambda, u)$ on the image.

2.2 Target Function

With the dichromatic model above, we proceed to define our target function for purposes of optimisation. At input, our algorithm takes a multispectral image whose pixel values correspond to the measurements of the spectral radiance $I(\lambda, u)$ over the bands indexed to the wavelengths $\lambda \in \{\lambda_1, \ldots \lambda_n\}$. In general, the dichromatic cost function of a multispectral image $\mathcal{I}$ is the weighted sum of its dichromatic error and a geometric regularisation term $R(u)$ for each image location.

$$F(\mathcal{I}) \triangleq \sum_{u \in \mathcal{I}} \left[\sum_{i=1}^{n} \left[I(\lambda_i, u) - L(\lambda_i)(g(u)S(\lambda_i, u) + k(u)) \right]^2 + \alpha R(u) \right] \tag{2}$$

In equation 2, α is a constant that acts as a balancing factor between the dichromatic error and the regularisation term $R(u)$ on the right-hand side. The wavelength-independent regularisation term $R(u)$ is related to the scene geometry and will be elaborated upon later.

For now, we focus our attention on the solution space of Equation 2. Note that minimising the cost $F(\mathcal{I})$ without further constraints is an underdetermined problem. This is due to the fact that, for an image with n spectral bands containing m pixels each, we would have to minimise over $2m + n + m \times n$ variables while having only $m \times n$ terms in the summation of Equation 2. However, we notice that this problem can be further constrained if the model is applied to smooth surfaces made of the same material, i.e. the albedo is regular across the patch under consideration. This imposes two constraints. Firstly, all locations on the surface share a common diffuse reflectance. Therefore, a regular surface P is assumed to have the same reflectance for each pixel $u \in P$, $S(\lambda_i, u) = S_P(\lambda_i)$. Note that this constraint significantly reduces the number of unknowns $S(\lambda_i, u)$ from $m \times n$ to $n \times N$, where N is the number of surface albedos in the scene. In addition, the smooth variation of the patch geometry allows us to formulate the regularisation term $R(u)$ in equation 2 as a function of the shading factor $g(u)$. In brief, smooth, regular surface patches naturally provide constraints so as to reduce the number of unknowns significantly, while providing a plausible formulation of the geometric regularisation term $R(u)$.

Following the rationale above, we proceed to impose constraints on the minimisation problem at hand. For smooth, regular surface patches, we consider the cost function for a surface patch $P \in \mathcal{I}$ and write

$$F(P) \triangleq \sum_{u \in P} \sum_{i=1}^{n} [I(\lambda_i, u) - L(\lambda_i)(g(u)S_P(\lambda_i) + k(u))]^2 + \alpha \sum_{u \in P} R(u)$$

As before, we have $S(\lambda_i, u) = S_P(\lambda_i)$, for all $u \in P$. Furthermore, the smoothness constraint on the patch implies that the shading factor $g(u)$ should vary smoothly across the pixels in P. This constraint can be effectively formulated by minimising the variation in the magnitude of the shading map. This, effectively, precludes discontinuities in the map of P via the regularisation term

$$R(u) \triangleq \left[\frac{\partial g(u)}{\partial x(u)}\right]^2 + \left[\frac{\partial g(u)}{\partial y(u)}\right]^2 \tag{3}$$

where the variables $x(u)$ and $y(u)$ are the x and y coordinates at pixel location u.

Thus, by making use of the set $\mathcal{P}$ of regular patches in the image $\mathcal{I}$, we can recover the dichromatic model parameters by minimising the target function

$$F^*(\mathcal{I}) \triangleq \sum_{P \in \mathcal{P}} \sum_{u \in P} \left[\sum_{i=1}^{n} [I(\lambda_i, u) - L(\lambda_i)(g(u)S_P(\lambda_i) + k(u))]^2 + \alpha R(u)\right] \tag{4}$$

2.3 Homogeneous Patch Selection

In the previous section, we formulated the recovery of the dichromatic model parameters as an optimisation procedure over the surface patch-set $\mathcal{P}$. In this section, we describe our method for automatically selecting regular surface patches for the minimisation of the cost function in Equation 4. The automatic patch selection method presented here allows the application of our method to arbitrary images. It is worth noting that this contrasts with other methods elsewhere in the literature [13,14], which are only applicable to pre-segmented images.

Our patch selection strategy is performed in two steps. Firstly, we identify edge pixels by applying a traditional edge detector such as that in [15] or the one presented by Canny [16] to the mean radiance across all spectral bands. Secondly, we subdivide the mean-radiance image into patches of equal size in a lattice-like fashion. Those patches containing less than t_p edge pixels, where t_p is a global parameter, are deemed to have a regular albedo. Note that, however, not all these patches are useful for purposes of illumination spectrum recovery. This is due to the fact that perfectly diffuse surfaces do not provide any information on the illuminant spectrum. The reason being that, for every band, the radiance of perfectly diffuse patches is given by the product of both, the illuminant spectrum and the surface reflectance. In the other hand, the dichromatic model implies that the specularities have the same spectrum as the illuminant, where the specular term can be viewed as a scaling factor solely dependent on the surface geometry.

We take profit of this dependency for purposes of recovering the light spectrum. We do this based upon the observation that, for highly specular surfaces, the surface diffuse radiance vector and the illuminant vector span a plane in the radiance space. This is a well known property in colour constancy, where a number of approaches [17,18] have employed subspace projection for purposes of light power spectrum recovery. As a result, for the recovery of the dichromatic model parameters, we use patches with the highest contrast amongst those deemed as regular albedo ones. We estimate the contrast of each patch as the variance of the average spectral radiance at each pixel.

2.4 Optimisation Procedure

Making use of the notation in Figure 1, we now present the optimisation procedure employed in our method. Here, we adopt an iterative approach so as to find the variables $\mathbf{L}, \mathbf{S}_P, \mathbf{g}_P$ and $\mathbf{k}_P$ which correspond to the minimum of the cost function in Equation 4. At each iteration, we minimise the cost function with respect to $\mathbf{L}$ and the triplet $[\mathbf{S}_P, \mathbf{g}_P, \mathbf{k}_P]$ of each patch P.

The algorithm commences with initialising the unknown light spectrum $L(\lambda)$. We note that objects in the scene reflect less energy than they receive from the illuminant, i.e. the radiance of a surface cannot be greater than the irradiance of the illuminant, the highest radiance across image pixels for each wavelength can be used as a lower bound estimate of the light energy. Thus, as an initialisation for $\mathbf{L}$ we employ the light spectrum estimated by the brightest pixel per band.

Now we elaborate on the coordinate descent approach. This is comprised by two interleaved minimisation steps. At each iteration, we can index the dichromatic variables to iteration number t and optimise, in interleaved steps, the variables $\mathbf{g}_P$, $\mathbf{k}_P$ and $\mathbf{S}_P$ with respect to the image radiance and the illuminant $\mathbf{L}$, and once the former are at hand, obtain a closed form solution for the latter.

In the first step, we estimate the optimal surface reflectance and geometry given the light spectrum $\mathbf{L}^{t-1}$. Note that, at iteration t, we can solve for the unknowns $\mathbf{g}_P^t, \mathbf{k}_P^t$ and $\mathbf{S}_P^t$ separately for each surface patch P. This is because these variables correspond to separate terms in Equation 4. This step is, therefore, reduced to minimising

$$\begin{aligned} F(P)|_{\mathbf{L}^{t-1}} &= \sum_{u \in P} \left[\|\mathbf{I}(u) - < \mathbf{L}^{t-1}, g(u)\mathbf{S}_P + k(u) > \|^2 + \alpha R(u) \right] \\ &= \sum_{u \in P} \left[\|\mathbf{I}(u) - g(u)\mathbf{D}_P^{t-1} - k(u)\mathbf{L}^{t-1}\|^2 + \alpha R(u) \right] \end{aligned} \tag{5}$$

$\mathbf{I}(u)$	: the spectral radiance vector at image pixel u, $\mathbf{I}(u) = [I(\lambda_1, u), \ldots I(\lambda_n, u)]^T$
$\mathbf{L}$	: the spectral power vector of the illuminant, $\mathbf{L} = [L(\lambda_1), \ldots L(\lambda_n)]^T$
$\mathbf{S}_P$	: the common spectral reflectance vector for each patch P, $\mathbf{S}_P = [S_P(\lambda_1), \ldots S_P(\lambda_n)]^T$
$\mathbf{g}_P$	: the shading map of all pixels in patch P, $\mathbf{g}_P = [g(u_1), \ldots g(u_l)]^T$ with $u_1, \ldots u_l$ being all the pixels in the patch P
$\mathbf{g}$	: the shading map of the whole image, $\mathbf{g} = [\mathbf{g}_{P_1}, \ldots \mathbf{g}_{P_r}]^T$ where $P_1, \ldots P_r$ are all patches in $\mathcal{P}$
$\mathbf{k}_P$	: the specularity map of all pixels in patch P, $\mathbf{k}_P = [k(u_1), \ldots k(u_l)]^T$
$\mathbf{k}$	: the specularity map of the whole image, $\mathbf{k} = [\mathbf{k}_{P_1}, \ldots \mathbf{k}_{P_r}]^T$

Fig. 1. Notations used in Section 2.4

where $\|.\| >$ and $< .,. >$, respectively, denote the inner product and the L_2-norm of the argument vectors and $\mathbf{D}_P^{t-1} \triangleq < \mathbf{L}^{t-1}, \mathbf{S}_P >$ is the diffuse radiance component for the surface patch P.

Note that the minimisation above involves $2|P| + n$ unknowns, where $|P|$ is the number of pixels in patch P. Hence, it becomes computationally intractable when the surface area is large. In practice, the selected patches need only be large enough so as to gather useful statistics from the radiance information. Moreover, as mentioned earlier, we can further reduce the degrees of freedom of the unknowns by noting that the spectral radiance vectors at all pixels in the same patch lie in a 2-dimensional subspace Q of $\mathbb{R}^n$ spanned by the diffuse radiance vector $\mathbf{D}_P^{t-1}$ and the light vector $\mathbf{L}^{t-1}$. This is a characteristic of the dichromatic model that has been widely utilised by prior work on colour constancy [13,14].

Having all the pixel radiance vectors $\mathbf{I}(u)$ at hand, one can obtain the 2-dimensional subspace Q via Single Value Decomposition (SVD). Denote the two basis vectors resulting from this SVD operation z_1 and z_2 and, accordingly, let the subspace $Q = span(z_1, z_2)$. Since $\mathbf{D}_P^{t-1} \in Q$, we can parameterise $\mathbf{D}_P^{t-1}$ up to scale as $\mathbf{D}_P^{t-1} = vz_1 + z_2$. Likewise, the light vector $\mathbf{L}^{t-1} \in Q$ can also be decomposed as $\mathbf{L}^{t-1} = w_1 z_1 + w_2 z_2$, where the values of w_1 and w_2 are two known scalars. Moreover, $\mathbf{D}_P^{t-1}$ is a linear function with respect to v. We also observe that the term $R(u)$ is only dependent on $g(u)$. Therefore, the objective function in Equation 5 can be reduced to a univariate rational function of v with degree 4, and can be minimised efficiently making use of the Levenberg-Marquardt algorithm [19]. Having obtained the optimal value of v, we can recover $\mathbf{g}_P$, $\mathbf{k}_P$ and $\mathbf{S}_P$, knowing the illuminant spectrum $\mathbf{L}$.

In the second step of each iteration t, we solve for $\mathbf{L}^t$ given $\mathbf{g}_P^t, \mathbf{k}_P^t$ and $\mathbf{S}_P^t$. Since the second term $R(u)$ in Equation 4 is independent of the illuminant, whereas the first term is a quadratic function of $\mathbf{L}$. Therefore, the illuminant spectrum can be solved in closed form as

$$\mathbf{L}^t(\lambda) = \frac{\sum_{P \in \mathcal{P}} \sum_{u \in P} I(\lambda, u)(g^t(u) S_P^t(\lambda) + k^t(u))}{\sum_{P \in \mathcal{P}} \sum_{u \in P} (g^t(u) S_P^t(\lambda) + k^t(u))^2}$$

The closed form solution above can be easily obtained by differentiating Equation 4 with respect to the illuminant irradiance and equating the resulting expression to zero.

3 Experiments

Here, we present results on the recovery of the illuminant spectra and skin recognition. To this end, we have acquired an image database of 51 human subjects, each captured under one of 10 light sources with varying directions and spectral power. The imagery has been acquired using a pair of OKSI Turnkey Hyperspectral Cameras. The light sources are divided into two rows, one placed above the camera system and one at the same height as the cameras. The light positions are evenly distributed in each row so that the illumination directions are towards the centre of the scene. The camera system is equipped with Liquid Crystal Tunable Filters which allow multi-spectral images to be resolved up to $10nm$ in both the visible (430–720nm) and near infrared (650–990nm) wavelength ranges. To obtain the ground truth for the illuminant spectrum, we have

Table 1. The accuracy of our illumant estimation method, in degrees, compared to Finlayson & Schaefer's method [13]

	Visible spectrum		Near-infrared spectrum	
Number of patches	Our method	Finlayson & Schaefer	Our method	Finlayson & Schaefer
1	10.87 ± 5.70	17.34 ± 9.04	5.98 ± 4.38	24.67 ± 9.46
2	10.77 ± 5.22	11.20 ± 6.18	5.06 ± 3.69	6.81 ± 5.57
4	10.94 ± 4.92	11.09 ± 5.32	4.47 ± 3.03	6.21 ± 4.89
5	11.26 ± 4.35	11.59 ± 4.16	4.21 ± 2.86	5.96 ± 4.15
10	11.26 ± 4.35	11.59 ± 4.16	4.21 ± 2.86	5.96 ± 4.15
20	11.51 ± 4.04	11.92 ± 3.65	4.08 ± 2.59	5.86 ± 3.81
30	11.62 ± 3.93	12.03 ± 3.52	4.08 ± 2.57	5.87 ± 3.65

measured the average radiance reflected from a white calibration target, i.e. a LabSphere Spectralon, illuminated by the light sources under consideration.

3.1 Illumination Spectrum Recovery

For our experiments on illumination spectra recovery, we compare the performance of our method with that yielded by the method of Finlayson and Schaefer [13]. We choose the work in [13] since it is based on the dichromatic model and can be adapted from trichromatic imagery to multispectral data in a straightforward manner.

In our experiment, we run our illuminant spectrum estimation algorithm on 1, 2, 4, 5, 10, 20 and 30 most constrasting patches of size 16×16 pixels from each image. The accuracy of light spectrum recovery is measured as the Euclidean angle between the estimated and ground truth spectrum in n dimensions, where n is the number of sampled bands for both spectra. Table 1 shows the mean and standard deviations of this angle, in degrees, over all images in the database versus the number of selected patches, in both the visible and infrared spectrum.

As shown in Table 1, our algorithm consistently achieves a higher accuracy than the alternative method. This is across all the number of patches for both spectral ranges. In general, the reported results match the expectation that both methods gain better accuracy with an increasing number of patches. However, it is noticeable that with a single patch, our method significantly outperforms the alternative method. This observation is consistent with the well-known fact that the dichromatic plane method and its variants require at least two homogeneous surfaces to compute the intersection between the dichromatic planes. Moreover, the deviation angle of the estimated illuminant decreases slowly with the growing number of patches, which indicates that our method is stable and does not require many homogeneous surfaces to achieve a reasonable estimate.

To illustrate the statistics in Table 1, we plot in Figure 2 both the visible and infrared spectra of the ground truth and the illuminant spectra of a sample light estimated from the image of a human subject in the database. All of the spectra are normalised by the highest spectral power. The ground truth spectra, the one estimated by our method and that recovered by Finlayson and Schaefer's algorithm, are drawn in red, blue and green, respectively. This illustrates the robustness our method, even when a single patch is used. Note that, this contrasts with the results delivered by the alternative method.

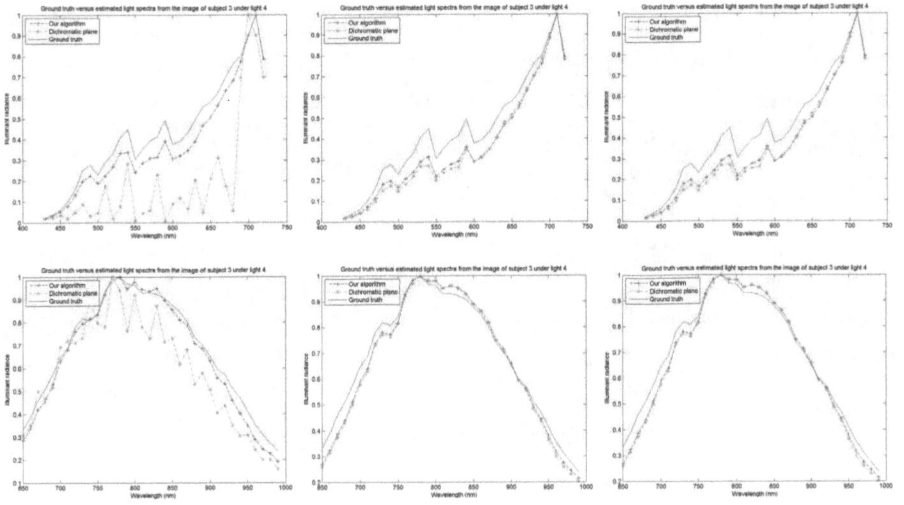

Fig. 2. Ground truth illuminant spectra and that delivered by our method and the alternative as estimated from the image of one subject in the database for both, the visible (top row) and near-infrared ranges (bottom row). From left to right: the spectra estimated from 1, 5 and 10 patches in the image.

3.2 Skin Recognition

We now turn our attention to the illumination invariance of the spectral reflectance recovered by our algorithm and its applications to recognition tasks. Specifically, we focus on using the surface reflectance extracted from an image for skin segmentation. This task can be viewed as a classification problem where the skin and non skin spectra comprise positive and negative classes. To perform this task, we select skin and non skin regions from the image of a subject taken under one of the illumination conditions in the data set and store the normalised reflectance spectra of these pixels as training data. Subsequently, we train a Support Vector Machine (SVM) classifier with a polynomial kernel of degree 2 on the training set. The resulting SVM model is applied for classifying skin versus non skin pixels in the remaining images of the subject, which have been acquired under diverse illuminant spectral conditions. In this manner, we can assert how robust are both, the illuminant spectrum recovered by our algorithm at training time and the normalised reflectance yielded by the method for skin recognition.

In our experiments, we also compare the results yielded by the SVM trained on the normalised reflectance spectra delivered by our method with those recovered using alternative recognition methods. To this end, we have used raw radiance spectra and principal components resulting from Linear Discriminant Analysis (LDA) projection of the original radiance spectra. The raw spectra and the LDA components are then used as input to SVMs whose parameters have been recovered using cross validation.

The top row of Figure 3 contains the training images for a number of sample subjects. Each training image shows the pixel regions from which skin and non-skin training data are collected. These have all been illuminated by the light source placed in a

Fig. 3. Skin segmentation results. Top row: sample training images for skin recognition; Second row: Skin probability map yielded by the recovered spectral reflectance on the image of the respective subject taken under another lighting condition; Third row: Skin probability map recovered making use of the normalised raw radiance data; Fourth row: Skin probability map recovered making use of 20 top principal components resulted from LDA projection on raw radiance data.

high-oblique position in front of the subject. For purposes of training, in all cases, we have employed 5 skin and 18 non-skin rectangular regions, of sizes 25×17 and 24×16, respectively. The second row shows the skin probability maps yielded by our method on images of the same subjects under a different illuminant in the dataset. The third and fourth rows, respectively, show the skin probability map for the input image used in the second row as recovered by making use of the normalised raw radiance spectra and the 20 top principal components obtained from LDA as the classification feature vectors. In these skin maps, lighter pixels are classified as more likely to be skin.

Figure 3 demonstrates that the recovered skin reflectance spectra are invariant to illuminant variations and light source direction changes. Indeed, the skin segmentation maps in the second row show a substantial proportion of skin pixels correctly classified. In addition, non-skin face details such as eye brows and mouth are correctly distinguished from skin. On the contrary, using normalised raw radiance data and LDA components for skin recognition, as shown in the third and fourth rows, causes much more false negatives in skin areas and false positives in other materials. The poor classification results in these cases can be explained by the variation induced by the illuminant spectrum and the surface geometry.

4 Conclusions

This paper has discussed an optimisation approach to the photometric invariance problem from a single multispectral image. The recovery process is based on optimising a cost function that imposes the smoothness constraint on uniform dichromatic reflectance surfaces. The method serves several purposes, including scene illuminant estimation, surface geometry recovery and illumination-invariant reflectance extraction. Firstly, the resulting illuminant spectra are showned to outperform the competitive method making use of the dichromatic plane hypothesis [13]. This is particularly evident when the number of homogeneous surface patches is limited. Secondly, we have shown how the invariance of the reflectance spectra recovered by our algorithm can be used for purposes of skin recognition.

References

1. Nayar, S., Bolle, R.: Reflectance based object recognition. International Journal of Computer Vision 17(3), 219–240 (1996)
2. Nayar, S.K., Bolle, R.M.: Computing reflectance ratios from an image. Pattern Recognition 26, 1529–1542 (1993)
3. Dror, R.O., Adelson, E.H., Willsky, A.S.: Recognition of surface reflectance properties from a single image under unknown real-world illumination. In: Proc. of the IEEE Workshop on Identifying Objects Across Variations in Lighting (2001)
4. Horn, B.K.P., Brooks, M.J.: The variational approach to shape from shading. CVGIP 33(2), 174–208 (1986)
5. Chang, J.Y., Lee, K.M., Lee, S.U.: Shape from shading using graph cuts. In: Proc. of the Int. Conf. on Image Processing, vol. I, pp. 421–424 (2003)
6. Brelstaff, G., Blake, A.: Detecting specular reflection using lambertian constraints. In: Int. Conf. on Comp. Vision, pp. 297–302 (1988)
7. Ragheb, H., Hancock, E.R.: A probabilistic framework for specular shape-from-shading. Pattern Recognition 36(2), 407–427 (2003)
8. Lin, S., Shum, H.: Separation of diffuse and specular reflection in color images. In: IEEE Conf. on Computer Vision and Pattern Recognition, vol. I, pp. 341–346 (2001)
9. Klinker, G., Shafer, S., Kanade, T.: A physical approach to color image understanding. Intl. Journal of Computer Vision 4(1), 7–38 (1990)
10. Suen, P.H., Healey, G.: Invariant mixture recognition in hyperspectral images. In: Int. Conference on Computer Vision, pp. 262–267 (2001)
11. Angelopoulou, E.: Objective colour from multispectral imaging. In: European Conf. on Computer Vision, pp. 359–374 (2000)
12. Shafer, S.A.: Using color to separate reflection components. Color Research and Applications 10(4), 210–218 (1985)
13. Finlayson, G.D., Schaefer, G.: Convex and Non-convex Illuminant Constraints for Dichromatic Colour Constancy. CVPR 1, 598–604 (2001)
14. Tominaga, S., Wandell, B.A.: Standard surface-reflectance model and illuminant estimation. Journal of the Optical Society of America A 6, 576–584 (1989)
15. Marr, D., Hildreth, E.: Theory of edge detection. In: Proceedings of Royal Society, London, vol. B–207, pp. 187–217 (1980)
16. Canny, J.: A computational approach to edge detection. IEEE Trans. Pattern Anal. Mach. Intell. 8(6), 679–698 (1986)
17. Klinker, G.J., Shafer, S.A., Kanade, T.: The measurement of highlights in color images. International Journal of Computer Vision 2, 7–32 (1988)
18. Lee, H.C.: Method for computing the scene-illuminant chromaticity from specular highlights. Journal of the Optical Society of America A 3, 1694–1699 (1986)
19. Marquardt, D.W.: An algorithm for least-squares estimation of nonlinear parameters. SIAM Journal on Applied Mathematics 11(2), 431–441 (1963)

Ensemble Combination for Solving the Parameter Selection Problem in Image Segmentation

Pakaket Wattuya and Xiaoyi Jiang

Department of Mathematics and Computer Science
University of Münster, Germany
{wattuya,xjiang}@math.uni-muenster.de

Abstract. Unsupervised image segmentation is of essential relevance for many computer vision applications and remains a difficult task despite of decades of intensive research. In particular, the parameter selection problem has not received the due attention in the past. Researchers typically claim to have empirically fixed the parameter values or train in advance based on manual ground truth. These approaches are not optimal and lack an adaptive behavior in dealing with a particular image. In this work we adopt the ensemble combination principle to solve the parameter selection problem in image segmentation. It explores the parameter space without the need of ground truth. The experimental results including a comparison with ground truth based training demonstrate the effectiveness of our framework.

Keywords: Image segmentation, parameter selection, ensemble combination concept, multiple segmentation combination.

1 Introduction

Segmentation algorithms mostly have some parameters and their optimal setting is not trivial. One fundamental problem is in fact to find suitable parameter values, preferably on a per-image basis. This need can be illustrated by the two pairs of images shown in Figure 1 and Figure 2. Each pair is segmented using the FH algorithm [3] based on exactly the same parameter set (see Section 3.3 for the meaning of these parameters). However, while Figure 1(a) and 2(a) show a nearly perfect segmentation, we obtain a very bad segmentation in Figure 1(b) and 2(b). Note that in this work we use the average normalized mutual information (ANMI) [9] (to be defined later) to quantitatively evaluate the quality of a segmentation against corresponding human ground truth segmentations.

Despite of its importance, the parameter selection problem has not received the due attention in the past. Researchers typically claim to have empirically determined the parameter values (in an ad-hoc manner). More systematically, the optimal parameter values can be trained in advance based on manual ground truth. A subspace of the parameter space is explored to find out the best parameter setting (with the largest average performance measure). Since fully exploring

N. da Vitora Lobo et al. (Eds.): SSPR&SPR 2008, LNCS 5342, pp. 392–401, 2008.

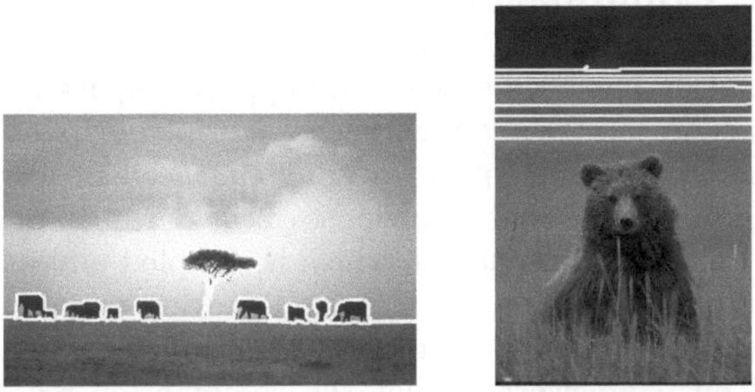

(a) ANMI = 0.8435 (b) ANMI = 0.5731

Fig. 1. Segmentations produced by the FH algorithm [3] given the same set of parameter values ($\sigma = 0.7$, $k = 700$, $min = 1500$)

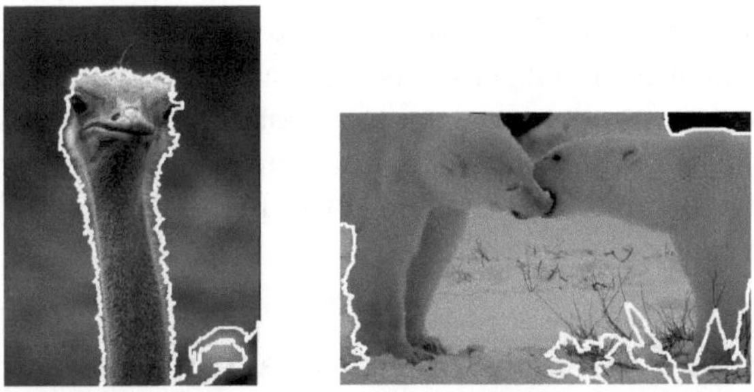

(a) ANMI = 0.6501 (b) ANMI = 0.1924

Fig. 2. Segmentations produced by the FH algorithm [3] given the same set of parameter values ($\sigma = 0.5$, $k = 700$, $min = 1500$)

the subspace can be very costly, space subsampling [8] or genetic search [2] has been proposed. While this approach is reasonable and has been successfully practiced in several applications, its fundamental disadvantage is the assumption of ground truth segmentation. The manual generation of ground truth is always painful and thus a main barrier of wide use in many situations. Another class of methods assumes a segmentation quality measure, which is used to control a parameter optimization process [1].

In this work we propose a novel framework of parameter handling based on ensemble combination. No ground truth is assumed in our framework. The fundamental idea is not to specify a particular parameter set. Instead, we compute a set of segmentations according to a subspace sampling of the parameter space and then to combine them into a final segmentation. This framework will be detailed in the next section. Then, we report the results of a case study based on a particular segmentation combination method in Section 3. Afterwards, comparison experiments are described in Section 4 to demonstrate that our framework even outperforms the ground truth based training approach. Finally, some discussions conclude this paper.

2 Ensemble Combination Framework

Ensemble combination has been successfully applied to pattern classification [7] and prototype learning [5]. We adopt this general principle to implicitly exploring the (segmentation) parameter space without the need of ground truth. It is assumed that we know a reasonable subspace of the parameter space (i.e. a lower and upper bound for each parameter), which is sampled into a finite number $\mathcal{N}$ of parameter settings. Then, we run the segmentation procedure for all the $\mathcal{N}$ parameter settings and compute a combined result of the $\mathcal{N}$ segmentations. The rationale behind our framework is that the combination result tends to be a good segmentation within the explored parameter subspace.

In this framework we do not explicitly determine the optimal parameter setting for a particular image. Instead, we try to reach an optimum out of the segmentation ensemble. One possibility is to compute an average, or more formally generalized median [5]. To some extent our combination method presented in the next section follows this line. The experimental results reported there show that the combined segmentation outperforms the majority of the input segmentations and is in many cases even superior to the best input segmentation. Given the fact that the optimal parameter setting may substantially vary among different images, our framework intends to achieve the highly desired adaptive behavior in dealing with a particular image.

3 Case Study: Multiple Segmentation Combination

For this case study the multiple segmentation combination algorithm in [11] is applied, which is briefly described in the following. The basis of this algorithm is the random walker algorithm for image segmentation [4]: Given (i) a graph $G = (V, E, w)$, where a node v_i corresponds to a pixel p_i, an edge e_{ij} connects two neighboring pixels in 4-neighborhood and a weight w_{ij} indicates similarity between the neighboring pixels v_i and v_j and (ii) a small number of K pixels with user-defined labels (or seeds), the algorithm labels unseeded pixels by resolving the probability that a random walker starting from each unseeded pixel will first reach each of the K seed points. A final segmentation is derived by selecting for each pixel the most probable seed destination for the random walker.

In principle there exists a natural link of our problem at hand to the random walker based image segmentation. The input segmentations provide strong hints about where to *automatically* place some seeds. Given such seed regions, we are then faced with the same situation as image segmentation with manually specified seeds and can thus apply the random walker algorithm [4] to achieve a quality final segmentation.

3.1 Multiple Segmentation Combination Algorithm

To develop the multiple segmentation combination algorithm [11] based on the random walker we need three components: (i) Generating a graph to work with,

(ii) extracting seed regions, and (iii) computing a final combined segmentation result.

Component 1: Graph Generation

A weight w_{ij} indicates how probably the two pixels p_i and p_j belong to the same image region, similar to the co-association weight in [12]. Clearly, this can be guided by counting the number n_{ij} of initial segmentations, in which p_i and p_j share the same region label. Thus, we define the weight function as a Gaussian weighting:

$$w_{ij} = \exp\left(-\beta \cdot \left(1 - \frac{n_{ij}}{\mathcal{N}}\right)\right) \tag{1}$$

where β is only a single parameter of an algorithm.

Component 2: Seed Region Generation

We describe a two-step strategy to automatically generate seed points: (i) extracting candidate seed regions from G; (ii) grouping them to form final seed regions to be used in the combination step.

Candidate seed region extraction: We build a new graph G^* by preserving those edges with weight $w_{ij} = 1$ only (p_i and p_j have the same label in all $\mathcal{N}$ initial segmentations) and removing all other edges. This step basically retains those edges between two neighboring nodes (pixels) which are most likely to belong to the same region. Then, we detect all *connected subgraphs* in G^* and regard them as a set of initial seed regions which are further reduced in the next step.

Grouping candidate seed regions: The number of candidate seed regions from the last step is typically higher than the true number of regions in an input image. Thus, a further reduction is performed by iteratively selecting the two candidate seed regions with the highest similarity value and merging them to build one single (possibly spatially disconnected) candidate seed region. Then, the similarity values between a new merged region and all remaining candidate seed regions are recomputed. The similarity $s(C_i, C_j)$ between candidate seed regions C_i and C_j is defined as following:

$$s(C_i, C_j) = \overline{\{n_{ij}/\mathcal{N} \mid (p_i, p_j) \in C_i \times C_j\}}$$

where $\overline{A}$ denotes the average of the set A.

A strategy for determining the final number K of seeds is to generate a series of results with $K \in [K_{\min}, K_{\max}]$ seed regions. Then, each initial result is fed to the ensemble combination part of our algorithm (Component 3) to achieve a total of $K_{\max} - K_{\min} + 1$ combination segmentations. Finally, we select an optimal one as the final combination segmentation (Section 3.2).

Component 3: Segmentation Ensemble Combination

Given the graph G constructed from the initial segmentations and K seed regions, we apply the random walker algorithm [4] to compute the combination segmentation. The computation of random walker probabilities can be exactly performed without the simulation of random walks, but by solving a sparse, symmetric, positive-definite system of equations. Each unseeded pixel is then assigned a K-tuple vector, specifying the probability that a random walker starting from that pixel will first reach each of the K seed regions. A final segmentation is derived by assigning each pixel the label of the largest probability.

3.2 Optimization of *K*

The quality of combination results can be evaluated in terms of consistency with the input ensemble. A good combination should share as much information as possible with the given $\mathcal{N}$ segmentations [9]. We follow the concept of mutual information to quantify the statistical information shared between two segmentations. An optimality criterion is proposed to determine the optimal K.

The *normalized mutual information* $\phi^{(\mathrm{NMI})}$ between two segmentations S_a and S_b is defined by

$$\phi^{(\mathrm{NMI})}(S_a, S_b) = \frac{\displaystyle\sum_{h=1}^{|S_a|}\sum_{l=1}^{|S_b|} |R_{h,l}| \log \frac{n \cdot |R_{h,l}|}{|R_h| \cdot |R_l|}}{\sqrt{\displaystyle\sum_{h=1}^{|S_a|} |R_h| \log \frac{|R_h|}{n} \sum_{l=1}^{|S_b|} |R_l| \log \frac{|R_l|}{n}}}$$

where R_h and R_l are regions from S_a and S_b, respectively, $R_{h,l}$ denotes the common part of R_h and R_l, and n is the image size. The value domain of $\phi^{(\mathrm{NMI})}$ is $[0, 1]$. Then, an information-share measure between a *single* segmentation $\hat{S}$ and a *set* $\Lambda = \{S_1, \ldots, S_N\}$ of $\mathcal{N}$ input segmentations can be defined as the average normalized mutual information (ANMI):

$$\phi^{(\mathrm{ANMI})}(\hat{S}, \Lambda) = \frac{1}{\mathcal{N}} \sum_{q=1}^{\mathcal{N}} \phi^{(\mathrm{NMI})}(\hat{S}, S_q)$$

We propose to select the *optimal* combination segmentation $S^{k_{\mathrm{opt}}}$ as the one with maximal average mutual information:

$$S^{k_{\mathrm{opt}}} = \arg\max_{\hat{S}} \phi^{(\mathrm{ANMI})}(\hat{S}, \Lambda) \tag{2}$$

where $\hat{S}$ covers all possible segmentations with $K \in [K_{\min}, K_{\max}]$.

If we replace $\hat{S}$ by a universe $\mathcal{U}$ of all possible segmentations of an image, then $S^{k_{\mathrm{opt}}}$ would represent the optimal segmentation in accordance with the generalized median concept of the input ensemble [5]. Therefore, our approach can be regarded as an approximation of generalized median segmentation by investigating the subspace of $\mathcal{U}$ consisting of the combination segmentations for all possible $K \in [K_{\min}, K_{\max}]$ only.

3.3 Experimental Results

We tested our approach on 300 natural images of size 481×321 from the Berkeley segmentation dataset [6], each having multiple manual segmentations. We apply $\phi^{(\mathrm{NMI})}$ to quantitatively evaluate the segmentation quality against the ground truth. One segmentation result is compared to all manual segmentations and the ANMI is reported. Larger ANMI values indicate better combination results that share more information with the ground truth.

The graph-based image segmentation algorithm [3] was used as the baseline segmentation method. This decision was made because of its competitive segmentation performance and high computational efficiency. The algorithm has three parameters: smoothing parameter (σ), a threshold function (k), and a minimum component size (min_area). We obtained 24 segmentations of an image by varying the parameter values (fixing $min\_area = 1500$ and varying $\sigma =$ 0.4, 0.5, 0.6, 0.7, 0.8, 0.9 and $k =$ 150, 300, 500, 700).

Figure 3 shows seven examples of combined segmentations produced by our method: column (a) using the optimality criterion in Eq. (2) to automatically determine K. For comparison purpose we also show the input segmentation with the worst, median and the best ANMI values (column (b)-(d)). Generally, we can observe a substantial improvement of our combination compared to the median input segmentation. These results demonstrate that we can obtain an "average" segmentation which is superior to the - possibly vast - majority of the input ensemble. Figure 4(a) shows a per-image performance of the first 50 images in the dataset, compared with the worst/best/average inputs. The combination performance tends to reach the best input segmentation. In some cases the combined segmentation even outperforms the entire input ensemble. This case is confirmed by Figure 4(b), which shows a statistic $f(n)$, indicating the number of images among the 300 test images, for which the combination segmentation is worse than the n best input segmentations. Remarkably, the combination segmentation outperforms all 24 input segmentations in $f(0) = 78$ cases. For 70% (210) of all 300 test images, the goodness of our solution is beaten by at most 5 input segmentations only. This statistic is a clear sign of combination quality of our approach.

Another perspective is given in Figure 5, showing the average performance of all 300 images with regard to each of the 24 individual configurations (parameter settings). This implies that for all 24 parameter settings the combination approach always achieved improved results in average.

There is a single parameter of our combination algorithm, β in Eq. (1). A value of β affects the smoothness of contours in a combination result. A result with a lower β value is a smoother version of that with a higher value. A systematic study of the effect of β values over combination results was undertaken by running our algorithm with $\beta = [10, 30, 60, 90, 120]$ on 100 images. The average standard deviation of performance of combination results is 0.0050. We thus conclude that the influence of β is rather small and set β to 30 for all our experiments.

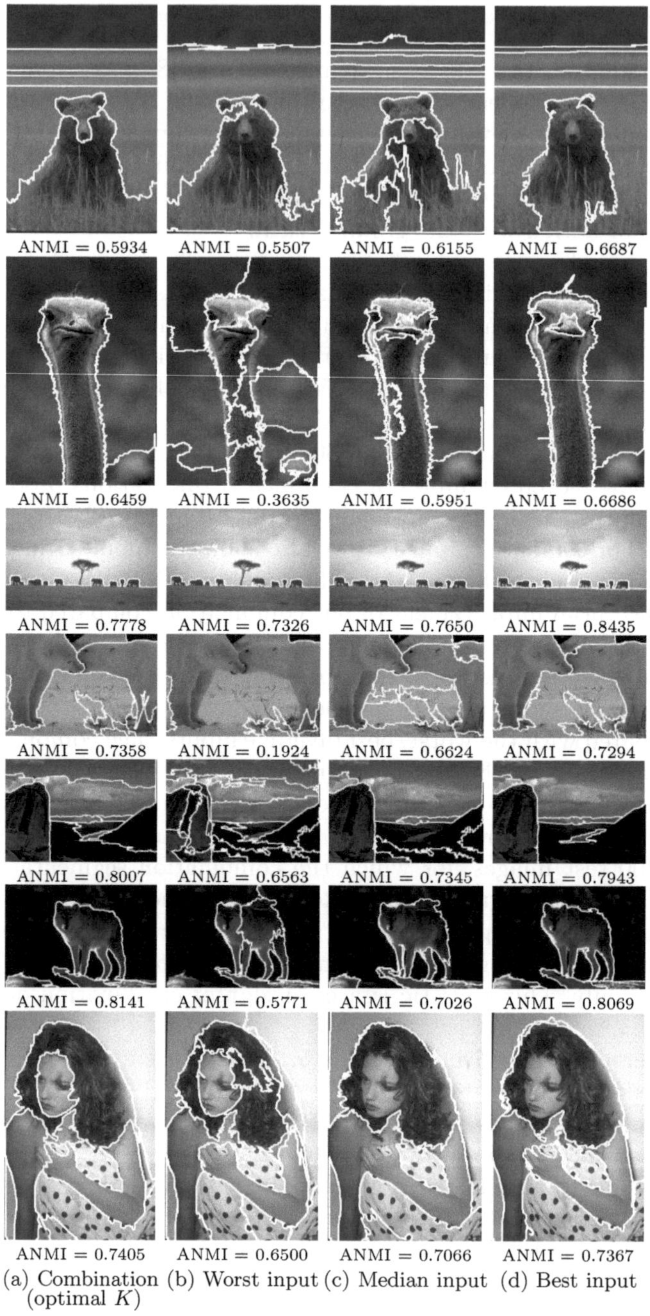

Fig. 3. Exploring parameter space: (a) Combined segmentation with optimal K; (b)-(d) show three input segmentations with the worst, median and the best ANMI scores, respectively

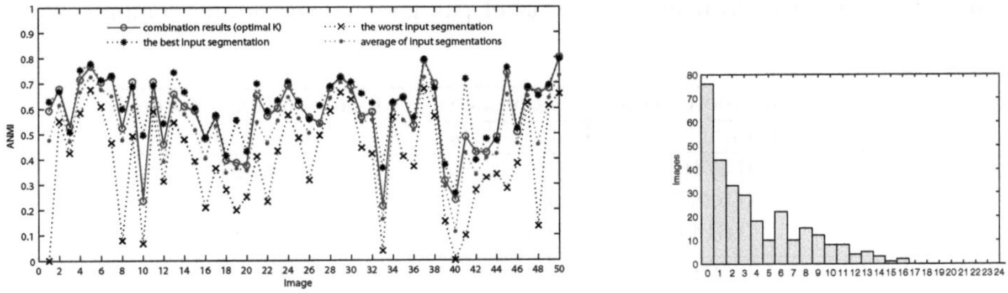

(a) (b)

Fig. 4. Comparison (per image): Worst/best/average inputs & combination result. Combined using an optimality criterion (Eq. (2)). (b) $f(n)$: Number of images for which the combination result is worse than the best n input segmentations.

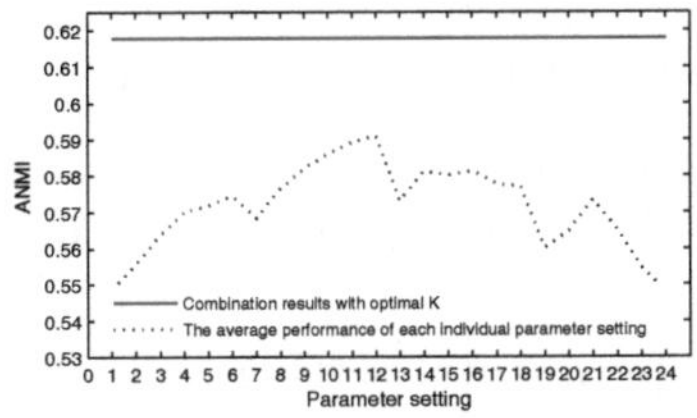

Fig. 5. Average performance of combined results over 300 images for each individual parameter setting

4 Combination Approach vs. Parameter Training

4.1 Automated Parameter Training

In recent years automated parameter training has become popular, mainly by probing a subspace of the parameter space by means of quantitatively comparing with a training image set with (manual) ground truth segmentation [2,8]. Assume that a reasonable parameter subspace is specified and sampled into a finite number $\mathcal{N}$ of parameter settings. For each parameter setting candidate a performance measure is computed in the following way:

- Segment each image of the training set based on the parameter setting;
- Compute a performance measure by comparing the segmentation result with the corresponding ground truth;
- Compute the average performance measure over all training images.

The optimal parameter setting is given by the one with the largest average performance measure.

4.2 Experimental Results

In this experiment series we investigate if our combination approach remains advantageous even if ground truth is available and the parameter training is

Table 1. Average performance measures of parameter training and combination approach on 3 test sets

Test set	Parameter training approach: Training data	Parameter training approach: Testing data	Combination approach (Optimal K)
1	0.5716	0.5936	0.6252
2	0.5887	0.5921	0.6208
3	0.6144	0.5793	0.6078
average	0.5916	0.5883	0.6179

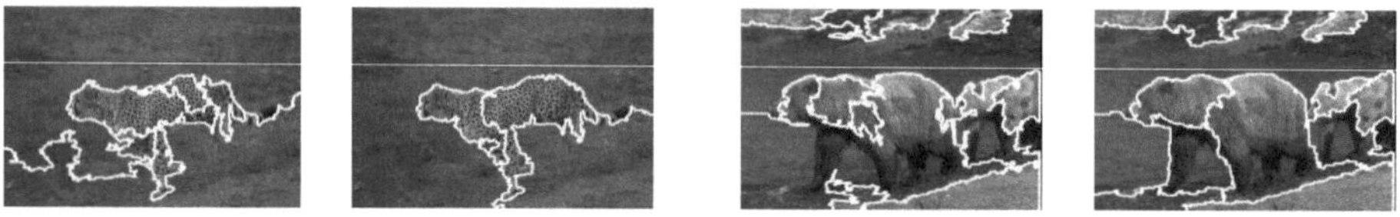

(1a) ANMI = 0.4244 (1b) ANMI = 0.4881 (2a) ANMI = 0.5038 (2b) ANMI = 0.5934

Fig. 6. Comparison of segmentation results produced by (a) trained parameters and (b) combination algorithm with optimality criterion in Eq. (2)

thus possible. The same experimental setting as in the combination approach is used (described in Section 3.3). The parameter subspace is sampled into 24 parameter settings. We apply a 3-fold cross validation in the training process described above. The dataset is randomly partitioned into 3 groups (100 images each). One group forms a 100-images training set while the rest two groups form a 200-images test set. By this way we have 3 different training sets with their corresponding test sets. The training procedure is then run 3 times on each training set to find its optimal parameter setting among the 24 parameter setting candidates.

The average performance measure over 100 images of each training set and 200 images of each test set are listed in the second and third column of Table 1, respectively. The forth column of Table 1 shows the average performance of combination approach on each test set. It clearly demonstrates that the combination approach is even superior to automated parameter training. Firstly, the combination approach needs no ground truth. Secondly, even in case of ground truth, the combination approach is able to produce segmentations (on test data) with higher average performance than those of the training approach. This is an indication that the trained parameters based on manual ground truth lack an adaptive ability for dealing with the variation of an input image. Figure 6 shows comparison of two examples of segmentation results produced by (a) trained parameters and (b) combination algorithm with optimality criterion in Eq. (2).

5 Conclusion

In this work we have taken a step towards solving the parameter selection problem in image segmentation. Since empirically fixing the parameter values or training in advance based on manual ground truth are not optimal and lack of adaptive behavior for dealing with the problem in a more general context, we have proposed to apply the concept of ensemble combination for exploring the (segmentation) parameter space without the need of ground truth. We verified our framework in a case study of segmentation combination. The experimental results confirm our expectation. Without using any ground truth information, our technique is able to produce segmentations with higher average quality than the training approach.

The focus of our current work is region-based image segmentation. It should be mentioned that our concept of ensemble combination is a general one. Given the demonstrated power we expect that it will be helpful towards solving the parameter selection problem in numerous other contexts. One such example is to explore the parameter space in a double contour detection problem [10]. We will consider further application scenarios in future.

References

1. Abdul-Karim, M.-A., Roysam, B., Dowell-Mesfin, N.M., Jeromin, A., Yuksel, M., Kalyanaraman, S.: Automatic selection of parameters for vessel/neurite segmentation algorithms. IEEE Trans. on Image Processing 14(9), 1338–1350 (2005)
2. Cingue, L., Cucciara, R., Levialdi, S., Martinez, S., Pignalberi, G.: Optimal range segmentation parameters through genetic algorithms. In: Proc. of 15th Int. Conf. on Pattern Recognition, Barcelona, vol. 1, pp. 474–477 (2000)
3. Felzenszwalb, P., Huttenlocher, D.: Efficient graph-based image segmentation. Int. Journal of Computer Vision, 167–181 (2004)
4. Grady, L.: Random walks for image segmentation. IEEE Trans. Patt. Anal. Mach. Intell. 28(11), 1768–1783 (2006)
5. Jiang, X., Münger, A., Bunke, H.: On median graphs: Properties, algorithms, and applications. IEEE Trans. Patt. Anal. Mach. Intell. 23(10), 1144–1151 (2001)
6. Martin, D., Fowlkes, C., Tal, D., Malik, J.: A database of human segmented natural images and its application to evaluating segmentation algorithms and measuring ecological statistics. In: Proc. 8th Int. Conf. Computer Vision, vol. 2, pp. 416–423 (2001)
7. Kuncheva, L.I.: Combining Pattern Classifiers. Wiley Interscience, Hoboken (2004)
8. Min, J., Powell, M., Bowyer, K.W.: Automated performance evaluation of range image segmentation algorithms. IEEE Trans. on SMC – Part B 34(1), 263–271 (2004)
9. Strehl, A., Ghosh, J.: Cluster ensembles – A knowledge reuse framework for combining multiple partitions. J. Machine Learning Research 3, 583–617 (2002)
10. Wattuya, P., Jiang, X.: A class of generalized median contour problem with exact solution. In: Yeung, D.-Y., Kwok, J.T., Fred, A., Roli, F., de Ridder, D. (eds.) SSPR 2006 and SPR 2006. LNCS, vol. 4109, pp. 109–117. Springer, Heidelberg (2006)
11. Wattuya, P., Jiang, X., Praßni, S., Rothaus, K.: A random walker based approach to combining multiple segmentations. In: Proc. of 19th Int. Conf. on Pattern Recognition, Florida (2008)
12. Fred, A., Jain, A.K.: Combining multiple clusterings using evidence accumulation. IEEE Trans. Patt. Anal. Mach. Intell. 27(6), 835–850 (2005)

Semantic Scene Classification for Image Annotation and Retrieval*

Özge Çavuş and Selim Aksoy

Department of Computer Engineering, Bilkent University, Ankara, 06800, Turkey
{cavus,saksoy}@cs.bilkent.edu.tr

Abstract. We describe an annotation and retrieval framework that uses a semantic image representation by contextual modeling of images using occurrence probabilities of concepts and objects. First, images are segmented into regions using clustering of color features and line structures. Next, each image is modeled using the histogram of the types of its regions, and Bayesian classifiers are used to obtain the occurrence probabilities of concepts and objects using these histograms. Given the observation that a single class with the highest probability is not sufficient to model image content in an unconstrained data set with a large number of semantically overlapping classes, we use the concept/object probabilities as a new representation, and perform retrieval in the semantic space for further improvement of the categorization accuracy. Experiments on the TRECVID and Corel data sets show good performance.

1 Introduction

Image annotation and content-based retrieval have been very active research areas with open problems due to the constant increase in the richness of the available image content. Contextual information plays a very important role for characterizing such content. A promising method for modeling context in images is scene classification because associating scenes with semantic labels has a high potential for providing a natural grouping of images instead of relying only on low-level features.

Scene classification and the related retrieval problems have two critical components: representing scenes and learning models for associating labels to these scenes. Given the difficulty of image segmentation in unconstrained data sets, most of the recent work use histograms of local features [1] or fixed grid layouts [2,3,4] for image representation but region-based models [5] can also be found. Intermediate semantic models that make use of the occurrence of common concepts, such as sky, water, grass, snow, have also been shown to improve the classification of natural scenes [4]. Although, recent work was mostly limited in terms of grid-based representations [2,3,4] or the small number of classes used [4], contextual modeling of image scenes using combinations of concepts and objects looks promising for decreasing the semantic gap.

* This work was supported in part by the TUBITAK Grant 104E077.

N. da Vitora Lobo et al. (Eds.): SSPR&SPR 2008, LNCS 5342, pp. 402–410, 2008.

In this paper, we describe an image annotation and retrieval framework that uses scene classification to detect high-level concepts for image indexing. First, images are segmented using clustering of color features and grouping of line structures (Sections 2 and 3). Next, each image is modeled using the histogram of quantized region types, and Bayesian classifiers are used to obtain the occurrence probabilities of high-level concepts and objects in images using these histograms (Section 4). An important observation is that a single class with the highest probability is not always sufficient to model image content in an unconstrained data set with semantically overlapping classes [2]. Therefore, we use the concept/object probabilities as a new representation, and perform retrieval in the semantic space for further improvement of the categorization accuracy (Section 5). Performance of the proposed models is evaluated using the TRECVID and Corel data sets.

2 Segmentation Using Color Information

Image segmentation is still an unsolved problem in computer vision. Although numerous algorithms have been shown to work well for images with only a few objects and a simple background, it seems impossible to find a fixed set of parameters that produces reasonable results in a large unconstrained data set. In this paper, we assume that a very precise segmentation of an image is not required for the scene classification and retrieval problem. Therefore, our goal is to obtain a rough estimate of important regions using unsupervised classification of color features and grouping of line segments.

For segmentation using color, we use the combined classifier approach in [6] because it fuses color and spatial information, and does not require the number of regions as an input parameter. First, an initial labeling of an image is done using k-means clustering of only the HSV values of pixels. Next, these pixel labels are used to train a multi-class nearest mean classifier on the HSV color features and a Parzen window classifier with a Gaussian kernel using the pixel positions as spatial features. The nearest mean classifier is selected for its simplicity and the Parzen window classifier is selected for its nonparametric nature for modeling a distribution with an indefinite number of modes (each mode corresponds to a segment). Then, the posterior probability outputs of each classifier for each pixel are combined using the product rule, and the pixels are assigned to the class with the largest probability. A new pair of classifiers are trained using these new pixel labels and the iterations continue until the pixel labels stabilize. Note that the number of clusters in the initial k-means clustering does not directly correspond to the number of segments, and can be empirically estimated using the number of dominant colors that can be found in the images in the selected data set.

Figure 1 shows example segmentations. The regions that are smaller than an area threshold are removed from the final segmentation where the results contain only contiguous sets of pixels that have a relatively uniform color distribution and are large enough.

Fig. 1. Segmentation using color. Top row: color images; bottom row: segmentation results in pseudocolor (pixels marked as white are not segmented).

3 Segmentation Using Line Structure

Not all objects/regions of interest can be characterized using uniformity of colors. Li and Shapiro [7] showed that features of line segments can be exploited for building recognition. We also use properties of nearby line segments to extract regions.

A line segment is found either at the boundary of two different regions or within a highly textured region. We expect that nearby line segments that have a similar color distribution around them may belong to the same object. Given the line segments in an image, they are grouped using average linkage hierarchical clustering according to the average color values on opposite sides of each segment. The average link criterion is used because we want all line segments that are selected as belonging to the same object to have similar color values.

The resulting clusters can contain line segments that have similar color pairs but belong to different objects. Therefore, a second level of clustering is performed to select the neighboring ones. This is done separately for each cluster using single linkage hierarchical clustering according to the distances between the end points of line segments. The single link criterion is used because we want to merge two groupings of line segments only by considering their nearest pair of end points. In both clustering steps, the number of clusters is automatically determined from the dendrogram [8].

Li and Shapiro [7] also required the line segments to have similar orientations in their building recognition algorithm. We do not consider orientation here because we observed that clustering of line segments can also detect highly textured regions such as trees where the tree branches have an almost random orientation. However, we perform a final post-processing step to eliminate the clusters with a very small number or a very sparse spatial distribution of line segments where the spatial coverage of a line cluster is computed as the ratio

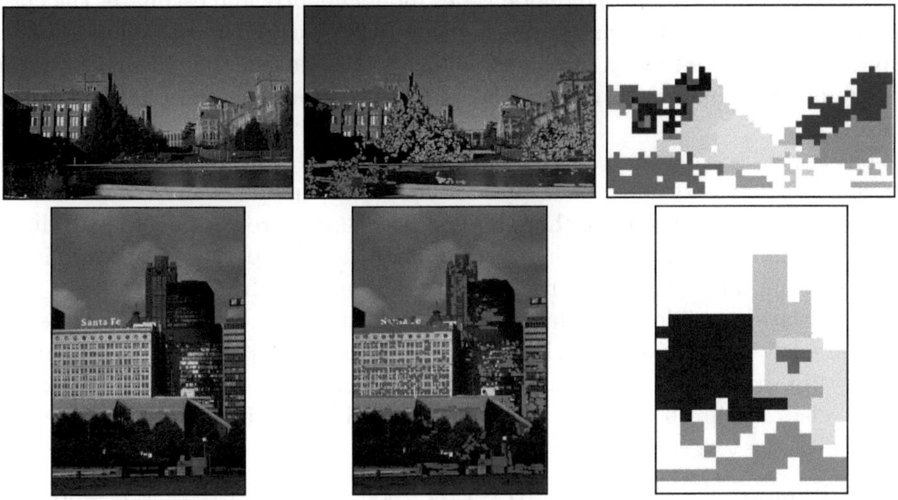

Fig. 2. Segmentation using line structure. First column: color images; second column: line clusters in pseudocolor; third column: final regions (pixels marked as white are not segmented).

of the area of the convex hull formed by the corresponding line segments to the number of line segments.

Finally, the resulting line segment clusters are converted to region representations by partitioning an image into non-overlapping grid cells and labeling each grid cell with the label of the cluster whose lines most frequently intersect with the cell. Figure 2 shows example segmentations.

4 Scene Classification

In this paper, a scene's content is represented as a collection of its regions. The regions that are extracted using pixel-based color information (Section 2) are modeled using their HSV histograms with 8 bins used for the H channel and 3 bins for each of S and V channels. Then, the k-means algorithm is used to create a codebook of k_1 region types. The regions that are extracted using line structure information (Section 3) are modeled using a 10-bin histogram of the orientation of their line segments. Similarly, the k-means algorithm is used to create a codebook of k_2 region types for the whole data set. After each region is assigned a type label, as the final representation, each scene is modeled using the histogram of $k_1 + k_2$ region types.

We use a Bayesian framework for scene classification and investigate two different settings for probability estimation. In both settings, the goal is to estimate the posterior probabilities $p(w_j|\mathbf{x})$, $j = 1, \ldots, c$, where w_j represents the j'th class, c is the number of classes, and $\mathbf{x}$ is the histogram of the quantized region types.

The first setting treats the region types independently and estimates the class-conditional probabilities using the multinomial model. The parameters of the model are computed using maximum likelihood estimates that involve counting the number of times each region type is observed for each class in the training data [9].

The second setting treats each class separately and trains a one-class Gaussian classifier using the training examples for each class independently. The one-class setting is particularly suitable here because the classes are not mutually exclusive and the commonly used two-class (target vs. others) approach is not applicable because sampling a sufficient number of training data from the "others" class is not always possible. One-class classifiers model only the "target" class and assume a low uniform distribution for the "others" class [10]. After a probability density (in this case, a Gaussian) is estimated using the training examples of the target class, a threshold is set on the tails of this distribution and a specified amount of the target data is rejected.

Experiments: We used the TRECVID 2005 and Corel data sets and their ground truth to evaluate the algorithms proposed in this paper. The TRECVID data set contains 24517 video shots in 18 classes and the Corel data set contains 4999 images in 21 classes. Two thirds of the images were used for training and the remaining one third for testing. Both k_1 and k_2 were set as 1000.

The Bayesian classifier assigns each image to the class with the highest posterior. Table 1 shows the confusion matrices when the multinomial model was used. The matrices show that the error rates were rather high (12.83% accuracy for the TRECVID data set and 34.76% accuracy for the COREL data set) but most of the misclassifications occurred among the classes that have a semantic overlap (e.g., boat–water, bus–road, sky–sunset). The confusions for the one-class Gaussian model were worse as most of the images were assigned to the outdoor class for the TRECVID data set and to the vegetation class for the Corel data set when the highest posterior was used. We also ran the bag-of-words model with probabilistic latent semantic analysis [1] on the Corel data set and obtained 19.58% accuracy (compared to 34.76% by the multinomial model in Table 1(b)).

An important observation was that, when all probabilities were considered, the image content can be modeled by multiple classes instead of a single class that has the highest probability. Figure 3 shows the posterior probability values for example images. These examples also confirm that the class models and the resulting probabilities are not necessarily wrong but a single class assignment is not appropriate when an unconstrained data set with a diverse set of classes is considered [2].

5 Image Annotation and Retrieval

The classification results show that an image cannot always be assigned to a single class if the categories are not mutually exclusive. Furthermore, different users may have different requirements where one may be interested in urban scenes but another may specifically look for vehicles. In image annotation, a

Table 1. Confusion matrices for Bayesian classification using the multinomial model

(a) TRECVID

	Airplane	Boat/shi	Building	Bus	Car	Desert	Fire	Mountai	Disaster	Outdoor	Road	Sky	Snow	Sports	Truck	Urban	Vegetati	Watersc
Airplane	0.19	0.07	0.07	0.04	0.00	0.04	0.07	0.07	0.00	0.00	0.04	0.04	0.00	0.04	0.04	0.00	0.00	0.30
Boat/ship	0.00	0.07	0.07	0.00	0.04	0.00	0.04	0.00	0.00	0.00	0.07	0.07	0.07	0.07	0.00	0.00	0.04	0.44
Building	0.01	0.01	0.14	0.02	0.09	0.05	0.04	0.02	0.02	0.08	0.07	0.08	0.02	0.04	0.03	0.12	0.09	0.06
Bus	0.00	0.00	0.27	0.00	0.07	0.00	0.00	0.07	0.00	0.00	0.27	0.07	0.00	0.00	0.07	0.00	0.20	0.00
Car	0.01	0.02	0.06	0.00	0.20	0.02	0.03	0.00	0.03	0.06	0.12	0.04	0.01	0.05	0.04	0.12	0.11	0.08
Desert	0.01	0.01	0.04	0.00	0.03	0.31	0.12	0.01	0.03	0.03	0.09	0.06	0.03	0.07	0.03	0.04	0.03	0.04
Fire	0.01	0.01	0.10	0.01	0.06	0.16	0.22	0.00	0.01	0.04	0.07	0.07	0.00	0.00	0.01	0.03	0.03	0.14
Mountain	0.02	0.07	0.04	0.02	0.00	0.04	0.02	0.22	0.02	0.07	0.17	0.02	0.02	0.00	0.00	0.02	0.04	0.20
Disaster	0.00	0.00	0.17	0.00	0.13	0.04	0.04	0.04	0.00	0.00	0.13	0.04	0.00	0.04	0.00	0.17	0.13	0.08
Outdoor	0.02	0.02	0.09	0.02	0.07	0.05	0.04	0.02	0.02	0.09	0.07	0.05	0.02	0.08	0.03	0.12	0.12	0.08
Road	0.01	0.01	0.10	0.03	0.14	0.05	0.05	0.01	0.03	0.05	0.09	0.04	0.04	0.06	0.03	0.08	0.09	0.08
Sky	0.03	0.03	0.08	0.01	0.06	0.08	0.05	0.04	0.02	0.05	0.08	0.10	0.03	0.04	0.03	0.07	0.10	0.10
Snow	0.04	0.00	0.07	0.00	0.04	0.07	0.07	0.11	0.00	0.04	0.04	0.00	0.29	0.00	0.07	0.00	0.04	0.14
Sports	0.01	0.02	0.04	0.03	0.06	0.02	0.01	0.02	0.00	0.06	0.06	0.01	0.03	0.38	0.03	0.06	0.14	0.06
Truck	0.00	0.00	0.08	0.02	0.06	0.10	0.06	0.04	0.00	0.04	0.18	0.04	0.04	0.00	0.00	0.10	0.08	0.14
Urban	0.01	0.01	0.12	0.02	0.09	0.04	0.04	0.02	0.01	0.10	0.09	0.06	0.01	0.06	0.03	0.13	0.09	0.07
Vegetation	0.01	0.02	0.09	0.02	0.07	0.04	0.03	0.03	0.01	0.07	0.07	0.04	0.02	0.11	0.03	0.08	0.20	0.06
Waterscape	0.03	0.07	0.05	0.01	0.03	0.04	0.02	0.02	0.01	0.03	0.05	0.05	0.05	0.02	0.04	0.05	0.05	0.38

(b) Corel

	Airplane	Boat/shi	Building	Bus	Car	Castle	Coast	Desert	Harbor	Mountai	Night	Road	Rock	Ruin	Sky	Snow	Sunset	Surfing	Train	Vegetati	Waterfa
Airplane	0.21	0.08	0.02	0.00	0.06	0.02	0.08	0.03	0.00	0.06	0.00	0.00	0.02	0.06	0.02	0.08	0.02	0.05	0.06	0.11	0.03
Boat/ship	0.02	0.38	0.04	0.01	0.01	0.02	0.12	0.01	0.02	0.00	0.01	0.00	0.00	0.02	0.04	0.08	0.01	0.11	0.08	0.00	0.01
Building	0.00	0.06	0.48	0.04	0.05	0.06	0.01	0.00	0.01	0.00	0.01	0.00	0.04	0.07	0.01	0.02	0.00	0.03	0.05	0.04	0.01
Bus	0.00	0.00	0.15	0.56	0.03	0.06	0.00	0.00	0.00	0.00	0.06	0.00	0.00	0.06	0.00	0.00	0.00	0.00	0.09	0.00	0.00
Car	0.00	0.03	0.05	0.01	0.38	0.07	0.03	0.02	0.00	0.01	0.01	0.00	0.02	0.02	0.04	0.00	0.14	0.02	0.04	0.10	0.00
Castle	0.00	0.03	0.08	0.00	0.02	0.20	0.03	0.02	0.03	0.07	0.01	0.00	0.08	0.16	0.00	0.04	0.01	0.01	0.06	0.14	0.00
Coast	0.01	0.15	0.02	0.00	0.00	0.03	0.29	0.02	0.01	0.08	0.00	0.01	0.02	0.05	0.02	0.05	0.03	0.11	0.04	0.04	0.01
Desert	0.02	0.02	0.05	0.02	0.03	0.00	0.03	0.17	0.02	0.03	0.02	0.03	0.25	0.05	0.05	0.03	0.03	0.03	0.02	0.11	0.02
Harbor	0.00	0.21	0.18	0.06	0.03	0.03	0.06	0.00	0.15	0.00	0.00	0.00	0.00	0.12	0.00	0.03	0.03	0.03	0.03	0.03	0.00
Mountain	0.01	0.02	0.01	0.00	0.00	0.03	0.13	0.00	0.00	0.32	0.00	0.02	0.06	0.03	0.01	0.09	0.00	0.12	0.03	0.07	0.04
Night	0.00	0.00	0.00	0.00	0.04	0.04	0.00	0.00	0.09	0.00	0.52	0.00	0.04	0.00	0.00	0.00	0.17	0.04	0.00	0.04	0.00
Road	0.00	0.00	0.00	0.00	0.00	0.09	0.12	0.15	0.00	0.00	0.00	0.03	0.09	0.00	0.00	0.09	0.03	0.00	0.12	0.26	0.03
Rock	0.00	0.03	0.00	0.00	0.01	0.03	0.01	0.10	0.00	0.05	0.03	0.00	0.47	0.09	0.01	0.02	0.01	0.02	0.00	0.10	0.01
Ruin	0.01	0.02	0.07	0.01	0.01	0.15	0.09	0.02	0.01	0.04	0.00	0.01	0.09	0.17	0.00	0.02	0.01	0.02	0.03	0.17	0.03
Sky	0.02	0.04	0.00	0.00	0.02	0.02	0.07	0.04	0.01	0.04	0.01	0.01	0.02	0.02	0.19	0.11	0.24	0.07	0.02	0.00	0.02
Snow	0.04	0.04	0.02	0.00	0.00	0.03	0.08	0.03	0.01	0.10	0.00	0.01	0.04	0.04	0.04	0.29	0.02	0.13	0.00	0.05	0.02
Sunset	0.00	0.00	0.02	0.00	0.00	0.00	0.02	0.05	0.03	0.02	0.03	0.00	0.05	0.00	0.13	0.00	0.62	0.02	0.02	0.00	0.00
Surfing	0.03	0.07	0.00	0.02	0.01	0.04	0.04	0.00	0.00	0.04	0.03	0.00	0.01	0.02	0.01	0.07	0.00	0.58	0.00	0.02	0.01
Train	0.00	0.11	0.09	0.05	0.01	0.12	0.03	0.02	0.01	0.00	0.01	0.02	0.01	0.07	0.00	0.02	0.00	0.02	0.30	0.09	0.01
Vegetation	0.01	0.00	0.00	0.00	0.01	0.08	0.02	0.01	0.00	0.01	0.00	0.00	0.08	0.08	0.00	0.04	0.00	0.02	0.05	0.58	0.01
Waterfall	0.00	0.03	0.00	0.00	0.00	0.03	0.03	0.00	0.00	0.09	0.00	0.00	0.06	0.03	0.00	0.21	0.00	0.03	0.00	0.21	0.27

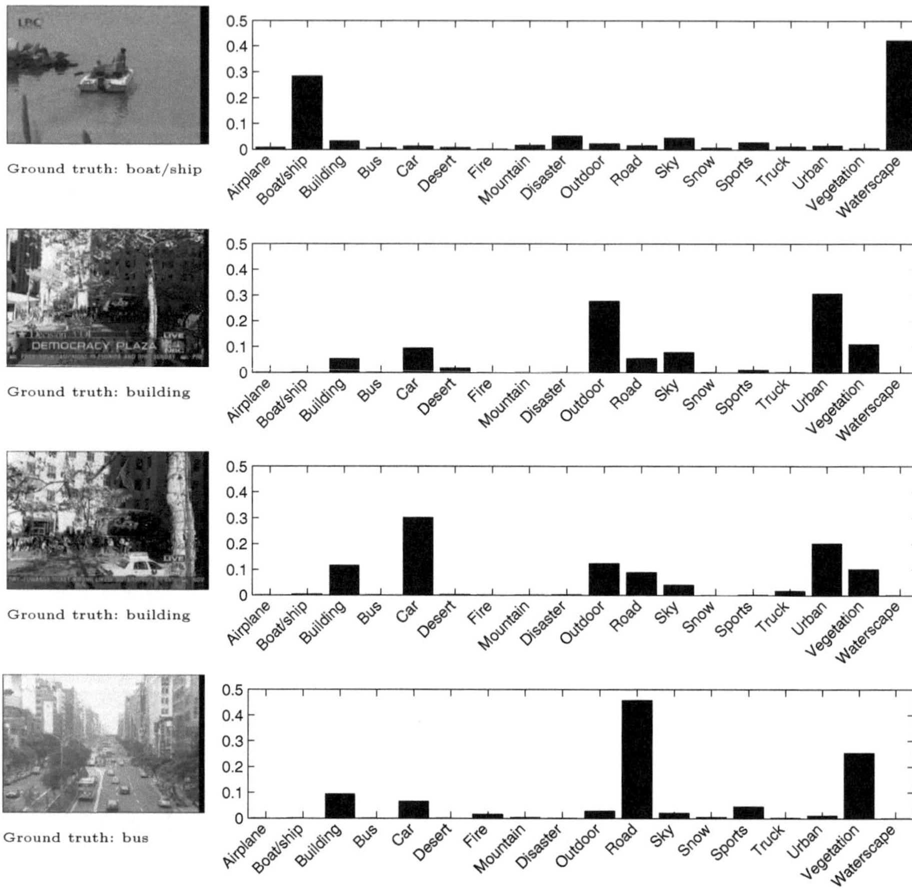

Fig. 3. Posterior probability examples. Although the ground truth class is not the one that received the highest posterior (i.e., counted as error in Table 1), the probability values are semantically correct

threshold is often applied to the posterior probabilities, and the concepts/classes with probabilities higher than this threshold are assigned to the image. To further improve the classification accuracy in an adaptive way regarding different user interests, we use the probabilities obtained from the Bayesian classifier as new features for each scene. This corresponds to a new representation in terms of a feature vector of length c for c classes.

This representation maps images to a semantic space where each component of the new feature vector corresponds to the probability of observing a particular concept/object in an image. Furthermore, the features complement each other and provide contextual information. For example, a car detector that uses low-level features may not be very reliable when used alone but the occurrence of a region that exhibits feature characteristics of a car becomes more relevant if that image also has a high probability of containing a road and having an

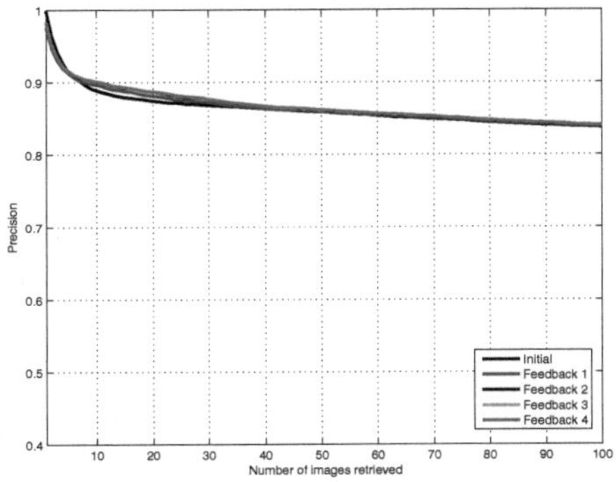

(a) TRECVID

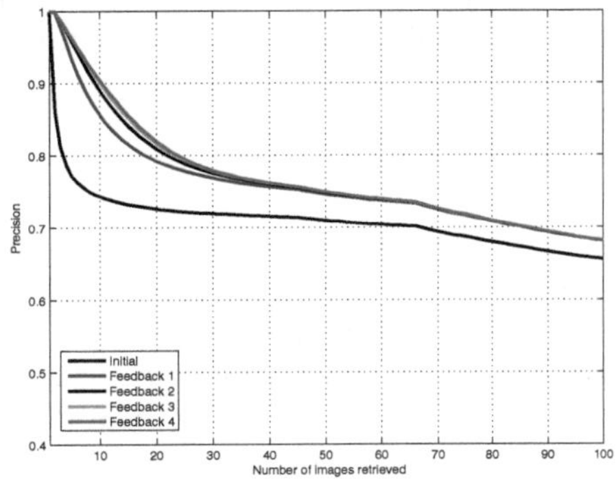

(b) Corel

Fig. 4. Precision vs. the number of images retrieved for the initial query and four feedback iterations

urban context. Therefore, this new representation compensates the limitations of individual object and scene detectors that use low-level features, and brings us one more step closer to bridging the semantic gap.

We use this representation in a retrieval framework. First, the user selects a category and the images that have a high probability of belonging to that category are retrieved. Next, the user selects one of these images as the query and an initial retrieval is performed.

The result of this query is a ranked list of the images in the database. If the user marks some of the images as relevant or irrelevant, this relevance feedback information can be used for further improvements. In this paper, we use the support vector data description (SVDD) model [11] as a one-class classifier. The one-class framework is intuitively applicable to the feedback problem because the images that are labeled as relevant provide a good sample for the target class of interest whereas the images that may be labeled as irrelevant often do not provide sufficient training data to learn a separate class. The SVDD model uses the relevant examples to learn a hypersphere that encloses the target class, and uses the irrelevant examples to minimize the volume of this sphere. Once the classifier is trained for a given iteration, the images can be re-ranked according to their distances to the resulting hypersphere. These results can also be used for annotating the query image using the most common concept/object labels among the images that fall into the hypersphere.

Experiments: The ground truth was used to automatically generate queries and provide feedback using the top 30 images by automatically labeling them as relevant or irrelevant at each iteration. Figure 4 shows the average precision results when the one-class Gaussian model was used for the Bayesian classifier. The first iteration gave the largest increase. Following iterations provided minor improvements. Although the classification accuracy was low when only the class with the highest posterior was considered, feedback iterations successfully converged to the true set of images by making use of the concept/object occurrence probabilities as semantic features.

6 Summary

We described an image annotation and retrieval framework that used scene classification for extracting semantic features for image representation. Scene classification was performed using Bayesian modeling of histograms of regions segmented using clustering of color features and line structures. Given the observation that multiple classes can describe the image content, a new semantic representation was constructed by contextual modeling of images using the occurrence probabilities of concepts and objects. The experiments showed that this new representation provided an increased precision in an unconstrained data set with a large number of semantically overlapping classes.

References

1. Quelhas, P., Monay, F., Odobez, J.M., Gatica-Perez, D., Tuytelaars, T.: A thousand words in a scene. IEEE Transactions on Pattern Analysis and Machine Intelligence 29(9), 1575–1589 (2007)
2. Boutell, M.R., Luo, J., Shen, X., Brown, C.M.: Learning multi-label scene classification. Pattern Recognition 37(9), 1757–1771 (2004)
3. van Gemert, J.C., Geusebroek, J., Veenman, C.J., Snoek, C.G.M., Smeulders, A.W.M.: Robust scene categorization by learning image statistics in context. In: CVPR (2006)
4. Vogel, J., Schiele, B.: Semantic modeling of natural scenes for content-based image retrieval. International Journal of Computer Vision 72(2), 133–157 (2007)
5. Li, Y., Shapiro, L.G., Bilmes, J.A.: A generative/discriminative learning algorithm for image classification. In: ICCV (2005)
6. Paclik, P., Duin, R.P.W., van Kempen, G.M.P., Kohlus, R.: Segmentation of multispectral images using the combined classifier approach. Image and Vision Computing 21(6), 473–482 (2003)
7. Li, Y., Shapiro, L.G.: Consistent line clusters for building recognition in CBIR. In: ICPR (2002)
8. Mojena, R.: Hierarchical grouping methods and stopping rules: An evaluation. The Computer Journal 20(4), 359–363 (1977)
9. Gokalp, D., Aksoy, S.: Scene classification using bag-of-regions representations. In: Proceedings of IEEE Conference on Computer Vision and Pattern Recognition, Beyond Patches Workshop, Minneapolis, Minnesota, June 23 (2007)
10. Tax, D.M.J.: One-Class Classification. PhD thesis, Delft University of Technology, Delft, The Netherlands (2001)
11. Tax, D.M.J., Duin, R.P.W.: Support vector data description. Machine Learning 54(1), 45–66 (2004)

Multi-angle View, Illumination and Cosmetic Facial Image Database (MaVIC) and Quantitative Analysis of Facial Appearance

Yen-Wei Chen[1], Takayuki Fukui[1], Xu Qiao[1], Takanori Igarashi[2], Keisuke Nakao[2], and Akio Kashimoto[2]

[1] Graduate School of Science and Engineering, Ritsumeikan University, Japan
[2] Beauty Cosmetic Research Lab, Kao Corporation, Japan
chen@is.ritsumei.ac.jp

Abstract. We constructed a multi-angle view, illumination and cosmetic facial image database (MaVIC) for quantitative characterization of facial appearance. The facial images in the database are captured from various angles of view and of illumination with a multi-angle image capturing system (MICS). In MaVIC, there are 170 Japanese female's natural facial images and 250 Japanese female's cosmetic facial images. Each subject was photographed in 13 different view-directions under 14 illuminations. Quantitative analysis has been done for the images in MaVIC by the use of principal component analysis (PCA).

Keywords: Multi-angle view, multi-angle illumination, cosmetic, facial image database, principal component analysis, opto-chromatic image.

1 Introduction

Quantitative characterization of facial appearance is an important issue in many fields and applications including cosmetic foundation design [1,2]. Two of the most important requirements for facial appearance studies are: (1) large database of facial images with multi-angle view, illumination and various types of facial appearance; and (2) a model describing the facial appearance with fewer parameters quantitatively.

To date, many face databases have been constructed for face recognition, face detection and face modeling, such as FERET [3], CMU PIE [4], ORL [5], MIT-CBCL[6], Yale[7], CAS-PEAL[8], AR[9], UMIST[10], Oulu[11], JAFFE[12] and so on. Especially, CMU PIE and CAS-PEAL are multi-angle view and illumination facial image databases. Unfortunately, since these face database are designed for face recognition and face detection, the image resolution is not enough for quantitative studies of facial appearance. There are also no any variations on facial appearance in CMU PIE or CAS-PEAL. Researchers at Rutgers University constructed a Skin Texture Database [13] with various variations of skin appearance, but it is not a facial image database.

We have constructed a facial image database for quantitative studies of facial appearance. We called this database the KAO-Ritsumeikan Multi-angle View, Illumination and Cosmetic Facial Image Database (MaVIC). The facial images in the

N. da Vitora Lobo et al. (Eds.): SSPR\&SPR 2008, LNCS 5342, pp. 411–420, 2008.

database are captured from various angles of view and of illumination with a multi-angle image capturing system (MICS) [14]. MICS is a system that consists of 20 CCD cameras and 40 flashes that are multi-directionally attached to a hemispherical gantry. In MaVIC, there are 170 Japanese female's natural facial images and 250 Japanese female's cosmetic facial images. Each subject was photographed in 13 different viewing directions under 14 illuminations. There are more than 76000 facial images in MaVIC. The unique features of our database are: (1) the multi-angles of view and of illumination are well controlled and multiple-view images under one illumination can be captured simultaneously in one shot; (2) the image resolution is very high (6M pixels) so that the image can also be used as skin texture image; and (3) various types of facial appearances of Japanese females in all ages, which are created by makeup artists, are available.

In order to characterize the opto-chromatic features of the facial appearance, each color facial image in MaVIC is transformed to L* image and C* image by the L*C*h transform, which represent the distribution mapping of brightness and chroma, respectively [15]. Principal Component Analysis (PCA) [16] is performed to L* images and C*images, respectively in order to find main components contributing to the facial appearance and the dependence on the viewing and illumination directions. Significant difference of PCA coefficients have been found for different facial appearances which are created by makeup artists such as "soft", "wet", "sharp" and "heavy".

The paper is organized as follows: the multi-angle image capture system is described in Section 2; the captured facial images and database MaVIC are described in Section 3; the quantitative analysis are presented in Section 4; finally the conclusion is given in Section 5.

2 Multi-angle Image Capture System

In order to capture facial images of a person from multi-angle, we built a multi-angle image capturing system (MICS)[14] in Beauty Cosmetic Research Lab, Kao Corporation. The setup of MICS is shown in Figure 1. MICS is a system that consists of 20 CCD cameras and 40 flashes that are multi-directionally attached to a hemispherical gantry with a radius of 1.5m. The subject sits in a chair with her head in the center. All cameras and flashes are pointed to the center. The relative locations of cameras and flashes are illustrated in Figure 2. The red one is a camera and the yellow one is a flash. The 20 CCD cameras are each separated by 15 degree. The 40 flashes are also each separated by 15 degree. The 13 cameras among 20 are set in a horizontal plane from full left to full frontal and on to full right. The horizontal plane is in equal height with the subject's head. The 7 other cameras include 3 above the center camera, 2 upper right and 2 upper left (see Figure 2). The images that the 13 cameras in the horizontal plane took from each position are shown in Figure 3. In our system, high resolution CCD cameras with 6M pixels are used. So the detailed skin texture can be imaged, which will be shown in Figure 4 later. All of the cameras and flashes are connected to and controlled by a computer. So we can capture the 20 images from different viewing angles simultaneously in one shot. The system is set in a darkroom resulting in a uniform dark background in the image.

Fig. 1. The setup of MICS

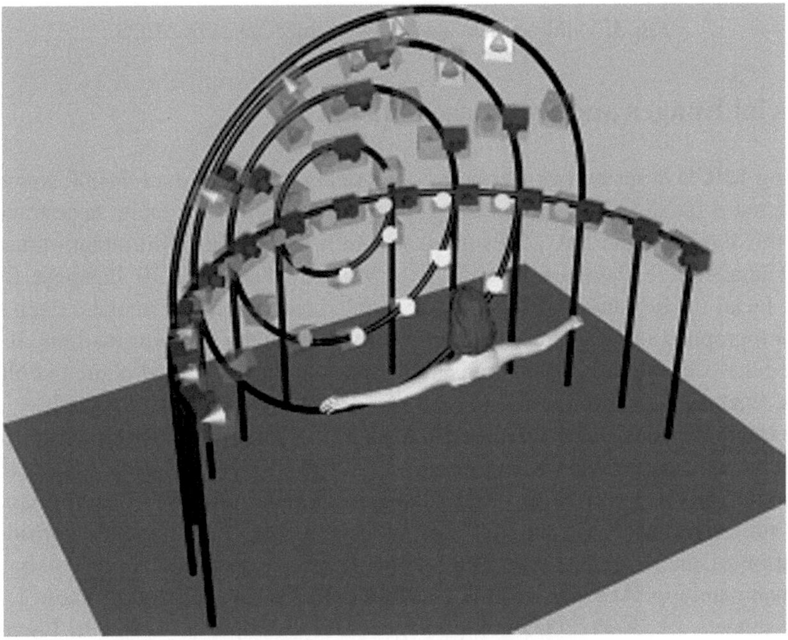

Fig. 2. Configuration of cameras and flashes

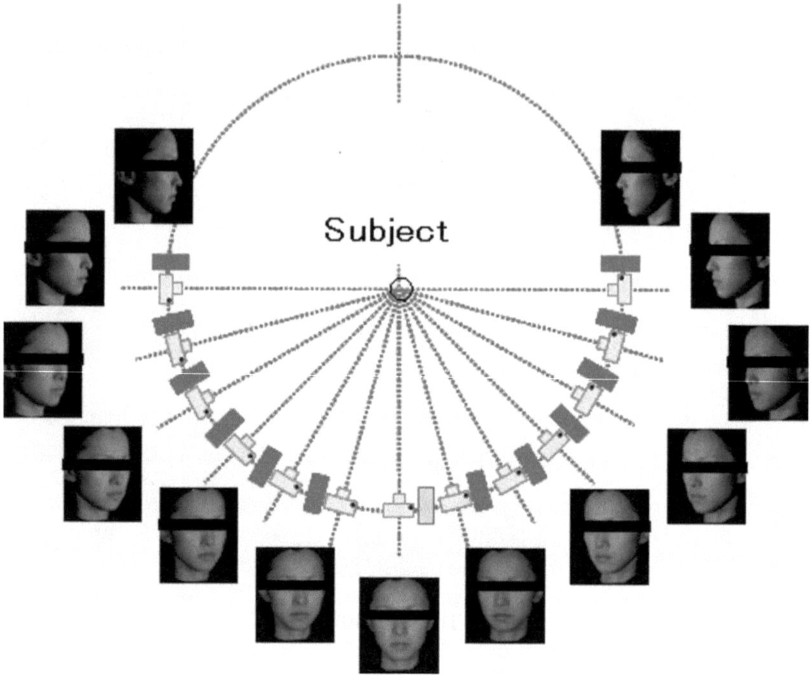

Fig. 3. Multi-angle images of one subject taken by MICS

3 Facial Images and Database: MaVIC

By using MICS described in Section 2, we collected more than 76000 images and constructed a facial image database for quantitative studies of facial appearance. We called this database the KAO-Ritsumeikan Multi-angle View, Illumination and Cosmetic Facial Image Database (MaVIC). In MaVIC, there are 170 Japanese female's natural facial images and 250 Japanese female's cosmetic facial images. Each subject was photographed in 13 different viewing directions under 14 illuminations. It should be noted that we only use the cameras and flashes in the horizontal plane to collect the images. Each subject is requested to shut eyes, sit straight and turn forward.

The database consists of three parts: MaVIC1 consisting of 70 Japanese female's natural facial images; MaVIC2 consisting of 100 Japanese female's natural facial images and MaVIC3 consisting of 250 Japanese female's cosmetic facial images. 100 of 250 subjects in MaVIC3 are the same subjects as MaVIC2. It can be considered as a variation of facial appearance. This is one of unique features of our database. The relationship among MaVIC1, MaVIC2 and MaVIC3 is summarized in Table 1.

The images are 3008x2000 color images. One example is shown in Figure 4(a). Each image is first geometrically normalized. The image is translated so that the eyes are positioned in line. Then the face image is cropped to include only the face with little hair and background. The size of the cropped face image is 800x800, which is

Table 1. Contents of MaVIC

Subjects	MaVIC1 (natural)	MaVIC2 (natural)	MaVIC3 (cosmetic)
70	○		
100		○	○
150			○

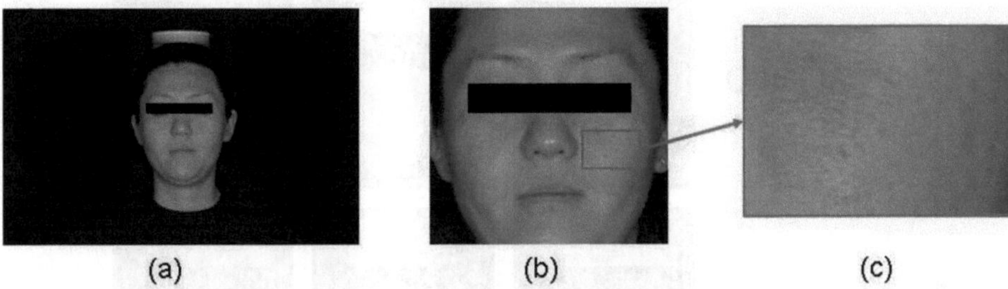

Fig. 4. (a) Original Image; (b) cropped face image; (c) skin texture image

The examples of multi-angle view and illumination facial images are shown in Figure 5 and four types (heavy, sharp, soft and wet) of cosmetic facial images created by makeup artists are shown in Figure 6.

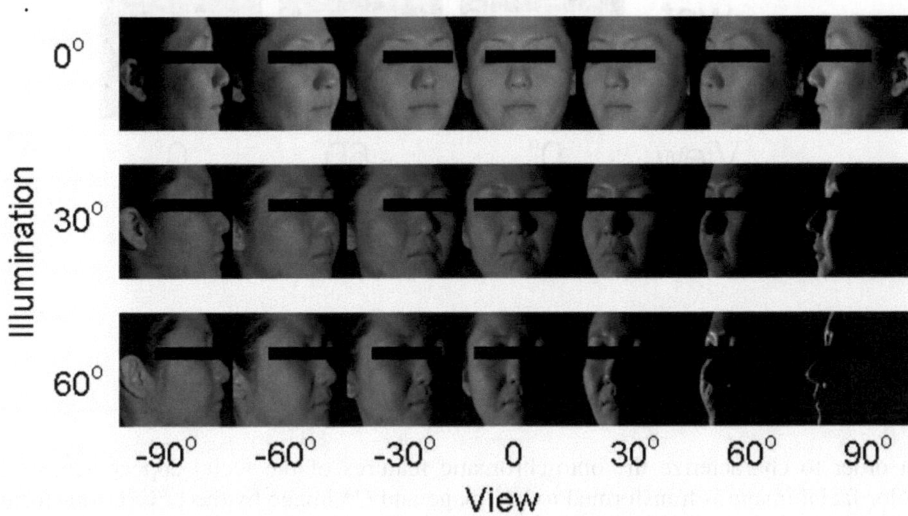

Fig. 5. Multi-angle view and illumination facial images

shown in Figure 4(b). The image resolution of our database is much higher than that of existing databases such as PIE [4], CAS-PEAL[8]. So detailed information about skin texture is also available as shown in Figure 4(c), which is another unique feature of our database.

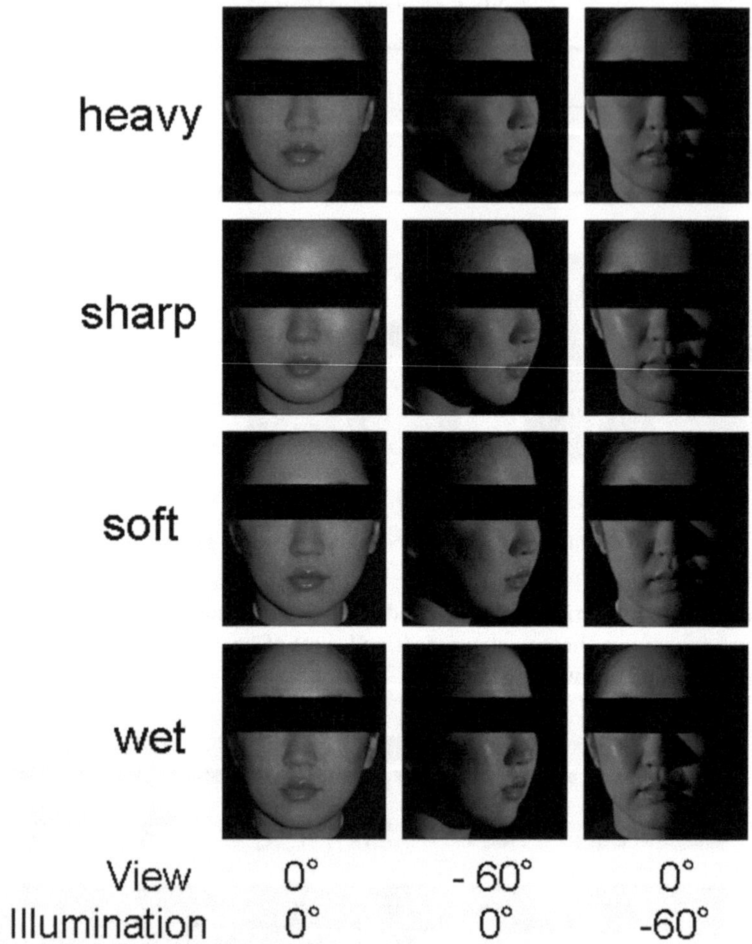

Fig. 6. Four types of cosmetic images created by makeup artists

4 Quantitative Analysis

4.1 Opto-chromatic Images

In order to characterize the opto-chromatic features of the facial appearance, each color facial image is transformed to L* image and C* image by the L*C*h transform, which represent the distribution mapping of brightness and chroma, respectively [11]. Examples of chromatic distribution maps for different types of facial appearance (Figure 6) created by makeup artists are shown in Figure 7. It can be seen that the facial appearance varied significantly depending on the directions from which they were viewed and illuminated. Specifically, we found that the patterns of the directional variations differed widely depending on the type of facial appearance, and this reflected the opto-chromatic features of their appearances.

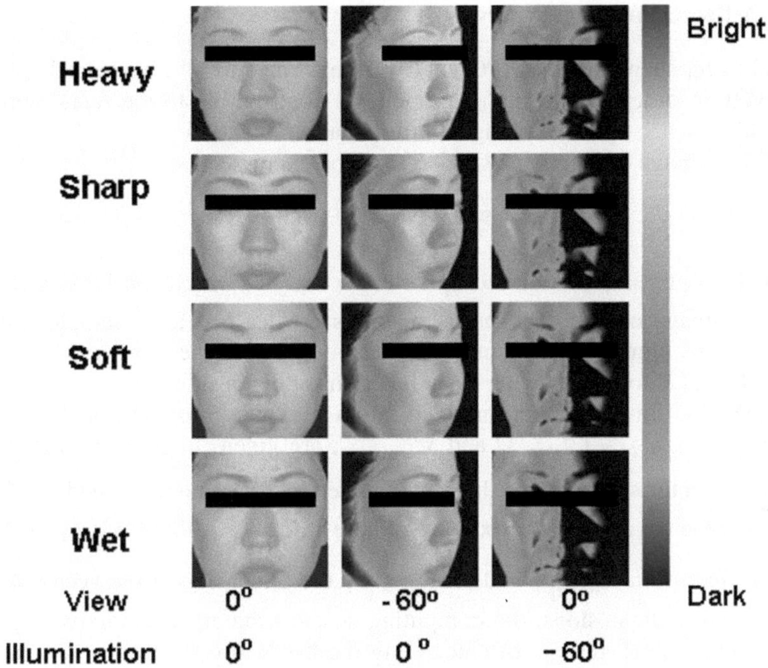

(a)

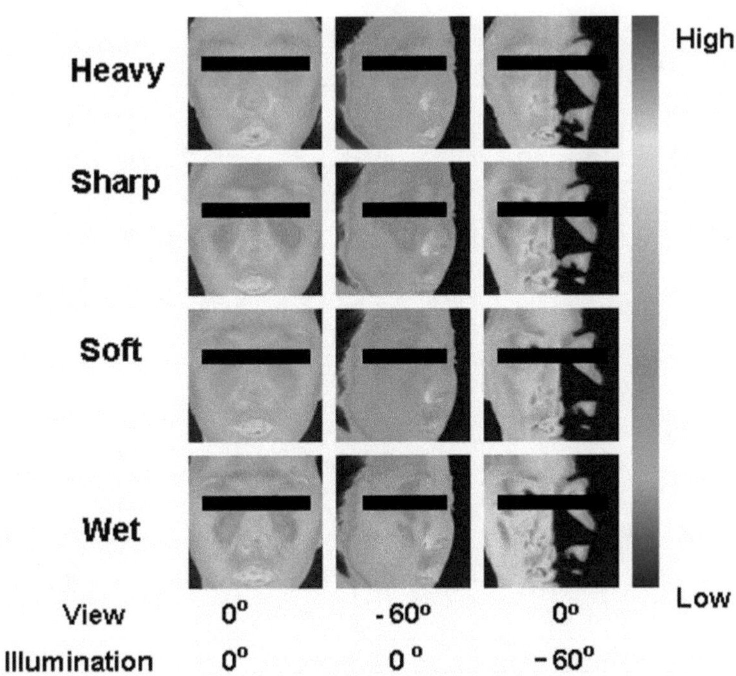

(b)

Fig. 7. (a) L* images; (b) C* images

4.2 PCA Based Subspace Analysis

Principal Component Analysis (PCA) [16] is performed to L* images and C* images, respectively in order to find main components contributing to the facial appearance and the dependence on the viewing and illumination directions.

The facial image or facial appearance **x** can be expressed as

$$\mathbf{x} = s_0\mathbf{v}_0 + s_1\mathbf{v}_1 + \cdots + s_n\mathbf{v}_n \tag{1}$$

where $\mathbf{v}_i$ is the basis function and s_i is the coefficient, which can be used as image features or image coding. The basis functions are learned from sample images by PCA. Since the dimension of sample images is large, we use a method called eigenfaces method [12] to calculate the basis functions or eigen subspace.

The 2D facial image is first unfolded into a vector **a**. Suppose a series of vectors with zero-mean, $\mathbf{a}_i, i = 1,2,\cdots,M$ and $\sum_{i=1}^{M}\mathbf{a}_i = 0$, are given. In conventional PCA, the bases of the vector space can be directly obtained by selecting the J leading eigenvectors, $\mathbf{v}_j$, of the covariance matrix $\mathbf{Cov}_1$, where $\mathbf{Cov}_1 = \mathbf{A}\cdot\mathbf{A}^T$ and $\mathbf{A} = [\mathbf{a}_1, \mathbf{a}_2, \cdots, \mathbf{a}_M]$. Since the dimension of $\mathbf{Cov}_1$ is determined by the dimension of the vector $\mathbf{a}_i$, if the vector has large dimensions, the computing cost for the eigenvectors is huge. In eigenface method [12], instead of calculating the bases directly from $\mathbf{Cov}_1$, the eigenvectors $\mathbf{u}_j$ is first calculated from $\mathbf{Cov}_2 = \mathbf{A}^T\cdot\mathbf{A}$. Then the basis of the unfolding

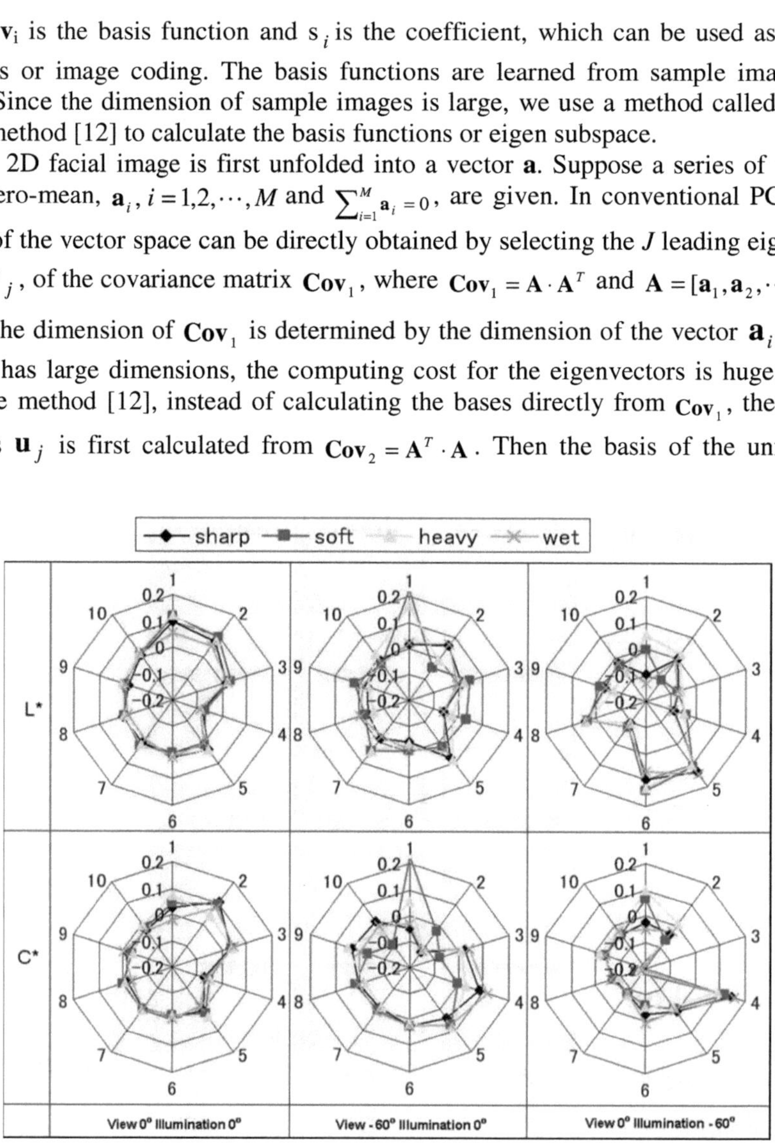

Fig. 8. Comparison of PCA coefficients

vector space $\mathbf{v}_j$ can be calculated by $\mathbf{v}_j = \mathbf{A} \cdot \mathbf{u}_j$. The dimension of $\mathbf{Cov}_2$ is depended on the number of training samples.

We first calculated the L* eigenface (brightness subspace) and C* eigenface (chromatic subspace) of each viewing angle and illumination condition using 300 corresponding L* facial images and C* images (150 natural images + 150 cosmetic images), respectively. Then we projected the L* image and C* image of various types of facial appearance (Figure 6) to the corresponding subspace, and calculated their coefficients. The 10 leading coefficients are plotted in Figure 8. It can be seen that the PCA coefficients are different (especially in the lower mode) for different facial appearances. When the viewing angle and illumination condition are different, the coefficients are also different. That means the PCA coefficients may be used as a quantitative measure of facial appearance. Detailed discussion about the analysis results is beyond the scope of this paper, it will be presented in [15].

5 Conclusion

We have constructed a multi-angle view, illumination and cosmetic facial image database (MaVIC), which is only one and useful database for appearance studies. By combining the MaVIC with chromatic distribution mapping and PCA based subspace analysis, we can understand the quantitative characterization of facial appearance, which will be very useful for designing and controlling various types of facial appearance using cosmetic foundations [11]. A more accurate statistical analysis in which all the face shape is normalized by warping is under research.

References

1. Igarashi, T., et al.: The Appearance of Human Skin: A Survey. Foundations and Trends in Computer Graphics and Visions 3, 1–95 (2007)
2. Tsumura, N., et al.: Image-based skin color and texture analysis/synthesis by extracting hemoglobin and melanin information in the skin. ACM Transactions on Graphics 22, 770–779 (2003)
3. Phillips, P.J., et al.: The FERET Evaluation Methodology for Face-Recognition Algorithms. IEEE Trans. Pattern Analysis and Machine Intelligence 22, 1090–1104 (2000)
4. Sim, T., Baker, S., Bsat, M.: The CMU Pose, Illumination, and Expression Database. IEEE Trans. Pattern Analysis and Machine Intelligence 25, 1615–1618 (2003)
5. http://www.cl.cam.ac.uk/research/dtg/attarchive/facedatabase.html
6. http://cbcl.mit.edu/software-datasets/heisele/facerecognitiondatabase.html
7. http://cvc.yale.edu/projects/yalefaces/yalefaces.html
8. Gao, W., et al.: The CAS-PEAL Large-Scale Chinese Face Database and Baseline Evaluation. IEEE Trans. on System Man, and Cybernetics (Part A) 38, 149–161 (2008)
9. http://rvl1.ecn.purdue.edu/~aleix/aleix_face_DB.html
10. http://images.ee.umist.ac.uk/danny/database.html
11. http://www.ee.oulu.fi/research/imag/color/pbfd.html

12. http://www.irc.atr.jp/Emlyons/jaffe.html
13. http://www.caip.rutgers.edu/rutgers_texture/
14. Igarashi, T.: Recent technical trends on cosmetic foundations. Fragrance Journal 34, 17–28 (2006)
15. Igarashi, T., et al.: Face Designing: Technology to Control Facial Appearance Based on Multi-angle Image Analysis and Electromagnetic Simulation. In: Proc. IFSCC 2008 (to be published, 2008)
16. Turk, M., Pentland: Eigenfaces for Recognition. Journal of Cognitive Neuroscience 3(1), 71–86 (1991)

Shape Representation Robust to the Sketching Order Using Distance Map and Direction Histogram

Kiwon Yun, Junyeong Yang, and Hyeran Byun

Dept. of Computer Science, Yonsei University, Seoul, Korea, 120-749
{wallace,gundid,hrbyun}@cs.yonsei.ac.kr

Abstract. We propose a powerful shape representation to recognize sketches drawn on a pen-based input device. The proposed method is robust to the sketching order by using the combination of distance map and direction histogram. A distance map created after normalizing a freehand sketch represents a spatial feature of shape regardless of the writing order. Moreover, a distance map which acts a spatial feature is more robust to shape variation than chamfer distance. Direction histogram is also able to extract a directional feature unrelated to the drawing order by using the alignment of the spatial location between two neighboring points of the stroke. The combination of these two features represents rich information to recognize an input sketch. The experiment result demonstrates the superiority of the proposed method more than previous works. It shows 96% recognition performance for the experimental database, which consists of 28 freehand sketches and 10 on-line handwritten digits.

Keywords: Shape representation, Sketch recognition, Sketching order, On-line handwriting recognition.

1 Introduction

Recently, as pen-based input devices became popular in the education area by the advance of tablet PC, it is increasing the needs to recognize not only handwritten characters, but also sketches with various shapes. The spatial information is used as an important factor in recognizing handwritten sketches because it illustrates the shape of the input object. However, contrary to printed characters, it has a large shape variation in terms of each individual's in handwritten characters [1]. Therefore, the variation of input must be strongly considered when describing the features of a sketch in the handwriting recognition; especially for writing independent sketch. Fortunately, the additional information of the writing, such as the number of strokes or the end points for each stroke, can be used to recognize on-line handwriting. In particular, the directional information of the writing for each stroke is another significant factor, in addition to the spatial information of writing. However, when the system tries to represent the directional feature of a sketch, it must consider that users can draw each sketch regardless of some

N. da Vitora Lobo et al. (Eds.): SSPR&SPR 2008, LNCS 5342, pp. 421–430, 2008.

decisive factors for directional feature, such as the order of input or the number of strokes.

A lot of study groups have over the past decades researched how to illustrate the feature information of freehand writing [2]. Also, their research has a widespread application in various environments [3], [4], [5]. Chan and Yeung [5] proposed a feature representation method to recognize handwriting by using fundamental properties of symbol consisting of a primitive set. This method has a weakness, which makes it difficult to recognize complex symbols or general sketches. Matsakis [6] represented shapes consisting of multi-strokes using aligned points that are invariant to the order of writing. However, this method does not consider the different number of strokes that can be drawn by different writers. It is also impossible to apply this method to symbols having loop like 0, 8 and ∞. Belongie [7] proposed shape context method using distance and direction between sampling points as feature information. It is less accurate in recognizing shapes that have large variations, but it can represent sketches as rich features simply by their distance and direction. Laviola [8] proposed a symbol representation method combined with fourteen different features such as the number of strokes, aspect ratio, intersection points, angle histogram and so on. It demonstrates that a combination feature consisting of several different kinds of slight features has enough accuracy to recognize a symbol.

In this paper, we propose a shape representation method with a high discriminating power using a combination of spatial and directional feature to recognize input sketch on Tablet PC. First of all, we use distance map, given the distance to the nearest edge at every pixel in the shape to represent spatial information. By the way, chamfer matching using distance map cannot handle large shape variation like handwriting because it just uses the edge of strokes as a measure [9]. In order to solve this problem, we use a distance map as a robust concerning large variations of on-line handwriting. This is due to the fact that it has more detail and rich spatial information of strokes. Moreover, we add the direction of strokes into feature data of input to increase recognition accuracy. In this work, we propose the representation method of directional features to be invariant to the order and the number of strokes using direction histogram.

After this section, we show how the proposed shape representation method can be robust in regards to writing order and variation by using distance map and direction histogram in section 2. In section 3 we show how to make features using the proposed method and to recognize the new input using them. Section 4 shows the results of our recognition experiments on a database including handwritten digits and freehand sketches. Finally, we present conclusions in section 5.

2 Shape Representation

In this chapter, we propose a shape representation method to be robust in its sketching order. The proposed method consists of two steps. One is to normalize freehand sketch and another is to extract features from normalized sketch data. Fig. 1 shows the conceptual diagram of the proposed method. For the the normalization step, control points are extracted using curvature properties of

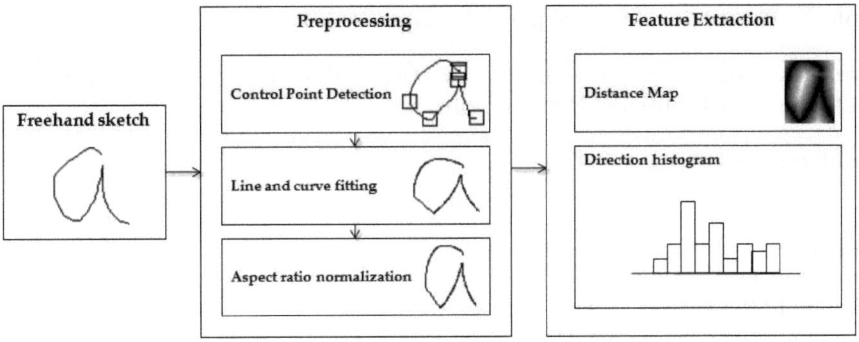

Fig. 1. System overview. After a couple of preprocessing from freehand sketch, the feature of sketch can be represented with distance map and direction histogram.

input sketch. Then, noises are removed by line and curve fitting using detected control points. Finally, the result of fitting process is interpolated into regular size, $m \times n$, to normalize aspect ratio. The second step is to extract spatial and directional features from normalized sketch data. In order to represent spatial feature, we use distance map computed by distance transform. This method helps the system reduce errors which can occur within the same category because features made solely by edges can be sparse. Next, in order to represent directional feature, we have to consider a variation of strokes drawn by different people for the same shape. This is because the order of drawing for a shape can not be the same for different writers and its effects on directional features for the shape. In order to solve this problem, we proposed directional feature which is invariant toward the order of drawing by the process of alignment between neighboring sampling points.

2.1 Sketch Normalization

A stroke is defined as a set of point, $\mathbf{x} = (x, y)^T$, on 2D plane. The set consists of p points between pen-down point and pen-up point. A stroke, $\mathbf{s}_i$, is defined as $\mathbf{s}_i = \{\mathbf{x}_1, \ldots, \mathbf{x}_p\}$. A sketch or online handwritten symbol is consisted of a set $\{\mathbf{s}_1, \ldots, \mathbf{s}_s\}$ having above one stroke. We define the set of the control point, $\mathbf{P}$, as the part of input points in which dramatically change the direction, such as the corner, a high curvature, and junction. By the definition, $\mathbf{P}$ satisfies $\mathbf{p}_i \in \{\mathbf{s}_1, \ldots, \mathbf{s}_s\}, \forall \mathbf{p}_i \in \mathbf{P}$. About a point $\mathbf{x}_i$, we define the nearest control point on a spatial domain as $\hat{\mathbf{p}}$ and define two vectors $\mathbf{d}_{xp}$ and $\mathbf{d}_{xx}$ where $\mathbf{d}_{xp} = \mathbf{x}_i - \hat{\mathbf{p}}$ and $\mathbf{d}_{xx} = \mathbf{x}_i - \mathbf{x}_{i+1}$. Now, the dot product, $\mathbf{D}_p(\mathbf{x}_i)$, of two vectors is expressed as $\mathbf{D}_p(\mathbf{x}_i) = (\mathbf{d}_{xp}^T \mathbf{d}_{xx})/(\|\mathbf{d}_{xp}\| \|\mathbf{d}_{xx}\|)$. If $\mathbf{D}_p(\mathbf{x}_i)$ is more than the threshold, it means that the curvature of $\mathbf{x}_i$ is small, i.e., the point is not a control point. We apply this rule to all input points and the point that satisfies the rule is added at a set of control points. This algorithm is summarized in Table 1. However, inadequate points remain around the corner of the stroke. We simply eliminate

Table 1. The proposed feature point detection algorithm based on the dot product

A stroke $\mathbf{X} = [\mathbf{x}_1, \cdots, \mathbf{x}_k], \mathbf{x}_i \in \Re^2$ is given. Define $\mathbf{P}$ as the set of control points and ϕ as the threshold. Initially, $\mathbf{P}$ is the empty set.

Initialization:
$\mathbf{P} \leftarrow \mathbf{P} \bigcup \mathbf{x}_0$

Iteration:
For i = 1 to k
    Compute the nearest control point of $\mathbf{x}_i$.
    Compute the dot product $\mathbf{D}_p(\mathbf{x}_i)$.
    If $\mathbf{D}_p(\mathbf{x}_i) < \phi$
        $\mathbf{P} \leftarrow \mathbf{P} \bigcup \mathbf{x}_i$

control points which are close together because they have a tendency of being together.

When we extracted control points from an stroke by using the proposed feature point detection algorithm, we classify each control point pair as a line or a curve. To this work, we used $\mathbf{L}_2/\mathbf{L}_1$ proposed in [10] where $\mathbf{L}_1(i,j) = \|\mathbf{x}_i - \mathbf{x}_j\|$ and $\mathbf{L}_2(i,j) = \sum_{l=i}^{j} \|\mathbf{x}_l - \mathbf{x}_{l+1}\|$, given the control point pair consisting of two neighbors, $\mathbf{P}_i$ and $\mathbf{P}_j$. By $\mathbf{L}_1$ and $\mathbf{L}_2$, each control point pair is classified as a curve, if $\mathbf{L}_2/\mathbf{L}_1$ is above the threshold, ϕ_c, otherwise it is classified as a line. The control point pair, which is classified as a line is beautified from the line, which pass through the start and end point of the pair. In the case of a curve, we expressed the curve as a quadratic Bezier curve by using the proposed algorithm in [10].

Lastly, the sketch which is fitted by line and curve is interpolated into regular size to reduce the effect of different aspect ratio. A sketch is interpolated with the scale ratio which adjusts the width and height of sketch into m and n. In this paper, we define the interpolated sketch as $\mathbf{I}(x,y) \in \Re^{m \times n}$.

2.2 Spatial Information Using Distance Map

We propose a spatial feature representation method using a distance map to be robust to shape variation as well as sketching order in handwritten sketches. The proposed spatial feature is trivially robust to sketching order because a distance map represents a captured image of sketch. In chamfer matching, first proposed by Barrow [11], distance transform creates a distance map, which represents how far each pixel is from the edges. Lower value is assigned to pixels near edges and higher value is assigned to pixels far from edges. By chamfer matching, the edge of new input is superimposed on the distance maps of each template. Then, the input is recognized as the category of the template with the lowest sum of the edge hits. Here, the sum of the edge hits is called chamfer distance. However, original chamfer matching using chamfer distance lacks robustness concerning the environment with a large shape variation like on-line handwriting [9]. On the

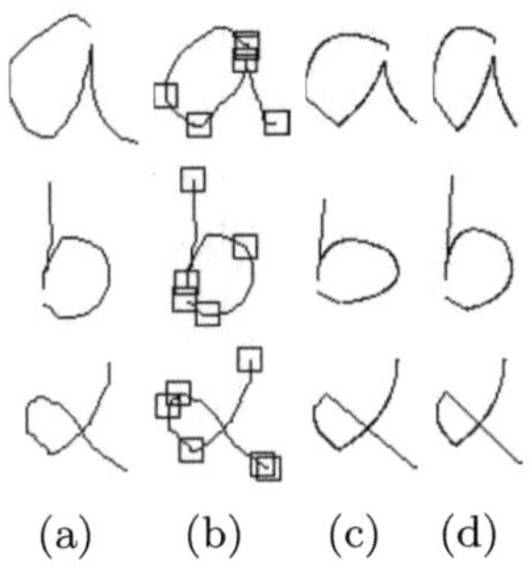

Fig. 2. Examples of the normalization result. (a) Freehand sketch. (b) Control point detection. (c) Line and curve fitting. (d) Aspect ratio normalization.

other hand, the case of using a distance map for spatial information of sketch gives us a richer spatial information because each pixel on distance map has more detail for each spatial feature. Given normalized sketch $\mathbf{I}(x, y)$ in regular size , $m \times n$, a distance map of the sketch is defined as $\mathbf{C}(x, y) \in \Re^{m \times n}$ by the proposed algorithm in [12]. Fig. 3 shows examples of a distance map.

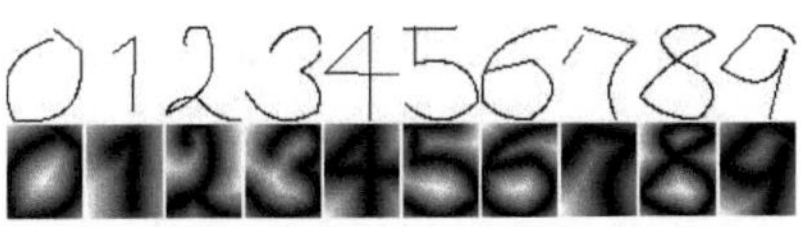

Fig. 3. Examples of a distance map. Input sketches are on top, and distance maps created by input sketches are on bottom. Black area is more closed to the edges, and white area is farther from edges in distance maps.

As we proposed, feature information using distance map can solve the another problem of original chamfer matching which is that chamfer distance considers edge pixels just as meaningful information. In this case, this problem is shown in Fig. 4. In contrast, our proposed approach not only solves this problem, but it also creates a robust feature toward shape variation. This is because all pixels on distance map can be meaningful to represent the shape.

2.3 Direction Histogram

We propose invariant directional feature using a histogram toward sketching order or the number of strokes. The directional information of input strokes is one of the significant features, in addition to the spatial information of writing. Thus, this factor is often used to recognize shape in several research area [9], [5], [8] [6]. When feature of shape is represented using directional information, one significant factor has to be considered. The same shape can be drawn in a different order by different persons. As a result, feature can describe the same sketch in a different way. For example, if two people try to draw input '1' on screen, one person can draw it from top to bottom, while another person can

14	10	7	5	3	0	3	7	14	21
10	7	5	3	0	0	3	7	14	21
7	5	3	0	3	0	3	7	14	15
5	3	0	3	3	0	3	7	10	14
3	0	3	5	3	0	3	5	7	7
0	3	3	3	3	0	3	3	3	3
0	0	0	0	0	0	0	0	0	0
3	3	3	3	3	0	3	3	3	3
7	7	7	5	3	0	3	7	7	7
14	14	10	7	3	0	3	7	14	14

(a)

28	21	14	7	3	0	3	7	14	21
28	21	14	7	3	0	3	7	14	21
28	21	14	7	3	0	3	7	14	21
28	21	14	7	3	0	3	7	14	21
28	21	14	7	3	0	3	7	14	21
28	21	14	7	3	0	3	7	14	21
28	21	14	7	3	0	3	7	14	21
28	21	14	7	3	0	3	7	14	21
28	21	14	7	3	0	3	7	14	21
28	21	14	7	3	0	3	7	14	21

(b)

Fig. 4. The problem of original chamfer matching. The number of each pixel is distance to the nearest edge and the dotted line is the edge of new input. (a) is the comparison between a template for digit 4 and a new input for digit 1, and (b) is the comparison between a template for digit 1 and a new input for digit 1. The chamfer distance is 0 in both cases, so this input can be recognized as both 4 and 1.

draw it from bottom to top for the same shape. In order to solve this problem, we propose a direction invariant method by using the process of alignment between neighbors of sampling points

Given normalized sketch $\mathbf{I}(x, y)$, the set of re-sampling points is define as $\{\mathbf{X}_i ... \mathbf{X}_k\}$, where $\mathbf{X}_i = (x_i, y_i)^T$ and the number of points in the set of normalized strokes is defined as k. Then direction of input points is given as :

$$\mathbf{D}(i) = \tan^{-1}(\frac{y_{j+1} - y_j}{x_{i+1} - x_i}) \tag{1}$$

In order to make direction invariant feature, we modified the proposed algorithm in [6]. Two neighbor points are sequentially selected from first two points in the set of re-sampling points. A primitive stroke is made by these two selected points. Then, each primitive stroke can be one of three types : horizontal, vertical, diagonal. Given two points, $\mathbf{X}_i$ and $\mathbf{X}_j$, a primitive stroke is horizontal if $x_i = x_j$, and vertical if $y_i = y_i$, and diagonal if $x_i \neq x_j$ and $y_i \neq y_j$. A primitive stroke is aligned by the regular order. In other words, if $x_i < x_j$ in horizontal strokes, or if $y_i < y_j$ in vertical and diagonal strokes, then let them leave. In other cases, two points of the primitive stroke swap their positions. As a result, the direction of a primitive stroke made by two points is always same. By this directional alignment, $\mathbf{D}(i)$ is $0 \leq \mathbf{D} \leq \pi$ as Eq. 1, while if the directional alignment process is not performed, $\mathbf{D}(i)$ is $-\pi \leq \mathbf{D} \leq \pi$.

By the way, if these direction are used as feature without any process, then feature information can be sparse. For this reason, direction angle quantize as n level to create direction histogram. In conclusion, this direction histogram represents invariant directional information toward the order of strokes and the number of strokes.

3 Sketch Recognition

In this paper, the spatial feature using distance map and directional feature using direction histogram are used to represent the features of shape. In order to use these two pieces of information as a feature vector, we first normalize them within an uniform range. First of all, the spatial feature is normalized in the following way. We first compute the maximum value of $\mathbf{C}(x, y)$, $\mathbf{C}_{max}$, for each distance map. Then, all the value of $\mathbf{C}(x, y)$ divided by $\mathbf{C}_{max}$ are normalized within $0 \sim 1$ value. On the other hand, the directional feature using direction histogram is also normalized within $0 \sim 1$ divided by the number of points in an entire stroke. Finally, these two feature vector can be used as the feature vector to represent sketch because they are normalized within an uniform range. In this paper, we define the size of distance map, m and n, as m=50 and n=40. These values are from the result of experiments to decide which value is the best way to illustrate a shape. Also, the quantization level for a direction histogram is defined as the most efficient value, 10, by the experiment. Eventually, the final feature vector of shape is combined by the spatial vector and the directional vector. The spatial vector is combined by each row of $\mathbf{C}(x, y)$ in distance map and the directional vector is quantized as 10 direction. In other words, the final feature vector of shape is defined as $\mathbf{F} = (\mathbf{C}^T\mathbf{D}^T)^T \in \Re^{(m\times n+10)\times 1}$. After all, templates can be made using this feature vector for each sketch. At this time, the feature vector is computed by the mean of total training data. In other words, given m training data and feature vector $\mathbf{F}_i$ for a sketch, the feature vector of a template, $\mathbf{T}$, is defined as $\mathbf{T} = \sum_{i=1}^{m} \mathbf{F}_i/m$. In conclusion, the new input sketch is recognized as the category of the template with the lowest Euclidian distance value by comparisons between the feature vector of input and the feature vector of templates.

4 Experiment

For an experiment, we collected a sketch database while writers draw given data-set with random sketching order. First, we gave 28 predefined shapes to 12 person and let them draw the shapes by using the pen of Tablet PC. Each person repeatedly drew each shape 3 times. Therefore, we collected a total of $1008(28 \times 12 \times 3)$ shapes. We utilized 336 sketches among the 1008 sketches for a training database, while others were utilized for an experimental database. Second, for an experiment to recognize the online handwritten digits, we asked 40 persons draw $0 \sim 9$ digits by using the pen of Tablet PC and they were also drawn 3 times for each digit. Therefore, we collected a total of $1200(40 \times 10 \times 3)$ shapes for digits. We utilized 900 sketches among the 1200 sketches for a training database, while others were utilized for an experimental database. Fig. 5 shows examples of two kinds of experimental databases.

We built templates according to each subject by using two training databases and evaluated the accuracy of sketch recognition about two experimental databases. The first experiment compares the accuracy between the proposed method using distance map and original chamfer matching. It is shown that the accuracy of original chamfer matching is 83.5% over 636 digits. On the other hand, it is shown that the accuracy of the case using the distance map is 94.2%. This result demonstrates how spatial feature representation using a distance map is a better method than original chamfer matching. The second experiment is

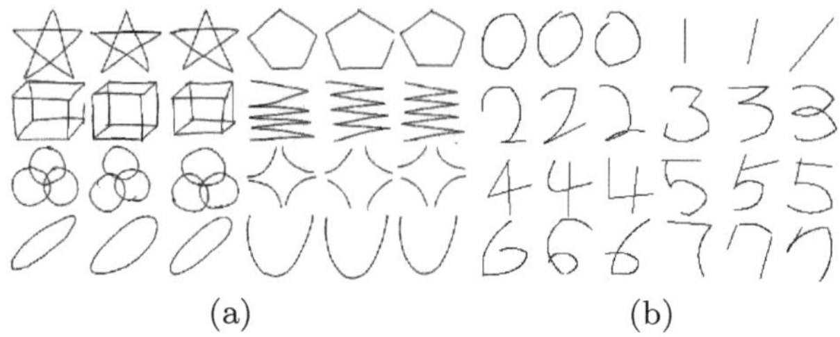

Fig. 5. Examples of sketch in our database. (a) figure sketches (b) numeric symbols.

the comparison experiment concerning the combination of two different features. We experiment with three different cases. One case uses only a directional feature, another uses only a spatial feature, and the other uses both together. It is shown that the accuracy of the case only using directional feature is 52.5%, while another case only using distance map as spatial feature is 87.7%. However, in the case of using both features together, the accuracy is highly increased to 93%. It demonstrates that it is a powerful feature for a shape representation which combines a spatial feature using a distance map and a directional feature using direction histogram. Fig. 6 shows the experimental result of the two experiments with numeric symbols in our database.

Table 2 shows that differences of the accuracy by giving different weight to spatial and directional features in the proposed shape representation method. We use both sketch database and online handwritten digit database in this experiment. It is shown that our proposed method with both spatial feature and directional feature has a high accuracy rate. In particular, the case which is given 0.4 weight to distance map and 0.6 weight to directional feature has the highest accuracy, which is the average rate 96%, on our database. It proves that the shape representation method using distance map and direction histogram is robust in regards to shape variation and invariant toward the direction of strokes.

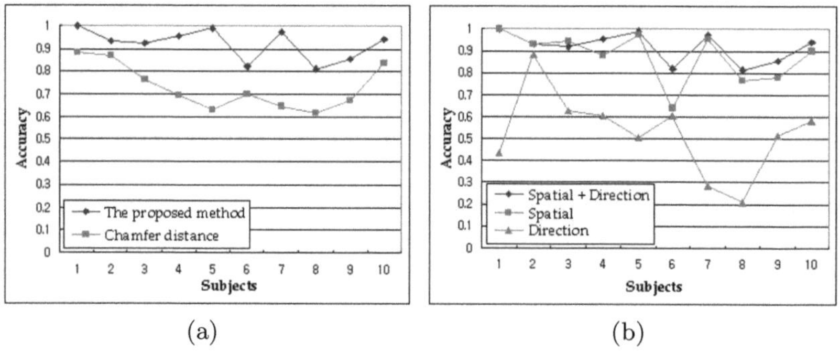

Fig. 6. The experimental result using numeric symbols. (a) The difference of accuracy between the proposed method and original chamfer matching. (b) The result of combination of spatial feature and directional feature.

Table 2. The result of experiment about all sketches in our database by different weight. In this table, w1 is the weight to spatial feature, and w2 is the weight to directional feature. Also, all values are the average of recognition accuracy in each database. Bold text is the highest accuracy rate for different weight.

Weight	Freehand Sketches	Numeric Symbols	Both Sketches
$w1 : 1.0, w2 : 0.0$	0.93	0.85	0.89
$w1 : 0.8, w2 : 0.2$	0.97	0.87	0.92
$w1 : 0.6, w2 : 0.4$	0.98	0.90	0.94
$w1 : 0.5, w2 : 0.5$	0.98	0.92	0.95
w1 : 0.4, w2 : 0.6	**0.98**	**0.93**	**0.96**
$w1 : 0.2, w2 : 0.8$	0.97	0.88	0.93
$w1 : 0.0, w2 : 1.0$	0.64	0.52	0.58

5 Conclusion and Future Work

In this paper, in order to recognize a sketch, we proposed the shape representation method being the combination of spatial and directional features on Tablet PC environment. It is obvious that the sketching order did not have to be considered in the spatial feature using distance map. Also, the experimental result demonstrated that the proposed template based on distance map is superior to the original chamfer matching. Furthermore, a directional feature is a good supplement feature to improve discriminant power. The proposed directional feature is more powerful on real application because it is unrelated to the drawing order. The experiment result showed 96% recognition performance about the experimental database which consist of 28 freehand sketches and 10 online handwritten digits.

In future work, we will study for making a classifier to recognize handwriting sketches more accurately by using this proposed feature representation method. Moreover, we will add more feature information to improve accuracy by giving weight to a discrimination area against other sketches. At this time, we will add alphanumeric symbols and complex mathematical expression symbols to experimental database with previous database. Through this research, we will make an unitive module to recognize various sketches, enabling us to go beyond characters, such as mathematical expression, geometrical figures and graph.

Acknowledgments. This research was supported by MIC, Korea under ITRC IITA-2008-(C1090-0801-0046).

References

1. Connell, S.D., Jain, A.K.: Template-based online character recognition. Pattern Recognition 34(1), 1–14 (2001)
2. Plamondon, R., Smithies, S.: On-line and off-line handwriting recognition: A comprehensive survey. IEEE Trans. on Pattern Analysis and Recognition 22(1), 63–84 (2000)

3. LaViola, J.J., Zeleznik, R.C.: MathPad2: a system for the creation and exploration of mathematical sketches. ACM Trans. on Graphics 23(3), 432–440 (2004)
4. Smithies, S., Novins, K., Arvo, J.: A handwriting-based equation editor. In: Proc. of Graphics Interface, Kingston, pp. 84–91 (1999)
5. Chan, K.F., Yeung, D.Y.: PenCalc: A Novel Application of On-Line Mathematical Expression Recognition Technology. In: Proc. of the Sixth International Conf. on Document Analysis and Recognition, Seattle, p. 774 (2001)
6. Matsakis, N.E.: Recognition of handwritten mathematical expression. Master's thesis, Massachusetts Institute of Technology (1999)
7. Belongie, S., Malik, J., Puzicha, J.: Shape matching and object recognition using shape contexts. IEEE Trans. Pattern Analysis and Machine Intell. 24(4), 509–522 (2002)
8. LaViola, J., Zeleznik, R.: A Practical Approach for Writer-Dependent Symbol Recognition Using a Writer-Independent Symbol Recognizer. IEEE Trans. on Pattern Analysis and Machine Intell., 1917–1926 (2007)
9. Thayananthan, A., Stenger, B., Torr, P., Cipolla, R.: Shape context and chamfer matching in cluttered scenes. In: Proc. IEEE Conf. on Computer Vision and Pattern Recognition, Madison, vol. 1, pp. 127–133 (2003)
10. Sezgin, T.M., Stahovich, T., Davis, R.: Sketch Based Interfaces: Early Processing for Sketch Understanding. In: Proc. 2001 Workshop Perceptive User Interfaces. ACM Press, Orlando (2001)
11. Barrow, H.G., Tenenbaum, J.M., Bolles, R.C., Wolf, H.C.: Parametric correspondence and chamfer matching: Two new techniques for image matching. In: Proc. 5th Int. Joint Conf. Artificial Intell., Cambridge, Massachusetts, pp. 659–663 (1977)
12. Borgefors, G.: Hierarchical chamfer matching: A parametric edge matching algorithm. IEEE Trans. Pattern Analysis and Machine Intell. 10(6), 849–865 (1988)

A Stochastic Approach to Median String Computation

Cristian Olivares-Rodríguez and Jose Oncina

Departamento de lenguajes y sistemas informáticos,
Universidad de Alicante
{colivares,oncina}@dlsi.ua.es

Abstract. Due to its robustness to outliers, many Pattern Recognition algorithms use the *median* as a representative of a set of points. A special case arises in Syntactical Pattern Recognition when the points (prototypes) are represented by strings. However, when the edit distance is used, finding the median becomes a NP-Hard problem. Then, either the search is restricted to strings in the data (*set-median*) or some heuristic approach is applied. In this work we use the (conditional) stochastic edit distance instead of the plain edit distance. It is not yet known if in this case the problem is also NP-Hard so an approximation algorithm is described. The algorithm is based on the extension of the string structure to multistrings (strings of stochastic vectors where each element represents the probability of each symbol) to allow the use of the Expectation Maximization technique. We carry out some experiments over a chromosomes corpus to check the efficiency of the algorithm.

Keywords: Median String, (Multi)string, Stochastic Edit Distance.

1 Introduction

Given a set of data points and a dissimilarity function, the *median* (also called *generalized median* or *geometric median*) point is defined as the point, in the whole space of the data points, that minimizes the sum of the dissimilarities with respect to the points of the set. It is easy to see that the customary definition used in statistics is a special case of this definition when the points are real numbers and the dissimilarity function is the absolute value of their differences. In a similar way, the mean can be defined as the point that minimizes the sum of the squared dissimilarities.

Due to the fact that the median is less sensitive to outliers than the mean (because of the squares), many Pattern Recognition techniques relies in computing the median instead of the mean when robustness is required. A special and very common case in Syntactical Pattern Recognition arises when the points are strings of features (symbols) and the dissimilarity function is the edit distance [1]. Unfortunately, in this case, the problem becomes NP-Hard [2]. Then, although some effectiveness is lost, it is customary to recur to the so called *set median* [3], that is, the median is restricted to be a point in the data set.

N. da Vitora Lobo et al. (Eds.): SSPR&SPR 2008, LNCS 5342, pp. 431–440, 2008.

Kohonen, in 1985 [4], proposed an approximation algorithm to find the generalized median based on a process of perturbation of the set median in order to modify it in the direction of the generalized median string. This results was improved in 2000 by Martínez-Hinarejos *et al.* [5][6]. Moreover, their experiments showed that using the generalized median instead of the set median improved their results.

In this work we are going to study the special case when the dissimilarity measure is the stochastic edit distance [7][8]. This distance is a reinterpretation of the classical edit distance where the edition costs are interpreted as the probabilities of inserting, deleting or substituting a symbol. The stochastic distance is then the minus logarithm of the probability that a string is a transformation of the other. The stochastic edit distance has the quality that the problem of learning the costs can be stated as a maximum likelihood problem and then, the Expectation Maximization technique [9] can be used [7] [8]. Unfortunately, although it is an open problem there are strong evidences that the problem of finding the median using the stochastic edit distance is also an NP-Hard problem.

In this work we propose an approximation algorithm to find the median. The proposed algorithm relies on an extension of the string data structure (a concatenation of symbols), to the multistring data structure, (a concatenation of stochastic vectors where each element represents the probability of each symbol). This pass to the continuum in the symbol space is seized to apply the expectation maximization technique and find the multistring that minimizes the sum of the distances to the set. After that, two possibilities are offered: the first is to recover a string by thresholding the multistring and the second is to transform the Pattern Recognition algorithm that uses the median to use multistrings instead of strings. In this paper we are going to explore the second one.

In the experimental section, since, to our knowledge, no other technique to find the generalized median string exists, our algorithm is compared with a straight forward adaptation of the Martínez-Hinarejos one.

2 Stochastic String Edit Distance

Let $X = \{a_1, \dots, a_n\}$ be a finite set (*alphabet*). A string x is any finite concatenation of symbols in X. Let X^* denote the set of all the strings that can be made using symbols in X and X^n the set of all the strings with exactly n symbols. The empty string is represented by ϵ. Let x be a string, x_i denotes the i-th symbol of x, then $x = x_1 \dots x_m$ where $m = |x|$. On the following we are going to use $a, b, \dots$ to denote symbols and $x, y, \dots$ to denote strings.

Classic String Edit Distance [1] is a dissimilarity measure between strings defined as the minimum number of *edit operations* needed to transform one string into the other, where an edit operation is an *insertion*, *deletion*, or *substitution* of a single symbol. This distance can be extended to use edit operations costs instead of simple edit operation counts. The edit distance can also be viewed as a model of the modifications suffered by a sequence of symbols when traversing a noisy channel.

More formally, let X (Y) be the alphabet of the input (output) strings. Let $E_d = \{(a,\epsilon) : a \in X\}$, $E_i = \{(\epsilon,b) : b \in Y\}$, $E_s = \{(a,b) : a \in X, b \in Y\}$ represent the deletion, insertion and substitution edit operations, and let $E = E_i \cup E_s \cup E_d$. The string Edit Distance is defined by a triple (X, Y, c_e) where $c_e : E \rightarrow \mathbf{R}$ is the primitive cost function. This structure induces a dissimilarity function d over pairs of strings $d : X^* \times Y^* \rightarrow \mathbf{R}$ as:

$$d(x,y) = \min \begin{cases} [c_e(a,b) + d(x',y')]_{x=x'a \wedge y=y'b} \\ [c_e(a,\epsilon) + d(x',y)]_{x=x'a} \\ [c_e(\epsilon,b) + d(x,y')]_{y=y'b} \end{cases} \tag{1}$$

Where $[x]_p$ returns x if predicate p is true and zero otherwise. (Note that $d(x,y)$ can be computed in $O(|x| \cdot |y|)$ time using dynamic programming techniques.)

Following the noisy channel model, if it is assumed the edit operations are based on a random phenomenon, a dissimilarity edit distance like measure can be defined as the probability of having string y in the output provided string x is in the input of the channel $(p(y|x))$[8][1]

Suppose the edit operations are independent. Let $c : E \rightarrow \mathbf{R}$ the cost function where:

- $c(a,\epsilon)$ is interpreted as the probability of deleting the symbol a provided the symbol a is the next symbol in the input string.
- $c(\epsilon,b)$ is interpreted as the probability of inserting the symbol b
- $c(a,b)$ is interpreted as the probability of substituting the symbol a by b provided the symbol a is the next symbol in the input string.

In this framework, a new probability should be introduced to represent the probability of stopping making insertions at the end of the string. Let we call γ this probability.

Then the probability of obtaining the string y provided the string x is in the input of the noisy channel, is the sum of the probabilities of all the possible ways of transforming the string x into y.

More formally, given a pair of strings $(x,y) \in X \times Y$, we denote by $E(x,y)$ the set of all the edit operation sequences that can transform x into y, that is $E(x,y) = \{(x_1,y_1)\ldots(x_n,y_n) : (x_i,y_i) \in E, x_1 \ldots x_n = x, y_1 \ldots y_n = y\}$.

Let $z = (x_1,y_1)\ldots(x_n,y_n)$ be a sequence of edit operations, we define

$$p(z) = \prod_{i=1}^{n} c(x_i,y_i)$$

the probability of generating y given x is defined as:

$$p(y|x) = \sum_{z \in E(x,y)} p(z)\gamma$$

[1] A similar approach was previously used by Ristad *et al.* [7] but based in a joint probability distribution. The results in this paper can be easily extended to the Ristad *et al.* approach.

This can be computed by means of an auxiliary function α as:

$$\begin{aligned}\alpha(x,y) = &[1]_{x=\epsilon\wedge y=\epsilon}\\ &+ [c(a,b)\cdot\alpha(x',y')]_{x=x'a\wedge y=y'b}\\ &+ [c(a,\epsilon)\cdot\alpha(x,y')]_{x=x'a}\\ &+ [c(\epsilon,b)\cdot\alpha(x',y)]_{y=y'b}\end{aligned}$$

And then,

$$p(y|x) = \alpha(x,y)\gamma$$

Of course, in order to have a well defined probability we have to assure the normalization condition:

$$\sum_{y\in Y^*} p(y|x) = 1 \;\forall x \in X^*$$

It can be seen [8] that the following condition over the cost function assures that the stochastic edit distance is well defined:

$$\begin{aligned}\gamma > 0, \quad c(a,b), c(a,\epsilon), c(\epsilon,b) \geq 0 & \qquad \forall a \in X, b \in Y\\ \sum_{b\in Y} c(\epsilon,a) + \sum_{b\in Y} c(a,b) + c(a,\epsilon) = 1 & \qquad \forall a \in X\\ \sum_{b\in Y} c(\epsilon,b) + \gamma = 1 &\end{aligned}$$

Finally, in order to have a dissimilarity measure, the *stochastic edit distance* is defined as:

$$d(x,y) = -\log p(y|x)$$

Oncina and Sebban, in 2006 [8], proposed an expectation-maximization based algorithm [9] to learn the probabilities of the cost function from a training set.

3 Median String and Set Median

Given a set $S \subset M$ and a dissimilarity function $d : M \times M \to \mathbf{R}$. The (*geometric* or *generalized*) *median* element of the set S is defined as the point in M that minimizes the sum of distances to the elements in S. That is,

$$m = \operatorname*{argmin}_{y\in M} \sum_{x\in S} d(y,x) \tag{2}$$

When the set M is an Euclidean space there exists fast iterative algorithms, like the Weiszfeld's algorithm, to find it.

Given an alphabet X, when the set M is X^* and the dissimilarity function is the edit distance it has been show that the problem becomes NP-Hard[2]. In such cases it is costumary to restrict ourself to search the median in the set S, the obtained element is then called the set-median.

Unfortunately, Martínez-Hinarejos *et al.* [5] [10] showed that the median string is better representative of a given set than the set median. In their work [11], they proposed several approximation methods to find the median. These methods are based on an iterative process of perturbation (see [4]) over an initial string. The most successful one, the joined iterative approach, applies each possible edit operation to each position of a string u (initially ϵ). From these strings, the one that minimizes the sum of the distances to the elements of S is selected, for the next iteration. The process is repeated until no changes are obtained.

In our case we are interested in the stochastic distance. It is not known if, using this distance, the problem of finding the median string of a set is also a NP-Hard problem.

In the next section an Expectation Maximization based method to found the median is proposed. In the experiments section it is compared with a straight forward adaptation to stochastic distances of the Martínez-Hinarejos *et al.* technique.

4 Stochastic Approach to Median String

Given an alphabet X of size n we are going to represent a multisymbol $\mathbf{a} = (\mathbf{a}_1, \ldots, \mathbf{a}_n)$ as a stochastic vector in $\mathbf{R}^n$, that is, $0 \leq \mathbf{a}_i \leq 1$ and $\sum_{i=1}^{n} \mathbf{a}_i = 1$. A multisymbol associates a probability to any symbol of X. Let $X = \{a_1, \ldots, a_n\}$ and a multisymbol $\mathbf{a}$ we say that $\mathbf{a}_i$ is the probability associated to the symbol $a_i \in X$ ($p_{\mathbf{a}}(a_i) = \mathbf{a}_{a_i}$).

Let $M_X = \{(\mathbf{a}_1, \ldots, \mathbf{a}_n) : |X| = n, 0 \leq \mathbf{a}_i \leq 1, \sum_{i=1}^{n} \mathbf{a}_i = 1\}$ be a multisymbol alphabet. In the same way as in the case of symbol strings, a multisymbol string is any finite concatenation of elements in M_X. We are going to represent by $\mathbf{x}_i$ the i-th multisymbol in the string $\mathbf{x}$ and by $\mathbf{x}_{i,j}$ the j-th component of the multisymbol $\mathbf{x}_i$. On the following we are going to use $\mathbf{a}, \mathbf{b}, \ldots$ to denote multisymbols, and $\mathbf{x}, \mathbf{y}, \ldots$ to denote multisymbol strings.

Note that a multistring $\mathbf{x}$ of length n ($|\mathbf{x}| = n$) defines a distribution probability $p_{\mathbf{x}}$ over X^n (where $\forall x \in X^n$, $p(x_i) = \mathbf{x}_{i,x_i}$). That is, $p_{\mathbf{x}}(x) \geq 0$ $\forall x \in X^n$ and $\sum_{x \in X^n} p_{\mathbf{x}}(x) = 1$.

Like in the plain string case, The (conditional) stochastic edit distance is given by a 4-tuple (X, Y, c, γ) where X and Y are the input and output alphabet respectively, $c : E \to \mathbf{R}$ is the cost function and $\gamma \in \mathbf{R}$.

The probability of generating the multistring $\mathbf{y}$ from the multistring $\mathbf{x}$ is defined as:

$$p(\mathbf{y}|\mathbf{x}) = \sum_{y \in \Sigma^{|\mathbf{y}|}} \sum_{x \in \Sigma^{|\mathbf{x}|}} p_{\mathbf{y}}(y) p_{\mathbf{x}}(x) p(y|x)$$

Note that the distance in plain string is a special case of this one when all the probabilities of each multisymbol are zero except the corresponding to the actual symbol in the string.

Similarly to the string case, this probability can be computed recursively as follows:

$$p(\mathbf{y}|\mathbf{x}) = \alpha(\mathbf{x}, \mathbf{y})\gamma$$

where

$$\begin{aligned}\alpha(\mathbf{x}, \mathbf{y}) = {} & [1]_{\mathbf{x}=\epsilon, \mathbf{y}=\epsilon} \\ & + [c(\mathbf{a}, \mathbf{b}) \cdot \alpha(\mathbf{x}', \mathbf{y}')]_{\mathbf{x}=\mathbf{x}'\mathbf{a} \wedge \mathbf{y}=\mathbf{y}'\mathbf{b}} \\ & + [c(\mathbf{a}, \epsilon) \cdot \alpha(\mathbf{x}', \mathbf{y})]_{\mathbf{x}=\mathbf{x}'\mathbf{a}} \\ & + [c(\epsilon, \mathbf{b}) \cdot \alpha(\mathbf{x}, \mathbf{y}')]_{\mathbf{y}=\mathbf{y}'\mathbf{b}}\end{aligned} \tag{3}$$

and where

$$c(\mathbf{a}, \epsilon) = \sum_{i=1}^{n} c(a_i, \epsilon)\mathbf{a}_i \tag{4}$$

$$c(\epsilon, \mathbf{b}) = \sum_{j=1}^{n} c(\epsilon, b_j)\mathbf{b}_j \tag{5}$$

$$c(\mathbf{a}, \mathbf{b}) = \sum_{i=1}^{n}\sum_{j=1}^{n} c(a_i, b_j)\mathbf{a}_i\mathbf{b}_j \tag{6}$$

This probability can be computed in $O(|\mathbf{x}||\mathbf{y}|)$ using dynamic programming techniques.

In a symetric way $p(\mathbf{y}|\mathbf{x})$ can be recursively computed by means of an auxiliary function β as:

$$p(\mathbf{y}|\mathbf{x}) = \beta(\mathbf{x}, \mathbf{y})\gamma$$

where

$$\begin{aligned}\beta(\mathbf{x}, \mathbf{y}) = {} & [1]_{\mathbf{x}=\epsilon, \mathbf{y}=\epsilon} \\ & + [c(\mathbf{a}, \mathbf{b}) \cdot \beta(\mathbf{x}', \mathbf{y}')]_{\mathbf{x}=\mathbf{a}\mathbf{x}' \wedge \mathbf{y}=\mathbf{b}\mathbf{y}'} \\ & + [c(\mathbf{a}, \epsilon) \cdot \beta(\mathbf{x}', \mathbf{y})]_{\mathbf{x}=\mathbf{a}\mathbf{x}'} \\ & + [c(\epsilon, \mathbf{b}) \cdot \beta(\mathbf{x}, \mathbf{y}')]_{\mathbf{y}=\mathbf{b}\mathbf{y}'}\end{aligned} \tag{7}$$

The stochastic edit distance can be extended to multistrings as:

$$d(\mathbf{x}, \mathbf{y}) = -\log p(\mathbf{y}|\mathbf{x})$$

Given a set of strings S we are interested in finding the median multistring $\mathbf{x}$, that is:

$$\begin{aligned}\mathbf{x} &= \operatorname*{argmin}_{\mathbf{x}\in M_X} \sum_{y\in S} d(\mathbf{x}, y) \\ &= \operatorname*{argmin}_{\mathbf{x}\in M_X} - \sum_{y\in S} \ln p(y|\mathbf{x}) \\ &= \operatorname*{argmax}_{\mathbf{x}\in M_X} \prod_{y\in S} p(y|\mathbf{x})\end{aligned}$$

Then the problem of finding the median multistring can be stated as a maximum likelihood problem.

Now, we are going restrict the problem to search the multistring $\mathbf{x}$ of a fixed length that maximizes the likelihood and then iterate for each possible length. To do that the standard expectation maximization algorithm is used.

It is easy to see in equation 3 that the parameters of $\mathbf{x}$ are only used when deleting (eq. 4) and substituting (eq. 6). Then the expectation $\bar{\mathbf{x}}_{i,j}$ of each parameter of $\mathbf{x}$ can be computed as:

$$\begin{aligned}\bar{\mathbf{x}}_{i,k} = & \sum_{j=1}^{|y|} \frac{\alpha(\mathbf{x}_{1\ldots i}, y_{1\ldots j})\mathbf{x}_{i,k} c(a_k, \epsilon)\beta(\mathbf{x}_{i+1\ldots n}, y_{j\ldots|y|})\gamma}{p(y|\mathbf{x})} \\ & + \sum_{j=1}^{|y|} \frac{\alpha(\mathbf{x}_{1\ldots i}, y_{1\ldots j})\mathbf{x}_{i,k} c(a_k, y_j)\beta(\mathbf{x}_{i+1\ldots n}, y_{j+1\ldots|y|})\gamma}{p(y|\mathbf{x})}\end{aligned}$$

And the maximization consists in renormalizing. That is,

$$\mathbf{x}_{i,j} = \frac{\bar{\mathbf{x}}_{i,j}}{\sum_{k=1}^{|X|} \bar{\mathbf{x}}_{i,k}}$$

As usual, both steps are repeated until a convergence criterion is reached.

5 Experiments and Results

5.1 *Copenhagen* Corpus

The database used to develop the experiments is the *Copenhagen chromosome dataset*. Each chromosome is depicted by a digitized image which was automatically transformed into a string through the procedure illustrated in figure 1. This procedure begins with the transformation of the images into its idealized profiles. Then, each profile is mapped into strings over the alphabet {1,2,3,4,5,6}. After that, these strings are coded in order to represent signed differences of successive symbols over the alphabet Σ = {e,d,c,b,a,=,A,B,C,D,E}. Taking into account that "a" correspond to difference of -1, "A" of +1, "=" of 0, and so on. For a complete reference to this procedure see [12].

The dataset has 200 strings per class and there are 22 non-sex chromosome types, so the dataset is formed by 4400 samples. Moreover, the behavior of the

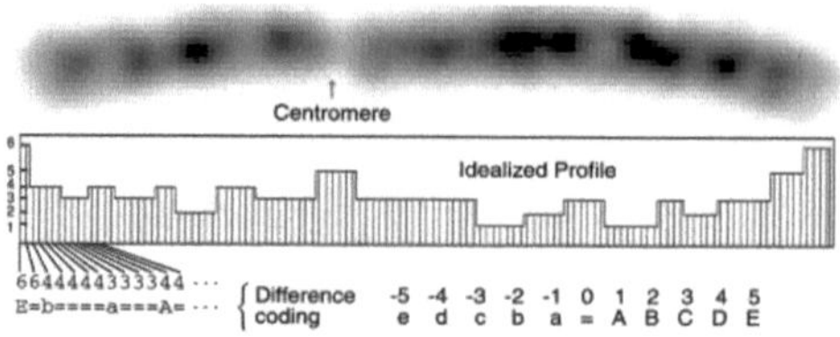

Fig. 1. Image of the chromosomes preprocessing

algorithms is evaluated based on a two-fold cross-validation. Accordingly, each class is divided into two equal-size sets with 100 samples each, then the F_j folds are formed by 2200 samples, where $j \in \{1, 2\}$.

5.2 Experiment 0: Learning the Stochastic Edit Distance

In order to use the stochastic edit distance we need to fix the edit operation probabilities. A similar approach to the one used in [8] was followed. Let P_j be a set of $(x, \mathrm{NN}(x))$ chromosomes pairs. Where each pair into the P_j set is formed by a sample x from F_l and its nearest neighbor $\mathrm{NN}(x)$. The P_j set was used as training set to learn Stochastic Distance. A different stochastic edit distance was learned for each set P_j.

5.3 Experiment 1: Clustering with k-Medians

Following the guidelines made in Martínez-Hinarejos *et al.* work we begin by making some experiments to know how much better are our medians with respect to the set-median and the medians found by the algorithm of Martinez-Hinarejos *et al.*. To do that the k-*medians* clustering algorithm along with the *minmax* initialization was use [13].

In the experiments, the median computation in k-medians was replaced by each of the three median computation algorithms that we are comparing. The number of clusters was increased from 1 to 100. As a quality measure the sum of the distances from medians to the elements in its cluster was used.

It can be observed in figure 2 that the worst results are obtained when the set-median is used. Of course, the sum of distances for the three algorithms converges as the number of prototypes grows.

5.4 Experiment 2: Classification with k-Nearest Neighbor

The main goal of this experiment is the evaluation of the k-Nearest Neighbor (k-NN) classification algorithm. According to this, the three versions of the c-medians algorithms were applied to each class to obtain c representative of the class (c varying from 1 to 100). Then the k-NN classification algorithm (for $k \in \{1, 3, 6\}$) was applied to classify the prototypes of an independent fold.

The classification error rates for each algorithm plus the k-NN using the whole set of prototypes are shown in figure 3.

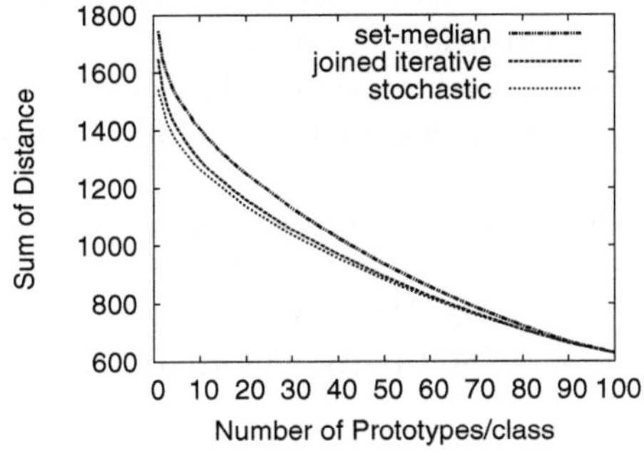

Fig. 2. Minimum sum of distances from the prototypes to its represented class

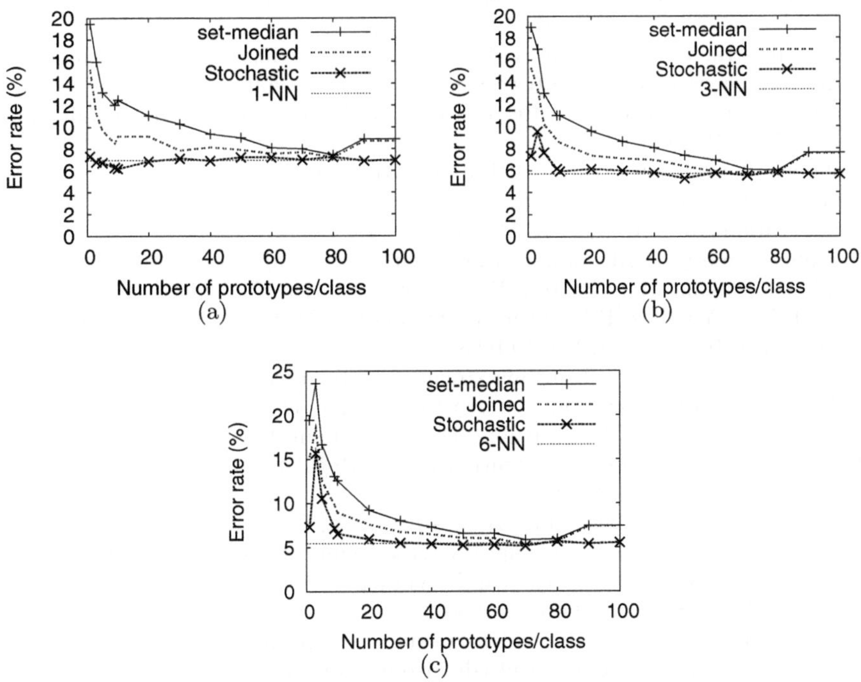

Fig. 3. Error rate of classification using 1-NN (a), 3-NN (b) and 6-NN (c)

Of course, using the whole set of prototypes obtains the lowest error rates, but, among the algorithms that are performing a prototype selection the one based in multistring obtains the lowest error rates.

6 Conclusions

In this work we have used the concept of a multistring in order to compute the median of a set of strings when using the stochastic edit distance. We have show that the problem can be stated as a maximum likelihood search and then, we used the Expectation Maximization algorithm.

In the experimental section we have shown that this approach is quite effective when comparing to it the set median or the generalized median.

Acknowledgments. The authors thank the Spanish CICyT for partial support of this work through projects DPI2006-15542-C04-01, the IST Program of the European Community, under the PASCAL Network of Excellence, IST–2002-506778, and the program CONSOLIDER INGENIO 2010 (CSD2007-00018).

References

1. Levenshtein, V.I.: Binary codes capable of correcting deletions, insertions and reversals. Soviet Physics Doklady 10, 707–710 (1966)
2. de la Higuera, C., Casacuberta, F.: Topology of strings: Median string is NP-complete. Theoretical Computer Science 230, 39–48 (2000)
3. Fu, K.S.: Syntactical Pattern Recognition and Applications. Prentice-Hall, Englewood Cliffs (1982)
4. Kohonen, T.: Median Strings. PRL 3, 309–313 (1985)
5. Martínez-Hinarejos, C.D., Juan, A., Casacuberta, F.: Use of Median String for Classification. In: ICPR, pp. 2903–2906 (2000)
6. Martínez-Hinarejos, C.D., Juan, A., Casacuberta, F.: Median Strings for k-nearest neighbour classification. Pattern Recog. Lett. 24, 173–181 (2003)
7. Ristad, E.S., Yianilos, P.N.: Learning String-Edit Distance. IEEE Trans. Pattern Anal. Mach. Intell. 20, 522–532 (1998)
8. Oncina, J., Sebban, M.: Learning stochastic edit distance: Application in handwritten character recognition. Pattern Recognition 39, 1575–1587 (2006)
9. Dempster, A.P., Laird, N.M., Rubin, D.B.: Maximum Likelihood from Incomplete Data via the EM Algorithm. Journal of the Royal Statistical Society. Series B (Methodological) 39, 1–38 (1977)
10. Martínez-Hinarejos, C.D., Juan, A., Casacuberta, F., Mollineda, R.A.: Reducing the Computational Cost of Computing Approximated Median Strings. In: IAPR International Workshop on Structural, Syntactic, and Statistical Pattern Recognition, pp. 47–55. Springer, London (2002)
11. Martínez-Hinarejos, C.D.: La cadena media y su aplicación en reconocimiento de formas. Phd. Thesis. Departamento de Sistemas Informáticos y Computación, Universidad Politécnica de Valencia (2003)
12. Granum, E., Thomason, M.G.: Automatically inferred markov network models for classification of chromosomal band pattern structures. Cytometry 11, 26–39 (1990)
13. Juan, A., Vidal, E.: Comparison of Four Initialization Techniques for the K - Medians Clustering Algorithm. In: Joint IAPR International Workshops on Advances in Pattern Recognition, pp. 842–852. Springer, London (2000)

Tracking by Hierarchical Representation of Target Structure*

Nicole M. Artner[1], Salvador B. López Mármol[2], Csaba Beleznai[1], and Walter G. Kropatsch[2]

[1] Austrian Research Centers GmbH - ARC, Smart Systems Division, Vienna, Austria
[2] PRIP, Vienna University of Technology, Vienna, Austria

Abstract. Tracking of spatially extended targets with variable shape, pose and appearance is a highly challenging task. In this work we propose a novel tracking approach using an incrementally generated part-based description to obtain a specific representation of target structure. The hierarchical part-based representation is learned in a generative manner from a large set of simple local features. The spatial and temporal density of observed part combinations is estimated by performing statistics over temporally aggregated data. Detected stable combinations consisting of multiple simpler parts encompass local, specific structures, which can efficiently guide a spatio-temporal association step of coherently moving image regions, which are part of the same target. The concept of our approach is proved and evaluated in several experiments.

Keywords: hierarchical representation, edge segments, spatial statistics, temporal statistics, tracking.

1 Introduction

Variable structure is hard to represent using models based on image statistics. Therefore, many tracking algorithms encounter difficulties in scenes with articulated objects, targets undergoing pose variations and partial occlusions. Part-based approaches [1,2] avoid some of these problems by decomposing variable structure into simpler parts, but often lack the flexibility to represent complex deformable structures, given that they either rely on an *a priori* model or the part-based model is too general (for example defined by a uniform grid of blocks).

This paper proposes an approach, which (i) builds a target-specific part-based model using a compositional framework [3] and (ii) employs the generated model for tracking. Our main motivation is to achieve reliable data association and thus robust tracking using a hierarchical system combining simple features into increasingly more complex parts and capturing the statistics of parts. Obtained parts with growing complexity represent specific entities occurring in images more rarely. Partitioning of the parts into trajectories based on structural similarity and motion coherence becomes feasible, since the combinatorial problem of matching involves fewer parts with less ambiguity across frames of an sequence.

* Supported by the Austrian Science Fund under grants P18716-N13.

N. da Vitora Lobo et al. (Eds.): SSPR&SPR 2008, LNCS 5342, pp. 441–450, 2008.

Hierarchical systems ensure an efficient way to represent exponential variability present in the visual data. Recent works [3,4] in visual object recognition demonstrate that hierarchical grouping of simple features generates a finite set of increasingly specific groups, able to represent structure of many object categories in a compact form. By applying the same concept to images of a video sequence, tracking can be formulated in a statistically-driven manner recovering specific parts occurring across multiple frames.

The paper is organized as follows. Sec. 2 introduces related work. Sec. 3 describes the basic concepts behind the proposed approach and Sec. 4 explains the algorithmic steps generating the hierarchical representation. Sec. 5 describes the association step enabling tracking. In Sec. 6 we discuss experimental results for several image sequences and in Sec. 7 we draw conclusions.

2 Related Work

Finding compact and informative representations able to handle the large complexity encountered when representing image content (structure, texture, appearance, etc.) has been an active field of research for many decades. Hierarchical representations organizing low-level visual information into more complex parts offer means to accomplish this task. Edge segments have been attractive low-level features offering a high degree of geometric and photometric invariance and encompassing a rich pool for building potential parts. Early works (e.g. [5]) employ pre-designed rules to partition and group edge segments into more complex entities. Partially driven by significant advances in visual object recognition recent frameworks propose statistical, learning-based methodologies. Opelt et al. [6] use object boundary fragments to detect multiple object classes in presence of clutter and partial occlusions, where class-discriminative boundary fragments are extracted and selected by Boosting. Agarwal and Triggs [7] introduce a novel multilevel visual representation called "hyperfeatures", where local sets of image descriptors are gradually organized into more complex and spatially sparse parts resulting in a system localizing several object categories. Crandall et al. [8] propose a recognition approach where appearance models of parts and spatial relations between parts are simultaneously estimated and used to localize objects. Bouchard et al. [9] present a hierarchical part-based description encoding geometry and appearance of object parts, where learned models vote in a bottom-up manner for object locations.

Hierarchical representation of structure for spatially extended targets in the context of tracking has not been investigated in great detail yet. Ommer et al. [10] present an approach where simple interest points are tracked in a frame-by-frame manner. Interest points as simplest parts are represented by local descriptors, which are used to analyze spatio-temporal relations between parts to learn compositional structured object models. In contrast to our work, Ommer et al. do not use a hierarchical representation and propagate hypothesized part compositions over consecutive frames. In our approach we apply similar concepts as presented by Fidler et al. [4,11] to hierarchically organize spatio-temporally aggregated low-level features in order to create compositions, which can be associated and tracked in an unambiguous manner.

3 Concept of Our Approach

Our approach starts with extracting simple oriented edge segments from a sequence of images. A local spatial configuration of multiple oriented edge segments – denoted further on as *combination* – encodes local structure. The main objective of the proposed approach is to select a set of temporally invariant combinations by statistical analysis, and to reliably associate them to form trajectories. Association exploits the *stability* (temporally invariant structure) and the high *specificity* (low occurrence frequency) properties of selected combinations. As can be seen in Fig. 1 the hierarchical representation is built in a bottom-up process, while the motion estimation (tracking) is carried out in a top-down manner.

The input of the bottom-up process are oriented edge segments detected in several consecutive frames of a video sequence. Starting at level 1, the edge segments are grouped together in combinations of two. By examining the temporal statistics of combinations, their number is reduced by retaining only the stable combinations. Then spatial statistics of edge segments for each remaining combination is examined, and combinations whose segments represent stable structures (i.e. occurring multiple times at the same relative positions) are kept. After the steps of temporal and spatial statistics, the stable combinations for level 1 are found and the process continues with level 2. Each stable combination is extended by an additional segment (selected from the pool of neighboring segments one-by-one) and the statistical analysis (temporal and spatial) is repeated. The incremental generation of edge combinations is carried out until a desired top level is reached, where only few, specific combinations remain, e.g.: level 3 in Fig. 1, or no more stable combinations are found (see Sec. 4).

The generated set of specific combinations at the top level is used in a top-down process to estimate the motion model of foreground objects. Combinations of segments at the top level are distinctive and in the best case they appear only once in an image. Hence, an association of combinations occurring at different time instances becomes feasible, providing an estimate of the underlying motion model. The motion models estimated by associating combinations at the highest level of the hierarchy, can be used to guide the association step of combinations at lower levels (with less specificity) yielding a dense structural description of foreground objects. Foreground objects are delineated by grouping stable combinations, which obey the same motion model (for details see Sec. 5).

4 Building the Hierarchical Representation

Each level L_i of the hierarchical representation contains M_{L_i} stable combinations $C = \{c_1, c_2, \dots, c_M\}$ of $i+1$ segments. A segment s_l represents one of O possible

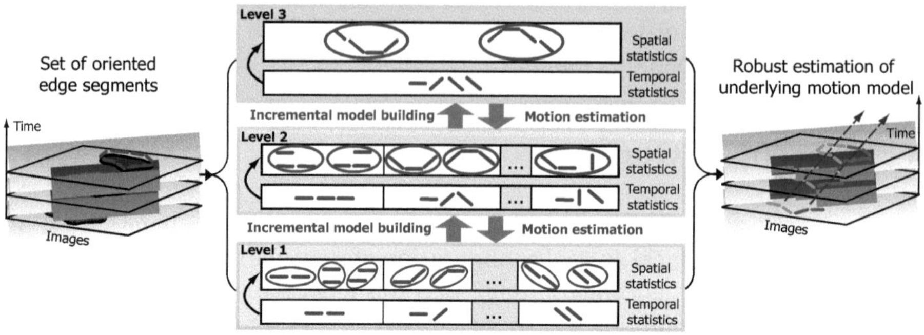

Fig. 1. Concept of our approach

edge orientations $S = \{s_1, s_2, \ldots, s_O\}$. To avoid an exponential complexity of O^N for N levels, the combinations of segments are built in the following way:

Each combination c_k of level L_i consists of i edge segments forming the primary part p_{prim} and one segment representing the marginal part p_{mar}. For example in level L_2 the primary part p_{prim} consists of a set of segments $\{s_1, s_2\}$ and the marginal part p_{mar} of the segment s_5. Except level L_1, the primary parts of the combinations are the stable combinations of level L_{i-1}. So the number of possible combinations for level L_i is $M_{L_{i-1}} \cdot O$.

4.1 Detection of Segments

The structure of foreground objects and background is described by oriented edge segments. A filter bank consisting of oriented Gabor filters (8 orientations in 0°-180°, $\sigma = 0.7$) is used for detecting local edge segments. Each frame is filtered using the oriented filter bank and the magnitude of filter responses (without sign) is computed for each orientation. The image is divided into a set of non-overlapping rectangular neighborhoods along a dense grid. In a given rectangular neighborhood of a size D – small enough to capture important shape details such as the shoulder silhouette of a human – the locally dominant orientation is determined by analyzing all filter responses within the neighborhood for all orientations and finding the orientation with the maximum response. Local maxima with a value smaller than T_m are considered as noise and are ignored. In our approach we used the values T_m=15 and D=10 pixels, the latter being approximately $\frac{1}{10}$ of the foreground object's height.

4.2 Building Levels

As mentioned in Sec. 3 the hierarchical representation is built in a bottom-up process. The stable combinations for each level are selected by temporal and spatial statistics as described in the following.

Level i: For each of the F frames all possible combinations of p_{prim} and p_{mar} are enumerated within local windows B consisting of multiple local neighborhoods D. In order to avoid prohibitive complexity due to the combinatorial nature of the enumeration task, the size of the local analysis window is defined to be small at lower levels of the hierarchy and increased at higher levels (see Tab. 2).

The local window is centered over each stable combination of level L_{i-1}, representing the primary part p_{prim}. Then all possible combinations of p_{prim} with an additional segment p_{mar} are formed. After all new combinations are built, temporal and spatial statistics are applied to select the stable combinations of level L_i.

Level 1: As there are no stable configurations from a previous level for level L_1 at the bottom of the hierarchy, local analysis is performed differently than in the other $N-1$ levels. The local analysis window is slid over all local neighborhoods in all frames starting in the top left corner with a step size equal to D. Within the sliding window at the actual position, all possible combinations are built, where the edge segment with the lower index defines p_{prim} (e.g.: s_1) and the segment with the higher index becomes p_{mar} (e.g.: s_5).

Temporal statistics: The task of temporal statistics is to estimate the density of combination occurrences over F frames by binning and to retain most frequently occurring combinations.

The density of combination occurrences is captured in form of a 3D histogram H, where the histogram is spanned by the primary part indices p_{prim}, the marginal part indices p_{mar} and the frame numbers f. Combinations frequently appearing across multiple frames are selected by a threshold T_f – representing a certain percentage of frames – and retained. The result of temporal statistics is a set of temporally stable combinations consisting of edge segments in an arbitrary spatial arrangement (see Fig. 1).

Spatial statistics: The combinations remaining after temporal analysis form the input for spatial statistics. Fig. 2 visualizes the creation of a co-occurrence histogram of segments, which encodes spatial relations. For each temporally stable combination the primary part p_{prim} of the combination is centered at $(0, 0)$ and the spatial distribution of corresponding marginal part p_{mar} – relative to p_{prim} – is built in form of a two-dimensional histogram.

The obtained set of spatial distributions is used to select combinations with frequently occurring spatial edge configurations within F frames. Mean shift mode seeking is employed to locate the most (up to four) significant modes of the distribution. Modes with densities below a threshold T_p and corresponding combinations are discarded. The outcome of spatial statistics is a set C of combinations corresponding to spatially stable edge configurations.

Fig. 3 shows bar diagrams displaying the effect of temporal and spatial statistics on the number of stable combinations and all occurrences of those combinations. Tab. 1 complements the information from Fig. 3(a) with the number of all possible combinations. For all experiments in Sec. 6 the parameters described in this section are set to the values of Tab. 2.

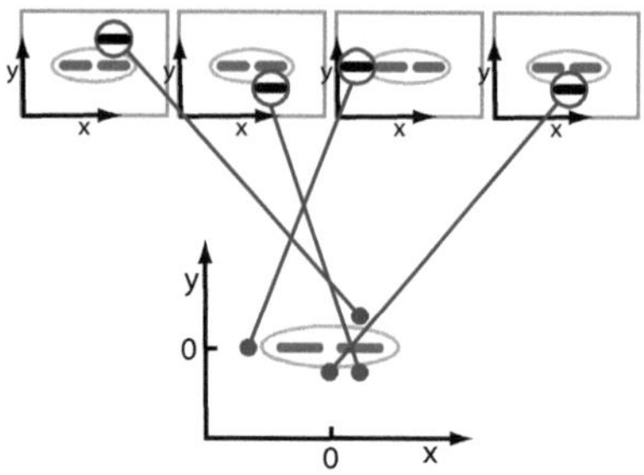

Fig. 2. The concept of spatial statistics. At the top different spatial edge configurations of a given combination are shown. At the bottom the obtained spatial distribution (relative to the centered primary part) of marginal parts is shown.

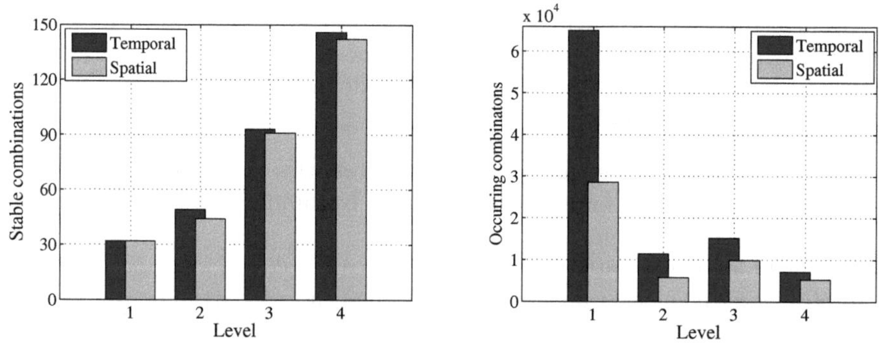

(a) Stable combinations. (b) All occurrences of combinations.

Fig. 3. Example numerical results of statistics. (a) Number of stable combinations after temporal (blue, dark) and spatial statistics (yellow, bright). (b) All occurrences of stable combinations after temporal (blue, dark) and spatial statistics (yellow, bright).

Table 1. The numerical data of the combinations of bar diagram (a) in Fig. 3 (third and forth column) and the number of all possible combinations (second column)

Level	All combinations	After temporal statistics	After spatial statistics
1	36	32	32
2	288	49	44
3	2304	93	91
4	18432	146	142

Table 2. Values of parameters for experimental results for each level

Level	D	B	T_f	T_p
1	10	$3 \cdot D \times 3 \cdot D$	70%	1.0
2	10	$3 \cdot D \times 3 \cdot D$	70%	1.0
3	10	$5 \cdot D \times 5 \cdot D$	50%	1.0
4	10	$7 \cdot D \times 7 \cdot D$	50%	1.0

5 Tracking Using the Built Hierarchy

Our proposed object tracking approach involves three steps. First, a temporal association between the obtained combinations is carried out using robust statistical estimation (see Sec. 5.1). Secondly, combinations following the same motion model are grouped together spatially, thus delineating the tracked object in each image frame (see Sec. 5.2). Thirdly, association of trajectory segments in overlapping space-time volumes (see Sec. 5.3).

5.1 Temporal Association of Combinations

Reliable association of combinations requires that combinations are (i) stable and (ii) highly specific. The previously described statistical analysis captures combinations which occur in multiple image frames of the analyzed spatio-temporal volume. Combinations with an increasing number of segments are more specific and occur less frequently. Given the set of formed combinations, we consider the estimation task of an underlying motion model as a regression problem. In order to keep the complexity of motion model estimation step low, we assume a linear motion model within the analyzed time span (typically 20 frames).

Robust regression is performed using the RANSAC algorithm [12]. Estimation is started at the highest level of hierarchy, where combinations are the most specific and their spatio-temporal distribution best exhibits the underlying linear structure. Typically, despite of the high specificity of combinations, the space-time distribution of a given combination contains multiple structures, therefore the regression task is challenging.

For each combination at the highest level we estimate the best fitting motion model. The slope of motion estimate encodes motion direction and magnitude in the image space. Motion vector estimates are accumulated – in a similar manner to layered motion representations [13] – in a two-dimensional vector space spanned by velocity components along x and y. Mode seeking is performed to find the underlying trend, i.e. peaks defined by velocity components of frequently occurring motion models. Detected peaks encompass combinations belonging to stationary background and moving foreground objects. Since the number of stable combinations at the highest level is low, the obtained set of coherently moving combinations defines a spatially sparse object description. In order to obtain a spatially more dense set of stable combinations, we perform RANSAC estimation also at lower levels of the hierarchy. Sampling is guided by the motion model estimated at the highest level: motion model estimates, which do not belong to any of the previously detected modes of accumulated motion vectors are discarded. In this manner, the estimation step is able to recover motion paths of less stable and less distinctive combinations at lower levels of the hierarchy and to provide a dense structural representation of targets (see example in Fig. 4(a)).

5.2 Spatial Grouping of Associated Combinations

The obtained set of combinations obeying the same motion model is used to spatially delineate the tracked object. Since a given stable combination is not

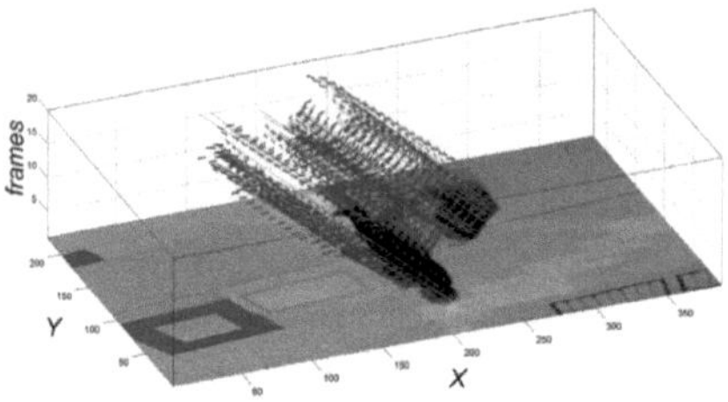

(a) Space-time plot.

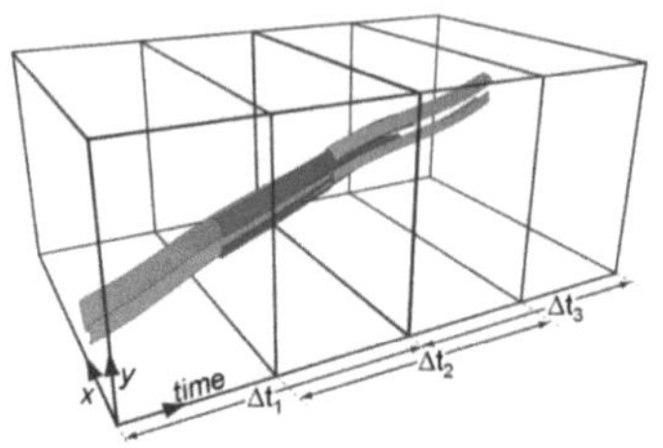

(b) Incremental spatio-temporal tracking.

Fig. 4. (a) Space-time plot showing all stable combinations describing a moving object and obeying the same motion model. (b) Illustration depicting the incremental spatio-temporal object tracking. Different colors indicate distinct stable combinations defining object trajectory segments in consecutive overlapping space-time volumes.

necessarily present in every frame, missing instances of combinations are generated by interpolating the location of each segment using the underlying motion model. Due to the interpolation step, the tracked object is described in each frame by the same number of stable combinations and spatial grouping can be carried out for each frame. Spatial delineation is performed by computing the convex hull of centroid locations of segments belonging to stable combinations.

5.3 Incremental Space-Time Processing

As the shape of a tracked target can change, usually varying sets of stable combinations represent the target structure at different time instances. Therefore, we perform the described hierarchical construction of combinations in an incremental manner. We aggregate edge segments from overlapping space-time volumes (see Fig. 4(b)) and form stable combinations in a given volume independently. Assuming kinematic smoothness of target motion, trajectory segments obtained for individual volumes are associated using the estimated motion models and spatial proximity.

6 Experiments and Discussion

We used three video sequences to prove our concept and qualitatively evaluate the performance of our approach. Video sequences 1 and 2 are parts from PETS 2006 and 2000 dataset.

In sequence 1 a moving pedestrian is segmented and tracked as foreground object. The obtained convex hull – spanned by segments belonging to stable combinations – covers image regions, where spatio-temporal stability was found, therefore heavily articulated parts, such as the feet are not part of the obtained target representation (see Fig. 5). Stationary structures (tiling on the ground) belonging to the background are delineated as well (not shown) explaining recovered motion with a zero velocity model.

Frame 1 Frame 15 Frame 30 Frame 45 Frame 60

Fig. 5. Convex hull of the tracked object in sequence 1

Sequence 2 shows a car tracking example. At lower levels of the hierarchy combinations formed in the image region of the moving target have similar structures as combinations detected at stationary objects in the background. Nevertheless, stable combinations constructed at higher levels are target-specific and delineate coherently moving parts in an unambiguous manner. Fig. 6 shows some frames of the sequence with the tracking results.

Frame 1 Frame 15 Frame 30 Frame 45 Frame 60

Fig. 6. Convex hull of the tracked object in sequence 2

Sequence 3 depicts two pedestrians where one partially occludes the other for a certain number of frames. Target-specific stable combinations found before and after the occlusion event follow the same motion model and thus they become associated with the same trajectory. The interpolation step of part locations generates observations, which are missing during occlusion. Both targets are segmented correctly and tracked in a stable manner, as shown in Fig. 7.

As the proposed representation for tracking employs no prior model, it can be applied to track arbitrary targets, rigid and non-rigid objects. In absence of a prior model, feature selection for the tracking task is completely data-driven and unbiased, implying that all detected features and their combinations – when stable and following a common motion model – are used to estimate the structure and motion of a target.

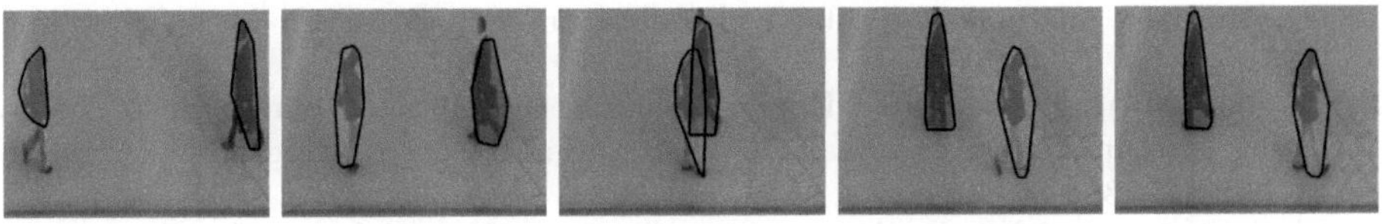

Frame 5 Frame 22 Frame 51 Frame 72 Frame 81

Fig. 7. Convex hull of tracked object of sequence 3

7 Conclusion

This paper introduces first concepts of a novel tracking approach, where the structure of a target is represented by edge segment combinations, which are formed in a hierarchical analysis framework. The obtained structural models represent specific entities, which can be reliably associated between frames of a space-time volume. Combinations are formed in a fully data-driven manner and they integrate all available low-level features representing temporally invariant target structure and coherent motion. The framework is applicable to multiple interacting targets and presented grouping and tracking results show promising performance. In future we plan to achieve rotation invariance of our description to improve the performance with articulated objects and change the temporal association in a way so that nonlinear motion models are possible.

References

1. Adam, A., Rivlin, E., Shimshoni, I.: Robust fragments-based tracking using the integral histogram. In: CVPR, vol. 1, pp. 798–805. IEEE Computer Society, Los Alamitos (2006)
2. Ramanan, D., Forsyth, D.A.: Finding and tracking people from the bottom up. In: CVPR, vol. 2, pp. 467–474. IEEE Computer Society, Los Alamitos (2003)
3. Geman, S., Potter, D., Chi, Z.: Composition systems. Quarterly of Applied Mathematics 60, 707–736 (2002)
4. Fidler, S., Leonardis, A.: Towards scalable representations of object categories: Learning a hierarchy of parts. In: CVPR, pp. 1–8. IEEE Computer Society, Los Alamitos (2007)
5. Gao, Q.: Perceptual tracking of edge features. In: ICIP, vol. 1, pp. 958–962 (1994)
6. Opelt, A., Pinz, A., Zisserman, A.: A boundary-fragment-model for object detection. In: Leonardis, A., Bischof, H., Pinz, A. (eds.) ECCV 2006. LNCS, vol. 3952, pp. 575–588. Springer, Heidelberg (2006)
7. Agarwal, A., Triggs, B.: Hyperfeatures multilevel local coding for visual recognition. In: Leonardis, A., Bischof, H., Pinz, A. (eds.) ECCV 2006. LNCS, vol. 3951, pp. 30–43. Springer, Heidelberg (2006)
8. Crandall, D.J., Huttenlocher, D.P.: Weakly supervised learning of part-based spatial models for visual object recognition. In: Leonardis, A., Bischof, H., Pinz, A. (eds.) ECCV 2006. LNCS, vol. 3951, pp. 16–29. Springer, Heidelberg (2006)
9. Bouchard, G., Triggs, B.: Hierarchical part-based visual object categorization. CVPR 1, 710–715 (2005)
10. Ommer, B., Buhmann, J.M.: Compositional object recognition, segmentation, and tracking in video. In: Yuille, A.L., Zhu, S.-C., Cremers, D., Wang, Y. (eds.) EMMCVPR 2007. LNCS, vol. 4679, pp. 318–333. Springer, Heidelberg (2007)
11. Fidler, S., Skočaj, D., Leonardis, A.: Combining reconstructive and discriminative subspace methods for robust classification and regression by subsampling. PAMI 28, 337–350 (2006)
12. Fischler, M.A., Bolles, R.C.: Random sample consensus: a paradigm for model fitting with applications to image analysis and automated cartography. Commun. ACM 24(6), 381–395 (1981)
13. Black, M.J., Anandan, P.: The robust estimation of multiple motions: Parametric and piecewise-smooth flow fields. Computer Vision and Image Understanding 63, 75–104 (1996)

Using the Shape Characteristics of Rain to Identify and Remove Rain from Video

Nathan Brewer[1,2] and Nianjun Liu[1,2]

[1] RSISE, The Australian National University
nathan.brewer@rsise.anu.edu.au
[2] NICTA Canberra Lab
nianjun.liu@nicta.com.au

Abstract. Dynamic weather effects such as rain cause rapid, distracting motion in a video sequence. This paper aims to remove rain and similar effects from video footage using a multi-step approach; Regions are identified as being potentially affected by rain if they exhibit a short-duration intensity spike. Falling rain drops are imaged by a video camera in a predictable way, as a streak with a consistent range of possible aspect ratios. To preserve scene motion, regions identified by this criterion are investigated, and those that do not fit into the expected range of aspect ratios are ignored. Information about the direction of rainfall is also used to reduce false detections. The effectiveness of this technique is shown on a number of video sequences. The method presented provides advantages over existing techniques.

1 Introduction

As outdoor surveillance becomes increasingly prevalent in society, vision algorithms that can handle the complex and unpredictable behaviour caused by changing light conditions and weather effects are becoming more important. This paper aims to detect and reduce the visibility of dynamic weather, specifically rain from video sequences. Rain has a negative effect on the performance of other vision algorithms, and can be distracting for a human observer.

Reducing or removing the effects of rain while preserving scene information is a difficult task, as rain streaks move very quickly through a scene and are difficult to separate from other motion in a scene, such as that caused by camera motion. Additionally, the visual appearance of rain depends both on the background of the streak and other scene lighting conditions, which makes it difficult to build a general appearance model.

Exploration of the physical and statistical properties of rain have been explored numerous times in the past, largely in an atmospheric science context. In recent times, a number of groups have approached the problem of reducing the appearance of rain from video in a variety of different ways.

Early methods for reducing the visibility of dynamic weather effects used a temporal median filter for each pixel [1,2]. While this approach was successful at reducing the effects of rain and snow in the video sequence, this type of filter

N. da Vitora Lobo et al. (Eds.): SSPR&SPR 2008, LNCS 5342, pp. 451–458, 2008.

resulted in blurring artifacts in footage with a moving camera or scene motion. More recent work [3,4,5] utilises additional properties of rain to identify and suppress only rain streaks, reducing the number of blurring artifacts generated.

Garg and Nayar [4] present a method that uses the optical properties of a falling rain drop to segment a video scene into rain and non-rain regions. Zhang et al [5] make use of the temporal property of rain, that any given pixel is not always covered by rain, together with the chromatic property of rain, that a rain drop will affect the red, green and blue channels of a pixel in a predictable fashion to improve segmentation. Finally, Barnum et al [3] use a global appearance model to identify the typical frequency-domain behavior of rain and snow and present a frequency-domain filter to reduce its visibility.

Each of these methods uses a significant number of frames in order to function optimally. The correlation filter presented by Garg and Nayar [4] uses 30 frames to identify rain-affected regions. The temporal property of Zhang et al [5] is applied across the entire video sequence. Finally, the spatio-temporal frequency method of Barnum et al [3] requires $2N - 1$ frames, where N is the number of desired removal cycles. This dependence on multiple frames makes these types of algorithms difficult to apply to short-duration sequences with rain or to apply in a short-delay online process.

The approach outlined in this paper uses only the local frame neighborhood to identify and remove the effects of rain based on both the optical and physical properties of a rain streak. In this way, we present a system that is able to reduce the visual effects of rain using fewer frames than similar techniques.

Section two of this paper outlines the algorithm that has been developed to reduce the appearance of rain. In section three we present qualitative results of the algorithm on a number of video sequences and a discussion of the effectiveness of the algorithm, and section four provides a brief summary and introduces potential future work.

2 Identifying and Removing Streaks

In this section, we describe the visual appearance of a typical rain streak as imaged by a video camera and a method for identifying and removing such streaks while preserving as much scene information as possible. We do this in a three step process, with an optional fourth step. The algorithm involves first identifying candidate patches, then eliminating those that do not meet the shape properties of rain. If the scene allows, we then identify the likely direction of rainfall to further eliminate false detections. Finally, we replace remaining candidate pixels with an approximation of their background.

2.1 Identification of Potential Rain-Affected Regions in Video

To isolate regions in a video which may contain rain, we can make use of Garg and Nayar's [4] photometric model to identify pixels which have potentially been affected by rain. If we identify pixels which undergo a short, positive intensity spike, we can build a set of candidate pixels corresponding to pixels that have potentially been affected by rain. As we expect the spike to be short duration, we need only compare the current frame to the previous and following frames. Given that real video is noisy, we select candidate pixels which exhibit a gray level that is higher by at least some threshold in the current frame than in the previous and following frames.

A threshold value of 3 gray levels was found to produce good results. This method allows for intensity changes in the background, provided that the rain streak is brighter than the background in both the previous and following frames. If the same pixel is affected by rain in more than one consecutive frame, this criterion fails to identify a pixel as being affected by rain. In practice, this is rarely a problem, as the velocity of a raindrop is high enough that it is only observable in a given pixel for a single frame, and the probability of different drops being present in the same pixel in consecutive frames is quite low.

In order to investigate the physical properties of each individual streak, contiguous blocks of candidate rain pixels are identified and labelled. As the presence of noise can cause pixels within a streak to be misclassified, a simple morphological fill operation is performed. To reduce noise, all identified regions with fewer than a small threshold Y pixels are removed from the set of candidate rain pixels. A Y value of 5 was experimentally found to be effective.

2.2 Shape Properties of a Raindrop

The terminal velocity of a raindrop depends on its diameter, and Foote and DuToit [6] developed an equation which describes the approximate velocity of a raindrop in metres per second as:

$$v(d) = -0.2 + 5.0d - 0.9d^2 + 0.1d^3 \tag{1}$$

where d is the drop diameter in millimetres.

Barnum et al [3] described the length l and breadth b of a rain streak in terms of the diameter of the drop, its distance from the camera z and various camera properties as:

$$b(d, z) = \frac{df}{z} \tag{2}$$

$$l(d, z) = \frac{v(d)ef}{z} + \frac{df}{z} \tag{3}$$

where f is the focal length of the camera and e is the exposure time.

In order to identify rain streaks in video, we can exploit these parameters. If we define the minimum and maximum sizes for a raindrop visible to a camera in the same fashion as [3] at 0.1mm and 3mm respectively, we can calculate a range of aspect ratios which correspond to those of visible rain streaks.

$$AR(d) = \frac{l(d, z)}{b(d, z)} \tag{4}$$

$$= \frac{v(d)e}{d} + 1 \tag{5}$$

We can see that the aspect ratio of a rain streak depends only on the diameter of the raindrop and the exposure time of the camera. As expected, this shows that a shorter exposure time on the camera will result in a less stretched visible streak regardless of the distance of a drop from the camera or other camera parameters.

Using an exposure time of 16ms, we then get an aspect ratio range of $47.58 \leq AR \leq 51.13$ for a drop of water. If the exposure time of the camera is unknown, we can assume that it lies in the range of a typical consumer video camera, between 1 and 40ms. Given this, for an unknown camera we have a range of possible rain streak aspect ratios of 3.9 to 95. Using such a large range of values, however, results in less effective separation of rain and non-rain regions than when the exposure time is known.

Calculating the aspect ratio of a streak in an image is quite challenging, as they are often deformed and, in most rain storms, the outline of a rain streak is difficult to prcisely determine. Because of this, instead of working directly with identified streaks, we find the ellipse with the equivalent second moments as the extracted blob. Assuming that the blob identified is entirely due to a single rain streak, this ellipse will have the properties expected of a falling rain drop. The aspect ratio of such an ellipse is taken as the major axis length divided by the minor axis length.

2.3 Finding the Direction of Rainfall

If a video sequence contains only small amounts of non-rain motion, we can also calculate a direction of rainfall based on the orientation of identified streaks, which can help to eliminate candidate blobs that are not due to rain.

We can easily find the orientation of the ellipses identified as having an appropriate aspect ratio in the previous section, and thus the orientation of the blob they are approximating. From this, we construct a histogram of streak orientation, with bins 5° wide. Assuming that rain falls in an approximately constant direction within any given frame, we can assume that all blobs with orientations that occur in bins with a low proportion of the total scene motion are not rain. Experimentally, we found that discarding bins with less than 10% of the total was effective.

This additional step further reduces the amount of non-rain scene motion that is erroneously removed, although it is only applicable in sequences with a high volume of rain motion.

2.4 Correcting Rain-Occluded Streaks

In order to remove the visual effects of a rain streak, we want to replace the information in each pixel in a particular streak with an estimate of what the pixel information would have been had it not been raining. As the identification of a rain streak relies on a relatively constant background, we can substitute the mean of the pixel values at the previous and following frame for each pixel identified as being affected by rain:

$$P_t = \frac{P_{t-1} + P_{t+1}}{2} \tag{6}$$

As the edges of identified blobs can be noisy, a morphological closing is performed to improve the visual output.

2.5 Summary of Rain Removal Procedure

In order to accurately identify and reduce the visible effects of rain in a video sequence, we apply a relatively simple algorithm. First, we identify which pixels may have been affected by rain by locating regions which display a short-duration intensity spike. We then investigate each blob of candidate pixels by observing its shape and orientation characteristics, discarding those that do not meet aspect ratio or orientation conditions, described above. Finally, for pixels that are identified as rain, we replace the affected pixel with the average of the pixel value in the previous and following frames.

3 Experimental Results and Discussion

The rain identification and removal method presented in section 2 was applied to a number of sequences featuring rain that have been used in previous work on rain removal. The algorithm performs reasonably well, removing a significant proportion of rain from a sequence with minimal distortion of scene information, as shown in Figure 1. While some rain streaks are missed, a significant proportion of rain has been identified and removed without camera stabilisation and with only three consecutive frames required.

Due to the difficulty of identifying and hand-labelling rain streaks, we were unable to develop a quantitative method for determining rain removal quality for all images. Presented in Figure 2 is a 100×100 pixel region of Figure 1 in which we hand-marked all pixels determined to be affected by rain.

Initial segmentation of the video is the major cause of misclassified rain streaks. Typically, misclassifications occur when multiple rain streaks intersect at some point, generating an unusual shape which does not meet the aspect ratio

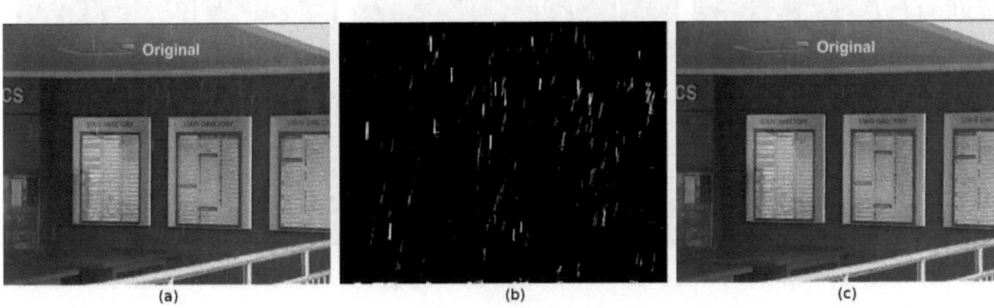

Fig. 1. A frame from a scene with a shaking camera and falling rain, from [5]. (a) is the original image, (b) the identified rain streaks and (c) the image after rain removal.

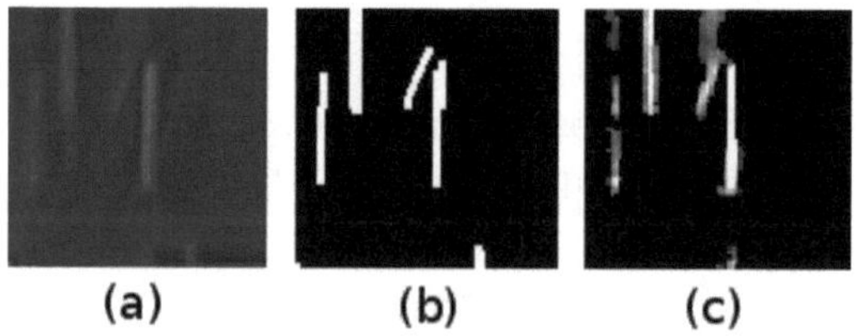

Fig. 2. A zoomed in section of Figure 1 (a) showing pixels identified as rain by hand (b) and by the proposed algorithm (c)

criteria. The other major cause of misclassifications occur when a rain streak intersects some other scene motion. Severely out of focus or distant rain streaks are difficult to detect using the intensity spike criterion. Additionally, water streaks which are caused by dripping water that is not rain can often have different orientations or sizes that do not correspond to the properties expected of rain.

In videos in which the direction of rain was seen as constant within a frame, and rain motion was the primary cause of initial detection, the direction of rainfall was taken into account. This further increased the accuracy of classification, as shown in Figure 3. As can be clearly seen, the streaks corresponding to the man's shirt that are changed in Figure 3 (b) are not changed when the direction parameter is applied, as they are not oriented in the primary rain direction.

In practice, particularly in sequences taken from movies, the aspect ratio of some raindrops does not fall in the expected range. This is most likely because much rain in movies is created using a rain tower, and so has not reached terminal velocity which results in a shorter streak. As such, the aspect ratio range for identifying a streak had to be broadened to identify these artificial rain drops.

This model can be applied in addition to other methods to reduce false detections. For example, when combined with the rain removal algorithm presented by Garg and Nayar in [4] as an additional step following the segmentation of rain regions. Doing so allows for identification and removal of blurring artifacts that appear as scene motion intersects with an area identified as being rain. Figure 4

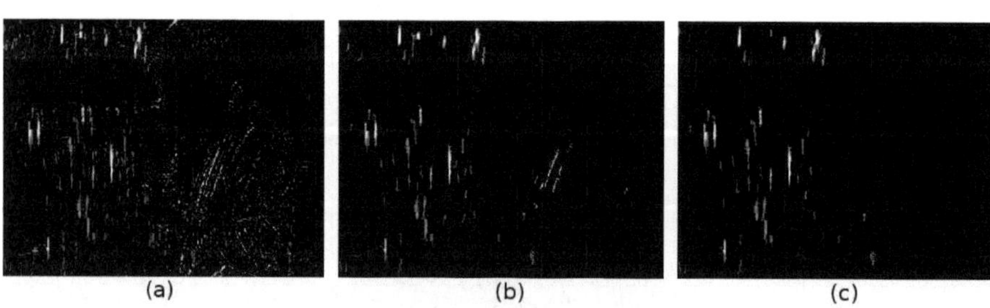

Fig. 3. A difference image from a frame from the film Magnolia showing pixels changed by the algorithm in white. (a) shows all potential pixels, (b) those identified by the shape property and (c) those identified by both the shape and direction properties.

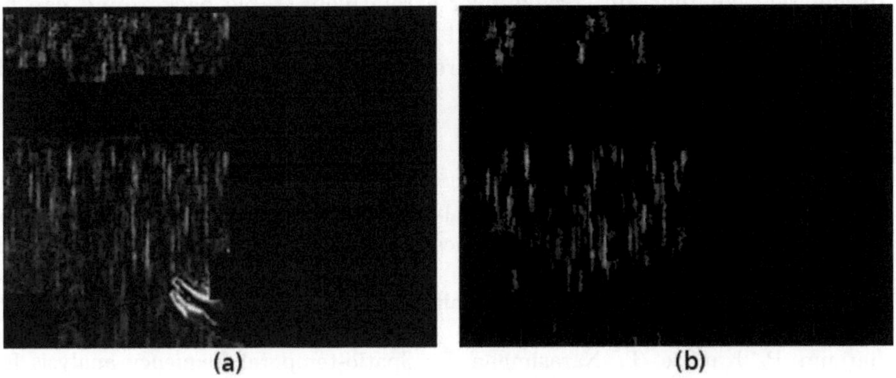

Fig. 4. A frame from the film Magnolia. (a) shows the pixels identified as rain by Garg and Nayar's method from [4], and (b) shows the streaks that are still identified after application of the presented algorithm.

shows rain identified in a frame from Magnolia using the method presented in [4] and the streaks that are identified as rain by the presented algorithm. As can clearly be seen, the detection that results from the man moving his hand is no longer identified as a rain streak in 4(b), resulting in significantly less blurring of scene motion.

Non streak-like background rain is not removed by this algorithm, as it often does not meet the segmentation criteria. However, the clear, bright foreground rain streaks which cause the majority of distraction to a human observer are detected and removed by the algorithm. This significantly improves the visual appearance of rain affected video sequences.

4 Conclusions and Future Work

This algorithm provides a method for reducing the amount of rain present in a section of video footage while preserving scene motion and background textures. By focusing on the physical properties of a rain streak, we are able to separate intensity changes due to rain from those due to scene motion. Other approaches require more frames to identify and remove rain from a video than the approach illustrated here, resulting in longer unavoidable lag times. The algorithm is able to be easily adapted to different scene types and camera parameters.

In order to further improve the algorithm, background subtraction methods may be useful to help isolate and segment rain streaks and scene motion, which can then be used to augment the intensity spike criterion and provide a better segmentation. Stabilisation techniques may also be useful to reduce camera motion and improve segmentation. Camera stabilisation techniques may also improve initial segmentation.

In general, isolated rain streaks over a relatively stable background can be readily identified as such and removed from a video sequence with minimal disruption of scene motion using only three frames with the algorithm presented.

References

1. Hase, H., Miyake, K., Yoneda, M.: Real-time snowfall noise elimination. In: Proceedings of 1999 International Conference on Image Processing, ICIP 1999, vol. 2, pp. 406–409 (1999)
2. Starik, S., Werman, M.: Simulation of rain in videos. In: Texture Workshop, ICCV, vol. 2, pp. 406–409 (2003)
3. Barnum, P., Kanade, T., Narasimhan, S.: Spatio-temporal frequency analysis for removing rain and snow from videos. In: Workshop on Photometric Analysis For Computer Vision (2007)
4. Garg, K., Nayar, S.: Vision and rain. International Journal of Computer Vision 75, 3–27 (2007)
5. Zhang, X., Li, H., Qi, Y., Leow, W.K., Ng, T.K.: Rain removal in video by combining temporal and chromatic properties. In: IEEE International Conference on Multimedia and Expo 2006, pp. 461–464 (July 2006)
6. Foote, G.B., Du Toit, P.S.: Terminal Velocity of Raindrops Aloft. Journal of Applied Meteorology 8, 249–253 (1969)

A Fast Approach to Improve Classification Performance of ECOC Classification Systems

Paolo Simeone[1], David M.J. Tax[2], Robert P.W. Duin[2], and Francesco Tortorella[1]

[1] DAEIMI, Università degli Studi di Cassino
Via G. Di Biasio 43, 03043 Cassino (FR), Italy
{paolo.simeone,tortorella}@unicas.it
[2] Information and Communication Theory Group
Delft University of Technology
Mekelweg 4, 2628 CD Delft, The Netherlands
{D.M.J.Tax,R.P.W.Duin}@tudelft.nl

Abstract. Error correcting output coding is a well known technique to decompose a multi-class classification problem into a group of two-class problems which can be faced by using a combination of binary classifiers. Each of them is trained on a different dichotomy of the classes. The way the set of classes is mapped on this set of dichotomies may essentially influence the obtained performance. In this paper we present a new tool, the k-NN lookup table to optimize this mapping in a fast way and a fast procedure to change the dichotomies in a proper way. Experiments on artificial and public data sets show that the proposed procedure may significantly improve the ECOC performance in multi-class problems.

Keywords: multiple classifiers systems, ECOC, k-NN .

1 Introduction

A common technique to solve an M class problem is to break the problem into dichotomies. Error Correcting Output Coding (ECOC) has arisen as one of the most eligible candidates to substitute monolithic classifier with multiple outputs by an ensemble of binary classifier. The reasons are mainly based on the stronger theoretical roots and the good understanding we have of some binary classifiers. The method, as originally conceived by Dietterich and Bakiri[1], consists of associating a binary string of length L to each of the M classes of the problem. This collection of strings is arranged into a $M \times L$ *coding matrix*. Each column in this matrix defines a binary classification problem. When a new sample is classified by the L dichotomizers, a new binary string is obtained, which has to be matched with the existing M binary class codes, using a suitable decoding technique. This approach has proved to provide a reliable probability estimation and a concurrent reduction of both bias and variance[2][3].

The original motivation for this method is based on the error correcting capabilities of the codes and a great effort has been devoted to improve the coding effectiveness. Allwein et al.[4] introduced a random technique for distributing the labels into the coding matrix and defined a weighted decoding technique to improve the performance in terms

N. da Vitora Lobo et al. (Eds.): SSPR&SPR 2008, LNCS 5342, pp. 459–468, 2008.

of general classification error. The random codes proved to perform not worse than the full code (the code containing all possible dichotomies). This leaded to the conclusion that it is not possible to define a coding strategy which can fit every problem.

Different codes applied to the same data set can yield different results as the classes may have an asymmetric influence on the problem. Some combinations of classes can better not be used for a dichotomy as they demand a too complicated base classifier. So changing the mapping of the classes on the set of dichotomies defined by the coding matrix may influence the performance of the base classifiers and thereby their combination.

Although Crammer et al.[5] demonstrated that designing the best output codes for a data set, given the set of base classifiers, is a NP-complete problem, approximate solutions can be practised. The proposal is to define a general framework to relate the data distribution in the space to the dichotomies and to use it to simplify the binary subproblems defined by each columns of the code. The class distributions in the feature space mainly determine the complexity of a classification problem. Some classes are probably closer than others as well as overlapping in some areas. ECOC provides a new labeling for these classes for each dichotomy, so the idea is to benefit of those which distribute binary labels in a proper way, such that a given binary classifier easily solves the new two class problem.

This target can be reached by permuting the rows of the code and thus changing the association class - binary string. An improvement on the error rate of the classification system is then expected by reducing the binary errors for each dichotomizer of the system. In any case, the idea is extremely powerful, because it will turn useful to analyze particular coding schemes for ECOC, like the LDPC codes[6] for example, which can take advantage of some error correcting techniques of general coding theory.

The optimization of the mapping of classes on the set of dichotomies can be prohibitively slow if it is based on retraining the base classifiers for different mappings or on training a multiple outputs classifier[7]. Moreover, it needs a separate validation set. We propose a much faster tool, the k-NN look-up table which simulates by k-NN classification results the use of more advanced base-classifiers. This table looks at the classes of the nearest k samples for each samples of the set and provides an evaluation of how easily a sample can be misclassified.

2 ECOC Framework

The ECOC approach consists of associating an L-length binary string (*codeword*) to each of the M class of the problem. This collection of strings is organized as an $M \times L$ *coding matrix*. Every column of the code defines a different binary subproblem that has to be solved by a dichotomizer as shown in Table 1. In the training phase, the new binary labels are fed to the binary classifiers. In the operating phase, the trained dichotomizers provide a string $\mathbf{o}(\mathbf{x})$ of L outputs for each sample which has to be classified.

Single errors on some of the L outputs not necessarily lead to the wrong final decision. Some of them can be recovered during the decoding phase performed on the outputs. Consider the Hamming Distance between two words of the code, i.e. the number of labels they differ from each other:

Table 1. Example of an exhaustive coding matrix for a 5 classes problem

classes	codewords														
	$\mathbf{f}_1$	$\mathbf{f}_2$	$\mathbf{f}_3$	$\mathbf{f}_4$	$\mathbf{f}_5$	$\mathbf{f}_6$	$\mathbf{f}_7$	$\mathbf{f}_8$	$\mathbf{f}_9$	$\mathbf{f}_{10}$	$\mathbf{f}_{11}$	$\mathbf{f}_{12}$	$\mathbf{f}_{13}$	$\mathbf{f}_{14}$	$\mathbf{f}_{15}$
ω_1	1	1	1	1	1	1	1	1	1	1	1	1	1	1	1
ω_2	0	0	0	0	0	0	0	0	1	1	1	1	1	1	1
ω_3	0	0	0	0	1	1	1	1	0	0	0	0	1	1	1
ω_4	0	0	1	1	0	0	1	1	0	0	1	1	0	0	1
ω_5	0	1	0	1	0	1	0	1	0	1	0	1	0	1	0

$$D_H(\mathbf{c}_i, \mathbf{c}_j) = \sum_{h=1}^{L} |c_{ih} - c_{jh}|. \tag{1}$$

A decoding technique is applied at the output of the dichotomizers and a sample is assigned to the k-th class if the Hamming distance between the output word $\mathbf{o}(\mathbf{x})$ and the k-th codeword is minimum:

$$\omega_k = \arg\min_h D_H(\mathbf{c}_h, \mathbf{o}). \tag{2}$$

The minimum Hamming distance (MHD) between two codewords of a defined code $d = \min_{i,j} D_H(\mathbf{c}_i, \mathbf{c}_j)$ is the most important parameter to take into account, because it tells how many binary errors can be corrected by the decoding rule. If the MHD is d, the code is able to correct $\lfloor (d-1)/2 \rfloor$ single errors from the base classifiers. A good coding matrix should therefore have a large MHD as possible between the rows of the code. A second demand, as observed by Dietterich himself, is to have also a good distance between the columns of the code. Similar columns produce high correlated classifiers that can easily provide the same errors on the same sample.

It is worth noting that MHD approach ignores the magnitude of prediction. Another approach based on computing the loss function for each base classifier, combining these measures into a total loss and minimizing the value has been successfully proposed in [4]. Notice that the choice of the decoding approach is not important for this paper and it has been analyzed in other papers [8] so we opted for the easiest criterion defined in eq. 2.

3 Reducing the Complexity of the Binary Problems

ECOC operates by binarization of multi class labels. Our aim is to estimate how good these binarizations/groupings of labels are in the feature space. It is easy to imagine that some dichotomies lead to simpler binary problems, separable by less complex classifiers, and some other to more complex ones that are hard to solve and quite complicated for a linear classifier.

It is worth noting that, once a random coding matrix C for a multi class problem is defined, the association between class label and binary codeword has an important role in the final performance of the system for the above given reason. The problem can be seen as a matter of finding the best grouping out of all possible labelings. The main

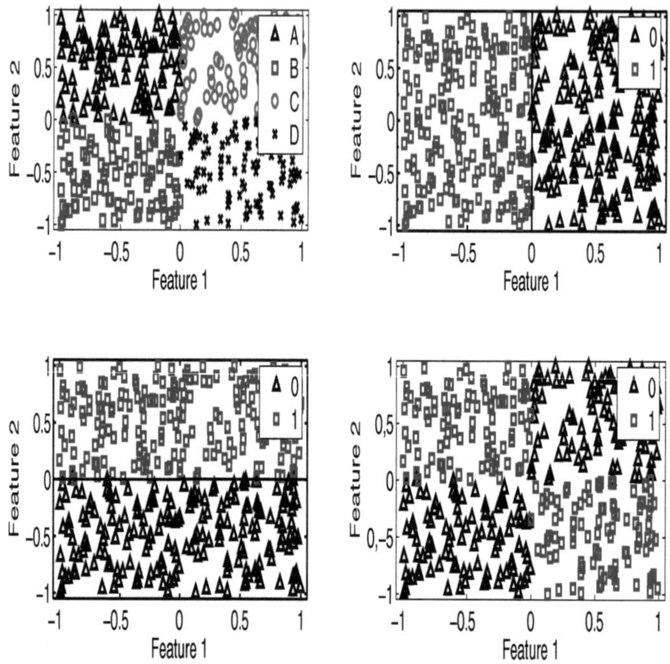

Fig. 1. An example of a 4 classes problem and some possible kind of binary coding for classes

idea is that a good relabeling is the one that groups together close classes in the feature space, so as to define the simplest possible binary problems.

In Fig. 1 a four class problem is shown in the top left corner. Each class is perfectly separable from the others. It is possible to solve this problem using the ECOC technique and given a low complexity classifier (a logistic one for example). Intuitively some grouping of labels are easier to solve. Putting together classes AB vs CD (top right), adopting a $[0011]^T$ dichotomy, or AC vs BD (bottom left), adopting a $[0101]^T$ dichotomy, the logistic classifier easily solves the problem, while it is not able to come up with a solution for the dichotomy $[1001]^T$ (bottom right).

What is needed is a tool to evaluate the difference between these dichotomies and possibly rank them. Notice that we assume that the coding matrix is given, and that we only want to optimize the association of the rows with the classes. We do not consider the problem of finding the best coding matrix as in some previous works[1][4].

Moreover it is useful to move toward an optimized view of the problem. If it is possible to define a tool to evaluate the re-labeling effects of the matrix, it will be possible to understand which dichotomies are good and which are not for a certain data distribution. The general problem of ECOC can be somewhat reformulated. It can be no more necessary to train a lot of classifiers to have very long codewords to be safe against binary errors, but, employing a suitable relabeling, few classifiers can grant the same performance obtained from a longer code. On the other hand simpler dichotomies imply the possibility to use simpler base classifiers. This is an additional reason to choose a linear classifier in the experiments.

Once fixed a code C for a M classes problem a fast way to change groupings of labels is to permute the rows to change the association multi class label - binary label. We look for the best of this association class-row by evaluating a measure to rank all the possible matrices obtained by permutations of the rows. A measure $\overline{\mathrm{m}}$ to distinguish the "quality" of the binary subproblems will be defined in the next section.

Moreover, it is worth noting that not all the possible permutations for a fixed coding matrix need to be examined, because a lot of them are effectively the same matrix with the same columns/dichotomies, but with different indexes. In order to simplify the search, we use a simple procedure of switching only two rows and analyze the resulting coding matrix. This means that, on the first stage of the algorithm, $M \times (M - 1)$ possible codes are created by switching two rows and the top one in terms of the quality parameter $\overline{\mathrm{m}}$ is picked. The procedure which switches two rows is repeated on the new candidate code until there is no improvement. The following algorithm summarizes our procedure:

Algorithm 1. Permute and Search C

```
Compute m̄ for C
while m̄ is not improved do
  for i = 1 to M do
    for j = 1 to M, j ≠ i do
      Swap i-th and j-th rows generating C*
      Compute the new quality measure m̄*
      if m̄* < m̄ then
        Set C = C*, m̄ = m̄*
      end if
    end for
  end for
end while
```

4 Measuring the Quality of the Coding

The next step is to define how to measure the parameter $\overline{\mathrm{m}}$ for algorithm 1. To this aim, a *k-NN look-up table* (example shown on the left of fig. 2) is built, where, for each sample of the set, all the labels of the first k nearest neighbors are stored.

The defined coding matrix C gives the correspondence between the multi class labels and the binary ones for each dichotomy. Consider a training set $S = \{(\mathbf{x}_1, y_1), ..., (\mathbf{x}_N, y_N)\}$, where $\mathbf{x}_i$ is a sample and $y_i \in \{1, .., M\}$ is the class label. As shown in fig. 2, for each dichotomy j of the coding matrix, the multi class label of the k-NN look-up table can be substituted by the binary one and the k-NN rule (f_j) can be evaluated for each sample. The error rate is then computed by:

$$\varepsilon_j = \frac{1}{N} \sum_{i=1}^{N} I(f_j(\mathbf{x}_i) \neq c_{y_i j}) \text{ for } j = 1, ..., L. \quad (3)$$

It is worth noting that this procedure does not go through massive training and combining of base classifiers, but the most expensive computations are done on the look-up

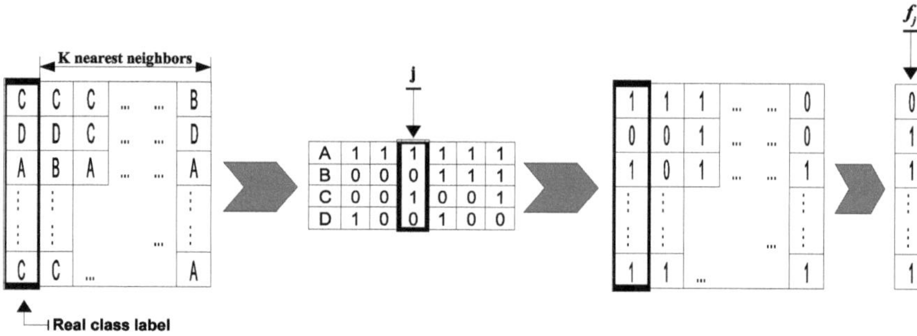

Fig. 2. k-NN look-up table

table itself. The value we wish to minimize is then the mean of this error rate over the L dichotomies:

$$\overline{\mathrm{m}} = \frac{1}{L}\sum_{j=1}^{L} \varepsilon_j. \tag{4}$$

It could be argued that also the error rate on computing the multi class labels could be computed at this stage. By evaluating the Hamming distance between the code and the collection of outputs of the binarized look-up tables a decoding technique has to be applied similarly to what is done with the ECOC itself. This measure has not been employed for two reasons. First the computation of the Hamming distance strongly enlarges the complexity of the framework, reducing all the benefits we had for the evaluation of many different codes in a fast way by replacing multi class labels with binary labels on the table. Second it is far too optimistic to think that this error rate computed at this stage and on the training set will behave like a typical ECOC scheme, which usually employs dichotomizers other than k-NN classifiers.

Attention must be paid to the choice of k. A big k can generate a lot of errors for the minority classes, while a little k can not be helpful to understand if some dichotomies have better performances. Experiments showed that a good rule of thumb, for now, is to choose a k next to the number of samples of the minority class. This way it should be avoided the possibility that some classes are encapsulated by others, while the effectiveness of the method is still maintained.

5 Experiments

In order to validate our tool, experiments have been made for an artificial data set and for two different multi class data sets from the UCI repository [9]. Notice that the given search algorithm outputs a temporary best code for every *while* iteration which evaluates $M \times (M-1)$ different matrices per time. Every output code for each of these iterations has been used as coding matrix for an ECOC system. We employed a logistic linear classifier as base classifier embedding the ECOC framework in the PR tools suite (`http://www.prtools.org`). Then, we observed the evolution of the test error at each step and studied how effective our tool is in providing a reliable evaluation for

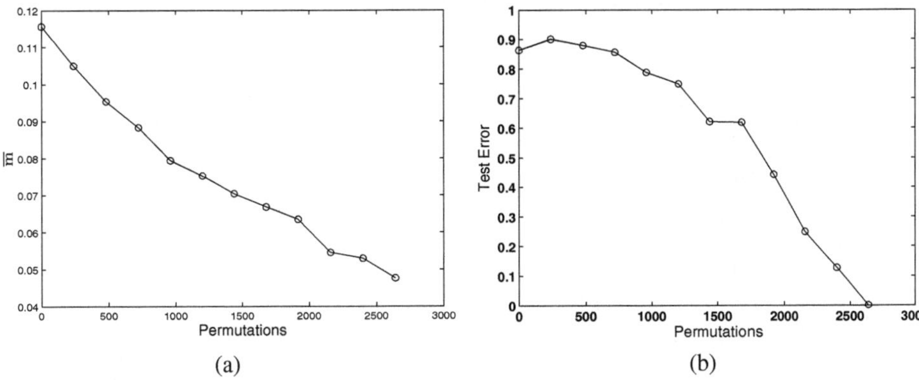

Fig. 3. Concurrent reduction of $\overline{\mathrm{m}}$ and of the test error for the artificial dataset

coding and dichotomies. In fig. 3 an example of the concurrent reduction of the quality parameter $\overline{\mathrm{m}}$ and of the test error for each *while* iteration of the algorithm 1 is shown for one run of experiments on the artificial data set. On the rest of the experimental section the results of training and test error are shown only at the starting and at the ending point of this algorithm.

5.1 Artificial Data Set

The artificial data set is formed by sixteen different classes uniformly distributed in a square region of a two dimensional feature space, which can be divided in a 4×4 pattern of squares as in fig. 4 (left). The ideal coding for this data set is represented by columns c_1 to c_6 of the code in fig. 4 (right), where c_1 to c_3 lead to three horizontal classifier and c_4 to c_6 to the three vertical, always shown on the same figure.

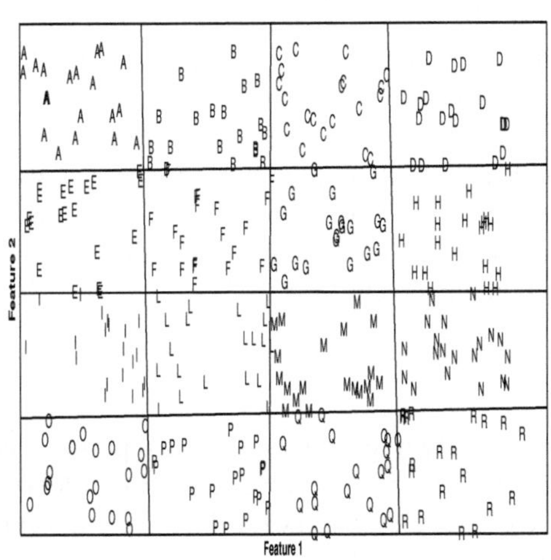

classes	$\mathbf{c_1}$	$\mathbf{c_2}$	$\mathbf{c_3}$	$\mathbf{c_4}$	$\mathbf{c_5}$	$\mathbf{c_6}$	$\mathbf{r_1}$	$\mathbf{r_2}$
A	1	1	1	0	0	0	1	1
B	1	1	1	1	0	0	1	1
C	1	1	1	1	1	0	1	1
D	1	1	1	1	1	1	1	1
E	0	1	1	0	0	0	0	1
F	0	1	1	1	0	0	1	1
G	0	1	1	1	1	0	1	1
H	0	1	1	1	1	1	1	0
I	0	0	1	0	0	0	0	1
L	0	0	1	1	0	0	0	1
M	0	0	1	1	1	0	1	0
N	0	0	1	1	1	1	1	0
O	0	0	0	0	0	0	0	1
P	0	0	0	1	0	0	0	0
Q	0	0	0	1	1	0	0	0
R	0	0	0	1	1	1	1	0

Fig. 4. Artificial data set in the two features space and a possible binary coding of the 16 classes

Table 2. Training and Test error for the Artificial Data Set: (a) ideal code (b) redundant code

(a)

Artificial Data set			
Training Error		Test Error	
0.7289	0.5045	0.7239	0.5042
0.9629	0.6869	0.9589	0.6815
0.8253	0.5453	0.8240	0.5546
0.9008	0.4865	0.9002	0.4998
0.8652	**0.0000**	0.8634	**0.0007**
0.9038	0.7737	0.9055	0.7719
0.9049	0.6224	0.9058	0.6254
0.8818	0.7569	0.8811	0.7583
0.9283	**0.0000**	0.9269	**0.0013**
0.8467	0.6858	0.8484	0.6875

(b)

Artificial Data set			
Training Error		Test Error	
0.8278	0.5298	0.8267	0.5289
0.9109	0.7355	0.9098	0.7339
0.8124	0.5449	0.8133	0.5519
0.9004	0.5656	0.8998	0.5436
0.8654	0.5553	0.8638	0.5656
0.8600	0.6303	0.8628	0.6384
0.8176	**0.3927**	0.8123	**0.3863**
0.8424	0.4178	0.8457	0.4030
0.8738	0.4381	0.8741	0.4310
0.8467	0.4637	0.8484	0.4711

We scramble ten times these multi class labels in order to have a casual row associated to a class and for each of these random initialization of the labels we run the search algorithm. We then wish to process and analyze different codings by permuting rows of these matrixes through the algorithm 1. Every time the optimum will be to reach the association multi class labels - binary labels shown in the figure, which correspond to an error rate of 0.0017.

Results in terms of error rate on the training and the test set are shown on table 2-(a): each row is one of the ten random initialization, while the train and test error column presents the values of the error rate, before and after running the algorithm 1. Other experiments were done with a redundant code, where two other columns are added to the ideal coding and corresponding to a diagonal split of the feature space. This coding is accomplished by the dichotomies r_1 and r_2 of the code in fig. 4 (right). Results of these experiments are on table 2-(b). The best error rate attainable with the entire code (c_1 - c_6 and r_1 - r_2) is 0.0785.

Observing the tables it is clear how the procedure always performs a reduction of both the error rates and we obtain the best performance on two runs for the code $c_1 - c_6$. In order to obtain the best association class-row we suggest then to carry out the random initialization procedure. The redundant code results also shows how the method always reduce the error rate, but the best performance is not reached because of the algorithm, which is only suboptimal and does not analyze a sufficient number of permutations to succeed. In both tables the best error rate attainable is bolded.

5.2 Public Data Sets

Differently from the previous set of experiments two real data sets from a public repository are chosen (characteristics showed in tab. 3) and ten cross validation folders are created for them. To fix the perfect choice of a coding matrix for the ECOC is beyond the purposes of this paper. We are interested in demonstrating that the tool is useful in understanding the best coding scheme for the ECOC. A random and short matrix

Table 3. Public data sets and coding matrices used in the experiments

Data Sets	# Classes	# Features	Coding Matrix	Length (L)	# Samples
SatImage	6	36	Random	15	5470
Letter	26	16	Die. 26-15	15	2970

is fixed for the SatImage data set, while for the Letter data set a code is selected from the set provided by Dietterich on his web page, just to make even more clear that the performance improvement does not involve the choice of the coding matrix. Even in this case, a multiple initialization procedure is done, randomly associating the classes to the rows of the coding, providing this way a different starting point for each run of the experiments.

Each row of table 4 is one of the initialization of the multi class labels, while train and test error columns are split into two columns: the first indicates the original error rate and the second is the error rate reached by the suboptimal procedure operated by the algorithm 1. It is clear that on SatImage data set (table 3-(a)) the error rate is always reduced as it happens for the parameter $\overline{\mathrm{m}}$, while this is not always true for the Letter data set (table 3-(b)).

These experiments show how the procedure is only suboptimal, because it always analyzes a number of coding matrices equal to $M \times (M-1) \times W$, where W is the number of $while$ cycles of the algorithm 1. For the Letter data set , for example, the total number of possible permutation of rows is 26!, while the algorithm tests $26 \times 25 \times 17$ different codes, i.e. a limited number of all the possible configurations. SatImage data set instead was analyzed for a number of combinations which was over the half of the number of permutations (6!) and the probability to find the best combination of rows is certainly higher. Another problem is in the choice of the base classifier, because it is probably hard for the logistic classifier to solve the binary problems defined by the dichotomies. This explains how the reduction of the quality parameter could not be in accordance with the reduction of the error rate sometimes. In both tables the best error rate attainable is bolded.

Table 4. Training and Test error for the for the Public Data Sets:(a) SatImage and (b) Letter

(a)

Satimage Data set			
Training Error		Test Error	
0.1956	**0.1710**	0.2041	**0.1786**
0.1925	**0.1710**	0.1960	**0.1786**
0.1962	0.1910	0.2014	0.1946
0.2005	0.1910	0.2053	0.1946
0.2230	**0.1710**	0.2306	**0.1786**
0.2003	0.1910	0.2057	0.1946
0.1725	**0.1710**	0.1805	**0.1786**
0.2018	**0.1710**	0.2047	**0.1786**
0.1889	**0.1710**	0.1940	**0.1786**
0.1926	0.1910	0.2005	0.1946

(b)

Letter Data set			
Training Error		Test Error	
0.6079	0.6188	0.6188	0.6289
0.6119	0.6124	0.6281	0.6239
0.6350	0.6150	0.6487	0.6217
0.6499	0.6285	0.6595	0.6333
0.6214	0.6235	0.6289	0.6373
0.6310	**0.5985**	0.6441	**0.6135**
0.6635	0.6302	0.6675	0.6439
0.6189	0.6352	0.6319	0.6492
0.6689	0.6284	0.6738	0.6412
0.6636	0.6117	0.6789	0.6304

6 Conclusions and Future Works

In this paper we proposed a fast and flexible tool to optimize the mapping between classes and the dichotomies defined in a chosen coding matrix. Both artificial and public data sets experiments validated our hypotheses. Using the random initialization procedure, on many cases we experienced the best performance. This suggests that many random initializations can be advantageous in order to find the best settings. A more appropriate and smart algorithm will be employed in the future works in order to fix the problems experimented with a huge data set like Letter.

Another known problem to fix is how to use the tool for an unbalanced data set. Future work will focus also on finding the optimal value of k in order to solve it. Moreover we will verify if the tool can be used to build a proper coding matrix from the data starting from the k-NN look-up table.

References

1. Dietterich, T.G., Bakiri, G.: Solving multiclass learning problems via error-correcting output codes. Journal of Artificial Intelligence Research 2, 263–286 (1995)
2. Kong, E.B., Dietterich, T.G.: Error-Correcting Output Coding Corrects Bias and Variance. In: International Conference on Machine Learning, pp. 313–321 (1995)
3. Kong, E.B., Dietterich, T.G.: Probability estimation via error-correcting output coding. In: Int. Conf. of Articial Inteligence and soft computing, Ban, Canada (1997)
4. Allwein, E.L., Schapire, R.E., Singer, Y.: Reducing multiclass to binary: A unifying approach for margin classifiers. In: Proceedings of the Seventeenth International Conference on Machine Learning, pp. 9–16 (2000)
5. Crammer, K., Singer, Y.: On the Learnability and Design of Output Codes for Multiclass Problems. Machine Learning 47 (2002)
6. Marrocco, C., Simeone, P., Tortorella, F.: Embedding Reject Option in ECOC Through LDPC Codes. In: Haindl, M., Kittler, J., Roli, F. (eds.) MCS 2007. LNCS, vol. 4472, pp. 333–343. Springer, Heidelberg (2007)
7. Pujol, O., Radeva, P., Vitrià, J.: Discriminant ECOC: A heuristic method for application dependent design of error correcting output codes. IEEE Trans. On Pattern Analysis And Machine Intelligence 28(6), 1001–1007 (2006)
8. Windeatt, T., Ghaderi, R.: Coding and decoding strategies for multi-class learning problems. Information Fusion 4(1), 11–21 (2003)
9. Blake, C., Keogh, E., Merz, C.J.: UCI Repository of Machine Learning Databases (1998), `http://www.ics.uci.edu/~mlearn/MLRepository.html`

Partially Corrective AdaBoost

Yan Jiang and Xiaoqing Ding

Department of Electronic Engineering, Tsinghua University, Beijing, China, 100084
{jyan,dxq}@ocrserv.ee.tsinghua.edu.cn

Abstract. We propose a novel algorithm to improve the ensemble performance of AdaBoost. The main contribution in our algorithm includes two aspects: (1) we aim to generate a distribution at each step that has less correlation with the previous classifiers so as to enhance the searching efficiency for new classifiers; (2) the classifiers weights can be iteratively modified along with the training process. In the proposed algorithm, the distribution is required to be corrective to some previous classifiers or some previous classifiers' linear combinations. Experiments on UCI Repository have validated the new algorithm's effectiveness.

1 Introduction

AdaBoost is a meta-algorithm, proposed by Y. Freund and R. E. Schapire ([1]) as an example to aggregate weak classifiers into a strong one in probably approximately correct (PAC) learning framework. A single weak classifier is only required to perform slightly better than random guess to assure the combined classifier is able to achieve arbitrarily good bounds on its training error.

Theoretical analysis reveals that the AdaBoost can be regarded as a gradient descent process in functional space ([2]) that minimizes a cost function, which is the upper bound of the training error. Based on the analysis, some researchers have suggested different ways to construct a more effective descent direction in order to get a faster convergence performance. For example, Friedman et al. ([3]) use Newton step in LogitBoost and GentleBoost and Li Ling pursues a conjugate gradient direction in his CGBoost to accelerate the training ([4]).

On the other hand, AdaBoost can also be regarded as an entropy projection process, which means the new distribution should be very 'close' to the the last in the sense of Kullback-Leibler divergence subjected to the constraint that requires the new distribution to be 'corrective' to the current classifier ([5]).

There are two main problems lies in AdaBoost: (1) the constraint for the next distribution is only designed to reduce the correlation between two adjacent member classifiers; (2) the member classifiers' coefficients can not be modified once assigned. The first problem results in the fact that the newly added member classifier, although uncorrelated to the last, is very similar to some previous classifiers. It may potentially counteract the searching efficiency for new member classifiers. Intuitively, the second problem will always weaken the ensemble performance since some previously assigned coefficients are no longer suitable when a new member is added. This can also be understood from a statistical view. In

N. da Vitora Lobo et al. (Eds.): SSPR&SPR 2008, LNCS 5342, pp. 469–478, 2008.

[6], AdaBoost is extended to regression model and regarded as a typical forward stage-wise variable selection method. However, this variable selection strategy is proved to be inferior to forward step-wise variable selection method, which iteratively modifies all the model coefficients once a new variable is selected.

2 Preliminaries

We are give N labeled training samples $\{(\mathbf{x}_i, y_i)\}_{1\leq i\leq N} \subset \mathcal{X} \times \{-1,+1\}$. AdaBoost selects one classifier $h_t : \mathcal{X} \to \{-1,+1\}$ at t-th round and assign a coefficient α_t to it. After T rounds of training, AdaBoost finally outputs a classifier ensemble $H_T = \mathrm{sgn}(F_T(\mathbf{x}))$, where $F_T(x) = \sum_{t=1}^{t=T} \alpha_t h_t(\mathbf{x})$.

2.1 Boosting Is a Gradient Descent Process

AdaBoost is analyzed as a gradient descent process to minimize a cost $C(F_t)$([2]).

$$C(F_t) = \log \sum_{i=1}^{i=N} \exp\big(-y_i F_t(\mathbf{x}_i)\big) \tag{1}$$

The gradient of $C(F_t)$ is denoted by $\nabla C(F_t)$ and its p-th component can be calculated as Eq.(2), where $d_t(p)$ is the p-th sample weight at t-th iteration.

$$\nabla C(F_t)_p \triangleq \frac{dC(F_t + s1_{\mathbf{x}_p})}{ds}\bigg|_{s=0} = -y_p d_{t+1}(p) \tag{2}$$

A new member classifier is required to be as 'close' as possible to the negative gradient direction of $C(F_t)$ in the sense of inner product (Eq.(3)). This process is equivalent to minimize the error rate according to the new distribution.

$$\big\langle -\nabla C(F_t), h_{t+1}\big\rangle = \sum_{i=1}^{i=N} h_{t+1}(\mathbf{x}_i) y_i d_{t+1}(i) = 1 - 2\varepsilon_{t+1} \tag{3}$$

h_{t+1} is selected as the new descent direction and we use exact line search to determine the step size (Eq.(4)), which is also the coefficient assigned to h_{t+1}.

$$\alpha_{t+1} = \arg\min_{\alpha\in\mathbb{R}} C(F_t + \alpha h_{t+1}) \tag{4}$$

2.2 Boosting as Entropy Projection and Totally Corrective AdaBoost

In [5], at each iteration, the new distribution is regarded as the 'projection' onto the hyperplane Δ_t (Eq.(5)). This actually gives the Lagrange dual of Eq.(4). The 'corrective' constraint of the hyperplane requires the new distribution makes the current weak hypothesis uncorrelated with the training samples. Therefore, the new classifier h_{t+1} must have some new classification behaviors compared with

h_t. It guarantees that AdaBoost can generate a sequence of diverse classifiers that is essential to the performance of the ensemble. However, as we have discussed, one 'corrective' constraint is not enough to adequately reduce the correlations between the new member classifier and all the previous ones.

$$\Delta_t \triangleq \left\{ \mathbf{u} \in \mathbb{R}^N \middle| \sum_{i=1}^{i=N} y_i h_t(\mathbf{x}_i) u_i = 0, \sum_{i=1}^{i=N} u_i = 1, u_i \geq 0, \forall 1 \leq i \leq N \right\} \quad (5)$$

$$\mathbf{d}_{t+1} = \arg\min_{\mathbf{u}\in\Delta_t} \sum_{i=1}^{i=N} u_i \log \frac{u_i}{d_t(i)} \quad (6)$$

Intuitively, we can obtain a more effective classifier that has less correlation with the already available classifiers by simply extending the 'corrective' requirement to all the previous classifiers (Eq.(7)). This is exactly the basic idea of totally corrective AdaBoost (TC-AB) ([5]). The dual problem of Eq.(6) subjected to the 'totally corrective' constraint Eq.(7) is Eq.(8). Consequently, we must modify the classifiers' weight as $F_t = F_{t-1} + \sum_{\tau=1}^{\tau=t} \alpha_{t,\tau} h_\tau$. TC-AB actually uses all the previous classifiers to construct the descent direction to minimize the same cost function as AdaBoost, where a 'good' descent direction is adopted.

$$\sum_{i=1}^{i=N} u_i y_i h_\tau(x_i) = 0, \qquad \forall 1 \leq \tau \leq t \quad (7)$$

$$\boldsymbol{\alpha}_t = \arg\min_{\boldsymbol{\alpha}=(\alpha_1,\alpha_2,\ldots,\alpha_t)\in\mathbb{R}^t} C(F_{t-1} + \sum_{\tau=1}^{\tau=t} \alpha_\tau h_\tau) \quad (8)$$

There are two main problems lie in TC-AB. Firstly, it is always not practical that the 'totally corrective' constraints have feasible solution; secondly, the number of the 'corrective' constraints continuously increases when more member classifiers are aggregated. So TC-AB is computationally expensive and its complexity can not be controlled. Inspired by the above discussion, we mainly focus on how to improve the cost function minimization by incorporating more information from previous member classifiers. Besides, the computational complexity of the developed algorithm can be adequately controlled.

3 Partially Corrective AdaBoost

In this section, we build a 'partially corrective' distribution instead by relaxing the 'totally corrective' constraints. We provides two different implementations for partially corrective AdaBoost (PC-AB).

3.1 Two Types of Partially Corrective AdaBoost

In the first case (PC-AB1), the new distribution is only required to be corrective to some of the previous classifiers(Eq.(9)). In the second case (PC-AB2), we construct a set of linear combinations of all the previous classifiers in advance and force the new distribution to be 'corrective' to these combinations(Eq.(10)).

The dual problems of PC-AB1 and PC-AB2 can be summarized in Eq.(11a) and Eq.(12b) respectively. PC-AB1 selects a subset from the previous classifiers to construct a new descent direction and the combination coefficients are determined by exact line search. PC-AB2 constructs a set of candidate directions by linearly combining all the classifiers (Eq.(12a)) and also uses exact line search to decide how to combine these directions (Eq.(12b)). It's easy to check that the distribution calculated from Eq.(9) satisfies Eq.(11b) and the distribution calculated from Eq.(10) also satisfies Eq.(12c). Consequently, member classifiers' coefficients must be backward adjusted respectively (Eq.(11c) and Eq.(12d)). Thus we can see that the requirement for a new distribution that is less correlated to the previous member classifiers is intrinsically consistent with the adjustment of member classifiers' coefficients from Lagrange duality.

PC-AB is regarded as a tradeoff between the requirement of time complexity and the adequate consideration of the previous classifiers' information. It is easy to see that AdaBoost and TC-AB can be summarized as special cases of PC-AB1; further, PC-AB2 implements a more general idea of PC-AB1. Actually, CGBoost is categorized as a special form of PC-AB2's idea, but it utilizes only one linear combination as the descent direction. Since our attempts in PC-AB1 and PC-AB2 aim to adopt a more effective descent direction in minimizing the same cost function as AdaBoost(Eq.(1)), it is easy to understand that we have the same training error bound as AdaBoost. But we can always achieve faster decreasing training errors because we optimize the cost function more aggressively.

$$\sum_{i=1}^{i=N} d_{t+1}(\mathbf{x}_i) y_i h_{t_k}(\mathbf{x}_i) = 0, 1 \leq t_k \leq t, 1 \leq k \leq m \tag{9}$$

$$\sum_{i=1}^{i=N} d_{t+1}(\mathbf{x}_i) y_i \Big(\sum_{\tau=1}^{\tau=t} \omega_{t,k,\tau} h_\tau(\mathbf{x}_i) \Big) = 0, 1 \leq k \leq m \tag{10}$$

3.2 Diversity and Heuristic Classifier Subset Selection for PC-AB1

The problem in PC-AB1 is how to choose a representative classifier subset in order to achieve a good descending performance. An intuitive idea is that the selected classifiers should have the following properties: (1) the selected classifiers should have good classification performance considering the current distribution; (2) the correlations among the selected classifiers should be small. These two points can be explained from a computational view: the first requires that all the selected directions are 'close' to the negative gradient and the second requires that these directions are also nearly orthogonal. Due to great time requirement, it is not practical to try all the classifier subsets to find the best choice. So we propose a heuristic classifier selection strategy for PC-AB1(Fig.1).

The discrepancy measure to evaluate two classifiers' similarity here is 'diversity'. Diversity and classification rate are regarded as two important factors in classifier ensemble. The latter is always measured by the error rate while

Input: t available classifiers $h_1, h_2, \ldots, h_t$ and a positive integer m, where $1 \le m \le t$.
Initialization: Initialize the candidate set as $C = \{h_t\}$.
Repeat: Choose a classifier $h_{\tilde{k}}$ from $\{h_1, h_2, \ldots, h_t\} \backslash C$ that satisfies Eq.(13) and add it to C. $\gamma > 0$ is a constant, diver$(\bullet, \bullet)$ is a diversity function and $\varepsilon_t(h) = \sum_{h(\mathbf{x}_i) \neq y_i} d_t(i)$.

$$h_{\tilde{k}} = \underset{h \in \{h_1, h_2, \ldots, h_t\} \backslash C}{\arg\max} \frac{\gamma}{\# C} \sum_{g_j \in C} \text{diver}(h, g_j) - \varepsilon_t(h) \tag{13}$$

Until: the size of C is m
Output: The selected classifier subset C

Fig. 1. Heuristic classifier subset selection

there is no straightforward criterion to measure the former. Presently, various pairwise diversity measures have been proposed, such as Disagreement measure, Q-statistics, Correlation coefficient and Kappa degree of agreement ([7]). We can choose any one of them and different diversity functions. On the whole, different diversity functions lead to similar performance in the experiments, although they may generate slightly different performance on different data sets. However, randomly selected classifier subset may counteract the training performance.

$$\boldsymbol{\alpha}_t = \underset{\boldsymbol{\alpha} = (\alpha_1, \alpha_2, \ldots, \alpha_m) \in \mathbb{R}^m}{\arg\min} C\Big(F_{t-1} + \sum_{k=1}^{k=m} \alpha_k h_{t_k}\Big) \tag{11a}$$

$$d_{t+1}(i) = d_t(i) \exp\big(-y_i \sum_{k=1}^{k=m} \alpha_{t,k} h_{t_k}(\mathbf{x}_i)\big) \Big/ \sum_{j=1}^{j=N} d_t(j) \exp\big(-y_j \sum_{k=1}^{k=m} \alpha_{t,k} h_{t_k}(\mathbf{x}_j)\big) \tag{11b}$$

$$F_t(\mathbf{x}) = F_{t-1}(\mathbf{x}) + \sum_{k=1}^{k=m} \alpha_{t,k} h_{t_k}(\mathbf{x}) \tag{11c}$$

$$g_{t,k} = \sum_{\tau=1}^{\tau=t} \omega_{t,k,\tau} h_\tau \tag{12a}$$

$$\boldsymbol{\alpha}_t = \underset{\boldsymbol{\alpha}(= \alpha_1, \alpha_2, \ldots, \alpha_m) \in \mathbb{R}^m}{\arg\min} C\Big(F_{t-1} + \sum_{k=1}^{k=m} \alpha_k g_{t,k}\Big) \tag{12b}$$

$$d_{t+1}(i) = d_t(i) \exp\big(-y_i \sum_{k=1}^{k=m} \alpha_{t,k} g_{t,k}(\mathbf{x}_i)\big) \Big/ \sum_{j=1}^{j=N} d_t(j) \exp\big(-y_j \sum_{k=1}^{k=m} \alpha_{t,k} g_{t,k}(\mathbf{x}_j)\big) \tag{12c}$$

$$F_t(\mathbf{x}) = F_{t-1}(\mathbf{x}) + \sum_{k=1}^{k=m} \alpha_{t,k} g_{t,k}(\mathbf{x}) \tag{12d}$$

3.3 A New Criterion for PC-AB

In this section, we discuss a more accurate evaluation for effective choice of classifiers. The basic idea is to approximate the cost function using a simple function and try to minimize the new function instead. For the classifiers $h_{t_1}, h_{t_2}, \ldots, h_{t_m}$, we define a $N \times m$ matrix $\mathbf{H} = \{y_p h_{t_q}(\mathbf{x}_p)\}_{1 \le p \le N, 1 \le q \le m}$. Its combination coefficients are represented by a vector $\mathbf{c}$. By Taylor expansion, we approximate the cost function as Eq.(14). The minimum of $K_t(\mathbf{H}, \mathbf{c})$ (denoted by $K_t(\mathbf{H})$, Eq.(16)) is achieved only when $\mathbf{c} = -(\mathbf{H}^T \nabla^2 C(F)\mathbf{H})^{-1}\mathbf{H}^T \nabla C(F)$.

$$C(F_t + \mathbf{Hc}) \approx C(F_t) + K_t(\mathbf{H}, \mathbf{c}) \tag{14}$$

$$K_t(\mathbf{H}, \mathbf{c}) = \mathbf{c}^T \mathbf{H}^T \nabla C(F_t) + \frac{1}{2}\mathbf{c}^T \mathbf{H}^T \nabla^2 C(F_t)\mathbf{Hc} \tag{15}$$

$$K_t(\mathbf{H}) \triangleq \min_{\mathbf{c} \in \mathbb{R}^m} K(\mathbf{H}, \mathbf{c}) = -\frac{1}{2}\nabla C(F_t)^T \mathbf{H}(\mathbf{H}^T \nabla^2 C(F_t)\mathbf{H})^{-1}\mathbf{H}^T \nabla C(F_t) \tag{16}$$

$\nabla^2 C(F_t)$ (Eq.(17)) is the Hessian matrix of $C(F_t)$ and it is both symmetric and semi-definite. But $\mathbf{H}^T \nabla^2 C(F_t)\mathbf{H}$ is always non-invertible. We slightly modify our criterion to $Q_t(\mathbf{H})$ by introducing a regularization item and discarding the constant factor. Eq.(18) corresponds to a modified goal function $\tilde{K}_t(\mathbf{H}, \mathbf{c})$ shown in Eq.(19). On the other hand, regularization is also helpful to control $\mathbf{c}$ in order to assure the validity of Taylor expansion. Next, we focus on how to minimize $Q_t(\mathbf{H})$ in order to achieve very promising descending potentialities.

$$\nabla^2 C(F_t)_{p,q} = \begin{cases} d_t(p)(1 - d_t(p)) & p = q \\ -y_p y_q d_t(p) d_t(q) & p \ne q \end{cases} \tag{17}$$

$$Q_t(\mathbf{H}) \triangleq \nabla C(F_t)^T \mathbf{H}(\mathbf{H}^T \nabla^2 C(F_t)\mathbf{H} + \lambda \mathbf{I})^{-1}\mathbf{H}^T \nabla C(F_t) \tag{18}$$

$$\tilde{K}_t(\mathbf{H}, \mathbf{c}) = \mathbf{c}^T \mathbf{H}^T \nabla C(F_t) + \frac{1}{2}\mathbf{c}^T \mathbf{H}^T \nabla^2 C(F_t)\mathbf{Hc} + \frac{\lambda}{2}\|\mathbf{c}\|_2^2 \tag{19}$$

First we can also replace the criterion Eq.(13) by our new criterion and use the same heuristic classifier subset selection strategy. Besides, the new criterion provides us a way to implement PC-AB2. Now we put all the member classifiers into $\mathbf{H}$. Since $\mathbf{H}^T \nabla^2 C(F_t)\mathbf{H} + \lambda \mathbf{I}$ is symmetric and positive-definitive, we have orthogonal matrix $\boldsymbol{\Sigma}$ that satisfies $\boldsymbol{\Sigma}^T(\mathbf{H}^T \nabla^2 C(F_t)\mathbf{H} + \lambda \mathbf{I})\boldsymbol{\Sigma} = \boldsymbol{\Lambda}$, where $\boldsymbol{\Lambda}$ is a diagonal matrix of the eigenvalues of $\mathbf{H}^T \nabla^2 C(F_t)\mathbf{H} + \lambda \mathbf{I}$. Therefore $Q_t(\mathbf{H})$ can be simplified as Eq.(20), where $\boldsymbol{\Lambda} = \mathrm{diag}\{\lambda_1, \lambda_2, \ldots, \lambda_m\}$ and $\boldsymbol{\Sigma}^T \mathbf{H}^T \nabla C(F_t)$ is a m-dimensional vector represented by $[l_1, l_2, \ldots, l_m]^T$. The detailed algorithm to construct classifier combinations at each iteration is shown in Fig.2, then follow Eq.(12)(a)-(d) to adjust sample distribution and member classifiers' coefficients.

$$Q_t(\mathbf{H}) = \nabla C(F_t)^T \mathbf{H}\boldsymbol{\Sigma}\boldsymbol{\Lambda}^{-1}\boldsymbol{\Sigma}^T \mathbf{H}^T \nabla C(F_t) = \sum_{k=1}^{k=m} \frac{l_k^2}{\lambda_k} \tag{20}$$

Input: t available classifiers $h_1, h_2, \ldots, h_t$ and a positive integer m, where $1 \leq m \leq t$.
Step1: Calculate $\nabla^2 C(F_t)$ as Eq.(17).
Step2: Find a orthogonal matrix $\boldsymbol{\Sigma}$ and a diagonal matrix $\boldsymbol{\Lambda} = \mathrm{diag}\{\lambda_1, \lambda_2, \ldots, \lambda_m\}$ that satisfies $\boldsymbol{\Sigma}^T(\mathbf{H}^T\nabla^2 C(F_t)\mathbf{H} + \lambda\mathbf{I})\boldsymbol{\Sigma} = \boldsymbol{\Lambda}$ and let $[l_1, l_2, \ldots, l_t]^T = \boldsymbol{\Sigma}^T\mathbf{H}^T\nabla C(F_t)$.
Step3: Sort $\left\{\frac{l_k^2}{\lambda_k}\right\}_{1\leq k\leq t}$ in descending order as $\frac{l_{j1}^2}{\lambda_{j_1}} \geq \frac{l_{j2}^2}{\lambda_{j_2}} \geq \cdots \geq \frac{l_{jt}^2}{\lambda_{j_t}}$.
Output: $g_{t,k} = \sum_{\tau=1}^{\tau=t} \sigma_{\tau,j_k} h_\tau, 1 \leq k \leq m$, where $\boldsymbol{\Sigma} = \{\sigma_{p,q}\}_{1\leq p\leq t, 1\leq q\leq t}$.

Fig. 2. Generating classifier linear combinations for PC-AB2

4 Experiments

4.1 Experiment Configurations

We performed the experiments on 12 UCI machine learning data sets ('balance-scale', 'breast-cancer-w', 'credit-a', 'credit-g', 'diabetes', 'heart-c', 'heart-statlog', 'ionosphere', 'kr-vs-kp', 'sonar', 'splice', 'waveform5000'). The base learner is decision stump implemented by WEKA. We use a 10-fold cross validation to compare the training and testing performance of standard AdaBoost and three proposed implementation strategies: 1)**PCAB1D**: PC-AB1 with heuristic classifier selection based on the criterion described in Eq.(13) ($\gamma = 1, m = 5$); 2)**PCAB1Q**: PC-AB1 with heuristic classifier selection based on the criterion described in Eq.(18) ($m = 5$); 3)**PCAB2Q**: PC-AB2 with given classifier combinations as described in section 3.3 ($\lambda = 0.001, m = 5$). m controls algorithm's complexity, so larger m will lead to faster training error convergence but more time consumption. γ and λ are both set by experience without optimization step.

4.2 Experiment Results

The results are shown in Fig.3(a)-(p) (Due to the page limitation, only part of the results are shown). In each graphic, we compare the ensemble error variations of AdaBoost (blue), PCAB1D (red), PCAB1Q (green) and PCAB2Q (black).

From the experiments, we find that PCAB1D, PCAB1Q and PCAB2Q all outperform AdaBoost and PCAB2Q always has the overall best performance. The diversity based classifier selection, as we have discussed, uses an empirical evaluation to select a classifier subset, so PCAB1D performs worse than PCAB1Q, which adopts a more accurate criterion to measure the cost function descent potentialities. Although PCAB1Q and PCAB2Q both select descent directions by the same criterion Q_t, PCAB2Q develops a more aggressive way by using the linear combinations of all the previous classifiers, it is believed that PCAB2Q can achieve better performance than the other two strategies, which is also substantiated by the experiments. Generalization performance is also improved in PC-AB, it can substantiate that the cost function not only bounds the training error, but also potentially controls the generalization risk.

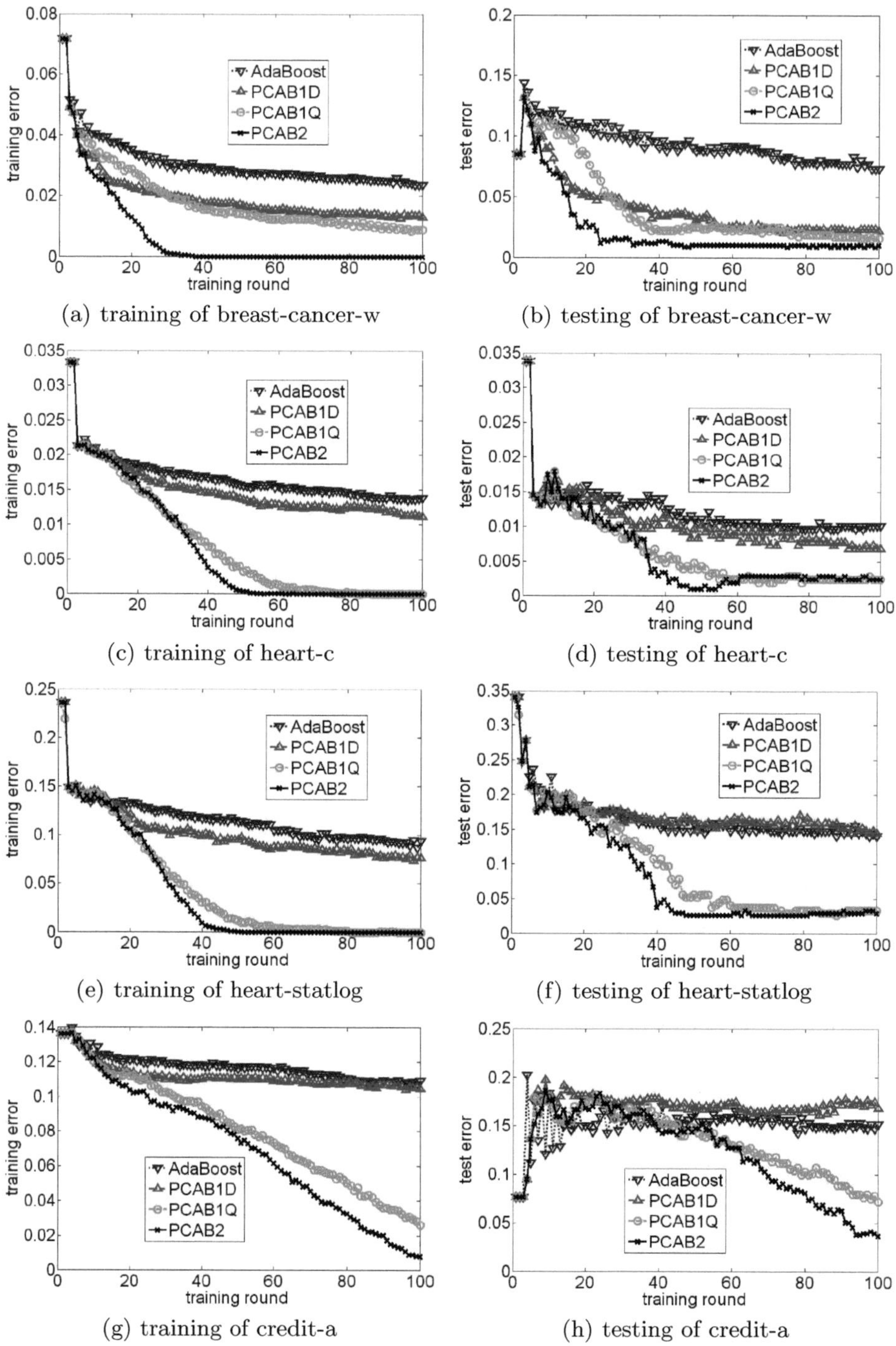

(a) training of breast-cancer-w (b) testing of breast-cancer-w

(c) training of heart-c (d) testing of heart-c

(e) training of heart-statlog (f) testing of heart-statlog

(g) training of credit-a (h) testing of credit-a

Fig. 3. Experiment results on selected UCI data sets

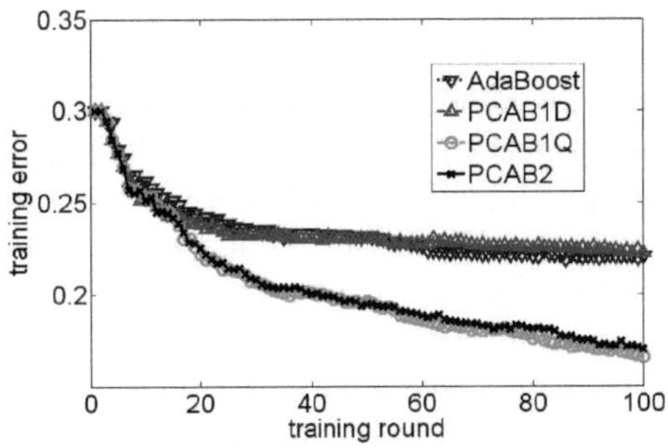

(a) training of credit-g

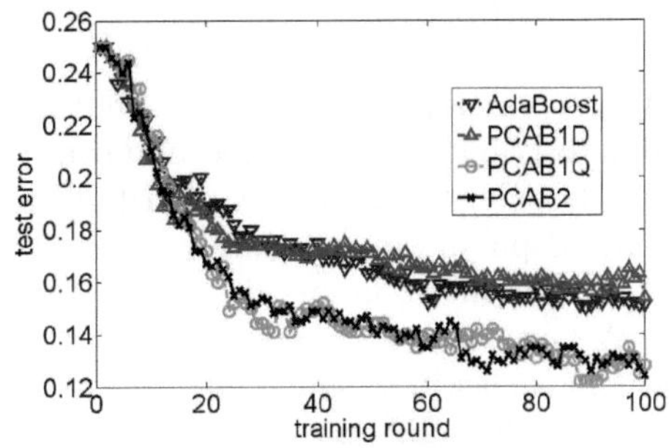

(b) testing of credit-g

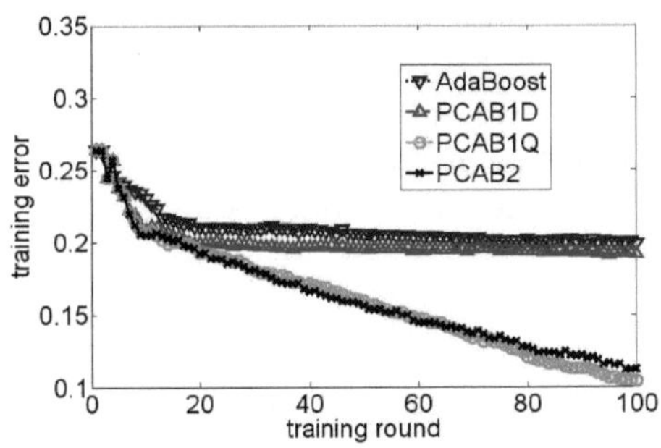

(c) training of diabetes

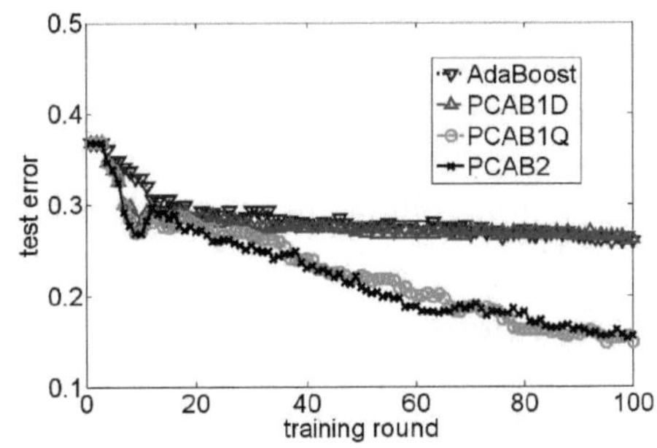

(d) testing of diabetes

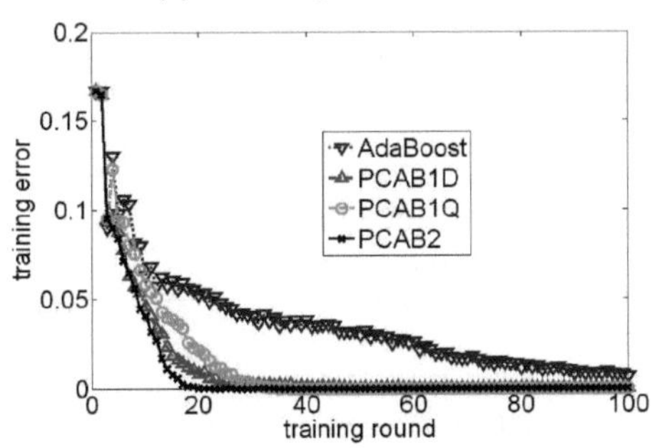

(e) training of ionosphere

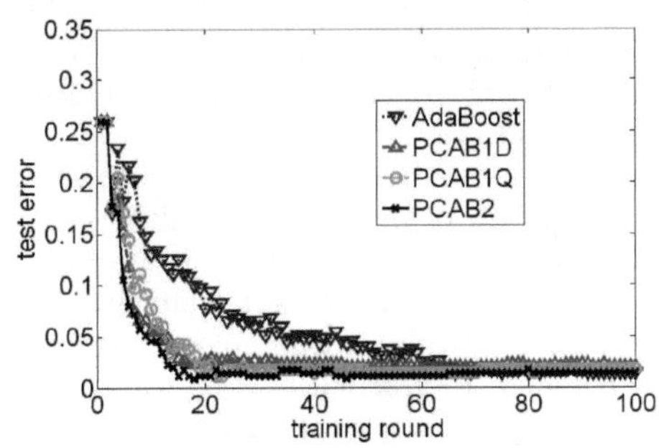

(f) testing of ionosphere

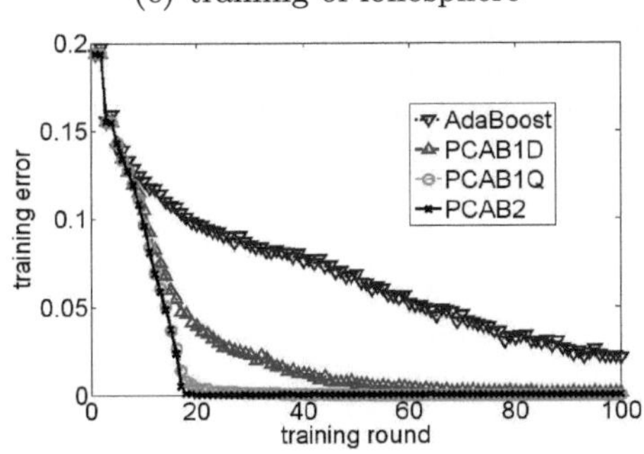

(g) training of balance-scale

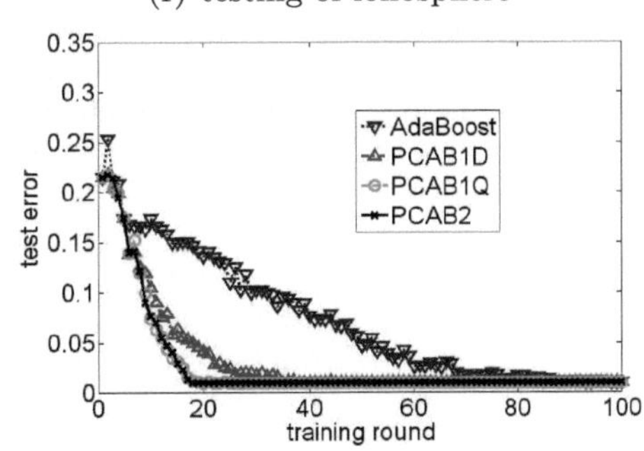

(h) testing of balance-scale

Fig. 3. (*continued*)

5 Conclusion

We propose a practical way to improve the convergence of AdaBoost by seeking for a distribution update method that reduces the correlations between the previous classifiers and the new distribution. Comparing with 'totally corrective', our method pursues a 'partially corrective' distribution to balance the requirement of considering more previous classifiers' information and time complexity.

We propose two types of PC-AB: PC-AB1 aims to select a subset from the previous classifiers and make the new distribution uncorrelated to these classifiers; in PC-AB2, the new distribution is required to be corrective to a set of classifier linear combinations. Lagrange duality theory shows that the proposed methods actually focus on how to construct a more effective descent direction by adequately considering the previously adopted directions. Thus the previous member classifiers' coefficients can be iteratively modified to achieve better ensemble performance. Experiments show that our method can significantly improve both training and generalization performance of standard AdaBoost.

References

1. Freund, Y., Schapire, R.E.: A Decision-Theoretic Generalization of On-line Learning and an Application to Boosting. In: Vitányi, P.M.B. (ed.) EuroCOLT 1995. LNCS, vol. 904, pp. 23–37. Springer, Heidelberg (1995)
2. Mason, L., Baxter, J., Bartlett, P.L., et al.: Boosting Algorithms as Gradient Descent. In: Advances in Neural Information Processing Systems (NIPS 1999), pp. 512–518 (1999)
3. Friedman, J.H., Hastie, T., Tibshirani, R.: Additive Logistic Regression: A Statistical View of Boosting. Annals of Statistics 28(2), 337–407 (1998)
4. Li, L., Abu-Mostafa, Y.S., Pratap, A.: CGBoost: Conjugate Gradient in Function Space. Technical Report CaltechCSTR: 2003.007, Learning Systems Group, California Institute of Technology (August 2003)
5. Kivinen, J., Warmuth, M.K.: Boosting as Entropy Projection. In: 12th Annual Conference on Computational Learning Theory (COLT 1999), pp. 134–144 (1999)
6. Bühlmann, P., Yu, B.: Boosting with the L_2-Loss: Regression and Classification. Journal of American Statistical Association 98, 324–339 (2003)
7. Kuncheva, L.I., Whitaker, C.J.: Measures of Diversity in Classifier Ensembles and Their Relationship with the Ensemble Accuracy. Machine Learning 51(2), 181–207 (2003)

Selection and Fusion of Similarity Measure Based Classifiers Using Support Vector Machines

Mohammad T. Sadeghi[1], Masoumeh Samiei[1], and Josef Kittler[2]

[1] Signal Processing Research Laboratory
Department of Electrical and Computer Engineering
Yazd University, Yazd, Iran
[2] Centre for Vision, Speech and Signal Processing
School of Electronics and Physical Sciences
University of Surrey, Guildford GU2 7XH, UK
{M.Sadeghi,J.Kittler}@surrey.ac.uk

Abstract. In this paper, we address the problem of selecting and fusing similarity measures based classifiers in LDA face space. The performance of a face verification system in an LDA feature space using different similarity measure based classifiers is experimentally studied first. The study is performed for both manually and automatically registered face images. A sequential search approach which is in principle similar to the "plus L and take away R" algorithm is then applied in order to find an optimum subset of the adopted classifiers. The selected classifiers are combined using the SVM classifier. We show that although, individually, one of the adopted scoring functions, the Gradient Direction distance outperforms the other metrics, by fusing different similarity measures using the proposed method, the resulting decision making scheme improves the performance of the system in different conditions.

1 Introduction

In spite of the rapid advances in machine learning, in many pattern recognition problems the decision making is based on simple concepts such as distance from or similarity to some reference patterns. This type of approach is particularly relevant when the number of training samples available to model a class of objects is very limited. Examples of such situations include content based retrieval from image or video databases, where the query image is the only sample at our disposal to define the object model, or biometrics where only one or a few biometric traits can be acquired during subject enrolment to create a reference template. In biometric identity verification, a similarity function measures the degree of similarity of an unknown pattern to the claimed identity template. If the degree exceeds a prespecified threshold, the unknown pattern is accepted to be the same as the claimed identity. Otherwise it is rejected.

The similarity score is computed in a suitable feature space. Commonly, similarity would be quantified in terms of a distance function, on the grounds that

N. da Vitora Lobo et al. (Eds.): SSPR&SPR 2008, LNCS 5342, pp. 479–488, 2008.

similar patterns will lie physically close to each other. Thus smaller the distance, the greater the similarity of two entities.

The role of the feature space in similarity measurement is multifold. First of all the feature space is selected so as to maximise the discriminatory information content of the data projected into the feature space and to remove any redundancy. However, additional benefit expected from mapping the original pattern data into a feature space is to simplify the similarity measure deployed for decision making. Linear Discriminant Analysis (LDA) is a powerful feature extraction tool for pattern recognition in general and for face recognition in particular. It was introduced to this application area by Belhumeur in 1996 [1]. An important contributing factor in the performance of a face authentication system within the LDA space is the metric used for defining a matching score. Theoretically, Euclidean distance provides an optimal measure in the LDA space. However, it has been demonstrated that it is outperformed by the Normalised Correlation (NC) and Gradient Direction (GD) [12]. Also, in [13], the performance of the NC scoring function was compared with the GD metric. The study was performed on the BANCA database [1] using internationally agreed experimental protocols by applying a geometric face registration method based on manually or automatically annotated eyes positions. It was concluded that overall the NC function is less sensitive to miss-registration error but in certain conditions GD metric performs better. However, in [14], it has been further demonstrated that by optimising the GD metric, this metric almost always outperforms the NC metric for both manually and automatically registered data. A variety of other metrics has been adopted as the similarity measure in different pattern recognition applications. Experimental results confirm that, individually other metrics, overall, do not perform as well as the NC and GD metrics in the LDA space for face verification. However, in different conditions, different classifiers can deliver better performance.

In this work we adopted 8 classifiers based on different similarity measures including Euclidean, City-block, Chebyschev, Canberra, Chi-square (χ^2) , NC, GD and Correlation coefficient-based distance. It is well known that a combination of many different classifiers can improve classification accuracy. The main idea behind the current study is to select and fuse the best classifiers in different conditions.

A variety of schemes have been proposed for combining multiple classifiers. Fusion rules can be divided into two main categories: fixed rules such as the sum, minimum, maximum and median rule [6] [7] [17] and trained rules like the weighted averaging of classifiers outputs[16] [11], Support Vector Machines (SVM)[14], bagging and boosting [4]. In [8], AdaBoost has been adopted for combining unimodal features extracted from face and speech signals of individuals in multimodal biometrics. Overall, the fixed rules are most often used because of their simplicity and do not requiring to a training data. In such a way, equal weights are used for all the classifiers [6] [3]. However, in different applications it has been demonstrated that trained approaches such as Support

[1] http://www.ee.surrey.ac.uk/banca/

Vector Machines (SVMs) have the potential to outperform the simple fusion rules, especially when a large enough training data is available.

Our approach to the classifier combination is at the decision-level. The individual classifiers are based on different similarity measures and each classifier outputs a score. The scores are treated as features and a second level classifier such as support vector classifiers, neural networks, Parzen classifier, etc., is constructed by training on these scores. Therefore, it is pertinent to ask which classifiers provide complementary information and how the expert scores should be fused to achieve the best possible performance of the face verification system. In [13] the fusion problem was solved by selecting the best classifier or a group of classifiers dynamically with the help of a gating function learnt for each similarity measures. In the present study we formulate the similarity measures fusion problem as a feature selection problem. The similarity measures are selected using a sequential search approach, the Plus L and Take Away R algorithm [2].

An important contributing factor in the selection algorithm is the "fusion rule" used for combining the similarity measure based classifiers. The experimental results in [14] confirm that, by fusing experts employing diverse similarity measures using the SVM classifier, the performance of the system is superior to that of any single similarity measure. It also improves compared to classical non trainable fusion strategies.

The main aim of this paper is to study the performance of the proposed similarity measure selection algorithm using the SVM classifiers. Surprisingly good results are obtained using the proposed method. The paper is organised as follows. In the next section the adopted scoring functions are introduced. The proposed method of classifier selection is described in Section 3. A description of the experimental design including the face database used in the study, the experimental protocols and the experimental setup are given in Section 4. The experimental results using the adopted scoring functions and the fusion results are presented and discussed in Section 5. Finally a summary of the main findings and conclusions can be found in Section 6.

2 Similarity Functions

In a similarity measure based face verification system, a matching scheme measures the similarity or distance of the test sample, $\mathbf{x}$ to the template of the claimed identity, $\boldsymbol{\mu}_i$. Note that $\mathbf{x}$ and $\boldsymbol{\mu}_i$ are the projections of the test sample and class mean into the feature space respectively. The general form of a group of similarity measures which is called *Minkowski Distance* or *power norm metrics* (L_p) is defined as:

$$s_M = \left| \sum_{j=1}^{m} (\boldsymbol{\mu}_{ij} - \mathbf{x}_j)^p \right|^{1/p} \tag{1}$$

where m is the dimensionality.

The most commonly used similarity measures, *Manhattan* or *City-block* metric, *Euclidean* Distance (ED) and *Chebyschev* Distance are derived from the above definition considering $p = 1$, $p = 2$ and $p \to \infty$ respectively , i.e. L_1, L_2 and L_∞ metrics:

$$s_{City} = \sum_{j=1}^{m} \left|\boldsymbol{\mu}_{ij} - \mathbf{x}_j\right|, \quad s_{ED} = \sqrt{(\mathbf{x} - \boldsymbol{\mu}_i)^T(\mathbf{x} - \boldsymbol{\mu}_i)}, \quad s_{Cheb} = \max_j \left|\boldsymbol{\mu}_{ij} - \mathbf{x}_j\right| \tag{2}$$

The *Canberra* Distance is also given by

$$s_{Canb} = \sum_{j=1}^{m} \frac{\left|\boldsymbol{\mu}_{ij} - \mathbf{x}_j\right|}{\left|\boldsymbol{\mu}_{ij}\right| + |\mathbf{x}_j|} \tag{3}$$

This can be considered as the normalised Manhattan Distance. The *Chi-squared* (χ^2) Distance is defined by

$$s_{\chi^2} = \sum_{j=1}^{m} \frac{(\boldsymbol{\mu}_{ij} - \mathbf{x}_j)^2}{\left|\boldsymbol{\mu}_{ij}\right| + |\mathbf{x}_j|} \tag{4}$$

which is basically a relative Euclidean squared distance and is usually meant for non negative variables only.

In different applications, it has been shown that a matching score based on *Normalised Correlation* (NC) scoring function, defined by Equation 5, is more efficient.

$$s_{NC} = \frac{||\mathbf{x}^T \boldsymbol{\mu}_i||}{\sqrt{\mathbf{x}^T \mathbf{x} \boldsymbol{\mu}_i^T \boldsymbol{\mu}_i}} \tag{5}$$

Another similarity measure which is conceptually same as the NC function is Distance based *Correlation Coefficients*. For more details, the reader is referred to [9].

The *Gradient Direction* (GD) metric measures the distance between a probe image and a model in the gradient direction of the aposteriori probability function associated with the hypothesised client identity i [12]. A mixture of Gaussian distributions with isotropic covariance matrix has been assumed as the density function representing the anticlass (world population) estimated from the data provided by all the other users ($\forall j \neq i$). The diagonal elements of the isotropic covariance matrix are assumed to have values related to the magnitude of variation of the image data in the feature space. It was demonstrated that applying GD metric is even more efficient than the NC function. The matching score is defined as

$$s_{GD} = \frac{||(\mathbf{x} - \boldsymbol{\mu}_i)^T \nabla_O P(i|\mathbf{x})||}{||\nabla_O P(i|\mathbf{x})||} \tag{6}$$

where $\nabla_O P(i|\mathbf{x})$ refers to the optimal gradient direction. For the isotropic structure of the covariance matrix, i.e. $\boldsymbol{\Sigma} = \sigma\mathbf{I}$,the optimal direction would be

$$\nabla_I P(i|\mathbf{x}) = \sum_{j \neq i, j=1}^{m} p(\mathbf{x}|j)(\boldsymbol{\mu}_j - \boldsymbol{\mu}_i) \tag{7}$$

Note that the magnitude of σ will affect the gradient direction through the values of density $p(\mathbf{x}|j)$.

3 Similarity Measure Based Classifiers Selection

One of the very promising research directions in the field of pattern recognition and computer vision is classifier fusion. It has been recognised that the classical approach to designing a pattern recognition system which focuses on finding the best classifier has a serious drawback. Any complementary discriminatory information that other classifiers may capture is not tapped. Multiple expert fusion aims to make use of many different designs to improve the classification performance. In the case considered here, as different metrics span the feature space in different ways, it seems reasonable to expect that a better performance could be obtained by combining the resulting classifiers.

Multiple expert fusion aims to make use of many different designs to improve the classification performance. The approach we adopted for selecting the best set of classifiers is similar in principal to the sequential feature selection methods [10]. In this study, the Sequential Forward Selection (SFS) and Plus'L' and Take away 'R' algorithms were examined for selecting an optimum subset of similarity measures. A trained fusion rule, realised by a Support Vector Machine, is used to combine the scores of the selected similarity based classifiers. The selection procedure keeps adding or taking away features (similarity measures in our case) until the best evaluation performance is achieved. Only the selected similarity measures are then used for decision making in the final multiple classifier system.

A Support Vector Machine is a two-class classifier showing superior performance to other methods in terms of Structural Risk [15]. In this study, decision level fusion approach drawing on SVMs has been adopted for combining the similarity measure based classifiers. For the face verification problem, the size of the training set for clients is usually less than that for impostors. In such a case, the class of impostors is represented better. Therefore, it is necessary to shift the optimal hyperplane of the SVM classifier towards the better represented class. In this work, the size of the shift is determined in the evaluation step using the Equal Error Rate criterion.

4 Experimental Design

In this section the face verification experiments carried out on images of the BANCA database are described. The BANCA database is briefly introduced first. The main specification of the experimental setup is then presented.

4.1 BANCA Database

The BANCA database has been designed in order to test multi-modal identity verification systems deploying different cameras in different scenarios (Controlled, Degraded and Adverse). The database has been recorded in several

Table 1. The usage of the different sessions in the BANCA experimental protocols

Name	Symbol	1	2	3	4	5	6	7	8	9	10	11	12
Matched Controlled	Mc	TI	CI	CI	CI								
Matched Degraded	Md					TI	CI	CI	CI				
Matched Adverse	Ma									TI	CI	CI	CI
Unmatched Degraded	Ud	T				I	CI	CI	CI				
Unmatched Adverse	Ua	T								I	CI	CI	CI
Pooled test	P	TI	CI	CI	CI	I	CI	CI	CI	I	CI	CI	CI
Grand test	G	TI	CI	CI	CI	TI	CI	CI	CI	TI	CI	CI	CI

languages in different countries. Our experiments were performed on the English section of the database. Each section contains 52 subjects (26 males and 26 females).

Each subject participated to 12 recording sessions in different conditions and with different cameras. Sessions 1-4 contain data under *Controlled* conditions whereas sessions 5-8 and 9-12 contain *Degraded* and *Adverse* scenarios respectively. In order to create more independent experiments, images in each session have been divided into two groups of 26 subjects (13 males and 13 females). Experiments can be performed on each group separately. In the BANCA protocol, 7 different distinct experimental configurations have been specified. Table 1 describes the usage of the different sessions in each configuration. "T" refers to the client training while "C" and "I" depict client and impostor test sessions respectively.

4.2 Experimental Setup

The performance of different decision making methods discussed in Section 2 is experimentally evaluated on the BANCA database using the configurations discussed in the previous section. The evaluation is performed in the LDA space. The original resolution of the image data is 720×576. The experiments were performed with a relatively low resolution face images, namely 64×49. The results reported in this article have been obtained by applying a geometric face normalisation based on the eyes positions. The eyes positions were localised either manually or automatically. A fast method of face detection and eyes localisation was used for the automatic localisation of eyes centre [5]. The XM2VTS database [2] was used for calculating the LDA projection matrix.

The thresholds in the decision making system have been determined based on the Equal Error Rate criterion, i.e. by setting the operating point so that the false rejection rate (FRR) is equal to the false acceptance rate (FAR). The thresholds are set either globally (*GT*) or using the client specific thresholding (*CST*) technique [12]. In the training sessions of the BANCA database 5 client images per person are available. In the case of global thresholding method, all these images are used for training each client template. The other group data is then used to set the threshold. In the case of the client specific thresholding

[2] http://www.ee.surrey.ac.uk/Research/VSSP/xm2vtsdb/

strategy, only two images are used for the template training and the other three along with the other group data are used to determine the thresholds. Moreover, in order to increase the number of data used for training and to take the errors of the geometric normalisation into account, 24 additional face images per each image were generated by perturbing the location of the eyes position around the annotated positions.

In the previous studies [12], it was demonstrated that the Client Specific Thresholding (CST) technique is superior in the matched scenario (Mc, Md, Ma and G) whereas the Global Thresholding (GT) method gives a better performance on the unmatched protocols. The results reported in the next section using thresholding were acquired using this criterion.

Also, the SVM classifier has been used in order to fuse the selected classifiers.

5 Experimental Results and Discussion

Tables 2 contains a summary of the results obtained using the individual scoring function on the evaluation and test sets when manually annotated eyes position were used for the face geometric normalisation. The values in the table indicate the Total Error rates in the Evaluation (TEE) and Test (TET) stages respectively. These results clearly demonstrate that among the adopted metric, individually, the GD metric is the outright winner. In a few cases, the NC results are comparable to the GD one. The performance of the other metrics is much worse. Results of the similar experiments considering automatically registered data demonstrate that in such a condition also the GD function deliver a better or at least comparable performance [14].

In the next step, the problem of combining classifiers derived from different similarity measures using the SVMs was considered. Table 3 contains the combined verification results using manually and automatically registered data. These results demonstrate that a better performance is achieved using the combined method especially on Mc and G protocols.

Table 2. ID verification results obtained using different scoring functions with Global and Client Specific Thresholding techniques for unmatched and matched protocols respectively. TEE: Total Error rate Evaluation and TET: Total Error rate Test.

	Mc		Md		Ma		Ud		Ua		P		G	
	TEE	TET	TEE	TET	TEE	TET	TEE	TET	TEE	TET	TEE	TET	TEE	TET
NC	1.93	8.08	3.57	13.36	3.79	14.61	24.81	25.93	37.63	38.81	27.69	28.01	7.26	9.75
GD	0.60	4.87	1.77	7.18	1.55	8.03	26.09	24.74	27.5	27.40	19.56	19.64	2.43	4.12
ED	7.97	25.89	17	32.34	25.06	38.62	52.37	51.15	59.26	60.42	47.12	48.22	46.33	54.3
City	11.6	29.65	22.9	37.4	34.17	43.71	57.82	58.4	66.44	67.3	54.25	54.25	57.24	62.26
Cheb	8.2	31.73	16.22	39.23	16	35.86	56.44	56.3	58.94	57.41	51.56	51.85	32.54	43.79
χ^2	7.49	20.41	14.88	28.88	22.99	34.17	48.17	47.15	56.35	60.48	44.46	45.45	42.91	48.12
Corr	2.25	11.22	4.74	15.6	4.54	17.43	22.66	26.25	36.57	37.44	34.44	34.54	8.02	10.85
Canb	5	13.85	8.69	20.25	12.01	24.2	34.26	33.5	51.54	52.37	26.74	27.69	22.54	24.04

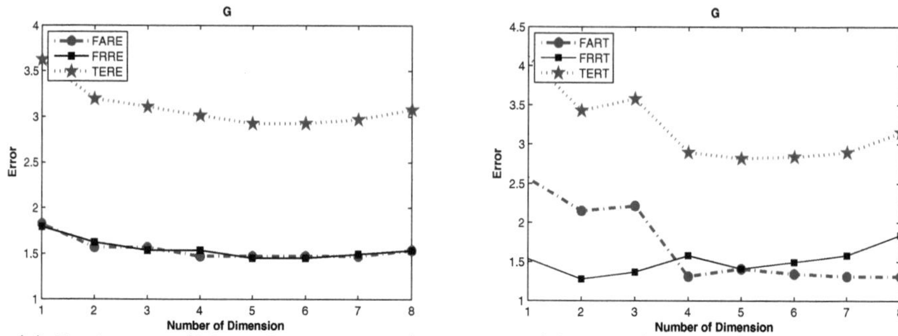

(a) Evaluation (Manual registration) (b) Test (Manual registration)

Fig. 1. ID verification results obtained when a subset of similarity measure based classifiers is selected using the SFS algorithm. SVM has been used for classifier fusion.

Then, the sequential search approaches, SFS and Plus 'L' and Take away 'R' algorithms with $L = 2$ and $R = 1$ were used for selecting a subset of similarity measure based classifiers. For this purpose, the scores associated to each similarity measure were appropriately normalised first. The normalised scores were then considered as a feature vector. The SVM classifier was finally used for decision making.

Figure 1 shows the resulting error rates for different number of similarity measures considering the G protocol using SFS algorithm. Figures 1-(a) and (b) contain the results considering the evaluation and test data respectively. Please note that based on the SFS algorithm, in each step, an optimum subset of the classifiers is selected considering the performance of the system using the evaluation data. The selected classifiers are then applied to the test data. As the evaluation plots show, when the number of the classifiers increases, the resulting error rate decreases first, then the error rate increases gradually. Moreover, the test results are almost consistent with the evaluation ones. Therefore, it is possible to find an optimum subset of the classifiers. Similar results were obtained using the Plus 'L' and Take away 'R' algorithm.

Table 3. ID verification results considering SVMs for combining classifiers derived from different metrics, manual registration(left) and automatic registration (right)

	Manual Registration						Automatic Registration					
	Evaluation			Test			Evaluation			Test		
	FAR	FRR	TER	FAR	FRR	TER	FAR	FRR	TER	FAR	FRR	TER
Mc	0.96	1.03	1.99	0.86	1.54	2.4	8.17	8.2	16.37	5.96	9.36	15.32
Md	3.75	3.85	7.6	2.98	3.97	6.95	8.36	8.46	16.82	12.79	6.28	19.07
Ma	2.4	2.31	4.71	3.65	5.27	8.92	5.57	5.51	11.08	2.02	15.9	17.92
Ud	9.61	9.61	19.22	10.38	13.46	23.84	13.46	13.46	26.92	15.09	15	30.09
Ua	12.88	13.08	25.96	16.63	10.39	27.02	17.88	17.82	35.7	28.17	23.08	51.25
P	8.87	8.85	17.72	9.29	9.06	18.35	15.64	15.6	31.24	14.33	16.62	30.95
G	1.54	1.54	3.08	1.31	1.84	3.15	7.05	7.05	14.1	7.53	8.76	16.29

Table 4. ID verification results using the proposed similarity measures selection method considering SVMs for combining classifiers derived from different metrics

	Manual Registration				Automatic Registration			
	SFS		+2-1		SFS		+2-1	
	TEE	TET	TEE	TET	TEE	TET	TEE	TET
Mc	1.99	2.14	1.99	2.14	16.14	23.88	16	22.17
Md	7.37	6.86	7.37	6.86	16.82	19.07	16.82	19.07
Ma	4.71	8.9	4.71	8.9	11.08	17.92	11.08	17.92
Ud	18.97	21.31	18.97	21.31	26.92	30.09	26.92	30.09
Ua	25.93	26.22	25.7	26.22	35.7	51.25	35.25	40.86
P	17.44	18.29	17.32	18	31.24	30.95	31.24	30.95
G	2.92	2.82	2.86	3.19	13.95	14.49	13.95	14.49

Table 4 contains the fusion results obtained using the SFS and "Plus 2 and take away 1" algorithms considering the manually and automatically registered data. As mentioned, based on the search algorithm, similarity measures are selected from the evaluation data for the whole data set. However, the algorithm is flexible so that, in different conditions, different classifiers can be adopted adaptively. For example considering the SFS algorithm, in the case of the G protocol the GD, Chebyschev, Chi-square (χ^2) ,Canberra and Manhattan distances were selected for manually registered data, whereas the GD, NC and Correlation Coefficient metrics were adopted in the Ua protocol. A comparison of the SFS and $+2-1$ results demonstrate that in most of the cases a similar performance is obtained using these algorithms. Our investigations confirm that a same subset of the scoring functions is selected using both methods when a same performance is achieved. As the result, since the SFS method is simpler than $+2-1$ method, applying this algorithm is sufficient in order to find the optimum metrics.

Overall, these results clearly demonstrate that the proposed similarity measures selection and fusion algorithms considerably improves the performance of the face verification system. The only exception is for the manually registered data of the MA protocol where the GD metric individually outperforms the combined system. It should be noted that in these experiments our thresholding technique has been applied for decision making when a metric is used individually while the SVM classifier has been used in the selection and fusion procedure.

6 Conclusions

We addressed the problem of selecting and fusing similarity measure based classifiers for face authentication in the LDA face space. An SVM-based sequential search approach similar to the "plus L, and take away R" algorithm was applied in order to find an optimum subset of similarity measures to be fused as a basis for decision making. Using the proposed method, the performance of the verification system can be considerably improved.

Acknowledgements. The financial support from Iran Telecommunication Research Centre is gratefully acknowledged. A partial support from the EU Network of Excellence Biosecure is also gratefully acknowledged.

References

1. Belhumeur, P.N., Hespanha, J.P., Kriegman, D.J.: Eigenfaces vs. Fisherfaces: Recognition using class specific linear projection. IEEE Trans. on Pattern Recognition and Machine Intelligence 19(7), 711–720 (1997)
2. Devijver, P., Kittler, J.: Pattern Recognition: A Statistical Approach. Prentice-Hall, Englewood Cliffs (1982)
3. Duin, R.P.W.: The combining classifier: to train or not to train? In: Inter. Conf. on Pattern Recognition (2002)
4. Freund, Y., Schapire, R.E.: A decision-theoretic generalization of on-line learning and an application to boosting. Journal of Computer and System Sciences 55 (August 1997)
5. Hamouz, M., Kittler, J., Kamarainen, J.-K., Paalanen, P., Kalviainen, H., Matas, J.: Feature-based affine-invariant localization of faces. IEEE Transactions on Pattern Analysis and Machine Intelligence 27(9), 1490–1495 (2005)
6. Kittler, J., Hatef, M., Duin, R.P.W., Matas, J.: On combining classifiers. IEEE Trans. on Pattern Recognition and Machine Intelligence, PAMI 20(3), 226–239 (1998)
7. Kittler, J., Roli, F.: Multiple classifier systems, vol. 2096. Springer, Heidelberg (2001)
8. Maghooli, K., Moin, M.S.: A new approach on multimodal biometrics based on combining neural networks using adaboost. In: ECCV Workshop BioAW (2004)
9. Perlibakas, V.: Distance measures for pca based face recognition. Pattern Recognition letters 25(6) (2004)
10. Pudil, P., Novovicova, J., Kittler, J.: Floating search methods in feature selection. Pattern Recognition Letters 15, 1119–1125 (1994)
11. Roli, F., Fumera, G.: Analysis of linear and order statistics combiners for fusion of imbalanced classifiers. In: 3rd Int. Workshop on Multiple Classifier Systems, Cagliari, Italy. Springer, Heidelberg (2002)
12. Sadeghi, M., Kittler, J.: Decision making in the LDA space: Generalised gradient direction metric. In: the 6th International Conference on Automatic Face and Gesture Recognition, Seoul, Korea, pp. 248–253 (May 2004)
13. Sadeghi, M., Kittler, J.: Confidence based gating of multiple face authentication experts. In: Joint IAPR International Workshops on Syntactical and Structural Pattern Recognition and Statistical Pattern Recognition, S+SSPR 2006, Hong Kong, China, August 17-19, pp. 667–676 (2006)
14. Sadeghi, M., Samiei, M., Kittler, J., Almodarresi, M.: Similarity measures fusion using svm classifier for face authentication. In: Proceedings of the third International Conference on Computer Vision Theory and Applications, VISAPP 2008, Portugal, vol. 2, pp. 105–110 (January 2008)
15. Vapnik, V.: The Nature of Statistical Learning Theory. Springer, New York (1995)
16. Verikas, A., Lipnickas, A., Malmqvist, K., Bacauskiene, M., Gelzinis, A.: Soft combining of neural classifiers: A comparative study. Pattern Recognition Letters 20, 429–444 (1999)
17. Xu, L., Krzyzak, A., Suen, C.Y.: Methods of combining multiple classifiers and their applications to handwriting recognition. IEEE Trans. on Systems, Man, and Cybernetics 22, 418–435 (1992)

On Mixtures of Linear SVMs for Nonlinear Classification

Zhouyu Fu[1] and Antonio Robles-Kelly[1,2]

[1] RSISE, Bldg. 115, Australian National University, Canberra ACT 0200, Australia
[2] National ICT Australia (NICTA)*, Locked Bag 8001, Canberra ACT 2601, Australia

Abstract. In this paper, we propose a new method for training mixtures of linear SVM classifiers for purposes of non-linear data classification. We do this by packaging linear SVMs into a probabilistic formulation and embedding them in the mixture of experts model. The weights of the mixture model are generated by the gating network dependent on the input data. The new mixture of linear SVMs can be then trained efficiently using the EM algorithm. Unlike previous SVM-based mixture of expert models, which use a divide-and-conquer strategy to reduce the burden of training for large scale data sets, the main purpose of our approach is to improve the efficiency for testing. Experimental results show that our proposed model can achieve the efficiency of linear classifiers in the prediction phase while still maintaining the classification performance of nonlinear classifiers.

1 Introduction

Support Vector Machines (SVMs) [1] have been widely used in different application areas including pattern recognition, computer vision and machine learning. SVMs seek to recover an optimal separating hyperplane in the feature space such that the margin between two classes, i.e. the minimum distance between data points with different labels, is maximised. By mapping the input data onto a higher-dimensional feature space in a nonlinear fashion and seeking an optimal separating hyperplane in the feature space, SVMs can generalise so as to handle linearly inseparable data sets. Moreover, the feature mapping can be done implicitly through the kernel trick.

Despite all their merits, nonlinear SVMs can carry a heavy computational burden for label prediction at testing time. To classify a single novel testing point, one needs to evaluate the nonlinear kernel function many times depending on the size of the set of support vectors extracted in the training phase. This can be very computationally intensive for large-scale prediction tasks. One such example is image classification, where feature vectors can be extracted at pixel

* NICTA is funded by the Australian Government as represented by the Department of Broadband, Communications and the Digital Economy and the Australian Research Council through the ICT Centre of Excellence program.

N. da Vitora Lobo et al. (Eds.): SSPR&SPR 2008, LNCS 5342, pp. 489–499, 2008.

level and an SVM applied to each of the pixels in the image. In contrast with their non-linear counterparts, linear SVMs can perform testing in linear time in the number of testing samples, whereas they are limited to those cases where the samples are linearly separable.

To combine the advantages of both, linear and nonlinear SVMs, and overcome their limitations, in this paper, we aim at developing a novel classification method with linear time complexity in the prediction stage and the classification performance comparable to the nonlinear classifiers. This is achieved by combining the mixture of experts model proposed by Jordan and Jacobs [2,3] with the use of linear SVM classifiers. Since the mixture of experts model is inherently a maximum likelihood probabilistic model, the trick here is to view the linear SVM from a probabilistic point so as to cast the problem into a maximum likelihood estimation (MLE) framework. We do this by viewing the classifiers as experts in a mixture model.

The work presented here differs from that elsewhere in the literature in a number of ways. Competing approaches [4,5] employ a divide-and-conquer strategy to partition the input space into disjoint regions. Once the space has been partitioned, a local non-linear SVM classifier is trained on samples falling in each region. Again, this may greatly reduce the training time but does not improve the efficiency for prediction due to the fact that non-linear SVMs are still used for the expert components. A straightforward remedy might be to replace the nonlinear SVMs with linear counterparts, but it is impossible, in general, to guarantee the training data in each region are linearly separable. This limits the use of linear classifiers in these settings. With our probabilistic formulation, we can sidestep this problem and formulate the problem in a more principled way. Moreover, the mixture of linear SVMs model can be trained using an EM algorithm by casting the problem into a MLE setting.

The remainder of the paper is organised as follows. Section 2 provides the background on the SVM classifier model. Section 3 details the proposed mixture of linear SVMs, including the model formulation and the EM algorithm employed for training. Experimental results are given in Section 4, followed by conclusions in Section 5.

2 The SVM Classifier Model

In this paper, we focus our attention in binary classification problems. Multiclass classification can be handled by converting to a number of binary classification problems via the pairwise coupling framework [7]. Given the labelled binary data set $(\mathbf{x}_i, y_i)$, where $i = \{1, \ldots, N\}$, and $y_i \in \{-1, 1\}$, the linear SVM classifier recovers an optimal separating hyperplane which maximises the margin of the classifier. This can be formulated as the following constrained quadratic optimisation problem

$$\begin{cases} \text{minimise} & \dfrac{||\mathbf{w}||^2}{2} + C\sum_i \epsilon_i \\ \text{subject to} & y_i(\mathbf{w}^T\mathbf{x}_i + b) \geq 1 - \epsilon_i \qquad i = \{1, \ldots, N\} \end{cases} \tag{1}$$

where $\mathbf{w}$ and b are the slope vector and the intercept of the hyperplane $y = \mathbf{w}^T\mathbf{x} + b$ respectively. The first term on the RHS of the above cost function is related to the classifier margin, which is the distance between the marginal hyperplanes $\mathbf{w}^T\mathbf{x} + b = +1$ and $\mathbf{w}^T\mathbf{x} + b = -1$. The second term on the RHS is related to the classification error. The parameter C controls the relative importance of the regularisation term and the error term.

Equation 1 is the primal form of the SVM formulation. It can be translated into the following dual form

$$\begin{cases} \text{maximise} & \sum_{i=1}^{N} \alpha_i - \frac{1}{2}\sum_{i=1}^{N}\sum_{j=1}^{N} \alpha_i\alpha_j y_i y_j \mathbf{x}_i^T\mathbf{x}_j \\ \text{subject to} & \sum_{i=1}^{N} \alpha_i y_i = 1 \text{ and } 0 \le \alpha_i \le C \qquad i = \{1, \ldots, N\} \end{cases} \tag{2}$$

The transformation above is important since it permits the original problem of recovering the hyperplane parameters to be reduced to the solution of the Lagrange multipliers α_i's, which is a standard QP problem. The condition for the primal problem in Equation 1 also implies that $w = \sum_{\alpha_i} \alpha_i y_i \mathbf{x}_i$. Hence, the optimal separating hyperplane is represented by

$$f(\mathbf{x}) = \sum_{\alpha_i} w^T x + b = \sum_{\alpha_i} \alpha_i y_i \mathbf{x}_i^T \mathbf{x} + b \tag{3}$$

Thus, testing points are classified according to the sign of the function above. The points $\mathbf{x}_i$ for which the associated $\alpha_i > 0$ are called support vectors. Geometrically, these are the points lying on or outside the marginal hyperplanes with $y_i(\mathbf{w}^T\mathbf{x_i} + b) \le 1$.

The linear SVM can be generalised to the nonlinear one by exploiting a nonlinear mapping from the input space to the high dimensional feature space known as the kernel trick. This is done by using an implicit mapping, i.e. a mapping whose closed form $x \rightarrow \Phi(x)$ is usually unknown. This mapping can be characterised by the inner product between any two vectors in the nonlinear feature space, which can be computed efficiently via the kernel function $K(x_i, x_j) = \Phi_i(x)^T\Phi_j(x)$. Compared to the dual form of the linear SVM in Equation 2, the dual problem for the nonlinear SVM is defined by substituting the inner product $\mathbf{x}_i^T\mathbf{x}_j$ with $K(\mathbf{x}_i, \mathbf{x}_j)$ as follows

$$\begin{cases} \text{maximise} & \sum_{i=1}^{N} \alpha_i - \frac{1}{2}\sum_{i=1}^{N}\sum_{j=1}^{N} \alpha_i\alpha_j y_i y_j K(\mathbf{x}_i, \mathbf{x}_j) \\ \text{subject to} & \sum_{i=1}^{N} \alpha_i y_i = 1 \text{ and } 0 \le \alpha_i \le C \qquad i = \{1, \ldots, N\} \end{cases} \tag{4}$$

The optimal separating hyperplane is then represented by

$$f(\mathbf{x}) = \sum_{\alpha_i} \alpha_i y_i K(\mathbf{x}_i, \mathbf{x}) + b \tag{5}$$

A novel testing point is then assigned to the positive class if $f(\mathbf{x}) > 0$ and otherwise to the negative class.

There are a number of alternatives for kernels used in nonlinear SVMs. These define the form of nonlinear mapping. The most common ones include Radial

Basis Functions (RBFs), polynomial and the Sigmoidal kernels. In this paper, we adopt the Radial Basis Functions (RBFs) with only one free parameter to tune. The kernel function is hence given by $\mathbf{K}(\mathbf{x}_j, \mathbf{x}_i) = \exp\left(-\beta\|\mathbf{x}_j - \mathbf{x}_i\|^2\right)$ where β is a scale parameter.

It is worth noting that, compared to their non-linear counterparts, linear SVMs cannot handle data sets that are linearly inseparable. Despite this apparent disadvantage, there are two major reasons to be in favor of the use of linear SVMs. Firstly, they are much more efficient at training time for large data sets. This is because training of linear SVMs can be achieved by minimising the cost function directly in the primal form in Equation 1, while nonlinear SVMs must be trained by minimising the cost function in the dual form as given in Equation 4. Efficient algorithms exist for training linear SVMs in linear time, respective of the size of training data set [8]. Secondly and more importantly, in general, the prediction time of linear SVMs is much lower than that of nonlinear SVMs. The efficiency of a nonlinear SVM in testing depends much on the number of support vectors. For a novel testing point, the kernel function, which is highly non-linear for the RBF function, needs to be evaluated m times to calculate the value of $f(\mathbf{x})$ in Equation 5. This can be very time consuming if there are thousands or millions of data points to classify in the testing set. In the other hand, the linear SVM only involves one inner product operation and one summation at testing time. This is since the weight vector $\mathbf{w}$ in Equation 3 can be explicitly expressed as the linear combination of support vectors. Hence it is desirable to develop an SVM based model with both the efficiency of the linear SVM at the prediction stage and the classification accuracy of the nonlinear SVM classifier.

3 Mixture of Linear SVMs

3.1 Model Formulation

As mentioned earlier, here we present a mixture of linear SVMs for classification making use of a mixture of experts model [2,3]. The architecture of the model is illustrated in Figure 1. The model consists of two parts. The experts at the bottom level are the linear SVM classifiers. The gating network produces a data-dependent weight distribution to combine the output of the linear SVM classifiers. The larger the weight, the more the corresponding component classifiers contribute to the classification of data.

Compared to conventional probabilistic classifier models like Linear Discriminant Analysis (LDA) or the logistic regressor, linear SVMs have better generalisation performance due to their large margin nature and are much cheaper to train thanks to the availability of efficient training methods [8]. However, to be able to use linear SVMs within the mixture of experts framework, we must be able to interpret it from a probabilistic point of view. To this end, we view the constrained quadratic optimisation problem in Equation 1 as that corresponding to the negative log-likelihood of an energy functional whose first term on the RHS corresponds to the prior probability of the parameters $P(w, b)$ and the

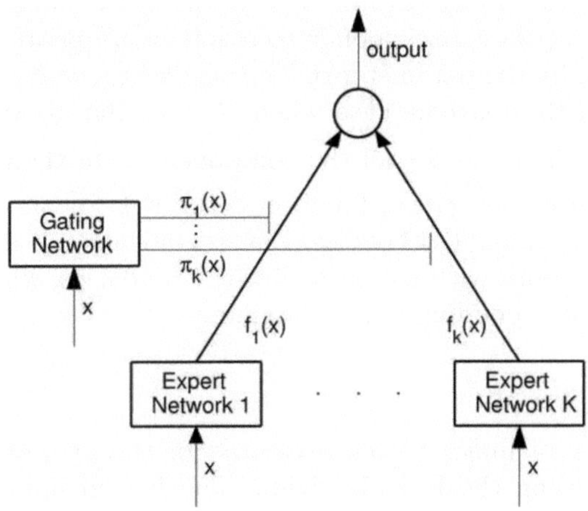

Fig. 1. Framework of Mixture of Experts

second term corresponds to the conditional probability $P(Y|X, w, b)$ related to classification errors. These are given by

$$P(w, b) = e^{-\gamma||w||^2} \tag{6}$$
$$P(Y|X, w, b) = \prod_i P(y_i|\mathbf{x}_i, w, b) = \prod_i e^{-\max(1-y_i(\mathbf{w}^T\mathbf{x}_i+b),0)}$$

where $\gamma = \log(\frac{1}{2C})$ and $\max(1, y_i(\mathbf{w}^T\mathbf{x}_i + b), 0)$ is the Hinge loss of the classification error. The loss is 0 for datapoint $\mathbf{x}_i$ if and only if it has a margin greater or equal to unity, i.e. $y_i(\mathbf{w}^T\mathbf{x}_i + b) \geq 1$. Here we assume the classification errors are independent and identically distributed for the conditional probability term.

With the above probabilistic interpretation of linear SVMs, we now introduce the mixture model. Consider K linear SVMs whose parameters are denoted by $\theta_k = \{\mathbf{w}_k, b_k\}$. By combining these SVMs, we can then obtain the log-likelihood function of the linear SVM mixture model by

$$l(\theta) = \sum_{i=1}^{N} \log \sum_{j=1}^{K} \pi_j(\mathbf{x}_i) P(y_i|\mathbf{x}_i, \theta_j) + \sum_{j=1}^{K} P(\theta_j) \tag{7}$$

where $\pi_j(\mathbf{x}_i|\mathbf{v}_j)$ is the jth output of the gating network for datapoint $\mathbf{x}_i$ with parameter $\mathbf{v}_j$. Unlike the conditional mixture of Gaussian models, the mixing weights π here are data-dependent. This allows for more flexibility in the modeling. Following Jordan and Jacobs [3], we specify the function $\pi_j(\mathbf{x})$ with the following multinomial log-linear model given by

$$\pi_j(\mathbf{x}|\mathbf{v}_j) = \frac{e^{\beta \mathbf{v}_j^T \mathbf{x}}}{\sum_j e^{\beta \mathbf{v}_j^T \mathbf{x}}} \tag{8}$$

The reasons for the above formulation of the gating function are twofold. Firstly, the above function corresponds to a soft-max operator, where the soft nature is controlled by the parameter β. The smaller the value of β, the "softer" the assignment. In the extreme case when $\beta = 0$, the above gating function will return equal weights $\frac{1}{K}$ for all the components. On the other hand, when β approaches infinity, the gating function returns 1 for the component with maximum value of $\mathbf{v}_j^T\mathbf{x}$ and 0 otherwise. This is equivalent to dividing the whole space into several regions with piecewise linear boundaries, where each region is processed by the corresponding linear classifier.

3.2 The EM Algorithm

The mixture model of linear SVMs presented in the previous section can be trained efficiently using the EM algorithm. The E-step updates the posterior probabilities of assigning each datapoint to the experts. The M-step involves simultaneously updating the gate networks and SVM parameters so as to solve two independent optimisation problems. The parameters for the gate networks are updated by solving a convex optimisation problem. This can be done efficiently by a quasi-Netwon method. The expert components, which are comprised by the linear SVMs, are retrained with sample weights specified by the posterior probabilities computed in the E-step.

Suppose the parameter estimate at iteration t is given by $\theta^{(t)} = \{\mathbf{w}^{(t)}, b^{(t)}\}$. In the E-step, the posterior probabilities of assigning the datapoint $\mathbf{x}_i$ to the jth expert, i.e. the jth linear SVM classifier, is given by

$$q_{i,j}^{(t)} = \frac{\pi_j(\mathbf{x}_i)e^{-\max(1-y_i(\mathbf{w}^T\mathbf{x}_i+b),0)}}{\sum_j \pi_j(\mathbf{x}_i)e^{-\max(1-y_i(\mathbf{w}^T\mathbf{x}_i+b),0)}} \tag{9}$$

With the posterior probabilities, we are able to derive a lower bound on the log-likelihood in the following fashion

$$Q(\theta, \theta^{(t)}) = \sum_i \sum_j q_{i,j}^{(t)} \log \pi_j(\mathbf{x}_i) P(y_i|\mathbf{x}_i, \theta) - \sum_i \sum_j q_{i,j}^{(t)} \log q_{i,j}^{(t)} + \sum_{j=1}^{K} P(\theta_j) \tag{10}$$

where $\theta^{(t)}$ on the LHS of the equation indicates the dependency of the lower bound function Q on the current parameter estimate, which directly influences the posterior probability $q_{i,j}^{(t)}$.

It can be shown that the gap between the true log-likelihood function and its corresponding lower bound is given by the sum of Kullback-Leibler divergences between $q_i^{(t)}$ and q_i as follows

$$l(\theta) - Q(\theta, \theta^{(t)}) = \sum_i \sum_j q_{i,j}^{(t)} \log \frac{q_{i,j}^{(t)}}{q_{i,j}} \tag{11}$$

$$q_{i,j} = \frac{\pi_j(\mathbf{x}_i) P(y_i|\mathbf{x}_i, \theta)}{\sum_j \pi_j(\mathbf{x}_i) P(y_i|\mathbf{x}_i, \theta)}$$

where q_i gives the posterior probability distribution of point i estimated from the parameters θ. The gap is non-negative and varnishes if and only if $\theta = \theta^{(t)}$.

Moreover, the observation above implies that $l(\theta) \geq Q(\theta, \theta^{(t)}$ and $l(\theta^{(t)}) = Q(\theta^{(t)}, \theta^{(t)})$. In the statistical literature, Equation 10 is also referred to as the complete log-likelihood function by viewing the classifier membership as hidden variables whose distributions are given by the posterior probabilities in Equation 9. Minimising the log-likelihood in Equation 7 directly is a hard optimisation problem. It is usually more tractable to minimise the complete log-likelihood as we do in the M-step. We need to update both, the parameters of the gating function π_j and each linear SVM classifier. Moreover, since the gating unit and the linear SVM experts are decoupled in the complete log-likelihood cost function in Equation 10, we are able to solve them individually through two separate optimisation procedures.

For the update of the parameters v_j in the gating function, we only need to maximise the following function

$$\max \sum_i \sum_j q_{i,j}^{(t)} \log \pi_j(\mathbf{x}_i | v_j) \tag{12}$$

where π_j is the multinominal log-linear model as defined in Equation 8. This is a weighted logistic regression problem with concave log-likelihood function, which can be efficiently maximised via standard convex optimisation routines. Here, we employ a limited memory BFGS quasi-Newton method [9], which can be applied to large scale data sets.

To update the SVM classifiers, we need to retrain the linear SVMs with samples weighted by the corresponding posterior probability. Specifically, for the jth linear classifier, we aim at maximising the following log-likelihood function

$$\begin{aligned} &\max \sum_i q_{i,j}^{(t)} \quad \log P(y_i | \mathbf{x}_i, \theta_j) + \log P(\theta_j) \\ &= \max - \sum_i q_{i,j}^{(t)} \max(1 - y_i(\mathbf{w}_j^T \mathbf{x}_i + b_j)) - \lambda \frac{1}{2} ||\mathbf{w}_j||^2 \end{aligned} \tag{13}$$

where $\lambda = e^{\gamma}$. This is equivalent to the following minimisation problem for training linear SVMs over weighted samples

$$\begin{aligned} &\min C \sum_i q_{i,j}^{(t)} \epsilon_i + \frac{1}{2} ||\mathbf{w}_j||^2 \qquad C = \frac{1}{2\lambda} \\ &\text{s.t. } \; y_i(\mathbf{w}_j^T \mathbf{x}_i + b_j) \geq 1 - \epsilon_i \end{aligned} \tag{14}$$

It is quite straightforward to observe that, at each EM iteration, the log-likelihood increases by the chain of relations $f(\theta^{(t)}) = Q(\theta^{(t)}, \theta^{(t)}) \leq Q(\theta^{(t+1)}, \theta^{(t)}) \leq f(\theta^{(t+1)})$. Hence, by repeating the EM steps we can obtain a convergent solution of the original maximum likelihood estimation problem.

4 Experimental Results

In this section, we demonstrate the performance of our algorithm with two experiments. The first experiment was performed on synthetic data. The second experiment was on pixel-level skin classification in hyperspectral images. For both experiments, we compared the proposed mixture of linear SVM classifiers (MixLSVM) with two alternative methods, the conventional linear SVM (LSVM) and nonlinear SVM classifiers with radial basis function (RBF) kernel. We set the number of linear SVMs to be 3 empirically for our mixture model in both experiments. The code was implemented in Matlab and the libsvm [10] package was used for training SVMs. The SVM parameters are selected via 5-fold cross validation.

First we focus on the synthetic data. A two-class data set was constructed comprising 200 points in each class uniformly drawn from two shifted sine signals in the two-dimensional space and perturbed with Gaussian noise in the vertical direction. We then applied the classification methods under study to the synthetic data. The resulting decision boundaries corresponding to kernel SVM, linear SVM and our approach were shown in Figure 2 respectively, superimposed on the generated synthetic data, where samples from different classes were plotted in different colors. We can see that our linear SVM mixture model is able to capture nonlinear information in the data for classification by locally combining linear classifiers. On the other hand, a globally linear classifier like the linear SVM fails to separate linearly inseparable data.

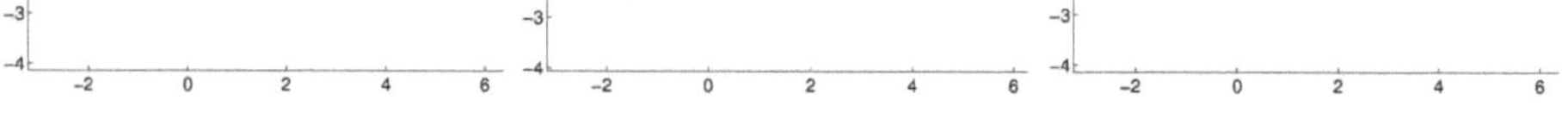

Fig. 2. From left to right, decision boundaries yielded by SVM, linear SVM and our approach superimposed on the synthetic data

Table 1. Testing result on the synthetic data

Data Size	Testing Error (%)			Testing Time		
	MixLSVM	SVM	LSVM	MixLSVM	SVM	LSVM
1K	**0.02**	0.01	9.28	0.011	0.022	0.006
10K	**0.02**	0.00	9.24	0.058	0.162	0.030
100K	**0.01**	0.00	9.26	0.562	1.687	0.323
1M	**0.01**	0.00	9.25	6.234	20.103	3.969

Table 2. Average cross validation error and testing time for skin classification

MixLSVM		SVM		LSVM	
Error (%)	Time (s)	Error (%)	Time (s)	Error (%)	Time (s)
4.95	**5.35**	4.54	76.25	8.66	3.65

Next, we generated testing data sets of different sizes from the same distribution and applied the trained classifiers for label prediction. The testing results are reported in Table 1. It can be seen that our algorithm achieves a desirable trade-off between classification performance and efficiency. It is more accurate than linear SVM with a much lower error rate and more efficient than kernel SVM, especially for large data sets containing millions of samples. Furthermore, the efficiency of kernel SVM for prediction replies highly on the number of support vectors. Since we only used a training set of 400 instances and about 1/3 of them are support vectors, it could even be worse if the kernel SVM is trained on a large training set. Both our algorithm and the linear SVM does not have such problem and the prediction efficiency only depends on the size of testing set.

We now turn our attention to skin detection in hyperspectral images. A set of hyperspectral images of human subjects was captured against a cluttered background by using a hyperspectral camera system comprised by a broad band monochrome camera and a Varispec liquid crystal tunable filter (LCTF). The spectral resolution of the LCTF is of 10nm between the 400nm to 720nm wavelengths, i.e. 33 bands covering the visible spectrum. The image spatial resolution is 696 by 520 pixels. The pseudo-colour images of some subjects are shown in the left-most column of Figure 3. For training, we have selected 2229 pixels in

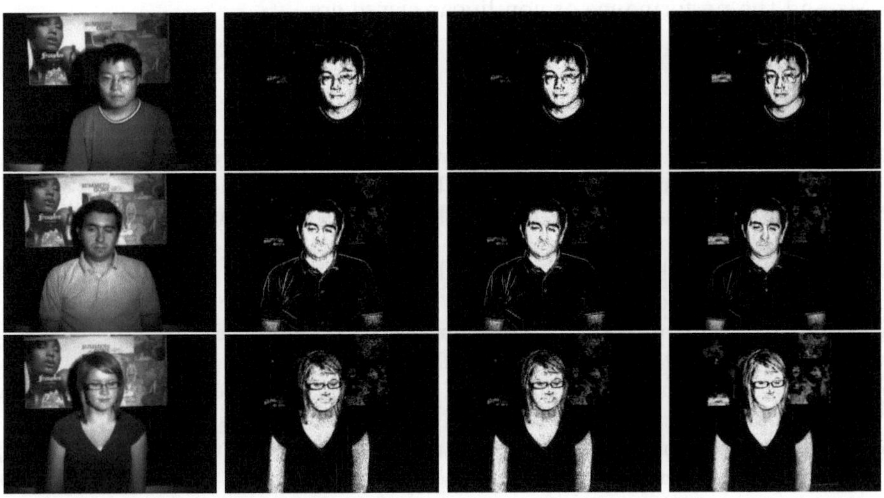

Fig. 3. Example skin classification results. Left-most column: pseudo-color images of sample human subjects; Second, third and fourth columns: classification results delivered by our algorithm, kernel SVM and linear SVM

the skin region and 6808 pixels in the background region from the images. In order to obtain the feature vectors used at input for our mixture of experts, we have applied the photometric invariant preprocessing method in [11].

The performance of the three algorithms under study are reported in Table 2, including the average cross validation error on the training set as well as the testing time in seconds for each image. From the table, we again see that our algorithm achieves a similar performance as compared to kernel SVM while being much more efficient. The linear SVM, on the other hand, is quite efficient for testing, yet it is unable to separate nonlinear data in the feature space and has a larger classification error than that yielded by non-linear SVMs and our approach.

Some example skin detection results are shown in the last three columns of Figure 3. These correspond, from left-to-right to the results delivered by our approach, the kernel and linear SVMs. We can see that both, our approach and the kernel SVM yield skin maps with similar visual quality. These compare favourably to the results yielded by the linear SVM, which recovers more false positives than the other two methods.

5 Conclusions

In this paper we have proposed a novel method to assemble linear SVM classifiers. Here, we have used the mixture of experts framework and re-interpreted linear SVMs in a probabilistic framework. We have adopted the EM algorithm to learn the parameters of the classifiers. The mixture of classifiers can handle nonlinear data in an efficient manner. This is evidenced by the experiments on skin classification in hyperspectral imaging. As compared to non-linear and linear SVMs, our method provides a means to the computational efficiency of linear SVMs and the performance of non-linear classifiers.

References

1. Boser, B., Guyon, I., Vapnik, V.: A training algorithm for optimal margin classifiers. In: ACM Conf. on Computational Learning Theory (1992)
2. Jacobs, R.A., Jordan, M.I., Nowlan, S.J., Hinton, G.E.: Adaptive mixtures of local experts. Neural Computation 3, 79–87 (1991)
3. Jordan, M.I., Jacobs, R.A.: Hierarchical mixtures of experts and the em algorithm. Neural Computation 6, 181–214 (1994)
4. Kwok, J.T.: Support vector mixture for classification and regression problems. In: Intl. Conf. on Pattern Recognition (1998)
5. Collobert, R., Bengio, S., Bengio, Y.: A parallel mixture of svms for very large scale problems. In: Advances in Neural Information Processing Systems (2002)
6. Kruger, S., Schaffoner, M., Katz, M., Andelic, E., Wendemuth, A.: Mixture of support vector machine for hmm based speech recognition. In: Intl. Conf. on Pattern Recognition (2006)
7. Hastie, T., Tibshirani, R.: Classification by pairwise coupling. In: Advances in Neural Information Processing Systems, vol. 10 (1998)

8. Joachims, T.: Training linear svms in linear time. In: SIGKDD (2006)
9. Nocedal, J., Wright, S.: Numerical Optimization. Springer, Heidelberg (2000)
10. Fan, R.E., Chen, P.H., Lin, C.J.: Working set selection using the second order information for training svm. Journal of Machine Learning Research 6, 1889–1918 (2005)
11. Fu, Z., Robles-Kelly, A., Tan, R., Caelli, T.: Invariant object material identification via discriminant learning on absorption features. In: Proc. of IEEE Conf. on Computer Vision and Pattern Recognition Workshop on Object Tracking and Classification Beyond the Visible Spectrum, pp. 140–147 (2006)

Adversarial Pattern Classification Using Multiple Classifiers and Randomisation

Battista Biggio, Giorgio Fumera, and Fabio Roli

Dept. of Electrical and Electronic Eng., University of Cagliari,
Piazza d'Armi, 09123 Cagliari, Italy
{battista.biggio,fumera,roli}@diee.unica.it
http://prag.diee.unica.it

Abstract. In many security applications a pattern recognition system faces an *adversarial classification* problem, in which an intelligent, adaptive adversary modifies patterns to evade the classifier. Several strategies have been recently proposed to make a classifier harder to evade, but they are based only on qualitative and intuitive arguments. In this work, we consider a strategy consisting in hiding information about the classifier to the adversary through the introduction of some randomness in the decision function. We focus on an implementation of this strategy in a multiple classifier system, which is a classification architecture widely used in security applications. We provide a formal support to this strategy, based on an analytical framework for adversarial classification problems recently proposed by other authors, and give an experimental evaluation on a spam filtering task to illustrate our findings.

1 Introduction

Pattern recognition techniques are currently applied to several security applications, like biometric personal authentication, intrusion detection in computer networks and spam filtering [1,2,3,4,5]. However, these kinds of application do not fit the standard pattern classification model [6]. The main reason is that, in these applications, a pattern classification system faces an intelligent, adaptive adversary who engineers patterns to defeat the system itself. The machine learning community is becoming aware of the relevance of this problem, as a recent workshop held as part of the NIPS 2007 conference shows.[1] To date, few works have explicitly addressed this problem [7,8,9,10]. In [9], the general issue of how to design machine learning and pattern classification systems which are hard to evade for an adaptive adversary was discussed. A taxonomy of attack types for different security applications was proposed and some potential approaches to design "evade hard" classification systems were suggested for future work. Practical solutions for securing classification systems have been also proposed in applications like intrusion detection in computer systems [3]. However, such

[1] Workshop on Machine Learning in Adversarial Environments for Computer Security, http://nips.cc/Conferences/2007/Program/event.php?ID=615

N. da Vitora Lobo et al. (Eds.): SSPR&SPR 2008, LNCS 5342, pp. 500–509, 2008.

solutions are indeed based on qualitative and intuitive arguments, besides the experimental evidence. In [7], the first attempt of developing an analytical framework for these kinds of problem (named *adversarial classification* problems) was made. An adversarial classification problem was formalised as a two-player game between a classifier and an adversary, in which both players try to maximise their expected utility by exploiting the knowledge they have about each other. In [7] it was *formally* shown that an adversary-aware classifier can perform much better than traditional adversary-unaware classifiers.

The aim of this work is to investigate and to provide a formal support to a new technique for designing hard to evade classifiers in adversarial classification problems, using the analytical framework developed in [7] (summarised in Sect. 2). Besides making a classifier adversary-aware, we consider the possibility of introducing some randomness in the placement of the classification boundary, which is a possible implementation of a general strategy suggested in [9], based on hiding information about the classifier to the adversary. We analyse the randomisation strategy in Sect. 3. In Sect. 4 we discuss a possible implementation using multiple classifier systems, which are widely used in security applications. We finally give an experimental validation of our findings in Sect. 5.

2 A Framework for Adversarial Pattern Classification

In [7], adversarial classification problems were formalised as two-player games. Each player chooses in his ply the move which maximises his expected utility (computed for that ply only, so using a greedy strategy[2]), based on the knowledge he has about the other player. The classifier faces a two-class problem in which patterns can be either malicious (if they are generated by the adversary) or innocent. Patterns are represented as n-dimensional feature vectors x in a feature space $\mathcal{X}$, and are viewed as instances of a random variable X generated i.i.d. from a given distribution $P(X)$ (without loosing generality, $\mathcal{X}$ is assumed to be discrete in [7]). Class labels y are denoted with $+$ (malicious) and $-$ (innocent). It is assumed that the adversary strategy to defeat the classifier consists in defining at each move a function $a : \mathcal{X} \rightarrow \mathcal{X}$ which modifies malicious patterns at operation phase, with the aim of making them being misclassified as innocent by the decision function constructed by the classifier at its previous move. The adversary can not modify malicious patterns at training phase, nor any innocent pattern. We point out that this assumption holds in several, but not all real applications (for instance, it holds in spam filtering tasks in which classifier training is made offline on a controlled training set, while it does not hold in some intrusion detection systems trained online). The classifier strategy consists instead in constructing at each move a decision function $y_C : \mathcal{X} \rightarrow \{+, -\}$ which discriminates malicious and innocent patterns, taking into account the knowledge he has about modifications introduced by the adversary in his previous move. For given $a(x)$ and $y_C(x)$, the utility function of the adversary is defined

[2] As explained in [7], computing the optimal strategies according to standard game theory is intractable for a classification problem.

as $U_A(x, y, a, y_C) = G_A(y_C(x), y) - W(x, a(x))$, where $G_A(y_C(x), y)$ is the gain accrued when a pattern (x, y) is labelled as $y_C(x)$ (with $G_A(-, +) > G_A(+, +)$, since the adversary's gain is higher when a malicious pattern is mislabelled as innocent than when it is correctly classified), and $W(x, a(x))$ is the cost faced to modify a pattern x to $a(x)$ ($W(x, a(x)) > 0$, if $a(x) \neq x$, $W(x, a(x)) = 0$ otherwise). The corresponding expected utility is:

$$E[U_A(x, y, a, y_C)] = \textstyle\sum_{(x,y) \in \mathcal{X} \times \mathcal{Y}} P(x, y)[G_A(y_C(a(x)), y) - W(x, a(x))]. \quad (1)$$

Analogously, the utility function of the classifier is defined as $U_C(x, y, a, y_C) = G_C(y_C(a(x)), y) - \sum_{i=1}^{N} V_i$, where $G_C(y_C(x), y)$ is the gain defined similarly as G_A (with $G_C(+, +) > G_C(-, +)$, and $G_C(-, -) > G_C(+, -)$), and $V_i > 0$ is the cost of measuring the i-th feature. The expected utility is thus:

$$E[U_C(x, y, a, y_C)] = \textstyle\sum_{(x,y) \in \mathcal{X} \times \mathcal{Y}} P(x, y)[U_C(y_C(a(x)), y) - \sum_{i=1}^{N} V_i]. \quad (2)$$

If the adversary has complete knowledge of the decision function constructed by the classifier at the previous move [7], its optimal strategy at any given move is the function $a(x)$ which maximises his expected utility (1):

$$a(x) = \text{argmax}_{x' \in \mathcal{X}}[G_A(y_C(x'), +) - W(x, x')], \quad (3)$$

if $y = +$, while $a(x) = x$, if $y = -$ (since he can modify only malicious patterns). Note that the above $a(x)$ amounts to choose between two alternatives. The first one is to keep x unchanged, and the corresponding utility is $G_A(+, +)$ (x is a malicious pattern correctly classified, and there is no cost since it is not modified). The second one is to modify it to some $x' \neq x$. The modification with the highest gain, denoted with x^*, is the one which leads to a pattern x^* misclassified as legitimate, with the minimum cost $W(x, x^*) = \text{argmin}_{x' \in \mathcal{X}} W(x, x')$. Such modification results in a gain equal to $G_A(-, +) - W(x, x^*)$. It follows that the adversary will keep a malicious pattern x unchanged, if $G_A(+, +) > G_A(-, +) - W(x, x^*)$, namely if the minimum cost required to evade the classifier is higher than the corresponding increase in the gain, $G_A(-, +) - G_A(+, +)$. Otherwise x would be modified to the x^* defined above. Analogously, under the assumption that the classifier knows the modification function $a(x)$ devised by the adversary at his previous move, the optimal strategy of the classifier is to compute at each move the decision function $y_C(\cdot)$ which maximises its expected utility (2).

Algorithms for computing the optimal strategies of adversary and classifier under the above framework strongly depend on the kind of classifier, on the kinds of feature used (for instance, on whether they are continuous or discrete), on the kinds of modification the adversary can make to malicious patterns and on the modification cost $W(\cdot, \cdot)$. In [7], specific algorithms were derived for the case in which the classifier is the naive Bayes. Experiments were also carried out on a spam filtering task, using features corresponding to words in the e-mail body, and assuming that the adversary can modify his e-mails by replacing words with synonyms or adding new words (these are real tricks used by spammers). These experiments showed that adversary-aware classifiers can perform much better than adversary-unaware ones designed with the traditional approach.

In the above analytical framework the hardness of evasion of a classifier is improved by making it adversary-aware, namely modifying it according to the knowledge on the strategy currently used by the adversary to defeat it. Although in real tasks the assumption that both players have complete knowledge about each other is likely to be violated, it is plausible that they can make at least some guesses about each other parameters [7] (for instance, the adversary can try to reverse engineer the classifier [10]). Using the same framework, in the next section we will formally show that the hardness of evasion can also be improved by hiding information about the decision function to the adversary.

3 Hardening Classifiers by Randomisation

An intuitive strategy for securing a classifier (namely, making it harder to evade) is to hide information about it to the adversary. A possible implementation of this strategy was suggested with qualitative arguments in [9]: introducing some randomness in the placement of the classification boundary. This implementation was not further investigated in [9] or in other works, to our knowledge. The goal of this section is to provide a more formal support to it, based on the analytical framework of [7]. To this aim, we consider a *single-shot* version of the game: first, the classifier computes its decision function $y_C(\cdot)$ assuming that the adversary does not modify his patterns. Then the adversary computes the optimal modification function $a(\cdot)$. Our aim is to compare the performance of both players (their expected utility) under two conditions: when the adversary has complete knowledge about the decision function of the classifier, as in [7] (named from now on the *deterministic* case), and when the classifier introduces some randomness in it (the *non-deterministic* case). In the following we will denote quantities related to the two cases with the superscripts "det" (for "deterministic") and "rnd" (for "randomisation").

In the deterministic case the optimal strategy $a^{\text{det}}(x)$ is the one which maximises the expected value of the utility function $U_A^{\text{det}}(x, y, a, y_C)$ over the distribution $P(x, y)$, for given $y_C(\cdot)$ and $a(\cdot)$. In the considered framework, introducing randomness in the decision function means that $y_C(\cdot)$ becomes a *random variable* $Y_C(\cdot)$ for the adversary. We denote with $\mathcal{Y}_C$ the domain of $Y_C(\cdot)$ (namely, the set of decision functions which the classifier can implement). Accordingly, in the non-deterministic case the expected utility has to be computed with respect to the distribution $P(x, y, y_C(\cdot))$. This distribution can be rewritten as $P(y_C(a(x))|x, y)P(x, y)$. Note that $P(Y_C(a(x)) = +|x, y)$ (written in the rest of the paper as $P(+|x, y)$ for the sake of brevity) is the probability that the classifier chooses a decision function $y_C(\cdot)$ such that $y_C(a(x)) = +$ (analogously for $P(Y_C(a(x)) = -|x, y)$). Note that such probability is given by $\sum_{y_C(\cdot)\in\mathcal{Y}_C} P(Y_C(\cdot) = y_C(\cdot)|x, y) \times I[y_C(a(x)), +]$, where $I[a, b] = 1$, if $a = b$, $I[a, b] = 0$ otherwise. Now, taking into account that the term $W(x, a(x))$

in (1) does not depend on $y_{\mathrm{C}}(a(x))$, the expected utility $E[U_{\mathrm{A}}]$ in the non-deterministic case can be written as:

$$\begin{aligned}&\textstyle\sum_{(x,y,y_{\mathrm{C}}(a(x)))\in\mathcal{X}\times\mathcal{Y}\times\mathcal{Y}} P(x,y,y_{\mathrm{C}}(a(x)))[G_{\mathrm{A}}(y_{\mathrm{C}}(a(x)),y) - W(x,A(x))]\\ &= \textstyle\sum_{(x,y)\in\mathcal{X}\times\mathcal{Y}} P(x,y)[P(+|x,y)G_{\mathrm{A}}(+,y) + P(-|x,y)G_{\mathrm{A}}(-,y) - W(x,a(x))].\end{aligned} \tag{4}$$

The term between square brackets in the second row of (4) is the conditional *expected* utility for a given pattern (x,y), taken over the distribution $P(y_{\mathrm{C}}(a(x))|x,y)$ (namely, over the possible realisations of the class boundary). In the determination of the optimal strategy it plays the same role as the utility function $U_{\mathrm{A}}^{\mathrm{det}}(x,y,a,y_{\mathrm{C}})$ in the deterministic case, and can be denoted here as $U_{\mathrm{A}}^{\mathrm{rnd}}(x,y,a)$: the optimal strategy $a^{\mathrm{rnd}}(x)$ is the one which maximises $U_{\mathrm{A}}^{\mathrm{rnd}}(x,y,a)$.

The two strategies $a^{\mathrm{det}}(x)$ and $a^{\mathrm{rnd}}(x)$ could be different, since the latter is chosen by maximising not the adversary's utility function for the decision function chosen by the classifier (which we denote now as $y_{\mathrm{C}}^*(\cdot)$), but instead its *expected* utility over all the possible decision functions. This means that the adversary could make sub-optimal choices in the non-deterministic case, with respect to the true but unknown $y_{\mathrm{C}}^*(\cdot)$, namely choices which would lead to a lower overall expected utility. More precisely, it can be easily shown (see below) that in the non-deterministic case the adversary *underestimates* the actual utility of each modification which evades the classifier, while he *overestimates* the actual utility of each modification which does not evade the classifier. This means that the adversary could make too conservative choices with respect to the optimal ones, to the extent that $a^{\mathrm{rnd}}(x)$ could keep unchanged some malicious pattern x correctly classified by $y_{\mathrm{C}}^*(\cdot)$, while $a^{\mathrm{det}}(x)$ would allow to evade the classifier. A simple example is shown in Fig. 1. Even more, $a^{\mathrm{rnd}}(x)$ could modify some malicious pattern x which evades $y_{\mathrm{C}}^*(\cdot)$ (and is thus kept unchanged by $a^{\mathrm{det}}(x)$) to a pattern x' which is correctly classified by $a^{\mathrm{rnd}}(x)$.

To prove the above assertion, consider any malicious pattern x $(y = +)$ and any modification x' which evades the classifier, namely $y_{\mathrm{C}}^*(x') = -$. The adversary's expected gain for x' is $P(+|x,+)G_{\mathrm{A}}(+,+) + P(-|x,+)G_{\mathrm{A}}(-,+)$, while the true gain is $G_{\mathrm{A}}(-,+)$. Since $G_{\mathrm{A}}(-,+) > G_{\mathrm{A}}(+,+)$, it follows that $U_{\mathrm{A}}^{\mathrm{rnd}}(x,+,x') < U_{\mathrm{A}}^{\mathrm{det}}(x,+,x',y_{\mathrm{C}}^*(x'))$, namely the adversary underestimates the true utility of modifying x to x'. Instead, if x' is correctly classified ($y_{\mathrm{C}}^*(\cdot) = +$), then the expected gain is the same as above, while the true gain is $G_{\mathrm{A}}(+,+)$. It follows that $U_{\mathrm{A}}^{\mathrm{rnd}}(x,+,x') > U_{\mathrm{A}}^{\mathrm{det}}(x,+,x',y_{\mathrm{C}}^*(x'))$, namely the adversary overestimates the true utility of modifying x to x'.

To sum up, we formally showed under the framework of [7] that randomising the decision boundary makes a classifier harder to evade, since the lack of information about the exact position of the decision boundary leads the adversary to make too conservative or too risky choices when deciding whether and how to modify a malicious pattern. One could raise the objection that in real tasks the adversary hardly ever has complete knowledge about the classifier decision function, and thus in practice it always acts in the non-deterministic case. However the above analysis can be easily extended to this case, to show more generally that the classifier can benefit by increasing the level of uncertainty of the

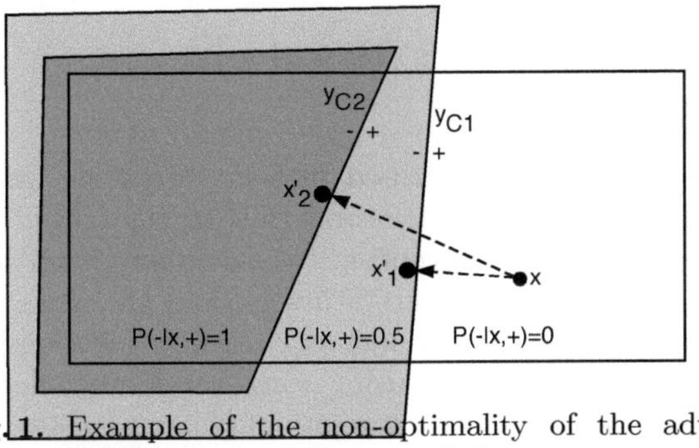

Fig. 1. Example of the non-optimality of the adversary's strategy in the non-deterministic case, for a given malicious pattern x in a 2-D feature space, assuming that the adversary's cost function is proportional to the Euclidean distance: $W(x, x') = |x - x'|$. The classifier chooses between two equiprobable decision functions y_{C_1} and y_{C_2} (both labelling patterns lying on the right as malicious), in a 2-D feature space. According to (3), in the deterministic case (when the adversary knows which decision function is used), the optimal strategy for x requires to choose between keeping x unchanged (with a gain of $G_A(+,+)$) and modifying it to x'_1 (if y_{C_1} is used, with a gain $G_A(-,+) - W(x, x'_1)$) or to x'_2 (if y_{C_2} is used, with a gain $G_A(-,+) - W(x, x'_2)$), being x'_1 and x'_2 the minimum cost modifications of x which allow to evade the corresponding decision function. If the adversary's gain and cost functions are the ones shown on the right, the optimal strategy against y_{C_1} is to modify x to x'_1, while against y_{C_2} is to keep x unchanged. In the non-deterministic case, patterns in the dark gray area are labeled as $-$ by both y_{C_1} and y_{C_2}, while patterns in the white area are labeled as $+$: accordingly, $P(-|, x, +)$ equals 1 in the former area and 0 in the latter. Patterns in the light gray area are labeled as $-$ by y_{C_1} only. Since y_{C_1} and y_{C_2} are equiprobable, $P(-|, x, +) = 0.5$ in that area. The optimal strategy in the non-deterministic case is to modify x to the pattern x' which maximises the conditional expected utility $P(+|x, +)G_A(+,+) + P(-|x, +)G_A(-,+) - W(x, x')$. This requires to choose between keeping x unchanged, modifying it to x'_1 or modifying it to x'_2 (any other choice has a higher cost but the same gain as one of these three choices). The best choice is to keep x unchanged. This means that the adversary's strategy in the non-deterministic case is optimal, only if the decision function is y_{C_2}. Otherwise the non-deterministic strategy does not allow to evade the classifier, while the deterministic strategy would have evaded it.

adversary, even starting from a non-zero uncertainty level (that is, by making $y_C(\cdot)$ still "more random"). On the other hand, it should be taken into account that an excessive randomisation could lead to select a classifier with a poor performance, to the extent that the advantage attained by hiding information to the adversary is lost. A trade-off is thus necessary between these two factors. Finally, we point out that randomisation can be viewed as a strategy to further improve the hardness of evasion of an adversary-aware classifier. Indeed, the analysis of the randomisation strategy can be extended to the repeated version of the game (when players continue to move indefinitely), allowing the classifier to retrain his set of decision functions according to the knowledge about the adversary's strategy $a^{\mathrm{rnd}}(x)$ (not reported here due to lack of space).

4 Evade Hard Multiple Classifiers Based on Randomisation Strategy

In real tasks the randomisation approach described in Sect. 3 can be implemented in several ways. It is important to understand that in the considered framework introducing randomness means preventing the adversary from having exact knowledge about one or more parameters involved in the design of the decision function. This can be achieved, for instance, tuning the classifier to a training set unknown by the adversary, or using some randomly selected parameter in the training phase (like the initial weights of a neural network). We propose here a specific implementation based on the use of MCSs. The reason is that MCSs turn out to be particularly suited in security applications as an alternative to the approach based on a "monolithic" classifier [2,3,1]. They are typically used for fusing classifiers each trained on a distinct feature-vector representation of patterns. The main motivations are derived from the field of classifier ensembles. It is indeed known that, if different sets of heterogeneous and loosely correlated features are available (as happens for IDSs), combining the outputs of different classifiers trained on different feature sets can be more effective than designing a single classifier in the feature space made up of all the available features (see for instance [11]). Moreover, if the overall number of features is large, such a single classifier would be more prone than a classifier ensemble to the so-called curse of dimensionality problem. Besides this, the MCS architecture is exploited (although often implicitly) in many commercial and open source spam filters and intrusion detection systems. Such kinds of system are usually made up of a set of independent modules which process a given subset of features, often focused on specific characteristics of malicious or innocent patterns, and provide a score denoting the likelihood that the input pattern is malicious. The final decision is then taken by fusing the scores with some combining function.

In this context, we point out that randomisation can be easily implemented exploiting one of the several techniques for constructing classifier ensembles, which are indeed based on randomisation like, for instance, bagging [12] and the random subspace method [13]. A concrete example of a security system with an MCS architecture is the SpamAssassin spam filter, in which filtering modules are provided with a default scoring system and the final decision is taken by thresholding the sum of the scores. However the user can modify the weights assigned to each module, for instance by tuning them to a given training set of legitimate and spam e-mails. Randomisation can be easily implemented here as the choice of a particular set of weights different than the default one. In sect. 5 we will present an experimental evaluation of the randomisation approach based on MCSs on a spam filtering task, using the SpamAssassin filter as the classifier.

5 Experimental Results

In this section we give an experimental evaluation on a spam filtering task of the effectiveness of class boundary randomisation for making a classifier harder to evade. We used the publicly available TREC 2007 data set,[3] made up of 25,220 legitimate and 50,199 spam e-mails. We used SpamAssassin as the classifier in its default configuration (all the available 619 different filtering rules were used, with their default scoring, and the decision threshold on the sum of the scores was set to 5), except for the embedded naive Bayes classifier, which was trained on the first 10,000 e-mails of the data set (in chronological order). We assumed that the adversary knew the decision threshold and the score produced by each module for any malicious pattern x. The randomisation was implemented as described in Sect. 3, by assigning different weights to each module. The weights were obtained using a support vector machine (SVM) classifier with linear kernel, trained on the scores produced by each filtering rule on 1,000 e-mails randomly drawn using the bootstrap technique from the 10,001st to 20,000th e-mails of the data set (the training set size of 1,000 was chosen since it provided a good trade-off between the uncertainty level of the adversary and the discriminant capability of the classifier). We computed 100 different sets of weights, and assumed that the classifier can choose one of them with identical probability. We also assumed that the adversary knew each set of weights and the fact that they are equiprobable.

Given that it was unfeasible to devise real modifications to the e-mails, tailored to each of the 619 filtering rules, we choose to simulate the effects of modifications by assuming that the adversary can arbitrarily reduce the score of each module. More precisely, let us denote with x_i the feature-vector representation of a given e-mail used by the i-th filtering rule, and with $s_i(x_i)$ the corresponding default score. We assumed that the adversary can modify any given spam e-mail x_i to any x'_i corresponding to any feasible value of the score such that $s_i(x_i) < s_i(x'_i)$[4] (note, however, that a modification is actually made by the adversary only if its cost is lower than the utility gain, as explained in Sect. 2). This simplifying assumption is in favour of the adversary, since we are not setting any constraint on how it can modify real spam e-mails. Given that it was also difficult to estimate the cost of modifications (or at least their relative difficulty) which allow to reduce the scores of the different modules, the modification cost $W_i(x_i, x'_i)$ for each module was assumed to be equal to the difference between the scores $s_i(x_i)$ and $s_i(x'_i)$ (which is always non-negative). Thus the total cost $W(x, x')$ is given by $\sum_{i=1}^{619}[s_i(x_i) - s_i(x'_i)]$. This is based on the reasonable assumption, in absence of more precise information, that attaining a higher score reduction requires a higher cost.

The experiments were made using the single-shot game setting explained in Sect. 3. Training the classifier corresponds to choose one out of the 100 weight sets. In the deterministic case, the adversary knows which weight set was chosen, and thus the corresponding decision function, and computes its optimal strategy $a^{\text{det}}(\cdot)$ against that function. In the non-deterministic case, he has to compute the optimal strategy $a^{\text{rnd}}(x)$ based on the distribution $P(y_C|x, y)$ (see (4)). He evaluates such distribution based on the knowledge that one of the 100 possible

[3] http://plg.uwaterloo.ca/~gvcormac/treccorpus07/

[4] Note that almost all SpamAssassin filtering modules output discrete scores, often binary-valued.

Table 1. First row: SpamAssassin results on the 55,419 testing e-mails are shown for reference. Rows 2 and 3: average and standard deviation (between parentheses) of the expected utility of the adversary ($E[U_{\mathrm{A}}]$) and of the classifier ($E[U_{\mathrm{C}}]$), and of the percentage of false positive (FP) and false negative (FN) error rates. The superscripts 'det' and 'rnd' denote respectively the deterministic and non-deterministic case. Second row: the adversary does not modify malicious instances. Third row: the adversary modifies malicious instances according to the optimal strategy.

$E[U_{\mathrm{A}}]^{\mathrm{det}}$	$E[U_{\mathrm{A}}]^{\mathrm{rnd}}$	$E[U_{\mathrm{C}}]^{\mathrm{det}}$	$E[U_{\mathrm{C}}]^{\mathrm{rnd}}$	$FN^{\mathrm{det}}(\%)$	$FN^{\mathrm{rnd}}(\%)$	FP (%)
0.49	-	0.65	-	9.78	-	0.15
0.05 (0.04)	-	1.67 (0.38)	-	0.97 (0.81)	-	0.69 (0.37)
0.98 (0.13)	0.56 (0.05)	1.30 (0.41)	1.46 (0.38)	19.55 (2.56)	11.21(1.05)	0.69 (0.37)

and equiprobable weight sets was chosen. Note that, in our experimental setting, computing the optimal strategy amounts to evaluating, for each set of scores $s = (s_1(x_1), \ldots, s_{619}(x_{619}))$ corresponding to a given e-mail, the maximum of $U_{\mathrm{A}}^{\mathrm{det}}(s, s')$ and $U_{\mathrm{A}}^{\mathrm{rnd}}(s, s')$ (defined in Sect. 3) for all possible combinations of score values of the filtering rules.[5] We then computed the performance of both players both in the deterministic and non-deterministic case as their average expected utility over the 100 possible decision functions, evaluated on the last 55,419 e-mails of the data set (the ones not used for training). The optimal strategy of the adversary was computed on the same 55,419 e-mails. We used the following utility functions: $G_{\mathrm{A}}(-,+) = 5, G_{\mathrm{A}}(+,+) = G_{\mathrm{A}}(-,+) = G_{\mathrm{A}}(-,-) = 0$ (namely, the adversary gains 5 if it evades the classifier, and 0 if a spam e-mail is correctly classified, or whatever happens to legitimate e-mails), $G_{\mathrm{C}}(-,+) = -1, G_{\mathrm{C}}(+,-) = -100, G_{\mathrm{C}}(+,+) = G_{\mathrm{C}}(-,-) = 1$ (namely, the classifier looses 1 for misclassifying a spam e-mail as legitimate, he looses 100 for misclassifying a legitimate e-mail as spam, and gains 1 for correct classifications, taking into account the fact that in spam filtering false positive errors are much more costly than false negative ones. According to the above G_{C}, the SVM classifier used for determining the weights of the filtering rules was trained with imbalanced error costs (1 for false negative and 100 for false positive errors). We did not consider the cost the classifier faces for measuring features, given that it was a constant term in these experiments.

The expected utilites and the false positive and false negative error rates are reported in Table 5, together with the results of the default SpamAssassin configuration for reference. The second row shows values corresponding to the deterministic case (average and standard deviation over the 100 sets of weights), when the adversary does not modify malicious patterns. The third row shows the values corresponding to the deterministic case (averaged as explained above) and the non-deterministic case (average and standard deviation over the 100 sets of weights). It can be seen that, in the deterministic case, the expected utility of the adversary exhibits a very large increase attained by modifying malicious pattern

[5] It was possible to carry out an exhaustive search, as most SpamAssassin filtering modules have binary scores.

at operation phase. This is mirrored by a reduction in the expected utility of the classifier. The average false negative error rate gives a more clear picture: it raises from 0.97% to 19.55% (note that the false positive error rate does not change, since the adversary modifies only malicious patterns, and the classifier is never retrained). Instead, when the classifier randomises the decision function, the expected utility of the adversary drops to about half the value attained in the deterministic case, and the same happens to the false negative error rate, while the expected utility of the classifier increases to a value in the middle between the ones achieved in the deterministic case, with and without the adversary reaction. We point out that this are *pessimistic* results for the classifier, for two reasons. First, we assumed that the adversary knew the exact distribution of the decision function, $P(y_C|x, y)$, while in a real setting he can only estimate it. Second, we computed the adversary's optimal strategy on testing e-mails.

We can conclude that these experiments support the analytical results derived in Sect. 3, based on the analytical framework in [7], showing that hiding information to the adversary through the randomisation of the decision function can improve the hardness of evasion of a classifier. Our results also show that randomisation can be naturally implemented in a MCS architecture.

References

1. Ross, A.A., Nandakumar, K., Jain, A.K.: Handbook of Multibiometrics. Springer, Heidelberg (2006)
2. Giacinto, G., Roli, F., Didaci, L.: Fusion of multiple classifiers for intrusion detection in computer networks. Pattern Recognition Letters 24, 1795–1803 (2003)
3. Perdisci, R., Gu, G., Lee, W.: Using an ensemble of one-class svm classifiers to harden payload-based anomaly detection systems. In: Proc. Int. Conf. Data Mining (ICDM), pp. 488–498. IEEE Computer Society, Los Alamitos (2006)
4. Sahami, M., Dumais, S., Heckerman, D., Horvitz, E.: A bayesian approach to filtering junk e-mail. AAAI Tech. Rep. WS-98-05, Madison, Wisconsin (1998)
5. Haindl, M., Kittler, J., Roli, F. (eds.): MCS 2007. LNCS, vol. 4472. Springer, Heidelberg (2007)
6. Duda, R.O., Hart, P.E., Stork, D.G.: Pattern Classification. Wiley, Chichester (2000)
7. Dalvi, N., Domingos, P., Mausam, Sanghai, S., Verma, D.: Adversarial classification. In: Proc. ACM Int. Conf. Knowledge Discovery Data Mining, pp. 99–108 (2004)
8. Globerson, A., Roweis, S.T.: Nightmare at test time: robust learning by feature deletion. In: Proc. Int. Conf. Mach. Learn., vol. 148, pp. 353–360. ACM, New York (2006)
9. Barreno, M., Nelson, B., Sears, R., Joseph, A.D., Tygar, J.D.: Can machine learning be secure? In: ASIACCS 2006: Proc. ACM Symp. Information, computer and communications security, pp. 16–25. ACM, New York (2006)
10. Lowd, D., Meek, C.: Adversarial learning. In: Proc. ACM Int. Conf. Knowledge Discovery Data Mining (KDD), pp. 641–647 (2005)
11. Kittler, J., Hatef, M., Duin, R.P., Matas, J.: On combining classifiers. IEEE Trans. Pattern Analysis and Machine Intelligence 20(3), 226–239 (1998)
12. Breiman, L.: Bagging predictors. Machine Learning 24(2), 123–140 (1996)
13. Ho, T.K.: The random subspace method for constructing decision forests. IEEE Trans. Pattern Analysis and Machine Intelligence 20(8), 832–844 (1998)

Adaptive Learning Rate for Online Linear Discriminant Classifiers

Ludmila I. Kuncheva and Catrin O. Plumpton

School of Computer Science,
Bangor University, Dean Street, Bangor Gwynedd LL57 1UT, UK
{l.i.kuncheva,c.o.plumpton}@bangor.ac.uk

Abstract. We propose a strategy for updating the learning rate parameter of online linear classifiers for streaming data with concept drift. The change in the learning rate is guided by the change in a running estimate of the classification error. In addition, we propose an online version of the standard linear discriminant classifier (O-LDC) in which the inverse of the common covariance matrix is updated using the Sherman-Morrison-Woodbury formula. The adaptive learning rate was applied to four online linear classifier models on generated and real streaming data with concept drift. O-LDC was found to be better than balanced Winnow, the perceptron and a recently proposed online linear discriminant analysis.

1 Introduction

In online classification, the data points[1] arrive one at a time. The classifier predicts a label, and immediately after that receives the correct label. The new data point is used to update the classifier. Although such instant feedback may not be available in real-life scenarios, this is a standard assumption that underpins incremental learning. The major advantage of this feedback is that the classification error can be monitored and changes in the classification environment (concept drift or concept shift) can be detected almost at their onset. A good online classifier should be able to self-tune to respond to changes. The classical online models, such as the perceptron [1] and the Winnow family [2,3,4,5] use a constant learning rate parameter to guide the classifier updates in case of a misclassification. Typically the value of the learning rate is fixed in advance through cross-validation experiments on a training data set. Using a fixed learning rate acts as a "forgetting mechanism" for the classifier because the new-coming points are given more weight in the online training. Here we propose to modify the learning rate as a function of a running estimate of the classification error. An increasing error rate may herald the onset of a change in the classification environment. In this case, an increased learning rate is desirable, so that the classifier "forgets more quickly" old observations. The larger the jump in the error is, the larger the increase of the learning rate should be.

[1] The term is used here as a synonym of *observations, examples, instances.*

N. da Vitora Lobo et al. (Eds.): SSPR&SPR 2008, LNCS 5342, pp. 510–519, 2008.

A notable omission from the toolbox of online linear classifiers is the classical Linear Disctiminant Classifier (LDC) [6] where the classes are assumed to be Gaussian with a common covariance matrix. The problem with the online implementation of LDC is that the update requires the inverse of the new covariance matrix. While the covariance matrix itself can be updated easily with a new observation, its inverse will have to be computed anew. Even though this can be done in constant time, running LDC online may become prohibitively expensive, especially for large number of features. Hence we propose here an Online Linear Discriminant Classifier (O-LDC) which avoids the matrix inverse at each step. A learning rate, λ is introduced to weight the contribution of the new data point.

The rest of the paper is organized as follows. Related work is discussed in Section 2. Section 3 introduces O-LDC. Experimental results with generated and real data for changing environments are presented in Section 6. Section 7 concludes the study.

2 Online Linear Classifiers

Consider data points in the n-dimensional real space, $\mathbf{x} \in \Re^n$. Let $\{\mathbf{x}_1, \ldots, \mathbf{x}_N\}$ be a data set, $\mathbf{x}_j \in \Re^n$, labelled in c classes.

2.1 Perceptron for Streaming Data (2 Classes)

The value for the learning rate η is chosen first. The coefficients of the linear discriminant function between the two classes, $\mathbf{w} = [w_0, \ldots, w_n]^T$, are initialised as small random numbers. Let $\mathbf{x}$ be the new observation, and $y(\mathbf{x}) \in \{-1, +1\}$ be its class label. Denote by $\mathbf{z} = [1 \quad \mathbf{x}^T]^T$ the augmented data vector. The first element, 1, multiplies the bias coefficient w_0. The predicted label (+1 or −1) for data point $\mathbf{x} \in \Re^n$, is calculated as

$$y_{\text{predicted}} = \text{sign}\left(\mathbf{z}^T \mathbf{w}\right),$$

where 'sign' is the signum function[2]. If $y_{\text{predicted}} \neq y(\mathbf{x})$, the weights are updated by $\mathbf{w} \leftarrow \mathbf{w} - \eta\, \mathbf{z}\, y_{\text{predicted}}$. Otherwise, the weights do not change.

2.2 Balanced Winnow for Streaming Data (2 Classes)

Balanced Winnow shares the same online update concept with the perceptron in that the weights are updated only upon a wrong prediction. First, a value for the learning rate β is chosen, $0 < \beta < 1$. There are two sets of weights, a positive set $\mathbf{w}^+$ and a negative set $\mathbf{w}^-$, all of which are initialized as positive random numbers. Using the augmented vector $\mathbf{z}$, the prediction formula is

$$y_{\text{predicted}} = \text{sign}\left(\mathbf{z}^T (\mathbf{w}^+ - \mathbf{w}^-)\right).$$

[2] Sign(a) = 1, if $a \geq 0$ and sign(a) = −1, if $a < 0$ The value of the function for $a = 0$ is irrelevant here and can be taken to be either +1 or −1.

When $y_{\text{predicted}} \neq y(\mathbf{x})$, the weights are updated by the following clause
if $y(\mathbf{x}) = +1$ then

$$w_i^+ \leftarrow \beta^{-z_i} w_i^+, \quad w_i^- \leftarrow \beta^{z_i} w_i^-, \quad i = 0, 1, \ldots, n. \tag{1}$$

else

$$w_i^+ \leftarrow \beta^{z_i} w_i^+, \quad w_i^- \leftarrow \beta^{-z_i} w_i^-, \quad i = 0, 1, \ldots, n. \tag{2}$$

2.3 Online Linear Discriminant Analysis (LDA) (c Classes)

The original data space is transformed as $\mathbf{y} = A^T\mathbf{x}$ where A is an $n \times m$ transformation matrix such that separability between the classes is maximised. The nearest mean classifier is applied in the new space. The procedure is equivalent to calculating linear discriminant functions in the original space. Fisher's linear discriminant analysis uses the Rayleigh quotient to measure separability. This criterion has the between-class scatter matrix, B, in the numerator and the within-class scatter matrix, W in the denominator. To make LDA suitable for online updates, Hiraoka et al. [7] propose an alternative criterion. The transformation matrix A is found by an iterative maximisation of the following potential function

$$\phi(A) = tr\left(A^T BA\left(I - \frac{1}{2}A^T WA)\right)\right),$$

assuming that the number of features in the new space, m, is chosen and fixed. It is stated in Ref [7] that the number of features m in the new space should be less than the number of classes c. In our case, two-class comparisons are carried out, which leaves only one dimension for the new space. In a set of pilot experiments we found that this seems to hamper Hiraoka's online LDC, hence we will use here $m = n$ (assuming non-singular covariance matrix). The learning rate parameter, ζ, for this classifier is a part of the iterative algorithm for computing A. Large learning rate places more weight on the new objects, as in the perceptron algorithm.

3 Online Linear Discriminant Classifier (O-LDC) (c Classes)

3.1 Online LDC for Static Environments

Let c be the total number of classes, $P^{(i)}$ be the prior probability for class i, $\mu^{(i)} \in \Re^n$ be the class mean, and Σ be the $n \times n$ covariance matrix, common for all classes. The linear discriminant classifier calculates c discriminant functions

$$g_i(\mathbf{x}) = \ln\ P^{(i)} - \frac{1}{2}{\mu^{(i)}}^T \Sigma^{-1} \mu^{(i)} + {\mu^{(i)}}^T \Sigma^{-1}\mathbf{x}, \quad i = 1, \ldots, c. \tag{3}$$

The class with the largest $g_i(\mathbf{x})$ is assigned to $\mathbf{x}$. This classifier guarantees minimum classification error (Bayes error) when the classes have normal distributions and equal covariance matrices. In an online setting the classifier is initialised by calculating estimates for the priors, the class means and the common covariance matrix from a training data set. A stream of data points is then submitted for classification. In order to update the discriminant functions so as to correspond to the new training set after each new observation, we need to update the class means and the *inverse* of the covariance matrix.

Let $\{\mathbf{x}_1, \ldots, \mathbf{x}_N\}$ be a data set, $\mathbf{x}_j \in \Re^n$, and $\mathbf{m}_{N_i}^{(i)}$ be the maximum likelihood estimate of the mean for class i, where N_i is the number of points from class i $(N_1 + N_2 + \ldots + N_c = N)$. Upon receiving data point $\mathbf{x}_{N+1}$ with label k, we update of the mean for class k using

$$\mathbf{m}_{N_k+1}^{(k)} = \frac{1}{N_k+1}(N_k \mathbf{m}_{N_k}^{(k)} + \mathbf{x}_{N+1}). \tag{4}$$

Denote by S_N the maximum likelihood estimate of the covariance matrix Σ.

$$S_N = \frac{1}{N} \sum_{i=1}^{c} \sum_{\mathbf{x} \in \text{ class } i} \left(\mathbf{x} - \mathbf{m}_{N_i}^{(i)}\right) \left(\mathbf{x} - \mathbf{m}_{N_i}^{(i)}\right)^T .$$

Knowing that $\mathbf{x}_{N+1}$ comes from class k, denote $\mathbf{z} = \mathbf{x} - \mathbf{m}_{N_k+1}^{(k)}$. Then

$$S_{N+1} = \frac{N}{N+1} \left(S_N + \frac{1}{N} \mathbf{z}\mathbf{z}^T \right) . \tag{5}$$

To apply the online updates, we reorganize (5) as

$$S_{N+1} = \frac{N}{N+1} \left(S_N + \sqrt{\frac{1}{N}}\, \mathbf{z} \, \sqrt{\frac{1}{N}}\, \mathbf{z}^T \right) \tag{6}$$

If A is an invertible matrix and $\mathbf{v}$ is a vector, the Sherman-Morrison-Woodbury formula states that

$$\left(A + \mathbf{v}\mathbf{v}^T\right)^{-1} = A^{-1} - \frac{A^{-1}\, \mathbf{v}\, \mathbf{v}^T\, A^{-1}}{1 + \mathbf{v}^T A^{-1} \mathbf{v}} . \tag{7}$$

Applying (7) to (6),

$$S_{N+1}^{-1} = \frac{N+1}{N} \left(S_N^{-1} - \frac{S_N^{-1}\, \mathbf{z}\, \mathbf{z}^T\, S_N^{-1}}{N + \mathbf{z}^T\, S_N^{-1}\, \mathbf{z}} \right) , \tag{8}$$

which is the update of the covariance matrix for $\mathbf{x}_{N+1}$.

Finally, the priors are estimated as $P_N^{(i)} = N_i/N$ and updated as

$$P_{N+1}^{(i)} = \begin{cases} \frac{N_i}{N+1}, & i \neq k \\ \frac{N_i+1}{N+1}, & i = k \end{cases} . \tag{9}$$

With $P_{N+1}^{(i)}$, $\mu_{N+1}^{(i)}$ and S_{N+1}^{-1} in place, the new discriminant functions $g_i(\mathbf{x})$ can be calculated.[3]

[3] In training O-LDC, when the covariance matrix of the training data appeared to be singular, we replaced it by the identity matrix.

3.2 Online LDC for Changing Environments

To enable updates that will accommodate changing environments, we introduce a learning rate λ to account for the weight of the new data point, $0 < \lambda < 1$. For $\lambda \to 0$, there is no update, while for $\lambda \to 1$ the classifier forgets everything except the new data point. When $\lambda = 1/2$, the update is exactly the one given by equations (4), (8) and (9). For values from 0 to 0.5 the classifier is "reluctant" to train while with $\lambda > 1/2$ it is "eager" to train. To accomplish this, the update formulas for the means are

$$\mathbf{m}^{(k)}_{N_k+1} = \frac{1}{(1-\lambda)N_k + \lambda} \times \left((1-\lambda)N_k \mathbf{m}^{(k)}_{N_k} + \lambda\, \mathbf{x}_{N+1}\right). \tag{10}$$

The prior probabilities are updated as

$$P^{(i)}_{N+1} = \begin{cases} \frac{(1-\lambda)N_i}{(1-\lambda)N+\lambda}, & i \neq k \\ \frac{(1-\lambda)N_i+\lambda}{(1-\lambda)N+\lambda}, & i = k \end{cases}. \tag{11}$$

Finally, by weighting the **z**-term in (5) by λ, and S_N by $(1-\lambda)$, while keeping the overall "soft" count at $(1-\lambda)N+\lambda$, we derive the update for the covariance matrix

$$S^{-1}_{N+1} = \frac{(1-\lambda)N+\lambda}{(1-\lambda)N} \times \left(S^{-1}_N - \frac{S^{-1}_N\, \mathbf{z}\,\mathbf{z}^T\, S^{-1}_N}{\frac{(1-\lambda)N}{\lambda} + \mathbf{z}^T\, S^{-1}_N\, \mathbf{z}} \right), \quad \lambda \neq 0, \lambda \neq 1. \tag{12}$$

4 Adaptive Learning Rate

4.1 Fixed Learning Rate - An Illustration

The effect of the learning rate on the classification accuracy is demonstrated below on two synthetic data sets simulating abrupt changes (STAGGER data) and gradual changes (moving plane data).

STAGGER data. We tested O-LDC, Perceptron, Winnow and Hiraoka's LDA on the popular STAGGER data used in [8]. Each data point is described by 3 features, each with three possible categories: size $\in$ {small, medium, large}, colour $\in$ {red, green, blue} and shape $\in$ {square, circular, triangular}. Three classification tasks were to be learned in a course of 120 points. From point 1 to point 40, the classes to be distinguished are [size = small AND colour = red] vs all other values; from 41 to 80, [colour = green OR shape = circular] vs all other values; and from 81 to 120, [size = small OR size = large] vs all other values. For each data point submitted as a part of the streaming data, an independent testing set of 100 objects was generated and labelled according to the current class description. The classifier was tested after each submission.

Here we are interested in the error rates of the four online linear classifiers with a *pre-fixed* parameter value that does not change with time. Therefore we varied the parameters that control the learning speed and then fine-tuned them

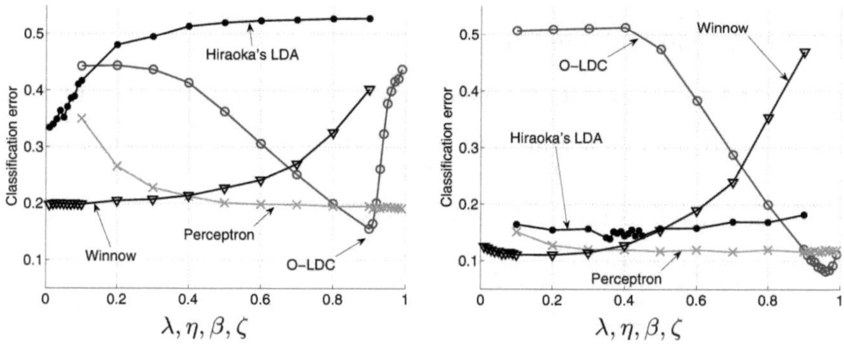

(a) STAGGER data (b) moving plane data

Fig. 1. Classification error versus learning rate

around the best value, all on the *testing* set. With each parameter value and each method we carried out 100 runs. The classification error was calculated as the mean error across the whole online training. Figure 1 (a) plots the error versus the parameter values.

Although O-LDC gives the best error rate, the range of values for the optimal λ is small. On the other hand, the perceptron and Winnow are fairly robust with respect to their learning rate. As expected, larger values of η and smaller values of β lead to faster learning, hence better overall results. The dip in the O-LDC error curve shows a balance between memorising past data and learning new data.

Moving plane data. A gradual changing environment was simulated by rotating the linear class boundary about the origin in a 2d space [9]. All data points came from a uniform distribution in the unit square. The class definition (concept) was changed with each new data point by rotating the boundary at a further angle of $1°$. Starting with $\theta = 0$ and finishing with $\theta = 359°$, we had a sequence of 360 data points. All classifiers were initialized at $\theta = 0$ using a random set of 10 points. An additional testing data set of 100 points was generated for each angle θ and used to evaluate the classifier performance. Figure 2 gives four snapshots of the moving-plane class configurations for $0 \leq \theta \leq 90°$.

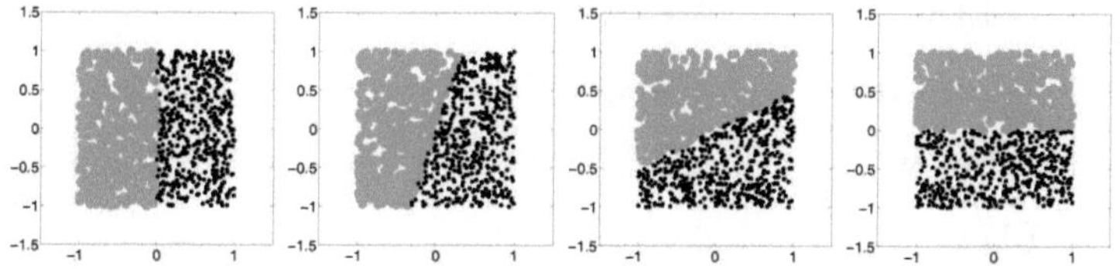

Fig. 2. Snapshots of the moving-plane data for angle $0 \leq \theta \leq 90°$

Table 1. Average error for the *best* choice of parameter values, the 95% confidence intervals and the optimal parameter values

Dataset	O-LDC	Perceptron	Winnow	Hiraoka's LDA
STAGGER	**0.156**±0.024	0.191±0.023	0.197±0.025	0.334±0.029 •
	($\lambda = 0.90$)	($\eta = 0.91$)	($\beta = 0.01$)	($\zeta = 0.01$)
Moving plane	**0.082**±0.005	0.116±0.003 •	0.110±0.009 •	0.139±0.005 •
	($\lambda = 0.96$)	($\eta = 0.70$)	($\beta = 0.20$)	($\zeta = 0.36$)
ELEC2.2	**0.162**±0.0342	0.169±0.035 •	0.171±0.035 •	0.190±0.036 •
	($\lambda = 0.50$)	($\eta = 0.90$)	($\beta = 0.10$)	($\zeta = 0.50$)

We varied the parameters of the algorithms, λ, η, β and ζ in the same way as with the STAGGER data. The results are presented in Figure 1 (b). A pattern similar to that with the STARGGER data is observed. O-LDC reaches the best error rate but only for a small range of λ. This time Hiraoka's LDA shows a good and stable performance across all tested values of ζ but is dominated by the perceptron. To evaluate the statistical significance of the differences, Table 1 shows the errors averaged across the length of the runs and then across the 100 runs. Only the *best* case is shown. The table also contains the 95% confidence intervals. The best results are indicated in boldface and the statistically significant differences in comparison with O-LDC are marked with '•'.

5 Variable Learning Rate

Let $W(t)$ be a window containing the past M objects in the streaming data, $W(t) = \{\mathbf{x}_{t-M+1}, ..., \mathbf{x}_t\}$. Denote by E_t the error rate of the objects in W, calculated in the course of the online classification. The error difference

$$\Delta_e = E_{t-M} - E_t$$

can be construed as a measure of the change in the environment in the past M steps. Large negative Δ_e will signify an abrupt deterioration in the classification accuracy. We propose to modify the learning rates of the four classifiers by taking the magnitude of Δ_e into account. The larger the change in the error, the larger the change in the learning rate in the direction of forgetting old data.

O-LDC	$\lambda \leftarrow \lambda^{(1+\Delta_e)}$
Perceptron	$\eta \leftarrow \eta^{(1+\Delta_e)}$
Winnow	$\beta \leftarrow \beta(1 + \Delta_e)$
LDA	$\zeta \leftarrow \zeta^{(1+\Delta_e)}$

For λ, η and ζ, the larger the learning rate, the more responsive the classifier. As all learning rates are within the interval [0,1], a power of $(1 + \Delta_e)$ will lead to an increase when $\Delta_e < 0$ and a decrease when $\Delta_e > 0$. The Winnow learning rate, β, on the other hand, already is used as the base for an exponent in the weights update ((1) and (2)). Thus it receives a more "gentle" update by only multiplying the old value by $(1 + \Delta_e)$. Negative Δ_e will derease β as smaller β makes the balanced Winnow more responsive. In all four models, we only update the learning rate if the current data point is misclassified.

At first glance it seems that we merely replace one parameter choice (the learning rate) with another (the moving window size). However, additional experiments with different window sizes showed that all four algorithms are much less sensitive to the window size than to the choice of their respective learning rates.

6 Experimental Results

The purpose of the experiment was to find out whether the automatic adaptation of the learning rate can lead to results similar to these with the optimal parameter values. The behaviour of the four online *linear* classifier models: O-LDC, perceptron, Winnow and Hiraoka's LDA was examined on the two synthetic data sets STAGGER and moving plane, and also on a real data set (ELEC2) used in other studies of classification in changing environments [10,11,12].

Figures 3 (a) and (b) show the patterns of the error progression along the online training for the two synthetic data sets. The two error peaks for the STAGGER data correspond to the concept shifts at observation 40 and 80. The "wavy" pattern for the moving plane data is caused by the inertia in the reaction of the classifier to the changes. Table 2 shows the mean classification errors for the four classifiers *with the adaptive learning rate.* In all cases the classifiers achieve error rates similar to these with the optimal choice of their respective learning rates (note that the optimal learning rate values were found on the *testing* data). O-LDC gives best overall results, where the difference is statistically signifiant ($p < 0.05$).

Electricity Market data set. This data set is one of the few publically available benchmark data sets for changing environments [11]. The version named Elec 2.2. was used here. It consists of 45,312 data points, each represented by three fatures: day of the week, time of the day and electricity demand of New South Wales, Australia at the time. The data set is a collection of successive measurements every 30 minutes, spanning the period from May 1996 to December 1998. The class label of each point is either UP or DOWN, referring to whether the electricity price at the specified time is higher or lower than the average price of the preceding 24 hours. In our experiments we used the error of the new data point in the sequence as the testing error before retrieving the correct label and adding the data point to the training set. Thus the overall error from an experiment is the average of correct/wrong predictions on the whole data set. Table 1 shows the error rates with the optimal parameter choice and Table 2, with the adaptive leanring rate. The error progression is displayed in Figure 3 (c)

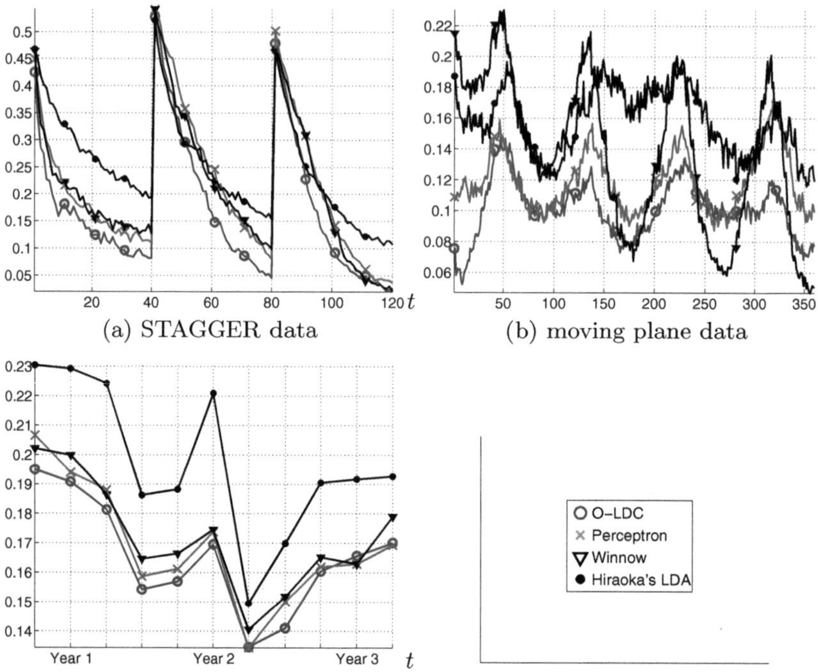

(a) STAGGER data (b) moving plane data

(c) ELEC2.2 data

Fig. 3. Classification error vs number of online observations t

Table 2. Average error with *adaptive learning rate* and 95% confidence intervals

Dataset	O-LDC	Perceptron	Winnow	Hiraoka's LDA
STAGGER	**0.171**±0.022	0.216±0.023	0.211±0.022 •	0.251±0.017 •
Moving plane	**0.101**±0.002	0.119±0.002 •	0.138±0.005 •	0.158±0.002 •
ELEC2.2	**0.165**±0.035	0.169±0.035	0.172±0.035	0.198±0.036 •

for the four online linear classifiers. Again, O-LDC shows superior performance compared to the perceptron, balanced Winnow and the LDA.

7 Conclusions

Linear online classifiers were found to be sensitive to the choice of their learning rate parameter. We propose a strategy for automatic updating of the learning rate based upon the magnitude of the error change during the online training. We also propose an online linear discriminant classifier (O-LDC) that is able to work with streaming data and changing environments. The inverse of the covariance matrix is updated through the Sherman-Morrison-Woodbury formula. The experiments demonstrated that the adaptive learning rate leads to error rates very close to those with the optimal learning rates. The results favoured

O-LDC in comparison with the perceptron, the balanced Winnow and a recently proposed online linear discriminant classifier, which we refer to as Hiraoka's LDA.

References

1. Rosenblatt, F.: Principles of Neurodynamics: Perceptrons and the Theory of Brain Mechanisms. Spartan Books, Washington (1962)
2. Littlestone, N.: Learning quickly when irrelevant attributes abound: A new linear threshold algorithm. Machine Learning 2, 285–318 (1988)
3. Grove, A., Littlestone, N., Schuurmans, D.: General convergence results for linear discriminant updates. Machine Learning 43(3), 179–210 (2001)
4. Wang, B., Jones, G.J.F., Pan, W.: Using online linear classifiers to filter spam emails. Pattern Analysis and Applications 9(4), 339–351 (2006)
5. Carvalho, V.R., Cohen, W.W.: Notes on single-pass online learning. Technical Report CMU-LTI-06-002, Carnegie Mellon University (2006)
6. Duda, R.O., Hart, P.E., Stork, D.G.: Pattern Classification, 2nd edn. John Wiley & Sons, Chichester (2001)
7. Hiraoka, K., Yoshizawa, S., Hidai, K., Hamahira, M., Mizoguchi, H., Mishima, T.: Convergence analysis of online linear discriminant analysis. In: Proc. Int. Joint Conf. on Neural Networks, vol. 3, pp. 387–391 (2000)
8. Widmer, G., Kubat, M.: Learning in the presence of concept drift and hidden contexts. Machine Learning 23, 69–101 (1996)
9. Narasimhamurthy, A., Kuncheva, L.I.: A framework for generating data to simulate changing environments. In: Proc. IASTED, Artificial Intelligence and Applications, Innsbruck, Austria, pp. 384–389 (2007)
10. Gama, J., Medas, P., Castillo, G., Rodrigues, P.: Learning with drift detection. In: Bazzan, A.L.C., Labidi, S. (eds.) SBIA 2004. LNCS, vol. 3171, pp. 286–295. Springer, Heidelberg (2004)
11. Harries, M.: Splice-2 comparative evaluation: Electricity pricing (1999)
12. Bifet, A., Gavaldà, R.: Learning from time-changing data with adaptive windowing. In: Proceedings of the Seventh SIAM International Conference on Data Mining, Minneapolis, Minnesota, USA, pp. 443–448 (2007)

Combining Online Classification Approaches for Changing Environments

Juan J. Rodríguez[1] and Ludmila I. Kuncheva[2]

[1] Lenguajes y Sistemas Informáticos, Universidad de Burgos, Spain
jjrodriguez@ubu.es
[2] School of Computer Science, Bangor University, UK
l.i.kuncheva@bangor.ac.uk

Abstract. Any change in the classification problem in the course of online classification is termed changing environments. Examples of changing environments include change in the underlying data distribution, change in the class definition, adding or removing a feature. The two general strategies for handling changing environments are (i) constant update of the classifier and (ii) re-training of the classifier after change detection. The former strategy is useful with gradual changes while the latter is useful with abrupt changes. If the type of changes is not known in advance, a combination of the two strategies may be advantageous. We propose a classifier ensemble using Winnow. For the constant-update strategy we used the nearest neighbour with a fixed size window and two methods with a learning rate: the online perceptron and an online version of the linear discriminant classifier (LDC). For the detect-and-retrain strategy we used the nearest neighbour classifier and the online LDC. Experiments were carried out on 28 data sets and 3 different scenarios: no change, gradual change and abrupt change. The results indicate that the combination works better than each strategy on its own.

1 Introduction

A change in a real-life classification problem may arise from demographic fluctuations, economic trends, seasonal drifts, change of measuring equipment, etc. Here we use the term *changing environments* to denote any change in the classification problem (also called concept drift, population drift, concept shift). Two general strategies to ensure that the classifier is up-to-date are the constant update of the classifier and update upon change detection. Various methods for classification in changing environments have been proposed [6]. Within the constant-update strategy group, many methods use a fixed window of objects or a fixed learning rate. Within the detect-and-retrain strategy group, changes are detected explicitly [3]. Ensembles of classifiers for changing environments have also been considered [4, 7, 9, 10], mostly following the constant-update strategy. Which method will be adequate for a particular problem depends on the types and frequencies of the changes. These are rarely known in advance. Hence, this work proposes to combine methods coming from the two strategies, using an ensemble of classifiers.

N. da Vitora Lobo et al. (Eds.): SSPR&SPR 2008, LNCS 5342, pp. 520–529, 2008.

The rest of the paper is organised as follows. Section 2 describes the online classification methods used in the ensemble. Section 3 contains the experiment, and Section 4 gives the concluding remarks.

2 Methods

2.1 Classification Methods

Three families of methods are considered:

- The nearest neighbour (1-NN) classifier with a fixed window. The reference set of the nearest neighbour are the latest M objects in the streaming data. Each new object is added to the reference set and the oldest one is discarded.
- The perceptron with a fixed learning rate η. The coefficients of a linear function are updated if the current object is misclassified. This method is designed for two-class problems. For multiclass problems, pairwise classification can be used [2] (either one versus all or all pairs of classes).
- An online version of the linear discriminant classifier (O-LDC). The classifier is updated with each new object by recalculating the means and the inverse of the covariance matrix using a shortcut formula. The strength of the updates is governed by a learning rate parameter, λ, with a value between 0 and 1. For $\lambda = 0.5$, all the objects have the same importance; if $\lambda > 0.5$, the new objects are more important than the older ones.

2.2 Change Detection Methods

The two detection methods considered here monitor the probability of error of the streaming objects. When a change in the error is detected (significant raise) the current classifier is abandoned and a new one is trained.

- The Drift Detection Method (DDM) proposed in (Gama et al., 2004) operates with the help of two thresholds, t_1 and t_2. When the average error rate e goes above t_1, a warning mode is activated and the time of the activation, t, is recorded. If e falls back bellow t_1, the warning mode is disabled. The second threshold is used for detecting the change. As soon as $e > t_2$, a change is detected. The new classifier is constructed using the objects that arrived after the latest recorded onset of change, t.
- The Sequential Probability Ratio Test (SPRT) charts are used in industrial engineering for process quality control [8]. Instead of the error rate e, the cumulative sum of errors, e_c, is being monitored. There are three thresholds, t_0, t_1, t_2. An SPRT is started, and at each step the cumulative error is compared with a reference value t_0. If the difference $e_c - t_0$ is greater than t_1, then change is signalled, the current SPRT is terminated, and a new SPRT is started. Otherwise, if $e_c - t_0 < t_2$, then the error rate is considered to be steady, the current SPRT is terminated, and a new SPRT is started. If $t_2 \leq e_c - t_0 \leq t_1$, SPRT continues with the next object.

Note that the change detection is not tied to the classifier model. Most often an online classifier model is used (constant-update), and upon detection of change, the window size or the learning rate are corrected to account for the change (detect-and-retrain). In our study, a classifier that does not have a forgetting mechanism is used. The current classifier is discarded, a new empty classifier is constructed and initialised with the object that triggered the detection.

2.3 Winnow

Winnow is a method for combining the votes of several experts using streaming data [5]. Substituting 'classifier' for 'expert', Winnow becomes an online strategy for combining classifiers [4]. Each classifier has a weight. Initially all the weights are set to 1. Given an object, the support for class k is calculated as the sum of the weights of the classifiers that predict that class. The prediction of the ensemble is the class with the largest support. If the prediction of the ensemble is wrong, the weights are updated. If the base classifier is correct, its weight is multiplied by α (a parameter of the method). If the classifier is incorrect, its weight is divided by α.

Winnow can be applied to classifiers of any type, both online and static. The update of the weights itself takes care of the changing environments. The novel idea in this study is to use Winnow with online classifiers in order to combine the benefits of the two strategies for handling changing environments.

3 Experiments

The question we seek to answer is whether combining classifiers with different abilities to react to changes is better than using a single online classifier.

The 28 real data sets used in the experiments are shown in Table 2. The features in all data sets are numerical and there are no missing values. Each dataset was normalized once, before starting the experiments. After the normalization, all the numeric features had mean 0 and standard deviation 1. A random subsample of the dataset was taken to be the testing set (10%), and the remaining data (90%) was the training set. One hundred such splits were made, and the classification error was estimated as the average of the 100 testing errors.

3.1 Scenarios

Three scenarios are considered: no change, gradual change and abrupt change.

No change. An initial classifier is constructed using a sample of 50 objects from the training data. Then, for 1000 iterations, a training object is selected randomly, the classifier is updated using this object and the error of the current classifier is estimated using all the testing data. The error of the method is the average error of the classifier for the 1000 iterations. The reported results are obtained as the average for 100 different random partitions of the data.

Table 1. Datasets used in the experiments. Notes: #F is the number of features, #O is the number of objects, #C is the number of classes.

Dataset	#F	#O	#C	Source
breast	9	277	2	UCI
german	24	1000	2	UCI
glass	9	214	6	UCI
heart	13	297	5	UCI
image	19	2310	7	UCI
ionosphere	34	351	2	UCI
iris	4	150	3	UCI
laryngeal1	16	213	2	Collection
laryngeal2	16	692	2	Collection
laryngeal3	16	353	3	Collection
liver	7	344	2	UCI
palynomorphs	31	609	3	Private
pendigits	16	10992	10	UCI
phoneme	5	5404	2	UCI

Dataset	#F	#O	#C	Source
pima	8	768	2	UCI
sat	36	6435	6	UCI
scrapie	14	3113	2	Private
sonar	60	208	2	UCI
spam	57	4601	2	UCI
spect	44	349	2	UCI
thyroid	5	215	3	UCI
vehicle	18	846	4	UCI
voice3	10	238	3	Collection
voice9	10	428	9	Collection
votes	16	232	2	UCI
vowel	10	990	11	UCI
wbc	30	569	2	UCI
wine	13	178	3	UCI

Gradual change. For each training/test partition the numeric features are paired randomly. If the number of features is odd, one of them will be unpaired. In each iteration, the values of the training and test data sets are rotated in the plane defined by each feature pair with an angle of $\pi/1000$. This means that in the last iteration the numeric values of the features will be the original ones with a sign change.

Abrupt change. For each training/test partition a pair of classes is randomly selected. At iteration 400, these two labels are swapped for the training and test data sets. The labels are swapped back at iteration 800.

3.2 Parameter Settings

The following parameter values were used in the experiments.

– Four moving window sizes were used with the nearest neighbour (NN) classifier: ∞ (the classifier does not forget), 50, 100 and 150. With smaller fixed window (used as the reference set), the classifier adapts quicker to changes than with larger window. On the other hand, with larger windows the classifier is likely to be more accurate.

– The learning rate η for the perceptron classifier was $\{0.1, 0.3, 0.5, 0.7, 0.9\}$. Smaller values of the learning rate mean slower training, hence slower adaptation to changes, and vice versa.

– The learning rate for O-LDC was $\{0.5, 0.7, 0.9, 0.95\}$. For $\lambda = 0.5$ O-LDC does not "forget" any observation, and at time t the classifier is equivalent to a linear discriminant classifier (assuming Gaussian densities) trained on all observations up until t. For $\lambda < 0.5$, O-LDC is "reluctant" to learn: new examples are considered less important than previous examples, and with $\lambda = 0$ O-LDC does not update at all. For $\lambda > 0.5$ the new example outweigh the old examples and the classifier can respond to changes. The higher the value, the more flexible the classifier.

Ideally, the parameter values should be tuned to suit the expected type of change in the environment. If the type and magnitude of the changes can not be evaluated in advance, a meta heuristic should be used to tune the parameters online. Instead of doing this, we propose to use an ensemble of classifiers that consists of classifiers suitable for different types of changes. We chose Winnow as the combination method because it largely follows the classifier selection strategy by constantly revising the weights of the classifiers in the ensemble. Thus, for any type of change, the classifiers (parameter values) that respond best to it will be awarded highest weights. This gives the ensemble a chance to outperform any single online classifier, specialised for a single type of change.

The change detection methods (DDM and SPRT) were combined with the two classifier methods that do not have a forgetting mechanism: NN_∞ and O-$LDC_{0.5}$.

The 17 classifiers considered in the experiment are

NN window	Perceptron η	O-LDC λ	Detection-based method-classifier
∞	0.1	0.5	SPRT-NN_∞
50	0.3	0.7	SPRT-O-$LDC_{0.5}$
100	0.5	0.9	DDT-NN_∞
150	0.7	095	DDT-O-$LDC_{0.5}$
	0.9		

For Winnow, α was set to 2, because good results were reported with this value [5]. Winnow was applied separately to the 4 NN models, 5 Perceptron variants and 4 O-LDC variants and the 4 detection-based classifiers. Finally Winnow was applied to all 17 classifiers.

3.3 An Illustration

The example below illustrates the effect of the parameter values on the classification. Consider a two-dimensional data set as shown in figure 1. There are four equiprobable Gaussian classes, each one with identity covariance matrix. The centres for the four classes are at $x = \pm 2$, $y = \pm 2$. The training data consists of 1000 objects, 250 from each class. Another set of 1000 objects (250 per class) is generated for testing. The results reported below are calculated on the testing set.

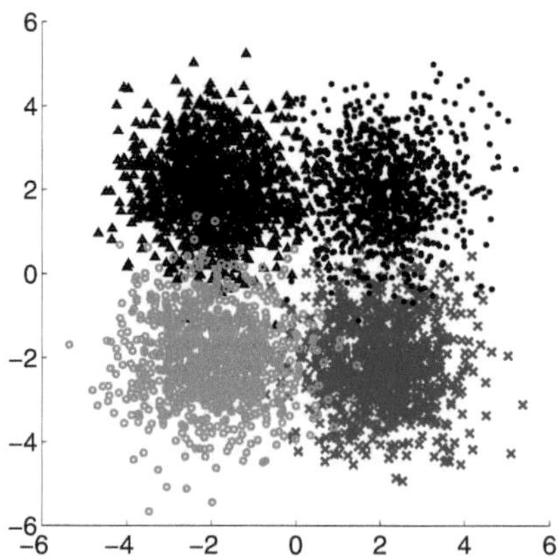

Fig. 1. 4-Gaussians dataset

Figure 2 shows the progression of the testing error for the different scenarios for O-LDC with $\lambda \in \{0.5, 0.7, 0.9, 0.95\}$. On each plot we also give the progression of the ensemble error where Winnow is applied to the 4 O-LDCs.

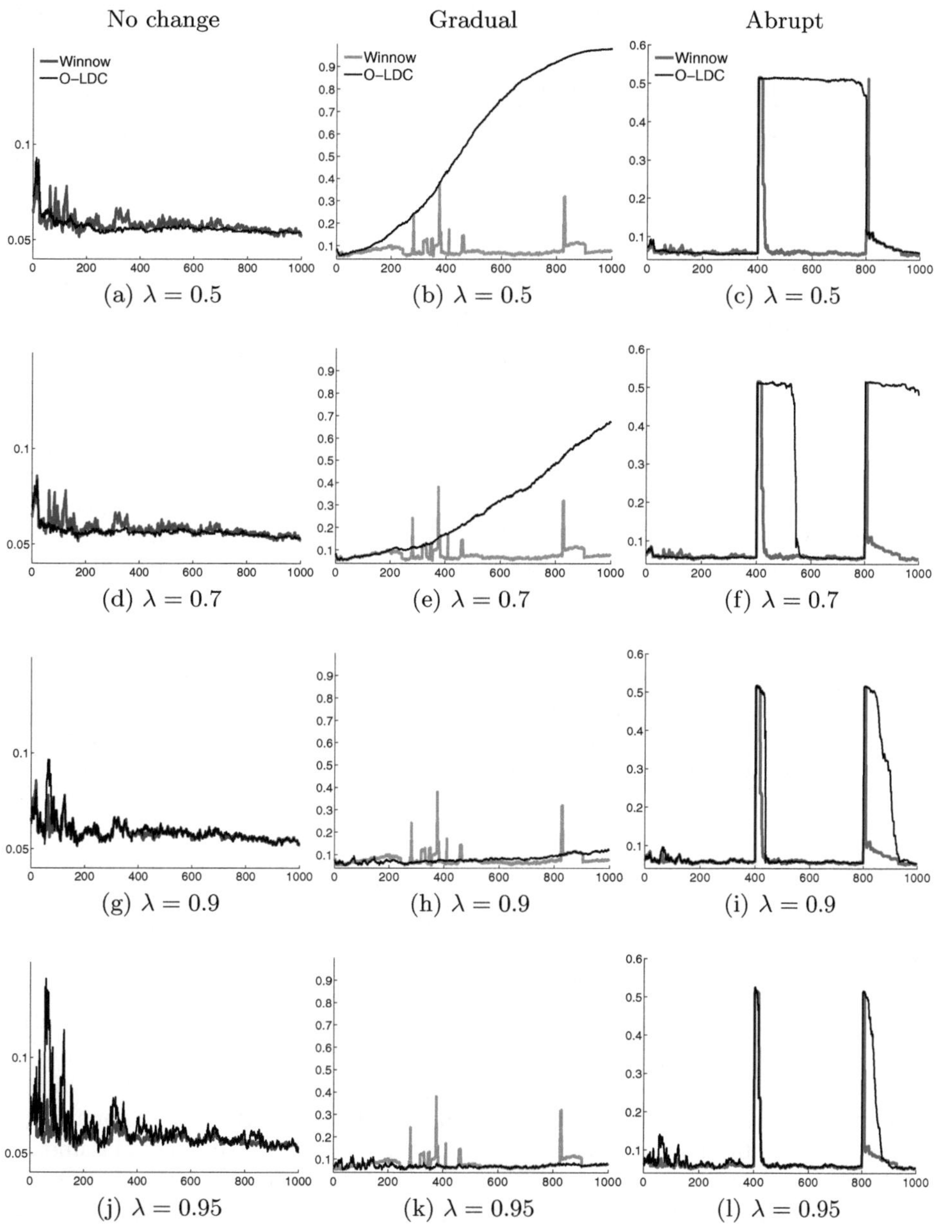

Fig. 2. Classification errors for LDC with different rates for the three scenarios. All the graphs include the errors for the method Winnow O-LDC.

For the no-change scenario, as the learning rate increases, the error of O-LDC increases too. The reason is that the learning rate acts as a soft window. Large λ corresponds to a small window of recent objects, and the training is limited

Table 2. Average ranks for the 17 individual classifiers and the 5 ensembles for the three scenarios

(a) No change		(b) Gradual change		(c) Abrupt change	
Rank	Method	Rank	Method	Rank	Method
5.96	Winnow All	4.43	Winnow All	3.04	Winnow All
6.96	O-LDC, rate=0.5	6.89	Winnow O-LDC	3.75	Winnow Detection
7.14	O-LDC, rate=0.7	7.25	Winnow Detection	5.43	SPRT NN
7.48	NN	7.79	O-LDC, rate=0.7	6.91	DDM NN
7.50	Winnow Detection	7.89	O-LDC, rate=0.9	7.86	Winnow Perceptron
7.82	DDM O-LDC	8.14	NN, window=150	8.43	SPRT O-LDC
8.79	Winnow O-LDC	9.75	Winnow NN	8.68	DDM O-LDC
8.82	Winnow NN	10.93	Winnow Perceptron	10.07	Winnow NN
8.88	DDM NN	11.00	NN, window=100	11.11	Perceptron, rate=0.3
10.79	Winnow Perceptron	12.04	DDM O-LDC	11.14	Perceptron, rate=0.7
11.04	SPRT NN	12.09	DDM NN	11.21	Perceptron, rate=0.5
11.93	SPRT O-LDC	12.23	NN	11.79	Perceptron, rate=0.9
12.54	O-LDC, rate=0.9	12.29	SPRT NN	11.93	Winnow O-LDC
13.43	NN, window=150	12.75	SPRT O-LDC	12.93	NN, window=50
14.32	Perceptron, rate=0.3	13.57	O-LDC, rate=0.95	13.18	Perceptron, rate=0.1
14.46	Perceptron, rate=0.5	13.61	Perceptron, rate=0.9	13.32	NN, window=100
14.68	Perceptron, rate=0.1	14.25	Perceptron, rate=0.7	14.43	NN, window=150
14.71	Perceptron, rate=0.9	14.32	Perceptron, rate=0.5	15.89	O-LDC, rate=0.95
14.82	Perceptron, rate=0.7	14.46	Perceptron, rate=0.3	15.93	O-LDC, rate=0.9
15.46	NN, window=100	15.54	NN, window=50	18.05	NN
16.82	O-LDC, rate=0.95	15.86	Perceptron, rate=0.1	18.14	O-LDC, rate=0.7
18.68	NN, window=50	15.93	O-LDC, rate=0.5	19.79	O-LDC, rate=0.5

to that. On the other hand, for $\lambda = 0.5$ the training set increases progressively with the new objects, which benefits the classifier. If a single classifier in the ensemble is the best method all the time, Winnow will not be able to improve on the best method as shown in plots (a), (d), (g) and (j). Winnow manages to get a similar error rate to O-LDC$_{0.5}$, which is the best classifier in the ensemble.

When the change is gradual, the results improve as the learning rate increases as seen in plots (b), (e), (h) and (k). Winnow's error follows a pattern similar to that of O-LDC with $\lambda = 0.95$, choosing again the one best classifier in the ensemble. The peaks in the Winnow error progression signify occasional slips where Winnow briefly fails to give the maximum weight to the best classifier.

For the abrupt change, the results are more interesting because at different times, different classifiers are the best ones. In the interval [400-800] the best results are for $\lambda = 0.95$ (plot(l)), while for [800-1000] the best results are for $\lambda = 0.5$ (plot(c)). In this case Winnow follows the best classifier rather well.

3.4 Results with the Real Datasets

Using Winnow, 5 ensemble designs were considered: the nearest neighbour classifiers, the Perceptrons, the O-LDC classifiers, the classifiers based on change detection and, finally, all the classifiers together. The average ranks [1] of the 17 individual classifiers and the 5 ensemble classifiers are shown in table 2.

The average ranks of the individual classifiers differ dramatically for the different scenarios. For instance, while O-LDC$_{0.5}$ is the best individual classifier for the no change scenario, it is the worst classifier for abrupt change. This variability indicates the dependency of the methods on the values of their parameter, as also illustrated in the previous subsection. In general, it is not possible to say that one method is better than another one with one exception: for the abrupt change scenario the methods based on change detection are the best.

The Perceptron is least sensitive to its parameter values; the ranks for the different η are close for all scenarios.

The Perceptron is designed for two classes. For c-class problems, $c(c-1)/2$ perceptron classifiers were trained. On the other hand, O-LDC can deal directly with multiclass datasets. Perhaps the results for O-LDC would be even better if a classifier was constructed for each pair of classes.

If we want a single method that could be used in the three scenarios, good options are DDM NN, DDM O-LDC and SPRT NN, because they have the lowest sum of ranks among the individual classifiers across the three scenarios.

The ensemble approach using Winnow outperforms the individual classifiers of all 4 types. Winnow with all the classifiers has the best average rank for the three scenarios. Winnow with the detection methods is among the 4 best methods for the three scenarios.

Figure 3 shows the error progression of Winnow All. For each scenario we plot also the error progression of the *best* individual classifier from table 2: O-LDC$_{0.5}$ (static classifier, no update) for the no-change scenario; O-LDC$_{0.7}$ for the gradual change scenario, and SPRT-NN (detect-and-update) for the abrupt change scenario. The errors are the averages across all the data sets. The graphs are meant to give us a general idea of the behaviour of the error rates for the different scenarios. Winnow All is better than the best individual classifiers for the three scenarios.

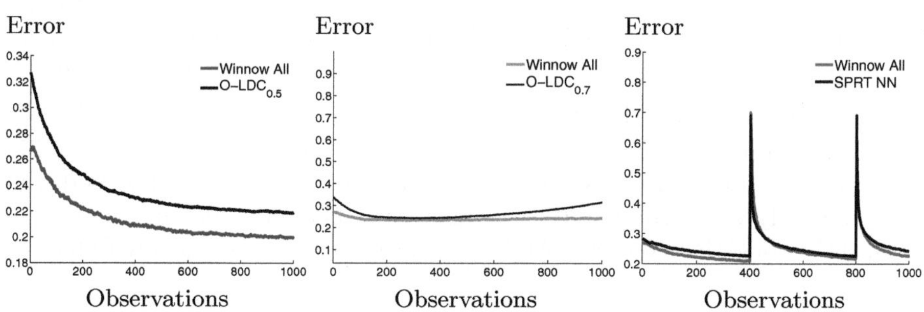

(a) No change (b) Gradual change (c) Abrupt change

Fig. 3. Progression of the classification error with the online observations for Winnow All and the best individual classifier from table 2

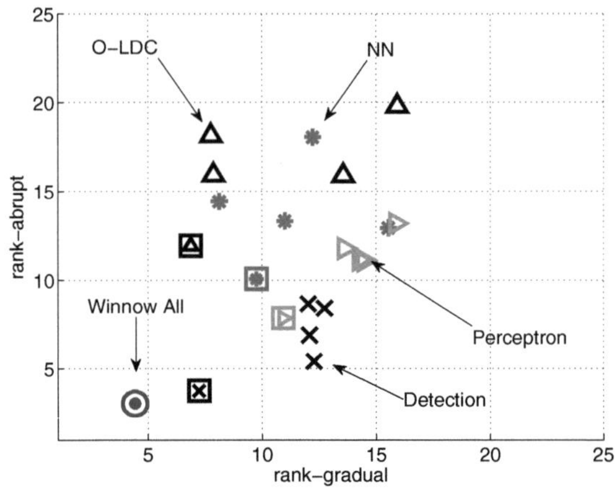

Fig. 4. Scatterplot of the 17 individual classifiers and the 5 Winnow ensembles

Figure 4 displays a scatterplot of the 22 classifiers in the 2-d space of the ranks for the gradual and the abrupt change scenarios. If there was an ideal classifier that was always the best in all experiments, it would be displayed as a point at (1,1). The closer the point to that ideal pair of ranks, the better the classifier, compared to the other classifiers in the experiment. The individual classifiers of the same types are shown with the same marker. The Winnow ensembles are indicated by a box. The figure reconfirms the finding that all 4 classifier models benefit from ensembles with Winnow. The best classifier is Winnow All, followed by Winnow Detection.

We apply the statistical analyses proposed in [1] on the ranks in table 2. Friedman's ANOVA followed by calculating the Iman and Davenport statistic F_F indicate that the entries in the table are statistically significant at significance level 0.05.

To determine which classifiers are significantly different from Winnow All (the best method for the three scenarios), the two-tailed Bonferroni-Dunn test was used as a post-hoc test. A classifier is significantly different from Winnow All if its average rank is greater than the average rank of Winnow All plus the CD. The critical difference for this test is defined as $CD = q_\alpha\sqrt{\frac{k(k+1)}{6N}}$ where N is the number of data sets. For $k = 22$ classifiers and $\alpha/(k-1) = 0.05/22$, $CD = 5.273$. The horizontal double lines are used in table 2 to delimit the classifiers that are significantly worse compared to Winnow All. The only classifier that is not significantly worse than Winnow All for the three scenarios is Winnow Detection.

4 Conclusion

The performance of different methods depends greatly on the type and frequencies of changes. If only one method had to be selected, we propose to choose one of the methods based on change detection, because they give the most consistent results. Our experiments showed that the combination of methods through Winnow leads to a significant improvement on the error rate of the individual methods.

Acknowledgements. This work has been supported in part by the "Junta de Castilla y León" project BU007B08 and the Spanish Ministry of Education and Science grant PR20060503.

References

1. Demšar, J.: Statistical comparisons of classifiers over multiple data sets. Journal of Machine Learning Research 7, 1–30 (2006)
2. Fürnkranz, J.: Round robin classification. Journal of Machine Learning Research 2, 721–747 (2002)
3. Gama, J., Medas, P., Castillo, G., Rodrigues, P.: Learning with drift detection. In: Bazzan, A.L.C., Labidi, S. (eds.) SBIA 2004. LNCS, vol. 3171, pp. 286–295. Springer, Heidelberg (2004)
4. Kuncheva, L.I.: Classifier ensembles for changing environments. In: Roli, F., Kittler, J., Windeatt, T. (eds.) MCS 2004. LNCS, vol. 3077, pp. 1–15. Springer, Heidelberg (2004)
5. Littlestone, N.: Learning quickly when irrelevant attributes abound: A new linear threshold algorithm. Machine Learning 2(4), 285–318 (1988)
6. Núñez, M., Fidalgo, R., Morales, R.: Learning in environments with unknown dynamics: Towards more robust concept learners. Journal of Machine Learning Research 8, 2595–2628 (2007)
7. Polikar, R., Udpa, L., Udpa, S.S., Honavar, V.: Learn++: An incremental learning algorithm for supervised neural networks. IEEE Transactions on Systems, Man and Cybernetics 31(4), 497–508 (2001)
8. Reynolds, M.R., Stoumbos, Z.G.: The SPRT chart for monitoring a proportion. IIE Transactions 30(6), 545–561 (1998)
9. Stanley, K.O.: Learning concept drift with a committee of decision trees. Technical Report AI-03-302, Computer Science Department, University of Texas-Austin (2003)
10. Wang, H., Fan, W., Yu, P.S., Han, J.: Mining concept-drifting data streams using ensemble classifiers. In: KDD 2003: Proceedings of the ninth ACM SIGKDD international conference on Knowledge discovery and data mining, pp. 226–235. ACM, New York (2003)

Optimal Kernel in a Class of Kernels with an Invariant Metric

Akira Tanaka, Hideyuki Imai, Mineichi Kudo, and Masaaki Miyakoshi

Division of Computer Science,
Graduate School of Information Science and Technology, Hokkaido University,
Sapporo, 060-0814, Japan
{takira,imai,mine,miyakosi}@main.ist.hokudai.ac.jp

Abstract. Learning based on kernel machines is widely known as a powerful tool for various fields of information science such as pattern recognition and regression estimation. One of central topics of kernel machines is model selection, especially selection of a kernel or its parameters. In this paper, we consider a class of kernels that forms a monotonic classes of reproducing kernel Hilbert spaces with an invariant metric and show that the kernel corresponding to the smallest reproducing kernel Hilbert space including an unknown true function gives the optimal model for the unknown true function.

1 Introduction

Learning based on kernel machines[1], represented by the support vector machine [2] and the kernel ridge regression[3,4], is widely known as a powerful tool for various fields of information science such as pattern recognition, regression estimation, and density estimation. In general, appropriate model selection is required in order to obtain a desirable learning result by kernel machines. There exists two classes of model selection. One is selection of a model space to which a learning machine belongs. The other is selection of a learning machine in a fixed model space. The latter, such as selection of a regularization parameter under a fixed kernel, is sufficiently investigated in terms of theoretical and practical senses (See [5,6] for instance). On the other hand, the former, that is, selection of a kernel or its parameters, is not discussed sufficiently in terms of theoretical sense although practical algorithms for selection of a kernel (or its parameters) such as cross-validation are revealed. The difficulty of the theoretical analyses for selection of a kernel lies on the fact that the metrics of two reproducing kernel Hilbert spaces (RKHS)[7,8] corresponding to two different kernels may differ in general, which means that we do not have a unified framework to evaluate learning results obtained by different kernels. In order to avoid this difficulty, we considered a class of kernels that forms a monotonic classes of functions with an invariant metric and introduced a method for selecting an optimal kernel from the class of kernels in the previous work[9]. However, mathematical properties of the selected kernel or the corresponding RKHS are not discussed sufficiently in [9].

N. da Vitora Lobo et al. (Eds.): SSPR&SPR 2008, LNCS 5342, pp. 530–539, 2008.

In this paper, we consider a class of kernels that forms a monotonic classes of functions with an invariant metric as the same with [9] and show that the kernel corresponding to the smallest RKHS including an unknown true function gives the optimal model for the true function among the class of kernels.

2 Mathematical Preliminaries for The Theory of Reproducing Kernel Hilbert Spaces

In this section, we prepare some mathematical tools concerned with the theory of reproducing kernel Hilbert spaces[7,8].

Definition 1. *[7] Let $\mathbf{R}^n$ be an n-dimensional real vector space and let $\mathcal{H}$ be a class of functions defined on $\mathcal{D} \subset \mathbf{R}^n$, forming a Hilbert space of real-valued functions. The function $K(\boldsymbol{x}, \tilde{\boldsymbol{x}})$, $(\boldsymbol{x}, \tilde{\boldsymbol{x}} \in \mathcal{D})$ is called a reproducing kernel of $\mathcal{H}$, if*

1. *For every $\tilde{\boldsymbol{x}} \in \mathcal{D}$, $K(\boldsymbol{x}, \tilde{\boldsymbol{x}})$ is a function of $\boldsymbol{x}$ belonging to $\mathcal{H}$.*
2. *For every $\tilde{\boldsymbol{x}} \in \mathcal{D}$ and every $f \in \mathcal{H}$,*

$$f(\tilde{\boldsymbol{x}}) = \langle f(\boldsymbol{x}), K(\boldsymbol{x}, \tilde{\boldsymbol{x}}) \rangle_{\mathcal{H}}, \tag{1}$$

where $\langle \cdot, \cdot \rangle_{\mathcal{H}}$ denotes the inner product of the Hilbert space $\mathcal{H}$.

The Hilbert space $\mathcal{H}$ that has a reproducing kernel is called a reproducing kernel Hilbert space (RKHS). The reproducing property Eq.(1) enables us to treat a value of a function at a point in $\mathcal{D}$. Note that reproducing kernels are positive definite [7]:

$$\sum_{i,j=1}^{N} c_i c_j K(\boldsymbol{x}_i, \boldsymbol{x}_j) \geq 0, \tag{2}$$

for any N, $c_1, \ldots, c_N \in \mathbf{R}$, and $\boldsymbol{x}_1, \ldots, \boldsymbol{x}_N \in \mathcal{D}$. In addition, $K(\boldsymbol{x}, \tilde{\boldsymbol{x}}) = K(\tilde{\boldsymbol{x}}, \boldsymbol{x})$ for any $\boldsymbol{x}, \tilde{\boldsymbol{x}} \in \mathcal{D}$ is followed[7]. If a reproducing kernel $K(\boldsymbol{x}, \tilde{\boldsymbol{x}})$ exists, it is unique[7]. Conversely, every positive definite function $K(\boldsymbol{x}, \tilde{\boldsymbol{x}})$ has the unique corresponding RKHS [7].

Next, we introduce the Schatten product [10] that is a convenient tool to reveal the reproducing property of kernels.

Definition 2. *[10] Let $\mathcal{H}_1$ and $\mathcal{H}_2$ be Hilbert spaces. The Schatten product of $g \in \mathcal{H}_2$ and $h \in \mathcal{H}_1$ is defined by*

$$(g \otimes h) f = \langle f, h \rangle_{\mathcal{H}_1} g, \quad f \in \mathcal{H}_1. \tag{3}$$

Note that $(g \otimes h)$ is a linear operator from $\mathcal{H}_1$ onto $\mathcal{H}_2$. It is easy to show that the following relations hold for $h, v \in \mathcal{H}_1$, $g, u \in \mathcal{H}_2$.

$$(h \otimes g)^* = (g \otimes h), \quad (h \otimes g)(u \otimes v) = \langle u, g \rangle_{\mathcal{H}_2} (h \otimes v), \tag{4}$$

where the superscript $*$ denotes the adjoint operator.

3 Formulation of Learning Problems

Let $\{(y_i, \boldsymbol{x}_i) | i = 1, \ldots, \ell\}$ be a given training data set with $y_i \in \mathbf{R}$, $\boldsymbol{x}_i \in \mathbf{R}^n$, satisfying

$$y_i = f(\boldsymbol{x}_i) + n_i, \tag{5}$$

where f denotes the unknown true function and n_i denotes a zero-mean additive noise. The aim of machine learning is to estimate the unknown function f by using the given training data set and statistical properties of noise.

In this paper, we assume that the unknown function f belongs to the RKHS $\mathcal{H}_K$ corresponding to a certain kernel function K. If $f \in \mathcal{H}_K$, then Eq.(5) is rewritten as

$$y_i = \langle f(\boldsymbol{x}), K(\boldsymbol{x}, \boldsymbol{x}_i) \rangle_{\mathcal{H}_K} + n_i, \tag{6}$$

on the basis of the reproducing property of kernels. Let $\boldsymbol{y} = [y_1, \ldots, y_\ell]'$ and $\boldsymbol{n} = [n_1, \ldots, n_\ell]'$ with the superscript $'$ denoting the transposed matrix (or vector), then applying the Schatten product to Eq.(6) yields

$$\boldsymbol{y} = \left(\sum_{k=1}^{\ell} [\boldsymbol{e}_k^{(\ell)} \otimes K(\boldsymbol{x}, \boldsymbol{x}_k)] \right) f(\boldsymbol{x}) + \boldsymbol{n}, \tag{7}$$

where $\boldsymbol{e}_k^{(\ell)}$ denotes the k-th vector of the canonical basis of $\mathbf{R}^\ell$. For a convenience of description, we write

$$A_K = \left(\sum_{k=1}^{\ell} [\boldsymbol{e}_k^{(\ell)} \otimes K(\boldsymbol{x}, \boldsymbol{x}_k)] \right). \tag{8}$$

The operator A_K is linear one that maps an element of $\mathcal{H}_K$ onto $\mathbf{R}^\ell$ and Eq.(7) can be written by

$$\boldsymbol{y} = A_K f(\boldsymbol{x}) + \boldsymbol{n}, \tag{9}$$

which represents the relation between the unknown true function f and an output vector $\boldsymbol{y}$. The information of input vectors is integrated in the operator A_K. Therefore, a machine learning problem can be interpreted as an inversion problem of the linear equation Eq.(9)[11]. In general, an estimated function $\hat{f}(\boldsymbol{x})$ is represented as

$$\hat{f}(\boldsymbol{x}) = L\boldsymbol{y}, \tag{10}$$

where L denotes a learning operator specified by a learning criterion such as those of the support vector machines and kernel ridge regression.

4 Kernel Specific Generalization Ability

In general, a learning result by kernel machines is represented by a linear combination of $K(\boldsymbol{x}, \boldsymbol{x}_i)$, which means that the learning result is an element in $\mathcal{R}(A_K^*)$ (the range of the linear operator A_K^*) since

$$\hat{f}(\boldsymbol{x}) = A_K^* \boldsymbol{\alpha} = \left(\sum_{k=1}^{\ell} [K(\boldsymbol{x}, \boldsymbol{x}_k) \otimes \boldsymbol{e}_k^{(\ell)}] \right) \boldsymbol{\alpha} = \sum_{k=1}^{\ell} \alpha_k K(\boldsymbol{x}, \boldsymbol{x}_k) \tag{11}$$

holds, where $\boldsymbol{\alpha} = [\alpha_1, \ldots, \alpha_\ell]'$ denotes an arbitrary vector in $\mathbf{R}^\ell$. The point at issue of this paper is to discuss selection of a model space, that is, the generalization ability of $\mathcal{R}(A_K^*)$ which is independent from criteria of learning machines. Therefore, we define the generalization ability of kernel machines specified by a kernel K as the distance between the unknown true function f and $\mathcal{R}(A_K^*)$ written as

$$J(K, f) = ||f - P_K f||_{\mathcal{H}_K}^2, \tag{12}$$

where P_K denotes the orthogonal projector onto $\mathcal{R}(A_K^*)$ and $||\cdot||_{\mathcal{H}_K}$ denotes the induced norm of $\mathcal{H}_K$. Note that the orthogonality of P_K is also defined by the metric of $\mathcal{H}_K$. Selection of an element in $\mathcal{R}(A_K^*)$ as a learning result is out of the scope of this paper since the selection depends on learning criteria. We also ignore the observation noise in the following contents since the noise does not affect Eq.(12).

Here, we give some propositions as preparations to evaluate Eq.(12).

Lemma 1

$$P_K = \sum_{i=1}^{\ell}\sum_{j=1}^{\ell}(G_K^+)_{i,j}\left[K(\boldsymbol{x}, \boldsymbol{x}_i) \otimes K(\boldsymbol{x}, \boldsymbol{x}_j)\right], \tag{13}$$

where G_K denotes Gram's matrix of K with input vectors $\boldsymbol{x}_i$, $(i = 1, \ldots, \ell)$ and the superscript $^+$ denotes the Moore-Penrose generalized inverse[12].

Proof. The orthogonal projector onto $\mathcal{R}(A_K^*)$ is given by $P_K = A_K^+ A_K$ and $A_K^+ = A_K^*(A_K A_K^*)^+$ holds. Since

$$\begin{aligned}
&A_K A_K^* \\
&= \left(\sum_{i=1}^{\ell}\left[\mathbf{e}_i^{(\ell)} \otimes K(\boldsymbol{x}, \boldsymbol{x_i})\right]\right)\left(\sum_{i=1}^{\ell}\left[\mathbf{e}_i^{(\ell)} \otimes K(\boldsymbol{x}, \boldsymbol{x_i})\right]\right)^* \\
&= \sum_{i=1}^{\ell}\sum_{j=1}^{\ell}\left[\mathbf{e}_i^{(\ell)} \otimes K(\boldsymbol{x}, \boldsymbol{x_i})\right]\left[K(\boldsymbol{x}, \boldsymbol{x_j}) \otimes \mathbf{e}_j^{(\ell)}\right] \\
&= \sum_{i=1}^{\ell}\sum_{j=1}^{\ell}\langle K(\boldsymbol{x}, \boldsymbol{x_j}), K(\boldsymbol{x}, \boldsymbol{x_i})\rangle_{\mathcal{H}_K}(\mathbf{e}_i^{(\ell)} \otimes \mathbf{e}_j^{(\ell)}) \\
&= \sum_{i=1}^{\ell}\sum_{j=1}^{\ell} K(\boldsymbol{x}_i, \boldsymbol{x_j})(\mathbf{e}_i^{(\ell)}\mathbf{e}_j^{(\ell)*})
\end{aligned}$$

is reduced to G_K, the concrete form of P_K is written as

$$\begin{aligned}
P_K &= A_K^+ A_K = A_K^*(A_K A_K^*)^+ A_K \\
&= \left(\sum_{i=1}^{\ell}\left[\mathbf{e}_i^{(\ell)} \otimes K(\boldsymbol{x}, \boldsymbol{x_i})\right]\right)^* G_K^+ \left(\sum_{i=1}^{\ell}\left[\mathbf{e}_i^{(\ell)} \otimes K(\boldsymbol{x}, \boldsymbol{x_i})\right]\right) \\
&= \left(\sum_{i=1}^{\ell}\left[K(\boldsymbol{x}, \boldsymbol{x_i}) \otimes \mathbf{e}_i^{(\ell)}\right]\right) G_K^+ \left(\sum_{i=1}^{\ell}\left[\mathbf{e}_i^{(\ell)} \otimes K(\boldsymbol{x}, \boldsymbol{x_i})\right]\right)
\end{aligned}$$

$$= \sum_{i=1}^{\ell}\sum_{j=1}^{\ell}(G_K^+)_{i,j}\left[K(\boldsymbol{x},\boldsymbol{x}_i)\otimes K(\boldsymbol{x},\boldsymbol{x}_j)\right],$$

which concludes the proof. □

Lemma 2. *For any $f \in \mathcal{H}_K$ and input vectors $\boldsymbol{x}_i$, $(i \in \{1,\dots,\ell\})$,*

$$||P_K f||^2_{\mathcal{H}_K} = \boldsymbol{f}' G_K^+ \boldsymbol{f} \tag{14}$$

holds, where $\boldsymbol{f} = [f(\boldsymbol{x}_1),\dots,f(\boldsymbol{x}_\ell)]'$.

Proof. Since $f \in \mathcal{H}_K$,

$$\begin{aligned} &P_K f \\ &= \sum_{i=1}^{\ell}\sum_{j=1}^{\ell}(G_K^+)_{i,j}\left[K(\boldsymbol{x},\boldsymbol{x}_i)\otimes K(\boldsymbol{x},\boldsymbol{x}_j)\right] f \\ &= \sum_{i=1}^{\ell}\sum_{j=1}^{\ell}(G_K^+)_{i,j} f(\boldsymbol{x}_j) K(\boldsymbol{x},\boldsymbol{x}_i) \end{aligned}$$

holds. Thus

$$\begin{aligned} &||P_K f||^2_{\mathcal{H}_K} \\ &= \left\langle \sum_{i=1}^{\ell}\sum_{j=1}^{\ell}(G_K^+)_{i,j} f(\boldsymbol{x}_j) K(\boldsymbol{x},\boldsymbol{x}_i), \sum_{k=1}^{\ell}\sum_{l=1}^{\ell}(G_K^+)_{k,l} f(\boldsymbol{x}_l) K(\boldsymbol{x},\boldsymbol{x}_k) \right\rangle_{\mathcal{H}_K} \\ &= \sum_{i=1}^{\ell}\sum_{j=1}^{\ell}\sum_{k=1}^{\ell}\sum_{l=1}^{\ell}(G_K^+)_{i,j}(G_K^+)_{k,l} f(\boldsymbol{x}_j) f(\boldsymbol{x}_l) \langle K(\boldsymbol{x},\boldsymbol{x}_i), K(\boldsymbol{x},\boldsymbol{x}_k)\rangle_{\mathcal{H}_K} \\ &= \sum_{i=1}^{\ell}\sum_{j=1}^{\ell}\sum_{k=1}^{\ell}\sum_{l=1}^{\ell} f(\boldsymbol{x}_j)(G_K^+)_{j,i}(G_K)_{i,k}(G_K^+)_{k,l} f(\boldsymbol{x}_l) \\ &= \boldsymbol{f}' G_K^+ G_K G_K^+ \boldsymbol{f} = \boldsymbol{f}' G_K^+ \boldsymbol{f} \end{aligned}$$

holds since G_K is symmetric. □

5 Optimal Kernel in a Class of Kernels Forming a Monotonic Classes of Functions with an Invariant Metric

Let us consider a monotonic classes of functions $\mathcal{H}_{K_i}$, $(i \in \{1,\dots,N\})$ written as

$$\mathcal{H}_{K_1} \subset \mathcal{H}_{K_2} \subset \cdots \subset \mathcal{H}_{K_N}, \tag{15}$$

forming Hilbert spaces specified by a class of kernels K_i, $(i \in \{1,\dots,N\})$ with an invariant metric satisfying

$$\langle f,g\rangle_{\mathcal{H}_{K_1}} = \langle f,g\rangle_{\mathcal{H}_{K_2}} = \cdots = \langle f,g\rangle_{\mathcal{H}_{K_N}} \tag{16}$$

for any $f, g \in \mathcal{H}_{K_1}$. Note that

$$||f||^2_{\mathcal{H}_{K_1}} = ||f||^2_{\mathcal{H}_{K_2}} = \cdots = ||f||^2_{\mathcal{H}_{K_N}} \tag{17}$$

trivially holds for any $f \in \mathcal{H}_{K_1}$ from Eq.(16).

Theorem 1. *[7] If K is the reproducing kernel of the class F with the norm $||\cdot||$, and if the linear class $F_1 \subset F$ forms a Hilbert space with the norm $||\cdot||_1$, such that $||f||_1 \geq ||f||$ for any $f \in F_1$, then the class F_1 possesses a reproducing kernel K_1 such that $K^c = K - K_1$ is also a reproducing kernel.*

According to Theorem 1, there exists a kernel K_j^c such that

$$K_j = K_1 + K_j^c,$$

for any $j > 1$ since $||f||_{\mathcal{H}_{K_j}} = ||f||_{\mathcal{H}_{K_1}}$ for any $f \in \mathcal{H}_{K_1}$. Note that corresponding Gram's matrices satisfy

$$G_{K_j} = G_{K_1} + G_{K_j^c}$$

for a set of input vectors.

Lemma 3. *[12] Let H_1 and H_2 be non-negative definite (n.n.d.) Hermitian matrices, then there exists a non-singular matrix T that jointly diagonalizes H_1 and H_2 written as*

$$TH_1T^* = \Lambda_1, \quad TH_2T^* = \Lambda_2,$$

where Λ_1 and Λ_2 denote diagonal matrices.

Lemma 4. *Let H_1 and H_2 be n.n.d. Hermitian matrices and let T be a non-singular joint diagonalizer of H_1 and H_2 written as*

$$TH_1T^* = \Lambda_1, \quad TH_2T^* = \Lambda_2,$$

where Λ_1 and Λ_2 denote diagonal matrices. The matrix $T^*(\Lambda_1 + \Lambda_2)^+T$ is generalized inverse matrix of $H_1 + H_2$.

Proof. Since $H_1 + H_2 = T^{-1}(\Lambda_1 + \Lambda_2)(T^*)^{-1}$,

$$\begin{aligned}
&(H_1 + H_2)(T^*(\Lambda_1 + \Lambda_2)^+T)(H_1 + H_2) \\
&= (T^{-1}(\Lambda_1 + \Lambda_2)(T^*)^{-1})(T^*(\Lambda_1 + \Lambda_2)^+T)(T^{-1}(\Lambda_1 + \Lambda_2)(T^*)^{-1}) \\
&= T^{-1}(\Lambda_1 + \Lambda_2)(\Lambda_1 + \Lambda_2)^+(\Lambda_1 + \Lambda_2)(T^*)^{-1} \\
&= T^{-1}(\Lambda_1 + \Lambda_2)(T^*)^{-1} = H_1 + H_2
\end{aligned}$$

holds, which concludes the proof. □

Lemma 5. *Let H_1 and H_2 be n.n.d. Hermitian matrices, then $H_1(H_1+H_2)^-H_2$ is invariant for any $(H_1 + H_2)^-$, where the superscript $^-$ denotes a generalized inverse[12].*

Proof. Let $B = H_1 + H_2$ and let B_1^- and B_2^- be different generalized inverse matrices of B, then

$$B_2^- = B_1^- + Y - B_1^- BYBB_1^- \tag{18}$$

holds with a certain matrix Y [12]. Thus,

$$\begin{aligned} & H_1 B_1^- H_2 - H_1 B_2^- H_2 \\ &= H_1 B_1^- H_2 - H_1(B_1^- + Y - B_1^- BYBB_1^-)H_2 \\ &= -H_1 Y H_2 + H_1 B_1^- BYBB_1^- H_2 \\ &= -H_1 Y H_2 + H_1 Y H_2 = O \end{aligned}$$

holds since $\mathcal{R}(H_i) \subset \mathcal{R}(B)$, $(i = 1, 2)$. □

Theorem 2. *Let H_1 and H_2 be n.n.d. Hermitian matrices and let $\boldsymbol{y} \in \mathcal{R}(H_1)$, then*

$$J = \boldsymbol{y}^*(H_1^+ - (H_1 + H_2)^+)\boldsymbol{y} \geq 0 \tag{19}$$

holds.

Proof. Since $\boldsymbol{y} \in \mathcal{R}(H_1)$, there exists $\boldsymbol{z}$ such that $\boldsymbol{y} = H_1\boldsymbol{z}$. Thus,

$$\begin{aligned} J &= \boldsymbol{y}^*(H_1^+ - (H_1 + H_2)^+)\boldsymbol{y} \\ &= \boldsymbol{z}^* H_1(H_1^+ - (H_1 + H_2)^+)H_1\boldsymbol{z} \\ &= \boldsymbol{z}^* H_1 \boldsymbol{z} - \boldsymbol{z}^* H_1(H_1 + H_2)^+ H_1 \boldsymbol{z} \qquad (20) \\ &= \boldsymbol{z}^* H_1 \boldsymbol{z} - \boldsymbol{z}^*(H_1 + H_2 - H_2)(H_1 + H_2)^+(H_1 + H_2 - H_2)\boldsymbol{z} \\ &= \boldsymbol{z}^* H_1 \boldsymbol{z} - \boldsymbol{z}^*(H_1 + H_2)\boldsymbol{z} - \boldsymbol{z}^* H_2(H_1 + H_2)^+ H_2 \boldsymbol{z} + 2\boldsymbol{z}^* H_2 \boldsymbol{z} \\ &= \boldsymbol{z}^* H_2 \boldsymbol{z} - \boldsymbol{z}^* H_2(H_1 + H_2)^+ H_2 \boldsymbol{z}. \qquad (21) \end{aligned}$$

From Eqs.(20) and (21), we have

$$\begin{aligned} & 2J \\ &= \boldsymbol{z}^*(H_1 + H_2)\boldsymbol{z} \\ &\quad -\boldsymbol{z}^* H_1(H_1 + H_2)^+ H_1 \boldsymbol{z} - \boldsymbol{z}^* H_2(H_1 + H_2)^+ H_2 \boldsymbol{z} \\ &= \boldsymbol{z}^*(H_1 + H_2)\boldsymbol{z} \\ &\quad -\boldsymbol{z}^*[(H_1 + H_2)(H_1 + H_2)^+(H_1 + H_2) \\ &\qquad - H_1(H_1 + H_2)^+ H_2 - H_2(H_1 + H_2)^+ H_1]\boldsymbol{z} \\ &= \boldsymbol{z}^*(H_1 + H_2)\boldsymbol{z} \\ &\quad -\boldsymbol{z}^*[(H_1 + H_2) - H_1(H_1 + H_2)^+ H_2 - H_2(H_1 + H_2)^+ H_1]\boldsymbol{z} \\ &= \boldsymbol{z}^*[H_1(H_1 + H_2)^+ H_2 + H_2(H_1 + H_2)^+ H_1]\boldsymbol{z}. \qquad (22) \end{aligned}$$

From Lemma 3, there exists a non-singular matrix T such that $TH_1T^* = \Lambda_1$ and $TH_2T^* = \Lambda_2$. From Lemmas 4 and 5, Eq.(22) is reduced to

$$\begin{aligned} 2J &= \boldsymbol{z}^*[H_1(H_1 + H_2)^+ H_2 + H_2(H_1 + H_2)^+ H_1]\boldsymbol{z} \\ &= 2\boldsymbol{z}^*[T^{-1}\Lambda_1(\Lambda_1 + \Lambda_2)^+\Lambda_2(T^*)^{-1}]\boldsymbol{z} \\ &= 2\boldsymbol{z}^* T^{-1}[\Lambda_1(\Lambda_1 + \Lambda_2)^+\Lambda_2](T^{-1})^*\boldsymbol{z}. \end{aligned}$$

On the basis of Sylvester's law of inertia, Λ_1 and Λ_2 are *n.n.d.*, which immediately yields $J \geq 0$. □

The following theorem is the main result of this paper.

Theorem 3. *For any $f \in \mathcal{H}_{K_1}$ and input vectors $\boldsymbol{x}_i$, $(i \in \{1, \ldots, \ell\})$,*

$$||f - P_{K_1} f||^2_{\mathcal{H}_{K_N}} \leq ||f - P_{K_j} f||^2_{\mathcal{H}_{K_N}} \tag{23}$$

holds for any $1 < j \leq N$.

Proof. For any $j \in \{1, \ldots, N\}$ and any $f \in \mathcal{H}_{K_1}$,

$$\begin{aligned}
&||f - P_{K_j} f||^2_{\mathcal{H}_{K_N}} = ||f - P_{K_j} f||^2_{\mathcal{H}_{K_j}} \\
&= \langle f, f\rangle_{\mathcal{H}_{K_j}} + \langle P_{K_j} f, P_{K_j} f\rangle_{\mathcal{H}_{K_j}} - 2\langle f, P_{K_j} f\rangle_{\mathcal{H}_{K_j}} \\
&= \langle f, f\rangle_{\mathcal{H}_{K_j}} + \langle P_{K_j} f, P_{K_j} f\rangle_{\mathcal{H}_{K_j}} - 2\langle f, P^2_{K_j} f\rangle_{\mathcal{H}_{K_j}} \\
&= \langle f, f\rangle_{\mathcal{H}_{K_j}} + \langle P_{K_j} f, P_{K_j} f\rangle_{\mathcal{H}_{K_j}} - 2\langle P_{K_j} f, P_{K_j} f\rangle_{\mathcal{H}_{K_j}} \\
&= \langle f, f\rangle_{\mathcal{H}_{K_j}} - \langle P_{K_j} f, P_{K_j} f\rangle_{\mathcal{H}_{K_j}} \\
&= \langle f, f\rangle_{\mathcal{H}_{K_N}} - \boldsymbol{f}' G^+_{K_j} \boldsymbol{f},
\end{aligned}$$

from Lemma 2 and the fact that P_{K_j} is an orthogonal projector. It is obvious that $\boldsymbol{f} \in \mathcal{R}(A_{K_1}) = \mathcal{R}(A_{K_1} A^*_{K_1}) = \mathcal{R}(G_{K_1})$. Therefore, Theorems 1 and 2 yield

$$\begin{aligned}
&||f - P_{K_j} f||^2_{\mathcal{H}_{K_N}} - ||f - P_{K_1} f||^2_{\mathcal{H}_{K_N}} \\
&= (\langle f, f\rangle_{\mathcal{H}_{K_N}} - \boldsymbol{f}' G^+_{K_j} \boldsymbol{f}) - (\langle f, f\rangle_{\mathcal{H}_{K_N}} - \boldsymbol{f}' G^+_{K_1} \boldsymbol{f}) \\
&= \boldsymbol{f}'(G^+_{K_1} - G^+_{K_j})\boldsymbol{f} \\
&= \boldsymbol{f}'(G^+_{K_1} - (G_{K_1} + G_{K^c_j})^+)\boldsymbol{f} \geq 0,
\end{aligned}$$

which concludes the proof. □

According to Theorem 3, given a class of kernels that forms a monotonic classes of functions with an invariant metric, it is concluded that the kernel corresponding to the smallest class of functions gives the best generalization ability among the given class of kernels.

6 Discussions

One of classes of kernels whose corresponding RKHSs have an invariant metric is the sinc kernel written as

$$K^{\alpha}(\boldsymbol{x}, \tilde{\boldsymbol{x}}) = \frac{\sin \alpha ||\boldsymbol{x} - \tilde{\boldsymbol{x}}||}{\pi ||\boldsymbol{x} - \tilde{\boldsymbol{x}}||},$$

where $\alpha > 0$ denotes the parameter. In [9], we proposed a method of selecting the best kernel from a class of kernels forming a monotonic classes of functions with an invariant metric and we observed that the RKHS corresponding to the selected kernel by the method was the smallest class of functions including the unknown true function in the numerical experiment using the sinc kernel. This result is given its theoretical ground in this paper.

As is well known, the best kernel that perfectly reconstructs the unknown true function f with a finite (at least one) number of training samples, is

$$K^P(\boldsymbol{x}, \tilde{\boldsymbol{x}}) = f(\boldsymbol{x})f(\tilde{\boldsymbol{x}}).$$

In fact,

$$\begin{aligned}
&P_{K^P} f(\boldsymbol{x}) \\
&= \sum_{i=1}^{\ell}\sum_{j=1}^{\ell}(G_K^+)_{i,j}\left[K^P(\boldsymbol{x},\boldsymbol{x}_i)\otimes K^P(\boldsymbol{x},\boldsymbol{x}_j)\right] f(\boldsymbol{x}) \\
&= \sum_{i=1}^{\ell}\sum_{j=1}^{\ell}(G_K^+)_{i,j}K^P(\boldsymbol{x},\boldsymbol{x}_i)f(\boldsymbol{x}_j) \\
&= \sum_{i=1}^{\ell}\sum_{j=1}^{\ell}(G_K^+)_{i,j}f(\boldsymbol{x})f(\boldsymbol{x}_i)f(\boldsymbol{x}_j) \\
&= f(\boldsymbol{x})\boldsymbol{f}'(\boldsymbol{f}\boldsymbol{f}')^+\boldsymbol{f} = f(\boldsymbol{x})\boldsymbol{f}^+\boldsymbol{f} = f(\boldsymbol{x})
\end{aligned}$$

holds since $\boldsymbol{f}^+\boldsymbol{f}$ is the one-dimensional projector with $\mathrm{rank}(\boldsymbol{f}^+\boldsymbol{f}) = 1$, that is, $\boldsymbol{f}^+\boldsymbol{f} = 1$. It is obvious that K^P corresponds to the smallest class of functions that includes f, which agrees with the analytical result of this paper.

As already mentioned, the scope of this paper is in selection of $\mathcal{R}(A_K^*)$. However, selection of $\mathcal{R}(A_K^*)$ does not always have precedence over selection of an element in $\mathcal{R}(A_K^*)$ since the total generalization ability, that is, minimization of $||f - \hat{f}||$ is required in practical problems.

In this paper, we discussed a class of kernels whose corresponding RKHSs have an invariant metric. However, this limitation is quite severe. In fact, this analysis cannot be applied to popular kernels such as Gaussian kernels and polynomial kernels. Thus, one of important future works is an extension of our discussions to a class of kernels with variant metrics.

7 Conclusion

In this paper, we considered a class of kernels that forms a monotonic classes of functions with an invariant metric and showed that the kernel corresponding to the smallest class of functions including the unknown true function gives the best model space for the true function. An extension of our result to a class of kernels corresponding to classes of functions with variant metrics is one of important future works that should be resolved.

Acknowledgement. This work was partially supported by Grant-in-Aid No. 18700001 for Young Scientists (B) from the Ministry of Education, Culture, Sports and Technology of Japan.

References

1. Muller, K., Mika, S., Ratsch, G., Tsuda, K., Scholkopf, B.: An introduction to kernel-based learning algorithms. IEEE Transactions on Neural Networks 12, 181–201 (2001)
2. Vapnik, V.N.: The Nature of Statistical Learning Theory. Springer, New York (1999)
3. Shawe-Taylor, J., Cristianini, N.: Kernel Methods for Pattern Recognition. Cambridge University Press, Cambridge (2004)
4. Cristianini, N., Shawe-Taylor, J.: An Introduction to Support Vector Machines and other kernel-based learning methods. Cambridge University Press, Cambridge (2000)
5. Sugiyama, M., Ogawa, H.: Subspace Information Criterion for Model Selection. Neural Computation 13, 1863–1889 (2001)
6. Sugiyama, M., Kawanage, M., Muller, K.: Trading variance reduction with unbiasedness: The regularized subspace information criterion for robust model selection in kernel regression. Neural Computation 16, 1077–1104 (2004)
7. Aronszajn, N.: Theory of Reproducing Kernels. Transactions of the American Mathematical Society 68, 337–404 (1950)
8. Mercer, J.: Functions of Positive and Negative Type and Their Connection with The Theory of Integral Equations. Transactions of the London Philosophical Society A, 415–446 (1909)
9. Tanaka, A., Sugiyama, M., Imai, H., Kudo, M., Miyakoshi, M.: Model selection using a class of kernels with an invarinat matric. In: Joint IAPR Internatioanl Workshops SSPR 2006 and SPR 2006, pp. 862–870. Springer, Heidelberg (2006)
10. Schatten, R.: Norm Ideals of Completely Continuous Operators. Springer, Berlin (1960)
11. Ogawa, H.: Neural Networks and Generalization Ability. IEICE Technical Report NC95-8, 57–64 (1995)
12. Rao, C.R., Mitra, S.K.: Generalized Inverse of Matrices and its Applications. John Wiley & Sons, Chichester (1971)

Chaotic Pattern Recognition: The Spectrum of Properties of the Adachi Neural Network

Ke Qin[1,⋆] and B. John Oommen[2,⋆⋆]

[1] School of Computer Science and Engineering
University of Electronic Science & Technology of China
Chengdu, China, 610054
thinker_qk@hotmail.com
[2] School of Computer Science, Carleton University,
1125 Colonel By Dr., Ottawa, ON, K1S 5B6, Canada
oommen@scs.carleton.ca

Abstract. Chaotic Pattern Recognition (PR) is a relatively new subfield of PR in which a system, which demonstrates chaotic behavior under normal conditions, resonates when it is presented with a pattern that it is trained with. The Adachi Neural Network (AdNN) is a classic neural structure which has been proven to demonstrate the phenomenon of Associative Memory (AM). In their pioneering paper [1,2], Adachi and his co-authors showed that the AdNN also emanates periodic outputs on being exposed to trained patterns. This was later utilized by Calitoiu *et al* [4,5] to design systems which possibly possessed PR capabilities. In this paper, we show that the previously reported properties of the AdNN do not adequately describe the dynamics of the system. Rather, although it possesses far more powerful PR and AM properties than was earlier known, it goes through a spectrum of characteristics as one of its crucial parameters, α, changes. As α increases, the AdNN which is first an AM become *quasi*-chaotic[1]. The system is then distinguished by two phases which really do not have clear boundaries of demarcation. In the former of these phases it is *quasi*-chaotic for some patterns and periodic for others. In the latter of these, it exhibits properties that have been unknown (or rather, unreported) till now, namely, a PR capability (which even recognizes masked or occluded patterns) in which the network resonates sympathetically for trained patterns while it is *quasi*-chaotic for untrained patterns. Finally, the system becomes completely periodic. All these results are, to the best of our knowledge, novel.

Keywords: Chaotic NNs, Chaotic Pattern Recognition, Adachi NN.

⋆ The work of this author was done while he was in Canada as a Visiting Scholar at Carleton University in 2008.

⋆⋆ *Chancellor's Professor* ; *Fellow : IEEE* and *Fellow : IAPR*. This author is also an *Adjunct Professor* with the University of Agder in Grimstad, Norway. The work of this author was partially supported by NSERC, the Natural Sciences and Engineering Research Council of Canada. The authors would like to thank Dr. Dragos Calitoiu for his time and his insight during our mutual discussions.

[1] A formal explanation of this expression is given in the body of the paper.

N. da Vitora Lobo et al. (Eds.): SSPR&SPR 2008, LNCS 5342, pp. 540–550, 2008.

1 Introduction

Although the field of PR has been extensively studied, the entire area of designing and implementing Neural Networks (NNs) which are inherently chaotic, and which simultaneously possess PR capabilities is relatively new[2]. As far as we know, the pioneering papers in this regard are those due to Calitoiu and his co-authors [4,5]. The aim of this paper is to demonstrate the fundamental properties of the NN used in [4,5], the AdNN, (which was earlier proposed by Adachi and his co-authors [1,2]). Quite simply, we shall show that the AdNN is an extremely powerful and efficacious system, and that its properties that have been reported in the literature are but a shadow of its true underlying capabilities. Indeed, we shall demonstrate that the same NN can demonstrate the phenomena of Associate Memory (AM), *quasi*-chaos, pattern recognition and periodicity for different values of a certain parameter[3].

An AM permits its user to specify part of a pattern, and to thereafter retrieve the values associated with that pattern. One of the limitations of most ANN models of AM is the dependency on an external input. Once an output pattern has been identified, the ANN remains in that state until the arrival of a new external input. This is in contrast to real biological neural networks which exhibit sequential memory characteristics. To be more specific, once a pattern is recalled from a memory location, the brain is not "stuck" to it, it is also capable of recalling other associated memory patterns without being prompted by any additional external inputs. This ability to "jump" from one memory state to another *in the absence of a stimulus* is one of the hallmarks of the brain, and *this is one phenomenon that a chaotic PR system has to emulate.*

The evidence that indicates the possible relevance of chaos to brain functions was first obtained by Freeman [7,13] through his clinical work on the large-scale collective behavior of neurons in the perception of olfactory stimuli. Based on his experiments, he conjectured that the quiescent state of the brain is chaos, while during perception, when attention is focused on any sensory stimulus, the brain activity becomes more periodic. The controlling of chaos gives rise to periodic behavior, culminating in the identification of the sensory stimulus that has been received. Thus, mimicing this identification on a neural network can lead to a new model of PR. encounters one of the memorized patterns as an input to the network, we want the network to *resonate* with that pattern, i.e., to generate *that* pattern with a certain periodicity. Between two consecutive appearances of the memorized pattern, the network can also be in an infinite number of states, but in none of the memorized ones. greater detail in [4,5], where we also show that in order to achieve recognition, one must decrease the level of chaos until a periodic behavior is obtained.

---

[2] The issue of how this sub-field of PR differs from the previously reported sub-fields and the sub-fields of NNs [3,8,9,11] (including those capable of associative memory and learning [6,12]), is described in [4,5,10].

[3] It is also very interesting to note that the parameter of interest in this paper, namely α, is one that has not been examined in the literature when it concerns any of these phenomena.

One of the primary differences between our results and the results reported earlier for the AdNN (and its variants) [1,2,4,5] is the demonstration that it is powerful enough to accomplish PR even if the pattern is masked or occluded. This phenomenon, which is almost "mystical" and inexplicable, has been both unknown and unreported. The second major difference between our results and the results reported earlier for the AdNN (and its variants) [1,2,4,5] is the concept of what we shall refer to as *quasi*-chaotic behavior, explained presently. The final difference of importance is that we show that all of the above-mentioned properties are a consequence of varying the parameter α and not the parameters k_r and k_f, as was previously anticipated.

1.1 Rationale and Contributions of the Paper

The entire field of Chaotic PR is in its infancy, and every step into the unknown leads to fascinating results. In particular, the papers of Adachi and his co-authors [1,2] first presented the AdNN with the conjecture that it demonstrated chaotic phenomena. Later, in [4,5], the authors showed by an analysis of the Lyapunov exponents, that the AdNN was *not* chaotic. But this was by virtue of only considering the coefficients, k_r and k_f. We emphasize here that no previous result delved into the importance of the parameter, α, the refractory scaling parameter[4].

The initial rationale for investigating the AdNN was to procure a clearer understanding of the role that α played. It turns out that unlike the coefficients k_r and k_f, α seems to have a more pronouncing effect on the chaotic, periodic and PR capabilities of the NN. Thus, by considering the range of values that α can assume, we encounter a fascinating spectrum of behavioral properties, which were previously unreported. Indeed, although the reason for the "peculiar" behavior is unknown, our experimental results (see Section 3.2) illustrate that the AdNN can truly be used to develop chaotic PR systems. Indeed, (a) We have shown that the chaotic behavior of the AdNN is not merely constrained by the coefficients k_r and k_f; (b) We have clearly demonstrated that unlike the parameters k_r and k_f, the refractory scaling parameter, α, plays a significant role in determining the phenomenon displayed by the NN. (c) As opposed to the previously claimed results, we show that the AdNN is not chaotic. Rather, it exhibits a behavior which is *quasi*-chaotic, implying that it is chaotic if the output is merely compared to the input, but if the output is examined as a stream in its own right, it is periodic all the same. Such a behavior was previously unknown; (d) We have shown that the same network, the AdNN can be made to demonstrate a wide spectrum of phenomena – Associative Memory, periodicity, pattern recognition and *quasi*-chaos. This fact was previously unknown too; and (e) Finally, and most importantly, we have shown that there is a "range" in the

[4] Adachi and his co-authors [1,2] set the value for α as 10 in their experiments, but this seemed rather arbitrary. No explanation was given as to how and why this value of α was used. The same comment can be made about the results of Calitoiu *et al.* [4,5].

value of the parameter α, in which the system is *quasi*-chaotic for untrained patterns, but periodic for trained patterns or their noisy incarnations.

We conclude this sub-section by observing that we are not aware of any comparable results in the field of NNs and CNNs.

2 The Adachi Neural Network Model and Its Variants

The AdNN[5] is a network of neurons with weights associated with the edges, a well-defined Present-State/Next-State function, and a well-defined State/Output function. It is composed of N neurons (Adachi set $N = 100$), topologically arranged as a completely connected graph, i.e, each neuron communicates with every other neuron, including itself. It is described by means of the following equations relating the two internal states $\eta_i(t)$ and $\xi_i(t)$, $i = 1...N$, and the output $x_i(t)$ as:

$$x_i(t+1) = f(\eta_i(t+1) + \xi_i(t+1)), \tag{1}$$

$$\eta_i(t+1) = k_f \eta_i(t) + \sum_{j=1}^{N} w_{ij} x_j(t), \tag{2}$$

$$\xi_i(t+1) = k_r \xi_i(t) - \alpha x_i(t) + a_i. \tag{3}$$

In the above, $x_i(t)$ is the output of the neuron i which has an analog value in [0,1] at the discrete time t. The internal states of the neuron i are $\eta_i(t)$ and $\xi_i(t)$, $f(\cdot)$ is the logistic function with the steepness parameter, ε, satisfying $f(y) = 1/(1+e^{-y/\varepsilon})$. k_f and k_r are the decay parameters for the feedback inputs and the refractoriness, respectively. $\{w_{ij}\}$ are the synaptic weights from the ith constituent neuron to the jth constituent neuron, and a_i denotes the temporally constant external inputs to the ith neuron. From the perspective of this paper, most importantly, α is the refractory scaling parameter, whose significance has, to date, not been investigated.

While the network dynamics are described by Equations (2) and (3), the outputs of the neurons are obtained by Equation (1). The feedback interconnections are determined according to the following symmetric auto-associative matrix of the p stored patterns by:

$$w_{ij} = \frac{1}{p} \sum_{s=1}^{p} (2x_i^s - 1)(2x_j^s - 1), \tag{4}$$

where x_i^s is the ith component of the sth stored pattern.

Calitoiu *et al* [5] proposed a model of CNNs which modifies the AdNN to enhance its PR capabilities. This NN, referred to as the Modified AdNN (M-AdNN), is actually also a Hopfield-like model, and manipulates the internal

[5] This review is necessarily very brief in the interest of space. However, we mention that the AdNN was initially modeled to serve as a dynamical Associative Memory. More details of the AdNN are found in [1,2,4,5,10].

structure and Present-State/Next-State equations of the original AdNN. Structurally, it is also composed of N neurons, topologically arranged as a completely connected graph. Again, each neuron i, $i = 1...N$, has internal states $\eta_i(t)$ and $\xi_i(t)$, and an output $x_i(t)$. Calitoiu *et al* [5] presented a brief rationale for each modification. The fundamental difference between the AdNN and the M-AdNN in terms of their Present-State/Next-State equations is that the latter has only a *single* global neuron (and its corresponding two global states) which is used for the state updating criterion for *all* the neurons.

Calitoiu *et al* [4] later enhanced the M-AdNN to yield an even more interesting NN, the Modified Blurring AdNN (Mb-AdNN). The latter was designed to solve the inverse PR problem, namely that of understanding how a neural system, which receives an exact pattern, can perceive it accurately in certain settings and yet see it as "blurred" in other settings. The authors of [4] accomplish this by forcing a set of neurons to have Present-State/Next-State functions which are updated using their current values, while the corresponding functions for the other neurons involve only a single global neuron as in the case of the M-AdNN. As we will not be investigating the Mb-AdNN in any detail here, the details of its design and analysis are omitted, but can be found in [4].

Since a detailed survey of the AdNN, the M-AdNN and the Mb-AdNN is not possible here, in the interest of completeness and continuity, we mention that its salient features are listed in [4,5,10]. However, we mention that the drawbacks of the current schemes are its excessive computational burden, the long transient phases, its imprecise PR capability, and the resonance for the untrained patterns. These have been tested for the data sets used by [1,2] and [4,5] given in Figure 1. Graphs demonstrating this are given in [10], and additional experimental results which clarify this phenomenon (which was earlier unreported) are listed in Table 1 for these data sets. We attempt to rectify these disadvantages here.

Table 1. Periodicities for trained and untrained patterns for the data set used by Adachi *et al* (the first two rows), and the digital data set used by Calitoiu *et al* (the last two rows). The entries of particular interest are in a **bold** font.

Pattern	1	2	3	4	With Noise	Random
Periodicity	20	20	39	39	39	**39**

Pattern	1	2	3	4	5	6	7	8	9	0	With Noise	Random
Periodicity	7,7,8	26	26	27	26	27	26	26	26	26	26	**25**

3 New Periodicity, *Quasi*-Chaotic and PR Properties of the AdNN

The goal of the field of Chaotic PR systems can be vocalized as follows: To be more specific, let us suppose that we want the chaotic PR system to recognize patterns P_i and P_j. To accomplish this, we train the system using both the patterns by a mere straightforward computation. This training phase assigns the weights between the neurons of the CNN, which effectively memorizes the

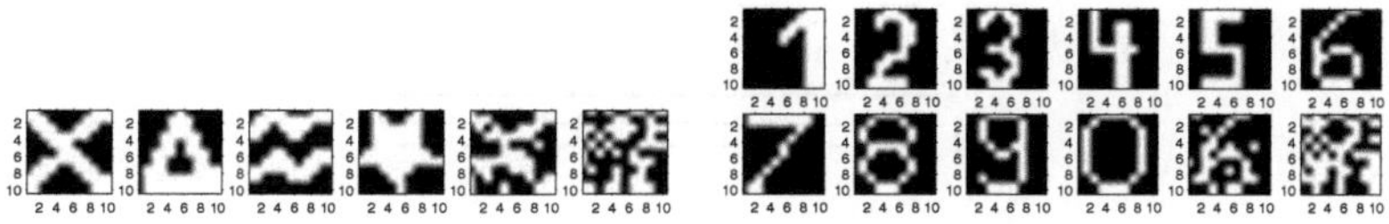

Fig. 1. (a) is the set of patterns used in Adachi *et al*'s and our experiments. The first four patterns constitute the set used by Adachi *et al.* Pattern5 is a noisy version of Pattern1 with 15% noise. Pattern6 is a randomly generated pattern. (b) is the set of patterns used in Calitoiu *et al*'s and our experiments. The first ten patterns constitute the set used by Calitoiu *et al.* Pattern11 is a noisy version of Pattern6 with 10% noise. Pattern12 is a randomly generated pattern, and is the reverse of Pattern6 in (a).

training patterns. Subsequently, on testing, if any pattern other than P_i or P_j is presented, the CNN must continue to be chaotic since it is not trained to recognize such a pattern. However, if P_i or P_j (or a pattern resembling either of them) is presented, the CNN must switch from being chaotic to being periodic. Note that as opposed to traditional PR systems, the output is not a single value. It is a sequence of values, which is chaotic (i.e., displays no periodicity) unless one of the trained patterns is presented.

We now present a sequence of results concerning the AdNN which were earlier unreported.

3.1 Periodicity Analysis Using Quasi-energy Functions

Although the AdNN was initially thought to be chaotic, Calitoiu *et al* [4] showed, by a Lyapunov analysis, that this was not the case. Rather, we present below an informal analysis for the observed periodicity.

As per the definition of Adachi *et al* [1], an alternate formulation for the dynamic equations of the AdNN is given by:

$$\begin{aligned} x_i(t+1) &= f(\sum_{j=1}^{N} w_{ij} \sum_{d=0}^{t} k_f^d x_j(t-d) - \alpha \sum_{d=0}^{t} k_r^d x_i(t-d) + a_i) \\ &= f(\sum_{j=1}^{N} w_{ij} \sum_{d=0}^{t} k_f^{t-d} x_j(d) - \alpha \sum_{d=0}^{t} k_r^{t-d} x_i(d) + a_i). \end{aligned} \tag{5}$$

Based on the above, we argue that the AdNN can never get "absorbed" into a fixed point. The details of the argument are rather lengthy and are omitted due to space limitations (they can be found in [10]) but can be summarized as:

1. If any neuron tends to converge towards the value '0' as a fixed point, it can be seen from the above equations that the dynamics of the system forces the output value of *that* neuron towards unity.

2. If any neuron tends to converge towards the value '1' as a forces the output value of *that* neuron towards '0'.

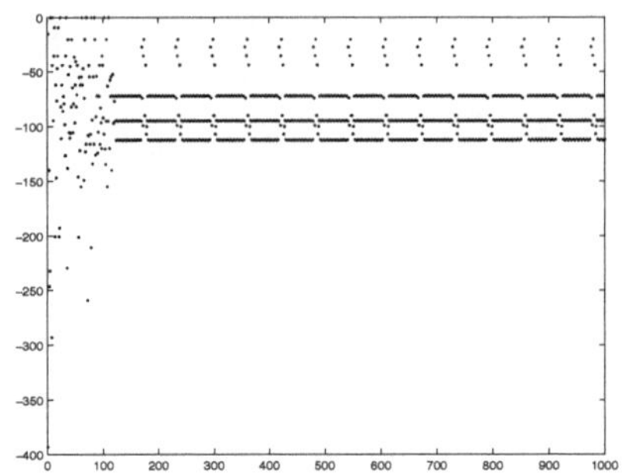

Fig. 2. A plot of the quasi-energy function with the parameters $\alpha = 17, k_f = 0.2, k_r = 0.9, \varepsilon = 0.00015$, and $a_i = 2$. The reader should observe the convergence and periodicity of the plot.

Since we have clearly argued that no neuron will be trapped at either fixed point '0' or '1', the question which has to be answered is that of knowing whether the system will be chaotic or periodic? Here, we believe that the critical parameter is the refractory parameter, α, in Equation (5), which completely describes the quasi-energy function of the dynamical AdNN.

Unfortunately, at this juncture, a formal analysis is unavailable due to the complexity of the overall network. Rather, we resort to an experimental analysis which shows that the system's behavior is periodic after a short transient phase, and that α is crucial to the length of this transient phase. As α increases, we observe that the transient phase becomes increasingly shorter. For example, if we set $\alpha \leq 15$, the transient phase is about $21,000$. However, the length of this phase reduces sharply to about $5,300$ when $\alpha = 16$. Thereafter, as α is increased (even incrementally, to $\alpha = 16.0001$), the quasi-energy function converges even faster, displaying a transient phase of only approximately $1,400$ iterations, which further reduces to only about 120 iterations when $\alpha = 17$. Indeed, the plots of the quasi-energy function (see for example, Figure 2) demonstrates that the AdNN is both convergent and periodic.

3.2 *Quasi*-Chaotic and PR Properties of the AdNN

We now report the PR properties of the AdNN. Although, in the ideal setting we would have preferred the AdNN to be chaotic (when the value of α does not lead to a periodic output), it turns out that, "unfortunately", its output is not really chaotic. Rather the AdNN demonstrates a phenomenon which we refer to as being *quasi*-chaotic since it is neither completely chaotic nor simply periodic. Indeed, this *quasi*-chaotic phenomenon implies that the system *appears* chaotic if the output is merely compared to the input. But if, on the other hand, the output is examined as a stream independently (i.e., by itself), *it is periodic all the same.* This unexplained behavior was previously unknown.

Table 2. The properties of the AdNN obtained for different values of α. The data set used is the one used by Adachi *et al* (see Figure 1), and the values of the parameters are $k_f = 0.2$, $k_r = 0.9$, and $\varepsilon = 0.00015$. The legend for the table is the following: AM — Associative Memory; QC — *Quasi*-Chaotic; C* — Periodic for a short time and then *quasi*-chaotic; P — Periodic. Observe the PR properties of the AdNN for $\alpha = 26$.

	P1	P2	P3	P4	P5	P6
$\alpha = 10$	A	A	A	A	A	A
$\alpha = 11$	A	A	A	A	A	A
$\alpha = 17$	QC	QC	QC	QC	QC	QC
$\alpha = 18$	QC	QC	QC	QC	QC	QC
$\alpha = 19$	QC	P	QC	QC	QC	QC
$\alpha = 25$	C*	C*	C*	C*	C*	C*
$\alpha = 26$	P	P	P	P	P	QC
$\alpha = 70$	P	P	P	P	P	P

Within this expanded view of perceiving things, we believe that it is still possible to design, develop and implement chaotic PR systems[6]. First of all, unlike the analysis done in [4,5] which concentrates on the Lyapunov analysis, it turns out that the parameters k_r and k_f do not play a crucial role to determine the overall phenomena – as was previously anticipated. Rather, it should be emphasized that the parameter of interest to achieve this, namely α, is one that has not been examined in the literature when it concerns a PR phenomena.

The highlights of the results obtained for the AdNN (for $a_i = 2$) is tabulated in Table 2. We summarize the results by stating that by increasing the value of α, the AdNN demonstrates the following amazing results:

1. **Associate Memory:** When $\alpha = 10$, the AdNN is merely an AM during its long transient phase. It can retrieve *all* the trained patterns non-periodically during its transient phase.
2. **Periodic but neither AM nor PR:** As α increases to 17, the AdNN begins to display some interesting properties. Initially, for $\alpha = 17$ or 18, the AdNN reproduces unrecognized patterns as can be seen from the detailed experimental figures included in [10] (omitted here in the interest of brevity). Observe that none of the trained pattern will be retrieved since the Hamming distance is never 0. Thus, amazingly, the AdNN is neither an AM nor a PR system when $\alpha = 17$!
3. **Periodic and *Quasi*-Chaotic:** If $\alpha = 19$ or 20, the output is a repeated version of unrecognized patterns for certain input patterns, while for other patterns, the system possesses the capability of recognizing them. This can also be observed from from the above-mentioned detailed experimental figures included in [10]. In this figure, the input is Pattern2, and the system yields as output, the same pattern (Pattern2) with a periodicity of 7.

[6] The detailed mechanics of such a system are, as yet, not fully explained. This is currently being undertaken, and will be, hopefully, reported in a future work.

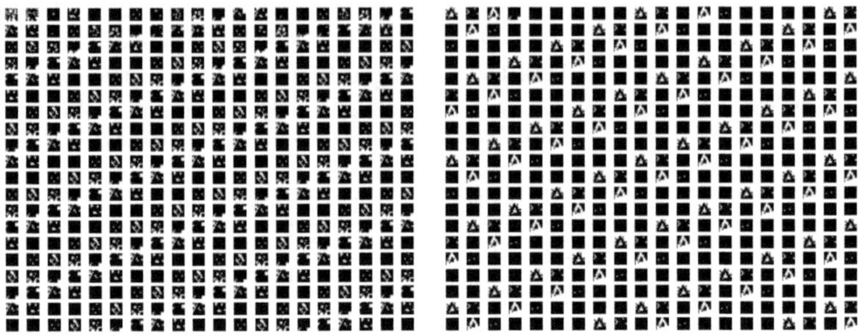

Fig. 3. Visualization of the output obtained from time t=1 to t=400. The input for the figure on the left is an untrained random pattern. The output yielded is neither chaotic nor simply periodic because the input pattern is never repeated, although a completely unknown pattern appears periodically at the output. This sequence of outputs given by the figure on the right demonstrates that the AdNN is capable of recognizing the input pattern correctly even if the percentage of noise is up to 90%.

4. **Short-term Periodic and *Quasi*-Chaotic:** If $\alpha = 21, 25, 29, 30$ or 35, the system has the ability of recalling the input patterns several times, and subsequently these samples from its memory.
5. **Pattern Recognition:** The most interesting scenario, the one which does yield PR, occurs when $\alpha = 26$. In this case, the system's output is periodic (repeating the input) as long as the input is one of the trained patterns. However, on the contrary, if the system is presented with untrained input patterns, it will result in unrecognized *quasi*-chaotic output. This interesting property can be seen from the figures in [10] and 3. The reader should observe the importance of this result: This implies that unlike the earlier claim of [4,5], the AdNN can be used to develop a chaotic PR system. Apart from the case when $\alpha = 26$, this powerful property is also exhibited at other values of α such as $32, 38, 44, 49$ and 61. Thus, it appears that part of the training phase of a chaotic PR system would involve determining the value of α that would lead such a behavior.
6. **Purely Periodic:** Finally, if α is increased to even larger values, for example, $\alpha = 70$, the output is a periodic version of the input independent of what the input is.
7. **Response to Weighted External Inputs:** Another fascinating feature of the AdNN is its behavior when external inputs are applied, that is, when $a_i = 2 + 6x_i$. From our preliminary experimental results it appears as if, in this case, the AdNN behaves like the Mb-AdNN [5].
8. **Recognition of *Inverted* Patterns:** Another property of the AdNN is that it is even capable of recognizing completely inverted patterns. Indeed, our experimental results demonstrate that the AdNN can recognize trained patterns even if the input is almost completely inaccurate. Thus, if the bit-wise percentage of noise is as high as **90%** (implying that the input is almost identical to a totally bit-wise inverted version of the original pattern), the AdNN recognizes the input pattern accurately by a periodic and

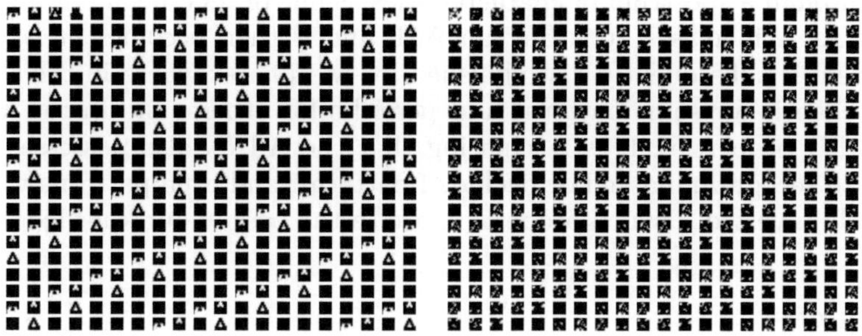

(a) (b)

Fig. 4. In the sequence of outputs on the left, (a), the input is a half-masked version of Pattern2. At the output, the AdNN recognizes Pattern2 correctly, by presenting the two halves in *periodic consecutive outputs*. Indeed, even if this masked pattern is garbled with noise up to **30**%, the AdNN continues to produce *quasi*-chaotic outputs, as seen in the sequence of outputs on the right, (b). Again, the parameters are $k_f = 0.2$, $k_r = 0.9$, and $\varepsilon = 0.00015$, with the crucial parameter, α being 26.

quasi-chaotic behavior. Detailed experiments demonstrating this property also are found in [10], omitted here due to space limitations.

9. **Recognition of *Masked* Patterns:** Perhaps the most fascinating unreported property of the AdNN is its ability to recognize masked or occluded patterns. In this case, the system is trained with a set of training patterns. On testing, however, the system is provided with a noisy version of one-half of a trained pattern, while the other half is completely random. In this case too, the network resonates sympathetically for trained patterns by presenting the two halves in *periodic consecutive outputs*[7], while it is *quasi*-chaotic for untrained patterns. This can be observed from Figure 4 where, in the first case, the masked image is generated as alluded to here, and in the second case, the input is further garbled with **30**% noise. In both of these cases, the system is capable of recognizing the true pattern with with a periodic and *quasi*-chaotic behavior.

4 Conclusions

In this paper we have concentrated on the field of Chaotic Pattern Recognition (PR), which is a relatively new sub-field of PR. Such systems, which have only recently been investigated, demonstrate chaotic behavior under normal conditions, and resonate when it is presented with a pattern that it is trained with. The reported systems work with the Adachi Neural Network (AdNN) [1,2] which has Associative Memory (AM) properties, and which also emanates periodic outputs on being exposed to trained patterns. In this paper, we have presented a collection of previously unreported properties of the AdNN. We have shown that

[7] There is no other expression to describe this phenomenon than *amazing*!

it goes through a spectrum of characteristics as one of its crucial parameters, α, changes. As α increases, it is first an AM, and it then becomes *quasi*-chaotic. The system is subsequently distinguished by two phases where in the former it is *quasi*-chaotic for some patterns and periodic for others, and in the latter, it exhibits PR properties. It is fascinating that the AdNN also possesses the capability to recognize masked or occluded patterns, and even patterns which are completely inverted – which properties, are to the best of our knowledge, novel.

References

1. Adachi, M., Aihara, K.: Associative dynamics in a chaotic neural network. Neural Networks 10(1), 83–98 (1997)
2. Aihara, K., Takabe, T., Toyoda, M.: Chaotic neural networks. Physics Letters A 144(6,7), 333–340 (1990)
3. Bishop, C.: Neural Networks for Pattern Recognition. Oxford University Press, Oxford (1995)
4. Calitoiu, D., Oommen, B., Nussbaum, D.: Desynchronzing of chaotic pattern recognition neural network to model inaccurate parception. IEEE Trans. on Systems, Man, and Cybernetics-part B: Cybernetics 37(3), 692–704 (2007)
5. Calitoiu, D., Oommen, B., Nussbaum, D.: Periodicity and stability issues of a chaotic pattern recognition neural network. Pat. Anal. Applic., 175–188 (October 2007)
6. Fausett, L.: Fundamentals of Neural Networks: Architectures, Algorithms And Applications. Prentice Hall, Upper Saddle River (1994)
7. Freeman, W.: Tutorial on neurobiology: from single neurons to brain chaos. International Journal of Bifurcation and Chaos in Applied Sciences and Engineering 2, 451–482 (1992)
8. Friedman, M., Kandel, A.: Introduction to Pattern Recognition, Statistical, Structural, Neural and Fuzzy Logic Approaches. World Scientific, Singapore (1999)
9. Fukunaga, K.: Introduction to Statistical Pattern Recognition. Academic Press, London (1990)
10. Qin, K., Oommen, B.J.: The Spectrum of Chaotic Pattern Recognition Properties Obtainable Using Adachi's Neural Network. Unabridged version of this paper (Submitted for Publication) (Can be made available if required)
11. Ripley, B.: Pattern Recognition and Neural Networks. Cambridge University Press, Cambridge (1996)
12. Schurman, J.: Pattern Classification, A Unified View of Statistical and Neural Approaches. John Wiley and Sons, New York (1996)
13. Skarda, C., Freeman, W.: How brains make chaos in order to make sense of the world. Behaviorai and Brain Sciences 10(2), 161–195 (1987)

On Euclidean Corrections for Non-Euclidean Dissimilarities

Robert P.W. Duin[1], Elżbieta Pękalska[2], Artsiom Harol[1], Wan-Jui Lee[1], and Horst Bunke[3]

[1] Faculty of Electrical Engineering, Mathematics and Computer Sciences, Delft University of Technology, The Netherlands
[2] School of Computer Science, University of Manchester, United Kingdom
[3] Department of Computer Science, University of Bern, Switzerland
r.duin@ieee.org, pekalska@cs.man.ac.uk, {a.harol,w.r.lee}@tudelft.nl, bunke@iam.unibe.ch

Abstract. Non-Euclidean dissimilarity measures can be well suited for building representation spaces that are more beneficial for pattern classification systems than the related Euclidean ones [1,2]. A non-Euclidean representation space is however cumbersome for training classifiers, as many statistical techniques rely on the Euclidean inner product that is missing there. In this paper we report our findings on the applicability of corrections that transform a non-Euclidean representation space into a Euclidean one in which similar or better classifiers can be trained. In a case-study based on four principally different classifiers we find out that standard correction procedures fail to construct an appropriate Euclidean space, equivalent to the original non-Euclidean one.

1 Introduction

In various pattern recognition applications the knowledge on a set of objects can be encoded with dissimilarity functions, which relate new objects to be classified to the training set. The main reason for such a preference is the difficulty of defining good features. In particular, structural descriptions of objects are an example of this. Instead of a feature-based approach, string matching or graph matching procedures can be applied [3,4,5,6], leading to (dis)similarity data.

In spite of the lack of a good feature representation, various tools have become available that build statistical pattern classifiers on dissimilarity data. This is possible thanks to the techniques that embed arbitrary dissimilarity functions into a fixed-dimensional vector space. Many successful examples are reported; see e.g. [7,8]. Some of these mappings are hampered by the use of non-Euclidean dissimilarity measures. These usually result from various matching procedures or when robustness or invariance are incorporated into the measure [9]. Consequently, a Euclidean embedding becomes imperfect and may loose information. In [1,2] we showed that classifiers based on non-Euclidean dissimilarity representations may lead to better results than those based on transformed dissimilarity measures that are either Euclidean or have reduced non-Euclidean component. In

N. da Vitora Lobo et al. (Eds.): SSPR&SPR 2008, LNCS 5342, pp. 551–561, 2008.

a series of 44 related experiments we found out that a significantly non-Euclidean measure performed best [2].

Non-Euclidean vector spaces, e.g. pseudo-Euclidean ones, are however not well equipped with the tools for training classifiers. Distances have to be computed in a specific way and are usually not invariant to orthogonal rotations. Densities may not be properly defined. Some density-based classifiers may still be used, albeit with some restrictions. So, Euclidean corrections become of interest. They have recently gained more attention, especially in relation to Support Vector Machines (SVMs) for indefinite kernels. Such kernels arise from similarity measures related to non-Euclidean dissimilarity ones. In general, they cannot guarantee a unique solution of SVM by quadratic programming [10]. Euclidean corrections may therefore be useful for SVMs if one deals with indefinite kernels, resulting e.g. from a kernel combining procedure or incorporation of invariance.

Many of the Euclidean corrections result in a continuous one-to-one correspondence between a non-Euclidean representation and the resulting Euclidean space. Consequently, vectors separating the classes in one space, separate the classes in the same way in the other space. This thereby gives hope that such corrections may determine a Euclidean space that can equally well (or even better) be used to train good classifiers.

In this paper we set out to study a set of Euclidean correction procedures on the basis of the performances before and after the correction. This topic is also discussed by [11]. We will present an experimental study based on four classifiers. These are local and global distance-based classifiers, such as 1-Nearest Neighbor rule (1-NN) and Nearest Mean Classifier (NMC), and local and global density-based classifiers, such as Parzen classifier with a small kernel width and Quadratic Discriminant Analysis (QDA). We focus on the same problem as studied in our previous papers on this topic [1,2]: the weighted edit distance between a set of contours obtained from the Chicken Pieces data [12].

The paper is organized as follows. Section 2 explains the Euclidean correction procedures. Section 3 summarizes the classifiers we use. Experiments are presented in Section 4, while conclusions are discussed in Section 5.

2 Euclidean Correction Procedures

Our starting point is a set of objects $\mathcal{X} = \{x_1, \ldots, x_m\}$ and a symmetric dissimilarity function that compares pairs of objects. This leads to a symmetric $m \times m$ dissimilarity matrix $D := (d_{ij})$, where $d_{ij} = d(x_i, x_j)$ and $d_{ii} = 0$. Such a matrix can be perfectly embedded in a pseudo-Euclidean space (PES) $\mathcal{E}$ by an isometric, distance-preserving, mapping [13,7]. $\mathcal{E} = \mathbb{R}^{(p,q)} = \mathbb{R}^p \oplus \mathbb{R}^q$ is a real vector space equipped with a non-degenerate indefinite inner product $\langle \cdot, \cdot \rangle_{\mathcal{E}}$ which is positive definite on $\mathbb{R}^p$ and negative definite on $\mathbb{R}^q$. $\mathcal{E}$ is characterized by the signature (p, q). This PES is determined by eigendecomposition of an (indefinite) Gram matrix $G = -\frac{1}{2} J D^{*2} J$ derived from D. J is the centering matrix, while $D^{*2} = (d_{ij}^2)$; see [13,7] for details. In this decomposition p positive and q negative eigenvalues arise, indicating the signature of $\mathcal{E}$. The axes of $\mathcal{E}$ are constituted by $\sqrt{|\lambda_i|}\mathbf{q}_i$, where $(\lambda_i, \mathbf{q}_i)$ is the corresponding eigenvalue-eigenvector pair of G. Note that the resulting configuration is uncorrelated and λ_i capture the variances of the embedded data.

Distances in $\mathcal{E}$ are defined as the square pseudo-Euclidean distance $d^2_{\mathcal{E}}(\mathbf{x},\mathbf{y}) = \langle \mathbf{x}-\mathbf{y}, \mathbf{x}-\mathbf{y}\rangle_{\mathcal{E}} = d^2(\mathbf{x}_p,\mathbf{y}_p) - d^2(\mathbf{x}_q,\mathbf{y}_q)$ for $\mathbf{x},\mathbf{y} \in \mathcal{E}$. The "-" sign distinguishes the pseudo-Euclidean distance from the Euclidean one. $p+q = m-1$ holds in a full embedding. It is perfect because the distances between the embedded PES vectors are equal to the original ones. In practice, the dimensions of PES related to small eigenvalues are discarded, which leads to more stable and better defined classifiers. Here we restrict ourselves to a full embedding because we want to make the comparisons with Euclidean corrections independent from the estimation of intrinsic dimension (i.e. number of significant eigenvalues). This has however consequences for some classifiers as appropriate regularization may be necessary. Euclidean correction procedures are described below.

The Positive part of the Pseudo Euclidean Space (PPES)
The most obvious correction for a pseudo-Euclidean space $\mathbb{R}^{(p,q)}$ is to neglect the negative definite subspace. This results in a p-dimensional Euclidean space $\mathbb{R}^p$ with many-to-one mappings to $\mathcal{E}$. Consequently, it is possible that the class overlap increases. It may, however, be worthwhile if the negative eigenvalues in the embedding procedure are mainly the result of noise and not informative for the class separation. In that case this correction may improve the classification.

The Associated Euclidean Space (AES)
Since $\mathbb{R}^{(p,q)}$ is a vector space, we can equip it with the traditional inner product, which leads to the so-called associated Euclidean space $\mathbb{R}^{p+q}$. It means that the vector coordinates are identical to those of PES, but now we use the norm and distance measure that are Euclidean. This is consistent with the natural topology of a vector space. This solution is identical to the one obtained by classical scaling based on the magnitudes of eigenvalues [14,7].

Dissimilarity Enlargement by a Constant (DEC)
Instead of modifying the embedding procedure, the dissimilarity matrix may be adapted such that it is embeddable into a Euclidean space. A simple way to avoid the negative eigenvalues is to increase all off-diagonal elements of the dissimilarity matrix such that $d^2_c(x_i,x_j) = d^2(x_i,x_j) + 2c, \forall_{i \neq j}$. The value of c is chosen such that $c \geq -\lambda_{min}$, where λ_{min} is the smallest negative eigenvalue in the pseudo-Euclidean embedding of D. As a result, all eigenvalues are increased by c [7]. In our experiments we set $c = -\lambda_{min}$. Since the eigenvalues reflect the variances of the embedded data, the dimensions of the resulting Euclidean space are unevenly scaled by $\sqrt{\lambda_i + c}$. Note that the dimension with the largest negative contribution in PES has now a zero variance. In this way, dimensions related to noisy negative eigenvalues are more pronounced. [7]

Relaxation by a Power Transformation (Relax)
Another way to adapt the dissimilarity matrix such that it is embeddable into a Euclidean space is to suppress the influence of large distances by a suitable concave transformation. Here, we relax the dissimilarity values by taking a small

power: $d_r(x_i, x_j) = d^\alpha(x_i, x_j), \forall_{i,j}$. For small values of α, $0 < \alpha < 1$, the dissimilarity values become increasingly more alike, by which the objects can eventually be embedded in a Euclidean space. We set $\alpha = 0.1$ in all experiments as it is just a sufficient value for the worst cases.

Laplace Transformation (Laplace)
We focus now on Euclidean corrections based on differential geometric properties of the given problem. We look for a smooth transformation T that changes large distances, but preserves the local structure as well as possible. Minimizing with respect to an objective functional, one may find a relation between the original distances, approximated by geodesics on a smooth manifold, and the Euclidean ones. For this purpose, we consider the following Hilbert-Schimdt operator $T\colon L_2 \rightarrow L_2$ acting in a Hilbert space of functions from L_2:

$$Tf(x) = \int_X \mathcal{D}(x, y) f(y) d\mu(y). \tag{1}$$

It is a compact linear operator associated to the kernel $\mathcal{D}(x, y)$, represented by the dissimilarity matrix D. T is defined on all configurations of our data points as specified by the function f. We assume that the data are distributed according to some unknown but finite probability measure $\mu(y)$. Hence, by defining a probability density function (pdf) $p(y)$, we have $d\mu(y) = p(y)dy$. Now a normalized kernel of the operator (1) can be constructed as $\mathcal{D}_p(x, y) = \frac{\mathcal{D}(x,y)}{p(x)p(y)}$. Consider now a normalization of T as

$$T_\nu f(x) = \frac{1}{\nu^2(x)} \int_X \mathcal{D}_p(x, y) f(y) dy, \tag{2}$$

with respect to the global scaling specified by the function $\nu^2(x) = \int_X \mathcal{D}_p(x, y)dy$, which is well defined iff $\mathcal{D}_p(x, y) \geq 0$. The resulting stochastic operator T_ν is widely used in spectral graph theory and usually referred to as a normalized graph Laplacian $\mathcal{L} = I - T_\nu$ [15]. $\mathcal{L}$ has a positive spectrum, by construction. Indeed, thanks to normalization, the spectrum of T_ν belongs to a sphere of radius 1. Moreover, the eigenfunctions of T_ν are identical to those of $\mathcal{L}$, leading to the same data configuration. T_ν has an asymmetric kernel due to the applied normalization. We can overcome this by working with its symmetric conjugate. For that, we conjugate $\mathcal{D}_p(x, y)$ with ν, and get a new normalized kernel $\mathcal{D}_\nu(x, y) = \frac{\mathcal{D}_p(x,y)}{\nu^2(x)} \cdot \frac{\nu(x)}{\nu(y)} = \frac{\mathcal{D}_p(x,y)}{\nu(x)\nu(y)}$. The associated operator is defined as

$$\widehat{T}_\nu f(x) = \int_X \mathcal{D}_\nu(x, y) f(y) dy, \tag{3}$$

and the Laplacian is $\widehat{\mathcal{L}} = I - \widehat{T}_\nu$. The optimal solution for f yields the largest correlation between f and $\widehat{T}_\nu$. As (3) is an averaging operator, this corresponds to the flattest configuration of the data in a Hilbert space. To obtain Euclidean distances from the Laplacian, we solve the following optimization problem

$$\begin{aligned} &\min\ \gamma \\ &\text{s.t. } \gamma I - \widehat{\mathcal{L}} \preceq 0,\ \gamma \geq 0, \end{aligned}$$

where the first constraint is equivalent to $(\gamma - 1)I + \widehat{T}_\nu \preceq 0$, while $\preceq$ denotes non-positive definiteness. This is equivalent to adding a constant to the spectrum of $\widehat{T}_\nu$ to make it non-positive definite, i.e. to represent Euclidean distances [14].

3 Classification Procedures

Generalization capabilities of classifiers in a vector space are based on two principles: densities and distances, related to generative and discriminative approaches, respectively. Objects are either classified into the most probable class derived from densities and prior probabilities, or into the class that is in some sense the nearest. Many classification algorithms rely on both principles, e.g. density estimators like Parzen and k-NN use distances between objects while distance-based classifiers like Fisher's Linear Discriminant and SVMs make use of parameter optimization procedures that somehow depend on the probability density of training vectors and not just their distances.

Both general densities and distances are not yet well formulated in PES, differently than in AES. E.g. we are not aware of the definition of a Gaussian distribution directly in PES, although the related quadratic classifier can easily be computed as it appears to be independent of the signature of the space. In fact, it can be computed directly in AES. Formally, the definition of a pdf does not make use of a metric, as the computation of integrals over volumes is sufficient. For the estimation of a pdf from a finite set of points, however, the metric may be very useful.

Pairwise distances between vectors are well defined in PES. Consequently, the 1-NN rule and also the Parzen classifier can be used. The interpretation of the latter as a consistent Bayes classifier for an appropriate asymptotic choice of its kernel function has still to be investigated. Distances between vectors and class boundaries, even if they are linear, are not well formulated in PES. Vectors on a linear boundary may have arbitrary large negative distances to the vectors on both sides of that boundary. An exception is the Nearest Mean Classifier, given that the contribution of the $\mathbb{R}^q$-subspace to the distances is smaller than of the $\mathbb{R}^p$-subspace. It produces a linear boundary, but its interpretation as the nearest class mean remains valid, because class means are properly computed.

We will use a consistent set of classifiers in order to compare different representation vector spaces. Four essentially different classifiers are selected in a Euclidean space such that they can also be computed in PES, but whose properties have to be still analyzed in the future. These are:

- 1-NN, the Nearest-Neighbor rule. It can directly be applied to a given dissimilarity matrix. Hence, embedding is not necessary.
- NMC, the Nearest Mean classifier. It is a global distance-based classifier (as it depends on all objects). Class means and distances of objects to these means have to be computed in the embedded PES.
- Parzen, a local-density based classifier. We use the same radial basis function as a kernel for all classes. The resulting classifier is thereby independent of

the dimensionality and is a function of the distances between training and test objects. Consequently, it can be computed directly on the dissimilarity matrix. In order to have a consistent choice for the kernel width over various representation spaces we set it to 0.1 of the 1% percentile of the distance distribution in the training set. This is a small value, in agreement with our intention to create a local density-based classifier.
- QDA, a global density-based classifier in a Euclidean space, equivalent to the Bayes classier assuming Gaussian class distributions. It is a global classifier as it everywhere depends on all training objects. As the dimension of PES is $m-1$ for a set of m objects it is necessary include some regularization. In order to compare over different representation spaces with different scalings we use QDA in a reduced PCA-space with a fixed dimension.

It is not our intention here to find the best classifier for a given problem. Instead, we want to investigate how different classifiers behave over several transformations, in order to characterize these transformations. We believe that the wide spectrum of the properties of the chosen classifiers serves thereby our target.

4 Experiments

We focus on a single problem, also studied by us before [1,2], based on the Chicken Pieces data [12]. It consists of 446 binary images representing five classes of chicken parts: wing (117 examples), back (76), drumstick (96), thigh and back (61), and breast (96). We estimate class prior probabilities by class frequencies here. Object contours are first approximated by straight line segments of a fixed length L; $5 \leq L \leq 40$ pixels. The sequence of angles between the neighboring segments becomes the initial string representation for which a family of edit distances [3] is derived. Insertion and deletion costs are set to 45 degrees, while the substitution cost is the magnitude of angle differences. The dissimilarity matrices are available from `http://www.iam.unibe.ch/fki/databases/string-edit-distance-matrices/`. The asymmetric dissimilarities are made symmetric by averaging, $d_{ij} = (d_{ij} + d_{ji})/2$.

Every dissimilarity matrix D_L from the family of edit-distances, $5 \leq L \leq 40$, is perfectly embedded in PES. The fraction of negative eigenvalues grows with increasing L indicating that the dissimilarity measure becomes increasingly non-Euclidean. After embedding (defined by all data), the performance of the four classifiers is computed for all values of L by averaging over ten runs of the 10-fold cross-validation. Experiments are handled with care; the same objects are used for training and testing in identical runs over different classifiers and different values of L. This results in smooth curves with reliable differences in classification performance for the given dissimilarity data.

The performance of four classifiers in six representation spaces is presented in Fig. 1 and 2. They show the same curves, organized either by classifier (Fig. 1) or by space (Fig. 2). The 1-NN results are essentially identical for the PES, DEC and Relax representations as monotonic transformations preserve the ranking of dissimilarity values. The 1-NN rule performs identically for the PPES and AES

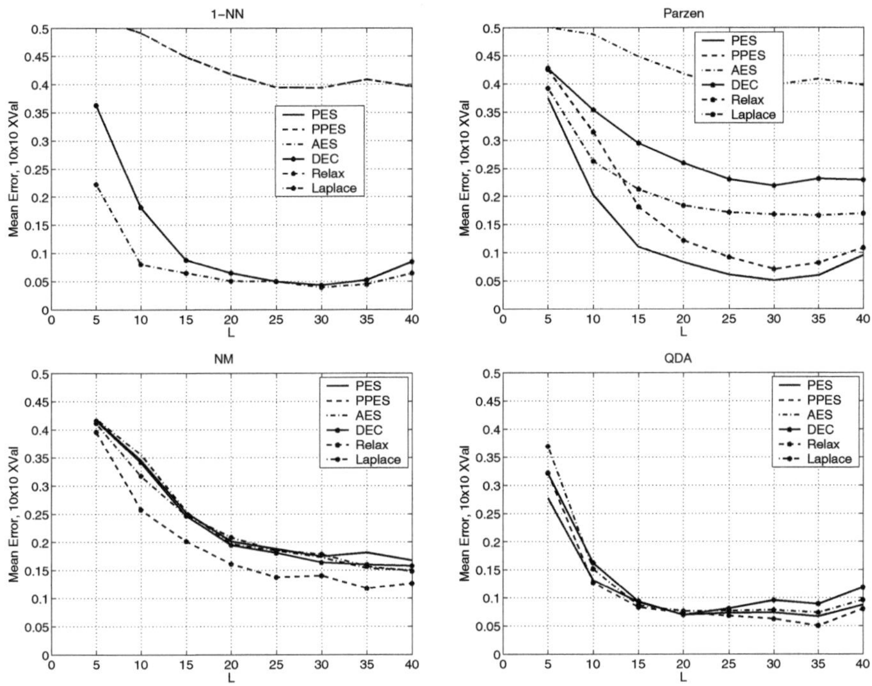

Fig. 1. Averaged 10-fold cross-validation performance of four classifiers in six representation vector spaces, shown by classifier. L (indicating different edit-distance data) is varied along the horizontal axis. In the top-left plot (1-NN), the curves for the PES, DEC and Relax cases are identical as well as the curves for AES and PPES. The curves for PES and PPES coincide in the top-right plot (Parzen). Standard deviations over the 10 repetitions are smaller than 0.01 everywhere.

cases and the Parzen classifier gives the same results for the PES and PPES cases. The following observations can be made w.r.t. various transformations:

- PPES (neglecting all 'negative' dimensions in PES) deteriorates the 1-NN results, but does not influence other classifiers. So it destroys very local distances without influencing the large ones. It also does not affect the densities.
- AES interprets PES as Euclidean, worsens the performance of locally-sensitive classifiers (1-NN, Parzen), but leaves the globally sensitive classification unaffected.
- DEC. Although this monotonic transformation preserves the ranking of dissimilarities, and so the 1-NN performance, it negatively influences the local densities, and thereby the Parzen results.
- Relax. The power transformation of the dissimilarities leads to better results than the DEC space. Parzen is less affected and NMC is even improved.
- Laplace. This shows an interesting result. The 1-NN performance improves but other classifiers deteriorate (Parzen and NM) or stay equal (QDA).

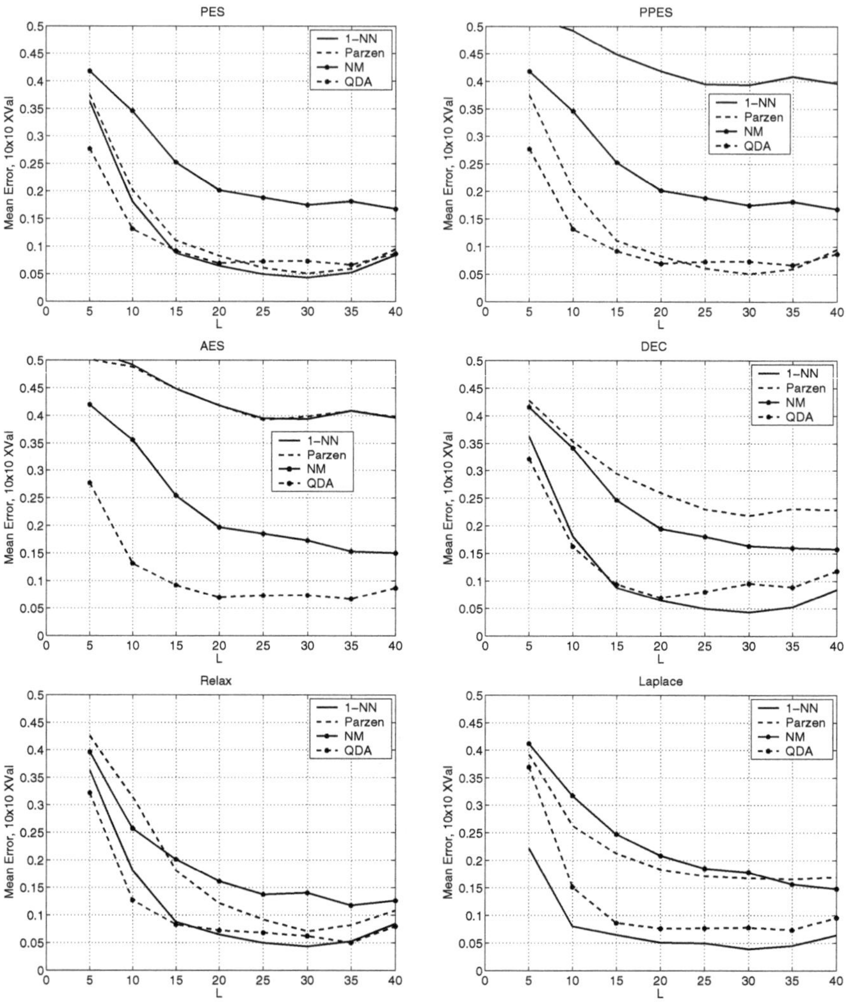

Fig. 2. Averaged 10-fold cross-validation performance of four classifiers in six representation vector spaces, shown by space. The curves for the 1-NN rule and Parzen (almost) coincide in AES. Standard deviations are smaller than 0.01 everywhere.

We observe that locally-sensitive classifiers are especially affected by the transformations. NMC and QDA remain stable or even improve (NMC by Relax). Parzen always deteriorates (except for PPES), and 1-NN deteriorates by PPES and AES, remains unaffected by DEC and Relax, and improves by Laplace.

Additional experiments are run for the best case, $L = 30$. The learning curves in Fig. 3 show no surprises: the relative performances of the classifiers are rather stable for the used sizes of the training set (90% from 445 objects). This indicates that the used data set is sufficiently large for the comparisons.

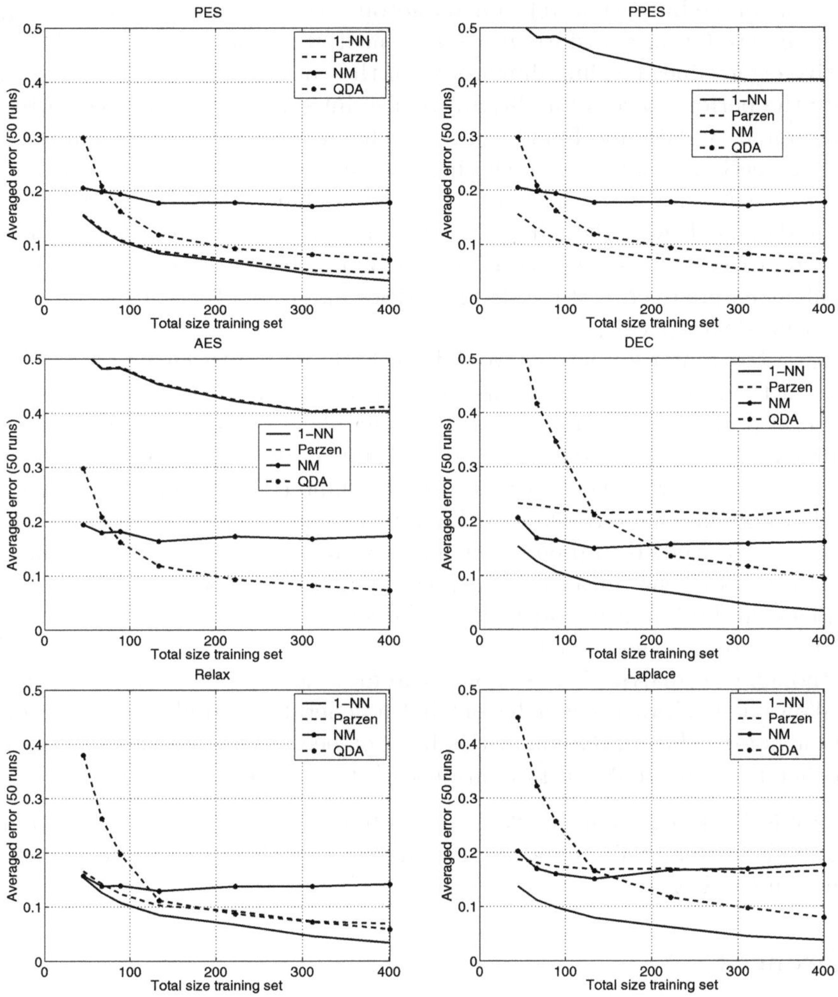

Fig. 3. Learning curves for the four classifiers in six representation spaces. Standard deviations are smaller than 0.002 everywhere.

5 Conclusions

In this paper we analyze the effect of Euclidean corrections for non-Euclidean dissimilarity data embedded in a pseudo-Euclidean space. We use a consistent and diverse set of classifiers that could be computed in the pseudo-Euclidean and Euclidean spaces. Our experiments rely on embedding both training and test data simultaneously (without using the labels of the test data). It is to be expected that all classifiers will perform worse if the test data are projected afterwards. We found that globally-sensitive classifiers are hardly affected by the correction procedures. The local distance-based 1-NN rule is insensitive to correction procedures that rely on monotonic transformations of the original dissimilarities. Other transformations that neglect or correct the contribution of negative eigenvalues (scaling dimensions of the embedded space), severely deteriorate the performance. In addition, we also introduced the Laplace correction for dissimilarity data and found that it might increase the 1-NN accuracy.

All correction procedures, including Laplace, deteriorate the results of the Parzen classifier for small kernels. This shows that the local structure in the data is damaged even though the nearest neighbor relations remain constant or are improved. The local separation capabilities based on densities may suffer from all attempts to correct the dissimilarities into a set of distances embeddable into a Euclidean space.

With a few minor exceptions we did not find any improvements of the classifiers after the Euclidean correction. So, such corrections do not seem worthwhile as general procedures for improving the generalization possibilities of the data in our example. Corrections need to be studied in relation with a classifier. Note that the above conclusions need to be handled with some care. There is a plethora of classifiers available in Euclidean spaces, much richer than the few ones currently studied in pseudo-Euclidean spaces, even though they still cover a wide range of generalization principles. It might be possible that very good classifiers can be found in corrected Euclidean spaces that have no counterpart (yet) in pseudo-Euclidean space.

Although our analysis is based on a careful study of just a single example, it still allows us to draw the conclusion that a general profitable correction procedure for non-Euclidean dissimilarities has not been found yet. Such procedures need thereby to be analyzed in relation with the classifier.

Acknowledgments. This work is supported by the Dutch Organization for Scientific Research (NWO) and the Engineering and Physical Sciences Research Council in the UK.

References

1. Pekalska, E., Duin, R., Günter, S., Bunke, H.: On not making dissimilarities Euclidean. In: S+SSPR, pp. 1145–1154 (2004)
2. Pekalska, E., Harol, A., Duin, R., Spillmann, B., Bunke, H.: Non-Euclidean or non-metric measures can be informative. In: S+SSPR, pp. 871–880 (2006)
3. Bunke, H., Sanfeliu, A. (eds.): Syntactic and Structural Pattern Recognition Theory and Applications. World Scientific, Singapore (1990)
4. Bunke, H., Shearer, K.: A graph distance metric based on the maximal common subgraph. Pattern Recognition Letters 19, 255–259 (1998)
5. Torsello, A., Hancock, E.: Computing approximate tree edit distance using relaxation labeling. Pattern Recognition Letters 24, 1089–1097 (2003)
6. Robles-Kelly, A., Hancock, E.: String edit distance, random walks and graph-matching. IJPRAI 18, 315–327 (2004)
7. Pekalska, E., Duin, R.: The Dissimilarity Representation for Pattern Recognition. Foundations and Applications. World Scientific, Singapore (2005)
8. Wilson, R., Luo, B., Hancock, E.: Pattern vectors from algebraic graph theory. IEEE Trans. on PAMI 27, 1112–1124 (2005)

9. Jacobs, D., Weinshall, D., Gdalyahu, Y.: Classification with Non-Metric Distances: Image Retrieval and Class Representation. IEEE TPAMI 22, 583–600 (2000)
10. Haasdonk, B.: Feature space interpretation of SVMs with indefinite kernels. IEEE TPAMI 25, 482–492 (2005)
11. Muñoz, A., de Diego, I.M.: From indefinite to positive semi-definite matrices. In: SSPR/SPR, pp. 764–772 (2006)
12. Andreu, G., Crespo, A., Valiente, J.M.: Selecting the toroidal self-organizing feature maps (TSOFM) best organized to object recogn. In: ICNN, pp. 1341–1346 (1997)
13. Goldfarb, L.: A new approach to pattern recognition. In: Kanal, L., Rosenfeld, A. (eds.) Progress in Pattern Recognition, vol. 2, pp. 241–402. Elsevier, Amsterdam (1985)
14. Gower, J.: Metric and Euclidean Properties of Dissimilarity Coefficients. J. of Classification 3, 5–48 (1986)
15. Chung, F.: Spectral Graph Theory. In: CBMS Regional Conference Series in Mathematics, vol. 92. American Mathematical Society (1997)

Feature and Classifier Selection in Class Decision Trees

Kazuaki Aoki and Mineichi Kudo

Division of Computer Science
Graduate School of Information Science and Technology
Hokkaido University
Kita-14, Nishi-9, Kita-ku, Sapporo 060-0814, Japan
{kazu,mine}@main.ist.hokudai.ac.jp
http://prml.main.ist.hokudai.ac.jp

Abstract. Feature selection is an important technique in pattern recognition. By removing features that have little or no discriminative information, it is possible to improve the predictive performance of classifiers and to reduce the measuring cost of features. In general, feature selection algorithms choose a common feature subset useful for all classes. However, in general, the most contributory feature subsets vary depending on classes relatively to the other classes. In this study, we propose a classifier as a decision tree in which each leaf corresponds to one class and an internal node classifies a sample to one of two class subsets. We also discuss classifier selection in each node.

1 Introduction

In pattern recognition, feature selection which chooses an effective feature subset for classification is an important technique. Generally, we tend to think that the greater the number of features is, the higher the recognition rate will be. However, when the number of features is large but the number of training samples is small, features that have little or no discriminative information weaken the performance of classifiers. This situation is typically called the *curse of dimensionality*. In that case, it is beneficial to choose a feature subset deriving the highest performance.

The benefits of feature selection are that it enables 1) improvement in the performance of classifiers, 2) reduction in the measurement cost of features and 3) reduction in the computational cost of classifiers in both the training phase and testing phase. Many feature selection methods have been proposed [1,2,3,4,5,6,7,8]. In all of these approaches, the same feature subset in common to classes is chosen. However, it is reasonable to assume that each feature subset has different discriminative information for different classes. For instance, a feature subset that has the most effective information in discriminating class ω_1 from classes ω_2 and ω_3 does not always work for discriminating ω_2 from ω_1 and ω_3. Therefore, in classification problems with many (at least three) classes, such as character recognition, feature selection depending on groups of classes

N. da Vitora Lobo et al. (Eds.): SSPR&SPR 2008, LNCS 5342, pp. 562–571, 2008.

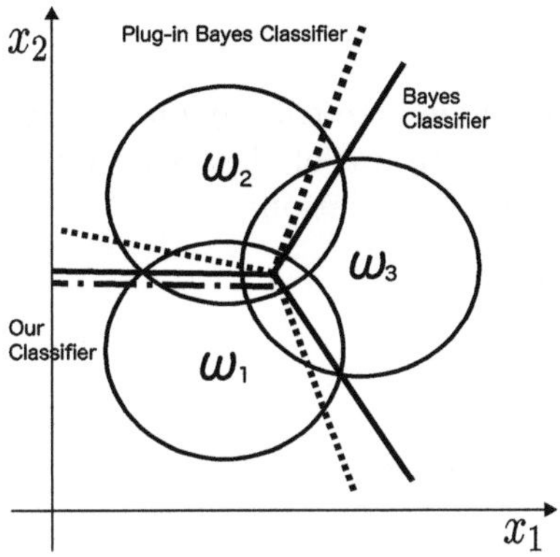

Fig. 1. Example

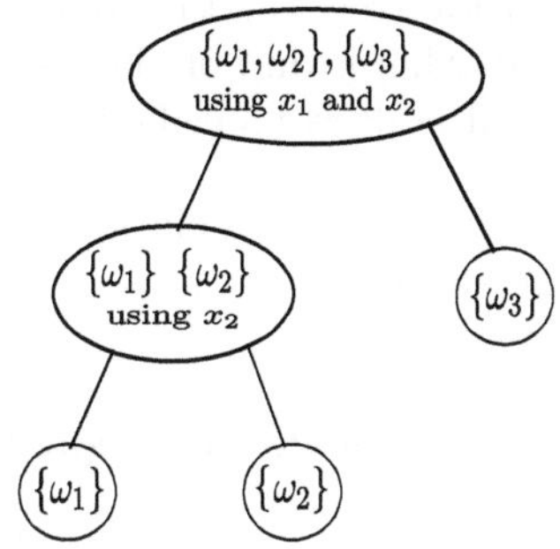

Fig. 2. Class decision tree for Fig.1

is a useful technique. We call such a feature subset a *"class-dependent feature subset."*

According to this idea, we have already proposed a class decision tree classifier using class-dependent feature subsets [9] and we have succeeded in showing the effectiveness in a Chinese character recognition problem of 30 classes. In [9], the classification rate was improved at 1.39% (on the average of four classifiers), and the validity of selected subsets was visually confirmed. In this paper, we investigate how degree such a trial is effective in a variety of problems. Especially, we make clear for which problems this way works most, with respect to the number of classes and the number of features. In addition, we try classifier selection in each internal node of the tree in addition to feature selection. It is well known that the best classifier depends on the problem, so that classifier selection, in each sub-problem corresponding to one internal node, is also promising for improving the performance of the decision tree classifier.

2 Class-Dependent Feature Subset

Our concept is explained by a simple example shown in Fig. 1. In Fig. 1, there are three classes according to normal distributions with the same covariance matrix and different means. The Bayes boundary, therefore, becomes linear. As long as a given training sample is finite, we cannot avoid misclassification in the plug-in Bayes classifier estimated from the training sample. However, we can reduce such misclassification using class-dependent feature subsets. Indeed, in this problem, only feature x_2 has discriminative information between ω_1 and ω_2. Thus, a classifier using only x_2 is expected to perform better than that using x_1 and x_2 in this case. A decision tree is designed naturally shown in Fig. 2.

3 Construction of Decision Tree

Here, we give a summary of the construction way of a class decision tree [9].

3.1 Class Decision Tree

In a class decision tree, the task of each internal node is to classify one group of classes from another group of classes. Here, let two groups of classes be Ω^1 and Ω^2. An internal node is identified by the following information:

1. (Ω^1, Ω^2): two groups of classes to be classified
2. F: feature subset
3. ϕ: classifier

Thus, an internal node t is denoted by four elements as

$$Node^t = \{\Omega_t^1, \Omega_t^2, F_t, \phi_t\}.$$

Our approach is different from conventional decision tree approaches [10,11] in the following two points:

1. In conventional decision tree approaches, each node is split in terms of *impurity* of two divided sample subsets in a feature, whereas in our approach, each node is split in terms of the *separability* between two groups of classes.
2. In conventional decision tree approaches, classification in each node is simple, and usually only one feature is used. In other words, the performance of classification in each node is not so good. Instead, those approaches complement the simplicity with splitting of data many times. In our approach, we split each node by a class-dependent feature subset. Thus, it is expected that the classification performance is improved in individual sub-problems.

3.2 Construction of Decision Tree

A decision tree is constructed in a bottom-up way like hierarchical clustering. The algorithm is as follows.

1. Initialization step: Set $\Omega_i = \{\omega_i\}, (i = 1, 2, \cdots, C)$, $c = C$,$t = 1$. Attach an *unprocessed* mark to all Ω_i. These Ω_i correspond to leaves.
2. Calculate the separability $S_{ij} = S(\Omega_i, \Omega_j)$ of pair (Ω_i, Ω_j) for all unprocessed nodes Ω_i and Ω_j, $(i, j = 1, \cdots, c)$.
3. Choose the pair $(\Omega_{i^*}, \Omega_{j^*})$ with the smallest separability $S_{i^*j^*}$. Let Ω_{i^*} be Ω_{c+1}^1 and Ω_{j^*} be Ω_{c+1}^2. Mark Ω_{i^*} and Ω_{j^*} as *processed.* Select a feature subset F_{c+1} that is effective for discrimination between Ω_{c+1}^1 and Ω_{c+1}^2.
4. Construct a classifier ϕ_{c+1} to classify Ω_{c+1}^1 and Ω_{c+1}^2 with feature subset F_{c+1}. In this step, we have a new node, $Node^{c+1} = \{\Omega_{c+1}^1, \Omega_{c+1}^2, F_{c+1}, \phi_{c+1}\}$.
5. $\Omega_{c+1} = \Omega_{c+1}^1 \bigcup \Omega_{c+2}^1$ and $c \leftarrow c + 1$, $t \leftarrow t + 2$.
6. Repeat steps 2-5 until $t = c$.

Here, choice of a similarity measure $S(\cdot, \cdot)$ between two class subsets is arbitrary, but some kind of estimated classification rate is preferred. In addition, any feature selection algorithm can be used for feature selection.

Table 1. Datasets

Name	#Class	#Feature	#Sample
waveform	3	40	337, 341, 322
glass	6	9	70, 17, 76, 13, 9, 29
mfeat-mor	10	6	200 per class
mfeat-zer	10	47	200 per class
mfeat-kar	10	64	200 per class
mfeat-fou	10	76	200 per class
mfeat-fac	10	216	200 per class
mfeat-pix	10	240	200 per class
letter-recognition	26	16	about 800 per class

In each node with (Ω_1, Ω_2), a sub-problem to be solved is now a two-class problem, but there exist more than two classes essentially. Therefore, the decision boundary is not so simple. For example, we cannot expect a linear boundary to work. Therefore, for classification, we solve the sub-problem as a multi-class problem for all classes included in either Ω_1 or Ω_2. Then we reassign the predicted class to Ω_1 or Ω_2. In this way, a linear classifier behaves like a piecewise linear classifier.

4 Experiments

We dealt with nine datasets from UCI Machine Learning Repository [12]. A summary of the datasets is shown in Table 1.

The separability measure, classifiers, and feature selection method used are as follows:

1. Separability: recognition rate estimated by the leave-one-out technique with 1-NN (the nearest neighbor) classifier (LOO_{1-NN}).
2. Feature selection method: an approach using normal vectors on the estimated Bayes discrimination boundary [8] (kNFS).
3. Classifiers: five classifiers of plug-in Bayes linear classifiers (L), plug-in Bayes quadratic classifier (Q), 1-nearest neighbor classifier (1NN), 3-nearest neighbor classifier (3NN), and support vector machine (S).

In the support vector machine, we used a RBF kernel function with soft margin parameter $C = 100$ and standard deviation $s = 10.0$. The recognition rate was calculated by 10-fold cross validation (5-fold CV only for the small "glass" dataset). Class decision trees were constructed from all training samples. In this first experiment, one classifier was used in common to all nodes, but feature subsets were chosen differently.

4.1 Plane vs. Tree

First, for examining basic performance, we compared two cases: the case in which a classifier is constructed in the usual way (which we call a "plane" approach)

Table 2. Comparison of plane and tree approaches using all features. The numbers are recognition rates in percentage.

Dataset		Classifier							
Name	#Feature	linear		quad		3-nn		svm	
		Plane	Tree	Plane	Tree	Plane	Tree	Plane	Tree
waveform	40	90.81	83.00	**94.75**	88.30	93.43	93.90	94.64	94.30
glass	9	55.50	53.75	52.50	54.23	56.00	63.55	54.00	**72.47**
mfeat-mor	6	46.15	**65.65**	55.25	58.15	44.50	44.80	46.75	43.95
mfeat-zer	47	81.40	78.85	80.60	78.55	79.00	79.10	**81.45**	80.85
mfeat-kar	64	95.10	93.65	96.65	93.65	97.75	97.90	**97.95**	97.45
mfeat-fou	76	80.60	81.70	82.15	78.00	83.75	**84.30**	81.40	82.00
mfeat-fac	216	96.94	97.15	95.90	68.75	95.25	96.85	**97.80**	97.70
mfeat-pix	240	95.65	92.25	94.80	52.45	97.75	**98.05**	84.95	97.90
letter	16	63.92	45.41	79.78	60.46	86.75	91.52	89.00	**91.66**
average	-	78.45	76.83	81.38	70.28	81.52	86.13	80.88	**87.53**

and the case in which the classifier is the same but a decision tree is used for classification (which we call "tree" approach). In this comparison, feature selection is not made. We omitted 1-NN from the comparison because the recognition rates for the plane classifier and tree classifier are the same. The results are shown in Table 2.

In Table 2, the recognition rate increases in some cases and decreases in other cases when the tree approach is used. The reason for the decrease in recognition rate is clear. This is because, in a class decision tree, an error occurred in an upper node cannot be recovered in the lower nodes. While, the improvement cases are quite interesting. The recognition rate was increased in all datasets when the 3-nearest neighbor was used and in 4 of the 9 datasets when SVM was used. In class decision trees, classification is easier in upper nodes and more difficult in lower nodes, so that classification rate is dominated in the lower nodes. It is obvious that, in a lower node, only a few classes and their training samples are concerned for classification. On the other hand, in the plane situation, all classes and all training samples are concerned with classification. In general, a flexible classifier such as k-NN or SVM is strongly affected by a small change in training samples, that is, they have a high variance. As a result, for these sensitive classifiers, class decision trees often work to reduce the sensitivity/variance. In other words, removing unrelated classes/samples is effective for these classifiers.

4.2 Class-Dependent Feature Subsets

Next, we examined the effectiveness of using class-dependent feature subsets. We used the kNFS [8] algorithm for feature selection. For comparison, we used the same feature selection in the corresponding plane approach. That is, a feature subset was chosen commonly to all classes by kNFS. The value of threshold θ needed for kNFS was set to 10.0%. The results are shown in Table 3.

Table 3. Comparison of plane and tree approaches with feature subsets selected by kNFS. The numbers are recognition rates in percentage.

Dataset			Classifier									
Name	#Feature		linear		quad		1-nn		3-nn		svm	
	Plane	Tree	Plane	Tree	Plane	Tree	Plane	Tree	Plane	Tree	Plane	Tree
waveform	34	35.0	91.01	90.60	94.95	94.90	92.83	93.70	94.55	94.60	**95.86**	94.70
glass	7	6.0	55.50	54.20	47.00	52.82	56.50	**64.96**	55.00	64.04	53.00	63.06
mfeat-mor	4	2.8	46.15	36.00	36.00	**47.05**	42.65	35.20	43.65	39.20	43.05	46.35
mfeat-zer	40	40.9	78.50	81.75	**82.50**	82.15	77.55	79.05	79.15	79.20	81.85	80.95
mfeat-kar	52	52.1	95.05	95.60	96.90	96.86	97.30	97.30	97.45	97.45	97.85	**98.05**
mfeat-fou	64	64.9	80.10	82.65	82.80	82.20	82.85	82.95	83.60	**84.15**	80.55	82.95
mfeat-fac	188	182.7	95.80	97.60	92.65	88.95	92.50	96.25	92.80	96.65	97.70	**97.85**
mfeat-pix	203	185.8	95.65	95.65	90.55	61.05	97.70	97.90	97.85	97.70	90.65	**98.40**
letter	13	14.0	61.72	69.47	76.23	80.44	83.91	88.21	83.63	90.70	88.98	**92.31**
average	67.22	64.91	79.93	79.49	81.33	76.27	80.42	81.25	80.85	82.63	81.05	**84.32**

Let us compare the plane approach using selected features (Plane_FS) and the tree approach using class-dependent features (Tree_FS). By using class-dependent features, recognition rates increased on average by 0.83% for 1NN, 1.78% for 3NN, and 3.27% for S but decreased by −0.44% for L and −5.06% for Q. Thus, class-dependent feature selection in class decision trees tends to be more effective for flexible classifiers, or classifiers with a low bias, than for stable classifiers, or classifiers with a low variance.

Fig. 3 shows the relationship between rate of improvement and number of classes. This is the ratio of recognition rate of class-dependent feature subsets to all features both in the class decision trees. Only the **mfeat-fou** dataset was used as a representative dataset of six datasets of the same size (10). From Fig. 3, the advantage of class-dependent feature subsets for a large number of classes is confirmed. The number of selected features is shown in Fig. 4. It is noted

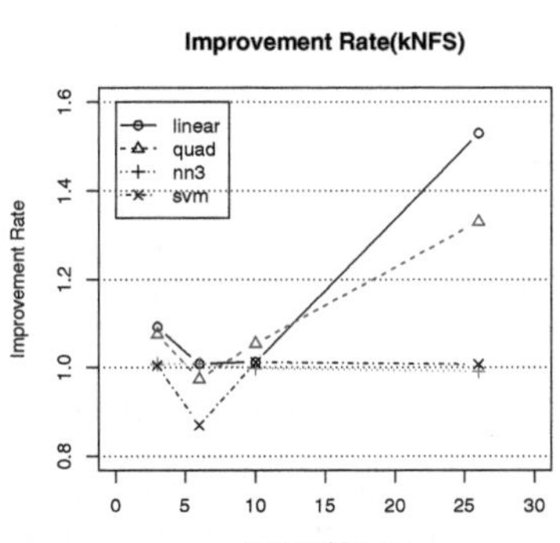

Fig. 3. Rates of Improvement by decision tree with kNFS

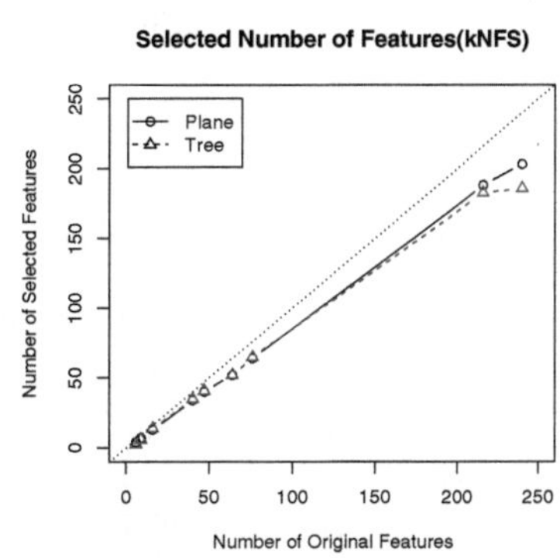

Fig. 4. Comparison of numbers of selected features in plane and tree approaches

that the number of class-dependent features on average is smaller than that of commonly selected features.

It should be noted that trees with class-dependent feature subsets are not always best among the four approaches of Plane, Plane_FS, Tree, and Tree_FS. It depends on datasets. Plane_FS was best for 2 of the 9 datasets, Tree was best for 3 datasets, and Tree_FS was best for 4 datasets with the best classifiers. However, for larger datasets in class, sample or dimensionality, Tree_FS is almost best.

5 Class-Dependent Classifier

Next, we examined the effectiveness of *class-dependent classifier selection*. We selected a *class-dependent classifier* in each internal node as the classifier to solve the corresponding sub-problem at the best recognition rate.

In this experiment, a class-dependent classifier was chosen from five previous classifiers abbreviated as L, Q, 1(NN), 3(NN) and S. In each node, a class-dependent feature subset was first chosen, and then a class-dependent classifier was chosen. Classifier selection is made by 10-fold CV in each node. The results are shown in Table 4 with the best single classifier for each approach.

A comparison with Tree_FS_CS and Tree_FS shows that the recognition rate was increased by using class-dependent classifier in 6 datasets and was decreased in 3 datasets. For degraded cases, it must be the case that CV did not work for classifier selection.

Table 4. Recognition rates with class-dependent classifiers. The characters in parentheses show the best single classifiers among L: linear, Q: quadratic, 1: 1-nearest neighbor, 3: 3-nearest neighbor, S: SVM. Approaches are named Plane: plane with all features, Tree: tree with all features, Plane_FS: plane with feature selection, Tree_FS: tree with feature selection, and Tree_FS_CS: tree with feature selection and classifier selection. In Tree_FS_CS, the characters in parentheses show the set of classifiers used in at least one node.

Dataset			Recognition rate (%)				
	#Class	#Feature	Plane	Tree	Plane_FS	Tree_FS	Tree_FS_CS
waveform	3	40	94.75(Q)	94.30(S)	**95.86(S)**	94.70(S)	95.50(Q,S)
glass	6	9	56.00(L)	**72.47(S)**	56.50(1)	64.96(1)	65.37(L,1,S)
mfeat-mor	10	6	55.25(Q)	**65.65(L)**	46.15(L)	47.05(Q)	65.60(L,Q,1,S)
mfeat-zer	10	47	81.45(S)	80.85(S)	**82.50(Q)**	82.15(Q)	79.75(L,Q,1,3)
mfeat-kar	10	64	97.95(S)	97.90(3)	97.85(S)	**98.05(S)**	94.70(L,Q,3,S)
mfeat-fou	10	76	83.75(3)	84.30(3)	83.60(3)	84.15(3)	**84.40(L,Q,1,3)**
mfeat-fac	10	216	97.80(S)	97.70(S)	97.70(S)	**97.85(S)**	97.30(L,Q,1,3,S)
mfeat-pix	10	240	97.75(3)	98.05(3)	97.85(3)	97.90(1)	**98.15(Q,1,3,S)**
letter	26	16	89.00(S)	91.66(S)	88.98(S)	92.31(S)	**95.05(L,Q,1,3,S)**
average	-	-	83.74	86.99	82.99	84.35	86.20
improvement	-	-	1.00	1.09	0.99	1.01	1.03

6 Total Evaluation

The recognition rates attained by five approaches are summarized in Fig. 5. The recognition rate was averaged over all datasets and all classifiers. From Fig. 5, we can conclude that both class-dependent feature selection and class-dependent classifier selection work well with class decision trees.

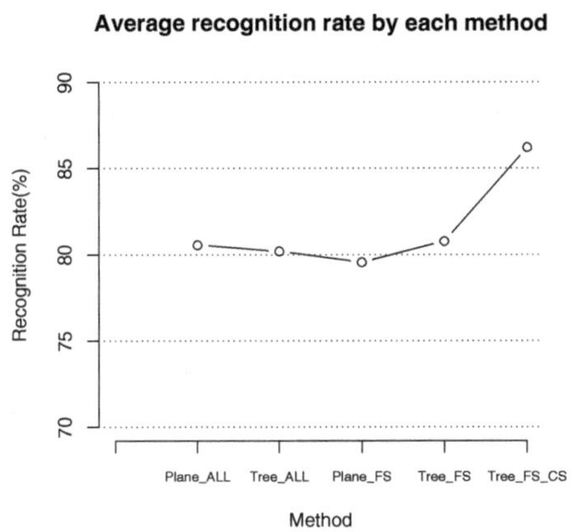

Fig. 5. Average recognition rate attained by each approach. The recognition rate was averaged over all datasets and all classifiers.

7 Discussion

The benefit of the proposed method is not only improvement of the recognition rate but also visualization of problem at hand. The decision tree constructed for **letter** dataset of 26 capital letters is shown in Fig. 6.

The number of selected features is almost 14 and the number of classifiers used is 12 (L or Q) : 4 (nearest neighbors) : 9 (SVM) on 25 internal nodes. It can be seen that k-NN and SVM are selected in lower nodes. Such flexible classifiers are suitable for these difficult sub-problems. In contrast, the linear and quadratic classifiers are selected in upper nodes. These observations imply that easier problems with many training samples should be solved by stable and low-variance classifiers and that more difficult problems with less samples should be solved by unstable but low-bias classifiers.

The features are 16 statistical moments and edge counts (Table 5). The selected features are almost the same in all nodes. Feature No.1 (horizontal position of the box) was removed from all nodes. This is obviously useless for classification. In the node classifying "M" and "W", No.1 and No.3 are removed. No.3 is the width of the box, and it does not contribute to this sub-problem because "M" and "W" have almost the same widths. We also notice that almost all features are invariant for upside down. A similar interpretation is possible for "H" and "K". The removed No.14 is not useful for classifying them.

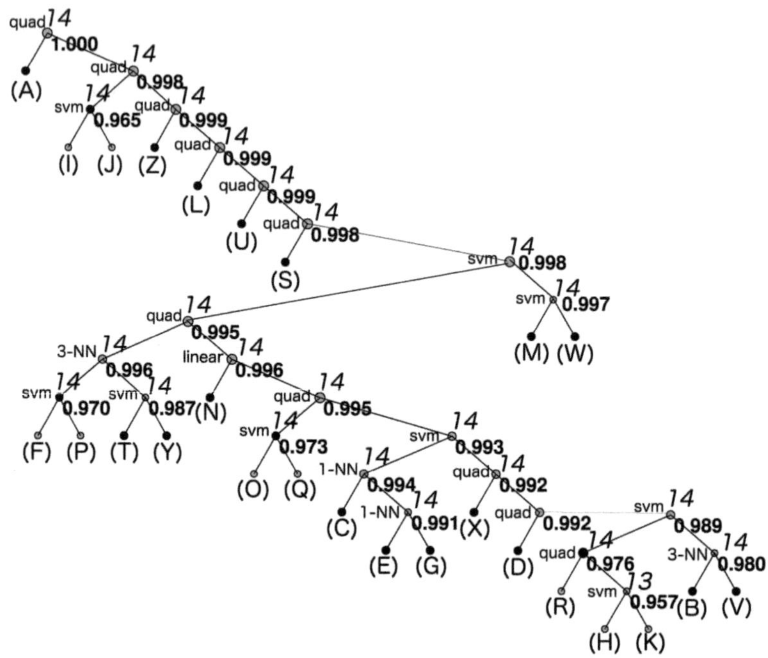

Fig. 6. Decision tree for **letter** dataset. As for a node, clockwise from top, the gray integer denotes the number of selected features, the decimal number denotes the separability, and the string denotes the classifier. In a leaf, the class is shown by the corresponding letter. The original number of features is 16.

Table 5. Details of features in **letter** dataset

No.	Name	Content	Type
1	x-box	horizontal position of box	integer
2	y-box	vertical position of box	integer
3	width	width of box	integer
4	high	height of box	integer
5	onpix	total # on pixels	integer
6	x-bar	mean x of pixels in box	integer
7	y-bar	mean y of pixels in box	integer
8	x2bar	mean x variance	integer
9	y2bar	mean y variance	integer
10	xybar	mean x y correlation	integer
11	x2ybr	mean of x * x * y	integer
12	xy2br	mean of x * y * y	integer
13	x-ege	mean edge count left to right	integer
14	xegvy	correlation of x-edge with y	integer
15	y-ege	mean edge count from bottom to top	integer
16	yegvx	correlation of y-edge with x	integer

8 Conclusion

We have experimentally studied the effectiveness of class-dependent feature subsets and class-dependent classifiers. Using class-dependent feature subsets and class-dependent classifiers at the same time resulted in significant improvement in class decision trees. In the future, better selection methods should be studied. Other ways to construct trees such as top-down construction and methods for choosing feature subsets should also be studied. If we determine a classifier first in each node, we can use classifier-specific feature selection algorithm, which is known to be superior to classifier-independent feature selection, as adopted in this paper.

References

1. Devijver, P.A., Kittler, J.: Pattern Recognition: A Statistical Approach. Prentice-Hall, Englewood Cliffs (1982)
2. Pudil, P., Novovičová, J., Kittler, J.: Floating search methods in feature selection. Pattern Recognition Letters 15, 1119–1125 (1998)
3. Somol, P., Novovičová, J., Paclík, P.: Adaptive floating search methods in feature selection. Pattern Recognition Letters 15, 1157–1163 (1998)
4. Ferri, F.J., Pudil, P., Hatef, M., Kittler, J.: Comparative study of techniques for large-scale feature selection. Pattern Recognition in Practice IV, 403–413 (1994)
5. Kudo, M., Sklansky, J.: A comparative evaluation of medium- and large-scale feature selection for pattern classifiers. In: 1st International Workshop on Statistical Techniques in Pattern Recognition, pp. 91–96 (1997)
6. Kudo, M., Sklansky, J.: Classifier-independent feature selection for two-stage feature selection. Advances in Pattern Recognition 1451, 548–554 (1998)
7. Zongker, D., Jain, A.: Algorithms for feature selection: An evaluation. In: 13th International Conference on Pattern Recognition, vol. 2, pp. 18–22 (1996)
8. Abe, N., Kudo, M.: Non-parametric classifier-independent feature selection. Pattern Recognition 39, 737–746 (2006)
9. Aoki, K., Kudo, M.: Decision tree using class-dependent feature subsets. In: Caelli, T.M., Amin, A., Duin, R.P.W., Kamel, M.S., de Ridder, D. (eds.) SPR 2002 and SSPR 2002. LNCS, vol. 2396, pp. 761–769. Springer, Heidelberg (2002)
10. Duda, R.O., Hart, P.E., Stork, D.G.: Pattern Classification: Second Edition. John Wiley & Sons, Chichester (2000)
11. Breiman, L., Friedman, J.H., Olshen, R.A., Stone, C.J.: Classification and Regresion Trees. Wadsworth & Brooks Cole Advanced Books & Software (1984)
12. Asuncion, A., Newman, D.: UCI machine learning repository (2007)

Soft Feature Selection by Using a Histogram-Based Classifier

Hiroshi Tenmoto[1] and Mineichi Kudo[2]

[1] Kushiro National College of Technology, Otanoshike Nishi 2-32-1, Kushiro, Hokkaido 084-0916, Japan
tenmo@kushiro-ct.ac.jp
[2] Division of Computer Science, Graduate School of Information Science and Technology, Hokkaido University, Sapporo 060-0814, Japan
mine@main.eng.hokudai.ac.jp

Abstract. Proposed is a histogram approach for feature selection and classification. The axes are divided into equally-spaced intervals, but the division numbers are different among axes. The main difference from similar approaches is that feature selection mechanism is embedded in the method. The optimal division is determined by an MDL criterion, so that the classifier is guaranteed to converge to the Bayes optimal classifier. We also introduce the concept of "soft feature selection" that is carried out by this method as an extension of traditional "feature selection."

Keywords: Histogram classifier, MDL criterion, Soft feature selection.

1 Introduction

Histogram classifiers are one of strong classifiers that are proved their convergence to the Bayes optimal classifier as the number of samples goes to infinity. Indeed, there are many histogram approaches (for example, [1,2,3]). Their main purpose is to speed up the convergence rate to the Bayes error. In this sense, the main attention is paid for large-scale cases. However, what we often have to face up is small- or medium-scale problems. Then one promising approach is to decrease the dimensionality by feature selection. Through feature selection stage, we can reduce a problem at hand to another problem in which a classifier with a high prediction performance can be constructed, e.g., by gaining a better estimation of the parameters of classifiers. Due to this virtue, many algorithms have been proposed for feature selection (for a comparison among them, see [4]). In this paper, we desire to attain both, that is, feature selection for small-scale and medium-scale cases and Bayes optimality for large-scale cases. This is done by thinking different-size division on feature axes. If no division is done, the feature is not useful for classification. On the other hand, for the axis to which the finest division is done, that axis (feature) is most important. In our approach, such an optimal division is obtained from an MDL criterion.

N. da Vitora Lobo et al. (Eds.): SSPR&SPR 2008, LNCS 5342, pp. 572–581, 2008.

2 MDL Coding of a Histogram-Based Classifier

Let us consider a problem to make a classifier under n training samples in m-dimensional Euclidean space $U = R^m$. Then, a sample is given by

$$x = (x_1, x_2, \ldots, x_m) \in R^m,$$

and a training sample sequence z^n is, with the class set $Y = \{1, 2, \ldots, c\}$, denoted by

$$z^n = (x, y)^n = \{(x_1, y_1), (x_2, y_2), \ldots, (x_n, y_n)\} \in (U \times Y)^n.$$

According to the MDL principle [5], we measure the cost of sending the class-label sequence y^n under the assumption that a receiver knows x^n, c and m. Let $L(\phi|x^n)$ be the bit length needed to send the classifier ϕ (the classification rule). In addition, let $L(y^n|\phi, x^n)$ be the cost of sending the class-label sequence information when ϕ is given. Then, the total cost is written as

$$L(y^n, \phi|x^n) = L(y^n|\phi, x^n) + L(\phi|x^n).$$

Of course, we would like to have $L(y^n|\phi, x^n)$ smaller than $n \log_2 c$ (the naive bit length of y^n). In our case, $L(\phi|x^n)$ is the bit length to send information of the histogram, that is, the division information of each axis, and $L(S|\phi, x^n)$ is the sum of the bit lengths of the class-label sequences in the cells forming the histogram. We assume that $x_1, x_2, \ldots, x_n$, as well as cells, can be ordered in some way, e.g., a dictionary order with numerical order.

What we think of as a classifier is a histogram. We divide ith axis into 2^{d_i} equally-spaced intervals. The ends of each axis are determined by the minimum and maximum values over training samples. Thus, a partition is expressed by m-tuple

$$\boldsymbol{d} = (d_1, d_2, \ldots, d_m).$$

By d we denote the sum of division indexes as $d = \sum_{i=1}^{m} d_i$. Then there are $\prod_{i=1}^{m} 2^{d_i} = 2^d$ cells.

We want to find the optimal division $\boldsymbol{d}$ in some sense. In our case, we use an MDL criterion for this goal. In the MDL criterion, a shorter length means a better partition for classification. We will describe in order the code length $L(y^n, \phi|x^n)$.

1. (Encoding the label sequence of samples in a cell). Let us consider a cell and let n be the number of samples falling in the cell. Let us denote the class-label sequence as

   $$y_1, y_2, \ldots, y_n.$$

   Then, for encoding this sequence, we need

   $$\log_2 \binom{n+c-1}{c-1} + \log_2 \binom{n}{n_1, n_2, \ldots, n_c} \text{ bits.}$$

Here, we do not need to inform the receiver the value of n. This is because once a partition was known, he/she can count the number of samples falling in each cell. By rewriting n by n^r for the rth cell, we have

$$L(y^n|\phi, x^n) = \sum_{r=1}^{R} \left\{ \log_2 \binom{n^r + c - 1}{c - 1} + \log_2 \binom{n^r}{n_1^r, n_2^r, \ldots, n_c^r} \right\} \text{bits},$$

where R is the number of non-empty cells.

It can happen that some cells have samples only from a single class. To think about this case, let us divide all non-empty R cells into R^P class-pure cells and R^M class-mixture cells. Therefore, for sending the value of R^M and the corresponding R^M positions, we need

$$\log_2 R + \log_2 \binom{R}{R^M} \text{ bits}.$$

In addition, we should send the class label of each class-pure cell. For this, we need

$$R^P \log_2 c \,\text{bits}.$$

As a result, we have

$$\begin{aligned} L(y^n|\phi, x^n) &= \log_2 R + \log_2 \binom{R}{R^M} + R^P \log_2 c \\ &+ \sum_{r=1}^{R^M} \left\{ \log_2 \binom{n^r + c - 1}{c - 1} + \log_2 \binom{n^r}{n_1^r, n_2^r, \ldots, n_c^r} \right\}. \end{aligned} \tag{1}$$

2. (Encoding the partition). We have to let the receiver know $\boldsymbol{d}$ too. First, we encode the number d and then $d_1, \ldots, d_m (d = \sum d_i)$. Thus, we need

$$L(\phi|x^n) = \log_2 \binom{d + m - 1}{m - 1} + \log_2^* d \,\text{bits}.$$

However, in many pattern recognition problems, only a few features are important for classification. In such a case, we can expect to economize the bit length as

$$\begin{aligned} L(\phi|x^n) &= \log_2 m + \log_2 \binom{m}{m - m^+} + \log_2^* d \\ &+ \log_2 \binom{d}{d_1, d_2, \ldots, d_m} + \log_2 d + \sum_{i=1}^{m^+ - 2} \log_2 d_i, \end{aligned} \tag{2}$$

where m^+ of m elements are non-zero d_i's.

In total, from (1) and (2), the bit length is evaluated by

$$L(y^n, \phi|x^n) = L(y^n|\phi, x^n) + L(\phi|x^n)$$

$$
\begin{aligned}
&= \log_2 R + \log_2 \binom{R}{R^M} + R^P \log_2 c \\
&+ \sum_{r=1}^{R^M} \left\{ \log_2 \binom{n^r + c - 1}{c - 1} + \log_2 \binom{n^r}{n_1^r, n_2^r, \ldots, n_c^r} \right\} \\
&+ \log_2 m + \log_2 \binom{m}{m - m^+} \\
&+ \log_*^2 d + \log_2 \binom{d}{d_1, d_2, \ldots, d_m} + \log_2 d + \sum_{i=1}^{m^+ - 2} \log_2 d_i
\end{aligned}
$$

We may use an entropy evaluation of this equation for large n. For this, we use the following two approximations

$$
\log_2 \binom{n + c - 1}{c - 1} \simeq (c - 1) \log_2 n,
$$

$$
\log_2 \binom{n}{n_1, n_2, \ldots, n_c} \simeq nH\left(\frac{n_1}{n}, \frac{n_2}{n}, \ldots, \frac{n_c}{n}\right) - \frac{c-1}{2} \log_2 n, \qquad (3)
$$

where

$$
H\left(\frac{n_1}{n}, \frac{n_2}{n}, \ldots, \frac{n_c}{n}\right) = -\sum_{i=1}^{c} \frac{n_i}{n} \log_2 \frac{n_i}{n}.
$$

Then, $L(y^n, \phi | x^n)$ is rewritten as

$$
\begin{aligned}
L(y^n, \phi | x^n) &\simeq \left(RH\left(\frac{R^P}{R}, \frac{R^M}{R}\right) + \frac{1}{2} \log_2 R + R^P \log_2 c \right) \\
&+ \left(n \sum_{r=1}^{R^M} \frac{n^r}{n} H\left(\frac{n_1^r}{n^r}, \frac{n_2^r}{n^r}, \ldots, \frac{n_c^r}{n^r}\right) + \frac{c-1}{2} \sum_{r=1}^{R^M} \log_2 n^r \right) \\
&+ \left(mH\left(\frac{m^+}{m}, \frac{m - m^+}{m}\right) + \frac{1}{2} \log_2 m \right) \\
&+ \left(\log_2^* d + dH\left(\frac{d_1}{d}, \frac{d_2}{d}, \ldots, \frac{d_m}{d}\right) + \frac{m-1}{2} \log_2 d \right) \\
&= \mathrm{I} + \mathrm{II} + \mathrm{III} + \mathrm{IV} \qquad (4)
\end{aligned}
$$

3 Evaluation of the MDL Coding

Let us examine how our criterion (4) works. First of all, it is noted that with this criterion our classification approaches to Bayes optimal classifier as n goes to infinity. In criterion (4), we can see more. First of all, the dominant terms are I, II and IV. When, the perfect classification on training samples is done by a certain $\boldsymbol{d}$, II vanishes because of $R^M = 0$ and $R^P = R$. Then, the problem reduces to minimize terms I and IV. Thus, what should be done is firstly to minimize the value of d and then to minimize the entropy of $\{d_i/d\}$. That is, this enhances feature selection in which some d_i's are expected to be zero to decrease the entropy. This is the biggest difference from previous similar MDL approaches [1,2,3]. It works often when the number n of training samples is small even when the classification on training samples is not perfect. This is because in such a case terms I and IV are dominant. In summary, our criterion works for feature selection when the sample number is small.

4 Classification of Unknown Samples by Subclass Method

Let us describe how to classify unknown samples in our approach. In a high-dimensional feature space, the regions including training samples may be very sparse. Therefore, we use the subclass method [6] with the histogram approach in order to interpolate the sample-containing regions. This method aims to find a set of hyper-rectangles in a class as its class region. Each hyper-rectangles is called "subclass" and satisfies the following two conditions:

1. Exclusiveness: The hyper-rectangle does not include any sample from negative class.
2. Maximalness: The hyper-rectangle is maximal with respect to the inclusion relation in the family of all subsets holding the above exclusiveness.

Since this problem requires a combinatorial examination with respect to the number of samples, we resort to a randomized algorithm [6] in order to obtain a suboptimal solution.

A concrete procedure for the classification is as follows:

1. Find a region including the unknown sample.
2. If the region contains training samples, determine the class label by applying majority rule.
3. If the region does not contain any training samples, determine the class label by using subclasses such that:
   (a) Find subclasses that include the sample-falling region.
   (b) If one or more subclasses are found, adopt the deepest subclass and determine the class label by that subclass.
   (c) If no subclass is found, adopt the nearest subclass and determine the class label by that subclass.

Example results on the first two features of Iris dataset are shown in Fig.1. The training samples are roughly separated by the regions, and subclasses form only on the pure regions. The resultant classification boundary appears based on the regions and subclasses.

5 Experiments

To solve the optimization/minimization problem of (4), we employed a GA. We encoded each d_i in gene and an m-tuple $(d_1, d_2, \ldots, d_m)$ in a chromosome, so that a chromosome shows $\boldsymbol{d}$.

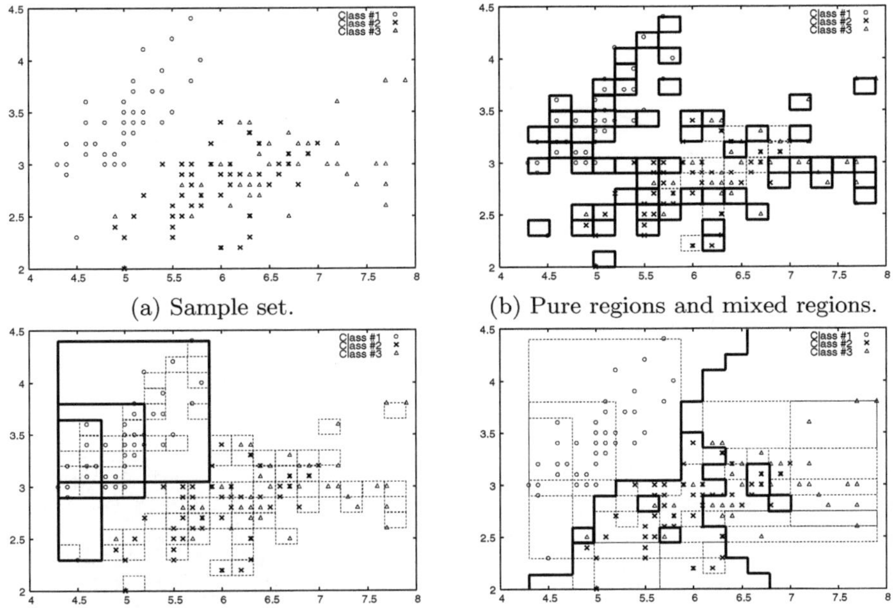

(a) Sample set. (b) Pure regions and mixed regions.

(c) Subclasses on class ω_1 regions. (d) Classification boundary.

Fig. 1. Classification boundary construction example for the first two features on Iris dataset

In the following experiments, the population size (the number of chromosomes) is 100, the crossover probability is 0.6, and the mutation probability is $1/m$. We used a roulette selection and elite strategy, and terminated the iteration with 100 generations.

5.1 Two-Dimensional Artificial Dataset

At first, we dealt with a 2-dimensional problem with zero Bayes error. In this dataset, the samples distribute uniformly within a circle with radius 0.5 centered at the origin. The two classes are separated by the curve of $y = 0.1 \sin(x\pi/0.5)$.

The true boundary, and the classification boundaries obtained by the proposed method are shown in Fig.2(a)–(d), varying the number n of samples. The change of the training and test error is also shown in Fig.3. The test errors are estimated by hold-out method. From these results, we can see: 1) the determined boundary approaches to the true one as n grows, 2) the division becomes finer as n increases and the ratio d_2/d_1 reflects faithfully the ratio of importance between the corresponding two features, and 3) feature selection ($2^{d_i} = 1$) is done when n is small. Especially, we can see that "soft feature selection" was done according to the increase of n. We can regard ith feature with a low ratio d_i/d as being not so important for classification. This is "soft feature selection" in which the importance of each feature is evaluated by this ratio and is an extension of

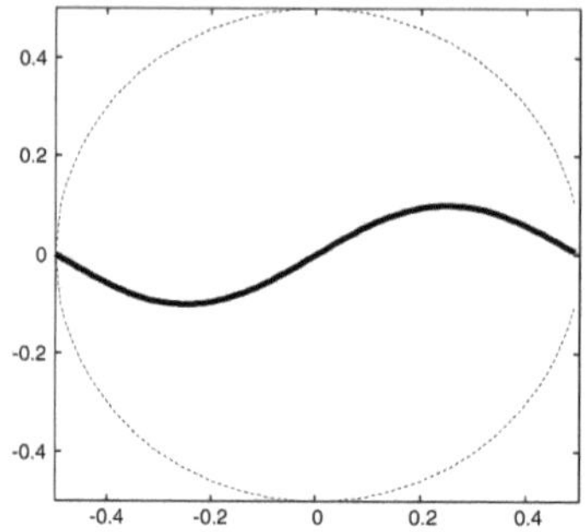

(a) True boundary (completely separable)

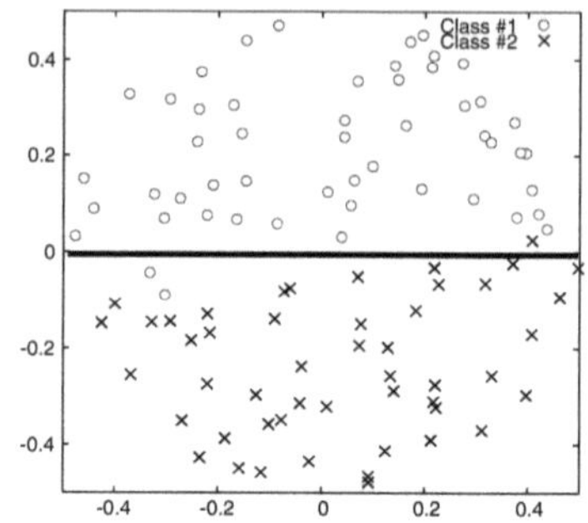

(b) $n = 100, 2^{d_1} = 1, 2^{d_2} = 2$

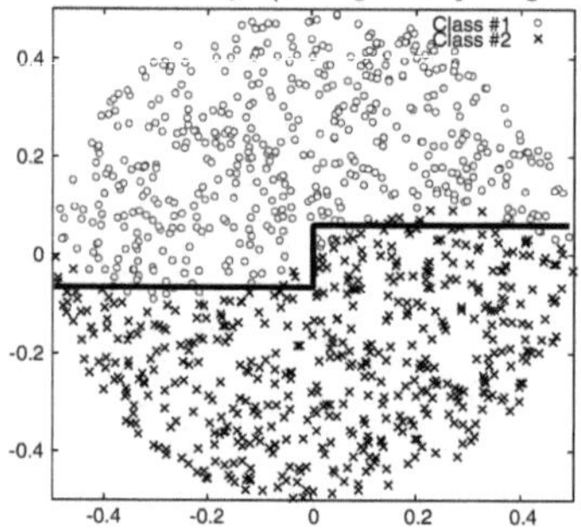

(c) $n = 1000, 2^{d_1} = 2, 2^{d_2} = 16$

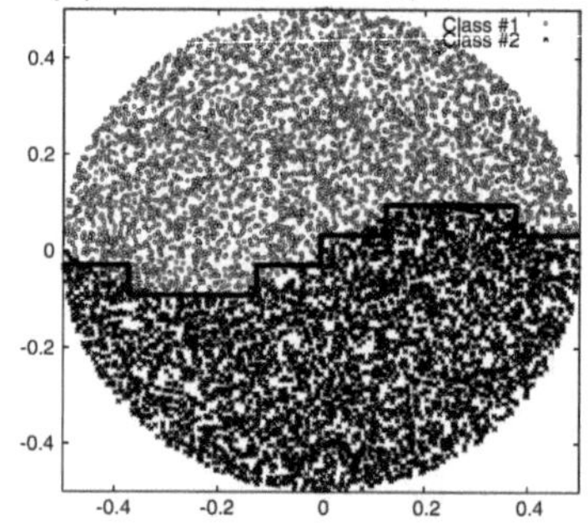

(d) $n = 10000, 2^{d_1} = 8, 2^{d_2} = 32$

Fig. 2. Refinement of classification boundary by the increase of samples. The symbols and lines denote the training samples and classification boundaries, respectively.

traditional use-or-not type feature selection. We might remove features with low ratios if we hope. It is reasonable to think of every feature as having at least some amount of contribution for classification. However, we cannot know this amount from a limited number of training samples. On the contrary, when an enough number of training samples is available, this amount would appear so that the value of d_i for such a feature takes a small but positive number. In the light of soft feature selection, we can deal with this amount properly.

To examine the performance as (soft) feature selection, we rotated the boundary so as to change the importance of individual features. The results are shown in Fig.5(a)–(d). From these results, we can see that the difference of the importance of features are captured well by the change of the ratio d_i/d.

5.2 Ten-Dimensional Artificial Dataset (Friedman Dataset)

To examine the performance of this method in high-dimensional cases, we used a famous two-class ten-dimensional artificial dataset "Friedman."

In this dataset, the discriminative features are leading four features, and the trailing six features are non-discriminative. The samples of class ω_1 simply distribute according to the 10-dimensional multivariate normal distribution with zero mean and the unit covariance matrix. The distribution of the samples of

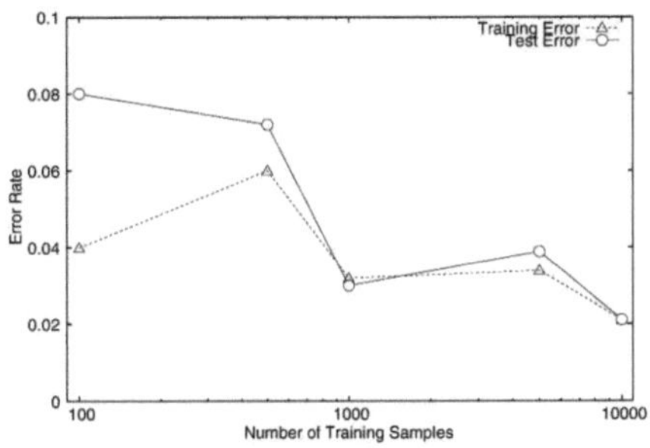

Fig. 3. Training and test errors for two-dimensional artificial dataset

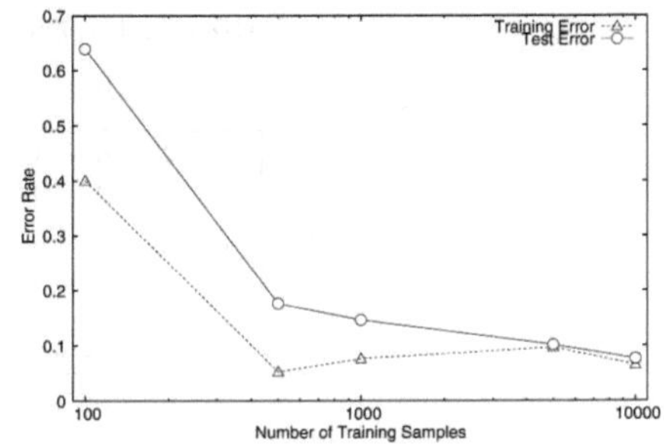

Fig. 4. Training and test errors for ten-dimensional artificial dataset (Friedman dataset)

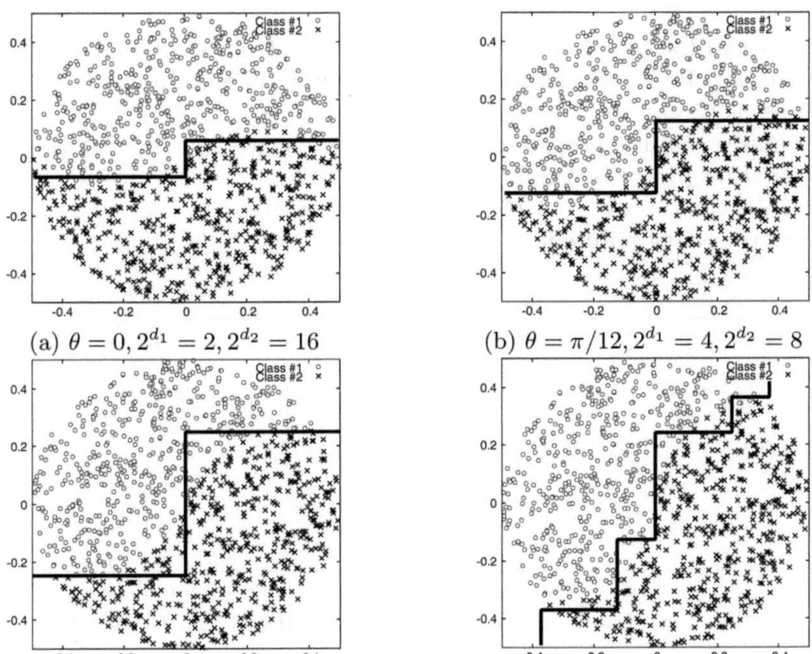

(a) $\theta = 0, 2^{d_1} = 2, 2^{d_2} = 16$

(b) $\theta = \pi/12, 2^{d_1} = 4, 2^{d_2} = 8$

(c) $\theta = 2\pi/12, 2^{d_1} = 4, 2^{d_2} = 4$

(d) $\theta = 3\pi/12, 2^{d_1} = 8, 2^{d_2} = 8$

Fig. 5. Change of division for the different rotations of true boundary. The number of samples n is fixed at 1000.

class ω_2 is as follows. For the first four features, the samples distribute uniformly between two hyper-spheres with radius 3.5 and 4.0, and for the last six features, the distribution is the same as that of ω_1.

The experimental result is shown in Table 1 and Fig. 4. When the number of samples is small, the training error is high, therefore the proposed method removes too many features. While, in medium- or large-size samples, only

Table 1. Division for Friedman data

n	2^{d_1}	2^{d_2}	2^{d_3}	2^{d_4}	2^{d_5}	2^{d_6}	2^{d_7}	2^{d_8}	2^{d_9}	$2^{d_{10}}$
100	1	1	4	1	1	1	1	1	1	1
500	4	4	4	4	1	1	1	1	1	1
1000	4	4	4	4	1	1	1	1	1	1
5000	4	4	4	4	1	1	1	1	1	1
10000	4	4	8	8	1	1	1	1	1	1
True	Necessary				Garbage					

necessary four features are chosen properly. Thus, feature selection were performed successfully.

5.3 Real Dataset (Ship Dataset)

We also tested the proposed method on real dataset called "Ship dataset" [7] in which 8 types of military ships are distinguished by 11 features, and the number of samples is 2545.

We can see that the proposed method works also well for the real dataset. As the generation proceeds, the MDL term IV decreases, so the feature selection is done. We can observe that the errors are not increasing while the number of features are getting reduced.

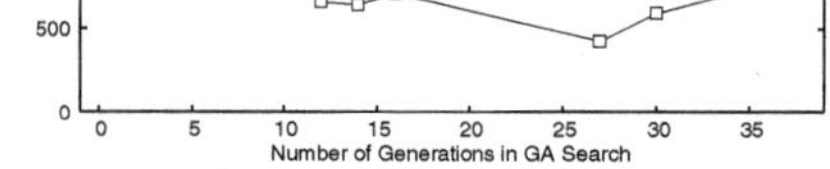

(a) MDL Term I and Term II.

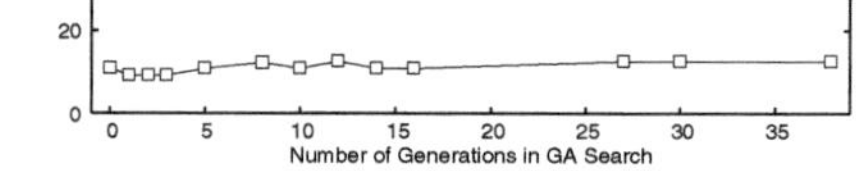

(b) MDL Term III and Term IV.

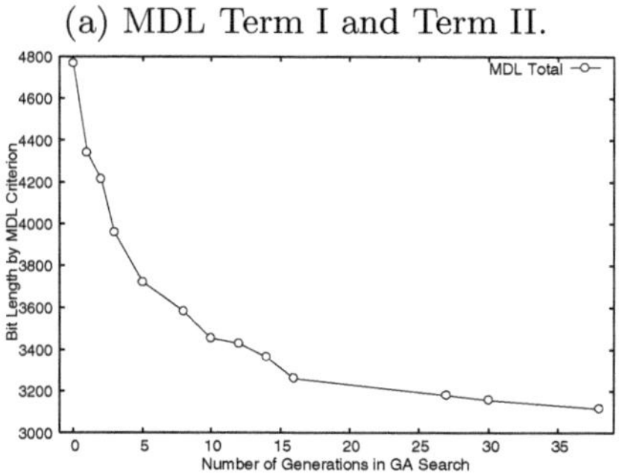

(c) MDL Total

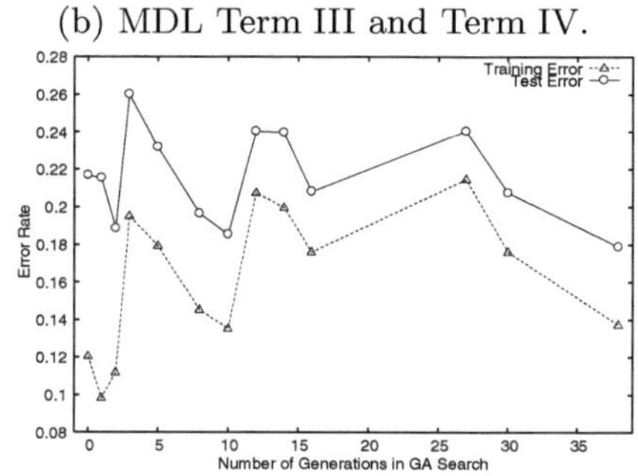

(d) Training error and test error.

Fig. 6. MDL value and error rate for Ship dataset

6 Conclusion

An MDL-based histogram classifier was proposed. In this method, feature selection mechanism is embedded in the optimization function which determines the optimal division of feature axes. This is a big difference from the previous similar approaches. It is shown that feature selection is done for small-scale (the number of training sample is small) problems and a quasi Bayes classification is done for large-scale problems. In addition, (soft) feature selection is done for medium-scale problems in which the degree of importance of each feature is evaluated by the ratio of the division number of the feature axis to the total number of division numbers. This means that use-or-not feature selection is not always appropriate when a rational number of training samples is available.

Although its convergence to the Bayes optimal classifier is guaranteed, the detailed analysis is needed for the convergence rate of this approach. In addition, we will reconsider our MDL criterion so that the convergence rate would be improved.

Acknowledgment

This research was partly supported by the Grant-in-Aid for Young Scientists (B) No. 20700203 of the Ministry of Education, Culture, Sports, Science and Technology, Japan.

References

1. Rissanen, J., Yu, B.: MDL Learning. In: Kueker, D.W., Smith, C.H. (eds.) Learning and Geometry: Computational Approaches, pp. 3–19. Birkhauser, Basel (1998)
2. Yamanishi, K.: Learning Non-Parametric Smooth Rules Using Stochastic Rules with Finite Partitioning. In: Proceedings of the Computational Learning Theory: EuroCOLT 1993, pp. 217–227 (1993)
3. Tsuchiya, H., Itoh, S., Mashimoto, T.: An Algorithm for Designing a Pattern Classifier by Using MDL Criterion. IEICE Transactions on Fundamentals, E79-A 6, 910–920 (1996)
4. Kudo, M., Sklansky, J.: Comparison of Algorithms that Select Features for Pattern Classifiers. Pattern Recognition 33(1), 25–41 (2000)
5. Rissanen, J.: Stochastic Complexity in Statistical Inquiry. Series in Computer Science, vol. 15. World Scientific, Singapore (1989)
6. Kudo, M., et al.: Construction of Class Regions by a Randomized Algorithm: A Randomized Subclass Method. Pattern Recognition 29, 581–588 (1996)
7. Park, Y., Sklansky, J.: Automated Design of Multiple-Class Piecewise Linear Classifiers. Journal of Classification 6, 195–222 (1989)

Embedded Map Projection for Dimensionality Reduction-Based Similarity Search

Simone Marinai, Emanuele Marino, and Giovanni Soda

Dipartimento di Sistemi e Informatica - Università di Firenze
Via S.Marta, 3 - 50139 Firenze - Italy

Abstract. We describe a dimensionality reduction method based on data point projection in an output space obtained by embedding the Growing Hierarchical Self Organizing Maps (GHSOM) computed from a training data-set. The dimensionality reduction is used in a similarity search framework whose aim is to efficiently retrieve similar objects on the basis of the Euclidean distance among high dimensional feature vectors projected in the reduced space. This research is motivated by applications aimed at performing Document Image Retrieval in Digital Libraries. In this paper we compare the proposed method with other dimensionality reduction techniques evaluating the retrieval performance on three data-sets.

1 Introduction

In Pattern Recognition, objects to be recognized are frequently represented by collections of features that, when organized in vectors, allow us to represent objects as points in a vector space. The use of a suitable distance (or similarity measure) among objects can give rise to significant differences in the recognition performance achievable, but in most cases the Euclidean distance is used.

In this work we deal with Euclidean vector spaces. Our main application domain is in the field of Document Image Retrieval (DIR) where the aim is to identify relevant documents relying on image features only (e.g. considering layout-based retrieval or word indexing [1]). In this paper we will consider three datasets. In each collection the objects are represented as n-dimensional points and we aim to retrieve objects on a query by example paradigm: given a query object (n-dimensional point) we identify the most similar objects by looking for nearest points in the feature space. In principle the approach is simply based on an exhaustive comparison of the query point with all the indexed points, followed by a sorting of the computed distances. This strategy has severe limits for the computational cost involved and several approaches have been proposed to alleviate this problem.

From one side various multidimensional indexing methods have been proposed as extensions of spatial indexes such as quad-trees and R-trees [2]. Multidimensional methods (e.g. X-tree) are aimed at indexing high-dimensional data more efficiently than the sequential scan. When dealing with "very high dimensional" data (hundreds or thousands of dimensions) many multidimensional indexes tend

N. da Vitora Lobo et al. (Eds.): SSPR&SPR 2008, LNCS 5342, pp. 582–591, 2008.

to degenerate and perform poorly than the sequential scan. Some studies have been performed to clarify these poor performances that are generally attributed to the so called *curse of dimensionality* [3]. The curse of dimensionality has several facets. One of the most known is the property that *independent and identically distributed* points are mostly concentrated on the sides of a unit cube as long as the number of dimensions grows [4]. These problems become evident for dimensions as low as 10-15. However, distributions of points representing real objects are unlikely to be uniformly distributed and therefore similarity methods still work for higher dimensions.

The latter consideration lead to a group of approaches that adopt a dimensionality reduction of the data as a preliminary processing step, before using a multidimensional index on the reduced space (e.g. [5]). Working on a reduced space the quality of the query results can be reduced, giving rise to wrong results both in terms of false positives and negatives [2]. If the general aim is to identify *some* relevant objects at the risk of loosing other positive hits (similarly to a Web search engine) then this approach can be considered.

In this paper we describe a dimensionality reduction technique that we designed to deal with an image based word indexing in a DIR application. The proposed method is based on the use of Growing Hierarchical Self Organizing Maps (GHSOM) that cluster input vectors into a hierarchy of multiple layers consisting of several independent SOMs. The hierarchy is deepest in correspondence with more complex clusters, that are represented with more details by lower level maps. The GHSOM has been mainly used as a visualization method and in this case the various maps are usually explored independently one to each other. The peculiarity of our approach is that we embed the lower level maps in the root one so as to obtain an unique low dimensional space where input patterns are projected by interpolation with respect to the cluster centers.

The paper is organized as follows. In Section 2 we summarize the previous work on dimensionality reduction and we describe the basic characteristics of Self Organizing Maps that are useful to understand the proposed method described in Section 3. The comparison of our method with other dimensionality reduction methods is analyzed in Section 4, whereas our final remarks are drawn in Section 5.

2 Related Work

The literature on dimensionality reduction is large, summarizing a longstanding research. In this section we concentrate only on some characteristics of the methods that we have considered for comparison with our approach.

Principal Components Analysis (PCA) is one of the most popular linear techniques. PCA performs dimensionality reduction by embedding the data into a lower dimensional space finding a linear basis in which the variance in the data is maximal. We omit here the details of PCA, however it is important to point out that PCA is based on the computation of the covariance matrix of the input data and subsequent evaluation of the principal eigenvectors of this matrix, that form the basis of the reduced space. After computing the PCA transformation, the mapping of points in the reduced space can be simply computed by means of a matrix multiplication and this is one of the main advantages of PCA-based dimensionality reduction.

In real data, often a linear mapping does not suffice to perform the dimensionality reduction and non-linear mappings should be considered. There are two main classes of algorithms: global and local techniques that attempt to preserve global (local) properties of the input data [6].

In the last few years deep architectures gained attention in the machine learning community. Autoencoders are a widely known architecture in this framework that have been used for global nonlinear dimensionality reduction since the 1990's [7]. Autoencoders are Multilayer perceptrons (MLP) having the same number of input and output units and a lower number of nodes in a hidden layer. The training is performed by means of the standard back-propagation algorithm where the network is forced to reproduce in the output layer the input patterns. After the training the hidden units are expected to describe the training data with a smaller representation, performing a non-linear dimensionality reduction. Similarly to other MLP-based architectures, autoencoders are prone to get stuck in local minima. To overcome this problem a new training strategy has been recently proposed [8]. The idea is to use Restricted Boltzmann Machines (RBM) for a preliminary unsupervised training of the recognition layers of the network. After the RBM training the reconstruction layers are formed by the inverse of the trained recognition layers. At the end of the process the overall autoencoder is fine tuned with the standard back-propagation algorithm. A trained autoencoder can be subsequently used to perform dimensionality reduction, by considering the output of the deeper recognition layer when presenting an input pattern in the original space.

Local Tangent Space Analysis (LTSA) is a local technique that is based on the representation of the local geometry of the manifold using tangent spaces learned by fitting an affine subspace in a neighborhood of each data point [9]. This local space is estimated by computing the PCA on the k nearest points of each input point. The tangent spaces are aligned so as to obtain the global coordinates of the data points with respect to the underlying manifold. This alignment is made by means of a partial eigendecomposition of the neighborhood connection matrix. One limitation of LTSA is the lack of an out-of-sample extension used to embed additional data points into an LTSA representation. It is therefore not possible to index additional objects or to perform queries with objects not indexed. This limit does not affect PCA, autoencoders and the method proposed in this paper.

2.1 Self Organizing Maps and GHSOM

The Self-Organizing Map (SOM) is a kind of artificial neural network that is based on unsupervised learning [10]. In the SOM the neurons are typically arranged in a two dimensional grid. Each neuron of the SOM is associated with a weight vector, or centroid, $w_i = [w_{i1}, w_{i2}, ..., w_{in}]^T \in \Re^n$. During learning, each input vector $x \in \Re^n$ is compared with all the weight vectors of the SOM and the Best Match Unit (BMU) is determined: the BMU is the weight vector closest to the input vector according to a given distance, for instance the Euclidean distance. The BMU c is therefore defined by

$$c = arg\min_i \parallel x - w_i \parallel \tag{1}$$

where x is an input vector and w_i is the weight vector associated with neuron i. The input vector x is then mapped to the cluster represented by neuron c, and its weight vector and those of its neighbors are updated. In the update process, two types of rules may be used: the *incremental* learning rule and the *batch* learning rule.

The need to predefine the SOM structure results in a significant limitation on the final mapping achievable . To avoid the problem of the static SOM structure various dynamic algorithms have been proposed. One example is the Growing Hierarchical Self-Organizing Map (GHSOM) that dynamically models the training data [11]. The GHSOM allows the network structure to grow in two dimensions, in width and in depth, so that it combines the advantages of the dynamic growth and of the hierarchical structure. The growing process of the GHSOM is regulated by two parameters. The parameter τ_1 regulates the growing process in width while τ_2 regulates the growing process in depth. Each layer of the trained GHSOM consists of several independent SOMs.

The GHSOM training starts from the first layer with an initial 2x2 grid. After a pre-defined number of training iterations, the unit E with the highest *quantization error* is identified, and a row or a column of neurons is inserted between the unit E and its most dissimilar neighboring unit. After this insertion, the training restarts with the standard SOM algorithm. When the training of a map ends, each unit satisfying a predetermined condition is subjected to hierarchical expansion, therefore producing the hierarchical structure of the GHSOM.

3 Embedded Map Projection

As already discussed, real patterns are unlikely to belong to a uniform distribution in the original vector space. On the opposite, the patterns can frequently be imagined as laying to low dimensional manifolds or, in other cases, as belonging to clusters, that can be more or less elongated in the input pattern space. The latter feature has been considered to speed-up the indexing and retrieval of objects in high dimensional spaces. For instance, the Cluster Tree [12] is an index structure that has been proposed to perform approximate search in high dimensional spaces on the basis of pattern clustering. In [13] we combined the SOM clustering with the PCA to efficiently index words represented by points in high dimensional spaces. Words in each cluster, that are expected to be more similar one to the others, are projected into a lower dimensional space by means of a local PCA, and this reduced representation is used to speed up the similarity search.

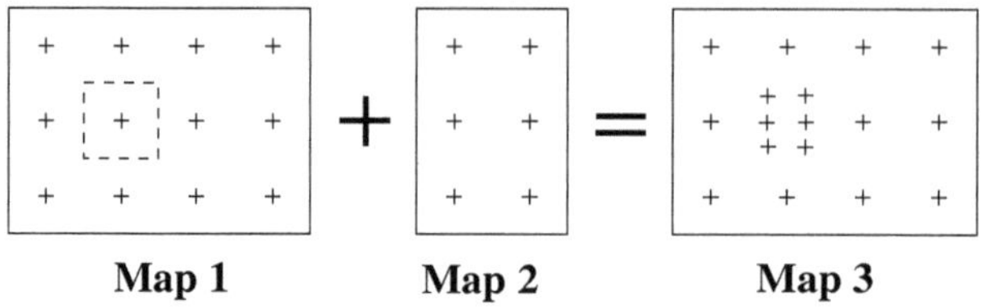

Fig. 1. Embedding of a second level SOM (Map2) into the parent one (Map1)

In [13] we did not consider the topological order of clusters because complex patterns, such as words, can not be easily modeled by a single map. One solution is to use a larger SOM, but training such map is not easy and the retrieval time risks to be very high, since the number of clusters quickly becomes very large. An alternative is to use a hierarchical map that is composed of several maps at different levels of detail such as the GHSOM. To have an idea of the type of maps that are built in the hierarchy, in the right part of Figure 3 we report the first level map and two sub-maps computed for the MNIST dataset.

The basic idea of the proposed approach is to embed lower maps in the output space that is implicitly defined by the first level map. Input points are then projected in this space. The dimensionality reduction is therefore a two step process: first, an embedding map is constructed starting from a trained GHSOM; second, input points are projected in this embedded map. Before describing the two steps, it is important to clarify that the GHSOM training, and subsequent embedded map building, is performed on a reduced number of points randomly selected from the collection to be indexed. The whole dataset is used in the projection step.

3.1 Embedded Map Building

To explain the concept we show in Figure 1 the embedding of a second order map (Map2) in the corresponding parent map (Map1). The SOM neurons (corresponding to cluster centers) can be considered as belonging to a two-dimensional grid, where each neuron is identified by two indexes. If we use the grid position as a low dimensional coordinate system, then the clusters correspond to uniformly distributed points represented by crosses in Map1. In the Figure, Map1 is a 3x4 map, whose cluster (1,1) (surrounded by a dashed box) is furtherly described with a 3x2 map (Map2). Embedding Map2 into Map1 we obtain Map3 that describes with more resolution the region in the output space corresponding to the neuron (1,1) in Map1. The actual embedding is obtained by a recursive linear scaling of lower level grids in the main one. A real embedded map computed with the MNIST dataset is shown in Figure 3.

3.2 Point Projection

As discussed in Section 2 the BMU for a pattern x represents the cluster having the highest similarity with x. By assigning x to this cluster we make a discrete

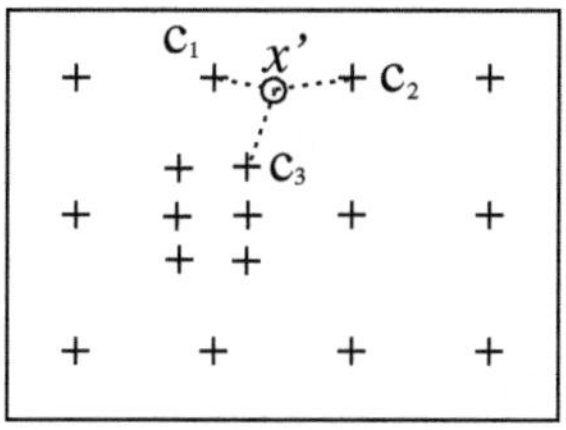

Fig. 2. Example of the projection of a point in the embedded map shown in Figure 1. Details are described in the text.

dimensionality reduction, where the output space is populated only in a limited number of fixed positions. If we aim at a less rigid projection, we can take advantage of the topological ordering of the map. Due to the curse of dimensionality, the quantization errors of nearest centroids have in general similar values, however we can consider these quantization errors as weights to be used to project the input point in a position in the embedded map that is between the closest centroids.

Previous approaches to perform this projection worked with single SOMs [14], or with each individual map in the GHSOM [15]. The projection described here deals with the embedded map centroids and projects each input point x as follows (see also Figure 2). Let c_1 be the BMU for point x and c_2 and c_3 be the next closest centroids with distances $d_i = \| x - c_i \|$ $(i = 1, 2, 3)$. The three distances are ordered so that $d_1 \leq d_2 \leq d_3$ and x should be placed somewhere between the three points, but closer to c_1. The projection is therefore defined by the following rule:

$$x' = \frac{d_1^{-1} c_1 + d_2^{-1} c_2 + d_3^{-1} c_3}{d_1^{-1} + d_2^{-1} + d_3^{-1}}. \tag{2}$$

It is important to remark that this projection has been already proposed [15], however in that case it was used for visualization of points in *separate* maps, whereas in our case we use it to project the data in a unique low dimensional space.

4 Experiments

In this section we describe the experiments performed on three datasets comparing four dimensionality reduction methods for similarity retrieval. The software used is implemented in Matlab. In particular the PCA, LTSA, and autoencoder software has been described in [16]. The proposed embedded map projection method has been implemented starting from the GHSOM Toolbox for Matlab [11].

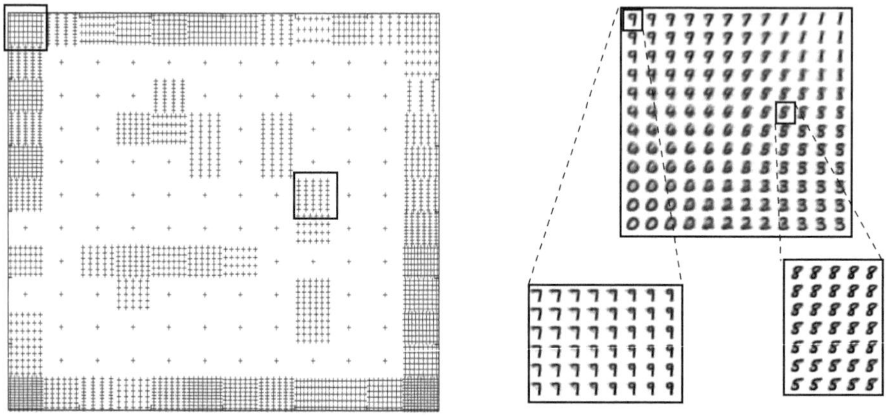

Fig. 3. The embedding obtained for the MNIST dataset. A part of the GHSOM is shown on the right. In particular we show the root map and two lower level maps (corresponding to the marked regions in the embedded map).

4.1 Datasets

The experiments have been performed on three datasets that represent various domains. The MNIST dataset is a widely used collection of handwritten digits containing 60,000 images with size of 28 x 28 pixels. The COIL20 dataset contains images of 20 objects, depicted from 72 different viewpoints each. Therefore, there are a total of 1,440 images that have a size of 32 x 32 pixels. The WORDS dataset is a collection of digitized printed words normalized to fit a 12 x 57 grid. In total we have 132,956 word images extrapolated from 1,302 pages that are part of an encyclopedia of the XIXth Century.

At first, we projected the input data of each dataset into a two dimensional space. In the PCA we only needed to define the output dimension (2 in this case). For the training we used 10,000 patterns for MNIST, all the patterns for COIL20, and 13,296 patterns for WORDS. For MNIST and WORDS we subsequently projected all the patterns in the datasets. For the autoencoders we used for all the datasets a 1000-500-250-2 units in the hidden layers and 50 epochs for RBM training, whereas the fine-tuning was made with 100, 200, 300 back-propagation epochs retaining the best network. For the GHSOM training we evaluated several combinations of parameters, but for all the datasets the best results have been achieved with $\tau_1 = 0.6$ and $\tau_2 = 0.005$. The resulting maps have the following features[1]: MNIST(2,60,2566), COIL20(4,70,395), WORDS(2,60, 2077). For LTSA we set $k = 12$.

Figure 5 shows the projections for MNIST of the whole dataset and of the first seven classes separately. From the global map we can notice that the proposed EMP method distributes the patterns in the output space more uniformly and,

[1] The notation used is DATASET(# levels, # maps, # clusters).

Table 1. Precision at 0 percent Recall with the four compared methods

Dataset	n	N	EMP	PCA	Autoencoder	LTSA
MNIST	784	10,000	87.14	55.17	66.14	77.25
		60,000	84.56	53.55	64.66	–
COIL20	1024	1,440	85.59	71.68	82.35	19.63
WORDS	684	132,956	79.80	9.91	30.29	–

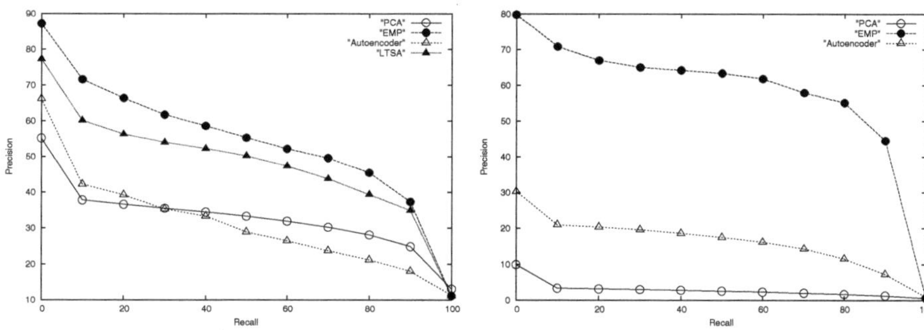

Fig. 4. Precision-Recall plot for the four compared methods working on the MNIST and WORDS data-sets

even if there are more outliers, in general the patterns of different classes are more separated with respect to the other methods.

4.2 Numerical Evaluation

The effectiveness of the methods has been measured in term of accuracy achieved by a query by example retrieval performed on the reduced space. We made several queries and we computed a Precision-Recall plot averaging each query. For the COIL20 dataset we used in turn each point as query evaluating the retrieval performance. For MNIST we used 10,000 queries randomly selected from the whole 60,000 patterns. In the case of the WORDS dataset there are only 576 labeled words that are used as queries.

Figure 4 shows the Precision-Recall plots for the MNIST and WORDS data-sets. In the MNIST dataset we projected only the 10,000 training patterns, and therefore we have also a plot for LTSA. In the WORDS dataset the LTSA plot is missing, since we indexed all the 132,956 points. To reduce the paper length we summarize in table 1 all the performed experiments reporting the Precision at Recall 0 for various experiments (this value is obtained by an interpolation of the Precision Recall plots). In the table, n is the dimension of the input space, N is the number of indexed objects.

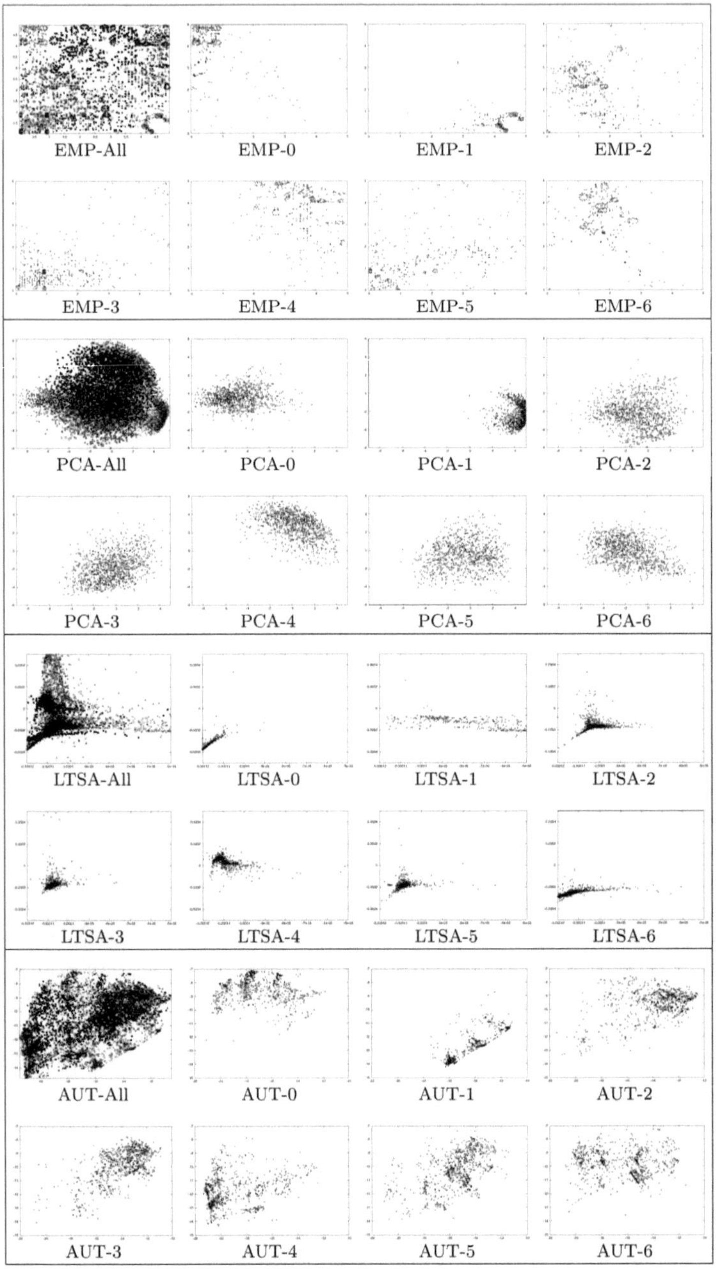

Fig. 5. Two dimensional projection of the MNIST dataset obtained with the four methods. For each method we report the distribution of points for the whole dataset, and for the first seven classes (0-6).

5 Conclusions

In this paper we propose a dimensionality reduction method that is based on the embedding of lower level maps of a GHSOM clustering of the input data. The method has been compared with other dimensionality reduction methods on a query by example retrieval application on three datasets. These preliminary results are encouraging, since the EMP method outperforms the compared ones.

References

1. Marinai, S., Marino, E., Soda, G.: Font adaptive word indexing of modern printed documents. IEEE Transactions on PAMI 28(8), 1187–1199 (2006)
2. Samet, H.: Foundations of multidimensional and metric data structures. Morgan Kaufmann, Amsterdam (2006)
3. Hastie, T., Tibshirani, R., Friedman, J.: The elements of statistical learning: data mining, inference, and prediction. Springer series in statistics, New York, NY, USA (2001)
4. Beyer, K., Goldstein, J., Ramakrishnan, R., Shaft, U.: When is "nearest neighbor" meaningful? In: Proc. 7th Int. Conf. on Database Theory, vol. 3, pp. 1763–1768 (1999)
5. Kanth, K.V.R., Agrawal, D., Singh, A.: Dimensionality reduction for similarity searching in dynamic databases. SIGMOD Rec. 27(2), 166–176 (1998)
6. van der Maaten, L., Postma, E., van den Herik, H.: Dimension reduction: A comparative review (preprint, 2007)
7. De Mers, D., Cottrell, G.: Nonlinear dimensionality reduction. In: NIPS-5 (1993)
8. Hinton, G.E., Salakhutdinov, R.R.: Reducing the dimensionality of data with neural networks. Science 313(5786), 504–507 (2006)
9. Zhang, Z., Zha, H.: Principal manifolds and nonlinear dimensionality reduction via local tangent space alignment. SIAM Journal of Scientific Computing 26(1), 313–338 (2004)
10. Kohonen, T., Kaski, S., Lagus, K., Salojarvi, J., Honkela, J., Paatero, V., Saarela, A.: Self organization of a massive document collection. IEEE Transactions on Neural Networks 11(3), 574–585 (2000)
11. Chan, A., Pampalk, E.: Growing hierarchical self organising map (ghsom) toolbox: visualisations and enhancements. In: Neural Information Processing, ICONIP 2002. Proceedings of the 9th International Conference, vol. 5, pp. 2537–2541 (2002)
12. Li, C., Chang, E., Garcia-Molina, H., Wiederhold, G.: Clustering for approximate similarity search in high-dimensional spaces. IEEE Transactions on Knowledge and Data Engineering 14(4), 792–808 (2002)
13. Marinai, S., Faini, S., Marino, E., Soda, G.: Efficient word retrieval by means of SOM clustering and PCA. In: Bunke, H., Spitz, A.L. (eds.) DAS 2006. LNCS, vol. 3872, pp. 336–347. Springer, Heidelberg (2006)
14. Wu, Z., Yen, G.: A som projection technique with the growing structure for visualizing high-dimensional data. In: Proceedings of the International Joint Conference on Neural Networks, vol. 3, pp. 1763–1768 (2003)
15. Yen, G., Wu, Z.: Ranked centroid projection: a data visualization approach with self-organizing maps. IEEE Transactions on Neural Networks 19(2), 245–258 (2008)
16. van der Maaten, L.: An introduction to dimensionality reduction using matlab. Technical Report Technical Report MICC 07-07 (2007)

Novel Incremental Principal Component Analysis with Improved Performance*

Myoung Soo Park and Jin Young Choi

School of Electrical Engineering and Computer Science, ASRI,
Seoul National University, Seoul 151-744, Korea
meister1@gmail.com, jychoi@snu.ac.kr

Abstract. In this paper, we present a novel incremental algorithm for principal component analysis (PCA). The proposed algorithm is a kind of covariance-free type algorithm which requires less computation and storage space in finding out the eigenvectors, than other incremental PCA methods using a covariance matrix. The major contribution of this paper is to explicitly deal with the changing mean and to use a Gram-Schmidt Orthogonalization (GSO) for enforcing the orthogonality of the eigenvectors. As a result, more accurate eigenvectors can be found with this algorithm than other algorithms. The performance of the proposed algorithm is evaluated by experiments on the data sets with various properties and it is shown that the proposed method can find out the eigenvectors more closer to those of batch algorithm than the others.

1 Introduction

In the studies of pattern recognition, principal component analysis (PCA) [1] has been extensively considered in many literatures. Many applications of pattern recognition suffer from the curse of dimensionalty, and one approach to resolve this problem is to reduce the original input dimensionality, by projecting data onto a linear subspace with small dimensionality. Among various methods for finding such linear subspace, PCA is most popular and has been widely used.

The conventional PCA are carried out in batch mode [2], that is, by using all data simultaneously. These methods involve computing the eigenvectors and eigenvalues of the sample covariance matrix, using numerical methods [3] such as the Jacobi method. However, there exist applications in which the data can be incrementally received and the number of data may be infinite. For such applications, it is inefficient or impossible to repeat batch PCA whenever a new datum arrives. When the original input dimension and the number of data are large, the computation and storage requirements make the batch PCA infeasible. This situation makes it necessary to study incremental algorithms for principal component analysis.

There have been many efforts devoted to incremental PCA (IPCA). Basically, there exist two types of IPCA: *covariance matrix-based methods* including the

* This research was supported by BK21 program and by the Ministry of Commerce, Industry and Energy of Korea.

N. da Vitora Lobo et al. (Eds.): SSPR&SPR 2008, LNCS 5342, pp. 592–601, 2008.

estimation of covariance matrix, and *covariance matrix-free methods* avoiding the explicit estimation and use of covariance matrix. There have been several covariance matrix-based incremental PCA algorithms, but these algorithms are generally similar in terms of accuracy and speed; the only difference is how to approximate the covariance matrix. To avoid the burden of covariance matrix estimation and storage and to speed up, a few covariance matrix-free IPCA methods [4][5][6][7][8], have been proposed. Among them, Candid Covariance free Incremental Principal Component Analysis (CCIPCA) [9] is a state-of-the art algorithm. CCIPCA algorithm has been reported to be much superior to other traditional algorithms for incremental principal component analysis.

However, CCIPCA has a few problems which should be overcome. [10] pointed out that despite the eigenvectors stasticially converging in CCIPCA, the higher-order eigenvectors converge very slow, and that the information in the previous eigenvecotrs may be lost in the calculation of new eigenvectors. To resolve this problem, a new covariance matrix-free algorithm named as largetst eigenvalue theory based IPCA (LET-IPCA) was introduced for eigenvector update. However, this has also important problems. When the data are given in incremental manner, the change of mean should be reflected in the update of eigenvectors. According to [11], it is especially important for image recognition, one of the major application areas of PCA, but it is ignored and the zero mean of data is assumed. Another problem is that the orthogonality of eigenvectors are not guaranteed and as a result the resultant eigenvectors of these methods become much different from those of batch PCA.

In this paper, we will propose a new incremental learning algorithm to solve the above problem of the changing mean and the orthogonality. After describing basic operations and problems of CCIPCA and LET-IPCA in the framework of power method, we will add an enforcement of eigenvector orthogonality to the update equation, then derive the update equation which can reflect the effect of changing mean. The performance of the proposed algorithm will be experimentally evaluated and compared with the performances of CCIPCA and LET-IPCA, by using various kinds of data in UCI Machine Learning repository and two face recognition data as benchmark data sets.

2 Proposed Incremental PCA

Tha main issues addressed by the proposed IPCA are to consider the change of mean and to enforce the orthogonality of eigenvectors. These two issues are not considered in the previous covariance matrix-free incremental PCA methods and we think that it is the major reason of their limited performance in extracting good features. This section will present the details of our method.

2.1 Problem Formulation

The goal of PCA is formulated as finding a basis vector $\mathbf{w}^*$ such that

$$\mathbf{w}^* = \arg\max_{\mathbf{w}} \mathbf{w}^T \mathbf{A}(n) \mathbf{w}, \tag{1}$$

where $\mathbf{w} \in R^d$ is a unit vector and $\mathbf{A}(n) = \frac{1}{n}\sum_{i=1}^{n}[\mathbf{x}(i) - \mathbf{m}(n)][\mathbf{x}(i) - \mathbf{m}(n)]^T$ is the auto-covariance matrix of each component in $\mathbf{x}(i) \in R^d$, and d is the input dimension, i.e. the number of components for representing $\mathbf{x}(i)$. $\mathbf{m}(n) = \frac{1}{n}\sum_{i=1}^{n}\mathbf{x}(i)$ is a mean vector of n data vectors, and n is the number of data for calculating $\mathbf{A}(n)$.

The $\mathbf{w}^*$ is called the 1st *principal axis* $\mathbf{w}_1(n)$ and $x_1(i) = \mathbf{w}_1(n)^T\mathbf{x}(i)$ is called the 1st *principal component* of $\mathbf{x}(i)$ for $i = 1, 2, ..., n$. Then, by removing the component of data in direction of $\mathbf{w}_1(n)$ and solving the above problem again, we can get a $\mathbf{w}^*$ which is orthogonal to the $\mathbf{w}_1(n)$. It is called the 2nd principal axis $\mathbf{w}_2(n)$ and $x_2(i) = \mathbf{w}_2(n)^T\mathbf{x}(i)$ is called the 2nd principal component of $\mathbf{x}(i)$ for $i = 1, 2, ..., n$. The other principal axes and principal components notified as $\mathbf{w}_k(n)$ and $x_k(i)$ for $k = 3, 4, ...$ are similarly achieved.

2.2 Update Equations of CCIPCA and LET-IPCA

According to [1], PCA can be carried out by solving the following eigenproblem,

$$\lambda_k(n)\mathbf{w}_k(n) = \mathbf{A}(n)\mathbf{w}_k(n), \tag{2}$$

where the kth principal axis $\mathbf{w}_k(n)$ corresponds to the eigenvector of $\mathbf{A}(n) = \frac{1}{n}\sum_{i=1}^{n}\mathbf{x}(i)\mathbf{x}(i)^T$ corresponding to the kth largest eigenvalue $\lambda_k(n)$. Power method is a simple method to solve this eigenproblem in iterative way; its basic procedure is to update $\mathbf{v}(n) \leftarrow \mathbf{A}(n)\frac{\mathbf{v}(n)}{||\mathbf{v}(n)||}$ with a nonzero initial value of $\mathbf{v}(n)$. It is shown that after a small number of iterations, $\mathbf{v}(n)$ converges to the true $\mathbf{v}_1(n)$.

The basic update equation of CCIPCA can be deduced from this. If we replace the $\mathbf{A}(n)$ with $\frac{n-1}{n}\mathbf{A}(n-1) + \frac{1}{n}\mathbf{x}(n)\mathbf{x}(n)^T$ and assume that $\mathbf{A}(n-1)\mathbf{v}_1(n)$ is almost the same as $\mathbf{A}(n-1)\mathbf{v}_1(n-1)$, $\mathbf{v}_1(n)$ can be determined by $\mathbf{v}_1(n) = \frac{n-1}{n}\mathbf{v}_1(n-1) + \frac{1}{n}\mathbf{x}(n)\mathbf{x}(n)^T$. This can be easily extended for higher eigenvectors.

$$\mathbf{v}_k(n) = \frac{n-1}{n}\mathbf{v}_k(n-1) + \frac{1}{n}\mathbf{x}_k(n)\mathbf{x}_k(n)^T, \tag{3}$$

where $\mathbf{x}_k(n)$ is $\mathbf{x}(n)$ whose components in the direction of $\mathbf{v}_j(n)$ where $j < k$ are removed, defined as

$$\mathbf{x}_k(n) = \mathbf{x}_{k-1}(n) - \mathbf{x}_{k-1}^T(n)\frac{\mathbf{v}_{k-1}(n)}{||\mathbf{v}_{k-1}(n)||}\frac{\mathbf{v}_{k-1}(n)}{||\mathbf{v}_{k-1}(n)||}, \tag{4}$$

and $\mathbf{x}_1(n) = \mathbf{x}(n)$ and $\mathbf{v}_0(n) = 0$. If necessary, this update equation for each k can be repeated more than one times. Then the eigenvector can be determined as $\mathbf{w}_k(n) = \frac{\mathbf{v}_k(n)}{||\mathbf{v}_k(n)||}$.

For more accurate calculation, the relations $\mathbf{A}(n) = \sum_k \lambda_k(n)\frac{\mathbf{v}_k(n)}{||\mathbf{v}_k(n)||}\frac{\mathbf{v}_k(n)^T}{||\mathbf{v}_k(n)||}$, where $\lambda_k = ||\mathbf{v}_k(n)||$, and $\mathbf{A}(n) = \frac{n-1}{n}\mathbf{A}(n-1) + \frac{1}{n}\mathbf{x}(n)\mathbf{x}(n)^T$ are used in LET-IPCA. Instead of calculating $\mathbf{x}_k(n)$, it is directly applied to the basic update procedure of the power method and the following update equation is achieved.

$$\mathbf{v}_k^t(n) = \frac{n-1}{n}\mathbf{v}_k^{\infty}(n-1) + \frac{1}{n}\mathbf{x}(n)\mathbf{x}(n)^T\frac{\mathbf{v}_k^{t-1}(n)}{||\mathbf{v}_k^{t-1}(n)||}$$

$$-\sum_{m=1}^{k-1} \lambda_m(n) \frac{\mathbf{v}_m^{\infty}(n)}{||\mathbf{v}_m^{\infty}(n)||} \frac{\mathbf{v}_m^{\infty}(n)^T}{||\mathbf{v}_m^{\infty}(n)||} \frac{\mathbf{v}_k^{t-1}(n)}{||\mathbf{v}_k^{t-1}(n)||}$$
$$+\frac{n-1}{n} \sum_{m=1}^{k-1} \lambda_m(n-1) \frac{\mathbf{v}_m^{\infty}(n-1)}{||\mathbf{v}_m^{\infty}(n-1)||} \frac{\mathbf{v}_m^{\infty}(n-1)^T}{||\mathbf{v}_m^{\infty}(n-1)||} \frac{\mathbf{v}_k^{t-1}(n)}{||\mathbf{v}_k^{t-1}(n)||}, \quad (5)$$

where $\mathbf{v}_k^0(n) = \mathbf{v}_k^{\infty}(n-1)$ and t is the number of iteration.

However, both update equations have a few problems. Both algorithms are based on the relation $\mathbf{A}(n) = \frac{n-1}{n}\mathbf{A}(n-1) + \frac{1}{n}\mathbf{x}(n)\mathbf{x}^T$, but in this case $\mathbf{A}(n)$ is different from a covariance matrix of $\mathbf{x}$ unless the mean of data is zero, which cannot be always satisfied. In fact, considering the change of mean is important for some applications such as image processing, and it needs to be reflected on the update equation. In addition, each eigenvector achieved from both algorithm are not guaranteed to be orthogonal to the others. It makes the achieved eigenvectors become different from the principal axes by batch PCA. To resolve these problems, we developed a new update equation starting from that of LET-IPCA.

2.3 A Modified Update Equation to Resolve Problems

After carefully checking out the derivation of the update equation of LET-IPCA, we found that it may be incorrect when $t \neq 1$. Therefore, we will use the update equation of LET-IPCA when $t = 1$ as our starting point. For finding the correct eigenvectors, each eigenvector needs to be made orthogonal to each other. The simplest ways to achieve this is to use the Gram-Schmidt Orthogonalization (GSO) procedure as follows.

$$\mathbf{v}_k(n) \leftarrow \mathbf{v}_k(n) - \sum_{m=1}^{k-1} \mathbf{v}_k(n)^T \frac{\mathbf{v}_m(n)}{||\mathbf{v}_m(n)||} \frac{\mathbf{v}_m(n)}{||\mathbf{v}_m(n)||}. \quad (6)$$

This use of GSO can also simplify the basic update equation. If the eigenvectors for $n-1$ data are already orthogonalized, that is, $\mathbf{v}_k(n-1)^T\mathbf{v}_m(n-1) = 0$ when $k \neq m$, the update equation can be simplified as follows,

$$\mathbf{v}_k(n) = \frac{n-1}{n}\mathbf{v}_k(n-1) + [\frac{1}{n}\mathbf{x}(n)\mathbf{x}(n)^T$$
$$-\sum_{m=1}^{k-1} \lambda_m(n) \frac{\mathbf{v}_m(n)}{||\mathbf{v}_m(n)||} \frac{\mathbf{v}_m(n)^T}{||\mathbf{v}_m(n)||}] \frac{\mathbf{v}_k(n-1)}{||\mathbf{v}_k(n-1)||}. \quad (7)$$

To account for the changing mean, a new relation between $\mathbf{A}(n)$ and $\mathbf{A}(n-1)$, which is matched to the correct covariance matrix, needs to be derived and used in the place of old relation. This new relation can be easily found as follows.

$$\mathbf{A}(n) = \frac{n-1}{n}\mathbf{A}(n-1) + \frac{1}{n-1}(\mathbf{x}(n) - \mathbf{m}(n))(\mathbf{x}(n) - \mathbf{m}(n))^T,$$

where $\mathbf{m}(n)$ is the mean vector and can be updated by the following equation.

$$\mathbf{m}(n) = \frac{n-1}{n}\mathbf{m}(n-1) + \frac{1}{n}\mathbf{x}(n). \quad (8)$$

If we use this relation in place of $\mathbf{A}(n) = \frac{n-1}{n}\mathbf{A}(n-1) + \frac{1}{n}\mathbf{x}(n)\mathbf{x}(n)^T$, the following new update equation can be achieved.

$$\mathbf{v}_k(n) = \frac{n-1}{n}\mathbf{v}_k(n-1) + [\frac{1}{n-1}[\mathbf{x}(n) - \mathbf{m}(n)][\mathbf{x}(n) - \mathbf{m}(n)]^T - \sum_{m=1}^{k-1} \lambda_m(n) \frac{\mathbf{v}_m(n)}{||\mathbf{v}_m(n)||} \frac{\mathbf{v}_m(n)^T}{||\mathbf{v}_m(n)||}] \frac{\mathbf{v}_k(n-1)}{||\mathbf{v}_k(n-1)||}. \tag{9}$$

After finding $\mathbf{v}_k(n)$, the eigenvector can be determined as $\mathbf{w}_k(n) = \frac{\mathbf{v}_k(n)}{||\mathbf{v}_k(n)||}$.

Combining the described modifications, the overall proposed Incremental PCA can be summarized as Algorithm 1.

Input: k (eigenvector, eigenvalue) pairs $\{(\mathbf{w}_j^{old}, \lambda_j^{old})|j = 1, 2, ..., k\}$
The number of given data n (including current data $\mathbf{x}$)
A mean vector data $\mathbf{m}^{old} \in R^d$
A new data $\mathbf{x}$
Output: k (eigenvector, eigenvalue) pairs $\{(\mathbf{w}_j^{new}, \lambda_j^{new})|j = 1, 2, ..., k\}$
A new mean vector of n data $\mathbf{m}^{new} \in R^d$

begin
  $\mathbf{m}^{new} \leftarrow \frac{n-1}{n}\mathbf{m}^{old} + \frac{1}{n}\mathbf{x}$
  $\mathbf{x}_c \leftarrow \mathbf{x} - \mathbf{m}^{new}$
  for $j \leftarrow 1$ **to** k **do**
    $\mathbf{v}_j^{old} \leftarrow \lambda_j^{old}\mathbf{w}_j^{old}$
    $\mathbf{v}_j^{new} \leftarrow \frac{n-1}{n}\mathbf{v}_j^{old} + \frac{1}{n-1}\mathbf{x}_c(n)\mathbf{x}_c(n)^T \frac{\mathbf{v}_j^{old}}{||\mathbf{v}_j^{old}||}$
    for $m \leftarrow 1$ **to** $j-1$ **do**
      $\mathbf{v}_j^{new} \leftarrow \mathbf{v}_j^{new} - \mathbf{v}_m^{new} \frac{\mathbf{v}_m^{new\,T}}{||\mathbf{v}_m^{new}||} \frac{\mathbf{v}_j^{old}}{||\mathbf{v}_j^{old}||}$
    end
  end
  for $j \leftarrow 1$ **to** k **do**
    for $m \leftarrow 1$ **to** $j-1$ **do**
      $\mathbf{v}_j^{new} \leftarrow \mathbf{v}_j^{new} - \mathbf{v}_j^{new} \frac{\mathbf{v}_m^{new\,T}}{||\mathbf{v}_m^{new}||} \frac{\mathbf{v}_j^{new}}{||\mathbf{v}_j^{new}||}$
    end
    $\lambda_j^{new} \leftarrow ||\mathbf{v}_j^{new}||$
    $\mathbf{w}_j^{new} \leftarrow \mathbf{v}_j^{new}/\lambda_j^{new}$
  end
end

Algorithm 1. Proposed Incremental PCA algorithm. The second for-loop implements the Gram-Schmidt orthogonalization procedure, and the others implement the basic update equation reflecting the change of mean.

Table 1. Properties of data sets used in the experiments

Data Set	input dimension	number of data
ECOLI	7	336
PIMA	8	768
GLASS	9	214
WBCancer	9	699
TICTACTOE	9	958
CMC	9	1473
WINE	13	178
IONOSHPERE	34	351
SONAR	60	208
YALE	630	165
ATnT	644	400

2.4 Computational Complexity

For a newly arrieved datum, the computational (or time) complexity of each method to determine k eigenvectors is as follows. Let n and d be the number of total data and the input dimension of each datum, respectively. Then, the computational complexities of batch PCA are $O((n+d)d^2)$ for Jacobi method [12], and $O((n+k)d^2)$ for Power method [3], while the computational complexities of CCIPCA, LET-IPCA, and the proposed IPCA are $O(kd)$, $O(k^2d)$ and $O(k^2d)$, respectively. Generally $k \ll d$ and $k \ll n$, and therefore the complexities of CCIPCA, LET-IPCA and the proposed method become less than that of batch PCA. The complexity of CCIPCA is better than those of LET-IPCA and the proposed method, but the improved performance of the proposed method can compensate its computational complexity. This can be clarified in the experimental study on the performance comparison.

3 Experiment

In the following experimental result, we will show that the result of proposed IPCA is closer to the result of batch PCA than other incremental algorithms, and therefore the proposed IPCA can generate similar eigenvectors with those of batch PCA.

3.1 Data Set

In the experiments, we used the data sets in UCI Machine Learning Repository and two face data sets. The UCI Machine Learning Repository includes various data sets consisting of artificial and real data, and can be accessed from the web site. Among them, we selected 9 data sets and used: Ecoli Database (ECOLI), Pima Indians Diabetes Database (PIMA), Glass Identification Database (GLASS), Wisconsin Breast Cancer Databases (WBCancer), Tic-Tac-Toe Endgame Database (TICTACTOE), Contraceptive Method Choice

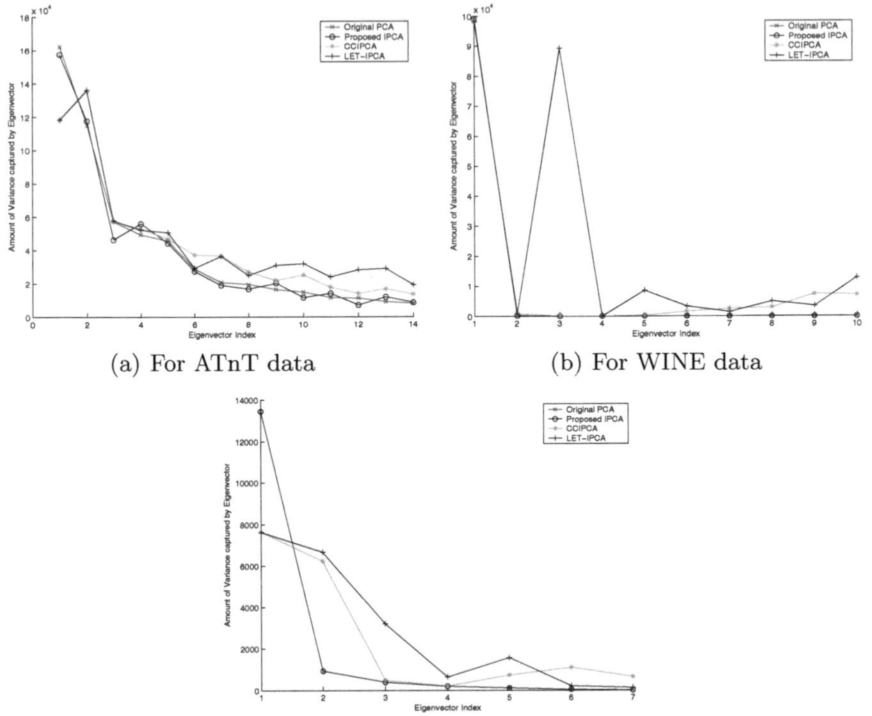

(a) For ATnT data

(b) For WINE data

(c) For PIMA data

Fig. 1. Distribution of variance captured with eigenvectors, obtained by batch and incremental PCA methods. x-axis represents the number of eigenvectors and the y-axis represents the amount of captured variance.

(CMC), Wine Recognition Database (WINE), Ionosphere Database (IONOSPHERE), Sonar data (SONAR). In addition, YALE and ATnT face data sets are used. The properties of the data used in experiments are summarized in Table 1. These data sets have various number of class labels, data, and attributes, so the experimental results using these data sets can cover the various cases.

3.2 Variance Capture Ability

We evaluated the quality of the extracted principal components (PCs) by different methods with two criterion. First, the distribution of captured variance by PCs are observed. As the distribution is more similar with that of batch PCA, the IPCA performance is considered to be good. Next, we evaluated the amount of total captured variance by PCs with the following performance index.

$$r(k, \cdot) = \frac{V(k)|_{\cdot}}{V(k)|_{PCA}} \times 100(\%) \tag{10}$$

where k is the number of eigenvectors extracted by batch or incremental PCA. and $V|\cdot$ is the variance captured by k eigenvectors, calculated by

$$V(k)|\cdot = \sum_{j=1}^{k} \left(\frac{1}{n} \sum_{i=1}^{n} [(\mathbf{w}_j(n)|\cdot)^T \mathbf{x}(i) - \mathbf{m}(n)]^2 \right) \quad (11)$$

where n is the number of data and $\mathbf{w}_j(n)|\cdot$ is the jth eigenvector extracted by incremental PCA method noted by $(\cdot)$.

Figure 1 shows the distribution of variance, captured by eigenvectors of batch PCA and incremental PCAs (CCIPCA, LET-IPCA, and proposed IPCA). Variance distribution is investigated for different numbers of eigenvectors, according to the input dimension of original data. For ATnT, WINE, and PIMA data, the variances captured by up to 14, 10, and 7 eigenvectors are investigated, respectively. The line with mark indicates the distribution of variance according to the principal component index; red line with cross mark, blue line with circle mark, green line with star mark, and black line with bar correspond to the results of batch PCA, proposed IPCA, CCIPCA, and LET-IPCA, respectively. It is clear that the result of proposed IPCA is closest to the result of batch PCA. We investigate the results for other data and achieve similar result.

Table 2 shows the values of $r(k,\cdot)$ in equation (10) for different data sets. $r(k,\cdot)$ is investigated for different number of eigenvectors k since the maximum number of eigenvectors are limited by the input dimension and the number of data. For 6 data sets (from ECOLI to CMC), variance capautre ratio is calculated for 7 eigenvectors. For 1 data set (WINE), variance capautre ratio is calculated for 10 eigenvectors. For 6 data sets (from IONOSPHERE to ATnT), variance capautre ratio is calculated for 14 eigenvectors.

Table 2. Variance Capture Ratio

(a) $k = 7$ case

Data Set	Proposed IPCA	CCIPCA	LET-IPCA
ECOLI	**100.00**	216.92	217.23
PIMA	**100.00**	112.71	132.40
GLASS	**99.47**	125.83	133.11
WBCancer	**99.93**	101.32	101.41
TICTACTOE	**99.33**	95.87	95.33
CMC	**99.94**	100.34	101.10

(b) $k = 10$ case

Data Set	Proposed IPCA	CCIPCA	LET-IPCA
WINE	**100.00**	122.35	224.70

(c) $k = 14$ case

Data Set	Proposed IPCA	CCIPCA	LET-IPCA
IONOSPHERE	**95.11**	225.00	230.47
SONAR	**80.11**	126.98	126.06
YALE	**97.82**	113.46	126.76
ATnT	**97.90**	109.03	117.45

Table 3. Classification Accuracy with the features extracted by batch PCA and different incremental PCA methods

Data Set	Batch PCA	Proposed IPCA	CCIPCA	LET-IPCA
ECOLI	81.25	**81.25**	75.89	75.00
PIMA	68.49	**68.10**	69.40	70.05
GLASS	75.23	**75.23**	72.43	72.90
WBCancer	96.71	**96.85**	**96.85**	97.14
TICTACTOE	73.17	**71.61**	97.70	97.91
CMC	47.79	48.47	**48.07**	**48.07**
WINE	76.97	**76.97**	**76.97**	**76.97**
IONOSPHERE	90.03	**89.74**	87.18	86.04
SONAR	83.65	**84.61**	76.44	73.56
ATnT	98.50	**98.25**	97.75	97.75
YALE	84.85	**84.24**	82.42	**84.24**

From the definition, it is clear that the $r(k, \cdot)$ goes close to 100.0% as the amount of captured variance by $(\cdot)$ is similar with that of batch PCA. The result shows that in most cases the amount of captured variance by the proposed IPCA is closer to 100.0%, compared to those of other methods, and we can concluded that the proposed method can extract better principal components than the other IPCA methods. In this result, note that $r(k, \cdot)$ of CCIPCA and LET-IPCA becomes larger than 100.0% for some cases. Since the eigenvectors of batch PCA maximally capture the variance when the same number of basis vectors are used, it may seem to be a weired result. But it is the result of broken orthogonality in eigenvectors of CCIPCA and LET-IPCA; the variance captured by higher eigenvectors are repeatedly captured by lower eigenvectors.

3.3 Classification Accuracy

In addition to the variance capture ability, the quality of features is evaluated from the classification accuracy achieved by using these features. The same data sets are used for experiments. Among various models, a simplest 1-Nearest Neighbor classifier is selected as classifiers so as to exposure the quality of features. The experimental results are as shown in Table 3.

In this table, the case of the incremental PCA whose classification accuracy is closest to that of batch PCA is notified by using the bold character. For most cases, the features extracted by the proposed method shows the closest result to that of batch PCA. It shows that the proposed IPCA method can extract the features most similar to that of batch PCA.

4 Conclusion

In this paper, we present a novel incremental algorithm for principal component analysis. Like the previous algorithms such as CCIPCA and LET-IPCA, this algorithm is a covariance-free type algorithm which requires less computation and storage for finding out the eigenvectors than the previous IPCA method using a covariance matrix. The major contribution of this paper is to explicitly deal with the changing mean and to combine a Gram-Schmidt Orthogonalization (GSO) for enforcing the orthogonality of the eigenvectors. As a result, more accurate eigenvectors can be found with this algorithm than the previous algorithms.

The performance of the proposed algorithm is evaluated by the experimental studies. The data sets of UCI Machine Learning repository, covering various different cases, are selected and used for the performance benchmark. The ratio of captured variance and the classification accuracy with the extracted features by CCIPCA, LET-IPCA and the proposed methods are compared. As a result, it is shown that the proposed method can find out the eigenvectors more closer to that of batch algorithm than other IPCA algorithms.

References

1. Jolliffe, I.: Principal Component Analysis. Springer, Heidelberg (1986)
2. Li, Y., Xu, L., Morphett, J., Jacobs, R.: An integrated algorithm of incremental and robust pca. In: Proceedings of International Conference on Image Processing, pp. 245–248 (2003)
3. Golub, G., Van Loan, C.: Matrix Computations, 3rd edn. The Johns Hopkins Press (1989)
4. Oja, E.: Neural networks, principal components, and subspace. International Journal of Neural Systems 1, 61–68 (1989)
5. Abbas, H., Fahmy, M.: Neural model for karhunen-loeve transform with application to adaptive image compression. IEE Proceedings I: Communications, Speech, Vision 140, 135–143 (1993)
6. Bannour, S., Azimi-Sadjadi, M.: Principal component extraction using recursive least squares learning. IEEE Transactions on Neural Networks, 457–469 (1995)
7. Diamantaras, K., K.S.Y.: Principal component neural networks: Theory and applications (1996)
8. Cichocki, A., Kasprzak, W., Skarbek, W.: Adaptive learning algorithm for principal component analysis with partial data. In: Proceedings of Cybernetics System, pp. 1014–1019 (1996)
9. Weng, J., Zhang, Y., Hwang, W.: Candid covariance-free incremental principal component analysis. IEEE Transactions on Pattern Analysis and Machine Intelligence 25(8), 1034–1040 (2003)
10. Yan, S., Tang, X.: Largest-eigenvalue-theory for incremental principal component analysis. In: Proceedings of IEEE International Conference on Image Processing, pp. 1181–1184 (2005)
11. Hall, P., Marshall, D., Martin, R.: Incremental eigenanalysis for classification. In: Proceedings of British Machine Vision Conference, pp. 286–295 (1998)
12. Press, W., Flannery, B., Teukolsky, S., Vetterling, W.: Numerical Recipes in C. Cambridge University Press, Cambridge (1993)

An Efficient Algorithm for Optimal Multilevel Thresholding of Irregularly Sampled Histograms

Luis Rueda

Department of Computer Science, University of Windsor, 401 Sunset Ave.,
Windsor, ON, N9B 3P4, Canada
Phone: +1 519 253-3000 x 3002; Fax: +1 519 973-7093
lrueda@uwindsor.ca

Abstract. *Optimal* multilevel thresholding is a quite important problem in image segmentation and pattern recognition. Although efficient algorithms have been proposed recently, they do not address the issue of irregularly sampled histograms. A *polynomial-time* algorithm for multilevel thresholding of irregularly sampled histograms is proposed. The algorithm is polynomial not just on the number of bins of the histogram, n, but also on the number of thresholds, k, i.e. it runs in $\Theta(kn^2)$. The proposed algorithm is general enough for a wide range of thresholding and clustering criteria, and has the capability of dealing with irregularly sampled histograms. This implies important consequences on pattern recognition, since optimal clustering in the one-dimensional space can be obtained in polynomial time. Experiments on synthetic and real-life histograms show that for typical cases, the proposed algorithm can find the optimal thresholds in a fraction of a second.

1 Introduction

Multilevel thresholding is an important technique that has many applications in image processing, including segmentation, classification, clustering and object discrimination. Various parametric and non-parametric thresholding methods and criteria have been proposed, being the three most important streams the following: Otsu's method [8], which aims to maximize the separability of the classes measured by means of the sum of between-class variances, Kittler and Illingworth's criterion [4], which minimizes the error between a mixture of Gaussians and the actual histogram, and the criterion that uses information-theoretic measures to maximize the separability of the classes [3]. Variations of these criteria include a bi-level method that correlates the discrepancy between the original and thresholded image and the number of bits used to represent the segmented image [14], the use of an information-theoretic criterion based on Renyi's entropy [10], and a unified method introduced in [13]. Other criteria that have been proposed use spatial information about the pixels, also known as 2D thresholding [11,17]. All these criteria proposed efficient algorithms for bi-level thresholding.

For *optimal* multilevel thresholding, various algorithms have been proposed. A fast multilevel thresholding algorithm was proposed in [5], and efficient implementations of the three criteria were proposed in [12], in which a reduced number

N. da Vitora Lobo et al. (Eds.): SSPR&SPR 2008, LNCS 5342, pp. 602–611, 2008.

of subsets are evaluated. Other efficient, but yet *sub-optimal*, thresholding algorithms have been proposed and which use different kinds of heuristics, including a method based on genetic algorithms [15], a hybrid method that combines Gaussian curve fitting and particle swarm optimization [16], and a hybrid approach that combines particle swarm optimization and expectation maximization [2]. All these approaches suffer from the lack of either optimality or running time efficiency. Polynomial-time algorithms have been recently proposed , which solve the problem efficiently [6]. However, these algorithms are for specific thresholding methods and for regularly sampled histograms only (i.e., the bins are consecutive over a certain range). Some image segmentation problems contain sparse histograms, or in some cases only a few gray levels are available. These can be represented by means of irregularly sampled histograms, which do not necessarily have consecutive bins, and which can be sampled in other domains (e.g. real numbers). In this paper, a polynomial-time algorithm for optimal multilevel thresholding of irregularly sampled histograms is proposed. This algorithm has important applications and consequences in pattern recognition, since optimal clustering in the one-dimensional space can be efficiently obtained.

2 Exhaustive Search Multilevel Thresholding

Although various thresholding criteria and methods have been proposed, three main streams of thresholding criteria are the most-widely used: maximizing the sum of between-class variances [8], maximizing the sum of entropies of each individual class [3], and minimizing the error between the actual histogram and a mixture of Gaussians [4]. A general framework for the thresholding problem on irregularly sampled histograms is discussed here.

Definition 1 (Histogram). *An irregularly sampled histogram H is an ordered set*[1] $\{h_1, h_2, \ldots, h_{n-1}, h_n\}$, $h_1 < h_2 < \ldots < h_{n-1} < h_n$, *where h_i has a probability, p_i, associated with it.* □

If the probabilities are unknown, and given in terms of frequencies, they can be computed as follows. If $F = \{f_1, f_2, \ldots, f_n\}$ are the frequencies of a histogram H, $P = \{p_1, p_2, \ldots, p_n\}$ can be computed as $p_i = f_i/N$, where $N = \sum_{i=1}^{n} f_i$.

Definition 2 (Threshold set). *A threshold set T is defined as an ordered set $T = \{t_0, t_1, \ldots, t_k, t_{k+1}\}$, where $0 = t_0 < t_1 < \ldots < t_k < t_{k+1} = n$ and $t_i \in \{0, 1, \ldots, n\}$.* □

For an ordered subset of T, the notation $T_{i,j} = \{t_i, t_{i+1}, \ldots, t_j\}$ is used, where $i, j = 0, 1, \ldots, k, k+1$, $i < j$, and $T = T_{0,k+1}$. The problem of multilevel thresholding consists of finding a threshold set T^*, in such a way that a function $\Psi : \{1, 2, \ldots, n\}^k \times [0, 1]^n \to \mathbb{R}^+$ is maximized/minimized. Using this threshold set, H is divided into $k+1$ classes: $\zeta_1 = \{h_1, h_2, \ldots, h_{t_1}\}$, $\zeta_2 = \{h_{t_1+1}, h_{t_1+2}, \ldots, h_{t_2}\}$, $\ldots$, $\zeta_k = \{h_{t_{k-1}+1}, h_{t_{k-1}+2}, \ldots, h_{t_k}\}$, $\zeta_{k+1} = \{h_{t_k+1}, h_{t_k+2}, \ldots, h_n\}$.

[1] In a b-bit grayscale image, the corresponding histogram is $H = \{0, 1, \ldots, 2^b - 1\}$.

In the original works of Otsu [8], Kapur *et al.* [3], and Kittler-Illingworth [4], efficient algorithms for finding bi-level thresholding were proposed. For optimal multilevel thresholding, different ways of implementing exhaustive-search algorithms were proposed in [5,12], which search for combinations of $t_1 < t_2 < \ldots < t_k$. This implies a number of subsets reduced to $\binom{n}{k}$ different subsets, or equivalently $\Omega\left(\left(\frac{n}{k}\right)^k\right)$ subsets, and subsequently an *exponential* complexity on the number of thresholds.

3 A Dynamic Programming Scheme

Dynamic programming algorithms for *optimal* multilevel thresholding have been designed specifically for the between-class, minimum error and entropy-based criteria, for regularly sampled histograms [6]. A generalized algorithm and framework for a wide range of criteria and for irregularly sampled histograms are given here.

Definition 3 (function ψ). *Let $P = \{p_1, p_2, \ldots, p_n\}$ be the probabilities of a histogram $H = \{h_1, h_2, \ldots, h_n\}$. The function $\psi_{l,r}$, where $l \leq r$, is a real, positive function of $p_l, p_{l+1}, \ldots, p_r$, $\psi_{l,r} : \{1, 2, \ldots, n\}^2 \times [0,1]^{l-r+1} \to \mathbb{R}^+ \cup \{0\}$.* □

Either $\psi(l, l+1, \ldots, r, p_l, p_{l+1}, \ldots, p_r)$, $\psi(l, r, p_l, p_{l+1}, \ldots, p_r)$ or for short, $\psi_{l,r}$, will be used to denote the function ψ for $l, r, \{p_l, p_{l+1}, \ldots, p_r\}$. Also, if $r < l$, by definition $\psi_{l,r} \triangleq 0$, and if $r = l = 0$, $\psi_{0,0} \triangleq 0$.

Definition 4 (function Ψ). *Let $T = \{t_0, t_1, \ldots, t_k, t_{k+1}\}$ be a threshold set, where $k \geq 1$. The function Ψ is defined for all m, where $1 \leq m \leq k+1$, as a function $\Psi : \{1, 2, \ldots, n\}^m \times [0,1]^n \to \mathbb{R}^+$ as follows:*

$$\Psi(T_{0,m}) = \Psi(\{t_0, t_1, \ldots, t_m\}) \triangleq \sum_{j=1}^{m} \psi_{t_{j-1}+1, t_j} , \tag{1}$$

where $1 \leq m \leq k+1$. If $m = 0$, then $\Psi(\{t_0\}) = \psi_{t_0,t_0} = \psi_{0,0} = 0$. □

As a consequence of this, if $k = -1$, $\Psi(\{t_0\}) = \psi_{t_0,t_0} = 0$, and if $k = 0$, $\Psi = \Psi(\emptyset) = \psi_{1,n}$, where $\emptyset$ is the empty thresholding set.

The optimal solution for multilevel thresholding can be characterized in terms of solutions to sub-problems. For this, Ψ and $\psi_{l,r}$ must satisfy positivity, decomposition into sum of independent terms, and a third condition on computing $\psi_{l,r}$.

Condition 1. *For any probability set P and any threshold set T, $\Psi > 0$ and $\psi \geq 0$.* □

Condition 2. *For any integer $m = 1, 2, \ldots, k$, $\Psi(T_{0,m})$ can be expressed as follows:*

$$\Psi(T_{0,m}) = \Psi(\{t_0, t_1, \ldots, t_{m-1}\}) + \psi_{t_{m-1}+1, t_m} . \tag{2}$$

□

Condition 3. *If $\psi_{t_{j-1}+1,t_j}$ is known, then $\psi_{t_{j-1}+2,t_j}$ can be computed in $\Theta(1)$ time.* □

The polynomial-time algorithm for optimal thresholding resorts on characterizing the optimal solution in terms of the solutions to sub-problems. Suppose that the optimal solution for $T_{0,j-1} = \{t_0, t_1, \ldots, t_{j-1}\}$, $\Psi^*(T_{0,j-1})$, is known. Then, the optimal solution for $T_{0,j}$, $\Psi^*(T_{0,j})$, is computed as:

$$\Psi^*(T_{0,j}) = \begin{cases} 0 & \text{if } j = 0 \\ \max_{\min\{t_{j-1}\} \leq t_{j-1} \leq \max\{t_{j-1}\}} \Psi^*(T_{0,j-1}) + \psi_{t_{j-1}+1,t_j} & \text{if } 1 \leq j \leq k+1 \end{cases} \quad (3)$$

where

$$\min\{t_j\} = \begin{cases} j \text{ if } 0 \leq j \leq k \\ n \text{ if } j = k+1 \end{cases} \quad (4)$$

and

$$\max\{t_j\} = \begin{cases} 0 & \text{if } j = 0 \\ n - k + j - 1 & \text{if } 1 \leq j \leq k+1 \end{cases} \quad (5)$$

The optimal solution for T is obtained by computing $\Psi^*(T) = \Psi^*(T_{0,k+1})$, and taking $T^* = \arg\max_T \Psi^*(T)$. The optimality is stated in the following theorem, whose proof follows by contradiction. Note that the above characterization of the optimal solution can also be used for a *minimization* criterion by substituting "$\max_{\min\{t_{j-1}\} \leq t_{j-1} \leq \max\{t_{j-1}\}}$" for "$\min_{\min\{t_{j-1}\} \leq t_{j-1} \leq \max\{t_{j-1}\}}$" in (3).

Theorem 1. *For any value of $t_j = j, j+1, \ldots, n$, $\Psi^*(T_{0,j})$ computed as in (3) is optimal by knowing the optimal value of Ψ for $T_{0,j-1}$, $\Psi^*(T_{0,j-1})$, for $\min\{t_{j-1}\} \leq t_{j-1} \leq \max\{t_{j-1}\}$.* □

Due to the recursive nature of the optimal solution for $T_{0,j}$, it is not difficult to see that a recursive algorithm can be derived by considering the base case for $j = 0$ and the general case for values of $j \geq 1$. Such an algorithm would run in the worst case in $\tau(n, j) = \sum_{i=j}^{n-1} \tau(i, j-1) + \Omega(n)$ time, where i is for computing the maximum and $\Omega(n)$ is needed to compute $\psi_{t_{j-1}+1,t_j}$ in the worst case. However, Equation (1) suggests a dynamic programming solution. This algorithm can be implemented for *any* criterion[2], whenever the objective function satisfies the underlying definitions and conditions. We show here how it can be implemented for the between-class variance criterion. The between-class variance criterion, function $\Psi_{\mathrm{BC}}(T)$, can be expressed as follows:

$$\Psi_{\mathrm{BC}}(T) = \sum_{j=1}^{k+1} \omega_j \mu_j^2 = \sum_{j=1}^{k+1} \psi_{t_{j-1}+1,t_j}\,, \quad (6)$$

where $\psi_{t_j+1,t_{j+1}} = \omega_j \mu_j^2$, $\omega_j = \sum_{i=t_{j-1}+1}^{t_j} p_i$, and $\mu_j = \frac{1}{\omega_j} \sum_{i=t_{j-1}+1}^{t_j} h_i p_i$.

As seen below, it is not difficult to show that $\Psi_{\mathrm{BC}}(T)$ as defined in (6) satisfies Conditions 1 to 3.

[2] The implementations of the entropy-based and minimum error criteria are not difficult to be derived. For the interested reader, they can be found in [9].

Condition 1: For all i, $p_i \geq 0$, and at least one $p_j > 0$, yielding $\omega_j \geq 0$, and at least one $\omega_j > 0$. Also, since at least one $\mu_j^2 > 0$, then at least one $\psi_{t_j+1,t_{j+1}} = \omega_j \mu_j^2 > 0$. This implies from (6) that $\Psi_{\text{BC}}(T) > 0$. □

Condition 2: Assume that $\sum_{j=l}^{r} \omega_j \mu_j^2 = 0$ if $l > r$. Then, for $m = 1, \ldots, k$, $\Psi_{\text{BC}}(T_{0,m})$ can be expressed as follows:

$$\Psi_{\text{BC}}(T_{0,m}) = \left(\sum_{j=1}^{m-1} \omega_j \mu_j^2 \right) + \omega_m \mu_m^2 = \Psi_{\text{BC}}(\{t_0, t_1, \ldots, t_{m-1}\}) + \psi_{t_{m-1}+1,t_m} . \tag{7}$$

□

Condition 3: If $\psi_{t_{j-1}+1,t_j}$ is known, then $\psi_{t_{j-1}+2,t_j}$ can be computed in $\Theta(1)$ time. From (6), and using the definitions of ω_j and μ_j, $\psi_{t_{j-1}+1,t_j}$ can be expressed as follows:

$$\psi_{t_{j-1}+1,t_j} = \omega_j \mu_j^2 = \frac{1}{\sum_{i=t_{j-1}+1}^{t_j} p_i} \left[\sum_{i=t_{j-1}+1}^{t_j} h_i p_i \right]^2 \tag{8}$$

$$= \frac{1}{p_{t_{j-1}+1} + \sum_{i=t_{j-1}+2}^{t_j} p_i} \left[h_{t_{j-1}+1} p_{t_{j-1}+1} + \sum_{i=t_{j-1}+2}^{t_j} h_i p_i \right]^2 . \tag{9}$$

Let $a \leftarrow p_{t_{j-1}+1} + \sum_{i=t_{j-1}+2}^{t_j} p_i$, and $b \leftarrow h_{t_{j-1}+1} p_{t_{j-1}+1} + \sum_{i=t_{j-1}+2}^{t_j} h_i p_i$. Since a and b are known from the previous step, then a, b and $\psi_{t_{j-1}+2,t_j}$ can be computed for the current step in $\Theta(1)$ as: $a \leftarrow a - p_{t_{j-1}+1}$, $b \leftarrow b - h_{(t_{j-1}+1)} p_{t_{j-1}+1}$, and $\psi_{t_{j-1}+2,t_j} \leftarrow \frac{b^2}{a}$. □

The algorithm, depicted in Fig. 1, starts by setting the ranges for t_0, t_1, ..., t_k, t_{k+1} as in (4) and (5), which is done by invoking findThresholdRanges(k). It uses a table C to store the solution to smaller values of k, and the table is incrementally filled by column, for $k = 1, 2, 3$, and so on. Thus, $C(t_j, j)$ contains the optimal solution for $T_{0,j} = t_0, t_1, \ldots, t_j$, $\Psi^*(T_{0,j})$, for $\min\{t_j\} \leq t_j \leq \max\{t_j\}$. The cell at position $(0, 0)$ is set to 0, while the cell at position $(n, k+1)$ contains the optimal value of $\Psi^*(T)$. As stated in Condition 3, ψ_{i+1,t_j} is computed in $\Theta(1)$ by knowing the value of ψ_{i,t_j}, and stored in variable "psi". The index 'i' in the inner for loop is used to represent t_{j-1}. Similarly, table D stores the thresholds for all the optimal solutions. Thus, $D(t_j, j)$ contains the value of t_{j-1} for which $\Psi^*(T_{0,j})$ is optimal. A detailed example can be found in [9]. The worst-case time complexity of Algorithm *Multilevel_Thresholding* is $\Theta(kn^2)$. For the between-class variance and minimum error criteria, it has been shown that the worst-case time complexity can be even reduced to $\Theta(kn)$, which implies linear time on the number of bins, and this, assuming regularly sampled histograms. For the entropy-based criterion, the worst-case time complexity is still $\Theta(kn^2)$.

Algorithm *Multilevel_Thresholding*
Input: A histogram, $H = \{h_1, \ldots, h_n\}$ with $P = \{p_1, p_2, \ldots, p_n\}$. Number of thresholds, k.
Output: A threshold set, $T = \{t_0, t_1, t_2, \ldots, t_k, t_{k+1}\}$.

```
minTj, maxTj ← findThresholdRanges(k)
C(0,0) ← 0; D(0,0) ← 0
for j ← 1 to k+1 do
  for t_j ← minTj(j) to maxTj(j) do
    C(t_j, j) ← 0;  psi ← ψ_{j,t_j}
    for i ← minTj(j−1) to min{maxTj(j−1), t_j − 1}  do
      if C(i, j−1) + psi > C(t_j, j) then
        C(t_j, j) ← C(i, j−1) + psi; D(t_j, j) ← i
      end if
      psi ← Compute ψ_{i+2,t_j} from psi, h_{i+1} and p_{i+1}
    end for
  end for
end for
return findThresholds(D)
procedure findThresholdRanges(k: integer)
for j ← 0 to k+1 do
  if j = k+1 then minTj(j) ← n else minTj(j) ← j end if
  if j = 0 then maxTj(j) ← 0 else maxTj(j) ← n − k + j − 1 end if
end for
return minTj, maxTj
end procedure
procedure findThresholds(D: table)
T(k+1) ← n
for j ← k downto 0 do
  T(j) ← D(T(j+1), j+1)
end for
return T
end procedure
```

Fig. 1. General algorithm for multilevel thresholding based on dynamic programming

4 Applications

The salient features of the proposed algorithm and general framework for optimal multilevel thresholding is that it can be applied to any criterion and irregularly sampled histograms. This provides a much more general framework than the previous approaches and algorithms. The implications of this generalization are in various aspects and applications of pattern recognition and image segmentation, and also, in general, in signal processing. Three of them are discussed below.

Segmentation of images using multilevel thresholding, in some cases, involves sparse histograms (i.e., histograms that with nonzero probabilities for a reduced number of bins, and spread out over a large domain). This is a typical case in segmentation of images representing the biofilm physical structure, and which are extracted from three-dimensional confocal laser scanning microscopy [1].

These images have a very high resolution and lead to sparse histograms on large domains.

Another important application of the proposed framework is to use it to finding the optimal number of clusters or classes, and still maintaining the optimality of the clustering and/or thresholding. Many indices of validity have been proposed, namely the I index, Davies Bouldin index, Dunns index, and the Xie-Beni index, among others [7]. It is not difficult to see that most of these indices satisfy the three conditions and can be used (i) to find the best number of clusters or classes, and (ii) as multilevel thresholding criteria on irregularly sampled histograms.

Lastly, an extremely important consequence of the proposed framework and algorithm in pattern recognition is in clustering. By modeling a one-dimensional clustering problem as an irregularly sampled histogram, the optimal clustering and optimal number of clusters can be obtained as follows. Assume that a given dataset, $D = \{x_1, x_2, \ldots, x_n\}$, is to be clustered, where $x_i \in \mathbb{R}$. By sorting D in increasing order, obtaining a new dataset, say for convenience in notation, $H = \{h_1, h_2, \ldots, h_n\}$, and assigning each probability $p_i = 1/n$, it is straightforward to see that the proposed framework and algorithm can optimally cluster D using any criteria, and even find the best (optimal, based on a certain criterion) number of clusters.

5 Experimental Results

A few experiments were performed on randomly generated histograms and on multilevel thresholding segmentation of real-life images. For the ease of notation, MTBC is used to refer to the implementation of the between-class variance criterion. The first set of experiments was performed on randomly generated histograms of various sizes and thresholded for various levels (values of k). The algorithm was implemented in Matlab and the experiments were run on a PC workstation at 3.4Ghz with 1GB of RAM. Nine different values of n, $n = 256, 512, \ldots, 65536$. For each value of n, nine different values of k were considered: $k = 2, 3, \ldots, 10$. Thus, for each value of n, nine different histograms were generated by a using a distribution that obeys a mixture of $k + 1$ Gaussians, to which white noise with a $N(0, 0.005)$ was incorporated in order to make it as real as possible. One such histogram generated with a mixture of three Gaussians ($n = 256$ and $k = 2$) is depicted in Fig. 2. The two optimal threshold levels were obtained by MTBC using only 0.005 seconds of CPU time.

The CPU time (in seconds) for all values of n and k are listed in Table 1. It can be observed that increasing n, the time taken by MTBC increases but polynomially. That is, for $2n$ bins, it takes approximately 3-4 times more than the time taken for n bins. This behavior is followed in general for all values of k. On the other hand, for a fixed value of n, increasing the value of k, again, increases the number of CPU seconds but linearly. It is interesting to see that for $n = 65536$ bins (or gray levels in image segmentation), MTBC can find $k = 10$ thresholds in only 2117 seconds, or approximately 35 minutes. Note that

Table 1. CPU times (in seconds) taken by MTBC to find k thresholds, $k = 2, 3, \ldots, 10$ on randomly generated histograms of sizes $n = 2^b$, where $b = 8, 9, \ldots, 16$

$n_\downarrow$ $k^\rightarrow$	2	3	4	5	6	7	8	9	10
256	0.005	0.007	0.011	0.015	0.020	0.022	0.025	0.029	0.033
512	0.014	0.029	0.043	0.056	0.069	0.085	0.099	0.111	0.126
1024	0.052	0.106	0.164	0.220	0.277	0.331	0.394	0.483	0.508
2048	0.214	0.419	0.655	0.873	1.112	1.340	1.583	1.815	2.064
4096	0.810	1.672	2.510	3.438	4.521	5.341	6.271	7.261	8.147
8192	3.312	6.878	10.152	19.283	17.742	20.966	25.013	29.195	32.668
16384	13.324	26.734	41.047	56.019	69.896	89.723	98.922	116.602	130.022
32768	51.376	105.440	159.379	231.875	287.749	351.783	404.136	469.089	527.862
65536	212.084	429.374	647.685	913.694	1167.983	1400.655	1628.987	1874.508	2116.873

an exhaustive-search multilevel thresholding algorithm (cf. [12]) would need to test $\frac{65536!}{65526!10!} \approx 4.02 \times 10^{41}$ different subsets of thresholds, which implies a huge number of computations. A practical comparison was done with an exhaustive search algorithm on finding 2 and 3 thresholds on a histogram of 2048 bins and took 0.616 and 185.967 seconds for $k = 2$ and 3 respectively, compared with the fractions of seconds taken by MTBC.

Other experiments were conducted with Lena and Peppers, for which the original images are depicted in Figs. 3 (a) and (d). The images segmented using the thresholds found by MTBC are depicted in Figs. 3 (b) and (e). The histogram was constructed for each of these images, and the optimal thresholds were found by using MTBC, where $k = 5, 4$ respectively. The histograms along with the optimal thresholds are depicted in Figs. 3 (c) and (f). All these images are 8-bit graylevel, and the segmentation of the images by computing the optimal thresholds were performed in 0.0224 and 0.0194 CPU seconds respectively. For all the images, an appropriate number of thresholds was found manually, and the histograms reported correspond to the best number of thresholds based on visually analyzing the histograms. However, it is not difficult to see that finding the best number of thresholds automatically is not a difficult task, as it would

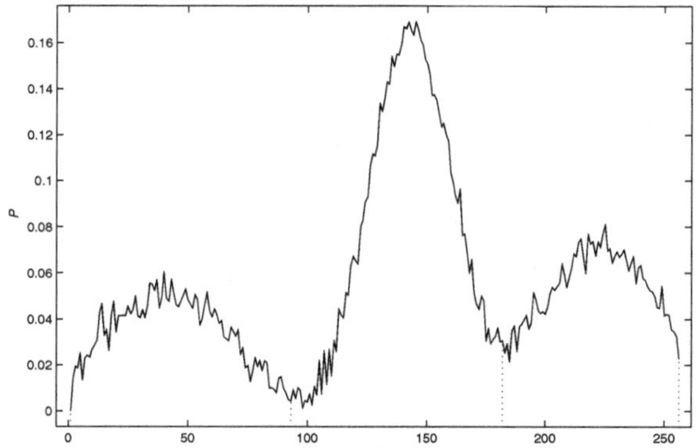

Fig. 2. An example of a randomly generated histogram using a mixture of three Gaussians. The two optimal thresholds were detected using MTBC.

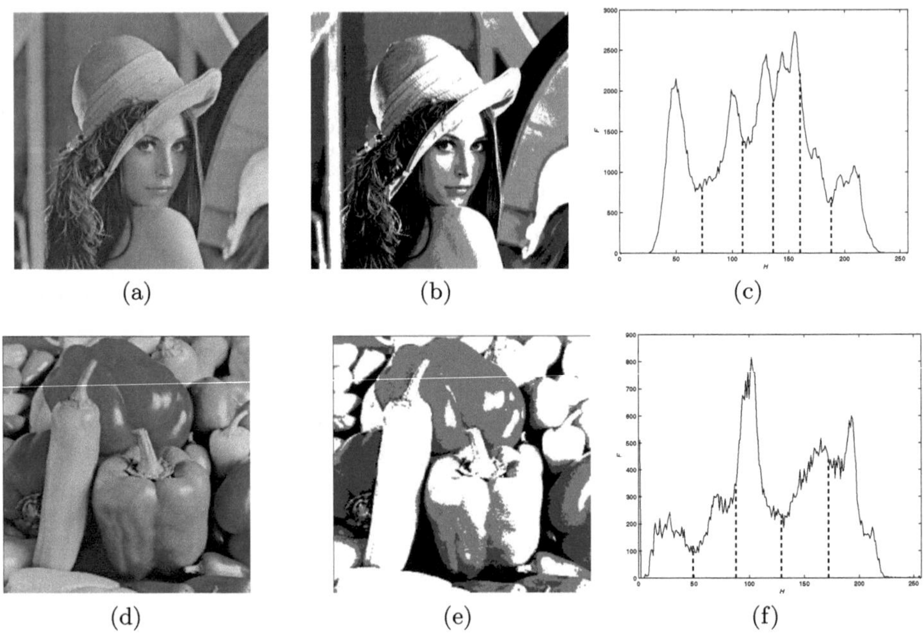

Fig. 3. Lena and Peppers segmented with the MTBC algorithm and the corresponding histograms with the optimal thresholds

imply to test an index of cluster validity after each column of table C is completed – thus, the complexity would remain the same.

6 Conclusions

An efficient algorithm for optimal multilevel thresholding of irregularly sampled histograms is proposed, which runs in polynomial time. A general algorithm based on dynamic programming is discussed, for which the optimality of the solution is also stated. Experiments have been carried out on randomly generated histograms and on segmentation of real-life grayscale images showing the efficiency of the algorithm. Applications of the proposed framework to different pattern recognition problems have been discussed, being the most important optimal one-dimensional clustering in polynomial time. As a consequence of this, clustering and thresholding problems for more than one dimension are topics that deserve further investigation.

References

1. Beyenal, H., Donovan, C., Lewandowski, Z., Harkin, G.: Three-dimensional Biofilm Structure Quantification. Journal of Microbiological Methods 59(3), 395–413 (2004)
2. Fan, S., Lin, Y.: A Multi-level Thresholding Approach Using a Hybrid Optimal Estimation Algorithm. Pattern Recognition Letters 28, 662–669 (2007)

3. Kapur, J., Sahoo, P., Wong, A.: A New Method for Gray-level Picture Thresholding Using the Entropy of the Histogram. Computer Vision Graphics and Image Processing 29, 273–285 (1985)
4. Kittler, J., Illingworth, J.: Minimum Error Thresholding. Pattern Recognition 19(1), 41–47 (1986)
5. Liao, P., Chen, T., Chung, P.: A Fast Algorithm for Multilevel Thresholding. Journal of Information Science and Engineering 17, 713–727 (2001)
6. Luessi, M., Eichmann, M., Shuster, M., Katsaggelos, A.: New results on efficient optimal multilevel image thresholding. In: Proc. of the IEEE International Conference on Image Processing, Atlanta, USA, pp. 773–776. IEEE Press, Los Alamitos (2006)
7. Maulik, U., Bandyopadhyay, S.: Performance Evaluation of Some Clustering Algorithms and Validity Indices. IEEE Trans. on Pattern Analysis and Machine Intelligence 24(12), 1650–1655 (2002)
8. Otsu, N.: A Threshold Selection Method from Gray-level Histograms. IEEE Trans. on Systems, Man and Cybernetics SMC-9, 62–66 (1979)
9. Rueda, L.: A Polynomial-time Algorithm for Optimal Multilevel Thresholding (submitted, 2008), `http://www.cs.uwindsor.ca/~lrueda/papers/PolyThresJnl.pdf`
10. Sahoo, P., Wilkins, C., Yeager, J.: Threshold Selection Using Renyi's Entropy. Pattern Recognition 30(1), 71–84 (1997)
11. Wang, S., Chung, F., Xiong, F.: A Novel Image Thresholding Method Based on Parzen Window Estimate. Pattern Recognition 41(1), 117–129 (2008)
12. Wu, B., Chen, Y., Chiu, C.: Efficient Implementation of Several Multilevel Thresholding Algorithms Using a Combinatorial Scheme. International Journal of Computers and Applications 28(3), 259–269 (2006)
13. Yan, H.: Unified Formulation of a Class of Optimal Image Thresholding Techniques. Pattern Recognition 29(12), 2025–2032 (1996)
14. Yen, J., Chang, F., Chang, S.: A New Criterion for Automatic Multilevel Thresholding. IEEE Trans. on Image Processing 4(3), 370–378 (1995)
15. Yin, P.: A Fast Scheme for Optimal Thresholding Using Genetic Algorithms. Signal Processing 72, 85–95 (1999)
16. Zahara, E., Fan, S., Tsai, D.: Optimal Multi-thresholding Using a Hybrid Optimization Approach. Pattern Recognition Letters 26, 1082–1095 (2005)
17. Zhou, Y., Yang, K.: New 2D Adaptive Image Thresholding Method Based on Within and Between Cluster Scatter. In: Proc. of the 27th International Congress on High-Speed Photography and Photonics, SPIE, vol. 6279, p. 62793N (2007)

A Family of Cluster Validity Indexes Based on a l-Order Fuzzy OR Operator

Hoel Le Capitaine and Carl Frélicot

MIA Laboratory, University of La Rochelle, La Rochelle 17042 Cedex, France
hoel.le_capitaine@univ-lr.fr, carl.frelicot@univ-lr.fr

Abstract. Clustering is one of the most important task in pattern recognition. For most of partitional clustering algorithms, a partition that represents as much as possible the structure of the data is generated. In this paper, we adress the problem of finding the optimal number of clusters from data. This can be done by introducing an index which evaluates the validity of the generated fuzzy c-partition. We propose to use a criterion based on the fuzzy combination of membership values which quantifies the l-order overlap and the intercluster separation of a given pattern.

1 Introduction

The objective of fuzzy clustering is to partition the data set into c distinct clusters. The fuzzy c-means (FCM) algorithm proposed by Bezdek [3] and its variations [8] are probably the most commonly used fuzzy clustering methods. However, these algorithms require the user to set the number c of clusters although the user do not always know it. A fuzzy c-partition obtained by FCM has to be validated because its quality depends on this number. Many cluster validity indexes have been proposed that evaluate each fuzzy c-partition and determines the optimal number $c^\star$ of clusters allowing to obtain the optimal partition of the data.

Compactness and separation of clusters are often considered to validate a partition [4,12,7]. In this paper, we propose a new family of indexes that combine an overlap measure and a separation measure both based on aggregation of membership values. Section 2 provides background information on basic fuzzy operators and the l-order fuzzy OR operator our work is based on. Next, in section 3, we briefly describe the FCM algorithm and recall some well-known cluster validity indexes we will use for comparison. The measures we propose to use and the definition of the new index are described in section 4. Experimental results on both synthetic and real data sets that show its efficiency and concluding remarks are given in section 5 and 6 respectively.

2 Mathematical Background

For the applications we have in mind, e.g. cluster validity, we are particulary interested in aggregation functions that map a collection $\mathbf{u}$ of c values in $[0, 1]$

N. da Vitora Lobo et al. (Eds.): SSPR&SPR 2008, LNCS 5342, pp. 612–621, 2008.

Table 1. Basic t-norms and t-conorms couples

Standard	$u_1 \top_S u_2 = \min(u_1, u_2)$
	$u_1 \bot_S u_2 = \max(u_1, u_2)$
Algebraic	$u_1 \top_A u_2 = u_1\, u_2$
	$u_1 \bot_A u_2 = u_1 + u_2 - u_1\, u_2$
Łukasiewicz	$u_1 \top_L u_2 = \max(u_1 + u_2 - 1, 0)$
	$u_1 \bot_L u_2 = \min(u_1 + u_2, 1)$

(i.e. a vector $\mathbf{u} = [u_1, \cdots, u_c]^T$) to a value in $[0, 1]$. Among the frequently used aggregation operators, one can cite the classes of triangular norms (t-norms) and triangular conorms (t-conorms). They have been introduced to characterize the general multivalued logic *AND* and *OR* operations and are widely used in fuzzy logic and fuzzy set theory to implement conjunctive and disjunctive operator respectively. A triangular norm is a commutative, associative and monotone function $\top$ having for neutral element 1. Alternatively, its dual operator, the triangular conorm, is a commutative, associative and monotone function $\bot$ having for neutral element 0. Examples of triangular norms couples are given in Table 1 for two operands, see [10] for a large survey. Note that various parameterized families have been introduced, e.g. the Hamacher family defined by $u_1 \top_H u_2 = \frac{u_1\, u_2}{\gamma+(1-\gamma)\,(u_1+u_2-u_1\, u_2)}$ and $u_1 \bot_H u_2 = \frac{u_1+u_2-u_1\, u_2-(1-\gamma)\, u_1\, u_2}{1-(1-\gamma)\, u_1\, u_2}$ where $\gamma \in [0, +\infty[$. The dual couple is generally associated with a fuzzy negation defined as $N(u_1) = 1 - u_1$ and mentionned as the triple $(\top, \bot, N)$.

Assume that the values u_i $(i = 1, \cdots, c)$ to be aggregated represent the degree to which an object $\mathbf{x}$ satisfies each group description, i.e. its similarity to the prototypes describing each group. Using this knowledge contained in $\mathbf{u}$, clustering consists in selecting the most appropriate group that the objects will be assigned to. The maximum operator is commonly used in this situation, but we may be interested in the lower values, which interact with the greatest value. In particular, if an object satisfies more than one group description, such an exclusive partitioning is not efficient. A fundamental issue becomes the determination of the overall degree of exclusive belongingness to a group or cluster. In [11], the authors define the l-order fuzzy OR operator. This operator evaluates degrees of satisfaction at a given order by combination of triangular norms. Let $\mathcal{P}$ be the powerset of $C = \{1, 2, ..., c\}$ and $\mathcal{P}_l = \{A \in \mathcal{P} : |A| = l\}$ where $|A|$ denotes the cardinality of subset A, then the fOR-l is defined by:

$$\overset{l}{\underset{i=1,\cdots,c}{\bot}} u_i = \underset{A\in\mathcal{P}_{l-1}}{\top} \left(\underset{j\in C\setminus A}{\bot} u_j \right) \tag{1}$$

It must be viewed as some kind of generalization of the notion of "l^{th} highest" value, with l in C. In particular, with standard triangular norms, $\overset{l}{\bot}(\mathbf{u})$ is exactly the "l^{th} highest" element of $\mathbf{u}$. For instance, let us take $C = \{1, 2, 3\}$, $l = 2$ and use standard triangular norms. We have $\mathcal{P}_1 = \{\{1\}, \{2\}, \{3\}\}$ and

$\overset{2}{\underset{i=1,\cdots,3}{\perp}} u_i = \min(\max(u_2, u_3) \max(u_1, u_3) \max(u_1, u_2))$ so that if $u_2 < u_1 < u_3$, $\overset{2}{\underset{i=1,\cdots,3}{\perp}} u_i = u_1$. This operator satisfies nice mathematical properties such as monotony, symmetry, see [11] for proofs and details.

3 Cluster Validity for Fuzzy Clustering

3.1 Fuzzy c-Means Algorithm

Clustering is an instance of unsupervised classification which aims at finding a structure of groups in set of n p-dimensional patterns $X = \{\mathbf{x}_1, ..., \mathbf{x}_n\}$. In this framework, the label vectors $\mathbf{u}_k = u(\mathbf{x}_k)$ do not exist and clustering algorithms can be used to obtain them from X. For instance, the *fuzzy c-means* (FCM) algorithm partitions X into $c > 1$ clusters by minimizing the following objective function [3]:

$$J_m(U, V) = \sum_{k=1}^{n} \sum_{i=1}^{c} u_{ik}^m \, ||\mathbf{x}_k - \mathbf{v}_i||^2 \tag{2}$$

where u_{ik} is the membership degree of $\mathbf{x}_k$ to the ith cluster represented by its centroid $\mathbf{v}_i \in \mathbb{R}^p$. Centroids are gathered into a $(c \times p)$ matrix $V = [\mathbf{v}_1, ..., \mathbf{v}_c]$. Degrees u_{ik} are subject to $\sum_{i=1}^{c} u_{ik} = 1$ for all $\mathbf{x}_k$ in X and to $0 < \sum_{k=1}^{n} u_{ik} < n$ $(\forall i = 1, \cdots, c)$. In addition, they are elements of the fuzzy $c-$partition matrix $U = [\mathbf{u}_1, \cdots, \mathbf{u}_n]$ of size $c \times n$. The so-called *fuzzifier* $m > 1$ is a weighting exponent which makes the resulting partition more or less fuzzy [12]. The higher m is, the softer the clusters' boundaries are. Minimization of (2) is obtained by iteratively updating (U, V) as follows:

$$u_{ik} = 1 \Bigg/ \sum_{j=1}^{c} \left(\frac{||\mathbf{x}_k - \mathbf{v}_i||}{||\mathbf{x}_k - \mathbf{v}_j||} \right)^{2/(m-1)} \tag{3}$$

$$\mathbf{v}_i = \frac{\sum_{k=1}^{n} u_{ik}^m \, \mathbf{x}_k}{\sum_{k=1}^{n} u_{ik}^m} \tag{4}$$

The usual euclidian norm $||.||$ induces hyperspherical clusters, hence FCM can only detect clusters with the same shape and orientation.

3.2 Classical Indexes

Validating the provided clustering of X consists in assessing whether the resulting partition reflects the data structure or not. Since c is a user-defined parameter of clustering algorithms such as FCM, most of works on cluster validity focus on the number of clusters problem. Many validity indexes have been proposed for fuzzy clustering (refer to [4,9,14,13,16] for comparative studies). They can be classified in two main categories. The first one is composed of indexes that only use membership degrees (U). Let us cite the *Partition Coefficient* [3], taking values in $[\frac{1}{c}, 1]$:

$$PC(c) = \frac{1}{n} \sum_{k=1}^{n} \sum_{i=1}^{c} u_{ik}^2 \tag{5}$$

or the *Partition Entropy* [2], taking values in $[0, \log(c)]$:

$$PE(c) = -\frac{1}{n} \sum_{k=1}^{n} \sum_{i=1}^{c} u_{ik}\, log(u_{ik}) \tag{6}$$

Both PC to be maximized and PE to be minimized are monotonic with c, as well as their bounds. Normalized versions have been proposed to reduce this monotonic tendency, *e.g.* in [6]. We will use these normalized versions in the experiments: $NPC(c) = \frac{cPC(c)-1}{c-1}$ and $NPE(c) = \frac{nPE(c)}{n-c}$. The second category consists of indexes that use membership degrees but also some information about the geometrical structure of the data (U, V, X), *e.g.* the Xie-Beni index [12,15]:

$$XB(c) = \frac{J_m(U, V)\,/n}{\min_{i,j=1,\cdots,c;j\neq i}\, ||\mathbf{v}_i - \mathbf{v}_j||^2} \tag{7}$$

or the Fukuyama-Sugeno index [7]:

$$FS(c) = J_m(U, V) - \sum_{k=1}^{n} \sum_{i=1}^{c} u_{ik}^m\, ||\mathbf{v}_i - \overline{\mathbf{v}}||^2 \tag{8}$$

where $\overline{\mathbf{v}}$ is the mean of centroids. Both XB and FS combine the FCM objective function (2) which measures the degree of compactness of the clusters and an additional term which measures the degree of their separation. Combination indicates that both indexes are to be minimized. The more compact and separated the clusters are, the more optimal c is.

4 The Proposed Index

Let $U = [\mathbf{u}_1, ..., \mathbf{u}_n]$ be a fuzzy c-partition provided by a fuzzy clustering algorithm, e.g. FCM. An overlap measure between l fuzzy clusters for each point $\mathbf{x}_k$ in X described by its membership degrees can be obtained by Eq. (1). By successively computing $\overset{l}{\perp}(\mathbf{u}_k)$ for different values of c, we get a combination of l-order overlap measure for $\mathbf{x}_k$. The most satisfied order is obtained by the disjunction of these measures, and we define the overall overlap measure as:

$$O(\mathbf{u}_k, c) = \overset{1}{\underset{l=2,c}{\perp}} \left(\overset{l}{\underset{i=1,\cdots,c}{\perp}} u_{ik} \right) \tag{9}$$

In [4], Bezdek and Pal show that intercluster separation plays a more important role in cluster validity than cluster diameters. We propose to introduce such a measure by quantifying the separation of each point $\mathbf{x}_k$ with $\overset{1}{\perp}(\mathbf{u}_k)$ which is the overlap measure within one fuzzy cluster, i.e. its separation from the other fuzzy clusters, since $\mathbf{u}_k$ sums up to one. Note that if standard triangular norms are used, $\overset{1}{\perp}(\mathbf{u}_k)$ is the maximum coordinate of $\mathbf{u}_k$. Finally, we define the family of l-order Fuzzy OR Indexes as:

$$lFORI(c, \top, \perp) = \frac{1}{n} \sum_{k=1}^{n} \frac{O(\mathbf{u}_k, c)}{\overset{1}{\underset{i=1,\cdots,c}{\perp}} u_{ik}} \tag{10}$$

Given U, the less overlaping and separated clusters are, the lower the value of $lFORI$ is expected to be and minimizing Eq. (10) will give the optimal number $c^\star$ of clusters.

If U is hard, i.e. $u_{ik} \in \{0, 1\}$, then one value equals 1 while the others are 0, say $u_{1k} = 1$ and $u_{2k} = \cdots = u_{ck} = 0$. Since 0 is the absorbing element of $\top$, it is easy to verify that $O(\mathbf{u}_k, c) = 0$ for all $\mathbf{u}_k$, whatever $(\top, \perp)$, therefore $lFORI(c, \top, \perp) = 0$:

$$\overset{1}{\underset{l=2,c}{\perp}} \left(\overset{l}{\underset{i=1,\cdots,c}{\perp}} u_{ik} \right) = \left(\overset{2}{\underset{i=1,\cdots,c}{\perp}} (1, 0, \cdots, 0) \right) \overset{1}{\perp} \cdots \overset{1}{\perp} \left(\overset{c}{\underset{i=1,\cdots,c}{\perp}} (1, 0, \cdots, 0) \right) \tag{11}$$

$$= \overset{1}{\perp} (\underbrace{0, \cdots, 0}_{c-1 \text{ times}}) = 0. \tag{12}$$

and $\overset{1}{\perp}(1, 0, \cdots, 0) = 1$, since 1 is the absorbing element of $\perp$.

On the other hand, if U is totally fuzzy, i.e. $u_{ik} = \frac{1}{c}$ $(\forall i = 1, \cdots, c)$, the resulting $lFORI$ value depends on the couple $(\top, \perp)$ because Eq. (9) only reduces to

$$\overset{1}{\underset{l=2,c}{\perp}} \left(\overset{l}{\underset{i=1,\cdots,c}{\perp}} u_{ik} \right) = \left(\overset{2}{\underset{i=1,\cdots,c}{\perp}} \left(\frac{1}{c}, \cdots, \frac{1}{c} \right) \right) \overset{1}{\perp} \cdots \overset{1}{\perp} \left(\overset{c}{\underset{i=1,\cdots,c}{\perp}} \left(\frac{1}{c}, \cdots, \frac{1}{c} \right) \right) \tag{13}$$

in the general case. If standard t-norms are used, we have:

$$\overset{1}{\underset{l=2,c}{\perp}} \left(\overset{l}{\underset{i=1,\cdots,c}{\perp}} u_{ik} \right) = \overset{1}{\perp} \left(\underbrace{\frac{1}{c}, \cdots, \frac{1}{c}}_{c-1 \text{ times}} \right) \tag{14}$$

$$= \frac{1}{c} \tag{15}$$

which is the value of $\overset{1}{\perp}(\mathbf{u}_k)$ for all $\mathbf{u}_k$, therefore $lFORI(c, \top_S, \perp_S) = 1$ for all c. Unfortunately, it is not possible to give a simple value for $lFORI$ for other couples $(\top, \perp)$, but an upper bound can be found. Due to lack of space, the proof is postponed to a forthcoming long paper, as well as properties that could help the user to choose the couple $(\top, \perp)$.

5 Experiments

5.1 Behavior According to Clusters' Separability

First, we generated a serie of 10 data sets, each composed of 800 points drawn from a mixture of $c = 4$ bivariate normal distributions. The covariance matrix of each component is the same $\Sigma_i = I$ ($\forall i = 1, \cdots, c$) and the mean vectors are: $\mu_1 = \alpha \begin{pmatrix} 1 \\ 1 \end{pmatrix}$, $\mu_2 = \alpha \begin{pmatrix} 1 \\ -1 \end{pmatrix}$, $\mu_3 = \alpha \begin{pmatrix} -1 \\ -1 \end{pmatrix}$ and $\mu_4 = \alpha \begin{pmatrix} -1 \\ 1 \end{pmatrix}$, for increasing values of $\alpha = 1, 2, \ldots, 10$. This successively moves the clusters in opposite directions, creating less overlap as the clusters become more and more separated. The first and last data sets are shown in Figure 1-*left*. Each data set was then clustered using FCM with $c = 4$, providing a fuzzy partition matrix U_α. Corresponding values of $lFORI$ for the different basic norms are plotted in Figure 1-*right* as a function of α. As expected, the proposed validity index decreases towards 0 as α increases whatever the couple $(\top, \bot)$.

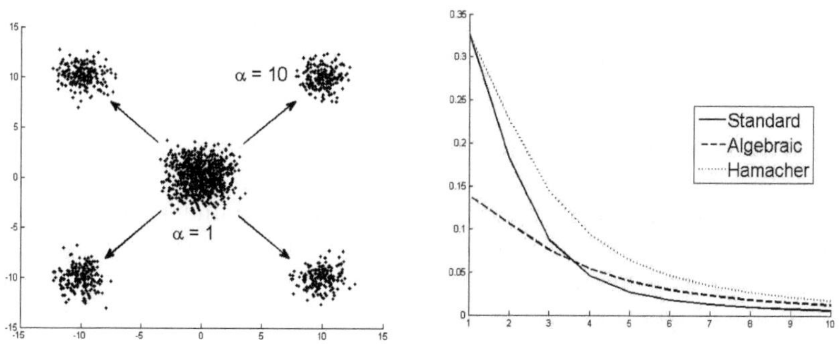

Fig. 1. α-separated data sets – $\alpha = 1$ and 10 (*left*) and values of $lFORI(c = 4)$ as a function of α for various t-norms (*right*)

5.2 Robustness to Outliers

In order to compare the proposed index to the classical ones, we generated a data set D_1 containing $n = 200$ points consisting of 50 points each drawn from a mixture of $c = 4$ bivariate normal distributions, see Figure 2-*left*. FCM was used with $m = 2$ for c varying from $c_{min} = 2$ to $c_{max} = 10$. A second artificial data set D_2 was generated. It is similar to D_1 except that 100 points drawn from a uniform distribution were added, as shown in Figure 2-*right*. These additional points act as noise and can make the FCM algorithm partitioning the data set into three clusters because the less separated groups in the right-lower area tend to become only one cluster. Values of the tested validity indexes on D_1 and D_2 are given in Table 2 and Table 3 respectively, where optimal values are bold faced and acceptable ones are italicized. Even if the classical indexes are known to be efficient, most of them fail in giving the right number of clusters on D_1 and with even stronger reason in presence of noise (D_2), whereas $lFORI$ always gives the

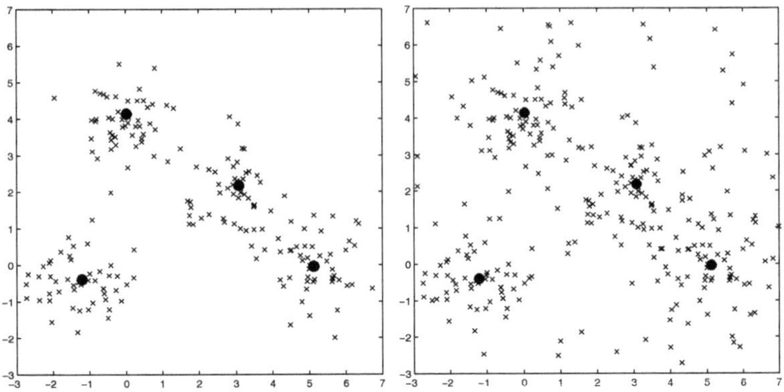

Fig. 2. Centroids of optimal clusters for data sets D_1 (*left*) and D_2 (*right*)

right number $c^\star = 4$ whatever the couple $(\top, \bot)$ for both data sets. Moreover, multiple runs of FCM with random initializations on data set D_2 showed us that the proposed index gives more stable results compared to the others, thus demonstrating a significantly higher robustness to noisy data.

5.3 Sensitivity to the Fuzzifier m

The FCM objective function depends on the fuzzy exponent m, see Eq. (2). Since the resulting fuzzy c-partition U is sensitive to this parameter, a cluster validity index can also be analysed with respect to m. In [4], Pal and Bezdek have shown that FCM provides best results for m lying in $[1.5, 2.5]$. We compare the different indexes in this range on data sets D_1 and D_2. Again, c is varying from $c_{min} = 2$ to $c_{max} = 10$. The selected number of clusters are reported in Table 4. As it can be seen, the proposed index $lFORI$ is less sensitive to m whatever the couple $(\top, \bot)$ for both data sets.

Table 2. Validity indexes on data set D_1

c	NPC	NPE	XB	FS $\times 10^{-3}$	$lFORI$ $(\top,\bot)_S$	$(\top,\bot)_A$	$(\top,\bot)_H$
2	0.57	**0.52**	0.13	-1.57	0.17	0.14	0.15
3	0.70	0.56	**0.07**	-3.39	0.12	0.12	0.14
4	**0.74**	0.59	*0.08*	-0.74	**0.07**	**0.07**	**0.12**
5	0.67	0.80	0.31	**-4.25**	0.12	0.11	0.17
6	0.61	1.00	0.45	-2.98	0.18	0.13	0.18
7	0.56	1.15	0.37	-1.53	0.19	0.10	0.20
8	0.530	1.27	0.30	-1.60	0.23	0.09	0.19
9	0.532	1.37	0.31	-2.43	0.29	0.10	0.19
10	0.50	1.47	0.28	-0.15	0.28	0.11	0.17

Table 3. Validity indexes on data set D_2

c	NPC	NPE	XB	FS $\times 10^{-3}$	$lFORI$ $(\top,\bot)_S$	$(\top,\bot)_A$	$(\top,\bot)_H$
2	0.49	**0.58**	0.23	-0.85	0.25	0.17	0.19
3	**0.60**	0.71	**0.11**	-2.33	0.23	0.11	0.20
4	*0.59*	0.87	0.13	-0.75	**0.17**	**0.09**	**0.15**
5	0.55	1.08	0.24	**-5.21**	0.18	0.13	0.21
6	0.50	1.26	0.35	-0.06	0.21	0.14	0.22
7	0.50	1.36	0.36	-0.12	0.21	0.17	0.22
8	0.46	1.50	0.44	-2.41	0.26	0.16	0.21
9	0.45	1.61	0.46	-2.61	0.30	0.16	0.19
10	0.44	1.70	0.36	-1.81	0.30	0.15	0.18

Table 4. Selected number of clusters in data sets D_1 and D_2 for different values of m

data set	m	NPC	NPE	XB	FS	$lFORI$ $(\top,\bot)_S$	$(\top,\bot)_A$	$(\top,\bot)_H$
D_1	1.5	4	4	3	5	4	4	4
	1.7	4	4	4	4	4	4	4
	1.9	4	2	3	4	4	4	4
	2.1	4	2	3	5	4	4	4
	2.3	4	2	3	5	4	4	4
	2.5	3	2	4	6	4	3	4
D_2	1.5	5	4	3	5	4	4	4
	1.7	4	4	3	3	4	4	4
	1.9	4	2	3	4	4	4	4
	2.1	3	2	3	5	4	4	4
	2.3	4	2	3	3	4	3	4
	2.5	4	2	4	4	4	3	4

5.4 Benchmark Data Sets

We finally compare the different indexes on benchmark data sets:

- *Iris* [5], composed of three classes of 50 flowers each described by 4 physical attibutes. Two classes have a substantial overlap in the feature space and the optimal number of clusters to be found is debatable: 2 or 3, see [4].
- *Wine* [5], which consists of 13 chemical attributes for $n = 178$ italian wines, divided into three classes. Classes are well separable, so the indexes found in the literature generally give the right number of clusters.
- *Wisconsin Breast Cancer* [5], composed of $n = 699$ malignant/benign cells described by 9 features computed from digitized images.
- The artificial data set $X30$ introduced in [4], consisting in $n = 30$ observations in $\mathbb{R}^2$, for which three clusters are expected.
- The bidimensional artificial data set *Bensaid* [1] characterized by 3 classes of very different cardinalities (6, 3 and 40).

Table 5. Selected number of clusters in benchmark data sets

Data set	$c^\star$	NPC	NPE	XB	FS	$IFORI$ $(\top,\bot)_S$	$IFORI$ $(\top,\bot)_A$	$IFORI$ $(\top,\bot)_H$
Iris	2 or 3	2	2	2	3	2	2	2
Wine	3	3	2	3	5	3	3	3
Breast	2	2	2	2	3	2	2	2
X30	3	3	2	3	7	3	3	3
Bensaid	3	3	2	3	7	3	3	3
Starfield	8 or 9	2	2	6	7	8	8	8

- The original *Starfield* [15], which contains the position and light intensity of $n = 51$ bright stars near Solaris. The expected number of clusters is 8 or 9, depending on the papers.

Table 5 summarizes the results obtained on these data sets. The $c^\star$ column gives the expected number of clusters and the other columns show the optimal number of clusters obtained using the validity indexes. The proposed index always finds the optimal number of clusters whatever the couple $(\top, \bot)$ while some others do not.

6 Conclusion

A new family of cluster validity indexes for fuzzy partitions has been proposed. These indexes combine a new measure of overlap of clusters and a separation measure. The novelty of the approach is that, for each data point to be clustered, the relative importance of each membership degree and the relationship of the degrees are taken into account through a combination of triangular norms $(\top, \bot)$. Results obtained on artificial and benchmark data sets have shown that the proposed family of indexes is most of time more efficient than well-known ones, less sensitive to the fuzzifier exponent and particularly robust in noisy environments.

Further results about properties of the proposed indexes based on the mathematical properties of the agregation operators involved as well as guidelines to choose a member $(\top, \bot)$ of the family appropriated to specific situations will come soon.

References

1. Bensaid, A.M., Hall, L.O., Bezdek, J.C., Clarke, L.P., Silbiger, M.L., Arrington, J.A., Murtagh, R.F.: Validity-guided (re)clustering with applications to image segmentation. IEEE Transactions on Fuzzy Systems 4(2), 112–123 (1996)
2. Bezdek, J.C.: Cluster validity with fuzzy sets. Journal of Cybernetics 3, 58–72 (1974)
3. Bezdek, J.C.: Pattern Recognition with fuzzy objective function algorithm. Plenum Press (1981)

4. Bezdek, J.C., Pal, N.R.: Some new indexes of cluster validity. IEEE Transactions on Systems, Man and Cybernetics 23(3), 301–315 (1998)
5. Blake, C., Merz, C.: Uci repository of machine learning databases (1998), `http://www.ics.uci.edu/mlearn/MLRepository.html`
6. Dave, R.N.: Validating fuzzy partitions obtained through c-shells clustering. Pattern Recognition Letters 17(6), 613–623 (1996)
7. Fukuyama, Y., Sugeno, M.: A new method for choosing the number of clusters for the fuzzy c-means method. In: Proc. 5th Fuzzy Systems Symposium, pp. 247–250 (1989)
8. Gustafson, D.E., Kessel, W.C.: Fuzzy clustering with fuzzy covariance matrix. In: Proc. IEEE Conference on Decision and Control, San Diego, California, pp. 761–766 (1979)
9. Kim, D.-W., Lee, K.H., Lee, D.: On cluster validity index for estimating the optimal number of fuzzy clusters. Pattern Recognition 37(3), 2009–2025 (2004)
10. Klement, E.P., Mesiar, R.: Logical, Algebraic, Analytic, and Probabilistic Aspects of Triangular Norms. Elsevier, Amsterdam (2005)
11. Mascarilla, L., Berthier, M., Frélicot, C.: A k-order fuzzy or operator for pattern classification with k-order ambiguity rejection. Fuzzy Sets and Systems 159(15), 2011–2029 (2008)
12. Pal, N.R., Bezdek, J.C.: On cluster validity for the fuzzy c-means model. IEEE Transactions on Fuzzy Systems 3(3), 370–379 (1995)
13. Wang, W., Zhang, Y.: On fuzzy cluster validity indices. Fuzzy Sets and Systems 158(19), 2095–2117 (2007)
14. Wu, K.L., Yang, M.S.: A cluster validity index for fuzzy clustering. Pattern Recognition Letters 26(9), 1275–1291 (2005)
15. Xie, X.L., Beni, G.: A validity measure for fuzzy clustering. IEEE Transactions on Pattern Analysis and Machine Intelligence 13(8), 841–847 (1991)
16. Zarandi, M.H.F., Neshat, E., Türksen, I.B.: A new cluster validity index for fuzzy clustering based on similarity measure. In: 11th International Conference on Rough Sets, Fuzzy Sets, Data Mining and Granular Computing, pp. 127–135 (2007)

Fast Multivariate Ordinal Type Histogram Matching

Sung-Hyuk Cha

Department of Computer Science, Pace University
861 Bedford rd, Pleasantville, NY 10570
scha@pace.edu

Abstract. When patterns are represented as histograms, the *earth mover's distance*, *EMD* has been considered an excellent metric between two distributions. EMD is formulated as the transportation problem which is a hard optimization problem. In similarity based pattern retrieval problems, computing *EMD*s for all histograms in the database against a query histogram would take too long time for users to wait for the output. Hence, the candidate selection technique is presented to speed up the *EMD* based multivariate ordinal type histogram retrieval problem. It guarantees to find all similar histograms while achieving significant speed up. Theoretical relationships between other metrics for multivariate histograms and their usages are presented as well.

1 Introduction

Distance between histograms plays important role in classification, clustering, and retrieval problems [1]. Among numerous metrics, the *Earth Mover's Distance*, or *EMD* in short, has received great attention in computer vision and pattern recognition [2-4]. It is the minimal amount of work that must be performed to transform one distribution into the other by moving "distribution mass" [2]. This concept was called the *Mallows* distance in Statistics earlier [3]. Other conceptually similar metrics includes the *edit distance* [5,6] which appeared in literature simultaneously as *EMD*. Only *univariate* histograms were considered and depending on the measurement types, $\Theta(b)$, $\Theta(b)$, and $O(b^2)$ efficient solutions for computing *nominal*, *ordinal* and *modulo* type histogram *edit distances* were discovered [5-7] where b is the number of bins.

When histograms are *multivariate*, computing the cost minimization can be formulated as a *transportation problem* [2] which is a hard optimization problem. One of the solutions is the *Simplex method* [9] and the efficient algorithm takes $O(b^3 \log b)$ [9]. Consider the *similarity based histogram retrieval* problem, *i.e.*, retrieving similar histograms out of w histograms in the database to a query histogram, q using the *EMD*. A naïve way to solve the problem would be running the *simplex method* or equivalent algorithm w times which takes unbearably long time for users to wait for the output.

For this reason, a fast method which quickly eliminates most histograms from consideration as possible matches is presented. Let a match be a similar histogram h whose $EMD(h,q) \leq t$, a threshold value. Suppose we have another distance measure, d_x such that $d_x(h,q) \leq EMD(h,q)$ where d_x can be computed very fast. Then we can use

N. da Vitora Lobo et al. (Eds.): SSPR&SPR 2008, LNCS 5342, pp. 622–632, 2008.

d_x to eliminate h from consideration if $d_x(h,q) > t$. This *candidate selection* method guarantees to find all matches while achieving significant speed up. The *candidate selection* technique was applied successfully for univariate *ordinal* [6] and *modulo* [10] type histograms. Here, two efficient multivariate histogram metrics whose computational complexities are linear are introduced for the multivariate ordinal type histogram retrieval problem. Theoretical relationships between metrics are presented.

The rest of the paper is organized as follows. In section 2, the *edit distance* between univariate histograms and the *EMD* between 2D cases are reviewed. Section 3 introduces two other metrics which can be computed very fast for multivariate histograms. Section 4 contains use of two efficient metrics in the similarity based histogram retrieval problem. Finally, section 5 concludes this work.

2 Preliminary: Ordinal Type Histograms

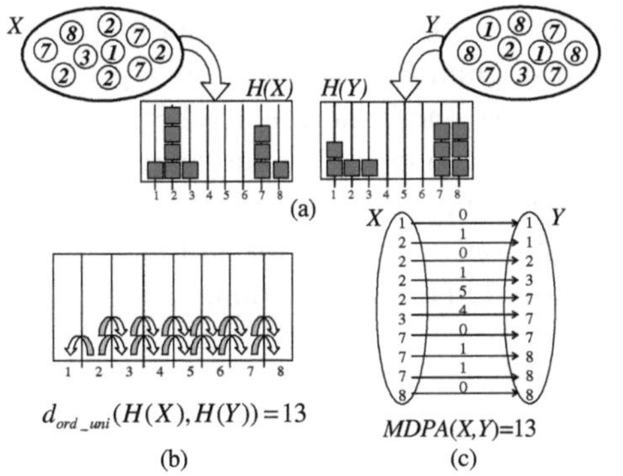

Fig. 1. Univariate ordinal type histogram and its metric

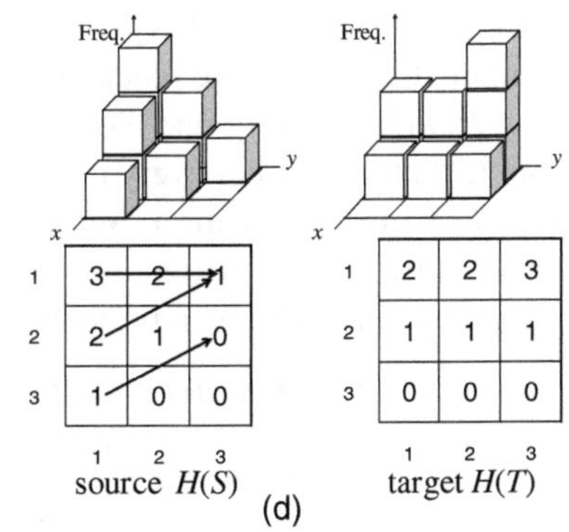

Fig. 2. 2-dimensional 3×3 ordinal histograms

2.1 Univariate Ordinal Type Histograms

Let X be a set of n elements whose possible values are finite. A histogram $H(X)$ of a set X represents the frequency of each value as shown in Figure 1 (a). The frequency value of the ith bin is denoted as $H_i(X)$, e.g., $H_2(X) = 4$ and $H_1(Y) = 2$ in Figure 1 (a). Let b be the number of bins and if b is strictly ordered and increases linearly, then the histogram is *univariate ordinal* type where values are simple scalar values.

Considering a histogram as a block world, the *edit distance* is the minimum amount of necessary movements to build one source histogram to the other target histogram by moving blocks left or right [7] as illustrated in Figure 1 (b) and the formula is:.

$$d_{ord\_uni}(H(X),H(Y)) = \sum_{i=1}^{b} \left| \sum_{j=1}^{i} H_j(X) - H_j(Y) \right| \qquad (1)$$

The *edit distance* is calculated by taking the sum of absolute values of prefix sum of difference for each level. Its computational complexity is $\Theta(b)$. Proofs for the correctness and complexity of the algorithm can be found in [7~9]. Note that the *edit distance* is one of the special cases of the *EMD* in section 2.2. It can be computed as a *transportation problem* [2] or the *minimum difference of pair assignment* problem (*MDPA*) [7] which can be computed in $O(n^3)$ using the *Hungarian method* [11] as depicted in Figure 1 (c). The minimum amount of necessary movements in histograms is the same as the minimum pair assignments in sets. It is important to conceive the eqn (2) because later arguments can be easily proven by the *MDPA* concept.

$$EMD(H(X),H(Y)) = MDPA(X,Y) \tag{2}$$

2.2 Two Dimensional Case

In higher dimensional histograms, values of elements are vectors rather than scalars. If each axis in the vector is *ordinal* and has a finite value, then the histogram is a *d*-dimensional ordinal type histogram. A 2-D ordinal type histogram has the number of bins, $b = r \times c$ and each axis is ordinal, *i.e.*, $x = (1,\ldots,c)$ and $y = (1,\ldots,r)$.

Consider 3x3 ordinal type histograms where $n = 10$ in Figure 2 as an illustrative example. One can move blocks in the source histogram $H(S)$ to make the target histogram $H(T)$ by moving one block from (1,1) to (1,3), one from (2,1) to (1,3), and one from (3,1) to (2,3). Let every rectangular coordinate bin in $H(S)$ be the source and that in $H(T)$ be the destination. Supply and demand values are bounded (0~n). Now the metric between two histograms can be formulated as a *standard form* of *Linear Programming* although it can be shortened as discussed in the later section.

$$\begin{aligned}
& d_{cm}(H(S),H(T)) = Z \\
& \text{Minimize} \quad Z = \sum_{i=(1,1)}^{(r,c)} \sum_{j=(1,1)}^{(r,c)} c_{ij} x_{ij} \\
& \text{s. t.} \quad \sum_{j=(1,1)}^{(r,c)} x_{ij} = H_i(S), \text{ for } i = (1,1),\ldots,(r,c) \\
& \qquad \sum_{i=(1,1)}^{(r,c)} x_{ij} = H_j(T), \text{ for } j = (1,1),\ldots,(r,c) \\
& \text{and} \quad x_{ij} \geq 0, \qquad \text{for all } i \text{ and } j
\end{aligned} \tag{3}$$

Let d_{ord_euc} and d_{ord_cbd} be the *EMD*s with the cost matrix $c_{i,j}$ using the Euclidean, L_2 and the city block L_1 distances between two grid indices, respectively. Two feasible solutions for $d_{ord\_euc}(H(S),H(T)) = 6.48$ and one for $d_{ord\_cbd}(H(S),H(T)) = 8$ are given in Figure 3.

Lemma 1: $d_{ord\_euc}(H(X),H(Y)) \leq d_{ord\_cbd}(H(X),H(Y))$.

Proof: $MDPA_{ord\_cbd}(X,Y) = d_{ord\_cbd}(H(X),H(Y))$. Let each pair assignment in the $MDPA_{ord\_cbd}(X,Y)$ setting be a_i. Then $MDPA_{ord\_cbd}(X,Y) = \sum_{i=1}^{n} L_1(a_i)$. Let $TEMP(X,Y) = \sum_{i=1}^{n} L_2(a_i)$. Then $TEMP(X,Y) \leq MDPA_{ord\_cbd}(X,Y)$ because $L_2 \leq L_1$. $MDPA_{ord\_euc}(X,Y) \leq TEMP(X,Y)$ by definition. $MDPA_{ord\_euc}(X,Y) \leq MDPA_{ord\_cbd}(X,Y)$ ■

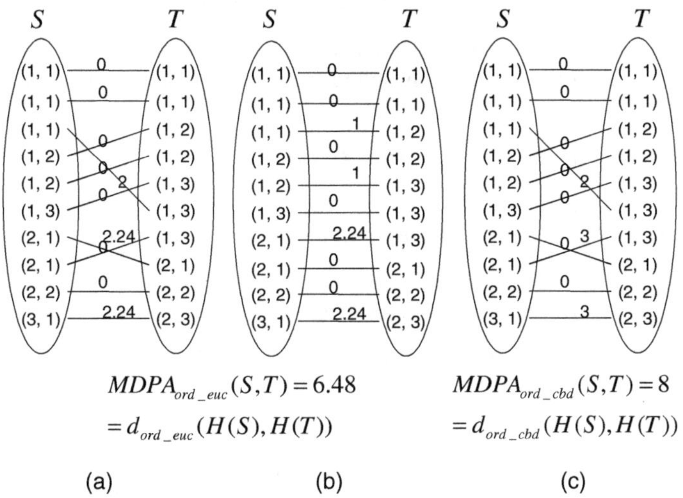

Fig. 3. (a) (b) 2 feasible solutions for d_{ord_euc} (c) d_{ord_cbd}

3 Efficient Metrics

This section introduces two metrics with linear time or lower complexities which shall be used to speed up the *EMD* based histogram retrieval.

3.1 Sum of Direct Subtractions (SDS)

Consider the cost matrix such that the diagonal cells have all 0's and all others have 1's. Histograms with such a cost matrix were referred to *nominal* type histograms [7]. *MDPA* for our examples is given in Figure 4 (a). Let d_{sds}, the *sum of direct subtractions*, *SDS* in short, be the *EMD* with such a cost matrix and its equation is:

$$d_{sds}(H(S),H(T)) = n - \sum_{i=1}^{b} \min(H_i(S),H_i(T)) = \frac{\sum_{i=1}^{b} |H_i(S)-H_i(T)|}{2} \tag{4}$$

The eqn (4) is often known as the non-intersection. Its computational complexity is linear: $\Theta(b) = \Theta(r\times c)$. The term, $\Sigma|H(S)-H(T)|$ is the *city block* distance, $L_1(H(S),H(T)) = 2 \times d_{sds}(H(S),H(T))$ [7]. It should be noted that $L_1(H(S),H(T)) \neq d_{ord\_cbd}(H(S),H(T))$ although the term '*city block distance*' appears in both metrics.

Let D be the direct subtraction of $H(T)$ from $H(S)$. $D=H(S)-H(T)$ contains positive and negative numbers as shown in Figure 4 (b). Let *DP* and *DN* be the sets of elements in positive and negative positions in D, respectively. Every element in *DP* must be assigned to a certain element in *DN* as shown in Figure 4 (c). Each assignment cost is 1. *SDS* is simply the cardinality of either *DP* or *DN*; $d_{sds}(H(S),H(T)) = |DP| = |DN|$.

Readers might easily notice that the constraints in eqn (3) can be reduced. $H(S)$ and $H(T)$ can be reduced to $H(DP)$ and $H(DN)$, respectively and non-zero cells in $H(DP)$ and $H(DN)$ are only necessary sources and destinations as shown in Figure 4 (d).

$$MDPA(DP,DN) = MDPA(S,T) \tag{5}$$

$$EMD(H(DP),H(DN)) = EMD(H(S),H(T)) \tag{6}$$

EMD may be computed efficiently by the constraint reduction but there is no asymptotical impact on average. It is stated because it simplifies proofs for following lemmas.

Lemma 2: $d_{sds}(H(S),H(T)) \leq d_{ord\_euc}(H(S),H(T))$

Proof: Let $a \in DP$ and $b \in DN$. $c_{a,b} = 1$ for $MDPA_{sds}(DP,DN)$ whereas $c_{a,b} \geq 1$ for $MDPA_{ord\_euc}(DP,DN)$. Hence, $MDPA_{sds}(DP,DN) \leq MDPA_{ord\_euc}(DP,DN)$. Thus $MDPA_{sds}(S,T) \leq MDPA_{ord\_euc}(S,T)$. ■

Lemma 3: $d_{sds}(H(S),H(T)) \leq d_{ord\_cbd}(H(S),H(T))$.

Proof: the similar argument to that of Lemma 2.

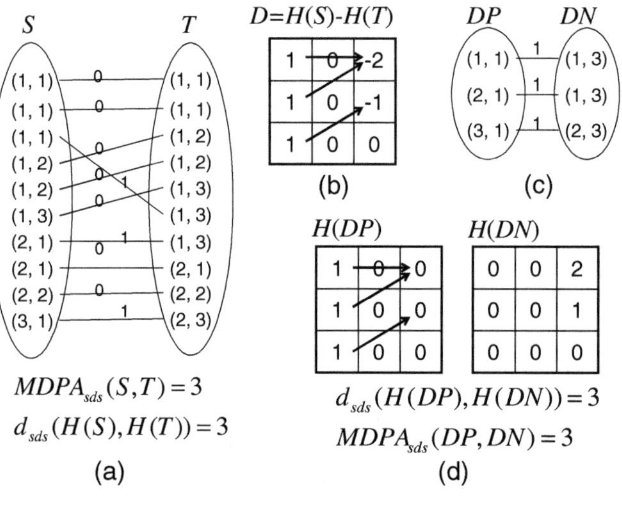

Fig. 4. *SDS* for nominal type histograms

3.2 Sum of Grid Differences (SGD)

Every vertical or horizontal grid dissects the histogram into two parts as shown in Figure 5 (a). In order to edit a source histogram to a target histogram, at least eqns (7) or eqn (8) amount of movements must pass the horizontal or vertical grid.

$$gd(k,*) = \left|\sum_{i=1}^{k} H_{(i,*)}(S) - \sum_{i=1}^{k} H_{(i,*)}(T)\right| \tag{7}$$

$$gd(*,k) = \left|\sum_{i=1}^{k} H_{(*,i)}(S) - \sum_{i=1}^{k} H_{(*,i)}(T)\right| \tag{8}$$

Let the *sum of grid differences* (*SGD*) be the sum of the eqn (7) for every horizontal grid and the eqn (8) for every vertical grid.

$$d_{sgd}(H(S),H(T)) = \sum_{i=1}^{r-1} gd(i,*) + \sum_{i=1}^{c-1} gd(*,i) \tag{9}$$

It can be very efficiently computed. Let $H^1(X),\ldots,H^d(X)$ be projection histograms the histogram is d-dimensional histogram. Note that each H^i is an univariate ordinal type histogram. Thus, *SGD* can be defined using the eqn (1):

$$d_{sgd}(H(S),H(T)) = \sum_{i=1}^{d} d_{ord\_uni}(H^i(S),H^i(T)) \tag{10}$$

In 2D histograms,

$$= \sum_{i=1}^{r} \left| \sum_{j=1}^{i} H_j^x(S) - H_j^x(T) \right| + \sum_{i=1}^{c} \left| \sum_{j=1}^{i} H_j^y(S) - H_j^y(T) \right| \tag{11}$$

where H^x and H^y are vertical and horizontal projection histograms.

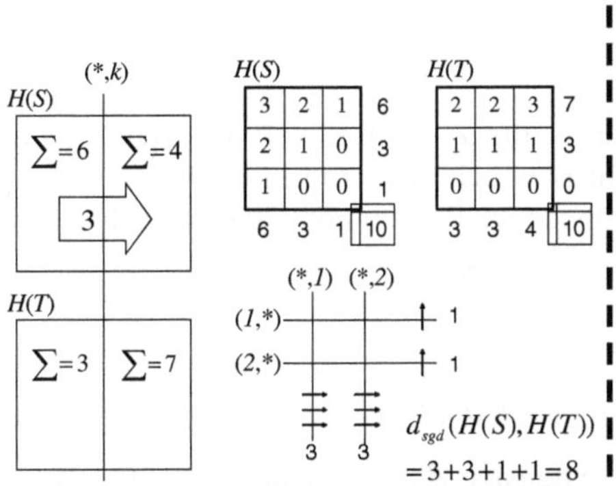

Fig. 5. Grid difference distance

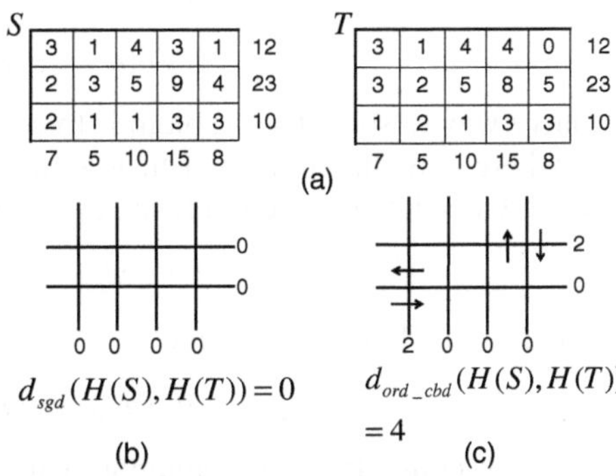

Fig. 6. Example for Lemma 4

The computational complexity is clearly $\Theta(d \times b)$ as one must build the projection histograms for each axis. When d is a small constant value like 2 or 3, $\Theta(b)$. Note that in many applications, the projection histograms might be pre-computed before matching. In this case, the complexity for *SGD* is $\Theta(r+c) \approx \Theta(\sqrt{b})$ for 2D and $\approx \Theta(d \times \sqrt[d]{b})$ for d-dimensional histograms).

Lemma 4: $d_{sgd}(H(S),H(T)) \leq d_{ord\_cbd}(H(S),H(T))$

Proof: Let $ag(*,k)$ and $ag(k,*)$ be the number of blocks that actually pass the kth vertical and horizontal grid, respectively when the histogram is edited using the city block distance cost matrix. Then

$$d_{ord\_cbd}(H(S),H(T)) = \sum_{i=1}^{r-1} ag(i,*) + \sum_{i=1}^{c-1} ag(*,i) \tag{12}$$

ag cannot be less than gd for every grid. If so, there is no way that the edited source histogram can be the same as the target histogram. However, $ag \geq gd$ because certain blocks can move in different positions on a certain grid in the handshaking manner as shown in Figure 6 (c). ∎

Fallacy 1: $d_{sgd}(H(S),H(T)) \leq d_{ord\_euc}(H(S),H(T))$.

Proof: a counter example is given in Figure 2, $d_{sgd}(H(S),H(T)) > d_{ord\_euc}(H(S),H(T))$, *i.e.*, 8 > 6.48 while there are other examples like in Figure 6, $d_{sgd}(H(S),H(T)) \leq d_{ord\_euc}(H(S),H(T))$, *i.e.*, 0 < 4 ■

Although the 2D cases are used for the simplicity sake, all claims are valid for any dimensional ordinal type histograms. It should be noted that *SGD* is only for multivariate *ordinal* type histograms but not for any other type histograms.

4 Multivariate Histogram Retrieval

Consider a query, "Find all similar histograms in a database D to a given histogram q." This problem can be tackled by defining metrics between histograms, preferably d_{ord_euc}. When a threshold value t is given, the problem can be formalized as:

$$S^t_{ord\_euc} = \forall h(d_{ord\_euc}(q,h) \leq t \wedge h \in D) \tag{13}$$

A naïve way to solve the problem is to compute the d_{ord_euc} for every histogram $\{h_1,\ldots, h_w\}$ where w is the number of histograms in D which takes too long. A better way is using the candidate selection technique. Suppose we first compute the candidate set.

$$C^t_{sds} = \forall h(d_{sds}(q,h) \leq t \wedge h \in D) \tag{14}$$

Then the eqn (13) is equivalent to:

$$S^t_{ord\_euc} = \forall h(d_{ord\_euc}(q,h) \leq t \wedge h \in C^t_{sds}) \tag{15}$$

Instead of computing the computationally expensive d_{ord_euc} for all h's in D, one can only compute h's in the candidate set in the eqn (15) to find matches.

In order to utilize *candidate selection* technique, the following necessary conditions must be met.

- C must be computed fast, *i.e.*, $o(b^3\log b)$
- C must contain all matches, *i.e.*, $C \supset S^t_{ord\_euc}$

The first condition is met as stated in the previous section; computing $d_{sds}(q,\forall h)$ takes $\Theta(wb)$. Theorems of the present section are for the second necessary condition.

Theorem 1: $S^t_{ord\_euc} \subset C^t_{sds}$

Proof: Consider an h whose $d_{ord\_euc}(q,h) \leq t$. From the Lemma 2, $d_{sds}(q,h) \leq d_{ord\_euc}(q,h) \leq t$. Hence, $S^t_{ord\_euc} \subset C^t_{sds}$ as shown in Figure 7 (a) ■
Suppose one is looking for all histograms in eqn (16).

$$S^t_{ord\_cbd} = \forall h(d_{ord\_cbd}(q,h) \leq t \wedge h \in D) \tag{16}$$

Then, the *SDG*.can be utilized to produce a candidate set:

$$C^t_{sgd} = \forall h(d_{sgd}(q,h) \leq t \wedge h \in D) \tag{17}$$

Moreover, one can also utilize the *SDS* as another candidate set. Hence the eqn (16) is equivalent to:

$$S^t_{ord\_cbd} = \forall h(d_{ord\_cbd}(q,h) \leq t \wedge h \in (C^t_{sds} \cap C^t_{sgd})) \tag{18}$$

Figure 7 (b) illustrates the eqn (18).

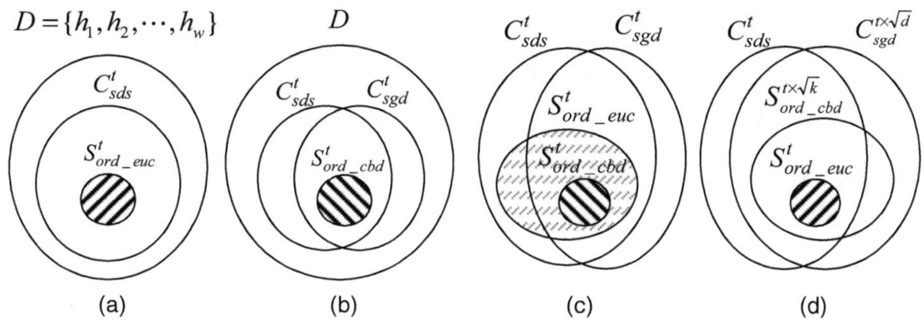

Fig. 7. Relationship between candidate sets

Theorem 2: $S^t_{ord\_cbd} \subset (C^t_{sds} \cap C^t_{sgd})$

Proof: $S^t_{ord\_cbd} \subset C^t_{sds}$ by Lemma 3 and $S^t_{ord\_cbd} \subset C^t_{sgd}$ by Lemma 4. ■

Unfortunately as depicted in Figure 7 (c), C^t_{sgd} cannot be a candidate set for $S^t_{ord\_euc}$; $S^t_{ord\_euc} \not\subset C^t_{sgd}$ because of Fallacy 1. Nevertheless, there is a way to utilize the *SGD* for d_{ord_euc}. Let p and q be points in d-dimensional *Euclidean space* and let L_1 and L_2 be the *city block* and *Euclidean* distances between two points, respectively.

Axiom 1: $L_2(p,q) \le L_1(p,q) \le \sqrt{d} \times L_2(p,q)$.

This bound can be shown by a simple geometry.

Lemma 5: $d_{ord\_cbd}(H(X),H(Y)) \le \sqrt{d} \times d_{ord\_euc}(H(X),H(Y))$

Proof: Let each pair assignment in $MDPA_{ord\_euc}$ be a_i. $MDPA_{ord\_euc}(X,Y) = \sum_{i=1}^{n} L_2(a_i)$. Let $TEMP(X,Y) = \sum_{i=1}^{n} L_1(a_i)$, then $\sum_{i=1}^{n} L_2(a_i) \le \sum_{i=1}^{n} L_1(a_i) \le \sum_{i=1}^{n} \sqrt{d} L_2(a_i)$ by Axiom 1. $TEMP(X,Y) \le \sum_{i=1}^{n} \sqrt{d} L_2(a_i) = \sqrt{d} \times \sum_{i=1}^{n} L_2(a_i)$.

$TEMP(X,Y) \le \sqrt{d} \times MDPA_{ord\_euc}(X,Y)$ and $MDPA_{ord\_cbd}(X,Y) \le TEMP(X,Y)$ by definition. $\therefore$ $MDPA_{ord\_cbd}(X,Y) \le \sqrt{d} \times MDPA_{ord\_euc}(X,Y)$ ■

$$S^{t\times\sqrt{d}}_{ord\_cbd} = \forall h(d_{ord\_cbd}(q,h) \le t \times \sqrt{d} \wedge h \in D) \quad (19)$$

Lemma 6: $S^t_{ord\_cbd} \subset S^t_{ord\_euc} \subset S^{t\times\sqrt{d}}_{ord\_cbd}$

Proof: by Lemma 1 and Lemma 5 ■

$$\text{Let} \quad C^{t\times\sqrt{d}}_{sgd} = \forall h(d_{sgd}(q,h) \le t \times \sqrt{d} \wedge h \in D) \quad (20)$$

Theorem 3: $S^t_{ord\_euc} \subset (C^{t\times\sqrt{d}}_{sgd} \cap C^t_{sds})$

Proof: $S^t_{ord\_euc} \subset S^{t\times\sqrt{d}}_{ord\_cbd}$ by Lemma 6. $S^{t\times\sqrt{d}}_{ord\_cbd} \subset C^{t\times\sqrt{d}}_{sgd}$ by Theorem 2. ■

In all, computing $S^t_{ord\_euc}$ by eqn (13) setting would take too long for users to wait for the output. Some speed up can be achieved in eqn (15) setting. Further speed up can be achieved by the following eqn (21) setting.

$$S^t_{ord\_euc} = \forall h(d_{ord\_euc}(q,h) \le t \wedge h \in (C^{t\times\sqrt{d}}_{sgd} \cap C^t_{sds})) \tag{21}$$

Corollary 1: $\exists h(h \in (C^t_{sds} - C^{t\times\sqrt{2}}_{sgd}))$

Corollary 2: $\exists h(h \in (C^{t\times\sqrt{2}}_{sgd} - C^t_{sds}))$

Proof: Let q, a, and b be 3x3 histograms with $n = 20$.

$$q = \begin{matrix} 2 & 3 & 2 \\ 1 & 2 & 3 \\ 2 & 2 & 3 \end{matrix} \qquad a = \begin{matrix} 1 & 3 & 2 \\ 2 & 1 & 1 \\ 3 & 3 & 4 \end{matrix} \qquad b = \begin{matrix} 0 & 1 & 2 \\ 3 & 3 & 4 \\ 1 & 4 & 2 \end{matrix}$$

If $t = 4$, a and b are examples of Corollary 1 and 2, respectively. $d_{sds}(q,a) = 4 \le t$ where $d_{sgd}(q,a) = 6 > t\times\sqrt{2}$. $d_{sds}(q,b) = 6 > t$ where $d_{sgd}(q,b) = 5 \le t\times\sqrt{2}$. ■

Corollaries 1 and 2 assure that using the eqn (21) is better than using either one of candidate sets alone. In other words, there is no need to compute $d_{ord\_euc}(q,a) = 4.8284 > t$ and $d_{ord\_euc}(q,b) = 6.4142 > t$ because their values are guaranteed to be larger than t by Theorem 3.

For the experiments, 101 histograms were randomly generated. Using the leave one out method [1], the table 1 shows the average percentage of retrieved similar histograms. For instance of a 3×5 histogram of n=70 and t=15, only 16.79% of histograms are needed to be validated for matches. The smaller t, the smaller the candidate set size.

Table. 1. Average candidate and match set sizes in %

$r\times c$	N	t	$\lvert C^t_{sds} \rvert$	$\lvert C^{t\times\sqrt{2}}_{sgd} \rvert$	$\lvert C^{t\times\sqrt{2}}_{sgd} \cap C^t_{sds} \rvert$	$\lvert S^t_{ord\_euc} \rvert$
3×5	50	10	7.9%	21.4%	4.65%	0.38%
3×5	50	12	23.9%	29.7%	13.60%	2.75%
3×5	70	12	6.8%	18.3%	3.45%	0.55%
3×5	70	15	27.3%	41.1%	16.79%	3.11%
5×7	100	28	16.2%	30.8%	7.94%	0.12%
5×7	100	30	30.3%	37.6%	15.43%	0.30%

The algorithm using the candidate selection technique for the d_{ord_euc} based d-dimensional ordinal type histogram retrieval problem is:

Input: q, D, and t.

Output: $S^t_{ord\_euc} = M$.

Procedure:

```
M = Ø
for i = 1 to w
    if d_sgd(q,h_i) ≤ √d × t
        if d_sds(q,h_i) ≤ t
            if d_ord_euc(q,h_i) ≤ t
                h_i ∈ M
return M
```

But for lines 3 and 4, it would take too long. The order of these lines matters. If the projection histograms for every histogram in D are pre-computed, $d_{sgd}(q,\forall h)$ takes $\Theta(w\times(r+c))$ much faster than $d_{sds}(q,\forall h)$ which takes $\Theta(w\times(r\times c))$ for 2D histograms and thus *SGD* should precede *SDS*. Then the running time complexity for the algorithm is $\Theta(w\times(r+c))+\Theta(|C_{sgd}^{t\times\sqrt{2}}|\times(r\times c))+O(|C_{sgd}^{t\times\sqrt{2}}\cap C_{sds}^{t}|\times b^3\log b)$. With a little modification, the above algorithm can be used for the *instance based classifier* [1]. However, in most classification problems, the threshold value is not given. One may set $t = \min(d_{sgd}(q,h))$ and slowly increase t to find the nearest one.

5 Conclusion

Similarity based multivariate histogram retrieval problem using the *EMD* as a metric was considered. The candidate selection method which quickly eliminates most histograms from consideration as possible matches was presented. All matches, *i.e.*, similar histograms to a query histogram are still guaranteed to be found while achieving significant speed up. Main contribution of this article is introducing two metrics which satisfy the two necessary conditions to produce a candidate set C for the *EMD* based multivariate histogram retrieval problem.

In addition to aforementioned two necessary conditions for the candidate selection method, a desirable condition is that the expected $|C|$ must be small, *i.e.*, $|C| \ll |D|$. Although Theorems in this paper imply $|C| \leq |D|$, finding the expected $|C|$ theoretically which requires probability analysis is one of the open problems. Finding a new *candidate selection* function such that the expected $|C|$ is small is also a challenging problem, e.g., an easy one might be finding a tighter bound in Lemma 5.

An example of 3D ordinal type histograms is the *RGB* histogram because each component histogram is ordinal type. Consider *HSV* histogram which is a multivariate heterogeneous type histogram. The *Hue* component histogram is *modulo* type whereas *Saturation* and *value* components are ordinal type. Dealing with heterogeneous type multivariate histograms is another open problem.

References

1. Duda, R.O., Hart, P.E., Stork, D.G.: Pattern Classification, 2nd edn. Wiley, New York (2000)
2. Rubner, Y., Tomasi, C., Guibas, L.J.: A metric for distributions with applications to image databases. In: Proc. of ICCV, pp. 59–66 (1998)
3. Levina, E., Bickel, P.: The Earth Mover's distance is the Mallows distance: some insights from Statistics. In: Proc. Of ICCV, pp. 251–256 (2001)
4. Puzicha, J., Rubner, Y., Tomasi, C., Buhmann, J.M.: Empirical evaluation of dissimilarity measures for color and texture. In: Proc. of ICCV, pp. 1165–1173 (1999)
5. Cha, S.-H.: Fast Image Template and Dictionary Matching Algorithms. In: Chin, R., Pong, T.-C. (eds.) ACCV 1998. LNCS, vol. 1351, pp. 370–377. Springer, Heidelberg (1997)
6. Cha, S.-H.: Efficient Algorithms for Image Template and Dictionary Matching. Journal of Mathematical Imaging and Vision 12(1), 81–90 (2000)

7. Cha, S.-H., Srihari, S.N.: On Measuring the Distance between Histograms. Journal of Pattern Recognition 35(6), 1355–1370 (2002)
8. Hiller, F.S., Lieberman, G.J.: Introduction to mathematical programming, 2nd edn. McGraw-Hill, New York (1995)
9. Ahuha, R.K., Magnanti, T.L., Orlin, J.B.: Network Flows. Prentice-Hall, Englewood Cliffs (1993)
10. Cha, S.-H.: A fast hue-based colour image indexing algorithm. Machine GRAPHICS & VISION 11(2/3), 285–295 (2002)
11. Kuhn, H.W.: The Hungarian method for the assignment problem. Naval Research Logistics Quarterly 2, 83–97 (1955)

Distance between Histograms with Shuffled Cost Matrix

Sung-Hyuk Cha

Department of Computer Science, Pace University
861 Bedford rd, Pleasantville, NY 10570
scha@pace.edu

Abstract. Comparing two distributions is of fundamental importance to statistics and pattern recognition. The *earth mover's distance* (*EMD*) has been considered an excellent distance measure between two distributions. It is defined to minimize the cost using a given cost matrix and formulated as the transportation problem which is a hard optimization problem. There are three special type cost matrices where efficient algorithms are known: nominal, ordinal, and modulo. Here the problem of identifying whether a given cost matrix has the shuffled ordinal property is considered and if so, the linear time complexity algorithm can be applied to compute the *EMD* efficiently.

1 Introduction

Comparing two distributions plays important role in pattern classification, clustering, and statistics problems [1]. Among numerous distance or similarity measures, the *Earth Mover's Distance*, or *EMD* in short, has received great attention in computer vision and pattern recognition. It is the minimal amount of work that must be performed to transform one distribution into the other by moving *distribution mass* [2]. The similar concept was called the *Mallows* distance in Statistics earlier [3].

Let X be a set of n elements whose possible values are finite. A histogram $H(X)$ of a set X represents the frequency of each value in N, an ordered list of bin names. Let b be the number of bins. The frequency value of the ith bin is denoted as $H_i(X)$. For examples in Figure 1 (a), $n = 10$, $b = 4$, $N = [\mathrm{A\ C\ G\ T}]$, $N_4 = \mathrm{T}$, $H_2(S) = 3$, and $H_1(T) = 2$, etc. $H(S)$ and $H(T)$ in Figure 1 (a) shall be used as illustrative examples throughout the rest of this paper.

Computing the *EMD* between two histograms can be formulated as a *transportation problem* which is a special linear programming problem to distribute any commodity from any group of supply centers, called sources, to any group of receiving centers, called destinations, in such a way as to minimize the total distribution cost [4]. In general, source $i = (1,2,\ldots,b)$ has a supply of $H_i(S)$ units to distribute to the destinations, and destination $j = (1,2,\ldots,b)$ has a demand for $H_i(T)$ units to be received from the sources. Let Z be the total distribution cost. Let $x_{i,j}$ and $c_{i,j}$ denote the number of units to be distributed and its cost from a source i to a destination j. Then the *linear programming* formulation of the *EMD* becomes

N. da Vitora Lobo et al. (Eds.): SSPR&SPR 2008, LNCS 5342, pp. 633–643, 2008.

$$
\begin{aligned}
&\text{Minimize} && Z=\sum_{i=1}^{b}\sum_{j=1}^{b}c_{ij}x_{ij} \\
&\text{subject to} && \sum_{j=1}^{b}x_{ij}=H_i(S), \quad \text{for } i=1,2,\dots,b \\
& && \sum_{i=1}^{b}x_{ij}=H_j(T), \quad \text{for } j=1,2,\dots,b \\
&\text{and} && x_{ij}\ge 0, \quad \text{for all } i \text{ and } j
\end{aligned}
\tag{1}
$$

$$
C_{i,j}=\begin{bmatrix}
A & c_{A,A} & c_{C,A} & c_{G,A} & c_{T,A} \\
C & c_{C,A} & c_{C,C} & c_{G,C} & c_{T,C} \\
G & c_{G,A} & c_{G,C} & c_{G,G} & c_{T,G} \\
T & c_{T,A} & c_{T,C} & c_{T,G} & c_{T,T} \\
 & A & C & G & T
\end{bmatrix}
\tag{2}
$$

In order to have any feasible solutions, a necessary and sufficient condition for a transportation problem is $\Sigma^{b}_{i=1}s_i = \Sigma^{b}_{j=1}d_j$ [4]. Since histograms to be compared are of the same size, $\Sigma^{b}_{i=1}H_i(S) = \Sigma^{b}_{j=1}H_j(T) = n$, the feasible solutions property is met.

The *EMD* value between two histograms varies depending on the cost matrix. Figure 1 (b), (c), and (d) show feasible solutions in bipartite graphs with edges as (quantity × cost) and *EMD* values between *H*(*S*) and *H*(*T*) using the following C_x, C_y, and C_z cost matrices, respectively.

$$
C_x=\begin{bmatrix}0&5&1&7\\5&0&4&6\\1&4&0&2\\7&6&2&0\end{bmatrix}\quad
C_y=\begin{bmatrix}0&4&5&1\\4&0&9&3\\5&9&0&6\\1&3&6&0\end{bmatrix}\quad
C_z=\begin{bmatrix}0&4&2&2\\4&0&2&2\\2&2&0&0\\2&2&0&0\end{bmatrix}
$$

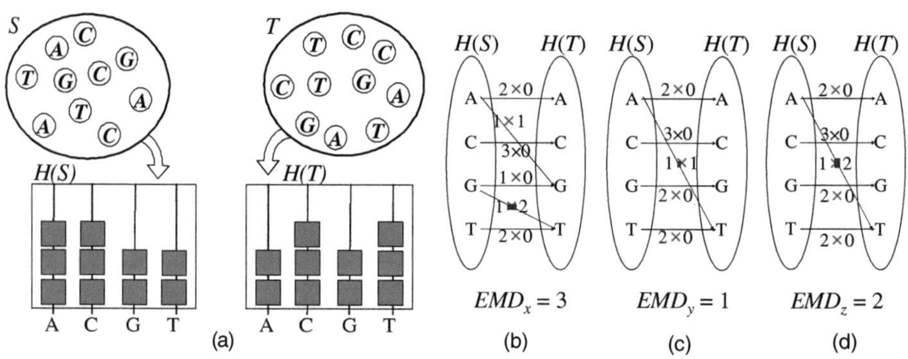

Fig. 1. Histogram representation and distance between two distributions

The traditional *Simplex method* [4] can solve the *EMD* problem and the efficient algorithm takes $O(b^3\log b)$ [2,5]. However, there are special histogram types where the *EMD* can be computed very fast. The metric was originally called the *edit distance* [6]. Depending on the measurement types, $\Theta(b)$, $\Theta(b)$, and $O(b^2)$ efficient solutions for computing *nominal*, *ordinal* and *modulo* type histogram *edit distances* were discovered [6-8]. These special type histograms have corresponding special cost matrices. Here the problem of determining whether a cost matrix is one of these special type cost matrices is considered. If so, the efficient algorithm can be applied.

The rest of the paper is organized as follows. In section 2, previously known cost matrices and their algorithms are reviewed and generalized. Section 3 introduces the algorithm to determine whether a given cost matrix has the *shuffled-ordinal* property. Finally, section 4 concludes this work with open problems.

2 Previously Known Cost Matrices with Fast Algorithms

Consider following nominal, ordinal, and modulo cost matrices.

$$C_{nom} = \begin{bmatrix} 0 & 1 & 1 & 1 \\ 1 & 0 & 1 & 1 \\ 1 & 1 & 0 & 1 \\ 1 & 1 & 1 & 0 \end{bmatrix} \quad C_{ord} = \begin{bmatrix} 0 & 1 & 2 & 3 \\ 1 & 0 & 1 & 2 \\ 2 & 1 & 0 & 1 \\ 3 & 2 & 1 & 0 \end{bmatrix} \quad C_{mod} = \begin{bmatrix} 0 & 1 & 2 & 1 \\ 1 & 0 & 1 & 2 \\ 2 & 1 & 0 & 1 \\ 1 & 2 & 1 & 0 \end{bmatrix}$$

In a nominal cost matrix, C_{nom}, the diagonal cells have all 0's and all others have 1's. Histograms with such a cost matrix were referred to as *nominal* type histograms [8] and EMD_{nom} can be computed using the following eqn (3):

$$EMD_{nom}(H(S), H(T)) = \frac{\sum_{i=1}^{b} |H_i(S) - H_i(T)|}{2} \tag{3}$$

It is nothing but a half of the *city block* distance and its computational complexity is clearly linear, $\Theta(b)$.

Next, if bin index value is ordered and increases linearly, then these histograms were referred to *univariate ordinal* type histograms [8]. Let the bin index be denoted as I. If an $I = [3,4,5,6]$ for example, then the cost matrix is C_{ord} and let's denote this special index be I_{ord}. Considering a histogram as a block world, the EMD_{ord} is the minimum amount of necessary movements to build one source histogram to the other target histogram by moving blocks left or right [6] as illustrated in Figure 2 (a) and can be computed in linear time $\Theta(b)$ using the following formula:

$$EMD_{ord}(H(X), H(Y)) = \sum_{i=1}^{b} \left| \sum_{j=1}^{i} H_j(X) - H_j(Y) \right| \tag{4}$$

$$= \sum_{i=1}^{b-1} \left| \sum_{j=1}^{i} H_j(X) - H_j(Y) \right| \tag{5}$$

It is calculated by taking the sum of absolute values of prefix sum of difference for each level. Proofs for the correctness and complexity of the algorithm can be found in [6]. Figure 2 (b) shows a couple of feasible solutions for two histograms in Figure 1 (a). The original formula in [6,8] is the eqn (4) but it is equivalent to the eqn (5).

When I values are angular values like north, east, south, and west directions, the first bin and the last bin are neighbors. These kinds of histograms were referred to as angular [7] or modulo type histograms [8] and Figure 2 (c) illustrates the edit distance between two modulo type histograms with I_{mod} and Figure 2 (d) shows a feasible solution for EMD_{mod} using the cost matrix C_{mod}. The algorithm and proofs for its correctness and computational time complexity, i.e., $O(b^2)$ can be also found in [8].

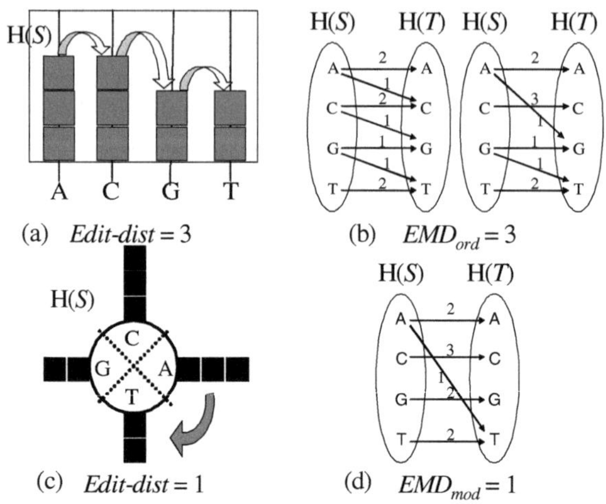

Fig. 2. Ordinal and modulo edit distance between histograms in Figure 1

When a constant value c is multiplied to nominal, ordinal, and modulo cost matrices, the new cost matrices also belong to the nominal, ordinal, and modulo type cost matrices, respectively, e.g,

$$C_{nom\times2}=\begin{bmatrix}0&2&2&2\\2&0&2&2\\2&2&0&2\\2&2&2&0\end{bmatrix}\quad C_{ord\times4}=\begin{bmatrix}0&4&8&12\\4&0&4&8\\8&4&0&4\\12&8&4&0\end{bmatrix}\quad C_{mod\times3}=\begin{bmatrix}0&3&6&3\\3&0&3&6\\6&3&0&3\\3&6&3&0\end{bmatrix}$$

The bin index values in Figure 3 (a) increase by one whereas those in Figure 3 (b) increase by two. Both are ordinal type histograms. Bin values in Figure 3 (c) increase by $2\pi/b$ and the histogram is a modulo type histogram whose cost matrix is $2\pi/b \times C_{mod}$. If $I_x = c\times I_{ord}$, then a cost matrix $C_x = c\times C_{ord}$

$$EMD_x(H(X), H(Y)) = c \times EMD_{ord}(H(X), H(Y)) \tag{6}$$

The same arguments for nominal and modulo types apply.

So far, bin index values, I in ordinal and modulo histograms increase uniformly by a certain constant value c. Consider three histograms in Figure 3 (d~f). Their indices and cost matrices are as follows:

$$I_{F3(d)}=\begin{bmatrix}1&4&5&7\end{bmatrix}\quad I_{F3(e)}=\begin{bmatrix}2&4&4&8\end{bmatrix}\quad I_{F3(f)}=\begin{bmatrix}0&3&6&7\end{bmatrix}$$

$$C_{F3(d)}=\begin{bmatrix}0&3&4&6\\3&0&1&3\\4&1&0&2\\6&3&2&0\end{bmatrix}\quad C_{F3(e)}=\begin{bmatrix}0&2&2&6\\2&0&0&4\\2&0&0&4\\6&4&4&0\end{bmatrix}\quad C_{F3(f)}=\begin{bmatrix}0&3&2&1\\3&0&3&4\\2&3&0&1\\1&4&1&0\end{bmatrix}$$

$$I'_{F3(d)}=\begin{bmatrix}3&1&2\end{bmatrix}\quad I'_{F3(e)}=\begin{bmatrix}2&0&4\end{bmatrix}\quad I'_{F3(f)}=\begin{bmatrix}3&3&1&1\end{bmatrix}$$

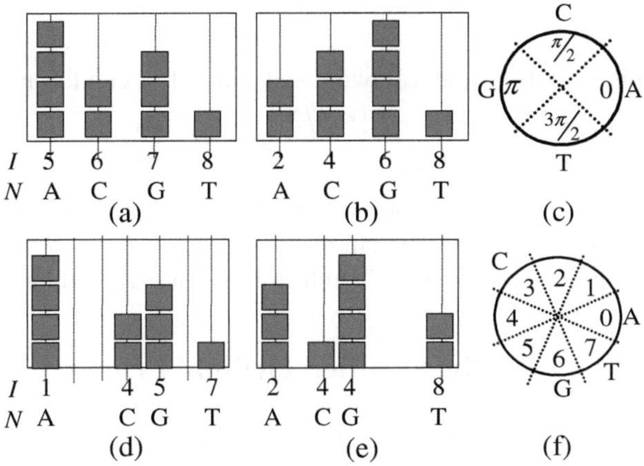

Fig. 3. Examples of ordinal and modulo type histograms

Albeit these indices do not increase uniformly, the concept of the edit distance [8] can be applied to compute the *EMD* with these kinds of cost matrices. For example of the ordinal case, the eqn(4) can be generalized to be

$$EMD_{ord}(H(X),H(Y))=\sum_{i=1}^{b-1}\left((I_{i+1}-I_i)\left|\sum_{j=1}^{i}H_j(X)-H_j(Y)\right|\right) \tag{7}$$

The term, $(I_{i+1} - I_i)$ is the interval difference I'_i between two adjacent bin index values. $I'_i = (I_{i+1} - I_i) = C_{i,i+1} = C_{i+1,i}$. In the *c*-uniform ordinal case, $I_{i+1} - I_i = c$ for all bins. Given an I_{ord}, a C_{ord} can be generated by taking absolute differences. Two different I_{ord1} and I_{ord2} have the same C_{ord} as long as $I'_{ord1} = I'_{ord2}$. To include the non-uniformly increasing ordinal histograms, the original algorithm in [6] for computing the *EMD* for ordinal histograms is generalized as follows.

Algorithm 1: Computing the *EMD* for ordinal histograms	
Input: *H*(*S*), *H*(*T*), and *I*	**Output:** hdist

```
prefixsum = 0
hdist = 0
for i = 1 to b − 1
    prefixsum += H_i(S) − H_i(T)
    hdist += (I_{i+1} − I_i) × |prefixsum|
```

Similarly, the algorithm in [8] for computing the distance between two modulo type histograms can be easily generalized to include the non uniform modulo histograms.

Definition 1: A histogram *H* is nominal iff .

$$C_{i,j}=\begin{cases}0, & \text{if } i=j\\ c, & \text{otherwise}\end{cases}$$

Definition 2: A histogram *H* is ordinal iff the following 3 conditions are met.

1. Elements in *I* must have scalar numeric values.

2. $I_i \le I_j$ if $i < j$.
3. $C_{i,j} = | I_i - I_j |$.

Definition 3: H is modulo iff the following and above 1~2 conditions are met.

3. $$C_{i,j} = \begin{cases} |I_i - I_j|, & \text{if } |I_i - I_j| \le \beta/2 \\ \beta - |I_i - I_j|, & \text{otherwise} \end{cases} \quad \text{where } \beta = \sum_{i=1}^{b} I'_i$$

$I'_b = (I_b - I_1) = C_{b,1}$ since the last bin and the first bin are neighbors. If all bins in a modulo histogram is crammed in the circular segment s.t. $I'_b \ge \beta/2$, it can be realized as an ordinal histogram. Hence, all ordinal histograms are special modulo histograms.

3 Cost Matrices with the Shuffled Ordinal Property

Suppose a histogram is given with C and N only without I. Determining whether C is nominal is trivial. In this section, a method to determine whether C is ordinal is presented. Let R^i be the ith row of C, e.g., R^1 is the first row of C.

Definition 4: A cost matrix C is ordinal iff

1. I $(= R^1)$ must satisfy condition 2 in Definition 2.
2. $C = C^1$ where C^1 is generated by the definition 2 with R^1 as I.

For example, C_y's $R^1{}_y = [0\ 4\ 5\ 1]$ violates the first condition and thus C_y is not ordinal. However, $C_{\text{F3(d)}}$'s $R^1{}_{\text{F3(d)}} = [0\ 3\ 4\ 6]$ and $C^1{}_{\text{F3(d)}} = C_{\text{F3(d)}}$ and thus $C_{\text{F3(d)}}$ is ordinal. In other words, *EMD* with $C_{\text{F3(d)}}$ can be solved in linear time using Algorithm 1 whereas *EMD* with C_y may take long time. Here, the concept of shuffling cost matrix to further include more C's like C_y to the linear time complexity cost matrix.

Definition 5: C_π is a *shuffled cost matrix* of C if N_π is a permutation of N and C_π is generated according to the N_π order.

For example, some C_π's of $C_{\text{F4(d)}}$ with $N_{\text{F4(d)}}$ = [A C G T] include

$$N_{\pi 1} = [A\ \ G\ \ T\ \ C] \quad N_{\pi 2} = [A\ \ G\ \ C\ \ T] \quad N_{\pi 3} = [G\ \ C\ \ T\ \ A]$$

$$C_{\pi 1} = \begin{bmatrix} 0 & 4 & 6 & 3 \\ 4 & 0 & 2 & 1 \\ 6 & 2 & 0 & 3 \\ 3 & 1 & 3 & 0 \end{bmatrix} \quad C_{\pi 2} = \begin{bmatrix} 0 & 4 & 3 & 6 \\ 4 & 0 & 1 & 2 \\ 3 & 1 & 0 & 3 \\ 6 & 2 & 3 & 0 \end{bmatrix} \quad C_{\pi 3} = \begin{bmatrix} 0 & 1 & 2 & 4 \\ 1 & 0 & 3 & 3 \\ 2 & 3 & 0 & 6 \\ 4 & 3 & 6 & 0 \end{bmatrix}$$

EMD using $C_{\text{F4(d)}}$ = *EMD* using any C_π. While $C_{\text{F4(d)}}$ is ordinal, $C_{\pi 1}$, $C_{\pi 2}$, and $C_{\pi 3}$ are not by definition 4. *EMD* using any C_π can be solved in linear time if N_π is reordered.

Definition 6: A cost matrix C is *shuffled-ordinal* if C is a C_π of C_{ord}.

Algorithm 2: *shuffled-ordinality* testing for C
1 R^m = any row of C which contains the maximum value in C.
2 R^s = sort R^m in the ascending order
3 $C^s{}_{ord}$ = the ordinal cost matrix by definition 2 with R^s as I.
4 $C^s{}_\pi$ = the shuffled cost matrix of C according to R^s order
Definition 7: A cost matrix C is *shuffled-ordinal* iff $C^s{}_{ord} = C^s{}_\pi$.

For earlier example of C_x, the maximum value 7 is located at first and last rows. If the last row is selected, $R^m{}_x = [7\ 6\ 2\ 0]$, $R^s{}_x = [0\ 2\ 6\ 7]$ and

$$\left(C^{s}_{x_{ord}} = \begin{bmatrix} 0 & 2 & 6 & 7 \\ 2 & 0 & 4 & 5 \\ 6 & 4 & 0 & 1 \\ 7 & 5 & 1 & 0 \end{bmatrix} \right) \neq \left(C^{s}_{x\pi} = \begin{bmatrix} 0 & 2 & 6 & 7 \\ 2 & 0 & 4 & 1 \\ 6 & 4 & 0 & 5 \\ 7 & 1 & 5 & 0 \end{bmatrix} \right)$$

Hence, C_x is not *shuffled-ordinal.* C_y and C_z are *shuffled-ordinal* though. If cost matrices are symmetric, two rows contain the maximum value. If cost matrices such as C_z violate the metric separation property, more than two rows may contain the maximum value. Whichever row with the maximum is chosen, the *shuffled-ordinal* property testing holds. For the example of C_y, the maximum value 9 appears in the second and third row. Let R^{s1} and R^{s2} be the sorted rows using the second and third rows which contain the maximum value, 9, respectively.

$$N_y^{s1} = [G\ \ A\ \ T\ \ C] \qquad N_y^{s2} = [C\ \ T\ \ A\ \ G]$$

$$R_y^{s1} = [0\ \ 5\ \ 6\ \ 9] \qquad R_y^{s2} = [0\ \ 3\ \ 4\ \ 9]$$

$$C^{s1}_{y_{ord}} = C^{s1}_{y_\pi} = \begin{bmatrix} 0 & 5 & 6 & 9 \\ 5 & 0 & 1 & 4 \\ 6 & 1 & 0 & 3 \\ 9 & 4 & 3 & 0 \end{bmatrix} \qquad C^{s2}_{y_{ord}} = C^{s2}_{y_\pi} = \begin{bmatrix} 0 & 3 & 4 & 9 \\ 3 & 0 & 1 & 6 \\ 4 & 1 & 0 & 5 \\ 9 & 6 & 5 & 0 \end{bmatrix}$$

The algorithm 2 finds that C_y is *shuffled-ordinal* using either R^{s1} or R^{s2}. The new histograms H^{s1} and H^{s2} have different orders though. Indeed, N^{s1} is the reverse order of N^{s2} and C^{s1} is the 180° rotated matrix of C^{s2}. As depicted in Figure 4, H^{s1} and H^{s2} are mirror images of each other and ordinal. Figure 4 assures that

$$EMD_y(H(S), H(T)) = EMD_{s1}(H^{s1}(S), H^{s1}(T)) = EMD_{s2}(H^{s2}(S), H^{s2}(T)) = 1.$$

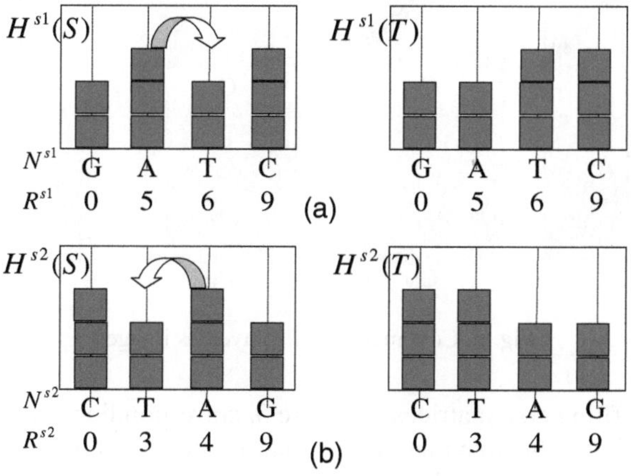

Fig. 4. Edit distances between *Shuffled ordinal* histograms

In order to visualize cost matrices, twelve 144×144 cost matrices are generated randomly as described in Table 1. Figure 5 shows the cost matrices as images; the darker means the closer to zero and the brighter means the closer to the maximum.

Table 1. Twelve 144×144 cost matrices with their metric properties

C	Generation method	Iden tity	sepa-ration	Non -neg	Sym-metry	Tri Ineq
C_1	Ordinal ($I_{i+1} - I_i = c$).	1	1	1	1	1
C_2	Ordinal ($I_{i+1} \geq I_i$).	1	0	1	1	1
C_3	Ordinal ($I_{i+1} > I_i$).	1	1	1	1	1
C_4	Nominal	1	1	1	1	1
C_5	Modulo ($I_{i+1} - I_i = c$).	1	1	1	1	1
C_6	Modulo ($I_{i+1} > I_i$).	1	1	1	1	1
C_7	*Euclidean* b/w 12×12 2D bins.	1	1	1	1	1
C_8	*Manhattan* b/w 12×12 2D bins.	1	1	1	1	1
C_9	Symmetric ($C_{i,j+1} > C_{i,j}$) for $j \geq i$.	1	1	1	1	0
C_{10}	Symmetric random with 0	1	0	1	1	0
C_{11}	Symmetric random with Separation	1	1	1	1	0
C_{12}	Totally random	0	0	1	0	0

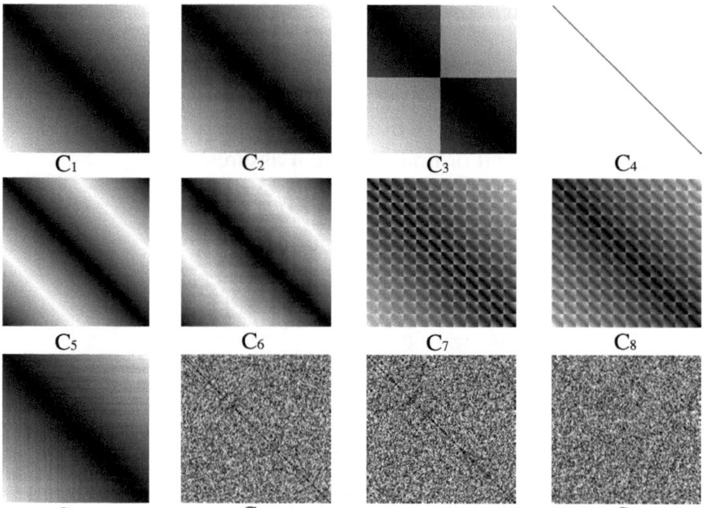

Fig. 5. Cost matrices displayed as images

Randomly shuffling cost matrices in Figure 5, corresponding shuffled cost matrices are displayed in Figure 6. Note that the cost matrix properties in Table 1 are preserved even after shuffling. Identity and symmetry properties are still visible in shuffled cost

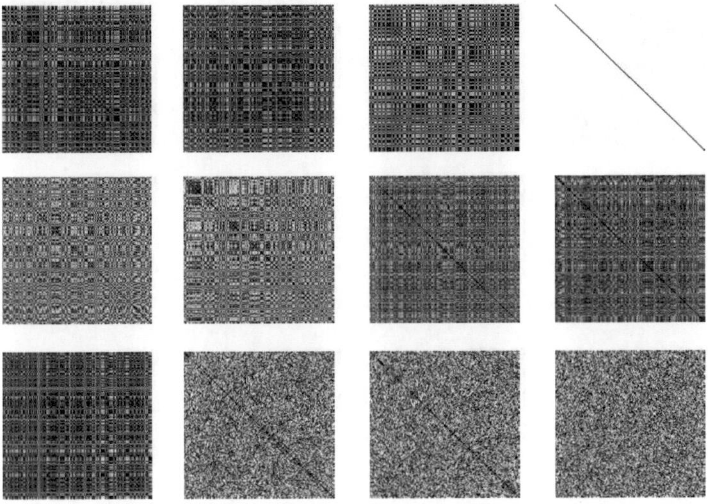

Fig. 6. Randomly shuffled cost matrices of those in Figure 5

matrices C_1~C_{11}. Note that a nominal type cost matrix is *shuffling invariant* because shuffling does not change the shape of the cost matrix.

Given twelve cost matrices in Figure 6, this section is concerned with the question of which cost matrices are *shuffled-ordinal*. According the algorithm 2, Figure 7 and Figure 8 shows the C^s_π's and C^s_{ord}'s for cost matrices in Figure 6, correspondingly. Only C_1, C_2, and C_3 have $C^s_\pi = C^s_{ord}$.

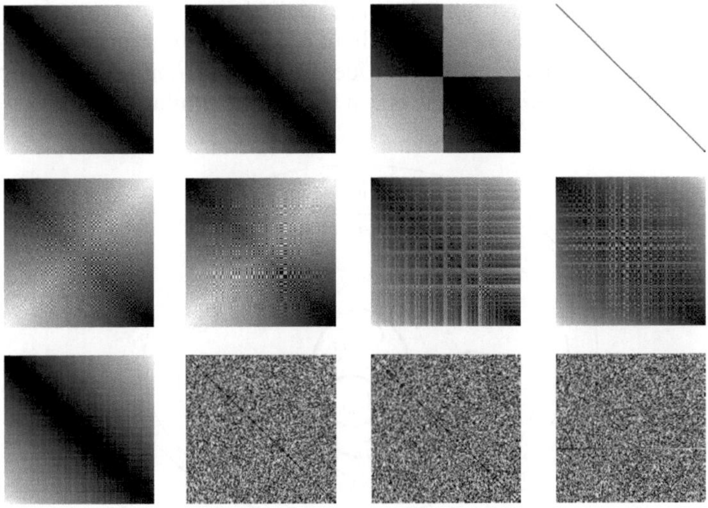

Fig. 7. C^s_π's for cost matrices in Figure 6

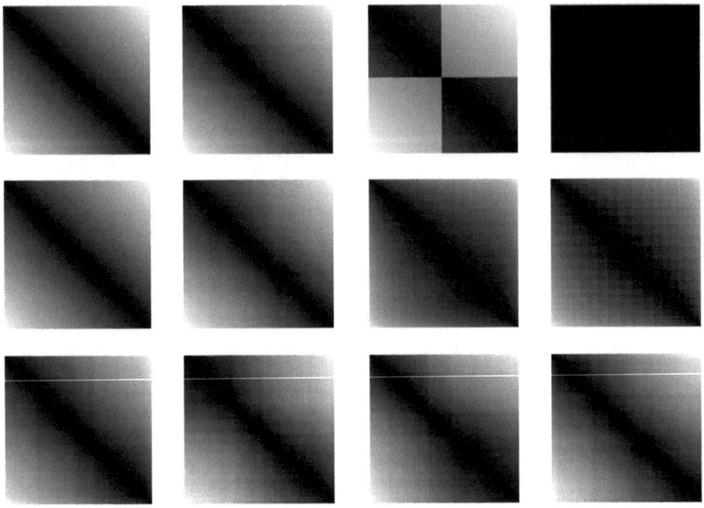

Fig. 8. C^{s}_{ord}'s for cost matrices in Figure 6

4 Discussions

In this paper, three special types of cost matrices are formally defined: nominal, ordinal, and modulo. Computing *EMS* with these types of cost matrices can be computed very efficiently: $\Theta(b)$, $\Theta(b)$, and $O(b^2)$, respectively. This paper considered the problem of identifying types of cost matrices. A method to identify whether a given cost matrix is *shuffled-ordinal* was presented. If so, the histograms can be reordered and the linear time complexity algorithm 1 can be applied.

When only the cost matrix is given without types of histogram information, identifying the nominal property is trivial but identifying the *shuffled-modulo* property is non-trivial. *Shuffled-modulo* cost matrix can be defined similar to definition 5.

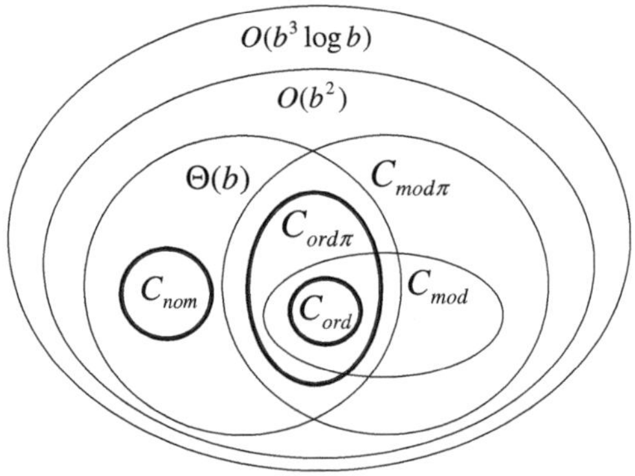

Fig. 9. Sets and their histograms

Definition 8: A cost matrix C is *shuffled-modulo* if C is a C_π of C_{mod}.

Finding the algorithmic definition for $C_{mod\pi}$ similar to definition 7 is an open problem.

Figure 9 depicts the cost matrices with their best known algorithm complexities. Lower bound algorithms are known only for nominal and ordinal types. Defining new types of cost matrices and finding better algorithms are open problems.

Cost matrices such as *BLOSUM* [9] and *PAM* matrices [10] in bio-informatics are often indexed by the alphabetical order of *N*. Indeed, *BLOSUM* had been the inspiration for this article although it does not have identity, non-negativity properties. Analyzing cost matrices in the real world applications is future works.

References

1. Duda, R.O., Hart, P.E., Stork, D.G.: Pattern Classification, 2nd edn. Wiley, New York (2000)
2. Rubner, Y., Tomasi, C., Guibas, L.J.: A metric for distributions with applications to image databases. In: Proc. of ICCV, pp. 59–66 (1998)
3. Levina, E., Bickel, P.: The Earth Mover's distance is the Mallows distance: some insights from Statistics. In: Proc. of ICCV, pp. 251–256 (2001)
4. Hiller, F.S., Lieberman, G.J.: Introduction to mathematical programming, 2nd edn. McGraw-Hill, New York (1995)
5. Ahuha, R.K., Magnanti, T.L., Orlin, J.B.: Network Flows. Prentice Hall, Englewood Cliffs (1993)
6. Cha, S.-H.: Fast Image Template and Dictionary Matching Algorithms. In: Chin, R., Pong, T.-C. (eds.) ACCV 1998. LNCS, vol. 1351, pp. 370–377. Springer, Heidelberg (1997)
7. Cha, S.-H., Srihari, S.N.: Distance between Histograms of Angular Measurements and its Application to Handwritten Character Similarity. In: Proceedings of 15th ICPR 2000, September 3-8, 2000, pp. 21–24. IEEE Computer Society, Barcelona (2000)
8. Cha, S.-H., Srihari, S.N.: On Measuring the Distance between Histograms. Journal of Pattern Recognition 35(6), 1355–1370 (2002)
9. Henikoff, S.: Amino Acid Substitution Matrices from Protein Blocks. In: PNAS 1989, pp. 10915–10919 (1992)
10. Dayhoff, M.O., Schwartz, R.M., Orcutt, B.C.: A model of evolutionary change in proteins, matrices for detecting distant relationships. In: Atlas of Protein Sequence and Structure, vol. 5, pp. 345–358. National Biomedical Research Foundation, Washington DC (1978)

Using Non Local Features for 3D Shape Grouping

Antonio Adán[1], Miguel Adán[1], Santiago Salamanca[2], and Pilar Merchán[2]

[1] Universidad de Castilla La Mancha. Escuela Superior de Informática. Ciudad Real. Spain
[2] Universidad de Extremadura. Escuela de Ingenierías Industriales. Badajoz. Spain
antonio.adan@uclm.es, miguel.adan@uclm.es, ssalaman@unex.es, pmerchan@unex.es

Abstract. This work faces the problem of 3D shape clustering when the whole surface information is available. The key of our method is to use a flexible feature, called Cone-Curvature, which provides local and extended information around every node of the mesh that represents the object. Thus as we increase the region around a node a new order of CC can be calculated. This feature, which was originally defined on spherical representation, has been adapted to work with standard triangular meshes and it is used for defining a similarity measure between shapes. Through a PCA technique, the dimensionality of the shape representation is drastically reduced and the hierarchical grouping can be efficiently carried out. This method has been tested under real conditions for a wide set of free shapes yielding promising results. We present a discussion of the clustering comparing human and computer results.

Keywords: Object clustering, Shape similarity, 3D representation, 3D models.

1 Introduction

3D shape clustering is an old topic on which many researchers continue working. From a human viewpoint the similarity is an ambiguous and relative concept. Intuitively, we put two objects in the same group taking into account the external appearance or some local/global features, but we can not be sure if we are right. Thus, different results depending on the perception of each person may be obtained [1]. Anyway, the similarity between two shapes is usually determined as a distance between some extracted features.

Roughly speaking, the clustering's goal is to achieve the best partition over a set of objects stored on a database in terms of similarity. For partitioning we need to extract or define features of each object in such a way as to be well characterized.

Similarity-base clustering is a simple technique that uses a similarity measure to guarantee if two objects are similar enough to put them in the same cluster. The similarity measure is usually defined through features of the object. In this sense, a good similarity measure is essential to carry out further clustering tasks.

As we know, there is a wide variety of approaches that propose different solutions to compare objects. In 3D environments, most of them use representation models based on geometrical descriptors and establish similarity relationships by means of feature matching procedures.

N. da Vitora Lobo et al. (Eds.): SSPR\&SPR 2008, LNCS 5342, pp. 644–653, 2008.

Local features describe the geometry of little regions close to a set of nodes (or vertices) of the mesh that represent the object. Thus, every point of the surface of the object has knowledge of the features of its immediate neighbours. Geometrical features such as simplex angle [2], curvature index [3] and integral Gaussian curvature [4], [5] correspond to these kinds of features. For a mesh representation, this means that a given node is handled as a single item isolated from the complete mesh and without any relationship to the remaining nodes. Usually these techniques are addressed to solve recognition and pose problems in occlusion circumstances more than for clustering purposes. Some of the most known techniques are as follows. Dorai and Jain [6] use a local curvature feature called shape index and define a disimilarity measure by comparing their histograms for several moments. Johnson and Hebert [7] compare two objects through spin image concept. In [8,9] (Yamany and Farag) a representation that stores surface curvature information from certain points, produces images, called surface signatures, at these points. As a result of this, a standard Euclidean distance to match objects is presented. Chua and Jarvis code surroundings information at a point of the surface through a feature called Point Signature [10].

On the other hand, methods that consider global characteristics are more suitable for carrying out clustering solutions because, contrary to local strategies, they define the whole shape and are not correlated with the observer viewpoint. Vandeborre, Couillet and Daoudi [11] obtain three distances by comparing local and global feature histograms. Adán et. al. [12], propose a recognition process where several distances, based on invariant global features, sequentially reduce the original model database until the object is identified. Liu, Sun, Kang and Shum propose the Directional Histogram Model for objects that have been completely reconstructed [13]. For each viewpoint, they obtain the histogram of thickness distribution that is later formulated with spherical harmonics.

2 A Flexible 3D Shape Descriptor

Adaptable strategies are rarely considered in literature. In this case, every node of the model has an extended knowledge of its surroundings and all or part of it can be used for further tasks. The main difficulty in using that idea, in a solid modeling environment, is to define a model with an appropriate topology and connectivity.

The models MWS (*Modeling Wave Set*) [14] have been designed to maintain new and wider relationships among subsets of mesh nodes. Contrary to most strategies, MWS structures allow us to compare two objects taking any part of information from the model. So, local, half, global, contiguous or discontiguous spatial information could be chosen to recognize an object. In order to express this concept of adaptability to different specifications, we have labelled our method as 'flexible'. Consequently, an adaptable (or flexible) similarity measure contrary to previous fixed similarity measures is defined in our case.

MWS was originally defined on a mesh of nodes with a 3-connectivity relationship from the tessellation of the unit sphere. This representation, with a constant number of nodes, has been efficient in solving recognition problems in medium database but with some shape limitations. This is due to the fact that a poor realistic geometric representation is yielded with such a mesh. In this work, we have extended MWS structure to a semi-regular triangular mesh which involves three evident improvements with respect to the original version. Firstly, we can use the standard topology in most of the CAD tools and 3D scanner software. Secondly, we achieve a realistic representation where the appearance of the object is as accurate as possible. As a consequence of this, old shape restrictions are now avoided and 3D clustering can be carried out more efficiently.

Let T be a triangular mesh fitted to the object and let N be a vertex of T. Note that T has been previously regularized and resampled to a fixed number of nodes *h*. *Modeling Wave* (MW) organizes the rest of the nodes of the mesh in concentric groups spatially disposed around N. Since this organization resembles the shape of a wave it is called *Modeling Wave (MW)*. Consequently, each of the disjointed subsets is known as *Wave Front (WF)* and the initial node N is called *Focus*. Let us call all the possible MWs that can be generated over T *Modeling Wave Set (MWS)*. Figure 1 illustrates the mesh model of an object and several *wave fronts* plotted for different *focuses*.

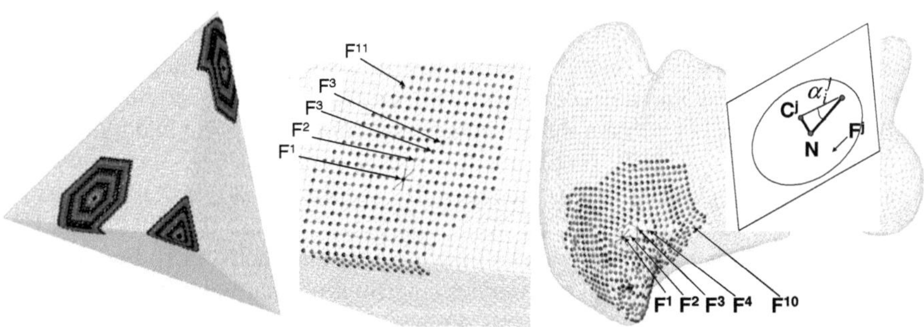

Fig. 1. *Wave fronts* in different *focuses* printed over the mesh model (left). Representation of j-th cone curvature in a point of the object (right).

Cone Curvature is defined taking into account MWS structure. We define *Cone Curvature* α^j of N∈T, the angle of the cone with vertex N whose surface inscribes the *j*th *Wave Front* of the *Modeling Wave* associated to N.

The range of CC values is [-π/2, π/2], being the sign assigned taking into account the relative location C^j with respect to T, C^j being the barycentre of the *j*th WF. Figure 1 right illustrates this definition and shows examples for positive and negative cases.

Note that a set of values $\{\alpha^1, \alpha^2, \alpha^3, \ldots \alpha^q\}$ gives an extended curvature information around N until the *q*th WF, where the word 'curvature' has a non-local meaning. So for each node N a set of *q* values could be used for exploring its surroundings

On the other hand, it can be said that a vector

$$V_{CC} = \{c^1, c^2, c^3 \ldots, c^q\},\ c^j : T \rightarrow [-\pi/2, \pi/2] \quad j = 1,\ldots q \tag{1}$$

q being the number of *Wave Front* considered, is established for all the nodes of T. Then the whole Cone Curvature information is stored in a CC-matrix M_{CC}, of *hxq* dimension. Note that M_{CC} is invariant, unless row permutations, to changes in the pose of the object. A wider explanation of CC concept can be found in [15].

The main properties are briefly mentioned as follows. Figure 2 illustrates a visualization of those, where the CC value is plotted over the mesh in a specific color code.

- CC is robust to noise. Small variations in the location of the nodes in the mesh, involve high variations in first orders of CC whereas for upper orders such variations do not meaningfully affect it. It can be said that an error filtering effect appears as the order increases. Keeping this in mind, local features are not suitable enough for characterizing an object unless errors are minimized.

- Spatial continuity of CC. It means that CC along any path of nodes is a continuous function for any CC-order considered. A visualization of this property, in the discrete sense, can be appreciated in Figure 2 where the CC value is plotted over the mesh in a specific code colour.

- Continuity on the CC-order. It can be said that there is also continuity in the CC vector $\{\alpha^1, \alpha^2, ...\alpha^q\}$. In other words, CC of every node of the mesh smoothly changes as the order grows.

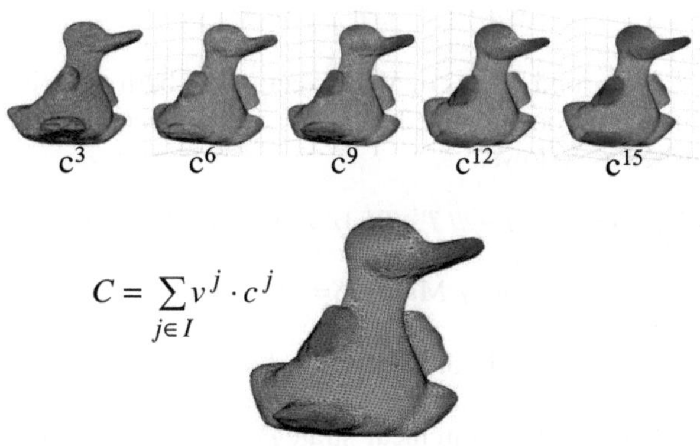

Fig. 2. Illustration of several CC orders (above) and the weighted cone curvature after considering from third until the 15-th CCs

3 Hierarchical Grouping

The hierarchical clustering algorithm is carried out after reducing the CC-matrix M_{CC} to a single vector. Our purpose is to obtain a single value for each row of the matrix M_{CC} from the analysis of the principal components performed on all the rows, so that each row is reduced to a single representative value. Then, by means of the adequate linear combination, for each node N, a single variable C will fuse the q values provided by its CCs. Therefore, every node will have an associated variable C that is called Weighted Cone Curvature (WCC). Consequently, the matrix M_{CC} is reduced to a vector of *hx1* dimension.

Formally, the variable Weighted Cone Curvature can be defined as a weighted combination of the different orders of CCs given by the expression:

$$C = \sum_{j \in I} v^j \cdot c^j \tag{2}$$

where j is extended to any set of CC-orders. Thus, we can choose contiguous or not contiguous *wave fronts,* near or far from the *focus* depending on the specifications of the experiment and the kind of shapes. Therefore, if we choose j-th CC we know exactly what it is because it is inside the whole structure MW. Therefore CC is not an isolated concept. This makes our representation more compact than others ([7,8,9,10].

Coefficients v^j in expression (2) are the coordinates of the eigenvector associated to the highest eigenvalue of the covariance matrix for the set of indexes I considered. These coefficients have been empirically determined by evaluating the principal component analysis over the models of our object database in our lab.

We define the distance d between two 3D shapes through their models T^i and T^j as follows:

$$d(T^i, T^j) = \left| C^i - C^j \right| = \sqrt{\sum_{k=1}^{h} (C^i(k) - C^j(k))^2} \tag{3}$$

where C^i and C^j are sorted values of WCC vectors for both models.

Once the distance d has been defined and considering a model database a *similarity function* is established as follows.

$$s(T^i, T^j) = s_{ij} = 1 - d(T^i, T^j) / max\{ d(T^i, T^j), \forall i, j \} \tag{4}$$

Finally we can define a Similarity Matrix $S = (s_{ij})$ which stores the whole similarity information for a database.

There are two major methods of clustering: hierarchical clustering and k-means clustering. We have applied the hierarchical clustering method based on the Similarity Matrix S. As it is known, in hierarchical strategies the data (in our case S) are not partitioned into a set of groups in a single step. Instead, a series of partitions takes place, which may run from a single cluster containing all objects to m clusters each containing a single object. Hierarchical Clustering is subdivided into agglomerative methods, which proceed by series of fusions of the m objects into groups, and divisive methods, which separate m objects successively into finer groupings. The result may be represented by a dendogram which illustrates the fusions or divisions made at each successive stage of analysis.

Among the different ways of measuring the distance between two clusters of objects, we have chosen the Ward's linkage which uses the incremental sum of squares; that is, the increase in the total within-cluster sum of squares as a result of joining two clusters. The within-cluster sum of squares is defined as the sum of the squares of the distances between all objects in the cluster and the centroid of the cluster. For clusters r and s the equivalent distance is given by:

$$d^2(r,s) = n_r n_s \frac{\left| \bar{x}_r - \bar{x}_s \right|^2}{n_r + n_s} \tag{5}$$

where |.| is Euclidean distance, $\overline{x}_r$ and $\overline{x}_s$ are the centroids, n_r and n_s are the respective number of objects in cluster r and s.

4 3D Shape Clustering Experimental Test

The hierarchical method proposed in this article has been verified using synthetic as well as real range data from a 3D laser scanner. We have built a database of 115 objects where 40 of them are models corresponding to real objects which have been built using a MINOLTA 910 scanner, 40 correspond to synthetic shapes which have been generated from CAD software and the rest were obtained deforming all those through smoothing, sharpening and bending processes. In this paper we present an experiment in which the clustering result is seemingly predictable for a human (see Figure 3).

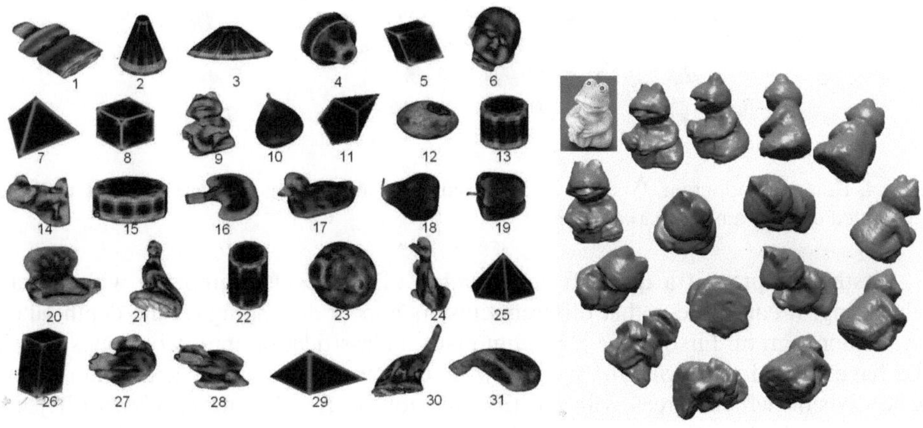

Fig. 3. Left: Objects database B. The colour code printed in the model mesh corresponds to the *weighted cone curvature* values. Right: synthetic model explored from multiple viewpoints.

Ten people were asked to perform the clustering procedure together over a selected database of 31 objects. A 3D visualization tool helped people to see the objects from any viewpoint (Figure 3 right). In order to compare the results of this experiment to those computed by our method, we asked them to carry out a splitting procedure in which the group of objects was iteratively separated into two parts until the end. So, after a consensus between all persons, a human dendogram Λ' was built. This is shown in figure 4. No conditions were imposed upon the number, the names of the groups and the belonging of each object to a specific group. The only requirement was to apply shape criteria.

On the other hand, we obtained the hierarchical cluster tree Λ using the minimum variance criteria within-clusters according to the Ward's linkage (see Figure 6). In this paper we only show the results obtained for *weighted cone curvatures* corresponding to orders 1 to 13.

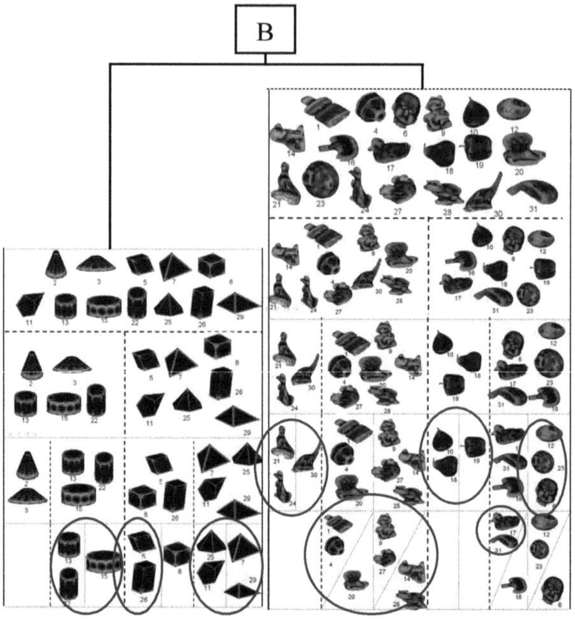

Fig. 4. Human dendogram Λ' built after consensus of ten persons. Coherent groups with the computed dendogram are marked on it.

Firstly, we present a qualitative evaluation of the results compared with human clustering. It can be seen that different clusters appear depending on the depth taken in the dendogram. Firstly, our algorithm gives a classification into two groups which we have labeled as *polyhedral* and *free-form* objects. This is an expected and reasonable division which agrees with the first division in Λ'. Among *polyhedral* objects we discussed what groups could be similar in Λ' and what label should be assigned to them. Finally we found Pyramids (I), tetrahedrons (II), paralelepipeds (III) and cylinders (IV). Curiously the cube (object 8) is more similar to cylinders than to paralelepideps and also two cones are not put in the same groups in the last level. Concerning *free-form* group we distinguish five groups, some of them appear in the tree Λ'. For instance we find *round* objects (X) and *elongated* objects (IX). Group VIII are also rounded but with a point at its end. Finally, groups VI and VII are difficult to understand and to label.

In order to provide a quantitative analysis of our approach, a concordance measure between groups of Λ and Λ' has been defined using a vote strategy. Let *a* and *a'* be two groups, $a \in \Lambda$ and $a' \in \Lambda'$. The concordance of group *a* in the tree Λ' is defined as:

$$H(a,a') = e^{-\frac{|V(a,a')-n_a|}{n_a}} \tag{6}$$

$$V(a,a') = max\{\sum_{o \in a} \delta(o,a') + \sum_{o \in a'} \delta(o,a)\} \tag{7}$$

where n_A is the number of objects in *A* and δ is the voting function

$$\delta(o,a')=\begin{cases}1 & o\in a'\\ -1 & o\notin a'\end{cases} \qquad \delta(o,a)=\begin{cases}1 & o\in a\\ -1 & o\notin a\end{cases} \tag{8}$$

Obviously, H is not a symmetrical matrix where the i-th column represents the concordance of the i-th group of Λ in Λ'. H is a normalized function in the range [0,1], where $H(a,a')=1$ if $a\equiv a'$ and $H(a,a')\rightarrow 0$ if any object of a is in a'.

In order to graphically illustrate H, we have presented it on a two dimension grey map where the grey level goes from the minimum value (0), plotted in black, to the maximum (1), plotted in white (Figure 5 left). In this figure, we have ranked H with respect to the computed clustering Λ. Note that we consider all subgroups that are generated throughout the whole splitting process and that the number of cluster generated in both dendograms is different (40 in Λ and 36 in Λ'). Looking at this figure we can appreciate high concordance between dendograms Λ and Λ' in several points. The clearer H, is the more concordance between Λ and Λ'.

Figure 5 right gives a more extended analysis taking into account the H value of the groups in which an object is classified along the tree Λ. In this figure we represent, for each one of the objects of the database, the average value of those values and the standard deviation. Finally, respective global mean values are calculated and pointed out in this figure superimposing a horizontal line. These measures provide a qualitative idea of whether an object is classified in more or less concordant groups throughout the whole clustering process. This means that, in a general sense, some objects are expected to be better classified than others. For instance, objects 11, 22 and 25 have average values above 0.8 whereas objects 2, 4 and 6 have values below 0.6. The global mean value was 0.72.

In figures 4 and 6 we have marked the groups with a high concordance measure. Groups with H values between 0.5 to 1 are pointed out by means of a coloured spot plotted on the division points. Note that the results in this experiment are surprising.

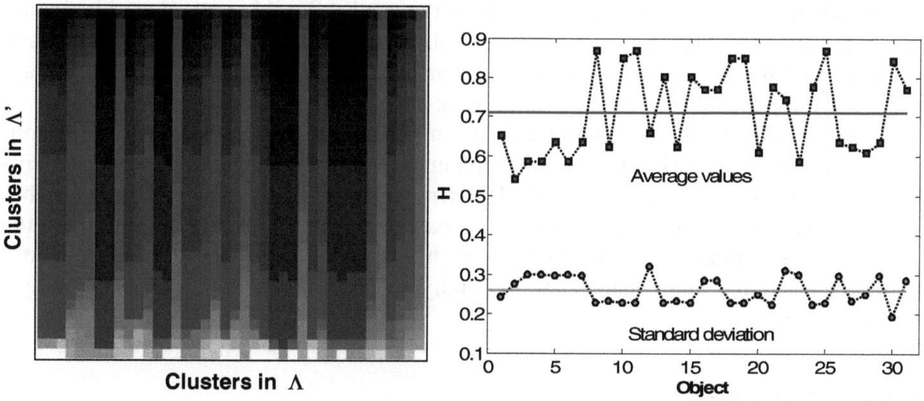

Fig. 5. Left: Comparison of clusters Λ and Λ' through H. Ordered values of H for each one of the groups generated in Λ. Right: Average values of H per object.

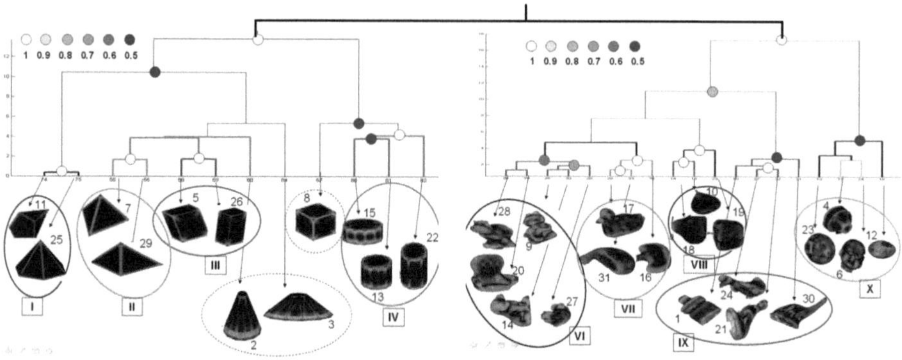

Fig. 6. Computed dendogram Λ. The groups with high concordance measure H are pointed out by means of coloured spots. Interesting groups in the last levels, with a notable value H, are also marked.

Thus 84% of the objects are classified in coherent groups with H values higher than 0.5, whereas only 16% (objects 2,3 ,8, 12 and 30) is put in unexpected groups or appear as isolated objects.

In conclusion we can say that clustering based on cone-curvature features can yield very good results.

5 Conclusions

Clustering is a well known topic in the image processing field. Roughly speaking, the clustering's goal is to achieve the best partition over a set of objects stored in a database in terms of similarity. This may become a tricky problem in 3D environments particularly when only shape information is considered.

For partitioning 3D shape databases we have used a flexible and adaptable feature which totally characterizes each object. This feature, called Cone Curvature, and originally defined on spherical representation models, has been adapted to standard triangular meshes and further used for solving clustering problems.

It can be said that the hierarchical clustering method has worked in a real environment which implies working with noise and error sources. In general, the experiments have yielded coherent and expected clustering results. Thus, high concordance between computed and human dendograms has been found. Nevertheless, we have started several initiatives to improve the method in two ways: firstly, to cope with a higher database and more shapes and, secondly, to cope with more exhaustive clustering evaluation.

Acknowledgment

This work is supported in the following projects: DPI2006-14794-C02 and PCI08-0052-1401 by the National and Regional Ministries of Education and Science.

References

[1] Santini, S., Jain, R.: Similarity measures. IEEE Trans. Pattern Anal. Mach. Intell 21(9), 871–883 (1999)

[2] Delinguette, H.: Simplex Meshes: A General Representation for 3D Shape Reconstruction, Technical Report 2214, INRIA, France (1994)

[3] Koenderink, J.J., Van Doorn, A.J.: Surface shape and curvature scales. Image and Vision Computing 10(8), 557–564 (1992)

[4] Dyn, N., Hormann, K., Kim, S.J., Levin, D.: Optimizing 3D Triangulations Using Discrete Curvature Analysis. In: Mathematical Methods for Curves and Surfaces, pp. 135–146. Vanderbilt University Press (2000)

[5] Alboul, L., Van Damme, R.: Polyhedral metrics in surface reconstruction. In: Mullineux, G. (ed.) The Mathematics of Surfaces VI, pp. 171–200. Clarendon Press, Oxford (1996)

[6] Dorai, C., Jain, A.K.: COSMOS - A Representation Scheme for 3D Free-Form Objects. IEEE Trans. PAMI 19(10), 1115–1130 (1997)

[7] Johnson, A.E., Hebert, M.: Recognizing Objects by Matching Oriented Points. In: IEEE Conf. Comp. Vision and Pattern Recogn, Puerto Rico, June, pp. 684–689 (1997)

[8] Yamany, S.M., Farag, A.A.: Free-Form Surface Registration Using Surface Signatures. In: Proc. IEEE Intern. Conf. on Computer Vision, vol. 2, pp. 1098–1104 (1999)

[9] Yamany, S.M., Farag, A.A.: Surfacing Signatures, An Orientation Independent Free-Form Surface Representation Scheme for the Purpose of Objects Registration and Matching. IEEE Trans. Pattern Anal. Mach. Intell 24(8), 1105–1120 (2002)

[10] Chua, C.S., Jarvis, R.: Point Signatures, A New Representation for 3D Object Recognition. International Journal of Computer Vision 25(1), 63–85 (1997)

[11] Vandeborre, J.-P., Couillet, V., Daoudi, M.: A practical approach for 3D model indexing by combining local and global invariants. In: 1st Int. Symp. On 3D Data Processing Visualization and Transmission, Padova, Italy, pp. 644–647 (2002)

[12] Adán, A., Cerrada, C., Feliu, V.: Global Shape Invariants: A Solution For 3D Object Discrimination/ Identification Problem. Pattern Recog. 34, 1331–1348 (2001)

[13] Liu, X., Sun, R., Kang, S.B., Shum, H.Y.: Directional Histogram Model for Three-Dimensional Shape Similarity. In: CVPR 2003, vol. 1, pp. 813–820 (2003)

[14] Adán, A., Cerrada, C., Feliú, V.: Modeling Wave Set: Definition and Application of a new Topological Organization for 3D Object Modeling. Computer Vision and Image Understanding 79, 281–307 (2000)

[15] Adán, A., Adán, M.: A Flexible Similarity Measure for 3D Shapes Recognition. IEEE Transactions on Pattern Analysis and Machine Intelligence 26(11), 1507–1520 (2004)

A Dynamic Programming Technique for Optimizing Dissimilarity-Based Classifiers*

Sang-Woon Kim[1] and Jian Gao[2]

[1] Senior Member, IEEE. Address: Dept. of Computer Science and Engineering, Myongji University, Yongin, 449-728 South Korea
kimsw@mju.ac.kr

[2] Address: Dept. of Computer Science and Engineering, Myongji University, Yongin, 449-728 South Korea
marsgao@mju.ac.kr

Abstract. The aim of this paper is to present a dissimilarity measure strategy by which a new philosophy for pattern classification pertaining to dissimilarity-based classifiers (DBCs) can be efficiently implemented. Proposed by Duin and his co-authors, DBCs are a way of defining classifiers among classes; they are not based on the feature measurements of individual patterns, but rather on a suitable dissimilarity measure among the patterns. The problem with this strategy is that we need to measure the inter-pattern dissimilarities for all the training samples to ensure there is no zero distance between objects of different classes. Consequently, the classes do not overlap, and therefore, the lower error bound is zero. In image classification tasks, such as face recognition, one of the most intractable problems is the distortion and lack of information caused by the differences in face directions and sizes. To overcome the above problem, in this paper, we propose a new method of measuring the dissimilarity distance between two images of an object when the images have different directions and sizes and there is no direct feature correspondence. In the proposed method, a dynamic programming technique, such as dynamic time warping, is used to overcome the limitation of one-to-one mapping. Furthermore, when determining the matching templates of two images in dynamic time warping, we use a correlation coefficient-based method. With this method, we can find an optimal warping path by surveying the images in a one-dimensional or two-dimensional way (that is, with vertical-only scanning or vertical-horizontal scanning). Our experimental results demonstrate that the proposed mechanism can improve the classification accuracy of conventional approaches for an artificial data set and two real-life benchmark databases.

1 Introduction

One of the most recent and novel developments in the field of statistical pattern recognition (PR) [1] is the concept of dissimilarity-based classifiers (DBCs) proposed by Duin and his co-authors (see [2] and [3]). Philosophically, the motivation for DBCs is as follows: If we assume that similar objects can be grouped together to form a class, the "class" is nothing more than a set of those similar objects [4]. On the basis of this idea, Duin and his colleagues argue that the notion of proximity (similarity or dissimilarity) is

* This work was supported by the Korea Research Foundation Grant (KRF-2007-313-D00714).

N. da Vitora Lobo et al. (Eds.): SSPR&SPR 2008, LNCS 5342, pp. 654–663, 2008.

actually more fundamental than that of a feature or a class. Indeed, it is more likely that the brain uses an intuitive DBC-based methodology rather than taking measurements, inverting matrices, and so on. Thus, DBCs are a way of defining classifiers among the classes; and the process is not based on the feature measurements of individual patterns, but rather on a suitable dissimilarity measure among the individual patterns [4]. The advantage of this method is that because it does not operate on class-conditional distributions, the accuracy can exceed the Bayes error bound; moreover, the method actually attempts to attain a zero-error bound [1], - which, in our opinion, is remarkable. Another salient advantage of such a paradigm is that it can avoid the problems associated with feature spaces, such as the "curse of dimensionality" [1], and the issue of estimating a large number of parameters.

In spite of the advantages of DBCs, several problems need to be resolved when they are applied to particular tasks, such as face recognition [5], [6]. The three major questions we encountered when designing the DBCs concern the selection of prototype subsets [3], [4], [7], [8]; the measurement of dissimilarities between object samples [9], [10], [11]; and the design of a classifier in the dissimilarity space [2], [12]. Various methods have been proposed in the literature [3], [4], [7] as a means of selecting a representative set of data that is both compact and capable of representing the entire data set. The details of these methods are omitted here because they are not directly related to the premise of our work. In the interest of completeness, however, we now attempt to explain the other two questions in the present paper.

Optimization of DBCs: There are a few ways by which the classification efficiency of DBCs can be optimized. The major task of this study is to deal with how the dissimilarity measure can be effectively computed. The reason we set this task as our goal comes from the necessity to measure the inter-pattern dissimilarities for all the training samples such that there is no zero distance between objects of different classes. Consequently, the classes do not overlap, and therefore, the lower error bound is zero. The application of DBCs for image classification, such as face recognition, requires accurate measurement of the dissimilarity between high-dimensional samples. Furthermore, in facial images, there are many kinds of variations based on such factors as pose (direction), facial expression, illumination, and distance. Thus, by simply measuring the differences in facial images for each class, we are unable to obtain a good representation. If a limited number of prototypes is available or the representational capability is insufficient to cover the possible variations of data, it is difficult to improve the performance of DBCs in the dissimilarity space.

To overcome this limitation and thereby improve the performance of DBCs, we propose another way of enriching the representational capability of dissimilarity measures. In particular, this goal can be achieved by using a dynamic programming technique [2] to

[1] The idea of the zero-error bound is based on the fact that dissimilarities may be defined such that there is no zero distance between objects of different classes. Consequently the classes do not overlap, and so the lower error bound is zero. We are grateful to Bob Duin for providing us with insight into this.

[2] Dynamic programming, in which the word programming means finding an acceptable plan of action, can deal with the local deformation or warping of the sequences by finding an optimal alignment (warping path) between two input sequences.

overcome the limitation of one-to-one mapping; the technique involves a mathematical procedure of solving sequential decision problems in operations research. In PR applications, various dynamic programming techniques have been proposed in the literature to solve the problems of time alignment, normalization, localization, and so on [13], [14]. Dynamic programming techniques that could be used include dynamic template warping [15], dynamic space warping [16], and affine invariant dynamic time warping [17]. However, for the purpose of simplifying the classification problem, we used dynamic time warping (DTW) [14], one of the most well-known dynamic programming techniques. In DTW, one-to-many points (vectors) mapping is performed after two respective subregions have been determined from the corresponding images; the purpose of the mapping is to find the optimal alignment between two image samples that have different directions or sizes. For greater accuracy of the classification, we propose a correlation coefficient-based DTW method in which matching templates from the two images are found by using a correlation coefficient based method with vertical and horizontal scanning. In the proposed method, the dissimilarity is measured in two steps: first we adjust the faces by finding the best warping path, w, with the correlation coefficient-based DTW method; we then compute the dissimilarity distance between the adjusted faces with conventional measures.

Contributions of the Paper: Our study makes two modest contributions to this field:

1. Our experimental results are the first to optimize [3] dissimilarity-based classifiers by resorting to a DTW dynamic programming technique. Although our results apply only to a case of appearance-based face recognition, the proposed approach can also be used in other high-dimensional classification tasks, such as multimedia retrieval, bioinformatics, computer vision, and text classification.

2. The proposed approach combines general distance measures traditionally associated with dissimilarity-based classifiers with the ability of the dynamic programming technique to accommodate variations in direction and size. The dynamic programming technique is used to establish one-to-many (or many-to-one) mapping as well as one-to-one mapping between the deformable images of an object class.

To the best of the our knowledge, these two contributions are novel to the field of high-dimensional image classification. The remainder of the paper is organized as follows: In Section 2, after providing a brief introduction to the dissimilarity representation, we present an algorithm that improves the classification accuracy of DBCs by utilizing dissimilarity representation and a dynamic programming technique. In Section 3, we present the experimental results of an artificial image data set and two real-life benchmark databases. In Section 4, we present our concluding remarks.

2 Optimizing DBCs with DTW

Dissimilarity Representation: A dissimilarity representation of a set of samples, $T = \{\boldsymbol{x}_1, \cdots, \boldsymbol{x}_n\} \in \Re^d$, is based on pairwise comparisons, and is expressed, for example, as an $n \times m$ dissimilarity matrix $D_{T,Y}[\cdot, \cdot]$, where $Y = \{\boldsymbol{y}_1, \cdots, \boldsymbol{y}_m\}$, a prototype set, is extracted from T, and the subscripts of D represent the set of elements, on which the

[3] Technically, the word "optimizing" in this paper is used in the context of "improving".

dissimilarities are evaluated. Thus, each entry $D_{T,Y}[i,j]$ corresponds to the dissimilarity between the pairs of objects $\langle \boldsymbol{x}_i, \boldsymbol{y}_j \rangle$, where $\boldsymbol{x}_i \in T$ and $\boldsymbol{y}_j \in Y$. Consequently, an object $\boldsymbol{x}_i$ is represented as a column vector as follows:

$$[d(\boldsymbol{x}_i, \boldsymbol{y}_1), d(\boldsymbol{x}_i, \boldsymbol{y}_2), \cdots, d(\boldsymbol{x}_i, \boldsymbol{y}_m)]^T, 1 \leq i \leq n. \quad (1)$$

Here, the dissimilarity matrix $D_{T,Y}[\cdot,\cdot]$ is defined as a *dissimilarity space*, on which the d-dimensional object, $\boldsymbol{x}$, given in the feature space, is represented as an m-dimensional vector $\delta(\boldsymbol{x}, Y)$, where if $\boldsymbol{x} = \boldsymbol{x}_i$, $\delta(\boldsymbol{x}_i, Y)$ is the i-th row of $D_{T,Y}[\cdot,\cdot]$. In this paper, the column vector $\delta(\boldsymbol{x}, Y)$ is simply denoted by $\delta_Y(\boldsymbol{x})$, where the latter is an m-dimensional vector, while $\boldsymbol{x}$ is d-dimensional [4]. Two factors to consider for a dissimilarity representation are to select a prototype subset from T, and to quantify the dissimilarity between two vectors. To do these things, various representative selection methods and dissimilarity measures have been proposed in [2], [3], and [4].

Fundamental to DBCs is the measure used to quantify the dissimilarity between two vectors[5]. Extensive experiments on various dissimilarity measures are reported in [3] (see Table 2 of [3]). The measures tested in [3] essentially fall into three categories: (a) the city block, $L_{0.8}$, the Euclidean, and the max norm, which are special cases of the L_p metric for $p = 1, 0.8, 2$, and ∞, respectively; (b) the Hausdorff norm and its variants, which involve *max-min* computations; and (c) traditional pattern recognition norms, such as the template matching norm and the correlation norm. The details of these methods are omitted here in the interest of brevity, but can be found in [2].

Schema for the Proposed Solution: On the basis of what we have briefly discussed, a conventional dissimilarity-based classification is summarized in the following:

1. Select the representative set, Y, from the training set, T, by resorting to one of the prototype selection methods as described in [3] and [4].
2. Using Eq. (1), compute the dissimilarity matrix, $D_{T,Y}[\cdot,\cdot]$, in which each individual dissimilarity is computed on the basis of the measures described in [2] and [4].
3. For a testing sample, $\boldsymbol{z}$, compute a dissimilarity column vector, $\delta_Y(\boldsymbol{z})$, by using the same measure used in Step 2.
4. Achieve the classification by invoking a classifier built in the dissimilarity space, and by operating the classifier on the dissimilarity vector, $\delta_Y(\boldsymbol{z})$. For example, if the nearest neighbor classifier is used and $\delta_Y(\boldsymbol{x}_i)$ that lies closest to $\delta_Y(\boldsymbol{z})$ is determined, $\boldsymbol{z}$ is assigned to the class of $\boldsymbol{x}_i$.

From the above four steps, we can see that the performance of the DBCs relies heavily on how well the dissimilarity space, which is determined by the dissimilarity matrix, $D_{T,Y}[\cdot,\cdot]$, is constructed. To improve the performance, we need to ensure that the dissimilarity matrix is well designed. With regard to measuring the dissimilarity of the sample points, we prefer not to directly measure the dissimilarity from the object points; rather, we propose a way of using the DTW dynamic programming technique to *adjust* or *scale* the object samples. This measure of dissimilarity effectively serves

[4] From this perspective, it becomes clear that the dissimilarity representation can be considered as a way of solving the small sample size problem [12].

[5] The details of the binary, categorical, ordinal, symbolic and quantitative features are omitted here, but can be found in [3].

as a new "feature" component in the dissimilarity space. In our method of optimizing DBCs, we use DTW [6] to align two face images that have different directions and sizes. The rationale for the proposed method is conceptually explained in Figure 1.

In Figure 1 we can see how DTW helps align two artificial faces, *FaceA* and *FaceB*. These faces have different directions: *FaceA* is facing left and *FaceB* is facing right. Thus, the two faces cannot be matched perfectly. To solve this problem, we propose a correlation coefficient-based DTW method. In this method, the dissimilarity is measured in an *adjustment* and a *measurement* step. In the *adjustment* step, the faces are aligned by matching the best warping path, w, with the correlation coefficients (see Figure 1). In the *measurement* step, the dissimilarity distance between the adjusted faces is then obtained with conventional measurements [7]. Theoretically, the stronger the linear relation between two vectors, the greater the similarity of their structures. On the basis of this idea, we use the right essence of the correlation coefficient as a measurement of the structural similarity between two face vectors [8].

To obtain an optimal warping path for two objects, we first compute the correlation coefficient between two subregions (a subsequence of vectors) selected from the object matrices by scanning through them in a vertical-only way or a vertical-horizontal way. Figure 1 shows a plot that represents the process of the dynamic alignment of DTW and explains how the DTW aligns *FaceA* and *FaceB* to make them overlap completely. In Figure 1, the diagonal line of the box is an optimal warping path that records the mapping between the two faces. The mapping indicated as $M1$ in the picture involves the mapping of "several" column vectors in *FaceB* to an "a" column vector in *FaceA*. In contrast, the mapping indicated as $M2$ represents the mapping of an "a" column vector in *FaceB* to "several" column vectors in *FaceA*. On the basis of the alignment process (the optimal warping path) in Figure 1, we measured the dissimilarity of *AdjustedFaceA* and *AdjustedFaceB*, rather than *FaceA* and *FaceB*.

Optimized Dissimilarity-Based Classification: As mentioned earlier, a good representation cannot be obtained by simply measuring the dissimilarity of facial images for each class. To overcome this limitation, we used the DTW algorithm. The basic strategy of the technique is to solve the classification problem by first adjusting the faces, and then constructing a dissimilarity matrix with the correlation coefficient-based DTW method; finally, DBCs are designed on the dissimilarity space to reduce the classification error rates. The proposed approach, which is referred to as an Optimized Dissimilarity-Based Classifier (ODBC), is summarized in the following:

1. This step is the same as Step 1 in the conventional DBC.
2. Using Eq. (1), compute the dissimilarity matrix $D_{T,Y}[\cdot,\cdot]$, in which each individual dissimilarity is computed using the correlation coefficient-based DTW method, rather than using one of the measures described in [2] and [4].
3. This step is the same as Step 3 in the conventional DBC.

[6] The details of DTW is omitted here in the interest of compactness, but can be found in [14].

[7] An algorithmic description of these *two* steps will be provided in the unabridged version of this paper (journal paper).

[8] The correlation coefficient, $\rho_{x,y}$, between x and y with their expected values, μ_x and μ_y, and standard deviations, σ_x and σ_y, is defined as: $\rho_{x,y} = E((x-\mu_x)(y-\mu_y))/\sigma_x\sigma_y$, where E is the operator for expected values.

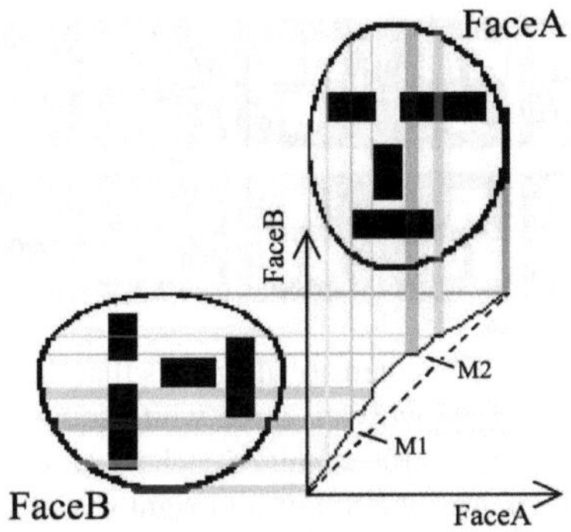

Fig. 1. The process of the DTW dynamic alignment. The DTW aligns *FaceA* and *FaceB* to make them overlap completely. The line in the picture is an optimal warping path, which records the mapping between the two faces. The details of the picture can be found in the text.

4. This step is the same as Step 4 in the conventional DBC.

As in the case of DBC, almost all the processing CPU-time of ODBC is utilized in executing Step 2. The time complexities and the space complexities of DBC and ODBC are, respectively, all the same as $O(dn^2 + \gamma)$ and $O(n(n + d))$, where γ is the time for doing classification with a classifier designed in the dissimilarity space.

3 Experimental Results

Experimental Data: The proposed method was tested and compared with conventional methods by conducting experiments on an artificial image data set and the two well-known benchmark face databases, namely, "AT&T" [9] and "UMIST" [10] .

Figure 2 shows an artificial data set of two-dimensional images that the proposed method can handle. This data set was included in the experiments as a baseline data set. In Figure 2, ten copies of each picture are generated with random noise, where the intensity of the added random noise varies from 3% to 21% at increments of 2% per generation. Thus, the data set consists of 90 sample pictures per class and 180 sample pictures in total. In addition, the size of each binary picture is 64×64 pixels for a total dimensionality of 4096. The face database of AT&T, formerly known as the ORL database of faces, consists of ten different images of 40 distinct subjects for a total of 400 images. The size of each image is 112×92 pixels for a total dimensionality of 10304 pixels. The UMIST repository is a multi-view face database, consisting of 564 images of 20 people, each covering a range of poses from profile to frontal views.

Experimental Method: All our experiments were performed with a "leave-one-out" strategy. To classify an image of an object, we removed that image from the training set and computed the dissimilarity matrix with the $n - 1$ images. This process was repeated n times for every sample, and a final result was obtained by averaging the results

[9] http://www.uk.research.att.com/facedatabase.html

[10] http://images.ee.umist.ac.uk/danny/database.html

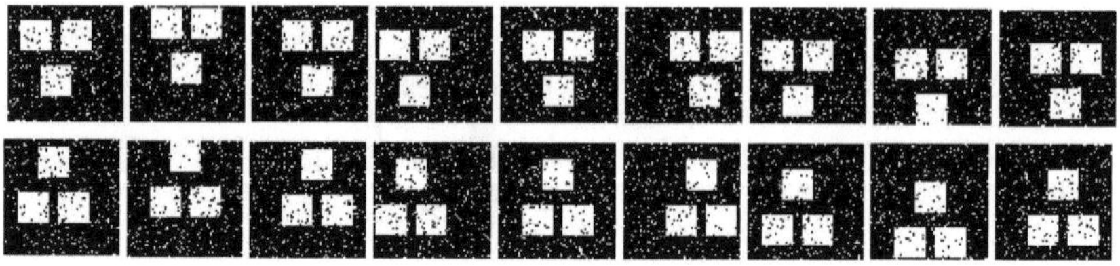

Fig. 2. Plots of an artificial data of two-dimensional images that the proposed method can handle. The top row shows a class of images involving three white squares, and the bottom row shows another class. The two pictures placed in the center of each row are the prototypes of their classes. The rest of the pictures illustrate warps due to left-and-right shifts and up-and-down shifts.

of each sample. To construct the dissimilarity matrix, then we used the proposed correlation coefficient-based DTW method in conjunction with a conventional measurement system, such as Euclidean distance (ED) measuring or Hamming distance (HD) measuring [11]. In the notations of ED1D and ED2D, the first two letters refer to the method of measuring the dissimilarity, namely ED measuring; the next two letters refer to the dimensionality; hence, 1D and 2D indicates that to obtain the optimal warping path the correlation coefficient is computed in a one-dimensional way or a two-dimensional way, respectively. In the one-dimensional way, the computation was performed solely with column vectors; in the two-dimensional way, the computation was performed with raw vectors and column vectors.

Because the construction of the dissimilarity matrix is time-consuming, we randomly selected five objects from the AT&T database and then constructed a 50×50 dissimilarity matrix instead of a full 400×400 matrix. This process was repeated 20 times for all objects of the database and a final result was obtained by averaging the results of each repetition [12]. On the other hand, to maintain the diversity between the dissimilarity-based classifications, we designed different classifiers, such as the k-nearest neighbor classifiers (k = 1, 3, 5, 7), the nearest mean classifiers, the support vector classifier, and the regularized normal density-based linear/quadratic classifiers. These classifiers, which were implemented with PRTools [13], are denoted in the next section as 1-NN, 3-NN, 5-NN, 7-NN, NMC, SVC, RLDC, and RQDC, respectively.

Experimental Results: The classification performance of the DBCs for the artificial data set is first illustrated in graphs. A numerical comparison of the two benchmark

[11] Here, we experimented with only two measures, namely, the ED and HD measuring methods. However, other measuring methods, such as the modified Hausdorff distances, the blurred Euclidean distance, etc., could also be employed. From this perspective, the question "what is the best measure ?" is an interesting issue for further study.

[12] We experimented with the simpler 50×50 matrix here. However, it should also be mentioned that we can have numerous solutions, depending on the representative selection, the dissimilarity measure, and the design of classifiers in the dissimilarity space. Especially, regarding the dissimilarity measure, it is possible to construct the dissimilarity matrix rapidly by employing a computational technique. From this perspective, we are currently investigating how the experiment with the full 400×400 matrix can be performed at a high speed.

[13] PRTools is a MATLAB toolbox for pattern recognition (refer to http://www.prtools.org/).

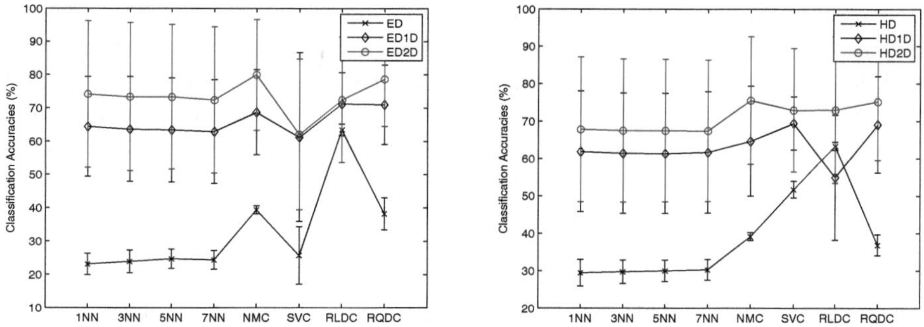

Fig. 3. A comparison of the classification accuracy rates (%) for the artificial data set: (a) left and (b) right. The dissimilarity matrices of (a) and (b) are obtained with the ED, HD, and their modified measuring methods, respectively.

databases is then made in relation to the conventional methods and the proposed scheme. Figure 3 shows a comparison of the classification accuracy rates (%) for the artificial data set with different dissimilarity measures.

These figures confirm the possibility of improving the performance of DBCs by effectively measuring the dissimilarity. The improvement can be seen by observing how the classification accuracy rates (%) change. For example, in Figure 3, for all the non-parametric classifiers, namely k-NN, NMC, and SVC, the classification results of the proposed measuring methods of ED1D and ED2D (which are marked as $\circ$ and $\diamond$) are *significantly* more accurate than those of the conventional ED measuring method (which is indicated as $\times$). For the parametric classifiers RLDC and RQDC, the classification results of the proposed methods are also more accurate than those of the conventional methods. The problem of theoretically analyzing this observation remains unresolved.

Table 1. A comparison of the averaged classification accuracy rates (%) of DBCs for AT&T and UMIST databases. The details of the table are discussed in the text.

Experimental Databases	Experimental Methods	Non-parametric Classifiers						Parametric Ones	
		1-NN	3-NN	5-NN	7-NN	NMC	SVC	RLDC	RQDC
AT&T	ED	89.00	76.80	69.70	66.70	73.90	93.60	**95.30**	84.40
	ED1D	89.20	84.40	81.80	78.50	79.20	94.10	88.20	**86.40**
	ED2D	**92.80**	**87.40**	**85.60**	**80.60**	**84.70**	**96.90**	76.70	84.70
	HD	**92.60**	86.10	78.10	74.00	47.50	63.40	33.90	87.90
	HD1D	91.80	81.10	77.10	74.00	43.70	60.90	48.50	**92.20**
	HD2D	90.70	**87.10**	**79.40**	**75.90**	**63.10**	**74.50**	**67.50**	91.40
UMIST	ED	99.27	99.07	99.00	98.33	99.07	**99.27**	**99.26**	99.26
	ED1D	**99.40**	**99.33**	**99.27**	**99.27**	**99.27**	**99.27**	**99.26**	**99.53**
	ED2D	99.33	**99.33**	**99.27**	**99.27**	**99.27**	**99.27**	98.33	99.33
	HD	99.93	**100**	**100**	99.8	87.53	**99.00**	90.00	99.13
	HD1D	**100**	**100**	**100**	**100**	91.33	**99.00**	93.53	99.20
	HD2D	**100**	**100**	**100**	**100**	**92.13**	**99.00**	**96.33**	**100**

To further investigate the advantage of using the proposed method, we conducted the following experiments for the real benchmark databases, namely AT&T and UMIST. First, the 50 facial images of five individuals were randomly selected from AT&T (and UMIST). Next, for the images, we evaluated the classification performance of the DBCs, which were designed in dissimilarity matrices constructed with conventional measures (of ED and HD) and the proposed methods (of ED1D, ED2D, HD1D and HD2D). After repeating the above two steps 20 times, we averaged the performances. (The best results of these evaluations are printed in bold in Table 1.) In comparing the conventional and new schemes, Table 1 shows a comparison of the averaged classification accuracy rates (%) for the facial images of AT&T and UMIST. We observed the same characteristics in Table 1 as in Figure 3 (see the bold-faced numbers). These results confirm that the classification performance of *almost* all classifiers was improved with the ED1D method and further improved with the ED2D method.

In review, it is not easy to crown one particular measuring method with superiority over the others in terms of solving the dissimilarity measuring problem. However, in terms of classification accuracies, the proposed measuring method is clearly more useful for simple pixel-based measurements, such as ED and HD measuring.

4 Conclusions

In our efforts to optimize DBCs, we used the DTW dynamic programming technique. To construct a dissimilarity matrix of DBCs, we measured the dissimilarity directly from the input images with one-to-many (or many-to-one) mapping as well as one-to-one mapping. This method overcomes the problems caused by variations in facial directions and sizes. Furthermore, instead of measuring the dissimilarity distances on the basis of the entire image, we used the correlation coefficient-based method where the entire image region is divided into several subregions according to their distances. The proposed method was tested on an artificial image data set and two well-known benchmark databases, and the results were compared with the results of conventional methods. The experimental results demonstrate that the proposed method is better than conventional methods in terms of classification accuracy. Although we have shown that DBCs can be optimized with our proposed measuring scheme, many tasks remain open. One of them is to improve the classification efficiency by combining the optimized classifiers in the dissimilarity space. Our aim is to conduct further research on this subject in the future.

References

1. Jain, A.K., Duin, R.P.W., Mao, J.: Statistical pattern recognition: A review. IEEE Trans. Pattern Anal. and Machine Intell. PAMI 22(1), 4–37 (2000)
2. Pekalska, E., Duin, R.P.W.: The Dissimilarity Representation for Pattern Recognition: Foundations and Applications. World Scientific Publishing, Singapore (2005)
3. Pekalska, E., Duin, R.P.W., Paclik, P.: Prototype selection for dissimilarity-based classifiers. Pattern Recognition 39, 189–208 (2006)
4. Kim, S.-W., Oommen, B.J.: On using prototype reduction schemes to optimize dissimilarity-based classification. Pattern Recognition 40, 2946–2957 (2007)

5. Belhumeour, P.N., Hespanha, J.P., Kriegman, D.J.: Eigenfaces vs. Fisherfaces: Recognition using class specific linear projection. IEEE Trans. Pattern Anal. and Machine Intell., PAMI 19(7), 711–720 (2007)
6. Zhou, W., Chellappa, R., Rosenfeld, A., Phillips, P.J.: Face recognition: a literature survey. ACM Compt. Surveys 35(4), 399–458 (2003)
7. Lozano, M., Sotoca, J.M., Sanchez, J.S., Pla, F., Pekalska, E., Duin, R.P.W.: Experimental study on prototype optimisation algorithms for prototype-based classification in vector spaces. Pattern Recognition 39, 1827–1838 (2006)
8. Orezco-Alzate, M., Duin, R.P.W., Castellanos-Dominguez, C.G.: Generalizing dissimilarity representations using feature lines. In: Rueda, L., Mery, D., Kittler, J. (eds.) CIARP 2007. LNCS, vol. 4756, pp. 370–379. Springer, Heidelberg (2007)
9. Paredes, R., Vidal, E.: A class-dependent weighted dissimilarity measure for neareast neighbor classification problem. Pattern Recognition Letters 21(12), 1027–1036 (2000)
10. Wang, J., Neskovic, P., Cooper, L.N.: Improving neareast neighbor rule with a simple adaptive distance measure. Pattern Recognition Letters 28(2), 207–213 (2007)
11. Kim, S.-W.: Optimizing dissimilarity-based classifiers using a newly modified Hausdorff distance. In: Hoffmann, A., Kang, B.-h., Richards, D., Tsumoto, S. (eds.) PKAW 2006. LNCS, vol. 4303, pp. 177–186. Springer, Heidelberg (2006)
12. Kim, S.-W., Duin, R.P.W.: On combining dissimilarity-based classifiers to solve the small sample size problem for appearance-based face recognition. In: Kobti, Z., Wu, D. (eds.) Canadian AI 2007. LNCS, vol. 4509, pp. 110–121. Springer, Heidelberg (2007)
13. Sakoe, H., Chiba, S.: A dynamic programming algorithm optimization for spoken word recognition. IEEE Trans. Acoust. Speech Signal Process. 26(1), 43–49 (1978)
14. Rabiner, L., Jung, B.H.: Fundamentals of Speech Recognition. Prentice Hall PTR, New Jersey (1993)
15. Ratan, A.L., Grimson, W.E.L., Wells, W.M.: Object detection and localization by dynamic template warping. International Journal of Computer Vision 36(2), 131–147 (2000)
16. Sahbi, H., Boujemaa, N.: Robust face recognition using dynamic space warping. In: Tistarelli, M., Bigun, J., Jain, A.K. (eds.) ECCV 2002. LNCS, vol. 2359, pp. 121–132. Springer, Heidelberg (2002)
17. Qiao, Y., Yasuhara, M.: Affine invariant dynamic time warping and its application to online rotated handwriting recognition. In: Proceedings of International Conference on Pattern Recognition, ICPR 2006, vol. 2, pp. 905–908. IEEE Computer Society, Los Alamitos (2006)

Supervised Principal Geodesic Analysis on Facial Surface Normals for Gender Classification

Jing Wu, William A.P. Smith, and Edwin R. Hancock

Department of Computer Science, The University of York, York, YO10 5DD, UK
{jwu,wsmith,erh}@cs.york.ac.uk

Abstract. In this paper, we perform gender classification based on facial surface normals (facial needle-maps). We improve our previous work in [6] by using a non-Lambertian Shape-from-Shading (SFS) method to recover the surface normals, and develop a novel supervised principal geodesic analysis (PGA) to parameterize the facial needle-maps. Experimental results demonstrate the feasibility of gender classification based on facial needle-maps, and shows that incorporating pairwise relationships between the labeled data improves the gender discriminating powers in the leading PGA eigenvectors and gender classification accuracy.

1 Introduction

Although the determination of gender from facial images has been the focus of sustained activity in both computer vision [1], [2], [3] and psychology [4], [5], most of the existing work is based on 2D images. Although it has been shown that gender is revealed by both the 2D texture and the 3D shape of the human face [4], few studies have investigated the role of 3D shape in gender classification [2]. To some extent this is due to the more complex computations required in 3D face shape analysis, and the limited effectiveness and high cost of the current 3D sensors. In [6], we have investigated the feasibility of gender classification based on 2.5D facial surface normals (or facial needle-maps), which exhibits the shape information from a fixed viewpoint and can be recovered from single images using shape-from-shading (SFS). In [6], an iterative principal geodesic SFS method, which combines the facial texture and shape information implicitly, is applied to facial images. However, this SFS method is only suitable for surfaces which exhibit Lambertian reflectance. Human face surfaces, on the other hand, do not strictly obey Lambert's law due to specularities caused by perspiration etc. In this paper, we apply an improved SFS method [7], which is suitable to non-Lambertian surfaces, to facial images. This method satisfies the brightness constraint using the Torrance and Sparrow reflectance model [8], and guarantees the accuracy of the recovered needle-maps by using a statistical height model.

Another improvement to [6] is the parameterisation of the needle-maps. Facial needle-maps can be considered as points lying on the $S^2(N)$ manifold. To parameterize data lying on this Riemmanian manifold, in [6], we use principal geodesic analysis (PGA) [9], [10]. However PGA locates the projections that maximize the

N. da Vitora Lobo et al. (Eds.): SSPR&SPR 2008, LNCS 5342, pp. 664–673, 2008.

variance of the data. Usually the projections that maximize the variance are not those that separate the data into distinct clusters. As a result the PGA parameters do not usually reveal cluster structure. In [11], the authors view PCA as locating projection that maximizes the sum of squared pairwise distances between the projected data. By associating a dissimilarity weight with each pairwise distance, they obtain a supervised PCA projection that emphasizes the inter-cluster separation. In this paper, we develop a novel supervised variant of PGA by extending this idea to the $S^2(N)$ space for the analysis of surface normals. By incorporating pairwise gender relationships, supervised PGA makes use of the interclass structure, and improves the gender discriminating power of the leading principal geodesics. We also show that the supervised PGA method can be viewed as applying standard PGA on the weighted facial needle-maps with the applied weight matrices specific to each facial needle-map. This links supervised PGA with the weighted PGA, we proposed before [12]. However the weight matrix is weighted PGA is a common one which can be obtained from a priori knowledge.

The outline of the paper is as follows. Section 2 reviews the PGA technique, and gives a detailed description of supervised PGA. Section 3 describes the non-Lambertian SFS method used to generate the needle-maps. Section 4 describes the classifier used for gender classification. Experiments on ground-truth needle-maps and brightness images are presented in Section 5. Finally, Section 6 concludes the paper and offers directions for future investigation.

2 Supervised Principal Geodesic Analysis

2.1 Principal Geodesic Analysis

A surface normal n may be considered as a point residing on a spherical manifold $n \in S^2$. Facial needle-maps, which are fields of N surface normals, may be considered as a point on the manifold $S^2(N) = \prod_{i=1}^{N} S^2$. For the analysis of data on this manifold, we turn to the intrinsic mean and PGA [9], [10].

Given $u \in T_nS^2$, a non-zero vector on the tangent plane to S^2, the exponential map, denoted Exp_n, maps u to the point $Exp_n(u)$ along the geodesic in the direction of u at distance $\| u \|$ from n. This is illustrated in Fig. 1. The log map, denoted Log_n is the inverse of the exponential map. In $S^2(N)$ space, the exponential and log maps are the products of N copies of the maps for S^2.

Intrinsic means minimize the sum-of-squared Riemannian distances on M: $\mu = \arg\min_{x \in M} \sum_{i=1}^{K} d(x, x_i)^2$. For a spherical manifold, the intrinsic mean of a set of surface normals $n_1, ..., n_K \in S^2$ can be found using the gradient descent method of Pennec [10]: $\mu^{(t+1)} = \mathrm{Exp}_{\mu^{(t)}}(\frac{1}{K}\sum_{i=1}^{K} \mathrm{Log}_{\mu^{(t)}}(n_i))$.

For data on a connected, complete manifold, PGA finds a sequence of nested geodesics that maximize the projected variance. In the spherical case, a geodesic corresponds to a great circle. The projection $\pi_G : S^2 \to G$ is defined as: $\pi_G(n_1) = \arg\min_{n \in G} d(n_1, n)^2$. For a geodesic G passing through the intrinsic mean μ, this projection can be approximated linearly in the tangent space $T_\mu S^2$:

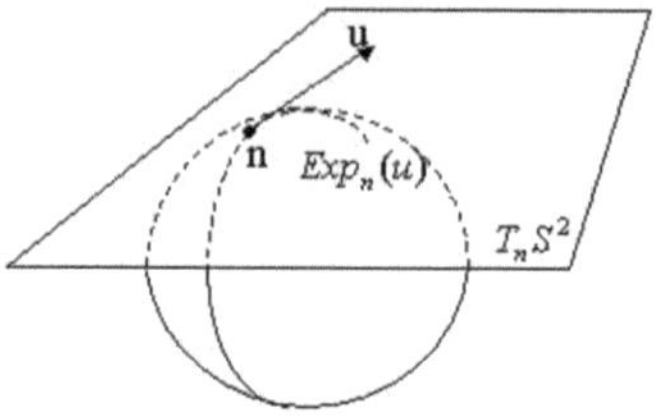

Fig. 1. The exponential map

$$Log_\mu(\pi_G(n_1)) \approx \sum_{i=1}^{m} v_i \cdot Log_\mu(n_1), \tag{1}$$

where $v_1, ...v_d$ is an orthonormal basis for $T_\mu S^2$ obtained through standard PCA.

2.2 Supervised Principal Geodesic Analysis

Supervised PCA is developed by Koren and Carmel [11]. They view PCA as computing the p-dimensional projection that maximizes $\sum_{i<j}(dist^p_{ij})^2$, where $dist^p_{ij}$ is the Euclidean distance between the projected data i and j. By incorporating a symmetric and nonnegative pairwise weight matrix $\{d_{ij}\}^n_{i,j=1}$, where d_{ij} is the dissimilarity of data i and j, they generalized PCA to seek the projection that maximizes the sum of weighted squared pairwise distances

$$dist^p_{sum} = \sum_{i<j} d_{ij}(dist^p_{ij})^2. \tag{2}$$

Koren and Carmel [11] proved that the p-dimensional projection that maximizes (2) is obtained by taking the p principal eigenvectors of the matrix $X^T LX$, where L is the Laplacian associated with the dissimilarities, and X is the row-wise data matrix. For labeled data, this weighted PCA turns to be supervised PCA when underweights the intra-cluster pairwise dissimilarities:

$$d^{labeled}_{ij} = \begin{cases} t \cdot d_{ij}, & \text{i and j have the same label;} \\ d_{ij}, & \text{otherwise.} \end{cases}$$

where $0 \leq t \leq 1$ is a decay factor. In this way, the supervised PCA constructs the projection which emphasizes inter-cluster separation.

Supervised PGA for Facial Needle-maps. As mentioned above, the approximation (1) enables us to compute the principal geodesics by applying standard PCA in the tangent plane $T_\mu S^2$. A straightforward generalization is to replace the standard PCA by the supervised PCA. This has the effect of enhancing the inter-cluster separability of the obtained projection.

Suppose there are K gender-labeled facial needle-maps, and each having N pixel locations. The surface normal at location l in the kth needle-map is n^k_l. We calculate the intrinsic mean μ_l at each pixel location l. n_l^k is then represented by its log map position $u_l^k = \text{Log}_{\mu_l}(n_l^k)$ in the tangent plane. $u^k = [u_1^k, \ldots, u_N^k]^T$ is the log mapped long vector of the kth training needle-map in the tangent plane $T_\mu S^2(N)$. The K long vectors form the column-wise data matrix $U = [u^1|\ldots|u^K]$. For gender classification, we construct the dissimilarity matrix:

$$d_{ij} = \begin{cases} 0, \text{ i and j are of same gender;} \\ 1, \text{ otherwise.} \end{cases}$$

and the corresponding Laplacian matrix is L. According to [11], the p leading eigenvectors with the highest eigenvalues of the matrix ULU^T form the projection Φ in the long vector space. Given a facial needle-map, its long vector $u = [u_1, \ldots, u_N]^T$ can be obtained through log-map of the surface normals, and the corresponding supervised PGA parameters can be calculated as $b = \Phi^T u$.

For face images, the dimension N is usually large. The memory storage and manipulation of the $N \times N$ matrix ULU^T are usually unfeasible. Since the Laplacian L is an $K \times K$ symmetric positive semi-definite matrix, its eigenvalues are non-negative. Through eigen-decomposition, L can be rewritten as $L = \Phi_L\sqrt{\Lambda}\sqrt{\Lambda}\Phi_L^T = (\Phi_L\sqrt{\Lambda})(\Phi_L\sqrt{\Lambda})^T$, where Φ_L and Λ are the eigenvectors and eigenvalues of L. Accordingly,

$$ULU^T = U(\Phi_L\sqrt{\Lambda})(\Phi_L\sqrt{\Lambda})^T U^T = (U\Phi_L\sqrt{\Lambda})(U\Phi_L\sqrt{\Lambda})^T.$$

Then, the numerically efficient snap-shot method of Sirovich [13] can be used to compute the eigenvectors and eigenvalues of ULU^T.

Representing ULU^T as $(U\Phi_L\sqrt{\Lambda})(U\Phi_L\sqrt{\Lambda})^T$, the supervised PGA can be viewed as standard PGA on weighted long vectors. The weight matrices for each data is the component-wise division of $(U\Phi_L\sqrt{\Lambda})$ and U. Some examples of the weight matrices for the ground-truth facial needle-maps are shown in Fig. 2. From the figure, the calculated weight matrices have higher weights in the areas which are used for gender discrimination by human observers. Explained as standard PGA on weighted data, the supervised PGA method is related with the weighted PGA proposed in [12]. The difference is the weight matrix adopted in weighted PGA is a common one which can be obtained from a priori knowledge, while in supervised PGA, the weight matrices are different for each data, and they are obtained from the labeled data's pairwise relationships.

3 Non-Lambertian Shape from Shading

Faces do not strictly reflect light according to Lambert's law. Therefore, we turn to a new shape-from-shading framework for non-Lambertian facial reflectance developed by Smith and Hancock [7]. This method not only imposes a hard non-Lambertian image brightness constraint, but uses a statistical constraint to impose the correct overall shape on the recovered facial needle-maps.

Fig. 2. Examples of the computed weight matrices. The two rows are according to the two directions. The first two columns are females, the last two columns are males.

The brightness constraint uses Torrance and Sparrow reflectance model [8]:

$$g_{\mathrm{T\&S}}(\mathbf{N}, \mathbf{L}, \mathbf{V}, \{\rho_d, \rho_s, \nu, L\}) = L\rho_d \mathbf{N} \cdot \mathbf{S} + \frac{L\rho_s e^{-\nu^2 \arccos\left(\mathbf{N} \cdot \frac{\mathbf{L}+\mathbf{V}}{\|\mathbf{L}+\mathbf{V}\|}\right)^2}}{\mathbf{N} \cdot \mathbf{V}},$$

where ρ_s is the specular coefficient, ν the surface roughness, L the light source intensity and, ρ_d is the diffuse albedo. The brightness error can be defined as a function of $n \in S^2$: $f(n) = (g(\Phi(n), \mathbf{L}, \mathbf{V}, \{\rho_d, \rho_s, \nu, L\}) - I)^2$, where $\Phi : S^2 \to R^3$ is an embedding. Making use of the exponential map, the local gradient of f can be approximated in terms of a vector on the tangent plane $T_n S^2$:

$$\nabla f(n) = \left(\frac{f\left(\mathrm{Exp}_n\left[(\epsilon, 0)^T\right]\right) - f(n)}{\epsilon}, \frac{f\left(\mathrm{Exp}_n\left[(0, \epsilon)^T\right]\right) - f(n)}{\epsilon} \right)^T.$$

Gradient descent is used to locate the local minimum of the brightness error function: $n_{t+1} = \mathrm{Exp}_{n_t}(-\gamma \nabla f(n_t))$, where γ is the step size.

The new SFS method also makes use of a surface height model constructed by applying PCA to a set of facial surface samples. Suppose Ψ_i is the ith principal eigenvector of samples' covariance matrix. The surface normals satisfying the statistical model can be expressed in terms of the parameter vector b:

$$\mathbf{n_b}(x, y) = \left(\sum_{i=1}^m b_i \partial_x \Psi_i(x, y), \sum_{i=1}^m b_i \partial_y \Psi_i(x, y), 1\right)^T.$$

In our experiments, we choose m to be 50 to maintain shape recovery accuracy while reducing noise. In order to apply this statistical constraint to $n(x, y)$ satisfying brightness constraint, the parameter vector $\mathbf{b}^*$ must be located minimizing the distance between $n(x, y)$ and $n_{b^*}(x, y)$. In [7], this minimization task is poses as that of minimizing the squared error between the surface gradients of $n(x, y)$ and $n_{b^*}(x, y)$.The optimal solution is: $\mathbf{b}^* = \left(\mathbf{\Psi}^T \mathbf{\Psi}\right)^{-1} \mathbf{\Psi}^T \mathbf{G}$, where G are the surface gradients of $n(x, y)$, and Ψ is the surface gradient of the eigenvectors.

A detailed description of this non-Lambertian SFS method is given in [7].

4 Classification

With a set of gender-labeled facial needle-maps, we obtain the projection matrix using supervised PGA. Both training and test needle-maps are represented by supervised PGA parameters. A Bayes classifier is used for gender classification.

Let C_f and C_m denote the female and male gender classes, x denote the p dimensional supervised PGA parameters of a test facial needle-map. Then according to the Bayes law, the probability that x is of class C_i is:

$$P(C_i|x) = \frac{P(x|C_i)P(C_i)}{\sum_{i\in\{f,m\}} P(x|C_i)}. \tag{3}$$

We assume x follow a Gaussian distribution for the different genders, and that the parameter vector mean and covariance of class C_i are μ_i and Σ_i. Then,

$$P(x|C_i) = \frac{1}{\sqrt{(2\pi)^p det(\Sigma_i)}} \exp[-\frac{1}{2}(x-\mu_i)^T \Sigma_i^{-1}(x-\mu_i)]. \tag{4}$$

The a posteriori probability $P(C_f|x)$ and $P(C_m|x)$ are computed by applying (4) and substituting equal a priori probability into (3). If $P(C_f|x) > P(C_m|x)$, then the face is classified as female. Otherwise, the face is classified as male.

5 Experimental Results

In this section, we report experiments on both ground-truth face needle-maps and needle-maps recovered from images using SFS. The ground-truth data are from the Max-Planck Face Database [14], [15]. It comprises 200 normalized 3D range images of human heads (100 females and 100 males), which are used to construct the statistical model for non-Lambertian SFS. The 168 (94 females and 74 males) brightness images with neutral expression and no glasses are from the AR Face Database [16]. These images are aligned using the centers of the eyes, the tip of the nose, and the middle of the mouth.

5.1 Experiments on Ground-Truth Data

We first evaluate the performance of supervised PGA by examining the gender discriminating powers of the eigenvectors. Then, we compare classification results using supervised PGA parameters with those using PGA parameters.

Discriminating Powers are calculated using the criterion function introduced in [17]: $J(\xi) = tr(S_w^{-1}S_b) = \sum_{k=1}^{d} \lambda_k$. The discriminating power of each of the 10 leading supervised PGA eigenvectors are shown in Fig. 3, compared with those using standard PGA. From the figure, it is clear that there is greater discriminating power in the first supervised PGA eigenvector. By comparison, the discriminating power of standard PGA eigenvectors are more widely distributed. This confirms our assumption that by making use of the interclass structure, supervised PGA improves the gender discriminating power of the leading principal geodesics.

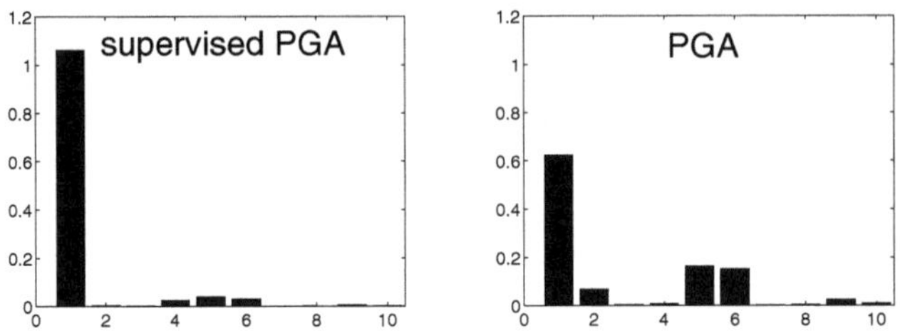

Fig. 3. Discriminating powers of the 10 leading eigenvectors

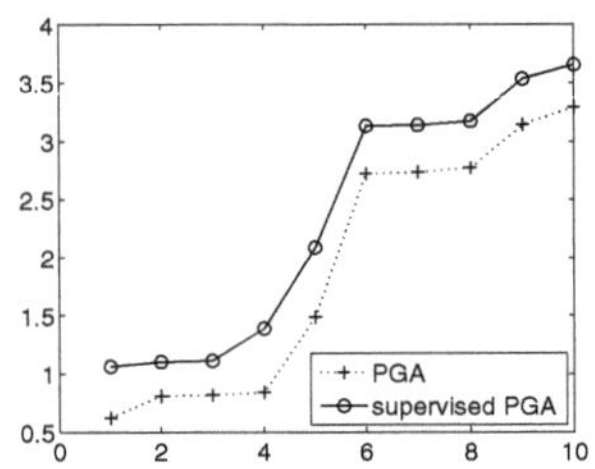

Fig. 4. Cumulative discriminating powers

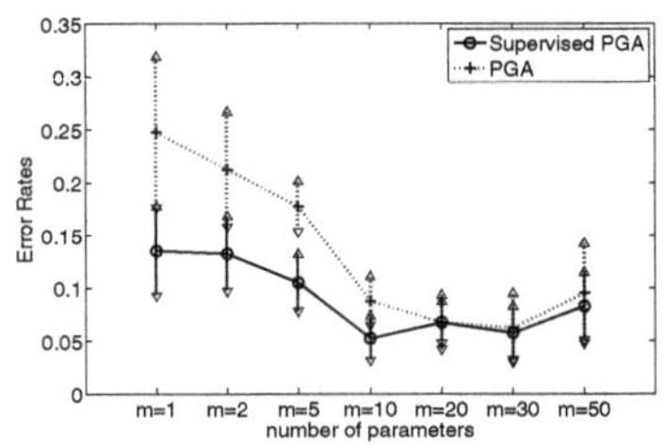

Fig. 5. Classification error rates for ground-truth data

Fig. 4 shows the cumulative gender discriminating powers. From the figure, it is clear that supervised PGA has significantly larger cumulative discriminating powers than standard PGA in the first 10 dimensions. This offers the expectation of higher classification accuracy using leading supervised PGA parameters.

Classification Results. 160 needle-maps are randomly selected from the 200 for use as training data to obtain the projection matrix, the remaining 40 as test data. Gender classification is performed by applying the Bayes classifier to the leading p (p=1,2,5,10,20,30,50) parameters for both supervised PGA and standard PGA. We repeat the randomization 10 times for each p. The average error rates and variance are shown in Figure 5, which exhibits interesting effects. First, when $p \leq 10$, using supervised PGA achieves significantly better accuracies than using standard PGA. This is expected from the cumulative discriminating powers. Second, the lowest error rate is 5.25% and 6.25% for the two methods, which indicate the feasibility of accurate gender classification based on facial shape information revealed by facial needle-maps. Finally, for both methods, the classification accuracy has an inflexion with increasing of p. It is consistent with the 'curse of dimensionality' phenomenon, as well as human classification which is based on just a few important attributes [17]. The redundant or irrelevant information in higher dimensions causes deteriorate of the classification accuracy.

5.2 Experiments with Brightness Images

Results of SFS. Four recovered needle-maps and surfaces are shown in Fig. 6. It is clear that the recovered needle-maps and the surfaces give realistic shape, overcoming the well-known local convexity-concavity instability problem.

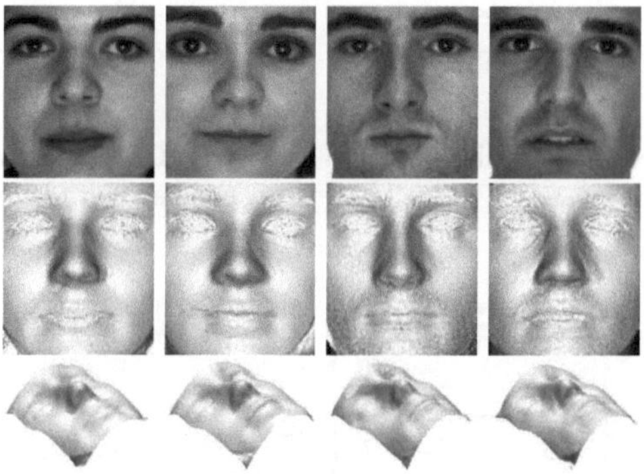

Fig. 6. Recovered needle-maps. First two columns are females, last two are males. From top to bottom are the input images, the recovered needle-maps, the recovered surfaces.

Discriminating Powers. We repeat the analysis of the discriminating powers of the 10 leading eigenvectors and the cumulative discriminating powers, and plot them in Figs. 7 and 8. From Fig. 7, it is clear that the first supervised PGA eigenvector encodes significantly more gender discriminating power. It is consistent with those of ground-truth data. However, the cumulative discriminating powers in Fig. 8 exhibit some systematic differences. The discriminating power of supervised PGA is superior to that of standard PGA when using only the first parameter. As the number of parameters increases, there is no significant difference between the two methods.

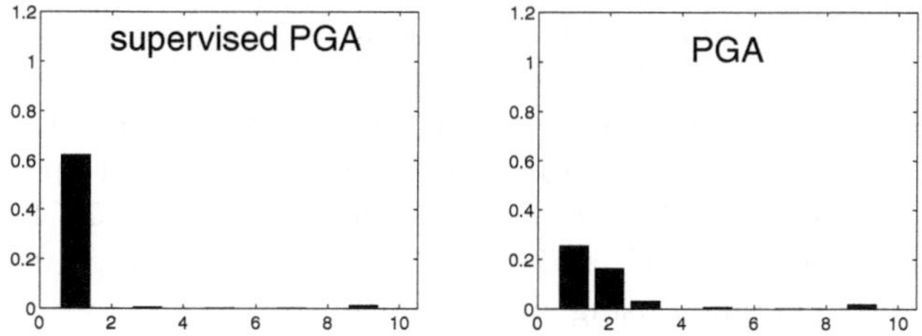

Fig. 7. Discriminating powers of the 10 leading eigenvectors

Classification Results. We randomly select 134 from the 168 brightness images as training data, and the remaining 34 as test data. The procedure described above is repeated. The classification results are shown in Figure 9. As expected

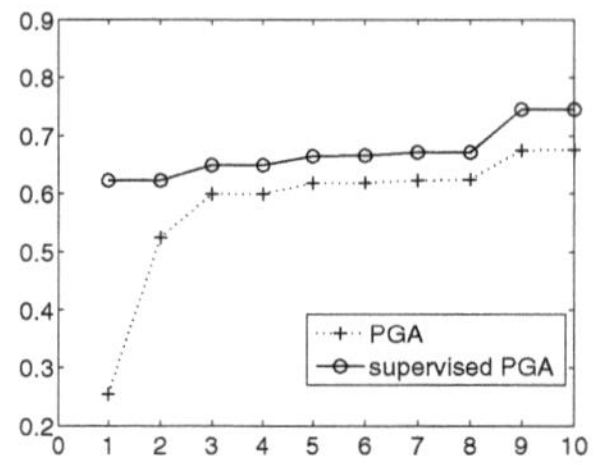

Fig. 8. Cumulative discriminating powers

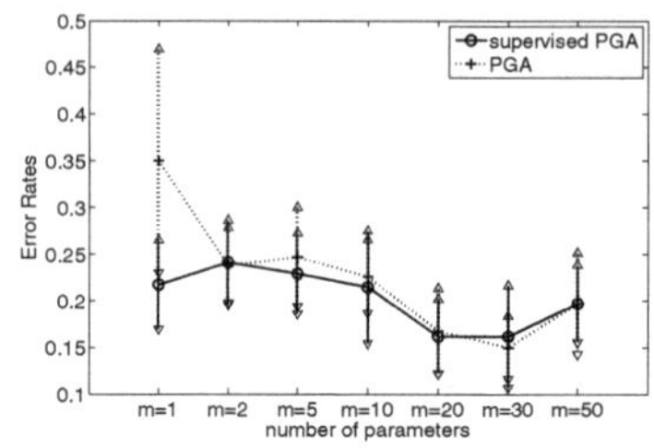

Fig. 9. Classification error rates for the recovered needle-maps

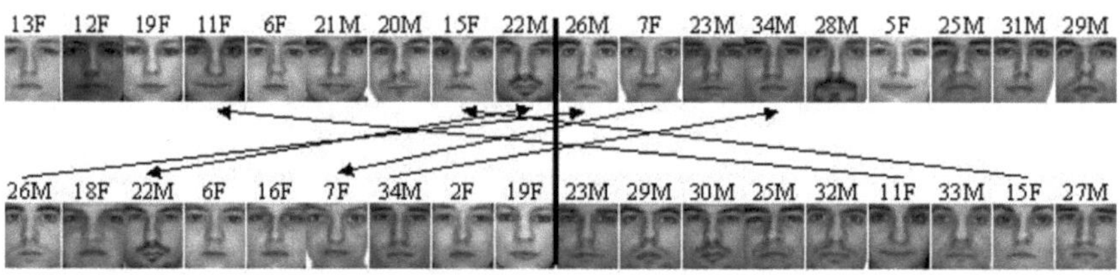

Fig. 10. Visualization of the classification

from the cumulative discriminating power, the classification results using supervised PGA are only superior to using standard PGA when $p = 1$. With increasing p, the advantage is not obvious. The best classification accuracy is about 84% for both methods, which confirms that the recovered facial needle-maps convey gender information. As the results on ground-truth data, the results on recovered needle-maps also confirm the 'curse of dimensionality' phenomenon.

For one classification, we arrange the test images into series according to their a posteriori classification probabilities. The middle portion of the series corresponding to gender overlap are shown in Fig. 10. The first row is using supervised PGA, and the second row is using standard PGA. The line in the middle is the Bayes decision boundary between the two genders (females to the left and males to the right). Each image is labeled with its index and actual gender. From the figure we can see, using supervised PGA, the misclassified faces (21M,20M,22M,7F,5F) are more concentrated in the overlap area, while in the case of standard PGA, the misclassified (26M,22M,34M,11F,15F) are more dispersed. Faces 26M, 34M, 11F and 15F which are misclassified using standard PGA, are correctly classified using supervised PGA.

6 Conclusion

In this paper, we perform gender classification on 2.5D facial surface normals. A recently reported non-Lambertian SFS method is used to recover facial needle-maps from brightness images. And a novel supervised PGA technique is developed to parameterize needle-maps so as to maximize gender discrimination. Experimental results reveal that supervised PGA improves gender discriminating powers of the leading eigenvectors. The high classification accuracy supports the feasibility of gender discrimination based on facial needle-maps.

Supervised PGA only appears to improve the classification on the recovered needle-maps using the first dimension of parameters. Establishing the reason for this is our first future work. Another topic is to investigate the distribution of the weight matrices. This may have some potential for unsupervised learning.

References

1. Buchala, S., Davey, N., Gale, T., Frank, R.: Principal component analysis of gender, ethnicity, age, and identity of face images. In: IEEE ICMI 2005 (2005)
2. Lu, X., Chen, H., Jain, A.: Multimodal facial gender and ethnicity identification. In: ICB 2006, pp. 554–561 (2006)
3. Moghaddam, B., Yang, M.: Learning gender with support faces. IEEE Trans. on PAMI 24(5), 707–711 (2002)
4. Bruce, V., Burton, A., Hanna, E., Healey, P., Mason, O., Coombes, A., Fright, R., Linney, A.: Sex discrimination: how do we tell the difference between male and female faces? Perception 22, 131–152 (1993)
5. Burton, A., Bruce, V., Dench, N.: What's the difference between men and women? evidence from facial measurement. Perception 22, 153–176 (1993)
6. Wu, J., Smith, W., Hancock, E.: Gender classification using shape from shading. In: BMVC 2007, pp. 499–508 (2007)
7. Smith, W., Hancock, E.: A new framework for grayscale and colour non-lambertian shape-from-shading. In: ACCV 2007, pp. 869–880 (2007)
8. Torrance, K., Sparrow, E.: Theory for off-specular reflection from roughened surfaces. Journal of the Optical Society of America 57, 1105–1114 (1967)
9. Fletcher, P., Joshi, S., Lu, C., Pizer, S.: Principal geodesic analysis for the study of nonlinear statistics of shape. IEEE Trans. on Medical Imaging 23, 995–1005 (2004)
10. Pennec, X.: Probabilities and statistics on riemannian manifolds: A geometric approach. Technical Report RR-5093, INRIA (2004)
11. Koren, Y., Carmel, L.: Robust linear dimensionality reduction. IEEE Trans. on VCG 10(4) (2004)
12. Wu, J., Smith, W., Hancock, E.: Weighted principal geodesic analysis for facial gender classification. In: Rueda, L., Mery, D., Kittler, J. (eds.) CIARP 2007. LNCS, vol. 4756, pp. 331–339. Springer, Heidelberg (2007)
13. Sirovich, L.: Turbulence and the dynamics of coherent structures. Quart. Applied Mathematics XLV(3), 561–590 (1987)
14. Troje, N., Bulthoff, H.: Face recognition under varying poses: The role of texture and shape. Vision Research 36, 1761–1771 (1996)
15. Blanz, V., Vetter, T.: A morphable model for the synthesis of 3d faces. In: SIGGRAPH 1999 Conference Proceedings, pp. 187–194 (1999)
16. Martinez, A., Benavente, R.: The ar face database. CVC Technical Report (1998)
17. Devijver, P., Kittler, J.: Pattern Recognition: A Statistical Approach. PrenticeHall (1982)

Exploring Margin Maximization for Biometric Score Fusion

Claudio Marrocco, Maria Teresa Ricamato, and Francesco Tortorella

DAEIMI - Università degli Studi di Cassino, Cassino, Italy
{c.marrocco,mt.ricamato,tortorella}@unicas.it

Abstract. Biometric systems are automated methods based on physical or behavioral characteristics of an individual for determining her/his identity. An important aspect of these systems is the reliability against forgery that is surely improved when using multiple sources of biometric information. In such cases combination rules can be applied to fuse the different scores thus obtaining a *multibiometric system.*

In this paper we analyze a method based on margin maximization for building a linear combination of biometric scores. The margin is a central concept in machine learning research and several theoretical results exist which show that improving the margin on the training set is beneficial for the generalization error of an ensemble of classifiers. Experiments performed on real biometric data and comparisons with other commonly employed fusion rules show that a combination based on margin maximization is particularly effective with respect to other established fusion methods.

Keywords: multiple classifiers systems, multibiometrics, margins, linear programming.

1 Introduction

Biometric systems aim to identify an individual by analyzing some physical characteristics such as fingerprints, iris, face, hand geometry, voice, etc. When dealing with a single source of biometric information, such systems could not ensure a sufficient degree of reliability because of the intrinsic noise affecting biometric data (e.g. a face can be partially occluded by sunglasses, fingerprints could vary due to different temperature and humidity conditions, etc.). Another problem is the vulnerability provided by only one characteristic which can be more easily counterfeited. An effective way to improve the accuracy is to combine multiple sources of biometric information: in this way more information is provided to the recognition/verification procedure and the whole process is more reliable and robust against spoofing attempts. Even though such combination could be performed at different levels, the most popular approach is to fuse the scores provided by a pool of classifiers (a.k.a.*matchers*) which are obtained by employing different feature extraction and/or matching techniques to different biometric characteristics. It is easy to see that such approach can be arranged in

N. da Vitora Lobo et al. (Eds.): SSPR&SPR 2008, LNCS 5342, pp. 674–683, 2008.

a Multiple Classifier System (MCS) framework, where the classification problem consists in discriminating between two categories: genuine and impostor. To this aim, several approaches have been proposed, based on nontrained combination rules [14], trained rules [14,15], classifiers or density estimation [8].

Like all the other classification problems, even in multibiometrics a central problem is to guarantee a good generalization degree of the recognition/verification system. In the area of the ensemble learning, the issue of generalization has been particularly considered in several papers aimed at explaining the reduction of the generalization error achieved by Adaboost. As it is well known, this algorithm generates weak classifiers via dinamically reweighting training samples on the basis of previous classification results; the final classifier is obtained by a convex combination of the constructed weak classifiers. In their analysis, Schapire et al. [12] introduced the concept of margin, which can be considered as a measure of the classification confidence, and showed that improving the margin on the training set guarantees an improvement in the upper bound on the generalization error of the ensemble of classifiers.

However these studies have been focused on Boosting algorithms which construct from scratch an ensemble of classifiers as different instances of a same base learning algorithm. The aim of our paper, instead, is to experimentally verify if the margin maximization can be beneficial also in biometric score fusion, where the classifiers in principle have different architectures, work on different sets of features and have been trained independently. To this aim, we have employed an algorithm for calculating the weights of a convex combination of matchers so as to maximize the margin of the combination on the training set. Several experiments performed on publicly available datasets have shown that this method is particularly effective if compared with other established fusion methods.

The rest of the paper is organized as follows: in sect. 2 the concept of margin and its characteristics are briefly recalled while in sect. 3 the method for maximizing the margin is described. In sect. 4 the results obtained on some publicly available data sets are presented and discussed. Sect. 5 concludes the paper with a summary and an outline of further research.

2 Margins and Generalization

Let us consider a two-class classification problem defined on an instance space X, in which the classes are identified by the two labels $\{-1, +1\}$. A classifier f can be described as a mapping from X to the interval $[-1, +1]$ such that a sample $\mathbf{x} \in X$ is assigned to one of the classes according to $\mathrm{sgn}(f(\mathbf{x}))$. If we assume that y is the correct label of $\mathbf{x}$, the *sample margin* associated to $\mathbf{x}$ is given by $yf(\mathbf{x})$. As a consequence, f provides a wrong prediction for $\mathbf{x}$ if the sample margin is negative. More precisely, while the sign of the margin indicates the correctness of the classifier, the magnitude estimates the confidence of the classifier in making its prediction on the sample $\mathbf{x}$. In this way, a large negative margin means that the classifier is wrong but highly confident.

Let us now consider a training set S containing N samples $\mathbf{x}_i$ where $i = 1, \cdots, N$, each associated to a label $y_i \in \{-1, +1\}$. The *minimum margin* or *classifier margin* for f over the set S is defined as: $\mu(f) = \min_i y_i f(\mathbf{x}_i)$ The classifier margin has an intuitive meaning [4]: it measures the distance that the classifier can travel in the feature space without changing the way it labels any of the sample points. Moreover, it has a direct connection with the generalization capability of the classifier as proved in several papers [2,3,13] which suggested upper bounds on generalization error expressed in terms of the classifier margin.

The margin theory has been also used in ensemble learning to analyze the generalization effectiveness of Adaboost. In particular, in [12] it has been proved that the error of a convex combination of classifiers is bounded by $\Pr_S[yf(\mathbf{x}) \leq \theta]$, i.e. the part of training examples with margin below a threshold θ, plus a "complexity penalty" term which is $O(1/\theta)$. In other words, the size of the minimum margin (i.e. the margin related to the worst result on the training set) appears to be the most relevant factor for improving generalization. On this basis, some experiments have been done aimed at improving the generalization performance of Adaboost by using a combination of the generated classifiers explicitly based on minimum margin maximization [5]. However, the results showed that the modified version of Adaboost performs generally worse than the original one. A possible reason is that the boosting approach focuses on the most difficult samples and tries to improve their margins, but with a significant possibility of overfitting when dealing with very difficult samples [11]. This problem could be further emphasized by a minimum margin maximization combination which would give higher weights to the classifiers with higher minimum margins, but more prone to overfitting.

However, as far as we know, the potential effectiveness of such a combination has not yet been examined when the classifiers of the ensemble are built independently and not according to a boosting approach and this becomes particularly interesting in the case of biometric score fusion.

3 Margin Maximization Via Linear Programming

In this section we describe the algorithm used for estimating the weights of a convex combination of classifiers which maximizes the minimum margin of the resulting classifier. As said in the previous section, the algorithm has been already used in [5] for experiments on Adaboost.

Let us consider again the training set S containing N samples $(\mathbf{x}_i, y_i)$. Let us moreover assume that K already trained classifiers are available $f_j(\mathbf{x}) \rightarrow [-1, +1]$ with $j = 1, \cdots, K$. The margin provided by the j-th classifier over the i-th sample is defined as:

$$\mu_{ij} = y_i \cdot f_j(\mathbf{x}_i) \quad (1)$$

so that the f_j correctly classifies $\mathbf{x}_i$ iff $\mu_{ij} > 0$. Let us now consider the linear combination of the K classifiers:

$$f_c(\mathbf{x}) = \sum_{j=1}^{K} w_j f_j(\mathbf{x}) \tag{2}$$

with $w_j \geq 0$ and $\sum_{j=1}^{K} w_j = 1$. The margin provided by f_c over the i-th sample is thus

$$\mu_i = y_i \cdot \left(\sum_{j=1}^{K} w_j f_j(\mathbf{x}_i) \right) = \sum_{j=1}^{K} w_j \mu_{ij} \tag{3}$$

while the margin of f_c is $\mu = \min_i \mu_i$. Actually the margin μ depends on the weights $\mathbf{w} = \{w_1, w_2, \cdots, w_K\}$ and thus such weights can be chosen to make the margin as large as possible. In this way we have a max-min problem which can be written as:

$$\begin{aligned} \text{maximize} \quad & \left(\min_i \sum_{j=1}^{K} w_j \mu_{ij} \right) \\ \text{subject to} \quad & \sum_{j=1}^{K} w_j = 1 \\ & w_j \geq 0 \qquad j = 1, 2, \cdots, K \end{aligned}$$

The problem can be recast as a linear problem [16] if we introduce the margin μ as a new variable:

$$\begin{aligned} \text{maximize} \quad & \mu \\ \text{subject to} \quad & \\ & \sum_{j=1}^{K} w_j \mu_{ij} \geq \mu \; i = 1, 2, \cdots, N \\ & \sum_{j=1}^{K} w_j = 1 \\ & w_j \geq 0 \quad j = 1, 2, \cdots, K \end{aligned}$$

If we collect the margins in a $N \times K$ matrix $\mathbf{M} = \{\mu_{ij}\}$, the weights in a vector $\mathbf{w}$ and define $\mathbf{e}_t$ the column vector consisting of t ones and $\mathbf{z}_t$ the column vector consisting of t zeros, the problem can be written in block-matrix form:

$$\begin{aligned} \text{maximize} \quad & \begin{bmatrix} \mathbf{z}_K^T \; 1 \end{bmatrix} \begin{bmatrix} \mathbf{w} \\ \mu \end{bmatrix} \\ \text{subject to} \quad & \\ & \begin{bmatrix} -\mathbf{M} \; \mathbf{e}_N \\ \mathbf{e}_N^T \quad 0 \end{bmatrix} \begin{bmatrix} \mathbf{w} \\ \mu \end{bmatrix} \begin{matrix} \leq \\ = \end{matrix} \begin{bmatrix} \mathbf{z}_N \\ 1 \end{bmatrix} \\ & \mathbf{w} \geq \mathbf{z}_K \end{aligned} \tag{4}$$

In principle, we could use any one of the numerous linear programming solving methods available, but it should be taken into account that the number of constraints could be very large since it equals the number of training samples. Therefore algorithms specially suited for large size problems should be used.

Table 1. Summary of biometric databases used in the experiments

Database	Biometric Traits	Matchers	Users	Training Set	Test Set
NIST Multimodal	Fingerprint (Left and Right) Face (2 matchers)	4	517	133645	133644
XM2VTS	Face (5 matchers) Speech (3 matchers)	8	295	40600	112200

4 Experiments

The effectiveness of the margin-based fusion rule has been evaluated on two publicly available biometric databases, namely, the partition NIST-Multimodal of the NIST-BSSR1 [9] and XM2VTS Benchmark [10]. The characteristics are summarized in table 1. All the data have been normalized using the min-max normalization [6]. For each experiment on NIST-Multimodal, half of the genuine and half of the impostor match scores were randomly selected to form the training set to train the fusion rules and the test set to evaluate their performance. Moreover, to avoid any bias in the following comparison, the hold out procedure has been repeated 20 times and all the reported results have been obtained as the mean values over each trial. For experiments on XM2VTS database the hold out procedure has not been implemented since we used the training-test partitioning suggested in [10].

In table 1 we also report the dimensions of training and test sets to show that in multibiometric problems the number of matchers is significantly less than the number of samples. This implies that our approach is usually involved in a linear maximization problem that faces with a low number of variables (i.e., a low-dimensional weight vector) but with a huge number of constraints (i.e., a high dimensionality of training data). In particular, on NIST-Multimodal we have only 4 matchers but 133645 samples while in XM2VTS the matchers are 8 and the number of constraints is equal to 40600. Therefore we employed the MOSEK Optimization Software [1] that provides specialized functions to solve large-scale linear programming problems.

In our first experiment we analyze the behavior of the proposed approach (hereafter called LPmax, i.e. Linear Programming maximization) on margins evaluation compared with the single matchers. To this aim, we consider the cumulative distributions of margins on the training set provided by LPmax and the single matchers. In figure 1 we report the margin cdfs for the proposed approach in comparison with the 4 matchers of the NIST-Multimodal database (figure 1.a) and the 8 matchers of XM2VTS benchmark (figure 1.b). Looking at these graphs, it could seem that some matchers exhibit higher margins. Actually, if we look at figure 2 (that reports the same plots but in logarithmic scale for the y-axis) we can note that the minimum margin of LPmax is in every case higher than all the matchers.

This agrees with the good results provided by LPmax and shown in figure 3. Here the performance on the test set have been evaluated in terms of Receiver

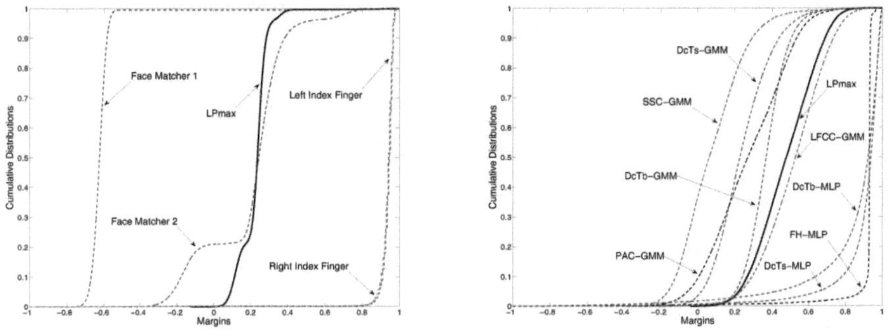

(a) (b)

Fig. 1. Margin distributions graphs for the LPmax approach and the single matchers on the training set of the NIST-Multimodal database (a) and the XM2VTS benchmark database (b)

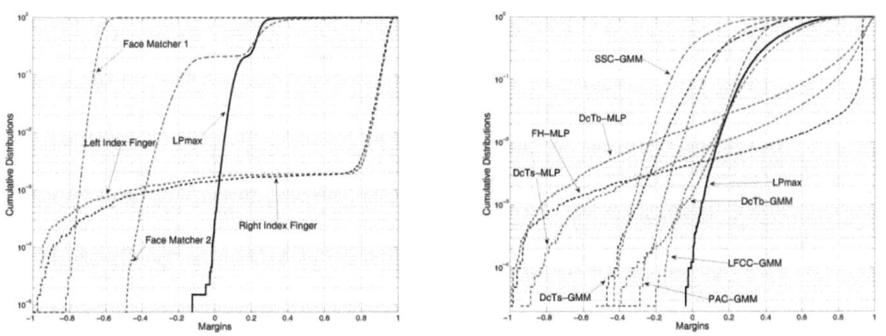

(a) (b)

Fig. 2. The same distribution graphs of figure 1 but with logarithmic scale on y-axis

Operating Characteristic (ROC) curves that plot the Genuine Acceptance Rate (GAR) vs. the False Acceptance Rate (FAR). The graphs show the curves only with a FAR between 0.01% and 0.1% since this is the only region useful to characterize the accuracy of multibiometric systems. Graphs show that the performance of LPmax are clearly better than all the single matchers on both data sets. It is worth noting that some matchers have margin distributions sensibly more concentrated on high values than LPmax (e.g. Right Index Finger on NIST and FH-MLP on XM2VTS); however, the better performance of LPmax suggests that a high value of minimum margin is the most relevant factor for increasing the performance.

In the second experiment we have compared LPmax with other fusion techniques. In particular, we considered two classifiers as combiners, i.e., AdaBoost and Support Vector Machine (SVM) and a parametric fusion model recently proposed in [8], hereafter called GMM-LR (Gaussian Mixture Model - Likelihood

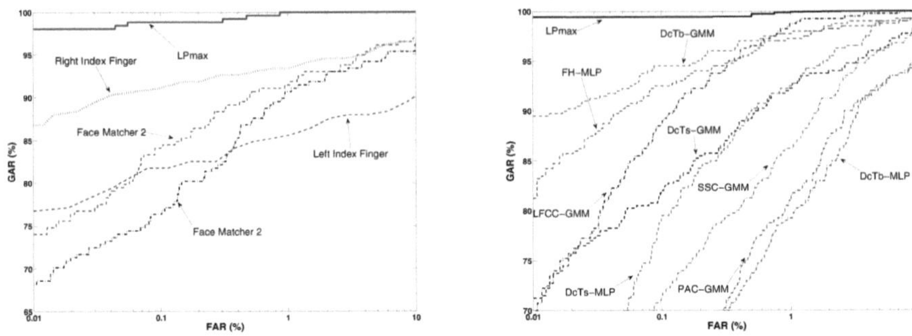

(a) (b)

Fig. 3. Performance of LPmax compared with the matchers on the test set of the NIST-Multimodal database (a) and the XM2VTS benchmark database (b)

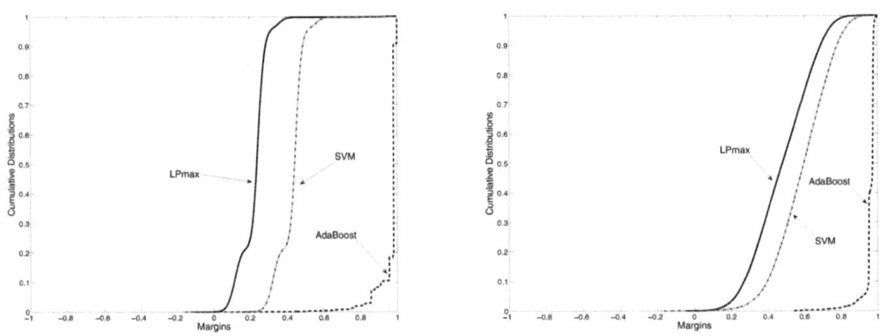

(a) (b)

Fig. 4. Margin distributions graphs for LPmax, AdaBoost and SVM on the training set of the NIST-Multimodal database (a) and the XM2VTS benchmark database (b)

Ratio). SVM has been used since it represents a linear classifier directly built for the margin maximization while AdaBoost, even if it is a combination of linear classifiers, could be considered as the state of the art for margin-based classifiers. AdaBoost has been implemented by means of Modest Adaboost [17], a more robust and stable version of standard AdaBoost, with 300 iterations of a simple decision tree used as weak learner while SVMs have been implemented by means of SVM^{light} software [7]. Following [8], linear SVMs have been used for both XM2VTS and NIST-Multimodal database. In all the experiments the C parameter of SVM has been put to infinite to keep fair the comparison with LPmax which does not use slack variables in the optimization algorithm.

Figure 4 shows this comparison respectively on the training set of the NIST-Multimodal and XM2VTS data sets[1]. In this case, the performance of SVM

[1] GMM-LR is not based on a margin fusion approach and thus we did not consider it for margin evaluation.

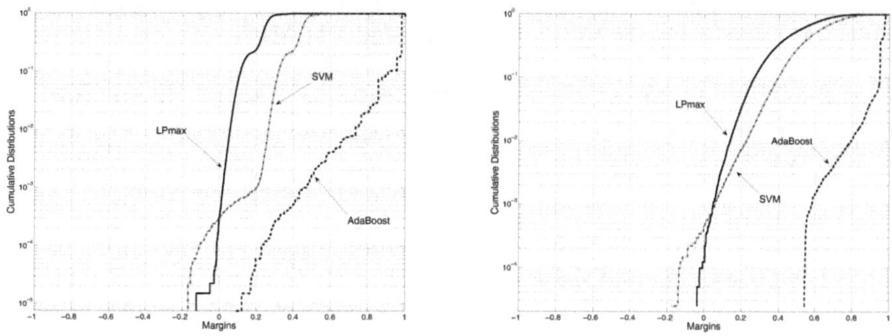

(a) (b)

Fig. 5. The same distribution graphs of figure 4 but with logarithmic scale on y-axis

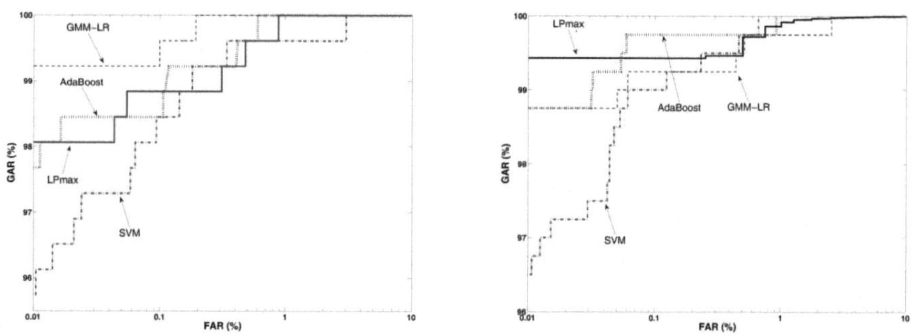

(a) (b)

Fig. 6. Performance of the compared approaches on the test set of the NIST-Multimodal database (a) and the XM2VTS benchmark database (b)

are quite good on margin distributions that have generally higher values than LPmax. However, if we look at the logarithmic graphs reported in figure 5 we can note that in both cases the minimum margin of LPmax is higher than SVM. AdaBoost, instead, exhibits clearly higher performance than the other approaches in terms of minimum margin on both data sets. As said in sect. 2, this is due to the boosting approach that focuses on the most difficult samples of the training set to be classified, giving almost perfect results on them.

Nevertheless, if we look at the performance on the test set it is possible to note that LPmax exhibits a better generalization capability. To this aim, in figure 6 we report the ROC curves of the different approaches. Moreover, to have a single value for the performance evaluation we also report in table 2 the mean GAR obtained at a FAR of 0.01% that is the point on the y-axis of the reported ROC curves. For NIST-Multimodal we provide the mean and the standard deviation obtained from the 20 runs of the hold out procedure.

Table 2. Performance on the test set in terms of mean GAR percentage at 0.01% FAR achieved by the different approaches. The best matchers are Right Finger and DcTb-GMM respectively for NIST-Multimodal and XM2VTS.

Database	LPmax	GMM-LR	AdaBoost	SVM	Best Matcher
NIST-Multimodal	98.5 ± 0.1	99.1 ± 0.4	98.4 ± 0.4	95.4 ± 1.9	85.3 ± 1.8
XM2VTS	99.5	98.7	98.8	96.2	89.5

In this case, the comparison shows different results on the two data sets. On NIST-Multimodal, GMM-LR has a higher ROC curve than all the other approaches while LPmax is comparable with AdaBoost and is clearly superior to SVM and the best matcher. The better performance of GMM-LR are probably due to the score distribution of the matchers that, as shown in [8], are well fitted by the Gaussian mixture model. On XM2VTS, instead, LPmax exhibits better performance than all the other methods even if for some FAR values the ROC of AdaBoost is higher than LPmax curve. However, if we focus on table 2, we can note that, when fixing the threshold at 0.01% FAR, LPmax has higher performance on XM2VTS and only GMM-LR exhibits better performance on NIST-Multimodal even if in this case the standard deviation of LPmax is tighter than GMM-LR, thus leading to a quite small difference between the two approaches.

Some final considerations could be made about the comparison with the SVM based combination which, in some regards, provides the most similar approach to LPmax: in fact both of them are linear combination and are based on margin maximization. The only difference is that SVM provide an optimal separating hyperplane whose margins from the two classes are equal; in other words, LPmax does not have the constraint of simmetrycal margins affecting the SVM learning algorithm and this could lead to a more effective minimum margin maximization.

5 Conclusions and Future Work

We have illustrated a framework for the fusion of match scores in a multibiometric system based on the margin maximization. Our approach relies on the evaluation of the weights of a convex combination of matchers such that the margin of the combination on the training set is maximized. The obtained results in comparisons with other recent and commonly employed approaches both on margin distribution and on classification performance encourage a deeper analysis of the proposed method. In particular, a relaxation of the constraint in the margin maximization could be adopted by introducing slack variables to face with noisy data and a more extensive experimentation considering other biometric datasets could be employed to better verify the effectiveness of this approach.

References

1. Mosek 2005. The MOSEK optimization tools version 5.0. User's manual and reference. MOSEK ApS, Denmark (2005)
2. Anthony, M., Bartlett, P.L.: Neural Network Learning: Theoretical Foundations. Cambridge University Press, Cambridge (1999)
3. Bartlett, P.L.: The sample complexity of pattern classification with neural networks: the size of the weights is more important than the size of the network. Information Theory, IEEE Transactions 44(2), 525–536 (1998)
4. Crammer, K., Gilad-Bachrach, R., Navot, A., Tishby, N.: Margin analysis of the lvq algorithm. In: Thrun, S., Becker, S., Obermayer, K. (eds.) Advances in Neural Information Processing Systems, vol. 15, pp. 462–469. MIT Press, Cambridge (2003)
5. Grove, A.J., Schuurmans, D.: Boosting in the limit: maximizing the margin of learned ensembles. In: AAAI 1998/IAAI 1998: Proceedings of the fifteenth national/tenth conference on Artificial intelligence/Innovative applications of artificial intelligence, Menlo Park, CA, USA, pp. 692–699. American Association for Artificial Intelligence (1998)
6. Jain, A.K., Nandakumar, K., Ross, A.: Score normalization in multimodal biometric systems. Pattern Recognition 38(12), 2270–2285 (2005)
7. Joachims, T.: Making large-scale svm learning practical. In: Schölkopf, et al. (eds.) Advances in Kernel Methods-Support Vector Learning, pp. 169–184. MIT Press, Cambridge (1999)
8. Nandakumar, K., Chen, Y., Dass, S.C., Jain, A.K.: Likelihood ratio-based biometric score fusion. Pattern Analysis and Machine Intelligence, IEEE Transactions 30(2), 342–347 (2008)
9. National Institute of Standards and Technology. Nist biometric scores set - release 1 (2004), `http://www.itl.nist.gov/iad/894.03/biometricscores`
10. Poh, N., Bengio, S.: Database protocols and tools for evaluating score-level fusion algorithms in biometric authentication. Pattern Recognition 39(2), 223–233 (2006)
11. Ratsch, G., Onoda, T., Muller, K.-R.: Soft margins for adaboost. Machine Learning 42(3), 287–320 (2001)
12. Schapire, R.E., Freund, Y., Bartlett, P., Lee, W.S.: Boosting the margin: A new explanation for the effectiveness of voting methods. The Annals of Statistics 26(5), 1651–1686 (1998)
13. Shawe-Taylor, J., Bartlett, P.L., Williamson, R.C., Anthony, M.: Structural risk minimization over data-dependent hierarchies. Information Theory, IEEE Transactions 44(5), 1926–1940 (1998)
14. Snelick, R., Uludag, U., Mink, A., Indovina, M., Jain, A.K.: Large-scale evaluation of multimodal biometric authentication using state-of-the-art systems. Pattern Analysis and Machine Intelligence, IEEE Transactions 27(3), 450–455 (2005)
15. Toh, K.-A., Kim, J., Lee, S.: Biometric scores fusion based on total error rate minimization. Pattern Recognition 41(3), 1066–1082 (2008)
16. Vanderbei, R.J.: Linear Programming: Foundations and Extensions, 2nd edn. Springer, Heidelberg (2001)
17. Vezhnevets, A., Vezhnevets, V.: Modest adaboost - teaching adaboost to generalize better. Graphicon-2005, Novosibirsk Akademgorodok, Russia (2005), `http://research.graphicon.ru/machine-learning/gml-adaboost-matlab-toolbox-3.html`

Biometric Template Update: An Experimental Investigation on the Relationship between Update Errors and Performance Degradation in Face Verification

Gian Luca Marcialis, Ajita Rattani, and Fabio Roli

University of Cagliari – Department of Electrical and Electronic Engineering
Piazza d'Armi – 09123 Cagliari (Italy)
{marcialis,ajita.rattani,roli}@diee.unica.it

Abstract. Current methods for automatic template update are aimed at capturing large intra-class variations of input data and at the same time restricting the probability of impostor's introduction in client's galleries. These automatic methods avoid the costs of supervised update methods, which are due to repeated enrollment sessions and manual assignment of identity labels. Most of state-of-the-art template update approaches add input patterns to the claimed identity's gallery on the basis of their matching score with the existing templates, which must be above a very high "updating" threshold. However, regardless of the value of such updating threshold, update errors do exist and impact strongly on the effectiveness of update procedures. The introduction of impostors into the galleries may degrade the performance quickly. This effect has not been studied in the literature so far. Therefore, a first experimental investigation is the goal of this paper, with a case study on a face verification system.

Keywords: Biometrics, Template update, Intra-class variations, Face biometrics, template update errors, impostors, threshold selection.

1 Introduction

The performance of a typical biometric system depends on the "representativeness" of the templates collected during the "enrollment" phase. Concisely,"representativeness" means the capability of enrolled template to recognize new input biometric data. In general, enrolled templates are captured in controlled environments, while input biometric data exhibit "substantial" intra-class variations. These may be (a) temporary variations like for face, changes in environment, expression, pose etc., or (b) permanent variations like for face, aging, accidental cuts or scars on subject's face, changing the subject's appearance.

To increase the templates representativeness, one can enroll multiple instances of each identity representing possible variations in input data, and repeating the process of enrollment over time [1]. However, this approach is based on supervised learning as the human supervisor assigns "identity" labels manually to the data used for updating, making the process very expensive, time consuming and inefficient [1]. Moreover, supervisor must be able to estimate possible intra-class variations in input data.

N. da Vitora Lobo et al. (Eds.): SSPR&SPR 2008, LNCS 5342, pp. 684–693, 2008.

Limitations of the supervised methods are addressed by introducing automatic template update methods based on semi-supervised learning [2-5]. These methods use both the initial set of enrolled and labelled templates and a set of unlabelled data acquired during the normal system's operation, for updating. The concept is that those unlabelled data classified (pseudo-labelled) as highly confident genuine users, i.e., genuine users with the matching score exceeding a given updating threshold, are used for updating the clients' galleries. The majority of these state-of–the-art "semi-supervised methods" [2-4] can be termed as "self update" methods, as they incrementally update themselves using their own knowledge.

It is worth noting that the threshold value selected for updating plays a vital role in the performance of these template update process. As only those input data, whose matching score is above the selected threshold, are added to the claimed identity's gallery. Thus, the efficiency of method and the security level of threshold selection has a direct bearing on the performance of the template update algorithms. The operation at very high or stringent threshold, may result in a very limited capability to capture input samples characterising significant intra-class variations [5]; on the other hand, relaxing the updating threshold may cause much introduction of impostors into the template gallery thus weakening the system's performance and security [5].

Although Ref. [5] pointed out the issue of threshold selection and impostor introduction, to the best of our knowledge, no other works in the literature investigated to which extent the updating threshold selection impacts on the performance degradation, and how relevant is the introduction of impostors into the template galleries. Most of the proposed methods accept the input data at a fixed non–user specific high threshold, corresponding to a certain working point in the ROC curve of the system. Threshold is estimated by using initial enrolled templates [2-3] and is not re-estimated after iterations of updating process. In this paper, we evaluate the performance of self update at different threshold setting methods, i.e., fixed non-user specific threshold ("same for each user") and user-specific threshold for constant and threshold re-estimated after each iteration of updation process at different security levels: zeroFAR, 1%FAR and 5%FAR. In particular, its effect on the verification performance degradation and introduction of impostors into the client's galleries has been investigated. We also draw attention on the vulnerability of self update methods to impostors even at very stringent threshold due to presence of "difficult" clients which can be easily confused with others, and clients, whose appearance intrinsically "imitates" others. These subjects have been named "lambs" and "wolves", according to the "Doddington's zoo" [6], respectively. In particular, these so called "wolves" can be classified as genuine users even at very stringent threshold settings and can intrude into galleries of another identity, in particular, the ones of lambs, hence gradually making the template self update counter productive.

In order to advance the state of art related to automatic template update procedures and to study the behaviour of the biometric template update methods from the stated viewpoint, a case study has been done on face authentication system based on EBGM [7]. Equinox Face Database [8] is used for experimental evaluation. The paper is organized as follows: Section 2, describes the self update systems in detail. Section 3, discusses the experiments and results. Conclusions are drawn in Section 4.

2 The Template "Self Update" Algorithm

In the "online" template self update algorithms [2-3], a matcher operating in mono-modal environment adapts itself to the variations of the input data available during the normal system's operation. The aim of these methods is to capture the temporal and temporary intra-class variations of the input data by modifying the templates and hence enhancing the generalization performance on the novel unseen data. The procedure to update the galleries is usually one of the two commonly used techniques, a) by adding the novel sample as another instance in the gallery set of the respective client ("instance based") [4] or b) by fusing the sample with the template to form a "super template", thus embedding the information into a single template [2-3] ("prototype based"). References [4-5] perform off-line updating: input data are collected during system's operations, but the updating process is performed "offline" that is, when the batch has been collected over time and system is not operating [4-5].

The general "online" self update method can be summarized as follows:

Initialization:
Each user is enrolled with its templates to build the initial gallery and the initial system parameter ("updating threshold" (*th*)) is set. The details about enrollment, feature extraction and matching processes of face verification can be found in [7].

Updating Process:
Repeat for each input data arrival

a) Compare the input data with the template(s) of the claimed identity's gallery to compute the matching score (*MS*).
b) **If** *MS>=th*
Update the template set of the respective client by either fusing with the current template or adding the input data as a separate instance into the gallery set of claimed identity
else
Reject the data
end

Specifically, the self update's internal operation can be represented by using a threshold graph (θ–graph) drawn on enrolled templates and input data. The graph gradually expands on each input sample arrival by connecting the template to the sample if the matching score between them is above the threshold θ. Thus, the self update graph is initialized with vertices representing enrolled templates and it gradually grows on each input data arrival by connecting the highly classified input data to current templates. The edge between two nodes represents that matching score among them is above θ. Fig. 1 shows an example of θ–graph build by the self updating process. Edges between vertices represent the matching score obtained, by comparing input data to current template set which is always above the acceptance threshold. The updating threshold is assumed to be 75 in this case (score normalized in [0,100] interval). The initial template (left side) is connected to an input due to the score which is above the threshold (see the edge label, namely, 85). This input is added to the gallery. A novel arrival is connected to such input, and/or connected to the initial template (current templates on the basis of score value), and so on.

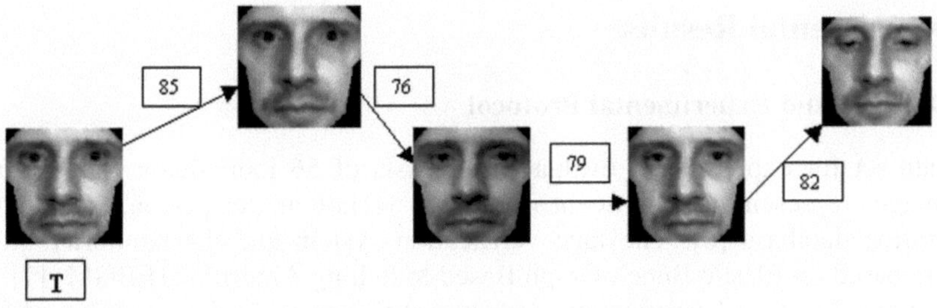

Fig. 1. The hypothetical theta graph (θ) maintained by self update's internal operation which keeps growing on the arrival of input data. The image marked **T** represents the enrolled template. The graph expands progressively by connecting the current templates to the input data, if the matching score is above acceptance threshold, which is assumed **75** in this example. All the input data connected to the initial template T (directly or through other input sample) are added to the template gallery for updating. Note that a new input data node can be linked to more than one node in the current template set on the basis of similarity above (θ). The present case only shows the connection of the input node to most similar node in template set.

Thus the connection between initial enrolled templates to highly confident examples and then the enrolled template together with the already classified examples are used to connect other samples whose scores are above θ, hence the name "θ–graph". Worth noting, if the same input data sequence is observed but the order is different, the corresponding θ–graph is also different.

Worth noting, the out-degree of θ-graph varies with θ, which corresponds to the updating threshold of self update. The more is θ, the less is the out-degree. Note that out-degree relates to number of images in the template gallery set which matches input sample at specific threshold. Obviously, at low threshold more images in template set will match input sample and vice-versa. Since highly genuine users are introduced in the clients' galleries for most of the template update procedures [2-5], operation at very stringent threshold, say zeroFAR, may be successful in avoiding impostors to certain extend due to very low out-degree of resulting θ-graph. However, this assumption completely neglects the typical characteristics of certain clients. Doddington et al. [6] introduced some peculiar terms to indicate clients wrongly classifiable even at high thresholds (θ), namely, "lambs", and clients which can "imitate" other clients, namely, wolves. Therefore, it is reasonable to argue that the existence of wolves and lambs among clients can have an adverse impact on the self update procedure even at stringent threshold condition. In particular, a wolf into a lamb's gallery may attract other wolves, so compromising seriously the representativeness of the obtained gallery. This effect may be accentuated by the low out-degree of θ-graph. These issues have not yet been studied in the literature. For many state-of-the-art methods, a fixed non-user specific threshold is set; although the access by impostors is assumed, their creeping in the templates galleries and the effect of impostors introduction has not been stated at all [1-3]. The threshold value fixed by using the initial enrolled is not re-estimated after updating iterations. To focus attention on these drawbacks, several experiments are conducted in this paper.

3 Experimental Results

3.1 Data Set and Experimental Protocol

The data set for experimental evaluation consists of 56 individuals with 91 frontal face images representing significant intra-class variations per person, from Equinox Corporation database [8]. The face verification system and the similarity measure used are based on Elastic Bunch Graph Based Matching Algorithm (EBGM) [7].

In a typical personal verification system, a different input data owing to different access attempt, is available for each client. In order to follow this evidence, the following protocol has been adopted: one image is selected as initial template for each client. In other words, the initial gallery of each subject is made up of only one template. Threshold values are always evaluated on template set by comparing each template to the templates of all the other clients thus estimating the impostor distribution and selecting threshold at desired FAR value, being the only set available in real environments where very less labelled data are present to set the system parameters.

Remaining client images i.e., ninety, are subdivided in two sets made up of thirty (input unlabelled set of genuine users) and sixty images per client (test set of genuine users). Both unlabelled and test sets are then randomly partitioned into 56 sets, such that the c-th partition does not contain images of c-th client. Each of these partitions, consisting of thirty and sixty images, represents the "impostor set" for c-th client: thirty images are added to the unlabelled set and sixty to the test set. The unlabelled set contains the images which are assumed to be the input data and each image in the unlabelled set is passed to the system for verification and accordingly on the basis of matching score between input image and the templates, the self update is performed. The test set evaluates the actual improvement in generalization of the system reached by self update techniques. Equal probability of genuine and impostors attempts are assumed by keeping equal number of genuine and impostors in unlabelled and test sets. Each experiment is repeated five times for different order of sequence of input data arrival (from "unlabelled set") to minimize the effect of order of sequence of input data arrival and related results are averaged.

The experiments are carried out using the above database characteristics and experimental protocol. In order to better study the impact of threshold selection on the capture of impostors or the difficult pattern representing intra-class variations, updating process in the current experiments is carried out by adding the input data as the separate instance in the gallery set of claimed identity. During updating, the input sample score is evaluated as the maximum value among scores computed between the sample and each template into the current gallery.

3.2 Results

The goal of our experiments is to evaluate the reliance of impostors introduction on various threshold selection conditions like fixed non-user specific vs. user specific. Each selection type is evaluated at different updating thresholds: very stringent; zeroFAR, stringent; 1% FAR, and relaxed; 5% FAR. Threshold can also be updated at each iteration, or kept fixed after the initial estimation on the available galleries. Thus, to show differences in the performance improvement and the progressive presence of impostors into the gallery by varying the initial experimental settings of template update, four threshold settings reported in Table 1 are investigated.

Table 1. Threshold settings adopted in reported experiments

Threshold Settings	
Along the algorithm iterations	**Along clients**
Fixed	Non user specific
Updated	User specific

For each experiment, two curves are reported and compared: Equal Error Rate (EER) on the test set at each updating iteration; % impostors into the galleries at each iteration. This is also an approximated average evaluation of the θ-graph out-degree. These curves are drawn for each threshold setting in combination (Table 1) in a plot, at three security levels: zeroFAR, 1%FAR, 5%FAR.

Figs. 2(a-f) show these plots. In particular:

1. Figs. 2(a-b). At zeroFAR updating threshold, there is no appreciable performance enhancement of face verification system for all the threshold settings methods (Table 1). Benefits of user-specific threshold appeared at the late iterations, as more amount of samples have been exploited in comparison to other threshold settings. The minimum EER value can be clearly noticed as a function of certain number of samples exploited for updating. With the increase in the number of samples exploited after certain iterations, EER starts increasing. By comparing Fig. 2(a) and Fig. 2(b), some correspondences can be found with the percentage impostors into the gallery, whose trend is always increasing, and the performance degradation depicted in Fig. 2(a). It can be hypothesised that, till the decreasing trend is noticed in EER value of Fig. 2(a), the genuine users are added into the gallery, thus the effect of impostors is not evident. This effect becomes prominent after the gradual accumulation of impostors, due, probably, to the cumulative "attraction" of impostors from other ones previously introduced ("wolves"?) in the same galleries. From the performance degradation viewpoint, it is worth noting that the difference in the impostor percentage of 1% (see values from 100 to 150 exploited samples in Fig. 2(b)) caused a remarkable decrease of EER (Fig. 2(a)).
2. Figs. 2(c-d). Observations similar to the previous ones can be drawn here also regarding impostors introduction. The minimum value of EER is lower than that of Fig. 2(a). This is due to the number of exploited samples, as can be seen by comparing Figs. 2(b, d). Moreover, the differences among threshold settings are more remarkable. Adopting an user-specific updated threshold appeared to be the best strategy in this case because it has apparently allowed to reduce the number of impostors introduced into the galleries. However, the trend of impostors introduction in Fig. 2(d) has increased very fast: from 4% impostors introduced in Fig. 2(b) to 20% impostors introduced there.
3. Figs. 2(e-f). For all threshold settings, the downward trend in performance is noticed even at the initial stages of updating. The worst performance is obtained by adopting a fixed and non user specific threshold, but, from the viewpoint of

impostors introduced, substantial differences with other settings cannot be appreciated. The amount of impostors introduction exceeds 50%, thus making self update very counter productive in this case. For both points 2 and 3, the θ–graph analysis is in our opinion superfluous, since the related out-degree is large even at the initial iterations of the algorithm, and impostors can be easily attracted.

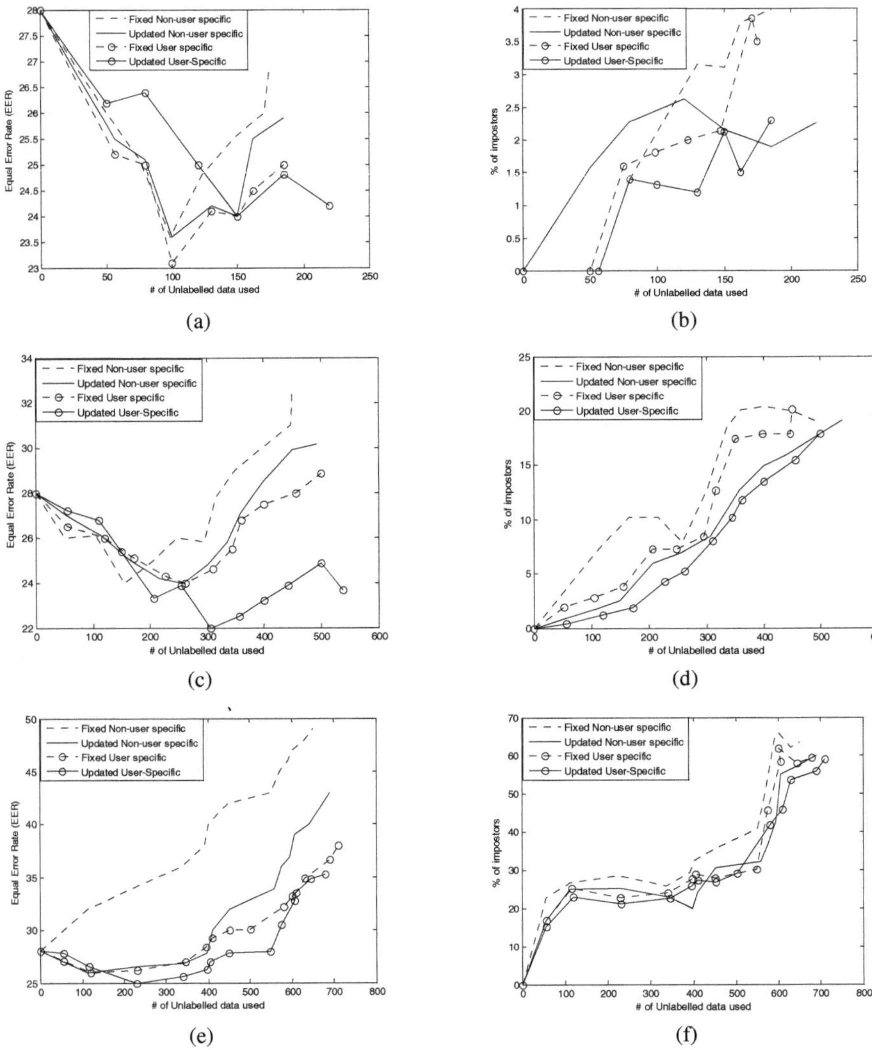

Fig. 2. Performance obtained in terms of %EER and %impostors encountered as a cumulative amount of unlabelled input data added at each iteration by self update for different threshold settings. Plots are reported as follows: (a-b) zeroFAR. (c-d) 1%FAR. (e-f) 5%FAR.

3.3 The Effect of Wolves and Lambs

The experimental results reported in Figs. 2(a–f) confirm the accumulative presence of impostors even at the stringent threshold selection conditions. The probabilistic presence of impostors in galleries during automatic self update process is due to basic FAR of the system, which increased very fast as shown in Section 3.3. Although this can be expected at relaxed updating thresholds, but operation at stringent threshold, say zeroFAR, the impostor introduction is still not avoidable. Obviously, a wrong estimation of zeroFAR threshold is to be taken into account. Beside this effect, which causes the introduction of "occasional" impostors, in our opinion the performance degradation can also be due to presence of wolves and lambs, which have so intrinsically overlapping nature (a wolf can be a lamb for others and vice-versa) that an estimation error, even very small, in the threshold at zeroFAR, causes a large number of impostors introduction, despite the low out-degree of the related θ-graph..

Accordingly, we analyzed the updated template galleries at zeroFAR. Some clients were found that acted as wolves, that is, these identities were very frequent among the impostors encountered in the template galleries. According to the Doddington's theory, some wolves were also lambs for other clients as shown by the same person which has been classified as a lamb in the same Table 2. Table 2 shows some impostors identities namely wolves which were always present in most of the updated gallery for all threshold settings. Some corresponding lambs are also reported.. For sake of clarity, values reported in third and fourth columns are related to the case of fixed threshold, as it is the most commonly adopted in literature [2-3]. The third column reports the overall percentage of clients affected by the related wolf, and the

Table 2. Some identities acted as wolves (present in updated templates galleries) for all the threshold selection method (Table 1) at zeroFAR. The current table shows the corresponding lambs, which are always intruded by impostors for all threshold settings (Table 1) at zeroFAR. The statistics related to %clients affected by the corresponding wolf (column 3) and % occurrence of samples of the coressponding wolf in updated galleries are presented at fixed-non user specific threshold. Note that at 0% FAR, 16.5% identities were not at all updated.

Wolves	Some of lambs affected by these wolves	%clients affected	%samples of these wolves in client's gallery
		18.2%	30%
		26.67%	34%
		23.33%	19%
		16.33%	16%

fourth one reports the average percentage of related wolf images among impostors introduced in the gallery of clients they have affected. This motivates why the above clients can be classified as wolves and lambs. The presence of wolves in several clients' galleries pointed out that, even at zeroFAR, the avoidance of intrusions by such identities is crucial for the self update effectiveness.

The effect of wolves can be modelled as in the θ–graph (as discussed in section 2) of Fig. 3. The θ–graph is taken from one of our experiments. For sake of space, it is not complete. The first image on the left-side is the initial template. It can be easily seen again that inclusion of an impostor during self-update process weakens the systems security and make the system vulnerable to others impostors too making the whole process counter-productive. In particular, the initial template of Fig. 3 is in the set of lambs of the fourth wolf of Table 2. It can be noticed that such wolf attracts quickly other samples similar to itself even after few iterations of self update. Finally, another wolf attack can also be seen by the last image (third subject of Table 2).

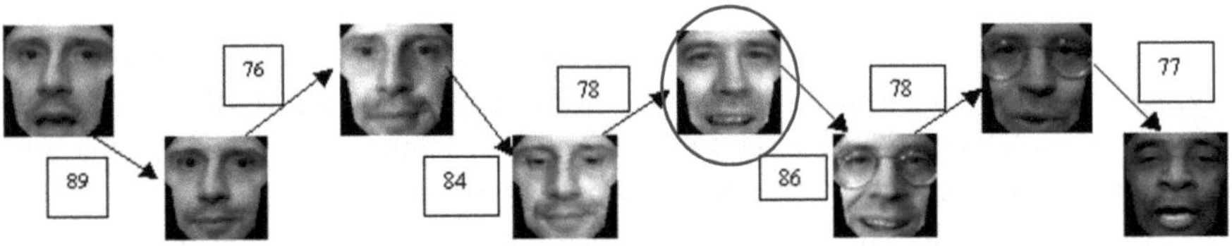

Fig. 3. θ–graph drawn for the selected identity at 0% FAR at fixed non-user specific threshold. It can be seen that introduction of an impostor makes the system vulnerable to same impostors and other impostors too. Red encircled identity is a wolf (Table 2, last row) which attracts more difficult samples of the same wolf with intra-class variations (spectacles and illumination invariance) and subject the system to other wolf too (last image; Table 2, third row).

Thus, the above experiments highlight that impostor inclusion is a serious issue related to automatic template self-update systems. So far, no method has been proposed to predict or avoid it. The only way to limit impostors introduction, in particular at stringent threshold, where a minimum of EER enhancement has been found, is the appropriate choice of the early stopping criterion of impostors. Since using a cross-validation methodology is not realistic in the case of self update, this criterion to avoid impostors must pass through an appropriate modelling of the impostor's introduction probability. In our opinion, some statistical solutions to predict wolves, lambs, goats, already proposed in the literature, could help to reduce or minimize this difficult task of modelling impostor introduction. For example, in [6] various tests like F-test, Kruskal-Wallis and Durbin tests have been proposed to model these peculiar characteristics of clients. Some criteria to rank the users in a database based on their variability in performance and wolves attack probabilities estimation have been also recently proposed [9-10]. These approaches has not yet been applied to self update algorithms, and may help in designing an appropriate early stopping criterion of impostor's introduction by modelling wolves and lambs probabilities, specially in the case of very stringent updating threshold. However, it is worth remarking that the problem is still an open issue.

4 Conclusions

In this paper, we clearly showed that the self update process is very much dependent on the threshold type settings and the security level for acceptance of input data which have the direct bearing on percentage of impostor introduction and performance degradation. This analysis has not yet been done in the literature on semi-supervised template update methods.

In particular, our investigation was related to self update methods. It has also been found that the impostors inclusion cannot be completely avoided even at strict threshold settings as apart from basic FAR of the system, identities so called wolves and lambs do have adverse effect on the self update process and may be introduced in template galleries of the client. Introduction of impostors has an accumulative adverse effect as they may attract more samples of them and other impostors too, especially if they act as wolves. The operation at optimal threshold value (user specific with updating) together with an appropriate early stopping criterion based on the "wolves and lambs" modelling may lead to reduce or minimize this evident drawback.

References

1. Uludag, U., Ross, A., Jain, A.: Biometric template selection and update: a case study in fingerprints. Pattern Recognition 37(7), 1533–1542 (2004)
2. Jiang, X., Ser, W.: Online Fingerprint Template Improvement. IEEE Trans. PAMI 24(8), 1121–1126 (2002)
3. Ryu, C., Hakil, K., Jain, A.: Template adaptation based fingerprint verification. In: Proc. of ICPR, Hong Kong, vol. 4, pp. 582–585 (2006)
4. Roli, F., Marcialis, G.L.: Semi-supervised PCA-based face recognition using self-training. In: Yeung, D.-Y., Kwok, J.T., Fred, A., Roli, F., de Ridder, D. (eds.) SSPR 2006 and SPR 2006. LNCS, vol. 4109, pp. 560–568. Springer, Heidelberg (2006)
5. Rattani, A., Marcialis, G.L., Roli, F.: Capturing large intra-class variations of biometric data by template co-updating. In: IEEE Workshop on Biometrics, Int. Conf. on Vision and Pattern Recognition CVPR 2008, Anchorage, Alaska, USA (in press, 2008)
6. Doddington, G., Liggett, W., Martin, A., Przybocki, M., Reynolds, D.: Sheeps, Goats, Lambs and Wolves: A Statistical Analysis of Speaker Performance in the NIST 1998 Speaker Recognition Evaluation. In: Proc. ICSLD, Sydney, Australia (1998)
7. Wiskott, L., Fellows, J.M., Kruger, N., von der Malsburg, C.: Face recognition by elastic bunch graph matching, Technical Report IR-INI, 96-08 (1996)
8. Equinox Face Database, http://www.equinoxsensors.com/products/HID.html
9. Poh, N., Bengio, S., Ross, A.: Revisiting Doddington's Zoo: A Systematic Method to Assess User-dependent Variabilities. In: Proc. of 2nd Work. on Multimodal User Authentication (MMUA), Toulose, France (May 2006)
10. Une, M., Otsuka, A., Imai, H.: Wolf Attack Probability: A New Security Measure in Biometric Authentication Systems. In: Lee, S.-W., Li, S.Z. (eds.) ICB 2007. LNCS, vol. 4642, pp. 396–406. Springer, Heidelberg (2007)

Automatic Mutual Nonrigid Registration of Dense Surfaces by Graphical Model Based Inference

Xiao Dong and Guoyan Zheng

MEM Research Center, University of Bern, Stauffacherstr. 78, CH-3014, Switzerland
guoyan.zheng@ieee.org

Abstract. This paper addresses the problem of fully automatic matching two triangulated surface meshes. In this paper, a similarity measurement is constructed to measure the consistency of the constraints among the correspondent landmarks, which is rigid transformation immune and robust to nonrigid deformations. The matching problem is then solved by directly finding correspondence between the landmarks of the two surfaces by graphical model based Bayesian inference. In order to reduce the computational complexity and to accelerate the convergence, a hierarchical graphical model is constructed which enables mutual registration and information exchange between the two surfaces during registration. Experiments on randomly generated instances from a PCA based statistical model of proximal femurs verified the proposed approach.

1 Motivation

Fully automatic nonrigid registration of shapes is highly desired for computer vision and medical image processing. For a nonrigid registration between two shapes $Shape_A$ and $Shape_B$, the ultimate goal should be to find the correspondence between subparts of the two shapes instead of to achieve a *perfect* match by applying a forced deformation without constraints on the objects. Such a correspondence assignment is usually defined and achieved as an optimization problem with a cost function abstracted as $H(S_A(\mathbf{Q}_A), S_B(Q_B), \mathbf{T}_{A\to B}, \mathbf{C}_{A\to B})$ to measure the difference of the two matched shapes, where $S_A(\mathbf{Q}_A)$ and $S_B(\mathbf{Q}_B)$ are the shape descriptors of S_A and S_B with shape parameters $\mathbf{Q}_A$ and $\mathbf{Q}_B$ in their own representation spaces, $\mathbf{T}_{A\to B}$ is a nonrigid transformation between the spatial coordinate spaces of the two shapes, and $\mathbf{C}_{A\to B}$ is the correspondence assignment between the subparts of the two shapes.

The cost function $H(S_A(\mathbf{Q}_A), S_B(Q_B), \mathbf{T}_{A\to B}, \mathbf{C}_{A\to B})$ can be defined in two ways. In the *robust point matching* (RPM), *active shape model*(ASM), *active appearance model* (AAM)[1], and nonrigid shape matching algorithms using the level set based contour and surface representation, shape signature harmonic embedding[3] and Gaussian mixture based shape models[4][5], the cost function is defined as

$$\begin{aligned} &H(S_A(\mathbf{Q}_A), S_B(Q_B), \mathbf{T}_{A\to B}, \mathbf{C}_{A\to B}) \\ &= E_{diff}(\mathbf{Q}_A), S_B(Q_B), \mathbf{T}_{A\to B}, \mathbf{C}_{A\to B}) + \gamma E_{deform}(\mathbf{T}_{A\to B}) \end{aligned} \tag{1}$$

N. da Vitora Lobo et al. (Eds.): SSPR&SPR 2008, LNCS 5342, pp. 694–704, 2008.

where $E_{diff}(S_A(\mathbf{Q}_A), S_B(Q_B), \mathbf{T}_{A\to B}, \mathbf{C}_{A\to B})$ is a measurement of the shape difference between the *aligned* shapes with the nonrigid transformation $\mathbf{T}_{A\to B}$ and the correspondence assignment $\mathbf{C}_{A\to B}$, $E_{deform}(\mathbf{T}_{A\to B})$ defines the *bending energy* of the nonrigid deformation $\mathbf{T}_{A\to B}$ and γ is a factor to balance the two terms. The advantage of this approach is that both the shape difference and the bending energy are explicitly expressed and we have a wide freedom to select their definitions with different underlying physical pictures. The disadvantage is that the explicit representations of $\mathbf{T}_{A\to B}$ and $\mathbf{C}_{A\to B}$ are iteratively defined on each other, i.e., a $\mathbf{T}_{A\to B}$ can only be defined on a given $\mathbf{C}_{A\to B}$ and vice versa. This usually results in that $\mathbf{T}_{A\to B}$ and $\mathbf{C}_{A\to B}$ can only be updated iteratively during the optimization. Such an optimization strategy can only converge to local optima and it usually asks for a proper initialization.

In the shape context[6], the spectral decomposition [7] and graphical model based registration algorithms [8][9][10], both the shape difference and the bending energy of the nonrigid deformation are defined implicitly, where both of them are encoded in the relative positions between correspondent landmarks of the shapes. In such an approach, the optimization procedure aims to directly find the optimal correspondence assignment $\mathbf{C}_{A\to B}$ and the nonrigid transformation $\mathbf{T}_{A\to B}$ never appears during the optimization. The advantage of this approach is that it can lead to a fully automatic registration since the optimization does no rely on an explicit transformation $\mathbf{T}_{A\to B}$ between the two shape. The disadvantage is that usually this approach is computational expensive.

In [8] and [9] graphical model based Bayesian inference was used for the purpose of point set matching. In [13] this idea was also explored for the purpose of surface matching. One limitation of the work of [13] is that it only matches shapes with the same number of landmarks and assumed that a correspondence assignment between the landmarks of these two shapes can always be achieved. The optimization procedure to find the optimal correspondence assignment was achieved by Gibbs sampling, which may not be efficient to explore the whole configuration space [14] and therefore apt to local optima.

In this paper, we explore the idea of graphical model based approach. Firstly, a cost function which is immune to rigid transformations and robust to nonrigid shape deformations is designed. Belief propagation is then used to find the optimal correspondence assignment. A hierarchical matching strategy is adopted to reduce the computational complexity and to ensure the correspondence assignment. To accelerate the convergence of the optimization procedure and to overcome local optima, we construct a parallel graphical model so that the two surfaces will mutually register to each other and also exchange local assignment information during the registration

2 Graphical Model Based Nonrigid Surface Matching

2.1 The Cost Function for Nonrigid Surface Matching

The inputs of our system are two dense triangulated surfaces $Surf_A$,$Surf_B$ with N_A and N_B ($N_A \approx N_B$) triangle faces respectively. The surfaces can be described as $\mathbf{S}_A = [\mathbf{s}_A^0, \mathbf{s}_A^1, \ldots, \mathbf{s}_A^{N_A-1}]$ and $\mathbf{S}_B = [\mathbf{s}_B^0, \mathbf{s}_B^1, \ldots, \mathbf{s}_B^{N_B-1}]$ respectively, where $\mathbf{s}_A^i = [\mathbf{x}_A^i, \mathbf{n}_A^i]$ ($\mathbf{s}_B^i = [\mathbf{x}_B^i, \mathbf{n}_B^i]$) stands for the 3d position of the center and the normal direction of the ith face of $Surf_A(Surf_B)$.

For each landmark $\mathbf{s}_A^i$ on the surface $Surf_A$, we can compute a *shape signature* $\mathbf{d}_A^i$ of the landmark at the ith face of $Surf_A$ as

$$\mathbf{d}_A^i = [\mathbf{d}_A^{i,0}, \ldots, \mathbf{d}_A^{i,j}, \ldots, \mathbf{d}_A^{i,N_A-1}]_{j \neq i}, \mathbf{d}_A^{i,j} = [D_A^{i,j}, A_A^{i,j}, AL_A^{i,j}], \quad (2)$$

where $D_A^{i,j} = \|\mathbf{x}_A^i - \mathbf{x}_A^j\|$ is the spatial distance of landmark $\mathbf{s}_A^i$ and $\mathbf{s}_A^j$, $A_A^{i,j}$ is the angle between the normal directions $\mathbf{n}_A^i, \mathbf{n}_A^j$ of the two faces and $AL_A^{i,j}$ is the angle between the normal direction $\mathbf{n}_A^i$ and the vector connecting $\mathbf{x}_A^i$ and $\mathbf{x}_A^j$ as shown in Fig. 1(a). This shape descriptor measures the distances and angles between landmarks and vectors, therefore a spatial shape $Surf_A(Surf_B)$ can be fully determined by the collection of all these shape signatures $\{\mathbf{d}_A^i\}(\{\mathbf{d}_B^i\})$. This shape descriptor is immune to rigid transformations. In the case of a nonrigid deformation, the landmarks and normal vectors can be regarded as connected by elastic strings and the deviation of these values can then be understood as the bending energy of the deformation. Therefore this descriptor can also handle nonrigid shape deformations.

Without losing any generality, we consider the case to register S_A to S_B. Given the shape descriptors $\{\mathbf{d}_A^i\}$ and $\{\mathbf{d}_B^i\}$, the difference between two surfaces with a given correspondence assignment between the landmarks of the surfaces $C_{A \to B}(i) = j, i = 0, 1, \ldots, N_A, j = 0, 1, \ldots, N_B$ can then be written as

$$H(S_A, S_B, C_{A \to B}) = \sum_i (\mathbf{d}_A^i - \widetilde{\mathbf{d}}_B^i) \Lambda (\mathbf{d}_A^i - \widetilde{\mathbf{d}}_B^i)^T \quad (3)$$

where $\widetilde{\mathbf{d}}_B^i = [\mathbf{d}_B^{C_{A \to B}(i), C_{A \to B}(0)}, \ldots, \mathbf{d}_B^{C_{A \to B}(i), C_{A \to B}(j)}, \ldots, \mathbf{d}_B^{C_{A \to B}(i), C_{A \to B}(N_B-1)}]$, $\Lambda = Diag(\sigma_D^2, \sigma_A^2, \sigma_{AL}^2, \ldots, \sigma_D^2, \sigma_A^2, \sigma_{AL}^2)^{-1}$ is the covariance matrix to balance the strengths of the elastic strings that connect the landmarks. The cost function measures the difference between the shape descriptor of S_A and the shape descriptor of S_B at the correspondent landmarks.

2.2 Graphical Model Based Correspondence Assignment Optimization

For the purpose of defining a 3-dimensional surface, the cost function in (3) is over-determined since its underlying graph G is a complete graph. In [9] the authors indicated that to determine a rigid shape in $\mathbb{R}^d$, the minimal underlying graph G is *k-tree* in $\mathbb{R}^d$. A denser graph topology will improve the robustness of the representation to keep the shape with the expense of a higher computational cost. Define a graph $G = \{V, E\}$ with N_A vertices, $V = \{v_i\}$ with each vertex correspondent to one landmark in S_A and $E = \{e_{ij}\}$ with $e_{ij} = 1$ if there is an edge connecting v_i and v_j. Then a generalized representation of (3) can be given as

$$\begin{aligned} H_G(S_A, S_B, C_{A \to B}) &= \textstyle\sum_{i,j,e_{ij}=1} E_A^{i,j}(C_{A \to B}) \\ &= \textstyle\sum_{i,j,e_{ij}=1} (\mathbf{d}_A^{i,j} - \widetilde{\mathbf{d}}_B^{i,j}) \Lambda' (\mathbf{d}_A^{i,j} - \widetilde{\mathbf{d}}_B^{i,j})^T \end{aligned} \quad (4)$$

where $\widetilde{\mathbf{d}}_B^{i,j} = \mathbf{d}_B^{C_{A \to B}(i), C_{A \to B}(j)}$, $\Lambda' = Diag(\sigma_D^2, \sigma_A^2, \sigma_{AL}^2)^{-1}$.

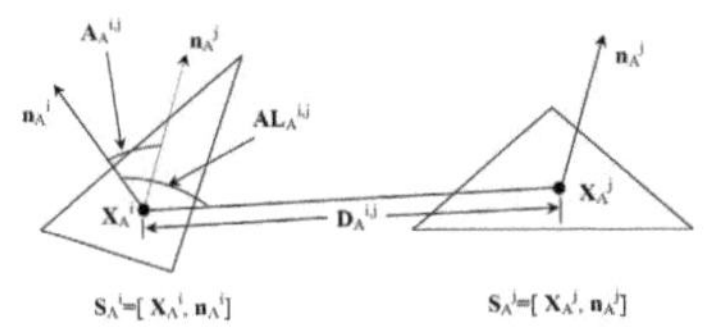

(a) The definition of the shape signature at each face of a mesh

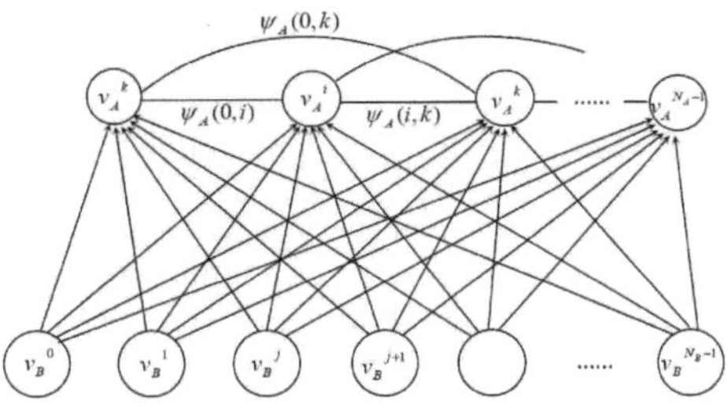

(b) The graphical models for belief propagation based nonrigid surface matching

Fig. 1. The graphical models for the registration of two surfaces taking $Surf_A$ as the template

Accordingly a distribution $p(H_G) = \prod_{i,j,e_{ij}=1} \exp(-E_A^{i,j}(C_{A\to B}))$ can be defined and represented by a graphical model as shown in Fig. 1(b). Each vertex v_A^i represents a landmark $\mathbf{s}_A^i$ of S_A and its state is a vector $sv_A^i = [p_A^{i,0}, \ldots, p_A^{i,N_B-1}]$ representing the probabilities that $\mathbf{s}_A^i$ is assigned as the correspondent landmark of the landmarks of S_B. Vertices $\{v_B^j\}$ stands for the the landmarks in S_B and its state is a vector $sv_B^j = [p_B^{j,0}, \ldots, p_B^{j,N_A-1}]$ representing that $\mathbf{s}_B^j$ should be the correspondence of each landmark in S_A. The vertices $\{v_A^i\}$ are connected according to the topology of G. The potentials between $\{v_A^i\}$ encode probabilities as $\psi_A(i,j) = \exp(-\beta E_A^{i,j}(C_{A\to B}))$, where $\beta = 1/\tau$ is the inverse of the annealing temperature τ. The optimization to find the optimal correspondence assignment can then be formalized as a ML estimation to find the peak of $p(H_G)$, which can be achieved by graphical model based Bayesian inference[8][9]. In our method, the Bayesian inference is achieved by running the belief propagation[8] on the graphical model as given in Fig. 1(b). When the belief propagation converges, the state sv_A^i is then the marginal distribution of the assignment of vertex v_A^i. If the annealing template τ is low enough, then $\{sv_A^i\}$ will give a unique peak which indicates the global optimum correspondence assignment for the matching.

3 Graphical Model Based Hierarchical Mutual Surface Registration

It can be observed that the above described graphical model based nonrigid surface matching faces three difficulties

- The existence of correspondence assignment: In our algorithm description, we assume that the correspondent landmarks for each landmark of $Surf_A$ can always be (*uniquely*) found on $Surf_B$. For a general case where $N_A \approx N_B$, this can not always be satisfied.
- The computational complexity: To matching two surfaces with N_A and N_B faces, the computational complexity of each belief propagation iteration is

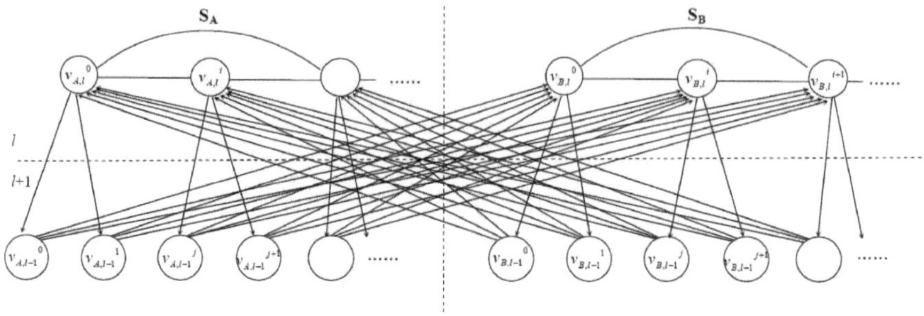

Fig. 2. The graphical model for parallel mutual registration of two hierarchically represented surfaces

$\mathcal{O}(\kappa N_A N_B^2)$, where κ is a factor related with the topology of the graphical model. The proposed algorithm is quite computational expensive for practical cases.
- The unbalanced information processing structure: In the graphical model based surface matching, the two surfaces play different roles. The graphical model is built on $Surf_A$ and the shape descriptors of the two surfaces are explored in different ways. Such an unbalanced information processing structure may lead to local optima as in the classic ICP based registration, where the correspondence is achieved by finding closest points from one data set to another. In an ideal surface matching solution, the two surfaces should be dealt with equivalently.

To overcome the above mentioned difficulties, we proposed a hierarchical mutual registration strategy, in which

- A hierarchical shape descriptor is used to describe each surface. In such a hierarchical structure, we can always match a sparse surface to a dense surface so that the existence of correspondence can be ensured.
- The computational cost on the hierarchical shape descriptor can be greatly reduced due to the fact that the matching results at coarse levels can be used as initializations for the finer levels so that the possible configurations of each landmark at finer levels can be greatly reduced.
- A parallel graphical model is used to achieve mutual matching of the two surfaces. During the matching procedure, the local matching result of each surface can be exchanged between surfaces so that it's more robust to local optima.

3.1 Hierarchical Shape Descriptors of Triangulated Surface Meshes

To reduce the computational cost, a hierarchical coarse-to-fine strategy is used to describe each surface as $\{\mathbf{S}_{A,L-1}, \ldots, \mathbf{S}_{A,1}, \mathbf{S}_{A,0}\}$ and $\{\mathbf{S}_{B,L-1}, \ldots, \mathbf{S}_{B,1}, \mathbf{S}_{B,0}\}$, were L is the levels of the hierarchical representation. In our system a simple k-mean clustering method is used to accomplish this task. $Surf_A$ and $Surf_B$

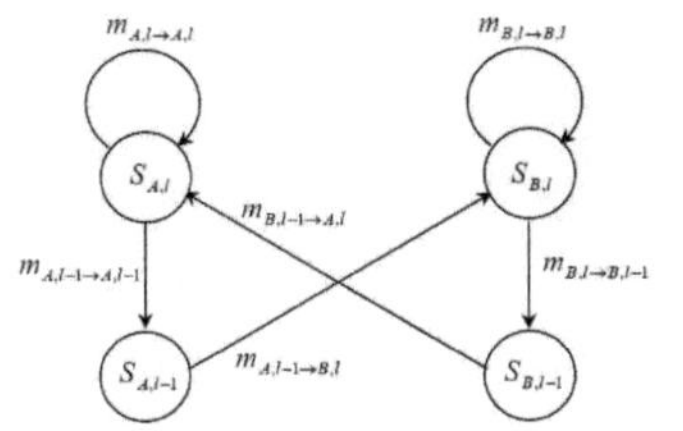

Fig. 3. Information transfer flow in the graphical model for the parallel mutual registration

are first segmented into $N_{A,L-1}$ and $N_{B,L-1}$ ($N_{A,L-1} \approx N_{B,L-1}$) clusters. Each cluster of $S_{A,L-1}$ ($S_{B,L-1}$) is represented as $\mathbf{s}^i_{A,L-1} = [\mathbf{x}^i_{A,L-1}, \mathbf{n}^i_{A,L-1}]$ ($\mathbf{s}^i_{B,L-1} = [\mathbf{x}^i_{B,L-1}, \mathbf{n}^i_{B,L-1}]$), where $\mathbf{x}^i_{A,L-1}$ ($\mathbf{x}^i_{B,L-1}$) is the center of the central triangle of this cluster and $\mathbf{n}^i_{A,L-1}$ ($\mathbf{n}^i_{B,L-1}$) is the mean normal direction of the faces in this cluster. Lower level representations are then obtained by further subdivide each cluster in higher level representations by k-mean clustering and each subdivided cluster can still be described by its center and normal direction. The finest level descriptors $S_{A,0}$ ($S_{B,0}$) are then S_A (S_B). To facilitate the correspondence assignment on each landmark $\mathbf{s}^i_{A,0}$ ($\mathbf{s}^i_{B,0}$)at the finest level, we build an extra representation level $S_{A,-1} = \{\mathbf{s}^i_{A,-1}\}, i = 0, 1, \ldots, KN_{A,0} - 1$ ($S_{B,-1} = \{\mathbf{s}^i_{B,-1}\}, i = 0, 1, \ldots, KN_{B,0} - 1$) by uniformly sampling K samples from each face in $S_{A,0}$ ($S_{B,0}$).

With such a hierarchical structure of each surface, we can construct hierarchical shape descriptors at each representation level as described in the previous part. The graphical model based surface matching can then be carried out to match a surface at a coarse level l to another surface at a finer level $l-1$. Due to the fact that $N_A \approx N_B$ and $N_{A,l} << N_{B,l-1}$ ($N_{B,l} << N_{A,l-1}$), the existence of correspondence can then be ensured.

3.2 Graphical Model Based Parallel Mutual Registration

In the graphical model based strategy to match S_A with S_B, only the shape information of S_A encoded in $\{\psi_A(i,j)\}$ is explicitly explored. This may lead to a slow convergence and it is apt to local optima.

The idea of parallel mutual registration is stimulated from the Turbo code in communication field, which indicates that an error-correction code which satisfies two sets of constraints simultaneously can greatly improve the its performance[15]. For our purpose of nonrigid shape matching, simultaneously exploring the different shape information of the two surfaces, may also improve the robustness and accelerate the convergence of our algorithm.

The graphical model for parallel mutual registration of $Surf_A$ and $Surf_B$ between representation level $l-1$ and l is a duplication of the graphical model of Fig. 1(b) as shown in Fig. 2 with extra directional edges to connect vertices representing landmarks of the same surface but at different representation levels.

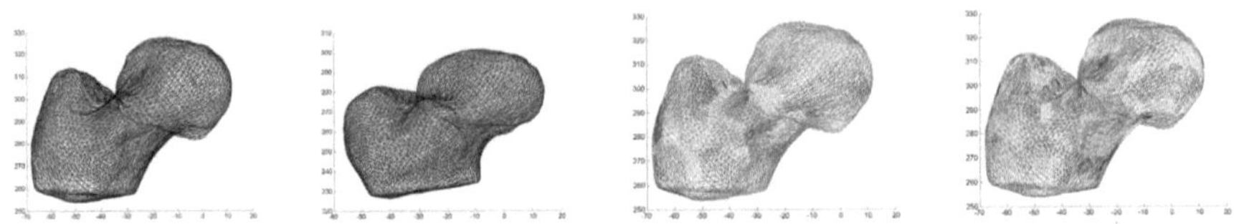

(a) Surface A (b) Surface B (c) Surface A, L=3 (d) Surface A, L=2

Fig. 4. Randomly generated surfaces from a PCA based statistical model of proximal femurs and their hierarchical representations

These edges only exist if a patch at a finer level is a subset of a patch at a coarser level. The equivalent cost function of the parallel mutual registration scheme can be written as $H_G(S_{A,l}, S_{B,l-1}, C_{A,l\rightarrow l-1}) + H_G(S_{B,l}, S_{A,l-1}, C_{B,l\rightarrow l-1})$ with a consistency requirement on $C_{A,l\rightarrow l-1}$ and $C_{B,l\rightarrow l-1}$ encoded in the message exchange between the two subparts of Fig. 2. In the graphical model for the mutual registration, the correspondence assignment information is transferred in the graphical model as shown in Fig. 3.

3.3 Hierarchical Mutual Registration

The parallel registration can be carried out at each successive level of the hierarchical representation of the surfaces, where the result of belief propagation at coarse levels can be used as the initialization at finer levels. The graphical model based hierarchical mutual surface matching is summarized as in Algorithm 1.

Algorithm 1. Graphical model based hierarchical automatic surface matching

---

- 1. $l = L - 1$, initialize all the messages as equal distributions at all configurations.
- 2. Run the parallel mutual matching between level l and level $l - 1$ with $\beta = \beta_l$ till it converges.
- 3. Using the correspondence assignment result between level l and $l - 1$, initialize the correspondence assignment between level $l-1$ and $l-2$. Increase β_l to β_{l-1}, go to 2 till $l = 0$.

---

4 Experiments

We verified the principle of the proposed matching algorithm on a PCA based statistical model of proximal femurs including 4098 vertices and 8192 faces. Two randomly generated instances are obtained from the statistical model and one of them is further scaled by a uniform scaling of 0.9 as shown in Fig. 4(a) and

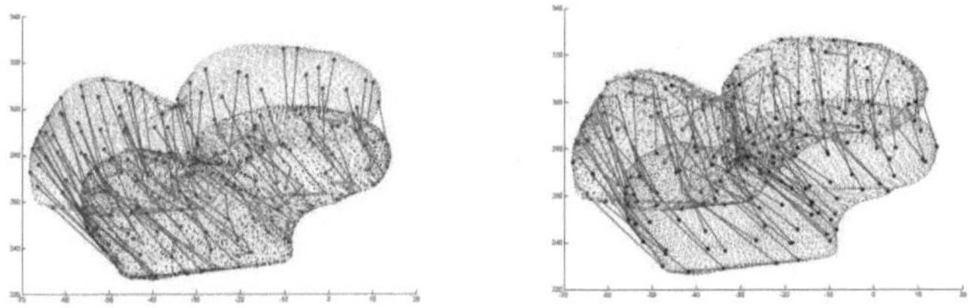

(a) Correspondence assignment between $S_{A,3}$ and $S_{B,2}$

(b) Correspondence assignment between $S_{B,3}$ and $S_{A,2}$

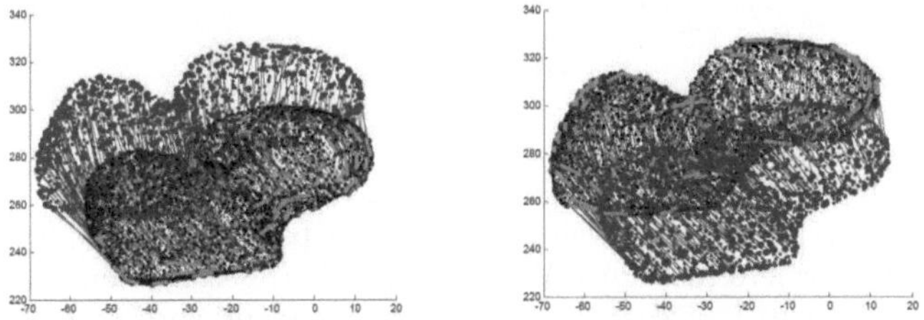

(c) Correspondence assignment between $S_{A,2}$ and $S_{B,1}$

(d) Correspondence assignment between $S_{B,2}$ and $S_{A,1}$

Fig. 5. The results of mutual registration at $l = 3$ and $l = 2$, the red lines show the correspondent landmarks on the two surfaces and the green lines show the distance between the assignment results of our algorithm and the correspondence assignment encoded in the PCA model

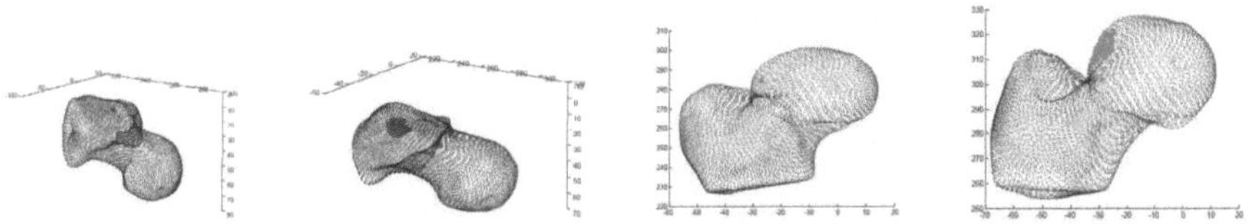

(a) The center of one patch in $S_{A,2}$

(b) Initialization of its correspondence assignment in $S_{B,1}$

(c) The center of one patch in $S_{B,2}$

(d) Initialization of its correspondence assignment in $S_{A,1}$

Fig. 6. The initialization for the mutual registration between level 2 and 1 using the assignment results at level 3

4(b). Since our matching algorithm is immune to rigid transformation, applying any rigid transformation to these surfaces will not influence the matching result. Hierarchical representations of the surfaces are constructed with $L = 3$, the first two levels ($L = 3$ and $L = 2$) of clustering results of $Surf_A$ are shown in Fig. 4(c) and 4(d).

The correspondence assignment using the graphical model based mutual registration is then carried out on them. The underlying graph topology G is selected as a random graph, in which the degree of each node is at least 5

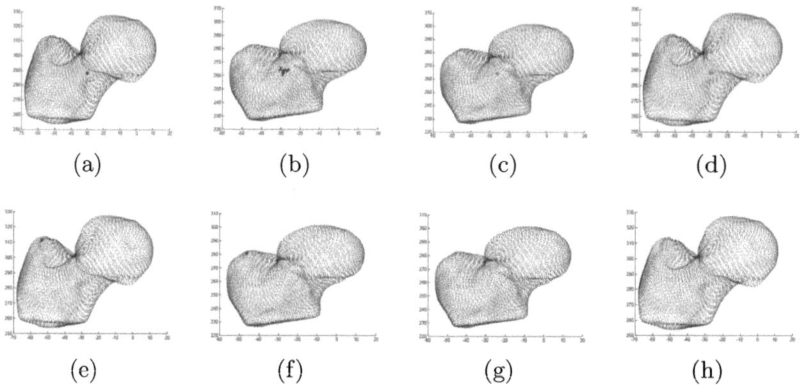

Fig. 7. The result of mutual registration between level 2 and 1, (a)(e): centers of two patches in $S_{A,2}$; (b)(f): centers of their correspondent patches in $S_{B,1}$; (c)(g): centers of two patches in $S_{B,2}$ where most probable correspondence assignment in $S_{B,1}$ of (a)(e) are their subpatches; (d)(h): centers of the correspondent patches of (c)(g) in $S_{A,2}$; the size of the dots in (b)(d)(f)(h) represents the probability of the correspondence assignment

(connected with 5 nearest neighbours) and the average degree of one node is 10. $\Lambda' = Diag(\sigma_D^2, \sigma_A^2, \sigma_{AL}^2)^{-1}$ in (4) is selected as $\sigma_D = \mathbf{d}_{A,l}^{i,j}/4$, $\sigma_A = \sigma_{AL} = 20°$.

- The results of the mutual registration at the coarsest level $l = 3$ and $l = 2$ are shown in Fig. 5. Obviously the two surfaces are already roughly aligned at the coarsest level and this shows that the proposed algorithm does not ask for any initialization.
- The registration result at coarser level is then used to initialize the correspondence assignment at finer level as shown in Fig. 6. It can be observed that the number of candidates of each landmark at a finer level is greatly reduced due to the initialization.
- The advantage of the parallel mutual registration between the two surfaces can be observed from Fig. 7. The mutual registration can explore the shape information of two surfaces and exchange correspondence assignment information between them. This strong constraint forces the correspondent landmarks on two surfaces to mutually assign to each other quickly. It's observed during the experiment the belief propagation at each level can converge in less than 10 iterations.
- As mentioned before, the goal of a nonrigid registration is to find optimal correspondence assignment between shapes. Since both surfaces are generated from the same PCA model, each landmark on a surface carries a point index in the PCA model. We take the landmarks with the same point index on the two surfaces as the ground truth of the correspondence assignment and then compute the registration error as the distance between the correspondent landmark obtained by our registration algorithm and the ground

truth position from prior knowledge of the PCA model. The registration error is evaluated on both shapes as $2.7 \pm 2.3mm$. Of course the prior correspondence knowledge may not be the ground truth but it can be regarded as a proper reference.

5 Conclusions

In this paper we proposed a fully automatic scheme for nonrigid surface matching. The nonrigid surface matching is formalized as a graphical model based Bayesian inference and the belief propagation is used to achieve the optimization to find the optimal correspondence assignment between shapes. To further reduce the computational cost and enhance the robustness to noise and local optima, a hierarchical mutual registration strategy is implemented so that the shape information of the two surfaces can be simultaneously explored. Experiments on randomly generated surfaces from a PCA based statistical model showed the capability of the proposed algorithm to achieve an automatic nonrigid surface registration.

The proposed scheme can also be extended to incorporate other shape descriptors such as the Gaussian curvature as used in [13] and the shape context [6] since they can be easily modeled as local believes of each vertex in our graphical model based scheme.

References

1. Cootes, T., Taylor, C.: Statistical models of appearance for computer vision. Technical report, University of Manchester, U.K (2004)
2. Xu, C., Yezzi, A., Prince, J.: A summary of geometric level-set analogues for a general class of parametric active contour and surface models (2001)
3. Lee, S.M., Abbott, A.L., Clark, N.A., Araman, P.A.: A shape representation for planar curves by shape signature harmonic embedding. In: CVPR 2006, pp. 1940–1947 (2006)
4. Roy, A.S., Gopinath, A., Rangarajan, A.: Deformable density matching for 3d non-rigid registration of shapes. In: MICCAI 2007, pp. 942–949 (2007)
5. Jiang, Y.F., Xie, J., Sun, D.Q., Tsui, H.: Shape registration by simultaneously optimizing representation and transformation. In: MICCAI 2007, pp. 809–817 (2007)
6. Belongie, S., Malik, J., Puzicha, J.: Shape matching and object recognition using shape contexts. IEEE Transactions on Pattern Analyis and Machine Intelligence 24, 509–522 (2002)
7. Jain, V., Zhang, H.: Robust 3d shape correspondence in the spectral domain. In: International Conference on Shape Modeling and Applications, SMI (2006)
8. Coughlan, J., Ferreira, S.: Finding deformable shapes using loopy belief propagation. In: ECCV 2003, pp. 453–468 (2002)
9. Caetano, T.S., Caeli, T., Barone, D.A.C.: An optimal probabilistic graphical model for point set matching. Technical Report Technical Report TR 04-03, University of Alberta, Edmonton, Alberta Canada (2004)
10. Rangarajan, A., Coughlan, J., Yuille, A.L.: A bayesian network framework for relational shape matching. In: ICCV 2003, pp. 671–678 (2003)

11. Zhang, L., Seitz, S.M.: Parameter estimation for mrf stereo. In: CVPR 2005, pp. 288–295 (2005)
12. Sun, J., Zheng, N.N., Shum, H.Y.: Stereo matching using belief propagation. IEEE Transactions on Pattern Analysis and Machine Interlligence 25, 1–14 (2003)
13. Xiao, P.D., Barnes, N., Caetano, T., Lieby, P.: An MRF and Gaussian curvature based shape representation for shape matching. In: CVPR 2007, pp. 17–22 (2007)
14. Gibbs, A.L.: Bounding the convergence time of the gibbs sampler in bayesian image restroation. Biometrika 87(4), 749–766 (2000)
15. McEliece, R.J., MacKay, D.J.C., Cheng, J.F.: Turbo decoding as an instance of pearl's "beliefpropagation" algorithm. IEEE Journal on Selected Areas in Communications 16, 140–152 (1998)

A Novel Video Classification Method Based on Hybrid Generative/Discriminative Models

Zhi Zeng, Wei Liang, Heping Li, and Shuwu Zhang

Digital Content Technology Research Center, Institute of Automation, Chinese Academy of Sciences, Beijing, China
{zzeng,wliang,hpli,swzhang}@hitic.ia.ac.cn

Abstract. We consider the problem of automatically classifying videos into predefined categories based on the analysis of their audio contents. In detail, given a set of labeled videos (such as news, sitcoms, sports, etc.), our objective is to classify a new video into one of these categories. To solve this problem, a novel audio features based video classification method combining an unsupervised generative model named probabilistic Latent Semantic Analysis (pLSA) with a multi-class discriminative classifier is proposed. Since general audio signals usually show complicated distribution in the feature space, k-means clustering method is firstly used to group temporal signal segments with similar low-level features into natural clusters, which are adopted as "audio words". Then, the audio stream of a video is decomposed into a bag of "audio words". To classify those bags of "audio words" which extracted from videos, latent "topics" are discovered by pLSA, and subsequently, training a multi-class classifier on the "topic" distribution vector for each video. Encouraging classification results have been achieved in our experiments.

Keywords: Video classification, pLSA, audio content mining.

1 Introduction

Due to the rapid development in various areas of computer infrastructure such as increased processing power, bigger and less expensive capacity of storage devices, and faster networks, the amount of video data has grown enormously in recent years. For a video viewer, the amount of video that he has to choose from is so large that it is infeasible for him to go through all these videos to find the video of interest. Classifying these videos into several categories or genres can narrow his choices. For a TV station or digital video library with a huge amount of digital video, video classification is an indispensable procedure in their management systems.

Many literatures [1-15] have addressed the video classification techniques. In general, for the purpose of video classification, features are drawn from three modalities: text, audio, and visual. So these methods could be divided into four groups: text-based methods [1-2], audio-based methods [3-7], visual-based methods [8-12] and those that used some combination of text, audio and visual features [13-14]. Regardless of which of features are used, there are some standard classifiers such as Support Vector

N. da Vitora Lobo et al. (Eds.): SSPR&SPR 2008, LNCS 5342, pp. 705–713, 2008.

Machine (SVM), Gaussian Mixture Models (GMMs), and Hidden Markov Models (HMMs) to be adopted. A comprehensive survey about this area can be found in [15].

In our method, only audio features are used to classify videos into predefined categories due to the following two reasons: (1) extracting audio features requires fewer computational resources than visual methods; (2) audio in video sequence has remarkable semantic content discrimination capability [3-7]. Audio classification is an active research topic in the fields of audio content analysis; L. Lu *et al.* [16] proposed a robust approach that is capable of classifying and segmenting an audio stream into speech, music, environment sound, and silence, G. Tzanetakis *et al.* [17] explored the automatic classification of audio signals into a hierarchy of musical genres.

In recent years, inspired by classical text document analysis techniques, some researchers [18-19] proposed an unsupervised approach to discover key audio elements in general audio documents. In [18], an iterative spectral clustering method is firstly used to group temporal signal segments with similar low-level features into natural clusters, which are analog to words in text documents and named audio elements, and then a TFIDF-like audio element weighting schema is presented to spot key audio elements, which is an analog to keyword extraction technique using term frequency (TF) and inverse document frequency (IDF) in text document analysis field. Motivated by these works, we try to utilize probabilistic Latent Semantic Analysis (pLSA) [20] to model video categories based on audio features..

The pLSA model was originally developed for topic discovery in a text corpus, where each document is represented by its word frequency. A. Bosch *et al.* [22] applied this model to images represented by the frequency of "visual words", and subsequently, training a multi-class classifier on the topic distribution vector for each image. This hybrid generative/discriminative approach also can be used to classify audio.

In this paper, we utilize the hybrid generative/discriminative approach in [21] to classify audio documents extracted from given videos into one of four predefined categories, namely news, sitcoms, sports, and talk shows. Firstly, the audio document is divided into frames, and a number of audio features are computed to characterize each audio frame. Then, a classical k-means clustering method is used to group similar frames into "audio words", and all the "audio words" form an "audio vocabulary". Subsequently, pLSA model is used to discover latent "topics" in an audio document using the bag of "audio words" representation. Finally, a multi-class classifier is trained on the "topic" distribution vector for each audio document. Our experiments show the effective of this method.

The rest of this paper is organized as follows. Section 2 describes in detail the features used to form the "audio vocabulary". Then, in Section 3, we briefly give an overview of the pLSA model and describe the hybrid classification algorithm based on applying pLSA to audio documents followed by discriminative classification. In Section 4, empirical experiments and performance evaluation of the proposed algorithm are presented. Finally, Section 5 concludes our work.

2 Audio Words and Audio Vocabulary

To utilize the pLSA model, a generative model from the statistical text literature, we need to compute a co-occurrence table, where each audio document is represented as a collection of "audio words", provided from an "audio vocabulary". This means that audio documents should be decomposed into segments, and these audio segments should be grouped into clusters. "Audio vocabulary" is formed by these cluster centers, and each audio segment could be denoted by an "audio word" in the "audio vocabulary", which is the most similar cluster center.

In this section, we first present the audio features we used, and then address how to form the "audio vocabulary" by classical k-means clustering algorithm.

2.1 Feature Extraction

Inspired by previous work on content-based audio classification and audio effect detection [16-19], in order to decompose an audio document into "audio words", the audio document is first down-sampled at 8000 Hz and 16 bits/sample, and divided into frames of 25 ms with 50% overlap. Then, a number of audio features are computed to characterize each audio frame. These audio features used in this work are modified from those described in [19, 20], which are following:

1) Root mean square volume (RMS);
2) Zero-crossing rate (ZCR);
3) Subband energy ratios (BER);
4) Brightness;
5) Bandwidth;
6) 8-order Mel-frequency cepstral coefficient (MFCCs);
7) Subband Spectral Flux.

These 7 kinds of audio features are collected into a 28-dimensional feature vector per audio frame. Moreover, the first two kinds of audio features are the temporal features, and the others are the spectral features. Following the work in [19], the spectral domain is equally divided into eight subbands in Mel-scale and then the subband features are extracted.

It's well-known that silence information in audio segments is useful to audio classification work. But the 28-dimensional feature vector we extracted can not efficiently recognize silence frame. So, due to the lack of ability to discriminate between silence frame and non-silence frame, we set a threshold on RMS of each frame to detect silence before feature vector extraction. If the RMS of a frame is below the threshold, we treat this frame as silence frame, and set its feature vector to $\mathbf{0}$ which is a 28×1 vector of all zero.

In order to reduce the computational complexity of the proposed method, we group audio frames into longer temporal audio segments, and use these longer segments as the basis for the next audio processing steps. We use a sliding window of 0.5s with non-overlap to segment the frame sequence, and the mean of the frame-based features at each window are used to represent the corresponding 0.5-second-long audio segment. Especially, if the silence frame ratio in a window is more than 60%, vector 0 is used to represent this audio segment.

2.2 Clustering

After audio feature extraction, we need to group the feature vectors extracted from the training audio set into V clusters. V is the total number of cluster, and also the amount of the “audio vocabulary”. In order to group similar feature vectors into V clusters, we use the traditional k-means clustering algorithm and adopt the Euclidean distance to measure the distance between two feature vectors.

By using clustering algorithm, every audio segment in the training audio set could be represented as a word in the “audio vocabulary”. Suppose we have a collection of audio Documents $D = d_1, ..., d_N$ with words from audio vocabulary $W = w_1, \ldots, w_V$, where N is total number of audio documents in training set and V is total number of “audio words” in the “audio vocabulary”. The training audio set could be denoted by a $V \times N$ co-occurrence table of counts $N_{ij} = n(w_i, d_j)$, where $n(w_i, d_j)$ denotes how often the word w_i occurred in an audio document d_j. And the audio segments in a new audio document could also be represented as a word index by choosing the nearest cluster center. This means that a test audio document could be represented by a V×1 vector of counts $N'_i = n(w_i, d_{test})$.

3 Video Classification

Since we classify videos based on audio features only, audio tracks will be extracted from videos both in training set and in testing set. And we just process these audio documents.

When the audio documents both in training set and in testing set are represented as vectors of counts of word occurrence, which are called bag-of-word in [21-22], we can use the pLSA to model these audio documents, and utilize the model results to classify these audio documents.

In Section 3.1, we give an overview of the pLSA model. Then we describe the hybrid classification algorithm in Section 3.2. Finally, our discussion is given in Section 3.3.

3.1 pLSA Model

The probabilistic Latent Semantic Analysis (pLSA) model [21] is a latent variable model for co-occurrence data which associates an unobserved topic variable $z \in Z = z_1, \ldots, z_Z$ with each observation, an observation being the occurrence of a word in a particular document. Here, we have audio documents as text documents, and we discover audio elements as topics (for example, speech and music) so that an audio document containing instances of several audio elements is modeled as a mixture of topics.

Let us introduce the following probabilities: $P(d_j)$ is used to denote the probability of observing a particular audio document d_j. $P(w_i|z_l)$ denotes the conditional probability of a specific word conditioned on the unobserved topic variable z_l, and finally $P(z_l|d_j)$ denotes an audio document specific probability distribution over the latent variable space. Using these definitions, one may define a generative model by the following scheme:

1) Select an audio document d_j with probability $P(d_j)$,
2) Pick a latent class z_l with probability $P(z_l|d_j)$,
3) Generate an audio word w_i with probability $P(w_i|z_l)$.

As a result, one obtains an observation pair (d_j, w_i), while the latent topic variable z_l is discarded. A representation of the pLSA model in terms of a graphical model is depicted in Figure 1.

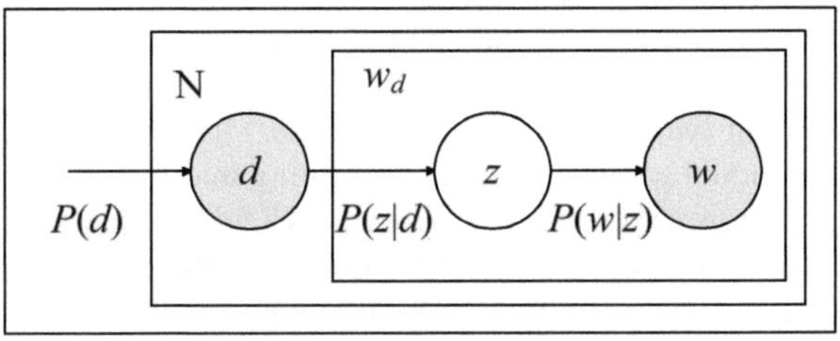

Fig. 1. Graphical model representation of the pLSA model

Translating the data generation process into a joint probability model results in the expression:

$$P(w_i, d_j) = P(d_j)P(w_i \mid d_j) \tag{1}$$

$$P(w_i \mid d_j) = \sum_{l=1}^{Z} P(w_i \mid z_l)P(z_l \mid d_j) \tag{2}$$

This amounts to a matrix decomposition with the constraint that both the word-topic matrix $P(w|z)$ and mixture coefficients matrix $P(z|d)$ are normalized to make them probability distributions. Following the likelihood principle, one can use a maximum likelihood formulation of the learning problem to determines ($V*Z+Z*N$) parameters: $P(w|z)$ and $P(z|d)$, i.e., one has to maximize the log likelihood function [21-22]:

$$\begin{aligned} L &= \log P(W, D) = \sum_{i=1}^{V}\sum_{j=1}^{N} n(w_i, d_j)\log P(w_i, d_j) \\ &= \sum_{j=1}^{N} n(d_j)\left[\log P(d_j) + \sum_{i=1}^{V}\frac{n(w_i, d_j)}{n(d_j)}\log\sum_{l=1}^{Z} P(w_i \mid z_l)P(z_l \mid d_j)\right] \end{aligned} \tag{3}$$

Here, $n(d_j) = \sum_{i=1}^{V} n(w_i, d_j)$ refer to the audio document length.

This is equivalent to minimizing the Kullback-Leibler divergence between the measured empirical distribution and the fitted model. As described in [21], the pLSA model is fitted by using the Expectation Maximization (EM) algorithm [23]. The goal of fitting this model is to determine the model that gives high probability to the distribution of "audio words" that appear in the corpus.

3.2 Hybrid Classification

Following the work in [22], we adopt the hybrid classification method with the following steps: firstly, discovering latent topics using pLSA model, and subsequently, training a multi-class classifier on the topic distribution vector for each audio document. Figure 2 shows graphically the hybrid generative/discriminative process for both training and testing.

Training proceeds in two stages: First, both $P(w|z)$ and $P(z|d_{training})$ are determined by fitting the pLSA model to the entire set of training audio documents. Each training audio document is then represented by a Z-vector $P(z|d_{training})$, where Z is the number of latent topics learned. Second, a multi-class discriminative classifier is trained given the vector $P(z|d_{training})$ of each training audio document and its class label. In this paper, we adopt the KNN classifier, which selects the K nearest neighbors of the new audio document within the training set (using Euclidean distance), and classifies the testing audio document to the most represented category label within the K nearest neighbors.

Classification of an unseen audio document similarly proceeds in two stages: First, $P(z|d_{testing})$ is computed by running EM algorithm in a similar manner to that used in learning, but only $P(z|d_{testing})$ are updated in each M-step with the learned $P(w|z)$ kept fixed. The testing audio document is represented by a Z-vector. In the second stage,

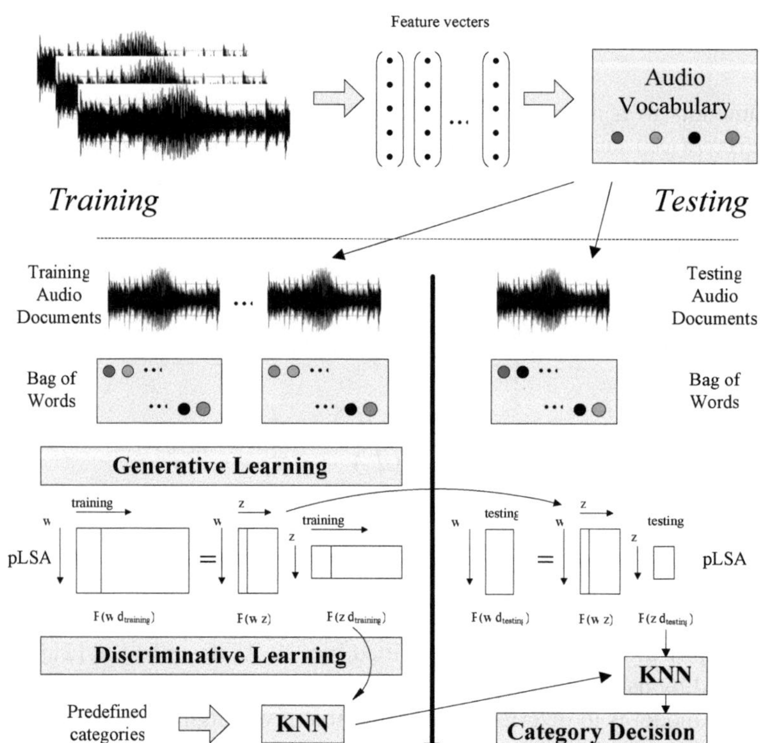

Fig. 2. Overview of audio vocabulary formation, learning, and classification stages

the testing audio document is then classified by the KNN multi-class classifier mentioned above.

3.3 Discussion

Unlike other audio content analysis techniques [16-17] or audio based video classification [3-7] approaches, which directly insert the low-level audio features into some kinds of statistical models or classifiers, our method discovers the semantic content from audio documents by 2-stage of abstraction of the low-level audio features. In the first stage, a group of low-level audio features of an audio segment is mapped into an "audio word", which is the nearest cluster center. Then, an audio document can be represented by an "audio word" set, which is called bag of words. In the second stage, latent topics are discovered within the bags of words by the pLSA model, and an audio document is modeled as a mixture of topics.

Since topics include many semantic contents of audio documents, using the topic distribution in an audio document as a feature vector in a classifier should be more effective than using low-level audio features directly. In the next section, experiments will validate this supposal.

We choose K nearest neighbors (KNN) classifier as the supervised discriminative classifier in our scheme, because of its simpleness. Compared to other classifier such as SVM, its performance is not too bad. Results from experiment in [22] show that SVM just performs around 1 percent better than KNN.

4 Experiment Results

In this section, the performance of the proposed hybrid approach to video categories classification is evaluated.

We collected video sequences from TV programs containing the following four classes: news, sitcoms, sports (basketball matches) and talk shows. These video sequences vary in duration from 15 minutes to 1 hour. These data is collected from different Chinese TV channels on different dates and at different times to ensure the variety of data. We divide the data into two sets: training and testing data sets. The training data set includes 100 video sequences, which are consisting of 24 news, 43 sitcoms, 15 basketball matches and 18 talk shows. And the testing data set include 70 video sequences, which are consist of 20 news, 40 sitcoms, 5 basketball matches and 5 talk shows. To apply out approach, audio tracks are extracted from these video sequences. The experiment in this section is on these audio documents.

In our classification approach, there are several parameters needed adjust: the number of audio words in the audio vocabulary (V), the number of latent topics (Z), and the number of neighbors (K in KNN). In the experiment, we set V equal to 500, Z equal to 10 and K equal to 5. The performance of our classification approach is given in Table 1.

The testing set contains 70 videos different from the training set. Out of these, 64 videos are correctly labeled while 6 are misclassified. Therefore, an accuracy of 91.4% was achieved. Most errors originate from non-news videos labeled as news

Table 1. Confusion matrix of video classification results

	News	Sitcom	Basketball	Talk Show
News	95%	5%	0%	0%
Sitcom	7.5%	92.5%	0%	0%
Basketball	0%	20%	80%	0%
Talk Show	20%	0%	0%	80%

videos, this is partially because that monologues in sitcoms or talk shows make them seem like news.

5 Conclusions

We have proposed a video classification approach that using audio feature only. In this approach, an audio vocabulary is first obtained by vector quantizing descriptors computed from training audio documents using k-means clustering, and an audio document is represented by a bag of words. Then latent topics and their distribution are trained using pLSA and used as a feature vector in the KNN classifier. In experiment, we try to classify videos into 4 categories namely news, sitcoms, sports and talk shows, and the results are encouraging.

Acknowledgments. This work has been supported by the National Science and Technology Supporting Program of China under Grant No. 2006BAH02A13 and No. 2006BAH02A36.

References

1. Zhu, W., Toklu, C., Liou, S.-P.: Automatic news video segmentation and categorization based on closed-captioned text. In: Proc. IEEE Int. Conf. Multimedia Expo (ICME), pp. 829–832 (2001)
2. Brezeale, D., Cook, D.J.: Using closed captions and visual features to classify movies by genre. In: Poster Session 7th Int. Workshop Multimedia Data Min (MDM/KDD), San Jose, CA (2006)
3. Liu, Z., Wang, Y., Chen, T.: Audio feature extraction and analysis for scene segmentation and classification. J. VLSI Signal Process. Syst. 20(1/2), 61–79 (1998)
4. Huang, J., Liu, Z., Wang, Y., Chen, Y., Wong, E.K.: Integration of multimodal features for video scene classification based on HMM. In: Proc. 3rd IEEE Workshop Multimedia Signal Process, pp. 53–58 (1999)
5. Roach, M., Mason, J.: Classification of video genre using audio. Eurospeech 4, 2693–2696 (2001)
6. Dinh, P.Q., Dorai, C., Venkatesh, S.: Video genre categorization using audio wavelet coefficients. In: 5th Asian Conf. Comput.Vis., Melbourne, Australia (2002)
7. Moncrieff, S., Venkatesh, S., Dorai, C.: Horror film genre typing and scene labeling via audio analysis. In: Proc. Int. Conf. Multimedia Expo (ICME), vol. 1, pp. 193–196 (2003)

8. Iyengar, G., Lippman, A.: Models for automatic classification of video sequences. In: Sethi, I.K., Jain, R.C. (eds.) Proc. SPIE Storage Retrieval Image Video Databases VI, vol. 3312, pp. 216–227 (1997)
9. Girgensohn, A., Foote, J.: Video classification using transform coefficients. In: Proc. IEEE Int. Conf. Acoust. Speech Signal Process (ICASSP), vol. 6, pp. 3045–3048 (1999)
10. Wei, G., Agnihotri, L., Dimitrova, N.: TV program classification based on face and text processing. In: Proc. IEEE Int. Conf. Multimedia Expo., vol. 3, pp. 1345–1348 (2000)
11. Truong, B.T., Dorai, C., Venkatesh, S.: Automatic genre identification for content-based video categorization. In: Proc. 15th Int. Conf. Pattern Recognit., vol. IV, pp. 230–233 (2000)
12. Lu, C., Drew, M.S., Au, J.: Classification of summarized videos using hidden Markov models on compressed chromaticity signatures. In: Proc. 9th ACM Int. Conf. Multimedia, pp. 479–482 (2001)
13. Qi, W., Gu, L., Jiang, H., Chen, X.-R., Zhang, H.-J.: Integrating visual, audio and text analysis for news video. In: Proc. 7th IEEE Int. Conf. Image Process (ICIP), pp. 520–523 (2000)
14. Wang, P., Cai, R., Yang, S.Q.: A hybrid approach to news video classification multimodal features. In: Proc. Joint Conf. 4th Int. Conf. Inf., Commun. Signal Process. 4th Pacific Rim Conf. Multimedia, pp. 787–791 (2003)
15. Brezeale, D., Cook, D.J.: Automatic Video Classification: A Survey of the Literature. IEEE Trans. Systems, Man, and Cybernetics-Part C: Applications and Reviews 38(3), 416–430 (2008)
16. Lu, L., Zhang, H.J., Jiang, H.: Content analysis for audio classification and segmentation. IEEE Trans. Speech Audio Process 10(7), 504–516 (2002)
17. Tzanetakis, G., Cook, P.: Musical genre classification of audio signals. IEEE Trans. Speech Audio Process 10(5), 293–302 (2002)
18. Cai, R., Lu, L., Hanjalic, A.: Unsupervised content discovery in composite audio. In: Proc. ACM Multimedia 2005, pp. 628–637 (2005)
19. Lu, L., Hanjalic, A.: Audio Keywords Discovery for Text-Like Audio Content Analysis and Retrieval. IEEE Trans. Multimedia 10(1), 74–85 (2008)
20. Cai, R., Lu, L., Hanjalic, A., Zhang, H.J., Cai, L.-H.: A flexible framework for key audio effects detection and auditory context inference. IEEE Trans. Audio, Speech, Lang, Process 14(3), 1026–1039 (2006)
21. Hofmann, T.: Unsupervised Learning by Probabilistic Latent Semantic Analysis. Machine Learning 41(2), 177–196 (2001)
22. Bosch, A., Zisserman, A., Munoz, X.: Scene classification using a hybrid generative /discriminative approach. IEEE Trans. PAMI 30(4), 712–727 (2008)
23. Dempster, A.P., Laird, N.M., Rubin, D.B.: Maximum likelihood from incomplete data via the EM algorithm. J. Royal Statist. Soc. B 39, 1–38 (1977)

Scale-Space Kernels for Additive Modeling

Chandan K. Reddy[1] and Jin-Hyeong Park[2]

[1] Department of Computer Science, Wayne State University,
Detroit, MI-48202, USA
reddy@cs.wayne.edu

[2] Integrated Data Systems Department, Siemens Corporate Research,
Princeton, NJ-08540, USA
jin-hyeong.park@siemens.com

Abstract. Various forms of additive modeling techniques have been popularly used in many pattern recognition and machine learning related applications. The efficiency of any additive modeling technique relies significantly on the choice of the weak learner and the form of the loss function. In this paper, we propose a novel scale-space kernel based approach for additive modeling. Our method applies a few insights from the well-studied scale-space theory for choosing the optimal learners during different iterations of the boosting algorithms, which are simple yet powerful additive modeling methods. For each iteration of the additive modeling, weak learners that can best fit the current resolution of the data are chosen and then increase the resolution systematically. We demonstrate the results of the proposed framework on both synthetic and real datasets taken from the UCI machine learning repository. Though demonstrated specifically in the context of boosting algorithms, our approach is generic enough to be accommodated in general additive modeling techniques. Similarities and distinctions of the proposed algorithm with the popularly used radial basis function networks and wavelet decomposition method are also discussed.

Keywords: additive modeling, boosting, weak learner, wavelet decomposition, scale-space theory.

1 Introduction

In statistical pattern recognition, additive modeling methods have been proven to be very effective for not only improving the classification accuracies but also in reducing the bias and variance of the estimated classifier. We choose to demonstrate our framework using '*boosting*' methods, which is a standard additive modeling algorithm popular in pattern recognition applications. In spite of its great success, boosting algorithms still suffer from a few open-ended issues such as the choice of the parameters for the weak learner. In this paper, we propose a novel scale-space based scheme for choosing optimal weak learners at multiple resolutions during the iterations in boosting or generic additive modeling. The proposed framework can model any arbitrary function using the boosting

N. da Vitora Lobo et al. (Eds.): SSPR&SPR 2008, LNCS 5342, pp. 714–723, 2008.

methodology at different resolutions of the data. In the scale-space based approach to statistical additive modeling, the weak learners are determined by analyzing the data at different resolutions. Our algorithm provides the flexibility of choosing the weak learners dynamically which will improve the modeling performance compared to using the static learners. For every learner in the additive model, the resolution is either maintained or doubled and a weak learner is modeled for fitting the data. To begin with, complete input data is considered for fitting a Gaussian model. As the iterations progress, the number of data points to be considered for fitting is reduced and a Gaussian model is fit to the data for every boosting iteration. We obtain an optimal Gaussian model that can fit the given number of data points at that particular resolution of interest in that iteration.

The scale-space concept allows for effective modeling of the dataset at any given resolution [6]. Scale-space based techniques have also been widely used for effective clustering [5]. In terms of analyzing a dataset at different resolutions, our approach closely resembles wavelet decomposition techniques which are effective tools in the field of multi-resolution signal analysis [4]. The main contributions of our work is: *Multi-resolution additive modeling by fitting the data using the concepts of scale-space theory.* This is an extended version of our previous work on scale-space based regression [7]. Though our method can be potentially applied with any base (or weak) learner, we chose to have Gaussian model because of its nice theoretical properties in the scale-space framework [8]. The rest of the paper is organized as follows: Section 2 shows the problem formulation in detail and discusses the concepts necessary to comprehend our algorithm. Section 3 describes our scale-space based additive modeling framework. Section 4 gives the experimental results of our algorithm on both synthetic and real datasets. Finally, Section 5 concludes our discussion with future research directions.

2 Preliminaries

This section gives details on boosting algorithm, introduces scale-space kernels and wavelet decomposition methods.

2.1 Problem Formulation

Let us consider N i.i.d. training samples $\mathcal{D} = (\mathcal{X}, \mathcal{Y})$ consisting of samples $(\mathcal{X}, \mathcal{Y}) = (x_1, y_1), (x_2, y_2), .., (x_N, y_N)$ where $\mathcal{X} \in \mathbb{R}^{N \times d}$ and $\mathcal{Y} \in \mathbb{R}^{N \times 1}$. For the case of binary classification problems, we have $y_i \in \{-1, +1\}$ and for regression problems, y_i takes any arbitrary real value. In other words, the univariate response $\mathcal{Y}$ is continuous for regression problems and discrete for classification problems. The goal of a regression problem is to obtain the function $F(\mathcal{X})$ that can approximate $\mathcal{Y}$. We chose to demonstrate the power of scale-space kernels in the context of Logitboost algorithm because of its popularity and the power of additive modeling. Let us denote $p(x)$ is an observation for $\hat{P}[y = 1|X = x]$. Algorithm 1 gives generic Logitboost algorithm to solve two class classification problem.

Algorithm 1. Logitboost Algorithm

Initialize $F^{(0)}(x_i) = 0$ and probabilities $p^{(0)}(x_i) = \frac{1}{2}$, for $i = 1, 2, .., N$
for $t = 1$ to T **do**
1. Compute response and weights for all datapoints

$$w_i^{(t)} = p^{(t-1)}(x_i) \cdot \left(1 - p^{(t-1)}(x_i)\right)$$
$$z_i^{(t)} = \frac{y_i - p^{(t-1)}(x_i)}{w_i^{(t)}}$$

2. Fit a weak learner $f^{(t)}$ to the training data using weights w_i.

$$f^{(t)} = arg\min_f \sum_{i=1}^{n} w_i^{(t)} \left(z_i^{(t)} - f(x_i)\right)^2$$

3. Update the Classifier and the probabilities

$$F^{(t)}(x_i) = F^{(t)}(x_i) + \frac{1}{2} f^{(t)}(x_i)$$
$$p^{(t)}(x_i) = \left(1 + exp(-2F^{(t)}(x_i))\right)^{-1}$$

end for
Output $C(X) = sign(F^{(t)}(x_i))$.

2.2 Scale-Space Kernels

In scale-space theory, mapping $p(x)$ is embedded into a continuous family $P(x, \sigma)$. Our method starts with an approximation of the entire dataset with Gaussian kernel of $\sigma = \infty$. As the resolution (or scale) increases, the sigma value is reduced and eventually converges to zero. In our case, the highest frequency (or resolution) corresponds to fitting every datapoint with a Gaussian kernel. Several practical considerations were made to make the implementation more efficient. More study on the scale-space kernels was made in the image processing literature and not much focus is given by the pattern recognition community.

Gaussian kernels are a simple and trivial choice for scale-space kernels which follow all the scale-space axioms [1]. The main reasons for choosing Gaussian kernels in the context of additive modeling as weak learners are that they are powerful universal approximators and they allow systematic hierarchical modeling. We choose to reduce the width by halves using the concepts of wavelet decomposition methods which were again well studied concepts in the context of handling image operations efficiently. Be it a classification, regression or image operation, one can consider in a generic way by treating them as a signal and the processing of this signal has to be done efficiently in order to obtain some interestingness of the signal. In this way, one can provide a unification framework

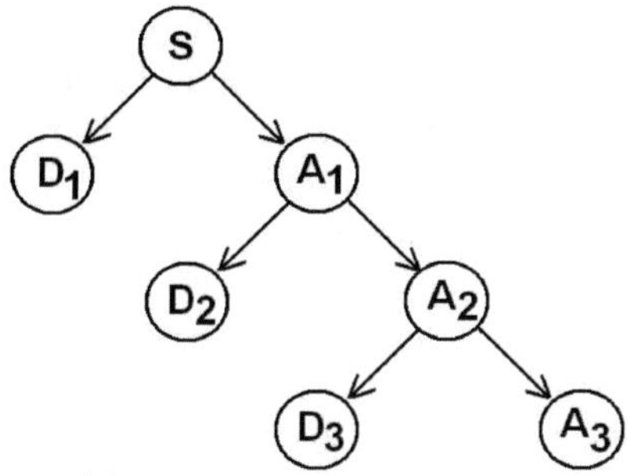

(a) Wavelet Decomposition

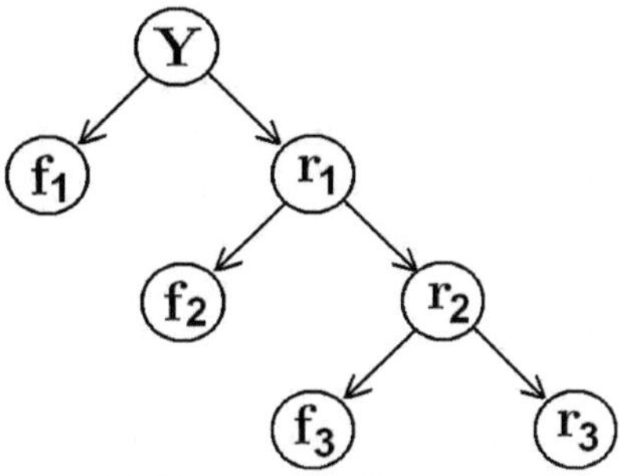

(b) Scale-space Boosting

Fig. 1. Similarities and differences between wavelet decomposition methods and scale-space boosting. (a) Wavelet techniques decompose the signal into approximations and details. (b) Scale-space based additive modeling decomposes the function by weak learners and the corresponding residuals.

that combines several well-studied concepts in a unique manner to handle some key challenges efficiently.

2.3 Wavelet Decomposition

In signal processing applications, wavelet transformation constructs a family of hierarchically organized decompositions. At a given level j, the j-level approximation is called A_j and a deviation signal called the j-level detail D_j. A_1 corresponds to the approximation of the original signal corresponding to the low frequencies of A_0, whereas the detail D_1 corresponds to the high frequency deviation. The selection of a suitable level for the hierarchy will depend on the user needs and is usually chosen based on a desired low-pass cutoff frequency. Fig. 1 shows a comparison of wavelet decomposition with scale-space additive modeling. Successive approximations A_1, A_2 and A_3 with lower resolution of a given signal are built. The original signal S can be built by summing up all the details with the approximation at the highest frequencies. i.e. $S = A_3 + D_3 + D_2 + D_1$. Our scale-space boosting algorithm works in a similar manner. The frequencies in the wavelet domain correspond to resolutions in our scale-space based algorithm. Basically, the low frequency components in wavelet decomposition correspond to fitting a Gaussian for the entire dataset and the high frequency components correspond to fitting fewer datapoints. The original target function ($\mathcal{Y}$) is decomposed using weak learners(f) and residuals(r). The final model at any given resolution is obtained by a weighted linear combination of the weak learners obtained so far. It should be noted the residual indicates the difference between the target function and the final learner (F) at that particular resolution. i.e. $r = |\mathcal{Y} - F|$.

3 Scale-Space Based Additive Modeling

Algorithm 2 describes our scale-space based approach for additive modeling. We demonstrate our approach on Logitboost algorithm, though our algorithm can be used to any additive modeling framework. The initial model is set to null or the mean value of the target values. Initially, the number of datapoints to be modeled will be the entire dataset. The feature values are first sorted (column-wise) independently which will facilitate the scale-space kernel based modeling that will be performed later on. As the iterations progress, the number of datapoints considered for fitting the weak regressor is retained or halved depending on the error of the model. For every iteration, the residual r is set to the absolute difference between the target value ($\mathcal{Y}$) and the final learner (F). The best multivariate Gaussian model will be fitted to this data at a given resolution. The details of the procedure *bestgaussianfit* which obtains the best Gaussian regressor at a particular resolution of the data is described later. The best Gaussian model corresponding to the resolution is obtained at every iteration and the model with the least error is chosen for that particular iteration.

Algorithm 2. Scale-space based Boosting

Input: Data ($\mathcal{D}$), No. of samples (N), No. of iterations (T).
Output: Final Regressor (F)
Algorithm:
set $n = N, \quad F = \emptyset$
for $i = 1 : d$ **do**
  $[\hat{\mathcal{X}}, idx(:, i)] = \text{sort}(\mathcal{X}(:, i))$
end for
for $t = 1 : T$ **do**
  $r = |\mathcal{Y} - F|$
  $[\hat{f}_0, err_0] = bestgaussfit(\hat{\mathcal{X}}, r, N, d, n, idx)$
  $[\hat{f}_1, err_1] = bestgaussfit(\hat{\mathcal{X}}, r, N, d, n/2, idx)$
  if $err_0 < err_1$ **then**
    $F = F + \hat{f}_0$
  else
    $F = F + \hat{f}_1$
    $n = n/2$
  end if
end for
return F

Our algorithm obtains the weak regressors and models the data in a more systematic hierarchical manner. Most importantly, the increase in the resolution is monotonically non-decreasing, i.e. the resolution either remains the same or increased. For every iteration, the best weak Gaussian regressor is fit to the data based on a single feature value at a particular resolution. The basic idea of *bestgaussianfit* is to slide a Gaussian window across all the datapoints corresponding to each feature at a given resolution. Depending on the given resolution

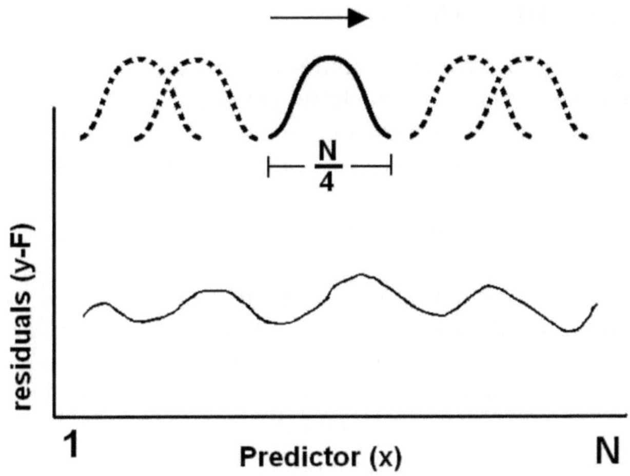

Fig. 2. Demonstration of the sliding Gaussian kernel. There are N datapoints and each has its corresponding response as a residual (y-F). A Gaussian kernel of resolution $\frac{N}{4}$ is slided across the entire range of data points to find the optimal fit. The optimal fit is obtained by minimizing the L_2 norm between the residual and the weighted Gaussian fit. The dotted Gaussians represent the sliding windows while the dashed Gaussian represent the active window.

(indicated by n datapoints), a Gaussian kernel containing n datapoints is slid across all the data points and the location where the minimal residual error is obtained. The concept of sliding kernels and active window is clearly illustrated in Fig. 2. Within each active window, the weight of individual data point is computed based on the normalized regression value.

The error is computed between the weak learner and the target regression values. It should be noted that the error is not computed using the given window alone. In other words, Gaussian kernel that will fit the data is computed based on the parameters computed within the window and the final error is computed based on the deviation in the entire dataset (not just the active window). This is a nontrivial aspect in our algorithm and failing to perform this might lead to erroneous results that might fit the data locally. This is due to the fact that every iteration, the result will be a local fit to the data and each of this weak regressor might not perform well globally. Finally, the best weak regressor across each feature that has the minimum error is returned.

The resolution is increased at a logarithmic scale. i.e. every time the resolution is doubled or the number of datapoints considered to fit a Gaussian is halved. In fact, we can use any other heuristic to change the resolution more efficiently. Experimental results indicated that this change of resolution is optimal and also this logarithmic change of resolution has nice theoretical properties (they mimic wavelet decomposition methods).

3.1 Relation to Other Models

Our method appears to have close connection to popularly used Radial Basis Function (RBF) Networks [2]. The RBF networks rely on the fact that any continuous function can be approximated by a sum of appropriately chosen Gaussian functions. These networks are one form of neural networks which have a static Gaussian function for the hidden layer processing elements to attain non-linearity. The RBF networks compute the input to output map using local Gaussian approximators. During the training phase the centers and width of the Gaussians in the hidden units along with the weights of the connections is determined. In spite of their success, the RBF networks suffer from some of the popular drawbacks of neural networks like fixing the topology, including the number of hidden nodes and the active connections between the layers. Our algorithm contains very minimal user-defined parameters and doesn't suffer from these problems. It gives a systematic approach to automatically model the regression function as a linear combination of Gaussian kernels. Being complex models, the time taken for the RBF networks during the training and testing phases are very high compared to our boosting algorithm. Being powerful local approximators, RBF networks perform poorly on noisy data since the coefficients are adjusted so as to accommodate local data and these networks do not average out noise over the entire data space. Using our approach, we avoid the local approach to regression of RBF networks and at the same time benefit from other nice properties of these networks.

4 Experimental Results

All programs were written in MATLAB 7.0 and run on pentium IV 2.8 GHz machines. Experiments were performed on both synthetic and real datasets.

4.1 Synthetic Datasets

To demonstrate the robustness of the proposed scale-space boosting algorithm, we performed experiments on two synthetic regression datasets. Regression datasets were chosen to demonstrate the capability of our algorithm to obtain smooth functions. Using, boost stumps or trees, one might be able to obtain more accurate models in terms of metrics like mean square error. But, this will come at a cost of obtaining highly discontinuous models which are not desirable. Hence, we demonstrated the behaviour of algorithm on two synthetic sine wave datasets: 1) noise-free sine wave and 2) the sine wave with Gaussian noise. For the first dataset, 241 samples were generated from a sine wave (with x from 0 to 12 uniformly and $y = \sin(x)$). For the second dataset, Gaussian random noise with zero mean and 0.2 standard deviation is overlaid onto all the samples of the first dataset. Figure 3 depicts the experiment results using the two synthetic sine wave datasets: 1) noise-free sine wave in the first row and 2) sine wave with Gaussian noise in the second row. To understand the difference between the original and the final data, we overlaid the noise-free sine wave in the second

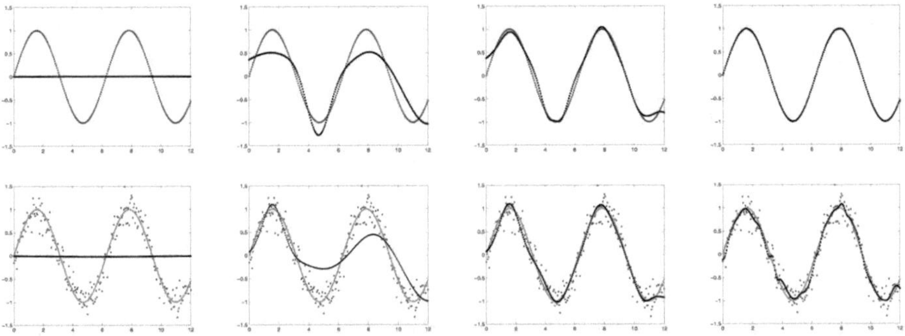

(a) Iteration 1 (b) Iteration 5 (c) Iteration 10 (d) Iteration 50

Fig. 3. Experimental results along with the number of iterations using two synthetic sinewave datasets. The first row is the sine wave without noise and the second row is the sine wave with Gaussian noise with μ=0 and σ=0.2. ($N(0, 0.2^2)$). The red color and blue color represent the original data and our algorithm results respectively. The green color in the second row represents the original sine wave without the Gaussian noise.

row in order to compare the system output results not only with the given noisy data but also with the original noise-free data.

4.2 Real-World Datasets

Different real-world datasets were chosen from the UCI machine learning repository [3] to demonstrate our scale-space based modeling technique. Since, the basic algorithm is for binary classification, we demonstrate the power of using scale-space kernels on datasets with 2 classes. One has to realize that the main claim of the paper is not about producing the best classification or regression algorithm. Rather, this paper tries to give a systematic understanding of different kinds of kernels that can be used in the additive modeling techniques. To the best of our knowledge, this is the first effort to combine the concepts of scale-space theory in the context of additive modeling. Additive modeling with smooth and continuous kernels will result in smooth functions for classifier boundary and regression functions. Hence, we compare the scale-space kernels with other static and dynamic kernels. Dynamic kernel (or random kernel) fits a kernel of random width during the boosting process. Static kernels will have static widths that do not change during the boosting process. We used kernel width of $n/8$ for demonstrating the performance of static kernels.

The scale-space kernel doesn't suffer from the generalization problem as clearly illustrated by the results on the test data shown in Table 1. Experiments were conducted using five-fold cross validation method and the test results are reported for the algorithm using various kernels. Mean squared error is reported between both training and test cases. Scale-space kernels are competitive with the best possible kernels and can be generically be used for any dataset. Since,

Table 1. Performance of scale-space kernels with other kernels on different real-world datasets. Test error along with the standard deviation using five-fold cross validation is reported.

Dataset	Cancer	Ionosphere	Sonar	Bupa
static kernel -n/8	0.244± 0.0869	0.4118± 0.1024	0.9148± 0.1267	0.9453± 0.0607
Dynamic kernel	0.1898± 0.0338	0.3553 ± 0.063	0.7543± 0.0885	0.869 ±0.053
Scale-space kernel	0.1895 ± 0.043	0.3371± 0.092	0.7125± 0.1433	0.8962 ± 0.1291

obtaining the width of the kernel during the additive modeling process can be a challenging task, the use of scale-space kernels can resolve the problem by using adaptive step-sizes. The fact that the scale-space kernels converge much faster than static kernels make them much suitable for additive modeling algorithms. Also, one can see that the results of the scale-space kernels are fairly robust compared to other kernels.

5 Conclusions and Future Research

Recently, additive modeling techniques have received a great attention from several researchers working in a wide variety of applications in science and engineering. In this paper, we proposed a novel additive modeling framework that uses scale-space theory to obtain the optimal weak learner at every iteration. Advantages of our method compared to other popular models proposed in the literature are clearly demonstrated. As a continuation of this work, we would like to explore the idea of using other scale-space kernels (different from Gaussian) as our weak learners. Optimal choice of the n value to be chosen adaptively during each boosting iteration must be investigated more thoroughly depending on individual dataset to be modeled. More theoretical insights related to combination of additive models and scale-space kernels are yet to be investigated.

References

1. Babaud, J., Witkin, A.P., Baudin, M., Duda, R.O.: Uniqueness of the gaussian kernel for scale-space filtering. IEEE Transactions on Pattern Analysis and Machine Intelligence 8(1), 26–33 (1986)
2. Bishop, C.M.: Neural Networks for Pattern Recognition. Oxford University Press, Oxford (1995)
3. Blake, C.L., Merz, C.J.: UCI repository of machine learning databases. University of California, Irvine, Dept. of Information and Computer Sciences (1998), `http://www.ics.uci.edu/~mlearn/MLRepository.html`
4. Graps, A.L.: An introduction to wavelets. IEEE Computational Sciences and Engineering 2(2), 50–61 (1995)
5. Leung, Y., Zhang, J., Xu, Z.: Clustering by scale-space filtering. IEEE Transactions on Pattern Analysis Machine Intelligence 22(12), 1396–1410 (2000)

6. Lindeberg, T.: Scale-Space Theory in Computer Vision. Kluwer Academic Publishers, Dordrecht (1994)
7. Park, J.-H., Reddy, C.K.: Scale-space based boosting for weak regressors. In: Kok, J.N., Koronacki, J., Lopez de Mantaras, R., Matwin, S., Mladenič, D., Skowron, A. (eds.) ECML 2007. LNCS, vol. 4701, pp. 666–673. Springer, Heidelberg (2007)
8. Sporring, J., Nielsen, M., Florack, L., Johansen, P.: Gaussian Scale-Space Theory. Kluwer Academic Publishers, Dordrecht (1997)

Learning Curves for the Analysis of Multiple Instance Classifiers

David M.J. Tax and Robert P.W. Duin

Delft University of Technology, Mekelweg 4, 2628 CD Delft, The Netherlands

Abstract. In Multiple Instance Learning (MIL) problems, objects are represented by a *set* of feature vectors, in contrast to the standard pattern recognition problems, where objects are represented by a single feature vector. Numerous classifiers have been proposed to solve this type of MIL classification problem. Unfortunately only two datasets are standard in this field (MUSK-1 and MUSK-2), and all classifiers are evaluated on these datasets using the standard classification error. In practice it is very informative to investigate their learning curves, i.e. the performance on train and test set for varying number of training objects. This paper offers an evaluation of several classifiers on the standard datasets MUSK-1 and MUSK-2 as a function of the training size. This suggests that for smaller datasets a Parzen density estimator may be preferrer over the other 'optimal' classifiers given in the literature.

1 Introduction

In many real-world classification problems objects cannot easily be represented by a single unique feature vector, because the objects are too rich and contain too many details and information. A typical example is the problem of image database retrieval. Each image can be of different size and can depict several physical objects in the same picture. In Multi(ple) Instance Learning (MIL) the standard pattern recognition assumption of having one feature vector per object is extended. The objects are represented by a collection (called a 'bag') of feature vectors (called 'instances'). In the training phase only bags are labeled (in a positive and negative class), but not the individual instances. A new bag has to be classified based on the collection of instances in that bag.

An example would be a medical classification problem: for the detection of abnormalities in the lung an X-ray image of the chest can be made. From this image small image patches can be extracted for which features can be derived. When each individual patch would have been labeled, an anomaly detector could have been trained. Although it is possible to classify a patient as healthy or ill, it is very hard to label each patch reliably. This therefore results in a multi-instance learning problem, where the collection (or bag) of patches (instances) is labeled in 'healthy' of 'ill', but where the individual instances are not. A patient is now classified as 'healthy' when none of the instances are classified as 'ill'.

Because objects are now represented by a collection of feature vectors, MIL datasets tend to be large and the classification problem difficult. Unfortunately,

N. da Vitora Lobo et al. (Eds.): SSPR&SPR 2008, LNCS 5342, pp. 724–733, 2008.

the effective sample size in MIL datasets is often limited. Not the instances have to be classified, but the bags. It is therefore not the number of instances, but the number of bags that is the determining factor for the performance of the classifier. Furthermore, these classifiers are often evaluated based on an error measure like classification error (or misclassification rate [11]). But for these low-sample-size problems, and problems where class priors and misclassification costs are unknown, the classification error may not be the most suitable. Performance measures like the Area under the ROC curve [2] have shown to be more reliable for small sample sizes.

Furthermore, the choice for the best classifier for a particular problem is often dependent on the sample size, i.e. the size of the training set. When more training samples are available, in general more complex classifiers can be trained. Reducing the training set size may result in a different (more simple) optimal classifier. It is therefore very informative to inspect the performance of classifiers for varying training set sizes.

In this paper we re-evaluate several MIL classifiers on the two standard datasets, MUSK-1 and MUSK-2. These two datasets from [6] are basically the only datasets that are consequently used in benchmarking of all MIL classification systems ([13,16,10,19,1,20,17,3,9,4] to name a few). To assess the relative performance of the methods over a wide range of sample sizes, we use the learning curve with the Area under the ROC curve. In section 2 we discuss a range of MIL classifiers, followed by a short introduction on learning curves and performance measures in section 3. In section 4 the results of the MIL classifiers is shown, and we summarize the conclusions in section 5.

2 Multi-instance Learning, Classifiers and Datasets

Assume that object i is represented by a bag B_i, containing a set of m_i instances $\mathbf{x}_j \in \mathbb{R}^p$: $B_i = \{\mathbf{x}_{ij}, j = 1, .., m_i\}$ (i.e. feature vectors of length p). Assume further that each bag is labeled with a positive or negative label: $Y_i \in \{\omega^+, \omega^-\}$. The label Y_i that will be assigned to a bag B_i is in principle determined by the number of instances $\mathbf{x}_{ij}$ that is positive. In the most extreme case a bag may be labeled positive when a single instance is classified as positive. When the classification is noisy, or a very large number of instances is present in each bag, a minimum number or a certain fraction of the total number of instances may be chosen. During training the label for each instance y_{ij} is unknown. A classifier has to determine which instances in each bag are informative for the bag label. Finally, a trained classifier has to estimate the class label from a bag of instances:

$$\widehat{Y} = f(B) = f(\{\mathbf{x}_j\}) \tag{1}$$

For the experiments in this paper we use a selected set of multi-instance learners. Due to space constraints not all classifiers can be discussed. The classifiers that are used to classify the multi-instance datasets are

Simple classifiers with bag combiner can be used when the multi-instance learning aspect of the learning problem is ignored. In training all instances are labeled according to their bag label. In the evaluation the classification outcomes of all the instances of a bag are combined to a single bag-outcome. One possible combining rule is the quantile combining. Assume that the (real) outcomes $o_j = f(\mathbf{x}_{ij})$ are sorted: $\boldsymbol{o}_i = \{o_1, o_2, ..., o_{N_i}\}$, where $o_1 \geq o_2 \geq ...o_{N_i}$. The quantile combining rule selects the q-th quantile value of this set:

$$\mathcal{Q}(\boldsymbol{o}_i) = o_n, \quad n = \lfloor qN_i \rfloor + 1 \tag{2}$$

This idea is similar to [3] where the bag-level classifier f_g is this quantile function.

In this paper we applied this combining rule to the outputs of the linear discriminant analysis (LDA) and the Parzen density estimator. The LDA is regularized by adding a small constant $\lambda = 10^{-6}$ to the diagonal. The Parzen density estimator fits a single Gaussian distribution with a fixed width parameter on each individual training sample. The width parameter in the Parzen density is optimized in a leave-one-out fashion on the training set [8]. In the quantile combination $q = 0.01$ is used, which means that for small bags only the presence of a single positive object is sufficient to classify the bag as positive.

Axis-parallel Rectangle [6] constructs a rectangular decision boundary, aligned with the feature axis, that is optimized such that at least one instance of each positive bag falls inside this box but such that none of the instances of the negative bags is inside. Three optimization schemes have been proposed, and in this paper the 'inside-out' scheme is used: it starts from a seed point and then grows a rectangle until it covers at least one instance per positive bag and no instances from negative bags. It also includes the possibility to select features such that in some feature dimensions no box face is defined. The implementation of the APR in this paper is based on [18], and the parameters are tuned to get a good average performance on the MUSK datasets.[1]

Citation k-NN [16] is a variant of the standard k nearest neighbor classifier. The standard k-NN is extended by considering not only the nearest bags on the training set, but also the training bags for which the test bag is the nearest neighbor. These are the so-called 'referees', or 'citers'. (See for a more complete explanation [16].) In this paper the distances between bags is computed using the Hausdorff distance, and $k = 5$ nearest neighbors and $r = 5$ citers are used.

Diverse Density [13] estimates the density of the co-occurrence of instances of positive bags, and corrects for the density of the negative instances. Areas of high diverse density contain at least one instance of all positive bags, but do not contain any negative instances. The diverse density method optimizes a location and a radius (basically, it models a Gaussian distribution) for which the diverse density is highest. When a new bag contains instances in this high diverse density area, it is classified as a positive bag.

Although the reported performances are very good, the optimization procedure of finding the location of a high diverse density area is computationally very intensive, and often local optima are reached. Therefore the procedure is

[1] For a full description of the algorithm, please see [6]. The chosen parameters are: the threshold distance in selecting significant features=1, $\tau = 0.995$, $\epsilon = 0.02$ and step size expanding the APR=0.1.

rerun several times with different initializations. In this paper we tried 10 different random initializations and used the solution which yielded the highest likelihood on the training set. Further parameter settings include the maximum number of epochs that the diverse density is optimized ($4 \times p$), and the tolerance in the change in position or change in likelihood during the optimization (here we used 10^{-5} and 10^{-7}).

EM-Diverse Density [19] is an extension of the Diverse Density approach. Here the highest diverse density position is optimized in an iterative Expectation-Maximization scheme. At each step of the EM scheme only the instance with the highest diverse density per bag is used. This simplifies the estimation of the diverse density, but because the estimated location of the maximum diverse density changes during the optimization, also the instance with the highest diverse density per bag changes each iteration. In practice it appears that this scheme is a bit more insensitive to local minima. The settings of this classifier is identical to these of the Diverse Density method.

MIL-Support vector classifier [10] is a standard support vector classifier that uses a multi-instance kernel to define a similarity between two bags of instances. In this paper very simple bag statistics are computed:

$$\tilde{\mathbf{x}}_i = \tilde{\mathbf{x}}(B_i) = [\min_j(\mathbf{x}_{ij}), \max_j(\mathbf{x}_{ij})] \tag{3}$$

This represents a single bag by one feature vector of length $2p$, where the minimum and maximum feature values per bag are stored. Notice that (3) does not compute a kernel, strictly speaking. On this feature vector standard kernels are computed, for instance the standard linear kernel $K(B_i, B_j) = \tilde{\mathbf{x}}_i^T \tilde{\mathbf{x}}_j$. In this paper the linear kernel, with a trade-off parameter $C = 1$ is used.

Table 1. The MUSK datasets and some characteristics

name	pos. bags	neg. bags	# instances	# features
MUSK-1	47	45	476	166
MUSK-2	39	63	698	166

The standard datasets in multi-instance learning are the MUSK-1 and MUSK-2 datasets. The task is to predict if a certain molecule can bind to a target molecule. The molecule is in fact a 3D object, and it can have multiple shapes (or conformations). When a molecule binds to a target, we know that at least one of the conformations has the ability to bind. When a molecule does not bind, none of the conformations have the correct shape. Each molecule can therefore described by a bag of conformations. In these datasets the 3D shape of each conformation is described by 166 features, and conformations of around 80-100 molecules are available. Dataset MUSK-1 contains on average 6 conformations per bag, while MUSK-2 has more than 60 conformations per bag. The task for these datasets is to predict on the basis of the collection of conformations if the molecule smells 'musky' or not. More specific information of the datasets are given in Table 1 and in [6].

3 Learning Curves and Performance Measures

Typically, a classification system is trained on a training set that is as large as possible, and evaluated on an independent test set [11]. It is expected that more training data results in a better performance of the classifier, because with more data the parameters of a classifier can be estimated more reliably. Often a cross-validation scheme is used where the data is split in M parts. $M-1$ parts are used for training, and the rest is used for testing. This is repeated M times and the performance is averaged. The classifier is thus trained M times on $M/(M-1)$ part of the data, and all data is used once for estimating the true error.

A learning curve shows the change in (true) classification error for a varying training set size. Often not only the true error is estimated, but also the apparent error, i.e. the error on the training set. When the difference between the apparent error and the true error is large, the classifier is called overtrained, or overfitted. The performance on the training set gives a too optimistic estimate on what can be expected in practice.

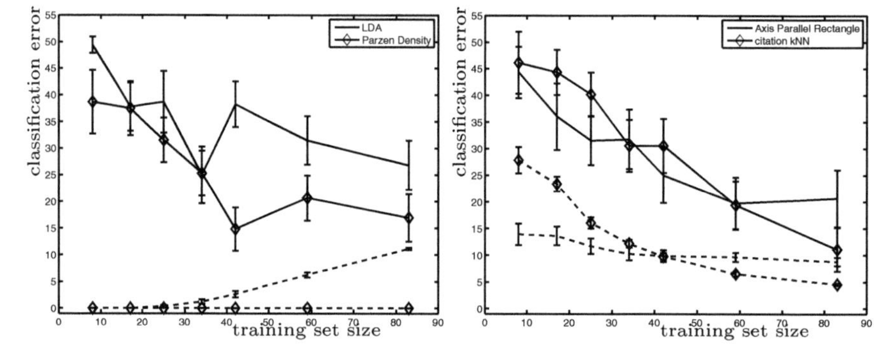

(a) Learning curves LDA and Parzen. (b) Learning curves APR and k-NN.

Fig. 1. Learning curves for the MUSK-1 dataset using the classification error

Figure 1(a) presents some typical learning curves, showing the true and apparent error rate for the MUSK-1 dataset by the solid and dashed lines respectively. (This dataset and the classifiers are explained in section 2.) The training set size is a fraction of the total training size, running from 10% (around 10 bags, smaller numbers caused problems for some classifiers) to 99% (84 bags). In the left subfigure the LDA and Parzen classifiers are shown. For both classifiers a significant difference between the apparent and true error is visible. Both classifiers perfectly fit the training set, and only the training error of the LDA deteriorates slightly for sample sizes larger than 40. This suggest that this dataset is too small for these classifiers and that the classifiers overtrain.

One can also extrapolate if classifiers may gain significantly in performance when more training data is added. A flat learning curve (like the one from the Parzen classifier, although it is a bit noisy) suggests that the classifier is already trained well, and that more training data will not help the classifier much. A more steeply decreasing learning curve (like the one from the LDA) suggests that a bit more data may be very beneficial for this classifier.

This general picture does not accommodate some special cases. For instance, the nearest neighbor classifier always has a positive bias in the evaluation of the training set. Because it is using the training set as the classifier model, evaluating the k nearest neighbors in the training set will always include a copy of the test object. This is visible in Figure 1(b). The training performance of the Citation k-NN (which is a special version of the standard k-NN classifier) *increases* for larger training set sizes. Still a positive bias is present over all training set sizes. The slope in the learning curve also suggests that a bit more data may be very beneficial for the k-NN performance.

The learning curves for the Axis Parallel Rectangle classifier shows a similar characteristic. Here the apparent error also decreases (slightly) with increasing sample size. This is caused by the fact that the parameter settings in this method are fixed to values that are more suitable for larger training sets. In particular the parameter τ, that specifies how far the rectangle boundaries should be placed around the training instances, is a sensitive parameter. For smaller sample sizes the decision boundary has to be set a bit wider than for larger sample sizes. Optimizing this parameter by cross-validation appears to be a very costly and gives very noisy results.

For the evaluation of a classifier often the classification error is used. It just counts the number of misclassifications in a (test) set. Assume a classifier f is trained for a two-class classification problem, and evaluated on a test set $\mathcal{X} = \{(\mathbf{x}_i, y_i), i = 1,.., N\}$ where $\mathbf{x}_i \in \mathbb{R}^p$ are p-dimensional feature vectors, and $y_i \in \{\omega^+, \omega^-\}$ are class labels. The classification error is estimated by: $\hat{\epsilon} = \frac{1}{N}\sum_{i=1}^{N} \mathcal{I}(f(\mathbf{x}) \neq y_i)$ where $\mathcal{I}(.)$ is the indicator function that outputs 1 when the statement is true and 0 otherwise.

A drawback of this measure is that it is sensitive to the class priors and that it does not take misclassification costs into account. Often the class priors and costs are unknown, and the empirical class priors are used. Especially when the data sampling is very skewed, deceiving performances are obtained. For these situations the area under the ROC curve (AUC) is more suitable (for MIL problems it is used in [14] for instance). It basically estimates the probability that a classifier gives a higher output for an object of the positive class than an object from the negative class [2]. It can be estimated by:

$$A\hat{U}C = \frac{1}{N^+N^-} \sum_{i=1}^{N^+} \sum_{j=1}^{N^-} \mathcal{I}(f(\mathbf{x}_i) > f(\mathbf{x}_j)) \tag{4}$$

where N^+ and N^- are the number of objects from the positive and negative class respectively. By this relative comparison this measure becomes independent of the class priors or class sampling. Furthermore, by the fact that it incorporates all possible pairs of positive and negative objects, the AUC tends to be a more stable performance estimator than the standard classification error [5] making it also easier to compare classifiers [12,15].

4 Experiments

In this section we evaluate all the classifiers on some datasets. We applied 10-fold stratified cross-validation on the bags, and varied the training set size from 10% to 99% of the total design set. All classifiers are implemented in the pattern recognition toolbox Prtools [7]. The implementations of the APR, the maximum Diverse Density and the EM-Diverse density are based on [18].

In figure 2 the learning curves for the same classifiers on *exactly* the same train and test sets as in figure 1 are shown, but instead of the classification error the AUC ($\times 100$) is used. First notice that a well performing classifier has a *low* classification error but a *high* AUC. The general trends are the same, but some subtle changes can be observed. Figure 1 using the classification error suggests that the performance for the Parzen classifier is slowly, but uniformly increasing. Figure 2 shows that the performance in terms of AUC is not improving significantly after $N = 30$. This suggests that around $N = 30$ the density of the target concept is estimated relatively reliably, but that the operating point is still uncertain. Because the AUC only estimates the ranking of the objects, it is not influenced by a poor operating point (or threshold on the density). Furthermore, the left graph in figure 1 suggests that for $N = 30$ bags per class the LDA and Parzen perform similarly. Figure 2 on the other hand suggests that averaged over all operating points the Parzen may still be preferred.

When we compare the final performances of the APR and the Citation k-NN in Figure 1, we observe another typical behavior: for smaller training sets the more simple APR classifier has smaller error. When the training set size is increased to over $N = 60$, the more complex Citation k-NN starts to win. This is even more prominent in figure 2 where the Citation k-NN is already better than the APR for $N = 25$.

In figures 3 and 4 all the learning curves for all classifiers are shown on datasets MUSK-1 and MUSK-2, respectively. From the results shown in figure 3 it can be concluded that for small sample sizes (upto $N = 50$) the Parzen density performs

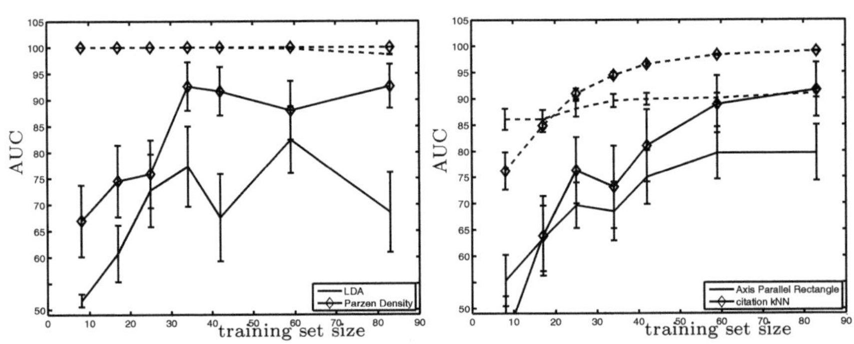

(a) Learning curves for LDA and Parzen density classifiers.

(b) Learning curves for the APR and k-NN classifiers.

Fig. 2. Learning curves for the MUSK-1 dataset using the AUC perf. measure ($\times 100$). The errorbars indicate one standard deviation of variation.

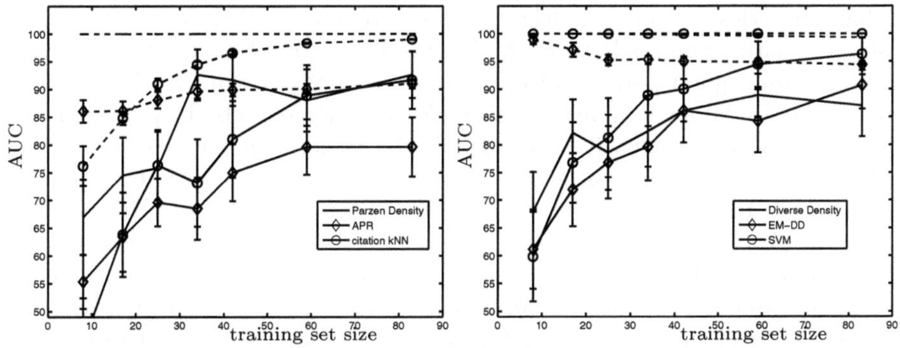

(a) Learning curves for Parzen density, APR and Citation k-NN

(b) Learning curves for Diverse Density, EM-DD and the support vector classifier

Fig. 3. Learning curves for MUSK-1 using the AUC performance measure

very well. Even for just 8 training bags per class, the AUC is around 0.70. For training sizes around $N = 30$ the Parzen reaches an AUC of more than 0.90, but it does not improve performance significantly for larger training sizes.

The Axis Parallel Rectangle performs relatively poorly in these experiments. The AUC for this method barely reaches 0.75 for the MUSK-1 dataset. The performance for the MUSK-2 dataset is slightly better, around 0.80.

The Citation k-NN and the Support vector classifier perform about equal. Both show promising performance, in that their learning curves are still increasing. This suggest that more training data may benefit the k-NN and SVM most. The very best performance on the MUSK-1 dataset is obtained by the SVM, with an AUC of more than 0.95. This suggest that the simple min, max-feature representation, given in equation (3), characterizes the problem very well.

In figure 4 the same classifiers are applied to the MUSK-2 dataset. This dataset has the same number of bags, but has far more instances per bag. Overall

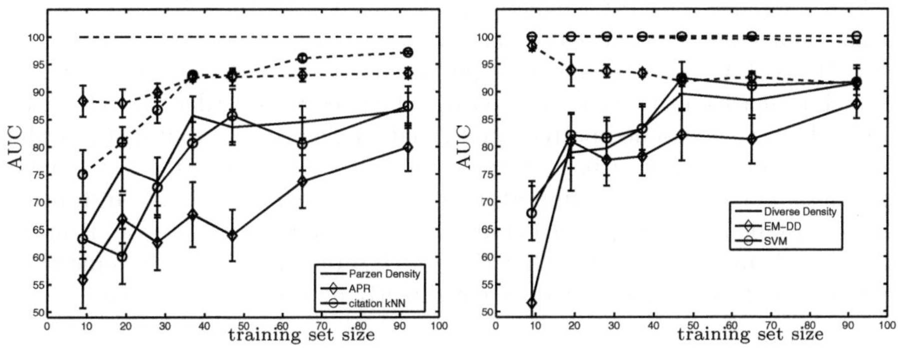

(a) Learning curves for Parzen density, APR and Citation k-NN

(b) Learning curves for Diverse Density, EM-DD and the support vector classifier

Fig. 4. Learning curves for MUSK-2 using the AUC performance ×100. Errorbars indicate one standard deviation over a 10-fold stratified cross-validation over the bags.

the characteristics are similar, although there are telling differences. The Parzen density estimator requires both in the MUSK-1 and MUSK-2 datasets about 30 bags to obtain a good performance, but Parzen finally obtains a slightly worse performance for larger training set sizes. Most other models, in particular the SVM and the Diverse Density, can exploit the larger number of instances per bag for the smaller sample sizes (around $N = 20$). For the SVM the minimum and maximum feature values can be estimated better, and for the Diverse Density a better estimate for the diverse density per bag can be obtained. For the higher sample sizes this advantage is lost, and the final best performance on the MUSK-2 dataset is just above 0.90, obtained again by the SVM. The EM-DD does not seem to gain too much by increasing the number of instances per bag, its performance increases just slightly after $N = 20$.

5 Conclusions

In real classification problems it is often interesting to see how classifiers behave for varying training set sizes. The resulting curve, the learning curve, shows which classifier is more suitable for small training set sizes, and which classifier has the most promising performance improvement when more data is available. For multi-instance learning problems this is very relevant, because the number of (in particular positive) bags is often limited. By comparing the learning curves of different classifiers for the MUSK-1 and MUSK-2 datasets, suitable models for different training set sizes can be obtained.

The experimental results in this paper suggest that simple classifiers like the Parzen density classifier, or the Support Vector Machine with a simple (but fitting) bag representation work well for both the MUSK datasets. The overall best performance for larger training set sizes are obtained using the SVM (over 0.95 AUC). Its learning curve suggest that even better performances can be obtained when more training data may be available.

The results also show that more instances per bag may not be advantageous. The best classification performances are obtained with the MUSK-1 dataset, that has on average 6 instances per bag, instead of the around 60 instances per bag in MUSK-2. When more instances per bag are present, the search problem to find the informative instance in each positive bag becomes harder. The classifiers therefore have to be trained better to find that single instance that distinguishes a positive from a negative bag. A topic for further research is if the number of instances per class is still important when not a single positive instance is required for labeling a bag positive, but a certain minimum *fraction* of instances. In that case it may be expected that the search problem does not become much harder and that more instances does not directly deteriorate the results.

References

1. Andrews, S., Hofmann, T., Tsochantaridis, I.: Multiple instance learning with generalized support vector machines. In: Proceedings of the AAAI National Conference on Artificial Intelligence (2002)

2. Bradley, A.P.: The use of the area under the ROC curve in the evaluation of machine learning algorithms. Pattern Recognition 30(7), 1145–1159 (1997)
3. Cannon, A., Hush, D.: Multiple instance learning using simple classifiers. In: Proceedings of the international conference on machine learning and applications, pp. 123–128 (2004)
4. Chen, Y., Bi, J., Wang, J.Z.: MILES: Multiple-instance learning via embedded instance selection. IEEE Transactions on Pattern Analysis and Machine Intelligence 28(12), 1931–1947 (2006)
5. Cortes, C., Mohri, M.: AUC optimization vs. error rate minimization. In: Advances in Neural Information Processing Systems (NIPS 2003) (2004)
6. Dietterich, T.G., Lathrop, R.H., Lozano-Perez, T.: Solving the multiple instance problem with axis-parallel rectangles. Artificial Intelligence 89(1-2), 31–71 (1997)
7. Duin, P., Juszcak, R.P.W., Paclik, P., Pekalska, E., de Ridder, D., Tax, D.M.J.: Prtools, a Matlab toolbox for pattern recognition, version 4.0 (January 2004)
8. Duin, R.P.W.: On the choice of the smoothing parameters for Parzen estimators of probability density functions. IEEE Transactions on Computers C-25(11), 1175–1179 (1976)
9. Gao, S., Sun, Q.: A generalized discriminative multiple instance learning for multimedia semantic concept detection. In: ICIP 2006, pp. 2901–2904 (2006)
10. Gärtner, T., Flach, P.A., Kowwalczyk, A., Smola, A.J.: Multi-instance kernels. In: Sammut, C., Hoffmann, A. (eds.) Proceedings of the 19th International Conference on Machine Learning, pp. 179–186. Morgan Kaufmann, San Francisco (2002)
11. Hand, D.J.: Construction and assessment of classification rules. Wiley, New York (1997)
12. Ling, C.X., Huang, J., Zhang, H.: AUC, a better measure than accuracy in comparing learning algorithms. In: Proceedings of the 2003 Canadian artificial intelligence conference (2003)
13. Maron, O., Lozano-Pérez, T.: A framework for multiple-instance learning. In: Advances in Neural Information Processing Systems, vol. 10, pp. 570–576. MIT Press, Cambridge (1998)
14. Ray, S., Craven, M.: Supervised versus multiple instance learning: an empirical comparison. In: ICML 2005: Proceedings of the 22nd international conference on Machine learning, pp. 697–704. ACM, New York (2005)
15. Rosset, S.: Model selection via the AUC. In: ICML 2004, pp. 703–710 (2004)
16. Wang, J., Zucker, J.D.: Solving the multiple-instance problem: A lazy learning approach. In: Proc. 17th International Conf. on Machine Learning, pp. 1119–1125. Morgan Kaufmann, San Francisco (2000)
17. Xu, X., Frank, E.: Logistic regression and boosting for labeled bags of instances. In: Proc. of the Pacific-Asia conference on knowledge discovery and data mining. Springer, Heidelberg (2004)
18. Zhang, M.-L.: Matlab toolbox: Mil learners and their ensemble versions
19. Zhang, Q., Goldman, S.: EM-DD: An improved multiple-instance learning technique. In: Advances in Neural Information Processing Systems, vol. 14. MIT Press, Cambridge (2002)
20. Zhou, Z.-H., Zhang, M.-L.: Ensembles of multi-instance learners. In: Lavrač, N., Gamberger, D., Todorovski, L., Blockeel, H. (eds.) ECML 2003. LNCS, vol. 2837, pp. 492–502. Springer, Heidelberg (2003)

Categorizing Perceptions of Indoor Rooms Using 3D Features

Agnes Swadzba and Sven Wachsmuth

Applied Computer Science, Faculty of Technology, Bielefeld University, Universitätsstraße 25, 33615 Bielefeld, Germany
{aswadzba, swachsmu}@techfak.uni-bielefeld.de

Abstract. In this paper, we propose a holistic classification scheme for different room types, like office or meeting room, based on 3D features. Such a categorization of scenes provides a rich source of information about potential objects, object locations, and activities typically found in them. Scene categorization is a challenging task. While outdoor scenes can be sufficiently characterized by color and texture features, indoor scenes consist of human-made structures that vary in terms of color and texture across different individual rooms of the same category. Nevertheless, humans tend to have an immediate impression in which room type they are. We suggest that such a decision could be based on the coarse spatial layout of a scene. Therefore, we present a system that categorizes different room types based on 3D sensor data extracted by a Time-of-Flight (ToF) camera. We extract planar structures combining region growing and RANSAC approaches. Then, feature vectors are defined on statistics over the relative sizes of the planar patches, the angles between pairs of (close) patches, and the ratios between sizes of pairs of patches to train classifiers. Experiments in a mobile robot scenario study the performance in classifying a room based on a single percept.

1 Introduction

Context is a rich source of information for interpreting tasks in complex scenes. This has already been recognized for a long time. Systems like CONDOR [1] or SPAM [2] coded explicit contextual rules that triggered other image operations. The main drawback of such kind of models was the complex knowledge engineering task and the semantic deficiencies of extensional systems dealing with uncertainty. More recently, graphical models have been applied in order to provide a more concise model relating objects and aspects of the scene [3,4]. Murphy, Torralba, and Freeman estimate global contexts, like persons, vehicles, furniture and vegetation from low-level image features [3]. These provide constraints on object categories and object scales. Hoiem, Efros, and Herbert first extract a 3D surface geometry from 2D images and relate the estimated local geometries to object classes predicted by a window-based object detector [4].

The work discussed so far mainly deals with 2D image information. Murphy et al. demonstrate that many different scene categories can be distinguished by

N. da Vitora Lobo et al. (Eds.): SSPR&SPR 2008, LNCS 5342, pp. 734–744, 2008.

purely considering 2D image statistics on texture and edges. There have also been other approaches using additional features like color histograms that successfully distinguished indoor/outdoor [5], sky/no-sky, vegetation/no-vegetation [6]. However, this does not necessarily extrapolate to more finely graded scene categories like different types of rooms, e.g. office or meeting room. Here, typical furniture like tables, chairs, and shelves reoccur, but in different layouts. Furthermore, furniture in the same type of room may have changing colors and textures or be viewed from different directions.

In these cases, a 3D description of the scene is much more invariant with regard to inner-class variations. However, strategies that provide a complete semantic interpretation of the 3D scene suffer from very constraint settings and extensive modeling efforts. Therefore, we aim at a more holistic 3D approach to scene classification in the spirit of the gist approach used by Torralba [7]. In the following, we describe the scene by a collection of planar structures and analyze whether it is possible to compute proper feature vectors for the classification of different room types (here: office, hall, and meeting room). The challenge faced is to categorize rooms only based on the information of one frame. Section 2 presents the 3D Time-of-Flight (ToF) sensor for acquiring 3D information in real time and introduces preprocessing steps specialized on this data. Section 3 presents necessary steps for determining sets of planar structures in this data, and in Section 4 and 5 features and classifiers are chosen and examined with regard to their performance in categorizing room percepts to room types.

2 Acquiring and Preprocessing 3D ToF Data

(a) (b) (c)

Fig. 1. (a) Swissranger SR3000, (b) example amplitude image, and (c) example 3D point cloud preprocessed

Our system uses the Swissranger SR3000 (Fig. 1(a)) provided by Swiss Center for Electronics and Microtechnology (CSEM) [8] delivering a matrix of distance measurements independent from texture and lighting conditions. It consists of 176 × 144 CMOS pixel sensors which are able to determine actively the distance between the optical center of the camera and the real 3D world point via measuring the time-of-flight of a near-infrared signal. Besides the distance value matrix (Fig. 1(c)), the camera provides a matrix containing amplitude values (Fig. 1(b)) for each frame. The amplitude value indicates the amplitude of the reflected near-infrared signal received by the sensor and implies therefore the amount of light reflected by a world point. A small amplitude corresponds to a small amount of light reflected indicating a weak signal.

To deal with noise arising from the different reflection properties several preprocessing techniques proposed in [9] are applied. The distance image is smoothed with a distance-adaptive median filter, which uses a different mask size (e.g. 3×3, 5×5, or 7×7) depending on the distance value for each pixel. As the amplitude value refers to the quality of the distance measurement, points with a small amplitude value are removed from the final 3D point cloud. The threshold needed adapts automatically to different reflection properties in different scenes. Further, edge points arising in the case when light from the fore- and the background hits the same pixel simultaneously are rejected. Finally, 3D coordinates are generated out of the distances with regard to a 3D camera coordinate system. With the assumption of ideal perspective projection, the known position of the principal point, pixel sizes, and focal length, the 3D coordinates can be computed from the distances via ray proportions in triangles. As a result the computed 3D points are organized regularly in a 2D matrix. This 2D arrangement enables us to apply 2D preprocessing and search methods nicely to 3D data saving computation time and complexity. Nevertheless, all methods proposed in the following are applicable on any type of 3D data.

3 Planes – Meaningful Structures

For many applications it is necessary to extract meaningful structures which enable a meaningful description of complex scenes. As the Swissranger camera provides $2\frac{1}{2}$D data from which 3D points are computed it can be focused on geometric aspects. Human-made environments – like walls, floors, and furniture – consist of large planar structures. Therefore, it is a reasonable step to find planar surfaces within a given 3D point set. It is assumed that preceptions of halls, offices, and meeting rooms can be categorized in a proper way using planar structures, because they provide more stable features compared to colors, textures, and materials occurring in different indoor scenarios.

In principal there are three possibilities to extract planes from a 3D point set. First, the Random Sample Consensus (RANSAC) algorithm [10] can be used to fit robustly plane models in 3D data, possibly combined with the Iterative Closest Points (ICP) algorithm [11] or SIFT features for refining the planes [12,13]. Second, the Expectation Maximization (EM) algorithm can be used to adjust the number of planes and estimates the locations and orientations by maximizing the expectation of a log-likelihood function [14,15]. Finally, region growing based approaches start from an initial triangle mesh and merge adjacent planar triangles iteratively [16]. These methods mentioned have several disadvantages. For example, RANSAC might lead to non-compact planes e.g. containing points of a table and a wall, simultaneously, or the EM algorithm has to run every time when the number of planes was updated.

In the following, a combination of seeded region growing [17] and RANSAC is introduced based on special values holding the correlating arrangement of points. This planar patch extraction works directly on 3D points instead on triangles as proposed by Hähnel saving computation time needed for triangle generation. The main idea is to decompose the point cloud into planarly connected regions and to extract planes in these regions for refinement.

Oriented Particles. First, oriented particles similar to Fua's approach [18] are defined for each point. A point's normal $\boldsymbol{n}_c$ is computed using a point set $\{\boldsymbol{p}_i \mid \boldsymbol{p}_i \in \mathcal{N}_{3\times 3} \text{ of } \boldsymbol{p}_c\}$ defined on the 8-neighborhood of the Swissranger image plane. The normal $\boldsymbol{n}_c$ of the current point $\boldsymbol{p}_c$ is determind by the principal component analysis of the points ($\{\boldsymbol{p}_i\} \in \mathcal{P}_c$). The deviation σ_c of the point $\boldsymbol{p}_c$ to the fitted plane can be used as a measure of the quality of the fit. Points with a deviation below a threshold θ_σ are classified as *locally planar*, otherwise as *nonplanar* [19]. Each plane $\{\mathcal{P}_c(\boldsymbol{n}_c, d_c)\}$ is described by the Hessian Normal Form

$$\mathcal{P}_c : \boldsymbol{n}_c \cdot \boldsymbol{x} - d_c = 0 \tag{1}$$

where d_c is equal to the Euclidean distance between the centroid $\boldsymbol{m}_c = \frac{1}{|\{\boldsymbol{p}_i\}|}\sum_i \boldsymbol{p}_i$ and the origin of the given world coordinate system.

Extracting Planar Patches. Here, the point cloud is decomposed into connected regions using region growing. Iteratively, points are selected randomly as seed of a region and extended with points of the 8-neighborhood $\mathcal{N}_{3\times 3}$ if four criteria are fulfilled. Two criteria are defined on the particles themselves, which are the validation of the points generated by the preprocessing and the local planarity as defined above. The other two criteria are computed on pairs of particles like the conormality and coplanarity measurement defined by Stamos and Allen [19]. Two points $\boldsymbol{p}_1$ and $\boldsymbol{p}_2$ are conormal, when their normals $\boldsymbol{n}_1$ and $\boldsymbol{n}_2$ hold:

$$\alpha = \cos^{-1}(\boldsymbol{n}_1 \cdot \boldsymbol{n}_2) < \theta_\alpha \tag{2}$$

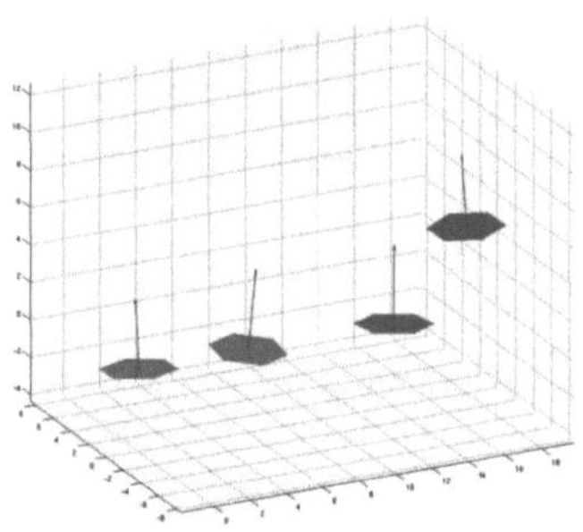

Fig. 2. (red) coplanar example of two patches, (blue) non coplanar pair of patches

Two points $\boldsymbol{p}_1$ and $\boldsymbol{p}_2$ are coplanar (Fig. 2) when the distance d

$$d = \max(\mid \boldsymbol{r}_{12} \cdot \boldsymbol{n}_1 \mid, \mid \boldsymbol{r}_{12} \cdot \boldsymbol{n}_2 \mid),$$
$$\boldsymbol{r}_{12} = \boldsymbol{p}_1 - \boldsymbol{p}_2 \tag{3}$$

is smaller than a threshold θ_d. The distance is computed with respect to the orientation and the distance of the oriented particles. As a result a set of mainly planar connected patches is provided. On each of these regions several runs of the RANSAC algorithm extract the largest and smoothest planes. This step can be seen as a postprocessing step where basically the parameters of the planes $\{\mathcal{P}_c \mid \boldsymbol{n}_c, d_c\}$ are refined.

Merging Planar Patches. Due to oversegmentation neighboring planar patches which are close to each other (so-called *close* patches) and belong to the same infinite plane have to be merged. In order to realize an efficient merging strategy, first, a patch is chosen randomly and related patches fulfilling the angle condition (Eq. 2) are determined. These selected patches are forwarded to the next step, where around the current patch a region of interest (ROI) is determined. Patch

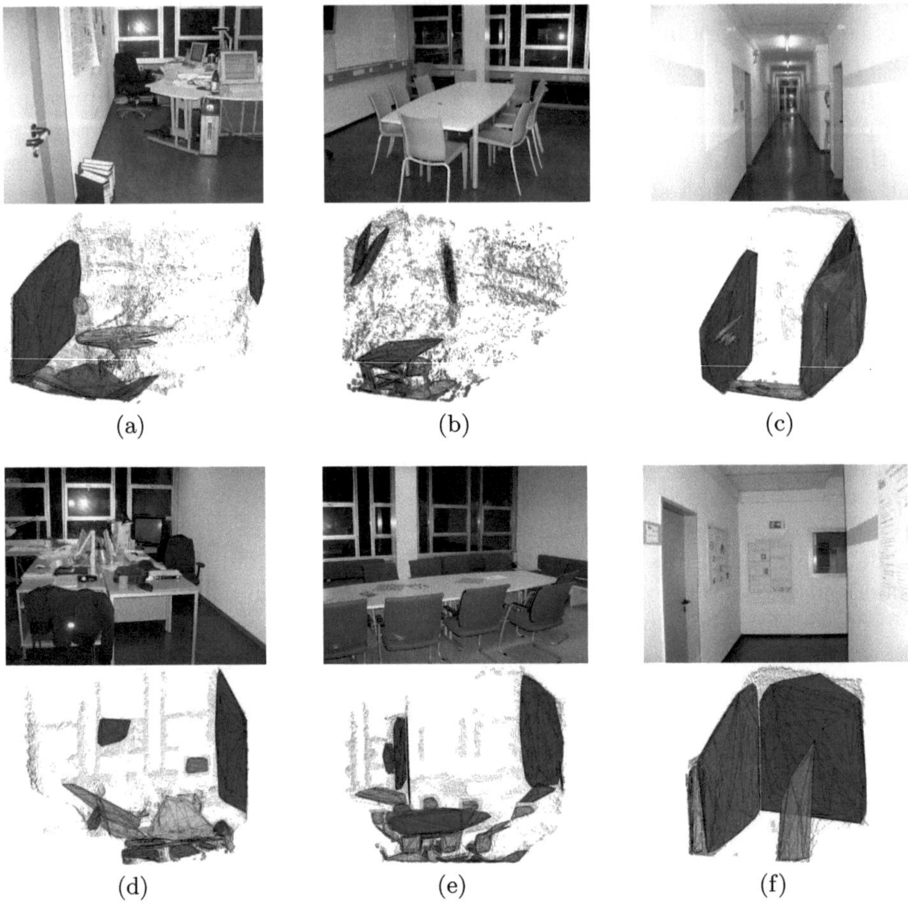

Fig. 3. First, exemplary photos and 3D point clouds of the training set are shown: the trained (a) office, (b) meeting room, and (c) hall. Second, the rooms for testing are displayed: the tested (d) office, (e) meeting room, and (f) hall.

candidates for merging are those patches which contain points that are inside this defined ROI. The candidates together with the current patch are merged to one plane containing all points of these patches. The parameters $(\boldsymbol{n}_c, d_c)$ of this new plane are recomputed using all 3D points. This merging procedure is repeated until the number of planes becomes stable.

Figure 3 presents exemplary photos of the six room scenarios – two office, two halls, and two meeting rooms – and planar patches produced by the algorithm introduced above using 3D point clouds provided by the Swissranger SR3000. The thresholds mentioned here were set to $\theta_\alpha = 10°$, $\theta_d = 0.2 \cdot z_c$ (the absolute Euclidean distances between neighboring pixel points vary over the distance to the camera), and $\theta_\sigma = \bar{\sigma} + \sqrt{\frac{1}{n}\sum_{c=1}^{n}(\sigma_c - \bar{\sigma})^2}$ with n is the number of valid points per frame and $\bar{\sigma} = \frac{1}{n}\sum_{c=1}^{n}\sigma_c$ the mean deviation.

4 Feature Extraction

For classification an extraction of meaningful features from the given planes is required. The aim is to classify a perception of a room (here: one frame of the 3D ToF sensor) while e.g. a robot enters the room. The result of the classification should be a hypothesis which room type was entered, even if the robot has not seen this specific room before.

As well defined feature vectors have to fulfill several conditions it is not suitable to use all plane parameters merged to one vector as features for classification as proposed by Lourenco [20]. The features should not only be independent from colors and textures in the scene, which is implemented by the extracted planar structures, but they should also be invariant with respect to changes in the absolute number of planes, changes in view angle and view direction of the camera, and invariant to in-class variation of the furniture configuration. In the following, different aspects of the planar patches $\{\mathcal{P}_i\}$ are examined as first simple features for classification concerning the conditions listed above.

(i) *Number of Points per Patch.* First, a feature vector is computed that encodes the size of patches in a frame, e.g. whether it contains large patches or a lot of small planar structures: $\forall i : n_i = \frac{|\mathcal{P}_i|}{\sum_j |\mathcal{P}_j|}$. The resulting terms have values between zero and one with a concentration in the region close to zero. As a feature vector (FV1) a histogram is computed using bins of different size – small close to zero and becoming large towards one.

(ii) *Angles between Normals of Patches.* Here, the orientation between patches is considered: $\forall i \neq j : \alpha_{ij} = \cos^{-1}(\boldsymbol{n}_i \cdot \boldsymbol{n}_j)$ normalized by the maximal possible value which is $\frac{\pi}{2}$. The feature vector (FV2) is created as a histogram with five intervals uniformly distributed over the values between zero and one. In Section 5, it is shown that for classification it is sufficient to compute the median of these angles to encode their information. Both, histograms over number of points per patch and angles between pairs of patches do not contain any structural information about the rooms. This information can be introduced by computing the feature histogram (FV3) over the angles α'_{ij} between pairs of close patches leading to better classification results.

(iii) *Ratios between Sizes of Patches.* This feature (FV4) encodes whether a frame contains a lot of patches of similar or different size: $\forall i \neq j : r_{ij} = \frac{\min(|\mathcal{P}_i|,|\mathcal{P}_j|)}{\max(|\mathcal{P}_i|,|\mathcal{P}_j|)}$, while the feature vector (FV1) over the number of points per patch refers to the absolute sizes of the patches.

The values in the bins of the feature histograms (FV1, FV2, FV3, FV4) are normalized to the range $[0, 1]$ by dividing the entries by the sum over all values in the bins per histogram.

5 Experiments and Discussion

For the following experiments 300 frames of two different offices, two halls, and two meeting rooms were acquired. The camera was positioned at a height of 145 cm (robot's camera head) and rotated horizontally 30° left/right and vertically 10° up/down in order to simulate a more or less random view on the room while entering. Also, the rooms chosen have significant differences in the layout within a room type as shown in Figure 3. The office tested is mirrored concerning the arrangement of furniture compared to the trained one, the table of the meeting room tested is rotated about 90 degree compared to the trained one, and the hall tested is not a straight corridor but a corridor following a corner. One office, one meeting room, and one hall (Fig. 3(a), 3(b), 3(c)) form the training set where 270 frames per room are used to train the classifiers and the remaining 30 frames to test the quality of the performance in recognizing an already seen room. 300 frames per room of the three other rooms (Fig. 3(d), 3(e), 3(f)) form the *main* test set for examining the performance of our system in categorizing percepts of rooms which have not be seen before. We intentionally started with a very small training set containing a single room per category in order to show the generalizability of the learned model.

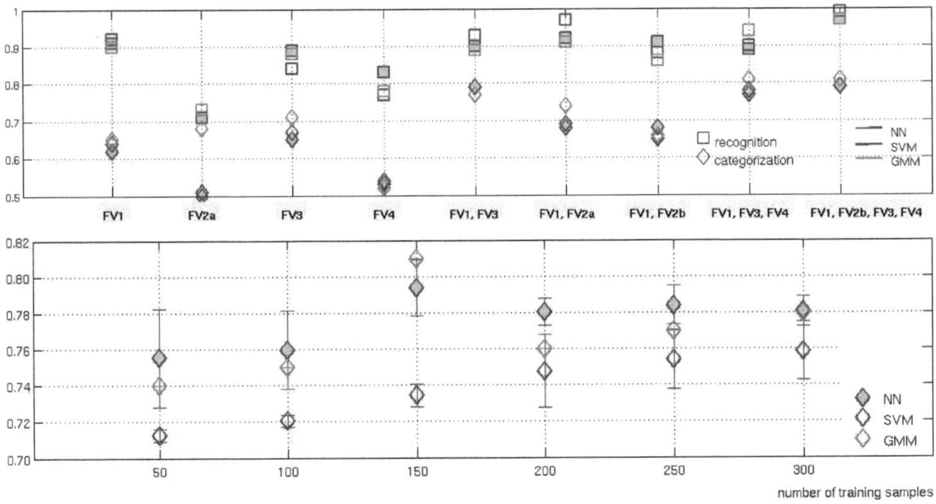

Fig. 4. (top) This plot presents results of the recognition and categorization using different combinations of the feature vectors (FV1, FV2, FV3, FV4: FV2a the histogram of angles between all patches, FV2b the median of these angles). Three classifiers are tested: a neuronal network (NN), a support vector machine (SVM), and a gaussian mixture model (GMM). (bottom) This plot shows the influence of the number of trainings data on the categorization result using [FV1, FV2b, FV3, FV4]. The vertical bars encode the standard deviation of the rates over 10 training runs of the NN, SVM or GMM with identical parameters (results using GMMs stay unchanged over 10 runs).

Three different classifiers are used to examine the proposed features in Section 4. The examined feature vectors are the number of points (FV1), the angles between patches (FV2a) and the median over these angles (FV2b), the angles between close patches (FV3), and the ratio of number of points between pairs of patches (FV4). The features are tested separately and in combination. A neural network (NN) with one hidden layer based on the Neural Network Toolbox of MatLab using backpropagation [21], the support vector machine SVM^{light} (SVM) [22,23] with a 5th-degree polynomial, and a gaussian mixture model (GMM) with five mixed distributions implemented in the toolkit ESMERALDA [24] are used for the classification task. Screening experiments provided five mixed distribution for GMM and a 5th-degree polynomial for SVM as quite suitable to deal with the proposed feature vectors.

Figure 4(top) presents all classification results from different feature vectors and combinations of them. The first four columns examine the feature vectors FV1, FV2a, FV3, and FV4 separated from each other. FV1 and FV3 turn out as features which contribute most to a good feature vector (recognition rate: 0.90, categorization rate: 0.65). The combination of these two features (FV1 and FV3) leads to a feature vector which provides promising categorization results up to 0.79 and recognition results up to 0.93. An additional test is executed to study the influence of FV2a compared to FV2b. FV2b performs similar to FV2a therefore it is assumed that the median of all angles can replace a histogram over all angles. The categorization can be improved up to 0.81 if the feature vector FV4 is added while the recognition rate stays on the level of 0.90. This rate can be further increased up to 0.99 using FV2b. As an assumption it can be stated that GMMs provide the most stable and proper classifiers using [FV1 FV2b FV3 FV4] as a feature vector. Round about 75% of the false classified vectors contains a mix up between meeting room and office. Since both room categories have analogies like a large table area in the middle of the room, this is an expected result.

Figure 4(bottom) shows the influence of the amount of training data on the classification rates. The vertical bars encode for each set of training data how reliable this classification rate is, since every trainings run with the same set of features and identical parameters leads to classifiers producing different classification rates on the same test data. Especially the NN and GMM classifiers seem to become saturated when using more than 150 training samples.

For additional experiments extra offices are recorded. Four of the now six different offices have a similar layout with two opposing work places while the other two rooms contain only a single work place. A categorization of the four double-place offices with the classifiers – trained with the train data introduced above – provides a rate of at least 0.69 right categorized percepts while only 0.34 to 0.51 of the percepts of the two single-place offices are classified correctly. If the train data is extended with frames of a single-place room the categorization rate of all offices can be increased to 0.88 on average.

Eighty percent of successful room categorization indicates that these planar structures extracted from the 3D point clouds provide meaningful information about categories of rooms whereon feature vectors can be defined suitable for classification. The categorization only based on the given 3D data provides promising first results that may be even further improved via applying more different statistics to the set of planar patches, like e.g. histograms over the smallest distances between pairs of patches, number of close patches, the shapes of patches, or the convex hulls.

6 Conclusion and Outlook

In this paper an approach for classifying and categorizing rooms is introduced. This is especially challenging as individual components reappear in different room categories and vary in the same category with regard to color and texture. We propose a holistic classification scheme of room types using 3D spatial information. The approach is designed to work in a real world setting combining different innovative techniques in a whole processing chain from sensor to classification result. First, planar structures on 3D ToF data are extracted. Oriented particles for each point over its 8-neighborhood are defined. Afterwards using region growing the current region is extended by points that are planar, valid with regard to the preprocessing, conormal and coplanar. On the resulting connected planar regions RANSAC refines smooth planar patches while a merging step fuses close patches that belong to the same infinite plane. Several statistics over these patches are computed like number of points per patch, angles between all pairs of patches, angles between pairs of close patches, and ratios between sizes of patch pairs. Histograms over these statistics define feature vectors that show a good performance in categorizing room percepts of offices, halls, and meeting rooms.

We show that the features defined can be utilized for room categorization providing context information important for a mobile robot acting in a home tour scenario [25,26]. To cope with such a scenario, the next step is to apply the proposed feature extraction and classification to data of other room types – like living rooms, kitchens, and bedrooms which are typical rooms for this scenario. The aim is to combine room hypotheses on 3D data with hypotheses of other sensors like speech or 2D data. There, it might be necessary to compute a huge amount of simple features and to extract the best ones using feature selection techniques like AdaBoost. As our approach is purely data driven and represents a bottom-up description of rooms, a second interesting field for further research is to generate models of different room categories out of planar structures. These models could be triggered by the holistic classification result and support the interpretation as top-down world knowledge. It further can help to build up a scene model of the robot's environment.

Acknowledgement

This work is funded by the Cooperative Research Center CRC673 "Alignment in Communication".

References

1. Strat, T.M., Fischler, M.A.: Context-based vision: Recognizing objects using both 2D and 3D imaging. PAMI 13, 1050–1065 (1991)
2. McKeown, D.M., Harvey, W.A., McDermott, J.: Rule-based interpretation of aerial imagery. In: Readings in Computer Vision: Issues, Problems, Principles, and Paradigms, pp. 415–430 (1987)
3. Murphy, K., Torralba, A., Freeman, W.T.: Using the forest to see the trees: A graphical model relating features, objects, and scenes. In: Advances in Neural Information Processing Systems, vol. 16 (2003)
4. Hoiem, D., Efros, A.A., Hebert, M.: Putting objects in perspective. CVPR 2, 2137–2144 (2006)
5. Szummer, M., Picard, R.W.: Indoor-outdoor image classification. In: CAIVD, pp. 42–51 (1998)
6. Paek, S., Chang, S.F.: A knowledge engineering approach for image classification based on probabilistic reasoning systems. ICME 2, 1133–1136 (2000)
7. Torralba, A.: Contextual priming for object detection. IJCV 53, 153–167 (2003)
8. Weingarten, J., Gruener, G., Siegwart, R.: A state-of-the-art 3D sensor for robot navigation. In: IROS (2004)
9. Swadzba, A., Liu, B., Penne, J., Jesorsky, O., Kompe, R.: A comprehensive system for 3D modeling from range images acquired from a 3D ToF sensor. In: ICVS (2007)
10. Fischler, M.A., Bolles, R.C.: Random sample consensus: A paradigm for model fitting with applications to image analysis and automated cartography. Commun. ACM 24, 381–395 (1981)
11. Besl, P.J., McKay, N.D.: A method for registration of 3D shapes. PAMI 14, 239–256 (1992)
12. Nüchter, A., Surmann, H., Hertzberg, J.: Automatic model refinement for 3D reconstruction with mobile robots. In: 3DIM, pp. 394–401 (2003)
13. Lee, S., Jang, D., Kim, E., Hong, S., Han, J.: A real-time 3D workspace modeling with stereo camera. In: IROS, pp. 2140–2147 (2005)
14. Liu, Y., Emery, R., Chakrabarti, D., Burgard, W., Thurn, S.: Using EM to learn 3D models of indoor environments with mobile robots. In: ICML (2001)
15. Lakaemper, R., Latecki, L.J.: Using extended EM to segment planar structures in 3D. In: ICPR, pp. 1077–1082 (2006)
16. Hähnel, D., Burgard, W., Thrun, S.: Learning compact 3D models of indoor and outdoor environments with a mobile robot. Robotics and Autonomous Systems 44, 15–27 (2003)
17. Adams, R., Bischof, L.: Seeded region growing. PAMI 16, 641–647 (1994)
18. Fua, P.: From multiple stereo views to multiple 3D surfaces. IJCV 24, 19–35 (1997)
19. Stamos, I., Allen, P.K.: Geometry and texture recovery of scenes of large scale. CVIU 88, 94–118 (2002)
20. Lourenco, A., Freitas, P., Ribeiro, M.I., Marques, J.S.: Detection and classification of 3D moving objects. In: Mediterranean Conf. on Control and Automation (2002)
21. Rumelhart, D.E., Hinton, G.E., Williams, R.J.: Learning internal representations by error propagation. Parallel Data Processing 1, 318–362 (1986)
22. Vapnik, V.N.: The Nature of Statistical Learning Theory (1995)
23. Joachims, T.: Learning to Classify Text Using Support Vector Machines. PhD thesis, Cornell University (2002)

24. Fink, G.A.: Developing HMM-based recognizers with ESMERALDA. In: Matoušek, V., Mautner, P., Ocelíková, J., Sojka, P. (eds.) TSD 1999. LNCS (LNAI), vol. 1692, pp. 229–234. Springer, Heidelberg (1999)
25. Haasch, A., Hohenner, S., Hüwel, S., Kleinehagenbrock, M., Lang, S., Toptsis, I., Fink, G.A., Fritsch, J., Wrede, B., Sagerer, G.: BIRON – The Bielefeld Robot Companion. In: ASER, pp. 27–32 (2004)
26. COGNIRON: The cognitive robot companion (FP6-IST-002020) (2004), http://www.cogniron.org

Template Selection by Editing Algorithms: A Case Study in Face Recognition

Biagio Freni, Gian Luca Marcialis, and Fabio Roli

University of Cagliari
Department of Electrical and Electronic Engineering
Piazza d'Armi - I-09123 Cagliari (Italy)
{biagio.freni,marcialis,roli}@diee.unica.it

Abstract. In this paper, we consider the problem of template selection in biometric systems as analogous to the problem, called "editing", of selecting representative prototypes when using the 1-Nearest Neighbour classifier (NN). Four editing algorithms are used and compared by experiments with state-of-the-art template selection algorithms. Experiments are performed on a benchmark face data set. Reported results show pros and cons of editing algorithms for template selection in biometric systems.

Keywords: Nearest neighbour, biometrics, face recognition, template selection.

1 Introduction

Biometrics are aimed to recognize or verify the personal identity by physiological or behavioral characteristics [1]. Among others, widely used biometrics are the fingerprint and the face. The core of a biometric system is the so-called gallery, made up of a set of at least one *template*, which is the feature set extracted, for example, from the face image of a certain person, collected during the enrolment session. Template(s) is stored into the system and it is used, by computing its "distance" from the input sample released by the person during the system operation, in order to authenticate, or recognize, his/her identity. As it can be easily noticed, the system performance (the recognition accuracy, or the trade-off between false and true authentications) is dependent on the score, that is, on the representativeness degree of the template for each identity (or client, as it is usually named in verification applications).

So far, only Uludag *et al.* addressed the problem of selecting templates for each client. They proposed some algorithms derived from the clustering theory. , they selected, from each gallery, the best samples according to a given criterion, which can be summarised in: maximum similarity among templates, in order to point out their common characteristics, and minimum similarity among them, in order to capture very different characteristics of the same client (intra-class variations). Worth noting, algorithms proposed by Uludag et al. selected the best templates by looking only at the instances collected of the given client,

N. da Vitora Lobo et al. (Eds.): SSPR&SPR 2008, LNCS 5342, pp. 745–754, 2008.

without analysing the representativeness of the template with respect to the others. Another issue for these algorithms is that they aren't able to find the number of templates for a certain client. This is a setting parameter given by the human expert, and does not take into account the intrinsic "difficulty" in classifying certain clients with respect to others. Furthermore, no generalization ability is estimated.

As alternative, we believe that the problem of template selection [2] can be dealt by considering its analogies with the "editing" of training sets in Nearest Neighoubour-based classification (NN) [3,5,6,7,8,9]. Editing consists in generating, from a given training set T, a subset E able to maintain the same classification accuracy of T on itself. In our opinion, the editing algorithms adopted for NN-based classification exhibit an interesting peculiarity: they focus on the whole collected training set T. Therefore, they are able to obtain the best galleries set E by analysing the whole structure of the data, i.e., all stored clients samples, by re-sizing the gallery for each client appropriately. This should have a twofold impact: (i) the overall analysis of the structure of T should help in finding templates (instances of T) exhibiting generalization abilities superior than that of algorithms proposed by Uludag et al [2]; (ii) some "difficult" clients could need galleries bigger than those of other "easy" ones, for which, on the contrary, related galleries could be small. In other words, the gallery size should be adapted as function of the client.

Therefore, in this paper, we consider the problem of the template selection as analogous to the problem of finding representative instances sets when using the NN classifier. It can be easily noticed that NN classifier and current approaches for person recognition are correlated, and share some aspects. For example, they are based on the same criteria of "distance from a template" and it is not difficult to see the set of galleries in biometrics as the training set used in NN-based classification. We experimentally compare such editing approaches with the state-of-the-art template selection algorithms proposed in [2]. Experimental results support our claims.

The paper is organized as follows. Section 2 introduces the common terminology we used both in general NN editing approaches and biometrics. Section 3 reviews the main general approaches to edit a given training set for NN-based applications, and summarizes the algorithms in [2]. Section 4 presents some experimental results by adopting a face data set. Section 5 concludes the paper.

2 Terminology and Notation Adopted

The following terminology has been used in this paper, which, in our opinion, can be commonly shared in NN-based classification and biometrics recognition. In brackets we pointed out the related application field:

- *class(NN)/client(biometrics)*: a set of objects with common characteristics [3,4]. In this paper, the number of classes is indicated with C. In personal recognition applications, each client identity corresponds to one class.

- *genuine users/impostors(biometrics)*: In personal verification applications, two subclasses are defined: the one of clients whose claimed and true identity correspond, named "genuine users", and the opposite one, named "impostors". In this paper, we only dealt with the face recognition application, thus these terms will not be used.
- *instance(NN)/template(biometrics)*: a certain feature set extracted from a given biometric. Although in the NN literature [5], the "instance" and "template" terms do not mean the same thing, in this paper we always considered the template and instance terms with the same substantial meaning. In particular, we indicated as instances the members of the set called T, that is, the initial training set of instances, collected during the enrolment session. At least one instance per class is present in T. We used the term "template" after the editing process is completed. In other words, with the terms "templates" we referred to the instances selected from T by the editing algorithm. We referred to the unknown input sample to the system with the term *sample*. The sample is represented in the same feature space of instances.
- *label(NN,biometrics)*: simply the class identifier. For example, if we have C classes, we can indicate the set of labels as $\Omega = \{\omega_1, ..., \omega_C\}$. In other words, a labeling function $l(*)$ is given such that, for each $x \in T$, $l(x) = \omega$, with $\omega \in \Omega$. In biometrics, the label for each class can be the user's name or identification number.
- *gallery(biometrics)*: is the set of possible instances (templates) for a certain client, stored into the memory of the system or embedded into a smart-card. The term "gallery" is typical in biometric applications.
- *distance(NN)/score(biometrics)*: they are quite similar, and exhibit non-negative output values. The only difference is that a small value of the distance function means that the instance/sample are "close", thus quite similar. Instead, a small value of the similarity function means that they are "far", so quite different. It will be possible to use indifferently distance or similarity function after an appropriate normalization [10]. In biometrics, the matching score can be derived from a given distance.

3 K-NN Editing for Biometrics and State-of-the-Art

The K-Nearest Neighbour classifier (KNN) belongs to the category of non-parametric classifiers [4]. Given a sample x, this classifier exploit a training set T of templates (instances) in order to find the K nearest neighbours templates on the basis of a pre-defined distance (or similarity) function.

The weak point of the KNN classifier is the high computational complexity. The more is the size of T, the more is the computation amount required. This is an issue in many real classification problems [3]. If a large amount of data are available in T, many of them could be redundant or outliers. Therefore, a possible solution to reduce the complexity of KNN is to find the most *representative* instances in the training set. This approach is called "editing". Several editing algorithms have been proposed so far [5,6,7,8,9].

Result of these algorithms is the so-called "edited" set $E \subseteq T$, satisfying some requirements given by the algorithm applied. This edited set must be representative of T. The manner to reach this aim can be roughly described as *incremental*, if E is initially empty and starts growing according to some specific criteria; *decremental*, if E is initially equal to T and instances are iteratively deleted until a certain stop criterion is met.

In this paper, we only focus the case of K=1, thus adopting the simple Nearest Neighbour algorithm.

Key points of the editing algorithms are: the distance or similarity function used, the selection criterion of the "best" instances in T, the stop criterion adopted. The distance or similarity function in this kind of application have a dramatic effect. In principle, such function impact on the classification accuracy, thus on the "nearest neighbours" of a given sample, or instance. In this paper we used the simple *Euclidean Distance*, but the selection of the best distance function is an open problem, although in biometrics it is simplified due to the fact that, often, a dedicated function is designed for the feature selected (e.g. String algorithm for fingerprint matching [12]). The selection and stop criteria depend on the specific algorithm as described in the following.

3.1 The Proposed Algorithms

CNN - Condensed NN: Aim of this incremental algorithm is to find an edited set $E \subseteq T$ such that the closest istance $y \in T$ to a given instance $x \in E$ exhibits the same class identifier of x. In this way, the maximum classification accuracy on T is reached [6]. Pseudo-code:

1. C instances $x_1, ..., x_C$ are initially selected, such that $l(x_i) = \omega_i$ (one instance per class).
2. $E \leftarrow \{x_1, ..., x_C\}$.
3. $T \leftarrow T - E$.
4. T is classified using E as training set.
5. Let Y the set of misclassified instances in T.
6. **if** $Y \neq \phi$ **then**
   (a) $E \leftarrow E \cup Y$.
   (b) $T \leftarrow T - Y$.
   (c) **Repeat** from 4.
7. Stop.

SNN - Selective NN: This algorithm is an extended version of the previous one [7]. In particular, the aim of this algorithm is to build an Edited set E as small as possible. According to this, not all the misclassified instances were added to E, but that who as much close to the own class than any other instance of different class. Therefore, SNN does not guarantee the 100% of classification accuracy on T. Since this algorithm is a clear extension of CNN, we omitted its pseudo-code for sake of space.

RNN - Reduced NN: This is a decremental algorithm. It starts with E set as T. Instances are iteratively deleted from E if their deletion does not imply a misclassification on T. This algorithm stops when no other instance can be removed from E [8]. Pseudo-code:

1. $E \leftarrow T$.
2. **for each** $x \in E$
   (a) $E \leftarrow E - \{x\}$
   (b) Classify T using E
   (c) Let Y be the set of misclassified samples in T
   (d) if $Y \neq \phi$ then $E \leftarrow E \cup \{x\}$

ENN - Edited NN: This algorithm also starts with an Edited set E equal to T [9]. Instances are removed from E, if do not agree with the majority of its "k-nearest neighbour" (in our case k=3). Pseudo-code:

1. $E \leftarrow T$.
2. **for each** $x \in E$
   (a) Let Y the set of "k nearest neighbour" instances of x.
   (b) Let m the majority class label of the instances in Y.
   (c) Let $l(x)$ the class label of x.
   (d) if $l(x) \neq m$ then $E \leftarrow E - \{x\}$

3.2 Editing Approaches vs. State-of-the-Art

The above algorithms have been experimentally compared with the ones proposed in [2], and named MDIST and DEND. These algorithms are described in the following.

MDIST: The rationale behind the MDIST algorithm comes from the distance-based clustering [11]. The aim is to build galleries made up of very close instances, for each client.

A very important parameter for this algorithm is the number of instances, namely, N, to select for each class. In [2] this number was an input parameter given by the user.

1. $E \leftarrow \phi$.
2. **for each** $\omega \in \Omega$
   (a) $T^{\omega} \leftarrow \{x \in T : l(x) == \omega\}$
   (b) $S \leftarrow \{S_1, ..., S_k\}$ such that $|S_i| == N$ and $S_i \leftarrow \{x \in T^{\omega}\}$
   (c) **for** $i = 0, ..., k$
      i. $d_i \leftarrow$ average distance among instances in S_i
   (d) $S_{min} \leftarrow \{S_i : i = argmin_j d_j\}$
   (e) $E \leftarrow E \bigcup S_{min}$

DEND: The basic idea of this algorithm comes also from the clustering theory (in particular, from the hierarchical clustering [11]). The selection criteria of this algorithm is opposite to the previous one: instead of the selection of very close instances, this algorithm selects the farest instances from a given gallery.

The rationale of DEND is to build galleries able to represent very large intra-class variations of each client.

1. $E \leftarrow \phi$.
2. **for each** $\omega \in \Omega$
   (a) $T^{\omega} \leftarrow \{x \in T : l(x) == \omega\}$
   (b) $S \leftarrow \{S_1, ..., S_k\}$ such that $|S_i| == N$ and $S_i \leftarrow \{x \in T^{\omega}\}$
   (c) **for** $i = 0, ..., k$
      i. $d_i \leftarrow$ average distance among instances in S_i
   (d) $S_{max} \leftarrow \{S_i : i = argmax_j d_j\}$
   (e) $E \leftarrow E \bigcup S_{max}$

A first critical analysis can be done after the description of state-of-the-art and proposed algorithms. Firstly, as stated in the Introduction, MDIST and DEND perform the search of best templates only by analysing the instances of each client. Proposed algorithms, instead, look for best templates by investigating the whole instance space. This can be an advantage, especially with respect to the DEND algorithm. Since it tries to find the largest intra-class variations, the result could be affected by some outliers, or noisy instances which could be very near to other clients. Secondly, the number of templates per gallery N is a fundamental parameter for MDIST and DEND. So far, no methods have been proposed to estimate such number. Therefore, a certain range of N values must be investigated. This problem does not affect the editing algorithms. On the other hand, it can be noticed that the result of *incremental* approaches (CNN and SNN) is quite dependent on the initial instances selected. Therefore, obtained gallery size can be unstable if outliers are selected. Moreover, *decremental* approaches starts from the whole T, and this could lead to a weak gallery size reduction. Finally, the stop criterion is strongly correlated with the effectiveness of both incremental and decremental algorithms. For example, CNN can reach 100% of accuracy on T only at the last iteration, before meeting the stop condition. If this one is altered, selected templates could be not representative enough.

4 Experimental Results

4.1 Data Set and Experimental Protocol

We adopted the Equinox Face Dataset [13]. In particular, our experiments are carried out on 50 clients randomly chosen, each one made up of 100 images. Therefore, the data set size is 5000 images.

Each image used was first cropped and, after histogram stretching and normalization, re-sized by keeping the intra-ocular distance on a constant value. This is a common practice in face recognition and verification applications [13]. The final size of each image is 100x100 pixels.

In our experiments we adopted the protocol below:

1. The overall data set has been randomly partitioned in two parts of the same size, which we named T and t. Thus, each client/class is made up of 50 instances collected during the so-called "enrolment" operation, that is, the registration of each client into the system.
2. PCA [3] was performed on T and t was transformed accordingly, in order to reduce the size of the feature space.

Table 1. Mean and, in brackets, standard deviation of number of templates per class in E (second column), and of the recognition accuracy of investigated editing approaches (third column). The first column reported the gallery related to the algorithm adopted. The term "TRAIN" means that the overall T is used for classification. Performance of MDIST and DEND has been reported in the case of six and nine instances per client. The best trade-off between recognition accuracy and galleries size has been highlighted in bold.

Gallery	#instances×class	TEST
TRAIN	50 (0)	99.62 (0.14)
CNN	**7 (3)**	**97.6 (0.45)**
SNN	4 (3)	73.66 (3.31)
RNN	17 (9)	98.43 (0.53)
ENN	49 (1)	99.35 (0.27)
MDIST	6 (0)	94.15 (0.68)
MDIST	9 (0)	96.56 (0.58)
DEND	6 (0)	89.11 (1.39)
DEND	9 (0)	94.03 (0.70)

3. T has been used as training set to be edited, by adopting proposed algorithms, MDIST and DEND. All algorithm led to a different edited set E, containing the "best" galleries according to the related method.
4. Performance of each approach has been estimated on t set, thus evaluating the generalization ability of the obtained gallery.
5. Items 1-3 were repeated for six runs.
6. Results have been averaged and reported in the following tables and plots.

The performance parameter adopted is the recognition accuracy when the 1-Nearest Neighbour is selected for classification.

4.2 Results

First of all, we verified if it is possible, by selecting templates with proposed editing algorithms, to obtain a generalization accuracy better than that achieved with state-of-the-art algorithms based on clustering, as MDIST and DEND.

In table 1, the average size of galleries in E, and the mean and standard deviation of the recognition accuracy were shown. Performance of MDIST and DEND has been reported in the case of six and nine instances per client.

From Table 1 the following observations can be drawn:

1. *Incremental* approaches (SNN and CNN) allowed a galleries size reduction much superior than decremental ones, without significant loss in accuracy. The exception is for SNN algorithm which exhibited an accuracy worse than that of other algorithms, due to its stringent working conditions: in fact, the galleries found by SNN have the smallest size.
2. Best accuracy is guarantee by *decremental* approaches (RNN and ENN) but at the expense of the reduction of the galleries size.

Table 2. Average recognition accuracy for classes for which the editing has built galleries larger than that of MDIST and DEND. It can be noted that the average accuracy is lower than that in Table 1 for such classes (MIDST and DEND columns), thus they can be classified as "difficult". On the other hand, the editing algorithms allowed to found a gallery large enough, thus allowing a better accuracy.

Gallery	#classes	#instances×class	TEST	MDIST 9	DEND 9	MDIST 6	DEND 6
CNN	8	12	96.18	93.78	87.56	89.41	77.26
RNN	41	20	98.58	96.05	93.25	93.20	87.89
ENN	50	49	99.35	96.56	94.03	94.15	89.11

Table 3. Average recognition accuracy for classes for which the editing has built galleries smaller than that of MDIST and DEND. It can be noted that the accuracy is about the same as the one reported by MDIST and DEND, thus those classes can be classified as intrinsically "easy".

Gallery	#classes	#instances×class	TEST	MDIST 9	DEND 9	MDIST 6	DEND 6
CNN	21	4	98.04	98.46	96.88	96.89	93.67
SNN	31	3	60.70	96.86	95.06	94.33	90.89
RNN	4	3	98.08	99.60	98.00	98.96	95.76

3. MDIST and DEND confirmed their strong dependence of the number of istances set, which is not simple to find a priori, and can be given as a constraint, relying, for example, on the maximum recognition time or memory available for storing galleries.
4. On the overall, CNN appeared to be the best trade-off between performance achieved and size of the galleries.

We also investigated if the gallery size of editing algorithms can be considered as function of the instrinsic difficulty of clients. In order to confirm or deny this claim, we proceeded as follows: (i) we isolated all clients for which the editing algorithm found a gallery larger than that of MDIST and DEND with $N = 9$; (ii) we also isolated all clients for which a gallery smaller than that of MDIST and DEND with $N = 6$ have been found by the editing algorithms. We expected that, for "difficult" clients, related to item (i), the average accuracy is higher for the proposed approaches, whilst it is about the same of state-of-the-art algorithms for "easy" clients of item (ii). Results related to "difficult" clients have been reported in Table 2, whilst Table 3 shows accuracy of "easy" clients.

In the case of Table 2, the SNN performance has not been reported since this algorithm strongly reduce the size of the galleries: only one client following constraints of item (i) was found. The opposite holds for the ENN algorithm in Table 3. No clients with a gallery size less than six were found (item (ii)).

By observing the other algorithms, it can be easily noticed that results confirmed what we expected. For "difficult" clients, the editing algorithms exhibited an accuracy more than that of state-of-the-art algorithm. Both incremental and decremental approaches were effective. The same can be noticed for "easy" clients: in this case the editing algorithms pointed out two important advantages: the gallery size is small, and the accuracy is unaltered. This has a positive effect when stringent conditions are required for galleries, especially in terms of the memory available for storing galleries or the matching time.

4.3 Discussion on Results

Reported results clearly showed that editing algorithms can be considered as a good alternative to the state-of-the-art approaches to the template selection: they allowed to select templates with good generalization ability and their number is modelled according to the characteristics of each client.

We believe that both features are correlated to the fact that the editing algorithms analysed the overall structure of the data, so they are able to exploit an amount of information larger than that exploited by MDIST and DEND.

On the other hand, some issues must be considered when using these algorithms for the practical selection of templates. A noticeable problem is how to limit the gallery size when strong requirements are present. For example, if the number of templates per clients can not exceed a certain value due to memory size or matching time constraints. By adopting incremental approaches, this limitation can be easily overcame: the increase of the number of templates for a client may stop when the limit is reached. However, it is worth noting that performing such an early stopping could strongly impact on the obtained galleries. A possible solution, which could work when using both incremental or decremental approaches, could be the joint use of them and state-of-the-art ones. Since the template editing approaches allow to find "easy" and "difficult" clients, a first pre-processing of the training set could be done by editing algorithms. Next, the DEND algorithm could be applied to the obtained galleries whose size exceeded the limit. We suggested DEND because it selects the most representative intra-class variations among templates found by editing algorithms. This could also allow to overcome the problem of DEND of selecting outliers or instances too near to the ones of other clients, thanks to the preliminary analysis performed by editing. The above suggestions need further experimental investigations to be confirmed or denied, thus the problem is still open.

Another aspect to point out is that these algorithms can be used when the number of collected instances per class has become large enough, since a large analysis on the data structure is necessary. This may be problematic for certain biometric applications (as the personal verification), where usually a large number of samples can be collected only after long periods of time. A possible solution could be to exploit samples collected during the system operations, and classified with very high matching score. This "semi-supervised" way of improving galleries have been also proposed in the literature [14], but disregarding the editing approaches.

5 Conclusions

In this paper, the problem of selecting templates from samples collected during one or multiple enrolment sessions has been treated, by considering its analogies with the problem of "editing" a training set in Nearest Neighbour-based classification. We applied the editing algorithms in the context of face recognition.

In particular, four NN-based editing algorithms have been compared with two state-of-the-art algorithms based on the clustering theory. Preliminary results showed that reported algorithms can find a gallery size such that overall accuracy is kept high, whilst the number of comparison per person notably decreased. The reported performance, in some cases, is better than that achieved by state-of-the-art algorithms, especially by considering that the best number of templates per class is fixed in such approaches.

Although this limited set of experiments is not enough to draw definitive conclusions, we believe that editing approaches for the templates selection could be considered as a good alternative to, or used jointly with, the algorithms so far proposed in the literature.

References

1. Jain, A.K., Pankanti, S. (eds.): Biometrics: Personal Identification in Networked society. Kluwer Academic Publishers, Dordrecht (1999)
2. Uludag, U., Ross, A., Jain, A.K.: Biometric template selection and update: a case study in fingerprints. Pattern Recognition 37(7), 1533–1542 (2004)
3. Duda, R.O., Hart, P.E., Stork, D.G.: Pattern Classification. John Wiley & Sons, Chichester (2001)
4. Koutroumbas, K., Theodoridis, S.: Pattern Recognition. Academic Press, London (2006)
5. Randall Wilson, D., Martinez, T.R.: Reduction Techniques for Instance-Based Learning Algorithms, `http://citeseer.ist.psu.edu/wilson00reduction.html`
6. Hart, P.E.: The Condensed Nearest Neighbor Rule. IEEE Transactions on Information Theory 14, 515–516
7. Ritter, G., Woodruff, H., Lowry, S., Isenhour, T.: An algorithm for a selective nearest neighbor decision rule. IEEE Transactions on Information Theory 21(6), 665–669
8. Gates, G.W.: The Reduced Nearest Neighbor Rule. IEEE Transactions on Information Theory 18(3), 431–433
9. Wilson, D.L.: Asymptotic properties of nearest neighbor rules using edited data. IEEE Transactions on System, Man and Cybernetics 2, 408–421 (1972)
10. Jain, A.K., Nadakumar, K., Ross, A.: Score Normalization in Multimodal Biometrics Systems. Pattern Recognition Letters (2005)
11. Jain, A.K., Dubes, R.C.: Algorithms for Clustering Data. Prentice-Hall, Englewood Cliffs (1988)
12. Jain, A.K., Hong, L., Bolle, R.: On-line Fingerprint Verification. IEEE Transactions on Pattern Analysis and Machine Intelligence 19(4), 302–314 (1997)
13. `http://www.equinoxsensors.com/products/HID.html`
14. Roli, F., Marcialis, G.L.: Semi-supervised PCA-based face recognition using self-training. In: Yeung, D.-Y., Kwok, J.T., Fred, A., Roli, F., de Ridder, D. (eds.) SSPR 2006 and SPR 2006. LNCS, vol. 4109, pp. 560–568. Springer, Heidelberg (2006)

A Novel Coding Scheme for Indexing Fingerprint Patterns*

Aglika Gyaourova and Arun Ross

West Virginia University, Morgantown WV 26506, USA
agyaourova@mix.wvu.edu, arun.ross@mail.wvu.edu

Abstract. Indexing is the process of assigning a numerical value to a database entry in order to facilitate its rapid retrieval. Indexing a fingerprint database can reduce the search space and improve the response time of an identification system. We discuss a novel method for generating index codes for fingerprint images by using a small set of pre-determined *reference* fingerprints. In the proposed method, the match scores generated by comparing an input fingerprint with the reference fingerprints are subjected to a discretization function, which converts them into an index code. A search mechanism based on the Hamming distance identifies those index codes in the database that are similar to the code of the input image. The proposed technique has several advantages: it obviates the need to extract complex features from the fingerprint image; it utilizes the matcher that is already associated with a particular application; and it can be used to index any biometric database irrespective of the trait or matcher being used. Experimental results on two fingerprint databases (NIST-4 and WVU) indicate that the proposed encoding scheme generates index codes that are well-scattered thereby allowing noisy query images to be indexed correctly.

1 Introduction

In a fingerprint identification system, the goal is to assign an input (or query) fingerprint to one of several identities present in the database. Every identity is defined by a fingerprint image and the identification operation entails the comparison of the query image with the images residing in the database. Filtering is the process of reducing the number of fingerprints to be retrieved from a database for comparison during an identification operation. Two important considerations make filtering a desirable part of large-scale fingerprint identification systems: (a) invoking a matcher on the entire database can be computationally expensive, and (b) the number of false positives grows exponentially with the size of the database [7][12]. Filtering can be accomplished by using two distinct schemes: classification or indexing. A classification scheme partitions the database of fingerprints into several classes. The query image is compared only against those images in the database belonging to the *same* class as itself. Indexing schemes,

* This work was supported by US NSF CAREER grant number IIS 0642554.

N. da Vitora Lobo et al. (Eds.): SSPR&SPR 2008, LNCS 5342, pp. 755–764, 2008.

on the other hand, assign an index value to every fingerprint and the query image is compared only against those images in the database having a *comparable* index value.

While the Henry fingerprint classification system[1] [1] has been traditionally used to perform filtering, fingerprint researchers have also developed alternate methods to filter fingerprints [11]. A review of the literature indicates that these techniques either utilize image features implicit in a fingerprint [2,3,5,6,9] or employ a fingerprint matcher to describe the relationship between multiple fingerprints based on pairwise match scores [10]. While the former approach has been widely researched, the latter has received very little attention in the literature primarily due to the need for generating and processing score matrices. For example, consider the work in [10] where the query image is sequentially matched against fingerprints in the database. At each step in the sequence, a decision is made regarding the choice of the next fingerprint to be retrieved. This decision is based on finding correlations in the score matrix containing the pairwise match scores of all fingerprints in the database. While this method obviates the need to perform additional image processing to compute index values, storing the matrix of match scores for a database containing millions of fingerprints can be prohibitive.

In this work, we utilize match scores to index and retrieve fingerprint images. However, the proposed technique does not use a score matrix in order to guide its search. Rather, the method relies on comparing a fingerprint with a *small* set of reference images and using the resulting match scores to derive an index code. The proposed method can be applied to any biometric database irrespective of the biometric trait or matcher being used.[2] Furthermore, it generates a compact code based on the evidence of a single impression of a finger. Thus, the proposed method has modest storage requirements.

The rest of the paper is organized as follows. In section 2 we discuss the three separate stages of the proposed method: (a) selecting reference images, (b) creating index codes for the images in the database, and (c) retrieving pertinent fingerprints from the database based on a query image. Section 3 describes the experimental setup using ternary-valued index codes. Experimental results on two databases are presented in section 4. Section 5 summarizes the paper and describes ways to further improve the proposed method.

2 Indexing Methodology

The crux of the proposed method relies on the use of a small set of reference fingerprint images, $\boldsymbol{R} = \{r_1, r_2, ..., r_n\}$, that have good representative and discriminative power. The representative power ensures that there are a sufficient *number* of reference fingerprints, while the discriminative power ensures that

[1] The disadvantages of the Henry classification system to filter one-print databases (as opposed to ten-print databases) have been discussed in [11].

[2] Indeed, it may be applied to any pattern retrieval problem provided an appropriate pattern matcher is available.

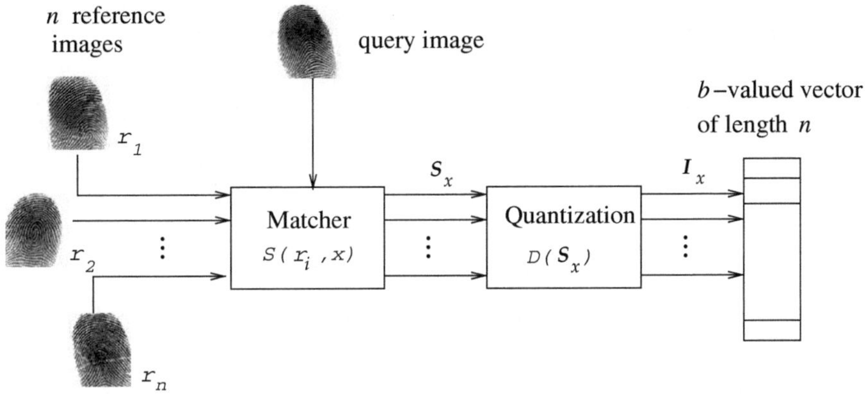

Fig. 1. Generating an index code using reference fingerprints

these prints exhibit sufficient *variability* amongst themselves. Rather than relying on implicit image features to select these reference images, the match scores generated by a fingerprint matcher are used.

The index of an arbitrary fingerprint image x is constructed in two steps. First, a set of match scores is computed by comparing x against the reference fingerprints, $r_1, r_2, ..., r_n$, in a fixed order. The result is the set $\boldsymbol{\mathcal{S}}_x = \{\mathcal{S}(r_1, x), ..., \mathcal{S}(r_n, x)\}$, where $\mathcal{S}()$ is the matching function. Next, a discretization function $\mathcal{D}_b : \mathbb{R}^n \rightarrow \mathbb{Z}_b^n$ is used to transform $\boldsymbol{\mathcal{S}}_x$ to an index code, i.e., $\boldsymbol{\mathcal{I}}_x = \mathcal{D}_b(\boldsymbol{\mathcal{S}}_x)$. Here, b is the number of distinct values that are possible for each element in the code. For example, $b = 2$ would imply that $\boldsymbol{\mathcal{I}}_x$ is a binary vector (see Figure 1).

During enrollment, the fingerprint image associated with an identity is subjected to the procedure above in order to generate its *model index code*. The model index code is stored in the fingerprint database along with the identity of the associated print. During identification, when a query image is presented to the system, a Hamming distance-based search mechanism is invoked to retrieve only those fingerprints in the database whose model index codes are probable candidates for a match.

Thus, the architecture of the proposed scheme is defined by the ordered set of reference fingerprint images ($\boldsymbol{R}$), the discretization function ($\mathcal{D}_b$), and the search mechanism used to identify potential model index codes. Figure 2 shows the framework and the information flow in the proposed indexing method.

2.1 Selecting Reference Images

Selection of the set of reference images is an important component of the proposed scheme. Successful indexing is possible only when the model index codes are well distributed over the index space. Therefore, the range of match scores obtained when comparing a reference image against different fingerprints in the database should be high. In the current implementation, the reference images are

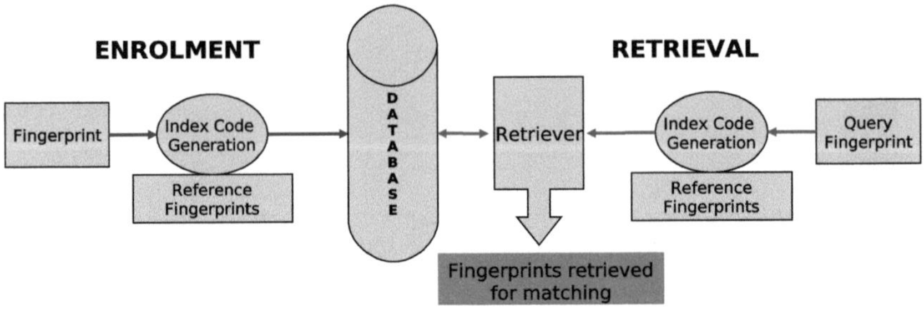

Fig. 2. Framework of the proposed method

selected from the fingerprint database itself (although they could be some type of digital artifacts as well, such as synthetic fingerprints). Fingerprint images whose impostor match scores exhibit a large variance are selected as reference fingerprints. While the entire database of prints may be viewed as a candidate pool for selecting reference images, practical considerations dictate the use of a small random subset of prints for this purpose.

Algorithm for selecting n reference fingerprints

Let $\boldsymbol{\mathcal{F}} = \{f_1, f_2, ..., f_P\}$ be the candidate pool of reference images, $\boldsymbol{\mathcal{D}} = \{d_1, d_2, ..., d_L\}$ be an "impostor" database of fingerprints, and $S()$ be the fingerprint matcher.

1. For each f_i compute $v_i = \text{Variance}\{\mathcal{S}(f_i, d_j)\}_{j=1}^{L}$.
2. Sort v_i in descending order.
3. Use the fingerprints corresponding to the top n v_i values as reference fingerprints.

There is a possibility of two reference images having very similar impostor match scores when compared against other images. Retaining both these reference images will not enhance the uniqueness of the index codes. Therefore, an additional constraint may be needed, which ensures that every pair of reference images are sufficiently dissimilar.

2.2 Creating the Index Code

The index code for an arbitrary image x is computed as follows. First, the vector $\boldsymbol{\mathcal{S}}_x$ is computed as $\boldsymbol{\mathcal{S}}_x = \{s_1^x, s_2^x, ..., s_n^x\}$, where $s_i^x = \mathcal{S}(r_i, x)$. Next, the index $\boldsymbol{\mathcal{I}}_x = \{i_1^x, i_2^x, ..., i_n^x\}$ of x is generated by using a discretization function $\mathcal{D}_b(\boldsymbol{\mathcal{S}}_x)$. In its simplest form, the discretization function applies thresholds on the match scores to convert them to a discrete domain. The choice of b, denoting the maximum number of output values of the discretization function, depends on the characteristics of the match scores generated by the matcher. For example, if the scores generated when comparing an arbitrary image with multiple impressions of a fixed impostor finger are stable, than b can be as large as the number of possible match score values. On the other hand, when these scores exhibit large variation, then a small value of b would be more appropriate.

2.3 Retrieving Fingerprints from the Database

Match score values may vary significantly depending on the quality of the fingerprint impressions. The quality is impacted by the photometric and geometric variations in the images. This variation can cause the index code of a query image to be different from its corresponding model index code. However, the quantization function ensures that a majority of the individual elements constituting these two index codes are the same. During identification, when a potential list of candidate matches has to be determined, the Hamming distance can be used to find model index codes located in the neighborhood of the query index code. The size t of the neighborhood depends on the expected error rate between the query and model index codes, and is estimated empirically. One straightforward implementation of the search process is presented below (note that step 1 adds to the computational overhead of the search process; however, Hamming distances can be computed rapidly thereby improving overall response time, since the number of fingerprint-to-fingerprint comparisons is reduced).

Algorithm for searching the model index codes

Let $\mathcal{I}_q$ be the index code of the query image and $\mathcal{I}_{x_i}$ be the index code of image x_i from the database.

1. Compute the Hamming distances $d_i = D_h(\mathcal{I}_q, \mathcal{I}_{x_i})$ for $i = 1, \ldots, M$.
2. Retrieve all x_i such that $d_i \leq t$.

More efficient algorithms for ordering and searching the model codes can be designed to minimize the number of computations. Clearly, if the model index codes are well scattered in terms of Hamming distance, the number of retrieved fingerprints will be small. In the ideal case, these fingerprints will pertain to the same identity as that of the query index code.

2.4 Properties of the Index Space

To ensure unique index codes and the ability to handle errors in coding, the index space has to be large and sparse. The size of the index space can be increased indefinitely by using a larger n (the number of reference fingerprints). However, a n value that is greater than, for example, 5% of the database will defeat the purpose of indexing. Larger value of b also lead to a larger index space. For example, for $n = 20$ and $b = 2$ the number of possible index codes is $2^{20} \approx 10^6$, while for $b = 3$ this number is $3^{20} \approx 3.5 \times 10^9$. However, the value of b cannot be changed arbitrarily because it depends on the distribution of the impostor match scores (see section 2.2).

The sparsity of the index space can be measured in terms of n, b, and t. The Hamming bound [8] provides a theoretical limit on the maximum number of unique model index codes such that the Hamming distance between an arbitrary query code and its associated model code is at most t, and all other model codes are at a distance greater than t. The Hamming bound is defined as:

$$A_b(n,t) \leq \frac{b^n}{\sum_{i=0}^{t} \binom{n}{i}(b-1)^i} .$$

For $n = 20, b = 3$, and $t = 6$, $A_3(20,6) \leq 1137$. For a database of size $M = 2,000$, if we were to assume that the model index codes achieve the Hamming bound and that no more than two fingerprints have the same index code, then only 0.1% of the database will be retrieved by the search process of the proposed method (this is called *penetration rate*). In practice, such an ideal distribution cannot be achieved. First, the distribution of the index codes created using the proposed method is "random" in nature and, therefore, the probability of achieving the Hamming bound is very small.[3] Second, the variance of the errors is large, which will further decrease the Hamming bound. For example, $A_3(20,10) \leq 10$ resulting in a 10% penetration rate.[4]

3 Experiments

In this section we present the implementation details of the proposed indexing method using ternary index codes (i.e., $b = 3$). Two thresholds, T_1 and T_2 are used to split the domain of the match score into three regions. The values of the thresholds are relative to the mean value, μ, and the standard deviation, σ, of the available impostor scores: $T_1 = \mu - \sigma/2$ and $T_2 = \mu + \sigma/2$ (Figure 3). Therefore, the index code $\boldsymbol{\mathcal{I}}_x = \{i_1^x, i_2^x, ..., i_n^x\}$ is obtained as $i_j^x = \begin{cases} 0, & \text{for } s_j^x < T_1 \\ 1, & \text{for } T_1 \leq s_j^x \leq T_2 \\ 2, & \text{for } s_j^x > T_2 \end{cases}$.
The search process extracts all model index codes whose Hamming distance from the query index code is smaller than a certain value t.

Two fingerprint data sets were used to test the proposed methods. The first data set is the NIST special database 4 (NIST-4) [13], which is the database of choice for evaluating fingerprint classification algorithms. It contains inked fingerprints and exhibits large variations in quality. The database consists of the images of 2,000 fingers with two impressions per finger. The second dataset was taken from the West Virginia University (WVU) multimodal biometric database [4]. A subset of 2400 images corresponding to 800 fingers (three impressions per finger) were used. This dataset contains impressions of the index and thumb

[3] Furthermore, an algorithmic way to design codes that achieve the Hamming bound does not exist [8].

[4] This is assuming that the fingerprints from the database are uniformly distributed over the space of unique model index codes.

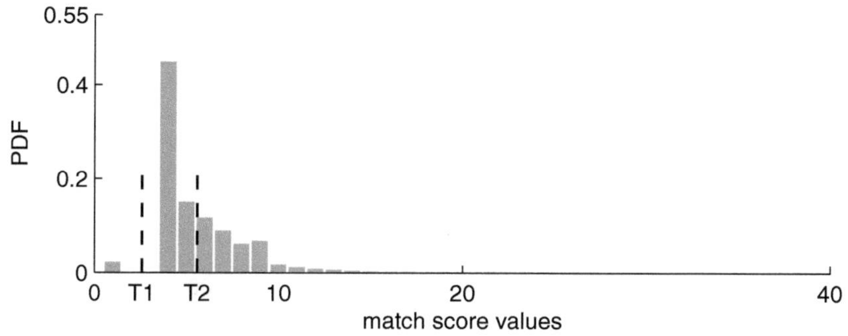

(a)

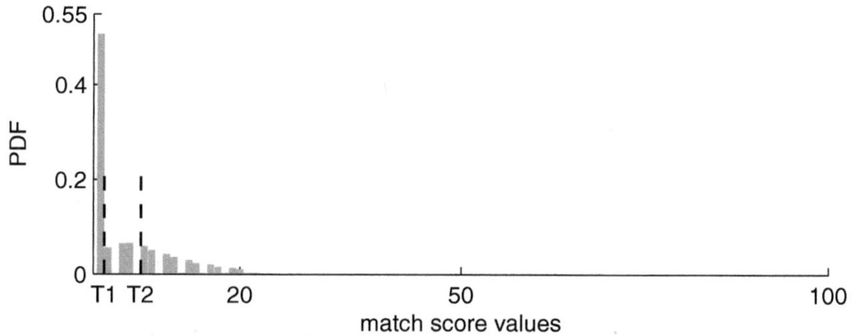

(b)

Fig. 3. Distribution of impostor scores and position of the thresholds. (a) NIST-4 data, $T_1 = 3, T_2 = 5$. (b) WVU data, $T_1 = 2, T_2 = 6$.

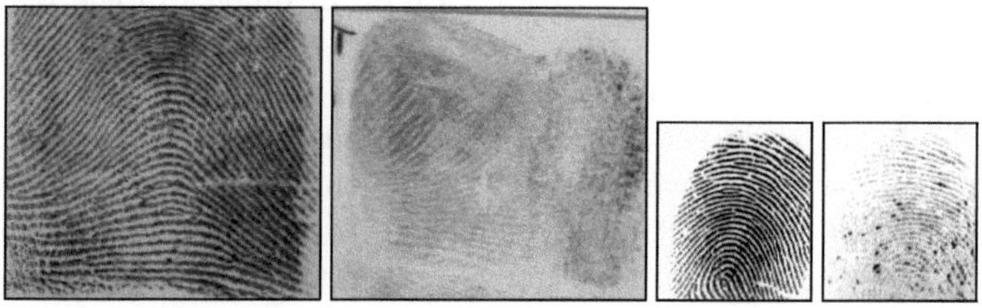

(a) (b) (c) (d)

Fig. 4. Sample images from the two datasets. (a) NIST-4, good quality, (b) NIST-4, poor quality, (c) WVU, good quality, (d) WVU, poor quality.

fingers of both hands of a subject obtained using a SecuGen Hamster scanner. Figure 4 presents examples of fingerprint images from the two datasets. The match scores were generated using the VeriFinger SDK by Neurotechnology. Several values of n were used in our experiments ranging from $n = 50$ to $n = 250$ in order to explore the amount of diversity needed to select "good" reference fingerprints.

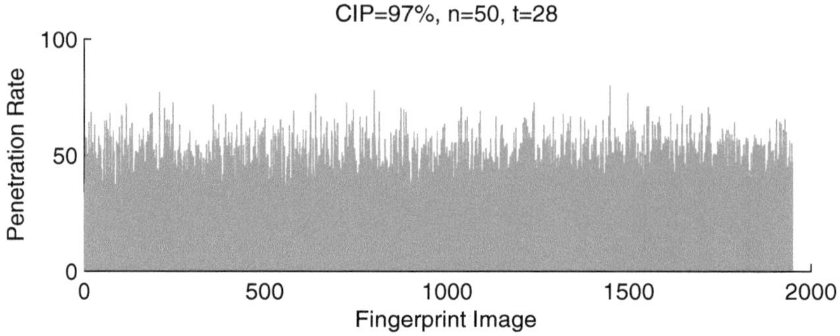

Fig. 5. Penetration rates of all images in the NIST-4 dataset for $n = 50$ and $t = 28$

4 Results

We evaluate the indexing performance in terms of Correct Index Power (CIP) [3] and penetration rate (R_p): $CIP = \frac{N_c}{M}$ and $R_p = \frac{N_a}{M}$, where N_c is the number of images that are correctly indexed, N_a is the number of images that are retrieved during a single identification operation, and M is the number of images in the data set. A query image is said to be correctly indexed, if one of the retrieved model index codes corresponds to the correct identity. In the experiments, one impression of each finger was used to populate the database (training set) while the other impressions of the finger were used to evaluate the performance of the proposed indexing method (test set). Cross-validation was performed by cycling through the multiple impressions of a finger whilst composing the training set. The penetration rate of all images in the test set of one particular experiment involving the NIST-4 database with $n = 50$ and $t = 28$ is shown in Figure 5. Figure 6 presents the CIP as a function of the penetration rate, which was averaged over all test images. For a 50% penetration rate the best CIP that was achieved was ~98% for the NIST-4 dataset (n=50) and ~99% for the WVU dataset (n=250). This is a promising result given that only match scores (rather than complex image features) have been used to facilitate indexing.

After a certain point, the indexing performance on the NIST-4 dataset decreases with increasing values of n. A possible explanation for this result is that the variability of the fingerprints in this dataset is so large that the use of a small set of reference fingerprints creates sparse index codes. In this case, increasing n beyond a certain value cannot add new variations and, therefore, will worsen the sparsity of the index space. Note that longer index codes usually leads to a larger number of errors. In contrast, the overall variability of the fingerprints in the WVU dataset may be small (due to the smaller size of this dataset and/or the nature of the fingerprints in it). Therefore, increasing the number of reference fingerprints (from 50 to 250) adds variability and a sparser index space is created.

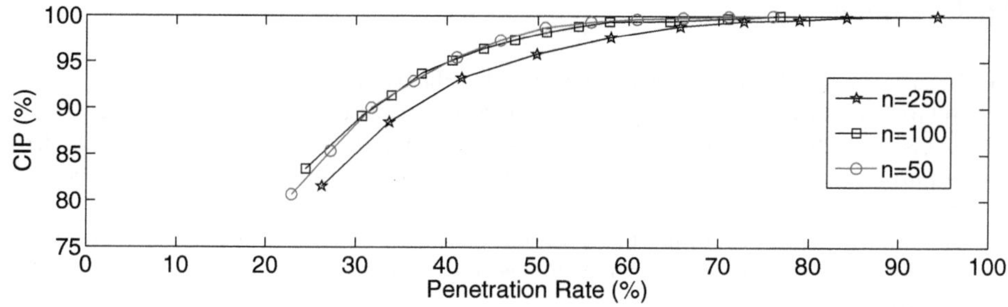

(a)

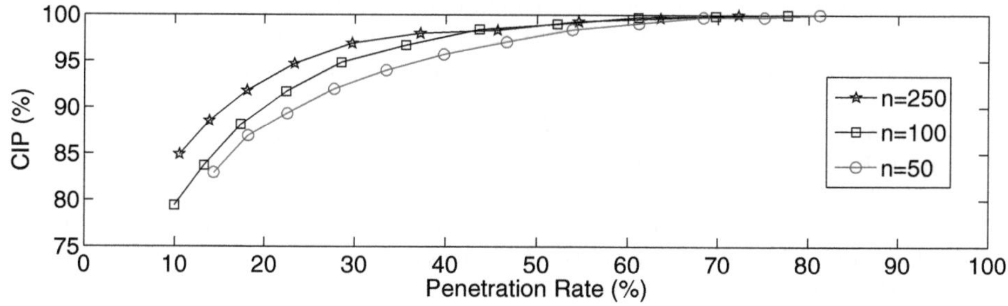

(b)

Fig. 6. Retrieval performance for different values of n. (a) NIST-4 dataset. (b) WVU dataset.

5 Summary and Future Work

In this paper, a new coding scheme for indexing fingerprint databases was presented. The proposed scheme relies on the use of a small set of reference fingerprints to generate a compact index code for labeling an arbitrary fingerprint. To handle noisy queries, candidate model index codes are identified using Hamming distances. The performance of the proposed scheme was evaluated on two fingerprint datasets. Experimental results demonstrate the efficacy of the scheme and its robustness to noisy query prints. However, there is plenty of room for improvement: (a) Estimating appropriate threshold values is critical for minimizing the number of errors in the index code caused by noisy images. We are exploring methods to automatically derive the discretization function based on the match score distributions associated with a matcher. (b) The selection of the reference fingerprints is important for obtaining a sparsely populated index space. Imposing further constraints on the selection mechanism for reference fingerprints may result in better sparsity. (c) The relationship between n and M for efficient indexing can be explored by conducting experiments on larger datasets. Indeed, it may be possible to derive upper bounds on indexing accuracy based on score distributions. We are currently evaluating the proposed scheme on other biometric traits.

References

1. International Biometric Group, Henry Classification System (2003), http://www.biometricgroup.com/HenryFingerprintClassification.pdf
2. Bebis, G., Deaconu, T., Georgiopoulos, M.: Fingerprint Identification Using Delaunay Triangulation. In: International Conference on Information Intelligence and Systems (ICIIS), pp. 452–459 (1999)
3. Bhanu, B., Tan, X.: Fingerprint Indexing Based on Novel Features of Minutiae Triplets. IEEE Trans. Pattern Analysis and Machine Intelligence 25(5), 616–622 (2003)
4. Crihalmeanu, S., Ross, A., Schuckers, S., Hornak, L.: A Protocol for Multibiometric Data Acquisition, Storage and Dissemination. Technical Report, WVU, Lane Department of Computer Science and Electrical Engineering (2007)
5. Feng, J., Cai, A.: Fingerprint Indexing Using Ridge Invariants. In: International Conference Pattern Recognition, vol. 4, pp. 433–436 (2006)
6. Germain, R.S., Califano, A., Colville, S.: Fingerprint Matching Using Transformation Parameter Clustering. IEEE Computational Science & Engineering 4(4), 42–49 (1997)
7. Jain, A., Pankanti, S.: Automated Fingerprint Identification and Imaging Systems. In: Lee, H.C., Gaensslen, R.E. (eds.) Advances in Fingerprint Technology, 2nd edn. CRC Press, Boca Raton (2001)
8. Lin, S., Costello, D.J.: Error Control Coding, 2nd edn. Prentice-Hall, Inc., Upper Saddle River (2004)
9. Lumini, A., Maio, D., Maltoni, D.: Continuous Versus Exclusive Classification for Fingerprint Retrieval. Pattern Recognition Letters 18(10), 1027–1034 (1997)
10. Maeda, T., Matsushita, M., Sasakawa, K.: Identification Algorithm Using a Matching Score Matrix. IEICE Trans. Information and Systems, Special Issue on Biometric Person Authentication, The Institute of Electronics, Information and Communication Engineers 84(7), 819–824 (2001)
11. Maltoni, D., Maio, D., Jain, A.K., Prabhakar, S.: Handbook of Fingerprint Recognition. Springer, New York (2003)
12. Mhatre, A., Palla, S., Chikkerur, S., Govindaraju, V.: Efficient Search and Retrieval in Biometric Databases. In: SPIE Biometric Technology for Human Identification II, pp. 265–273 (March 2005)
13. Watson, C.J., Wilson, C.L.: NIST Special Database 4, Fingerprint database: Users' Guide. U.S. National Institute for Standards and Technology (1992), http://www.nist.gov/data/WebGuide/SD_4/FingerprintDB_4.htm

A Theoretical and Experimental Analysis of Template Co-update in Biometric Verification Systems

Luca Didaci[1,*], Gian Luca Marcialis[2,*], and Fabio Roli[2,*]

[*] University of Cagliari, Italy
[1] Department of Pedagogical and Philosophical Sciences
[2] Department of Electrical and Electronic Engineering
Piazza d'Armi – I-09123 Cagliari (Italy)
{luca.didaci,marcialis,roli}@diee.unica.it

Abstract. Template update in biometric recognition system is aimed to improve the representativeness of available templates in order to make them adaptive to the large intra-class variations characterizing biometrics (e.g. fingerprints and faces). Among others, semi-supervised approaches to template update have been recently proposed. Since the lack of representativeness is due to the impossibility of sampling all possible variations of a given client biometric, these approaches exploit samples submitted during the recognition phase by adding the "highly genuine" ones to the related client gallery. In particular, the template co-update algorithm, which uses the mutual help of two complementary biometric matchers, has shown promising experimental results. However, no theoretical model has been proposed to explain the behaviour of the co-update algorithm and support the experimental results. This is the goal of this paper. Experimental results show the correctness of the proposed theoretical model.

1 Introduction

Biometric-based personal verification systems reached a good maturity degree, both scientifically and technologically. In particular, face and fingerprint are the most widely adopted biometrics [1-2].

However, advancements of the state of the art have been able only partially to solve crucial problems as the dependency of performance of the recognition algorithms from the behavioural and environmental conditions. In many cases, these variations are not easily predictable, because they can be related to short-time or medium-time causes. For example, a scratch on the finger skin or the face, or an unknown expression, or the so-called "aging" effect, which is crucial in performance of face recognition systems [1]. Although many efforts have been made to deal effectively with such problems, a well-founded and definitive solution has not yet been achieved.

With regard to recent works, the state of the art moved the above problems from the input patterns, which are difficulty predictable, to the patterns already stored into the system, which concur to define the client's "gallery": the so-called template(s). The template(s) is a sample, or a set of samples, which is considered "representative"

N. da Vitora Lobo et al. (Eds.): SSPR\&SPR 2008, LNCS 5342, pp. 765–774, 2008.

of the subject appearance. In order to capture such templates, the "enrolment" (or "registration") session is performed, under the supervision of a human expert, with the goal of selecting the most appropriate templates according to pre-defined qualitative parameters (e.g. neutral expression, light equally distributed along the face, fingerprint ridge and valleys visually clear, etc.). Thus, the problem of facing with short-time and medium-time intra-class variations can be formulated as the problem of collecting representative templates of these variations [1]. However, it is widely acknowledged that capturing all possible subject variations during an enrolment session is impossible. On the other hand, the system performance strongly depends on such representativeness.

The first solution proposed to store representative templates has been proposed in [3] in a completely supervised manner. A novel and interesting approach to template selection and update can be referred as "semi-supervised" [4-9]. In this approach, samples collected during the system operations, thus "labelled" neither genuine users nor impostors, are exploited. In particular, samples showing very strong similarity to existing templates are added into the gallery of the related subject. The small variations introduced by these samples can gradually capture larger variations from other ones, so "attracting" samples which should be considered "difficult" when using only existing templates. This iterative process can increase the system robustness, leading to galleries truly representative of the subject appearance [4-9].

In particular, the authors proposed the "co-update" algorithm, which showed that the mutual "help" of two weakly correlated biometrics (e.g., fingerprints and faces) can increase the gallery representativeness more quickly than systems adopting only one biometric [7-8], thus improving the overall performance. Despite of these promising results, the co-update algorithm still lacks of a theoretical framework able to explain the behaviour of the algorithm, and, in particular, which are the "hidden" parameters that rule such behaviour. To this aim, we investigate the co-update algorithm by proposing a statistical framework for modelling the gallery size increase. The proposed model starts from hypotheses which fit the practical scenarios involving fingerprint and face biometrics. The aim is to give a first analytical framework able to point out the relationship between the amount of unlabelled data collected and the error rate reduction required for a certain application.

Experimental results on a chimerical data set show the reliability of the framework by comparing the gallery sizes predicted by the model with the gallery sizes obtained on test data.

The paper is organized as follows. Section 2 describes the proposed model. Section 3 reports the experimental results which validate the model. Section 4 concludes the paper.

2 A Theoretical Framework for Template Co-update

2.1 The Template Co-update Algorithm

The key idea behind the biometric co-updating is the exploitation of the mutual complementary performance of multi-modal biometrics [7]. In fact, the "complementary" performances of biometric matchers using distinct biometric traits, for example, face and fingerprint, are one of the fundamental motivations for multi-modal biometrics [10]. Intuitively, each recognizer is expected to assign correct labels to certain input data which are difficult for the other. In the case of co-update, two matchers are used (bi-modal system).

Let us suppose that we have a bi-modal biometric system made up of a face and a fingerprint recognition system, and a small set D_l of biometric data, labelled with the users' identities, acquired during the enrolment session.

The set D_l is used to create the initial template galleries. During the on-line operation of the biometric system, a batch D_u (usually, much larger than D_l) of data is acquired. Each element of D_u is a pair of biometric samples, i.e. face and fingerprint, from the same individual, so each biometric sample of the couple is characterized by the same true identity and by the same claimed identity (it is worth noting that we are not considering the possibility of fraudulent accesses to the system with fake biometrics, that is an open issue left for future works).

During the off-line co-update phase, each matcher is applied to the batch D_u to verify the identity of each couple of biometric samples. A biometric sample is added to the set D_l if and only if the identity of the other biometric sample of the couple has been verified. In other words, a face sample is added to the set D_l if and only if the correspondent fingerprint sample's identity has been verified, and vice-versa.

```
In the enrolment (supervised) session, collect a set D_l of face and fingerprint images.
  A couple of face and fingerprint images is acquired for each user.
Create the fingerprint and face templates using the set D_l
Loop for H iterations:
    Collect a set D_u without supervision. Each element of D_u is a pair of face and
        fingerprint samples {x1, x2} from the same individual
    For each biometry i, i ∈ {0,1}
        Let i be the 'master' biometry and |1-i| be the 'slave' biometry
        For each couple of elements in D_u
            If the matcher 'master' verifies the claimed identity, the sample of the
            'slave' biometry is added to the set D_l
  Update face and fingerprint templates using the augmented labelled set D_l
```

Fig. 1. Co-updating algorithm. The main steps of the algorithm are shown.

It is worth noting that in the co-update process the two biometrics play complementary role. One biometry verifies the claimed identity, assuming the role of supervisor, and allowing the gallery of the other biometry to be augmented. Let us call "master" the biometry that assumes the supervisor's role, and "slave" the biometry whose gallery is augmented thanks to the master biometry. During the co-training process both biometrics assume, alternatively, master and slave roles. Both matchers are re-trained with this augmented data set (e.g., by updating the face and fingerprint template galleries), and the process is repeated a specified number of times.

Intuitively, co-updating is expected to work because each matcher may assign correct labels to certain examples, considered as difficult for the other one. Therefore, each matcher can increase the galleries size with examples which are very informative for the complementary one.

Insertion of new samples without external supervision might introduce incorrectly labelled samples in gallery, which becomes noise and will degrade the performance of the system. For this reason we need to choose "confident examples" using a very high co-update threshold value at which no false matches occur, namely, FAR = 0%. Figure 1 shows the main steps of the algorithm.

Some observations can be made about this version of the algorithm. First of all, it is possible to use a lower threshold, at the cost of introduce into the galleries a certain number of 'impostors' [8]. This is a common problem in all co-update and self-update algorithms [8]. Moreover, the genuine example added to the gallery could be characterized by a low quality, providing no informative contribution to the gallery. This issue will be addressed in future work. The update method used in this work consists into add a new template to the gallery, but more sophisticated methods are possible, i.e. to obtain a new template "fusing" different examples [14].

2.2 The Theoretical Framework

Let n be the number of different templates contained in a gallery, for a given identity. Let N_{TOT} be the maximum number of templates that a gallery can contain. The role, master or slave, of the two biometrics, is highlighted by superscripts M and S.

Some assumptions are made in order to propose a mathematically tractable model. The first assumption regards the finite discrete space used to represent different biometric impressions. Each identity can produce a maximum of $N_{TOT}^{(face)}$ and $N_{TOT}^{(fingerprint)}$ different observations of the face and fingerprint biometrics, respectively. This assumption can be motivated easily in strongly controlled environments, where a small amount of substantial intra-class variations is admitted, but not collectable by the supervised enrolling. The main consequence of this assumption is that a new example acquired by the system does not necessarily lead to a substantial performance improvement, because the new example can be considered already present in the gallery. In the literature, it has been suggested to "fuse" such similar variations in order to obtain a super-template [6, 9].

The second assumption states that all biometric impressions are equally probable (as usually stated in biometric applications), and the two traits are conditionally independent given the identity. This is motivated by the fact that no clear correlation has been found between the face appearance and the related client's fingerprint image.

In the co-update stage, a set D_u of $|D_u|=k$ pairs of biometric samples (i.e. face and fingerprint) are presented to the system. As described in §2.1, the 'master' biometry validates the claimed identity of correspondent samples of the 'slave' identity. Let us call k' the number of samples of the master biometry whose identity has been verified (i.e. is over the updating threshold in practice). Thus, k' is the number of samples of the slave identity that can be added to the slave gallery.

The value k' can be easily computed by considering that each of k biometric samples of the master biometry can be considered as drawn with replacement from a homogeneous population of $N_{TOT}^{(M)}$ elements. The probability that a sample will be verified is $p = 1 - FRR^{(M)}$, where $FRR^{(M)}$ is the False Rejection Rate for the master biometry. The problem can be modeled using a binomial distribution, so the expected value of verified samples is $E[\#\,verified\ samples] = k\,p = k\left(1 - FRR^{(M)}\right)$

Let us assume that for each co-update step the number of verified samples *k'* is equal to its expected value:

$$k' = k\left(1 - FRR^{(M)}\right). \tag{1}$$

It is worth noting that the effectiveness of the master biometry as supervisor towards the 'slave' biometry is related to its FRR value. If FRR=0 (*k'*=*k*) the master biometry is equivalent to a "true" supervisor because it will verify the claimed identity of all 'genuine' samples in D_u. If FRR>0, then only *k'*<*k* samples can be verified, thus only *k'*<*k* samples will be added to the slave gallery.

Since the two biometrics (face and fingerprint) are conditionally independent, from the point of view of the 'slave' biometry each of these *k'* elements can be considered as drawn with replacement from a homogeneous population of $N_{TOT}^{(S)}$ elements.

In order to calculate how many samples provide informative contribution to the gallery we are interested in the expected value of elements that are different among themselves and are not present in gallery.

Let us call *d* the random variable representing the number of elements in *k'* that are different among themselves, and *dn* the random variable representing the number of elements in *k'* that are different and not present in gallery.
The pdf $p_d(d)$ can be modeled using a multinomial distribution:

$$p_d(d) = \binom{N_{TOT}}{d} \sum_{i=0}^{d} (-1)^i \binom{d}{i} \left(\frac{d-i}{N_{TOT}}\right)^{k'}. \tag{2}$$

The conditional pdf $p_{dn/d}\,(dn/d)$ – the probability to have *dn* different samples not present in gallery if *d* samples are different among themselves – can be modeled using an hyper-geometric distribution:

$$p_{dn|d}(dn \mid d) = h\left(dn; N_{TOT}, (N_{TOT} - n), d\right). \tag{3}$$

where N_{TOT} is the number of items in the population, $(N_{TOT} - n)$ the number of items in the population that are classified as 'highly confident genuine users', *d* the number of items in the sample, *dn* the number of items in the sample that are classified as 'highly confident genuine users'. Its expected value is $E[dn] = d\left(N_{TOT} - n\right)/N_{TOT}$.

Recalling (2) and (3), the pdf $p_{dn}(dn)$ can be written as:

$$p_{dn}(dn) = \sum_{d=1}^{k'} p_{dn/d}(dn/d)\,p_d(d) =$$

$$= \begin{cases} \sum_{d=1}^{k'} \left[\binom{N_{TOT} - n}{dn}\binom{n}{d - dn}\right] \left[\sum_{i=0}^{d} (-1)^i \binom{d}{i}\left(\frac{d-i}{N_{TOT}}\right)^{k'}\right] & if\ dn \le d \\ 0 & if\ dn > d \end{cases} \tag{4}$$

The expected value $E[dn]$ represents the expected enlargement of the gallery due to the collection of *k'* elements. Let us assume that for each co-update step the number $\Delta n^{(S)}$ of new and distinct samples added to the slave gallery is about equal to its expected value $E[dn]$, since the correspondent variance var$[dn]$ is very small. Recalling eq. (1):

$$\Delta n^{(S)} \approx E[dn] = \sum_{dn=0}^{k'} \{dn \cdot p_{dn}(dn)\} = \left(1 - \left(1 - \frac{1}{N_{TOT}^{(S)}}\right)^{k\left(1-FRR^{(M)}\right)}\right)\left(N_{TOT}^{(S)} - n^{(S)}\right). \quad (5)$$

Let $n_0^{(S)}$ be the initial size of the gallery, and $n_{(i)}^{(S)}$ the size of the gallery at the *i*-th co-update step. The length of the gallery at the *i*-th step can be written as:

$$\begin{aligned} n_{(0)}^{(S)} &= n_0^{(S)} \\ n_{(i)}^{(S)} &= n_{(i-1)}^{(S)} w + N_{TOT}^{(S)} (1-w) \end{aligned}; \quad w = \left(1 - 1/N_{TOT}^{(S)}\right)^{k\left(1-FRR^{(M)}\right)}. \quad (6)$$

The weight *w* takes into account the effectiveness of the master gallery as a supervisor.

The $FRR^{(M)}$ value determine the system's dynamics, so it is worth to relate its value to the length of the master gallery. In order to model FRR, let m_i be the average number of examples that produce a 'score' over the updating threshold, if compared with a generic sample of the *i*-th client. Therefore, "difficult" clients can be modeled by a low value of m_i, whilst high value of m_i is adopted for "easy" clients. With regard to this value, we can make the following observations. Let $N_{TOT}^{(M)}$ be the size of the user population for the master biometric:

1. If $m_i > \left(N_{TOT}^{(M)} - n^{(M)}\right)$, at least one of these m_i examples is in the 'master' gallery, therefore, the claimed identity of the example *x* will be verified (that is, their matching is above the updating threshold), and $FRR^{(M)}=0$.
2. If $m_i \leq \left(N_{TOT}^{(M)} - n^{(M)}\right)$, the probability that exactly *r* examples among the m_i are in the gallery can be modeled using an hyper-geometric distribution $p(r) = h\left(r; N_{TOT}^{(M)}, m_i, n^{(M)}\right)$. The probability $p(r=0)$ that no examples among the m_i are in the gallery corresponds to the FRR value of the system:

$$FRR^{(M)} = \begin{cases} 0 & \text{if } m_i > \left(N_{TOT}^{(M)} - n^{(M)}\right) \\ h(0; N_{TOT}^{(M)}, m_i, n^{(M)}) & \text{if } m_i \leq \left(N_{TOT}^{(M)} - n^{(M)}\right) \end{cases}. \quad (7)$$

To sum up, eq. (6) allows us to model the gallery size increase during the co-update iterations, whilst eq. (7) models the FRR of the master biometrics. These relationships can be used to predict the behaviour of co-update. In the next Section the experimental validation of the proposed model by eqs. (6-7) is done by co-updating a face and a fingerprint matcher on real data.

3 Experimental Results

3.1 The Data Set

The data set adopted consists of 42 individuals composed of 20 face and fingerprint images for each individual, by keeping in mind the independence of face and fingerprint traits. The time span of both the collected data sets spans over one year. Forty-two frontal face images with 20 instances representing significant illuminations changes and variations in facial expressions per person were used from the Equinox corporation database [13]. The fingerprint data set has been collected by the authors using Biometrika Fx2000 optical sensor. The images are acquired with variations in pressure, moisture and time interval to represent large intra-class variations. The results are computed on five random coupling of face and fingerprint datasets and are averaged. Whilst minutiae are simply extracted from the fingerprint images after commonly used processing algorithms [2], PCA is computed on the whole data set and applied to face images in order to reduce the size of the overall feature space. 95% of energy is retained according to the current literature [1].

It is worth noting that fingerprint and face data sets are strongly different in terms of environmental conditions: the face one is notably "simpler" than the fingerprint one. We adopted so different data sets in order to show the effect of intra-class variations on the model prediction ability.

3.2 Experimental Protocol

In order to investigate the correctness of the proposed model, we first implemented a simple bi-modal identification system made up of a PCA-based face matcher and a fingerprint matcher using the "String" matching algorithm ("String" is based on minutiae points). We used the standard versions of these two recognition algorithms [11-12]. Then, we implemented the template co-update algorithm in Figure 1. Both the eigenspace of the PCA-based face matcher and the co-update threshold value at FAR = 0% are computed using the whole dataset, and has not been updated during the co-update process. This approach is common to some template update practices [4-5]. The update of face and fingerprint templates is performed simply by adding new examples to the user's gallery.

In the experiments and in the model m_i has been set as follows. We considered all samples for each client i, and computed m_i as the average number of samples exceeding the updating threshold. Then, we selected, as the initial template in D_l, the image "near" to m_i samples. The rationale for this choice is to exclude outliers from the initial gallery, likewise to what happens in real situations, where the initial template is chosen in completely supervised fashion.

In order to simulate the acquisition of a batch set D_u, several sets of $k=10$ couples – face and fingerprint – of 'genuine' examples are drawn with replacement from a homogeneous population of $N_{TOT}^{(face)} = N_{TOT}^{(fingerprint)} = 20$ samples, and used as batches collected during the system operations. We are aware that adopted database size may not be very well appropriate for the task, but it respects, on average, the size adopted in other template update works reported in literature [4-9].

3.3 Experimental Results

Figures 2-3 compare the gallery sizes predicted by the model (thin lines) with the gallery sizes obtained on test data (bold lines), for fingerprint and face biometrics, respectively. For each iteration of the algorithm depicted in Fig. 1, the size of the galleries is reported, whilst theoretical lines follow eqs. (6-7). Marked lines report gallery sizes averaged over all the identity. Largest and smallest gallery cases, related to clients exhibiting the largest and smallest m_i, are shown using dashed and solid lines.

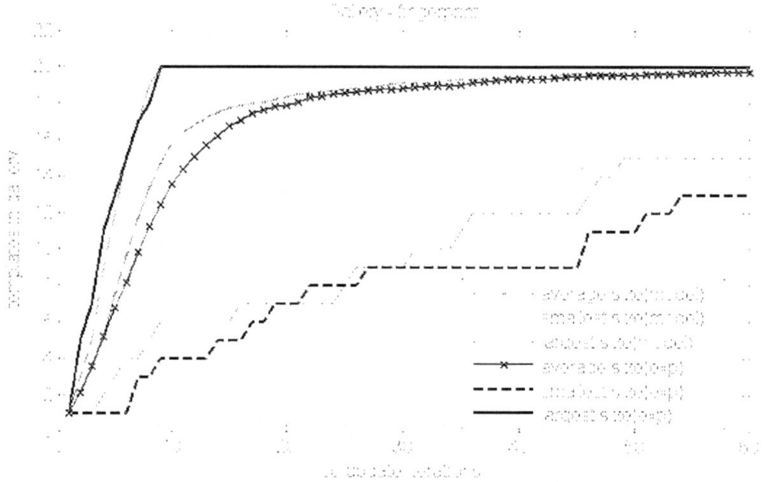

Fig. 2. Fingerprint galleries. Marked lines represent the gallery size averaged across all the identities. Dashed and solid lines represent the smallest and biggest galleries. Experimental trend (bold) and predicted values (thin) are shown.

The following observations can be drawn from Figures 2-3:

- the model is able to predict the experimental performance, with a negligible difference;
- prediction on the face matcher is the less accurate. This is due for sure to the strong intra-class variations which affect the fingerprint images. In other words, m_i exhibits a variance so high that, on average, prediction becomes unstable especially at the initial iterations of the co-update algorithm. On the other hand, it can be noticed that, when focusing on single clients, the prediction is good even for the face matcher (thin and bold dotted lines);
- prediction on the fingerprint matcher is very good, even when evaluating the average behaviour of co-update. This is thanks to the good estimation of FRR of related face matcher, which takes the role of master-supervisor during the iterations. In other words, m_i of clients faces exhibit a variance low enough. This is due to the fact that the face data set is "simpler" than the fingerprint one;

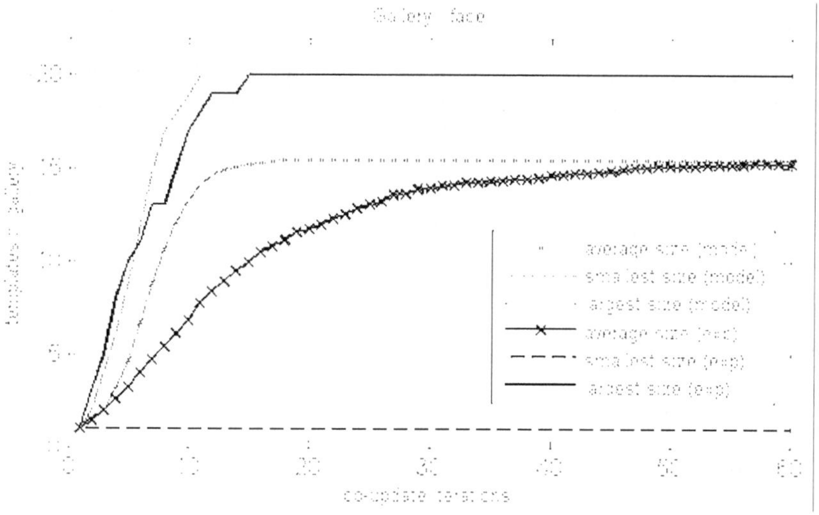

Fig. 3. Face galleries. Marked lines represent the galleries length averaged across all the identities. Dashed and solid lines represent the smallest and biggest galleries. Experimental trend (bold) and predicted values (thin) are shown.

– the saturation of theoretical and experimental curves is obviously due to the form of eqs.(6-7). In particular, the weight w tends to decrease as the gallery size of both matchers increase, depending on the value of their FRR, thus the size of galleries must converge to N_{TOT} if m_i is not zero.

It is worth to point out that for 10 clients (details are omitted for brevity) the 'face' gallery cannot be enlarged by using the co-update algorithm, both in the model and in the experiments, because no fingerprint examples produce a 'score' over the updating threshold (m_i=0). In this case, the reason lies in the high value of the global updating threshold that has been chosen. A user-specific threshold will permit to avoid this problem.

These good results must take into account some issues which have not yet been investigated here: (1) in real environments N_{TOT} is unknown, accordingly an approach to estimate it is necessary. This problem can be overcame by considering N_{TOT} as the number of samples in the batch exhibiting, for each client, a score above the zeroFAR threshold adopted. Worth noting, this assumption does not consider the problem of impostors insertion, a still open issue as remarked in [9]; (2) the same for the estimation of m_i. Therefore, in a future work, we will investigate the impact of N_{TOT} and m_i predicted on a batch of samples including even impostors, on a larger, non chimerical data set.

4 Conclusions

In this paper, we proposed a theoretical framework able to model the behaviour of the co-update algorithm. The co-update method appears very promising for galleries and performance improvement, because it exploits the complementary characteristics of two matchers.

The framework works on the basis of some assumptions which are, in our opinion, reasonable for biometrics in real applications. The reported experimental validation of the model showed that its prediction ability is good even in the case of a strong performance difference among matchers, due to intra-class variations in the data.

Although we pointed out some issues which will be clarified or solved in future works, we believe that this model can be considered as a good starting point for the theoretical investigation of the co-update algorithm behaviour.

References

1. Li, S.Z., Jain, A.K. (eds.): Handbook of face recognition. Springer, Heidelberg (2005)
2. Maltoni, D., Maio, D., Jain, A.K., Prabhakar, S.: Handbook of fingerprint recognition. Springer, Heidelberg (2003)
3. Uludag, U., Ross, A., Jain, A.K.: Biometric template selection and update: a case study in fingerprints. Pattern Recognition 37(7), 1533–1542 (2004)
4. Jiang, X., Ser, W.: Online Fingerprint Template Improvement. IEEE Trans. PAMI 24(8), 1121–1126 (2002)
5. Ryu, C., Hakil, K., Jain, A.: Template adaptation based fingerprint verification. In: Proc. of ICPR, vol. 4, pp. 582–585 (2006)
6. Roli, F., Marcialis, G.L.: Semi-supervised PCA-based face recognition using self-training. In: Yeung, D.-Y., Kwok, J.T., Fred, A., Roli, F., de Ridder, D. (eds.) SSPR 2006 and SPR 2006. LNCS, vol. 4109, pp. 560–568. Springer, Heidelberg (2006)
7. Roli, F., Didaci, L., Marcialis, G.L.: Template co-update in multimodal biometric systems. In: Lee, S.-W., Li, S.Z. (eds.) ICB 2007. LNCS, vol. 4642, pp. 1194–1202. Springer, Heidelberg (2007)
8. Roli, F., Didaci, L., Marcialis, G.L.: Adaptive biometric systems that can improve with use. In: Ratha, N., Govindaraju, V. (eds.) Advances in Biometrics: Sensors, Systems and Algorithms, pp. 447–471. Springer, Heidelberg (2008)
9. Rattani, A., Marcialis, G.L., Roli, F.: Capturing large intra-class variations of biometric data by template co-updating. In: IEEE Workshop on Biometrics, Int. Conf. on Vision and Pattern Recognition CVPR 2008, Anchorage, Alaska, USA (2008) (in press)
10. Ross, A., Nandakumar, K., Jain, A.K.: Handbook of Multibiometrics. Springer, Heidelberg (2006)
11. Jain, A.K., Hong, L., Bolle, R.: On-line Fingerprint Verification. IEEE Transactions on Pattern Analysis and Machine Intelligence 19(4), 302–314 (1997)
12. Turk, M., Pentland, A.: Eigenfaces for Face Recognition. Journal of Cognitive Neuroscience 3(1), 71–86 (1991)
13. `http://www.equinoxsensors.com/products/HID.html`
14. Freni, B., Marcialis, G.L., Roli, F.: Replacement algorithms for fingerprint template update. In: Campilho, A., Kamel, M.S. (eds.) ICIAR 2008. LNCS, vol. 5112. Springer, Heidelberg (in press, 2008)

A New Solution Scheme of Unsupervised Locality Preserving Projection Method for the SSS Problem

Yong Xu[1] and David Zhang[2]

[1] Harbin Institute of Technology, Shenzhen Graduate School
518055 Shenzhen, China
laterfall2@yahoo.com.cn
[2] The Biometrics Research Centre, The Hong Kong Polytechnic University
Kowloon, Hong Kong, China
csdzhang@comp.polyu.hk

Abstract. When locality preserving projection (LPP) method was originally proposed, it takes as the LPP solution the minimum eigenvalue solution of an eigenequation. After that, LPP has been used for image recognition problems such as face recognition. However, almost no researcher realizes that LPP usually encounters several difficulties when applied to the image recognition problem. For example, since image recognition problems are usually small sample size (SSS) problems, the corresponding eigenequation cannot be directly solved. In addition, it seems that even if one can obtain the solution of the eigenequation by using the numerical analysis approach, the obtained conventional LPP solution might produce the 'presentation confusion' problem for samples from different classes, which is disadvantageous for the classification to procedure a high accuracy. In this paper we first thoroughly investigate the characteristics and drawbacks of the conventional LPP solution in the small sample size (SSS) problem in which the sample number is smaller than the data dimension. In order to overcome these drawbacks, we propose a new LPP solution for the SSS problem, which has clear physical meaning and can be directly and easily worked out because it is generated from a non-singular eigenequation. Experimental results the proposed solution scheme can produce a much lower classification error rate than the conventional LPP solution.

Keywords: Locality preserving projection; Feature extraction; Face recognition.

1 Introduction

LPP is well-known as a linear graph embedding method. While LPP transforms different samples into their respective new representations using a same linear transform, it tries its best to preserve the local structure of the samples [1]--[12]. That it, after the LPP transformation, close samples in the original space is still close in the new space. LPP was firstly proposed as an unsupervised method that does not exploit the class-label information. So far a number of improvements of LPP such as kernel-based LPP [10], [13], two-dimensional LPP [14] and orthogonal LPP [15] have

N. da Vitora Lobo et al. (Eds.): SSPR&SPR 2008, LNCS 5342, pp. 775–781, 2008.

been developed. Recently, it has been demonstrated that LPP is theoretically related to other linear dimensionality reduction methods. In other words, these methods can be described as implementations of the linear graph embedding framework with different choices of the weight matrix and a related matrix [16].

Conventional LPP, which takes as the optimal solution of LPP the minimum eigenvalue solution of an eigenequation, will suffer from several difficulties in the SSS problem, in which the data dimension is much larger than the number of samples. The first difficulty is as follows: since the dimension of the sample is larger than the number of the samples, the eigenequation cannot be directly solved due to matrix singularity. An image recognition problem such as face recognition is a typical SSS problem. Note that image recognition covers a wide range of pattern recognition problems; therefore, the study to how to properly apply LPP to the SSS problem is very significant. But to our best knowledge, almost no researcher has worked for this and no satisfactory approach is proposed. It seems that existing LPP-based image recognition applications are usually interested in avoiding the SSS problem rather than finding a good solution to this problem. For example, most of face recognition applications of LPP firstly reduce the size of the face image and then implement the LPP algorithm based on the resized images. Theoretically, to make the LPP algorithm be workable, the dimension of the one-dimensional vector of the resized image should be smaller than the number of the training samples. Consequently, in order to make the dimension of the image not be larger than the sample number to avoid the SSS problem, the original image usually should be resized into a quite low size. Therefore, the process of reducing the size of the face image will case a large quantity of image information loss. This is disadvantageous for recognition to obtain a high accuracy.

The second difficulty of conventional LPP in the SSS problem is that the minimum eigenvalue solution seems not to be the genuine optimal solution for the purpose of locality preserving projection. The reason is as follows. According to the theory of eigenvalue, if a real symmetric matrix is not full rank, it has zero eigenvalues and the number of zero eigenvalues is the same as the result of its dimension subtracted by the rank. As a result, in the SSS problem, even if the eigenequation can be numerically solved by employing the numerical analysis approach, there will be a large number of zero eigenvalues. Conventional LPP will take as transforming axes the eigenvectors corresponding to the zero eigenvalues. Consequently, after conventional LPP transforms samples into new representations using these transforming axes, a sample statistically will have the same representation as its 'neighbors', as demonstrated in the context below. Indeed, in this case, even if the 'neighboring' samples are from different same classes, conventional LPP will still produce the same representation for them, which is harmful for the classification procedure.

In this paper, we propose a new solution scheme for LPP. The new solution scheme does not suffer from the same difficulties as conventional LPP. Moreover, the new solution scheme appears to be subject to the motivation of locality preserving projection and has clear justification. The remainder of the paper is organized as follows. In Section 2 we introduce conventional LPP and its difficulties in the SSS problem. In Section 3 we present our LPP solution scheme and show its advantages. In Section 4 we describe the experimental results. Finally we offer our conclusion in Section 5.

2 Algorithm and Analysis of Unsupervised LPP

LPP was proposed to transform samples into a new space by use a linear transform and to make close samples from the original space be still close in the new space. According to the objective function of LPP, its optimal transforming axis should be the minimum eigenvalue solution z of the following eigenequation:

$$(XDX^T)^{-1}XLX^Tz = \lambda z \tag{1}$$

where $L = D - W$, X stands for the matrix consisting of n training samples $x_1 \quad x_2 \ldots x_n$ i.e. $X = [x_1 \quad x_2 \ldots x_n]$. D is defined as $D_{ii} = \sum_j w_{ij}$. Note that D is a dialog matrix and W is a symmetric matrix and the element w_{ij} of W is defined as follows: if x_j (or x_i) is one of k neighbors of x_i (or x_j), then $w_{ij} = \exp\left(-\|x_i - x_j\|^2 / t\right)$; otherwise, $w_{ij} = 0$. Both L and D are positive semi-definitive matrices. For simplicity of presentation, hereafter we define that $D_1 = XDX^T$ and $L_1 = XLX^T$.

We call the eigenvector corresponding to the minimum eigenvalue of (1) conventional LPP solution and call the solving scheme conventional LPP. For real-world applications, conventional LPP usually firstly sorts the eigenvectors in increasing order of the corresponding eigenvalues and takes the first number of eigenvectors as transforming axes to implement the LPP transform. The dimension of the transform result of the sample is same as the number of the used transforming axes.

Conventional LPP may suffer from the following problem: in the case of the so-called SSS problem where the dimension of the sample is larger than the number of the samples, the eigenequation (1) cannot be directly solved since the matrix D is usually singular. Indeed, it is clear that the rank of D_1 must not be larger than the number of the samples. Consequently, if the dimension of the sample vector is larger than the number of the samples, (1) cannot be solved directly. On the other hand, if one modifies Eq. (1) into the form following form:

$$(D_1 + \mu I)^{-1}L_1 z = \lambda z \tag{2}$$

where μ is a small positive constant, conventional LPP solution can be directly worked out by solving (2). In practice, a similar procedure to solve singular eigenequation has been widely used in numerical computation [17], [18]. However, another problem arises in the above procedure for solving LPP. It is when the obtained minimum eigenvalue solution serves as LPP transforming axes of the SSS problem, the transform result will have poor data representation. This is because in the case of the SSS problem, there are a number of zero eigenvalues and conventional LPP will take as transforming axes the eigenvectors corresponding to zero eigenvalues of the eigenequation. Consequently, as shown in the following theorem, different samples may produce the same transform result.

Theorem 1. In the case where the minimum eigenvalue of the eigenequation is zero, after conventional LPP transforms samples into new representations by using the eigenvectors corresponding to the minimum eigenvalues as transforming axes, a sample will statistically have the same representation as its 'neighbors'.

Proof. Based on (2), we have $z^T L_1 z = \lambda z^T (D_1 + \mu I) z$. If (2) has zero eigenvalues and z_0 is the eigenvector corresponding to a zero eigenvalue, then we can obtain $z_0^T L_1 z_0 = 0$. We know that $\frac{1}{2}\sum_{ij}(y_i - y_j)^2 w_{ij} = z^T L_1 z$ [4]. As a result, if the eigenvector corresponding to a zero eigenvalue is taken as the transforming axis, the transform result will statistically satisfy the following condition $\frac{1}{2}\sum_{ij}(y_i - y_j)^2 w_{ij} = 0$, which implies that in the transform space the 'neighboring' samples must have the same representation. Indeed, because $(D_1 + \mu I)^{-1} L_1$ is positive semi-definitive, no negative eigenvalue exists and the minimum eigenvalue is of course zero. Theorem 1 shows that the conventional LPP solution is not quite subject to the purpose of local structure preservation, which requires that the transform results of 'neighbor samples' be also close rather than the same. Another drawback of conventional LPP is that the 'neighbor samples' from different same classes also statistically have the same representation in the transform space. We call this problem the presentation confusion problem, because it is disadvantageous for pattern classification tasks.

3 A New LPP Solution Scheme

3.1 Formal Description of the New Solution Scheme

In this subsection, we present a new LPP solution scheme that does not suffer from the same difficulties as conventional LPP. Suppose that $\alpha_1, \alpha_2, ..., \alpha_r$ are the eigenvectors corresponding to positive eigenvalues of D_1 and $\alpha_{r+1}, \alpha_{r+2}, ..., \alpha_N$ are the eigenvectors corresponding to the zero eigenvalues. It is easy to prove that valid LPP solution should be from the range space of D_1. Therefore, the new solution scheme is designed as follows. We first define that $R = [\alpha_1 \quad \alpha_2 ... \alpha_r]$. Using R, we respectively transform D_1 and L_1 into the following matrices: $\bar{D} = R^T D_1 R$, $\bar{L} = R^T L_1 R$. We construct the eigenequation $\bar{L} z = \lambda \bar{D} z$ (3) Since $\bar{D}$ is of full rank, we can directly solve this equation. Moreover, according to experimental results, the rank of $\bar{L}$ is usually close to that of $\bar{D}$. As a result, (3) almost has no zero eigenvalue. Therefore, the minimum eigenvalue solution of (3) usually transforms close samples into close rather than the same presentation in the transform space. This means that the new solution scheme is more subject to the goal of LPP than conventional LPP. Let $\beta_1, \beta_2, ..., \beta_r$ denote the eigenvectors corresponding to increasing eigenvalues $\lambda_1, \lambda_2, ..., \lambda_r$ of Eq.(3). If the sample is required to be transformed into an m-dimensional space, we should select $\beta_1, \beta_2, ..., \beta_m$ as the m transforming axes of LPP.

3.2 Transform Procedure

We summarize the transform procedure associated with the new LPP solution scheme as follows. Using the matrix R, we first produce $x_1 = x^T R$ for a sample x. Then we transform x_1 into y, where $y = x_1 \mathrm{H}$, $\mathrm{H} = [\beta_1 \quad \beta_2 ... \beta_m]$.

4 Experimental Design and Results

In this section, we test the new solution scheme (NSS) and the conventional LPP solution. The experiment is performed on the ORL database (http://www.cam-orl.co.uk). In this experiment the first five face images of all the subjects are used as training samples, and the corresponding remaining images are regarded as test samples. Conventional LPP is implemented using three schemes. The first scheme (FS) takes the minimum eigenvalue solutions of (2) as the LPP solution. The second scheme (SS) takes nonzero minimum eigenvalue solutions of (2) as the LPP solution. The third scheme (TS) first resizes each image and then solves Eq. (1) as previous literatures did. To obtain non-singular D_1 and make Eq. (1) be directly solvable, we reduce the size of each image to 7 by 23 by down-sampling [19]. Fig. 1 shows that NSS obtains a low classification error rate with the minimum being 17.5%, whereas the three schemes of conventional LPP have higher error rates. This clearly demonstrates that our solution to unsupervised LPP outperforms the conventional LPP solution. The reason why FS performs badly is that in the SSS problem there are a number of zero eigenvalue serving as the minimum eigenvalues and consequently

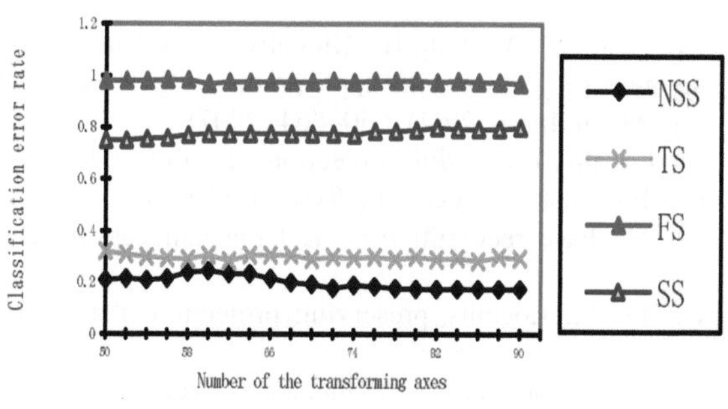

Fig. 1. Classification error rates of NSS, FS, SS and TS. The parameter k in the weight matrix W is set to $k = 1$.

the transform results of neighboring samples are the same even if they belong to different classes. This prevents the classification procedure from obtaining a high accuracy. Though SS and TS almost do not suffer from the zero-eigenvalue problem and outperform FS, their performances are still unsatisfactory. The following are the reasons. In SS, the minimum eigenvalues are quite small so that the transform results of two neighboring samples will be too close even if they are from different classes. Consequently, in SS weak 'presentation confusion' problem also occurs. For TS, the reason why it also does not obtain good result is that resizing images causes too much image information loss.

5 Conclusion

Solid theoretical analysis presented in this paper shows that the proposed LPP solution scheme is quite suitable for the SSS problem, because it does not suffer from the difficulties of conventional LPP such as the matrix singularity problem and the presentation confusion problem. The proposed scheme produces a well defined eigenequation that is directly solvable. Moreover, the obtained solution is more subject to the essence of locality preserving projection. Experimental results sufficiently show the validness and effectiveness of the solution proposed in this paper.

Acknowledgment

This work was partially supported by NSFC under Grant Nos.60602038, 60620160097, the National High-Tech Research and Development Plan of China (863) under Contract No.2007AA01Z195 as well as NSF of Guangdong Province, China under grant No.06300862.

References

1. Belkin, M., Niyogi, P.: Laplacian eigenmaps and spectral techniques for embedding and clustering. In: Proceedings of the Conference on Advances in Neural Information Processing Systems (2001)
2. Yang, J., Zhang, D., Yang, J.-Y., Niu, B.: Globally maximizing, locally minimizing: Unsupervised discriminant projection with applications to face and palm biometrics. IEEE Trans. Pattern Anal. Mach. Intell. 29(4), 650–664 (2007)
3. He, X., Niyogi, P.: Locality preserving projections. In: Proceedings of the Conference on Advances in Nerual Information Processing Systems (2003)
4. He, X., Yan, S., et al.: Face recognition using laplacianfaces. IEEE Trans. PAMI 27(3), 328–340 (2005)
5. Min, W., Lu, K., He, X.: Locality preserving projection. Pattern Recognition J. 37(4), 781–788 (2004)
6. He, X., Cai, D., Min, W.: Statistical and computational analysis of locality preserving projection. In: Proceedings of the 22nd International Conference on Machine Learning, pp. 281–288 (2005)

7. He, X., Yan, S., Hu, Y., Zhang, H.: Learning a Locality Preserving Subspace for Visual Recognition. In: ICCV 2003, pp. 385–393 (2003)
8. Zheng, Z., Yang, F., Tan, W., Jia, J., Yang, J.: Gabor feature-based face recognition using supervised locality preserving projection. Signal Processing 87, 2473–2483 (2007)
9. He, X.: Incremental semi-supervised subspace learning for image retrieval. ACM Multimedia, 2–8 (2004)
10. Cheng, J., Liu, Q., Lu, H., Chen, Y.-W.: Supervised kernel locality preserving projections for face recognition. Neurocomputing 67, 443–449 (2005)
11. Chen, H.-T., Chang, H.-W., Liu, T.-L.: Local discriminant embedding and its variants. In: Proc. CVPR (2005)
12. Yan, S., Xu, D., Zhang, B., Zhang, H.-J., Yang, Q., Lin, S.: Graph embedding and extension: A general framework for dimensionality reduction. IEEE Transactions on PAMI 29(1) (2007)
13. Feng, G., Hu, D., Zhang, D., Zhou, Z.: An alternative formulation of kernel LPP with application to image recognition. Neurocomputing 69(13-15), 1733–1738 (2006)
14. Hu, D., Feng, G., Zhou, Z.: Two-dimensional locality preserving projections (2DLPP) with its application to palmprint recognition. Pattern Recognition 40(1), 339–342 (2007)
15. Cai, D., He, X.: Orthogonal locality preserving indexing. In: SIGIR 2005, pp. 3–10 (2005)
16. Cai, D., He, X., Hu, Y., Han, J., Huang, T.S.: Learning a Spatially Smooth Subspace for Face Recognition. In: CVPR 2007 (2007)
17. Xu, Y., Zhang, D., Jin, Z., Li, M., Yang, J.-Y.: A fast kernel-based nonlinear discriminant analysis for multi-class problems. Pattern Recognition 39(6), 1026–1033 (2006)
18. Xu, Y., Yang, J.-y., Yang, J.: A reformative kernel Fisher discriminant analysis. Pattern Recognition 37(6), 1299–1302 (2004)
19. Xu, Y., Jin, Z.: Down-sampling face images and low-resolution face recognition. In: The third international conference on innovative computing, information and control (in Press)

High Occupancy Vehicle Detection

Alberto J. Pérez-Jiménez, Jose Luis Guardiola, and Juan Carlos Pérez-Cortés

Instituto Tecnológico de Informática*
Universidad Politécnica de Valencia
Camino de Vera s/n, 46022 Valencia, Spain
aperez@disca.upv.es, joguagar@iti.upv.es, jcperez@disca.upv.es

Abstract. High occupancy vehicles lanes (HOV) are highway lanes usually reserved to vehicles carrying at least two persons. They are designed to help move more people though congested areas. In this context, automatic passenger counting systems could be useful to grant access to or to control vehicles in those lanes.

In this work, we propose a real-time passenger detection system based on the analysis of visual images. Each person is detected by mixing the information from different types of classifiers in order to make the detection process faster and more robust.

1 Introduction

Detecting the number of passengers in vehicles can be useful to grant access to or to control vehicles in high occupancy lanes (HOV). Those lanes are usually reserved to vehicles carrying at least two users and they are designed to help move more people though congested areas. Variable highway toll fees depending on passenger number and traffic conditions (HOT) is another possible application.

The control of vehicles in those lanes is usually performed by humans, although some attempts to automate or ease this task have been performed. In [1] a video-based system operated by humans is proposed to assist in this task. A first attempt to automate the problem is proposed in [2]. In that work, authors propose to detect passenger faces by analyzing differences in visible and near infrared images by means of computers. Based on that technology, some commercial systems have appeared recently [3].

In order to work correctly, those visual systems require the occupants to explicitly make themselves visible. For this reason in [4] the use of internal passenger sensors already present in the vehicles (safety belt detection, airbags activation, etc.), is proposed along with transponder technology to count the number of passengers. The major drawback of this method is that it implies vehicle modification.

We propose a passenger detection system based on the analysis of visual images. The passengers are detected by combining the results of different types of classifiers, searching for different features that characterize people (faces, safety

* Work partially supported by the Spanish *Ministerio de Educación y Ciencia* under project DPI2006-15542-C04 and Consolider CSD2007-018.

N. da Vitora Lobo et al. (Eds.): SSPR&SPR 2008, LNCS 5342, pp. 782–789, 2008.

belts, etc.). A cascade of boosted classifiers [5,6] is implemented for fast feature detection. To minimize false detections, more powerful but slower classifiers, as k nearest neighbor (k-NN) [7], are used to filter the cascade classifier results.

This work is organized as follows: section 2 describes the proposed system, section 3 describes the experiments performed to test the system. section 4 shows the results obtained and, finally, a 5 section is included.

2 System Description

2.1 Faces Detector

In a first attempt to count passengers, a face detector classifier was used. This classifier analyzes subwindows of the image at different positions and sizes (see figure 2). The classifier used is in fact a cascade of boosted classifiers based on *haar-like* features [5,6]. As can be seen in figure 1, this classifier is formed by a set of classifiers arranged in a cascaded structure. Each classifier refuses the input data or allows them to pass to the next stage. Input windows not refused in any stage are detected as positives.

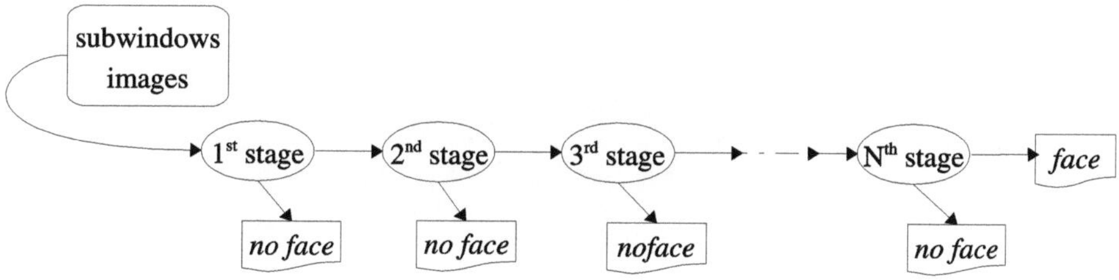

Fig. 1. Cascade classifier structure

At each stage, the classifier is designed to allow a maximum false positive rate (F_p), while having a minimum success rate (T_p). Doing so, the first stages generate very simple classifiers that refuse very quickly those subwindows of the image that are very different from the objects to detect (faces in our case), while more complex and costly last stages deal with the more ambiguous subwindows. This strategy allows to analyze images very fast but, due to the complexity of the captured images (bad contrast, highlights in the windshield, wildly changing illumination, etc., see figure 3), a too high false positive (false detection) rate is obtained.

To reduce the false positive rate, a k nearest neighbor classifier (k-NN) [7] is trained to filter the results of the cascade classifier. This type of classifier is simple to train and can be very powerful, although its computational cost is high, which restricts its use in this case to the analysis of a moderate number of selected hypothesis rather than the whole image.

The system works as follows: the subwindows detected as faces by the cascade classifier are then fed into the k-NN classifier to be classified again as *face* or

Fig. 2. Subwindows of different sizes are analyzed at different positions all over the input image

Fig. 3. Low contrast image with windshield highlights obtained by the acquisition system

false . To do so subwindows are first resized to 20×20 pixels to obtain a 400-dimensional vector representation. In order to simplify the k-NN task, and speed up the classification process, the dimensionality is reduced by means of principal component analysis (PCA) [7]. A fast implementation of the k-NN classifier based on *kdtrees* [8] has been implemented.

Finally, a join process is performed to merge face detections in the same area, since one face can be detected by the cascade classifier at several scales at slightly different positions.

2.2 Safety Belt Detectors

A significant number of occluded faces occurs in the captured images due to windshield highlights, car sunshades, etc. In order to be able to correctly account for the

Fig. 4. Right passenger face is hidden behind the car sunshade but the safety belt is perfectly visible. AOIs are drawn in the image.

real number of passengers, other features can be used. After studying the images, safety belts seem to be a good additional feature to look for (see figure 4).

Due to the simple geometry of the safety belts, the cascade classifier seems specially suitable for detecting them. Two new cascade classifiers were created for left and right safety belts, this simplified the training process and allowed us to filter out false detections that are at the wrong side of the car.

After safety belt detection, a join process is performed as before. In this case, an additional filter is used, taking into account the number of detections in a group, allowing us to easily filter out false detections.

2.3 Areas of Interest

The areas of interest (AOI) are rectangular areas defined to bound the position of each passenger in the image (see figure 4). The AOIs are defined relatively to the windshield location and allow us to filter out false detections in wrong positions, simplifying the merge of the information from faces and safety belt detectors.

The filtering process is done by excluding detected faces that are not completely inside the AOIs. In the case of safety belt detections, a certain amount of overlap is allowed. A passenger detection will be signaled in an AOI if either a face or a safety belt in the right position is detected.

Because the AOIs are defined relative to the windshield, they only have to be redefined when the camera view is changed.

2.4 Context Information

Finally, any information about the capture context that could simplify the problem could be added to the system. For instance, we can consider that the driver will always be in the car, this simplifies the problem since it is only necessary to identify the existence of the right passenger (or the left one some countries).

2.5 Final System

A graphical summary of the whole system can be seen in figure 5.

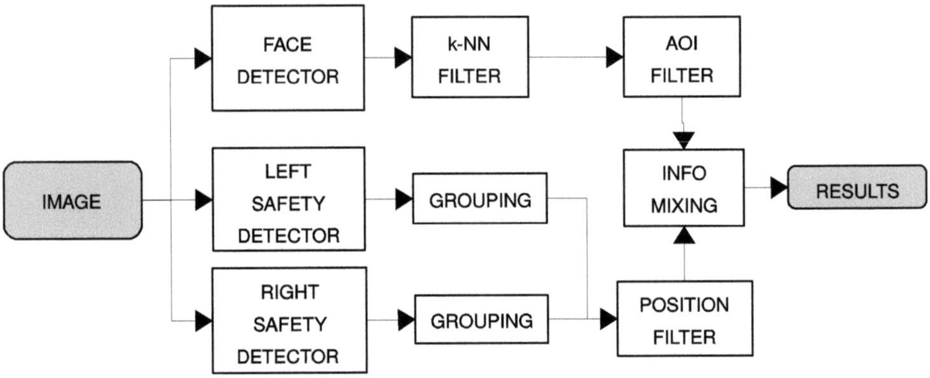

Fig. 5. Architecture of the system for passenger detection

3 Methods and Experiments

To develop and test the system, a set of 3357 images with a 1280×1024 pixel resolution was employed. The images were taken from a highway toll station using the same capture system throughout several days. This set was split randomly into a training set of 2347 images and a test set of 1210 images. Every image in the training and in test sets was manually labeled to mark the location of the faces and the windshield.

The cascade classifier for face detection was trained using the whole training set. The T_p and the F_P rates for every stage were chosen to be 0.999 and 0.5 respectively, obtaining a final classifier with 19 stages.

The k-NN classifier was trained using the subwindows detected as faces by the cascade classifier. Those subwindows were labeled as *face* or *false* according to the ground truth. In order to obtain a larger number of samples, the last 4 stages of the cascade classifier were not used to obtain the training corpus for the k-NN classifier. As stated in the previous section, each subwindows was rescaled to 20×20 pixels and the dimensionality reduced by means of PCA from 400 to 50. A total of 505000 training vectors were obtained. A value of 13 for the k parameter was estimated to give the best success rate.

For the safety belts detectors, a set of 200 images randomly selected from the training set were used. The images were labeled manually to indicate the safety belt positions. Two cascade classifiers were trained using the values of 0.99 and

0.5 for T_p and F_p respectively. After the training process, a number of 9 and 7 stages were obtained for the right and left safety belt detectors respectively.

After experimentation a group size parameter of 4 is selected for safety belts joining filtering. That is, after safety belts detection, groups are formed and those not having at least 4 members are eliminated. This helps to minimize the number of false detections.

The test set (1280 images) has been used to evaluate the performance of the developed system. In the next section, the obtained results are shown.

4 Results

4.1 Detection Results

Table 1 shows the results obtained over the whole test in different stages of the system development. The *ok* label refers to the percentage of images where the number of passengers was correctly detected, while the label *false detection* refers to the percentage of images where a number of passengers larger than the real one were detected.

Table 1. Passenger detection results obtained over the test set in the different stages of the system

	Faces	**Faces + *k*-NN**	**Faces + *k*-NN + safety belts**
Ok	79.35 %	77.66 %	84.42 %
False detection	12.48 %	1.61 %	1.92 %

We can see how the cascade classifier for faces presents a high false detection rate, this rate is significantly reduced by the *k*-NN classifier, losing only a small percentage of correctly identified images. After adding the safety belt detectors to the system, the *ok* success rate increased significantly without an important increase of the *false detection* percentage.

In many areas, where most vehicles have the wheel at the same side, we can improve the success rate considering that "context information". In table 2 the improvement obtained using this information is shown.

Finally, the detection results using only the safety belt detectors are shown in table 3. As we can see, the success rate is only moderate, but it is important

Table 2. Passenger detection results using context information (driver always present) or not

	No context	**Driver present**
Ok	84.42 %	89.60 %
False detection	1.92 %	1.92 %

Table 3. Passenger detection results based only in the safety belts detector

	Safety belts
Ok	66.27 %
False detection	0.51 %

to recall that safety belts can not be seen in all of test-set images. On the other hand, the *false detection* percentage is really small, and therefore this detector will not introduce a significant number of *false detections* in the final result (as can be seen in table 1).

4.2 Computational Cost

The system takes approximately 5 seconds to analyze a whole image of 1280×1024 pixels resolution in a computer with an Intel Core 2 Duo 2.0Ghz. If the windshield is pre-located (something easily done in most cases), an important reduction of the computational cost is obtained, taking about 0.5 seconds the image process with windshields of size around 600×200 pixels.

A multithreaded implementation was performed to take advantage of new multicore processors. The cascade classifier and the k-NN applications were parallelized using the support provided by the *openCV* libraries [9] and the *Open MultiProcessing* directives provided in the 4.2 GCC compiler.

With the parallelized code the computational cost can be reduced to 60 % in double core systems. Quadruple core systems only attain a reduction of 50 % due probably to the dual port memory bottleneck.

5 Conclusions and Future Works

The system described is able to detect the frontal passengers with a high success rate (89.60 %), while maintaining the false detection percentage relatively low (1.92 %). The system could be used to control vehicles in the HOV with a low probability of allowing unauthorized vehicles (i.e., with only one person) to stay in those lanes.

The use of different cascade classifiers to detect different passenger features has proved to be effective in our problem, maintaining a reasonable computational cost. Finally, combining cascade classifiers with k-NN classifiers can lead to fast and precise classification.

An important percentage of images from the set used show a very low contrast, windshield highlights and other acquisition problems. Improving the quality of images will lead to significantly better results.

A confidence rate measure of the hypothesis obtained for every image could be useful for practical purposes. Having this measure, images over a certain confidence threshold can be processed automatically, while images with a confidence value under the threshold will be supervised by human operators. This way, even if the succes rate is not close to 100%, the system can be used to effectively reduce the installation costs.

In our work the windshield location was indicated manually during the labelling process, a robust automatic system to detect the windshield based on the Hough transform and corner classification is already being tested.

Finally because our method only uses images in the visual range, the required image adquisition equipment presents a lower cost compared to IR-based methods.

References

1. Billheimer, J.W., Kaylor, K., Shade, C.: Use of videotape in hov lane surveillance and enforcemente. Technical Report 6429, California Department of Transportation (1990)
2. Pavlidis, I., Symosek, P., Fritz, B., Bazakos, M., Papanikolopoulos, N.: Automatic detection of vehicle occupants: the imaging problem and its solution. Machine Vision and Applications 11(6), 313–320 (2000)
3. Video-Occupancy, C.: Cyclops, occupancy counter system (2005), `http://www.vehicleoccupancy.com/products.htm`
4. Schijns, S., Matthews, P.: A breakthrough in automated vehicle occupancy monitoring sustems for hov/hot facilities. In: 12th HOV Systems Conference (2005)
5. Viola, P., Jones, M.: An extended set of haar-like features for rapid object detection. In: IEEE ICIP, vol. 1, pp. 900–903 (2002)
6. Lienhart, R., Maydt, J.: Rapid object detection using a boosted cascade of simple features. In: IEEE Conference on Computer Vision and Pattern Recognition, vol. 1, pp. 511–518 (2001)
7. Fukunaga, K.: Statistical Pattern Recognition, 2nd edn. Academic Press, London (1990)
8. Friedman, J., Bentley, J., Finkel, R.: An algorithm for finding best matches in logarithmic expected time. In: ACM Transactions on Mathemathical Software, vol. 3, pp. 209–226 (1977)
9. Davies, B., Bornet, O., Bradki, G.: altera: Open Computer Vision Library. Version 1.0 (2006), `http://sourceforge.net/projects/opencvlibrary/`

Automatic Classification of NMR Spectra by Ensembles of Local Experts*

Kai Lienemann, Thomas Plötz, and Gernot A. Fink

Dortmund University of Technology, Intelligent Systems Group, Germany
{Kai.Lienemann,Thomas.Ploetz,Gernot.Fink}@udo.edu

Abstract. A new approach for the automatic detection of drug-induced organ toxicities based on Nuclear Magnetic Resonance Spectroscopy data from biofluids is presented in this paper. Spectral data from biofluids contain information on the concentration of various substances, but the combination of only a small subset of these cues is putatively useful for classification of new samples. We propose to divide the spectra into several short regions and train classifiers on them, using only a limited amount of information for class discrimination. These *local experts* are combined in an ensemble classification system and the subset of experts for the final classification is optimized automatically. Thus, only local experts for relevant spectral regions are used for the final ensemble classification. The proposed approach has been evaluated on a real data-set from industrial pharmacology, showing an improvement in classification accuracy and indicating relevant spectral regions for classification.

1 Introduction

The early detection of drug-induced adverse effects being toxic for particular organs is pursued in safety pharmacology, a crucial step in industrial drug design. The effect of applied drugs on different (regions of) organs is typically determined by measurements of various pattern of marker enzymes in urine samples in clinical chemistry followed by an interpretation by an expert. Thus, the detection of drug-induced organ toxicities is an expensive and time-consuming process requiring expert knowledge. An automation of this process is highly desirable.

The analysis of the metabolic profile from experimental animals after drug application can be automated by ^{1}H *Nuclear Magnetic Resonance (NMR) spectroscopy* of urine samples. Each peak in the resulting spectrum (cf. figure 1) corresponds to certain molecules' hydrogen atoms with an equal magnetic environment, which is influenced by chemical bonds and surrounding atoms. The final peak position is normalized to an added standard substance and to the measurement frequency (cf. [1]). Since the peak intensity correlates with the molecules'

* Parts of this work have been funded by a grant from Boehringer Ingelheim Pharma GmbH & Co. KG., Genomics group. The authors would like to thank the General Pharmacology Group of the company for providing the sample set.

N. da Vitora Lobo et al. (Eds.): SSPR&SPR 2008, LNCS 5342, pp. 790–800, 2008.

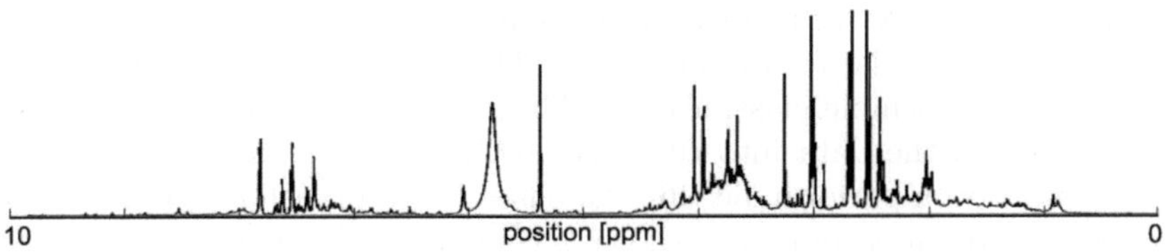

Fig. 1. Exemplary ^{1}H-NMR-spectrum from rat urine

concentrations, a change in the metabolism affects several peak intensities. For a full automation of the whole process, characteristic changes corresponding to a drug-induced organ toxicity have to be determined, facilitating the automatic classification of new samples – a classical pattern recognition task.

While in clinical chemistry the patterns or marker enzymes, with known relevance for the detection of adverse effects, are determined, NMR spectroscopy principally measures all molecules emitting signals of sufficient intensity. Since the underlying molecules from all peaks within an NMR spectrum are not known, background information on the relevance of certain molecules for the detection of adverse effects can not be introduced in the analysis and relevant peaks have to be selected in a data-driven way. This allows for the use of information from molecules not taken into account by clinical chemistry and possibly leading to a deeper understanding of metabolic changes caused by drug-induced organ-toxicities. Therefore, we propose to divide the classification of the whole spectrum into several local classifications in varying scale, incorporating a different amount of local information, and to combine these *local experts* in a multiple classifier system. Thereby, several cues from across the whole spectrum are combined for a final classification and regions not relevant are automatically excluded.

Related work aiming at an automatic classification of NMR spectra will be briefly reviewed in the next section. The proposed classification process for NMR spectra by an ensemble of local experts will be described in section 3. The results of the experimental evaluation are presented in section 4.

2 Related Work

The probably most comprehensive project within the field of Metabolomics, aiming at an automatic analysis and classification of NMR data w.r.t. drug-induced organ toxicities, was the *Consortium for Metabonomic Toxicology (COMET)* [2]. Toxicity prediction has been realized by an approach comparable to mixture density classification, namely *Classification of Unknowns by Density Superposition (CLOUDS)*. Alternative classification approaches using statistical models have also been investigated within and beyond this project (cf. [3,4,5]).

Although multiple classifier systems have been applied successfully in various fields of pattern recognition, they have been hardly used for the detection of drug-induced organ toxicities based on the analysis of NMR spectra. Basically, several base classifiers are trained on a single sample set created by modifying samples, or different data representations. The classifiers' predictions are combined to a final ensemble classification. Thereby, the incorporation of different experts' *views* on the data into the ensemble classification can lead to an improved classification performance [6]. An ensemble approach for classification of NMR spectra presented in [7] has shown a general improvement in classification performance by using Support Vector Machines (SVMs) [8] as base classifier and a modified version of Random Subspace Sampling [9] for ensemble creation. The application of spectral preprocessing and feature extraction methods in different parametrizations is also leading to different *views* on the sample set and aggregation of SVMs trained on these views leads to an improved ensemble classification in comparison to the best single SVM (cf. [10]).

3 Ensembles of Local Experts

Our first experiments on the use of ensemble methods for classification of NMR spectra have shown, that some spectral regions are more important for classification purposes than others (cf. [7]). Incorporating weights in a Random Subspace Sampling procedure for ensemble creation has improved classification performance and apparently relevant spectral regions were indicated by the highest weights. NMR spectra were preprocessed in these experiments by a bucketing procedure [11] for the reduction of dimensionality and peak-shifts, thereby, integrating all intensity values of short spectral regions into single values by summation. However, full compensation of peak shifts can not be guaranteed by bucketing without a considerable decrease in resolution and integration of several peaks into a single bucket value, thereby combining intensity information from different molecules (cf. [12]). Therefore, the analysis should be performed in a sufficiently high resolution in order to be able to infer a single peak's intensity. Thus, it is also possible to identify metabolites showing changes in intensity subject to the degree of an investigated organ toxicity (*Biomarkers*).

We developed a new approach for the automatic classification of high-dimensional NMR spectra by the combination of several classifiers trained on short spectral regions in a multiple classifier system, as illustrated in Fig. 2. Thereby, the classification of the whole spectrum is subdivided into several classification problems using short spectral regions of interest (SROIs) in different scales, incorporating different amounts of local information. An alignment procedure is applied to each SROI for compensation of peak shifts and a classifier is trained. These classifiers serve as *local experts*, modeling class discrimination based on a limited view on the data. Local experts individually leading to a sufficient classification performance are combined in a multiple classifier system. An optimized subset of local experts is used for the final classification, combining information on intensity changes from different spectral regions of the whole spectrum. Details on the selection of SROIs, spectral alignment and combination of a subset of local experts in a multiple classifier system will be given in the following.

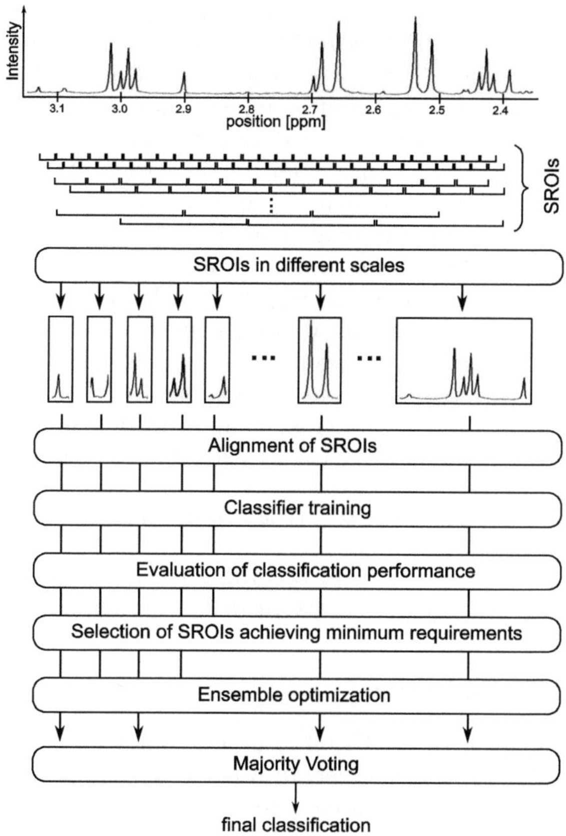

Fig. 2. Classification of NMR spectra by an ensemble of local experts

3.1 Spectral Regions of Interest

The initial selection of SROIs aims at the detection of all spectral regions putatively containing information beneficial for the further classification process. The main information within NMR spectra is principally contained in peak intensities, but the reliable detection of all peaks in a spectrum is a challenging task and their individual relevance for the current classification purposes is not known. Therefore, we propose to use a sliding window approach with overlap and varying scale, detecting SROIs among the whole spectrum including different amounts of local information within each region. Non-informative regions will be excluded from further analysis at a following stage according to their classification results. Thus the relevance of spectral regions is not defined by the pure presence of a peak, but by its local classification performance.

3.2 Alignment of SROIs

A reasonable analysis of NMR spectra in high resolution presumes an alignment procedure, normalizing the exact position of corresponding peaks among all spectra. These are modified by peak-shifts, primarily induced by varying ion concentration and pH value of the sample to be analyzed. Shifts of different peaks predominantly occur independent of each other, thus, a global shift of the whole spectrum is not feasible for peak-shift compensation. Alignment approaches either aim at the detection of corresponding peaks among all spectra (cf. [13]) or modify the exact position of peaks for maximization of similarity among all spectra (cf. [14]). Restricting the analysis on short spectral regions simplifies the alignment process due to the very limited amount of independently varying peaks present in each region. Therefore, peak-alignment can be realized by assigning a global shift factor to the particular region of each spectrum. Thus, the shape similarity within a particular SROI across all spectra is maximized. Maximum similarity is described within the proposed alignment procedure based on the minimization of the spectrum's reconstruction error ε. This similarity measure is defined according to the first principal component $\boldsymbol{\alpha}$ of a PCA model and the mean vector $\boldsymbol{\mu}$ determined on the data set excluding the spectrum $\boldsymbol{x}$.

$$\varepsilon = \| \boldsymbol{x} - \boldsymbol{\mu} - (((\boldsymbol{x} - \boldsymbol{\mu})\boldsymbol{\alpha}^T)\boldsymbol{\alpha}) \|$$

Thereby, the alignment is performed according to the shape of all spectra and is not dependent on peak-detection algorithms or a reference spectrum.

3.3 Ensemble Optimization

The initial determination of SROIs (cf. 3.1) selects numerous regions among the spectra in a non data-driven manner. In order to reduce the complexity of the subsequent ensemble optimization method, SROIs not individually leading to a sufficient classification accuracy are excluded from the further process. Subsequently, the predictions of all classifiers selected for further processing have to be aggregated to a final classification.

The most prominent and intuitive method for classifier aggregation is **majority voting**, classifying samples into the class with the maximum number of votes from all base classifiers. Furthermore, **decision templates** proposed by Kuncheva [15] are an alternative method for the combination of continuous-valued predictions of an ensemble of classifiers. Thereby, every classifiers' support for every class is organized in a decision profile matrix for every training sample. Subsequently, class-specific decision templates are obtained by averaging over all decision profile matrices of the corresponding training samples. Test samples are classified according to the class corresponding to the decision template with minimum distance to the sample's decision profile matrix.

The aforementioned combination strategies respect all classifiers' predictions within the ensemble for final classification. However, in case of redundancy or classifiers performing poorly within the ensemble, the selection of a subset of all available classifiers and the use of majority voting on their predictions can improve classification performance in the final ensemble. Especially in the proposed approach, the combination of certain local experts putatively leads to an improved classification performance due to the combination of several cues on relevant peak-intensity changes and exclusion of non-informative regions. This combination of local experts for ensemble optimization has to be determined automatically and two different approaches will be presented in the following.

Sequential optimization is a feasible approach for the determination of an expert selection by iteratively including / excluding the best / worst expert in an ensemble until all experts are included in or excluded from the ensemble. The optimal number of classifiers is used for final classification. The best individual expert is used as starting point in *best selection* and at every iteration the next best expert is included in the ensemble. Instead of respecting only the individual experts' classification performance within every iteration, the increase of ensemble classification performance is used in *forward selection* as criterion for the selection of the expert to be added to the ensemble. This procedure is modified in *backward selection* by starting with an ensemble of all available experts and excluding at every iteration the expert, leading to the best ensemble classification performance if removed from the ensemble.

Widely used approaches for solving optimization problems are **genetic algorithms** (cf. [16]), which can easily be adapted to the problem of expert selection for ensemble classification. Each individual within a population encodes an expert selection as a string of zeros and ones, representing the inclusion or exclusion of every expert for the final ensemble classification by majority voting. The initial population is initialized randomly and each individual is selected for the next generation with a probability relative to the ensemble classification performance of its encoded ensemble selection. The genetic operations mutation and cross-over can be applied by randomly including or excluding experts in the selection and by the exchange of selection information between two individuals. Furthermore, elite selection can be applied, selecting a defined number of best individuals for the next generation without application of genetic operations. The iterative process is stopped after a defined time-limit and the best selection is used for the final ensemble classification.

4 Experimental Evaluation

The effectiveness of the proposed ensemble of local experts for automatic classification of NMR spectra has been investigated based on a real-world set of ^{1}H NMR spectra from pharmaceutical industry. Details on the data-set, evaluation methodology and results are presented in the following.

4.1 Data-Set and Methodology

The basis for an experimental evaluation of the proposed approach was a set of 896 ^{1}H NMR spectra of urine samples from rats treated with pharmaceuticals currently investigated in industrial safety pharmacology. Overall 52 pharmaceuticals were applied, whereby 34 (= 637 samples) has been labeled as non-toxic and 18 (= 259 samples) as toxic by experts' predictions w.r.t. proximal tubule (kidney) toxicity based on literature investigations and histological judgment.

All spectra were initially down-sampled by a bucketing procedure with a bucket-width 0.001 ppm in the spectral region from 0.2 ppm up to 10 ppm, excluding the water and urea peaks from 4.5 ppm up to 6.0 ppm. The selected bucket-width preserves shapes of individual peaks and provides the same discretization for each spectrum. These bucket values serve as data representation for the further process. Additionally, all spectra were scaled to a total sum of 1000 units for compensation of changes in sample concentration (further details on spectra measurements, data treatment and histological judgment are given in [10]).

Training, parameter optimization and final test were performed by means of a five-fold cross-validation and test. Therefore, all substances were grouped according to their targets and indications and divided among 5 sets structurally preserving grouping and focussing on similar ratios of non-toxic and toxic substances among all sets. Three fifth of the sets were used for training (training set), one fifth for parameter optimization (cross-validation set) and the remaining fifth for test (test set) in every possible configuration. The final rates were averaged over the results on the particular five sets. The evaluation criterion for all training and optimization procedures is the *Matthews Correlation Coefficient* [17] - MC (normalized to $[-1 \ldots 1]$), due to its robustness to imbalanced data-sets:

$$MC = \frac{TP \times TN - FP \times FN}{\sqrt{(TP+FN)(TP+FP)(TN+FP)(TN+FN)}}$$

(TP: number of true positive predictions, FP: false positives, TN: true negatives, FN: false negatives). Classification accuracy, specificity and sensitivity will be shown additionally.

The major goal in the current application is the classification of pharmaceuticals at a certain dose as non-toxic or toxic, therefore, predictions of all samples corresponding to a single drug from different collection time-points and gender have to be aggregated to a final classification. Thus, majority voting among samples corresponding to a certain drug at each collection time-point is performed and classified as toxic if the current drug is classified as toxic at one time-point. False predictions caused by biological noise due to different individual responses to the applied pharmaceutical are, thereby, compensated by averaging over different experimental animals and genders.

4.2 Results

The initial determination of SROIs was performed by sliding windows of size 0.025 ppm, 0.05 ppm, 0.1 ppm and 0.2 ppm (25, 50, 100 and 200 bucket values, respectively) and an overlap of 50 % each, resulting in an overall amount of 1 160 SROIs. These SROIs were aligned by the procedure outlined in 3.2 and the classification performance evaluated by SVMs using a radial basis function kernel function [8]. All SROIs not achieving at least a Matthews of 0.5 have been excluded from the further evaluation, resulting in 147 SROIs for the final ensemble evaluation.

Table 1. Evaluation results for different types of base classifiers

Measure	**NN**	***k*NN**	**RF**	**LSVM**	**RSVM**
			cross-validation set		
Accuracy [%]	94.2	92.3	86.5	96.2	**98.1**
Specificity [%]	97.1	**100**	**100**	**100**	**100**
Sensitivity [%]	88.9	77.8	61.1	88.9	**94.4**
MC	0.872	0.834	0.712	0.916	**0.958**
			test set		
Accuracy [%]	82.7	86.5	82.7	86.5	**90.4**
Specificity [%]	85.3	97.1	**100**	97.1	94.1
Sensitivity [%]	77.8	66.7	50	66.7	**83.3**
MC	0.623	0.700	0.629	0.700	**0.785**

We first compared Nearest Neighbor classifier (NN), K Nearest Neighbor classifier (*k*NN) (cf. [18]), Random Forests (RFs) [19], and SVMs with linear (LSVMs) and radial basis kernel function (RSVMs) [8] for their use as base classifier in the ensemble. The optimal number of neighbors used for *k*NN is optimized according to a fix grid from three to 31 in steps of two on each SROI, and the model parameters for LSVMs and RSVMs are determined by a grid-search algorithm. RFs are parametrized dependent on the data dimensionality v, using $\lceil log_2(v) \rceil$ decision trees in the forest and selecting $\lceil \sqrt{v} \rceil$ variables randomly at each node. The evaluation results on the cross-validation and test set, using the previously selected SROIs and ensemble optimization by forward selection are shown in table 1. RSVMs show the best classification performance on the cross-validation and test set, and are used as base classifier in the further process.

A comparison of different ensemble optimization methods' performance for RSVMs as base classifier is shown in table 2. Decision templates (DT) were used in combination with the squared euclidean distance as similarity measure (cf. [15]) and genetic algorithm optimization (GA) was performed with a population size of 100, mutation rate of 0.05, cross-over rate of 0.6 and elite selection of the 10 best individuals. Best selection (select), forward selection (fselect) and backward selection (bselect) were used as described in 3.3 and the optimal number of experts determined. Forward selection performs identical to genetic algorithm optimization on the cross-validation set, but outperforms all optimization strategies on the test set and the corresponding expert selection is used for the final ensemble classification.

The classification performance of the proposed ensemble of local experts is compared to alternative classification approaches in Table 3. In order to apply single RSVM classification on the full spectra, the data-set was preprocessed by a bucketing procedure using a bucket-width of 0.2 ppm, scaled by standard normal variate transformation [20], and features extracted by partial least squares transformation [21] using 15 model components, resulting in 15-dimensional samples.

Table 2. Evaluation results for different ensemble optimization strategies for an ensemble of RSVMs

Measure	Majority	DT	GA	select	fselect	bselect
cross-validation set						
Accuracy [%]	90.4	90.4	**98.1**	94.2	**98.1**	92.3
Specificity [%]	**100**	97.1	**100**	97.1	**100**	**100**
Sensitivity [%]	72.2	77.8	**94.4**	88.9	**94.4**	77.8
MC	0.793	0.786	**0.958**	0.872	**0.958**	0.834
test set						
Accuracy [%]	84.6	82.7	84.6	86.5	**90.4**	84.6
Specificity [%]	**97.1**	**97.1**	94.1	91.2	94.1	**97.1**
Sensitivity [%]	61.1	55.6	66.7	77.8	**83.3**	72.2
MC	0.657	0.613	0.652	0.699	**0.785**	0.743

Table 3. Comparison of single SVM classification, ensemble classification based on variation of data preprocessing, and the proposed ensemble of local experts

Measure	single RSVM	preprocessing ensemble	local experts
cross-validation set			
Accuracy [%]	86.5	86.5	**98.1**
Specificity [%]	88.2	88.2	**100**
Sensitivity [%]	83.3	83.3	**94.4**
MC	0.707	0.707	**0.958**
test set			
Accuracy [%]	86.5	86.5	**88.5**
Specificity [%]	85.3	85.3	**94.1**
Sensitivity [%]	**94.4**	88.9	83.3
MC	0.768	0.719	**0.785**

An ensemble based on variation of data preprocessing is used according to [10], but using RSVMs as base classifier and forward selection for ensemble optimization in order to use the same classifier in all methods compared.

The proposed ensemble of local experts leads to a general improvement in classification performance on the cross-validation set and slightly outperforms single RSVM classification on the test set. While single RSVM classification assigns hard class labels to the samples, ensemble methods provide further information on the classification by the agreement of all experts' predictions on the final classification. Furthermore, the degree of the induced organ toxicity can be derived from the percentage of votes for the toxic class, showing a dose-dependent response. The spectral regions used for final classification can be derived from the subset of local experts, indicating regions of possibly relevant peaks for classification. Thus, an improved classification is achieved and the decision can be further interpreted, which is especially important for applications in drug design.

5 Summary

We presented an ensemble approach for the detection of drug-induced adverse effects based on the classification of NMR spectra. RSVMs are trained on short spectral regions, representing local experts for the discrimination between toxic and non-toxic samples with a limited view on the data. The combination of these local experts is realized by a multiple classifier system and the subset of experts, leading to an improved final ensemble classification, is determined automatically. Thus, new samples are classified based on a combination of classifiers, focussing on a limited amount of spectral information, showing an improvement in classification performance. Furthermore, the analysis of the ensemble classification allows for interpretation of the decision, especially indicating spectral regions of relevant information for classification purposes – a very important aspect in safety pharmacology.

References

1. Freeman, R.: Magnetic resonance in chemistry and medicine. Oxford University Press, New York (2003)
2. Lindon, J.C., et al.: Contemporary issues in toxicology the role of metabonomics in toxicology and its evaluation by the COMET project. Toxicology and Applied Pharmacology 187(3), 137–146 (2003)
3. Holmes, E., et al.: Development of a model for classification of toxin-induced lesions using ^{1}H NMR spectroscopy of urine combined with pattern recognition. NMR in Biomedicine 11(4-5), 235–244 (1998)
4. Beckonert, O., et al.: NMR-based metabonomic toxicity classification: hierarchical cluster analysis and k-nearest-neighbour approaches. Analytica Chimica Acta 490, 3–15 (2003)
5. Fieno, T., Viswanathan, V., Tsoukalas, L.: Neural network methodology for ^{1}H NMR spectroscopy classification. In: ICIIS 1999: Proc. Int. Conf. on Information Intelligence and Systems, pp. 80–85. IEEE Computer Society, Los Alamitos (1999)
6. Kittler, J., Hatef, M., Duin, R.P., Matas, J.: On combining classifiers. IEEE Transactions on Pattern Analysis and Machine Intelligence 20(3), 226–239 (1998)
7. Lienemann, K., Plötz, T., Fink, G.A.: On the application of SVM-Ensembles based on adapted random subspace sampling for automatic classification of NMR data. In: Haindl, M., Kittler, J., Roli, F. (eds.) MCS 2007. LNCS, vol. 4472, pp. 42–51. Springer, Heidelberg (2007)
8. Schölkopf, B., Smola, A.J.: Learning with Kernels. MIT Press, Cambridge (2002)
9. Ho, T.K.: The random subspace method for constructing decision forests. IEEE Transactions on Pattern Analysis and Machine Intelligence 20(8), 832–844 (1998)
10. Lienemann, K., Plötz, T., Pestel, S.: NMR-based urine analysis in rats: Prediction of proximal tubule kidney toxicity and phospholipidosis. Journal of Pharmacological and Toxicological Methods 58(1), 41–49 (2008)
11. Spraul, M., et al.: Automatic reduction of NMR spectroscopic data for statistical and pattern recognition classification of samples. Journal of Pharmaceutical & Biomedical Analysis 12, 1215–1225 (1994)
12. Torgrip, R.J.O., et al.: New methods of data partitioning based on pars peak alignment for improvedmultivariate biomarker/biopattern detection in ^{1}H NMR spectroscopic metabolic profiling of urine. Metabolomics 2(1), 1–19 (2006)

13. Torgrip, R.J.O., Åberg, M., Karlberg, B., Jacobsson, S.P.: Peak alignment using reduced set mapping. Journal of Chemometrics 17, 573–582 (2003)
14. Skov, T., van den Berg, F., Tomasi, G., Bro, R.: Automated alignment of chromatographic data. Journal of Chemometrics 20(11-12), 484–497 (2006)
15. Kuncheva, L., Bezdek, J., Duin, R.: Decision templates for multiple classifier fusion: an experimental comparison. Pattern Recognition 34, 299–314 (2001)
16. Goldberg, D.E.: Genetic Algorithms in Search, Optimization and Machine Learning. Addison-Wesley Longman Publishing Co. (1989)
17. Matthews, B.W.: Comparison of the predicted and observed secondary structure of the T4 phage lysozyme. Biochimica et Biophysica Acta 405, 442–451 (1975)
18. Duda, R.O., Hart, P.E., Stork, D.G.: Pattern Classification, 2nd edn. John Wiley & Sons, Chichester (2001)
19. Breiman, L.: Random forests. Machine Learning 45(1), 5–32 (2001)
20. Barnes, R.J., et al.: Standard normal variate transformation and de-trending of near-infrared diffuse reflectance spectra. Applied Spectroscopy 43(5), 772–777 (1989)
21. Nord, L.I., Kenne, L., Jacobsson, S.: Multivariate analysis of ^{1}H NMR spectra for saponins from quillaja saponaria molina. Anal. Chim. Acta 446, 197–207 (2001)

Bagging, Random Subspace Method and Biding

Satoshi Shirai, Mineichi Kudo, and Atsuyoshi Nakamura

Division of Computer Science
Graduate School of Information Science and Technology
Hokkaido University
Kita-14, Nishi-9, Kita-ku, Sapporo 060-0814, Japan
{satoshi,mine,atsu}@main.ist.hokudai.ac.jp
http://prml.main.ist.hokudai.ac.jp

Abstract. In recent years, many approaches for achieving high performance by combining some classifiers have been proposed. We exploit many random replicates of samples in the bagging, and randomly chosen feature subsets in the random subspace method. In this paper, we introduce a method for selecting both samples and features at the same time and demonstrate the effectiveness of the method. This method includes a parametric bagging and a parametric random subspace method as special cases. In some experiments, this method and the parametric random subspace method showed the best performance.

1 Introduction

Many classifiers have been proposed for pattern recognition, including plug-in Bayes classifiers, nearest neighbors, support vector machines, etc. It is well known that no classifier keeps the best position for every problem, so that we have to choose one of them depending on the problem at hand. When the number of training samples is small, an appropriate classifier with a complexity suitable for the problem size must be chosen. One attractive approach is to combine several classifiers to bring stability over them. This approach is called *classifier fusion* and has been widely studied [1,2,3,4,5]. Each such classifier is called a *component classifier*. This approach aims at decreasing the generalized error by producing many weak classifiers in one way or another from a provided training dataset and by consolidating them in majority voting (the case of classification) or averaging (the case of regressive prediction).

Two methods have been widely used: bagging [1] and random subspace method [4]. In bagging, many different training sets are generated by resampling and fed to a *base classifier* (a designing algorithm). On the other hand, in the random subspace method, many different feature subsets are randomly chosen for producing component classifiers. As a natural extension, simultaneous selection of sample and feature subsets has been studied [6] and has been shown to often outperform one-side selections.

N. da Vitora Lobo et al. (Eds.): SSPR&SPR 2008, LNCS 5342, pp. 801–810, 2008.

The advantageous effect of classifier fusion has been analyzed in different ways. In the case of regressive prediction, it has been revealed that independent, hopefully negatively-correlated, component classifiers are effective. Even in the case of classification, with some different definitions of correlation, use of negatively-correlated component classifiers is known to be effective [6,7,8]. On the other hand, diversity of component classifiers [9,10,11] is often a useful measure to estimate the performance. Typically, higher diversity of component classifiers means better performance in fusion.

In this paper, we focus on resampling rates of bagging and the random subspace method. Examining possible resampling rates in a systematical way is useful for revealing the potential abilities of bagging and the random subspace method in detail. In addition, we propose simultaneous selection of sample and feature subsets with optimized resampling rates.

2 Formulation of Classifier Fusion

Producing different component classifiers from a single training sample set by resampling of samples or by reselection of features or both is typical in classifier fusion. Let us divide a training set into form and contents. Here, the form is denoted by

$$U_{N\times M} = T_N \times F_M,$$

where T_N is the index set of samples, $T_N = \{1, 2, \ldots, N\}$, and F_M is the index set of features, $F_M = \{1, 2, \ldots, M\}$. The contents are specified by a sampling (an observation) function $x : U_{N\times M} \to R$ with $(i, j) \mapsto x_{ij}$. Hereafter, U and the relatives are considered along with the contents $\{x_{ij}\}$ unless no confusion occurs.

2.1 Number of Distinct Elements by Resampling

In some cases, it is better to consider the set of distinct samples instead of the bootstrap sample T^* itself. Indeed, duplication of samples does not cause any change in the rule of nearest neighbors and support vector machines without a soft margin.

When $N\alpha$ $(0 \leq \alpha \leq 1)$ elements are randomly and uniformly selected from N elements with replacement, the expected size $n = E\hat{n}$ of distinct elements is given by

$$n = (1 - (1 - \frac{1}{N})^{N\alpha})N \simeq (1 - e^{-\alpha})N.$$

Thus, $\theta = n/N = 0.63$ for $\alpha = 1$ (bagging case).

2.2 Bagging

In bagging, one resampling selects T_N^* of N samples from T_N of N samples, but some samples in T_N^* are duplicated. The average size n of distinct samples is

$n = 0.63N$. As a result, on average, we have the form $U_{n\times M} = T_n \times F_M$. The bagging method is a majority vote of predicted classes by $\phi(T_N^{*1}), \phi(T_N^{*2}), \dots, \phi(T_N^{*B})$, where ϕ is a base classifier and $T_N^{*1}, T_N^{*2}, \dots, T_N^{*B}$ are the B bootstrapped samples. The corresponding $T_{n_1}^1, T_{n_2}^2, \dots, T_{n_B}^B$ of distinct samples are different in size, but the basic performance for fixed $\theta_S^B = 0.63$ and $n = 0.63N$ can be analyzed.

2.3 Random Subspace Method

On the other hand, in the random subspace method [4], $F_{M/2}^*$ of $M/2$ distinct features are typically selected randomly from F_M of M features. As a result, we have the form $U_{N\times M/2}$ with $\theta_F^R = 0.5$ and $m = 0.5M$. Similarly to bagging, we use B bootstrapped features $F_{M/2}^1, F_{M/2}^2, \dots, F_{M/2}^B$ for designing component classifiers.

2.4 Parametric Bagging and Random Subspace Method

Let us analyze the performance of above two aggregating methods of component classifiers produced by random sampling with/without replacement. To make clear the difference in which direction sampling is made, we use θ_S for samples and θ_F for features. Then we have

$$U_{\theta_S N\times M} = T_{\theta_S N} \times F_M : \textit{(parametric) bagging},$$
$$U_{N\times \theta_F M} = T_N \times F_{\theta_F M} : \textit{(parametric) random subspace method}.$$

The question "what are the best values of θ_S and θ_F ?" naturally arises. To examine the relationship between θ_S (θ_F) and classification performance, we examined 10 real-world datasets. They are UCI datasets [12] that have wore 10 featuresand consist of continuous features only. The classification rate was estimated by 10-fold cross validation. In the following, the bagging is simulated by $\theta_S = 0.63$ and the random subspace method by $\theta_F = 0.5$. In this case, a certain subset of size $n(m)$ is chosen from a set of size $N(M)$ with a probability of $1/\binom{N}{n}$ $(1/\binom{M}{m})$. Here, $\theta_S(\theta_F)$ was changed in the range of 0.1 to 1.0 by step 0.1. The base classifier was the nearest-neighbor and $B = 100$. The results are shown in Table 1 and Table 2. In bagging, the parameter tuning did not greatly improve the performance. However, it did improve the performance when θ_S was low (for example, in **heart-statlog** and **waveform-5000**, the best value of θ_S is 10%). This is an astonishing result. This result implies the possibility of *sparse* sampling. Note that the nearest neighbor does not change its rule if $\theta_S > 0.5$, such as $\theta_S^B = 0.63$ for bagging, because the true nearest neighbor of a given sample appears over a half of $T_n^1, T_n^2, \dots, T_n^B$ with $n = \theta_S N$. The classification rates using the parametric random subspace method were improved at 3% on average from those using the original random subspace method. The optimal value of θ_F^* ranges from 0.2 to 0.7 but is mostly around 0.3. That is, the empirical selection $\theta_F = 0.5$ of a typical random subspace method is a little larger than the best choice.

Table 1. Recognition rates in bagging (Bag) with 1-NN. Here, θ_S^* is the sampling ratio bringing the best performance.

Dataname	#class	#sample	1-NN	Bag($\theta_S^B = 63\%$)	P-Bag($\theta_S^*\%$)	P-Bag/Bag
heart-statlog	2	270	59.62	59.62	68.14 (10)	1.148
ionosphere	2	351	86.92	86.92	87.76 (50)	1.009
optdigits	10	5620	98.75	98.75	98.84 (30)	1.001
page-blocks	5	5473	95.41	95.43	95.88 (30)	1.004
segment	7	2310	96.49	96.49	96.49 (60−100)	1.000
sonar	2	208	83.64	83.64	83.64 (60−100)	1.000
spambase	2	4601	82.41	82.37	82.78 (40)	1.004
vehicle	4	846	65.71	65.71	66.42 (50)	1.010
waveform-5000	3	5000	76.58	76.68	84.52 (10)	1.102
wine	3	178	74.73	74.73	75.26 (30)	1.007
Average			82.03	82.03	83.97 (41.0)	1.028

Table 2. Recognition rates in the random subspace method (RS) with 1-NN.Here, θ_F^* is the sampling ratio bringing the best performance

Dataname	#feature	SVM	RS($\theta_F^R = 50\%$)	P-RS($\theta_F^*\%$)	P-RS/RS
heart-statlog	13	59.62	66.67	73.33 (40)	1.100
ionosphere	34	86.92	90.61	94.02 (20)	1.037
optdigits	64	98.75	99.02	99.05 (70)	1.000
page-blocks	10	95.41	95.52	95.70 (60)	1.002
segment	19	96.49	97.74	97.74 (50)	1.000
sonar	60	83.64	85.07	89.90 (20)	1.056
spambase	57	82.41	91.72	93.21 (30)	1.016
vehicle	18	65.71	69.14	71.14 (20)	1.029
waveform-5000	40	76.58	84.60	84.82 (30)	1.002
wine	13	74.73	88.20	96.07 (30)	1.089
Average		82.03	86.83	89.50 (37.0)	1.033

2.5 Biding

Choice of the best pair $(\theta_S, \theta_F)^*$ is clearly more effective than a single choice of best θ_S^* or θ_F^*. Therefore, we consider simultaneous selection of (θ_S, θ_F) and call the aggregating method using the selected parameter pair "Biding" (abbreviation of BIDirectional aggregatING). In biding, a component classifier has a form

$$U_{\theta_S N \times \theta_F M} = T_{\theta_S N} \times F_{\theta_F M} : biding.$$

Parametric bagging (P-Bag) corresponds to the case of $(\theta_S, 1.0)$ and the parametric random subspace method (P-RS) corresponds to the case of $(1.0, \theta_F)$. The functional relation between performance and (θ_S, θ_F) is shown in Fig.1 concretely for **heart-statlog** and **segment** datasets. From Fig.1, we see the effectiveness of simultaneous selection. We conducted the same experiment with (θ_S, θ_F). The performance is shown in Table 3. Biding is naturally the best among all three

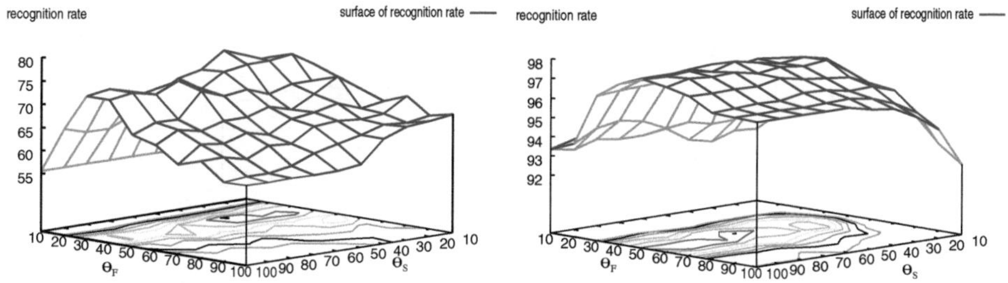

(a) heart-statlog (b) segment

Fig. 1. Recognition rate over (θ_S, θ_F): (a) the best pair $(\theta_S, \theta_F)^* = (40, 30)$ in heart-statlog, (b) the best pair $(\theta_S, \theta_F)^* = (60, 50)$ in segment

Table 3. Comparison of parametric bagging (P-Bag), parametric random subspace (P-RS) method and biding

Dataname	P-Bag(θ_S^*%)	P-RS(θ_F^*%)	Biding(θ_S%, θ_F%)*	Biding/P-Bag	Biding/P-RS
heart-statlog	68.14 (10)	73.33 (40)	78.51 (40,30)	1.152	1.070
ionosphere	87.76 (50)	94.02 (20)	94.02 (100,20)	1.071	1.000
optdigits	98.84 (30)	99.05 (70)	99.06 (60,80)	1.002	1.000
page-blocks	95.88 (30)	95.70 (60)	96.08 (40,90)	1.002	1.004
segment	96.49 (60−100)	97.74 (50)	97.92 (60,50)	1.015	1.002
sonar	83.64 (60−100)	89.90 (20)	89.92 (90,20)	1.075	1.000
spambase	82.78 (40)	93.21 (30)	93.21 (100,30)	1.125	1.000
vehicle	66.42 (50)	71.14 (20)	71.50 (60,30)	1.076	1.005
waveform-5000	84.52 (10)	84.82 (30)	85.84 (20.60)	1.015	1.012
wine	75.26 (30)	96.07 (30)	97.18 (40,20)	1.291	1.011
Average	83.97	89.50	90.32	1.082	1.010

variants including itself. The improvement ratios to the parametric bagging and parametric random subspace method with the optimal parameters were 1.08 and 1.01, respectively.

Table 3 also shows the difference between $(\theta_S, \theta_F)^*$ and (θ_S^*, θ_F^*). From Fig.2, we see that $(\theta_S, \theta_F)^*$ tends to be located at the right-hand or upper-right of (θ_S^*, θ_F^*). This is natural because either θ_S or θ_F is fixed to one when θ_S^* or θ_F^* is calculated. Combining results shown in Table 1, Table 2, and Table 3, biding improved the original (but simulated) bagging at ratio $0.3 - 31.7\%$ and the random subspace method at ratio $0.2 - 2.3\%$.

3 Estimation of $(\theta_S, \theta_F)^*$ by Optimization

For practical usage, we have to estimate the optimal parameters in biding and its two variants. It is natural to use a validation dataset cut from the training set. The problem is the cost of calculation. It costs the number of (θ_S, θ_F) pairs to be

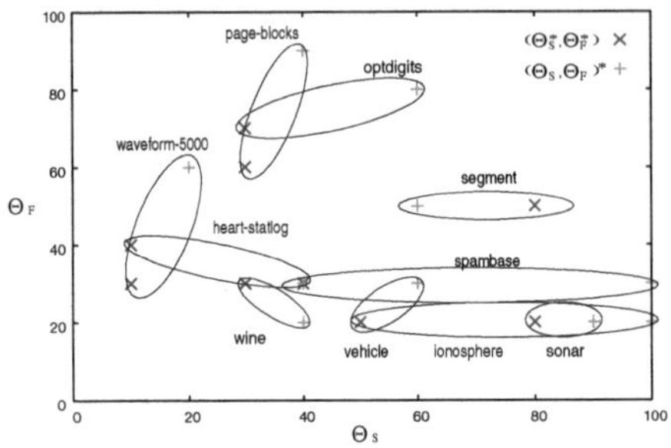

Fig. 2. Point diagram of (θ_S^*, θ_F^*) and $(\theta_S, \theta_F)^*$. An ellipsoid encloses the same dataset.

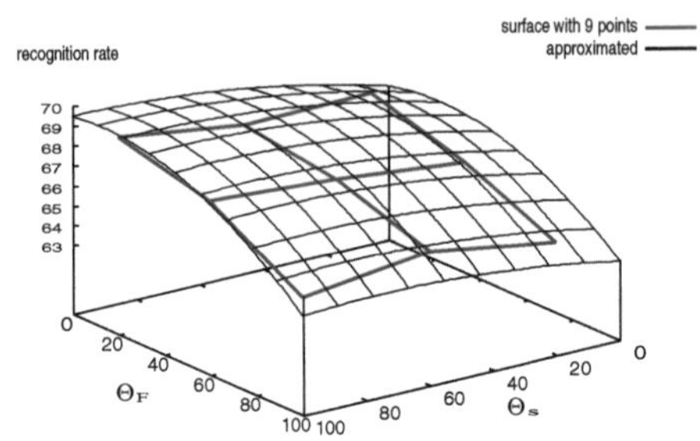

Fig. 3. Quadratic approximation by nine points

examined multiplied by the cost of one-time validation test and by the number B of bootstrapping. It is therefore crucial to reduce the size of the candidate set of (θ_S, θ_F)'s.

To do this, assuming smoothness of the target function (see Fig.1), we approximate it by a quadratic function that is estimated surrounding eight points and the center point (Fig.3). Then we narrow the search region by a factor of $1/4$ for once. By three times of narrowing, we find an estimation of $(\theta_S, \theta_F)^*$ with 27 values of the target function on $[0.2, 1.0] \times [0.2, 1.0]$. For P-Bag and P-RS, we use 20 points in $[0.05, 1.0]$ on each axis.

4 Experiments

We conducted some experiments to demonstrate the effectiveness of P-Bag, P-RS and biding. The base classifier is either of the nearest neighbor and the support vector machine (SVM) with a radial basis kernel, and $B = 100$. The result of

Table 4. Recognition rate by SVM method

Data	SVM	Bag	RS	P-Bag(θ_S^*%)	P-RS(θ_F^*%)	Biding($(\theta_S, \theta_F)^*$)
heart-statlog	62.22	62.22	73.70	59.63 (72.0)	80.37 (28.5)	**81.11** (33.2, 21.0)
ionosphere	93.46	92.60	**94.04**	92.31 (83.5)	92.90 (64.5)	92.61 (58.2, 49.6)
optdigits	96.49	96.21	**99.04**	96.37 (65.5)	**99.04** (50.0)	98.81 (58.3, 38.8)
page-blocks	92.42	92.51	94.06	92.44 (80.5)	**94.46** (33.5)	93.88 (57.2, 30.0)
segment	92.51	92.38	96.74	92.55 (93.5)	96.71 (43.0)	**97.10** (64.8, 35.9)
sonar	81.17	80.69	80.69	80.21 (72.0)	**81.69** (55.5)	76.68 (51.0, 44.7)
spambase	86.61	86.79	**93.74**	86.55 (65.0)	93.68 (43.0)	93.33 (51.8, 41.9)
vehicle	57.20	56.85	69.84	55.55 (62.0)	**73.87** (29.0)	72.21 (34.8, 21.9)
waveform-5000	83.08	85.08	85.78	**86.52** (21.0)	85.60 (45.5)	86.44 (25.0, 60.2)
wine	70.72	69.64	89.87	69.64 (61.5)	96.57 (32.5)	**97.71** (34.5, 24.2)
Average	81.59	81.50	87.75	81.18 (65.7)	**89.49** (42.5)	88.98 (46.9, 36.8)

SVM is shown in Table 4. We also obtained the result of 1-NN, but it is not shown here for economizing the space, but the tendency was similar. We have tried to use SVM in the experiments of the previous section, but gave up due to its high computation cost.

In both 1-NN and SVM, P-RS was best for many cases and biding followed. This is mainly because the number 27 of search points in biding was insufficient for 2-dimensional search space, compared to 20 for 1-dimensional search space in the other two variants. Indeed, it is noted that when (θ_S, θ_F) is chosen close to the optimal value in Table 3, the biding attained the best performance.

5 Analysis of Classifier Fusion in Classification

So far, many studies have been invoked to reveal when classifier fusion works best [1,3,4,5]. The standard ways to connect the generalized error possibly reduced by classifier fusion with the characteristics of component classifiers can be divided into three: 1) bias and variance decomposition, 2) dependency analysis, that is, what degree the component classifiers are correlated to each other, and 3) diversity measures among component classifiers.

Let us first use a disagreement measure. The disagreement measure $Dis_{b,b'}$ between component classifiers ϕ_b and ϕ'_b is calculated by N samples with N_1 outputs of both correct or both incorrect and N_0 outputs of one-sided correct such as

$$Dis_{b,b'} = \frac{N_0}{N_0 + N_1}, \quad Dis = \frac{1}{B(B-1)} \sum_b \sum_{b' \neq b} Dis_{b,b'}. \tag{1}$$

The relationship between improvement rate and diversity is shown in Fig.4. In Fig.4, we notice that there is a positive correlation between them, and biding and random subspace method brought the largest diversity. Note that $Dis_{b,b'}$ is a post-event evaluation, that is, this value is available only with true labels. On the other hand, we can use this measure without such true labels if the number of classes is two.

Next, we state the relationship among above 1), 2) and 3) in a theoretical way. Let us assume that a classifier (strictly speaking, a design algorithm) ϕ produces an estimate $\hat{\eta}(x)$, with a training sample set, for the posterior density $\eta(x) = P(Y=1|X=x)$ in a two-class problem ($Y \in \{1,0\}$). Then the assignment by $\hat{\eta}$ is given by $\hat{y}(x) = \mathrm{sign}\,(\hat{\eta}(x) - 1/2)$. Let $y_B(x)$ be the assignment by the Bayes classifier as $y_B(x) = \mathrm{sign}\,(\eta(x) - 1/2)$.

According to Friedman [13], the bias-variance decomposition at a certain point x for classification is given by a multiplicative form, unlike an additive form for regression, as

$$P(\hat{y}(x) \neq Y) - P(y_B(x) \neq Y) = |2\eta(x) - 1| \mathrm{Erf}\left(\frac{-b(\eta(x), \mathrm{E}\,\hat{\eta}(x))}{\sqrt{\mathrm{Var}\,\hat{\eta}(x)}} \right), \tag{2}$$

where $\mathrm{Erf}\,(x)$ is the error function defined by $\mathrm{Erf}\,(x) = 1/\sqrt{2\pi} \int_x^\infty e^{-u^2/2} du$ and the *boundary bias* b is given by

$$b(\eta(x), \mathrm{E}\,\hat{\eta}(x)) = \mathrm{sign}\,(\eta(x) - 1/2)\,(1/2 - \mathrm{E}\,\hat{\eta}(x)). \tag{3}$$

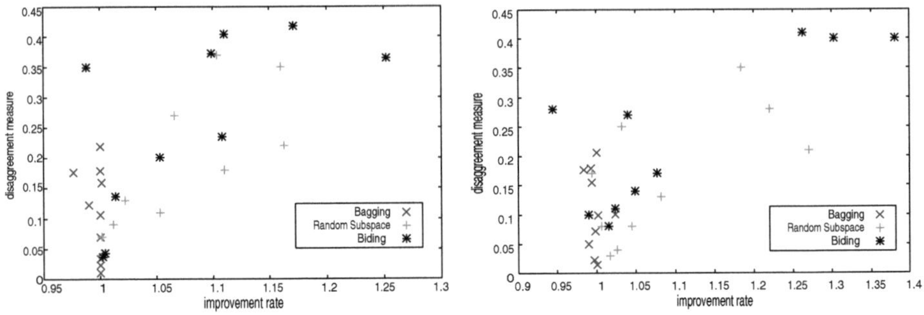

(a) 1-NN (b) SVM

Fig. 4. Relationship between improvement rate and diversity

That is, the generalized error $P(\hat{y}(x) \neq Y)$ at x is expressed by the Bayes error plus a multiplicative decomposition of the boundary bias $b(\eta(x), \mathrm{E}\,\hat{\eta}(x))$ and the square-rooted variance $\sqrt{\mathrm{Var}\,\hat{\eta}(x)}$. Here, the distribution of $\hat{\eta}(x)$ is assumed to be a Gaussian to exploit the error function.

Eq. (2) implies that if the boundary bias is negative, hence the assignment is agreed between the Bayes classifier and ϕ, then a smaller variance is effective for improving the performance, and, if not, a larger variance is effective for making $\mathrm{Erf}(x)$ approach 1/2.

From (2), we have the total generalized error $\mathrm{Err}(\phi)$ as

$$0 \leq \mathrm{Err}(\phi) - \mathrm{Err}_{\mathrm{Bayes}} = \int |2\eta(x) - 1| \mathrm{Erf}\left(\frac{-b(\eta(x), \mathrm{E}\,\hat{\eta}(x))}{\sqrt{\mathrm{Var}\,\hat{\eta}(x)}} \right) p(x) dx. \quad (4)$$

When aggregation is applied to B component classifiers with

$$\hat{y}_1, \hat{y}_2, \ldots, \hat{y}_b, \ldots, \hat{y}_B, \quad \forall b\,, \mathrm{E}\,\hat{y}_b = \mathrm{E}\,\hat{y},$$

we have an aggregated estimate and assignment as

$$\hat{\eta}_A = \frac{1}{B} \sum_b \hat{y}_b, \quad \hat{y}_A = \mathrm{sign}\,(\hat{\eta}_A - 1/2)\,.$$

Then,

$$\mathrm{E}\,\hat{\eta}_A = \mathrm{E}\,\hat{y}, \quad \mathrm{Var}\,\hat{\eta}_A = \mathrm{E} \left\{ \left(\frac{1}{B} \sum_b \hat{y}_b - \mathrm{E}\,\hat{y} \right)^2 \right\}$$

$$= \frac{1}{B} \mathrm{Var}\,\hat{y} + \frac{1}{B^2} \sum_b \sum_{b' \neq b} \mathrm{Cov}\,(\hat{y}_b, \hat{y}_{b'}), \quad (5)$$

where

$$\mathrm{Var}\,\hat{y} = \mathrm{E}\,\hat{y}(1 - \mathrm{E}\,\hat{y}), \quad \mathrm{Cov}\,(\hat{y}_b, \hat{y}_{b'}) = \mathrm{E}\,(\hat{y}_b - \mathrm{E}\,\hat{y})(\hat{y}_{b'} - \mathrm{E}\,\hat{y}). \quad (6)$$

That is, by aggregating, the variance is reduced by $1/B$ and a positive (negative) correlation increases (decreases) the variance.

Here, we notice that

$$\begin{aligned} \mathrm{E}\,(\hat{y}_b - 1/2)(\hat{y}_{b'} - 1/2) &= \mathrm{E}\,(\hat{y}_b - \mathrm{E}\,\hat{y})(\hat{y}_{b'} - \mathrm{E}\,\hat{y}) + (\mathrm{E}\,\hat{y} - 1/2)^2 \\ &= \mathrm{Cov}\,(\hat{y}_b, \hat{y}_{b'}) + (\mathrm{E}\,\hat{y} - 1/2)^2. \end{aligned}$$

Then, we can evaluate Eq. (5) as

$$\begin{aligned} \mathrm{Var}\,\hat{\eta}_A = \frac{1}{B}\mathrm{Var}\,\hat{y} - \frac{1}{B^2}\sum_b \sum_{b' \neq b} \mathrm{E}\,(\hat{y}_b - 1/2)(1/2 - \hat{y}_{b'}) \\ - (1 - \frac{1}{B})(\mathrm{E}\,\hat{y} - 1/2)^2. \end{aligned} \tag{7}$$

That is, the variance is reduced if the degree of disagreement between b and b' is (positively) high or if $\mathrm{E}\,\hat{y}$ is far from $1/2$, or both.

On the other hand, the *disagreement* measure *Dis* (1) is evaluated over $\{x_i\}$ as

$$\begin{aligned} Dis &= \frac{1}{B(B-1)}\sum_b \sum_{b' \neq b} \frac{1}{N}\sum_{i=1}^{N} \{2(\hat{y}_b(x_i) - 1/2)(1/2 - \hat{y}_{b'}(x_i)) + \tfrac{1}{2}\} \\ &\to \frac{2}{B(B-1)}\sum_b \sum_{b' \neq b} \int (\hat{y}_b(x) - 1/2)(1/2 - \hat{y}_{b'}(x))p(x)dx + \frac{1}{2} \\ &\quad as \;\; N \to \infty. \end{aligned} \tag{8}$$

Comparing Eqs. (7) and (8), we notice that *Dis* (1) shows to what degree the variance of the aggregated classifier is reduced on average as a sample estimation. The difference is that the expectation is taken for the training sets in (7) and is taken for $\{x\}$ in (8).

In total, (5), (7) and (8) reveal the relationship among the variance of an aggregating classifier, the correlation between component classifiers, and the disagreement between component classifiers. When the disagreement is high, then the correlation becomes negative and thus the variance of aggregation is reduced.

6 Conclusion

We have proposed a parametric bagging, a parametric random subspace method, and biding as a parent of them for classifier fusion. We confirmed that the parametric random subspace method and biding are both effective in most cases. We also found that 1) sparse sampling (a parametric bagging with a small sampling rate) may be useful for large datasets, 2) traditional random subspace method is reasonable but a little larger in the sampling rate, and 3) simultaneous random selection of samples and features is promising.

Among them, biding is the most powerful in theory. The main drawback of biding is the cost of estimating optimal parameters. In the experiments, a very simple technique for optimization was employed with only 27 function values. A more sophisticated optimization technique would be considered for improving the performance of biding.

References

1. Breiman, L.: Bagging predictors. Machine Learning Journal 2 24, 123–140 (1996)
2. Efron, B.: Bootstrap methods: Another look at the jackknife. Annals of Statistics 7, 1–26 (1979)
3. Freund, Y., Iyer, R., Shapire, E., Singer, Y.: An efficient boosting algorithm for combining preferences. The Jounal of Machine Learning Research 4, 933–969 (2003)
4. Ho, T.: The random subspace method for constructing decision forests. IEEE Trans. on Pattern Analysis and Machine Intelligence 20, 832–844 (1998)
5. Jiang, Y., Ling, J.J., Li, G., Dai, H., Zhou, Z.: Dependency Bagging. LNCS, pp. 491–500 (2005)
6. Skurichina, M., Duin, R.: Bagging, boosting and the random subspace method for linear classifiers. Pattern Analysis and Applications 5, 121–135 (2002)
7. Oh, S.: On the relationship between majority vote accuracy and dependency in multiple classifier systems. Pattern Recognition Letters 24, 359–363 (2003)
8. Ruta, D., Gabrys, B.: A theoretical analysis of the limits of majority voting errors for multiple classifier systems. Pattern Analysis and Applications 5, 333–350 (2002)
9. Brown, G., Wyatt, J., Harris, R., Yao, X.: Diversity creation methods: a survey and categorisation. Information Fusion 6, 5–20 (2005)
10. Kuncheva, L., Whitaker, C.: Measures of Diversity in Classifier Ensembles and Their Relationship with the Ensemble Accuracy. Machine Learning 51, 181–207 (2004)
11. Skalak, D.: The sources of increased accuracy for two proposed boosting algorithms. In: Proc. American Association for Artificial Intelligence, AAAI 1996, Integrating Multiple Learned Models workshop (1996)
12. Blake, C., Merz, C.: UCI repository of machine learning databases (1998), `http://www.ics.uci.edu/~mlearn/mlrepository.html`
13. Friedman, J.: On Bias, Variance, 0/1-Loss, and the Curse-of-Dimensionality. Data Mining and Knowledge Discovery 1, 55–77 (1997)

Use of Structured Pattern Representations for Combining Classifiers

Raisa Socorro[1] and Luisa Micó[2]

[1] Instituto Superior Politécnico Jose Antonio Echevarría
La Habana, Cuba
[2] Departamento de Lenguajes y Sistemas Informáticos
Universidad de Alicante
P.O. box 99, E-03080 Alicante, Spain

Abstract. In Pattern Recognition, there are problems where distinct representations can be obtained for the same pattern, and depending on the type of classifiers (statistical or structural) one type of representation is preferred versus the others. In the last years, different approaches to combining classifiers have been proposed to improve the performance of individual classifiers. However, few works investigated the use of structured pattern representations. In this paper combination of classifiers has been applied using tree pattern representation in combination with strings and vectors for a handwritten character classification task. In order to save computational cost, some proposals based on the use of both embedding structured data and refine and filter framework are provided.

Keywords: combining classifiers, tree representation, string representation, edit distance, k-NN rule.

1 Introduction

Multiple classifier systems have received much attention in the recent years. Its success is in the use of different decisions to obtain a more accurate classification than single classifiers. Classifier ensemble has mainly two advantages: on the one hand, the recognition performance is higher in opposite of single classifiers. On the other hand, the behavior of a multiple classifier system is more robust due to the fact that a final decision is taken on the basis of several decisions (critical individual decision can be avoided) [1]. As stated in a recent work by Duin [2], a consistent set of different classifiers should be used, but also they should be comparable. In the same work different sets of consistent combined classifiers are presented (using different initializations, parameters choice, classification schemes, etc) obtaining the best results when different feature sets have been used.

While regular vector-space represented data are widely used for combining classifiers, structured data have been rarely used. The main reason is the wide range of different classification methods that can be applied when patterns are represented as vectors. However, the number of classifiers to use is drastically

N. da Vitora Lobo et al. (Eds.): SSPR&SPR 2008, LNCS 5342, pp. 811–820, 2008.

reduced for structural pattern representations. For example, in the case of structured data (strings, trees, graphs ...), if a dissimilarity measure is defined, a k nearest neighbor (k-NN) classifier can be applied. Also, recently, the use of kernel functions based on edit distances [3] or the use of embedding for structured data in real vector spaces [4] have enabled the use of alternative statistical methods for classification.

Structural representation of data have been successfully used in many recognition tasks. In fact, it is the more natural representation of data in many applications, as for example in computational biology. The main advantages of using a structural representation is that the number of features is not fixed and the relationship between individual feature components appears explicitly in the representation (in contrast with the feature vector representation).

In this work we analyse how the combination of more than a structural representation can improve the classification error rate. As well as strings and vectors have been used previously in classifier ensembles, trees have not been used ever. The experiments performed show that the performance of the classification improves when trees are used in combination with vectors and strings. Moreover, experiments performed using embedding strings and vectors and a *"filter and refine"* classifier show that a significant speedup can be achieved at the expense of a decrease in the rate of success. In all the experiments, the classifier we adopted is the k-NN. As different representations are used, different dissimilarity measures are needed, associated to the type of data. For that reason, in this work the Euclidean distance (ED), the string-edit-distance (SED) and the tree-edit-distance (TED) have been used.

In the following sections, feature extraction methods for data and ensemble schemes used in this work are described. Then, the results for the ensembles are presented for binary images of handwritten characters from the NIST 3 Database National Institute of Standards and Technology [5]. Finally, the conclusions drawn from the results are discussed, pointing the research to further work lines.

2 Feature Extraction of the Data Set and Distances

Three different feature extraction methods have been used in this work, two of them are structural representations (trees and strings) and the third is a representation in a real vector space. Given a pattern image, the key idea is that each representation gives some type of information about the pattern that it is not given by the others. For example, strings represent the contour of the images, trees represent the skeleton, and each dimension in a n-dimensional vector represents the ratio of black pixels over the total in a region of the image.

2.1 Tree Data Representation

The first step before obtaining a tree data representation of patterns is the use of a thinning algorithm to eliminate redundant information in the image [6].

The image is scanned from left to right starting from the top looking for the first black pixel. This first pixel is marked with label "0". Every tree node has so many descendents as unmarked neighbors has the selected and marked pixel; each of the new branches is extended following their neighborhood until: a) the branch has a maximum size (parameter R of the algorithm [1]), b) is a terminal pixel or c) is an intersection pixel. After that, a new node is assigned to every end of branch. The labels of the nodes (except for the root) have values between "1" and "8" depending on the direction of the branch (see figure 1 on the left). More detailed information about this algorithm can be found in [7].

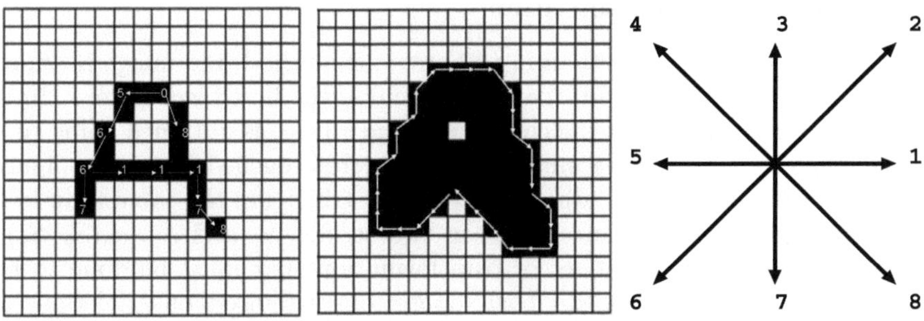

Fig. 1. On the left, example of a tree and a contour string. On the right, symbols used to encode strings and trees

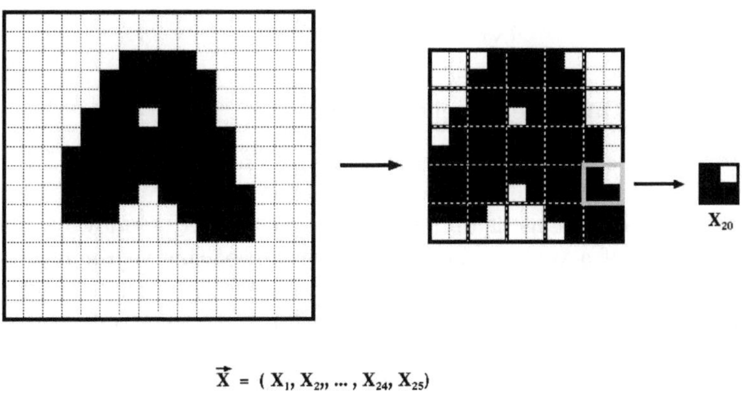

Fig. 2. Example of a vectorial data representation

A general TED is defined by Shasha and Zhang in [8]. Given trees T_1 and T_2, a dynamic programming algorithm with space complexity $O(|T_1| \times |T_2|)$ and $O(|T_1| \times |T_2| \times \min(depth(T_1); leaves(T_1)) \times \min(depth(T_2); leaves(T_2))$ in time

[1] In this work the parameter R=2, that means that the maximum length of each branch of the tree is 2 pixels.

complexity has been used in this work. For the edit basic operations (insertion, deletion and substitution of nodes) in the tree, the following weights were used:

- $w_i = w_d = 1$ (insertion and deletion weights)
- $w_{ij} = min(|i-j|; 8-|i-j|)$ (substitution weights)

The distance $d_t(T_1, T_2)$, is finally normalized (to avoid the size invariance problem) with the maximum size of the two trees [2]:

$$d_N(T_1, T_2) = \frac{d_t(T_1,T_2)}{\max(|T_1|,|T_2|)}$$

2.2 String Data Representation

Following the same strategy than in the previous method, the image is scanned from left to right starting from the top, searching the first black pixel [3]. In this case, the external contour of the characters (the string) is obtained going to the right (in the clockwise sense) until the first pixel is reached again (see figure 1 in the middle). The symbols (directions) in the strings are the same than in the case of trees. The edit distance between two strings S_1 and S_2, $d_s(S_1, S_2)$, is defined as the minimum-cost set of transformation that must be done to turn a string into the other. The edit basic operations are insertions, deletions and substitution of individual symbols in the string. The cost values for the operations are equal that those used in TED. This distance can be computed in time $O(|x| \times |y|)$ [9]. As in trees, the distance is finally normalized with the maximum of the length of the strings to avoid the size invariance problem:

$$d_N(S_1, S_2) = \frac{d_s(S_1,S_2)}{\max(|S_1|,|S_2|)}$$

2.3 Vectorial Data Representation

Firstly, to obtain a size invariance representation of characters, the smallest squared matrix that includes the character in the original image is selected (see Figure 2). The scaled image is divided in n fixed squared windows, where n will be the dimension of a vector, and each dimension of this vector represents the ratio of black pixels in relation to the total number of pixels in each window. After some preliminary experiments, the best results were obtained using 25-dimensional vectors.

3 Classifier Ensembles

Designing a suitable method of decision combinations is a key point for the ensemble's performance. In this paper, two different combination schemes have

[2] Different normalizations have been applied, but in this work only results with the best behavior are shown.

[3] In order to reduce the length of the strings, all the images were scaled from 128×128 to 64×64.

been employed, one of them is based on a confidence voting scheme (*sum rule*) [10] and the second is a scheme based on the *filter-and-refine* framework [11].

3.1 Confidence Voting Approach

In confidence voting methods, voters express the degree of their preference for a candidate. This is done by assigning a value (called the confidence value) to the candidate. The higher the confidence value, the more the candidate is preferred by the voter [1].

Although different confidence methods have been applied (some of them can be found in [12] [13]) in this work the best results have been obtained with the following confidence:

$$C_i(x) = \frac{\sum_i \frac{1}{d_i(x)^2}}{\sum_{j=1}^{k} \frac{1}{d_j(x)^2}} \qquad i : 1 \ldots C$$

where the k-nearest distances to the sample x have been used. In particular $\sum_i \frac{1}{d_i(x)^2}$ is the sum of the inverse squared distances of the nearest neighbors belonging to the class i among the k nearest neighbors.

Kittler et al. performed in [10] an experimental comparison of different classifier combination schemes. They show that the combination rule called *sum rule* outperforms others combination schemes when different pattern representation of data are used. For this reason in this work we have used this combination (even so, others combination schemes were also tested with worst results).

Sum rule: each voter has to give a confidence value for each candidate. Next, all confidence values are added for each candidate and the candidate (the class) with the highest value of the sum is selected.

3.2 Filter and Refine Schemes

In domains where the distance measure is computationally expensive, different approaches to accelerate the search have been proposed. For example, approximate cascade classifiers can been used [14]. In this scheme normally inaccurate but cheap classifiers are considered, followed by the more accurate and expensive, ie, the idea is to use at different levels different classifiers (starting by the less expensive). The number of levels depends on the problem and different possibilities can be build for the same problem.

On the other hand, embedding methods have been recently proposed for speeding up the search [11]. These methods embed the strings or trees (or any structural object for which a distance has been defined) in a vector space, so the distances of the embedded objects approximates the actual distances. Thus, the search can be performed on the embedded objects using, for example, the ED.

Embedding methods can be used in a "filter and refine" scheme to accelerate the search [14]. This method consists on:

– firstly, the sample in the embedding space is used as a *"filter"*, where a small set of candidates are selected (the most promising);
– secondly, a new classification in the original space is performed using only the selected set to *"refine"* the result [4].

In this work, two different two-level "filter and refine" schemes have been defined based on the use of these ideas.

4 Experiments

To perform the experiments, binary images of handwritten segmented uppercase characters from the NIST 3 Database National Institute of Standards and Technology were used (see some examples in figure 3):

a) number of classes in this task: 26
b) types of representation: 25-dimensional vectors, strings and trees
c) distances: ED, SED and TED

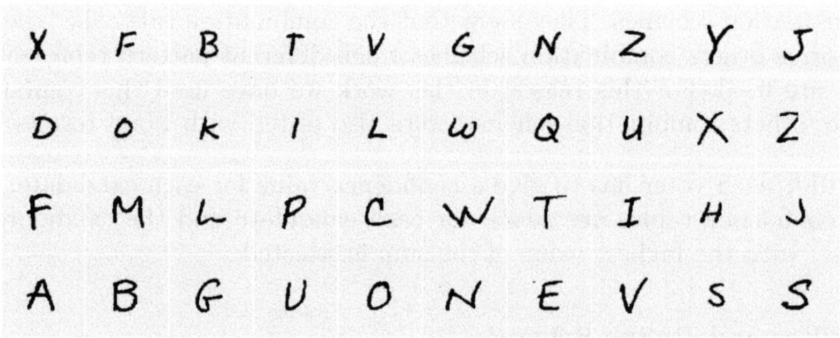

Fig. 3. Some examples of upper handwritten characters from the NIST Database 3

Each experiment were repeated 10 times using training and test sets with 1080 samples on average.

To perform a combination based on the confidence voting scheme, the results of the k-NN rule of individual classifiers with different representations of data has been used as input. In particular, the following combinations have been performed:

C1: strings and vectors
C2: strings and trees
C3: trees and vectors
C4: strings, trees and vectors

[4] In practice, this two steps can be interleaved.

Table 1. Average classification error using the k-NN rule (standard deviation are in brackets) using strings, trees and vectors and four combinations

	individual classifiers			classifier ensembles			
k	strings	trees	vector	C1	C2	C3	C4
1	**12.9(1.1)**	26.4(1.6)	27.4(1.4)	19.6(1.0)	19.4(1.4)	26.5(1.1)	13.3(1.1)
3	13.8(1.3)	27.4(1.8)	28.9(1.2)	11.0(1.0)	11.5(1.3)	16.7(1.3)	**08.9(1.1)**
5	15.9(1.6)	30.4(1.8)	32.4(1.0)	10.5(0.7)	10.9(1.4)	15.8(1.5)	09.1(1.0)
7	18.0(1.6)	33.2(2.0)	35.6(1.0)	10.5(0.6)	11.1(1.5)	16.2(1.7)	09.3(1.1)

Table 1 shows the classification error rate (in %) when the k-NN classifier is used for the three types of representation. The first three columns show the result of the individual classifiers for each representation. In this case, the best results are obtained when the k-NN classifier is used with $k = 1$, and strings outperforms significantly trees and vectors. The four columns on the right of table show the classification error rate for the ensembles. In this case, it can be observed that the classification error rate is influenced by the number of nearest neighbors k, and depends on the combination. Some interesting results are obtained: firstly, depending on the combination, the best results were obtained for k values greater than 1; secondly, any combination of two representation were strings are represented (C1 and C2) improves the results of any individual classifier; thirdly, the best results are obtained using combination C4 with $k = 3$ (using the three representations).

As the use of edit distance with strings or trees is very time consuming, other alternatives have been analyzed trying to reduce it. In particular, experiments with two schemes based on refine and filter classifiers and embedding methods were done.

Embedding strings and trees. Strings and trees were embedded as vectors using the edit distance to a selected number of prototypes. That is, an object (string or tree) can be transformed into a vector by calculating the edit distance to all the selected objects, where each distance represents one vector component. Formally, given a set of objects P and a subset $B = \{b_1, \dots, b_n\} \subseteq P$, the transformation $t_n^B : P \rightarrow \mathbb{R}^n$ is defined as a function where $t_n^B(x) = (d(x, b_1), \dots, d(x, b_n))$, and where $d(x, b_i)$ is the edit distance between the objects x and b_i.

Spillman et al. proposed in [4] some prototype selection methods to use [5]. The method used in this work was the following: the first one, b_1, is randomly selected among P and then, for $i = 2, 3 \dots n$

$$b_i = \text{argmax}_{p \in (P - B_i)} \min_{k=1}^{i-1} d(p, b_k),$$

where $B_i = \{b_1, \dots, b_{i-1}\}$.

For every $p \in P$, the edit distance from these n selected objects are computed, and these n distances are used as the n coordinates of each object p in a new

[5] These methods are similar than the used in some fast nearest search algorithms based on prototype selection [7].

Table 2. Average classification error (standard deviation are in brackets) using different number of reference points in the embedding for $k = 5$

	dimension			
type	**50**	**100**	**150**	**200**
strings	21.2(1.2)	19.8(1.1)	18.9(1.0)	18.7(0.9)
trees	30.0(1.3)	30.1(1.0)	30.3(1.0)	30.0(1.0)

Table 3. Average classification error (standard deviations are in brackets) using the k-NN rule with individual and combined embedded strings, trees and vectors

	individual classifiers			combining classifiers
k	**Emb_string**	**Emb_trees**	**vector**	**C5**
1	22.2(2.2)	31.4(1.4)	27.4(1.4)	18.0(1.2)
3	21.4(1.8)	30.5(1.1)	28.9(1.2)	12.6(1.3)
5	21.2(1.2)	30.0(1.3)	32.4(1.0)	**12.1(1.0)**
7	21.5(1.3)	30.3(1.2)	35.6(1.0)	12.4(1.3)

transformed n-dimensional space. The experiment presented in Table 2 were performed to select the dimensionality in the transformed vector space.

The combination with the best behavior in Table 1 (C4: combining strings, trees and vectors) were repeated using the embedding of trees and strings (for $n = 50$). The results can be shown in Table 3. Except for $k = 1$, for the other values of k we found a better result using the new combination (called C5) than the better individual classifier with the difference that now the classification is faster that in the individual classifiers.

Refine and filter classifiers. Two schemes were applied:

EmbS-OrS: Firstly, a k-NN rule is used with the embedded strings in a 50-dimensional vector space using the ED. Secondly, the 100 nearest samples to the test were used as training set for the second classifier. For this second classifier, the original strings using the SED were used.

Vec-OrS: Firstly, a k-NN rule is used with the 25-dimensional vector representation obtained in section 2.3 using the ED. Secondly, the 200 nearest samples to the test were used as training set for the second classifier. For this second classifier, the original strings using the SED were used.

Table 4 shown results with both schemes. It can be observed that the best results are obtained when the embedding method is used in the first classifier.

In this work, the main goal for using embedding and refine and filter classifiers were the reduction of the time complexity. Table 5 shows this reduction (see the last three columns) were for the combination C5 (using the embedding of strings, trees and the vector representation), slightly reduces the error rate obtained by the contour string, but also reducing up to 8.0% the classification time. However, if our main goal is to reduce the error rate, the combination C4 can be used. In our experiments, we have reduced the error from 12.9% to 8.9%.

Table 4. Average classification error (standard deviations are in brackets) for a filter and refine framework

Second class.	EmbS-OrS	Vec-OrS	Contour string
1	13.2(1.2)	13.6(1.2)	12.9(1.1)
3	12.3(0.9)	14.5(1.4)	13.8(1.3)
5	12.4(1.1)	16.5(1.7)	15.9(1.6)
7	12.9(1.2)	18.5(1.7)	18.0(1.6)

Table 5. Classification error and time for all combinations. Contour string error rate and time is used as baseline.

	Contourn string	C4	C5	EmbS-OrS	Vec-OrS
error (%)	12.9	**8.9**	12.1	12.3	13.6
time (%)	100	158	**8**	13.9	18.37

5 Conclusions

In this work classifier ensembles using structural information of data have been applied in a handwritten character recognition task. Different ensembles have been evaluated, some of them to improve the classification rate and others to reduce the time complexity. The proposed schemes improve the classification rate or the time consuming in relation to the individual classifiers. To reduce the time complexity a scheme based on the use of a refine and filter framework an embedding methods have been used, reducing this time until a 8% maintaining similar error rate that the best individual classifier. The combination of trees, strings and vectors reduce the error rate from 12.9% to 8.9% when a training set of 1080 samples on average is used (but at expense of consuming more time). In future work, new schemes using alternative embedding methods and fast approximated nearest neighbors algorithms will be analyzed to reduce both the error rate and the time complexity.

Acknowledgments

This work was supported by the Spanish project CICyT DPI2006-15542-C04-01 and the IST Programme of the European Community, under the Pascal Network of Excellence, IST-2002-506778.

References

1. Kuncheva, L.: Combining Pattern Classifiers: methods and algorithms. Wiley, Chichester (2004)
2. Duin, R.: The combining classifier: to train or not to train? In: Proceedings of the International Conference on Pattern Recognition ICPR 2002, Quebec Canada, vol. II, pp. 765–770 (2002)

3. Neuhaus, M., Bunke, H.: Edit distance-based kernel functions for structural pattern classification. Pattern Recognition 39(10), 1852–1863 (2006)
4. Spillmann, B., Neuhaus, M., Bunke, H., Pekalska, E., Duin, R.: Transforming strings to vector spaces using prototype selection. In: Yeung, D.-Y., Kwok, J.T., Fred, A., Roli, F., de Ridder, D. (eds.) SSPR 2006 and SPR 2006. LNCS, vol. 4109, pp. 287–296. Springer, Heidelberg (2006)
5. Grother, P.: Handprinted forms and character database. In: NIST Special Database 19 (2001)
6. Carrasco, R., Forcada, M.: A note on the Nagendraprasad-Wnag-Gupta thinning algorithm. Patter Recognition Letters 16, 539–541 (2003)
7. Rico, J., Micó, L.: Comparison of aesa and laesa search algorithms using string and tree edit distances. Pattern Recognition Letters 24(9), 1427–1436 (2003)
8. Zhang, K., Shasha, D.: Simple fast algorithms for the editing distance between trees and related problems. SIAM J. Comput. 18, 1245–1262 (1989)
9. Wagner, R., Fischer, M.: The string-to-string correction problem. Journal of the Association for Computing Machinery 211, 168–173 (1974)
10. Kittler, J., Hatef, M., Duin, R., Matas, J.: On combining classifiers. IEEE Trans. on Pattern Analysis and Machine Intelligence 20, 226–239 (1998)
11. Hjaltason, G., Samet, H.: Properties of embedding methods for similarity searching in metric spaces. IEEE Trans. on Pattern Analysis and Machine Intelligence 25, 530–549 (2003)
12. Duin, R.P.W., Tax, D.M.J.: Classifier conditional posterior probabilities. In: Amin, A., Pudil, P., Dori, D. (eds.) SPR 1998 and SSPR 1998. LNCS, vol. 1451, pp. 611–619. Springer, Heidelberg (1998)
13. Arlandis, J., Perez-Cortes, J.C., Cano, J.: Rejection strategies and confidence measures for a k-nn classifier in an ocr task. In: 16th International Conference on Pattern Recognition ICPR 2002, pp. 576–579. IEEE, Los Alamitos (2002)
14. Athitsos, V., Alon, J., Sclaroff, S.: Efficient nearest neighbor classification using a cascade of approximate similarity measures. In: Proceedings of the 2005 IEEE Computer Society Conference on Computer Vision and Pattern Recognition (CVPR 2005), vol. 1, pp. 486–493 (2005)

Combination of Experts by Classifiers in Similarity Score Spaces

Roberto Tronci[1], Giorgio Giacinto[2], and Fabio Roli[2]

[1] AmiLab Laboratorio Intelligenza d'Ambiente, Sardegna Ricerche, 09010 Pula, Italy
[2] DIEE Department of Electrical and Electronic Engineering, University of Cagliari, 09123 Cagliari, Italy
{roberto.tronci,giacinto,roli}@diee.unica.it

Abstract. The combination of different experts is largely used to improve the performance of a pattern recognition system. In the case of experts whose output is a similarity score, different methods had been developed. In this paper, the combination is performed by building a similarity score space made up of the scores produced by the experts, and training a classifier into it. Different techniques based on the use of classifiers trained on the similarity score space are proposed and compared. In particular, they are used in the framework of Dynamic Score Selection mechanisms, recently proposed by the authors. Reported results on two biometric datasets show the effectiveness of the proposed approach.

1 Introduction

The combination of ensemble of experts is largely used by the pattern recognition community to improve the performance and the reliability of the classification [1][2]. This combination can be performed at different levels, and with different types of information depending on the pattern recognition problem.

In a number of applications (e.g., biometric authentication, spam filtering, etc.) the task is either to assess if a given test pattern matches to a specific training pattern, or if it belongs or not to a class of patterns (i.e., one-class problems). The task is typically accomplished by "experts" that produce a similarity score. The decision is taken by setting a decision threshold on the score produced by the matching process, so that the outcome is *positive* if the score is larger than the threshold, while it is *negative* otherwise. As a consequence, the matching process subdivide patterns into two classes, namely the *positive* and *negative* classes. Within the *positive* and *negative* classes, the pattern can be further subdivided into "true" and "false" patterns, depending on the fact that the actual class of the pattern is the one assigned by the expert or not. The goal of such experts is to maximize the true positive rate (i.e. the number of *positive* patterns classified as *positive* over the total number of *positive* patterns), while minimizing the false positive rate (i.e. the number of *negative* patterns classified as *positive* over the total number of *negative* patterns). Typically this tradeoff is measured through the use of the Receiver Operating Characteristic curve (ROC),

N. da Vitora Lobo et al. (Eds.): SSPR&SPR 2008, LNCS 5342, pp. 821–830, 2008.

which plots the *true positive* rate against the *false positive* rate as the decision threshold th varies along the score range [3][4]. The Area Under the ROC Curve (AUC), and the Equal Error Rate (EER) are two performance measure derived from the ROC that are typically used to compare the performance of different experts. The AUC is the area under the ROC curve, the larger the AUC, the better the ROC. The EER represents the point where the *false positive* rate and the *false negative* rate are equal, and a good system should keep this value as small as possible.

It is easy to see that to increase the reliability of the matching process, a number of different matching techniques can be used, thus originating a set of similarity scores. Then, this set of scores can be exploited to produce an ensemble decision, through some functions designed to "fuse" the available scores (e.g., max, min, linear combination, etc.). The aim of these combination functions is to produce "new" scores whose distributions for the two classes (i.e., *positive* and *negative*) exhibit a larger degree of separation than those produced by the individual experts. As a consequence, it is possible to choose decision thresholds allowing to increase the true positive rate at a given false positive rate. In the literature, different types of score combination techniques have been proposed. In particular, some techniques focused on the optimization of the combination parameters in order to maximize some measures of class separability [5][6][7].

In this paper, we propose to build a similarity score space, where each component is related to a different expert. This space may allow a better discrimination between true matching pairs and false matching pairs, as different similarity measures can produce false matching on different pairs. A similar representation space has been proposed by the authors in the framework of dynamic classifier selection [8]. Other authors have used this space representation to focus the separation of the classes [9], while similar spaces have been used in specific recognition problems [10]. Finally, the formulation of the combination of experts has been also carried our in terms of the joint distribution of the scores produced by the ensemble of experts [7].

In the similarity score space, we can then train a classifier by using a set of labeled patterns, where the labels are the true outcomes of the matching process (i.e., "positive" or "negative"). When a test pattern is submitted to the system, the similarity score space is built from the ensemble of experts, and the output of the classifier is produced. We propose and compare four uses of the output of the classifier: "crisp" classification, as in a two-class problem; treating the output as a new score, so that the Receiver Operating Characteristic curve can be built; using the "crisp" output to estimate the true outcome in the Dynamic Score Selection (DSS) mechanism (see Section 3)[11]; using the continuous output in the generalized DSS mechanism (see Section 3). The aim of this paper is to assess the feasibility of this space for the combination of experts, and to compare the four solution designed to exploit the information provided by classifiers. The paper is organized as follows: Section 2 describes the use of a score space for classification. Section 3 describes the Dynamic Score Selection and its modification. Section 4 reports the experimental phase made to compare the different uses of the classifier. Finally the conclusion are outlined in Section 5.

2 The Similarity Score Space of an Ensemble of Two-Class Experts

As stated in the introduction, it has been proved that the combination of different experts can improve the performance of a pattern recognition system [1][2].

Let $E = \{E_1, E_2, \dots E_j \dots E_N\}$ be a set of N experts and $X = \{x_i\}$ be a set of patterns to be tested. Let also s_{ij} be the score produced by the expert E_j for the pattern x_i. Let S_j be the set of all the scores produced by an expert E_j for all the patterns X. Generally a combination function of experts takes as input the similarity scores estimated for a pattern and combines them in some way to obtain a combined score as shown in Figure 1.

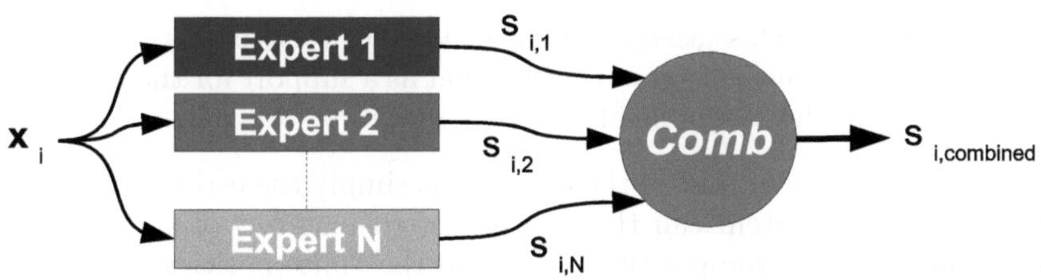

Fig. 1. General schema of an expert combination system

These combination functions produce a "new" score usually through a mathematical function of the set of scores estimated by the N experts, e.g., max, min, linear combination, etc.

In this paper we use these scores not as an input of a combination function, but as a "similarity" feature vector to represent each input pattern, whose components are the scores assigned to that pattern by the ensemble of experts E_j. Thus, for each pattern x_i the following feature vector $v_i = \{s_{i1}, \dots s_{ij} \dots s_{iN}\}$ can be built [8][9][10][11].

Using these vectors, we are able to represent the patterns in a similarity score space. An example of the use of this space is shown in Figure 2. In this figure a 2D and a 3D spaces are shown, where the 2D space is related to an ensemble of two experts, while the 3D space is obtained from the 2D space by adding another expert. In the figure, the distribution of *positive* (colored in blue) and *negative* (colored in red) patterns are reported. Thus it follows that we transformed a matching problem (or a one-class problem) into a two-class problem.

From the examples in the two figures it is easy to see that the increase of the number of experts produces a larger separation of patterns belonging to different classes. It is also evident that generally the separation surface of the two classes is not linear. To estimate this separation surface, a classifier can be trained, and its output can be used in two ways:

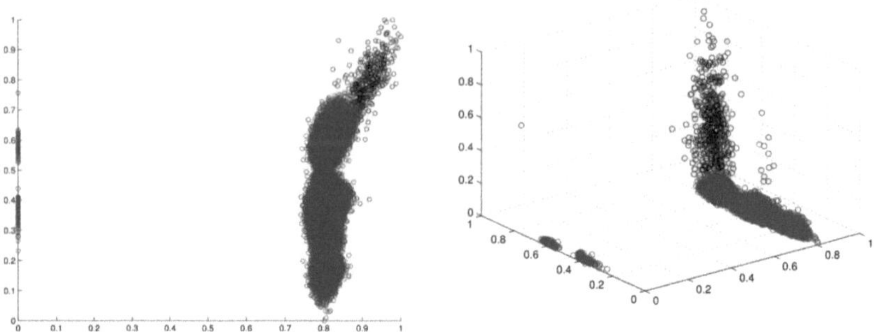

Fig. 2. Example of the distribution of scores in a space in a two-dimensional and three-dimensional space. The three-dimensional space is obtained by adding a dimension to the two-dimensional space. By increasing the number of dimensions, the separability increases. The *positive* class is represented in blue, while the *negative* class, in red.

1. Using directly the class assigned by the classifier ("crisp" classification)
2. Using the continuous output of the classifier as a support for the combination of the scores of the ensemble of experts

In the first case, the output of the classifier is simply the estimated class. The performance of the system can then be measured in terms of the accuracy of classification, i.e. the "true positive" rate, and the "false positive" rate.

In the second case, we can consider the output of a classifier as the results of the combination produced by the ensemble of experts, thus the performance can be expressed using the same measures used for the individual experts E_j, i.e. the Receiver Operating Characteristic curve (ROC), the Area Under the ROC Curve (AUC), and the Equal Error Rate (EER).

It is easy to see that if the "crisp" classification always exhibits an accuracy of 100%, then the direct use of the class assigned by a classifier is more advantageous than the use of the continuous output of the classifier. In practical application it is very hard to reach a 100% accuracy, thus it is worth investigating further exploitation of the output of the classifier. To this end, we propose two algorithms based on "crisp" outputs, and continuous outputs, respectively. These algorithms are based on the Dynamic Score Selection algorithm [11], whose basic concepts are recalled in the following Section.

3 Dynamic Score Selection Based on Classifiers in Similarity Spaces

Dynamic Score Selection (DSS) techniques are based on the concept of the ideal score selector [11] who aims to select one score from those assigned to the pattern by the ensemble of experts. In particular, ideal selection has been defined as the process of selecting for each pattern the maximum score among those assigned to the pattern by the ensemble if it is a *positive* pattern, and the minimum score if it is a *negative* pattern.

$$s_{i,*} = \begin{cases} \max\{s_{ij}\} \text{ if } x_i \text{ is a } \textit{positive} \text{ pattern} \\ \min\{s_{ij}\} \text{ if } x_i \text{ is a } \textit{negative} \text{ pattern} \end{cases} \tag{1}$$

It can be proved that the above ideal score selector exhibits a better ROC curve than every single ROC curve of the individual experts used in the combination, and consequently a larger AUC [11].

We propose to implement the DSS by exploiting either the "crisp" output of a classifier or the continuous output, when available. When a "crisp" output is available, it can be used as an estimation of the state of nature of a test pattern (i.e., if it is *positive* or *negative*), and the score is chosen according to Equation (1).

The proposed DSS algorithm is made up of the following steps:

1. Construct the N-dimensional vector space made up of the score outputs of the N available experts
2. Train a classifier C in the N-dimensional vector space using a set of labeled patterns
3. Classify test patterns with the classifier C
4. Select the score s_{sel} based on the output class of classifier C as follows

$$s_{sel} = \begin{cases} \max_j(s_{ij}) & \text{if } \textit{class} = \textit{positive} \\ \min_j(s_{ij}) & \text{if } \textit{class} = \textit{negative} \end{cases}$$

We call such a DSS algorithm as "crisp" DSS as it exploit only the "crisp" decisions made by the classifier.

If we consider the continuous outputs provided by the classifier, it is possible to develop a Dynamic Score Combination algorithm, that can be considered as a generalization of the basic DSS algorithm:

$$s_{i,dss} = \beta_i \cdot \max_j(s_{ij}) + (1 - \beta_i) \cdot \min_j(s_{ij}) \tag{2}$$

where $\beta_i \in [0, 1]$ represents the continuous outputs of a classifier. If $\beta = 1$, than the pattern is estimated as *positive*, otherwise if $\beta = 0$ the pattern is estimated as *negative*. It is easy to see that the formulation in Equation 2 embodies the "crisp" DSS.

Thus, the output of the classifier can be used either to perform a "hard" selection of one of the available scores (Equation (1)), or a "soft" selection, based on the linear combination of the max and min scores by a dynamic weight that is the output of the classifier (Equation (2)).

4 Experiments

The aim of the reported experiments is to show the capabilities of similarity score space for the combination of experts, and to compare the four different uses of classifiers trained on a similarity score space, i.e., the "crisp" output, the continuous output, the "crisp" DSS, and the generalized DSS. Experiments have been performed using two multimodal biometric authentication datasets.

Firstly, we will compare the performance of "crisp" classifiers with those of the DSS based on "crisp" outputs and the generalized DSS, in terms of false *positive* and false *negative* rates. The dataset used for these experiments is made up by the scores obtained during the third Fingerprint Verification Competition[1] (FVC2004) [12]. From the FVC2004, we use both the Open category (made up of 41 high performance experts), and the Light category (made up of 26 low performance experts). The fingerprint images were acquired using four sensors. For each sensor and for each expert, 2800 of positive and 4950 of negative scores are available. For the details on how the scores where obtained and normalized, the reader is referred to [12]. This database is not freely available, so our algorithms were executed at the Biometric Systems Lab (University of Bologna) which organizes the competition.

Then, the performances of the "crisp" DSS, the generalized DSS, and that of the continuous output of classifiers will be compared in terms of AUC and EER. In these experiments the dataset used is the Biometric Scores Set Release 1 of the National Institute of Standards and Technology[2] (NIST). The dataset contains similarity scores from two face recognition systems and one fingerprint system on left and right index fingerprints. For aim of this paper, the set containing the face and the fingerprint systems were used. In this multimodal dataset are present 517 subjects. The scores were normalized using the *Min-Max* rule.

In order to create a training set for the classifiers, we randomly divided the set of scores into four subsets of the same size for both the NIST and the FVC2004 datasets. Each of the four subsets has been used for training, while the remaining three subsets have been used for testing. Using this partitioning of the dataset, we performed an exhaustive multi-modal combination experiment for the NIST dataset, and an exhaustive multi-algorithm combination experiment for the FVC2004 dataset for each sensor.

In Figures (3-4), the graphs related to experiments carried out on the FVC2004 dataset are shown. For these experiments four classifiers have been used: the k-Nearest Neighbor (kNN), the Linear Discriminant Classifier (LDC), the Quadratic Discriminant Classifier (QDC), and a Parzen classifier. The comparison is made in the following way, as the "crisp" outputs of classifiers are obtained by fixing a decision threshold at 0.5 (the output of classifiers is normalized in the range $[0, 1]$):

- each classifier is trained on the four subdivision of the dataset described above.
- for each trained classifier C the FMR and $FNMR$ are computed on the test set as follows considering the estimated state of nature

$$FMR = negatives\ classified\ as\ positives\ /\ total\ negatives$$

$$FNMR = positives\ classified\ as\ negatives\ /\ total\ positives$$

[1] Web site: http://bias.csr.unibo.it/fvc2004/

[2] Web site: http://www.itl.nist.gov/iad/894.03/biometricscores/

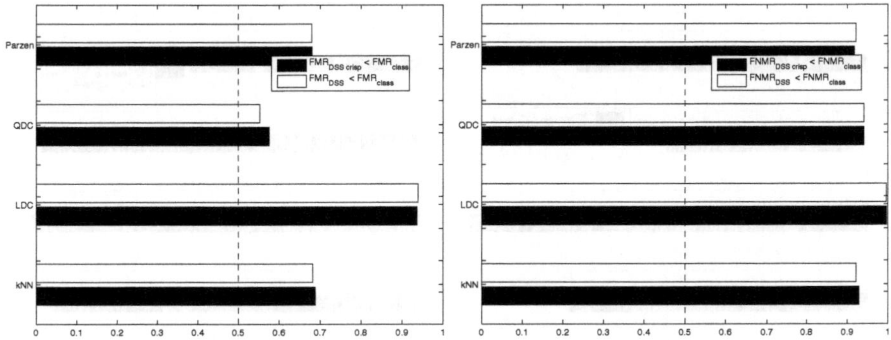

(a) (b)

Fig. 3. Comparison between the "crisp" classifier and the "crisp" and generalized DSS using the Open FVC2004 dataset. In *white* are reported the occurrences when the generalized DSS exhibits lower errors than the classifier. In *black* are reported the occurrences when the "crisp" DSS exhibits lower errors than the classifier. In (a) the threshold for the DSS is chosen to obtain $FNMR_{class} = FNMR_{DSS}$. In (b) the threshold is chosen to have $FMR_{class} = FMR_{DSS}$.

- for each classifier a "crisp" and generalized DSS is built using the output of the classifier
- for each classifier and subdivision, the value of FMR is compared to that produced by the related DSS when the value of $FNMR$ of the DSS is equal to the value of $FNMR$ produced by the classifier.
- for each classifier and subdivision, the value of $FNMR$ is compared to that produced by the related DSS when the value of FMR of the DSS is equal to the value of FMR produced by the classifier.

In Figure (3) the results on the Open dataset are shown, while in Figure (4) the results on the Light dataset are reported. The bars represent the percentage of cases in which the DSS outperformed the "crisp" classifier. From these figures it is clear that the use of the classifier as a component of a Dynamic Score Selector is generally preferable to the direct use of the "crisp" classifier. Moreover, the analysis of the figures clearly shows that this advantages depends on the classifier used, when the comparison is made by fixing the value of FNMR. The more simple the classifier, the better the performance of the DSS with respect to that of the classifier, e.g. the performance of LDC are very low with respect to the DSS based on LDC, while in the case of QDC the performance are quite similar. On the other hand, if the value of FMR is fixed, the DSS always outperforms the "crisp" classifier in terms of FNMR. Thus, the DSS algorithm allows attaining a better tradeoff of FMR and FNMR than that attained by a "crisp" classifier. The comparison between the two versions of the DSS shows that the "crisp" version slightly performs better than the generalized version when the FNMR value is fixed. When the value of FMR is fixed, the two versions of DSS show a quite similar performance. However, a better comparison of the performance of

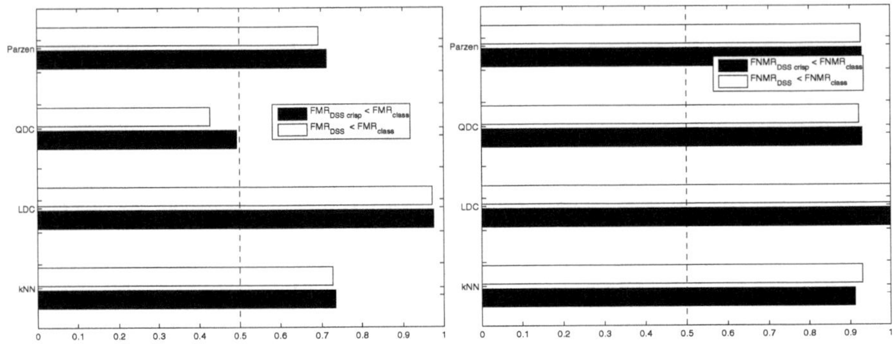

(a) (b)

Fig. 4. Comparison between the "crisp" classifier and the "crisp" and generalized DSS using the Light FVC2004 dataset. In *white* are reported the occurrences when the generalized DSS exhibits lower errors than the classifier. In *black* are reported the occurrences when the "crisp" DSS exhibits lower errors than the classifier. In (a) the threshold for the DSS is chosen to obtain $FNMR_{class} = FNMR_{DSS}$. In (b) the threshold is chosen to obtain $FMR_{class} = FMR_{DSS}$.

Table 1. Mean and standard deviation of AUC as the dimension of the similarity score space increases. All the possible combination of 2,3 and 4 experts are used.

AUC	**2 Experts**	**3 Experts**	**4 Experts**
LDA	0.9887(±0.0073)	0.9945(±0.0040)	0.9977(±0.0019)
DSS LDC "crisp"	0.9743(±0.0103)	0.9728(±0.0088)	0.9744(±0.0055)
DSS k-NN "crisp"	0.9859(±0.0068)	0.9917(±0.0050)	0.9955(±0.0023)
DSS QDC "crisp"	0.9833(±0.0112)	0.9919(±0.0086)	0.9966(±0.0050)
DSS SVM "crisp"	0.9789(±0.0094)	0.9799(±0.0077)	0.9799(±0.0000)
DSS LDC	0.9752(±0.0103)	0.9741(±0.0087)	0.9760(±0.0051)
DSS k-NN	0.9936(±0.0065)	0.9987(±0.0016)	**0.9998(±0.0001)**
DSS QDC	0.9901(±0.0083)	0.9964(±0.0039)	0.9991(±0.0008)
DSS SVM	0.9946(±0.0064)	**0.9996(±0.0004)**	**0.9998(±0.0002)**
LDC est	0.9891(±0.0076)	0.9947(±0.0038)	0.9973(±0.0017)
k-NN est	0.9699(±0.0226)	0.9928(±0.0075)	0.9973(±0.0051)
QDC est	**0.9953(±0.0047)**	0.9985(±0.0005)	0.9986(±0.0001)
SVM est	0.9049(±0.0600)	0.9360(±0.0191)	0.9996(±0.0003)

different versions of the DSS is made in the following Tables in terms of AUC and EER.

Tables (1-2) show the comparison of performances of the "crisp" DSS, the generalized DSS, the output of the classifiers treated as a "new" score, and a "trained fusion" rule based on a linear score combination whose weights are computed by Linear Discriminant Analysis (LDA) [9][13]. These results have been obtained on the NIST dataset, and the performance are reported in terms of AUC and EER, for different number of experts combined (i.e., 2, 3, and 4

Table 2. Mean and standard deviation of EER as the dimension of the similarity score space increases. All the possible combination of 2,3 and 4 experts are used.

EER	2 Experts	3 Experts	4 Experts
LDA	0.0431(±0.0166)	0.0296(±0.0123)	0.0200(±0.0068)
DSS LDC "crisp"	0.0634(±0.0162)	0.0647(±0.0148)	0.0632(±0.0098)
DSS k-NN "crisp"	0.0377(±0.0153)	0.0236(±0.0072)	0.0199(±0.0067)
DSS QDC "crisp"	0.0453(±0.0228)	0.0243(±0.0182)	0.0138(±0.0124)
DSS SVM "crisp"	0.0525(±0.0170)	0.0497(±0.0140)	0.0547(±0.0000)
DSS LDC	0.0629(±0.0164)	0.0642(±0.0149)	0.0632(±0.0098)
DSS k-NN	0.0261(±0.0175)	0.0104(±0.0053)	0.0071(±0.0025)
DSS QDC	0.0324(±0.0198)	0.0147(±0.0092)	0.0097(±0.0044)
DSS SVM	**0.0212(±0.0182)**	**0.0048(±0.0026)**	**0.0040(±0.0024)**
LDC est	0.0429(±0.0164)	0.0300(±0.0116)	0.0206(±0.0069)
k-NN est	0.0340(±0.0238)	0.0101(±0.0051)	0.0058(±0.0033)
QDC est	0.0214(±0.0161)	0.0094(±0.0050)	0.0070(±0.0013)
SVM est	0.0965(±0.0582)	0.0640(±0.0191)	**0.0040(±0.0024)**

experts). Four classifiers have been used, namely, the k-Nearest Neighbor (kNN), the Linear Discriminant Classifier (LDC), the Quadratic Discriminant Classifier (QDC), and the Support Vector Machine with a radial basis kernel. All the classifier are trained using the dataset subdivision described above, and all the possible combination of experts had been performed.

The results reported in Tables (1-2) confirm one of the initial intuition of the paper: the larger the dimension of the score space, the better the separability. In fact all the methodologies, but the LDC, increase their performance as the number of scores for pattern increases. This aspect is more evident when we take into account the results obtained with the estimation of the SVM, as the increasing number of scores per pattern leads to a high increase in performance. If we compare the performance of different version of DSS and those of the output of the classifier, it is easy to see that the generalized version of the DSS outperform the "crisp" DSS both in terms of AUC and EER. Moreover, the generalized version of the DSS outperforms the LDA in terms of AUC and EER when the kNN, the QDC, and the SVM are considered. The performances attained by the output of the classifier used as a "new" score are quite similar to those of the generalized DSS. However, the best results have been produced by the generalized DSS when the SVM classifier is used.

5 Conclusions

In this paper we showed the use of a similarity score space to combine different experts. In this space a classifier can be trained and its output used to improve the performance. In particular, we showed different ways of using the trained classifier: using directly its "crisp" classification, using the continuous output of the classifier either as a new score, or as a support for Dynamic Score Selection mechanisms. Reported experiments showed that the best use of the output of classifiers built in similarity score spaces is with Dynamic Score Selection mechanisms, as they allow obtaining lower errors than those obtained through the "crisp" classification. In particular, the best performances are provided by a generalized Dynamic Score Selection mechanism based on SVM. Moreover, reported results also show that the larger the dimension of the similarity score space (i.e., the larger the number of experts in the ensemble), the better the performance.

References

1. Kittler, J., Hatef, M., Duin, R., Matas, J.: On combining classifiers. IEEE Transactions on Pattern Analysis and Machine Intelligence 20(3), 226–239 (1998)
2. Kuncheva, L.: Combining Pattern Classifiers: Methods and Algorithms. John Wiley & Sons, Chichester (2004)
3. Bradley, A.: The use of the area under the ROC curve in the evaluation of machine learning algorithms. Pattern Recognition 30(7), 1145–1159 (1997)
4. Fawcett, T.: An introduction to roc analysis. Pattern Recognition Letters 27(8), 861–874 (2006)
5. Marrocco, C., Molinara, M., Tortorella, F.: Exploiting AUC for optimal linear combinations of dichotomizers. Pattern Recognition Letters 27(8), 900–907 (2006)
6. Ross, A., Nandakumar, K., Jain, A.: Handbook of Multibiometrics. Springer, Heidelberg (2006)
7. Nandakumar, K., Chen, Y., Dass, S., Jain, A.: Likelihood ratio-based biometric score fusion. IEEE Trans. on PAMI 30(2), 342–347 (2008)
8. Giacinto, G., Roli, F.: Dynamic classifier selection based on multiple classifier behaviour. Pattern Recognition 34(9), 179–181 (2001)
9. Ross, A., Jain, A.K.: Information fusion in biometrics. Pattern Recognition Letters 24, 2115–2125 (2003)
10. Lumini, A., Nanni, L.: Advanced methods for two-class pattern recognition problem formulation for minutiae-based fingerprint verification. Pattern Recognition Letters 29(2), 142–148 (2008)
11. Tronci, R., Giacinto, G., Roli, F.: Dynamic Score Selection for Fusion of Multiple Biometric Matchers. In: ICIAP 2007, pp. 15–20 (2007)
12. Maio, D., Maltoni, D., Cappelli, R., Wayman, J., Jain, A.: FVC2004: Third Fingerprint Verification Competition. In: Zhang, D., Jain, A.K. (eds.) ICBA 2004. LNCS, vol. 3072, pp. 1–7. Springer, Heidelberg (2004)
13. Duda, R., Hart, P., Stork, D.: Pattern Classification. John Wiley & Sons, Chichester (2001)

Improving the Performance of a NER System by Post-processing and Voting

Asif Ekbal and Sivaji Bandyopadhyay

Department of Computer Science and Engineering
Jadavpur University, Kolkata, India-700032
asif.ekbal@gamil.com, sivaji_ju@vsnl.com

Abstract. This paper reports about the development of a NER system in Bengali by combining outputs of the classifiers such as Maximum Entropy (ME), Conditional Random Field (CRF) and Support Vector Machine (SVM). The training set consists of approximately 250K wordforms and has been manually annotated with the four major named entity (NE) tags such as *Person*, *Location*, *Organization* and *Miscellaneous* tags. The classifiers make use of the different contextual information of the words along with the variety of features that are helpful in predicting the various NE classes. Lexical context patterns, which are generated from an unlabeled corpus of 1 million wordforms in a semi-automatic way, have been used as the features of the classifiers in order to improve their performance. In addition, we have used the second best tags of the classifiers and applied several heuristics to improve the performance. Finally, the classifiers are combined using a majority voting approach. Experimental results show the effectiveness of the proposed approach with the overall average recall, precision, and f-score values of 90.78%, 87.35%, and 89.03%, respectively, which shows an improvement of 11.8% in f-score over the best performing SVM based *baseline* system and an improvement of 15.11% in f-score over the least performing ME based *baseline* system. The proposed system also outperforms the other existing Bengali NER system.

Keywords: Natural Language Processing, Named Entity Recognition, Maximum Entropy, Conditional Random Field, Support Vector Machine, Majority Voting.

1 Introduction

Named Entity Recognition (NER) is an important tool in almost all Natural Language Processing (NLP) application areas including machine translation, question answering, information retrieval, information extraction, automatic summarization etc. The current trend in NER is to use the machine-learning approach, which is more attractive in that it is trainable and adoptable and the maintenance of a machine-learning system is much cheaper than that of a rule-based one. The representative machine-learning approaches used in NER are Hidden Markov Model (HMM) (BBN's IdentiFinder [1]), Maximum Entropy (ME) (New York University's MENE [2]) and Conditional Random Fields (CRFs) [3]. Support Vector Machines (SVMs) based

N. da Vitora Lobo et al. (Eds.): SSPR&SPR 2008, LNCS 5342, pp. 831–841, 2008.

NER system was proposed by Yamada et al. [4] for Japanese. The process of stacking and voting method for combining strong classifiers like boosting, SVM and TBL, on NER task can be found in [5]. Florian et al. [6] tested different methods for combining the results of four systems and found that robust risk minimization worked best. Munro et al. [7] employed both voting and bagging for combining classifiers. The work reported in this paper differs from that of Wu et al. [5], Florian et al. [6] and Munro et al. [7] in the sense that here, we have conducted a number of experiments to post-process the outputs of the classifiers with the lexical context patterns, which are generated in a semi-automatic way from an unlabeled corpus of 1 million wordforms, and used several heuristics to improve the performance of each classifier before applying majority voting.

Named Entity (NE) identification and classification in Indian languages in general and in Bengali in particular is difficult and challenging. Unlike English and European languages, Bengali lacks capitalization information, which plays a very important role in identifying NEs. Indian names are more diverse and a lot of these words can be found in the dictionary with specific meanings. Another problem in Indian language NER is the scarcity of NE annotated corpus. A pattern directed shallow parsing approach for NER in Bengali is reported in [8]. A HMM based NER system for Bengali has been reported in [9], where additional contextual information has been considered during emission probabilities and NE suffixes are kept for unknown word handling. More recently, the works in the area of Bengali NER can be found in [10] and [11] with the CRF, and SVM approach, respectively.

Other than Bengali, the works on Hindi can be found in [12] with CRF, [13] with a language independent method, [14] with a hybrid feature set based ME approach and in [15] using MEMM. As part of the IJCNLP-08 NER shared task, various works of NER in Indian languages using various approaches can be found in NERSSEAL[1].

2 Named Entity Recognition in Bengali

Bengali is the seventh popular language in the world, second in India and the national language of Bangladesh. We have used a Bengali news corpus [16], developed from the web-archive of a widely read Bengali newspaper for NER. Out of 34 million wordforms, 150K wordforms have been manually annotated with the four NE tags namely, *Person, Location, Organization* and *Miscellaneous*. We have also collected the unannotated corpus of around 100K wordforms, particularly from the literature, agriculture and scientific domains and manually annotated with the four NE tags. Both the annotations were carried out by a person and verified by an expert. In order to properly denote the boundaries of NEs, these four NE tags are further divided in the form:

B-XXX: Beginning of a multi-word NE, I-XXX: Internal of a multi-word NE, E-XXX: Ending of multi-word NE (e.g., *sachin ramesh tendulkar* is tagged as *sachin* /B-PER *ramesh*/I-PER *tendulkar*/E-PER). The single-word NE is tagged as PER/ LOC/ORG/MISC. The words that do not denote the NEs are tagged as NNE.

---

[1] http://ltrc.iiit.ac.in/ner-ssea-08/proc/index.html

3 Approaches to NER for Bengali

Incorporating diverse features in an HMM-based NE tagger is difficult and complicates the smoothing typically used in such taggers. In contrast, ME, CRF or SVM based method can deal with the diverse and morphologically complex features of the Indian languages.

In this paper, we have used ME, CRF and SVM frameworks in order to identify NEs from a Bengali text and to classify them into *Person, Location, Organization* and *Miscellaneous*. A semi-automatic context pattern induction method, second best tag of the classifier and several heuristics have been used in order to improve the performance of each of the classifiers. Finally, we have combined these classifiers with the help of a simple majority voting scheme.

We have used the C++ based ME package (http://homepages.inf.ed. ac.uk/ s0450736/software/maxent/maxent-20061005.tar.bz2)and C^{++} based OpenNLP CRF++ package (http://crfpp.sourceforge.net) for NER. Support Vector Machines (SVMs) have advantages over conventional statistical learning algorithms, such as Decision Tree, HMMs, ME in terms of its high generalization performance independent of dimension of feature vectors and its ability of learning with all combinations of given features without increasing computational complexity by introducing the kernel function. We have used *YamCha (*http://chasen-org/~taku/software/yamcha) toolkit, an SVM based tool for detecting classes in documents and formulating the NER task as a sequential labeling problem. Here, the pair wise multi-class decision method and *second degree polynomial kernel function* were used. We have used TinySVM-0.07 (http://cl.aist-nara.ac.jp/~taku-ku/software/TinySVM) classifier that seems to be the best optimized among publicly available SVM toolkits.

3.1 Named Entity Features

Experiments have been carried out in order to find out the most suitable features for NER in Bengali. Following are the details of the set of features that have been applied to the NER task:

- Context words: Preceding and following words of a particular word. This is based on the observation that the surrounding words are very effective in the identification of NEs.
- Word suffix and prefix: Word suffix and prefix information are helpful to identify NEs. A fixed length (say, *n*) word suffix/prefix of the current and/or the surrounding word(s) can be treated as feature(s). If the length of the corresponding word is less than or equal to *n-1* then the feature values are not defined and denoted by ND. The feature value is also not defined (ND) if the token itself is a punctuation symbol or contains any special symbol or digit. Another way to use the suffix information is to modify the feature as binary valued. Variable length suffixes of a word can be matched with predefined lists of useful suffixes (e.g., *-babu*, *-da*, *-di* etc. for persons and *-land*, *-pur*, *-lia* etc. for locations). These features are useful to handle the highly inflective Indian languages like Bengali.
- Named Entity Information (dynamic feature): NE tag(s) of the previous word(s).

- First word (binary valued): Current token is the first word of the sentence or not. First word is generally NE.
- Length of the word (binary valued): Length of the token is less than three or not. This is based on the observation that the short words are rarely NEs.
- Infrequent word (binary valued): A cut off frequency has been chosen in order to consider the infrequent words in the training corpus. Frequently occurring words are rarely NEs.
- Digit features: Several digit features have been considered depending upon the presence and/or the number of digit(s) in a token. These binary valued features are helpful in recognizing miscellaneous NEs such as time, monetary and date expressions, percentages, numerical numbers etc.
- Position of the word (binary valued): Position of the word in a sentence is a good indicator of NEs. Generally, verbs occur at the last position of the sentence. This feature is used to check whether the word is the last word in the sentence.
- Part of Speech (POS) Information: We have used a CRF-based POS tagger [18] that was originally developed with 26 POS tags, defined for the Indian languages. We have considered a coarse-grained POS tagger that has the following tags: Nominal, PREP (Postpositions) and Other. Postpositions are considered as these often appear after the NEs.
- Gazetteer Lists: Gazetteer lists, developed from the Bengali news corpus [11], have been used as the features in each of the classifiers. If the current token is in a particular list, then the corresponding feature is set to 1 for the current and/or the surrounding word(s); otherwise, it is set to 0. The following is the list of gazetteers along with the number of entries:

(1).Organization clue word (e.g., *kong, limited* etc): 94, Person prefixes (e.g., *sriman, sreemati* etc.): 245, Middle names: 1,491, Surnames: 5,288, Common location (e.g., *sarani, road* etc.): 547, Action verb (e.g., *balen*, *ballen* etc.): 241, Function words: 743, Designation words (e.g., *neta*, *sangsad* etc.): 947, First names: 72,206, Location names: 7,870, Organization names: 2,225, Month name (English and Bengali calendars): 24, Weekdays (English and Bengali calendars): 14

(2).Common word (521 entries): Most of the Indian languages NEs appear in the dictionary with some other meanings. For example, the word *kamol* may be the name of a person but also appears in the dictionary with another meaning *lotus*, the name of a flower; the word *dhar* may be a verb or also can be the part of a person name. We have manually created a list, containing the words that can be NEs as well as valid dictionary words.

(3).Lexicon (128,000 entries): We have developed a lexicon from the Bengali news corpus in an unsupervised way. The feature 'LEX' has value 0 for those words that appear in the lexicon; otherwise, the value is 1. This feature has been considered, as the words that appear in the lexicon are rarely NEs.

4 Semi-automatic Context Pattern Induction

For English NER, a context pattern induction method through successive learners is reported in [17]. Here, we have developed a semi-automatic context pattern induction method in a different way in order to use the contextual information of words as features of the classifiers as well as to post-process the outputs of each of the classifiers for Bengali NER. For this purpose, we have used an unlabeled corpus of 1000K wordforms of the Bengali news corpus [11]. We have collected the frequently occurring words from a part of this news corpus and the training set to use as the seeds. There are 123, 87, and 32 entries in the person, location, and organization seed lists, respectively. The unlabeled corpus is tagged with the elements from the seed lists. For example, *<Person> sonia gandhi </Person>*, *<Location> kolkata </Location>* and *<Organization> jadavpur viswavidyalya </Organization>*.

For each tag T inserted in the training corpus, the algorithm generates a *lexical* pattern *p* using a context window of maximum width 6 (excluding the tagged NE) around the left and the right tags, e.g., p = [$l_{-3}l_{-2}$ l_{-1} <T> ...</T> l_{+1} l_{+2} l_{+3}], where $l_{\pm i}$ are the *context* of p. Any of $l_{\pm i}$ may be a punctuation symbol. In such cases, the width of the lexical patterns will vary. All these patterns, derived from the different tags of the tagged training corpus, are stored in a Pattern Table (or, set P), which has four different fields namely, pattern *id* (identifies any particular pattern), pattern *example* (pattern), pattern *type* (*Person/Location/Organization*) and *relative frequency* (indicates the number of times any particular pattern appears in the entire training corpus relative to the total number of patterns generated). This table has 38,986 entries, out of which 21,031 patterns are distinct.

Every pattern p in the set P is matched against the same unannotated corpus of 1 million wordforms. In a place, where the context of p matches, p predicts the occurrence of the left or right boundary of name. The POS information of the words as well as well as some linguistic rules is used to detect the other boundary of the NE. In order to do this, the corpus is tagged with the CRF-based POS tagger [18] and then a regular expression Nominal* (NE is a sequence of nominals) has been used. Each new NE (not appearing in the seed lists) so obtained in the unlabeled corpus is manually checked for correctness. The corpus is further tagged with these newly acquired NEs to identify further lexical patterns. The bootstrapping is applied on the training corpus until no new patterns can be generated. The patterns are added to the pattern set P with the *type* and *relative frequency* fields set properly, if they are not already in the pattern set P with the same *type*. Any particular pattern in the set of potential patterns P may occur many times but with different *type* and with equal or different *relative frequency* values. For each pattern of the set P, the *relative frequencies* of its occurrences as person, location and organization names are calculated. For the candidate patterns acquisition, a particular threshold value of *relative frequency* is chosen. If the *relative frequency* for a particular pattern (along with the *type)* is less than this threshold value then this pattern (only for that *type)* is discarded otherwise, it is retained in the Pattern Table (or, set P). An additional 1,078 number of new patterns were generated through bootstrapping.

To extract the candidate patterns from the set of patterns of the Pattern Table, the NE annotated corpus of 250K wordforms have been used as the gold standard development data. Each pattern p of the set P is matched against the development set. The identified NE is checked for the correctness against the gold standard tag(s) of that entity/or the sequence of entities. A pattern is penalized if it extracts entities, which belong to the other classes. We consider the extracted entity of a particular class (*type*) to be an instance of positive example if it is identified by a pattern of that *type*.

The extracted entity is counted as an instance of the negative example if it is extracted by pattern of other *type* or it is not at all a named entity. Let positive (p) and negative (p) be respectively the number of distinct positive and negative examples extracted by a pattern p. Now, we calculate the following function for each pattern p:
precision(p)=positive(p)/[positive(p)+negative(p)]

We discard all the patterns p if the values of precision (p) are below a predetermined threshold value. The remaining patterns are called the *high precision patterns* and can be very effective for NE identification. We also discard the patterns whose total positive example (distinct) extraction is less than a certain threshold value. The remaining patterns in the set P form the set of final patterns to be applied during evaluation and this resultant set is denoted by *Accept Pattern*. There are 12,633 entries in this set.

5 Evaluation Results

A set of 250K wordforms, collected from the news corpus, has been manually annotated with the four NE tags with the help of *Sanchay Editor*[2], a text editor for the Indian languages. Out of 250K wordforms, 150K wordforms were collected from the news corpus [16] and the remaining 100K wordforms have been collected from the agriculture, literature and scientific domains. Around 30K NE tagged corpus is selected as the development set and the rest 220K wordforms are used as the training set of each of the classifiers.

A number of experiments have been conducted taking the different combinations of the available words, context, orthographic word level features and gazetteers to identify the best-suited set of features in the ME, CRF and SVM frameworks for NER in Bengali. These are defined as the *baseline* models.

Evaluation results of the development set are presented in Table 1 in terms of Recall (R), Precision (P) and F-Score (FS) that are defined as below:

Recall= No. of NEs retrieved correctly/Relevant NEs in the gold standard data; Precision= No. of NEs retrieved correctly /No. of NEs identified by the system; F-Score=(2*Recall*Precision)/(Recall + Precision).

Table 1. Results of the development set for the *baseline* models

Model	Recall (in %)	Precision (in %)	F-Score (in %)
ME	78.48	73.40	74.32
CRF	82.29	76.42	77.71
SVM	82.32	77.49	78.30

Results have demonstrated that the ME based *baseline* system performs best (f-score=74.32%) for the context window of size three (i.e., previous, current and next word), NE information of the previous word, POS information of the current word, prefixes and suffixes upto three characters of the current word along with the gazetteers and other features. The *baseline* CRF model has shown best performance

[2] Sourceforge.net/project/nlp-sanchay

(f-score=77.71%) for the context window of size five, POS information of the current and previous words along with the other set of features like ME. The SVM based *baseline* system has performed best among the three models and has demonstrated the f-score value of 78.3% for the context window of size six, NE information of the previous two words, POS information of the current, previous and the next words along with the other set of features as like ME and CRF. During all the experiments, we have observed that word context, prefixes, suffixes, POS information, dynamic NE tag(s) and digit features are the most effective features for NER. Other features are not so effective but helpful in increasing the overall performance of the NER system.

5.1 Use of Context Patterns as Features

The *Accept Pattern* set (discussed in section 4) is used to prepare three lists of the frequently occurring words in the left and/or the right contexts of person, location and organization names. These words are used as the *trigger words* and are very useful to identify the NEs. A feature 'ContextInformation' is set appropriately if the window [-3, 3] (three words spanning to left and right) of the current word contains any *trigger word* of a particular NE type. The use of this feature along with the previously discussed features increase the overall f-score values to **75.72%, 79.87%**, and **80.38%** in the *baseline* ME, CRF, and SVM based systems, respectively. Evaluation results are presented in Table 2.

5.2 Second Best Tag and Heuristics

If the best tag given by any of the classifiers is "NNE" (other than NE) and the confidence of the second best tag is greater than a particular threshold value then the second best tag is considered as the correct tag. We conducted a number of experiments to identify the threshold value for each of the classifiers. Finally, the threshold value is set to 0.19% for ME and 0.17% for CRF and SVM. Evaluation results of the system are presented in Table 2. Results show an improvement in the recall values by 9.05%, 9.31%, and 9.69% in the ME, CRF, and SVM, respectively. But, the choice of second tag decreases the precision values by 1.42%, 2.44%, and 2.65% in the ME, CRF, and SVM, respectively. This resulted in the overall performance improvement in terms of f-score values.

We have used the following heuristics to further improve the performance of each of the classifiers. Some of the rules are useful to improve the recall values, whereas

Table 2. Results with context features and second best tag

Baseline + Context features			Baseline + Context features + Second best tags			
Model	R (in %)	P (in %)	FS (in %)	R (in %)	P (in %)	FS (in %)
ME	79.24	72.50	75.72	85.10	74.75	79.59
CRF	82.97	77.00	79.87	88.72	76.59	82.21
SVM	83.07	77.86	80.38	90.17	77.13	83.14

some are effective to increase the precisions. Many of the heuristics are also helpful to identify the boundaries properly. Following are the set of heuristics.

(a). The NNE tag of a particular word is replaced by the appropriate NE tag, if that word appears somewhere in the output with that NE tag.
(b). If any word is tagged as B-XXX/I-XXX/E-XXX (XXX: PER/LOC/ORG/MISC) and the previous and next words are tagged as NNE then that word is assigned the NE tag of type XXX.
(c). The NNE tag of a word is replaced by the E-XXX if the previous word is already tagged as B-XXX.
(d). NNE tag of a word is replaced by B-XXX, if the next word is already tagged as E-XXX.
(e). If there is sequence B-XXX/I-XXX followed by XXX in the output, then the tag XXX is replaced by the E-XXX.
(f). If the sequence of tags is of the form XXX B-XXX1/I-XXX1/E-XXX1 NNE (XXX#XXX1) for three consecutive words in the output, then the tag B-XXX1/I-XXX1/E-XXX1 is replaced by the XXX1.
(g). If there is a word (current), tagged not as B-XXX/I-XXX/NNE followed by a word that is tagged as B-XXX/I-XXX/E-XXX then the current word is assigned the tag B-XXX.
(h). If the words, tagged as NNE, contain the variable length NE suffixes (used as the feature in the *baseline* models) then the words are assigned the NE tags. The *types* of the NE tags are determined by the *types* of the suffixes (e.g., *Person* tag is assigned if matches with the person name suffix).

We have presented the experimental results of the classifiers in Table 4 for each heuristics. It shows how each heuristics improve the performance.

Table 3. Results using each heuristics FS (Model): Denotes the f-score for a particular classifier, H[i,0]: Heuristics of the ith row and 0th column of the table

Heuristic	FS (ME)	FS (CRF)	FS (SVM)
a	80.65	83.23	84.02
b + H[2, 0]	80.98	83.47	84.56
c + H[3,0]	81.34	83.79	84.83
d + H[4,0]	81.62	84.22	85.25
e + H[5,0]	82.91	84.63	85.57
f + H[6,0]	83.12	84.89	85.82
g + H[7,0]	83.26	85.02	85.91
h + H[8,0]	83.42	85.23	86.29

5.3 Voting

Voting is a very widely used term and generally used to combine a set of classifiers together into a final system. It is better to give importance to the outputs of all the classifiers rather than giving importance to the output of any particular classifier. It is possible to assign varying weights to the models in order to give more priority to one model than the others. The voting scheme becomes effective in order to improve the overall performance of any system. Here, we have combined the three classifiers by simply assigning equal weight to each model and selecting the classifications, which were proposed by the majority of models. The final voted system has demonstrated the recall, precision, and f-score values of **90.23%, 86.07%,** and **88.1%**, respectively. Hence, we observe the improvement the f-score values of 4.68%, 2.87%, and 1.81% in the ME, CRF, and SVM, respectively.

5.4 Results of the Test Set

Three systems are tested with a gold standard test set of 40K wordforms. Approximately, 21% of the NEs are unknown in the test set. Experimental results are shown in Table 4. Meaning of the notations used in the experimental results are given below: FS (Bs): F-Score of the *baseline* model; FS (Bs + C): F-Score using context features; FS (Bs +C+ S): F-Score using context features and second best tag; FS (Bs + C+S+H): F-Score using context features, second best tag and heuristics.

Evaluation results show the f-score values of 73.92%, 76.35%, and 77.23% for the *baseline* models of ME, CRF, and SVM, respectively. Evaluation results have demonstrated the fact that the use of context features, second best tag and post-processing can improve the performance of each individual system by the impressive margins of 9.5%, 8.85%, and 9.06% in f-scores over the *baseline* ME, CRF, SVM, respectively. These post-processed systems are then combined together into a final system by using majority voting. Experimental results have shown the recall, precision, and f-score values of **90.78%, 87.35%,** and **89.03%**, respectively. Clearly, this is the improvement in performance compared to any individual system.

Table 4. Results for the test set

Model	FS (Bs)	FS (Bs+C)	FS(Bs + C+S)	FS (Bs + C+ S+ H)
ME	73.92	76.11	79.65	83.42
CRF	76.35	79.22	81.17	85.23
SVM	77.23	80.13	82.23	86.29

5.5 Comparison with Other NER Systems

The most recent existing Bengali NER systems, i.e., HMM based system [9], CRF based system 10 and SVM based system [11] have been trained and tested with the same datasets. Evaluation results are presented in Table 5. Results show the effectiveness of the proposed NER model that outperforms the other existing Bengali NER systems by an impressive margin. The system outperforms the HMM based system with **17.27%** f-sore, CRF based system with **9.22%** f-score and SVM based system with **9.07%** f-score. Thus, it can be decided that purely statistical approaches cannot always perform well across the domain (here, both training and test sets are collected from the mixed domains). So, post-processing the output of the statistical framework with the context patterns, second best tags and several heuristics can yield a reasonably good performance. Results also suggest that combination of several classifiers is more effective than the single classifier.

Table 5. Comparisons with other models

Model	Recall (in %)	Precision (in %)	F-Score (in %)
HMM	73.02	70.55	71.76
CRF	80.46	79.17	79.81
SVM	81.98	78.04	79.96
Voted System (proposed)	90.78	87.35	89.03

6 Conclusion

In this paper, we have developed a NER system for Bengali by combining the ME, CRF and SVM classifiers using a majority voting approach. Performance of each *baseline* classifier has been improved by the context pattern induction method, second best tag and several heuristics. It has been experimentally shown that post-processing the output of each classifier is effective and improves the performance significantly. The proposed approach shows the effectiveness with the reasonably high recall, precision and f-score values. We have shown that the reported system outperforms the other existing Bengali NER systems. This approach can be applicable for NER in other languages, especially for the Indian languages. Future works include investigating how the proposed approach effect to each of the NE classes. We would like to observe the effect by recognizing miscellaneous expressions using some regular expressions and adopting weighted voting scheme.

References

1. Bikel, D.M., Schwartz, R., Weischedel, R.M.: An Algorithm that Learns What's in Name. Machine Learning (Special Issue on NLP), 1–20 (1999)
2. Bothwick, A.: A Maximum Entropy Approach to Named Entity Recognition. Ph.D. Thesis, New York University (1999)
3. Lafferty, J., McCallum, A., Pereira, F.: Conditional Random Fields: Probabilistic Models for Segmenting and Labeling Sequence Data. In: Proc. of 18th ICML, pp. 282–289 (2001)
4. Hiroyasu, Y., Kudo, T., Matsumoto, Y.: Japanese Named Entity Extraction using Support Vector Machine. Transactions of IPSJ 43(1), 44–53 (2003)
5. Wu, D., Ngai, G., Carpuat, M.: A Stacked, Voted, Stacked Model for Named Entity Recognition. In: Proceedings of CoNLL 2003 (2003)
6. Florian, R., Ittycheriah, A., Jing, H., Zhang, T.: Named Entity Recognition through Classifier Combination. In: Proceedings of CoNLL 2003 (2003)
7. Munro, R., Ler, D., Patrick, J.: Meta-learning Orthographic and Contextual Models for Language Independent Named Entity Recognition. In: Proceedings of CoNLL 2003 (2003)
8. Ekbal, A., Bandyopadhyay, S.: Lexical Pattern Learning from Corpus Data for Named Entity Recognition. In: Proc. of 5th ICON, India, pp. 123–128 (2007)
9. Ekbal, A., Naskar, S., Bandyopadhyay, S.: Named Entity Recognition and Transliteration in Bengali. Named Entities: Recognition, Classification and Use, Special Issue of Lingvisticae Investigationes Journal 30(1), 95–114 (2007)
10. Ekbal, A., Haque, R., Bandyopadhyay, S.: Named Entity Recognition in Bengali: A Conditional Random Field Approach. In: Proc. of IJCNLP 2008, pp. 589–594 (2008)

11. Ekbal, A., Bandyopadhyay, S.: Bengali Named Entity Recognition using Support Vector Machine. In: Proc. of NERSSEAL, IJCNLP 2008, pp. 51–58 (2008)
12. Li, Wei, McCallum, Andrew: Rapid Development of Hindi Named Entity Recognition Using Conditional Random Fields and Feature Inductions. ACM TALIP 2(3), 290–294 (2003)
13. Cucerzan, S., Yarowsky, D.: Language Independent Named Entity Recognition Combining Morphological and Contextual Evidence. In: Proc. of the Joint SIGDAT Conference on EMNLP and VLC, pp. 90–99 (1999)
14. Saha, S., Sarkar, S., Mitra, P.: A Hybrid Feature Set based Maximum Entropy Hindi Named Entity Recognition. In: Proc. of IJCNLP 2008, pp. 343-349 (2008)
15. Kumar, N., Bhattacharyya, P.: Named Entity Recognition in Hindi using MEMM. Technical Report, IIT Bombay, India (2006)
16. Ekbal, A., Bandyopadhyay, S.: A Web-based Bengali News Corpus for Named Entity Recognition. Language Resources and Evaluation Journal 40 (2008)
17. Niu, C.g., Li, W., Ding, J., Srihari, R.: A Bootstrapping Approach to Named Entity Classification Using Sucessive Learners. In: Proc. of ACL 2003, pp. 335–342 (2003)
18. Ekbal, A., Haque, R., Bandyopadhyay, S.: Bengali Part of Speech Tagging using Conditional Random Field. In: Proc. of SNLP, Thailand (2007)

A Low Cost Incremental Biometric Fusion Strategy for a Handheld Device

Lorene Allano, Sonia Garcia-Salicetti, and Bernadette Dorizzi

Telecom & Management SudParis – Institut Telecom
9 rue Charles Fourier, 91011 Evry Cedex, France
{Lorene.Allano,Sonia.Salicetti,
Bernadette.Dorizzi}@it-sudparis.eu

Abstract. In this paper, we present some results on multimodality implementation resulting from the VINSI ("Vérification d'Identité Numérique Sécurisée Itinérante" for Secured Mobile Digital Identity Verification) French project. The VINSI handheld terminal allows identity verification in mobile conditions (airport gates) using two biometrics usable in future biometric passports (fingerprint and face). We propose an incremental fusion strategy aiming at improving the global performance of the combined system over each individual recognizer while optimizing the cost resulting from the fusion. Indeed, in this kind of application, time and complexity optimization is essential. To this aim, we split the fingerprint scores' range into different interest zones, on which we do not apply the same strategy depending on the relative quality of the modalities at hand. Results on a virtual database corresponding to VINSI applicative conditions (Combination of BIOMET fingerprints and FRGCv2 faces) show that this incremental fusion strategy allows the same improvement in performance as global fusion methods while significantly reducing the cost.

Keywords: Biometrics, Fusion, Incremental Strategy, Face, Fingerprint, Handheld Device, Cost Optimization, Multiple Classifiers System..

1 Introduction

The VINSI handheld terminal is a mobile secured terminal allowing the control of identity in mobile conditions. Two biometric modalities are implemented, namely face and fingerprint, on this mobile device which contains a highly secured part, named "security box", allowing the terminal itself to be safe and secured from the point of view of storage and processing of biometric data. This is in accordance with the specification of the new biometric passport which will contain, stored on a contactless chip, the fingerprint and face references of the owner of the passport. This way, this terminal could be used to verify a person's identity in airport gates at the exit of planes or even more generally outdoors, in the framework of mobile customs for instance. Note that the terminal embeds not only the softwares but also the sensors (camera, fingerprint sensor) necessary for the verification stage.

N. da Vitora Lobo et al. (Eds.): SSPR&SPR 2008, LNCS 5342, pp. 842–851, 2008.

In this article we want to explore the interest of using multimodality in this context. Indeed, multimodality is often presented as a way to increase the performance of the system, to provide an alternative when one modality alone cannot be used and to enhance robustness to forgeries.

However, this VINSI application deserves several remarks. In this mobility context where a lot of disturbance on images will come from illumination variations (indoor versus outdoor acquisitions), fingerprints will give much better results than face. Processing time and usage friendliness are also important issues. Indeed, the processing capabilities of such a mobile terminal are limited compared to those of a fixed PC and in verification applications using the terminal, acquisition of both face and fingerprints can appear fastidious both to the user and to the operator.

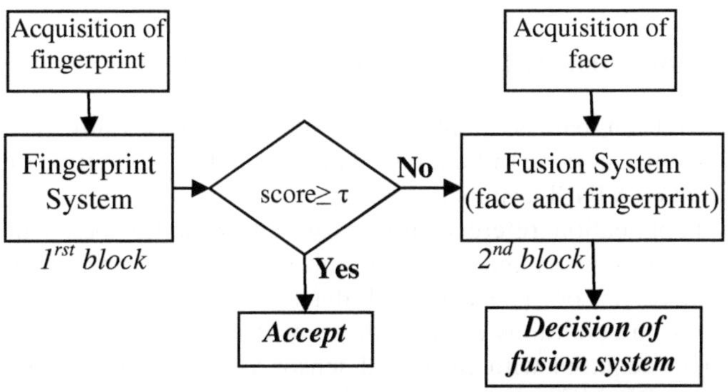

Fig. 1. The proposed incremental biometric fusion strategy composed of two blocks: the fingerprint system and the fusion system

For this reason, we propose a novel incremental fusion strategy, based on the multistage approach proposed in [1], that we exploit for fusion of biometric modalities. In [1] Pudil et al. proposed a multi-experts system based on a serial architecture with a reject option. At each stage, the reject option consists in an impossible decision of the expert due to a low confidence in its decision. This multistage approach with reject option was also used in [2, 3] applied to handwritten digit and word recognition. In biometrics, Cordella et al. proposed in [4] a cascaded multi expert system for signature verification based on the same approach. Our idea is to used a multi-stage approach applied to multimodal biometrics, that is the fusion of two different biometrics. In most of multimodal systems, experts are fused in a parallel way that we call “global” fusion methods, that is all the available scores are systematically used. The architecture that we propose is a sequential fusion strategy composed of two blocks (see Fig. 1). The first block corresponds to the fingerprint system alone, and the second, eventually used, is a fusion system of the two biometric scores (face and fingerprint). Therefore, the architecture has increasing complexity, since using the second block involves the acquisition of a second modality, face.

This strategy was also proposed by Marcialis et al. in [5] for a multimodal biometric system composed of fingerprint and face. The major difference between our architecture and the one in [5] is the second stage in which the authors use only the second system while we applied fusion of the two modalities. This guarantees better results, as performance of fusion when performed in an optimal way are better than those of any of the two modalities alone [11].

Our incremental fusion strategy will allow increasing system performance to the same level as the one reached by global fusion methods (which always use the two modalities) while using only the additional modality (in this case face) in the situations when this can bring an improvement over using fingerprint alone.

We first describe briefly our experimental monomodal context (systems and database), then we present the global fusion schemes that are commonly used in the literature before introducing our new incremental fusion strategy. Comparative results are finally presented allowing putting forward the interest of this incremental approach.

2 Bimodal Database and Evaluation Protocol

Experiments have been realized on publicly available databases suitable in the context of this application because we did not have at disposal during the time of the project a database acquired on VINSI terminal.

In the VINSI application, reference images stored on the ID document are acquired with an optical sensor while the platform embedded sensor which will be used for test, is a capacitive sensor. The BIOMET database allows therefore evaluating the fingerprint system in conditions close to those of the application because it includes fingerprint images acquired with an optical sensor as well as images acquired with a capacitive sensor. BIOMET database contains for each person 6 optical fingerprint images and 6 capacitive fingerprint images. Thus, we can compute in the best case 36 matching scores per person for the fingerprint system. However, as a quality control is applied in order to select images with good enough quality to be used by the system, we selected 124 persons among all BIOMET users having enough genuine scores (30 scores in average).

FRGCv2 [7] database has been chosen for the face system evaluation. This database includes important lighting variations and blurred images as will be the case in the VINSI application. We chose to use the "Experiment 4" protocol associated to FRGCv2. It is a degraded conditions protocol with controlled images (indoor, controlled light, controlled background) as references and uncontrolled images (outdoor, uncontrolled light, blurred images ...) for test. As for fingerprints, the face modality also has a quality control that selected images that have good enough quality to be used by the face system. We thus selected from FRGCv2, 124 persons with in average 30 genuine face matching scores per person, so that we have the same number of persons and the same number of matching scores per person for both biometric modalities.

In order to realize fusion experiments we considered a database of virtual persons obtained by combining a person of the BIOMET database with a person of the FRGCv2 database. This can be considered as valid as fingerprint and face are two uncorrelated modalities [8]. For each person, each of the 30 fingerprint matching scores is coupled with a face matching score. This permits to have 124 virtual persons with in average 30 pairs of genuine scores (one fingerprint score associated to one face score).

The fingerprint system used in this study is NFIS2 [9] (NIST Fingerprint Image Software) developed by NIST (National Institute of Standards and Technology). The Face system has been developed by our team and is based on PCA (Principal component analysis) [10].

Score distributions provided by both systems and presented in Fig. 2, allow noticing that performance of both systems are very different. Client and impostor scores distributions are well separated for the fingerprint system as shown in Fig. 2 (left).Because of this sharp separation, the fingerprint system considered individually can take the correct decision in most of the cases. On the contrary, scores distributions of clients and impostors for the face system have an important overlap, as shown in Fig. 2 (right). This will induce a lot of decision errors when considering the face system individually.

3 Global Fusion Methods

The aim of all fusion methods is to take a final decision, that is to separate the 2 classes (client and impostor) in the 2-dimensional space of scores (face and fingerprint scores).

In this section we will present global fusion methods, that is, methods that always fuse all the available scores. Most of these methods are described in [11].

We will present two types of global fusion methods:

- Score combination after a normalization step. Combination could be a simple average of the scores or a weighted average. Normalization of scores as well as of weighted coefficients if needed, requires a learning phase for the normalization parameters and the weights. Combination methods normalize each type of score separately before combining them.
- Classification with a learning phase. In this paper, we will present only two methods that model the bi-dimensional distribution of the scores provided by the fingerprint and face systems. These methods model the two types of scores together by estimating their joint distribution as already presented in [12].

We will now detail two score normalization schemes for combination methods:

- MinMax normalization method.
- Normalization method using posterior probabilities.

The MinMax score normalization method is very simple. The aim is to rescale each system's scores in the same interval. Indeed, as we can see in Fig. 2 (left and right), fingerprint scores vary from 0 to 1000 while face scores vary from -1 to 1. MinMax normalization permits to rescale each system's scores in the [0,1] interval. Normalization parameters are estimated on a dedicated training set.

Normalization using posterior probabilities consists in transforming each system's scores in posterior probabilities to be a client. To this aim, an assumption of normal distribution for the two classes (client and impostors) is made. This assumption permits to estimate the class conditional densities of each system by using a Gaussian component whose parameters are learnt on a dedicated training set. Bayes rule is then use to compute the posterior probabilities.

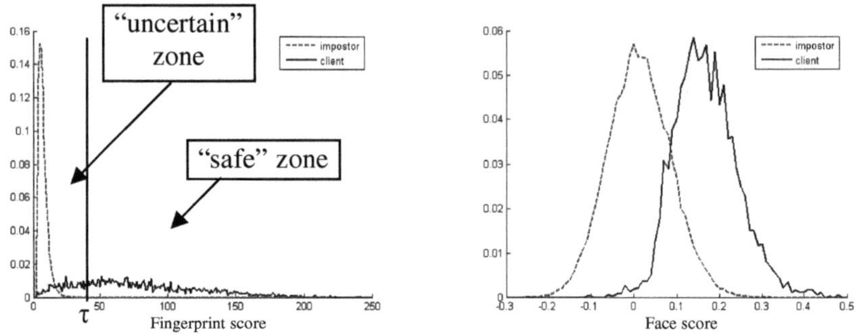

Fig. 2. Scores distributions of the fingerprint system (left) and face system (right). The "safe" and "uncertain" zones for the fingerprint system (first block of the architecture) are reported in Fig. 2 (left).

This normalization also allows rescaling each system's scores in the [0,1] interval. It generally turns out to be more precise than MinMax normalization presented previously because it relies on a model of each class and not only on a simple rescaling of scores. However, the model requires a training set representative of the application population and large enough to learn the Gaussian parameters accurately.

We will now present two fusion methods based on a bi-dimensional scores distribution model:

- Joint class conditional densities estimation with assumption of joint normal distribution.
- Estimation of joint class conditional densities based on Gaussian Mixture Model (GMM) [13].

These two bi-dimensional fusion methods consider face and fingerprint scores jointly and not separately as done by combination methods. Their objective is to estimate the joint class conditional densities in order to estimate the probability to be a client knowing the scores pair (face and fingerprint). This probability is the posterior probability given by the Bayes rule.

The difference between the two bi-dimensional fusion methods is the way the two class conditional densities are estimated. The first method is based on a normal assumption of the two-dimensional class conditional densities. The second method is less restrictive on the shape of the joint density of scores, since it is based on Gaussian Mixture Model (GMM) [13].

4 Performance of Global Fusion Methods

We run fusion experiments using the global fusion methods presented above. We also tested another combination method using a weighted sum. For this method, scores are normalized using MinMax method before being weighted. Weights are learnt in order to minimize the Half Total Error Rate (HTER) on the training set, by an exhaustive search of weights such as their sum equals 1.

In Table 1, the Equal Error Rate (EER) of each fusion method is reported as well as their relative improvement compared to face and fingerprint systems. The 90% confidence intervals for EER values are also indicated in square brackets. Confidence Intervals are estimated using a parametric method as described in [14]. Performance is estimated using a leave-one-out cross-validation protocol [15] on the persons. In practice, for each person, we learn the fusion parameters on the 123 remaining ones in order to produce the result associated to this person. All the global fusion methods improve the performance of the fingerprint system considered individually. Fusion by MinMax normalization and average of scores which is the simplest method does not improve significantly performance of the fingerprint system alone. All other schemes bring a big improvement on the performance compared to the best unimodal systems (fingerprint) by 42% to 45%. We notice that when scores are weighted before combining them (after normalization), performance is in the same range as those obtained with more complex methods involving density estimation (GMM, normalization using posterior probabilities, joint density model). It could be explained by the fact that the size of the training set is too small to correctly estimate densities for models like GMM.

Table 1. Performance of global fusion methods

	Average after MinMax normalization	Average of posterior probabilities	Client posterior probability after joint normal density estimation	Client posterior probability after joint density estimation by GMM	Weighted sum after MinMax normalisation (Fingerprint weight= 0.8 ; Face weight = 0.2)
EER (%)	4.33 [3.97;4.68]	2.65 [2.37;2.93]	2.57 [2.29;2.84]	2.57 [2.29 ;2.84]	2.49 [2.22 ;2.76]
Relative improvement of EER compared to fingerprint system (EER=4.59% [4.23;4.95])	6%	42%	44%	44%	45.7%
Relative improvement of EER compared to face system (EER=13.87% [13.27;14.47])	69%	81%	81.5%	81.5%	82%

5 Incremental Fusion Strategy

In Fig. 2 (left) that represents the client and impostor distributions of fingerprint scores on the test set, we can distinguish two zones where the fingerprint system has different behaviors and performance: a "safe" zone for high fingerprint scores, that corresponds only to client scores, and an "uncertain" zone for low fingerprint scores, where client and impostor scores overlap.

The implementation of our incremental fusion strategy is therefore dependent on a threshold τ on the fingerprint score (see Fig. 1). If the fingerprint score is higher than τ (belongs to the "safe" zone), the person is accepted, otherwise (the score is in the "uncertain" zone), we exploit the second block of our incremental fusion architecture, which is the fusion system combining both biometrics (see Fig. 1).

We define the cost of the fusion scheme as the average time (in seconds) for acquisition and processing of the biometric data of one user. The longer is the identity verification process, the higher is the cost. The total time is composed of 5 steps: fingerprint acquisition, fingerprint processing, face acquisition, face processing and fusion of fingerprint and face. In Table 2, we give approximations of the time needed for each of these 5 steps in the framework of the VINSI application on the handheld terminal.

Table 2. Time estimations of each step of the verification process on the VINSI terminal

Fingerprint acquisition	Fingerprint processing	Face acquisition	Face processing	Processing of scores fusion	Average time for one modality (fingerprint)	Average time for two modalities
10 s	1 s	10 s	4 s	0.001 s	11 s	25 s

Acquisition times are estimated empirically while processing times correspond to experiments on the mobile terminal. All the values given in Table 2 are average estimations and can vary from one user to the other and from one system to the other. Time for score fusion is not significant compared to all the other times. These values are only estimations allowing illustrating the cost gain using the incremental architecture.

The cost for global fusion methods is 25s. For the incremental fusion, for example if 30% of the users required two modalities, the cost will be 15,2s, that is a decrease of 40% compared to global fusion methods. Indeed, for these clients, the operator needs to acquire the face and process the fusion of both biometrics, that represents extra time cost mainly because of face acquisition, quality control and preprocessing. This extra time cost may be critical in certain applications where human flow is extremely important. The cost (c) is linearly related to the proportion (p) of clients requiring two modalities by the equation: c=11*(1-p)+25*p .

The objective of our strategy is to minimize the cost while still keeping a good performance (at least as good as the global fusion one). For this purpose, we propose to determine the acceptance threshold (τ) for fingerprint scores by the following formula:

$$\tau^* = \arg\min_{\tau}\left(FAR(\tau) \middle| FAR(\tau) > 0\right) \tag{1}$$

That is, τ^* corresponds to the threshold giving the lower non zero FAR (False Acceptance Rate). It follows that τ^* also corresponds to the maximum value of the impostor scores.

In order to determine this value, we used an independent development set of fingerprint data extracted from the same BIOMET database. On this development set, the corresponding acceptance threshold estimated by Eq. (1) is 35. It roughly corresponds to 70% of client scores and 0% of impostor scores in the "safe" zone

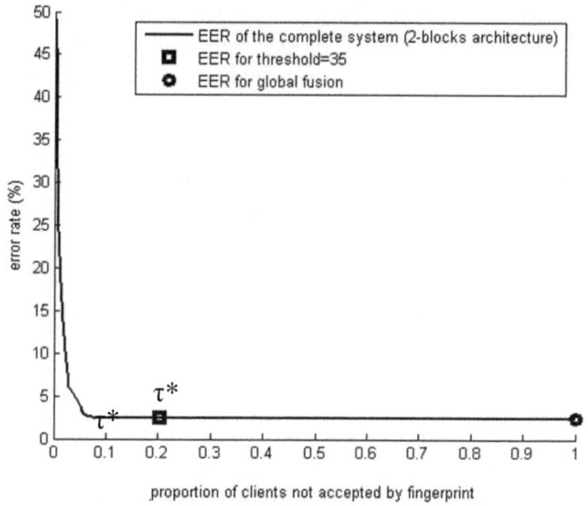

Fig. 3. Error rate of the complete fusion system (2-blocks architecture) vs. the proportion of clients not accepted by fingerprint only at the first stage, on the test set

(that is, 70% of clients are only processed by the first block of the architecture (fingerprint)) and 100% of impostor scores and 30% of client scores in the "uncertain" zone. It corresponds to an average cost of 15,2s (0,7*11+0,3*25) on the development set, that is a decrease of 45% of the cost compared to global fusion.

This value of the threshold corresponds on the test set to 80% of client scores and 0% of impostor scores in the "safe" zone and to 20% of client scores and 100% of impostor scores in the "uncertain" zone. It gives an average cost of 13,8s (0,8*11+0,2*25) on the test set.

With this strategy, we do not insure the optimality of the choice of the threshold (as defined in Eq. (1)) on the test database but due to the fact that the impostor score distribution drops sharply in its upper tail, the corresponding values of cost and error rates on the test set will be very similar to those obtained on the development set. The important point is that the strong decision power of the fingerprint system in the "safe" zone (no impostors) is kept on the test database when using this strategy for choosing the threshold on the development set.

More generally, we can study the variation of the Error Rate in function of the cost when the threshold τ is varying. In Fig. 3 the complete system (2-blocks architecture) error rate is presented depending on the proportion of clients needing to be processed by the second stage. Two special cases are considered, namely proportion=1, indicated in Fig. 3 by the mark (o). In this case, the two scores are always fused through a global fusion strategy, corresponding to a cost of 25s. The second value indicated in Fig. 3 by the mark (□) correspond to the proportion=0.2 (τ*=35). We can observe that for all the values in the interval [0.2, 1], the error rate roughly remains constant. It is even true for lower values (until p=0.08). Another way to determine the threshold could be to find the lowest cost associated to the incremental strategy whose performance is still equivalent to that of the global strategy on the development dataset (threshold for p=0.08). But the main drawback of this threshold optimization is that it is not robust in generalization. As we can see in Fig. 3, a small variation of threshold around 0.08 could imply a large degradation of performance. On the contrary, our choice of the threshold as defined in Eq. (1) is robust to small variations as explained above.

In the second block of our incremental architecture, the fusion method used to combine fingerprint and face can be of two sorts: it can be a global fusion method as described in Section 3; in this case, the parameters of the fusion scheme are learnt on the complete training dataset (all the available scores). But it could also be a local fusion method whose parameters are learnt only on the scores of the training set which belong to the "uncertain" zone. This direction will be studied in future work.

6 Conclusion

In this paper, we realized fusion experiments in the context of a mobile terminal aiming at performing identity verification using fingerprint and face. We implemented global fusion methods where fingerprint and face are combined for each test access. We noticed a significant improvement of performance due to fusion: 40% to 45% of relative improvement for global fusion methods compared to the best modality (fingerprint). We also implemented an incremental fusion strategy where the second biometric modality (face in this example) is used only when the fingerprint system alone is not sufficient to take a reliable decision. The aim of our incremental fusion strategy is to minimize the cost of using several modalities while keeping the same level of performance than global fusion methods. In the VINSI application, the cost is due to acquisition of biometric modalities and their processing. It is estimated in our strategy by the proportion of clients for which we need to acquire the second modality (face). With our incremental strategy we have shown that 80% of clients can be authenticated using fingerprint alone without the need of the face modality leading to a decrease of 55% of the cost compared to global fusion methods. This property is very important particularly in the context of a mobile identity verification application through a handheld device like the VINSI terminal, in which computational constraints are significant and real time response is mandatory. The only drawback of our incremental strategy is that if we do not use a second modality, we cannot guarantee that the fingerprint has not been forged, while it is possible at low effort as explained in [16]. Nevertheless, in this kind of application, a human operator is physically present and can therefore do a visual control thanks to the face image that is stored in the passport and can be visualized on the screen of the VINSI terminal. The big cost improvement that results from our incremental fusion strategy is due in particular to the fact that the client and impostor scores of the fingerprint system are very well separated and thus that fusion can be largely driven by the fingerprint modality. Anyhow, this incremental fusion strategy can also be applied with profit to other application contexts reflecting other behaviors of the modalities at hand.

Acknowledgments. This work was funded by the French ministry of Industry.

References

1. Pudil, P., Novovicoca, J., Blaha, S., Kittler, J.: Multistage Pattern Recognition with Reject Option. In: Proc. 11th IAPR Internat. Conf. on Pattern Recognition, The Hague, pp. 92–95. IEEE Computer Society Press, Los Alamos (1992)
2. Alpaydin, E., Kaynak, C.: Cascading Classifiers. Kybernetika 34(4), 369–374 (1998)
3. Mandhavanath, S., Govindaraju, V.: Serial Classifier Combination for Handwritten Word Recognition. In: Proceedings of the Third international Conference on Document Analysis and Recognition, August 14-15. ICDAR, vol. 2, p. 911. IEEE Computer Society, Washington (1995)
4. Cordella, L.P., Foggia, P., Sansone, C., Tortorella, F., Vento, M.: A Cascaded Multiple Expert System for Verification. Multiple Classifier Systems, 330–339 (2000)
5. Marcialis, G.L., Roli, F.: Serial Fusion of Fingerprint and Face Matchers. In: Haindl, M., Kittler, J., Roli, F. (eds.) MCS 2007. LNCS, vol. 4472, pp. 151–160. Springer, Heidelberg (2007)
6. Garcia-Salicetti, S., Beumier, C., Chollet, G., Dorizzi, B., Leroux-Les Jardins, J., Lunter, J., Ni, Y., Petrovska-Delacretaz, D.: BIOMET: a Multimodal Person Authentication Database Including Face, Voice, Fingerprint, Hand and Signature Modalities. In: Kittler, J., Nixon, M.S. (eds.) AVBPA 2003. LNCS, vol. 2688, pp. 845–853. Springer, Heidelberg (2003)
7. FRGCv2 database, `http://www.frvt.org/FRGC/`
8. Dorizzi, B., Garcia-Salicetti, S., Allano, L.: Multimodality in Biosecure: Evaluation on Real vs. Virtual Subjects. In: Proc. of ICASSP, Toulouse, France, May 14-19 (2006)
9. NIST Fingerprint system NFIS2, http://www.itl.nist.gov/iad/894.03/fing/fing.html
10. Turk, M., Pentland, A.: Eigenfaces for Recognition. Journal of Cognitive Neuroscience 3(1), 71–86 (1991)
11. Ross, A.A., Nandakumar, K., Jain, A.K.: Handbook of multibiometrics. Springer, Heidelberg (2006)
12. Allano, L., Morris, A.C., Sellahewa, H., Garcia-Salicetti, S., Koreman, J., Jassim, S., LyVan, B., Wu, D., Dorizzi, B.: Non intrusive multi-biometrics on a mobile device: a comparison of fusion techniques. In: Proc of SPIE 2006, conference on Biometric Techniques for Human Identification III, Orlando, Florida, USA, April 17-21 (2006)
13. Reynolds, D.A., Rose, R.C.: Robust Text-Independent Speaker Identification Using Gaussian Mixture Speaker Models. IEEE Transactions on Speech and Audio Processing 3(1) (January 1995)
14. Bolle, R.M., Ratha, N.K., Pankanti, S.: Error Analysis of pattern recognition systems – the subset bootstrap. Computer Vision and Image Understanding 93 (2004)
15. Devijver, P.A., Kittler, J.: Pattern Recognition: A Statistical Approach. Prentice-Hall, London (1982)
16. Matsumoto, T., Matsumoto, H., Yamada, K., Hoshino, S.: Impact of Artificial Gummy Fingers on Fingerprint Systems. In: Proceedings of SPIE. Optical Security and Counterfeit Deterrence Techniques IV, vol. 4677 (2002)

A Hidden Markov Model Approach to Classify and Predict the Sign of Financial Local Trends

Manuele Bicego, Enrico Grosso, and Edoardo Otranto

DEIR - University of Sassari
via Torre Tonda, 34 - 07100 Sassari (Italy)
{bicego,grosso,eotranto}@uniss.it

Abstract. In the field of financial time series analysis it is widely accepted that the returns (price variations) are unpredictable in the long period [1]; nevertheless, this unappealing constraint could be somehow relaxed if *sufficiently short* time intervals are considered. In this paper this alternative scenario is investigated with a novel methodology, aimed at analyzing short (local) financial trends for predicting their sign (increase or decrease). This peculiar problem needs specific models – different from standard techniques used for estimating the volatility or the returns – able to capture the asymmetries between increase and decrease periods in the short time. This is achieved by modeling directly the signs of the local trends using *two* separate Hidden Markov models, one for positive and one for negative trends. The approach has been tested with different financial indexes, with encouraging results also in comparison with standard methods.

1 Introduction

The analysis of financial signals represents a widely investigated research area, where many different methodologies have been proposed in the past, employing different techniques such as Neural Networks (see for example the special issue [2]), Support Vector Machines [3] or Hidden Markov Models (HMMs) [4,5,6].

Recently, the efforts of financial econometricians have been mainly concentrated on the prediction of volatility in financial time series [7], rather than on the price variations. In fact it has become widely accepted by that community that the price variations are unpredictable [1], following the idea that if a pattern exists it should be destroyed by the power of the market. Nevertheless this unappealing constraint could be somehow relaxed if *sufficiently short* periods are considered, as a couple of weeks: prior to the deletion of a pattern, the market needs a brief period (a local trend) to detect it and to change its dynamics. This idea is the basis of the technical analysis (see, for example, [8]), where different heuristics-driven graphical techniques and simple indicators (climate, flow-of-funds, market indicators) were developed to detect and forecast short trends in financial markets.

This paper addresses the aforementioned problem, proposing a methodology based on HMMs [9], aimed at recognizing and forecasting the sign of financial

N. da Vitora Lobo et al. (Eds.): SSPR&SPR 2008, LNCS 5342, pp. 852–861, 2008.

local trends. The direct forecasting of the sign – with respect to the real value – represents a quite novel trend in the econometric literature (for example in [10] a set of heuristics and trivial models are used to this aim). The main feature of the proposed approach is represented by the explicit encoding of the asymmetry present in the problem, obtained by employing *two different* HMMs to separately model the situations leading to increases or decreases. This is substantially different from standard financial approaches based on HMMs, where just one model is employed [4,11]; actually such single model is able to capture only the implicit level of asymmetry present in the signal (represented with the hidden states). On the contrary, our approach considers and models the separate signals using *two models*, highly and explicitly emphasizing the asymmetries typically present in this kind of data. Another key feature is that, instead of hypothesizing a distribution on the levels or the returns of the series (which is actually a crucial problem), we model directly the signs, transforming the data in discrete binary sequences (modeled by discrete HMMs) representing positive variations (increases) and negative variation (decreases) with respect to the previous period.

The models were trained with sequences that we trust be of increase (or decrease) — i.e. short sequences of T periods (few weeks of daily data), ending with τ increases (decreases). Each HMM training has been carefully initialized, while the number of states has been directly estimated from data using the Akaike Information Criterion (AIC – [12]). Recognition and prediction of signs are made on the basis of the likelihoods of the models; an automatic threshold-based reject rule has been also introduced, able to characterize "uncertainty" zones (zones neither of decrease nor of increase).

The proposed approach has been tested with encouraging results on the U.S.A. Dow Jones index and other shares of the same stock market with different risk; a comparison with alternative schemes has been also carried out.

2 The Proposed Approach

In this section the proposed approach is presented. Let us call F_+ the set of sequences ending with τ positive signs (sequences of "reasonable sure increase"), F_- the set of sequences ending with τ negative signs (sequences of "reasonable sure decrease") and $F_?$ the set of all the other sequences, which present an irregular alternation of the last τ signs. The choice of τ is quite subjective, depending on the number of consecutive positive days considered as an interesting gain by the investor, and could have a not negligible impact on the performances.

The main feature of the proposed methodology is that the F_+ and F_- sets are modeled separately. Actually many studies in financial econometrics (see, for example, [11,4]) stress the fact that the volatility of financial time series is characterized by asymmetric behavior, because the quiet and turmoil periods have different length and persistence; in this framework a *single* HMM seems suitable, each state representing a different regime of volatility. The asymmetric transition probability matrix provides the different length and persistence of the states along the full period studied. A similar reasoning could be applied when modeling the signs. Even more, the asymmetry could be furthermore stressed by explicitly emphasizing two degrees of asymmetry: first, the sequences belonging to F_+ and F_- have different behaviors (exploited by estimating two different independent models); second, each set of sequences can show asymmetries also in each proper dynamics (exploited by different states in each model).

2.1 HMM Definition

In this section, HMMs are briefly introduced, mainly to set up notation. A discrete-time hidden Markov model $\boldsymbol{\lambda}$ is defined by the following entities [9]:

- a set $S = \{S_1, S_2, \cdots, S_N\}$ of (hidden) states;
- a transition matrix $\mathbf{A} = \{a_{ij}\}$, where $a_{ij} \geq 0$ represents the probability of going from state S_i to state S_j;
- an emission matrix $\mathbf{B} = \{b(o|S_j)\}$, indicating the probability of emission of symbol o from state S_j;
- an initial state probability distribution $\boldsymbol{\pi} = \{\pi_i\}$, representing the probability of the first state $\pi_i = P[Q_1 = S_i]$

For a sequence $\mathbf{O}$ and an HMM $\boldsymbol{\lambda}$, there is a standard recursive procedure, called the *forward-backward* procedure [9], able to compute the probability $P(\mathbf{O}|\boldsymbol{\lambda})$. Given a set of observed sequences $\{\mathbf{O}_i\}$, the learning of the HMM parameters is usually performed using the well-known Baum-Welch algorithm [9], which is able to determine the parameters maximizing the loglikelihood $\log P(\{\mathbf{O}_i\}|\boldsymbol{\lambda})$.

2.2 Coding

When dealing with HMMs and financial series, a serious problem is to choose a proper distribution (namely a proper emission function) for the levels or the returns of the series, difficult task for the presence of high peaks, fat tails and higher kurtosis with respect to the Normal distribution; the typical solution is to increase the complexity of the distributions, assuming Student's t, GED, Gamma, etc. In our approach we overcame this critical choice by modeling directly the signs of the sequences, transforming the original series in binary strings (series of increase or decrease). Discrete data have been used in several works dealing with financial time series [13,14]. It should be noted that using more than two symbols could be also reasonable: in this last case a proper quantization mechanism should be inserted[1].

2.3 Training

The training phase is aimed at building two models, one devoted to modeling the sequences belonging to F_+ and one for the sequences of F_-. The two sets of

[1] A preliminary experimentation using *Vector Quantization* has been carried out, showing no significant variations.

sequences F_+ and F_- are used to train two discrete Hidden Markov Models (i.e. HMMs where the emission probability $b(o|S_j)$ is a binomial distribution), called $\boldsymbol{\lambda}_+$ and $\boldsymbol{\lambda}_-$ respectively. The training is performed using the standard Baum Welch re-estimation method [9], which is stopped at likelihood convergence. In order to deal with the initialization issue, the training has been repeated several times starting from different random conditions, choosing the model leading to the highest likelihood. Following findings in a related context [15], the number of states has been chosen with the *Akaike Information Criterion* (AIC – [12])[2]

3 Experimental Results

The proposed approach has been tested using the Dow Jones index (DJ) and some other shares with different risks, belonging to the components of the Dow Jones index: GE (General Electric) and IBM (International Machines Corp.) (low risk shares), BGP (Borders Group) and CGI (Commerc Group) (high risk shares). We employed daily close prices from 30 November 1995 to 5 February 2001, adjusted for dividends and splits — Yahoo Finance source. The models were trained using the period from December 1995 to June 1998, whereas the remaining part of the series has been used for testing.

A first large scope analysis (not shown here) was carried out with the Dow Jones index, in order to find the most suitable values of the parameters. The test was carried with 4 different sequence lengths (T=10, 15, 20, 25 – it has poor financial sense to perform the analysis outside the range two - five weeks) and 3 different τ (τ=3, 4, 5 — we stopped the analysis at $\tau = 5$ since no sequences with $\tau = 6$ were present in the dataset.): best results were obtained with $T = 10$ and $\tau = 5$. We noticed that in general the performances worsen when increasing the length of the sequence, especially with $\tau = 3$. This confirms the aforesaid intuition: longer sequences mislead the system, since the market is able to detect and correspondingly destroy the pattern. In table 1 the number of sequences used to build the models are displayed, together with the optimal number of states determined by the AIC criterion for all models. It is evident that the number of increases is often really different than the number of decreases, this suggesting an asymmetry between the two typologies of signals. This is confirmed by considering that the corresponding optimal models show almost always a different number of states.

In order to test both the recognition and the forecasting capabilities of the method, two different experiments have been carried out. The recognition capability of the proposed methodology was tested by classifying all the sequences of reliable increase and reliable decrease extracted from the testing set (disjoint from the training). Each sequence has been fed to the two trained HMMs, assigning it to the class whose model shows the highest log likelihood (Bayes rule). The number of correct classifications are reported in the second column of table 2.

[2] Actually, the proof, involving Monte Carlo simulation, was made for Autoregressive Markov Switching Models, that are dynamic models with parameters subject to Markov regime switching.

Table 1. Number of training sequences and best number of states for different indices

Index	Train seq F_+	Train seq F_-	States λ_+	States λ_-
DJ	28	8	3	2
GE	16	20	4	4
IBM	17	17	3	2
BGP	7	25	2	4
CGI	12	38	3	4

Table 2. Recognition and forecasting accuracies of the proposed approach for different indices (correct/total)

Index	Recogn. Accuracy	Forecast -1	Forecast -2	Forecast -3
DJ	26/26	26/26	26/26	26/26
GE	30/30	30/30	30/30	25/30
IBM	39/39	39/39	39/39	39/39
BGP	32/32	32/32	32/32	32/32
CGI	50/50	50/50	50/50	50/50

The second experiment was performed in order to assess the forecasting performances of the proposed approach, namely the capability of predicting the increase or decrease *before* its happening. To this end the sequences of reliable increase or decrease extracted from the testing set have been truncated and fed to the models to be classified; the classification result represents an attempt to predict in advance the behavior of the whole sequence (i.e. by just observing a part of it). Results are displayed in table 2 (last three columns), for different temporal horizons. More precisely, "Forecast -k" means that the forecasting is performed with a horizon of k days, namely by observing only $T - k$ signs. It is interesting to notice that the method shows an almost optimal performance in the recognition and forecast of the reliable sequences, and the different risk does not affect the forecast accuracy.

3.1 Comparative Analysis

The proposed approach was compared with two other techniques:

- Single Continuous HMM: the main features of the presented approach are the direct modeling of the sign (different than standard modeling of the return) and the employment of two HMMs (different than standard use of one HMM). So a single continuous HMM (Gaussian emission), directly modeling the returns and trained using all sequences, represents the natural competitor of our approach. The recognition / prediction is made by forecasting the next return using the Gaussian of the most probable state given the observed sequence (or forecasting the most probable state in k time steps).

– Local Polynomial Trend: this method is based on local polynomial trend (LP model – [16]):

$$y_t = \alpha + \sum_{i=1}^{p_h} \beta_i t^i \tag{1}$$

where y_t represents the observation (the level of the series) at time t and h refers to the model of increase (h=1) or decrease (h=2). This choice seems to be adequate for this kind of analysis: actually, the technical analysts do not employ statistical models but empirical evidence for their operations – see [8]. Different models have been estimated using different degrees p_1 and p_2 (ranging from 1 to T - 4), finally choosing the configuration maximizing the prediction accuracy. In this case, the coefficients α and β_i were estimated (via ordinary least squares) for each local trend to be recognized (or predicted). In practice, we have not utilized a training set, but the local trends of interest (eliminating the latest k observations in the forecasting experiments). As a matter, the information used in this analysis is more accurate with respect the HMM experiments. Then we used the estimated coefficients to estimate the observations of the local trend and verified if the estimated signs of the returns are equal to the true ones.

The analogous experiments described before have been repeated with these models (forecasting with single continuous HMM has been repeated 1000 times in order to ensure statistical significance of the sampling); the results are shown in Table 3 and 4. By comparing the results with Table 2, it could be noted the

Table 3. Recognition and forecasting accuracies for different indices (correct/total) when using one single continuous HMM

Index	Recogn. Accuracy	Forecast -1	Forecast -2	Forecast -3
DJ	22.3/26	21.9/26	21.6/26	21.7/26
GE	29.2/30	27.2/30	28.8/30	27.9/30
IBM	34.2/39	26.2/39	33.2/39	29.7/39
BGP	27.3/32	24.6/32	27.1/32	27.1/32
CGI	42.3/50	19.8/50	22.6/50	22.9/50

Table 4. Recognition and forecasting accuracies for different indices (correct/total) when using Polynomial Local Trends

Index	Recogn. Accuracy	Forecast -1	Forecast -2	Forecast -3
DJ	25/26	24/26	22/26	20/26
GE	23/30	26/30	23/30	17/30
IBM	36/39	34/39	31/39	24/39
BGP	15/32	21/32	13/32	10/32
CGI	33/50	38/50	29/50	23/50

great increase in accuracy gathered by the HMM modeling strategy proposed in this paper. In particular, for difficult shares (like CGI) the predictions of the other methods are almost completely random.

4 Detection of Fluctuant Sequences

In practical cases, given a generic sequence, the goal of an investor is to get some suggestions for the solution of the following problems: does the sequence present a clear and defined trend? If so, does the sequence represent a reliable increase or a reliable decrease? Further, can we detect such situation in advance? The main purpose of this paper was to answer to the second and third questions. Nevertheless, explicitly addressing the first issue is crucial for practical applications. In this section we provide a possible solution to this aspect by introducing in the proposed methodology a reject rule. The most widely employed rejection rule is the Chow rule [17], which indicates to reject a decision if the maximum posterior probability is under a given threshold. This rule is optimal in a Bayesian sense, leading to the optimal reject-error tradeoff, but assumes the complete knowledge of the posterior. Moreover it is defined in a multi-class context, and could not completely exploit the dichotomy of the problem. For these reasons a different rule is introduced here, which could be easily derived from our asymmetric modeling methodology. In particular the difference between the log likelihoods is considered

$$C(O) = |\log P(O|\boldsymbol{\lambda}_+) - \log P(O|\boldsymbol{\lambda}_-)| \tag{2}$$

If the system does not show a clear evidence in favor of a certain class, i.e. $C(O) < \theta$ (for a chosen threshold θ), the sequence is not assigned neither to F_+ nor to F_-, but it is rejected (namely it is assigned to $F_?$).

The choice of the threshold θ is obviously crucial, and it is linked to the level of confidence desired by the user. In our approach it is directly computed on the basis of the training set: in particular we divide the training set into two subsets, one containing the sequences with a clear trend (i.e. both the sequences of reliable increase and reliable decrease), and one with the others. Then the quantity $C(O)$ is computed for both the sets and the two empirical pdf's are estimated; the threshold is then computed in a Bayesian way, by considering the point where the two pdfs intersect. We assumed equal priors, even if a different choice based on prior knowledge of the analyzed data set is also possible. The idea is briefly sketched in Fig. 1: the continuous line represents the empirical distribution of the statistic (2) for the sequences of length $T = 10$ of the training set belonging to $F_?$ and the dot line the analogous distribution for the sequences with $T = 10$ and $\tau = 5$ of the training set belonging to F_+ and F_-. The intersection point of the two curves detects the threshold θ (equal to 3.1868). All the sequences of the testing set with $Confidence(O) < \theta$ will be considered as belonging to $F_?$, whereas the others will be classified in F_+ or F_-.

In Table 5 we show the percentages of the right rejections when trying to recognize all the possible sequences. These accuracies represent the number of

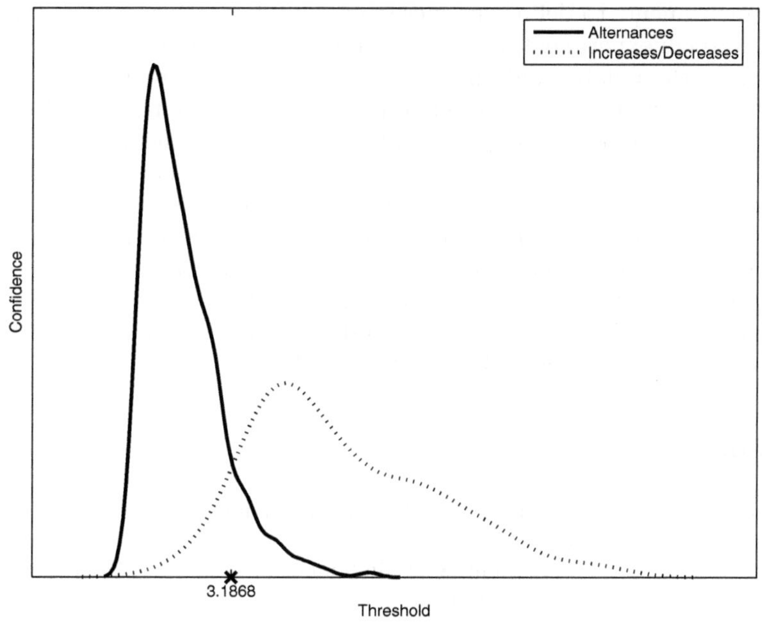

Fig. 1. Empirical distributions of the confidence statistics for $T = 10$ and $\tau = 5$ and detection of the threshold

Table 5. Percentages of correctly rejected sequences

Index	Correct
DJ	93.1%
GE	95.1%
IBM	85.8%
BGP	91.3%
CGI	93.3%

times the system is able to correctly rejecting a sequence belonging to $F_?$: we could notice that the percentage is generally high.

5 Conclusions

In this paper a novel approach for recognizing and forecasting brief sequences of series relative to financial markets has been proposed. The model explicitly and directly exploits the natural asymmetry present in the market by training two separate models, one for the increase situation and one for the decrease. Experiments on different indices show the feasibility of the proposed method.

Future efforts will be devoted to extend the presented model. An immediate attempt will deal with different definitions of local trends (namely what we define as sequences of reliable increase/decrease), for example using some technical analysis instruments to define the increasing and decreasing periods. Moreover, a more thorough experimental analysis will be made in order to test recognition accuracies on the reliable and unreliable sequences at different rejection levels.

Another interesting research direction is to try to enrich the model by linking some external information; e.g., it is plausible that the movements in a leading market (such as the U.S. market) influence the movements in other markets. There are many possibilities: inserting in the transition probabilities of the HMM a dependence on the value of the other markets, as in [18], or modeling multiple sequences (namely different indices or shares) with the same Markov Chain or a multiple Markov Chain, with a mutual influence among the states of different variables (as in [19]), or finally by modeling interactions between couple of processes (as in the Coupled Hidden Markov Models [20]).

References

1. Campbell, J., Lo, A., MacKinlay, A.: The Econometrics of Financial Markets. Princeton University Press, Princeton (1997)
2. IEEE: Special issue on neural networks in financial engineering. IEEE Transactions on Neural Networks 12(4) (2001)
3. Cao, L., Tay, F.: Support vector machine with adaptive parameters in financial time series forecasting. IEEE Trans. on Neural Networks 14(6), 1506–1518 (2003)
4. Dueker, J.: Markov switching in garch processes and mean-reverting stock market volatility. Journal of Business and Economic Statistics 15, 26–34 (1997)
5. Bengio, Y., Lauzon, V.P., Ducharme, R.: Experiments on the application of IOHMMs to model financial returns series. IEEE Transaction on Neural Networks 12(1), 113–123 (2001)
6. Liehr, S., Pawelzik, K., Kohlmorgen, J., Lemm, S., Müller, K.R.: Hidden markov mixtures of experts for prediction of non-stationary dynamics. In: IEEE Proc. of Neural Networks for Signal Processing IX, pp. 195–204 (1999)
7. Engle, R.: Autoregressive conditional heteroskedasticity with estimates of the variance of u.k. inflation. Econometrica 50, 987–1008 (1982)
8. Pring, M.: Technical Analysis Explained, 3rd edn. McGraw-Hill, New York (1991)
9. Rabiner, L.: A tutorial on Hidden Markov Models and selected applications in speech recognition. Proc. of IEEE 77(2), 257–286 (1989)
10. Christofferson, P., Diebold, F.: Financial asset returns, direction-of-change forecasting, and volatility dynamics. Management Science 52(8), 1273–1287 (2006)
11. Hamilton, J.: Time Series Analysis. Princeton University Press, Princeton (1994)
12. Akaike, H.: A new look at the statistical model identification. IEEE Trans. on Automatic Control AC-19, 716–723 (1974)
13. Frankel, J., Rose, A.: Currency crashes in emerging markets: an empirical treatment. Journal of International Economics 41, 351–366 (1996)
14. Holthausen, R., Larcker, D.: The prediction of stock returns using financial statement information. Journal of Accounting and Economics 15, 373–412 (1992)
15. Psaradakis, Z.: Spagnolo: On the determination of the number of regimes in Markov-Switching autoregressive models. Journal of Time Series Analysis 24, 237–252 (2003)
16. Kendall, G., Stuart, A., Ord, J.: The advanced Theory of Statistics. Griffin, London (1983)

17. Chow, C.: An optimum character recognition system using decision functions. IRE Trans. on Electronic Computers 6, 247–254 (1957)
18. Diebold, F., Lee, J., Weinbach, G.: Regime switching with time-varying transition probabilities. In: Hargreaves, P. (ed.) Nonstationary Time Series Analysis and Cointegration, pp. 283–302. Oxford University Press, Oxford (1994)
19. Otranto, E.: The multi-chain markov switching model. Journal of Forecasting 24(7), 523–537 (2005)
20. Brand, M., Oliver, N., Pentland, A.: Coupled hidden markov models for complex action recognition. In: Proc. of IEEE Conf. on Computer Vision and Pattern Recognition (1997)

Local Metric Learning on Manifolds with Application to Query–Based Operations

Karim Abou-Moustafa and Frank Ferrie

The Artificial Perception Laboratory
Centre for Intelligent Machines, McGill University
3480 University street, Montreal, QC, Canada H3A 2A7
{karimt,ferrie}@cim.mcgill.ca

Abstract. We first investigate the combined effect of data complexity, curse of dimensionality and the definition of the Euclidean distance on the distance measure between points. Then, based on the concepts underlying manifold learning algorithms and the minimum volume ellipsoid metric, we design an algorithm that learns a local metric on the lower dimensional manifold on which the data is lying. Experiments in the context of classification on standard benchmark data sets showed very promising results when compared to state of the art algorithms, and consistent improvements over the Euclidean distance in the context of query–based learning.

1 Introduction

The Euclidean distance between two points $\mathbf{x}$ and $\mathbf{y}$ in an Euclidean space $\mathbb{R}^d$ or finding the nearest neighbour(s) or matching points to a query point, are two basic operations for a plethora of algorithms in the literature on pattern recognition, machine learning, computer vision, image retrieval and data mining. Indeed, the solid roots of the Euclidean distance in geometry, its intuitive meaning and its simple computation make it unarguably a prime choice as a similarity measure or simply as the true distance between points. Here, our concern is focused on query–based operations where a typical scenario is to have a set $\mathcal{X}$ of high dimensional vectors (images, image patches, feature vectors, etc) and it is required to find a set of nearest neighbors or matches to a point that is either in the same set or a new incoming one. In such situations, the Euclidean distance is usually the measure of choice for assessing the similarity between points. Despite its success in many applications, and with a sober look at the Euclidean distance, there are few reasons that raise a question on the full validity of this metric when dealing with high dimensional real life data.

A first reason for that is the curse of dimensionality. That is, in high dimensional spaces, the distance between every pair of points is almost the same for a wide variety of data distributions and distance functions [5]. Hence, in such high dimensional spaces, the notion of similarity becomes less accurate. The second reason stems from the complexity of data arising from real life applications. Usually, such data are: (1) High dimensional, highly structured and

N. da Vitora Lobo et al. (Eds.): SSPR&SPR 2008, LNCS 5342, pp. 862–873, 2008.

nonlinear such as images, text documents, proteins, etc. (2) Measurements from various sources at different scales and with various degrees of variability and correlation, and (3) Prone to various sources of noise that may largely deviate measurements and raise outliers in the data. The third reason, which is the definition of the Euclidean distance, combines and builds over the aforementioned ones. By expanding the squared Euclidean norm $\|\mathbf{x}-\mathbf{y}\|_2^2$ to $(\mathbf{x}-\mathbf{y})^T(\mathbf{x}-\mathbf{y}) = (\mathbf{x}-\mathbf{y})^T\mathbf{I}(\mathbf{x}-\mathbf{y})$, where $\mathbf{I}$ is the identity matrix, one directly obtains an instance of the general family of Mahalanobis distances between points $\mathbf{x}$ and $\mathbf{y}$: $D_S(\mathbf{x},\mathbf{y}) = (\mathbf{x}-\mathbf{y})^T\mathbf{S}^{-1}(\mathbf{x}-\mathbf{y})$, where $\mathbf{S}$ is a symmetric and positive definite matrix. Replacing $\mathbf{S}$ by $\mathbf{I}$ implies that the Euclidean distance takes it for granted that all variables are independent, the variance across all dimensions is one and that covariances among all variables are zero, a situation that is hardly attained in real life data. Therefore, the Euclidean distance, by definition, ignores the structure, scale, variance and correlations in the data and consequently it is wise to say that *"in the absence of clear evidence of Euclidean geometry, the metric structure should be inferred from the data"* [13].

Contribution: We are interested in learning a local metric for query–based operations. We combine the concepts underlying manifold learning algorithms and the minimum volume ellipsoid metric (MVEM) [1] in a unified algorithm that tries to overcome the aforementioned problems with using the Euclidean distance as a metric or as similarity measure. That is, given a data set $\mathcal{X}$ of some high dimensional points and a query point $\mathbf{x}_q$ of similar dimensionality, we are interested in learning a similarity measure based on the information in $\mathbf{x}_q$, a small neighbourhood $\mathcal{N}_{\mathbf{x}_q}$ around $\mathbf{x}_q$ and the data set $\mathcal{X}$; i.e. the metric is adaptive for each query point (whether $\mathbf{x}_q \in \mathcal{X}$ or is a new incoming one), such that each $\mathbf{x}_q$ can better define its nearest neighbors from $\mathcal{X}$. To this end, our approach will rely on defining the local metric on the lower dimensional manifold on which $\mathbf{x}_q$ lies on. This is different from previous metric learning algorithms that focused on learning a metric specifically for k–NN (nearest neighbors) classification, exemplified by [19,12,22], in that the proposed algorithm is unsupervised, self–adaptive for each new query point, and define the metric on the lower dimensional manifold on which the query point is lying.

2 Related Work

The earliest work on metric learning is due to Short and Fukunaga [19] where they define an optimal distance measure to minimize the difference between the finite sample k-NN error and the asymptotic k-NN error. Recently, there are arguably three main streams for metric learning: global metric learning using labels or similarity constraints (side–information) [23,2,22], locally fully supervised metric learning similar to [19,12], and unsupervised metric learning exemplified by [20,17,14]. We briefly review some of these algorithms and the interested reader can see [24] for a comprehensive review on the topic.

Most of the algorithms in the first category learn a global metric through the general family of Mahalanobis distances $\mathrm{D}_A(\mathbf{x},\mathbf{y}) = (\mathbf{x}-\mathbf{y})^T\mathbf{A}(\mathbf{x}-\mathbf{y})$ and the differences between these algorithms are due to the constraints defining each metric. In metric learning using labels, [10] defines a differentiable probability function (softmax) using $D_A(\mathbf{x}, \mathbf{y})$ with $\mathbf{A}$ as its parameter. This function is optimized to maximize the probability of correct classification using the labels in the training set. In an extended work, [9] uses the same objective function to map all points that belong to the same class into a single point. Alternatively, [22] searches for a matrix $\mathbf{A}$ that defines a linear transformation such that k nearest neighbours of the same class are always kept together while samples from other classes are separated by a large margin.

In learning with similarity constraints, [18] uses similarity constraints in the form of triplet relative comparisons; i.e. for samples $\mathbf{x}$, $\mathbf{y}$ and $\mathbf{z}$ the relative comparison information is in the form of $D_A(\mathbf{x}, \mathbf{y}) > D_A(\mathbf{x}, \mathbf{z})$. Hence, the objective is to find a matrix $\mathbf{A}$ with minimum rank that respects such constraints. Using a different form of constraints, [23] tries to find a matrix $\mathbf{A}$ that will keep similar points close to each other while keeping dissimilar points far from other, i.e. using positive and negative similarity constraints. RCA [2] in a simpler setting uses only positive constraints and tries to find a matrix $\mathbf{A}$ that keeps similar points close to each other.

3 The Minimum Volume Ellipsoid Metric

In previous work [1], we have introduced the Minimum Volume Ellipsoid Metric (MVEM) as well as the Minimum Volume Ellipsoid of Nearest Neighbors (MiniVenn) algorithm to learn the proposed metric. The MVEM is a similarity measure that tries to mitigate the difficulties of data complexity and curse of dimensionality by using a parametrized Mahalanobis distance instead of the Euclidean distance. The MVEM is defined independently for each point (referred to as a query point) in a data set based on the information in a small neighborhood around it. Hence, the MVEM is locally defined for each query point. First, the MiniVenn is briefly reviewed to see how it can learn this local metric, then its limitations are addressed in the next subsection and finally the modified algorithm is presented with a discussion on its generalization.

Consider a data set $\mathcal{X} = \{\mathbf{x}_i \mid 1 \leq i \leq n\} \subset \mathbb{R}^d$ that is drawn from a probability distribution $p(\mathbf{x})$, and a query point $\mathbf{x}_q \in \mathbb{R}^d$ that is also assumed to be drawn from $p(.)$, where d is the dimensionality of the input space, and n is the number of samples. The first step in MiniVenn is to find the m nearest neighbors to $\mathbf{x}_q$ from the set $\mathcal{X}$ using the Euclidean distance. Under the concept of locality [3], it is assumed that in a small neighborhood $\mathcal{N}_{\mathbf{x}_q}$ around $\mathbf{x}_q$ (ϵ–ball or m nearest neighbors), points will tend to be similar and share some common properties. Such similarities can be characterized by means of the neighborhood's covariance matrix (or local covariance) $\mathbf{S}_q$. The induced Mahalanobis distance using $\mathbf{S}_q$ can measure the similarity between $\mathbf{x}_q$ and its neighbors while taking correlations and variances into considerations. In other words, the distance between $\mathbf{x}_q$ and any other point $\mathbf{x}_i \in \mathcal{X}$ will be based on the information in $\mathbf{x}_q$ and $\mathcal{N}_{\mathbf{x}_q}$ that is encoded in $\mathbf{S}_q$. However, due to the curse of dimensionality effect, the high nonlinearity of real life data and noise, estimating such a local covariance matrix using a Maximum Likelihood estimator (MLE) is hard and not reliable [7]. Therefore, instead of using a MLE, MiniVenn uses a robust estimator, the Minimum Volume Covering Ellipsoid (MVCE) estimator [16], to compute an accurate and a robust estimate for $\mathbf{S}_q$. Finally, the estimated $\mathbf{S}_q$ is used to define a more accurate Mahalanobis distance to the measure the similarity between $\mathbf{x}_q$ and any other point in $\mathcal{X}$.

Limitation of the MiniVenn algorithm: MiniVenn and consequently the MVEM suffered from three drawbacks. The first drawback arises from using the Euclidean metric in a high dimensional space to find the nearest neighbors of $\mathbf{x}_q$. As mentioned in the introduction, due to the curse of dimensionality effect, the notion of similarity in a small neighborhood around $\mathbf{x}_q$ will be inaccurate. Second, in order to compute the robust estimate using the MVCE estimator, MiniVenn has to solve a convex optimization problem that relies on a difficult formulation for the problem. In addition to using a general convex optimization package [11] that did not consider the special problem structure, this formulation led to a very slow algorithm for computing the MVCE. This, by its turn, hindered the usage of the MVEM in practical situations that require fast query–based operations. Finally, the literature on manifold learning algorithms assumes that the data actually lies on or near a lower dimensional nonlinear manifold that captures most of the data variability and is embedded in the high dimensional input space. MiniVenn in its current design state does not take this assumption into consideration and it is our objective to let MiniVenn defines the metric on the lower dimensional manifold on which $\mathbf{x}_q$ lies on.

Algorithm 1. Regularized Minimum Volume Ellipsoid of Nearest Neighbors: *Learns a local metric for query point* $\mathbf{x}_q$ *on the manifold on which* $\mathbf{x}_q$ *is lying*

Require: $\mathcal{X}_{n\times d}$, $\mathbf{x}_q$, m, τ and ρ where $\mathcal{X}_{n\times d}$ is the training set with n d-dimensional samples, $\mathbf{x}_q$ is the query point, $m \geq d+1$ is a user input that controls the size of the neighborhood, $\tau > 0$ is the threshold to select the leading (tangent) directions with large eigenvalues along the manifold and $\rho \in [0, 1]$ is the MVEM regularization parameter.

1: Find the set $\mathcal{N}_{\mathbf{x}_q}$ that has the m similar points to $\mathbf{x}_q$ using the similarity measure in Section 3.1.

2: Compute the robust estimate of $\mathbf{S}_q$ defined by the MVCE estimator for the set $\mathcal{N}_{\mathbf{x}_q}$ and centre $\mathbf{x}_q$ using Titterington algorithm [21].

3: Compute the eigen decomposition of $\mathbf{S}_q = \mathbf{V}\mathbf{L}\mathbf{V}^T$ where $\mathbf{V} = [V_1 \ \dots \ V_d]$, $\mathbf{L} = \text{diag}(\lambda_1, \dots, \lambda_d)$ are the matrices of eigenvectors and eigenvalues respectively and $\lambda_1 > \lambda_2 > \dots > \lambda_d$.

4: Select the d_0 leading eigenvalues such that $\lambda_{[1\ :\ d_0]} > \tau$ and form the matrix $\tilde{\mathbf{L}} = \text{diag}(\rho, \dots, \rho, \frac{1}{\lambda_{d_0+1}}, \dots, \frac{1}{\lambda_d})$

5: **return** $\tilde{\mathbf{S}}_q^{-1} = \mathbf{V}\tilde{\mathbf{L}}\mathbf{V}^T$

3.1 The Modified MiniVenn Algorithm

The modified MiniVenn algorithm, shown in Algorithm (1), tries to overcome the above mentioned limitations by modifying the definition of local neighborhoods, by modifying the computation of the MVCE estimator, and finally by adding two extra steps for manifold detection. The algorithm proceeds as follows. In step 1, the algorithm defines a local neighborhood $\mathcal{N}_{\mathbf{x}_q}$ for $\mathbf{x}_q$ based on a similarity measure, explained in the following subsection, different than the Euclidean distance. In step 2, similar to the original MiniVenn, the algorithm computes the robust estimate of the covariance matrix $\mathbf{S}_q$ using the MVCE of the set $\mathcal{N}_{\mathbf{x}_q}$. The difference here is in the formulation and algorithm used to compute the MVCE of $\mathcal{N}_{\mathbf{x}_q}$. Steps 3 and 4 are new steps in the algorithm concerned with manifold detection, estimation of the local intrinsic dimensionality at $\mathbf{x}_q$, and MVEM regularization. In the following, each modification and addition will be explained in more detail.

Redefining local neighbourhoods: The definition of $\mathcal{N}_{\mathbf{x}_q}$ for the query point $\mathbf{x}_q$ in the original MiniVenn used the Euclidean distance as a similarity measure to find the m nearest neighbors to $\mathbf{x}_q$. Since our major objective is to find the most similar points to $\mathbf{x}_q$, one can define a different similarity measure between points and use it to define local neighborhoods. Indeed, selecting appropriate neighbourhoods is a key factor to the success of local learning algorithms [14,20]. A flexible and easy to compute similarity measure is the dot product between two vectors. That is, let $s_i = \text{Sim}(\mathbf{x}_q, \mathbf{x}_i) = \langle \mathbf{x}_q^{'}, \mathbf{x}_i^{'} \rangle = \mathbf{x}_q^{' T} \mathbf{x}_i^{'}$, be the similarity measure between $\mathbf{x}_q$ and $\mathbf{x}_i$, where $\mathbf{x}_q^{'} = \mathbf{x}_q / \|\mathbf{x}_q\|_2$, $\mathbf{x}_i^{'} = \mathbf{x}_i / \|\mathbf{x}_i\|_2$ and $\mathbf{x}_i \in \mathcal{X}$. This is equivalent to finding the m nearest neighbors for $\mathbf{x}_q^{'}$ after normalizing all the points to lie on the unit sphere in $\mathbb{R}^d$. This similarity measure, formulated as a dot product, can be extended to accommodate different similarity measures using the kernel trick and hence can mitigate the curse of dimensionality effect.

Fast computation of the MVCE: Let $\mathcal{N}_{\mathbf{x}_q} = \{\mathbf{x}_j \mid 1 \le j \le m,\ \mathbf{x}_j \in \mathcal{X}\}$ be the set of similar neighbors to the point $\mathbf{x}_q$ using the above defined similarity measure. The MVCE of $\mathcal{N}_{\mathbf{x}_q}$ with centre $\mathbf{x}_q$ is denoted by $\mathcal{E}$ and is parameterized by a symmetric and positive definite matrix $\mathbf{S}_q \in \mathbb{R}^{d \times d}$ as follows [4]:

$$\mathcal{E} = \{\mathbf{x}_j \mid \|\mathbf{S}_q^{-\frac{1}{2}} \mathbf{x}_j - \mathbf{b}\|_2^2 \le 1,\ \forall j\} \tag{1}$$

where $\mathbf{b} = \mathbf{S}_q^{-\frac{1}{2}} \mathbf{x}_q$. Since $V(\mathcal{E}) \propto \det(\mathbf{S}_q)$, where $V(\mathcal{E})$ is the ellipsoid's volume, minimizing this volume can be formulated as follows:

$$\min_{\mathbf{S}_q} \ \log\det \mathbf{S}_q, \quad \text{s.t.} \quad \|\mathbf{S}_q^{-\frac{1}{2}} \mathbf{x}_j - \mathbf{b}\|_2^2 \le 1,\ \forall j \tag{2}$$

The objective and the constraints in (2) are convex in $\mathbf{S}_q$, therefore this optimization problem has a unique global optimal solution. However, as in [1], directly solving this optimization problem using standard convex optimization libraries such as CVX [11] showed to be computationally expensive and not efficient for practical situations. Alternatively, the dual of this optimization problem, thanks to Titterington [21], is easier to optimize and has a very fast and efficient algorithm for its computation (see [21] for algorithm details) :

$$\max_{\mathbf{S}_q, \Phi} \quad \log\det(\mathbf{S}_q) \quad \text{s.t.} \quad \mathbf{\Phi} \in \mathbb{R}^m, \quad \mathbf{\Phi} \geq 0, \quad \mathbf{\Phi}^T\mathbf{e} = 1 \tag{3}$$

$$\mathbf{S}_q = \sum_{j=1}^{m} \phi_j(\mathbf{x}_j - \mathbf{x}_q)(\mathbf{x}_j - \mathbf{x}_q)^T + \gamma\mathbf{I}$$

where Φ is the vector of dual variables ϕ_j, $\gamma \geq 0$ and $\gamma\mathbf{I}$ is an extra constraint that guarantees a minimal diameter of the ellipsoid in all directions. This would prevent the ellipsoid from collapsing to zero volume especially in large dimensional spaces [6].

Manifold detection: The literature on manifold learning algorithms assumes that despite the high dimensionality of the data in the input space, most of the data variability can be captured by far fewer dimensions known as the intrinsic dimensionality of the data. Accordingly, it is assumed that the data lies on or near (due to noise) a lower dimensional nonlinear manifold that is embedded in the high dimensional input space. Real life data, however, are usually highly structured, nonlinear and prone to various sources of noise. Consequently, the data might not actually lie on a single nonlinear manifold, but rather on or near several disconnected nonlinear manifolds [13]. It is this last observation that motivates our objective to let MiniVenn define the metric on the lower dimensional manifold on which $\mathbf{x}_q$ is lying. To detect this manifold, MiniVenn performs an eigen decomposition and a regularization step for the robust estimate $\mathbf{S}_q$. The benefit of the eigen decomposition is twofold: (1) It can estimate the intrinsic dimensionality of the data using Fukunaga's algorithm [8] by means of the number of dominating eigenvalues of $\mathbf{S}_q$ (which is the role of parameter τ), and (2) The orthogonal eigenvectors of $\mathbf{S}_q$ decide which vectors are tangent or normal to the underlying manifold. That is, the eigenvector associated with the smallest eigenvalue (or lowest variance in $\mathcal{N}_{\mathbf{x}_q}$) is normal to the manifold, while the eigenvector associated with the largest eigenvalue (or highest variance in $\mathcal{N}_{\mathbf{x}_q}$) is tangent to the manifold and the latter is the main direction of interest since it is the direction that goes along the manifold and contributes the most to the dissimilarity measure in the neighborhood of $\mathbf{x}_q$. Note that in a d–dimensional space and for a d_0–dimensional manifold with $d_0 \ll d$, there will be approximately d_0 tangent vectors associated with the largest eigenvalues.

The Mahalanobis distance, however, measures the similarity using $\mathbf{S}_q^{-1}$, i.e. by taking the inverse of the eigenvalues, thus assigning small weights to high variance components (tangent eigenvectors) and large weights to low variance components (normal eigenvectors). It is at this point that the regularization parameter ρ is needed to emphasize the contribution of the main tangent vectors over the contribution of normal and less significant tangent vectors. More specifically, ρ influences the notion of similarity of the MVEM, however this is task dependent since it can tune the MVEM according to the objective of the task under consideration.

3.2 Generalization of the MVEM

Generalization of the MVEM is controlled by the MiniVenn's four parameters: m, τ, ρ and implicitly γ to compute the optimization in (3). While m and τ reflect the topological properties of the data, ρ influences the notion of similarity of the obtained metric. Using [8], τ can be fixed for a data set since it is a threshold on the normalized eigenvalues. Similarly, γ can be fixed for each data set separately although it was fixed to either 0 or 0.1 in all our experiments. More attention however, is required to select m and ρ. A large value of m will over smooth the main tangent directions of the patch on which $\mathbf{x}_q$ is lying, while a very small value will lead to crude and rather fragile estimates of these directions. An intuitive approach is to select m and ρ via an optimization procedure. This can be achieved by linking the two parameters to an objective function that can be optimized. The optimal objective function in this case would be the objective function of the task under consideration. Implicitly, this means that the metric (or the MVEM) will be tuned to maximize or minimize this objective function. For instance, in the case of our experiments on query-based learning, m and ρ were optimized by a grid search to minimize the expected zero–one loss $E[L(Y, f(X))] = E[1 - \delta(Y, f(X))]$ (or miss-classification rate) on the available training set, where Y is the true label of the input X, $f(X)$ is the decision obtained from the classifier, and the $\delta(.,.)$ is the Kronecker delta function. Accordingly, since there is a training phase to optimize m and ρ directly on the task's objective function, the MVEM is expected to generalize well on unseen data sets.

4 Experimental Results

To assess the validity and generalization of the modified MiniVenn algorithm, we have conducted extensive experiments in the context of classification on a large variety of standard benchmark data sets. In our experiments, fifteen data sets were used from the UCI Machine Learning Repository [15], shown in Table 1,

Table 1. The fifteen UCI [15] data sets used in our experiments with the number of classes (C), size (S) and dimensionality (D). m_{opt} is the optimal neighbourhood size for each k–NN classifier ($k = 1$, $k = 3$, $k = 5$) using the MVEM.

ID	Dataset	C	S	D	m_{opt}	ID	Dataset	C	S	D	m_{opt}
bal	Balance	3	625	4	7,8,10	new	NewThyroid	3	215	5	11,6,6
bup	Bupa	2	345	6	15,8,16	pim	Pima	2	768	8	22,12,9
gla	Glass	7	214	9	11,16,20	seg	Segment	7	2086	18	33,21,20
hou	HouseVotes	2	341	16	43,19,19	tic	TicTacToe	2	958	9	10,10,19
ion	Ionosphere	2	350	33	42,35,85	wdb	WDBC	2	569	30	31,43,31
iri	Iris	3	150	4	8,7,9	win	Wine	3	168	13	14,35,31
lym	Lymphography	4	148	18	41,49,33	yea	Yeast	10	1484	6	8,17,7
pag	Pageblocks	5	5473	10	29,23,27						

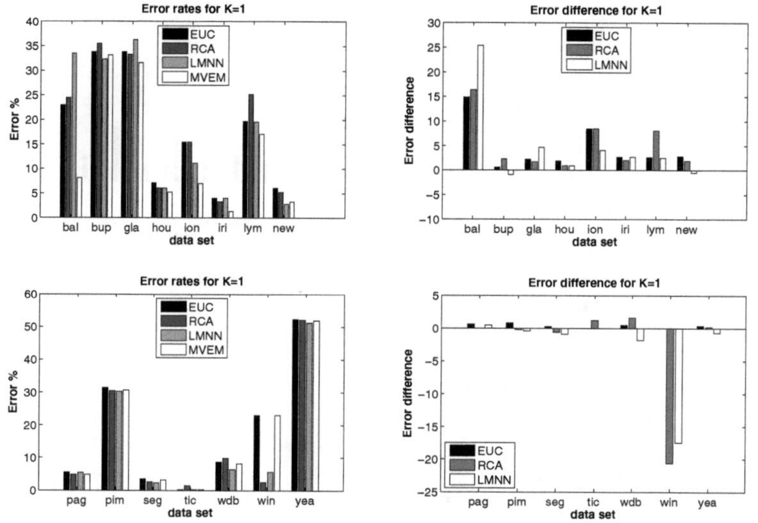

Fig. 1. (a) *Left* The error rates for k–NN classification ($k = 1$) using the four metrics: EUC, RCA [2], LMNN [22] and MVEM on the 15 UCI data sets. *Right* Difference in error for EUC, RCA and LMNN against the MVEM. Positive value implies the MVEM is better and a negative value implies the other metric is better. Ex.: Error(EUC) – Error(MVEM).

with various sizes, number of classes and dimensionalities. Since there is no explicit training and test sets for these data sets, 10 Folds Double Cross Validation (FDCV) were used to report the errors on all the data sets. For query–based learning, a k–Nearest Neighbour (k–NN) classifier was used with three different values for $k = (1, 3, 5)$. Only raw data was used in the experiments without any kind of preprocessing. Similar to all local learning algorithms [19,20,17,14], the parameter m plays a key role in the MVEM's performance. In all our experiments, the parameter m was set relative to the data dimensionality (d); i.e. $d + 1 \leq m \leq 3d$. The optimal value of m for each k–NN classifier using the MVEM, denoted by (m_{opt}), is shown Table (1).

In terms of comparisons, the k–NN classifier using the MVEM was compared with k–NN classifiers using three different metrics[1]: Euclidean (EUC), large margin nearest neighbour (LMNN) [22] and relevant component analysis (RCA) [2]. Since RCA depends on the availability and the amount of side–information (true labels in the context of classification), all the true labels for a given training set were provided to RCA in order to peel off any doubts about its performance.

Results analysis: Figures 1, 2 and 3 show the error rate and the difference in error for the three classifiers, $k = 1$, $k = 3$ and $k = 5$ respectively, using the four metrics on all data sets (Please see caption of Figure 1 for details). It

[1] The source code for LMNN and RCA was downloaded from the author's website.

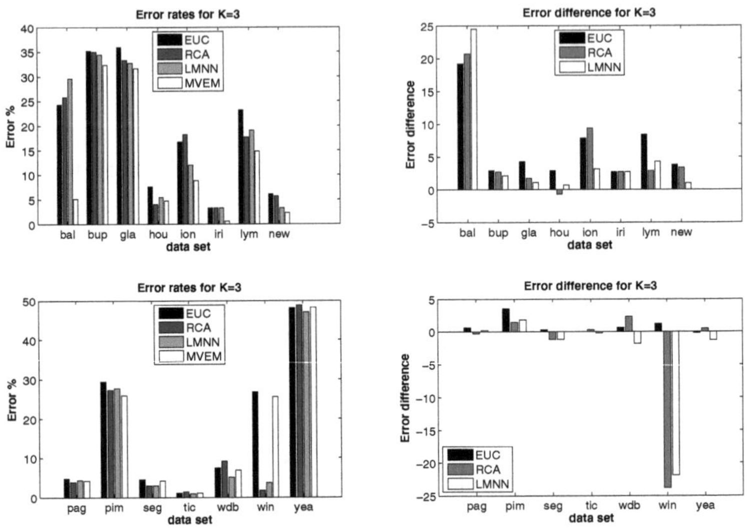

Fig. 2. Error rates and error differences for k–NN classification using $k = 3$. Please see caption of Figure (1) for explanation.

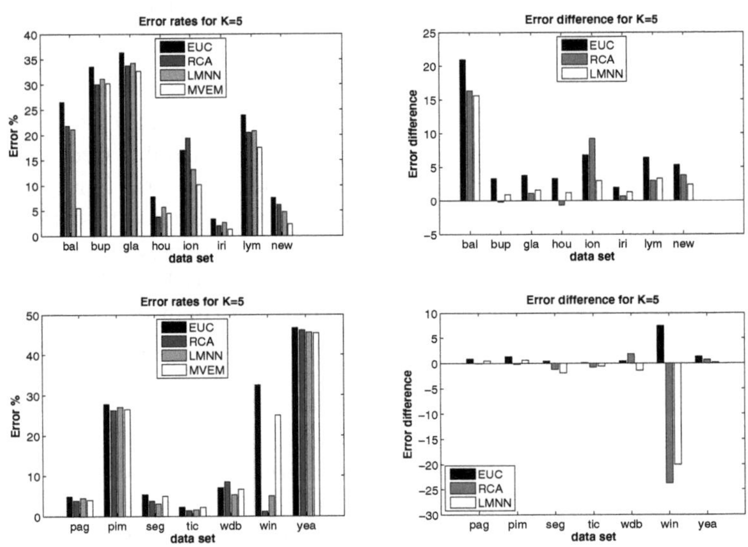

Fig. 3. Error rates and error differences for k–NN classification using $k = 5$. Please see caption of Figure (1) for explanation.

can be seen clearly that for most of the cases, the MVEM is placed first with a large margin in error difference, or placed second with a very small margin in error difference against the competing metric (except for the wine case, see

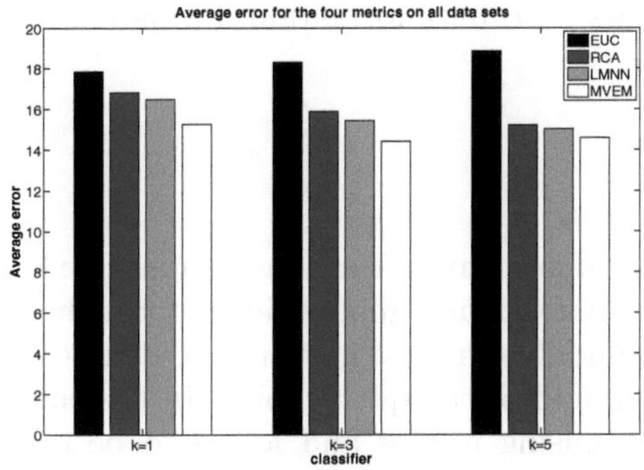

Fig. 4. Average error for the three classifiers using the four metrics on all data sets

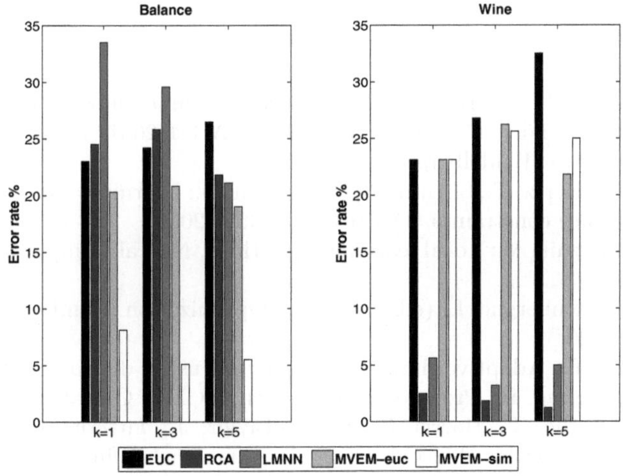

Fig. 5. The error rate using the Five metrics: EUC, RCA, LMNN, MVEM–euc (using Euclidean distance to define neighbourhoods) and MVEM–sim (using the similarity measure in Section 3.1). Note that the *wine* case reflects the power of the true labels exploited by RCA and LMNN. This is not observed in the *balance* case where the EUC error is significantly better than RCA and LMNN. The *balance* case also reflects the impact of well defined neighborhoods of a query point for the MVEM.

captions of Figure 5). The MVEM is consistently better than the Euclidean metric and very competitive with a more dedicated algorithm like LMNN which was specifically designed to learn a metric that minimizes the error of k–NN classification (i.e. discriminative training). Similar behaviour is observed when comparing between MVEM and RCA. Although, as shown in Figure 4, that the MVEM is slightly better as an overall performance, statistical significance tests showed that on average, the MVEM is not significantly different than the more dedicated algorithms LMNN and RCA. This is a very interesting result since MiniVenn had less *a priori* information during training and yet it showed similar performance.

5 Conclusion

We have introduced an algorithm for learning an adaptive local metric for query–based operations. The algorithm combines ideas from the minimum volume ellipsoid metric and manifold learning algorithms to define the local metric on the lower dimensional manifold of the query point. In the context of classification, the metric showed promising results and found to be competitive with other metric learning algorithms in the literature. These results motivate us to extend the proposed algorithm and metric to the domain of clustering and unsupervised learning with complete absence of side–information and labels.

References

1. Abou-Moustafa, K., Ferrie, F.: The minimum volume ellipsoid metric. In: Hamprecht, F.A., Schnörr, C., Jähne, B. (eds.) DAGM 2007. LNCS, vol. 4713, pp. 335–344. Springer, Heidelberg (2007)
2. Bar-Hillel, A., Hertz, T., Shental, N., Weinshall, D.: Learning a mahalanobis metric from equivalence constraints. JMLR 6, 937–965 (2005)
3. Bottou, L., Vapnik, V.: Local learning algorithms. Neural Computation 4(6), 888–900 (1992)
4. Boyd, S., Vandenberghe, L. (eds.): Convex Optimization. Cambridge Univ. Press, Cambridge (2004)
5. Ding, C., Li, T.: Adaptive dimension reduction using discriminant analysis and k–means clustering. In: Proceedings of the 24th ICML, Corvallis, OR (2007)
6. Dolia, A., De Bie, T., Harris, C., Shawe-Taylor, J., Titterington, D.: The minimum volume covering ellipsoid estimation in kernel-defined feature spaces. In: Fürnkranz, J., Scheffer, T., Spiliopoulou, M. (eds.) ECML 2006. LNCS, vol. 4212. Springer, Heidelberg (2006)
7. Friedman, J.: Regularized discriminant analysis. J. of the American Statistical Assoc. 84(405), 165–175 (1989)
8. Fukunaga, K., Olsen, R.: An algorithm for finding intrinsic dimensionality of data. IEEE Trans. on Computers 20(2), 176–183 (1971)
9. Globerson, A., Roweis, S., Hinton, G., Salakhutdinov, R.: Metric learning by collapsing classes. In: NIPS, vol. 18, pp. 451–458. MIT Press, Cambridge (2006)
10. Goldberg, J., Roweis, S.: Neighbourhood component analysis. In: NIPS, vol. 17, pp. 513–520. MIT Press, Cambridge (2005)
11. Grant, M., Boyd, S., Yinyu, Y.: Matlab software for disciplined convex programming (2005), `http://www.stanford.edu/~boyd/cvx`
12. Hastie, T., Tibshirani, R.: Discriminant adaptive nearest neighbour classification. IEEE Trans. PAMI 18(6), 607–615 (1996)
13. Lebanon, G.: Metric learning for text documents. IEEE Trans. PAMI 28(4), 497–508 (2006)
14. Lin, T., Zha, H.: Riemannian manifold learning. IEEE. Trans. PAMI 30(5), 796–809 (2008)

15. Newman, D., Hettich, S., Blake, C., Merz, C.: UCI repository of machine learning databases (1998)
16. Rousseeuw, P.: Multivariate estimation with high breakdown point. In: Proc. of the 4th Pannonian Symp. on Mathematical Statistics, vol. 3, pp. 283–297 (1983)
17. Roweis, S., Saul, L.: Nonlinear dimensionality reduction by locally linear embedding (LLE). Science 290(5500), 2323–2326 (2000)
18. Schultz, M., Joachims, T.: Learning a distance metric from relative comparisons. In: NIPS, vol. 16. MIT Press, Cambridge (2004)
19. Short, R., Fukunaga, K.: The optimal distance measure for nearest neighbour classification. IEEE Trans. on Information Theory 27(5), 622–627 (1981)
20. Tenenbaum, J., de Silva, V., Langford, J.: A global geometric framework for nonlinear dimensionality reduction. Science 290(5500), 2319–2323 (2000)
21. Titterington, D.: Estimation of correleation coefficients by ellipsoidal trimming. J. of Royal Statistical Society 27(3), 227–234 (1978)
22. Weinberger, K., Blitzer, J., Saul, L.: Distance metric learning for large margin nearest neighbor classification. In: NIPS, vol. 18, pp. 1473–1480. MIT Press, Cambridge (2006)
23. Xing, E., Ng, A., Jordan, M., Russell, S.: Distance metric learning with application to clustering with side-information. In: NIPS, vol. 15, pp. 505–512 (2003)
24. Yang, L.: Distance metric learning: A comprehensive review. Technical report, Dept. of Computer Science and Engineering, Michigan State University (2006)

Behavior Analysis of Volume Prototypes in High Dimensionality

Maiko Sato, Mineichi Kudo, and Jun Toyama

Division of Computer Science
Graduate School of Information Science and Technology
Hokkaido University
Kita-14, Nishi-9, Kita-ku, Sapporo 060-0814, Japan
{maico,mine,jun}@main.ist.hokudai.ac.jp
http://prml.main.ist.hokudai.ac.jp

Abstract. In these years, we often deal with an enormous amount of data or a data stream in a large variety of pattern recognition tasks. As a promising approach for economising the amount, we have previously defined *a volume prototype* as a geometric configuration that represents some data points inside and proposed a single-pass algorithm for finding them. In this paper, we analyze the convergence behavior of volume prototypes in high-dimensional cases. In addition, we show the applicability of volume prototypes to high-dimensional classification problems.

1 Introduction

In these years, we often deal with an enormous amount of data or a data stream in a large variety of pattern recognition tasks. Such data require a huge amount of memory space and computation time for processing. So it is preferable to use a more simple expression of data without losing characteristics of original datasets. We have proposed *volume prototypes* [1] as an extension of conventional point prototypes and defined each as a geometric configuration that represents many data points inside.

Volume prototypes can be generated by a single-pass algorithm, and thus they are significantly effective for streaming data that can be accessed only once. As another advantage, volume prototypes typically converge to modes of the underlying distribution, so that we do not need to care the number of prototypes. It suffices to take a sufficiently large number of initial prototypes, unlike mixture models.

In the initial study [1], we conducted some experiments and analyzed a typical behavior of volume prototypes in two-dimensional problems. However, it is still unclear about how they move and how they are effective in high dimensionality. In this paper, we clarify the behavior of volume prototypes in high-dimensional spaces and show the applicability of them to high-dimensional problems.

N. da Vitora Lobo et al. (Eds.): SSPR&SPR 2008, LNCS 5342, pp. 874–884, 2008.

2 Related Works and Achieved Status

The prototype approaches for massive data are divided into four groups: prototype selection, density estimation by mixture models, clustering and classification. The volume prototype approach is related to all of these groups of studies, especially it is strongly related to clustering and mixture models (Table 1). However, there are clear differences from them. A major difference is that volume prototype approach is at a lower level than clustering or mixture models. That is, volume prototypes are regarded as one of intermediate expressions, more close to data, between data points and components as seen in clustering or mixture models.

Table 1. Comparison of volume prototypes, clustering and mixture models

	Volume prototypes	Clustering	Mixture models
Goal	To condense (large) data appropriately	To analyze relative locations and distributions of clusters	To estimate mechanisms of data generation
Definition	Local prototypes which have volume	Partition of data without overlapping	Sum of component distributions to estimate original distributions
Example	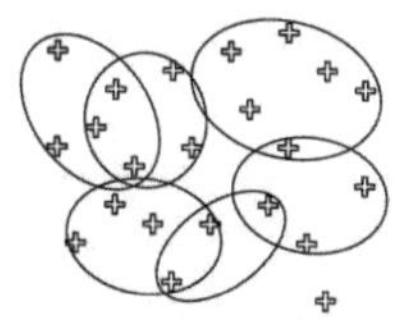	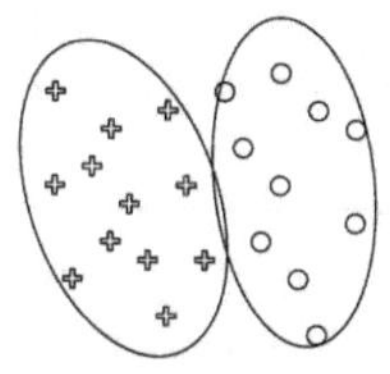	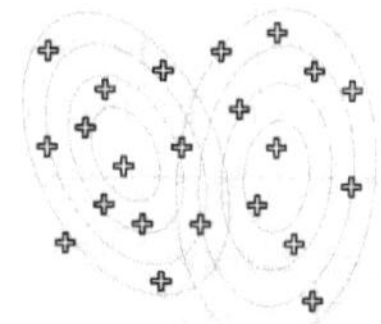

The name *volume prototype* is not new. Kaymak and Setnes [2] already used the terminology, but their volume prototypes are clusters with a boundary. On the contrary, our volume prototypes are closer to data themselves.

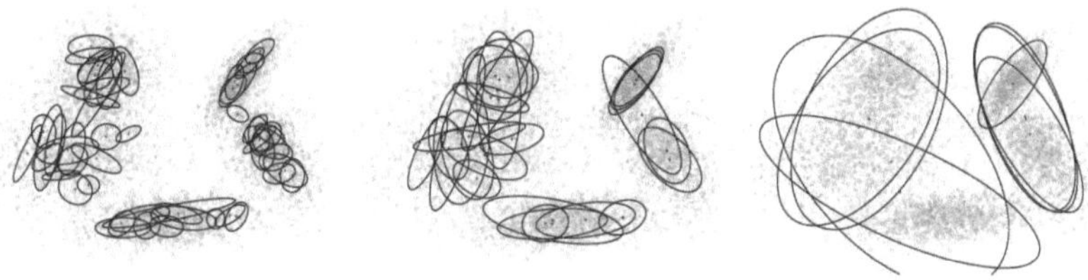

$\theta = 0.85$ $\theta = 0.90$ $\theta = 0.95$

Fig. 1. For one two-dimensional data (with five modes), volume prototypes with squared radius $r^2 = \chi_2^2(\theta)$. Here 10^5 artificial data were compressed into 100 or less volume prototypes.

An example of volume prototypes is shown in Fig. 1 for a two-dimensional data. From Fig. 1, we see that 1) the volume prototypes converge to the modes or ridge of distributions, 2) they approximate the distribution from inside if $\theta \leq 0.9$, and 3) only a few volume prototypes can summarize a massive data.

3 Algorithm

In this section, we show the concrete algorithm VP to generate volume prototypes. The single-pass algorithm is shown in Fig. 2.

Inputs:(Unlimited samples) $\boldsymbol{x}_1, \boldsymbol{x}_2, \ldots, \boldsymbol{x}_N, \ldots$

N is the number of samples for **ME Step**.

M is the number of seed prototypes.

N_v is the number of latest samples to be kept.

θ is the value to determine radius r of the acceptance region.

Procedure for obtaining volume prototypes

ME (mode estimation) step:

Repeat M times the following for the first N samples.

1. Choose a random permutation σ to reorder the samples to $\boldsymbol{x}_{\sigma 1}, \boldsymbol{x}_{\sigma 2}, \ldots, \boldsymbol{x}_{\sigma N}$.
2. Initialize a prototype by the way described in the detailed description.
3. Use $\boldsymbol{x}_{\sigma 2}, \ldots, \boldsymbol{x}_{\sigma N}$ in order to update the *seed prototype* by the same way as 2. (using only the samples in the acceptance region (1)).

C (Convergence) step:

For each sample $\boldsymbol{x}_i (i = N + 1, N + 2, \ldots)$ do the following.

1. Update the seed prototypes in which x_i falls by (2)–(4).
2. Keep the latest N_v samples.

Exploit time:

From M prototypes, select greedily $L(\leq M)$ prototypes until no improvement is found for covering N_v samples. Count the number of samples falling in each prototype in N_v samples. Here a sample is divided by k when k prototypes share it.

Fig. 2. Volume Prototype Algorithm (VP)

The outline is as follows. With multiple scans over the first N samples in different orders, we obtain M *seed prototypes* (ME Step). We bring up those seed prototypes by updating them using the samples fell in their acceptance regions (C Step). For *final prototypes*, we carry out prototype selection using a set covering criterion to have L final prototypes.

3.1 Details

Dataset: We consider a data stream $\boldsymbol{x}_1, \boldsymbol{x}_2, \ldots, \boldsymbol{x}_N, \boldsymbol{x}_{N+1}, \ldots$ of d-dimensional vectors.

Prototype: Let $\boldsymbol{p} = (\boldsymbol{\mu}, \boldsymbol{\Sigma}, r, n)$ be a prototype. Here, $\boldsymbol{\mu}$ is the prototype center, $\boldsymbol{\Sigma}$ is the covariance matrix, r is the Mahalanobis radius and n is the number of samples inside the prototype. Here, r is a significant parameter that affects the behavior of the prototype growth and is set to $\chi_d^2(\theta)$ ($100 \times \theta$ % point of chi-squared statistics with d degrees of freedom). With $\theta = 0.9$, we expect that 90% of surrounding points are covered.

Included sample set: Let $\boldsymbol{S}_p$ be the set of data points inside the prototype $\boldsymbol{p}$, which is specified by $\boldsymbol{S}_p = \{\boldsymbol{x} | (\boldsymbol{x} - \boldsymbol{\mu})' \boldsymbol{\Sigma}^{-1} (\boldsymbol{x} - \boldsymbol{\mu}) \leq r^2\}$. Then $n = |\boldsymbol{S}_p|$. We approximate n by the number of samples used for updating the prototype.

Acceptance region: We specify the acceptance region $\boldsymbol{A}_p$ of prototype $\boldsymbol{p} = (\boldsymbol{\mu}, \boldsymbol{\Sigma}, r, n)$, by the acceptance Mahalanobis radius R as

$$\boldsymbol{A}_p = \{\boldsymbol{x} | (\boldsymbol{x} - \boldsymbol{\mu})' \boldsymbol{\Sigma}^{-1} (\boldsymbol{x} - \boldsymbol{\mu}) \leq R^2\}. \tag{1}$$

Here, $R (\geq r)$ is determined from r and n by $R = r + \sqrt{\frac{\chi_d^2(0.95)}{n}}$. Here, 0.95 means to take the 95% confidence region of the sample mean in Mahalanobis radius. It is noted that R approaches to r as n goes to infinity.

Updating procedure: When a sample $\boldsymbol{x}$ falls into the acceptance region $\boldsymbol{A}_p$ of prototype $\boldsymbol{p}_t$ at time $t+1$, $\boldsymbol{p}_t$ is updated to $\boldsymbol{p}_{t+1}$ by

$$n_{t+1} = n_t + 1, \tag{2}$$

$$\hat{\boldsymbol{\mu}}_{t+1} = \frac{1}{n_{t+1}} \left(n_t \hat{\boldsymbol{\mu}}_t + \boldsymbol{x} \right), \tag{3}$$

$$\hat{\boldsymbol{\Sigma}}_{t+1} = \frac{1}{n_{t+1}} \left(n_t \hat{\boldsymbol{\Sigma}}_t + \boldsymbol{x}\boldsymbol{x}' + n_t \hat{\boldsymbol{\mu}}_t \hat{\boldsymbol{\mu}}_t' - n_{t+1} \hat{\boldsymbol{\mu}}_{t+1} \hat{\boldsymbol{\mu}}_{t+1}' \right). \tag{4}$$

In addition, radius R of the acceptance region is also updated.

Initialization of ME step: We initialize the center $\hat{\boldsymbol{\mu}}^{(0)}$ and the covariance matrix $\hat{\boldsymbol{\Sigma}}^{(0)}$ by $\hat{\boldsymbol{\mu}}^{(0)} = \boldsymbol{x}_1$(the first element after random reordering), and $\hat{\boldsymbol{\Sigma}}^{(0)} = \lambda^2 \boldsymbol{I}$. Here,

$$\lambda^2 = \frac{1}{dN} \sum_{i=1}^{N} \min_{j \neq i, j \in \{1,2,\ldots,N\}} \|\boldsymbol{x}_i - \boldsymbol{x}_j\|^2$$

($\|\cdot\|^2$ is the squared Euclidean norm) and $\boldsymbol{I}$ is the unit matrix. In addition, we set the initial value of $n^{(0)}$ to $n^{(0)} = d + 1$.

4 Behavior Analysis

Here, let us make clear the behavior of volume prototypes, especially where they move to, how fast they move and whether they shrink or not, as the number t of updatings increases.

The behavior of a volume prototype is very complicated. The center $\boldsymbol{\mu}_t$ and covariance matrix $\boldsymbol{\Sigma}_t$ are updated to $\boldsymbol{\mu}_{t+1}$ and $\boldsymbol{\Sigma}_{t+1}$ through (2) – (4). The factors $\boldsymbol{\mu}_t$, $\boldsymbol{\Sigma}_t$, r, n_t affect the updating as well as $\boldsymbol{X}_{t+1}$ generated from the underlying distribution f. Here let us derive the typical behavior under some strong but generally acceptable assumptions.

4.1 Movement of Prototype Centers

First, let us analyze the direction and speed of movement of a volume prototype center. Let a prototype after tth updating be $\boldsymbol{p}_t = (\boldsymbol{\mu}_t, \boldsymbol{\Sigma}_t, r, n_t)$. For simplicity, we assume that $\boldsymbol{\Sigma}_t = \boldsymbol{\Sigma}_0$ and $R^2 = r_\theta^2 = \chi_d^2(\theta)$ for $t = 1, 2, \ldots$ Here, $n_t = n_0 + t$. Then, the acceptance region $\boldsymbol{A}_{p_t}$ of $\boldsymbol{p}_t$ is given by

$$\boldsymbol{A}_{p_t} = \{\boldsymbol{x} \mid (\boldsymbol{x} - \boldsymbol{\mu}_t)'\boldsymbol{\Sigma}_0^{-1}(\boldsymbol{x} - \boldsymbol{\mu}_t) \le r_\theta^2\}.$$

With a sample $\boldsymbol{X}_{t+1} \in \boldsymbol{A}_{p_t}$, the center $\boldsymbol{\mu}_t$ is updated to $\boldsymbol{\mu}_{t+1}$ by

$$\begin{aligned}\boldsymbol{\mu}_{t+1} &= \frac{n_0 + t}{n_0 + t + 1}\boldsymbol{\mu}_t + \frac{1}{n_0 + t + 1}\boldsymbol{X}_{t+1} \\ &= \boldsymbol{\mu}_t + \frac{\boldsymbol{X}_{t+1} - \boldsymbol{\mu}_t}{n_0 + t + 1} = \boldsymbol{\mu}_t + \frac{\Delta\boldsymbol{X}_{t+1}}{n_0 + t + 1}.\end{aligned}$$

Then the expected updated center is given by

$$E_{\boldsymbol{X}_{t+1} \in \boldsymbol{A}_{p_t}}\boldsymbol{\mu}_{t+1} = \boldsymbol{\mu}_t + \frac{E_{\boldsymbol{X}_{t+1}}\Delta\boldsymbol{X}_{t+1}}{n_0 + t + 1}.$$

If $\boldsymbol{\Sigma}_0$ is small enough to allow f to be linearly approximated in $\boldsymbol{A}_{p_t}$ (Fig. 3) as $f(\boldsymbol{x}) = f(\boldsymbol{\mu}_t) + \nabla f(\boldsymbol{\mu}_t)'(\boldsymbol{x} - \boldsymbol{\mu}_t)$, we can have

$$E_{\boldsymbol{X}_{t+1}}\Delta\boldsymbol{X}_{t+1} = \frac{r_\theta^2}{d+2}\boldsymbol{\Sigma}_0\frac{\nabla f(\boldsymbol{\mu}_t)}{f(\boldsymbol{\mu}_t)},$$

(for the proof, see Fukunaga [3](pp.534–535)). Therefore, we have

$$E_{\boldsymbol{X}_{t+1} \in \boldsymbol{A}_{p_t}}\boldsymbol{\mu}_{t+1} = \boldsymbol{\mu}_t + \frac{1}{n_0 + t + 1}\frac{r_\theta^2}{d+2}\boldsymbol{\Sigma}_0\frac{\nabla f(\boldsymbol{\mu}_t)}{f(\boldsymbol{\mu}_t)}. \quad (5)$$

From Eq. (5), we see that 1) the direction of displacement is $\boldsymbol{\Sigma}_0\nabla f(\boldsymbol{\mu}_t)$, 2) the speed $E\boldsymbol{\mu}_{t+1} - \boldsymbol{\mu}_t$ is proportional to $\frac{r_\theta^2}{d+2}$ and $\frac{\nabla f(\boldsymbol{\mu}_t)}{f(\boldsymbol{\mu}_t)}$, therefore, 3) the speed is low when $\nabla f(\boldsymbol{\mu}_t)$ is small or $f(\boldsymbol{\mu}_t)$ is large depending on the previous position $\boldsymbol{\mu}_t$, and 4) the speed reduces at ratio $1/(n_0 + t + 1)$ for the increase of t, hence, the accumulated amount is approximately $\log t$. Note that $r_\theta^2/(d+2)$ does not

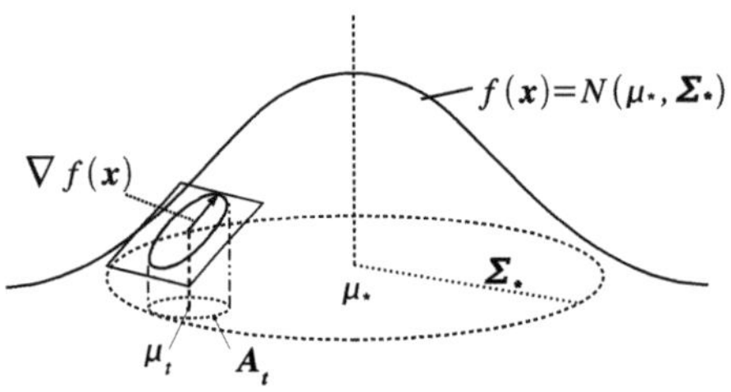

Fig. 3. Movement of a prototype center

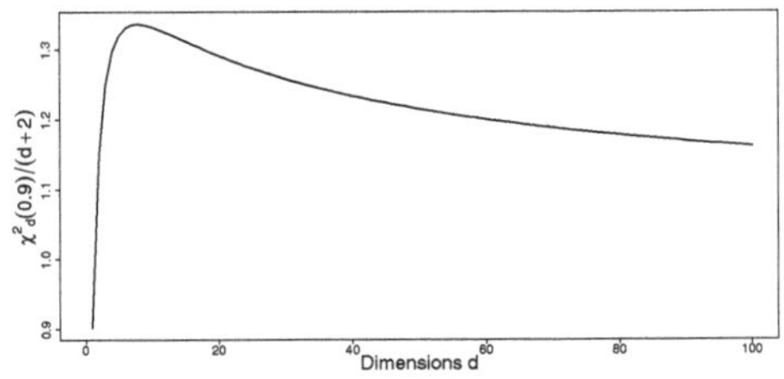

Fig. 4. $\chi_d^2(0.9)/(d+2)$ vs. dimensions

decrease linearly to $1/d$ because r_θ depends on both d and θ. Indeed, this value changes in a little strange way as shown in Fig. 4 ($\theta = 0.9$). In $\theta = 0.9$, up to around $d = 8$, the value increases, hence the movement speed of the center increases, but, after $d = 8$, the value decreases toward one, hence the speed decreases, as the dimensionality d increases.

We can obtain more intuition when $\boldsymbol{\Sigma}_0 = \alpha \boldsymbol{\Sigma}_*$ with a constant α (< 1), where $\boldsymbol{\Sigma}_*$ is the covariance matrix of the underlying normal distribution $f(\boldsymbol{x}) = N(\boldsymbol{x}; \mu_*, \boldsymbol{\Sigma}_*)$. Then, for Eq. (5), we have a simpler equation:

$$\begin{aligned} E_{\boldsymbol{X}_{t+1}} \boldsymbol{\mu}_{t+1} &= \boldsymbol{\mu}_t + \frac{1}{n_0 + t + 1} \frac{r_\theta^2}{d+2} \boldsymbol{\Sigma}_0 \frac{\nabla f(\boldsymbol{\mu}_t)}{f(\boldsymbol{\mu}_t)} \\ &= \boldsymbol{\mu}_t + \frac{1}{n_0 + t + 1} \frac{\alpha r_\theta^2}{d+2} \boldsymbol{\Sigma}_* \boldsymbol{\Sigma}_*^{-1} (\boldsymbol{\mu}_* - \boldsymbol{\mu}_t) \quad \text{(for Gaussian } f) \\ &= \boldsymbol{\mu}_t + \frac{1}{n_0 + t + 1} \frac{\alpha r_\theta^2}{d+2} (\boldsymbol{\mu}_* - \boldsymbol{\mu}_t) \end{aligned} \tag{6}$$

Therefore, the expected $\boldsymbol{\mu}_{t+1}$ is the point between the global mean $\boldsymbol{\mu}_*$ and $\boldsymbol{\mu}_t$ at ratio $\frac{1}{n_0+t+1} \frac{\alpha r_\theta^2}{d+2} : (1 - \frac{1}{n_0+t+1} \frac{\alpha r_\theta^2}{d+2})$ and climbs up to the steepest ascend direction.

Experiment 1

Data: $N(\mathbf{0}, \boldsymbol{I}_d), d = 2, 4, 8, 16, 32, 64$

Sample size: Out of 10^5 samples, the first $N = 10^3$ samples were used in ME step.

The number of prototype: $M = 10$

Radius: $R^2 = {r_\theta}^2 = \chi_d^2(\theta)$, $\theta = 0.85, 0.90, 0.95$

Procedure: Just after ME step, we have $\boldsymbol{p}_0 = (\boldsymbol{\mu}_0, \boldsymbol{\Sigma}_0, r, n_0)$, where n_0 depends on the prototype.

Results: The upper row of Fig. 5. Here, the horizontal axis is the logarithm of the number of samples.

From Fig. 5 (the upper row), we can confirm that the speed is linear to $\log t$ within a range in which $\frac{\nabla f(\boldsymbol{\mu}_t)}{f(\boldsymbol{\mu}_t)}$ is almost the same. The speed (the slope) increases up to $d = 8$ and then decreases for larger d. When θ, thus r_θ, increases, they converge more quickly. These observations show the validity of our analysis.

As a result, the centers of volume prototypes move at speed of the reciprocal of the number of samples and converge to one of local peaks of the underlying distribution, and they will stay there (because ∇f is close to zero) or approach to a more higher place (dense part) very slowly.

4.2 Convergence of Covariance Matrix

Next, let us analyze whether volume prototypes vanish or not with a sufficiently large number of samples.

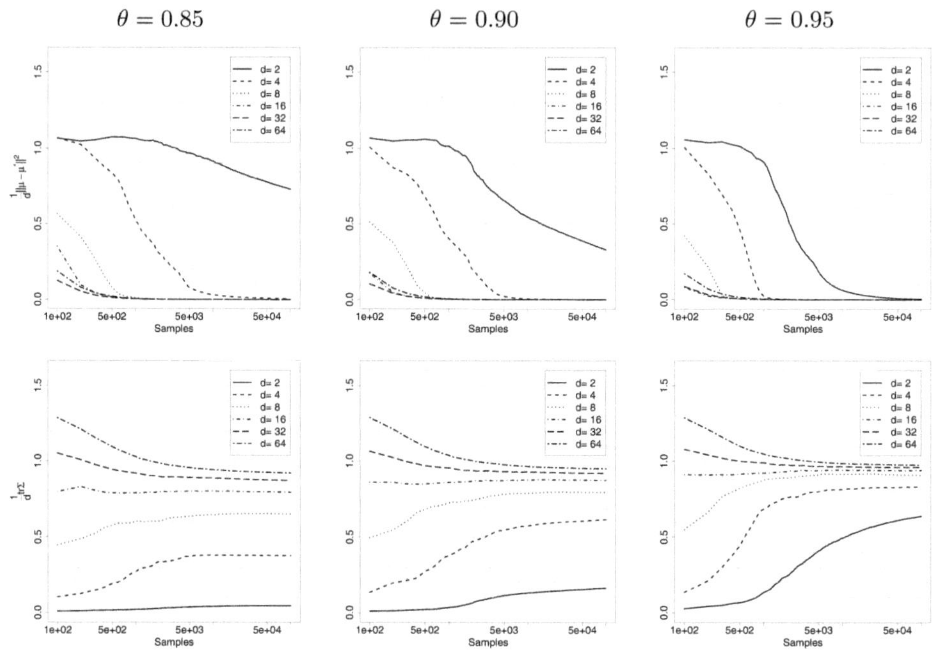

Fig. 5. The behavior of prototypes for $f(\boldsymbol{x}) = N(\boldsymbol{x}; \mathbf{0}, \boldsymbol{I}_d)$. The distance $\|\boldsymbol{\mu}_t\|^2/d$ (the upper row) and $\mathrm{tr}\boldsymbol{\Sigma}_t/d$ (the lower row) are plotted for increasing number of sample in log-scale in $d = 2, 4, 8, 16, 32, 64$. From left to right, $\theta = 0.85, 0.90, 0.95$.

For simplify, let us assume that $f(\boldsymbol{x}) = N(\boldsymbol{x}; \mathbf{0}, \boldsymbol{\Sigma}_*)$ and a prototype $\boldsymbol{p}_t = (\mathbf{0}, \boldsymbol{\Sigma}_t, r_\theta, n_t)$ as the one of which center has already converged to the origin. In addition, we assume that $\boldsymbol{\Sigma}_t = \alpha_t \boldsymbol{\Sigma}_*$, that is, the configuration is the same as $\boldsymbol{\Sigma}_*$ and only the scale is different by α_t.

Then the behavior of $\boldsymbol{\Sigma}_t$ is identical to that of α_t. Let $Y^2 = \boldsymbol{X}'\boldsymbol{\Sigma}_*^{-1}\boldsymbol{X}$. Then Y^2 obeys χ_d^2 and $EY^2 = d$. The acceptance region $\boldsymbol{A}_{p_t} = \{\boldsymbol{x} | \boldsymbol{x}'\boldsymbol{\Sigma}_t^{-1}\boldsymbol{x} \le r_\theta^2\}$ is rewritten by $\boldsymbol{A}_{p_t} = \{\boldsymbol{x} | \boldsymbol{x}'\boldsymbol{\Sigma}_*^{-1}\boldsymbol{x} \le \alpha_t r_\theta^2\}$. Then, with $f(y^2) = \chi_d^2(y^2)$, we have

$$E_{\boldsymbol{X} \in \boldsymbol{A}_{p_t}} Y^2 = \int_0^{\alpha_t r_\theta^2} y^2 f(y^2) dy^2 \Big/ \int_0^{\alpha_t r_\theta^2} f(y^2) dy^2$$
$$= \frac{2\gamma\left(\frac{d}{2} + 1, \alpha_t r_\theta^2\right)}{\gamma\left(\frac{d}{2}, \alpha_t r_\theta^2\right)}, \tag{7}$$

where $\gamma(\cdot, \cdot)$ is the incomplete gamma function defined by $\gamma(a, x) = \int_0^x e^{-t} t^{a-1} dt$. Note that $E_{\boldsymbol{X} \in \boldsymbol{A}_{p_t}} Y^2 \to d$ as $r_\theta^2 \to \infty$. From Eq. (7) and the fact that $E_{\boldsymbol{X} \in \boldsymbol{A}_{p_t}} Y^2 = \mathrm{tr}(\boldsymbol{\Sigma}_*)^{-1} E_{\boldsymbol{X} \in \boldsymbol{A}_{p_t}} \boldsymbol{X}\boldsymbol{X}'$, we have

$$E_{\boldsymbol{X} \in \boldsymbol{A}_{p_t}} \boldsymbol{X}\boldsymbol{X}' = \frac{2\gamma\left(\frac{d}{2} + 1, \alpha_t r_\theta^2\right)}{d\gamma\left(\frac{d}{2}, \alpha_t r_\theta^2\right)} \boldsymbol{\Sigma}_*. \tag{8}$$

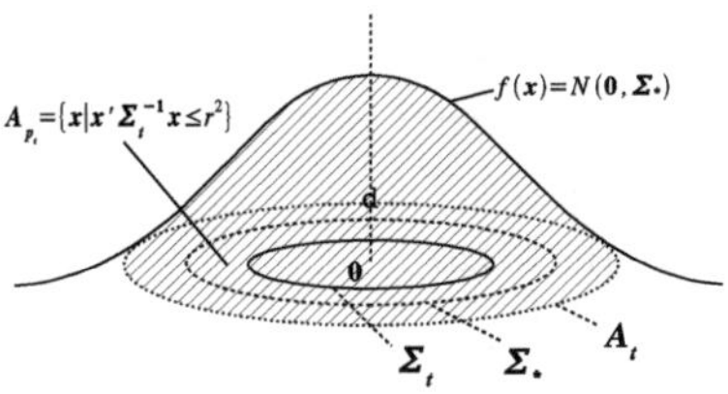

Fig. 6. Acceptance region of prototype $\boldsymbol{p}_t$

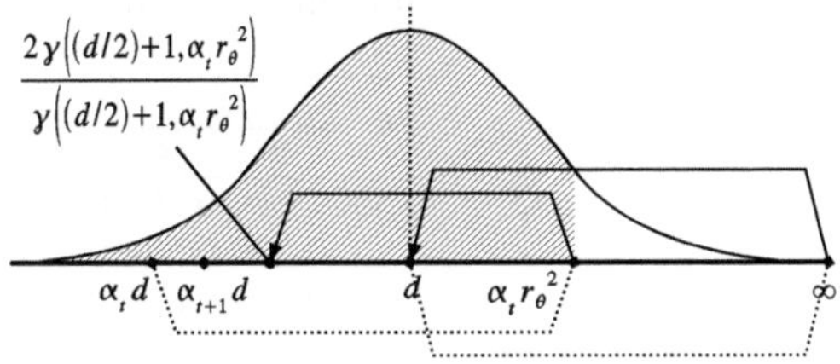

Fig. 7. Update scheme of covariance matrix

On the other hand, the covariance $\boldsymbol{\Sigma}_t$ is updated to $\boldsymbol{\Sigma}_{t+1}$ by $\boldsymbol{X}_{t+1}$ as

$$\boldsymbol{\Sigma}_{t+1} = \frac{n_0 + t}{n_0 + t + 1}\boldsymbol{\Sigma}_t + \frac{1}{n_0 + t + 1}\boldsymbol{X}_{t+1}\boldsymbol{X}'_{t+1}.$$

By taking the expectation over $\boldsymbol{X}_{t+1} \in \boldsymbol{A}_{p_t}$, from (8), we have

$$\alpha_{t+1}\boldsymbol{\Sigma}_* = E_{\boldsymbol{X}_{t+1}\in\boldsymbol{A}_{p_t}}\boldsymbol{\Sigma}_{t+1} = \boldsymbol{\Sigma}_t + \frac{E\boldsymbol{X}_{t+1}\boldsymbol{X}'_{t+1} - \boldsymbol{\Sigma}_t}{n_0 + t + 1}$$
$$= \alpha_t\boldsymbol{\Sigma}_* + \frac{\frac{2\gamma\left(\frac{d}{2}+1,\alpha_t r_\theta^2\right)}{d\gamma\left(\frac{d}{2},\alpha_t r_\theta^2\right)} - \alpha_t}{n_0 + t + 1}\boldsymbol{\Sigma}_*.$$

That is,

$$\alpha_{t+1} = \alpha_t + \frac{\frac{2\gamma\left(\frac{d}{2}+1,\alpha_t r_\theta^2\right)}{d\gamma\left(\frac{d}{2},\alpha_t r_\theta^2\right)} - \alpha_t}{n_0 + t + 1},$$

$$\alpha_{t+1}d = \left(1 - \frac{1}{n_0 + t + 1}\right)\alpha_t d + \frac{1}{n_0 + t + 1}\frac{2\gamma\left(\frac{d}{2}+1,\alpha_t r^2\right)}{\gamma\left(\frac{d}{2}+1,\alpha_t r^2\right)}. \tag{9}$$

At time t, $\boldsymbol{\Sigma}_t$ and $\boldsymbol{A}_t$ correspond to $\alpha_t d$ and $\alpha_t r_\theta^2$, respectively (Fig. 6). Note that $r_\theta^2 > d$ for large θ, say $\theta = 0.9$. Then the contribution by $\boldsymbol{X}_{t+1}$ corresponds to $2\gamma\left(\frac{d}{2}+1,\alpha_t r^2\right)/\gamma\left(\frac{d}{2},\alpha_t r_\theta^2\right)$. Thus a pair $(\alpha_t d, \alpha_t r_\theta^2)$ produces $2\gamma\left(\frac{d}{2}+1,\alpha_t r^2\right)/\gamma\left(\frac{d}{2},\alpha_t r_\theta^2\right)$.

When $\alpha_t > 1$, since $2\gamma\left(\frac{d}{2}+1,\alpha r_\theta^2\right)/\gamma\left(\frac{d}{2},\alpha r_\theta^2\right) < d < \alpha_t d < \alpha_t r_\theta^2$, we have $\alpha_{t+1}d < \alpha_t d$ from (9). On the other hand, when α_t is smaller than α^* defined below, we have $\alpha_t d < 2\gamma\left(\frac{d}{2}+1,\alpha_t r_\theta^2\right)/\gamma\left(\frac{d}{2},\alpha_t r_\theta^2\right) < \alpha_t r_\theta^2$, then $\alpha_{t+1}d > \alpha_t d$. The latter situation is illustrated in Fig. 7.

To find the convergence value α^*, it suffices to solve the equation

$$\alpha d = \frac{2\gamma\left(\frac{d}{2}+1,\alpha r_\theta^2\right)}{\gamma\left(\frac{d}{2},\alpha r_\theta^2\right)}.$$

Table 2. Equilibrium value α^* of $\alpha = 2\gamma\left(\frac{d}{2}+1, \frac{\alpha r_\theta^2}{2}\right)/d\gamma\left(\frac{d}{2}, \frac{\alpha r_\theta^2}{2}\right)$

θ	d					
	2	4	8	16	32	64
0.85	—	0.367	0.628	0.769	0.850	0.901
0.90	0.346	0.634	0.781	0.862	0.911	0.940
0.95	0.715	0.835	0.899	0.936	0.958	0.972

Let us denote $\alpha^* = \alpha^*(d, \theta)$ by making clear the dependency. Several values of $\alpha^*(d, \theta)$ are shown in Table. 2. Here $r_\theta^2 = \chi_d^2(\theta)$. As a result, the covariance matrix $\boldsymbol{\Sigma}_t$ approaches to $\alpha^*(d,\theta)\boldsymbol{\Sigma}_*$ as $t \to \infty$.

We conducted the same experiment as Experiment 1. The result is shown in Fig. 5. In Fig. 5 (the lower row), we plotted the value of $\text{tr}\boldsymbol{\Sigma}_t/d$ as an estimate of α_t because $\text{tr}\boldsymbol{I}_d = d$. From Fig. 5, we can see that α_t approaches to the theoretical equilibrium value $\alpha^*(d, \theta)$.

In total, we can say that volume prototypes converge to local peaks at a speed proportional to the reciprocal of the number of samples and then the covariance matrices converges to $\alpha^*(d, \theta)$ times the local covariance matrices.

5 Experiments

5.1 Classification by Volume Prototypes

Here, we present a simple but natural classifier constructed by volume prototypes. Each volume prototype can be regarded as a component of a mixture model, but is closer to a point as a prototype. Thus according to the following rule we classify a class-unknown sample $\boldsymbol{x}$ to c^*.

$$c^* = \arg\max_{c} \max_{k|c} P(c)P(k|c)f(\boldsymbol{x}|\boldsymbol{\theta}_{c,k}),$$

where $P(c)$ is the prior probability of class c, $P(k|c)$ is the probability of kth prototype in c, and $f(\boldsymbol{x}, \theta_{c,k})$ is the probabilistic distance of $\boldsymbol{x}$ to kth prototype of class c. In the estimation, we used $P(k|c) = n_{c,k}/n_c$ $(n_c = \sum_k n_{c,k})$ and $f(\boldsymbol{x}|\boldsymbol{\theta}_{\boldsymbol{c},\boldsymbol{k}}) = N(\boldsymbol{x}|\boldsymbol{\mu}_{c,k}, \boldsymbol{\Sigma}_{c,k})$ for $\boldsymbol{p} = (\boldsymbol{\mu}_{c,k}, \boldsymbol{\Sigma}_{c,k}, n_{c,k})$ where n_c is the total number of the latest samples of c that were kept in VP algorithm. Here $n_{c,k}$ was modified to satisfy $\sum_k n_{c,k} = n_c$ by taking into account the degree of overlapping among volume prototypes. To estimate the classification rate of the classifier, we used 10-fold cross validation.

We used a multi-featured dataset of 10 handwritten numerals represented in terms of four different feature sets [4]. The number of prototypes was set to $M = 10$, the first $N = 20$ samples were used for ME step and the remaining 180 samples were used for C step. Algorithm VP was invoked class by class. The radius was $r^2 = \chi_d^2(\theta)$ with $\theta = 0.9$. The latest sample number was $N_v = 180$ because no more sample was available.

Table 3. Recognition rate in each feature set. The figure of parentheses is the dimensionality.

Classifier	Recognition rate (%)			
	mfeat-zer (47)	mfeat-kar (64)	mfeat-fou (76)	mfeat-pix (240)
VP Clf ($\theta = 0.90$)	83.5	82.5	97.3	97.9
SVM	80.9	82.0	97.5	97.9

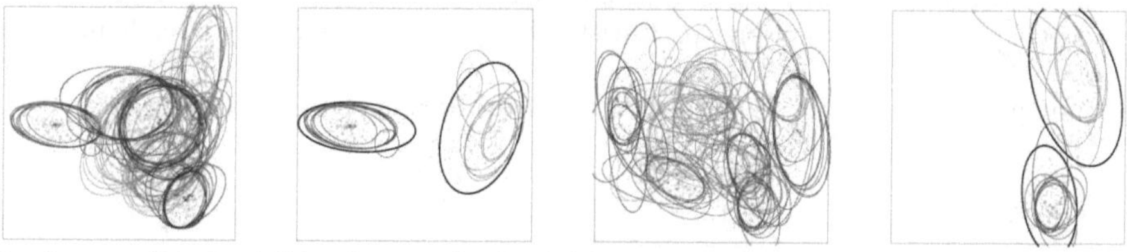

(a) mfeat-zer (10-class) (b) mfeat-zer (2-class) (c) mfeat-pix (10-class) (d) mfeat-pix (2-class)

Fig. 8. Volume prototypes in each dataset (10 classes). The number of volume prototypes is 10 in each class. The radius is $r^2 = \chi_d^2(0.9)$. The groups of various marks are the centers of prototypes and the group of ellipsoids with a same color is the volume prototypes of the class. The covariance of datapoints is drown as a black bold ellipsoid.

The result is shown in Table 3 and Fig. 8. For comparison, the results by SVM with a RBF kernel is also shown.

From Fig. 8, we see that 1) 10 prototypes are closely located but they represent well the spread of the class distribution (Fig. 8 (a), (c)), 2) the sizes are a little smaller than that of the covariance matrix of all the samples of the class (two of them are chosen for Fig. 8, (b), (d)). Note that the prototype sizes look larger than the real size when they are projected on a 2–dimensional space, because the volume of an ellipsoid increases exponentially in dimensionality. As a whole, volume prototypes well represent the configuration of dataset. The recognition rates are comparable to those of SVM. This means that the volume prototypes keep the characteristics of the original data.

6 Conclusion

We have clarified the behavior of volume prototypes toward large and high-dimensional datasets. A volume prototype moves toward a dense part of samples, e.g., a mode, and the covariance matrix of a prototype converges to a certain size depending on the local structure of the underlying distribution. The size of converged prototype is affected by the dimensionality and the user-specified radius for prototypes.

The efficiency of the classifier with volume prototypes is also shown. There are various possible applications of volume prototypes in pattern recognition tasks, in virtue of the high compression rate and the single-pass algorithm for generating them. In the future, we will particularly investigate the availability of volume prototypes to clustering and mixture model estimation.

References

1. Tabata, K., Kudo, M.: Information compression by volume prototypes. The IEICE Technical Report, PRMU 106(429), 25–30 (2006) (in Japanese)
2. Kaymak, U., Setnes, M.: Fuzzy clustering with volume prototypes and adaptive cluster merging. IEEE Transactions on Fuzzy Systems 10(6), 705–712 (2002)
3. Fukunaga, K.: Introduction to Statistical Pattern Recognition. Academic Press, London (1990)
4. Asuncion, A., Newman, D.: UCI machine learning repository (2007)

Pattern Recognition Approaches for Classifying IP Flows*

Alice Este[1], Francesco Gargiulo[2], Francesco Gringoli[1], Luca Salgarelli[1], and Carlo Sansone[2]

[1] DEA, Università degli Studi di Brescia
Via Branze, 38, 25123 Brescia, Italy
first.last@ing.unibs.it

[2] Dipartimento di Informatica e Sistemistica, Università degli Studi di Napoli Federico II,
Via Claudio, 21 I-80125 Napoli, Italy
{francesco.grg,carlosan}@unina.it

Abstract. The assignment of an IP flow to a class, according to the application that generated it, is at the basis of any modern network management platform. However, classification techniques such as the ones based on the analysis of transport layer or application layer information are rapidly becoming ineffective. Moreover, in several network scenarios it is quite unrealistic to assume that all the classes an IP flow can belong to are *a priori* known. In these cases, in fact, some network protocols may be known, but novel protocols can appear so giving rise to *unknown* classes.

In this paper we propose to face the problem of classifying IP flows by means of different pattern recognition approaches. They have been explicitly devised in order to effectively address the problem of the *unknown* classes, too. An experimental evaluation of the various proposal on real traffic traces is also provided, by considering different network scenarios.

1 Introduction

Accurate classification of traffic is at the basis of several techniques that are essential to provide Quality of Service (QoS) guarantees and to face attacks in IP based packet networks. The assignment of a packet flow to the class, from a set of classes that represent all the possible traffic crossing a network edge, is, in fact, requisite to allocate resources according to QoS and security policies. Classical approaches based on port analysis and Deep Packet Inspection are currently implemented on several commercially available products such as firewalls, routers and Application Level Gateways. These devices, that are widely used to engineer Internet traffic, assign flows to classes by thoroughly analyzing the payload of observed packets at the Network and Transport layers. The emergence of new network applications is, however, putting into discussion the reliability of these mechanisms: apart from dynamic port negotiation, several issues arise when existing protocols are used in a way they were not originally conceived for or when protocols

* This work has been partially supported by the Ministero dell'Università e della Ricerca in the framework of the PRIN Project "Robust and Efficient traffic Classification in IP nEtworks".

N. da Vitora Lobo et al. (Eds.): SSPR&SPR 2008, LNCS 5342, pp. 885–895, 2008.

are tunneled one inside another by ad-hoc encapsulation. Indeed the use of obfuscation techniques and encryption mechanisms definitely make payload based approaches completely ineffective: a clear example is given by the latest Peer–to–Peer applications that exploit all these tricks with the specific objective of bypassing the policies at the network boundary [1]. Research efforts have been unceasingly growing in the last few years to counter all these impairments and to restore traffic control capabilities.

Rather than concentrating on deterministic properties such as port numbers or packets payload, recently proposed approaches have been designed to statistically analyze features of network flows, such as combination of packet size, inter-arrival times, flow duration, etc. Several encouraging results have been reported on the application of Pattern Recognition (PR) or Machine Learning (ML) techniques to the problem of traffic classification. Recent works, such as [2,3,4], persuaded the scientific community to accept the idea that the behavior of network traffic is so heavily influenced by the underlying application protocols to enable the recognition of traffic classes by observing a very limited set of features using ML or PR techniques. Thanks to these works, several other approaches are now being considered to investigate traffic similarities and to determine by means of both supervised or unsupervised algorithms how to divide flows into classes or clusters depending on the application that has generated them: in particular, the work presented in [5] focused on investigating the performance of several supervised approaches (including Naïve Bayes, Bayesian Networks and Decision Trees) in IP flow classification. In [6] unsupervised approaches based on K-Means, Gaussian Mixture Models and Hidden Markov Model are applied for assigning flows to clusters.

From a PR point of view, the approach considered in [5] corresponds to a standard classification problem where one has to classify a pattern by assigning it to one of a set of known classes. A similar approach has been followed also in [7] where seven classes, each one made up of several protocol applications, have been considered during the training of a Support Vector Machine. In the problem presented in [6], instead, it is assumed that novel classes, which can be considered as *unknown*, can appear during the testing phase of the system. While the authors in [6] employ some heuristics for labeling a cluster as *unknown*, an alternative approach can be that of rejecting a pattern that does not fit to any of the known classes [8]. To this concern, it has to be noted that, even if several proposals have been made in the open literature for realizing a classifier system with a reject option, in most cases [9,10] authors consider the need of rejecting samples occurring in the overlap region between two known classes (the so-called *confusion-based* reject or *ambiguity-based* reject [11]). In the application context described here, instead, we are interested in the so-called *distance-based* reject, i.e. in a reject rule that is mainly devised for rejecting distant samples (with respect to the *known* classes) [11].

Therefore, in this paper we propose three PR approaches based on Support Vector Machines, Gaussian Mixture Models and Decision Trees respectively, that have been explicitly developed for coping with the presence of the *unknown* classes by means of a *distance-based* reject. Moreover, since it is also possible to define a network scenario in which users can only use a set of predefined applications, in this paper we also evaluate how the above classifiers perform when all the classes are *a priori* known. In order to compare the different approaches, we also took into account an aspect that seems to be disregarded in the open literature on IP flow classification. Different scenarios, in fact, exist in network management. Each of them corresponds to different costs that can be assigned to the different kinds of errors that can be made by the considered classification system. So, an effective comparison among different approaches can be made only after fixing a specific network scenario for which a cost matrix has been defined. To this aim, we present three different network scenarios, two of which refer to the presence of unknown classes, for IP flow classification and propose a cost matrix for each of them.

The rest of the paper is organized as follows: the three PR approaches proposed for IP flow classification are presented in Section 2. The considered network scenarios are illustrated in Section 3 together with some guidelines for defining the corresponding cost matrices. Then, an experimental comparative evaluation of the three approaches on real traffic traces is reported in Section 4. Finally, some conclusions are drawn.

2 Three Approaches for Classifying IP Flows

The basic traffic unit we consider in this paper is the *flow*, that we define as bi-directional ordered sequence of packets with the same IP addresses and TCP port numbers. The only flows we select have a complete and valid opening TCP three-way-handshake and terminate with a complete TCP shutdown or when a packet with RST flag set is seen.

For each flow we extract a *feature vector* $\mathbf{x} = (x_1, x_2, ...x_n)$ containing the payload sizes of the first packets in the same order as they are seen by the monitored network node at which we want to perform the classification. We leave off packets without payload because they are mostly used to exchange connection state information, for example to set up the connection or to acknowledge the reception of data, and their features give us less information on the specific application protocol the flow carries. To preserve the information of the direction in which a packet travels, the corresponding feature value is multiplied by -1 if the packet is moving from the server to the client, while it keeps its sign if it is sent by the client.

We choose the above features for the identification of these classes because payload size should only depend on the finite state machine that drives the application layer protocol, especially in the early steps of the exchange.

Finally, with the term *class* in this paper we refer to the set of flows generated by the same application–layer protocol, such as HTTP, POP3 or SMTP protocols.

In the following, for each considered classification technique, we present both an architecture able to handle the case in which all classes are *a priori* known and an architecture that can address the problem of the *unknown* classes, by using a suitably defined *distance-based* reject option. Note that differently from [6], in our approaches we consider only supervised techniques, since we can use a labeled dataset that can be divided in a training set and a test set.

2.1 SVM–Based Classifier

The first architecture we present is based on Support Vector Machines (SVM) [12]. As stated in the introduction, it has been yet fruitfully employed in IP flows classification when all classes are *a priori* known. For the applications that do not contemplate the presence of *unknown* protocols we choose to use a multi–class *one-against-all* SVM with a Gaussian kernel function. In this case, the regularization coefficient, the kernel parameter and the size of the feature vectors n have been chosen by maximizing the accuracy of the classification on the training samples.

To solve the problem of *unknown* protocol identification we have designed an alternative architecture, using the one–class SVM proposed by Schölkopf et al. in [13]. The classifier trains a one–class SVM model for each known protocol class building a surface that contains the training observations with a reliability level determined by the model parameters. The one–class models need as input only the samples of the corresponding training class. We use also the flows of the other known training classes for selecting the values of the parameters that allow to identify correctly the flows of the trained class and to rule out the samples of the other classes.

In order to classify a flow, our classifier finds what surfaces, among the ones calculated during the training phase, the corresponding feature values fall in: if there is only one candidate surface, then the flow is assigned to the corresponding class (i.e., application protocol). If, instead, the flow falls outside all the available surfaces, i.e. it is *distant* from the all the target classes, the flow is marked as not generated by any of the known applications. In some cases it may happen that the flow falls inside the surfaces of two or more different classes: to assign the flow in these cases, we use a multi-class SVM stage like that we described above.

2.2 GMM–Based Classifier

The second classifier we adopt is founded on Gaussian Mixture Models (GMM) [14]. This choice is related to the fact that it is quite simple to adapt the classifier so as to address the problem of *unknown* classes.

In the training phase this classifier trains a GMM for each protocol in the feature space. The size n of the feature vectors is chosen equal to four, following also the choice suggested in [6]. This model approximates the distribution that generates the training flows of a protocol as a linear combination of L basic Gaussian distributions.

The distribution parameters have been computed from the training set using the Expectation Maximization (EM) algorithm [14], while the number of Gaussians L has been determined for each protocol minimizing a parameter that represents the Minimum Description Length (MDL) [15]. This quantity depends on the number of model parameters and on the value of the maximum likelihood computed on the training data, for a given number of Gaussians.

During the test phase, the classifier selects the class that maximizes the *a posteriori* probability of the protocol that has likely generated the flow. When the classifier needs to detect *unknown* protocols we introduce a *threshold* parameter. The threshold is used during classification to assign a testing flow adopting a *distance-based* reject option: when all the *a posteriori* probability values calculated for the flow are lower than the threshold, we label it as *unknown*. The choice of the threshold value entails a trade-off between the outlier identification and the correct recognition of the samples belonging to the training classes: it is, in fact, set by maximizing the overall classification accuracy, including also the correct recognition of the flows that do not belong to any of the known classes. To this end we introduced a new class, that we name *other*, representing a subset of the *unknown* classes: it is, in fact, assembled using a little subset of the test set composed of only flows that do not belong to any of the known protocols (we cannot use the training set for this purpose as it lacks this kind of samples). As a consequence, the resulting *other* class represents samples of *unknown* protocols and is only used to tune the *threshold* value during the above described maximization step together with the other classes in the training set, using the same weight for each application protocol class.

2.3 Decision Tree–based Classifier

The third classifier model is based on a *Decision Tree* [14]. This model has been chosen since its classification speed makes it very effective for an on-line implementation [5]. To cope with the presence of the *unknown* class we also defined a suitably reject option, as it will be illustrated in the following. As in the previous case, the size n of the feature vectors was fixed equal to four.

In case of known classes we used a standard C4.5 model; more precisely, the J48 implementation from the Weka library[1]. When considering the presence of known classes only, we used a post pruning method to reduce the tree size and increase prediction accuracy. The pruning method is based on the MDL principle.

For addressing the presence of the *unknown* classes, we instead considered an unpruned version of the tree. We then added a *reliability* value to each leaf. Let us suppose that the DT attributes the *j-th* leaf to the *i-th* class: the reliability value of the *j-th* leaf is therefore calculated, in the training phase, as the ratio between the number of the training set samples of the *i-th* class that fall in the *j-th* leaf and the number of training set samples belonging to the *i-th* class.

In this way, a low reliability is associated to leafs which few samples of the correct class fall in. These leafs are then representative of regions of the feature space that should be quite distant from the ones in which the most part of the samples lies. So, after the training phase, we set a threshold in order to reject samples that fall in leafs whose reliability value is below such a threshold. These samples are considered as belonging to the *unknown* class since are reasonably distant from the *target* classes.

As in the GMM case, we used a little subset of the test set for setting the threshold so as to optimize the overall accuracy of the DT, including the recognition on the class *other*.

3 The Considered Network Scenarios

3.1 Known Classes

We evaluate the performance of the considered approaches in a first scenario where only a fixed set of known protocols have been allowed within the workstations connected by a Local Area Network (LAN) to the Internet. In this scenario each application protocol is described by a class and the union of the M allowed classes represent all the possible

[1] Weka is an open source collection of data-mining tools and is freely available at the website http://www.cs.waikato.ac.nz/ml/weka

Table 1. A possible protocol categorization in a QoS context

Priority class	Service	Applications/Protocols
1	online gaming	HalfLife, Counter-Strike, Age of Empires, etc.
2	voip	skype, WengoPhone, Ekiga, Asterisk
3	streaming	mms (wmp), real, quicktime, shoutcast, etc.
4	network mgmt	dns, netbios, smb, snmp, ntp
5	interactive	http, https, ssh, ftp(control)
6	chat	IRC, msn messenger, yahoo messenger, AIM
7	mail	smtp, pop3, pop3s, imap, imaps
8	news	nntp
9	bulk	ftp(data), sftp (data)
10	p2p	FastTrack, eDonkey2000, BitTorrent, Gnutella, etc.
11	other	

traffic that can cross the edge router of the LAN. This scenario appears in networks where the workstations are strictly controlled by administrators and users cannot install new software: the set of network software suites installed on these workstations determines the M allowed protocols. Even in this scenario, however, traffic classification could be essential for QoS deployment. Given, in fact, that users can configure applications for retrieving data from remote hosts, there could be no correspondence between Transport Layer ports and underneath application protocols: e.g. a user can configure its mail client to send messages through an external provider by connecting to the mail exchanger on port 587; another user can connect to its home gateway to retrieve web information on the same port. Administrators could however decide to give more priority to web traffic than email: to this end they can use one of the proposed classifiers to mark each of the above connections with a tag that identifies the generating application class. Given the nature of the proposed approaches, administrators have to first train classifiers on pre-recorded traffic traces, each representing one of the M allowed protocols: later, when a new flow comes, the deployed algorithm appends the class tag that is then used by the edge router to speed up the de-queuing of those packets whose class should be prioritized according to a given policy.

We report an example of a QoS policy in Table 1, where protocols have been grouped depending on their usage context. For example, since POP3 and IMAP are both used for message retrieval they are given the same priority. We assign then an higher priority to chat related protocols - one row above, and so on. Note that the last line of the table is not used within this scenario: we will consider it in the next one. To compare the classification performance of the proposed algorithms, we can then introduce a cost matrix: the cost of misclassifying protocol A as protocol B is obtained by simply computing, for each couple (A, B), the distance between their priorities.

3.2 Known+unknown Classes

When control on the hosts is released, protocols that have not been considered for training can unexpectedly appear: users can, in fact, upgrade applications or install new ones; mobile users coming from the outside can use the network to connect their

Table 2. A possible protocol categorization in a Security context

Class	Service	Applications/Protocols
accepted traffic	web interactive mail network management bulk	http, https, ssh, ftp(control), smtp, pop3, pop3s, imap, imaps, dns, netbios, smb, snmp, ntp, ftp(data), sftp (data)
denied traffic	chat p2p voip news streaming multimedia online gaming **other**	IRC, msn messenger, yahoo messenger, AIM, FastTrack, eDonkey2000, BitTorrent, Gnutella, OpenNap, Soulseek, Ares, MP2P, Dirrect Connect, GoBoogy, Soribada, PeerEnabler, skype, WengoPhone, Ekiga, Asterisk, nntp, mms (wmp), real, quicktime, shoutcast, vbrick streaming, logitech Video IM, HalfLife, Counter-Strike, Age of Empires

equipment such as laptop or organizers. In both cases a flow generated by a protocol for which no training data was available, could cross the edge router where classification algorithms, as those considered in the previous section, will inevitably assign it to one of the known classes – probably to the class whose behavior is closest to the protocol that is generating this flow.

To avoid to assign not characterized flows to any of the classes chosen for the training, algorithms should be hence partially modified to be given the ability to reject these flows, as explained in Section 2. It is worth noting that erroneous assignments can occur also in this scenario: in this case, however, misclassification not only determines impairments to QoS but can impact also the security of the network. Suppose, in fact, that the administrators of a company network are told to protect some business related information that are commonly accessed by inside users: a first mandatory measure is to prohibit P2P communications to prevent users to unconsciously export important information through file sharing. Misclassification could hence let a P2P flow pass opening up a hole in the security of the network.

To evaluate the performances of the proposed approaches in these two contexts we consider again error costs, separating the scenario in two sub-cases. In the first we will consider the QoS context: similarly to what described in Section 3.1 we take the same QoS policy as that reported in Table 1. This time the last row expresses the priority associated to flows that do not fall inside any of the classes depicted in the upper rows.

In the second sub-case we will examine the security context. Here we divide protocols in two classes, one composed of protocols that can pass the edge router, the other composed of forbidden protocols. The former class has to be completely formalized and each allowed protocol be described by a distinct subclass. The latter class, instead, is composed of a few protocols, that are explicitly described by separated subclasses, and of all the protocols for which no training data was available. A possible protocol categorization is reported in Table 2.

4 Experimental Results

4.1 Dataset Description

The packet traces used in our experimental evaluation were collected at the border router of *University of Brescia*'s network. For every packet we captured the first 400 bytes of payload and we can apply on these bytes pattern-matching mechanisms to assess the actual application that has generated each TCP flow, in some cases with the addition of manual inspection. The network infrastructure comprises several 1000Base-TX segments routed through a Linux-based dual-processor box and includes about a thousand workstations with different operating systems. All the traffic traces were gathered on the 100Mb/s link connecting the edge router to Internet during the first three weeks of June 2007. A total of 50GB of traffic was collected by running Tcpdump for fifteen minutes regularly every hour.

Both the training and the test sets are composed of six different protocol classes. These protocols belong to different application types: web browsing, mail services, P2P and interactive. The six protocols we select are responsible for the generation of the most part of the traffic, and, given their variety, they allow us to verify the general applicability of our classification techniques. In Table 3 [left] we report the number of flows for the training and the test set. We took care of choosing the training and the test set from traces collected in two separate and consecutive time frames.

In addition to the six protocols mentioned above, we captured another class of flows called *other*. This class was used during the test phase to verify the classifier's ability to recognize protocols different than those used during the training phase. This subset

Table 3. Left: number of flows included in the considered traces. **Right:** Cost matrix for the considered protocols. Values in parentheses refer to a Security context, while the other ones to a QoS application.

protocol	training set	test set	http	smtp	pop3	ftp	bitTorrent	msn	unknown
http	7063	4928	–	2(1)	2(1)	0(1)	5(10)	1(10)	6(10)
smtp	19427	19480	2(1)	–	0(1)	2(1)	3(10)	1(10)	4(10)
pop3	19611	19940	2(1)	0(1)	–	2(1)	3(10)	1(10)	4(10)
ftp	6296	14458	0(1)	2(1)	2(1)	–	5(10)	1(10)	6(10)
bitTorrent	5057	7412	5(10)	3(10)	3(10)	5(10)	–	4(1)	1(1)
msn	1024	1033	1(10)	1(10)	1(10)	1(10)	4(1)	–	5(1)
other		7912	6(10)	4(10)	4(10)	6(10)	1(1)	5(1)	–

Table 4. Comparison among the three different approaches in different scenarios

	Overall Accuracy		Cost (QoS)		Cost (Security)
	known	known+unknown	known	known+unknown	known+unknown
SVM	93.87%	94.72%	0.08	0.32	0.18
GMM	98.83%	97.07%	0.02	0.13	0.20
DT	98.27%	92.53%	0.03	0.29	0.38

is composed by flows that did not match the expressions of the training protocols but matched at least one of the patterns of all the other protocols in [16] (such as edonkey, ssl, smb, gnutella, imap, etc): we are hence relatively sure that they belong to other protocol classes than those listed in Table 3 [left].

By using the considerations made in Section 3, we defined the cost matrices for this dataset (see Table 3 [right]) in both security and QoS contexts. Note that the last row and the last column of this table are not used in the *known classes* scenario.

4.2 Analysis of the Classification Results

Table 4 reports a summary of the results we obtained both for the scenario when only known protocols are tested and in the case known and unknown protocols are classified. In the second case, we fixed to 1.5e-14 and 0.02 the thresholds for the reject option of the GMM and of the DT, respectively.

Generally speaking, we observe that the accuracy of correctly classifying known protocols gets worse when *unknown* classes are introduced. This surely happens with the GMM and the DT approaches since some of the flows that are correctly reported as their actual classes in the first scenario, can be mapped to the *unknown* class in the second: e.g., all flows that are correctly classified by a GMM when there are no *unknown* classes and that get a value of the *a posteriori* probability lower than the threshold, are misclassified. While this behavior is intrinsic for these two approaches, the recognition rate for the various protocols does not necessarily decrease when the SVM architecture is used. As explained, in fact, in Section 2.1 the SVM architecture completely changes when *unknown* classes are introduced.

We point out now some differences between the three approaches, considering the scenario with unknown protocols. In particular, the design of the SVM–based classifier presents characteristics differing from the other two architectures: i) it is the only classifier that does not require to be trained on samples coming from unknown classes; ii) while thresholds in the GMM and DT classifiers have the same value for all the training protocols, we find the one–class SVM parameters independently for each protocol and iii) finally the models of all application classes for both the GMM and DT approaches have been determined on the same n–dimensional space (a fourth dimensional space has been chosen in both approaches), while SVM builds each one–class model on the space that allows better classification results for that class (the value of n varied between two and six). This optimization process for the SVM approach leads to better results in classifying unknown flows when compared to the other two techniques even if the accuracy of all the proposed techniques are very similar as reported in the second column of Table 4.

Finally, a note on accuracy, which cannot be used as a single quality parameter for a classifier. Even if the GMM always exhibited the best accuracy (Table 4), in fact, there is a scenario (Security with *unknown* classes) in which the cost achieved by an SVM is lower. Furthermore, if we are interested in an on-line implementation, DT could be also a good choice since it performs acceptably, especially in a QoS scenario.

5 Conclusions

In this paper we presented three approaches to the classification of IP flows, based on Support Vector Machines, Gaussian Mixture Models and Decision Trees. One of the problems that arise in this context is that the classifiers need to deal with the sudden appearance of new application protocols. We proposed to address this problem by rejecting a pattern that does not fit to any of the known classes, by means of a suitably defined *distance-based* reject option. A further contribution of this work has been the definition of different network scenarios and the proposal of a cost matrix for each of them so as to weight in a different way the various classification errors. This permitted a more effective comparison of the proposed approaches on real traffic traces.

In the future we plan to implement the proposed classification architectures in an on-line system allowing the comparison of their computational requirements as well as their behavior when applied to different network scenarios.

References

1. Skype, `http://www.skype.com`
2. Karagiannis, T., Papagiannaki, K., Faloutsos, M.: BLINC: multilevel traffic classification in the dark. In: Proceedings of the 2005 conference on Applications, technologies, architectures, and protocols for computer communications (SIGCOMM 2005), Philadelphia, PA, USA, pp. 229–240 (August 2005)
3. Roughan, M., Sen, S., Spatscheck, O., Duffield, N.: Class–of–service mapping for QoS: a statistical signature–based approach to IP traffic classification. In: Proceedings of the 4th ACM SIGCOMM conference on Internet measurement (IMC 2004), Taormina, Sicily, Italy, pp. 135–148 (October 2004)
4. McGregor, A., Hall, M., Lorier, P., Brunskill, J.: Flow Clustering Using Machine Learning Techniques. In: Barakat, C., Pratt, I. (eds.) PAM 2004. LNCS, vol. 3015, pp. 205–214. Springer, Heidelberg (2004)
5. Williams, N., Zander, S., Armitage, G.: A Preliminary Performance Comparison of Five Machine Learning Algorithms for Practical IP Traffic Flow Classification. SIGCOMM Computer Communication Review 36(5), 7–15 (2006)
6. Bernaille, L., Teixeira, R., Salamatian, K.: Early Application Identification. In: Proceedings of the 2006 ACM CoNEXT conference (CoNEXT 2006), Lisboa, Portugal, pp. 1–12 (December 2006)
7. Li, R.Y.Z., Guan, X.: Accurate Classification of the Internet Traffic Based on the SVM Method. In: Proceedings of the 42th IEEE International Conference on Communications (ICC 2007), Glasgow, Scotland, pp. 1373–1378 (June 2007)
8. Tax, D.M.J., Duin, R.P.W.: Growing a multi–class classifier with a reject option. Pattern Recognition Letters 29, 1565–1570 (2008)
9. Fumera, G., Roli, F., Giacinto, G.: Reject option with multiple thresholds. Pattern Recognition 33(12), 2099–2101 (2000)
10. De Stefano, C., Sansone, C., Vento, M.: To reject or not to reject: That is the question – an answer in case of neural classifiers. IEEE Trans. on Systems, Man and Cybernetics 30(1), 84–94 (2000)
11. Landgrebe, T., Tax, D.M.J., Paclík, P., Duin, R.P.W.: The interaction between classification and reject performance for distance–based reject–option classifiers. Pattern Recognition Letters 27(8), 908–917 (2006)

12. Vapnik, V.N.: Statistical Learning Theory. John Wiley and Sons, New York (1998)
13. Schölkopf, B., Platt, J.C., Shawe–Taylor, J., Smola, A.J., Williamson, R.C.: Estimating the Support of a High–Dimensional Distribution. Neural Computation 13, 1443–1471 (2001)
14. Duda, R.O., Hart, P.E., Stork, D.G.: Pattern Classification, 2nd edn. Wiley, Chichester (2000)
15. Rissanen, J.: Modeling by shortest data description. Automatica 14, 465–471 (1978)
16. L7 Filter, `http://l7--filter.sourceforge.net`

Outlier Robust Gaussian Process Classification

Hyun-Chul Kim[1] and Zoubin Ghahramani[2]

[1] Yonsei University, 262 Seongsanno, Seodaemun-gu, Seoul 120-749, Korea
[2] University of Cambridge, Trumpington Street, Cambridge CB2 1PZ, UK

Abstract. Gaussian process classifiers (GPCs) are a fully statistical model for kernel classification. We present a form of GPC which is robust to labeling errors in the data set. This model allows label noise not only near the class boundaries, but also far from the class boundaries which can result from mistakes in labelling or gross errors in measuring the input features. We derive an outlier robust algorithm for training this model which alternates iterations based on the EP approximation and hyperparameter updates until convergence. We show the usefulness of the proposed algorithm with model selection method through simulation results.

1 Introduction

In many real-world classification problems the labels provided for the data are noisy. There are typically two kinds of noise in labels. Noise near the class boundaries often occurs because it is hard to consistently label ambiguous data points. Labelling errors *far* from the class boundaries can occur because of mistakes in labelling or gross errors in measuring the input features. We call labelling errors far from the boundary, *outliers*. While many methods have been proposed for dealing with noisy class boundaries, far less attention has been placed on dealing with classification outliers. In this paper we present a Bayesian kernel classification algorithm which is robust to outliers.

GPCs are a Bayesian kernel classifier derived from Gaussian process priors over functions which were developed originally for regression [1,2,3]. In classification, the target values are discrete class labels. To use Gaussian processes for binary classification, the Gaussian process regression model can be modified so that the sign of the continuous latent function it outputs determines the class label. Observing the class label at some data point constrains the function value to be positive or negative at that point, but leaves it otherwise unknown. To compute predictive quantities of interest we therefore need to integrate over the possible unknown values of this function at the data points. Exact evaluation of this integral is computationally intractable. However, several successful methods have been proposed for approximately integrating over the latent function values, such as the Laplace approximation [1], Markov Chain Monte Carlo [3], and variational approximations [2]. Opper and Winther (2000) used the TAP approach originally proposed in statistical physics of disordered systems to integrate over the latent values [4]. The TAP approach for this model is equivalent

N. da Vitora Lobo et al. (Eds.): SSPR&SPR 2008, LNCS 5342, pp. 896–905, 2008.

to the more general Expectation Propagation (EP) algorithm for approximate inference [5]. Minka's formulation [5] has a special hyperparameter which can potentially be used to deal with outliers. Since outliers can be a big obstacle to learning, in this paper we propose an outlier robust learning algorithm based on EP for Gaussian process classification. It turns out that the algorithm is useful with the model selection between it and the regular GPC.

The paper is organized as follows. Section 2 introduces Gaussian process classification. In section 3, we introduce the EP/TAP method for approximate inference. In section 4, we derive the algorithm for outlier treatment. In section 5, we present model selection method based on Bayesian model evidence. In section 6, we show experimental results on both synthetic and real data sets. In section 7, we discuss our approach and future work.

2 Gaussian Process Classification

Let us assume that we have a data set D of data points $\mathbf{x}_i$ with binary class labels $y_i \in \{-1, 1\}$: $D = \{(\mathbf{x}_i, y_i)|i = 1, 2, \ldots, n\}, X = \{\mathbf{x}_i|i = 1, 2, \ldots, n\}, Y = \{y_i|i = 1, 2, \ldots, n\}$. Given this data set, we wish to find the correct class label for a new data point $\tilde{\mathbf{x}}$. We do this by computing the class probability $p(\tilde{y}|\tilde{\mathbf{x}}, D)$.

We assume that the class label is obtained by transforming some real valued latent variable $\tilde{f}$, which is the value of some latent function $f(\cdot)$ evaluated at $\tilde{\mathbf{x}}$. We put a Gaussian process prior on this function, meaning that any number of points evaluated from the function have a multivariate Gaussian density (see [6] for a review of GPs). Assume that this GP prior is parameterized by Θ which we will call the hyperparameters. We can write the probability of interest given Θ as:

$$p(\tilde{y}|\tilde{\mathbf{x}}, D, \Theta) = \int p(\tilde{y}|\tilde{f}, \Theta) p(\tilde{f}|D, \tilde{\mathbf{x}}, \Theta)\, d\tilde{f} \tag{1}$$

The second part of Eq 1 is obtained by further integration over $\mathbf{f} = [f_1 f_2 \cdots f_n]$, the values of the latent function at the data points.

$$p(\tilde{f}|D, \tilde{\mathbf{x}}, \Theta) = \int p(\mathbf{f}, \tilde{f}|D, \tilde{\mathbf{x}}, \Theta)\, d\mathbf{f} = \int p(\tilde{f}|\tilde{\mathbf{x}}, \mathbf{f}, \Theta) p(\mathbf{f}|D, \Theta) d\mathbf{f} \tag{2}$$

where $p(\tilde{f}|\tilde{\mathbf{x}}, \mathbf{f}, \Theta) = p(\tilde{f}, \mathbf{f}|\tilde{\mathbf{x}}, X, \Theta)/p(\mathbf{f}|X, \Theta)$ and

$$p(\mathbf{f}|D, \Theta) \propto p(Y|\mathbf{f}, X, \Theta) p(\mathbf{f}|X, \Theta) = \{\prod_{i=1}^{n} p(y_i|f_i, \Theta)\} p(\mathbf{f}|X, \Theta). \tag{3}$$

The first term in 3 is the likelihood for each observed class given the latent function value, while the second term is the GP prior over functions evaluated at the data. Writing the dependence of $\mathbf{f}$ on $\mathbf{x}$ implicitly, the GP prior over functions can be written

$$p(\mathbf{f}|X, \Theta) = \frac{1}{(2\pi)^{N/2} |\mathbf{C}_\Theta|^{1/2}} \exp(-\frac{1}{2}(\mathbf{f} - \boldsymbol{\mu})^\top \mathbf{C}_\Theta^{-1} (\mathbf{f} - \boldsymbol{\mu})), \tag{4}$$

where the mean $\boldsymbol{\mu}$ is usually assumed to be the zero vector $\mathbf{0}$ and each term of a covariance matrix C_{ij} is a function of $\mathbf{x}_i$ and $\mathbf{x}_j$, i.e. $c(\mathbf{x}_i, \mathbf{x}_j)$.

One form for the likelihood term $p(y_i|f_i, \Theta)$, which relates $f(\mathbf{x}_i)$ monotonically to probability of $y_i = +1$, is

$$p(y_i|f_i, \Theta) = \frac{1}{\sqrt{2\pi}} \int_{-\infty}^{y_i f(\mathbf{x}_i)} \exp(-\frac{z^2}{2})\, dz = \mathrm{erf}(y_i f(\mathbf{x}_i)). \quad (5)$$

Other possible forms for the likelihood are a sigmoid function $1/(1+\exp(-y_i f(\mathbf{x}_i)))$, a step function $H(y_i f(\mathbf{x}_i))$, and a step function with a labelling error $\epsilon + (1 - 2\epsilon)H(y_i f(\mathbf{x}_i))$. In this paper we use the step function with a labelling error which was also used in [5].

Since $p(\mathbf{f}|D, \Theta)$ in Eq 3 is intractable due to the nonlinearity in the likelihood terms, we use an approximate method. Laplace method, variational method, Markov Chain Monte Carlo method was used in [1], [2], and [3], respectively. Expectation propagation, which is described in the next section, was used in [4,5]

3 EP for GPC

The Expectation-Propagation (EP) algorithm is an approximate Bayesian inference method [5]. We review EP in its general form before describing its application to GPCs.

Consider a Bayesian inference problem where the posterior over some parameter ϕ is proportional to the prior times likelihood terms for an i.i.d. data set

$$p(\phi|y_1, \ldots, y_n) \propto p(\phi) \prod_{i=1}^{n} p(y_i|\phi) \quad (6)$$

We approximate this by

$$q(\phi) \propto \tilde{t}_0(\phi) \prod_{i=1}^{n} \tilde{t}_i(\phi) \quad (7)$$

where each term (and therefore q) is assumed to be in the exponential family. EP successively solves the following optimization problem

$$\tilde{t}_i^{\text{new}}(\phi) = \arg\min_{\tilde{t}_i(\phi)} \mathrm{KL}\left(\frac{q(\phi)}{\tilde{t}_i^{\text{old}}(\phi)} p(y_i|\phi) \middle\| \frac{q(\phi)}{\tilde{t}_i^{\text{old}}(\phi)} \tilde{t}_i(\phi) \right) \quad (8)$$

Where KL is the Kullback-Leibler divergence and

$$\mathrm{KL}(p(x)||q(x)) = \int p(x) \log \frac{p(x)}{q(x)}\, dx \quad (9)$$

for the density functions $p(x)$ and $q(x)$. Since q is in the exponential family, this minimization is solved by matching moments of the approximated distribution. EP iterates over i until convergence. The algorithm is not guaranteed to converge although it did in practice in all our examples. Assumed Density Filtering (ADF) is a special online form of EP where only one pass through the data is performed $(i = 1, \dots n)$.

We describe EP for GPC referring to [5,4,7]. The latent function $\mathbf{f}$ plays the role of the parameter ϕ above. The form of the likelihood we use in the GPC is

$$p(y_i|f_i) = \epsilon + (1 - 2\epsilon)H(y_i f_i), \tag{10}$$

where $H(x) = 1$ if $x > 0$, and otherwise 0. The hyperparameter, ϵ in Eq 10 models labeling error outliers. The EP algorithm approximates the posterior $p(\mathbf{f}|D) = p(\mathbf{f})p(D|\mathbf{f})/p(D)$ as a Gaussian having the form $q(\mathbf{f}) \sim \mathcal{N}(\mathbf{m_f}, \mathbf{V_f})$, where the GP prior $p(\mathbf{f}) \sim \mathcal{N}(\mathbf{0}, \mathbf{C})$ has covariance matrix $\mathbf{C}$ with elements C_{ij} defined by the covariance function

$$C_{ij} = c(\mathbf{x}_i, \mathbf{x}_j) = v_0 \exp\{-\frac{1}{2}\sum_{m=1}^{d} l_m d_m(x_i^m, x_j^m)\} + v_1 + v_2\delta(i,j), \tag{11}$$

where x_i^m is the mth element of $\mathbf{x}_i$, and

$$d_m(x_i^m, x_j^m) = \begin{cases} (x_i^m - x_j^m)^2 & \text{if } x^m \text{ is continous;} \\ 1 - \delta(x_i^m, x_j^m) & \text{if } x^m \text{ is discrete,} \end{cases} \tag{12}$$

and $\delta(x_i^m, x_j^m)$ is a Kronecker delta function. The hyperparameter v_0 specifies the overall vertical scale of variation of the latent values, v_1 the overall bias of the latent values from zero mean, v_2 the latent noise variance, and l_m the (inverse) lengthscale for feature dimension m. The erf likelihood term 5 is equivalent to using the threshold function in Eq 10 with $\epsilon = 0$ and non-zero latent noise v_2.

EP tries to approximate $p(\mathbf{f}|D) = p(\mathbf{f})/p(D)\prod_{i=1}^{n} p(y_i|\mathbf{f})$, where $p(\mathbf{f}) \sim \mathcal{N}(0, \mathbf{C})$. Each term $p(y_i|\mathbf{f}) = t_i(\mathbf{f})$ is approximated by $\tilde{t}_i(\mathbf{f}) = s_i \exp(-\frac{1}{2v_i}(f_i - m_i)^2)$. From this initial setting, we can derive EP for GPC by applying the general idea described above. The resulting EP procedure is virtually identical to the one derived for BPMs in [5]. We define the following notation[1]: $\Lambda = \mathrm{diag}(v_1, \dots, v_n)$; $h_i = E[f_i]$; $h_i^{\setminus i} = E[f_i^{\setminus i}]$, where $h_i^{\setminus i}$ and $f_i^{\setminus i}$ are ones obtained from a whole set except for $\mathbf{x}_i$. The EP algorithm is as follows which we repeat for completeness—please refer to [5] for the details of the derivation. After the initialization $v_i = \infty, m_i = 0, s_i = 1, h_i = 0, \lambda_i = C_{ii}$, the following process is performed until all (m_i, v_i, s_i) converge:

Loop $i = 1, 2, \dots, n$:

1. Remove the approximate density $\tilde{t}_i$ (for ith data point) from the posterior to get an 'old' posterior: $h_i^{\setminus i} = h_i + \lambda_i v_i^{-1}(h_i - m_i)$
2. Recompute part of the new posterior: $z = \frac{y_i h_i^{\setminus i}}{\sqrt{\lambda_i}}$; $Z_i = \epsilon + (1 - 2\epsilon)\mathrm{erf}(z)$ $\alpha_i = \frac{1}{\sqrt{\lambda_i}} \frac{(1-2\epsilon)\mathcal{N}(z;0,1)}{\epsilon + (1-2\epsilon)\mathrm{erf}(z)}$; $h_i = h_i^{\setminus i} + \lambda_i \alpha_i$, where $\mathrm{erf}(z)$ is a cumulative normal density function.

[1] $\mathrm{diag}(v_1, \dots, v_n)$ means a diagonal matrix whose diagonal elements are $v_1, \dots, v_n$. Similarly for $\mathrm{diag}(\mathbf{v})$.

3. Get a new $\tilde{t}_i$: $v_i = \lambda_i(\frac{1}{\alpha_i h_i} - 1)$; $m_i = h_i + v_i\alpha_i$; $s_i = Z_i\sqrt{1 + v_i^{-1}\lambda_i}\exp(\frac{\lambda_i\alpha_i}{2h_i})$
4. Now that v_i is updated, finish recomputing the new posterior: $\mathbf{A} = (\mathbf{C}^{-1} + \boldsymbol{\Lambda}^{-1})^{-1}$; For all i, $h_i = \sum_j A_{ij}\frac{m_j}{v_j}$; $\lambda_i = (\frac{1}{A_{ii}} - \frac{1}{v_i})^{-1}$

Our approximated posterior over the latent values is:

$$q(\mathbf{f}) \sim \mathcal{N}(\tilde{\mathbf{C}}\boldsymbol{\alpha}, \mathbf{A}), \tag{13}$$

where $\tilde{C}_{ij} = y_j c(\mathbf{x}_i, \mathbf{x}_j)$ (or $\tilde{\mathbf{C}} = \mathbf{C}\mathrm{diag}(\mathbf{y})$). The approximate evidence can be obtained in the same way as for BPMs:

$$p(Y|X, \Theta) \approx \frac{|\Lambda|^{1/2}}{|\mathbf{C} + \Lambda|^{1/2}}\exp(B/2)\prod_{i=1}^{n} s_i \tag{14}$$

where $B = \sum_{ij} A_{ij}\frac{m_i m_j}{v_i v_j} - \sum_i \frac{m_i^2}{v_i}$. The approximate evidence in 14 can be used to evaluate the feasibility of kernels or their hyperparameters to the data. But, it is tricky to get a updating rule for the hyperparameters from Eq 14. In the following section, we derive the algorithm to find the hyperparameters automatically based not on Eq 14 but a variational lower bound of the evidence [8]. Classification of a new data point $\tilde{\mathbf{x}}$ can be done according to $\underset{\tilde{y}}{\mathrm{argmax}}\, p(\tilde{y}|\tilde{\mathbf{x}}) = \mathrm{sgn}(E[\tilde{f}]) = \mathrm{sgn}(\sum_{i=1}^{n} \alpha_i y_i c(\mathbf{x}_i, \tilde{\mathbf{x}}))$.

4 Outlier Robust Learning Algorithm

We derive an algorithm that learns the labelling error hyperparameter, ϵ, and is robust to outliers. Our algorithm is based on the EP approximation mentioned above, and described in more detail below. The goal is to find ϵ and other model hyperparameters to maximize the evidence $p(Y|X, \epsilon)$. We put $\Theta = \Theta_{\mathrm{cov}} \cup \{\epsilon\}$, and $\Theta_{\mathrm{cov}} = \{v_0, v_1, v_2\} \cup \{l_p | p = 1, 2, \ldots, d\}$ The following procedure is iterated until convergence after initializing ϵ with a small value:

1. EP iterations are performed given the hyperparameter ϵ (see section 3). As a result of EP, the posterior $p(\mathbf{f}|D, \Theta)$ is approximated as a Gaussian density

$$q(\mathbf{f}) \sim \mathcal{N}(\tilde{\mathbf{C}}\boldsymbol{\alpha}, \mathbf{A}) \tag{15}$$

   where $\tilde{C}_{ij} = y_j C_{ij}$ (or $\tilde{\mathbf{C}} = \mathbf{C}\,\mathrm{diag}(\mathbf{y})$), $\boldsymbol{\alpha}$ and $\mathbf{A}$ are obtained from EP.
2. Using $q(\mathbf{f})$, we form a lower bound for the log evidence by Jensen's inequality:

$$\log p(Y|X, \Theta) \geq \int q(\mathbf{f}) \log \frac{p(Y|\mathbf{f}, \epsilon)p(\mathbf{f}|X, \Theta_{\mathrm{cov}})}{q(\mathbf{f})}\, d\mathbf{f} \tag{16}$$

   The only term in the lower bound that depends on ϵ is $F_\epsilon \overset{\mathrm{def}}{=} \int q(\mathbf{f}) \log p(Y|\mathbf{f}, \epsilon)\, d\mathbf{f}$ so we optimize F_ϵ with respect to ϵ.

$$F_\epsilon = \sum_i \int q(f_i) \log p(y_i|f_i, \epsilon)\, df_i \tag{17}$$

$$= \sum_i \int q(f_i) \log(\epsilon + (1 - 2\epsilon)H(y_i f_i))\, df_i \tag{18}$$

Now, this is easy to compute in terms of erf functions (cumulative normal density functions). Define $\omega_i \stackrel{\text{def}}{=} \int q(f_i)H(y_i f_i)\, df_i = \text{erf}(y_i h_i/\sqrt{\lambda_i})$ where $f_i \sim \mathcal{N}(h_i, \lambda_i)$ then,

$$F_\epsilon = \sum_i (1 - \omega_i) \log \epsilon + \omega_i \log(1 - \epsilon) \tag{19}$$

Differentiating with respect to ϵ and solving we get:

$$\epsilon^* = \frac{1}{n} \sum_i (1 - \omega_i) \tag{20}$$

Although we have focused on learning the outlier model ϵ, in step 2, the other hyperparameters of the covariance function Θ_{cov} are also optimized by taking some gradient steps. In fact, the above algorithm is a form of approximate EM algorithm [9] with EP in the E-step and the labeling error hyperparameter updating in the M-step.

5 Model Selection Based on Evidence

In designing Gaussian process classifiers there is considerable flexibility both in the choice of the covariance function (i.e. kernel) and in the choice of the form of the likelihood term (labelling noise model). We have focused on the likelihood term and to model outliers we introduced the ϵ parameter in the likelihood term (Eq 10). However, for certain data sets an outlier model may not be necessary, i.e. we can set $\epsilon = 0$. One can do Bayesian model comparison between GPCs with and without an outlier model by evaluating the *evidence*, which is the probability of the data given the model [10]. While the exact evidence is intractable to compute, EP provides an approximation (Eq 14). Consider comparing two models, the regular GPC Θ_{regular} (learned with EP (ϵ=0)) and the robust GPCs Θ_{robust} (learned with the robust algorithm above). The model we would select based on this data is:

$$\hat{\Theta} = \underset{\Theta \in \{\Theta_{\text{regular}}, \Theta_{\text{robust}}\}}{\text{argmax}} \log p(Y|X, \Theta). \tag{21}$$

An alternative approach is to simply run the more complicated robust model and examine whether ϵ converges to exactly zero. However, there are two problems with this: first EM only finds local optima and the $\epsilon = 0$ solution may not be found because of this, even if it is a good solution. Second, we may actually believe that the labels are clean so it is useful to explicitly consider the hypothesis that $\epsilon = 0$.

6 Simulation Results

We tested the proposed algorithm for outlier treatment. We applied it to a synthetic data set, which can be visualized, and a real world data set.

We generated a simple 2-class 2-dimensional data set whose input features x_1 and x_2 are random numbers between -1 and 1. The class $+1$ or -1 is determined by the $x_1^2 + x_2^2 \geq 0.5$ decision rule. We generated 1000 data points and selected 40 data points as the training set and added labeling errors to two randomly selected data points in the training set. We compared a version of our algorithm with a fixed ($\epsilon = 0$) labeling error hyperparameter to one where ϵ was adapted. The hyperparameters are set to $v_0 = 1$, $v_1 = 10^{-8}$, $l_1 = l_2 = 0.5$. In the regular GPC ($\epsilon = 0$), v_2, which allows a soft decision boundary, was updated with the EM-EP algorithm [11]. In the robust GPC we used the proposed algorithm and also updated v_2 (as in [11]) in step 2 of the proposed algorithm.

Figure 1 shows the training points, test points, decision boundary from GPC and decision boundary from robust GPC. Obviously, robust GPC gives a better decision boundary than GPC.

Table 1 shows the classificaton error rates and the log of the approximate evidence (Eq 12). GPC with the adapted labeling error hyperparameter clearly outperforms the one with it fixed by both approximate evidence and classification test error rate. If we apply a model selection scheme based on an approximate evidence, we select robust GPC which is obviously a better model for this data set.

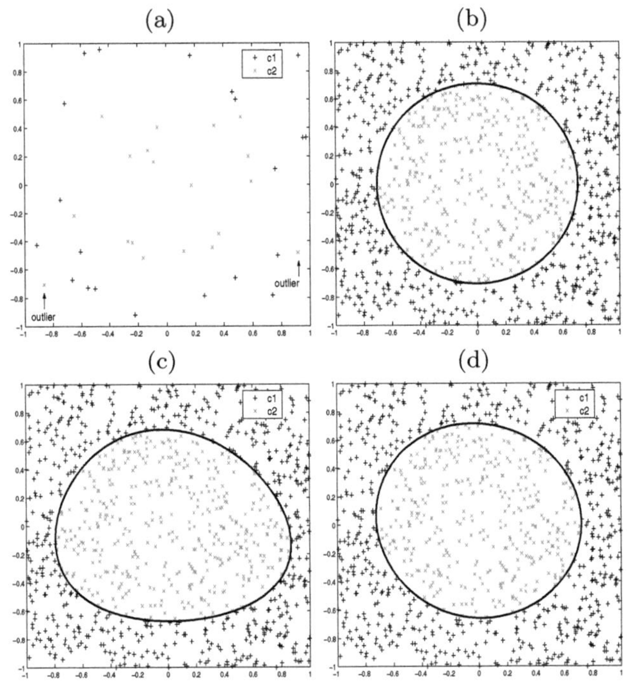

Fig. 1. Experiment results of the proposed outlier robust algorithm for the artificial data set with outliers: (a) Training set, (b) Test set, (c) Classification result with GPC ($\epsilon = 0$), (d) Classification result with robust GPC

Table 1. Comparison of GPCs with adpated and fixed label error hyperparamter

Methods:	GPC with $\epsilon = 0$	Robust GPC
$\log p(Y\|X, \Theta)$	-26.96	-25.31
Error rate (%)	7.08	2.40

To illustrate the proposed algorithm we applied it to a real world data set from the UCI Machine Learning Repository. We used the New Thyroid data set which has 2 classes ("normal" and "not normal", reduced from 3 classes) and 215 data points. We created 10 pairs of training and test sets by randomly dividing the whole data set into two. Outliers were generated in the training set by changing labels of 0, 5, 10, 15 % of the data points[2]. We applied GPC with $\epsilon = 0$ and the proposed robust GPC method, with ϵ updated. We also tested a model selection method which chose between GPC and robust GPC based on the model *evidence*.

Table 2 shows means of approximate log evidences (available only for GPCs), means of test set errors and standard deviations of those mean estimators computed averaged over the 10 runs. In Table 2, "label-change rate" is the rate of the data whose labels are changed in the training set,"error" is the test set error,"log-ev" is the approximate log evidence, "SVM" is a soft-margin support vector machine, "GPC" is the model with ($\epsilon = 0$), "robust-GPC" is the proposed model adapting ϵ, and "MS-robust-GPC" is the model selection method which selects between GPC and robust GPC based on which has higher evidence on the training data. The hyperparameters in the GPC covariance function were set to $v_0 = 1$,$v_1 = 10^{-8}$, $l_1 = 0.05$, and $l_2 = 0.05$. In the regular GPC ($\epsilon = 0$), v_2, which allows a soft decision boundary, was updated with the EM-EP algorithm [11]. The SVM used the same kernel as the covariance function of the GPC, except that v_2 was set to 0 and instead the regularization parameter C was chosen by cross-validation—a standard approach to SVMs. Note that C also allows a soft decision boundary, like v_2. In the robust GPC we used the proposed algorithm and also updated v_2 (as in [11]) in step 2 of the proposed algorithm.

The robust-GPC performed better than SVMs and GPCs for data with more outliers whose label-change rates are 10% and 15%, but the GPC performed better than SVMs and robust-GPC for less outliered data whose label-change rates are 0% and 10%. The reason why GPCs are better than robust-GPC for the data with 0% and 5% label-change rates seems to be that there are no outlier or only a relatively small number of outliers. MS-robust-GPC performed as well or better than SVMs and GPCs for all label-change rates studied (including 0%). The MS-robust-GPC model combined the best of both worlds, choosing the simpler GPC when there were no outlier and chosing the robust GPC when there were many outliers.

[2] This does not mean that the training set has 0, 5, 10, 15 % of its data as outliers, because the data whose labels are changed which are far from the decision boundary are outliers. However, the training set with more label-changed data tends to have more outliers.

Table 2. Comparison of SVM, the regular GPC, the proposed robust GPC, and the model selection based robust GPC

label-change rate(%)		0	5	10	15
SVM	error(%)	4.07±0.60	5.09±0.96	6.76±1.10	8.80±1.13
GPC	log-ev	-24.4±0.6	-41.8±0.8	-51.8±1.1	-60.2±1.0
	error(%)	**3.70±0.36**	4.90±0.77	7.50±0.88	8.15±1.42
robust GPC	log-ev	-27.8±0.5	-41.7±0.8	-50.9±0.7	-59.2±0.7
	error(%)	4.54±0.55	5.28±0.56	**6.30±0.89**	6.94±0.75
MS-robust-GPC	log-ev	-24.4±0.6	-41.0±0.7	-50.5±0.7	-58.6±0.8
	error(%)	**3.70±0.36**	**4.54±0.62**	6.76±0.76	**6.85±0.73**

7 Discussion

We have proposed an algorithm for outlier robust Gaussian process classification. This algorithm is an approximate EM algorithm which updates a labeling error hyperparameter. The experimental results show that having an outlier model is important for classification and that robust GPCs with model selection based on the Bayesian evidence provide a promising approach to robust classification.

The complexity of learning in the outlier robust GPC is not higher than that in normal GPC. Both of them go through the almost similar steps. The extension to multi-class classification is not straightforward to do. It is not straightforward to apply the EP algorithm to multi-class classification models (e.g. the multi-class classification model proposed in [12]). It is a challenging problem to get the EP algorithm for a multi-class GPC model.

The notion of an outlier is relative to the complexity of the model. If the model is very complicated, it may not have any outliers in the sense that the model can fit all data points easily. Since the complexity of the model and the need for an explicit outlier model are closely related, this poses a challenging set of issues for future work in model selection.

Acknowledgement

This work has been partially supported by the BK21 Research Center for Intelligent Mobile Software at Yonsei Univerisity in Korea.

References

1. Williams, C.K.I., Barber, D.: Bayesian classification with Gaussian processes. IEEE Transactions on Pattern Analysis Machine Intelligence 20, 1342–1351 (1998)
2. Gibbs, M., MacKay, D.J.C.: Variational Gaussian process classifiers. IEEE Transcations on Neural Networks 11(6), 1458 (2000)
3. Neal, R.: Monte Carlo implementation of Gaussian process models for Bayesian regression and classification. Technical Report CRG–TR–97–2, Dept. of Computer Science, University of Toronto (1997)

4. Opper, M., Winther, O.: Gaussian processes for classification: Mean field algorithms. Neural Computation 12, 2655–2684 (2000)
5. Minka, T.: A family of algorithms for approximate Bayesian inference. PhD thesis, MIT (2001)
6. Williams, C.K.I., Rasmussen, C.E.: Gaussian processes for regression. In: NIPS, vol. 8. MIT Press, Cambridge (1995)
7. Seeger, M., Lawrence, N., Herbrich, R.: Sparse representation for Gaussian process models. In: NIPS, vol. 15 (2002)
8. Jordan, M.I., Ghahramani, Z., Jaakkola, T.S., Saul, L.K.: An introduction to variational methods for graphical models. Machine Learning 37, 183–233 (1999)
9. Neal, R., Hinton, G.: A view of the EM algorithm that justifies incremental, sparse, and other variants. In: Jordan, M.I. (ed.) Learning in Graphical Models. Kluwer, Dordrecht (1998)
10. MacKay, D.J.C.: Bayesian interpolation. Neural Computation 4(3), 415–447 (1992)
11. Kim, H.C., Ghahramani, Z.: The EM-EP algorithm for Gaussian process classification. In: Proceedings of the Workshop on Probabilistic Graphical Models for Classification (ECML) (2003)
12. Kim, H.C., Ghahramani, Z.: Bayesian Gaussian process classification with the EM-EP algorithm. IEEE Transactions on Pattern Analysis and Machine Intelligence 28(12), 1948–1959 (2006)

2D Shape Classification Using Multifractional Brownian Motion

Manuele Bicego and Alessandro Trudda

DEIR - University of Sassari
via Torre Tonda, 34 - 07100 Sassari (Italy)
{bicego,atrudda}@uniss.it

Abstract. In this paper a novel approach to contour-based 2D shape recognition is proposed. The main idea is to characterize the contour of an object using the multifractional Brownian motion (mBm), a mathematical method able to capture the local self similarity and long-range dependence of a signal. The mBm estimation results in a sequence of Hurst coefficients, which we used to derive a fixed size feature vector. Preliminary experimental evaluations using simple classifiers with these feature vectors produce encouraging results, also in comparison with the state of the art.

1 Introduction

Object recognition represents an important research area in computer vision and pattern recognition. The classification of three-dimensional (3D) objects has been faced by different approaches [1,2], many of which are based on the analysis of two-dimensional (2D) aspects of objects (2D shapes). In this context, the analysis could be performed by analyzing the boundary (external approaches) or the whole shape (internal approach). In the former context, many contour descriptors have been proposed, like, for example, Fourier descriptors and chain code, whereas, in the latter context features like the medial axis, or moments, have been employed [2]. In particular, object contours proved to be very effective in many applications, with many different approaches being presented over the past years, each with different characteristics, like robustness to noise and occlusions, invariance to translation, rotation and scale, computational requirements, and accuracy [2,3].

In this paper a novel approach to contour-based 2D shape recognition is proposed. The main idea is to characterize the contour of each object using the multifractional Brownian motion (mBm) [4,5,6], a mathematical method employed in the finance community to characterize and model the financial series. The simplest way of introducing the multifractional Brownian motion is to consider it as an extension of the fractional Brownian motion (fBm) [7], a process characterized by a slowly decaying autocorrelation function depending on the parameter $H \in (0, 1]$ (named Hurst exponent). The fBm may model the self similarity and long-range dependence of a process. The Hurst exponent H has

N. da Vitora Lobo et al. (Eds.): SSPR&SPR 2008, LNCS 5342, pp. 906–916, 2008.

a clear interpretation: if $H = 1/2$, then the fBm represents the standard Brownian motion (namely the process has no memory); for $H > 1/2$, the process is characterized by a persistence (positive or negative) – namely the process has a clear trend; finally, for $H < 1/2$ the process is characterized by an antipersistence. Even if this method has been largely applied in the finance community (mainly to understand some characteristics of the financial series), some works applying the fBm function have also been presented in the pattern recognition context, like in medical imaging [8] or in speech recognition [9]. It is important to note that fBm is related to the fractal theory [10] – a theory largely employed in image analysis [11] or even object classification [12,13] – since it is modeling self-similarity. Actually it can be shown that, for fractal or self-similar processes, the Horst exponent H can be related to a fractal dimension D_h [10] through the equation $D_h = 2 - H$, where D_h is the fractal Hausdorff dimension [14].

Nevertheless, from a shape recognition point of view, we consider it too strong to assume contours showing a complete self similarity, and modeling them with a fBm process (or fractals) could be too approximate. Still, the description power of such models may be successfully applied if we model *local* self-similarity, that is by considering objects as self similar in limited region. This behavior could be modeled by the so called multifractional Brownian motion (mBm), an extension of fBm so that the Hurst exponent can vary over time. As a result of this process, a function $H(t)$ is obtained, describing the self similarity of the process in a window centered in t. The mBm process is no longer stationary nor self-similar; despite this, the process is extremely versatile since the time dependency of H is useful to model phenomena whose punctual regularity is time changing, this characteristic being possibly really useful in characterizing an object's shape.

In this paper we explore the use of the mBm function for characterizing the shape contours with the aim of 2D shape classification. In the proposed approach, the contour is modeled with the curvature (similarly to [15,16]), which was fitted with a mBm function. The sequence of Horst coefficients computed was subsequently employed to derive a fixed length vector, which characterizes each shape. Some experimental evaluations were performed using simple Nearest Neighbor classifier (with the Euclidean and the Minkowsky distance) on the Chicken database, showing promising results also in comparison with the state of the art.

2 The Multifractional Brownian Motion (mBm) Approach

This section will present the theory of the mBm approach, as well as the estimation methodology we used here.

2.1 The mBm Model

As reported in the introduction, a convenient way to introduce the mBm is to consider it as an extension of the fractional Brownian motion (fBm). Defined in a celebrated paper by Mandelbrot and Van Ness [7], the fBm is characterized by a slowly decaying autocorrelation function depending on the parameter $H \in (0,1]$, named *Hurst* exponent. Following the definition that can be found in [17], the process has the following moving average representation

$$B_H(t) = C\{\pi K(2H)\}^{1/2} \int_{\mathbb{R}} f_t(s) dB(s) \tag{1}$$

with

$$f_t(s) = \frac{1}{\Gamma\left(H+\frac{1}{2}\right)} \left\{ |t-s|^{H-\frac{1}{2}} 1_{]-\infty,t]}(s) - |s|^{H-\frac{1}{2}} 1_{]-\infty,0]}(s) \right\}$$

where C is a positive constant, K is the function defined on $]0,2[$ as $K(\alpha) = \Gamma(\alpha+1)\frac{\sin\frac{\alpha\pi}{2}}{\pi}$, and $B(\cdot)$ stands for the ordinary Brownian motion process – introduced by Robert Brown in 1872 to describe the random movements of particles suspended in a liquid or gas [18]. The process is self-similar[1] of parameter H and has stationary increments. Its covariance function reads as

$$E\left(B_H(t)B_H(s)\right) = \frac{c^2}{2}\left(|t|^{2H} + |s|^{2H} - |t-s|^{2H}\right) \tag{2}$$

The fBm can be generalized by allowing H to vary over time. This extension – known as *multifractional Brownian motion* (mBm) [4,5,6] – has the following representation

$$M_{H(t)}(t) = C\{\pi K(2H(t))\}^{1/2} \int_{\mathbb{R}} f_t(s) dB(s) \tag{3}$$

with

$$f_t(s) = \frac{1}{\Gamma\left(H(t)+\frac{1}{2}\right)} \left\{ |t-s|^{H(t)-\frac{1}{2}} 1_{]-\infty,t]}(s) - |s|^{H(t)-\frac{1}{2}} 1_{]-\infty,0]}(s) \right\}$$

where $H : [0,\infty) \rightarrow (0,1]$ is required to be a Hölder function of order $0 < \eta \leq 1$ to ensure the continuity of the motion.

Notice that since $H(t)$ is the punctual Hölder exponent of the mBm at point t, the process is locally asymptotically self-similar with index $H(t)$ (see, e.g. [20]) in the sense that, denoted by $Z(t, au) := M_{H(t+au)}(t+au) - M_{H(t)}(t)$ the increment process of the mBm at time t and lag au, it holds

$$\lim_{a \to 0^+} a^{-H(t)} Z(t, au) \stackrel{d}{=} B_{H(t)}(u),\ u \in \mathbb{R}. \tag{4}$$

[1] We recall that the process $\{X(t),\, t \in T\}$ is said *self-similar* with parameter H if for any $\alpha > 0$ $\{X(\alpha t)\} \stackrel{d}{=} \{\alpha^H X(t)\}$, where the equality holds for the finite-dimensional distributions of the process (see e.g. [19]).

The above distributional equality indicates that at any point t there exists an fBm with parameter $H(t)$ tangent to the mBm. Moreover, since $B_{H(t)}(u) \sim \mathcal{N}(0, C^2 u^{2H(t)})$, the infinitesimal increment of the mBm at time t, normalized by $a^{H(t)}$, normally distributes with mean 0 and variance $C^2 u^{2H(t)}$ $(u \in \mathbb{R}, a \to 0^+)$. The increments of the mBm are no longer stationary nor self-similar; despite this, the process is extremely versatile since the time dependency of H is useful to model phenomena whose punctual regularity is time changing.

2.2 Point-Wise Estimation of the Mbm

One of the main problems when using the mBm process is to estimate the function $H(t)$ from actual data. To face this problem one could think to adapt the traditional estimators of H available in literature in order to shadow the dynamics of $H(t)$. The weakness of this approach resides in the fact that very large samples are needed to get reliable estimates and over a long period of time H is likely to change even further. So, more efficient estimators are needed in the case of the mBm. An answer to this problem is provided by Bianchi in [21], who develops the work of [22] and defines a family of "moving-window" estimators of $H(t)$ based on the k-th absolute moment of a Gaussian random variable of mean zero and given variance V_H (the variance of the unit lag increment of a mBm). Given a series of length N and a window of length δ, the estimator has the form

$$H^k_{\delta,N}(t) = \frac{\log\left(2^{k/2}\Gamma\left(\frac{k+1}{2}\right)V_H^{k/2}\right) - \log\left(\frac{\sqrt{\pi}}{\delta}\sum_{j=t-\delta}^{t-1}|X_{j+1,N} - X_{j,N}|^k\right)}{k\log(N-1)} \tag{5}$$

$$\text{for } j = t-\delta, ..., t-1; \quad t = \delta+1, ..., N; \quad k \geq 1. \tag{6}$$

When $H \in \left(0, \frac{3}{4}\right)$, the estimators' rate of converge is $O\left(\delta^{-\frac{1}{2}}(\log N)^{-1}\right)$, what allows to get reliable estimates even for very short $\delta's$.

The family of estimators (5) was proved to be correct and normally distributed as

$$H^k_{\delta,N}(t) \sim \mathcal{N}\left(H(t), \frac{\pi}{\delta k^2 \ln^2(N-1)\, 2^k \left(\Gamma\left(\frac{k+1}{2}\right)\right)^2}\sigma^2\right). \tag{7}$$

σ^2 being the variance of a Gaussian random variable defined as a proper rescaled sum. Toilsome computations show that when $H = \frac{1}{2}$ the variance of the estimator reduces to

$$Var(H^k_{\delta,N}(t)) = \frac{\sqrt{\pi}}{\delta k^2 \ln^2(N-1)\left[\Gamma\left(\frac{k+1}{2}\right)\right]^2} \cdot \left(\Gamma\left(\frac{2k+1}{2}\right) - \frac{1}{\sqrt{\pi}}\left[\Gamma\left(\frac{k+1}{2}\right)\right]^2\right) \tag{8}$$

and the optimal value of k is deduced by minimizing the last relation. So one finds that the minimum of (8) takes place when $k = 2$, value which will be used in the experiments discussed below.

3 The Proposed Approach

In this section the proposed approach is summarized: first the contours are encoded in one-dimensional signals, then the mBm is estimated and a feature vector is extracted from $H(t)$.

3.1 Coding of Contours

Each object is represented by the sequence of the curvature coefficients, computed as described in [15,16]. Curvature has been chosen since it presents many different appealing characteristics, like invariance to translation and somehow to rotation (it results just in a shift of the signal). In order to compute curvature sequence, the contours are firstly extracted by using the *Canny* edge detector; the boundary is then approximated by segments of approximately fixed length d_L. Finally, the curvature value at point x is computed as the angle between the two consecutive segments intersecting at x. The initial point is the leftmost point lying on the horizontal line passing through the object centroid, following the boundary in a counterclockwise manner.

3.2 Feature Extraction

It should be noted that, in the finance community, the main goal is to approximate the series with the mBm process, in order to gain knowledge from the sequence of Hurst coefficients and understand the persistence or anti-persistence behavior of the financial series. Here the goal is different: we want to approximate the contour signal with the mBm function, subsequently extracting a feature vector from the sequence $H(t)$ used to characterize the shape.

In particular each object's curvature sequence is modeled with a mBm process and a function $H(t)$ is extracted. It is worthwhile to observe that the signal $H(t)$ could be used as it is as a signature of the shape, with many attractive characteristics (one for all: rotation of the object simply results in a shift of the curvature and consequently only a shift of the $H(t)$). Nevertheless, in this preliminary study, we opted for a more simple and compact solution: the feature vector was obtained by sub-sampling $H(t)$ with a regular step (this has a clear impact on the rotation-invariance); in order to obtain exactly the same number of points for each shape, the sub-sampling step varies with the length of the sequence. In order to increase the robustness, the sub-sampling was carried out by averaging the signal in a small interval around the sampling points.

To summarize, there are two crucial parameters in the proposed system: the window size δ (lag), defining the "scale" of the analysis, and the number of sampled points, defining the trade-off between the capacity of the descriptor

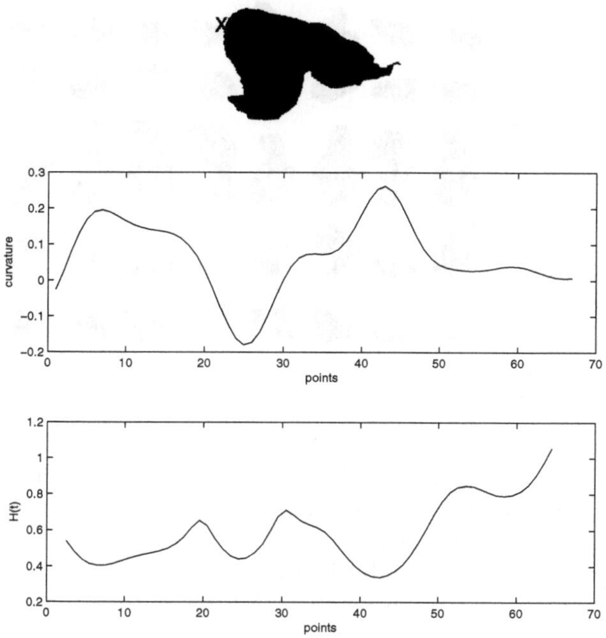

Fig. 1. (Top): one sample object; (center): curvature extracted from the object contour – starting from the "X" and following the boundary in a counterclockwise manner; (bottom) the estimated $H(t)$, computed with a lag $\delta = 5$. The value of $H(t)$ at a certain point x derives from the application of the estimator of H to the subwindow of the signal centered in x.

and the curse of dimensionality. In the experimental part, we made a thorough analysis of the accuracy of the descriptors when varying these parameters.

An example of the application of the above described approach is displayed in Fig. 1: the curvature signal (middle) of the contour of the object (top) is approximate with a mBm, which $H(t)$ is displayed in the bottom part. It is clear how this $H(t)$ encodes the regularities/indentations of the signal: where the signal has a change in the regularity (at points 25 and 45 – corresponding to the abrupt change in the contour) $H(t)$ suddenly decreases – modeling a somehow turbulent behavior. On the other side, when the behavior is regular (zone 50-70), $H(t)$ suddenly approaches 1.

4 Experimental Results

We chose to study the Chicken Pieces Database[2], denoted also as Chicken data [23]. This set consists of 446 binary images of chicken pieces, whose examples are shown in Fig. 2. Each piece belongs to one of the five classes, representing

[2] See `http://algoval.essex.ac.uk:8080/data/sequence/chicken/`.

Fig. 2. Examples from the Chicken data

specific parts of a chicken: wing (117 samples), back (76), drumstick (96), thigh and back (61), and breast (96). The shapes are usually first described by contours, which are further encoded by suitable sequences. This poses a difficult classification task. The results published in [24] report a baseline leave-one-out (LOO) accuracy of $\approx 67\%$ by using the 1-NN on the Levenshtein (non-cyclic) edit distance computed on the contour chain code.

The sequences of the dataset were encoded using the proposed approach in order to obtain a feature space. In the feature space any classifier may be used. In order to have a clear idea of the discriminative capability of the features we used a very simple one, namely the Nearest Neighbor. We computed the leave-one-out (LOO) accuracy using the proposed approach. Two different vector-based distances were employed in the feature space: the classic Euclidean distance and the Minkowsky distance. We let the parameter δ vary from 2 to 25, whereas the dimensionality of the feature vector was analyzed in the range [2-10]. The results are graphically shown in Fig. 3, while increasing window size δ. For each δ, we chose the dimensionality of the feature vector rising to the best accuracy.

From the figure it could be noted that results are quite satisfactory, even if the task is extremely difficult. Results with Euclidean Distance seem to be slightly better than those with Minkowski distance. It may also be noted that the results seem to improve when increasing the analysis window size (actually the best result was obtained with $\delta = 20$ with a 76.5% of accuracy). This is true up to a certain point: when increasing too much the δ, the system performances start to decrease, mainly because the mBm will consider larger parts of the shape as self similar (so, to some extent, leading to the fBm).

4.1 Comparative Analysis

This database has been widely used in the literature, with different representation and different classifiers. Since our goal is to compare the discriminative power of the proposed features, here we have reported results of other methods which only employ simple classifiers, like the Nearest Neighbor (or the K-Nearest

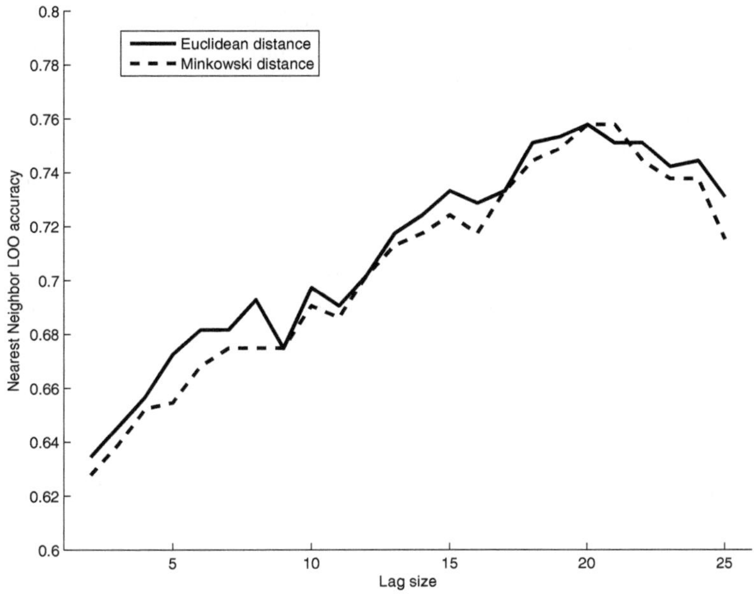

Fig. 3. LOO accuracy using the NN classifier in the proposed feature space

Neighbor). Some of them are taken from literature, some of them have been computed by the authors.

1. Fractional Brownian motion (fBm): naturally the first comparison is obtained by applying the simpler fBm to the sequences extracted from the shapes. As explained in the previous sections, this representation in too simplistic and not suitable for describing a shape. This is confirmed by experimental evidence, since the LOO accuracy using NN classifier (and Euclidean Distance) was really poor, about 38.79%.
2. In [24], a structural approach has been applied to this problem, with a cyclic string matching procedure. Many different techniques have been proposed: as said before, the Levenshtein (non-cyclic) edit distance computed on the contour chain code returns a LOO error (with Nearest Neighbor) of about 67%, whereas some extensions and improvements arrive up to 78%[3]
3. Another structural approach has been proposed by Neuhaus and Bunke in [16] where a kernel based method, based on a dissimilarity representation, was proposed. As a reference system, given the sequences encoded with curvature coefficients – similarly to what done here – they compute the accuracy of a K-Nearest Neighbor where the distances were determined using an

[3] Approximate values are shown since in the paper no numerical values were given, just a graphical table – Tab. 5(b).

efficient cyclic string edit distance algorithm [25]. The accuracy, computed by splitting the set in Training/Testing/Validation was 74.3%[4].

4. Hidden Markov Models: recognition of 2D shapes is a quite unconventional application of HMMs, even though promising results have been reported [26,27,15]. The conventional approach is train one HMM for each class, subsequently classifying with the standard Bayes rule. Nevertheless, since the tackled dataset presents a notably high intra-class variability – and a single HMM per class would be too general and so not able to solve the task – we trained one model for each training sequence, assign an unknown sequence to the class of the model showing the highest likelihood (similarly to what was done in [28]): clearly this may be seen as a 1-nearest-neighbor (1-NN) classifier, with the proximity measure defined by the likelihood function. After a preliminary evaluation, HMMs are trained with $K = 3$ states for all classes. The LOO error computed was of 73.77%.

From the previous results we may observe the suitability of the proposed representation technique, also in comparison with the state of the art.

5 Conclusions

In this paper a novel 2D shape descriptor has been proposed, which characterizes the contour of each object using the multifractional Brownian motion (mBm). This function represents a mathematical method able to capture the local self similarity and long-range dependence of the silhouette of an object. The mBm estimation results in a sequence of Hurst coefficients, which we used to derive a fixed size feature vector. Preliminary experimental evaluations using simple classifiers on these feature vectors produced encouraging results.

Current research directions are mainly directed toward two goals: first, the automatic selection of the dimension of the window size δ – this possibly being inferrable from a local R/S analysis; second, a more effective classification scheme able to exploit the description power of the proposed features. Clearly, a more thorough experimental evaluation is also needed, involving different datasets and different classifiers.

References

1. Campbell, R., Flynn, P.: A survey of free-form object representation and recognition techniques. Computer Vision and Image Understanding 81, 166–210 (2001)
2. Loncaric, S.: A survey of shape analysis techniques. Pattern Recognition 31(8), 983–1001 (1998)
3. Suetens, P., Fua, P., Hanson, A.: Computational strategies for object recognition. ACM Compuiting Surveys 24(1), 5–61 (1992)
4. Peltier, R., Véhel, J.L.: Multifractional brownian motion: definition and preliminary results. Technical Report Rapport de recherche no. 2645, INRIA (1995)

[4] With the kernel based method they were able to reach 81.1% of accuracy.

5. Véhel, J.L.: Fractal approaches in signal processing. Fractals 3, 755–775 (1995)
6. Ayache, A., Véhel, J.L.: The generalized multifractional brownian motion. Statistical Inference for Stochastic Processes 3, 7–18 (2000)
7. Mandelbrot, B., Ness, J.V.: Fractional brownian motions, fractional noises and applications. SIAM Review 10, 422–437 (1968)
8. Lundahl, T., Ohley, W., Kay, S., Siffert, R.: Fractional brownian motion: A maximum likelihood estimator and its application to image texture. IEEE Trans. on Medical Imaging 5, 152–161 (1986)
9. SantAna, R., Coelho, R., Alcaim, A.: Text-independent speaker recognition based on the hurst parameter and the multidimensional fractional brownian motion model. IEEE Trans. on Audio, Speech and Language Processing 14(3) (2006)
10. Mandelboot, B.B., et al.: The Science of Fractal Images. Springer, New York (1988)
11. Chen, S., Keller, J., Crownover, R.: On the calculation of fractal features from images. IEEE Trans. om Pattern Analysis and Machine Intelligence 15 (1993)
12. Taylor, R., Lewis, P.: 2d shape signature based on fractal measurements. IEE Proc. of Vision, Image and Signal Processing 141(6), 422–430 (1994)
13. Uemura, S., Haseyama, M., Kitajima, H.: Efficient contour shape description by using fractal interpolation functions. In: Proc. Int. Conf. Image Processing, vol. 1, pp. I–485–I–488 (2002)
14. Falconer, K.: Fractal Geometry: Mathematical Foundations and Applications. Wiley, Chichester (2003)
15. Bicego, M., Murino, V.: Investigating Hidden Markov Models' capabilities in 2D shape classification. IEEE Trans. on Pattern Analysis and Machine Intelligence - PAMI 26(2), 281–286 (2004)
16. Neuhaus, M., Bunke, H.: Edit distance-based kernel functions for structural pattern classification. Pattern Recognition 39, 1852–1863 (2006)
17. Coeurjolly, J.: Identification of the multifractional brownian motion. Bernoulli 11(6), 987–1008 (2005)
18. Brown, R.: A brief account of microscopical observations made in the months of June, July and August, 1827, on the particles contained in the pollen of plants; and on the general existence of active molecules in organic and inorganic bodies. Phil. Mag. 4(1828), 161–173
19. Samorodnitsky, G., Taqqu, M.: Stable non-Gaussian random processes. Chapman & Hall, Boca Raton (1994)
20. Benassi, A., Jaffard, S., Roux, D.: Elliptic gaussian random processes. Matematica Iberoamericana 13(1), 19–89 (1997)
21. Bianchi, S.: Pathwise identification of the memory function of the multifractional brownian motion with application to finance. Int. J. of Theoretical and Applied Finance 8(2), 255–281 (2005)
22. Peltier, R., Véhel, J.L.: A new method for estimating the parameter of fractional brownian motion. Technical Report Rapport de recherche no. 2396, INRIA (1994)
23. Andreu, G., Crespo, A., Valiente, J.: Selecting the toroidal self-organizing feature maps (tsofm) best organized to object recognitio. In: Proc. of IEEE ICNN 1997, vol. 2, pp. 1341–1346 (1997)
24. Mollineda, R., Vidal, E., Casacuberta, F.: Cyclic sequence alignments: Approximate versus optimal techniques. Int. Journal of Pattern Recognition and Artificial Intelligence 16(3), 291–299 (2002)
25. Peris, G., Marzal, A.: Fast cyclic edit distance computation with weighted edit costs in classification. In: Proc. of Int. Conf. on Pattern Recognition, vol. 4, pp. 184–187 (2002)

26. He, Y., Kundu, A.: 2-D shape classification using Hidden Markov Model. IEEE Trans. Pattern Analysis Machine Intelligence 13(11), 1172–1184 (1991)
27. Arica, N., Yarman-Vural, F.: A shape descriptor based on circular Hidden Markov Model. In: IEEE Proc. Int. Conf. Pattern Recognition, vol. 1, pp. 924–927 (2000)
28. Bicego, M., Murino, V., Figueiredo, M.: Similarity-based classification of sequences using hidden markov models. Pattern Recognition 37(12), 2281–2291 (2004)

A New Performance Evaluation Method for Two-Class Imbalanced Problems

Vicente García[1,2], Ramón A. Mollineda[2], and J. Salvador Sánchez[2]

[1] Lab. Reconocimiento de Patrones, Instituto Tecnológico de Toluca
Av. Tecnológico s/n, 52140 Metepec (México)
[2] Dept. Llenguatges i Sistemes Informàtics, Universitat Jaume I
Av. Sos Baynat s/n, 12071 Castelló de la Plana (Spain)

Abstract. In this paper, we introduce a new approach to evaluate and visualize the classifier performance in two-class imbalanced domains. This method defines a two-dimensional space by combining the geometric mean of class accuracies and a new metric that gives an indication of how balanced they are. A given point in this space represents a certain trade-off between those two measures, which will be expressed as a trapezoidal function. Besides, this evaluation function has the interesting property that it allows to emphasize the correct predictions on the minority class, which is often considered as the most important class. Experiments demonstrate the consistency and validity of the evaluation method here proposed.

Keywords: Imbalance, performance measure, learning.

1 Introduction

In the last years, the class imbalance problem has received considerable attention in areas such as Machine Learning and Pattern Recognition. A two-class data set is said to be imbalanced when one of the classes (the minority one) is heavily under-represented in comparison to the other class (the majority one). This issue is particularly important in real-world applications where it is costly to misclassify examples from the minority class, such as diagnosis of rare diseases, detection of fraudulent telephone calls, text categorization, information retrieval and filtering tasks. Because of examples of the minority and majority classes usually represent the presence and absence of rare cases, respectively, they are also known as positive and negative examples.

Most research on this topic has traditionally focused on a number of solutions both at the data and algorithmic levels. Data level methods for balancing the classes consists of resampling the original data set, either by over-sampling the minority class or by under-sampling and/or under-sampling the majority class, until the classes are approximately equally represented [2, 4, 10, 14, 23]. Solutions at the algorithmic level consist of adapting existing algorithms and techniques to the particular characteristics of the imbalanced data sets [5, 7, 12, 15, 16, 20].

Recently, some authors have also drawn attention to the fact that there does not exist a direct correlation between class imbalance and the loss of performance [3, 9, 12, 17, 18], suggesting that the class imbalance is not a problem by itself. From this point of view,

N. da Vitora Lobo et al. (Eds.): SSPR&SPR 2008, LNCS 5342, pp. 917–925, 2008.

the degradation of performance is also related to other data characteristics, such as the size of the data set, distribution of the data within each class, small disjuncts, data duplication, density and overlap complexity.

Within the context of class imbalance, a third area to investigate refers to measuring the classifier performance. Common metrics estimate classification accuracy and/or error rates. However, when the prior class probabilities are very different and the error costs are unequal, the use of these measures might produce misleading conclusions since they do not take into account misclassification costs, are strongly biased to favor the majority class, and are sensitive to class skews [19, 11, 6]. Under these circumstances, it can be more appropriate to employ the ROC curve or other alternative metrics, such as the geometric mean (g-mean) or the area under the ROC curve (AUC) [2,7,8].

The geometric mean, AUC and other measures give an estimate of the overall classifier performance, but they cannot reflect the contribution of each class to it. In some cases, it can be even more important to know whether the accuracies on each class are balanced and if not, to find out which is the 'dominant class' (the class with the highest accuracy rate). For this purpose we will here suggest a simple metric to reflect the 'dominance' relationship between the majority class and the minority class. After this, we will also introduce a new method to evaluate and visualize the classifier performance in class imbalance problems.

2 Evaluation of Classifier Performance in Imbalanced Domains

Typical metrics for measuring the performance of learning systems are classification accuracy and error rates, which can be easily derived from a 2×2 confusion matrix as that given in Table 1 (for a two-class problem).

Empirical evidence shows that most of these commonly used measures are biased with respect to the data imbalance and proportions of correct and incorrect classifications. Shortcomings of these evaluators have motivated search for new measures, such as the geometric mean of class accuracies or the AUC [2,7,8].

Given the True Positive rate, $TPrate = TP/(TP + FN)$, and the True Negative rate, $TNrate = TN/(TN + FP)$, the *geometric mean* of $TPrate$ and $TNrate$ is computed as $\sqrt{TPrate \cdot TNrate}$. This measure can be seen as a sort of correlation between both rates, because a high value occurs when they both are also high, while a low value is related to at least one low rate. An alternative is the ROC space, in which the classifier performance is characterized by a point that represents a trade-off between the $TPrate$ and the False Positive rate, $FPrate = FP/(FP + TN)$ (also known as

Table 1. Confusion matrix for a two-class problem

	Predicted positive	Predicted negative
Actual positive	True Positive (TP)	False Negative (FN)
Actual negative	False Positive (FP)	True Negative (TN)

false alarm rate). The performance can be finally computed by means of the AUC, which is a combined measure of $TPrate$ and specificity $(1 - FPrate)$.

These measures are sensitive to skewed distributions of positive and negative instances, but they cannot explain the contribution of each class to it, nor which is the prevalent class. For this reason, in this paper we propose a simple performance metric, here called *dominance*, which will be able to reflect the dominance or prevalence relationship between the positive class and the negative class. The dominance can be computed as follows:

$$Dominance = TPrate - TNrate \tag{1}$$

This measure can take on any value between -1 and $+1$, since both the $TPrate$ and the $TNrate$ are in the range $[0, +1]$. A dominance value of $+1$ represents a situation of perfect accuracy on the positive class, but failing on all negative cases; a value of -1 corresponds to the opposite situation. The closer the dominance is to 0, the more balanced both individual rates. In practice, the dominance can be interpreted as an indicator of how balanced the $TPrate$ and the $TNrate$ are. Obviously, any particular value of dominance could represent very different scenarios (for example, a dominance value of 0 could correspond to an optimal case with $TPrate = TNrate = +1$, but also to a dramatic situation with $TPrate = TNrate = 0$). In order to estimate the performance of a classifier, one should use the dominance together with other measures, such as the geometric mean or the accuracy.

2.1 The Accuracy-Dominance Space

The performance of a classification model on imbalance scenarios could be better evaluated by combining a suitable measure of its overall accuracy (g-mean, AUC or others) and the dominance.

This section introduces the $Accuracy - Dominance$ (AD) space as a tool to visualize and measure the behavior of a classifier from the joint perspective of global accuracy and dominance. Fig. 1 illustrates the AD space as a two-dimensional coordinate system where the X axis corresponds to dominance and the g-mean is plotted on the Y axis. An AD graph depicts relative trade-offs between dominance and g-mean. The performance of a classifier measured as a pair ($dominance$, g-mean) corresponds to a single point in this AD space. The upper central point $(0, +1)$ represents a perfect classifier where $TPrate = TNrate = 1$, while points $(-1, 0)$ and $(+1, 0)$ match with the useless cases of $TPrate = 0, TNrate = 1$ and $TPrate = 1, TNrate = 0$, respectively.

The upper left $(-1, +1)$ and upper right $(+1, +1)$ points correspond to the 'unfeasible cases' because when dominance is -1 or $+1$, one of the two class rates is 0 what makes impossible for the g-mean to achieve values greater than 0. Actually, there is an infinite number of points in the AD graph which represents unfeasible cases. Fig. 1(left) shows the bell-shaped region of probable cases, while the part out of this region corresponds to those unfeasible cases. The rationale behind this effect is that g-mean decreases when the absolute value of dominance increases.

Given a point (d, g) in an AD graph, it would be interesting to quantify the trade-off between dominance and g-mean represented by that point. A plain solution could

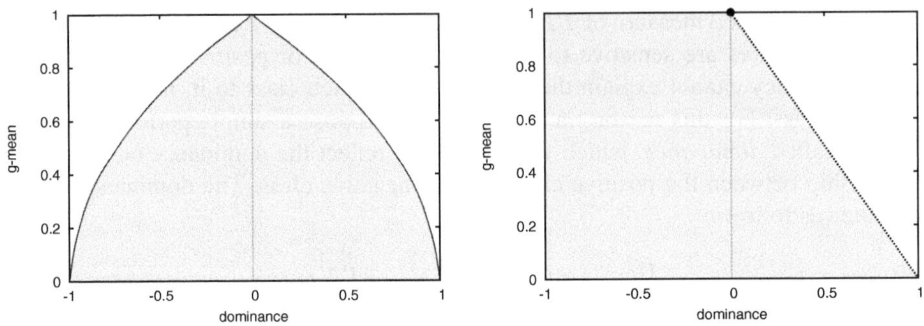

Fig. 1. Accuracy-Dominance Space. Left plot corresponds to the region of feasible values. Right plot represents the optimal point (maximum dominance and maximum g-mean), producing the maximum area.

be to measure the area of the triangle defined by $(-1, 0)$, (d, g) and $(+1, 0)$, which is maximized for $d = 0$, $g = +1$ representing a perfect classification. Nevertheless, the triangular function only depends on the value of the g-mean (the height of a triangle with base equal to 2), thus resulting in a constant area for an infinite number of pairs $(TPrate, TNrate)$ that produce the same g-mean value. Consequently, no information about the prevalence of a class accuracy over the other would be given by the triangular function. Instead, as we are seeking a function dependent on both dominance and g-mean, we will employ the area of a rectangular trapezium whose vertices are at points $(-1, 0)$, $(-1, g)$, (d, g) and $(+1, 0)$ (see Fig. 1(right)).

The area of such a trapezium will conveniently reflect the trade-off between dominance and g-mean. Besides, it is able to capture the fact that the minority class is usually deemed as the class of most importance and with a greater misclassification cost: for a given g-mean, the value of the trapezoidal function increases with the dominance. For two models A and B producing g-means $g_A = g_B$ and dominance values $d_A > d_B$, it can be easily proved that the trapezoidal function of model A will be greater than that of model B.

The trapezoidal function can take on any value between 0 and $+1.5$ (the area of the greatest trapezium, that is, the one corresponding to a point with dominance equal to 0 and g-mean equal to +1, which represents a perfect classification). Informally, one point in AD space is better than another if that is to the northeast (g-mean is higher, dominance is higher, or both) of the second.

3 Experimental Data Sets and Results

A number of experiments have been conducted to illustrate the performance evaluation method introduced in the previous section. Ten real data sets taken from the UCI Machine Learning Database Repository (`http://archive.ics.uci.edu/ml/`) have been employed in the experiments (see a summary in Table 2). All data sets were transformed into two-class problems. For example, in Vehicle database the objects of

Table 2. Data sets used in the experiment

	Positive Examples	Negative Examples	Classes	Majority Class
Breast	81	196	2	1
Ecoli	35	301	8	1,2,3,5,6,7,8
German	300	700	2	1
Glass	17	197	9	1,2,4,5,6,7,8,9
Haberman	81	225	2	1
Phoneme	1586	3818	2	1
Pima	268	500	2	1
Satimage	626	5809	7	1,2,3,5,6,7
Vehicle	212	634	4	2,3,4
Yeast	429	1055	10	1,3,4,5,6,7,8,9,10

classes 2, 3, and 4 were joined to form a unique majority class and the original class 1 was left as the minority class.

For each database, we have estimated the geometric mean, the dominance and the area by 5×10–fold cross–validation: each data set is divided into ten blocks of size $n/10$ (n is the total number of objects), using nine folds as the training set and the remaining block as the test set. This is repeated 5 times. Values in Table 3 correspond to the average results with the nearest neighbor (1-NN) classifier, a support vector machine (SVM) and a multi-layer perceptron (MLP). Each classifier has been applied to the original training set and also to sets whose training partitions have been preprocessed by SMOTE (Synthetic Minority Oversampling Technique) and by random under-sampling. These two techniques have been included in order for checking validity and consistency of the new performance measure: theory dictates that this sort of preprocessing techniques should balance the class distributions and correspondingly, the value of the area for the preprocessed sets should result greater than that for the original training set. All hyper–parameters of the classifiers have been set to the default values given in the Weka toolkit [21].

SMOTE and random under-sampling are two well-known data-driven strategies to tackle the class imbalance problem. These consists of resampling the training data by expanding the minority class (SMOTE) or by shrinking the majority class (under-sampling), thus obtaining an altered class distribution. SMOTE [4] adds artificial examples of the minority class by interpolating existing instances that lie close together. This algorithm first finds the k nearest neighbors belonging to the minority class for each positive example and then, the synthetic examples are generated in the direction of some or all of the nearest neighbors, depending on the amount of over-sampling required. In the case of random under-sampling, instances of the majority class are randomly discarded from the data set until the ratio between the minority and majority classes is at the desired level.

As already mentioned in Section 2, the purpose of the new performance evaluation method is to reflect a certain trade-off between the geometric mean of accuracies and the dominance, but emphasizing the true predictions on the minority class. This means that when comparing two classifiers with equal geometric means, the model with the

Table 3. Results for the three classifiers

	1-NN								
	Original			SMOTE			Random under-sampling		
	g-mean	dominance	area	g-mean	dominance	area	g-mean	dominance	area
Breast	0.59	-0.31	0.79	0.60	-0.24	0.83	0.60	0.05	0.92
Ecoli	0.70	-0.43	0.90	0.83	-0.14	1.19	0.85	0.04	1.30
German	0.59	-0.31	0.79	0.61	-0.16	0.86	0.61	0.02	0.93
Glass	0.55	-0.63	0.65	0.71	-0.32	0.95	0.69	0.12	1.08
Haberman	0.50	-0.44	0.64	0.55	-0.24	0.77	0.55	-0.07	0.81
Phoneme	0.87	-0.14	1.25	0.88	-0.09	1.28	0.87	0.00	1.30
Pima	0.64	-0.27	0.88	0.68	-0.11	0.98	0.68	-0.05	1.00
Satimage	0.82	-0.26	1.13	0.90	-0.03	1.33	0.87	0.06	1.34
Vehicle	0.62	-0.39	0.81	0.66	-0.19	0.94	0.71	-0.01	1.06
Yeast	0.63	-0.34	0.83	0.65	-0.20	0.91	0.66	-0.02	0.98
	SVM								
	Original			SMOTE			Random under-sampling		
	g-mean	dominance	area	g-mean	dominance	area	g-mean	dominance	area
Breast	0.55	-0.52	0.69	0.65	-0.02	0.97	0.64	-0.04	0.95
Ecoli	0.00	-1.00	0.00	0.88	0.04	1.33	0.85	0.19	1.35
German	0.65	-0.41	0.85	0.72	0.01	1.09	0.71	0.02	1.07
Glass	0.00	-1.00	0.00	0.55	0.60	1.00	0.55	0.59	0.98
Haberman	0.00	-1.00	0.00	0.56	-0.56	0.69	0.56	-0.49	0.70
Phoneme	0.70	-0.28	0.95	0.75	0.10	1.16	0.75	0.09	1.15
Pima	0.94	-0.36	1.24	0.74	-0.06	1.09	0.74	-0.07	1.09
Satimage	0.00	-1.00	0.00	0.70	0.43	1.20	0.68	0.47	1.19
Vehicle	0.00	-1.00	0.00	0.74	0.08	1.15	0.72	0.07	1.10
Yeast	0.42	-0.79	0.47	0.71	0.05	1.08	0.71	0.06	1.08
	MLP								
	Original			SMOTE			Random under-sampling		
	g-mean	dominance	area	g-mean	dominance	area	g-mean	dominance	area
Breast	0.56	-0.48	0.70	0.57	-0.30	0.77	0.58	-0.02	0.86
Ecoli	0.83	-0.24	1.14	0.88	-0.09	1.29	0.87	0.05	1.32
German	0.64	-0.31	0.86	0.66	-0.12	0.95	0.67	0.00	1.00
Glass	0.37	-0.83	0.40	0.74	-0.32	1.00	0.75	0.07	1.15
Haberman	0.49	-0.65	0.57	0.63	-0.01	0.95	0.63	-0.03	0.93
Phoneme	0.78	-0.13	1.12	0.81	0.10	1.26	0.81	0.11	1.26
Pima	0.68	-0.31	0.92	0.73	0.03	1.10	0.72	-0.04	1.07
Satimage	0.78	-0.33	1.05	0.86	-0.13	1.24	0.87	0.04	1.32
Vehicle	0.75	-0.29	1.01	0.78	-0.10	1.13	0.78	0.00	1.17
Yeast	0.67	-0.35	0.88	0.71	-0.04	1.05	0.69	-0.03	1.03

highest value of dominance will be selected as the "best", which in practice corresponds to the trapezium with largest area.

Table 3 reports the g-mean, the dominance and the trapezoidal area for the three classifiers in order to illustrate the behavior of the evaluation method here introduced.

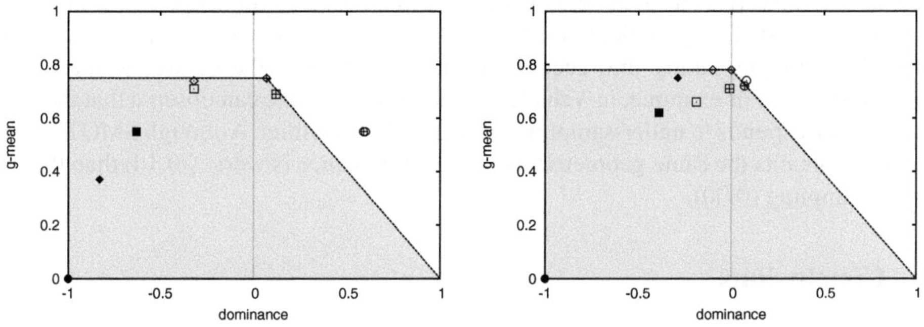

Fig. 2. Scatterplot of the results for Glass (left) and Vehicle (right) databases. Shaded is the area of the best approach. Squares are for 1-NN, circles represent SVM, and diamonds are for MLP. Black figures correspond to the results with the original training set, white are for SMOTE, and crossed figures are for random under-sampling.

The main question is whether the area can be considered as a consistent measure. To answer this question one can observe that in all cases, the area of the trapezium for the sets processed with SMOTE and under-sampling are greater than the value for the original training set.

In order to fully understand the meaning of the trapezoidal function, we can focus on the results for Pima domain. In this case, the geometric mean of both SMOTE and under-sampling with 1-NN is equal to 0.68, making it impossible to consider one technique better than the other. Nevertheless, it is clear that the cost of misclassifications on the minority class (presence of diabetes) is higher than that on the majority class (absence) and therefore, the under-sampling model should be preferable over SMOTE because it has a higher dominance. What is important to note here is that the value of the area for under-sampling (1.00) is also higher than that for SMOTE (0.98). Correspondingly, one could select the best model by looking at the value of the trapezoidal function.

It is also interesting to note the results for Glass database in Table 3 with the 1-NN classifier. In this case, SMOTE obtains a higher geometric mean than the under-sampling technique. By using only this metric, one should select SMOTE as the best model. Nevertheless, the value of the area gives the under-sampling as the best because it also looks at the true positive rate and the true negative rate by means of the dominance. In some way, the evaluation method here proposed prefers some loss in the geometric mean if it produces a higher $TPrate$.

Results on several databases (Ecoli, Glass, Haberman, Satimage, and Vehicle) with the SVM are worth to be mentioned. Here the g-mean for the original training set is equal to 0 and the dominance is -1, thus meaning that the $TPrate = 0$. Obviously, this corresponds to a very undesirable case because all predictions on the minority class are wrong. Such a situation has been conveniently reflected by the trapezoidal function, which is equal to 0. Futhermore, the behavior of the SVM when it ignores the positive instances, it suggests they have been treated as noise [1].

Figure 2 shows a scatterplot of the nine methods (combining the original training set and the two preprocessing techniques with the three classifiers) for Glass and Vehicle databases. The (greatest) area corresponding to the best pair (*dominance*, *g*-mean) is also represented. This graphical model allows to visualize in an easy way the behavior of each technique, taking into account a trade-off between the geometric mean and the dominance. For example, in Vehicle (on the right side) one can observe that the best method corresponds to under-sampling with the MLP classifier. Although SMOTE with the MLP obtains the same geometric mean, the dominance is worst (-0.10) than that of under-sampling (0.00).

4 Conclusions

In this paper, we have introduced a new method to evaluate and visualize the performance of learning algorithms in two-class imbalanced domains. This is based on the use of two metrics: the geometric mean of accuracies and the dominance (which is here defined as the difference between the true positive rate and true negative rate). The way in which we have defined the evaluation method allows to stress the importance of the true predictions on the minority class because in many practical problems, the minority class constitutes the class of primary interest. For these situations, the cost of misclassifications on the minority class is higher than that on the majority class.

The experiments carried out have been thought as a way of illustrating the behavior of this new evaluation method. To this end, we have used three classifiers and two well-known resampling techniques to balance the class distributions. We have empirically shown that the computation of the trapezoidal function allows to represent the best model in terms of trade-off between the geometric mean and the dominance. The new evaluation method has been demonstrated to be valid and consistent in all cases.

Future work will be addressed to relate the value of the trapezoidal area with the AUC, trying to complement the particular features of each one. Also, defining a three-dimensional space by the inclusion of some other metric could provide more accurate estimates of the classifier performance, especially in class imbalance domains with unequal misclassification costs.

Acknowledgment

This work has been partially supported by grants DPI2006–15542 and CSD2007–00018 from the Spanish Ministry of Education and Science, GV–2007–105 from Generalitat Valenciana and SEP-2003-C02-44225 from the Mexican CONACyT.

References

1. Akbani, R., Kwek, S., Japkowicz, N.: Applying support vector machines to imbalanced datasets. In: Proc. XVth European Conference on Machine Learning (ECML 2004), Pisa, Italy, pp. 39–50 (2004)
2. Barandela, R., Sánchez, J.S., García, V., Rangel, E.: Strategies for learning in class imbalance problems. Pattern Recognition 36, 849–851 (2003)
3. Batista, G.E., Prati, R.C., Monard, M.C.: Balancing Strategies and Class Overlapping. In: Proc. 6th Intl. Symposium on Intelligent Data Analysis, Madrid, Spain, pp. 24–35 (2005)

4. Chawla, N.V., Bowyer, K.W., Hall, L., Kegelmeyer, W.P.: SMOTE: synthetic minority over-sampling technique. Journal of Artificial Intelligence Research 16, 321–357 (2002)
5. Chawla, N.V., Lazarevic, A., Hall, L.O., Bowyer, K.W.: SMOTEBoost: Improving prediction of the minority class in boosting. In: Proc. 7th European Conf. on Principles and Practice of Knowledge Discovery in Databases, Cavtat–Dubrovnik, Croatia, pp. 107–119 (2003)
6. Daskalaki, S., Kopanas, I., Avouris, N.: Evaluation of classifiers for an uneven class distribution problem. Applied Artificial Intelligence 20, 381–417 (2006)
7. Domingos, P.: Metacost: a general method for making classifiers cost-sensitive. In: Proc. 5th Intl. Conf. on Knowledge Discovery and Data Mining, San Diego, CA, pp. 155–164 (1999)
8. Drummond, C., Holte, R.C.: Explicitly representing expected cost: an alternative to ROC representation. In: Proc. 6th ACM SIGKDD Intl. Conf. on Knowledge Discovery and Data Mining, Boston, MA, pp. 198–207 (2000)
9. García, V., Mollineda, R.A., Sánchez, J.S., Alejo, R., Sotoca, J.M.: When overlapping unexpectedly alters the class imbalance effects. In: Pattern Recognition and Image Analysis, pp. 499–506 (2007)
10. Han, H., Wang, W.Y., Mao, B.H.: Borderline-SMOTE: A new over-sampling method in imbalanced data sets learning. In: Proc. Intl. Conf. on Intelligent Computing, Hefei, China, pp. 878–887 (2005)
11. Huang, J., Ling, C.X.: Using AUC and accuracy in evaluating learning algorithms. IEEE Trans. on Knowledge and Data Engineering 17, 299–310 (2005)
12. Japkowicz, N., Stephen, S.: The class imbalance problem: a systematic study. Intelligent Data Analysis 6, 40–49 (2002)
13. Jo, T., Japkowicz, N.: Class imbalances versus small disjuncts. ACM SIGKDD Explorations 6, 40–49 (2004)
14. Kubat, M., Matwin, S.: Adressing the curse of imbalanced training sets: one-sided selection. In: Proc. 14th Intl. Conf. on Machine Learning, Nashville, TN, pp. 179–186 (1997)
15. Liu, X.Y., Zhou, Z.H.: The influence of class imbalance on cost-sensitive learning: an empirical study. In: Proc. 6th Intl. Conf. on Data Mining, pp. 970–974 (2006)
16. Maloof, M.A.: Learning when data sets are imbalanced and when costs are unequal and unknow. In: ICML–2003 Workshop on Learning from Imbalanced Data Sets II (2003)
17. Okamoto, S., Yugami, N.: Effects of domain characteristics on instance-based learning algorithms. Theoretical Computer Science 298, 207–233 (2003)
18. Prati, R.C., Batista, G.E., Monard, M.C.: Class imbalance versus class overlapping: an analysis of a learning system behavior. In: Proc. 3rd Mexican Intl. Conf. on Artificial Intelligence, Mexico City, Mexico, pp. 312–321 (2004)
19. Provost, F., Fawcett, T.: Analysis and visualization of classifier performance: Comparison under imprecise class and cost distributions. In: Proc. 3rd Intl. Conf. on Knowledge Discovery and Data Mining, Newport Beach, CA, pp. 43–48 (1997)
20. Raskutti, B., Kowalczyk, A.: Extreme rebalancing for svms: a case study. SIGKDD Explorations 6, 60–69 (2004)
21. Witten, I.H., Frank, E.: Data Mining: Practical Machine Learning Tools and Techniques. Morgan Kaufmann, San Francisco (2005)
22. Yen, S.H., Lee, Y.S., Lin, C.H., Ying, J.C.: Investigating the effect of sampling methods for imbalanced data distributions. In: Proc. IEEE Intl. Conf. on Systems, Man, and Cybernetics, Taipei, Taiwan, pp. 4163–4168 (2006)
23. Zhang, J., Srihari, R.K.: kNN approach to unbalanced data distributions: a case study involving information extraction. In: ICML 2003 Workshop on Learning from Imbalanced Data Sets II (2003)

A Pruning Rule Based on a Distance Sparse Table for Hierarchical Similarity Search Algorithms

Eva Gomez-Ballester, Luisa Micó, and Jose Oncina

Dept. Lenguajes y Sistemas Informáticos
Universidad de Alicante, E-03080 Alicante, Spain
{eva,mico,oncina}@dlsi.ua.es

Abstract. Nearest neighbour search is a simple technique widely used in Pattern Recognition tasks. When the dataset is large and/or the dissimilarity computation is very time consuming the brute force approach is not practical. In such cases, some properties of the dissimilarity measure can be exploited in order to speed up the search. In particular, the metric properties of some dissimilarity measures have been used extensively in fast nearest neighbour search algorithms to avoid dissimilarity computations. Recently, a distance table based pruning rule to reduce the average number of distance computations in hierarchical search algorithms was proposed. In this work we show the effectiveness of this rule compared to other state of the art algorithms. Moreover, we propose some guidelines to reduce the space complexity of the rule.

Keywords: fast nearest neighbour search, metric space, tree-based algorithms, pruning rules.

1 Introduction

Similarity search has been applied in many areas as pattern recognition, machine learning, computer vision, robotics or data mining were data has complex representations and then, complex functions are used to quantify their dissimilarities.

Some examples of proximity searches are range search, nearest neighbour search, k-nearest neighbours search, reverse nearest neighbour search or proximity join search [14]. In this work we focus on nearest neighbour search. That is, given a set P of *objects*, a *query* object q and a dissimilarity measure $d(\cdot,\cdot)$, the goal is to retrieve the object p_{nn} of the set P closest to the query object according to d:

$$p_{nn} = \operatorname*{argmin}_{\forall p \in P} d(p,q)$$

A brute force implementation is impractical when the dissimilarity measure is very time consuming and/or the dataset is very large. As a result, many techniques have been devised to speed up the search.

Some of this techniques relies on the fact that some of the dissimilarity measures are metrics and then, they fulfils the the triangle inequality. For example,

N. da Vitora Lobo et al. (Eds.): SSPR&SPR 2008, LNCS 5342, pp. 926–936, 2008.

to avoid distance computations a lower bound of the distance can be defined using some precomputed distances [3][12].

In the last few years, a large number of works appeared in the literature related with such type of methods. The interest of these methods relies on the possibility of providing high extensibility when comparing with methods that requires more specific information, such as a vector representation. Some interesting surveys have been published recently [1,7,10].

The use of hierarchical data structures is very common in most of the proposed algorithms [11,9,15,17,13], as these structures provide a simple way to avoid the exploration of some subsets of objects. Several searching strategies can be performed in these structures, the most popular are depth-first and best-first strategies (even a combination of both), but independently of the strategy, similar pruning rules can be applied.

One of the first hierarchical-based search algorithms was proposed by Fukunaga and Narendra (`FN`) [11]. This algorithm is very suitable for studying tree building strategies (based on different types of space partitioning) with minimum changes in the other properties that we want to keep (pruning rules, searching strategy, ...) and *vice versa*. The algorithm is also very suitable to apply new pruning rules as a previous step for extending it to other hierarchical search methods.

Recently a new pruning rule for hierarchical similarity search algorithms based on the use of a distance table was proposed [8]. To apply the rule, the distance table should be built up and stored in preprocessing time. This table stores the distances between the nodes of the hierarchical structure and the objects in the data set.

In this work some strategies are proposed for reducing the space complexity of the table (quadratic with the size of the dataset) and some experiments are performed to test their effectiveness. To perform the experiments some state of the art fast Nearest Neighbour Search (NNS) algorithms have been used: the `FN` (using the same hierarchical structure mentioned above) [11], the Spatial Approximation Tree (`SAT`) [6] and the vp-tree [15] algorithms. They will serve as a baseline for the comparisons.

The paper is organized as follows: we first introduce the basic algorithm and the pruning rules (section 2). In section 3, we provide experiments on artificial and real world data. We conclude suggesting some future works in section 4.

2 The Basic Algorithm

The `FN` is a NNS algorithm that uses a hierarchical data structure where objects are stored in nodes. With each node t is associated S_t, the set of the objects stored in the leaves of the t sub-tree. Each node stores M_t (a representative of S_t) and the radius R_t of a ball centered in M_t that includes S_t ($R_t = \max_{x \in S_t} d(M_t, x)$).

The tree is generally built using recursive calls to a clustering algorithm. In the original `FN` algorithm the c-means algorithm was used. In [2] other strategies were explored. The best strategy lied in using the representative of the father node as one of the representatives of the two child nodes and the other as its farthest object in the set. Thus, each time an expansion of the node is necessary, only one new distance (instead of two) should be computed during the search, reducing the number of distance computations.

In Algorithm 1, a simplified version of `FN` algorithm is presented. The `Prune_FNR` function call should be changed when considering other pruning rules. In order to make the pseudo-code simpler, p_{nn} (actual nearest neighbour to the query) and d_{nn} (distance from the query to p_{nn}, initially with $-\infty$ value) are considered global variable. Only binary trees with one object on the leaves are considered. Taken into account that leaf nodes only have one object, in the algorithm it is not necessary to define a special case pruning rule for leaves (this is why the leaves are not considered in the algorithm). When a new query is given, its nearest neighbour is searched in the tree using a depth-first strategy (the best-first strategy was also explored, but despite the results reported in [5], in this case, similar results were obtained. Due to lack of space we omit them in the experimental section).

Algorithm 1: search(t, q)

Data: t: a node tree ; q: query;
Result: p_{nn}: the nearest neighbour object; d_{nn}: the distance to q_{nn};
if *t is not a leaf* **then**
  $r = right\_child(t)$; $\ell = left\_child(t)$;
  $d_r = d(q, M_r)$; $d_\ell = d(q, M_\ell)$; $//d_\ell$ previously computed;
  update d_{nn} and p_{nn};
  if $d_\ell < d_r$ **then**
    if *not Prune_FNR(ℓ)* **then**
      search(ℓ, q);
    if *not Prune_FNR(r)* **then**
      search(r, q);
  else
    if *not Prune_FNR(r)* **then**
      search(r, q);
    if *not Prune_FNR(ℓ)* **then**
      search(ℓ, q);

Next, we present the two pruning rules used in this work. If the disimilarity function is a metric, (the triangle inequality is fulfilled) it is guaranteed that the nearest neighbour can't be found among the eliminated elements with these rules.

2.1 Fukunaga and Narendra Rule (`FNR`)

This rule only makes use of the information in the node t to be pruned (with representative M_t and radius R_t) and the hyperspherical volume centered in the query q with radius $d(q, p_{nn})$, where p_{nn} is the nearest object considered up to the moment. To apply this rule to a node t, the distance from the query q to the representative of the node should be available.

Proposition 1 ([8]). *No $y \in S_t$ can be the nearest neighbour to q if $d(q, p_{nn}) + R_t < d(q, M_t)$.*

2.2 Table Rule (TR)

In order to use the table pruning rule (TR), the distance from an object p to a node t is defined as $d(p,t) = \min_{e \in S_t} d(p,e)$ (this definition can be seen as a particular case of the distance between two sets [16]). At preprocessing time, the distances from each object to each node of the tree are computed and stored in a table. Note that the size of this table grows with the square of the number of objects in the database.

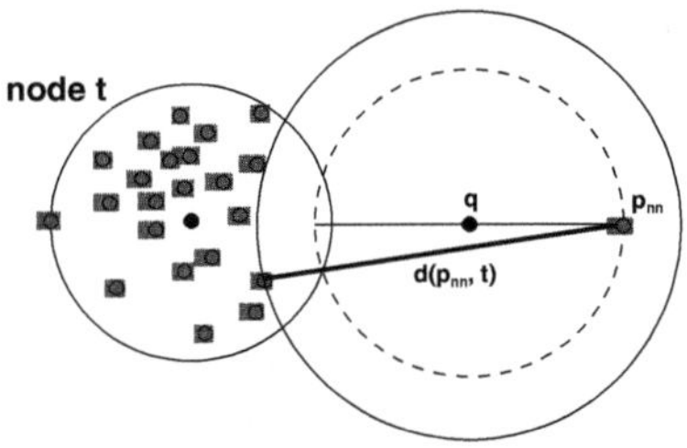

Fig. 1. Application of the table rule

The TR states that no object in a set can be the nearest neighbour to a query if the distance from the current nearest neighbour to the set is bigger than two times the distance from the current nearest neighbour to the query (see Figure 1).

Proposition 2 ([8]). *Let t be a node such that $d(p_{nn}, t) \geq 2d(p_{nn}, q)$. Then, $d(e, q) \geq d(p_{nn}, q)$ $\forall e \in S_t$*

2.3 Reducing the Table Size

As it was stated in the previous section, the distances between the objects in the dataset and all the nodes in the tree should be computed and stored in the table to use the TR. Then, as the size of the table grows with the square of the number of objects, the space complexity can be a bottleneck for using this rule. In this section we will present some ideas about how to detect, in preprocessing time, in which circumstances the application of the TR will be useless. In these cases, the computation and storage of $d(n,t)$ (where n is an object and t a node) can be avoided. It is evident that in order to have effective storage reductions, the table should be implemented as a sparse matrix.

Let n be an object and let t be a node of the tree. Let we denote by $\hat{d}_{nn}$ the expected value of the distance to the nearest neighbour ($\hat{d}(q, p_{nn})$) (of course this value is unknown in preprocessing time, we are going to use it as a parameter in the table reduction system). There are two circumstances where the TR is going to be probably useless:

1. When $\hat{d}_{nn} > \frac{d(n,t)}{2}$. In such case the TR is not going to prune and then the distance $d(n,t)$ is not stored in the table. This is the situation where TR is not fulfilled because the object n is very near to the node t.
2. When $\hat{d}_{nn} < d(M_t, n) - R_t$. This case arises when we expect that the FNR rule is going to prune and then, the result of the TR is useless. This formula derives from an approximation of the FNR where we suppose that n is very near to the query and then $d(M_t, n) \sim d(M_t, q)$ (this is usually the case after some steps of the algorithm [4]). In this case the FNR will be effective and then the TR is probably not needed to prune node t. In this case, the distance $d(n,t)$ is not stored in the table.

The experiment section analyses the price to pay, in distance computations, versus table size reduction obtained when our techniques are applied.

3 Experiments

Some experiments with synthetic and real data were carried out to evaluate our proposal. In the case of synthetic data, objects were obtained randomly from the unit hypercube with a uniform distribution and for several dimensions (5, 10, 15 and 20). In this case, the Euclidean distance was applied. As real data experiment, the Spanish dictionary in `http://www.sisap.org` using the Levenshtein distance has been used. In all the experiments sets of 1000 queries were used. Each experiment was repeated 10 times. The average and the deviations of each experiment are shown in the plots.

Moreover, comparison with other state of the art fast NNS methods was performed. In particular, the following algorithms were used: vp-tree (VPT) [15], SAT [6] and FN [11].

3.1 Experiments in the Unit Hypercube

In the first set of experiments, the results of the FN algorithm, using both FNR and TR (FNT in the figures), are compared with VPT, SAT[1] and the basic FN algorithm (using only FNR). Figure 2 shows the average number of distance computations by query when increasing the size of the dataset.

The results show that for lower dimensions, the average number of distance computations when FNR is used alone (FN curve) or in combination with TR (FNT curve) outperforms all the other methods, although this difference decreases when the dimension grows. This is an expected behaviour, because of the *curse of the dimensionality*: for high dimensions (30 or more) all the methods compute all the distances (see Figure 3) and then, all are equivalent to the brute force

[1] Code for VPT and SAT was obtained from `http://www.sisap.org`

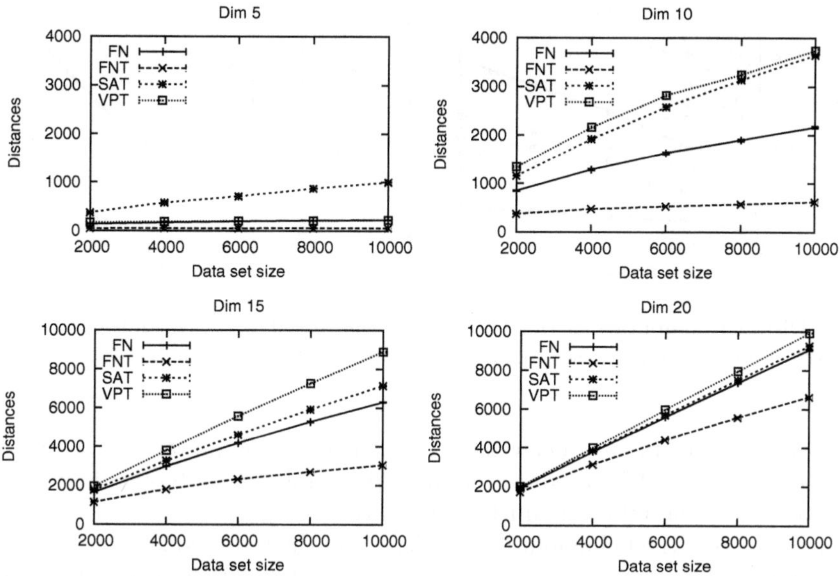

Fig. 2. Distance computations using increasing dataset size for several methods in the unit hypercube. In the particular case of 5-dimensional space, the best behaviour is obtained with the FNT method, and the worst for the SAT method.

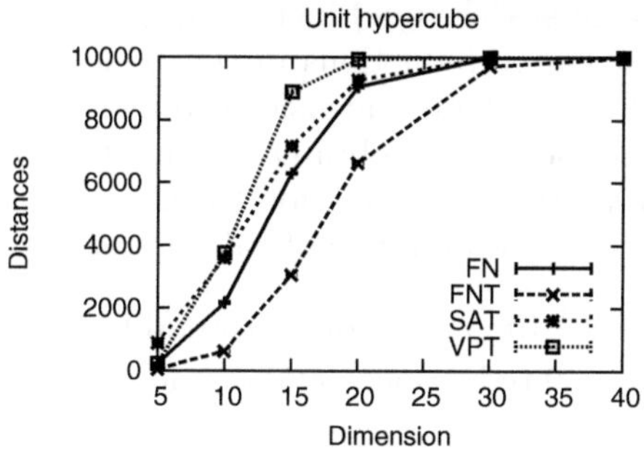

Fig. 3. Distance computations using a dataset of 10000 objects and different dimensions

algorithm. In this figure can be observed that for all dimensions the FNT method (use of both, FNR and TR rule) outperforms all the other methods. Moreover, although the behaviour of the FNT method also degenerates with the dimension, it degenerates for higher dimensions more slowly than the others. For example, the VPT, FN and SAT methods compute almost all the distances to the training set in a 20-dimensional space. To have a similar result with FNT method we have to go to a 30-dimensional space.

Reduction of the table. Due to the quadratic space complexity, the use of the TR pruning rule can be a problem when using very large data sets. A a second

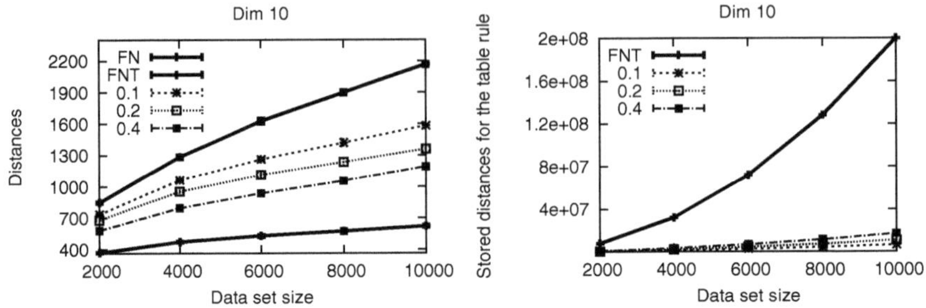

Fig. 4. Distance computations (left) and number of stored distances on the table (right) when the dataset size increases, using several values of the parameter $\hat{d}_{nn}$

group of experiments have been performed to study the power of the strategies described in section 2.2 to reduce the table size while trying to avoid an increase of the distance computations during the search.

Figure 4 shows the behaviour of `FNT` (left) and the reductions of the size of the table (right) when $\hat{d}_{nn}$ increases (from 0.1 to 0.8 depending on the dimension). This figure also shows the average number of distance computations when the table is full (FNT curve) and when is empty (FN curve). This behaviour is directly related with the number of distances stored in the table. As the number of distances stored (in preprocessing time) in the table decreases, the number of distance computations (in search time) increases.

In all the cases significative reductions in the table size are obtained. Saves of around 80% of the table in high dimensions (dimension 20) and around 98% in low dimensions (dimension 5) are obtained without increasing very much the number of distance computations in relation to the use of the whole table.

Some experiments using a wide range of values of $\hat{d}_{nn}$ show that:

1. using only $\hat{d}_{nn} > \frac{d(n,t)}{2}$ the reduction of the table increases with $\hat{d}_{nn}$ (in dimension 10, for $\hat{d}_{nn} = 0$ reductions are 0%, and for $\hat{d}_{nn} = 0.75$ the reduction is around 90%).
2. using only $\hat{d}_{nn} < d(M_t, n) - R_t$, the reduction of the table decreases with the increase of $\hat{d}_{nn}$ (in dimension 10, for $\hat{d}_{nn} = 0$ reductions are around 96%, and for $\hat{d}_{nn} = 0.75$ the reduction is around 60%).

If both criteria are used, a minimum in the number of distance computations for a specific value of $\hat{d}_{nn}$ can be found. This minimum is related with a minimum reduction in the size of the table. This behaviour can be observed in Figure 5. For very high values of $\hat{d}_{nn}$ the main reduction on the table is obtained using the first criterion. On the other hand, for very low values of d_{nn}, the main reduction on the table is obtained with the contribution of the second criterion.

These results confirm that a significative reduction in the number of computed distances can be obtained using only a small percentage of the table.

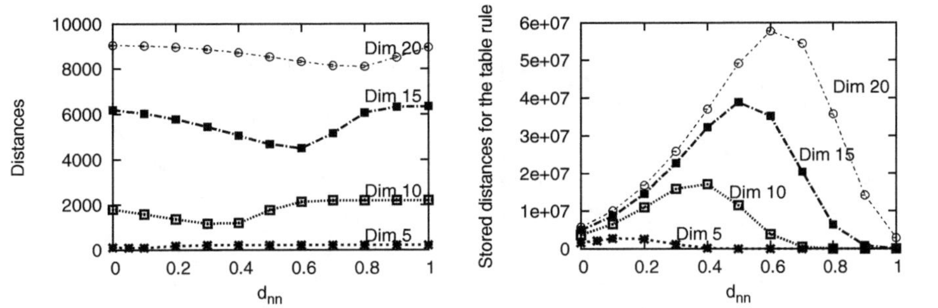

Fig. 5. Distance computations (left) and number of stored distances on the table (right) for several values of the parameter $\hat{d}_{nn}$ using a 10000 samples database

3.2 Experiments with the Spanish Dictionary

Similar experiments were repeated simulating a spell corrector task using a Spanish dictionary (`http://www.sisap.org`). The words to correct were obtained distorting the words by means of random insertion, deletion and substitution operations over the words in the original dictionary.

The edit distance was used to compare the words. In these experiments, the values of the weighting operation costs for the edit distance were fixed to 1 for insertions, deletions and substitutions. As, in this case, the edit distance is a metric (symmetry, nonnegativity and triangle inequality properties are fulfilled), it makes the FNR and TR rules applicable.

Figures 6 and 7 show similar behaviour that Figures 2 and 4 when strings are used: the average number of distance computations when FNR is used alone or in combination with TR outperforms all the other methods. Moreover, similar reduction of the table can be achieved (around 85%) maintaining a good reduction in the average number of computed distance when comparing with the basic FN algorithm (FNR only).

Also, a final experiment (similar to Figure 5) was repeated using the dictionary. It can be observed in Figure 8 a minimum number of distance computations

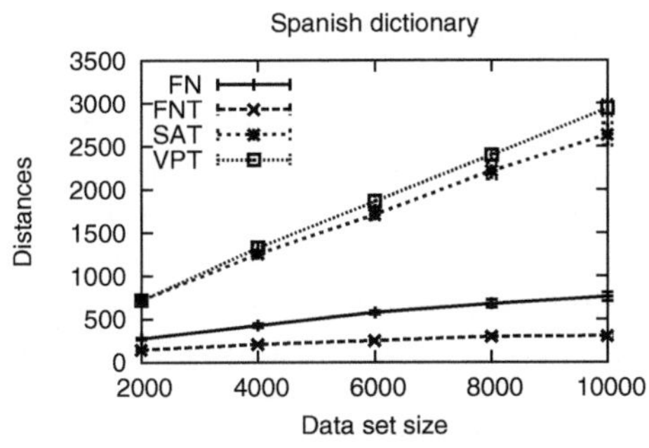

Fig. 6. Distance computations with increasing dataset size for different methods

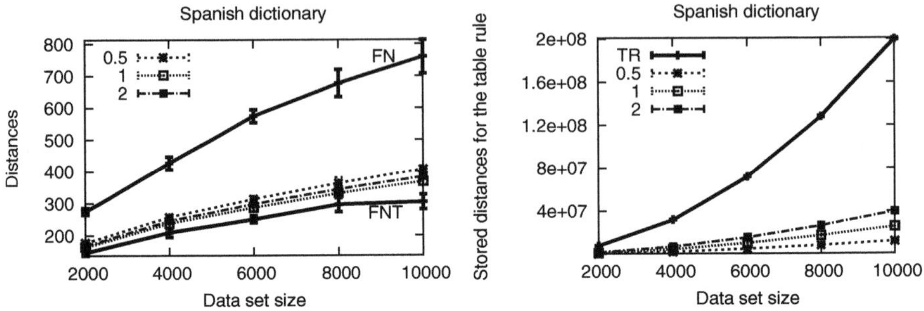

Fig. 7. Distance computations (on the left) and number of stored distances on the table (on the right) for different values of the parameter $\hat{d}_{nn}$

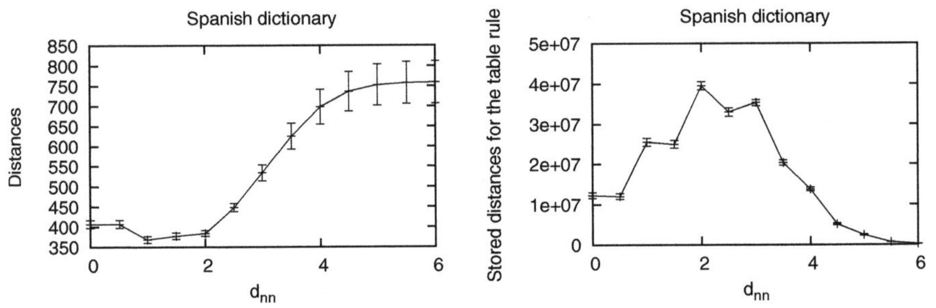

Fig. 8. Distance computations (left) and number of stored distances on the table (right) for different values of the parameter $\hat{d}_{nn}$ using 10000 samples are training set

for a specific value of $\hat{d}_{nn}$. This minimum is related with a minimum reduction in the size of the table. These results confirm that also for the dictionary a significative reduction in the number of computed distances can be obtained using only a small percentage of the table.

4 Conclusions and Further Works

A recently proposed pruning rule for hierarchical search algorithms based on the use of precomputed distances between nodes in the tree and the objects of the training set has been explored and compared with some state of the art methods. In this work some proposals to reduce the size of the table have also been presented. Two criteria, based on the use of a parameter ($\hat{d}_{nn}$) was defined. The key idea is to avoid the storage of distances that rarely would be used (TR is not fulfilled because objects are very close to nodes or FNR is fulfilled).

As the results suggest, significant reductions in distance computations can be obtained when comparing with the basic FN algorithm and other state of the art algorithms (SAT and VPT). Moreover, this criteria to reduce the table size has shown its effectiveness allowing significant table reductions while increasing moderately the number of distance computations.

In future works, more selective reduction of the table should be explored.

Acknowledgments. The authors thank the Spanish CICyT for partial support of this work through projects DPI2006-15542-C04-01, the IST Programme of the European Community, under the PASCAL Network of Excellence, IST–2002-506778, and the program CONSOLIDER INGENIO 2010 (CSD2007-00018).

References

1. Chávez, E., Navarro, G., Baeza-Yates, R., Marroquin, J.L.: Searching in metric spaces. ACM Computing Surveys 33(3), 273–321 (2001)
2. Gómez-Ballester, E., Micó, L., Oncina, J.: Some improvements in tree based nearest neighbour search algorithms. In: Sanfeliu, A., Ruiz-Shulcloper, J. (eds.) CIARP 2003. LNCS, vol. 2905, pp. 456–463. Springer, Heidelberg (2003)
3. Vidal, E.: New formulation and improvements of the nearest-neighbour approximating and eliminating search algorithm (aesa). Pattern Recognition Letters 15(1), 1–7 (1994)
4. Moreno-Seco, F., Micó, L., Oncina, J.: A modification of the laesa algorithm for approximated k-nn classification. Pattern Recognition Letters 22, 1145–1151 (2003)
5. Moreno-Seco, F., Micó, L., Oncina, J.: Approximate nearest neighbour search with the fukunaga and narendra algorithm and its application to chromosome classification. In: Sanfeliu, A., Ruiz-Shulcloper, J. (eds.) CIARP 2003. LNCS, vol. 2905, pp. 322–328. Springer, Heidelberg (2003)
6. Navarro, G.: Searching in metric spaces by spatial approximation. VLDB Journal 11(1), 28–46 (2002)
7. Hjaltason, G.R., Samet, H.: Index-driven similarity search in metric spaces. ACM Trans. Database Syst. 28(4), 517–580 (2003)
8. Oncina, J., Thollard, F., Gómez-Ballester, E., Micó, L., Moreno-Seco, F.: A tabular pruning rule in tree-based pruning rule fast nearest neighbour search algorithms. In: Martí, J., Benedí, J.M., Mendonça, A.M., Serrat, J. (eds.) IbPRIA 2007. LNCS, vol. 4478, pp. 306–313. Springer, Heidelberg (2007)
9. Friedman, J.H., Bentley, J.L., Finkel, R.A.: An algorithm for finding best matches in logarithmic expected time. ACM Transactions on Mathematical Software 3, 209–226 (1977)
10. Clarkson, K.: Nearest-neighbor searching and metric space dimensions. In: Nearest-Neighbor Methods for Learning and Vision: Theory and Practice, pp. 15–59. MIT Press, Cambridge (2006)
11. Fukunaga, K., Narendra, P.M.: A branch and bound algorithm for computing k-nearest neighbors. IEEE Transactions on Computers, IEC 24, 750–753 (1975)
12. Micó, L., Oncina, J., Vidal, E.: A new version of the nearest-neighbour approximating and eliminating search algorithm (aesa) with linear preprocessing time and memory requirements. Pattern Recognition Letters 15, 9–17 (1994)
13. Ciaccia, P., Patella, M., Zezula, P.: M-tree: An efficient access method for similarity search in metric spaces. In: Proceedings of the 23rd International Conference on VLDB, Athens, Greece, pp. 426–435. Morgan Kaufmann Publishers, San Francisco (1997)

14. Zezula, P., Amato, G., Dohnal, V., Batko, M.: Similarity Search: The Metric Space Approach. Springer, Heidelberg (2006)
15. Yianilos, P.N.: Data structures and algorithms for nearest neighbor search in general metric spaces. In: Proceedings of the ACM-SIAM Symposium on Discrete Algorithms, pp. 311–321 (1993)
16. Duda, R.O., Hart, P.E., Stork, D.G.: Pattern Classification, 2nd edn. Wiley, New York (2000)
17. Brin, S.: Near neighbor search in large metric spaces. In: Proceedings of the 21st International Conference on Very Large Data Bases, pp. 574–584 (1995)

Classification and Automatic Annotation Extension of Images Using Bayesian Network

Sabine Barrat and Salvatore Tabbone

LORIA-UMR 7503, University of Nancy 2,
BP 239, 54506 Vandœuvre-lès-Nancy, France
{barrat,tabbone}@loria.fr

Abstract. In many vision problems, instead of having fully annotated training data, it is easier to obtain just a subset of data with annotations, because it is less restrictive for the user. For this reason, in this paper, we consider especially the problem of classifying weakly-annotated images, where just a small subset of the database is annotated with keywords. In this paper we present and evaluate a new method which improves the effectiveness of content-based image classification, by integrating semantic concepts extracted from text, and by automatically extending annotations to the images with missing keywords. Our model is inspired from the probabilistic graphical model theory: we propose a hierarchical mixture model which enables to handle missing values. Results of visual-textual classification, reported on a database of images collected from the Web, partially and manually annotated, show an improvement by 32.3% in terms of recognition rate against only visual information classification. Besides the automatic annotation extension with our model for images with missing keywords outperforms the visual-textual classification by 6.8%. Finally the proposed method is experimentally competitive with the state-of-art classifiers.

Keywords: probabilistic graphical models, Bayesian networks, image classification, image annotation.

1 Introduction

The rapid growth of internet and multimedia information has shown a need in the development of multimedia information retrieval techniques, especially the image retrieval. We can distinguish two main trends.

The first one, called "text-based image retrieval", consists in applying text-retrieval techniques from fully annotated images. The text describes high-level concepts but this technique requires a tedious work of annotation. The second approach, called "content-based image retrieval" is a more young field. These methods rely on visual features (color, texture or shape) computed automatically, and retrieve images using a similarity measure. However the obtained performances are not really acceptable, except in the case of well-focused corpus.

In order to improve the recognition, a solution consists in combining visual and semantic informations. Some researchers have already explored this possibility [1,2,3,4].

N. da Vitora Lobo et al. (Eds.): SSPR&SPR 2008, LNCS 5342, pp. 937–946, 2008.

Finally, automatic image annotation can be used in image retrieval systems to organize and locate images of interest from a database, or to perform visual-textual classification. This method can be regarded as a type of multi-class image classification with a very large number of classes, as large as the vocabulary size. Typically, image analysis in the form of extracted feature vectors and the training annotation words are used by machine learning techniques to attempt to automatically apply annotations to new images. Many works have been proposed in this sense [5,6,7,8,9,10].

In this perspective, the contribution of this paper is to propose a scheme for image classification optimization, by using a joint visual-text clustering approach and automatically extending image annotations.

2 Motivations

The motivation of the model presented here is to use it for the both tasks of weakly-annotated image classification and annotation. In fact the classification methods before mentionned are efficient but they requires that all image, or image blobs are annotated. Moreover the annotation models are not enable to classify images.

The proposed model does not require that all images be annotated: when an image is weakly annotated, the missing keywords are considered as missing values. Besides it enables to automattically extend existing annotations to weakly-annotated images, without User intervention.

The model [5] is the most related to our approach, because it enables to classify images based on visual and textual features and to automatically annotate new images. However our model is less restrictive for the user. In fact our classifier does not need that all images be annotated, and the existing keywords are associated to the whole images, not to image regions.

The proposed approach is derived from the probabilistic graphical model theory. We introduce a method to deal with missing data in the context of text annotated images as defined in [5,4]. The uncertainty around the association between a set of keywords and an image is tackled by a joint probability distribution over the dictionary of keywords and the numerical features extracted from our collection of grey-level and color images.

The rest of this paper is organized as follows: Section 3 decribes the probabilistic model of weakly-annotated image representation and how to use it to classify and extend existing annotations to images. Section 4 presents the experimental results. Finally conclusions and future works are given in Section 5.

3 Representation and Classification of Weakly-Annotated Images

Our work is focused on weakly-annotated image modelling and classification. Now visual descriptors often provide vectors of continuous values, and the associated keywords often correspond to discrete variables. So we have chosen to construct a Bayesian classifier which allows for the discrete and continuous variable combination and the problem of missing values handling.

Let f_j be a query image characterized by a set of features F. F is composed of:

- m visual features, denoted $v_1, ..., v_m$,
- n possible keywords, denoted $KW\ 1, ..., KW\ n$.

The chosen visual features are issued from one color descriptor, a color histogram [11], and one shape descriptor based on the Fourier/Radon transform [12]. We are interested in the probability distributions of these features and their conditional dependence relations. Let us consider the visual features as continuous random variables and theirs associated keywords as discrete variables. This model is too big to be represented as a unique joint probability distribution, therefore it is required to introduce some sparse and structural *a priori* knowledge. The probabilistic graphical models, and especially Bayesian networks, are a good way to solve this kind of problem. In fact within Bayesian networks the joint probability distribution is replaced by a sparse representation only among the variables directly influencing one another. Interactions among indirectly-related variables are then computed by propagating inference through a graph of these direct connections. Consequently the Bayesian networks are a simple way to represent a joint probability distribution over a set of random variables, to visualize the conditional properties and to compute complex operations like probability learning and inference, with graphical manipulations. Then a Bayesian network seems to be appropriate to represent and classify images and associated keywords.

3.1 A Gaussian-Mixtures and Multinomial Mixture Model

We present a hierarchical probabilistic model of multiple-type data (images and associated keywords) in order to classify large annotated image databases. A Gaussian Mixtures and Multinomial Mixture model is proposed. In fact, the observation of some peaks on the different histograms of the feature variables, has led us to consider that the visual features can be estimated by mixtures of Gaussian densities. The discrete variables corresponding to the possible keywords are assumed to be distributed as a multinomial distribution over the vocabulary of keywords.

Now let F be the training set composed of m instances $f_{1_i}, ..., f_{m_i}, \forall i \in \{1, ..., n\}$, where n is the dimension of the signatures provided by the concatenation of the feature vectors issued from the computation of all the descriptors on each image on the training set. Each instance $f_j, \forall j \in \{1, ..., m\}$ is then characterized by n continuous variables. A supervised classification is considered then F instances are divided into k classes $c_1, ..., c_k$. Let $G_1, ..., G_g$ be g groups whose each has a Gaussian density with a mean $\mu_l, \forall l \in \{1, ..., g\}$ and a covariance matrix $\sum_l$. Besides, let $\pi_1, ..., \pi_g$ be the proportions of the different groups, $\theta_l = (\mu_l, \sum_l)$ the parameter of each Gaussian and $\Phi = (\pi_1, \pi_1, ..., \pi_g, \theta_1, ..., \theta_g)$ the global mixture parameter. Then the probability density of F conditionally to the class $c_i, \forall i \in \{1, ..., k\}$ can be defined by

$$P(f, \Phi) = \sum_{l=1}^{g} \pi_l p(f, \theta_l)$$

where $p(f, \theta_l)$ is the multivariate Gaussian defined by the parameter θ_l.

Then, we have one Gaussian Mixture Model per class. This problem can be represented by the probabilistic graphical model in Figure 1, where:

- The "Class" node is a discrete node, which can take k values corresponding to the pre-defined classes $c_1, ..., c_k$.
- The "Component" node is a discrete node which corresponds to the components (i.e the groups $G_1, ..., G_g$) of the mixtures. This variable can take g values, i.e the number of Gaussians used to compute the mixtures. It's an hidden variable which represents the weight of each group (i.e the $\pi_l, \forall l \in \{1, ..., g\}$).
- The "Gaussian" node is a continuous variable which represents each Gaussian $G_l, \forall \in \{l = 1, ..., g\}$ with its own parameter ($\theta_l = (\mu_l, \sum_l)$). It corresponds to the set of feature vectors in each class.
- Finally the edges represent the effect of the class on each Gaussian parameter and its associated weight. The green circle is just used to show the relation between the proposed probabilistic graphical model and GMMs: we have one GMM (encircled in green), composed of Gaussians and their associated weight, per class.

Now the model can be completed by the discrete variables, denoted $KW\ 1, ...,$ $KW\ n$, corresponding to the possible keywords associated to the images. Dirichlet priors [13], have been used for the probability estimation of the variables $KW\ 1, ..., KW\ n$. That is we introduce additional pseudo counts at every instance in order to ensure that they are all "virtually" represented in the training set. Therefore every instance, even if it is not represented in the training set, will have a not null probability. Like the continuous variables corresponding to the

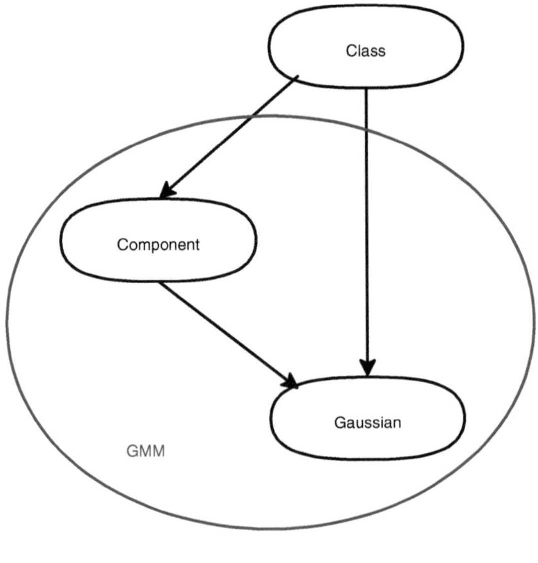

Fig. 1. A Probabilistic graphical model as GMMs

visual features, the discrete variables corresponding to the keywords are included in the graphical model by connecting them to the class variable.

Then our classifier can be depicted by the Figure 2. The hidden variable "α" shows that a Dirichlet prior is used. The box around the variable KW denotes n repetitions of KW, for each keyword. This Bayesian classifier means that each image and its keywords are assumed to have been generated conditional on the same class. Therefore the resulting multinomial and Gaussian mixture parameters should correspond: concretely if an image, represented by visual descriptors, has an high probability under a certain class, then its keywords should have an high probability under the same class.

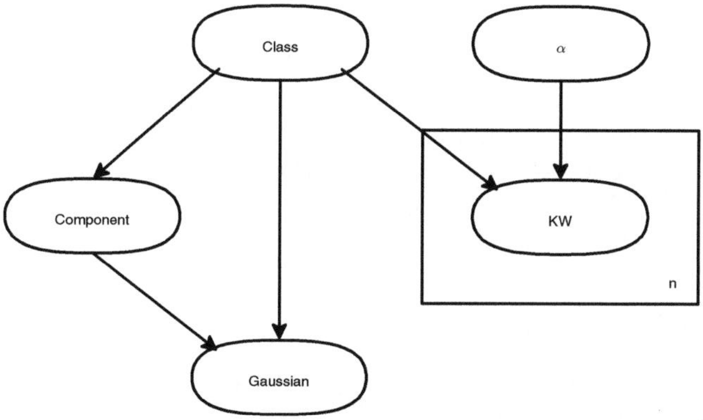

Fig. 2. The Gaussian-Mixtures and Multinomial mixture model

3.2 Parameter Learning and Inference

The EM algorithm has been used to learn the Gaussian mixture parameters. But our major problem deals with missing values. Indeed only some data are fully observed. It's the case of all visual features for color images or just shape features for grey-level images. On the contrary the color features for grey-level images, and especially some keywords for a large subset of images, are missing. Concerning the visual features, the missing values are clearly homogeneously distributed (because they correspond to grey-level images). But the missing values are randomly distributed for the variables $KW\ i, \forall i \in \{1, ..., n\}$. This kind of problem can be tackled with the EM algorithm too. The general purpose of this algorithm, explained in detail in [14] consists in computing, in an iterative way, the likelihood maximum when the instances can be viewed as incomplete data: each iteration consists of an Expectation computation step before a Maximization step. This algorithm has been chosen for its simplicity and generality. Applied to the parameter learning problem, the EM algorithm starts off with the random model initialisation, then, at each iteration, the parameters are updated in order to obtain a local optimal of the maximum-likelihood. An inference algorithm is required to compute the expected sufficient statistics in the expectation step. Then, in the maximization step, the parameters of the model are adjusted to fit the data.

An inference algorithm is also necessary to classify new images. Indeed, the inference process consists in computing posterior probability distributions of one or several other subsets of nodes. In the case of classification, the class node is inferred. According to our Bayesian network topology, the inference process propagates the values from the image feature level represented by the "Gaussian" node, trough the "Component" and Keyword nodes, until the "Class" node level. A message passing algorithm [15] is applied to the network. In this technique, each node is associated to a processor, which can send some messages to its neighbors, in an asynchronous way, until it reaches a stability.

Thus a query image f_j, characterized by its visual features $v_{j_1}, ..., v_{j_m}$ and its possible keywords $KW\ 1_j, ..., KW\ n_j$ is considered as an "evidence" represented by:

$$P(f_j) = P(v_{j_1}, ..., v_{j_m}, KW\ 1_j, ..., KW\ n_j) = 1$$

when the network is evaluated. Thanks to the inference algorithm, the probabilities of each node are updated in function of this evidence. After the belief propagation, we know, $\forall i \in \{1, ..., k\}$, the posterior probability $P(c_i|f_j) = P(c_i|v_{j_1}, ..., v_{j_m}, KW\ 1_j, ..., KW\ n_j)$. The query f_j is assigned to the class c_i which maximizes this probability.

3.3 Annotation Extension of Images

Given an image without keyword, or a weakly annotated image, the proposed Bayesian model before described can be used to compute a distribution over words conditionally to the image and its possible existing keywords. In fact, for a query image f_j annotated by $k, \forall k \in \{0, ..., n\}$ keywords, where n is the maximum keyword number per image, the inference algorithm enables to compute the posterior probability $P(KW_{i_j}|f_j, KW_1, ..., KWk), \forall i \in \{k+1, ..., n\}$. This distribution represents a prediction of the missing keywords for that image.

For example, let us consider Table 1 which presents 3 images with possible keywords and the keywords obtained after automatic annotation extension. The first image, without annotation, has been annotated by 2 suitable keywords. In

Table 1. Examples of images and possible keywords before and after annotation extension

image	initial possible keywords	keywords after annotation extension
		bridge water
	bridge cloud	bridge cloud sunset
	bridge	bridge cloud sunset

the same way, the second image annotation, composed at the beginning of 2 keywords, has been extended to 3 keywords. The added keyword, “sunset” is suitable. On the contrary, the third image, initially annotated by 1 keyword, has obtained 2 other keywords after the automatic extension. But the second new keyword, “sunset” is wrong. This mistake is probably due to the large number of database images annotated by these 3 keywords “bridge”, ”cloud” and “sunset”, and the inference algorithm.

4 Experimental Results

In this section, we present an evaluation of our model on more than 3000 free images collected from the Web, and kindly provided by Kherfi et al. [4]. These images are split up into 16 classes. For example, 4 images of the class “horse” are given in Figure 3.

Fig. 3. Examples of horse-class images

65% of the image database have been manually annotated by 1 keyword, 28% by 2 keywords and 6% by 3 keywords, using a vocabulary set of 39 keywords. For example, among the 4 images from the Figure 3, the first image is annotated by the 2 keywords “animal” and “horse”. The second is only annotated by one keyword: “animal”. The others have not been annotated. We have evaluated our method by performing 5 cross validations whose each proportion of the training set is 25%, 35%, 50%, 65% and 75% of the database, the remaining respectively 75%, 65%, 50%, 35% and 25% are hold for test set. In each case the tests are repeated 10 times in order that each database instance would be used for the training and the test. For each training set size, the recognition rate is obtained by taking the mean recognition rate of the 10 tests. Since we want to improve the recognition rate by introducing semantics, and automatically extending image annotations, we limit ourselves to the experiments comparing:

- the visual-textual classification to the only visual information classification,
- the visual-textual classification before and after automatic annotation extension of all images,
- the proposed model of visual-textual classification to two state-of-art classifiers

Let us consider Table 2. The notation “C + S” means that the color and shape descriptors (“C” for Color, “S” for Shape) have been combined. The notation “C + S + KW” indicates that both visual descriptors and textual information (“KW” for Keywords) have been combined. The recognition rates confirm that combining visual with semantic features performs always better than any of them alone. In fact we observe that the combination of our visual features and keywords (when they are available) increases the recognition rate by 38.5% compared to the results of color information alone, 58% compared to the shape information classification and 37% compared to the only textual information classification. Besides we can notice that for all experiments, combining the both visual descriptors is better, by 16% on average, to use just one of them. Finally, the visual-textual classification shows an improvement by 32.3% in terms of recognition rate against only visual information classification.

Table 2. Recognition rates (in %), obtained by the GM-M model, of only visual classification vs. visual-textual classification

Specifications		Color	Shape	Keywords	C + S	C + S + KW
training part	test part					
25%	75%	35	17.8	36.6	39.4	69.7
35%	65%	36.9	18.1	38.9	42.2	74.4
50%	50%	38.7	18.5	41.1	45	79.1
65%	35%	41.1	20.6	41.5	46.6	81.7
75%	25%	43.5	21.8	45.1	52.9	82.9

Then, Table 3 shows the effectiveness of our approach (GM-M mixture) compared to the SVM and FKNN classifiers. The results have been obtained by using the visual features and theirs possible associated keywords. It appears that the GM-M mixture results are always better than the ones of SVM and FKNN results.

Finally, annotations have been extended to all images of the database in order that each would be annotated by 3 keywords. Then, to evaluate the quality of this annotation extension, the visual-textual classification has been re-done, with the same specifications as in Table 2. Table 4 shows the efficiency of our automatic annotation extension. In fact, the recognition rates after automatic annotation extension are always better than before. Moreover the automatic annotation extension outperforms the recognition by 6.8% on average.

Table 3. Recognition rates (in %) of the SVM classifier and the FKNN vs. our Gaussian-Mixtures and multinomial mixture model

Specifications		SVM	FKNN $k = 1$	FKNN $k = m$	GM-M mixture
training part	test part				
25%	75%	38.3	59.1	58.5	69.7
35%	65%	41.3	62.3	58.3	74.4
50%	50%	39.9	68.2	58.2	79.1
65%	35%	40.5	72.9	67	81.7
75%	25%	41.9	73.2	69.3	82.9

Table 4. Recognition rates (in %), obtained by the GM-M model, of visual-textual classification before and after automatic annotation extension

Specifications		Before annotation extension	After annotation extension
training part	test part		
25%	75%	69.7	77
35%	65%	74.4	79.3
50%	50%	79.1	85.4
65%	35%	81.7	87.6
75%	25%	82.9	92.7

5 Conclusion and Future Works

We have proposed an efficient model which enables to combine visual and textual information, handle missing values and extend image annotations to other images. Our experiments have been done on a partially annotated Web image database. The results show that our visual-textual classification method improves the recognition rate compared to the only visual information classification. Moreover our Bayesian network can be used to extend image annotations, what outperforms the recognition rate. Finally the proposed method is competitive with state-of-art classifiers. Further works will be devoted to capture the user's preference by considering a relevance feedback process. More precisely, the user's preference can be represented by the network parameter update (i.e the probabilities of each variable in function of the new classified instance) during the inference process.

References

1. Barnard, K., Duygulu, P., Forsyth, D., De Freitas, N., Blei, D.M., Jordan, M.I.: Matching words and pictures. Journal of Machine Learning Research 3(6), 1107–1135 (2003)
2. Benitez, A., Chang, S.F.: Perceptual knowledge construction from annotated image collections. ICME 2002 1, 189–192 (2002)
3. Grosky, W.I., Zhao, R.: Negotiating the semantic gap: From feature maps to semantic landscapes. In: Pacholski, L., Ružička, P. (eds.) SOFSEM 2001. LNCS, vol. 2234, pp. 33–52. Springer, Heidelberg (2001)
4. Kherfi, M.L., Brahmi, D., Ziou, D.: Combining visual features with semantics for a more effective image retrieval. In: ICPR 2004, vol. 2, pp. 961–964 (2004)
5. Blei, D.M., Jordan, M.I.: Modeling annotated data. In: SIGIR 2003, pp. 127–134 (2003)
6. Gao, Y., Fan, J., Xue, X., Jain, R.: Automatic image annotation by incorporating feature hierarchy and boosting to scale up svm classifiers. In: ACM MULTIMEDIA 2006, pp. 901–910 (2006)
7. Yang, C., Dong, M., Hua, J.: Region-based image annotation using asymmetrical support vector machine-based multiple-instance learning. In: CVPR 2006, pp. 2057–2063 (2006)

8. Wang, C., Jing, F., Zhang, L., Zhang, H.J.: Image annotation refinement using random walk with restarts. In: ACM MULTIMEDIA 2006, pp. 647–650 (2006)
9. Rui, X., Li, M., Li, Z., Ma, W.Y., Yu, N.: Bipartite graph reinforcement model for web image annotation. In: ACM MULTIMEDIA 2007, pp. 585–594 (2007)
10. Feng, S., Manmatha, R., Lavrenko, V.: Multiple bernoulli relevance models for image and video annotation. CVPR 2004 2, 1002–1009 (2004)
11. Swain, M.J., Ballard, D.H.: Color indexing. International Journal of Computer Vision 7(1), 11–32 (1991)
12. Tabbone, S., Wendling, L.: Technical Symbols Recognition Using the Two-dimensional Radon Transform. In: ICPR 2002, vol. 2, pp. 200–203 (2002)
13. Robert, C.: A decision-Theoretic Motivation. Springer, Heidelberg (1997)
14. Dempster, A.P., Laird, N.M., Rubin, D.B.: Maximum likelihood from incomplete data via the em algorithm. Journal of the Royal Statistical Society. Series B (Methodological) 39(1), 1–38 (1977)
15. Kim, J.H., Pearl, J.: A computational model for combined causal and diagnostic reasoning in inference systems. In: IJCAI 1983, pp. 190–193 (1983)

Selection of Suitable Set of Decision Rules Using Choquet Integral

Laurent Wendling[1], Jan Rendek[1], and Pascal Matsakis[2]

[1]LORIA - ESIAL, Universit Henri Poincar
54506 Vandœuvre-ls-Nancy, France
[2]CIS Dpt, Universit de Guelph
ON N1G 2W1 Canada
{wendling,rendek}@loria.fr, matsakis@cis.uoguelph.ca

Abstract. An approach to automatically extract pertinent subsets of soft output classifiers, assumed to decision rules, is presented in this paper. They are aggregated into a global decision scheme using the Choquet integral. A selection scheme is defined that discards weak or redundant decision rules, keeping only the most relevant subset. An experimental study, based on real world data attest the interest of our method.

1 Introduction

A pattern recognition system can be roughly decomposed into three successive steps [6,18]. Firstly, the shapes are extracted from their surrounding background. This step, called segmentation, relies heavily on *a priori* knowledge of the documents to be processed. Secondly, a representation is built from the extracted patterns. This representation allows to computationally judge the (dis)similarity between two patterns. It can either be a set of measurements performed on the patterns, forming a vector of features, or a symbolic description of how the pattern can be divided into basic shapes. Thirdly, a decision rule is built, using the representation of the pattern. This rule predicts to which class an observed pattern most likely belongs. It can either be built by introducing some expert knowledge about the patterns, or learned on a representative subset of labeled patterns. Surveys on pattern representation and classification techniques over the last decade fail to conclude whether a set of generic methods performing best on any kind of document, can be found [6,3,11]. Rather, a collection of techniques has been developed to address domain specific issues. For a given application, the choice of a pattern representation or a decision scheme relies at best on the extensive testing of several combinations of techniques. It is not uncommon that a designer chooses by making an educated guess between the techniques he has at hand, or between those he is the most acquainted with. A tempting approach is to combine several decision rules, based on various representations and classification schemes, instead of electing only one. The expected outcome is a more robust final decision, taking advantage of all the decision rules qualities. This approach appears well suited to cope with cases where the

N. da Vitora Lobo et al. (Eds.): SSPR&SPR 2008, LNCS 5342, pp. 947–955, 2008.

available data is too scarce to determine the best method by thorough testing, or to build robust decision rules by learning. Many classifier combination systems have been proposed and compared in the literature [9,16,10,19]. In this paper, an improvement of the aggregation of decision rules using the Choquet integral is proposed. The Choquet integral is part of the aggregation techniques based on fuzzy integrals and have been successfully used as fusion operators in various applications.

2 Background on Choquet Integral Fusion

2.1 Decision Rules Fusion

Let us consider m classes, $\mathcal{C} = \{C_1, \ldots, C_m\}$, and n Decision Rules (DRs) $X = \{D_1, \ldots, D_n\}$. When a new pattern x^o is observed, we wish to find the class it most likely belongs to. Labeling this unknown pattern is a three-steps process. Firstly, for each decision rule j and each class i, we compute ϕ_j^i the degree of confidence in the statement "According to D_j, x^o belongs to the class C_i". Secondly, we combine all these partial confidence degrees into a global confidence degree by choosing a suitable aggregation operator $\mathcal{H}$. Thus, the global confidence degree in the statement "x^o belongs to C_i", noted $\Phi(C_i|x^o)$, is given by:

$$\Phi(C_i|x^o) = \mathcal{H}(\phi_1^i, \ldots, \phi_n^i)$$

Finally, x^o is assigned to the class for which the confidence degree is the highest.

$$label(x^o) = \overset{m}{\underset{i=1}{\arg}}\ max\ \Phi(C_i|x^o)$$

Many aggregation operators were introduced in the literature. If the classification issue implies more than two classes, two learning approaches can be followed. Either each class C_i is paired with his own aggregation operator $\mathcal{H}^i$, or a single global aggregation operator is learned. In the former case, the final decision is slightly modified as the global confidence degree depends on the operator associated with the class.

$$\Phi(C_i|x^o) = \mathcal{H}^i(\phi_1^i, \ldots, \phi_n^i)$$

2.2 Fuzzy Measures and the Choquet Integral

The Choquet integral was first introduced in the capacity theory [2,15]. Let us denote by $X = \{D_1, \ldots, D_n\}$ the set of n decision rules, and $\mathcal{P}$ the power set of X, i.e. the set of all subsets of X.

Definition 1. *A* fuzzy measure *or* capacity, μ, *defined on X is a set function* $\mu : \mathcal{P}(X) \longrightarrow [0, 1]$, *verifying the following axioms:*
$\mu(\emptyset) = 0$, $\mu(X) = 1$
$A \subseteq B \Longrightarrow \mu(A) \leqslant \mu(B)$

Fuzzy measures generalize additive measures, by replacing the additivity axiom by a weaker one (monotonicity). Fuzzy measures embed particular cases including probability measure, possibility and necessity measures, or belief and plausibility functions. In our context of decision rules fusion, $\mu(A)$ represents the importance, or the degree of trust in the decision provided by the subset A of DRs. The next step in building a final decision, is to combine the partial confidence degree according to each DR into a global confidence degree, taking those weights into account.

Definition 2. *Let μ be a fuzzy measure on X. The* discrete Choquet *integral of $\phi = [\phi_1, \ldots, \phi_n]^t$ with respect to μ, noted $\mathcal{C}_\mu(\phi)$, is defined by:*

$$\mathcal{C}_\mu(\phi) = \sum_{j=1}^{n} \phi_{(j)} [\mu(A_{(j)}) - \mu(A_{(j+1)})]$$

where $(.)$ is a permutation and $A_{(j)} = \{(j), \ldots, (n)\}$ represents the $[j..n]$ associated criteria in increasing order and $A_{(n+1)} = \emptyset$.

2.3 Determining the Fuzzy Measure

There are several methods to determine the most adequate fuzzy measure to be used for a given application and the most straightforward learning approach is based on optimization techniques. The aim is to find the fuzzy measure that minimizes best a criterion on the training set, such has the square error. Considering $(x^k, y^k), k = 1, \ldots, l$, l learning samples where $x^k = [x_1^k, \ldots, x_n^k]^t$ is a n-dimensional vector, and y^k the expected global evaluation of object k, the fuzzy measure can be determined by minimizing [5]: $E^2 = \sum_{k=1}^{l} (\mathcal{C}_\mu(x_1^k, \ldots, x_n^k) - y^k)^2$. This criterion can be put under a quadratic program form and solved by the Lemke method. Nevertheless the method requires at least $n!/[(n/2)!]^2$ learning samples. When little data is available, matrices may be ill-conditioned, causing a bad behavior of the algorithm. To cope with the above problems, "heuristic" algorithms have been developed. To our knowledge, the algorithm providing the best approximation was proposed by Grabisch in [4]. It assumes that in the absence of any information, the most reasonable way to aggregate the partial matching degrees is to compute the arithmetic mean on all the inputs.

2.4 Behavioral Analysis of the Aggregation

The importance index is based on the definition proposed by Shapley in game theory [17]. It is defined for a fuzzy measure μ and a rule i as:

$$\sigma(\mu, i) = \frac{1}{n} \sum_{t=0}^{n-1} \frac{1}{\binom{n-1}{t}} \sum_{\substack{T \subseteq X \setminus i \\ |T| = t}} [\mu(T \cup i) - \mu(T)]$$

It can be interpreted as a average value of the marginal contribution $\mu(T \cup i) - \mu(T)$ of the decision rule i alone in all combinations. The interaction index [14] represents the degree of interaction between two decision rules. If the fuzzy measure is non-additive then some sources interact. The *marginal interaction* between I and j, conditioned to the presence of elements of the combination $T \subseteq X \backslash ij$ is:

$$(\Delta_{ij}\mu)(T) = \mu(T \cup ij) + \mu(T) - \mu(T \cup i) - \mu(T \cup j)$$

Averaging this criterion over all the subsets of $T \subseteq X \backslash ij$ gives the interaction index of sources i and j.

$$I(\mu, ij) = \sum_{T \subseteq X \backslash ij} \frac{(n-t-2)!t!}{(n-1)!}(\Delta_{ij}\mu)(T)$$

A positive interaction index for two DRs i and j means that the importance of one DR is reinforced by the second. A negative interaction index indicates that the sources are antagonist, and their combined use impairs the final decision.

3 Extraction of Decision Rules

3.1 Handling with Learning Error

Lattices (associated to fuzzy measures) are initialized at the arithmetic mean, and are approximated using a training set via a gradient descent. From training pattern, m training samples are created $\Phi^1, \ldots, \Phi^m$, with $\Phi^i = (\phi_1^i, \ldots, \phi_m^i)$ where ϕ_j^i represents the confidence in the fact that the sample belongs to class i, according to DR j. Each of these samples is paired with a target value, i.e. the value an ideal operator would output using this sample as input. For techniques that use a single fuzzy measure no real formula exists and often the following one is used:

$$\mathcal{C}_\mu(\Phi^i) = \begin{cases} 1, \text{ if sample belongs to class } i, \\ 0, \text{ otherwise.} \end{cases}$$

For techniques that use a different fuzzy measure per class, the optimal target value minimizing the quadratic error is known for two classes [4]. Best criterion is defined as follows.

$$E^2 = \sum_i \sum_j |\Psi(\Delta\Phi_{12}(X_k^j)) - 1|^2$$

with X_k^j is the kth training data of class j, Ψ a sigmoid-type function and $\Delta\Phi_{12}$ is given by:

$$\Delta\Phi_{12} = \Phi_{\mu_1}(C_1|X_k^j) - \Phi_{\mu_2}(C_2|X_k^j)$$

The error used in the gradient descent algorithm is given by:

$$e = \Psi(\Delta\Phi_{12}(X_k^j)) - 1)$$

The criterion is modified as follows to process with N-classes:

$$\Delta\Phi_{qr_{r\in X-\{q\}}} = \Phi_{\mu_q}(C_q|X_k^j) - \bigotimes_{\{r\in X-\{q\}\}} \Phi_{\mu_r}(C_r|X_k^j)$$

with q and r two classes. Operator $\bigotimes$ can be a median, min... Here the max was kept to move away ambiguous classes and so to favor a discrimate behavior. Median is interesting to preserve acceptable results even if some sources may be contradictory. The sign of $\Delta\Phi_{qr_{r\in X-\{q\}}}$ is studied to set the error to be propagated in the lattice as follows.

$$sign(\Delta\Phi_{qr_{r\in X-\{q\}}}) = \begin{cases} +e = 1./f(\Delta\Phi_{qr_{r\in X-\{q\}}}) \\ -0,\ e = 0 \text{ for } q,\ 1 \text{ for others.} \end{cases}$$

with f an increasing function.

3.2 Weaker Decision Rule Extraction

Once the lattice is learned, the individual performance of each DR is analysed in the produced fuzzy measure. This analysis is performed using the importance and interaction indexes defined in 2.4. The aim is to track the DRs having the weak importance in the final decision, and that positively interact the least with the other rules. Such DRs are assuming to blur the final decision. First low significant rules S_L having an importance index lower than 1 are selected:

$$S_L = \{k \mid n \cdot \sigma(\mu, k) < 1\}$$

The set of rules to be removed MS_L is composed of the rules having an interaction index lower than the mean of the interaction indexes of S_L:

$$MS_L = \{k \mid \sum_{j=1,n} I(\mu, kj) < m\}_{k\in S_L}$$

with the global mean interaction index $m = \frac{1}{|S_L|}\sum_{k\in S_L}\sum_{j=1,n} I(\mu, kj)$

3.3 Extracting Decision Rules

We use the classic fuzzy pattern matching setup. Each class $C_1, \ldots, C_n$ is associated with a fuzzy measure $\mu_1, \ldots, \mu_n$. The evaluation measures are generally classifier dependant [13]. For their computation the classification of training data must be performed to identify the fuzzy measure for each class. The application of our learning algorithm followed by the extraction of the most relevant DRs forms a training epoch is quite similar to global scheme but regarding a set of classes aggregated at las using a $argmax$ criterion. As it is not easy to valuate each combinaison of classes. we consider them indepedently. A Greedy feature selection algorithm is used to ensure a continuous extraction of unexpected features per class. At each epoch the weakest descriptor calculated from indices

is removed and so on, while improving a gain (recognition rates). The overall algorithm is described below.
% Initialization %
For EachC_i $\mathcal{C}$
– Learned μ_i
– Extract weakest descriptor
End For Each
% Main %
While Minimization is on the way
– Replace old capacity with new capacity
– Evaluate gain (minimizing cost function)
– Keep the "best" new capacity
– Extract feature for that new capacity
End while

4 Experimental Results

The proposed approach being aimed at situations where little information is available for training, databases having a fair number of categories and a small number of samples have been used . For the experiments, several decision rules are set from different photometric descriptors associated to a basic similarity measure to belong in the same range. A set of nine pattern recognition methods $R = \{R_i\}_{i=1,9}$ is used here, most of them having a low processing time, easy to implement, and invariant to affine transforms such as translation, rotation, or scaling. The descriptors computed on the shapes are: $DC = \{$ ART [8], angular signature [1], GFD [22], Ellipticity,f_0 and f_2-histograms [12], moments of Zernike [7], Yang [21], Radon signature [20]$\}$. B1(9 classesx11samples) and B2(18,12) are well-known sharvit' databases (see figure 1).

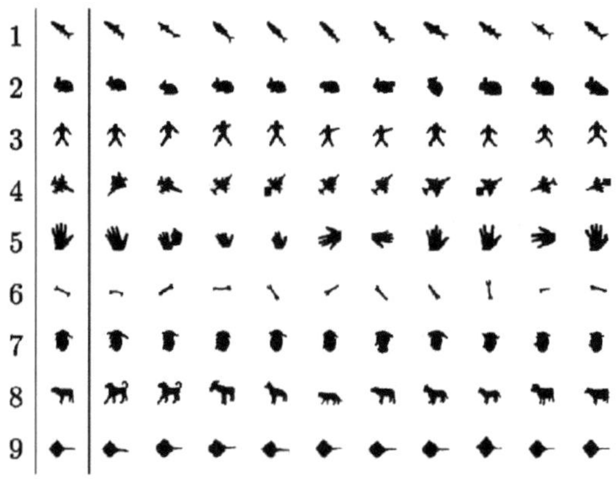

Fig. 1. The first Sharvit' database (B1)

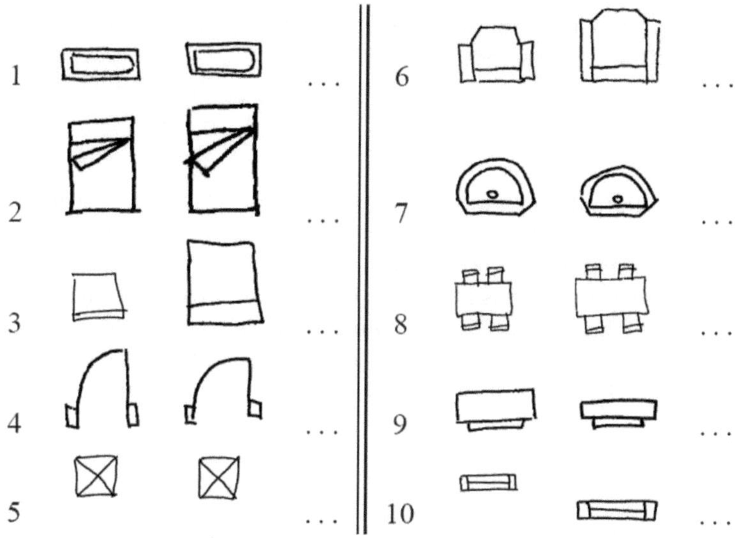

Fig. 2. Samples of the CVC database (B3)

Table 1. Recognition rates reached for each method

	R_1	R_2	R_3	R_4	R_5	R_6	R_7	R_8	R_9	Mi	Ma	Me	Ch
B1	93	59	98	48	74	50	96	85	66	43	99	78	100
B2	88	56	90	48	62	46	86	66	56	59	87	76	93
B3	80	30	38	22	20	46	80	76	56	46	82	78	89

B3(10x300) is a database kindly provided by the CVC Barcelone. Symbols have been drawn by ten people using "anoto" concept. Few samples are provided in figure 2.

An experimental study is carried out from each decision rules R_i, basic aggregation operators as Min, Max, $Median$ and the proposed method based on $Choquet$ integral.

For each test, a crossvalidation was applied (1/3 and 2/3). Table 1 shows the good behavior of the method on these databases. The results reached independently by each decision rule R_i are coherent with the amount of information processed by associated features. Simple aggregation operators: minimum, maximum and median exhibit various behavior depending on the database. In terms of recognition rates, only the maximum achieves better results than the best simple DR on two databases. In comparison to the simple DRs and the simple aggregation operators, our fusion operators based on the Choquet integral consistently achieve better results on each databases, in terms of recognition rates. On each dataset, our recognition method at worse improves a little the recognition rates; it never worsens them. Moreover a decreasing of the number of decision rules around 50% per class has been obtained. It allows to define a kind of identity map per class consisting in the most suitable decision rules following the application under consideration.

5 Conclusion

The algorithm appears well suited for the situation when no *a priori* knowledge is available about the relevance of a set of decision rules for a given dataset, or when the training set available is to small to build reliable decision rules, or to determine which method is the best. It finds the best consensus between the rules, taking their interaction into account, and discarding the undependable or redundant ones. A way to merge both numerical and expert decision rules in order to improve the recognition process is under consideration.

References

1. Bernier, T., Landry, J.-A.: A new method for representing and matching shapes of natural objects. Pattern Recognition 36(8), 1711–1723 (2003)
2. Choquet, G.: Theory of capacities. Annales de l'Institut Fourier 5, 131–295 (1953)
3. Cordella, L.P., Vento, M.: Symbol recognition in documents: a col of technics? International Journal of Document Analysis and Recognition 3(2), 73–88 (2000)
4. Grabisch, M.: A new algorithm for identifying fuzzy measures and its application to pattern recognition. In: Int. conf. FUZZ'IEEE 1995, pp. 145–150 (1995)
5. Grabisch, M., Nicolas, J.M.: Classification by fuzzy integral - performance and tests. Fuzzy Sets and Systems 65, 255–271 (1994)
6. Jain, A.K., Duin, R.P.W., Mao, J.: Statistical Pattern Recognition: A Review. IEEE Transactions on Pattern Analysis and Machine Intelligence 22(1), 4–37 (2000)
7. Khotanzad, A., Hong, Y.H.: Invariant Image Recognition by Zernike. IEEE Transactions on Pattern Analysis and Machine Intelligence 12(5), 489–497 (1990)
8. Kim, W.-Y., Kim, Y.-S.: A new region-based shape descriptor. In: TR 15-01, Pisa, Italy (1999)
9. Kittler, J., Hatef, M., Duin, R., Matas, J.: On combining classifiers. IEEE Transactions on Pattern Analysis and Machine Intelligence 20(3), 226–239 (1998)
10. Kuncheva, L.I., Whitaker, C.J.: Measures of diversity in classifier ensembles. Machine Learning 51, 181–207 (2003)
11. Lladós, J., Valveny, E., Sánchez, G., Martí, E.: Symbol Recognition: Current Advances and Perspectives. In: Blostein, D., Kwon, Y.-B. (eds.) GREC 2001. LNCS, vol. 2390, pp. 104–127. Springer, Heidelberg (2002)
12. Matsakis, P., Wendling, L.: A New Way to Represent the Relative Position Between Areal Objects. IEEE Transactions on Pattern Analysis and Machine Intelligence 21(7), 634–643 (1999)
13. Mikenina, L., Zimmermann, H.-J.: Improved feature selection and classification by the 2-additive fuzzy measure. Fuzzy Sets and Systems 107, 197–218 (1999)
14. Murofushi, T., Soneda, S.: Techniques for reading fuzzy measures(iii): interaction index. In: Proc. of the 9th Fuzzy Set System, pp. 693–696 (1993)
15. Murofushi, T., Sugeno, M.: A theory of fuzzy measures: representations, the Choquet integral, and null sets. Journal of Math. Anal. Appl. 159, 532–549 (1991)

16. Ruta, D., Gabrys, B.: An overview of classifier fusion methods. Comp. and Information System 7(1–10) (2000)
17. Shapley, L.: A value for n-person games. In: Khun, H., Tucker, A. (eds.) Annals of Mathematics Studies, pp. 307–317. Princeton University Press, Princeton (1953)
18. Smeulders, A.W.M., Worring, M., Santini, S., Gupta, A., Jain, R.: CB Image Retrieval at the End of the Early Years. IEEE Transactions on Pattern Analysis and Machine Intelligence 22(12), 1349–1380 (2000)
19. Stejic, Z., Takama, Y., Hirota, K.: Mathematical aggregation operators in image retrieval: effect on retrieval performance and role in relevance feedback. Signal Processing 85(2), 297–324 (2005)
20. Tabbone, S., Wendling, L.: Binary shape normalization using the Radon transform. In: Nyström, I., Sanniti di Baja, G., Svensson, S. (eds.) DGCI 2003. LNCS, vol. 2886, pp. 184–193. Springer, Heidelberg (2003)
21. Yang, S.: Symbol Recognition via Statistical Integration of Pixel-Level Constraint Histograms: A New Descriptor. IEEE Transactions on Pattern Analysis and Machine Intelligence 27(2), 278–281 (2005)
22. Zhang, D., Lu, G.: Shape-based image retrieval using generic fourier descriptor. Signal Processing: Image Communication 17, 825–848 (2002)

Evaluating the Stability of Feature Selectors That Optimize Feature Subset Cardinality

Petr Somol[1,2] and Jana Novovičová[1,2]

[1] Dept. of Pattern Recognition, Institute of Information Theory and Automation, Academy of Sciences of the Czech Republic, 182 08 Prague, Czech Republic
{somol,novovic}@utia.cas.cz
http://ro.utia.cz/
[2] Faculty of Management, Prague University of Economics, Czech Republic

Abstract. Stability (robustness) of feature selection methods is a topic of recent interest. Unlike other known stability criteria, the new consistency measures proposed in this paper evaluate the overall occurrence of individual features in selected subsets of possibly varying cardinality. The new measures are compared to the generalized Kalousis measure which evaluates pairwise similarities between subsets. The new measures are computationally very effective and offer more than one type of insight into the stability problem. All considered measures have been used to compare two standard feature selection methods on a set of examples.

Keywords: Feature selection, stability, relative weighted consistency measure, sequential search, floating search.

1 Introduction

Feature selection (FS) has been a highly active area of research in recent years due to its potential to improve both the performance and economy of automatic decision systems in various applicational fields. It has been pointed out recently that not only model performance but also stability (robustness) of the FS process is important. Domain experts prefer FS algorithms that perform stably when only small changes are made to the data set. However, relatively little attention has been devoted to the stability of FS methods so far. Recent works in the area of FS methods' stability mainly focus on various stability indices, introducing measures based on Hamming distance, Dunne et al. [1], correlation coefficients and Tanimoto distance, Kalousis et al. [2], consistency index, Kuncheva [3] and Shannon entropy, Křížek et al. [4]. Most of these recent works focus on the stability of single FS methods, while Saeys et al. [5] construct and study an ensemble of feature selectors. Stability of FS procedures depends on sample size, criteria utilized to perform FS and complexity of FS procedure (Raudys [6]).

To evaluate the stability of feature selectors we propose in this paper several new measures to be called the *consistency measure* (C), the *weighted consistency measure* (CW) and the *relative weighted consistency measure* (CW_{rel}). Unlike most other measures, these can be used for assessing FS methods that yield

N. da Vitora Lobo et al. (Eds.): SSPR&SPR 2008, LNCS 5342, pp. 956–966, 2008.

subsets of varying sizes. We compare the new measures to the generalized form of Kalousis measure (GK) [2]. All four measures have been used to compare the stability of Sequential Forward Selection (SFS) [7] and Sequential Forward Floating Selection (SFFS) [8] on a set of examples.

2 The Problem of Feature Selection Stability

It is common that classifier performance is considered the ultimate quality measure, even when assessing the FS process. However, misleading conclusions may be easily drawn when ignoring stability issues. Unstable FS performance may seriously deteriorate the properties of the final classifier by selecting the wrong features. In the following we focus on several new measures allowing to assess FS stability.

Let Y be the original set of features of size (cardinality) $|\mathrm{Y}|$. Following [2] we define the *stability* of the FS algorithm as the robustness of the feature preferences it produces to differences in training sets drawn from the same generating distribution. FS algorithms express the feature preferences in the form of a selected feature subset $\mathrm{S} \subseteq \mathrm{Y}$. Stability quantifies how different training sets drawn from the same generating distribution affect the feature preferences.

2.1 Considered Measures of Feature Selection Stability

Let $\mathrm{Y} = \{f_1, f_2, \ldots, f_{|\mathrm{Y}|}\}$ be the set of all features and let $\mathcal{S} = \{\mathrm{S}_1, \ldots, \mathrm{S}_n\}$ be a system of $n > 1$ ($n \in \mathbb{N}$) feature subsets $\mathrm{S}_j = \{f_i | i = 1, \ldots, d_j, f_i \in \mathrm{Y}, d_j \in \{1, \ldots, |\mathrm{Y}|\}\}, j = 1, \ldots, n$ obtained from n runs of the evaluated FS algorithm. Let F_f be the number of occurrences (frequency) of feature f in system $\mathcal{S}$. Let X be the subset of Y representing all features that appear anywhere in $\mathcal{S}$:

$$\mathrm{X} = \{f | f \in \mathrm{Y}, F_f > 0\} = \bigcup_{i=1}^{n} \mathrm{S}_i, \quad \mathrm{X} \neq \emptyset. \tag{1}$$

Let N denote the number of all features in system $\mathcal{S}$, i.e.,

$$N = \sum_{g \in \mathrm{X}} F_g = \sum_{i=1}^{n} |\mathrm{S_i}|, \quad N \in \mathbb{N}, \quad N \geq n. \tag{2}$$

Let us now introduce several measures usable for evaluating FS stability.

Definition 1. *The consistency $C(\mathcal{S})$ of system $\mathcal{S}$ is defined as*

$$C(\mathcal{S}) = \frac{1}{|\mathrm{X}|} \sum_{f \in \mathrm{X}} \frac{F_f - 1}{n - 1} \,. \tag{3}$$

Definition 2. *The weighted consistency $CW(\mathcal{S})$ of system $\mathcal{S}$ is defined as*

$$CW(\mathcal{S}) = \sum_{f \in \mathrm{X}} w_f \frac{F_f - 1}{n - 1} \,, \tag{4}$$

where $w_f = \frac{F_f}{\sum_{g \in \mathrm{X}} F_g}$, $0 < w_f \leq 1$, $\sum_{f \in \mathrm{X}} w_f = 1$.

Because $F_f = 0$ for all $f \in \mathrm{Y}\backslash\mathrm{X}$, the *weighted consistency* $CW(\mathcal{S})$ can be equally expressed using notation (2) as

$$CW(\mathcal{S}) = \sum_{f \in \mathrm{Y}} \frac{F_f}{N} \cdot \frac{F_f - 1}{n - 1} . \tag{5}$$

The main properties of both $C(\mathcal{S})$ and $CW(\mathcal{S})$ are:

1. $0 \leq C(\mathcal{S}) \leq 1$, $0 \leq CW(\mathcal{S}) \leq 1$.
2. $C(\mathcal{S}) = 1$, $CW(\mathcal{S}) = 1$ if and only if (iff) all subsets in $\mathcal{S}$ are identical.
3. $C(\mathcal{S}) = 0$, $CW(\mathcal{S}) = 0$ iff all subsets in $\mathcal{S}$ are disjunct from each other.

It is obvious that $CW(\mathcal{S}) = 0$ iff $N = |\mathrm{X}|$, i.e., iff $F_f = 1$ for all $f \in \mathrm{X}$. This is unrealistic in most of real cases. Whenever $n > |\mathrm{X}|$, some feature must appear in more than one subset and consequently $CW(\mathcal{S}) > 0$. Similarly, $CW(\mathcal{S}) = 1$ iff $N = n|\mathrm{X}|$, otherwise all subsets can not be identical.

Clearly, for any N, n representing some system of subsets $\mathcal{S}$ and for given Y there exists a system $\mathcal{S}_{min}$ with such configuration of features in its subsets that yields the minimal possible $CW(\cdot)$ value, to be denoted $CW_{min}(N, n, \mathrm{Y})$, being possibly greater than 0. Similarly, a system $\mathcal{S}_{max}$ exists that yields the maximal possible $CW(\cdot)$ value, to be denoted $CW_{max}(N, n)$, being possibly lower than 1.

It can be easily seen that $CW_{min}(\cdot)$ gets high when the sizes of feature subsets in system approach the total number of features $|\mathrm{Y}|$, because in such system the subsets get necessarily more similar to each other. Consequently, using measure (3) or (4) for comparison of various FS methods may lead to misleading results if the methods tend to yield systems of differently sized subsets. For this reason we introduce another measure, to be called the *relative weighted consistency*, which suppresses the influence of the sizes of subsets in system on the final value.

Definition 3. *The relative weighted consistency* $CW_{rel}(\mathcal{S}, \mathrm{Y})$ *of system* $\mathcal{S}$ *characterized by* N, n *and for given* Y *is defined as*

$$CW_{rel}(\mathcal{S}, \mathrm{Y}) = \frac{CW(\mathcal{S}) - CW_{min}(N, n, \mathrm{Y})}{CW_{max}(N, n) - CW_{min}(N, n, \mathrm{Y})} , \tag{6}$$

$$CW_{rel}(\mathcal{S}, \mathrm{Y}) = CW(\mathcal{S}) \quad \textit{for} \quad CW_{max}(N, n) = CW_{min}(N, n, \mathrm{Y}).$$

Remark: The values $CW_{min}(\cdot)$ and $CW_{max}(\cdot)$ will be derived in Section 3.

It can be seen that for any N, n representing some system of subsets $\mathcal{S}$ and for given Y it is true that $0 \leq CW_{rel}(\mathcal{S}, \mathrm{Y}) \leq 1$ and for the corresponding systems $\mathcal{S}_{min}$ and $\mathcal{S}_{max}$ it is true that $CW_{rel}(\mathcal{S}_{min}) = 0$ and $CW_{rel}(\mathcal{S}_{max}) = 1$. The measure (6) does not exhibit the unwanted behavior of yielding higher values for systems with subset sizes closer to $|\mathrm{Y}|$, i.e., is independent on the size of feature subsets selected by the examined FS methods under fixed Y. We can say that this measure characterizes for given $\mathcal{S}, \mathrm{Y}$ the relative degree of randomness of the system of feature subsets on the scale between the maximum and minimum values of the weighted consistency (4).

A conceptually different measure of FS stability can be derived from the *similarity* measure between two subsets of features S_i and S_j, $\mathrm{S}_K(\mathrm{S}_i, \mathrm{S}_j)$ of arbitrary cardinality introduced by Kalousis et al. [2] as a straightforward adaptation of the Tanimoto distance measuring the amount of overlap between two sets.

Definition 4. *The similarity measure (to be called generalized Kalousis) of system $\mathcal{S}$ is the average similarity over all pairs of feature subsets in $\mathcal{S}$:*

$$GK(\mathcal{S}) = \frac{2}{n(n-1)} \sum_{i=1}^{n-1} \sum_{j=i+1}^{n} S_K(\mathrm{S}_i, \mathrm{S}_j) = \frac{2}{n(n-1)} \sum_{i=1}^{n-1} \sum_{j=i+1}^{n} \frac{|\mathrm{S}_i \cap \mathrm{S}_j|}{|\mathrm{S}_i \cup \mathrm{S}_j|}. \tag{7}$$

$GK(\mathcal{S})$ takes values from $[0,1]$ with 0 indicating empty intersection between all pairs of subsets $\mathrm{S}_i, \mathrm{S}_j$ and 1 indicating that all subsets are identical.

The properties of all introduced measures are discussed further in Sect. 4.

3 Relative Weighted Consistency Measure

To obtain the explicit formula for the *relative weighted consistency* measure $CW_{rel}(\mathcal{S}, \mathrm{Y})$ of system $\mathcal{S}$ for given Y as defined in Eq.(6) we derive in this Section the minimum and the maximum values for the *weighted consistency* measure (5). First we introduce supporting concepts.

3.1 Compacted Form of Arbitrary System of Feature Subsets

It follows from Eq.(5) that $CW(\cdot)$ is constant for all systems of subsets with equal N, n and identical feature frequencies. Therefore for any system $\mathcal{S}$ we can derive its *compacted form* yielding equal $CW(\cdot)$ value.

Definition 5. *The compacted form of system $\mathcal{S}$ is system $\mathcal{S}^{com}$ with equal characteristics N, n and equal feature frequencies, but with features reordered among subsets so that subset sizes are maximally equalized.*

It can be seen that for each system $\mathcal{S}$ a compacted form $\mathcal{S}^{com}$ exists yielding equal $CW(\cdot)$ value. It can also be seen that $\mathcal{S}^{com}$ consists of n_1 subsets of size $(k+1)$ and $(n-n_1)$ subsets is of size k, where

$$n_1 = N \bmod n \ , \qquad k = \frac{N - N \bmod n}{n} \ . \tag{8}$$

A compacted form is illustrated in Fig. 1.

3.2 The Impact of Feature Replacement on Consistency Value

Consider a system of subsets $\mathcal{S}$ characterized by N, n. We will now investigate how the value $CW(\mathcal{S})$ changes if one instance of some feature f_i in some subset in $\mathcal{S}$ is removed and another instance of some other feature f_j is added instead, so that system characteristics N, n remain unchanged. Let F_i and F_j denote the frequency of features f_i and f_j for all $i, j = 1, \ldots, |\mathrm{Y}|, i \neq j$ in $\mathcal{S}$.

$\mathcal{S}$:			$\mathcal{S}^{com}$:	
{3, 2}			{1, 2, 3, 6}	
{1, 6, 3, 2, 5}			{1, 2, 4, 6}	
{1, 2, 3}	N=23		{1, 2, 4}	N=23 n_1=2
{6, 5, 3, 4}	n=7	⇨	{1, 3, 4}	n=7 k=3
{4, 2, 3}	|X|=6		{2, 3, 5}	|X|=6
{1, 2}			{2, 3, 5}	
{1, 4, 2, 5}	CW($\mathcal{S}$)= $\frac{38}{69}$		{2, 3, 5}	CW($\mathcal{S}^{com}$)= $\frac{38}{69}$

Fig. 1. Compacting system $\mathcal{S}$ to $\mathcal{S}^{com}$ does not change N, n, feature frequencies, nor the respective $CW(\cdot)$ value

Lemma 1. *Assume $F_i \leq F_j$. Replace one instance of the (equally or less frequent) feature f_i in system $\mathcal{S}$ by one new instance of the (equally or more frequent) feature f_j to obtain system $\mathcal{S}^{\mp}$. Then $CW(\mathcal{S}) < CW(\mathcal{S}^{\mp})$.*

Lemma 2. *Assume $F_i > F_j$. Replace one instance of the more frequent feature f_i in system $\mathcal{S}$ by one new instance of the less frequent feature f_j to obtain system $\mathcal{S}^{\mp}$. Then $CW(\mathcal{S}) \geq CW(\mathcal{S}^{\mp})$, with equality iff $F_i = F_j + 1$.*

Proof. (of Lemmas 1 and 2) Let us assume that $F_i = F_j + d$, where $d \in \mathbb{Z}$, $\mathbb{Z}$ is the set of integers. If $d \leq 0$ then $F_i \leq F_j$. If $d \geq 1$ then $F_i > F_j$. Let us denote by $\mathcal{S}^{\mp}$ the system in which one instance of feature f_i has been removed and one instance of feature f_j has been added, i.e., the frequency of feature f_i is $F_i - 1$ and the frequency of feature f_j is $F_j + 1$. Then we have

$$\begin{aligned}
&CW(\mathcal{S}) - CW(\mathcal{S}^{\mp}) \\
&= K \cdot \big\{F_i(F_i - 1) + F_j(F_j - 1) - [(F_i - 1)(F_i - 1 - 1) + (F_j + 1)(F_j + 1 - 1)]\big\} \\
&= K \cdot \big\{(F_j + d)(F_j + d - 1) + F_j(F_j - 1) \\
&- \big[(F_j + d - 1)(F_j + d - 1 - 1) + (F_j + 1)(F_j + 1 - 1)\big]\big\} = K \cdot 2(d - 1),
\end{aligned}$$

where $K = 1/\big(N(n-1)\big)$. It immediately follows that $CW(\mathcal{S}) < CW(\mathcal{S}^{\mp})$ iff $d \leq 0$, $CW(\mathcal{S}) > CW(\mathcal{S}^{\mp})$ iff $d > 1$ and $CW(\mathcal{S}) = CW(\mathcal{S}^{\mp})$ iff $d = 1$. □

To summarize less formally – we have shown that when replacing one feature instance in system $\mathcal{S}$ by another while keeping the system characteristics N, n unchanged, it is true that: a) increasing the difference between frequencies of two features increases the value of $CW(\mathcal{S})$ defined in Eq.(5), while b) decreasing the difference between frequencies of two features decreases the value of $CW(\mathcal{S})$.

3.3 Minimum Value of Weighted Consistency

Consider an arbitrary system of subsets $\mathcal{S}$ characterized by N, n and given Y. We will now focus on finding the lower bound on $CW(\mathcal{S})$.

Definition 6. *The minimal system of subsets $\mathcal{S}_{min}$ characterized by N, n and for given* Y *is such system, where* $\max_{i,j\in\{1,\ldots,|\mathrm{Y}|\}}(|F_i - F_j|) \leq 1$.

An example of a minimal system is given in Fig. 2(a).

a)			b)		
$\mathcal{S}_{min}$:	{1, 2, 3, 4}		$\mathcal{S}_{max}$:	{1, 2, 3, 4}	
	{5, 6, 1, 2}	N=23 n_1=2		{1, 2, 3, 4}	N=23 n_1=2
	{3, 4, 5}	n=7 k=3		{1, 2, 3}	n=7 k=3
	{6, 1, 2}	|Y|=6		{1, 2, 3}	|X|=4
	{3, 4, 5}			{1, 2, 3}	|X|≤|Y|
	{6, 1, 2}			{1, 2, 3}	
	{3, 4, 5}	$CW(\mathcal{S}_{min})=\frac{11}{23}$		{1, 2, 3}	$CW(\mathcal{S}_{max})=\frac{64}{69}$

Fig. 2. (a) The system expected to yield the lowest value of $CW(\cdot)$ given N, n, Y. (b) The system expected to yield the highest value of $CW(\cdot)$ for given N, n.

Lemma 3. *Let $\mathcal{S}$ be a system of subsets characterized by N, n and given* Y. *If $\mathcal{S}_{min}$ is the minimal system with equal N, n and* Y, *then $CW(\mathcal{S}_{min}) \leq CW(\mathcal{S})$.*

Proof. Taking use of Lemma 2 as long as there exist features $f_i, f_j \in \mathrm{Y}$ such that $F_i > F_j + 1$, modify $\mathcal{S}$ so that one instance of f_i is replaced by one new instance of f_j, what decreases $CW(\mathcal{S})$. No decrease is possible iff there is no chance to take use of Lemma 2, i.e., the system conforms to Definition 6. □

Consider now for fixed N, n and given Y the compacted form of system $\mathcal{S}_{min}$. Let us denote for simplicity

$$D = N \bmod |\mathrm{Y}| \, . \tag{9}$$

It can be seen that no feature frequency in $\mathcal{S}_{min}$ can be lower than $F^{'}$, where

$$F^{'} = (N - D)/|\mathrm{Y}| \, . \tag{10}$$

Lemma 4. *The minimum value $CW_{min}(N, n, \mathrm{Y})$ of the consistency measure $CW(\mathcal{S})$ for a system $\mathcal{S}$ with characteristics N, n and for given* Y *is*

$$CW_{min}(N, n, \mathrm{Y}) = \frac{N^2 - |\mathrm{Y}|(N - D) - D^2}{|\mathrm{Y}|N(n-1)} \, . \tag{11}$$

Proof. It is obvious that in the compacted form of $\mathcal{S}_{min}$ exactly $(|\mathrm{Y}| - D)$ features occur F' times and D features occur $(F' + 1)$ times. Substituting in (5) we obtain

$$CW_{min}(N, n, \mathrm{Y}) = (|\mathrm{Y}| - D)\left(\frac{F'}{(|\mathrm{Y}| - D)F' + D(F' + 1)} \cdot \frac{F' - 1}{n - 1}\right) \tag{12}$$

$$+ D\left(\frac{F' + 1}{(|\mathrm{Y}| - D)F' + D(F' + 1)} \cdot \frac{(F' + 1) - 1}{n - 1}\right) = \frac{N^2 - |\mathrm{Y}|(N - D) - D^2}{|\mathrm{Y}|N(n - 1)} \, .$$

□

3.4 Maximum Value of Weighted Consistency

Consider an arbitrary system $\mathcal{S}$ of subsets characterized by N, n and given Y. We will now focus on finding the upper bound on $CW(\mathcal{S})$. First denote for simplicity

$$H = N \bmod n \, . \tag{13}$$

Definition 7. *The maximal system of subsets $\mathcal{S}_{max}$ characterized by N, n is such system, where k features [defined in (8)] occur n times and, if $H > 0$, one feature occurs H times.*

An example of a maximal system is given in Fig. 2(b).

Lemma 5. *Let $\mathcal{S}$ be a system of subsets characterized by N, n. If $\mathcal{S}_{max}$ is the maximal system with the same characteristics, then $CW(\mathcal{S}) \leq CW(\mathcal{S}_{max})$.*

Proof. Taking use of Lemma 1 as long as there exist features $f_i, f_j \in \mathrm{Y}$ such that $F_i \leq F_j$, modify $\mathcal{S}$ so that one instance of f_i is replaced by one new instance of f_j, what increases $CW(\mathcal{S})$. No increase is possible only if there is no chance to take use of Lemma 1, i.e., the system conforms to Definition 7. □

Lemma 6. *The maximum value $CW_{max}(N, n)$ of the consistency measure $CW(\mathcal{S})$ for a system $\mathcal{S}$ with characteristics N, n is*

$$CW_{max}(N,n) = \frac{(N-H)}{n}\left(\frac{n}{N}\cdot\frac{n-1}{n-1}\right) + 1\cdot\left(\frac{H}{N}\cdot\frac{H-1}{n-1}\right) = \frac{H^2 + N(n-1) - Hn}{N(n-1)}\ . \quad (14)$$

Proof. Substitute to Eq.(5) the feature frequencies specified in Definition 7. □

3.5 Explicit Formula for Relative Weighted Consistency

Collecting the results from Sect. 3.3 and 3.4 we obtain the following proposition.

Proposition 1. *The relative weighted consistency measure of system $\mathcal{S}$ characterized by N, n and for given Y becomes*

$$CW_{rel}(\mathcal{S}, \mathrm{Y}) = \frac{|\mathrm{Y}|\left(N - D + \sum_{f\in \mathrm{Y}} F_f(F_f - 1)\right) - N^2 + D^2}{|\mathrm{Y}|\left(H^2 + n(N-H) - D\right) - N^2 + D^2}\ . \quad (15)$$

Proof. Substitute (5), (11) and (14) using (9) and (13) to Eq.(6). □

4 Practical Differences between the Discussed Measures

Assuming n is the number of subsets in $\mathcal{S}$, the GK time complexity is $O(n^2)$ as each pair of subsets is evaluated, while the complexity of C, CW and CW_{rel} is $O(n)$ as each subset is processed only once to collect feature occurrence statistics. The weighted consistency CW was defined to overcome the deficiency of consistency C which underrates systems of the type illustrated in Fig. 3. The measures C, CW and CW_{rel} all differ from GK in principle; GK evaluates pairwise similarities between subsets in system while measures C, CW and CW_{rel}

$\mathcal{S}_1$: { {1}, {1}, {1}, {1}, {1}, {1,2}, {1}, {1}, {1}, {1}, {1}, {1}, {3}, {1}, {1} }

$C(\mathcal{S}_1)=0.31$ $CW(\mathcal{S}_1)=0.813$ $CW_{rel}(\mathcal{S}_1)=0.806$ $GK(\mathcal{S}_1)=0.805$

Fig. 3. Illustrating the deficiency of the C measure on a system that clearly should not be rated as severely inconsistent

$\mathcal{S}_2$:
{1, 2, 3, 4, 5, 6, 7}
{1, 2, 3, 4, 5, 6}
{1, 2, 3, 4, 5}
{1, 2, 3, 4}
{1, 2, 3}
{1, 2}
{1}

$GK(\mathcal{S}_2)=0.5$
$C(\mathcal{S}_2)=C(\mathcal{S}_3)=0.5$
$CW(\mathcal{S}_2)=CW(\mathcal{S}_3)=0.\overline{66}$
$CW_{rel}(\mathcal{S}_2)= CW_{rel}(\mathcal{S}_3)=0.\overline{33}$

$\mathcal{S}_3$:
{1, 2, 3, 4}
{1, 2, 3, 4}
{1, 2, 3, 4}
{1, 2, 3, 4}
{1, 2, 3, 5}
{1, 2, 6, 5}
{1, 7, 6, 5} $GK(\mathcal{S}_3)=0.564$

Fig. 4. Comparing the behavior of the considered measures on synthetic example

evaluate the overall occurrence of features in the system as a whole. The difference between C, CW, CW_{rel} and GK is illustrated in Fig. 4. The measures C, CW and GK tend to yield the higher values the closer the sizes of subsets in system are to the size of Y. This property seriously hinders the usability of these measures for comparison of various FS methods, should the compared methods yield differently sized subsets. The measure CW_{rel} overcomes this deficiency, yielding values unaffected by feature subset size issues.

Note that the measure CW_{rel} cannot be interpreted simply as a measure evaluating how much the selected subsets overlap. Instead, it shows the relative amount of randomness inherent in the concrete FS process. For a given total number of features in evaluated system and given size of Y it yields values on a scale $[0, 1]$ where 0 represents the outcome of completely random occurrence of features in the selected subsets and 1 indicates the most stable FS outcome possible. Note that even completely random FS process will lead to positive CW and GK values in most cases. The CW_{rel} helps to indicate cases where seemingly consistent results (that may be evaluated as highly consistent by CW or GK) are not the result of consistent FS performance, but follow from the inherent characteristics of certain systems of subsets.

5 Experimental Evaluation

In order to illustrate the considered measures we have conducted a series of FS experiments on real data from the UCI Repository [9]: *wine* data (13-dim., 3 classes of 59, 71, 48 samples), *wdbc* data (30-dim., 2 classes of 357 and 212 samples) and *cloud* data (10-dim., 2 classes of 1024 and 1024 samples).

We focused on comparing the stability of two standard FS methods: SFS and SFFS in the Wrapper [10] setting that allows the methods to be used as both the optimizers of feature subset and of subset size (subsets of all sizes are selected, then the one with the highest classification accuracy is chosen; in case of ties the one with lower cardinality is preferred).

Table 1. Consistency of FS Wrappers evaluated on Wine data, 13-dim, 3-class

Wrap.	FS Meth.	Classif. rate Mean	Classif. rate *St.Dv.*	Subset size Mean	Subset size *St.Dv.*	C	CW	CW rel	GK	FS time h:m:s	*CW min*	*CW max*
Gauss.	SFS	0.590	*0.023*	3.73	*1.70*	0.310	0.519	0.353	0.379	00:02:20	*0.286*	*0.947*
	SFFS	0.625	*0.023*	3.58	*1.23*	0.298	0.514	0.365	0.389	00:14:30	*0.275*	*0.932*
3-NN	SFS	0.982	*0.004*	7.12	*1.47*	0.547	0.752	0.467	0.615	00:07:45	*0.547*	*0.985*
	SFFS	0.987	*0.003*	6.91	*1.60*	0.531	0.763	0.508	0.637	00:33:20	*0.531*	*0.988*
SVM	SFS	0.980	*0.005*	9.09	*1.92*	0.699	0.758	0.203	0.611	00:16:49	*0.699*	*0.991*
	SFFS	0.989	*0.003*	8.46	*1.36*	0.650	0.816	0.516	0.697	01:08:41	*0.650*	*0.971*

Table 2. Consistency of FS Wrappers evaluated on WDBC data, 30-dim, 2-class

Wrap.	FS Meth.	Classif. rate Mean	Classif. rate *St.Dv.*	Subset size Mean	Subset size *St.Dv.*	C	CW	CW rel	GK	FS time h:m:s	*CW min*	*CW max*
Gauss.	SFS	0.963	*0.003*	11.95	*5.30*	0.398	0.506	0.181	0.332	01:02:04	*0.397*	*0.996*
	SFFS	0.969	*0.003*	12.17	*4.66*	0.405	0.556	0.259	0.387	09:13:03	*0.405*	*0.988*
3-NN	SFS	0.976	*0.002*	15.45	*5.74*	0.514	0.584	0.148	0.401	07:27:39	*0.514*	*0.984*
	SFFS	0.979	*0.002*	17.96	*5.67*	0.598	0.658	0.149	0.481	33:53:55	*0.598*	*0.998*
SVM	SFS	0.982	*0.002*	9.32	*4.12*	0.310	0.433	0.185	0.283	07:13:02	*0.310*	*0.977*
	SFFS	0.983	*0.002*	10.82	*4.58*	0.360	0.472	0.179	0.310	30:28:02	*0.360*	*0.987*

Table 3. Consistency of FS Wrappers evaluated on Cloud data, 10-dim, 2-class

Wrap.	FS Meth.	Classif. rate Mean	Classif. rate *St.Dv.*	Subset size Mean	Subset size *St.Dv.*	C	CW	CW rel	GK	FS time h:m:s	*CW min*	*CW max*
Gauss.	SFS	0.998	*4e-4*	4.80	*1.09*	0.480	0.794	0.644	0.671	00:03:54	*0.480*	*0.967*
	SFFS	0.999	*3e-4*	5.02	*0.87*	0.501	0.839	0.682	0.737	00:17:42	*0.501*	*0.997*
3-NN	SFS	1	*0*	1	*0*	1	1	1	1	05:25:24	*0.099*	*1.0*
	SFFS	1	*0*	1	*0*	1	1	1	1	11:04:39	*0.099*	*1.0*
SVM	SFS	1	*0*	1	*0*	1	1	1	1	02:41:40	*0.099*	*1.0*
	SFFS	1	*0*	1	*0*	1	1	1	1	04:13:19	*0.099*	*1.0*

We used the classification accuracy of three conceptually different classifiers as FS criteria: *Gaussian classifier*, *3-Nearest Neighbor* (majority voting) and *Support Vector Machine* (with Radial Basis Function kernel [11]).

In each setup FS was repeated 1000× on randomly sampled 80% of the data (class size ratios preserved). In each FS run the criterion was evaluated using 10-fold cross-validation, with 2/3 of available data randomly sampled for training and the remaining 1/3 used for testing.

5.1 Results

The results of our experiments are collected in Tables 1 to 3. Note that *CW* and *GK* exhibit similar behavior (except the slightly higher *CW* values' level), while

C is to be considered less reliable (cf. Fig. 3). The measure CW_{rel}, however, reveals different properties of FS process – note in Table 2 that with the *wdbc* data both FS methods yield too random results (note the low CW_{rel} values and also the high deviations in subset size). This may indicate some pitfall in the FS process – either there are no clearly preferable features in the set, or the methods overfit, etc. Another notable result is the consistent tendency of SFFS to yield higher CW, CW_{rel} and GK values than SFS in most of the experiments.

6 Conclusions

We propose several new consistency measures especially suitable for evaluating the stability of FS methods that yield subsets of varying sizes (although they can be used in fixed subset size problems). The key new measures CW and CW_{rel} are compared to the generalized Kalousis measure GK. Both CW and CW_{rel} are computationally less demanding than GK. Each of the considered measures evaluates the FS process from a different perspective – consequently they complement each other well. GK evaluates pairwise similarities between selected feature subsets. CW and CW_{rel} evaluate the overall occurrence of individual features in selected feature subsets. Unlike CW, the CW_{rel} shows the relative amount of "randomness" of the FS process, independently on subset size. The measures have been used to compare two standard FS methods on a set of examples. The SFFS has been shown as more stable than the SFS.

Acknowledgements. The work has been supported by GAAV CR project 1ET400750407 and AV0Z1075050506, GAČR 102/08/0593, 102/07/1594 and CR MŠMT grants 2C06019 ZIMOLEZ and 1M0572 DAR.

References

1. Dunne, K., Cunningham, P., Azuaje, F.: Solutions to instability problems with sequential wrapper-based approaches to feature selection. Technical Report TCD-CD-2002-28, Dept. of Computer Science, Trinity College, Dublin, Ireland (2002)
2. Kalousis, A., Prados, J., Hilario, M.: Stability of feature selection algorithms: a study on high-dimensional spaces. Knowledge and Inf. Syst. 12(1), 95–116 (2007)
3. Kuncheva, L.I.: A stability index for feature selection. In: Proc. 25th IASTED Int. Multi-Conf. Artificial Intelligence and Applications, pp. 421–427 (2007)
4. Křížek, P., Kittler, J., Hlaváč, V.: Improving stability of feature selection methods. In: Kropatsch, W.G., Kampel, M., Hanbury, A. (eds.) CAIP 2007. LNCS, vol. 4673, pp. 929–936. Springer, Heidelberg (2007)
5. Saeys, Y., Abeel, T., de Peer, Y.V.: Towards robust feature selection techniques. In: Proceedings of Benelearn, pp. 45–46 (2008)
6. Raudys, Š.: Feature over-selection. In: Yeung, D.-Y., Kwok, J.T., Fred, A., Roli, F., de Ridder, D. (eds.) SSPR 2006 and SPR 2006. LNCS, vol. 4109, pp. 622–631. Springer, Heidelberg (2006)
7. Devijver, P.A., Kittler, J.: Pattern Recognition: A Statistical Approach. Prentice-Hall International, London (1982)

8. Pudil, P., Novovičová, J., Kittler, J.: Floating search methods in feature selection. Pattern Recognition Letters 15, 1119–1125 (1994)
9. Asuncion, A., Newman, D.: UCI machine learning repository (2007), http://www.ics.uci.edu/~mlearn/MLRepository.html
10. Kohavi, R., John, G.: Wrappers for feature subset selection. Artificial Intelligence 97, 273–324 (1997)
11. Chang, C.C., Lin, C.J.: LIBSVM: a library for support vector machines (2001), http://www.csie.ntu.edu.tw/~cjlin/libsvm

A Learning Scheme for Recognizing Sub-classes from Model Trained on Aggregate Classes

Ranga Raju Vatsavai[1], Shashi Shekhar[2], and Budhendra Bhaduri[1]

[1] Computational Sciences and Engineering Division,
Oak Ridge National Laboratory, Oak Ridge, TN 37831, USA
vatsavairr@ornl.gov, bhaduribl@ornl.gov
[2] Dept. of Computer Science, University of Minnesota
4-192 EE/CS Bldg., 200 Union Street SE, MN 55455
shekhar@cs.umn.edu

Abstract. In many practical situations it is not feasible to collect labeled samples for all available classes in a domain. Especially in supervised classification of remotely sensed images it is impossible to collect ground truth information over large geographic regions for all thematic classes. As a result often analysts collect labels for aggregate classes. In this paper we present a novel learning scheme that automatically learns sub-classes from the user given aggregate classes. We model each aggregate class as finite Gaussian mixture instead of classical assumption of unimodal Gaussian per class. The number of components in each finite Gaussian mixture are automatically estimated. Experimental results on real remotely sensed image classification showed not only improved accuracy in aggregate class classification but the proposed method also recognized sub-classes.

Keywords: Semi-supervised learning, EM, GMM, Remote Sensing.

1 Introduction

Remote sensing, which provides inexpensive, synoptic-scale data with multi-temporal coverage, has proven to be very useful in land cover mapping, environmental monitoring, forest and crop inventory, urban studies, natural and man made object recognition, etc. Thematic information extracted from remote sensing imagery is also useful in a variety spatio-temporal applications. For example, land management organizations and the public have a need for more current regional land cover information to manage resources and monitor land use changes. Likewise, intelligence agencies, such as, National Geospatial Intelligence Agency (NGA), and Department of Homeland Security (DHS), utilizes pattern recognition and data mining techniques to classify both natural and man made objects from large volumes of high resolution imagery.

Image classification, i.e., assigning class labels to pixels, using some discriminant function, is one of the fundamental analysis technique used in remote sensing to generate thematic information. Image classification can be formally

N. da Vitora Lobo et al. (Eds.): SSPR&SPR 2008, LNCS 5342, pp. 967–976, 2008.

defined as finding a function $g(x)$ which maps the input patterns x onto output classes y_i (some times y_i's are also denoted as ω_i or c_i). The main objective is to assign a label (e.g. Water, Forest, Urban) to each pixel in the image to be classified, given corresponding feature vector x_j's in the input image. Depending on the type of supervised learning method used, the objective of a supervised learning could be finding a function $g(x)$ (also called a discriminant function), that divides the input $d-$dimensional feature space into several regions, where each region corresponds to a thematic class y. One such simple function is given by: $x \in y_i$ if $p(y_i|x) > p(y_j|x)$ $\forall j \neq i$. That is, the feature vector x belongs to class y_i if $p(y_i|x)$ is the largest. Even though it sounds simple, this assignment problem is very difficult. There is no single algorithm which will correctly classify any given image. Multi-spectral image classification is still an open problem.

Collecting ground truth (labels) data over large geographic regions is costly, time consuming, and poses several other practical problems. Given these practical limitations, often the analyst groups the relevant classes and collects labels only for those aggregate (grouped) classes. Typical examples are forest, agriculture, urban, and other land-use classes. Usually, the analyst given forest class may contains samples from all types of forests, such as hardwoods, conifer, etc., each of which can be described by a unique statistical distribution. These distributions are clearly identifiable in many image classification problems (depending on the spectral resolution), though in some cases these finer classes may be highly overlapping (in low spectral resolution images).

Aggregating several classes into a single class such as forest poses two problems, first it violates the common assumption that each class is unimodal, secondly the estimated parameters could be wrong. Since we are combining distinctly identifiable classes, the estimated covariance matrix is large, as it accounts for both inter-class covariance and intra-class covariances (of component classes). It is very important to estimate covariance matrix accurately because even for a fixed means, it can be shown that increase in variance of any one class (keeping means fixed), leads to the increase in probability of error. This observation motivated us to develop a new classification scheme which relaxes the common assumption that the class has to be a unimodal distribution. Instead, we assume that each class is a finite mixture model.

Related Work and Our Contributions: Supervised methods are extensively used in remote sensing imagery classification [9,4]. Finite mixture model parameter estimation [5] for unlabeled samples is also investigated extensively under various disguises. For example, semi-supervised learning approaches are very close to the approach we presented in Section 3. Well-known studies in this area include, but not limited to [7,3,8,1,11]. Model selection approaches have also been extensively studied [12,6] and used in finding the number of clusters [2].

In this work we clearly identified an important practical problem in supervised classification of remotely sensed imagery. We developed a novel learning algorithm which takes user defined aggregate classes and automatically discovers sub-classes within each aggregate class. Each aggregate class is modeled as Gaussian mixture and is split into sub-classes (components) using a Gaussian splitting criteria. The resulting classifier showed not only improvement in overall classification accuracy but also recognized finer classes which analyst is always interested in, but sufficient ground truth data cannot be collected for the finer classes.

The rest of this paper is organized as follows. In Section 2, we provide a basic statistical framework for Bayesian classification and maximum likelihood based parameter estimation. We present our proposed learning scheme and parameter estimation in Section 3. Experimental results are given in Section 4. Related work is presented in Section 5, followed by conclusions and future directions in Section 6.

2 Statistical Classification Framework

In the classification of a remote sensing images, our objective is to assign a class label (y) to each pixel (x) based on certain decision criterion. Maximum likelihood classification (MLC) and maximum a posteriori (MAP) classification are two of the most widely used classifiers in remote sensing. Assuming the training samples were generated by a multivariate normal or Gaussian density, we can write the decision rule for maximum a posterior (MAP) classifier as following:

$$g_i(x) = \ln P(y_i) - \frac{1}{2}\ln|\Sigma_i| - \frac{-1}{2}(x-\mu_i)^t|\Sigma_i|^{-1}(x-\mu_i) \quad (1)$$

If we don't have *a priori* knowledge about the classes then we can drop $P(y_i)$ term in the above equation and the resulting decision rule is known as maximum likelihood classification (MLC). The covariance matrix Σ plays a key role in discriminant analysis. Covariance accounts for the shape (size and orientation) of classes in the feature space. The effectiveness of ML/MAP classification depends on the quality of the estimated parameter vector Θ (i.e., mean vector μ and the covariance matrix Σ for each class) from the training samples. Using a well-known parameter estimation technique, maximum likelihood estimation (MLE), we can obtain the parameters μ and Σ as following:

$$\hat{\mu} = \frac{1}{n}\sum_{k=1}^{n} x_k; \quad \hat{\Sigma} = \frac{1}{n}\sum_{k=1}^{n}(x_k-\hat{\mu})(x_k-\hat{\mu})^t. \quad (2)$$

2.1 Limitations of MLC and MAP

One of the classical assumptions in supervised (statistical) classification is that the classes are unimodal. We now test the impact of violation of this constraint through a simulated example. We generated bivariate Normal samples (150 per class) for three distinctly identifiable classes using the parameters given in the Table 1 and 2. Maximum likelihood estimates from the simulated samples were also summarized along with the original parameters.

Table 1. Simulation Parameters (Means)

Classes	Given X1	Given X2	No of Samples	Estimated (MLE) X1	Estimated (MLE) X2
C_1	55.00	25.00	150.00	54.93	24.50
C_2	80.00	50.00	150.00	79.58	50.23
C_3	50.00	40.00	150.00	49.74	39.55

Table 2. Simulation Parameters (Covariance)

Features	C_1 X1	C_1 X2	C_2 X1	C_2 X2	C_3 X1	C_3 X2
X_1	30.00	25.00	60.00	40.00	60.00	50.00
X_2	25.00	40.00	40.00	90.00	50.00	70.00
Estimated Parameters (MLE)						
X_1	27.03	19.10	57.66	43.43	56.81	40.60
X_2	19.10	31.58	43.43	100.03	40.60	55.84

Table 3. ML Estimates of Aggregated Class

C_{23}		X1	X2
Mean		64.66	44.89
Covariance	X1	280.49	121.80
	X2	121.80	106.26

Let us now assume that classes 2 and 3 are sub-classes of an aggregate class C_{23}, i.e., analyst gave a single label to all the samples generated from the classes C_2 and C_3. The new estimates of the aggregate class C_3 are given in the Table 3. For understanding the distribution (and interaction) of original classes (C_1, C_2, C_3) and aggregate classes (C_1, C_{23}), we have generated the bivariate density plots which are shown in Figure 1.

The overlap between classes (C_1, C_2, C_3) is almost negligible (see Figure 1(a)). However, aggregation of classes C_2, C_3 into C_{23} has greatly increased its overlap with class C_1 (see Figure 1(b)). We have seen previously that this overlap directly accounts for the probability of error, p_E. One can expect to improve the classification accuracy, if somehow the original classes (C_2, C_3) which gave raise to aggregate class C_{23} can be automatically discovered. Also, there is a great need for finer class discovery from remotely sensed images, and precisely this is what our proposed algorithm tries to accomplish.

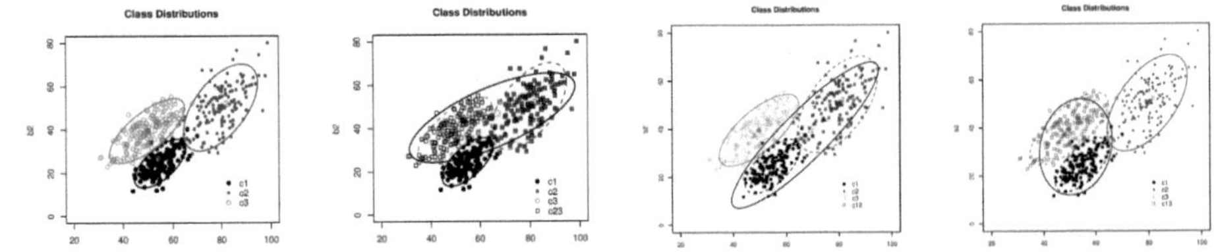

(a) Original Distribution (b) C_2, C_3 aggregated into C_{23} (c) C_1, C_2 aggregated into C_{12} (d) C_1, C_3 aggregated into C_{13}

Fig. 1. Interaction Between Finer and Aggregate Classes

3 Learning To Discover Sub-classes

Basic idea behind the proposed algorithm is very simple. Instead of assuming that each class is a unimodal multivariate Gaussian, we assume that the samples from each class are generated by finite Gaussian mixture. There are two sub-problems associated with this assumption: First, we don't have labels for any of the component (sub-class) so that we can employ regular MLE technique to estimate the parameters of each component; second, we don't know how many components (sub-classes) are there in this finite mixture model. We address these two problems in the following two sub-sections.

3.1 Estimating Finite Mixture Parameters

First problem is solved by assuming that the training samples in each (aggregate) class were generated a mixture density and then estimate parameters of mixture density for arbitrary number of components using the expectation maximization (EM) algorithm. The EM algorithm consists of two steps, called the E-step and and M-step as given below.

E-Step: For multivariate normal distribution, the expectation $E[.]$, which is denoted by p_{ij}, is the probability that Gaussian mixture j generated the data point i, and is given by:

$$p_{ij} = \frac{\left|\hat{\Sigma}_j\right|^{-1/2} e^{\left\{-\frac{1}{2}(x_i-\hat{\mu}_j)^t \hat{\Sigma}_j^{-1}(x_i-\hat{\mu}_j)\right\}}}{\sum_{l=1}^{M} \left|\hat{\Sigma}_l\right|^{-1/2} e^{\left\{-\frac{1}{2}(x_i-\hat{\mu}_l)^t \hat{\Sigma}_l^{-1}(x_i-\hat{\mu}_l)\right\}}} \tag{3}$$

M-Step: The new estimates (at the kth iteration) of the model parameters in terms of the old parameters are computed using the following update equations:

$$\hat{\alpha}_j^k = \frac{1}{n}\sum_{i=1}^{n} p_{ij} \tag{4}$$

$$\hat{\mu}_j^k = \frac{\sum_{i=1}^{n} x_i p_{ij}}{\sum_{i=1}^{n} p_{ij}} \tag{5}$$

$$\hat{\Sigma}_j^k = \frac{\sum_{i=1}^{n} p_{ij}(x_i - \hat{\mu}_j^k)(x_i - \hat{\mu}_j^k)^t}{\sum_{i=1}^{n} p_{ij}} \tag{6}$$

The EM algorithm iterates over these two steps until convergence is reached. We can now put together these individual pieces into the following algorithm (Table 4) which computes the parameters for each component in the finite Gaussian mixture model that generated our training data D (without any labels).

3.2 Estimating the Number of Component of a Finite Gaussian Mixture

We now address our second problem, i.e., we don't know how many components (sub-classes) are there in each (aggregate) class. As with estimating model parameter for finite Gaussian mixture model, we assume that the training dataset D is generated by a finite Gaussian mixture model, but we don't know either the number of components or the labels for any of the sub-component. In previous

Table 4. Algorithm for Computing Parameter of Finite Gaussian Mixture Model Over Unlabeled Training Data

Inputs: D_j, training dataset (no labels for sub-classes) for any aggregate class y_j; M, the number of sub-classes in the corresponding aggregate class.
Initial Estimates: Do clustering by K-Means, and estimate initial parameter using MLE, to find $\hat{\theta}$ (see Equations 2)
Loop: While the complete data *log-likelihood* improves:
E-step: Use current classifier to estimate the class membership of each unlabeled sample, i.e., the probability that each Gaussian mixture component generated the given sample point, p_{ij} (see Equation 3).
M-step: Re-estimate the parameter, $\hat{\theta}$, given the estimated Gaussian mixture component membership of each unlabeled sample (see Equations 4, 5, 6)
Output: Parameter vector Θ.

section, we devised an algorithm to find parameters by assuming a M-component finite Gaussian mixture model. In general, we can estimate parameters for any arbitrary M-component model, as long as there are sufficient number of samples available for each component and the covariance matrix does not become singular. Then the question remains, which M-component model is better? This question is addressed in the area of model selection, where the objective is to chose a model that maximizes a cost function. There are several cost functions available in the literature, most commonly used measures are Akaike's information criterion (AIC), Bayesian information criteria (BIC), and minimum description length (MDL). The common criteria behind these models is to penalize the models with additional parameters, so BIC and AIC based model selection criteria follows the principal of parsimony. In this study we considered BIC as a model selection criteria, which is also takes the same form as MDL. We also chose BIC, as it is defined in terms of maximized log-likelihood which any way we are computing in our parameter estimation procedure defined in the previous section. BIC can be defined as

$$BIC = MDL = -2 \log L(\Theta) + m \log(N) \quad (7)$$

where N is the number of samples and m is the number of parameters. We now describe our BIC based model selection criteria to determine the number components in each aggregate class. First, we take the aggregate class and split it into two Gaussians at a time using the Gaussian splitting criteria specified in [10]. Then the parameters of this new mixture model are estimated using the algorithm 4. This process is recursively applied for a fixed number times or BIC is minimized. We then repeat the algorithm for each (aggregate) class in the original classification problem. At the end of each iteration we have parameters for each sub-class within an aggregate class. We can now apply MLC/MAP in two ways. First, we modified MLC/MAP to output both aggregate classes (original analyst given classes) and as well sub-classes which were discovered automatically using the procedure just described. We can combine the finer classes (predicted) into the corresponding aggregate class in order to find the aggregate class classification accuracy.

4 Experimental Results

We have conducted several experiments using simulated and as well as the real dataset.

Dataset 1: The objective of first experiment on simulated data is to see performance of proposed method in aggregate class classification and as well as finer class classification. We used the parameters listed in Table 1 to generate two distinct datasets. First dataset consisted of 150 samples at 50 samples per class, and second dataset consisted of 450 samples at 150 samples per class. We used first dataset for training and the second dataset for testing. We conducted following three experiments.

Experiment 1: MAP classification was carried out using all three classes, whose distribution is shown in Figure 2(a). Test accuracy of MAP is shown in Table 5.

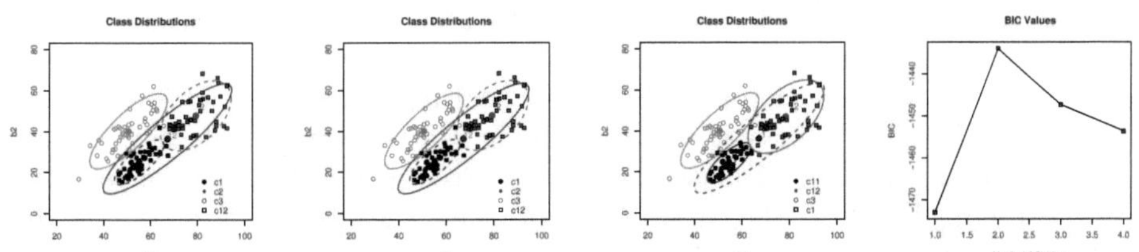

(a) Original Distribution (b) C_1, C_2 aggregated into C_{12} (c) Sub-classes C_{11}, C_{12} discovered from C_{12} (d) BIC Values For Different Number of Clusters

Fig. 2. Interaction Between Finer, Aggregate, and Newly Discovered Classes

Table 5. Accuracy (All Classes)

G.Truth	C_1	C_2	C_3	P.Acc
C_1	141.00	4.00	5.00	94.00
C_2	1.00	147.00	2.00	98.00
C_3	1.00	1.00	148.00	98.67
U.Acc	98.60	96.71	95.48	(OA) 96.89

Table 6. Accuracy ($C_1, C_2 \rightarrow C_1$)

G.Truth	C_1	C_3	P.Acc
C_1	281.00	19.00	93.67
C_3	2.00	148.00	98.67
U.Acc	99.29	88.62	(OA) 95.33

Table 7. Accuracy ($C_1 \rightarrow C_1, C_2$)

G.Truth	C_1	C_2	C_3	P.Acc
C_1	132.00	11.00	7.00	88.00
C_2	1.00	147.00	2.00	98.00
C_3	1.00	1.00	148.00	98.67
U.Acc	98.51	92.45	94.27	(OA) 94.89

Table 8. Accuracy ($C_1 \rightarrow C_1, C_2 \rightarrow C_1$)

G.Truth	C_1	C_3	P.Acc
C_1	291.00	9.00	97.00
C_3	2.00	148.00	98.67
U.Acc	99.32	94.27	(OA) 97.56

Table 9. Accuracy (Aggregate Classes)

G. Truth	1	2	3	4	P. Acc.
Forest(1)	1475.00	9.00	28.00	0.00	97.55
Ag.(2)	90.00	142.00	2.00	0.00	60.68
Urban(3)	0.00	0.00	45.00	0.00	100.00
Wetlands(4)	18.00	0.00	2.00	34.00	62.96
Users Acc.	93.18	94.04	58.44	100.00	91.92

Table 10. Accuracy (Sub-Classes)

GT	1	2	3	4	P. Acc.
(1)	1448.00	13.00	51.00	0.00	95.77
(2)	14.00	214.00	6.00	0.00	91.45
(3)	0.00	0.00	45.00	0.00	100.00
(4)	3.00	0.00	13.00	38.00	70.37
UA.	98.84	94.27	39.13	100.00	94.58

Experiment 2: Classes C_1, C_2 were combined into one aggregate class and class C_3 remained untouched. Resulting new class distributions were shown in Figure 2(b). Test accuracy of MAP using aggregated class is shown in Table 6.

Experiment 3: Our new algorithm was applied on the dataset generated in Experiment 2. We tested both aggregate classification performance and as well as the finer class performance. Newly discovered classes were shown in Figure 2(c). Test accuracy of MAP using newly discovered classes is shown in Table 7, and the corresponding aggregated class accuracy is shown in Table 8.

Dataset 2: In this experiment we used a spring Landsat 7 scene, taken on May 31, 2000 over the Cloquet town located in Carlton County of Minnesota state. The training dataset consisted of sixty plots and four aggregate classes, namely, Forest(1), Agriculture(2), Urban(3), and Wetlands(4). We have an independent test dataset consisting of 205 plots. Feature vectors were extracted from the Landsat image (6-dimensional) by placing a 3×3 window at each of these plots. Maximum likelihood classification is carried out using the conventional approach and as well as the proposed approach. The results were summarized in the following contingency tables (or error matrices).

Table 9 gives MLC accuracy using the standard approach and Table 10 provides the classification accuracy obtained by our proposed method. We regrouped the sub-classes into the corresponding aggregate classes for testing the accuracy using same test dataset (consisting of four aggregate classes).

4.1 Analysis

Accuracy assessment on simulated data shows interesting results. Bivariate density plot shown in Figure 2(a) and as well as hight test accuracy (Table 5) shows that the three classes were clearly separable. Aggregation of classes $C_1, C_2 \rightarrow C_1$ has increased the overlap between the aggregate class C_1 and C_3 (see Figure 2(b)), and this overlap has resulted in more classification error (Table 6) as compared to finer class classification error. Our proposed algorithm on this aggregate data has discovered two sub-classes in the aggregate class C_1 (see Figure 2(c)) and the corresponding BIC value (for number of clusters = 2) is maximum (Figure 2(d)). The test accuracy of MAP classifier trained on newly discovered classes is given in the Table 7 and the corresponding aggregated class accuracy is shown in Table 8. First, comparison of this accuracy table with original classification accuracy (Table 5) reveals that the sub-classes discovered closely corresponds to the original classes. Second, the aggregate (i.e., predicted sub-classes were merged) classification accuracy with our new scheme is higher than the MAP on original aggregate class classification (compare with Table 6). This study revels that our new algorithm not only discovered sub-classes that are close to the original fine classes (without providing any labeled training data) but also improved classification accuracy of original aggregate classes.

Let us now compare the classification error matrices (Table 9 and Table 10) obtained on the real dataset. From these two tables, we can see that our new procedure improved overall classification accuracy (OA) for the same training dataset without any additional (sub-class related training) information. In addition the new procedure automatically discovered sub-classes within each aggregate class. In this (aggregate class) training dataset, our new procedure discovered four additional component in the forest class, two additional components in the agriculture class, and two additional components in the urban class. Our preliminary investigation into the newly discovered sub-classes reveled very interesting information. The four classes discovered roughly corresponds to the following information classes: upland hardwood, lowland hardwood, upland conifer, and lowland conifer forests. There is a great demand for such additional information in many real world applications.

5 Conclusions

We identified an important practical classification problem that requires knowledge discovery approaches for automatically discovering the sub-classes from the aggregate classes. We developed a new classification scheme that automatically discovers the sub-classes from the user given aggregate classes, without any additional labeled training data for sub-classes. In addition, the procedure showed improvement in the classification of aggregate classes as well. This improvement can be attributed to the fact that the aggregate classes tend to increase the overlap between class distributions. Our preliminary investigation also showed a strong correspondence between sub-classes and true information classes. Further research is needed to automatically (or with minimal efforts) label these sub-classes. We are investigating the semantic relationships between various information classes that are common in this domain which might help to automatically label these sub-classes.

Acknowledgments

We would like to thank our former collaborators Thomas E. Burk, Jamie Smedsmo, Ryan Kirk and Tim Mack at the University of Minnesota for useful comments and inputs into this research. The comments of Eddie Bright, Phil Coleman, and Veeraraghavan Vijayraj, have greatly improved the technical accuracy and readability of this paper. Prepared by Oak Ridge National Laboratory, P.O. Box 2008, Oak Ridge, Tennessee 37831-6285, managed by UT-Battelle, LLC for the U. S. Department of Energy under contract no. DEAC05-00OR22725.

References

1. Cozman, F.G., Cohen, I., Cirelo, M.C.: Semi-supervised learning of mixture models. In: Twentieth International Conference on Machine Learning, ICML (2003)
2. Figueiredo, M.A.T., Jain, A.K.: Unsupervised selection and estimation of finite mixture models. In: Pattern Recognition, 2000. Proceedings. 15th International Conference on 2000, vol. 2, pp. 87–90 (2000)
3. Goldman, S., Zhou, Y.: Enhancing supervised learning with unlabeled data. In: Proc. 17th International Conf. on Machine Learning, pp. 327–334. Morgan Kaufmann, San Francisco (2000)
4. Jensen, J.R.: Introductory Digital Image Processing, A Remote Sensing Perspective. Prentice Hall, Upper Saddle River (1996)
5. Mclachlan: Mixture Models: Inference and Applications to Clustering. CRC, New York (1987)
6. Miloslavsky, M., van der Laan, M.J.: Fitting of mixtures with unspecified number of components using cross validation distance estimate. Comput. Stat. Data Anal. 41(3-4), 413–428 (2003)
7. Mitchell, T.: The role of unlabeled data in supervised learning. In: Proceedings of the Sixth International Colloquium on Cognitive Science, San Sebastian, Spain (1999)
8. Nigam, K., McCallum, A.K., Thrun, S., Mitchell, T.M.: Text classification from labeled and unlabeled documents using EM. Machine Learning 39(2/3), 103–134 (2000)
9. Richards, J.A., Jia, X.: Remote Sensing Digital Image Analysis. Springer, New York (1999)
10. Sankar, A.: Experiments with a gaussian merging-splitting algorithm for hmm training for speech recognition. In: Proceedings of the Broadcast News Transcription and Understanding Workshop, pp. 99–104 (1998)
11. Shahshahani, B., Landgrebe, D.: The effect of unlabeled samples in reducing the small sample size problem and mitigating the hughes phenomenon. IEEE Trans. on Geoscience and Remote Sensing 32(5) (1994)
12. Xuelei, Lei, X.U.: Investigation on several model selection criteria for determining the number of clusters. Neural Information Processing - Letters and Reviews 4(1), 139–148 (2004)

An Empirical Evaluation of Common Vector Based Classification Methods and Some Extensions*

Katerine Díaz-Chito, Francesc J. Ferri, and Wladimiro Díaz-Villanueva

Dept. Informàtica, Universitat de València. Spain
{Katerine.Diaz,Francesc.Ferri,Wladimiro.Diaz}@uv.es

Abstract. An empirical evaluation of linear and kernel common vector based approaches has been considered in this work. Both versions are extended by considering directions (attributes) that carry out very little information as if they were null. Experiments on different kinds of data confirm that using this as a regularization parameter leads to usually better (and never worse) results than the basic algorithms.

1 Introduction

Recent developments in pattern recognition methodology suggest that good trade-offs between intrinsic complexity of the problems and models used to represent them constitute one of the key factors that drive the good behavior of some approaches against others [5,7]. In particular, learning techniques that explicitly aim at reducing the complexity of the models either by reducing features, prototypes or both have experienced a deep development in the last decades [8,2]. Feature selection and extraction methods and dimensionality reduction in general are commonly used when there is enough a priori evidence that significant redundancy is present in the representation of data. Techniques that deal with high dimensional data by explicitly managing projections onto low dimensional subspaces are usually known as subspace methods [12]. In fact, many classical approaches to representation or classification such as principal component analysis (PCA) or Fisher or Linear Discriminant Analysis (LDA) can be properly included in this category.

The key idea of subspace methods consists of finding an appropriate subspace (and the corresponding projection operator) in which the original recognition problem becomes significantly easier or even trivial. Subspace methods in general have become an active field of research in recent times both because of the proliferation of practical problems exhibiting very high dimensional representations and also since the use of the kernel trick [15] has lead to a relatively easy way of extending the methods to the nonlinear case.

Subspace methods usually rely on the fact that the intrinsic dimensionality of the problem is much lower than the real dimension of the data. Moreover, usually

* This work has been partially funded by FEDER and Spanish MEC through projects DPI2006-15542-C04-04 and Consolider Ingenio 2010 CSD2007-00018.

N. da Vitora Lobo et al. (Eds.): SSPR&SPR 2008, LNCS 5342, pp. 977–985, 2008.

linear separability and dealing with relatively clean and well-behaved data (from a numerical stability point of view) needs to be assumed for its practical use. This is the case of appearance-based face recognition in which subspace methods have become increasingly popular [18].

Subspace methods based on the concept of common vectors have been recently proposed for classification problems in high dimensional spaces like speech and face recognition [6,4]. Common vector methods can be thought of as projection methods that make all samples of a particular class to collapse on a single point (the common vector). In this subspace, the classification of test samples can be based on their (usually Euclidean) distance to the common vector in the appropriate subspace. Recent works have proposed extensions of the basic methods aimed at making them applicable to broader domains [3,17].

This paper presents an empirical evaluation of common vector based methods and their extensions in a range of different situations ranging from face recognition to graph-based chemical compound analysis. Further extensions and alternatives are introduced and studied in this particular context.

2 The Discriminant Common Vector Method

Let us suppose we have given m samples in $\mathbb{R}^n$ (with $m < n$) where m_k of them correspond to the k-th class, $\{x_i^k\}_{i=1}^{m_k}$, $\sum_k m_k = m$ and $k = 1, \ldots, c$.

Let us define the mean vector for each class and the within and between class scatter matrices, S_W and S_B in the usual way [5]. We use $R(S)$ and $N(S)$ to refer to the complementary range and null spaces of a scatter matrix S. For example, $R(S_W)$ and $N(S_W)$ are the linear subspaces spanned by the eigenvectors of S_W corresponding to nonzero and zero eigenvalues, respectively.

It has been proved [4] that under certain assumptions all the samples from a particular class collapse into a unique vector when orthogonally projected onto $N(S_W)$. This consequently implies that the rank of the projected total scatter matrix, $S_T = S_W + S_B$ is $c-1$ (or less if two or more classes collapse together). If the c common vectors are distinct, a projection onto a $c-1$ dimensional linear subspace exists that maps all training data into these vectors. New data can be then classified by projecting it into the same subspace and measuring the (usually Euclidean) distance to the common vectors of each class. This is a particular case of a nearest mean (NM) classifier and we will use this terminology in the present work. This method is known as the Discriminative Common Vector (DCV) approach and has been proposed in the context of very high dimensional data as in appearance-based face recognition [4] being the limitation of having more dimensions than data the most serious one to make the method broadly applicable.

The DCV method can be extended to the nonlinear case by applying the kernel trick [14,3]. That is, by first projecting the data into a much higher, possibly infinite dimensional space implicitly using a given mapping, ϕ. First, the data is projected onto the range space $R(S_T^\phi) = R(S_W^\phi + S_B^\phi)$ where the superscript indicates the new representation space. As in this new space, scatter matrices can be analytically obtained in terms of kernel matrices (given by dot products among given data), then the final projection onto $N(S_W^\phi)$ can be obtained also in terms of kernel matrices. A further PCA step is applied to obtain a final projection onto a $c-1$ subspace as in the case of linear DCV. In the kernel DCV (kDCV) method, the distinctness of common vectors is guaranteed if the training kernel matrix is positive definite [3].

This nonlinear extension of the plain DCV has several practical advantages. First, more complex problems involving nonlinear dependences can be conveniently tackled. Also, the dependence of the cost on the number of dimensions is changed into dependence on the number of training samples. One practical drawback is that kernel parameters need to be tuned on the particular data. Also, care must be taken when applying kDCV because of overtraining that can appear more often than in the linear case.

3 Extended Discriminant Common Vectors

The Rough Common Vector method (rDCV) is a recent linear extension of the DCV method that has been proposed by relaxing the condition about the null space of S_W [17]. This consists basically in taking as zero the small enough eigenvalues from the eigendecomposition of S_W. This on one hand makes not possible the use of the differencing methods proposed in [4] but instead overcomes the limitation of $m < n$ and increases in some way the generalization ability of the method which is now controlled by the amount of extra dimensions which are included in the extended null space. The generalization ability of linear common vector based methods and how this behaves across different kinds of data has been studied only recently [1] and only from an empirical point of view.

In this paper, the same extension is included in the kDCV by setting a variance parameter, f, that controls the nonzero eigenvalues of S_W^ϕ that are taken as zero. In particular, if λ_i are the normalized and increasingly ordered eigenvalues, we choose as the extended null space the subspace spanned by the eigenvectors corresponding to eigenvalues $\lambda_1, \ldots, \lambda_s$ satisfying $\sum_{i=1}^{s} \lambda_i \leq f$. Note that this parameter is different from the one introduced in [17] but in this way the parameter has the meaning of suppressed variance and we have the original DCV and kDCV methods when $f = 0$ in both cases.

The effect of increasing f in both linear and kernel DCV methods has not been fully studied. The observable effect is that the classes do not collapse anymore into common vectors. In [17] the projected mean of each class is taken as common vector and the $(c-1)$-dimensional subspace spanned by them as the final projection where distances to common vectors (i.e. NM) is used to classify.

But in fact, as classes do not collapse, the rank of S_T or S_T^ϕ will be certainly greater than $c-1$ thus opening the possibility of considering final projected spaces of higher dimensions. Also and independently, the possibility of using all training data to classify instead of only common vectors can be considered by substituting the NM by the nearest(s) neighbor(s) (NN). These two possibilities are explored in this work using data from different domains.

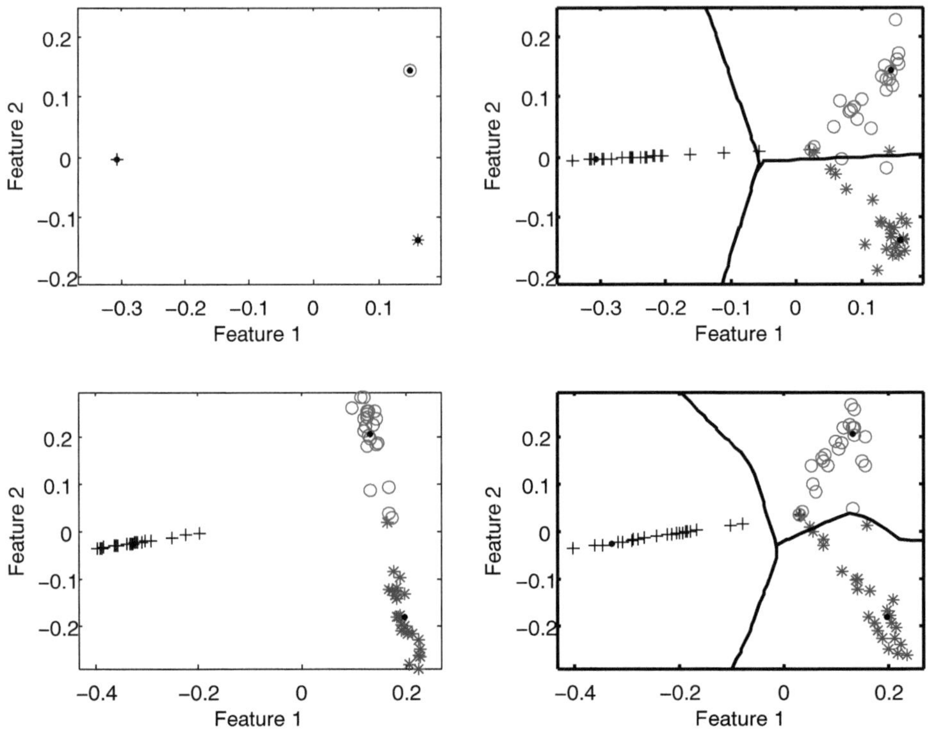

Fig. 1. Illustrative example of the Kernel DCV method and extension using the Iris database. The original method is used in the upper row and a variance parameter of 0.2 has been used for the lower row. Left column shows the projected training set and right column shows the projected test set along with the boundaries of the NM (upper) and NN (lower) classifiers.

Figure 1 contains an illustrative example of using kDCV on a random partition of the well-known Iris data into train and test sets. In the upper row, train (left) and test (right) sets are projected onto the final 2-dimensional space. Boundaries of the NM classifier are also shown along with the projected test set. In the lower row, kDCV with variance parameter of 0.2 is applied to the same data. NN boundaries are now shown instead. Although this is in fact a toy and very small example, it can be seen that the projected training set carries out significant information about how the (independent) test set will be projected onto the final space.

4 Experimental Results

A number of experiments have been conducted on several kinds of data of relatively high dimensionality in order to compare both linear and kernel versions of

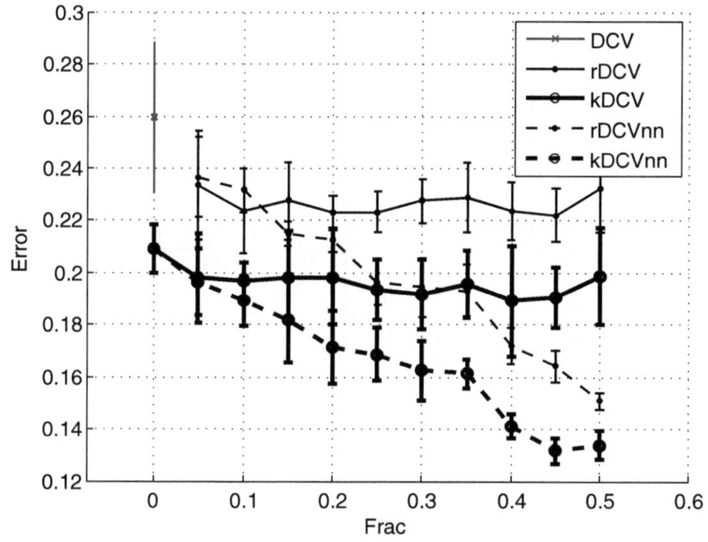

Fig. 2. Classification results using NM and NN classifiers on Coil dataset projected using DCV, rDCV, and kDCV using several values of the variance parameter f (Frac). The kernel parameter for this data is 1000.

the Discriminant Common Vector approach and their extensions using a variance parameter.

In all experiments using kernels, the Gaussian kernel with an appropriately tuned parameter has been considered. Values of the variance parameter ranging from zero (plain kDCV and DCV) to 0.5 (half the total variance of the range space of S_W^{ϕ} or S_W) have been considered.

The first dataset considered is made from the Coil-100 dataset [11] by taking a random subset of 30 images of each of 40 different objects. Original images have been downsampled to 40x40. The final dataset consists then of 1200 1600-dimensional objects from 40 different classes.

The results of using the considered methods on this data are shown in Figure 2. As can be seen, kDCV methods significantly improve the plain DCV results. Also, the consideration of the variance parameter improves the results with regard to the plain methods. But this improvement is really significant when NN classifier is used in the projected space instead of the NM classifier.

In this experiment 1/3 of each class is used for training and the remaining ones for test. Results are the average of 3 different and independent runs.

Given the nature of this dataset in which some of the objects are very different, the value of the variance parameter can be increased without producing high degrees of overlapping in the projected space.

The second dataset corresponds to a particular problem in the domain of classification of chemical compounds from their topological and structural description [10]. In particular, a dataset of 434 samples consisting of 62 molecular

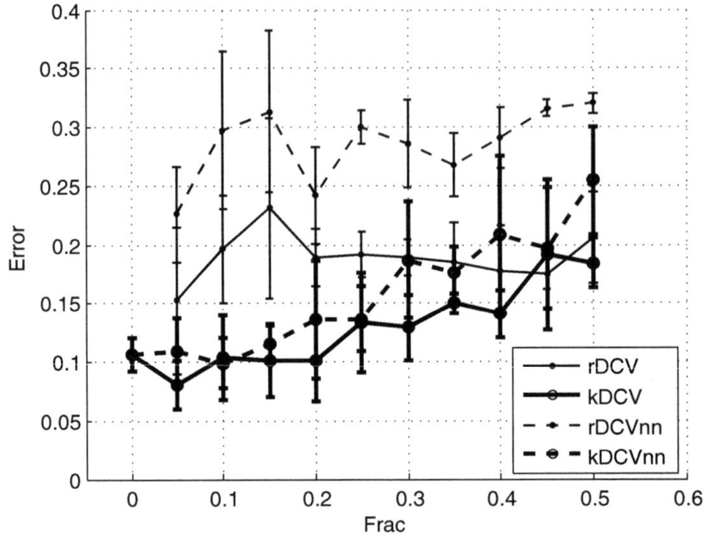

Fig. 3. Classification results using NM and NN classifiers on Antibacterial drug dataset projected using rDCV, and kDCV using several values of the variance parameter f. The kernel parameter for this data is 10.

descriptors computed from the graph corresponding to its chemical formula is considered. Roughly half of these compounds (218) are known to exhibit antibacterial activity. So the dataset consists of 434 62-dimensional objects from 2 classes. Note that in this case the plain DCV method cannot be used.

Experiments use 2/3 of the available data for training and the remaining 1/3 for testing in a classic 3-fold cross validation scheme. The final results are the corresponding average that takes into account all data.

The results using this data are shown in Figure 3. In this case, no improvement is observed when the variance parameter is increased. Also, using NN instead of NM is always equal or worse. At least, it can be seen that kDCV with variance values from 0 to 0.2 behaves equally well and constitutes the best option for this data.

The last experiment involves data from frontal face images taken from the well-know datasets AR [9], ORL [13] and YALE [16]. All images have been manually aligned and conveniently cropped and downsampled to 40x40. Moreover, to study the behavior of the methods considered when data is degraded with noise, corrupted versions of the faces using variable amounts of Gaussian additive noise have been added. A total of 95 different classes is considered and there are 50 different images of each person. Training sets of 10 random images per person have been considered and results on the remaining 40 are averaged over 5 different runs.

Figure 4 shows the results obtained using both linear and nonlinear methods considered. In this case, plain DCV gives a significantly worse result. Both linear and kernel versions using NM exhibit a similar behavior with the variance

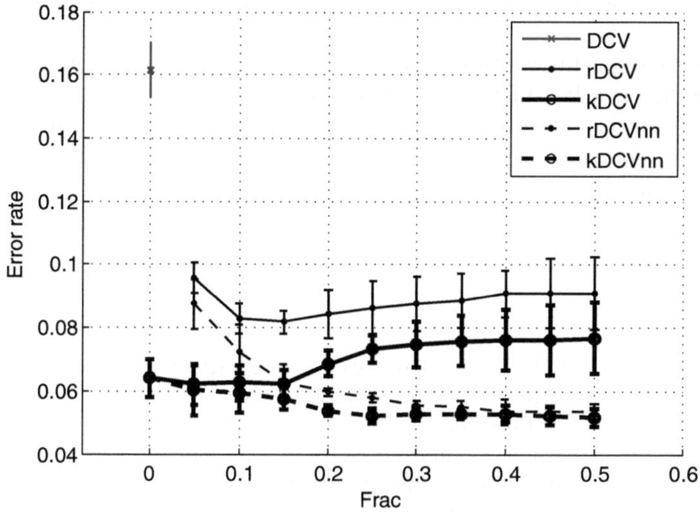

Fig. 4. Classification results using NM and NN classifiers on AR dataset (with Gaussian noise added) projected using rDCV, and kDCV using several values of the variance parameter f. The kernel parameter for this data is 100.

parameter. First a slight improvement or stability and then a slow decrease of performance from values of about 0.2. On the other hand, when using NN both linear and kernel methods are able to significantly improve the results obtained for relatively high values of the variance parameter.

5 Concluding Remarks and Further Work

An empirical evaluation of linear and nonlinear common vector based classification approaches has been carried out on several different data domains. The methods have been extended by considering pseudo null spaces in which ortogonal directions that are discarded by plain DCV and kDCV are taken into account if their corresponding variance is small enough.

The main conclusion is that usually the kernel version of DCV improves the results of the linear case if the kernel parameter is appropriately tuned.

More importantly, from the results shown, the variance parameter can be used in most cases as a convenient regularization which leads to a better generalization abilities. In particular, using the NN classification rule using the projected training prototypes looks as a significantly better option that largely improves both DCV and kDCV without extensions.

Although experiments in this paper give results whose absolute values can be far from the state of the art in their corresponding domain (i.e. there are better adhoc solutions possibly using more apriori knowledge about the problem), their relative values allow us to draw conclusions about the relative merits of the different options.

More realistic experimentation in a more restricted domain in which the generalization ability is a key factor should be considered and is now under investigation. Further mechanisms to construct robust classifiers in the finally projected space could also be considered together with alternative and more effective ways of controlling generalization in these models.

References

1. Armengot, M., Ferri, F.J., Diaz-Villanueva, W.: Experiments about the generalization ability of common vector based methods for face recognition. In: Proceedings of PRIS 2007, Madeira, Portugal, ICEIS, pp. 129–137 (August 2007)
2. Bezdek, J.C., Kuncheva, L.: Some notes on twenty one (21) nearest prototype classifiers. In: Proceedings of the Joint IAPR International Workshops on Advances in Pattern Recognition, pp. 1–16. Springer, London (2000)
3. Cevikalp, H., Neamtu, M., Wilkes, M.: Discriminative common vector method with kernels. IEEE Transactions on Neural Networks 17(6), 1550–1565 (2006)
4. Cevikalp, H., Neamtu, M., Wilkes, M., Barkana, A.: Discriminative common vectors for face recognition. IEEE Transactions on Pattern Analysis and Machine Intelligence 27(1), 4–13 (2005)
5. Duda, R.O., Hart, P.E., Stork, D.G.: Pattern Classification, 2nd edn. John Wiley and Sons, Chichester (2001)
6. Gulmezoglu, M.B., Dzhafarov, V., Keskin, M., Barkana, A.: A novel approach to isolated word recognition. IEEE Transactions on Speech and Audio Processing 7(6), 620–628 (1999)
7. Jain, A.K., Duin, R.P.W., Mao, J.: Duin, and Jianchang Mao. Statistical pattern recognition: A review. IEEE Trans. Pattern Anal. Mach. Intell. 22(1), 4–37 (2000)
8. Liu, H., Motoda, H.: Feature Extraction, Construction and Selection: A Data Mining Perspective. Kluwer Academic Publishers, Norwell (1998)
9. Martinez, A.M., Benavente, R.: The AR face database. Technical Report 24, Computer Vision Center, Barcelona (1998)
10. Murcia-Soler, M., Pérez-Giménez, F., García-March, F.J., Salabert-Salvador, M.T., Díaz-Villanueva, W., Castro-Bleda, M.J., Villanueva-Pareja, A.: Artificial neural networks and linear discriminant analysis: A valuable combination in the selection of new antibacterial compounds. J. Chem. Inf. and Comp. Sciences 3, 1031–1041 (2004)
11. Nene, S.A., Nayar, S.K., Murase, H.: Columbia object image library (coil-100). Technical Report CUCS-006-96, Columbia University (February 1996)
12. Oja, E.: Subspace Methods of Pattern Recognition. Research Studies Press (1983)
13. Samaria, F., Harter, A.: Parameterisation of a stochastic model for human face identification. In: 2nd IEEE Workshop on Applications of Computer Vision (1994)
14. Scholkopf, B., Smola, A.: Learning with Kernels. MIT Press, Cambridge (2002)
15. Schölkopf, B., Smola, A., Müller, K.-R.: Nonlinear component analysis as a kernel eigenvalue problem. Neural Comput. 10(5), 1299–1319 (1998)

16. Swets, D.L., Weng, J.: Using discriminant eigenfeatures for image retrieval. IEEE Trans. Pattern Anal. Mach. Intell. 18(8), 831–836 (1996)
17. Tamura, A., Zhao, Q.F.: Rough common vector: A new approach to face recognition. In: Proceedings of the 2007 SMC. Intl. Conf. on Syst., Man and Cyb., Montreal, Canada, pp. 2366–2371 (October 2007)
18. Turk, M., Pentland, A.: Eigenfaces for recognition. Journal of Cognitive Neuroscience 3(1), 71–86 (1991)

Data Complexity Analysis: Linkage between Context and Solution in Classification

Tin Kam Ho

Bell Labs, Alcatel-Lucent
tkh@research.bell-labs.com

Abstract. For a classification problem that is implicitly represented by a training data set, analysis of data complexity provides a linkage between context and solution. Instead of directly optimizing classification accuracy by tuning the learning algorithms, one may seek changes in the data sources and feature transformations to simplify the data geometry. Simplified class geometry benefits learning in a way common to many methods. We review some early results in data complexity analysis, compare these to recent advances in manifold learning, and suggest directions for further research.

1 Introduction

Challenges from practical problems represented by publicly shared data sets have contributed much to the wide participation and diverse advances in statistical pattern recognition. The data sets, and the baseline accuracies of standard algorithms on such, provided a reference frame for evaluating the effects and differential advantages of many new methods and their variants. Researchers of classification algorithms benefited from being isolated from the often tedious and resource-intensive stage of preparing training data. However, an undesirable side effect is that they were also isolated from the context of the recognition problem, driven to concentrate on obtaining incremental progresses in accuracies using minor tweaks on the learning algorithms. Such a narrow view of a pattern recognition problem resulted in many uninteresting experiments that do not lead to useful conclusions.

Data complexity analysis attempts to address this dilema. In developing a way for characterizing a data set by its geometrical and topological complexity, we study whether, and to what extent, the class boundaries are learnable. Detailed quantification of learnability provides operational guidance on the expected accuracies of a family of algorithms that share common geometrical behavior. It allows for the return of attention to many issues surrounding the compilation of the training data, and provides a chance to relate the classification task to its original context. In the cases where accuracy is constrained by the geometrical properties inherent in the data, one is prompted to look for improvements upstream, that is, improvements in the data collection, feature extraction, or feature transformation stages that will lead to reduction in data complexity.

N. da Vitora Lobo et al. (Eds.): SSPR&SPR 2008, LNCS 5342, pp. 986–995, 2008.

Early explorations in data complexity analysis have followed an empirical approach, and resulted in several discoveries: (1) a large collection of problems are shown to span a continuum in several measures of geometrical complexity, with no obvious gaps and clusters; (2) apparently there exist a small number of complexity measures that provide independent characterization of different important aspects relevant to classification; (3) there are identifiable domains of dominant competence for a set of standard classifiers; and (4) measurable complexity may change due to feature selection and feature transformation. In this article we review these early progresses, discuss related efforts in newer approaches, and suggest some future directions.

2 Parameterization of Data Complexity

Complexity Classes and Complexity Scales

The concern in data complexity analysis is on characterizing each instance of the classification problem. This is different from the worst case analysis that is common for combinatorial optimization problems and algorithms.

In classification, early discussions on data complexity focused on the split between linearly separable problems and those that are not. The discovery of the perceptron's inability to learn linearly nonseparable classes led to a long silence in activities in neural networks, until newer structures and learning algorithms were found to overcome the obstacle. In the meantime, efforts were made to stretch linear classifiers to adapt to complex class boundaries. Piecewise linear classifiers, polynomial discriminants, low-dimensional feature projections and high-dimensional feature transformations, nearest-neighbor classifiers, and decision trees were among the popular pursuits. Optimism built up with breakthroughs in multi-layer perceptron training, support vector machines, and ensemble learning methods. Many arguments were made on the generality of the ability of large-capacity learning methods on arbitrary problems. Accuracy limits observed in empirical studies are believed to be setbacks that can be addressed by careful control of overfitting.

Yet we believe that such practical limits deserve a deeper investigation.

In the pursuit of multiple classifier systems that may benefit from the merits of all the existing learning algorithms, and the pursuit of generative algorithms like random decision forests that may create as many classifiers as needed, it graduately became clear that some problems are intrinsically very difficult to learn under the lack of a complete sample along with class-labeling. It is difficult for a small-sample representation of the problem to contain complete knowledge of the classes, especially in high-dimensional spaces. There is a need for a better understanding of the class geometry to provide guidance to the creation of classifier components. Data complexity analysis was motivated by this need.

Continuum of Complexity Distribution

Early studies [14] revealed that a collection of many benchmarking data sets span a nearly continuous range in several complexity measures. The complexity measures include known descriptors of class geometry, such as Fisher's discriminant ratio, ratio of between-class and within-class distances, departure from convexity and smoothness of boundaries, fragmentation of the support of the classes, and length of the class boundary. In addition, the error rates of several simple classifiers, such as nearest-neighbor classifiers, linear classifiers induced by linear programming minimizing the sum of error distances, can also be considered useful parameters of data complexity.

For a standard problem, e.g., discrimination of two classes represented by two Gaussians of different means, it is understandable that changes in the distributional parameters may lead to differences in classification difficulty. Say, with the variances fixed, increasing distance between the means promotes separability, and such changes can be continuous.

The surprise in the discovery is that the observed continuum is spanned by a large but almost arbitrary collection of benchmarking data sets. This gives raise to a doubt that the common practice of arguing about relative merits of learning algorithms on a very limited collection of problem instances can be misleading. Simple counts of the number of times an algorithm "wins", or even numerical measures of the best, worst, average accuracies over the entire set, have little relevance to the algorithm's performance on a future problem of unknown complexity. Better characterization of the merits of learning algorithms should involve a careful, balanced selection of problem instances to cover differences in data complexity in different aspects.

Intrinsic Dimensionality of Complexity Distribution

Another observation from the early experiments is that the distribution of the problem instances in the complexity space is far from uniform. The distribution concentrates on a narrow band across the space, but there are outliers. Correlation between some measures explains the narrow distribution in the projection to certain dimensions. For example, it is observed that the error rate of a nearest-neighbor classifier is highly correlated with the fraction of points located on the class boundary. This can be explained by a close examination of how a boundary point contributes to nearest-neighbor classification error. On the other hand, fragmentation and sphericity of the classes have very low correlation with measures of separation distances and class convexity [15]. In one experiment, principal component analysis shows that six linear combinations of the measures would explain 94% of the variance [14].

If the behavior of other classifiers is used to parameterize data complexity, one can easily adopt too many measures for the analysis to be practical [23]. Examples include the size of a decision tree grown to fit the training data, number of nearest neighbors needed after condensation, or the type and parameters of kernels useful in support vector machines. Arguably many of these gauge the data complexity in some sense. However, even after removing those that do not provide good characterization for problems with known complexity, there are too many possibilities. Clustering the measures offers a solution [23], but a more interesting approach will be to uncover those measures that directly describe the essential geometrical and topological characteristics that determine the behavior of the more sophisticated classifiers. We shall return to this subject when we discuss the interplay between the learning algorithms and geometrical complexity.

Bayes Error and Uncertainty in Complexity Estimates

In classification, the analysis is complicated by the fact that while the application context can be reasonably specified, the full difficulty of a problem can only be understood through a sample data set. A consequence of such indirectness is that a fair analysis of the problem's complexity is affected by the sampling rate and sampling distribution. Only in very rare cases, a generative model is available that can produce an infinite amount of representative data [13].

In classical statistics, a classification problem's difficulty is quantified by the Bayes error, which can be calculated if the probability density function of each class is known. The Bayes error gives a lower bound on a classifier's asymptotic error rate. The lower bound is achievable by parametric classifiers suited to the distributional characteristics in some simple problems (e.g. two Gaussians). Some nonparametric methods like nearest-neighbor classifiers are also shown to be able to closely approach the Bayes error with infinite training samples [8].

When logistic regression is used for classification, the uncertainty of class prediction can be described by the confidence interval on the log-odds prediction. For other classifiers, a large literature exists on relating sample sizes to classification error [9][24], and relating generalization error to sample size and classifier capacity [7][27][28].

Given problems of the same Bayes error, differences in their geometrical complexity can cause large differences in classification accuracies because of the simple geometrical models used by many standard classifiers. However, the characterization of data geometry is subject to uncertainty due to small sample sizes. This issue is especially severe if the data are given as dissimilarity representations rather than feature vectors [10]. In lack of additional data, one may have to invoke assumptions on the problem domain, or general heuristics such as the compactness hypothesis [10].

3 Relating Classifier Behavior to Data Complexity

Domain of Classifier Competence

In an analysis of several classifiers' relative merits in the data complexity space, concentrations are observed in the distribution of problem instances where some classifiers have dominant advantage. Most notable is that the domain of dominant competence of the linear classifier and that of the nearest-neighbor classifier occupy opposite ends of the complexity distribution when projected to a measure of boundary length and a measure of boundary nonlinearity [19]. Similarly, one can specify the region where a certain classifier has advantages [20], or compare the competence regions of two similar classifiers [12]. One useful result from these studies is that several complexity measures, including the boundary length, boundary nonlinearity, and the ratio of intra-inter class nearest-neighbor distances, are the most relevant metrics for discriminating between competence domains of different classifiers [16].

However, while interesting observations like these have been made, the study is far from conclusive. On close examination of the competence domains, one can see substantial spread and overlap. Given the similarities of the classifiers under study, it is conceivable that their competence domains are intrinsically ambiguous. It is unclear to what extent the ambiguity can be alleviated by using additional complexity measures. The correlation between the complexity measures complicates the issue. Furthermore, one has to consider the uncertainty in the complexity estimates for each problem due to small sample sizes.

Another difficulty is that in our study using empirical data sets, certain regions in the complexity space are not well covered. The uneven coverage makes it difficult to obtain a comprehensive mapping from the complexity measures to classifier preferences. In this discrimination task, large uncertainty in the prediction can occur due to small sample sizes (few problem instances) in certain regions. To address this difficulty, a systematic scheme to select or synthesize sample problems will be needed.

Changes in Complexity Due to Feature Selection

Feature selection has long been an important theme in classification. Often, the goal is to reduce the demand on run time and computer memory, and to alleviate the effect of the curse of dimensionality, in other words, to avoid overfitting irrelevant noise contained in the data.

A study on selecting subsets of features using an LPSVM (Linear Programming Support Vector Machine) formulation and a Forward Feature Selection procedure shows that such selection can change the data complexity substantially [21]. A note of caution is that the changes may occur only in the *apparent* complexity, i.e., the complexity estimate calculated from the particular data set. If the sample size is too small for the problem, the changes may not be a reliable indicator of changes in the problem's intrinsic difficulty.

4 Experimental Design for Classifier Evaluation

Experiences from the early studies point to the need for further efforts in selecting appropriate complexity measures and designing a better classifier evaluation strategy. This can be compared to a conventional computational experiment, where critical issues include a proper parameterization of the input domain and a good way to generate samples to cover the domain.

In pattern recognition, this amounts to simulating classification problems arising from different contexts. The goal is to provide a comprehensive coverage of (or a close approximation to) all possible problems that may arise in a practical application. Going back to the application context, one may seek modifications of the problem formulation, so that the classification task can be matched to one with the lowest data complexity. A few approaches can be considered in designing the evaluation strategy.

Synthesis of Classification Problems with Target Complexity

In this approach, synthesizing a classification problem involves creating a suitable distribution of the training data points in the feature space and their class labeling. Ideally, the synthesis procedure provides an even cover of the chosen complexity measure. In [18] an interesting attempt is made on covering the range of a boundary length measure. The experiment first generates n data points that follow a uniform random distribution in an m-dimensional space. A minimum spanning tree (MST) is calculated for each realization. Different choices in designating some edges to be class-crossing result in data sets of different complexity according to this measure. The reference numbers for n, m may come from a real-world problem. To compare classifier accuracies on the synthetic data with those on real-world data, the same number of MST edges are made to be class-crossing. Therefore, the synthetic data are at the same apparent complexity with the real-world data under this measure. Observation of the classifier accuracies shows that this method provides a pessimistic estimate of achievable accuracy. A possible cause is that the real-world problems have a more compact geometry than the uniform distribution in the synthetic data, so that they are actually easier. To close this discrepancy, additional complexity measures are needed to guide the production of synthetic data towards similar compactness.

Synthesis of Problems with References to Natural Processes

A more plausible approach to create synthetic data is to follow certain well known models of stochastic spatial processes, e.g., the Neyman-Scott process, that represent certain natural processes. Using such models allows for formal inferences on the resulting data complexity. Furthermore, uncertainty in the complexity estimates can be evaluated in a principled way.

Systematic Degradation of Real-World Problems

One may also produce new classification problems by systematically degrading existing ones. An advantage of this approach is that the process starts with a realistic data geometry. The drawback is that the reference geometry also introduces a bias, which needs to be removed by sufficient coverage using a larger variety of real-world problems. This could be very difficult for high dimensional spaces.

A version of this method is to take a real-world problem that has a parameterized domain model. Perturbing the model parameters leads to different realizations of the problem with different complexity. For example, in [13] a document image defect model is used to create increasing difficult instances of a character recognition problem. Other variations include simulations of imperfect sampling conditions or class labeling errors.

5 Simplification of Class Geometry

A constructive procedure to improve classification performance is to seek to reduce data complexity by changing a problem's formulation, introducing clever feature extractors, and mapping the classification task to an easier space by feature selection and transformation. This is an age-old theme in pattern recognition. But we argue for more than just a return of attention to this theme.

Instead of relying on intuition and heuristics, we argue for a way to formalize this process and set goals for systematic optimization. Description of data complexity gives an operational definition of learnability, against which systematic optimization can be driven. An extreme example is that one may seek to extract enough features so that the classification task is reduced to be linearly separable in the lowest dimensionality. Another example is to include clever normalization procedures and observe problem-specific invariances that can compress the within-class scatter. In many engineering contexts this practice has been followed implicitly or explicitly. Ideally, more automation can be introduced into the process by linking the complexity measures to controllable processes in the upstream.

Manifold Learning and Dimensionality Reduction

Recent advances in the methods for manifold learning provide some interesting directions. Several methods have been proposed to fit localized linear models to the data manifold. The local models are embedded into a global model along the intrinsic dimensions. In the process the data cloud is unfolded to a smooth and flat surface. The emphasis of many manifold learning techniques is on obtaining a useful transformation that highlights the intrinsic dimensions, yields better clustering, and helps visualization. Some transformations can be applied to unseen data from the same source (out-of-sample extensions) [5].

Most manifold learning methods do not give explicit characterization of the data geometry. This is deferred to a subsequent clustering step. Also, little is said about the existence of holes, class fragmentations, etc. Recent efforts in topological data analysis may address these concerns [6]. Other interesting attempts include modeling entire data sets and their similarities by Grassmann manifolds [26], and applying the embedding methods to tensor representations [11].

A *supervised* manifold learning procedure highlights the geometry of the class boundary and promotes linearization of the separating surface. A pioneering work in this area is [25], where a supervised locally linear embedding (LLE) method is proposed. By applying LLE to a distance matrix where between-class distances are exaggerated, the authors show that superior classification accuracy can be achieved especially on high-dimensional data where the class structures contain curved low-dimensional manifolds. An symptom of such data is a high contrast between the global dimensionality and local intrinsic dimensionality. More recent work extends this to tensor representations [17].

In many of these studies, the merits of the entire approach, including both the feature transformation step and the classification and clustering step, are evaluated jointly as if the steps are parts of one complete method. This incurs a risk that the classification method chosen may not be the optimal companion to the feature transformation step. With data complexity analysis, an intermediate goal can be made explicit. One may seek a manifold learning/mapping technique that reduces the complexity of the data geometry, and let the next step be guided by a known match of data complexity to classifier choices.

6 Application Examples

The limit of achievable recognition accuracy remained a subject for the experts for a long time, until recently, a few high-profile projects brought this to public attention. A few examples are as follows.

A clever way to exploit the limits of mechanical pattern recognition is the CAPTCHA system [1]. The system challenges a user to enter the text that appears as an image that is distorted from its regular shape, in an arbitrary font, corrupted with noise and overprinting. It is believed that the recognition task is beyond machine reach, so that a user who enters the text correctly is trusted to be a living human. This is also known as a reverse Turing test, or human interactive proofs [3].

The Netflix challenge [22] was open to the public since 2006. The challenge is a public contest on improvements to the video rental company's system for predicting movie rating for its users based on the user's previous ratings on other movies. The project offers a prize to the first team with an algorithm that can achieve a 10% improvement on the test set over the company's in-house algorithm. While progress has been made since the beginning of the contest, at the time of this paper's writing, it remains unclear whether the project's goal is intrinsically achievable.

Another example is the list of "Human Intelligence Tasks" offered at the Amazon web service "Mechanical Turk" [2]. Many of these are small pattern recognition tasks that machines do not perform well, for example, identifying and tagging objects in an image sequence. Do these task represent the limit of power in automatic pattern recognition? Or do we expect that, someday in the future, a fair portion of these can be fully automated by a smarter classifier technology?

7 Conclusions

We reviewed the development of data complexity analysis over the last decade, and summarized the experiences from early explorations. We argued that data complexity provides an important focus for the optimization of the formulation and solution of a classification task. It provides a linkage between optimizations related to the problem context and those related to the learning algorithms.

Our discussion has focused primarily on the feature vector space representation of classification problems. We have not addressed challenges arising from alternative representations (e.g. similarity based representations, tensors, categorical variables, strings), data processing issues (unit and scale normalization, missing values, mislabeling), or multi-layered or meshed dependence structures. We also note that more can be explored on coupling data complexity analysis with active and sequential learning.

Acknowledgements

My collaborators on this endeavor have contributed many of the ideas summarized in this article: Mitra Basu, Ester Bernado-Mansilla, Richard Baumgartner, Martin Law, Erinija Pranckeviciene, Albert Orriols-Puig, and Nuria Macia. I would also like to thank Henry Baird, George Nagy, Anil Jain, Robert Duin, Josef Kittler, Ludmila Kuncheva, David Hand, Thomas Bengtsson, Peter Bickel, Bin Yu, Emina Soljanin, and participants of the Multiple Classifier Systems workshops for many stimulating discussions.

References

1. von Ahn, L., Blum, M., Hopper, N.J., Langford, J.: CAPTCHA: Telling Humans and Computers Apart. In: Biham, E. (ed.) EUROCRYPT 2003. LNCS, vol. 2656, pp. 294–311. Springer, Heidelberg (2003), `http://www.captcha.net/`
2. Amazon, Mechanical Turk (2005), `http://www.mturk.com/mturk/welcome`
3. Baird, H.S.: Complex Image Recognition and Web Security. In: [4], pp. 287–298
4. Basu, M., Ho, T.K. (eds.): Data Complexity in Pattern Recognition. Springer, London (2006)
5. Bengio, Y., Paiement, J.-F., Vincent, P.: Out-of-Sample Extensions for LLE, Isomap, MDS, Eigenmaps, and Spectral Clustering. In: NIPS 2003, pp. 177–184 (2003)
6. Carlsson, G.: Topology and Data, Dept of Math., Stanford Univ., August 10 (preprint, 2008), `http://comptop.stanford.edu/preprints/topologyAndData.pdf`
7. Cherkassky, V., Ma, Y.: Data Complexity, Margin-Based Learning, and Popper's Philosophy of Inductive Learning. In: [4], pp. 91–114
8. Cover, T.M., Hart, P.E.: Nearest Neighbor Pattern Classification. IEEE Trans. on Inf. Theory 13, 21–27 (1967)
9. Devroy, L.: Automatic Pattern Recognition: A Study of the Probability of Error. IEEE Trans. on Pat. Anal. and Mach. Intell. 10(4), 530–543 (1988)
10. Duin, R.P.W., Pekalska, E.: Object Representation, Sample Size, and Data Set Complexity. In: [4], pp. 25–58
11. He, X., Cai, D., Niyogi, P.: Tensor Subspace Analysis, NIPS 2005 (2005)
12. Ho, T.K.: A Data Complexity Analysis of Comparative Advantages of Decision Forest Constructors. Pattern Analysis and Applications 5, 102–112 (2002)
13. Ho, T.K., Baird, H.S.: Large-Scale Simulation Studies in Image Pattern Recognition. IEEE Trans. on Pat. Anal. and Mach. Intell. 19, 1067–1079 (1997)
14. Ho, T.K., Basu, M.: Complexity Measures of Supervised Classification Problems. IEEE Trans. on Pat. Anal. and Mach. Intell. 24(3), 289–300 (2002)
15. Ho, T.K., Basu, M., Law, M.H.C.: Measures of Geometrical Complexity in Classification Problems. In: [4], pp. 3–23

16. Ho, T.K., Mansilla, E.B.: Classifier Domains of Competence in Data Complexity Space. In: [4], pp. 135–152
17. Li, X., Lin, S., Yan, S., Xu, D.: Discriminant Locally Linear Embedding with Higher-Order Tensor Data. IEEE Trans. on Sys., Man, and Cyb., Part B: Cyb. 38(2), 342–352 (2008)
18. Macia, N., Mansilla, E.B., Orriols-Puig, A.: Preliminary Approach on Synthetic Data Sets Generation Based on Class Separability Measure. In: Proc. of the 19th Int'l. Conf. on Pat. Recog., Tampa, U.S.A, December 7-11 (2008)
19. Mansilla, E.B., Ho, T.K.: On Classifier Domains of Competence. In: Proc. of the 17th Int'l. Conf. on Pat. Recog., Cambridge, U.K, August 22-26, vol. 1, pp. 136–139 (2004)
20. Mansilla, E.B., Ho, T.K.: Domain of Competence of XCS Classifier System in Complexity Measurement Space. IEEE Trans. on Evol. Comp. 9(1), 82–104 (2005)
21. Pranckeviciene, E., Ho, T.K., Somorjai, R.: Class Separability in Spaces Reduced By Feature Selection. In: Proc. of the 18th Int'l. Conf. on Pat. Recog., Hong Kong, China, August 20-24, vol. 2 (2006)
22. Netflix Prize (2006), `http://www.netflixprize.com/`
23. Raudys, S.: Measures of Data and Classifier Complexity and the Training Sample Size. In: [4], pp. 59–68
24. Raudys, S., Jain, A.K.: Small Sample Size Effects in Statistical Pattern Recognition: Recommendations for Practitioners. IEEE Trans. on Pat. Anal. and Mach. Intell. 13(3), 252–264 (1991)
25. de Ridder, D., Kouropteva, O., Okun, O., Pietikainen, M., Duin, R.P.W.: Supervised Locally Linear Embedding. In: Kaynak, O., Alpaydın, E., Oja, E., Xu, L. (eds.) ICANN 2003 and ICONIP 2003. LNCS, vol. 2714, pp. 333–341. Springer, Heidelberg (2003)
26. Srivastava, A.: A Bayesian Approach to Geometric Subspace Estimation. IEEE Trans. Sig. Proc. 48(5), 1390–1400 (2000)
27. Vapnik, V.: Estimation of Dependences Based on Empirical Data. Springer, Berlin (1982)
28. Vapnik, V.: Statistical Learning Theory. Wiley, New York (1998)

Graph Classification Based on Dissimilarity Space Embedding

Horst Bunke and Kaspar Riesen

Institute of Computer Science and Applied Mathematics, University of Bern, Neubrückstrasse 10, CH-3012 Bern, Switzerland
{bunke,riesen}@iam.unibe.ch

Abstract. Recently, an emerging trend of representing objects by graphs can be observed. In fact, graphs offer a powerful alternative to feature vectors in pattern recognition, machine learning, and related fields. However, the domain of graphs contains very little mathematical structure, and consequently, there is only a limited amount of classification algorithms available. In this paper we survey recent work on graph embedding using dissimilarity representations. Once a population of graphs has been mapped to a vector space by means of this embedding procedure, all classification methods developed in statistical pattern recognition become directly available. In an experimental evaluation we show that the proposed methodology of first embedding graphs in vector spaces and then applying a statistical classifier has significant potential to outperform classifiers that directly operate in the graph domain. Additionally, the proposed framework can be considered a contribution towards unifying the domains of structural and statistical pattern recognition.

1 Introduction

The field of pattern recognition can be divided into two sub-fields, namely the statistical and the structural approach. In statistical pattern recognition, patterns are represented by feature vectors $(x_1, \ldots, x_n) \in \mathbb{R}^n$. The recognition process is based on the assumption that patterns of the same class are located in a compact region of $\mathbb{R}^n$. In recent years a huge amount of methods for the classification of patterns represented by feature vectors have been proposed, such as Bayes classifier, neural network, support vector machine, and many more [1]. Object representations given in terms of feature vectors have a number of useful properties. For example, object similarity, or distance, can easily be computed by means of Euclidean distance. Yet graph-based representations, which are used in the field of structural pattern recognition, have a number of advantages over feature vectors. Graphs are much more powerful and flexible than vectors, as feature vectors provide no direct possibility to describe structural relations in the patterns under consideration. Furthermore, while the size of a graph can be adjusted to the size and complexity of the underlying pattern, vectors are constrained to a predefined length, which has to be preserved for all patterns encountered in a particular application. On the other hand, a major drawback of

N. da Vitora Lobo et al. (Eds.): SSPR&SPR 2008, LNCS 5342, pp. 996–1007, 2008.

graph representations is their lack of suitable methods for classification. This is mainly due to the fact that some of the basic operations needed in classification are not available for graphs.

A promising direction to overcome the lack of algorithmic tools for graph classification is graph embedding. Basically, an embedding of graphs into a vector space establishes access to the rich repository of algorithmic tools developed in statistical pattern recognition. The present paper considers a new class of graph embedding procedures which are based on dissimilarity representation and graph edit distance computation. Originally the idea was proposed in [2] in order to map feature vectors into dissimilarity spaces. This idea was first generalized to string based object representation [3] and later to the domain of graphs [4,5,6,7,8,9]. Graphs from a given problem domain are mapped to feature spaces by computing the distance to some predefined prototype graphs, or sets of prototype graphs. The resulting distances can be used as a vectorial representation of the considered graph.

Note that our graph embedding approach can be applied to both directed and undirected graphs, as well as to graphs without and with labels on their nodes and/or edges. In case there are labels on the nodes and/or edges, these labels can be of any nature (discrete symbols, the set of integer or real numbers, or whole attribute vectors). Even hypergraphs can be embedded with the method described in this paper [10]. Hence, the proposed embedding approach is more general than other graph embedding techniques where sometimes restrictions on the type of underlying graph are imposed (e.g. [11,12,13]).

2 Dissimilarity Space Embeddings of Graphs

In [2] the authors claim that the concept of proximity is more fundamental than that of a feature or a class. Furthermore, it is pointed out that in the case of structural data (like graphs) the extraction of numerical features may be difficult or even intractable, while proximity can directly be derived from the data using an adequate dissimilarity model.

Assume we have a labeled set of sample graphs, $\mathcal{G} = \{g_1, \ldots, g_N\}$ and a graph dissimilarity measure $d(g_i, g_j)$. Note that $\mathcal{G}$ can be any kind of graph set. For the sake of convenience, however, in the context of the present work we define $\mathcal{G}$ to be the set of training graphs $\mathcal{T}$. After having selected a set of prototype graphs $\mathcal{P} = \{p_1, \ldots, p_n\} \subseteq \mathcal{T}$, we compute the dissimilarity of a given input graph g to each prototype $p \in \mathcal{P}$. Note that g can be an element of $\mathcal{T}$ or any other graph. This leads to n dissimilarities, $d_1 = d(g, p_1), \ldots, d_n = d(g, p_n)$, which can be arranged in an n-dimensional vector $(d_1, \ldots, d_n)$. In this way we can transform any graph from the training as well as any other graph set (for instance a validation or a test set of a classification problem), into a vector of real numbers.

Definition 1 (Graph Embedding). *Let $\mathcal{G}$ be a finite or infinite set of graphs and $\mathcal{P} = \{p_1, \ldots, p_n\} \subseteq \mathcal{G}$ be a set of prototypes. Then, the mapping $\varphi_n^{\mathcal{P}} : \mathcal{G} \rightarrow \mathbb{R}^n$ is defined as the function*

$$\varphi_n^P(g) \mapsto (d(g, p_1), \ldots, d(g, p_n)),$$

where $d(g, p_i)$ is some graph dissimilarity measure between graph g and the i-th prototype.

Though conceptually $d(g, p_i)$ can be any kind of dissimilarity measure, the embedding procedure proposed in this paper makes use of graph edit distance. The key idea of graph edit distance is to define the dissimilarity, or distance, of graphs by the minimum amount of distortion that is needed to transform one graph into another [14]. Compared to other approaches, graph edit distance is very flexible since it can handle arbitrary graphs and any type of node and edge labels. Furthermore, by defining costs for edit operations, the concept of edit distance can be tailored to specific applications.

Optimal algorithms for computing the edit distance of graphs are typically based on combinatorial search procedures that explore the space of all possible mappings of the nodes and edges of the first graph to the nodes and edges of the second graph [14]. A major drawback of those procedures is their computational complexity, which is exponential in the number of nodes of the involved graphs. However, a number of efficient suboptimal methods for graph edit distance computation have been proposed (e.g. [15] with cubic time complexity). Consequently, given n predefined prototypes the embedding of one particular graph is established by means of n distance computations with polynomial time.

Regarding the general embedding procedure established above, a strong relationship to graph kernel methods can be observed [16]. In fact, based on the graph embedding $\varphi_n^{\mathcal{P}}$ established above, one can define a valid graph kernel κ by applying valid kernel functions defined for vectors to two graph maps in the resulting vector space (e.g. Scalar Product, Radial Basis Function, Sigmoid Function, etc.) [17].

3 Survey of Work on Dissimilarity Based Graph Embedding

In this section we provide an overview of our recent work on graph embeddings using dissimilarity representation, i.e. we recapitulate several papers addressing different aspects of graph embedding and its applications [4,5,6,7,8,9].

3.1 Prototype Selection

The embedding method described in the previous section crucially depends on the prototypes. Therefore, an important problem to be solved is an appropriate choice of the prototype set $\mathcal{P} = \{p_1, \ldots, p_n\}$. A good selection of n prototypes seems to be crucial to succeed with the classification algorithm in the feature vector space. The prototypes should avoid redundancies in terms of selection of similar graphs, and prototypes should include as much information as possible.

In the following six different prototype selection strategies are discussed. For a more thorough review of the prototype selection methods we refer to [4], where these strategies are applied to several graph data sets.

Center. The Centers prototype selector selects prototypes situated in the center of the training set $\mathcal{T}$. This is achieved by iteratively taking the set median graph out of the set $\mathcal{T}$. The set median graph is the graph whose sum of distances to all other graphs in the set is minimal.

Border. The Border prototype selector selects prototypes situated at the border of the training set $\mathcal{T}$, i.e. it iteratively takes the set marginal graph out of the set $\mathcal{T}$. The set marginal graph is the graph whose sum of distances to all other graphs in this set is maximal.

Random. A random selection of n prototypes from $\mathcal{T}$ is performed.

Spanning. The first prototype selected is the set median graph. Each additional prototype selected by the Spanning prototype selector is the graph the furthest away from the already selected prototype graphs.

k-Centers. The k-Centers prototype selector tries to adapt to the graph distribution of set $\mathcal{T}$ and selects graphs that are in the center of densely populated areas. First a k-means clustering procedure is applied to set $\mathcal{T}$. The number of clusters to be found is equal to the number of prototypes to be selected. Once the clusters have been established, the set median of each cluster is selected as a prototype.

Targetsphere. The Targetsphere prototype selector first selects a graph g_c situated in the center of $\mathcal{T}$. Next the graph $g_f \in \mathcal{T}$ whose distance to g_c is maximum is located. The remaining prototypes are uniformly distributed from g_c to g_f. That is, the interval $[0, d_{max}]$ ($d_{max} = d(g_c, g_f)$) is divided into $m-1$ equidistant subintervals of width $w = \frac{d_{max}}{m-1}$. The $m-2$ graphs for which the corresponding distances to the center graph g_c are located nearest to the interval borders in terms of edit distance are selected as prototypes.

In [4] it has been our intention to improve the accuracy achieved by nearest-neighbor classifiers in the graph domain by the use of classifiers operating on graph maps, i.e. vectorial representations. We applied support vector machine (SVM) as a popular method from statistical pattern recognition and showed that this approach outperforms the nearest-neighbor classifiers in the graph domain. In the experiments reported in [4] it turns out that the quality of the prototype selectors introduced above depends on the application and the underlying data. Therefore, the question which of the prototype selectors is globally best cannot definitely be answered. Nevertheless, there is a clear tendency that prototype selectors that distribute the graphs more or less uniformely over the whole graph set (Spanning, k-Centers, Targetsphere) lead to higher recognition rates.

3.2 Prototype Reduction

The prototype selection strategies introduced above use some heuristics based on the underlying dissimilarities in the original graph domain. Another idea to select the members of set $\mathcal{P}$ is prototype reduction methods in conjunction with nearest neighbor classifiers [18]. These reduction schemes determine a subset $\mathcal{P} \subseteq \mathcal{T}$ such that the elements in $\mathcal{T}$ (or at least a considerable part of them) are still correctly classified using a nearest neighbor classifier.

We use *selective* prototype selectors where the number of prototypes is *uncontrollable* [19]. These two constraints are motivated through the following considerations. First, the fact that we are dealing with graphs makes the creation of new prototypes quite difficult. Therefore, we restrict set $\mathcal{P}$ to include only graphs from $\mathcal{T}$. Secondly, we want to bypass the time consuming validation of the dimensionality of the resulting embedding space by means of the target classifier. Hence, we leave the determination of the number of prototypes to the prototype selection algorithm.

Condensing (Cond). The idea of condensing a training set $\mathcal{T}$ is to iteratively select graphs $g_i \in \mathcal{T}$ as prototypes until all graphs from $\mathcal{T}$ are correctly classified using the respective prototypes. As a disadvantage, this procedure depends on the order in which the graphs are processed.

Modified Condensing (mCond). Modified condensing overcomes the limitation of order dependency. In this scheme we start with a basic set of prototypes containing the set center graph (*centroid*) of each class. The centroid of $\mathcal{G}$ is the graph for which the maximum distance to all other graphs in $\mathcal{G}$ is minimum. The graphs from $\mathcal{T}$ are classified by means of this initial set of prototypes. Using only the misclassified graphs, the class centroids are computed and subsequently added to the existing set of prototypes. This procedure is repeated until all graphs from $\mathcal{T}$ are correctly classified.

Editing (Edit). The basic idea of editing a training set $\mathcal{T}$ is to delete outliers from $\mathcal{T}$. For this purpose, we classify each graph g_i from $\mathcal{T}$ with a 3-NN classifier. If g_i is misclassified we assume that this particular graph is an outlier and therefore should not be included in the prototype set.

Reducing (Red). The idea of reducing is built up on condensing. First, the training set $\mathcal{T}$ is condensed to a prototype set $\mathcal{P}$. Next, each prototype p_i is iteratively removed from $\mathcal{P}$. The training graphs are then classified using the reduced prototype set $\mathcal{P} \setminus \{p_i\}$. If all graphs are classified correctly with this reduced prototype set, the respective prototype is useless and can therefore be omitted. Otherwise, the prototype is necessary and therefore kept in $\mathcal{P}$.

Merging (Merg). The basic idea of merging a training set is to define two graph sets $\mathcal{P}$ and $\mathcal{Q}$, where initially $\mathcal{P}$ is empty and $\mathcal{Q}$ contains all training graphs from $\mathcal{T}$. First, an arbitrary graph from $\mathcal{Q}$ is selected as prototype, i.e. moved from $\mathcal{Q}$ to

$\mathcal{P}$. Next, we consider the two closest graphs p and q from $\mathcal{P}$ and $\mathcal{Q}$, respectively. If the class of p is not the same as that of q, q is moved from $\mathcal{Q}$ to $\mathcal{P}$. Otherwise, p and q are merged to $p^* \in \mathcal{Q}$, where p^* minimizes the sum of distances to p and q. The accuracy of the NN classifier using $\mathcal{P} \cup \{q\}$ is then compared with the accuracy when $\mathcal{P} \setminus \{p\} \cup \{p^*\}$ is used as prototype set. Whenever the former outperforms the latter, q is moved from $\mathcal{Q}$ to $\mathcal{P}$. Otherwise, p and q are removed from $\mathcal{P}$ and $\mathcal{Q}$, respectively, and p^* is moved from $\mathcal{Q}$ to $\mathcal{P}$. This procedure is repeated until no graphs are left in $\mathcal{Q}$.

Selecting (Sel). Another algorithm for reducing the training set $\mathcal{T}$ is based on the idea of related neighbors. We define $g_j \in \mathcal{T}$ as a related neighbor to $g_i \in \mathcal{T}$ if g_i and g_j are out of the same class, and g_j is nearer to g_i than any other sample $g_k \in \mathcal{T}$ from another class. The selection of the prototypes is now stated as finding a small number of graphs such that each of these graph has at least one related neighbor. In the present paper a greedy algorithm is employed seeking for a small number of prototypes.

For a more detailed discussion on these prototype reduction schemes in conjunction with graph embeddings we refer to [9]. Furthermore, in Section 4.3, an experimental evaluation of this approach is given.

3.3 Dimensionality Reduction

In Section 3.1 and 3.2 a number of prototype selection strategies have been introduced. In the current section we describe an alternative approach where we use all available elements from the training set as prototypes, i.e. $\mathcal{P} = \mathcal{T}$ and subsequently apply dimensionality reduction methods. This process is more principled and allows us to completely avoid the problem of finding the optimal prototype selection strategy. For dimensionality reduction, we make use of Principal Component Analysis (PCA), Fisher's Linear Discriminant Analysis (LDA) [1], and kernel PCA [20].

Principal Component Analysis (PCA). PCA [1] is a linear transformation. It seeks the projection which best represents the data. PCA is an unsupervised method which does not take any class label information into consideration. We first normalize the data by shifting the mean to the origin of the coordinate system and making the variance of each feature equal to one. Then we calculate the covariance matrix of the normalized data and determine the eigenvectors $\mathbf{e}_i$ and the eigenvalues λ_i of the covariance matrix. The eigenvectors are ordered according to decreasing magnitude of the corresponding eigenvalues, i.e. $\lambda_1 \geq \lambda_2 \geq \ldots \geq \lambda_N \geq 0$. The data is then represented in a new coordinate system defined by the eigenvectors. For reducing the dimensionality of the transformed data we retain only the $n < N$ eigenvectors with the n highest eigenvalues.

Fisher's Linear Discriminant Analysis (LDA). LDA [1] is a linear transformation as well. In contrast with PCA, LDA takes class label information into account. In its original form, LDA can be applied to two-class problems only.

However, we make use of a generalization, called Multiple Discriminant Analysis (MDA), which can cope with more than two classes. In MDA, we are seeking the projection of the data which best separates the classes from each other. For all further details, we refer to [1]. Note that under this transformation the maximal dimensionality of the transformed feature space is $c-1$, where c is the number of classes.

Kernel PCA. The result of a kernel function $k(\mathbf{x},\mathbf{y})$, applied to two feature vectors $\mathbf{x}$ and $\mathbf{y}$, is equal to the result that one obtains by mapping those vectors to a possibly infinite dimensional feature space $\mathcal{F}$ and computing their dot product subsequently [17]. This procedure offers a very elegant way to construct non-linear extensions of linear algorithms in pattern recognition. The fundamental observation that makes kernel theory so interesting in the field of pattern recognition is that many of the algorithms (e.g. PCA [20]) can be *kernelized*, i.e. they can be formulated entirely in terms of dot products. Consequently, instead of applying PCA in the original vector space, the linear transformation is applied in an implicit existing feature space $\mathcal{F}$ by substituting the dot product by an appropriate kernel function κ. This procedure is commonly referred to as the *kernel trick* [17].

Experimental results and a more detailed description of these dimensionality reduction strategies applied to vector space embedded graphs can be found in [6,7]. The main finding in [6] is that the performance of a k-nearest neighbor classifier in the graph domain, used as a reference system, can be outperformed with statistical significance using this embedding approach. In case of classification problems with many classes, the MDA based system is preferable, while for a small number of classes the PCA based system is the method of choice. Moreover, in [7] we observe that the approach with kernelized PCA outperforms the former approach with linear PCA on most of the data sets.

3.4 Relationship to Lipschitz Embeddings

Our graph embedding approach can also be regarded as a special instance of a Lipschitz embedding [21,22] where a coordinate space is defined such that each axis corresponds to a reference set of objects. Formally, rather than a single prototype set $\mathcal{P} = \{p_1, \ldots, p_n\}$ we define a set $\mathcal{S} = \{\mathcal{P}_1, \ldots, \mathcal{P}_n\}$ that consists of n prototype sets. The n subsets $\mathcal{P}_i \subset \mathcal{T}$ define the *reference sets* of the Lipschitz embedding. The extended graph edit distance function between graphs and reference sets is defined as $d(g, \mathcal{P}_i) = \min_{p \in \mathcal{P}_i}\{d(g,p)\}$. The Lipschitz embedding with respect to $\mathcal{S} = \{\mathcal{P}_1, \ldots, \mathcal{P}_n\}$ is then defined as

$$\varphi_n^{\mathcal{S}}(g) = (d(g, \mathcal{P}_1), \ldots, d(g, \mathcal{P}_n)) \enspace .$$

Obviously, the range of function $\varphi_n^{\mathcal{S}}$ is a vector space where each dimension corresponds to a subset $\mathcal{P}_i \subset \mathcal{T}$ and the coordinate values of the embedded graph g are the distances from g to the nearest element in $\mathcal{P}_i$.

In [8] two different methods to define our reference sets are applied, viz. random selection and a more advanced technique based on k-Centers prototype selection. In the random method we randomly select n subsets $\mathcal{S} = \{\mathcal{P}_1, \ldots, \mathcal{P}_n\}$ each of size m. After drawing a graph g from $\mathcal{T}$, g is put back such that the same graph can have multiple occurences in $\mathcal{P}_i$, and can occur in different sets $\mathcal{P}_i$ and $\mathcal{P}_j$ of $\mathcal{S}$ as well.

For the second method k-Centers prototype selection is applied on $\mathcal{T}$. This results in n disjoint subsets $\mathcal{P}_i \subset \mathcal{T}$ of different size. Next, we iteratively remove the set marginal graphs of $\mathcal{P}_i$ until m graphs remain in $\mathcal{P}_i$. Note that no graphs are removed whenever the size of a set $\mathcal{P}_i$ is smaller than, or equal to, m. Note that in contrast with the random selection this method leads to disjoint subsets.

From the experimental evaluation conducted in [8] one can draw the following conclusions. A definition of the reference sets based on a more elaborated method than random selection is preferable. Classifiers using the Lipschitz embedded graphs outperform classification systems using the original graph edit distances. Finally, the generalization of the Lipschitz embedding to the case where the subsets are not necessarily singletons ($m \geq 1$) is clearly beneficial.

3.5 Multiple Classifier System

Recently, the field of multiple classifier systems has become a very active area of research. The fundamental observation that motivates the combination of classifiers is that the sets of patterns misclassified by different classifiers do not necessarily heavily overlap. Hence, errors of a single classifier can be compensated by the other classifiers of an ensemble [23]. In the case of statistical patterns, that is, patterns represented by feature vectors, a large number of methods for the creation and combination of classifiers have been developed over the past years [23,24,25].

Regarding the graph embedding procedure proposed in this paper, if we repeat the process of random prototype selection a number of times, we can derive different graph subsets that can be used to map a given population of graphs to various vector spaces. That is, we obtain n different vector sets all representing the same graph set. For each vector set an individual classifier is trained and thus one gets an ensemble of classifiers. Consequently, a number of methods become available for combining the results of the individual ensemble members. In [5] this method was tested on a number of graph data sets with different characteristics, comming from various application domains. From the results of our experiments, one can conclude that the classification accuracy can be enhanced by most ensemble methods on almost all data sets.

4 Experimental Results

Lacking space we present the results for one particular graph embedding setting only, namely the graph embedding in conjunction with the prototype reduction schemes introduced in Section 3.2. For further results achieved with the novel graph embedding methodology, we refer to [4,5,6,7,8,9].

4.1 Experimental Setup

The classifier applied in the graph domain is a k-nearest-neighbor classifier. Note that this specific classifier is basically the only classifier that is directly applicable in the graph domain. The classifier applied to the vector space embedded graphs $\varphi_n^{\mathcal{P}}(g)$ is the support vector machine (SVM) with RBF Kernel. Hence, the kernel values are given by

$$\kappa_{RBF}(g_i, g_j) = exp\left(-\gamma ||\varphi_n^{\mathcal{P}}(g_i) - \varphi_n^{\mathcal{P}}(g_j)||^2\right)$$

where $\gamma > 0$.

In each of our experiments we make use of three disjoint graph sets, viz. a *validation set*, a *test set* and a *training set*. The validation set is used to determine optimal parameter values for classification. They consist of parameter k for the nearest neighbor classifier and the different parameters for the SVM, viz. γ (RBF) and C (weighting of the maximization of the margin and the minimization of the error). The parameter combination that results in the lowest classification error on the validation set is finally applied to the independent test set.

The pattern classification tasks considered in this paper involves a total of five different graph data sets. Because of lack of space, we can give only a short description of the data. For a more thorough description we refer to [4] where the data sets are discussed in greater detail. Note that each of our graph sets is divided into three disjoint subsets, viz. a training, a validation, and a test set.

The first database used in the experiments consists of graphs representing distorted letter drawings out of 15 classes (Letter). The second data set is given by graphs representing fingerprint images of the NIST-4 database [26] out of the four classes *arch*, *left*, *right*, and *whorl* (Fingerprint). The third graph set is constructed from the AIDS Antiviral Screen Database of Active Compounds [27] (AIDS). Graphs from this data set represent molecules out of two classes (*active*, *inactive*). The last data set consists of graphs representing webpages [28] that originate from 20 different categories (*Business*, *Health*, *Politics*, ...) (Webgraph).

4.2 Reference Systems

Two reference systems are used to compare the performance of the proposed graph embedding procedure with. Similarly to our novel approach, both reference systems make use of an SVM. The first reference system uses a similarity kernel directly derived from the edit distances [29] (referred to as GED). That is, for this reference method no explicit graph embedding is conducted but the dissimilarites are merely turned into kernel values $\kappa(g_i, g_j) = -d(g_i, g_j)^2$. The second reference system interprets all distances to the whole training set $\mathcal{T}$ as a vectorial description, i.e. the graphs are explicitly embedded but no attempts are made to reduce the number of prototypes (referred to as All). For this reference system an RBF kernel is applied to the vector space embedded graphs, too.

4.3 Results and Discussion

In Table 1 the classification results of all reference systems and the proposed approach using all prototype reduction schemes are given. Comparing the results of our novel approach with the results achieved by the first reference system (GED), we observe the following. On the Webpage data the first reference method outperforms all other systems. On this particular data set, the transformation of the edit distances into kernel values seems to be the best choice. However, on the three remaining data sets it is beneficial to use the embedding approach rather than the direct transformation. On the Letter data the SVM based on the merged prototypes performes better than the first reference system and on the AIDS and Fingerprint data even all reduction schemes lead to better classification accuracies than the similarity kernel. Note that 12 out of 13 improvements, but only 5 out of 11 deteriorations, compared to the first reference system are statistically significant.

Regarding the results achieved by the second reference system (All), we observe that our approach using prototype reduction outperforms this reference system on all data sets (at least with one of the proposed reduction schemes). Hence, besides the speed-up in computation, it is beneficial for achieving a higher recognition rate to use prototype reduction for embedding rather than using the whole training set as prototypes.

Comparing the prototype reduction schemes against each other, one can conclude that the merging approach performs generally best. On three out of four data sets this reduction scheme leads to the overall best classification result (including the reference methods). On the other hand, condensing and modified condensing lead on three data sets to the lowest recognition rate among the prototype reduction approaches.

Table 1. Experimental Results

	Ref. System		Proposed Method					
Data Set	GED	All	Cond	mCond	Edit	Red	Merg	Sel
Letter	92.27	91.73	91.73	91.47	92.00	92.00	92.53	92.00
AIDS	93.60	97.20	97.13	97.53	97.20	97.20	98.27	97.20
Fingerprint	79.35	82.10	81.60	81.75	81.70	82.10	82.80	82.30
Webpage	84.62	82.44	82.44	83.21	76.92	82.95	81.15	81.03

5 Conclusions

Although graphs have a higher representational power than feature vectors, there is a lack of methods for pattern classification using graph representations. By contrast, a large number of methods for classification have been proposed for object representations given in terms of feature vectors. The present paper introduces a novel graph embedding procedure in order to bridge the gap between structural and statistical pattern recognition. These graph embeddings make explicit use of graph edit distance and can therefore deal with various kinds of graphs (labelled, unlabelled, directed, undirected, etc.). The basic idea of the embedding procedure is to describe a graph by means of n dissimilarities to a predefined prototype set or to sets of prototypes. That is, a graph g is mapped explicitly to the n-dimensional real space $\mathbb{R}^n$ by arranging the distances of g to all of the n prototypes, or n prototype sets, as a vector. We show that the embedding process can be controlled by different prototype selectors or by well-known dimensionality reduction algorithms. Furthermore, the proposed graph embedding process lends itself to a method for the automatic generation of classifier ensembles. From the results of our experiments presented in the present paper as well as in other contributions, one can conclude that the classification accuracy can be statistically significantly enhanced by all embedding methods compared to reference systems directly operating in the graph domain.

Acknowledgements

This work has been supported by the Swiss National Science Foundation (Project 200021-113198/1).

References

1. Duda, R., Hart, P., Stork, D.: Pattern Classification, 2nd edn. Wiley-Interscience, Hoboken (2000)
2. Pekalska, E., Duin, R.: The Dissimilarity Representation for Pattern Recognition: Foundations and Applications. World Scientific, Singapore (2005)
3. Spillmann, B., Neuhaus, M., Bunke, H., Pekalska, E., Duin, R.: Transforming strings to vector spaces using prototype selection. In: Yeung, D.-Y., Kwok, J.T., Fred, A., Roli, F., de Ridder, D. (eds.) SSPR 2006 and SPR 2006. LNCS, vol. 4109, pp. 287–296. Springer, Heidelberg (2006)
4. Riesen, K., Neuhaus, M., Bunke, H.: Graph embedding in vector spaces by means of prototype selection. In: Escolano, F., Vento, M. (eds.) GbRPR 2007. LNCS, vol. 4538, pp. 383–393. Springer, Heidelberg (2007)
5. Riesen, K., Bunke, H.: Classifier ensembles for vector space embedding of graphs. In: Haindl, M., Kittler, J., Roli, F. (eds.) MCS 2007. LNCS, vol. 4472, pp. 220–230. Springer, Heidelberg (2007)
6. Riesen, K., Bunke, H.: Reducing the dimensionality of dissimilarity space embedding graph kernels. Engineering Applications of Artificial Intelligence Engineering Applications of Artificial Intelligence (accepted, 2008)
7. Riesen, K., Bunke, H.: Non-linear transformations of vector space embedded graphs. In: 8th International Workshop on Pattern Recognition in Information Systems (accepted, 2008)
8. Riesen, K., Bunke, H.: On Lipschitz embeddings of graphs. In: 12th International Conference on Knowledge-Based and Intelligent Information & Engineering Systems (accepted, 2008)
9. Riesen, K., Bunke, H.: Dissimilarity based vector space embedding of graphs using prototype reduction schemes (submitted)
10. Bunke, H., Dickinson, P., Kraetzl, M.: Theoretical and algorithmic framework for hypergraph matching. In: Roli, F., Vitulano, S. (eds.) ICIAP 2005. LNCS, vol. 3617, pp. 463–470. Springer, Heidelberg (2005)

11. Luo, B., Wilson, R., Hancock, E.: Spectral embedding of graphs. Pattern Recognition 36(10), 2213–2223 (2003)
12. Wilson, R., Hancock, E., Luo, B.: Pattern vectors from algebraic graph theory. IEEE Trans. on Pattern Analysis ans Machine Intelligence 27(7), 1112–1124 (2005)
13. Robles-Kelly, A., Hancock, E.: A Riemannian approach to graph embedding. Pattern Recognition 40, 1024–1056 (2007)
14. Bunke, H., Allermann, G.: Inexact graph matching for structural pattern recognition. Pattern Recognition Letters 1, 245–253 (1983)
15. Riesen, K., Bunke, H.: Approximate graph edit distance computation by means of bipartite graph matching. Image and Vision Computing (accepted, 2008)
16. Gärtner, T., Lloyd, J., Flach, P.: Kernels and distances for structured data. Machine Learning 57(3), 205–232 (2004)
17. Shawe-Taylor, J., Cristianini, N.: Kernel Methods for Pattern Analysis. Cambridge University Press, Cambridge (2004)
18. Bezdek, J., Kuncheva, L.: Nearest prototype classifier designs: An experimental study. Int. Journal of Intelligent Systems 16(12), 1445–1473 (2001)
19. Kim, S., Oommen, B.: A brief taxonomy and ranking of creative prototype reduction schemes. Pattern Analysis and Applications 6, 232–244 (2003)
20. Schölkopf, B., Smola, A., Müller, K.R.: Nonlinear component analysis as a kernel eigenvalue problem. Neural Computation 10, 1299–1319 (1998)
21. Bourgain, J.: On Lipschitz embedding of finite metric spaces in Hilbert spaces. Israel Journal of Mathematics 52(1-2), 46–52 (1985)
22. Hjaltason, G., Samet, H.: Properties of embedding methods for similarity searching in metric spaces. IEEE Trans. on Pattern Analysis ans Machine Intelligence 25(5), 530–549 (2003)
23. Kuncheva, L.: Combining Pattern Classifiers: Methods and Algorithms. John Wiley, Chichester (2004)
24. Breiman, L.: Bagging predictors. Machine Learning 24, 123–140 (1996)
25. Freund, Y., Shapire, R.: A decision theoretic generalization of online learning and application to boosting. Journal of Computer and Systems Sciences 55, 119–139 (1997)
26. Watson, C., Wilson, C.: NIST Special Database 4, Fingerprint Database. National Institute of Standards and Technology (1992)
27. DTP, AIDS antiviral screen (2004), `http://dtp.nci.nih.gov/docs/aids/aids_data.html`
28. Schenker, A., Bunke, H., Last, M., Kandel, A.: Graph-Theoretic Techniques for Web Content Mining. World Scientific, Singapore (2005)
29. Neuhaus, M., Bunke, H.: Bridging the Gap Between Graph Edit Distance and Kernel Machines. World Scientific, Singapore (2007)

Author Index

Abou-Moustafa, Karim 862
Adán, Antonio 644
Adán, Miguel 644
Aksoy, Selim 402
Albarelli, Andrea 197
Allano, Lorene 842
Alquézar, René 298
Aoki, Kazuaki 562
Artner, Nicole M. 441
Atmosukarto, Indriyati 267

Balika, Anupriya 329
Bandyopadhyay, Sivaji 831
Barrat, Sabine 937
Battiato, Sebastiano 137
Beleznai, Csaba 441
Bhaduri, Budhendra 967
Bicego, Manuele 852, 906
Biggio, Battista 500
Brewer, Nathan 451
Brun, Luc 227, 350
Bunke, Horst 2, 15, 287, 551, 996
Byun, Hyeran 421

Çavuş, Özge 402
Cha, Sung-Hyuk 622, 633
Chen, Liming 277
Chen, Yen-Wei 411
Choi, Heeyoul 247
Choi, Jin Young 592
Cipolla, Roberto 137
Cisar, Petr 107

de Amicis, Raffaele 167
Dellandréa, Emmanuel 277
Delmas, Patrice 127
Díaz-Chito, Katerine 977
Díaz-Villanueva, Wladimiro 977
Didaci, Luca 765
Ding, Xiaoqing 469
Dinh, H. Quynh 187
Domingos, Pedro 3
Dong, Xiao 694
Dorizzi, Bernadette 842

Duin, Robert P.W. 35, 459, 551, 724
Dupé, François-Xavier 227

Ekbal, Asif 831
ElGhawalby, Hewayda 207
Emms, David 5
Escolano, Francisco 237
Este, Alice 885

Farinella, Giovanni Maria 137
Ferrer, Miquel 15
Ferri, Francesc J. 977
Ferrie, Frank 862
Fink, Gernot A. 790
Flach, Boris 177
Fothergill, Simon 372
Frélicot, Carl 612
Freni, Biagio 745
Fu, Huanzhang 277
Fu, Zhouyu 489
Fukui, Takayuki 411
Fumera, Giorgio 500

Gallo, Giovanni 137
Gao, Jian 654
García, Vicente 917
Garcia-Salicetti, Sonia 842
Gargiulo, Francesco 885
Ghahramani, Zoubin 896
Giacinto, Giorgio 821
Gimel'farb, Georgy 127
Gomez-Ballester, Eva 926
Gringoli, Francesco 885
Grosso, Enrico 852
Guardiola, Jose Luis 782
Gyaourova, Aglika 755

Habrard, Amaury 86
Hall, Peter M. 329
Hammond, Tracy 247
Hancock, Edwin R. 5, 117, 157, 207, 217, 237, 257, 308, 318, 664
Harle, Robert 372
Harol, Artsiom 551

Heesch, Daniel 97
Heroux, Pierre 45
Ho, Tin Kam 1, 986
Holden, Sean 372
Huynh, Cong Phuoc 382

Igarashi, Takanori 411
Imai, Hideyuki 530
Iñesta, José Manuel 55, 86

Jehan-Besson, Stéphanie 350
Jeudy, Baptiste 65
Jiang, Xiaoyi 392
Jiang, Yan 469

Kashimoto, Akio 411
Kasparis, Takis 147
Kim, Hyun-Chul 896
Kim, Sang-Woon 654
Kittler, Josef 479
Kok, Stanley 3
Kropatsch, Walter G. 441
Kudo, Mineichi 530, 562, 572, 801, 874
Kuncheva, Ludmila I. 4, 510, 520

Le Capitaine, Hoel 612
Lee, Wan-Jui 35, 551
Li, Heping 705
Liang, Wei 705
Lienemann, Kai 790
Liu, Nianjun 451
Lowd, Daniel 3
Lozano, Miguel A. 237

Ma, Yunqian 107
Marcialis, Gian Luca 684, 745, 765
Marinai, Simone 582
Marino, Emanuele 582
Mármol, Salvador B. López 441
Marrocco, Claudio 674
Matsakis, Pascal 947
Merchán, Pilar 644
Micó, Luisa 811, 926
Miyakoshi, Masaaki 530
Molina-Abril, Helena 361
Mollineda, Ramón A. 917
Morris, Jonh 127

Nakamura, Atsuyoshi 801
Nakao, Keisuke 411
Née, Guillaume 350
Novovičová, Jana 956

Olivares-Rodríguez, Cristian 431
Oncina, Jose 431, 926
Oommen, B. John 540
Otranto, Edoardo 852

Park, Jin-Hyeong 714
Park, Myoung Soo 592
Paulson, Brandon 247
Pękalska, Elżbieta 551
Pelillo, Marcello 197
Pérez-Cortés, Juan Carlos 782
Pérez-Jiménez, Alberto J. 782
Pérez-Sancho, Carlos 55
Petrou, Maria 97
Plötz, Thomas 790
Plumpton, Catrin O. 510
Poon, Hoifung 3
Pujol, Alain 277

Qiao, Xu 411
Qin, Ke 540

Ramel, Jean-Yves 45
Rattani, Ajita 684
Real, Pedro 361
Reddy, Chandan K. 714
Ren, Peng 257, 308
Rendek, Jan 947
Revenu, Marinette 350
Ricamato, Maria Teresa 674
Richardson, Matt 3
Riesen, Kaspar 2, 287, 996
Rizo, David 55, 86
Robles-Kelly, Antonio 382, 489
Rodríguez, Juan J. 520
Roli, Fabio 500, 684, 745, 765, 821
Ross, Arun 755
Rueda, Luis 602

Sadeghi, Mohammad T. 479
Salamanca, Santiago 644
Salgarelli, Luca 885
Samiei, Masoumeh 479
Sánchez, J. Salvador 917
Sanromà, Gerard 298
Sansone, Carlo 885
Sato, Maiko 874
Schlesinger, Dmitrij 177

Sebban, Marc 86
Serratosa, Francesc 15, 298
Shapiro, Linda G. 267
Sharpe, James 157
Shekhar, Shashi 967
Shibata, Tomoyuki 339
Shin, Dongjoe 25
Shirai, Satoshi 801
Shorin, Alexander 127
Shorter, Nicholas 147
Sidere, Nicolas 45
Simeone, Paolo 459
Singla, Parag 3
Smith, William A.P. 664
Socorro, Raisa 811
Soda, Giovanni 582
Somol, Petr 956
Song, Yi-Zhe 329
Sumner, Marc 3
Swadzba, Agnes 734
Symonova, Olga 167

Tabbone, Salvatore 937
Tan, Robby 97
Tanaka, Akira 530
Tax, David M.J. 459, 724
Tenmoto, Hiroshi 572
Thollard, Franck 65, 76
Tjahjadi, Tardi 25
Tortorella, Francesco 459, 674
Toyama, Jun 874
Tronci, Roberto 821
Trudda, Alessandro 906

Valveny, Ernest 15
van Leuken, Reinier H. 167
Vatsavai, Ranga Raju 967
Veltkamp, Remco C. 167
Viviani, Sebastiano 197

Wachsmuth, Sven 734
Wang, Jue 3
Wattuya, Pakaket 392
Wendling, Laurent 947
Wilson, Richard C. 5, 217, 257, 308
Wu, Jing 664

Xia, Shengping 117, 318
Xiao, Bai 217, 329
Xu, Liefei 187
Xu, Yong 775

Yamaguchi, Osamu 339
Yang, Junyeong 421
Yun, Kiwon 421

Zdziobeck, Arnaud 76
Zeng, Zhi 705
Zhang, David 775
Zhang, Shuwu 705
Zheng, Guoyan 694
Zliobaite, Indre 4

Lecture Notes in Computer Science

Sublibrary 6: Image Processing, Computer Vision, Pattern Recognition, and Graphics

For information about Vols. 1–4190
please contact your bookseller or Springer

Vol. 5359: G. Bebis, R. Boyle, B. Parvin, D. Koracin, P. Remagnino, F. Porikli, J. Peters, J. Klosowski, L. Arns, Y.K. Chun, T.-M. Rhyne, L. Monroe (Eds.), Advances in Visual Computing, Part II. XLIV, 1203 pages. 2008.

Vol. 5358: G. Bebis, R. Boyle, B. Parvin, D. Koracin, P. Remagnino, F. Porikli, J. Peters, J. Klosowski, L. Arns, Y.K. Chun, T.-M. Rhyne, L. Monroe (Eds.), Advances in Visual Computing, Part I. XLIV, 1173 pages. 2008.

Vol. 5342: N. da Vitora Lobo, F. Roli, J.T. Kwok, G.C. Anagnostopoulos, M. Loog (Eds.), Structural, Syntactic, and Statistical Pattern Recognition. XIX, 1021 pages. 2008.

Vol. 5305: D. Forsyth, P. Torr, A. Zisserman (Eds.), Computer Vision – ECCV 2008, Part IV. XIX, 893 pages. 2008.

Vol. 5304: D. Forsyth, P. Torr, A. Zisserman (Eds.), Computer Vision – ECCV 2008, Part III. XIX, 827 pages. 2008.

Vol. 5303: D. Forsyth, P. Torr, A. Zisserman (Eds.), Computer Vision – ECCV 2008, Part II. XIX, 851 pages. 2008.

Vol. 5302: D. Forsyth, P. Torr, A. Zisserman (Eds.), Computer Vision – ECCV 2008, Part I. XXXVII, 801 pages. 2008.

Vol. 5277: A. Egges, A. Kamphuis, M. Overmars (Eds.), Motion in Games. IX, 257 pages. 2008.

Vol. 5259: J. Blanc-Talon, S. Bourennane, W. Philips, D. Popescu, P. Scheunders (Eds.), Advanced Concepts for Intelligent Vision Systems. XXIII, 1138 pages. 2008.

Vol. 5256: S. Liu, T. Maibaum, K. Araki (Eds.), Formal Methods and Software Engineering. XI, 397 pages. 2008.

Vol. 5242: D.N. Metaxas, L. Axel, G. Fichtinger, G. Székely (Eds.), Medical Image Computing and Computer-Assisted Intervention – MICCAI 2008, Part II. LI, 1111 pages. 2008.

Vol. 5241: D.N. Metaxas, L. Axel, G. Fichtinger, G. Székely (Eds.), Medical Image Computing and Computer-Assisted Intervention – MICCAI 2008, Part I. LI, 1087 pages. 2008.

Vol. 5197: J. Ruiz-Shulcloper, W.G. Kropatsch (Eds.), Progress in Pattern Recognition, Image Analysis and Applications. XXI, 809 pages. 2008.

Vol. 5188: M. Sebillo, G. Vitiello, G. Schaefer (Eds.), Visual Information Systems. XIV, 340 pages. 2008.

Vol. 5166: A. Butz, B. Fisher, A. Krüger, P. Olivier, M. Christie (Eds.), Smart Graphics. XI, 277 pages. 2008.

Vol. 5158: S.N. Srihari, K. Franke (Eds.), Computational Forensics. IX, 229 pages. 2008.

Vol. 5128: T. Dohi, I. Sakuma, H. Liao (Eds.), Medical Imaging and Augmented Reality. XVI, 441 pages. 2008.

Vol. 5116: E.A. Krupinski (Ed.), Digital Mammography. XXVII, 769 pages. 2008.

Vol. 5112: A. Campilho, M.S. Kamel (Eds.), Image Analysis and Recognition. XXII, 1126 pages. 2008.

Vol. 5099: A. Elmoataz, O. Lezoray, F. Nouboud, D. Mammass (Eds.), Image and Signal Processing. XVI, 625 pages. 2008.

Vol. 5098: F.J. Perales, R.B. Fisher (Eds.), Articulated Motion and Deformable Objects. XIV, 458 pages. 2008.

Vol. 5096: G. Rigoll (Ed.), Pattern Recognition. XIII, 538 pages. 2008.

Vol. 5046: W. Liu, J. Lladós, J.-M. Ogier (Eds.), Graphics Recognition. IX, 336 pages. 2008.

Vol. 4992: D. Coeurjolly, I. Sivignon, L. Tougne, F. Dupont (Eds.), Discrete Geometry for Computer Imagery. XIII, 554 pages. 2008.

Vol. 4987: X. Gao, H. Müller, M.J. Loomes, R. Comley, S. Luo (Eds.), Medical Imaging and Informatics. XV, 388 pages. 2008.

Vol. 4958: V.E. Brimkov, R.P. Barneva, H.A. Hauptman (Eds.), Combinatorial Image Analysis. XVI, 446 pages. 2008.

Vol. 4931: G. Sommer, R. Klette (Eds.), Robot Vision. XI, 468 pages. 2008.

Vol. 4901: D. Zhang (Ed.), Medical Biometrics. XII, 324 pages. 2007.

Vol. 4889: A. Pasko, V. Adzhiev, P. Comninos (Eds.), Heterogeneous Objects Modelling and Applications. VII, 285 pages. 2008.

Vol. 4844: Y. Yagi, S.B. Kang, I.S. Kweon, H. Zha (Eds.), Computer Vision – ACCV 2007, Part II. XXVIII, 915 pages. 2007.

Vol. 4843: Y. Yagi, S.B. Kang, I.S. Kweon, H. Zha (Eds.), Computer Vision – ACCV 2007, Part I. XXVIII, 969 pages. 2007.

Vol. 4842: G. Bebis, R. Boyle, B. Parvin, D. Koracin, N. Paragios, S.-M. Tanveer, T. Ju, Z. Liu, S. Coquillart, C. Cruz-Neira, T. Müller, T. Malzbender (Eds.), Advances in Visual Computing, Part II. XXXIII, 827 pages. 2007.

Vol. 4841: G. Bebis, R. Boyle, B. Parvin, D. Koracin, N. Paragios, S.-M. Tanveer, T. Ju, Z. Liu, S. Coquillart, C. Cruz-Neira, T. Müller, T. Malzbender (Eds.), Advances in Visual Computing, Part I. XXXIII, 831 pages. 2007.

Vol. 4815: A. Ghosh, R.K. De, S.K. Pal (Eds.), Pattern Recognition and Machine Intelligence. XIX, 677 pages. 2007.

Vol. 4814: A. Elgammal, B. Rosenhahn, R. Klette (Eds.), Human Motion – Understanding, Modeling, Capture and Animation. X, 329 pages. 2007.

Vol. 4792: N. Ayache, S. Ourselin, A. Maeder (Eds.), Medical Image Computing and Computer-Assisted Intervention – MICCAI 2007, Part II. XLVI, 988 pages. 2007.

Vol. 4791: N. Ayache, S. Ourselin, A. Maeder (Eds.), Medical Image Computing and Computer-Assisted Intervention – MICCAI 2007, Part I. XLVI, 1012 pages. 2007.

Vol. 4781: G. Qiu, C. Leung, X.-Y. Xue, R. Laurini (Eds.), Advances in Visual Information Systems. XIII, 582 pages. 2007.

Vol. 4778: S.K. Zhou, W. Zhao, X. Tang, S. Gong (Eds.), Analysis and Modeling of Faces and Gestures. X, 305 pages. 2007.

Vol. 4768: D. Doermann, S. Jaeger (Eds.), Arabic and Chinese Handwriting Recognition. VIII, 279 pages. 2008.

Vol. 4756: L. Rueda, D. Mery, J. Kittler (Eds.), Progress in Pattern Recognition, Image Analysis and Applications. XXI, 989 pages. 2007.

Vol. 4738: A.C.R. Paiva, R. Prada, R.W. Picard (Eds.), Affective Computing and Intelligent Interaction. XVIII, 781 pages. 2007.

Vol. 4729: F. Mele, G. Ramella, S. Santillo, F. Ventriglia (Eds.), Advances in Brain, Vision, and Artificial Intelligence. XVI, 618 pages. 2007.

Vol. 4713: F.A. Hamprecht, C. Schnörr, B. Jähne (Eds.), Pattern Recognition. XIII, 560 pages. 2007.

Vol. 4679: A.L. Yuille, S.-C. Zhu, D. Cremers, Y. Wang (Eds.), Energy Minimization Methods in Computer Vision and Pattern Recognition. XII, 494 pages. 2007.

Vol. 4678: J. Blanc-Talon, W. Philips, D. Popescu, P. Scheunders (Eds.), Advanced Concepts for Intelligent Vision Systems. XXIII, 1100 pages. 2007.

Vol. 4673: W.G. Kropatsch, M. Kampel, A. Hanbury (Eds.), Computer Analysis of Images and Patterns. XX, 1006 pages. 2007.

Vol. 4642: S.-W. Lee, S.Z. Li (Eds.), Advances in Biometrics. XX, 1216 pages. 2007.

Vol. 4633: M.S. Kamel, A. Campilho (Eds.), Image Analysis and Recognition. XII, 1312 pages. 2007.

Vol. 4625: R. Stiefelhagen, R. Bowers, J.G. Fiscus (Eds.), Multimodal Technologies for Perception of Humans. XIII, 556 pages. 2008.

Vol. 4584: N. Karssemeijer, B. Lelieveldt (Eds.), Information Processing in Medical Imaging. XX, 777 pages. 2007.

Vol. 4569: A. Butz, B. Fisher, A. Krüger, P. Olivier, S. Owada (Eds.), Smart Graphics. IX, 237 pages. 2007.

Vol. 4538: F. Escolano, M. Vento (Eds.), Graph-Based Representations in Pattern Recognition. XII, 416 pages. 2007.

Vol. 4535: H. Ito, M. Kano, N. Katoh, Y. Uno (Eds.), Computational Geometry and Graph Theory. XII, 237 pages. 2008.

Vol. 4522: B.K. Ersbøll, K.S. Pedersen (Eds.), Image Analysis. XVIII, 989 pages. 2007.

Vol. 4485: F. Sgallari, A. Murli, N. Paragios (Eds.), Scale Space and Variational Methods in Computer Vision. XV, 931 pages. 2007.

Vol. 4478: J. Martí, J.M. Benedí, A.M. Mendonça, J. Serrat (Eds.), Pattern Recognition and Image Analysis, Part II. XXVII, 657 pages. 2007.

Vol. 4477: J. Martí, J.M. Benedí, A.M. Mendonça, J. Serrat (Eds.), Pattern Recognition and Image Analysis, Part I. XXVII, 625 pages. 2007.

Vol. 4472: M. Haindl, J. Kittler, F. Roli (Eds.), Multiple Classifier Systems. XI, 524 pages. 2007.

Vol. 4466: F.B. Sachse, G. Seemann (Eds.), Functional Imaging and Modeling of the Heart. XV, 486 pages. 2007.

Vol. 4418: A. Gagalowicz, W. Philips (Eds.), Computer Vision/Computer Graphics Collaboration Techniques. XV, 620 pages. 2007.

Vol. 4417: A. Kerren, A. Ebert, J. Meyer (Eds.), Human-Centered Visualization Environments. XIX, 403 pages. 2007.

Vol. 4391: Y. Stylianou, M. Faundez-Zanuy, A. Esposito (Eds.), Progress in Nonlinear Speech Processing. XII, 269 pages. 2007.

Vol. 4370: P.P. Lévy, B. Le Grand, F. Poulet, M. Soto, L. Darago, L. Toubiana, J.-F. Vibert (Eds.), Pixelization Paradigm. XV, 279 pages. 2007.

Vol. 4358: R. Vidal, A. Heyden, Y. Ma (Eds.), Dynamical Vision. IX, 329 pages. 2007.

Vol. 4338: P.K. Kalra, S. Peleg (Eds.), Computer Vision, Graphics and Image Processing. XV, 965 pages. 2006.

Vol. 4319: L.-W. Chang, W.-N. Lie (Eds.), Advances in Image and Video Technology. XXVI, 1347 pages. 2006.

Vol. 4292: G. Bebis, R. Boyle, B. Parvin, D. Koracin, P. Remagnino, A. Nefian, G. Meenakshisundaram, V. Pascucci, J. Zara, J. Molineros, H. Theisel, T. Malzbender (Eds.), Advances in Visual Computing, Part II. XXXII, 906 pages. 2006.

Vol. 4291: G. Bebis, R. Boyle, B. Parvin, D. Koracin, P. Remagnino, A. Nefian, G. Meenakshisundaram, V. Pascucci, J. Zara, J. Molineros, H. Theisel, T. Malzbender (Eds.), Advances in Visual Computing, Part I. XXXI, 916 pages. 2006.

Vol. 4245: A. Kuba, L.G. Nyúl, K. Palágyi (Eds.), Discrete Geometry for Computer Imagery. XIII, 688 pages. 2006.

Vol. 4241: R.R. Beichel, M. Sonka (Eds.), Computer Vision Approaches to Medical Image Analysis. XI, 262 pages. 2006.

Vol. 4225: J.F. Martínez-Trinidad, J.A. Carrasco Ochoa, J. Kittler (Eds.), Progress in Pattern Recognition, Image Analysis and Applications. XIX, 995 pages. 2006.

Vol. 4191: R. Larsen, M. Nielsen, J. Sporring (Eds.), Medical Image Computing and Computer-Assisted Intervention – MICCAI 2006, Part II. XXXVIII, 981 pages. 2006.

MIX
Papier aus verantwortungsvollen Quellen
Paper from responsible sources
FSC® C105338

If you have any concerns about our products,
you can contact us on
ProductSafety@springernature.com

In case Publisher is established outside the EU,
the EU authorized representative is:
Springer Nature Customer Service Center GmbH
Europaplatz 3, 69115 Heidelberg, Germany

Printed by Libri Plureos GmbH
in Hamburg, Germany